Desk Reference of
Functional Polymers

Desk Reference of
Functional Polymers
Syntheses and Applications

Reza Arshady, Editor

Imperial College of Science, Technology and Medicine
University of London

American Chemical Society, Washington, DC

Library of Congress Cataloging-in-Publication Data

Desk reference of functional polymers: syntheses and applications / Reza Arshady, editor.

    p.   cm.— (ACS Professional reference book)

Includes bibliographical references and index.

ISBN 0–8412–3469–8 (alk. paper)

1. Polymerization.   2. Polymers.

I. Arshady, R. (Reza)   II. Series: ACS professional reference book.

QD281.P6D47   1996
668.9—dc21   96–39093
                                                              CIP

The paper used in this publication meets the minimum requirements of American National Standard for Information Sciences—Permanence of Paper for Printed Library Materials, ANSI Z39.48–1984.   ∞

PRINTED IN THE UNITED STATES OF AMERICA

Advisory Board

About the Editor

Reza Arshady is a Senior Fellow at the Department of Chemistry, Imperial College of Science, Technology and Medicine, University of London, London SW15 2AY, United Kingdom. He was previously Professor and Chairman of the Department of Chemistry, and Chairman of Academic–Industrial Liaison, University of Kashan, Kashan, Iran. Before that he was Assistant Professor and Director of the Polymer and Industrial Chemistry Laboratory at the University of Tabriz, Tabriz, Iran, with sabbatical periods at the Medical Research Council (MRC), Cambridge (U.K.); Imperial College, London (U.K.); and Munich University of Technology, Munich, Germany.

He is a keen educationist and advocate of active learning. His main area of research is organic polymer chemistry, including polymer synthesis and functional polymers for emerging technologies in chemistry, engineering, biotechnology, and medicine. His other research interests include heterophase polymerization processes, microspheres, and microcapsules. He is the author or coauthor of more than 120 publications, including a number of course books, and is the editor of the forthcoming *Handbook of Microspheres, Microcapsules and Liposomes*.

He obtained his Ph.D. in organic polymer chemistry from the University of Liverpool, Liverpool, U.K., and B.Sc. from the University of Teachers in Tehran, Iran. He helped to organize the Microencapsulation Symposium of the ACS National Meeting in Washington, D.C., in August 1994, and is a frequent invited speaker at international symposia and conferences. He is on the editorial boards of *Trends in Polymer Science* and the *Journal of Microencapsulation,* and is listed in *Who's Who in Science & Engineering* and *Who's Who in the World.* His interests beyond chemistry include history, linguistics, and science and technology in society.

Contents

Contributors

Manfred Anders *page 227*
Institute of Organic Chemistry, University of Tuebingen, Auf der Morgenstelle 18, D–72076 Tuebingen, Germany

Joseph M. Antonucci *page 719*
Dental and Medical Materials Group, Polymers Division, National Institute of Standards and Technology, Gaithersburg, MD 20899

Reza Arshady *pages 133, 15*
Department of Chemistry, Imperial College of Science, Technology, and Medicine, South Kensington, London, SW7 2AY, England

Warren E. Baker *page 133*
Department of Chemistry, Queen's University, Kingston, Ontario K7L 3N6, Canada

Vincenzo Bertini *page 635*
Istituto di Analisi e Tecnologie Farmaceutiche ed Alimentari, Università di Genova, Via Brigata Salerno, I-16147 Genova, Italy

Karsten Blatter *page 73*
Freie Universität Berlin, Institut für Organische Chemie, Takustrasse 3, D–14195 Berlin, Germany

D. Bosc *page 489*
France Telecom, CNET/LAB, Département TEP, Route de Trégastel, F–22301, Lannion France

B. Boutevin *page 489*
Laboratoire de Chimie Appliquée, UPRESA 5076, Ecole Nationale Supérieure de Chimie de Montpellier, 8 rue de l'Ecole Normale, F–34053, Montpellier, France

L. B. Bridwell *pages 371, 387*
Department of Physics and Astronomy, Southwest Missouri State University, Springfield, MO 65804

D. J. Broer *page 451*
Philips Research Laboratories, Prof. Holstlaan 4, 5656 AA Eindhoven, The Netherlands

J. C. Brosse *page 201*
Laboratoire de Chimie et Physicochimie Macromoleculaire, Université du Maine, Avenue Olivier, Messiaen, 72000 Le Mans, France

Marcel Camps *page 3*
University of Provence, 13331, Marseille Cedex 3, France

P. Chandrasekhar *page 529*
Ashwin-Ushas Corporation, Inc., 206 Ticonderoga Blvd., Freehold, NJ 07728

Daniel Chappard *page 699*
L.H.E.A. Faculty of Medicine and Pharmacy, University of Angers, 49045 Angers Cedex, France

Michel Chatzopoulos   *page 3*
Faculty of Sciences and Techniques, University of
Saint-Etienne, 42023, Saint-Etienne Cedex 2, France

T. C. Chung   *page 35*
Department of Materials Science and Engineering,
The Pennsylvania State University, University
Park, PA 16802

Benedetto Corain   *page 151*
Dipartimento di Chimica Ingegneria Chimica e
Materiali, via Vetoio, Coppito Due, 67010 L'Aquila,
Italy

J. M. G. Cowie   *page 427*
Chemistry Department, Heriot-Watt University,
Riccarton, Edinburgh EH14 4AS, Scotland

Al Czanderna   *page 567*
National Renewable Energy Laboratory, 1617 Cole
Boulevard, Golden, CO 80401

Semih Erhan   *page 261*
Center for Protein and Polymer Research, Albert
Einstein Medical Center, Philadelphia, PA
19141–3098

Alan B. Fischer   *page 657*
Advanced Surface Technology, Inc., 9 Linnell
Circle, Billerica, MA 01821–3902

Marye Anne Fox   *page 549*
Chemistry Department, University of Texas at
Austin, Austin, TX 78712

Kurt E. Geckeler   *pages 227, 601*
Institute of Organic Chemistry, University of
Tuebingen, Auf der Morgenstelle 18, D–72076
Tuebingen, Germany

R. E. Giedd   *pages 371, 387*
Department of Physics and Astronomy, Southwest
Missouri State University, Springfield, MO 65804

Adelheid Godt   *page 73*
Freie Universität Berlin, Institut für Organische
Chemie, Takustrasse 3, D–14195 Berlin, Germany

Sylwester Gogolewski   *page 677*
Department of Polymers, AO/ASIF Research
Institute, CH-7270 Davos, Switzerland

W. Göpel   *page 601*
Institute of Physical and Theoretical Chemistry,
University of Tuebingen, Auf der Morgenstelle 8,
D–72076 Tuebingen, Germany

T. T. Hinchcliffe   *page 427*
Chemistry Department, Heriot-Watt University,
Riccarton, Edinburgh EH14 4AS, Scotland

Akira Hirao   *page 19*
Faculty of Engineering, Department of Polymer
Chemistry, Tokyo Institute of Technology, 2–12–1,
Ohokayama, Meguro-ku, Tokyo 152, Japan

Krister Holmberg   *page 739*
Institute for Surface Chemistry, P.O. Box 5607,
S114 86 Stockholm, Sweden

Sung Il Hong   *page 293*
Department of Fiber and Polymer Science, Seoul
National University, Seoul 151–742, Korea

Toru Ishiji   *page 589*
Riken Keiki Company Ltd., 2–7–6 Azusawa,
Itabashi, Tokyo 174, Japan

Hiroshi Ito   *pages 311, 341*
IBM Research Division, Almaden Research Center,
650 Harry Road, San Jose, CA 95120–6099

Satoru Iwata   *page 169*
Department of Molecular Chemistry and
Engineering, Faculty of Engineering, Tohoku
University, Aoba, Sendai 980, Japan

So Young Joo   *page 293*
Textile Division, National Institute of Technology
and Quality, Kwacheon-City 427–010, Korea

Gary Jorgensen   *page 567*
National Renewable Energy Laboratory, 1617 Cole
Boulevard, Golden, CO 80401

Jun-ichi Kadokawa   *page 93*
Department of Molecular Chemistry and
Engineering, Faculty of Engineering, Tohoku
University, Sendai 980, Japan

Jaroslav Kahovec   *page 699*
Institute of Macromolecular Chemistry, Academy
of Sciences of the Czech Republic, 16026 Prague 6,
Czech Republic

Masao Kaneko   *page 589*
Department of Chemistry, Ibaraki University,
2–1–1 Bunkyo, Mito 310, Japan

Doo Whan Kang   *page 293*
Department of Polymer Science and Engineering,
Dan Kook University, Seoul 140–714, Korea

Stephen Kelley *page 567*
National Renewable Energy Laboratory, 1617 Cole
Boulevard, Golden, CO 80401

Joseph P. Kennedy *page 57*
Institute of Polymer Science, University of Akron,
Akron, OH 44325–3909

Shiro Kobayashi *pages 93, 169*
Department of Molecular Chemistry and
Engineering, Faculty of Engineering, Tohoku
University, Sendai 980, Japan

Yuichiro Kojima *page 753*
Department of Microbiology, Kumamoto
University School of Medicine, 2–2–1 Honjo,
Kumamoto city, Kumamoto 860, Japan

Keisuke Kurita *page 239*
Department of Industrial Chemistry, Faculty of
Engineering, Seikei University, Musashino-shi,
Tokyo 180, Japan

Wayne W. Y. Lau *page 621*
Department of Chemical Engineering, National
University of Singapore, Republic of Singapore
11926

G. Legeay *page 201*
Centre de Transfert et de Technologie du Mans,
Departement Matériaux, Technopôle Université,
rue Thalès de Milet, 72000 Le Mans, France

Matthias Löffler *page 73*
Freie Universität Berlin, Institut für Organische
Chemie, Takustrasse 3, D–14195 Berlin, Germany

Ih-Houng Loh *page 657*
Advanced Surface Technology, Inc., 9 Linnell
Circle, Billerica, MA 01821–3902

Hiroshi Maeda *page 753*
Department of Microbiology, Kumamoto
University School of Medicine, 2–2–1 Honjo,
Kumamoto city, Kumamoto 860, Japan

Munmaya K. Mishra *page 57*
Research and Development, Ethyl Corporation,
500 Spring Street, Richmond, VA 23218–2158

Jean-Pierre Monthéard *pages 3, 699*
Faculty of Sciences and Techniques, University of
Saint-Etienne, 42023, Saint-Etienne Cedex 2,
France

Yukio Nagasaki *page 183*
Department of Materials Science and Technology,
Science University of Tokyo, Noda 278, Japan

Jan H. Näsman *page 115*
Department of Polymer Technology, Åbo Akademi
University, Porthansgatan 3-5, FIN-20500 Åbo,
Finland

Jan Olsson *page 739*
Department of Cariology, Faculty of Odontology,
University of Göteborg, Box 33070, S400 33
Göteborg, Sweden

John Pern *page 567*
National Renewable Energy Laboratory, 1617 Cole
Boulevard, Golden, CO 80401

Richard A. Pethrick *page 463*
Department of Pure and Applied Chemistry,
Thomas Graham Building, University of
Strathclyde, Cathedral Street, Glasgow G1 1XL,
United Kingdom

David Phillips *page 407*
Department of Chemistry, Imperial College,
London SW7 2AY, England

Marco Pocci *page 635*
Istituto di Analisi e Tecnologie Farmaceutiche ed
Alimentari, Università di Genova, Via Brigata
Salerno, I-16147 Genova, Italy

F. Poncin-Epaillard *page 201*
Laboratoire de Chimie et Physicochimie
Macromoleculaire, Université du Maine, Avenue
Olivier, Messiaen, 72000 Le Mans, France

Pan Jiang Qing *page 621*
Institute of Chemistry, Academia Sinica, Beijing
10080, China

Helmut Ritter *page 103*
Bergische Universität-GH Wuppertal, FB 9,
Macromolecular Chemistry, Gauß Strasse 20,
D–42097 Wuppertal, Germany

Douglas R. Robello *page 505*
Imaging Research and Advanced Development,
Eastman Kodak Company, Rochester, NY
14650–2110

A. Rousseau *page 489*
Laboratoire de Chimie Appliquée, UPRESA 5076,
Ecole Nationale Supérieure de Chimie de

Montpellier, 8 rue de l'Ecole Normale, F–34053, Montpellier, France

Paul Schissel   *page 567*
National Renewable Energy Laboratory, 1617 Cole Boulevard, Golden, CO 80401

Arnulf-Dieter Schlüter   *page 73*
Freie Universität Berlin, Institut für Organische Chemie, Takustrasse 3, D–14195 Berlin, Germany

Min-Shyan Sheu   *page 657*
Advanced Surface Technology, Inc., 9 Linnell Circle, Billerica, MA 01821–3902

Shin-ichiro Shoda   *page 169*
Department of Molecular Chemistry and Engineering, Faculty of Engineering, Tohoku University, Aoba, Sendai 980, Japan

Jeffrey W. Stansbury   *page 719*
Dental and Medical Materials Group, Polymers Division, National Institute of Standards and Technology, Gaithersburg, MD 20899

Mats J. Sundell   *page 115*
Department of Polymer Technology, Åbo Akademi University, Porthansgatan 3-5, FIN-20500 Åbo, Finland

Hans-Joachim Timpe   *page 273*
Research and Development Division, Polychrome GmbH, An der Bahn 80, D–37520 Osterode, Germany

Vladimir P. Torchilin   *page 769*
Center for Imaging and Pharmaceutical Research,

Massachusetts General Hospital and Harvard Medical School, Charlestown, MA 02129

Hiroshi Uyama   *page 93*
Department of Molecular Chemistry and Engineering, Faculty of Engineering, Tohoku University, Sendai 980, Japan

Y. Q. Wang   *pages 371, 387*
Acadiana Research Laboratory, University of Southwestern Louisiana, Lafayette, LA 70504
Former address: Department of Physics and Astronomy, Southwest Missouri State University, Springfield, MO 65804

Diana M. Watkins   *page 549*
Chemistry Department, University of Texas at Austin, Austin, TX 78712

Hai-Qi Xie   *page 133*
Department of Chemistry, Queen's University, Kingston, Ontario K7L 3N6, Canada

Marco Zecca   *page 151*
Centro per lo Studio della Stabilità e Reattività dei Composti di Coordinazione, Italian Research Council, via Marzolo 1, 35131 Padova - c/o Dipartimento di Chimica Inorganica Metallorganica e Analitica, via Marzolo 35131 Padova, Italy

R. Zhou   *page 601*
Institute of Organic Chemistry, University of Tuebingen, Auf der Morgenstelle 18, D–72076 Tuebingen, Germany

Preface

This book was conceived as a comprehensive reference volume on functional polymers for emerging technologies. The term "functional polymers" signifies a broad area of polymer science and engineering involving the design, synthesis, and study of macromolecular materials of increasing structural sophistication and complexity useful for a wide range of chemical, physicochemical, and biomedical processes. "Emerging technologies" are also broadly understood to include new technological developments, beginning at the forefront of conventional industrial practices and extending into anticipated and speculative industries of the future.

With these broad descriptions in mind, a glance through the pages of science and engineering literature shows that the use of functional polymers for emerging technologies represents one of the most active areas of research and development throughout the fields of chemistry, physics, life sciences, and related technologies. In addition to being of technological interest, the subject of functional polymers is a fascinating area of interdisciplinary research and a major source of inspiration and motivation in its own right.

The scope of the present book thus extends far beyond emerging technologies per se. It presents a wealth of new ideas in the design, synthesis, and study of sophisticated macromolecular structures. The conception of the book was based as much on the serendipity of these new ideas as on the need to offer blueprints for the design of functional polymer devices for emerging technologies.

The book contains 42 chapters, logically organized into five parts:

Part One: General Synthetic Methods
Part Two: Radiation Effects and Applications
Part Three: Optoelectronic Properties and Applications
Part Four: Chemical and Physicochemical Applications
Part Five: Biomedical Applications.

Part One introduces a wide selection of functional polymers, their syntheses, properties, and numerous current and potential applications that they relate to. Various topics covered in Part One include a Japanese open-and-shut case, a stretcher of ladders to ribbons, enzymic synthesis, Ziegler–Natta and living ionic polymerizations, as well as functional polymers based on chloromethylstyrene, germylene, stannylene, silicon, polyamines, polyacids, organometallics, and naturally occurring polymers (cellulose, chitin, chitosan, and gelatin).

Part Two is devoted to photochemistry and photocross-linking of functional polymers, microlithography, and ion implantation. Part Three relates to photophysics of functional polymers, chiral liquid crystalline (LC) polymers, LC networks, and polymers with electrical conductivity, optical conductivity, and nonlinear optical behavior. Part Four discusses the use of functional polymers for electron and energy transfer processes, solar energy utilization, chemical sensors, selective flocculation of minerals and chemical synthesis, and catalysis. Part Five focuses on the biomedical uses of polymers such as poly(hydroxyethyl methacrylate) (PHEMA), polyurethanes, plasma-treated commodity polymers, polymeric dental aids, polymer–drug conjugates, and liposomes.

The choice of material for a book of this nature is inevitably influenced by a complex set of judicial and practical factors, and hence the outcome could hardly be expected to be ideal. Thus, to the extent that the chosen topics in the book are consistent and cohesive, I am greatly indebted to the efforts and courtesy of all the contributors who keenly agreed to my revisions (and in some cases re-writing) their chapters. I am also grateful to David Phillips and Maurice H. George (Imperial College, London) for their encouragement and support, and to Gary Rumbles (Imperial College, London) for reading some of the chapters in Part Three. I also wish to thank Robert A. Weiss (University of Connecticut, Storrs, CT) for encouraging the conception of the book.

The publication of the book took somewhat longer than had been planned because of events beyond my control, but independent reviewers assure me that the book will be a rich source of information for some years to come. It had been anticipated that the book would contain a comprehensive overview, and shorter overviews for each of the five parts, presenting an overall picture of the subject of functional polymers at the forefront of emerging technologies, and outlining their general features, potentials, and limitations, but these were omitted because of unexpected processing difficulties. However, considerable effort has been made from the outset to ensure that each individual chapter is self-contained, as far as practicable, with the necessary background information about the functional polymers and emerging technologies discussed, and anticipating future potentials and prospects of these polymers and technologies.

The book contains contributions from polymer scientists from around the world. My hope is that these contributions represent the flavor and fascination of the whole field of functional polymers, their synthesis, properties, and current and potential uses in emerging technologies in chemistry, engineering, biotechnology, and medicine.

Reza Arshady
Imperial College of Science,
    Technology and Medicine
University of London
London, England

Part One

General Synthetic Methods

Functional Polymers via Free-Radical Polymerization of Chloromethylstyrene

Jean-Pierre Monthéard, Michel Chatzopoulos, and Marcel Camps

Free-radical polymerization, until recently, has been the method of choice for the synthesis of most functional polymers based on styrene, (meth)acrylates, acrylamide, and other monomers. Among these monomers, vinylbenzylchloride (VBC; chloromethylstyrene) is the most widely used and is now commercially available. Numerous nucleophilic substitutions provide new monomers that are generally suitable for the synthesis of functional polymers by polymerization or copolymerization. Applications of VBC polymers and copolymers are now well established in the field of fine chemistry as microlithographic agents, reactive polymers, polymers supports, and catalysts. Growing numbers of reactions and applications have been described yearly; and new developments such as optical lenses, optical materials, and dyes for nonlinear optics were found recently. This chapter provides an overview of the synthesis of VBC, its conversion to other functional monomers, and the synthesis and applications of its polymers and copolymers.

Free-radical polymerization provides the most versatile route for the synthesis of vinyl-type polymers, although until recently it lacked some of the favorable architectural features of anionic, cationic, and Ziegler–Natta polymerizations. Our chapter presents an overview of the synthesis, polymerization, copolymerization, modifications, and applications of vinylbenzylchloride (VBC; chloromethylstyrene) as one of the most widely studied functional monomers.

VBC (*1, 2*) is probably one of the most important functional monomers. Its first industrial synthesis was described in 1957 (*3*), but it was scarcely used before 1972. Since then, however, thousands of articles and patents have been presented about the synthesis, properties, and applications of this monomer and corresponding polymers or copolymers. Some factors may well explain this interest:

1. VBC is a functional monomer now commercially available (Dow Chemical, United States; or Seimi, Japan) in a mixture of meta and para isomers (ratio, 60/40). The pure para isomer also has been commercially available for some years (Kodak, United States; or Nagase, Japan).

2. Because of the benzylic chlorine, a great number of nucleophilic substitutions are possible and leave the vinylic bond undamaged; hence, new monomers are provided. These new monomers, even with bulky substituents, are generally able to be polymerized or copolymerized.

3. Poly(VBC) or copolymers including VBC units can also react with various nucleophilic groups.

Poly(VBC) or Poly(VBC-*co*-styrene) are different from the polymers resulting from chloromethylation of polystyrene. Chloromethylation reactions of polystyrene are generally carried out with a starting cross-linked polystyrene (*4*) synthesized from styrene and divinylbenzene (DVB). Even if these polymers are prepared starting from soluble polystyrene, the resulting chloromethylated polystyrene is generally insoluble due to cross-linking by the CH_2Cl groups, which are mainly attached in the para position of the aromatic nucleus (5). Cross-linked chloromethylated polystyrenes have numerous applications as ion-exchange resins, polymer supports, and catalysts (6–9). Therefore, to prevent confusion, we report only the main work and applications devoted to VBC polymers and copolymers but not products obtained by chloromethylation of polystyrene. To simplify writing, we shall use the symbols shown in Chart I.

Syntheses and Characterizations of VBC

Different methods for the preparation of VBC can be divided into industrial processes and laboratory processes. As previously explained, VBC, in meta–para mixture or para isomer, is a commercial product. Therefore, the industrial methods are briefly described and only a very short summary of the laboratory syntheses is given.

Industrial Preparations of VBC

The well-known chloromethylation method was applied to styrene to prepare, in one step, the VBC monomer. But unfortunately, this method only gave a mixture of 1-phenyl-3-chloropropane and 1-phenyl-3-chloroprop-1-ene (10). Chloromethylation of ethylbenzene (3) was more successful and gave ethylchloromethylbenzene, which was photobrominated and then thermally dehydrobrominated in VBC. Chloro-

ethylbenzylchloride prepared by a chloromethylation reaction of 2-chloroethylbenzene can be pyrolyzed in VBC (mixture of isomers) (11, 12).

Another approach to the synthesis of VBC was described via thermal chlorination of vinyl-toluene at 550–600 °C (13). The obtained VBC was a 60/40 mixture of meta and para isomers. A similar synthesis from pure para methylstyrene provided para isomer of VBC (14). Different schemes for industrial syntheses of VBC are given in Scheme I.

Laboratory Preparations of VBC

Because VBC is commercially available, the laboratory syntheses (previously reviewed) (2) are less useful today, except for the preparation of pure isomers. o-, m-, or p-VBC can be synthesized in three steps from commercial products (15). The meta isomer of VBC has been prepared from acetophenone by means of a chloromethylation reaction using the carcinogenic chloromethyl methyl ether (16) (Scheme II), but the experimental procedure has not been fully described.

Characterizations of VBC Monomers

VBC has been characterized mainly by spectroscopy to identify the mixture of meta and para (2) and the pure isomers. The vibrations of the double bonds are observed by IR spectroscopy (17) at 1630 cm^{-1} (ortho), 1637 cm^{-1} (meta), and 1631 cm^{-1} (para), and the absorptions of the vinylic hydrogens are in the ranges of 919–987 cm^{-1} (ortho), 907–987 cm^{-1} (meta), and 901–985 cm^{-1} (para). The vibrations of the carbon–chlorine (CH_2Cl) bond for the three isomers or the meta–para mixture are in the range of 1255–1265 cm^{-1} (15). Mass spectroscopy (18) does not allow a distinction between the various isomers of VBC. The main peak at m/z = 117 is due to the benzylic ion ($CH_2=CHC_6H_4CH_2{}^+$). ^1H and ^{13}C NMR spectra of the three isomers of VBC have been studied (15). The main physical properties (1) for the meta–para mixture are boiling point (at 760 mmHg), 229 °C and (at 1 mmHg) 56 °C; refractive index ($n^{25}{}_D$), 1.5702; and density, 1.083 g/mL at 20 °C.

Substitution of Chlorine on VBC

Because of the benzylic chlorine group, nucleophilic reactions of VBC are easy and leave the double bond undamaged when the experimental conditions are well chosen. A solvent for the reaction is usually necessary, and the duration of the reaction must be relatively short. An inhibitor of polymerization (substituted phenols or aromatic amines) must be used. The resulting monomers can generally be polymerized or

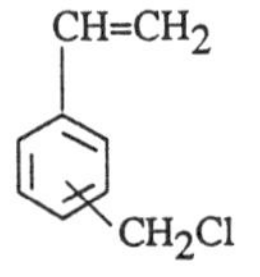

Vinylbenzylchloride (VBC)   Poly(vinylbenzylchloride)

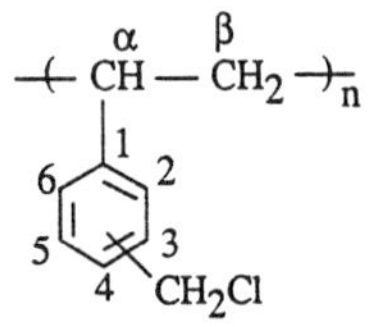

Mixture of isomers   P(VBC)

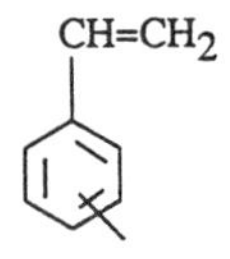

Styryl residue 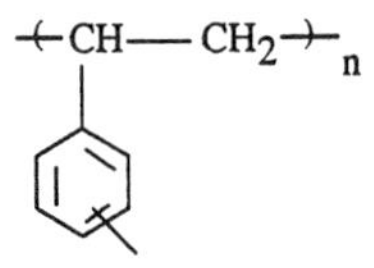   Polystyryl residue

Chart I. Chemical structures and commonly used abbreviations for VBC and VBC derivatives and polymers.

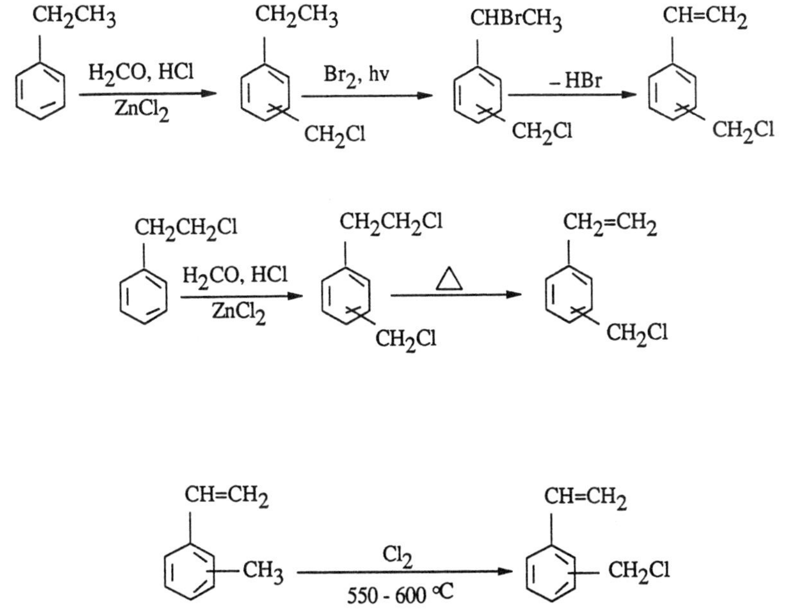

Scheme I. Industrial processes for preparation of VBC.

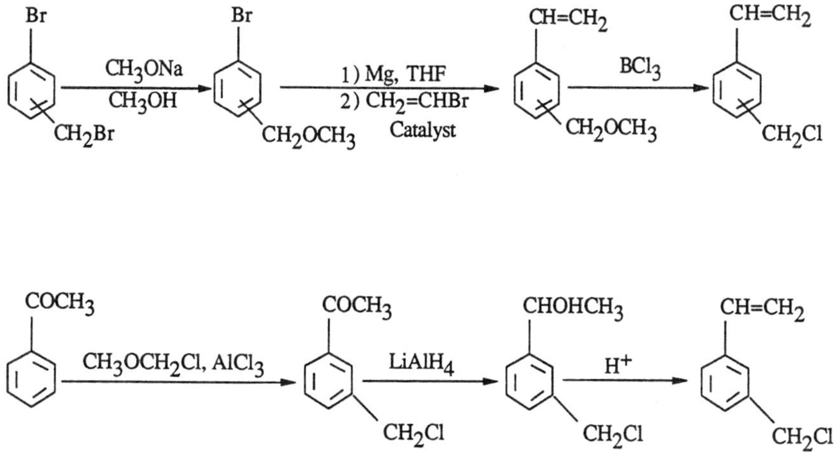

Scheme II. Two laboratory preparations of VBC.

copolymerized even when a bulky group has been attached to the benzylic carbon of VBC. Nucleophilic reactions can also be carried out with bifunctional compounds to give new monomers with two styrenic units (Scheme III).

The main nucleophilic substitution reactions of VBC can be divided as the formations of the following:

- ether functions
- ester functions
- amines and ammonium salts
- thioethers or thioesters
- carbon–carbon bonds by alkylation
- carbon–phosphorus bonds or phosphonium salts
- carbon–silicon bonds

Since the beginning of the work on VBC, several hundred substitutions have been described for this monomer. Therefore, they cannot all be reviewed in this chapter. However, very similar methods have been used for these substitutions.

The ether, ester, thioethers, and thioesters are, respectively, prepared by reaction of alcohols or phenols, acids, thiols, and thioacids in basic medium or with their corresponding metallized (Na or K) derivatives. Ether functions can also be prepared by a phase-transfer catalyzed reaction of alcohol with VBC.

Amines and ammonium salts and phosphonates and phosphonium salts are, respectively, synthesized by reaction of amines (primary, secondary, or tertiary) and trialkyl phosphine. The experimental conditions

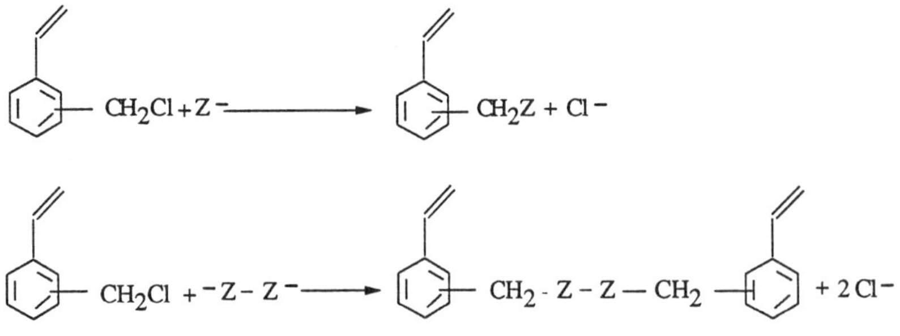

Scheme III. General schemes for nucleophilic substitution of VBC with mono- and bifunctional nucleophiles.

for the phosphorus derivatives generally need heating at 100–120 °C for 24–48 h.

The alkylation reaction is usually done with carbanions of organic compounds bearing withdrawing groups (e.g., diketones or ketoesters) or, for some examples, by reaction of the Grignard compound of VBC with dihalogenoalkanes.

Table I gives the most significant examples of nucleophilic transformations of VBC with references that provide experimental details. A detailed procedure for nucleophilic substitution using a phase-transfer catalysis etherification (19) of a perfluorinated alcohol with p-VBC is given as an example.

$1H,1H,2H,2H$-Perfluorohexanol [$CF_3(CF_2)_3CH_2CH_2$-OH; 80 mmols] is mixed with 160 mL of 50% aqueous NaOH. Then, 160 mL of methylene chloride and 8 mmoles of tetrabutylammonium hydrogen sulfate are added, and the resulting suspension is vigorously stirred. After addition of 88 mmoles of p-VBC, the reaction mixture turns brilliant yellow. After stirring

for 18 h at 40 °C, the organic layer is separated, washed once with diluted HCl and three times with water, and dried over sodium sulfate. A brown oily liquid is obtained after filtration and evaporation of the solvent. The para-substituted VBC is purified by distillation under high vacuum (yield, 76%). Because the reaction of secondary amines with VBC is very useful, a simple procedure for this reaction is also given with the pure para isomer of VBC (27).

Seventy grams of p-VBC (0.46 mol) are added to 67 g of diethylamine (0.92 mol) in 100 mL of methanol, and the resulting mixture is refluxed for 2 h. The solvent is removed by evaporation, and 100 mL of 6 M HCl is added to the residue, which is subsequently extracted three times with diethyl ether to remove any unreacted p-VBC. Fifty grams of sodium hydroxide dissolved in 50 mL of water is then carefully added to the aqueous layer, which is further extracted with diethyl ether. The ether extracts are combined and dried over magnesium sulfate, and the solvent is evaporated. Seventy-three grams of (4-vinylbenzyl)-diethylamine are obtained (yield, 84%).

Synthesis and Characterization of VBC Polymers

Because of the benzylic chlorine, VBC can be polymerized with radical and cationic initiators but not with anionic initiators. The cationic polymerization uses classical initiators such as Lewis acids or tertiary amines. These amines react with the benzylic chlorine and give a benzylic ammonium ion, which acts as an initiator of the cationic reaction. These initiators provide cross-linked products (2).

The most important polymerization reactions of VBC have been carried out by free-radical initiators. VBC can be polymerized in bulk, in solution, in suspension, or in emulsion. Each of these methods may be used successfully, depending on the requirements of the intended application. Suspension polymerization has been studied for application to the preparation of polymer support (35). Emulsion polymerization of VBC is the best process to have high molecular

Table I. Nucleophilic Substitutions of VBC with References Providing Experimental Details (R = Various Alkyl, St = Styryl, X = Halogen)

Nucleophilic Reagent	Product	Ref.
$CF_3(CF_2)_3CH_2CH_2OH$	St–$CH_2OCH_2CH_2(CF_2)_3CF_3$	19
$H(OCH_2CH_2)_nR$	St–$CH_2(OCH_2CH_2)_nR$	20
OH-crown ether	St–CH_2O crown ether	21
HO-C_6H_4-X	St–$CH_2OC_6H_4$-X	22
3-Methyl-3-hydroxy-methyloxetane	St–CH_2-O methyloxetane group	23
KOH	St–CH_2O-CH_2O-St	24
$HOOCCH_3$	St–CH_2OCOCH_3	25
$KOOCC(CH_3)=CH_2$	St–$CH_2OCOC(CH_3)=CH_2$	26
$HOOCC_6H_4R$	St–$CH_2OCOC_6H_4R$	22
$(C_2H_5)_2NH$	St–$CH_2N(C_2H_5)_2$	27
R_3N	St–$CH_2N^+(R)_3Cl^-$	28
CH_3SNa	St–CH_2-S-CH_3	29
$(C_2H_5)_2N$-CSS-Na	St–$CH_2SCSN(C_2H_5)_2$	30
$CH_2(COOC_2H_5)_2$	St–$CH_2CH(COOC_2H_5)_2$	31
$Br(CH_2)_5Br$	St–$(CH_2)_6Br$	32
PR_3	St–$CH_2P^+R_3$, Cl^-	33
$(CH_3)_3SiCl$	St–$CH_2Si(CH_3)_3$	34

weights, high rates of polymerization, and a good conversion in a short time (*1*). Microspheres of P(VBC) have been prepared by dispersion polymerization (*36*).

Pure isomers or *o-*, *m-*, or *p*-VBC have been polymerized in bulk (*37*). When a commercial mixture of meta and para isomers has been used, the resulting polymer is a copolymer of *m-* and *p*-VBC; but as the reactivities of the two isomers of VBC seem to be equal, the initial proportions are maintained in the resulting P(VBC) (*1*). The heat of polymerization of VBC (mixture of meta and para) is –64.37 kJ/mol at 195 °C, and the glass transition temperature (T_g) of the polymer is 82 °C (*1*).

P(VBC) has been characterized by means of ^{1}H and ^{13}C NMR spectroscopy (Table II). ^{13}C NMR spectra of cross-linked P(VBC) resulting from copolymerization of VBC with DVB (*38*) and spectra resulting from the polymerization of pure meta or para isomers (*37*) are given (Figures 1 and 2, respectively). Studies of the thermal stability of P(VBC) show that these polymers are stable up to 250–300 °C (*39*). The values of T_g for the poly(*o*-VBC), poly(*m*-VBC), and poly(*p*-VBC) are, respectively, 134 °C, 48 °C, and 100 °C (*37*).

A simple example of VBC polymerization in bulk without initiator is described (*1*). VBC is distilled to remove the inhibitors. A small amount of the monomer is placed in a glass tube, and nitrogen is allowed to bubble through the liquid for 2–3 min. After purging with nitrogen, the sample is placed under vacuum and sealed. The tube is placed in a controlled-temperature water bath at 80 °C until the monomer has solidified, after which it is transferred to an oven for 24 h at 115 °C. If it is desirable to remove the volatiles, the polymer can be dissolved in methyl ethyl ketone (5% solution) and reprecipitated with methanol. The polymer is then filtered and dried.

Copolymers of VBC with Other Monomers

VBC (commercial mixture of meta and para isomers or pure para isomer) has been copolymerized with a great number of comonomers chiefly in the presence of radical initiators. A listing of the copolymerizations until 1982 was previously given (*2*). Numerous new copolymerizations have been described since 1982. The new comonomers used are α-acetoxystyrene (*40*), 4-acetoxymethylstyrene (*25*), allylmaleimide (*41*), 3-vinylbenzylacetylacetone (*42*), ethyleneglycol dimethacrylate (*43*), *p*-trimethylsilane styrene (*44*), isoprene (*45*), *p*-methoxystyrene (*46*), *p*-pentamethyldisilylstyrene (*47*), 2-(2-propenylpentafluoro)triphosphazene (*48*), tetramethyl germanium methacrylate (*49*), 2-vinyl-4,4-dimethyl-5-oxazolone (*50*), and *N*-pyrrolylalkylstyrene (*51*). All these references do not give

Table II. Chemical Shifts in ^{13}C NMR Spectra of Polymers of *o-*, *m-*, and *p*-VBC

Carbon	Poly(o-VBC)	Poly(m-VBC)	Poly(p-VBC)
α	34.6	40.5	40.5
β	37–46	41–47	40.5–47
CH$_2$Cl	44	46.4	46.2
1	142–145	144–147	145.5–147
2	135.2	128	127.8
3	130.5–131	137.2	128.4
4	126.6	126	135
5	126.6	128.6	127.8
6	129.2	127.5	128.4

Note: See Chart I for carbon designations.

Source: Reproduced with permission from reference 37. Copyright 1987 Hüthig-Hepf Verlagwepf.

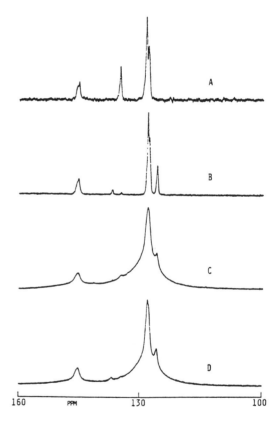

Figure 1. ^{13}C NMR spectra of the aromatic carbons of soluble poly(*p*-VBC) (A), poly(styrene-*co*-VBC) (25 wt% VBC) (B), 1 wt% DVB-cross-linked polystyrene chloromethylated to 1.6 mequiv/g (C), and 0.97 wt% DVB-cross-linked poly-(styrene-*co*-VBC) (1.6 mequiv/g VBC) (D).

details of the reactivity ratios. Table III gives a short listing of copolymerizations in which the reactivity ratios have been measured either for the meta–para mixture or for the pure para isomer. Because of the recent syntheses of pure isomers (ortho and meta) of VBC (*15, 16*), data about their copolymerization reac-

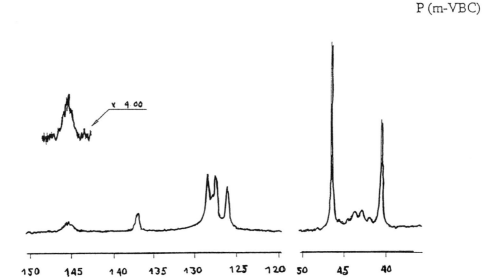

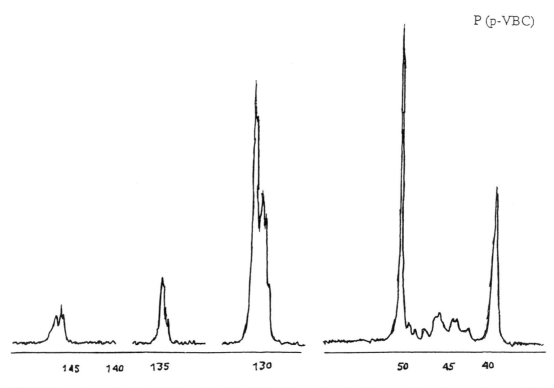

Figure 2. ^{13}C NMR spectra of poly(m-VBC) and poly(p-VBC). (Reproduced with permission from reference 37. Copyright 1987 Hüthig-Wepf Verlag.)

tions (reactivity ratios and Alfrey–Price parameters) are not yet available.

Various values of the Alfrey–Price parameters Q and e have been given for VBC. According to the first copolymerization report (1), the calculated values of Q and e are, respectively, 1.06 and –0.45 for the meta–para mixture. Similar results have been obtained for

the copolymerization of pure para isomer of VBC with styrene or methyl methacrylate (53) ($Q = 1.13$ or 1.16; $e = -0.58$ or –0.69). Therefore, the reactivities of the two isomers, meta and para, seem to be similar. But other authors have found different values for the copolymerization of VBC with DVB (54, 55) ($Q = 0.88$ and $e = -0.20$).

Table III. Copolymerization of VBC (M_1) with Various Comonomers (M_2) and Their Reactivity Ratios (r_1 and r_2, Respectively)

M_2	r_1	r_2	Ref.
Copolymerization with meta–para VBC			
Vinyl acetate	36.8	0.02	1
Ethyl acrylate	1.51	0.42	1
2-Hydroxyethyl acrylate	0.754	0.372	1
2-Hydroxypropyl acrylate	0.712	0.334	1
Maleic anhydride	1.08	0.02	2
2-Sulfoethyl methacrylate	0.47	0.70	1
Styrene	1.08	0.72	1
p-Methylstyrene	0.62	0.65	52
2-(2-Propenyl)pentafluorocyclo-triphosphazene	1.59	0.015	48
Copolymerization with p-VBC			
Styrene	1.37	0.72	53
α-Acetoxystyrene	1.33	0.67	40
p-Divinylbenzene	0.68	1.47	54
m-Divinylbenzene	1.20	0.83	54

A procedure for emulsion copolymerization of VBC with methyl acrylate described by Dow Chemical (*1*) is described. Emulsion copolymerization provides high-molecular-weight linear or lightly cross-linked copolymers of VBC. VBC (60 g), methyl acrylate (20 g), and varying percentages of DVB are weighed into clean bottles containing deionized water (175 mL), 20 mL of aqueous 20% sodium lauryl sulfate, 9.6 mL of aqueous 5% sodium hydrogen carbonate, and 9.6 mL of aqueous 5% potassium persulfate. The mixture is cooled in an ice bath for 1 h, and then 6.8 mL of aqueous 5% sodium metabisulfite is added. The bottle contents are purged with prepurified nitrogen for 20 min in an ice bath, and the bottle is sealed. The bottle is then agitated at 122 rpm for 18 h, and the polymer is recovered. However, the yield of the polymer is not given in reference 1.

Chemical modifications of cross-linked P(VBC) and related copolymers need the best accessibility of the reactants to the CH_2Cl groups (and its derivatives), especially for the polymers prepared by suspension to obtain beads or microspheres (*56, 57*). Gel-type or macroporous polymers are synthesized so as to obtain an insoluble product with a good accessibility to the CH_2Cl groups. The comonomers are often styrene and DVB, and the cross-linked copolymer is best represented by the structure shown in Chart IIA.

A similar structure can be obtained by means of a chloromethylation of the copolymer of styrene and DVB (*4*) but with additional cross-linking (Chart IIB). Therefore, the copolymerization route is preferred. Because copolymerization does not lead to additional

A)

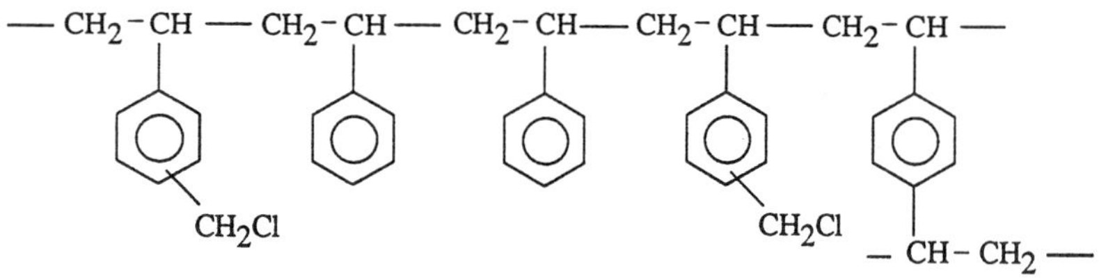

B)

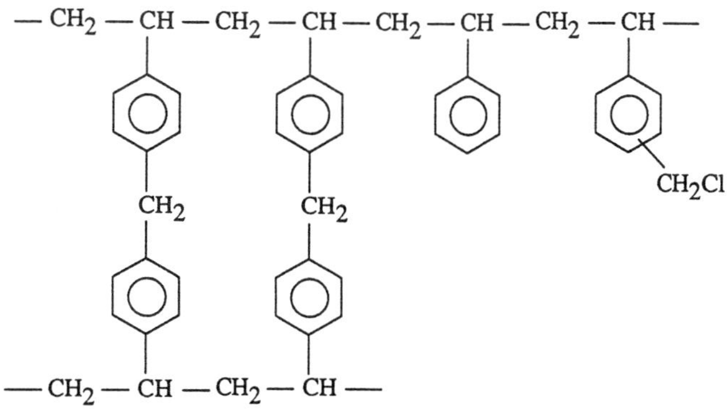

Chart II. Structure and cross-linked copolymer of VBC with styrene (A) and a possible structure obtained by chloromethylation of copoly(styrene-DVB) (B).

cross-linking, the composition of the copolymer can be calculated beforehand. With a low percentage of DVB (1–5%), the copolymer is of a gel type, which when swollen in a hydrophobic solvent allows the fast diffusion of the reactants. In macroporous-type resins, there are permanent pores (Figure 3) (58). Experimental procedures for preparation of the resins have been described (59), and the distribution of the pores in

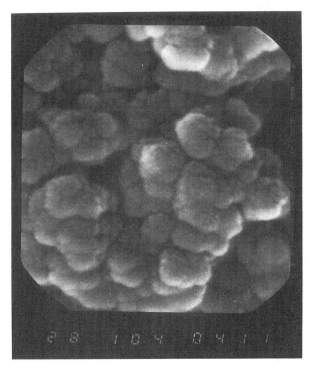

Figure 3. Scanning electron micrograph of internal structure of a poly(styrene–DVB–VBC) bead. (Reproduced with permission from reference 58. Copyright 1989 Elsevier.)

macroporous resins has been studied by nitrogen absorption–desorption (60) and thermoporometry (61).

Grafted copolymers of VBC are also synthesized by using a special procedure with a microdispersion of water in the styrene–DVB mixture followed by graft copolymerization of VBC (62) (Schemes IV and V). Another procedure uses functional azo initiators on which the VBC can be graft copolymerized (63). Polymerization of styrene in the presence of $Al(C_2H_5)_2Cl$ and a copolymer of isobutylene–VBC gives a grafted copolymer (64).

The microstructure of (VBC–styrene–DVB) terpolymer has been studied by ^{13}C NMR spectroscopy, and the assignment of the chemical shifts of aromatic carbons has been reported (37). By studies of their volatile fractions (pyrolysis–gas chromatography), the terpolymers of (VBC–DVB–styrene) can be distinguished from chloromethylated polystyrenes or from chlorinated poly(4-methylstyrene) (65).

Applications of Polymers and Copolymers of VBC

Because of the numerous chemical reactions of VBC and its (co)polymers, many applications have been found for these materials and their derivatives. But before giving an overview of these applications, we should distinguish between products resulting from chloromethylation of polystyrene with those of P(VBC) obtained by direct (co)polymerization (*see* previous sections). Substitution reactions of chloromethylated polystyrene are numerous and very similar to those of P(VBC) (66), but the cost of chloromethylated polystyrene is relatively low compared with that of

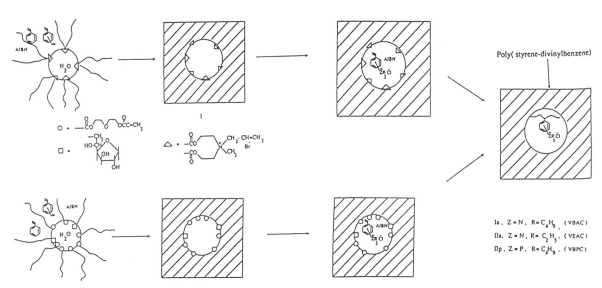

Scheme IV. Preparation of porous polymer-supported poly(quaternary onium salts) by using aqueous surfactants. (Reproduced from reference 62. Copyright 1992 American Chemical Society.)

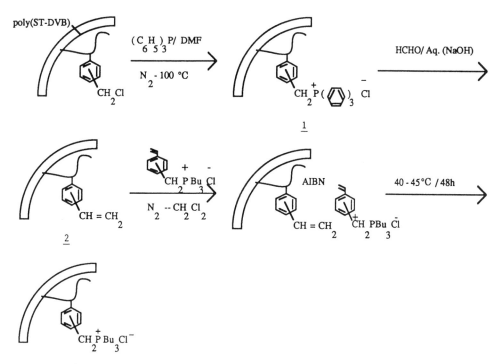

Scheme V. Preparation of porous polymer-supported poly(VBC) by using surface-grafted poly(VBC) as anchors. (Reproduced from reference 62. Copyright 1992 American Chemical Society.)

P(VBC). Therefore, modified polystyrene is used more generally than P(VBC) for applications.

Solid-phase peptide synthesis in which the CH_2Cl groups are used to anchor the growing peptide chains is, for example, based largely on chloromethylated polystyrene (Merrifield resin) rather than on P(VBC). Similarly, ion-exchange resins that are polyelectrolytes (prepared by sulfonation of aromatic rings or by quaternization of the CH_2Cl groups with tertiary amines) or other polymer reagents resulting from modification of the CH_2Cl groups are prepared from the cross-linked polystyrene. Therefore, the application of VBC polymers and copolymers is limited to scientific research in fine chemistry.

A typical procedure for the derivatization of P(VBC) in the presence of a phase-transfer catalyst is described as an example (67). P(VBC) (1.53 g; 10 mmol) is dissolved in 20 mL dimethylformamide, potassium 4-nitrophenoxide (0.36 g; 2 mmol) and tetrabutylammonium bromide are added, and the mixture is stirred at 30 °C for 24 h. The reaction mixture is poured in water, and the obtained polymer is purified by reprecipitating twice from tetrahydrofuran (THF) into water and from THF into methanol. The polymer is finally dried at 50 °C. The degree of substitution is 19.2 mol%, as calculated from chlorine analysis (9).

In addition to previous remarks about applications such as ion-exchange resins and phase-transfer catalysis, a summary of recent work dealing with P(VBC) is provided.

Microlithography

Poly(VBC) and some copolymers of VBC are used in high-resolution microlithography. The homolytic scission of the CH_2Cl group gives a Cl radical. Elimination of the hydrogen radical of the methine group on the polymer chain may also occur. Combination of free radicals gives, via a cross-linking reaction insoluble products (68), in the exposed areas; The developer only dissolves the protected areas. Sensitivity and contrast depend on the average molecular weights and on molecular weight distributions according to Charlesby–Pinner equations (69). The effect of chlorine content in copolymers of VBC and 4-methylstyrene has been studied (70). Copolymerization of VBC with monomers carrying silicium (71) or germanium (49) atoms increases the sensitivity to radiation. A similar phenomenon has been observed with a copolymer of VBC modified by tetrathiofulvalene groups (72). In this case, a charge transfer occurs and transforms the covalent species into ionic species. This charge transfer avoids the use of a slightly polar solvent, which swells the polymer and decreases the contrast (72).

Supported Polyelectrolytes

Polymers and copolymers of VBC are easily modified into polyelectrolytes (Scheme VI). The main modifications and applications of polyelectrolytes of P(VBC) are given in Table IV. Similar products can be synthe-

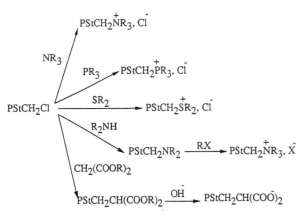

Scheme VI. Preparation of polyelectrolytes from P(VBC); X = halogen.

sized by chemical modification of the VBC monomer followed by polymerization or copolymerization. The main applications of these macromolecules are ion exchange of anions and cations, electron exchange, and selective permeation. When the tertiary amine is bipyridine, viologens (V^{++}) are obtained (Schemes VII and VIII and Figure 4). V^{++} is easily reduced to $V^{+\circ}$, and oxidation with oxygen gives V^{++}. These products have been used for solar energy transfer (90).

These copolymers also were used to study the kinetics of reduction of ferricyanide ions (86) with various membranes (A–E) (Figure 4). These membranes were prepared by the reaction of copolymer of VBC and acrylonitrile (AN) with various proportions of 4,4'-bipyridyl, 1-propyl-4-(4-pyridyl)pyridinium bromide, or diazabicyclo[2.2.2] octane (DABCO).

Table IV. Polyelectrolytes Prepared from (Co)Polymers of VBC

Comonomers	Pendant Groups	Use	Ref.
DVB	$N^+(CH_3)(CH_2)_7CH_3$, Cl^-	Oleophilic membrane for anion exchange	73
DVB	N^+R_3, Cl^-; (R = C_2H_5, C_3H_7, C_8H_{17})	Anion exchange	74
2M5VP–AN–B	2-Methylpyridinium	Anion exchange	75
—	Ammonium group of VP in poly(VP-S)	Anion exchange for Cr/Fe battery	76
St–DVB	$-(CH_2)_nN^+R_3$, Cl^- (n = 5,7; R = alkyl groups)	Anion exchange	77
St–DVB	$P^+(C_4H_9)_3Cl^-$	Catalysis of substitution of bromine by CN ions of $Br(CH_2)_7CH_3$	78, 79
St–DVB	$(CH_2)_nP^+(C_4H_9)$,Cl^-	Catalysis of substitution of bromine by CN ions of $Br(CH_2)_7CH_3$	80
St–DVB	N^+R_3,Cl^- (R = C_4H_9, C_8H_{17})	Catalyst for addition of dichlorocarbene onto cyclohexene	81
St	$N^+(C_2H_5)_3$,Cl^- and $N^+(C_2H_4OH)_3$,Cl^-	Coated electrode to increase the diffusion coefficient	83, 84
St	$N^+(CH_3)_2X=CH_2NH_2$, X = $(CH_2)_3$, N^+CH_3, Cl	Anti-corrosive for electrodes	85
AN; AN–VP	4,4'-Dipyridinium-N-propyldichloride and 1,4-diazabicyclo[2,2,2]octane	Electron-transfer membrane	86
St	$-CH_2R$ (R = phenothiazine group; N-methyl-4,4'-bipyridyl; 2,2'-bipyridyl derivatives	Electron-transfer decreasing of electron diffusion	87, 88
St	4,4'-Dipyridinium-N-hexadecyl	Electroactive emulsifant	89
NVP	4,4'-Bipyridyl derivatives	Bleaching with O_2	90
VMO	$N^+(CH_3)_3$,Cl^-	Opening of VMO cycle and immobilization of proteins	50
EGDMA	$N^+(CH_3)_3$,Cl^-	Protein chromatography	43
St	$N^+(CH_3)_2(CH_2)NCOO^-$, $N^+(CH_3)_2CH_2$-$P(C_6H_5)OO^-$	Betaine synthesis	91
—	$N^+R_1R_2R_3Cl^-$ (R = various alkyl radicals)	Antibacterial agent	28
Grafted VBC onto PP	$P^+R_1R_2R_3Cl^-$ (R = alkyl)	Antibacterial agent	33
—	$P^+(C_6H_5)_3Y$; Y^- = BH_4^-, SCN^-, NO_2^-	Nucleophilic reagents	92
DVB	$N^+(CH_3)_3[Mo_x^{VI}O_y]^n$	Epoxydation catalyst	93

Note: Abbreviations used for comonomers: DVB, divinylbenzene; St, styrene; 2M5VP, 2-methyl-5-vinylpyridine; AN, acrylonitrile; B, butadiene; NVP, N-vinyl-2-pyrrolidone; VMO, 2-vinyl-4,4-dimethyl-5-oxazolone; EGDMA, ethylene glycol dimethacrylate; VP, vinylpyridine; PP, polypropylene; — means the homopolymer of VBC was used.

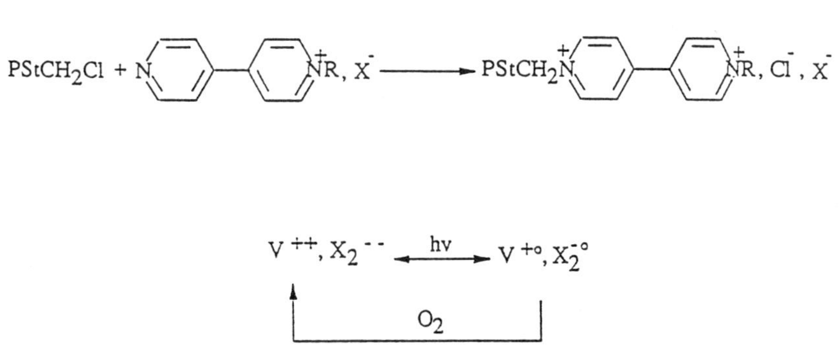

$$PStCH_2Cl + N\!\!\!\bigcirc\!\!\!-\!\!\!\bigcirc\!\!\!\overset{+}{N}R, X^- \longrightarrow PStCH_2\overset{+}{N}\!\!\!\bigcirc\!\!\!-\!\!\!\bigcirc\!\!\!\overset{+}{N}R, Cl^-, X^-$$

$$V^{++}, X_2^{--} \xrightarrow{\;\;hv\;\;} V^{+\circ}, X_2^{-\circ}$$

$$O_2$$

Scheme VII. Preparation of viologen derivative of P(VBC). (Reproduced with permission from reference 90. Copyright 1989 John Wiley and Sons.)

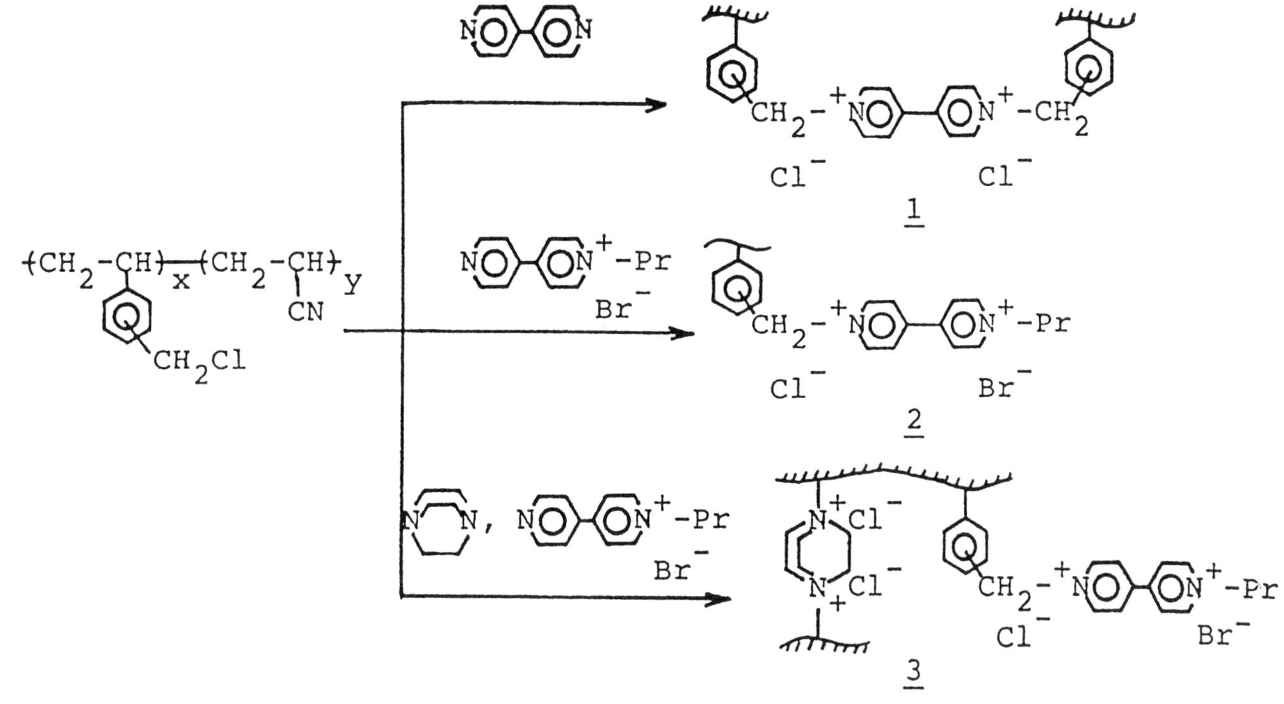

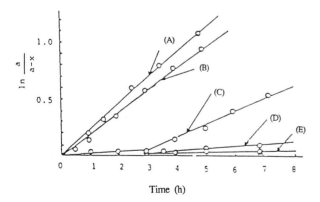

Scheme VIII. Schemes of reduction of ferricyanide with the use of polymeric membranes. (Reproduced from reference 86. Copyright 1983 American Chemical Society.)

Figure 4. Kinetics of reduction of ferricyanide with the use of polymeric membranes. A, 4.8 mol% of 4,4-bipyridyl; B, 5.5 mol% of 1-propyl-4-(4-pyridyl)pyridinium bromide; C, 5.2 mol% of DABCO; D, 5.3 mol% of 1-propyl-4-(4-pyridyl)pyridinium bromide; E, prepared by the reaction of membrane D with 2 mol% of DABCO. (Reproduced from reference 90. Copyright 1989 American Chemical Society.)

The permeability of polyelectrolyte membranes prepared from quaternized VBC has been studied to improve the efficiency of electrodialysis. Polyelectrolytes have been used as phase-transfer catalysts (triphase catalysis: solid catalyst and two liquid phases). Oxidation of CH_2Cl groups by dimethylsulfoxide, potassium dichromate, or potassium permanganate gives carboxylic resins useful for ionic chromatography (*94*).

Chloromethylated polystyrene is very useful for the preparation of polymer-supported reagents and catalysts; however, the activity of immobilized catalysts is lower than that of soluble catalysts. Therefore, the introduction of a spacer arm between the catalyst moiety and the polystyrene backbone has been carried out by modification of VBC (*32*). Subsequent copolymerization of this monomer with styrene and DVB affords a spacer-modified polymer bearing a phosphonium salt. The catalyst activity of these copolymers is increased by a factor of 2.5 compared with the catalyst prepared from chloromethylated polystyrene in the cyanation of bromoalkanes (80). New anion-exchange resins were also prepared by the same procedure, and these resins have better properties (77) compared with the resin synthesized from cross-linked chloromethylated polystyrene.

Complexing Agents

Chemical modifications of P(VBC) and related copolymers with formation of ether functions give products with surface-active and complexing properties. Ether functions carried by P(VBC) are synthesized by nucleophilic substitutions of the chlorine atom with crown-ethers (95), poly(ethylene glycol) (96), or oxetane (23). Sulfoxide groups were grafted onto P(VBC) to prepare phase-transfer catalysts (97). Other complexing and chelating resins are synthesized from P(VBC) according to Scheme IX. The resins from reaction I have been used to extract uranium salts from seawater (98).

The resins from reaction II bind various transition metal salts (99). Resins from reaction III have been used for complexing Cu^{++}, UO_2^{++}, and Ag^+ (100), and resins from reaction IV have been used for rare earth cations (55). Cyclic tetramines have been grafted onto VBC. After polymerization or copolymerization with styrene, the resulting resins have been used for extraction of heavy metals or transition metals (101).

Polymeric Reagents and Catalysts

Polymeric reagents and catalysts are discussed in another chapter, but three examples are given for the sake of reference. Tertiary amines are often used for catalysis of the esterification of alcohols. Catalysis also takes place when the amine function is attached to a styrenic matrix. Such polymers are prepared by reaction of a secondary amine with VBC (102). The com-

plexation of rhodium salts on the polyamine derived from the styrenic matrix provides a catalyst for the elimination of oxygen in N–O bonds (103).

Photosensitizers can produce singlet oxygen $^1O_2$, which starts photooxidation reactions. Photosensitizers such as substituted xanthene (a dye) can be linked to P(VBC) or a copolymer of VBC with, for example, the carboxylic group of the dye. The supported dye that has the highest quantum yield is the "rose bengal" attached onto P(VBC) (trade name, Sensivox 1) (104). Other photosensitizers of xanthene type, halogenated derivatives of succinylfluoresceine (105), and benzophenone derivatives (67) have also been attached to P(VBC).

Similar nucleophilic substitutions of P(VBC) with phthalocyanines carrying phenol function and porphyrines with amine functions give photosensitizers for visible radiation (106).

A copolymer of styrene and VBC reacts with triethylamine and flavin (Poly I); the resulting terpolymer is an oxidant of the 1-benzyl-1,4-dihydronicotinamide (107) (Scheme X). Quaternization of P(VBC) with (S)-(–)-N-(pyridin-3-oyl)-2-aminobutan-1-ol provides the corresponding quaternized polymer, which is reduced with sodium hydrosulfite to give polymer II in Scheme X. This polymer reduces methylbenzoylformate with the same enantiomeric excess as the free reagent (108).

Mercaptoacrylamide ($HS–C(SC_4H_9)=C(CN)CONH_2$) reacts easily with P(VBC) to give $PStCH_2–S–C(SH)=C(CN)CONH_2$. This product is a ketone carrier. It reacts with cyclohexanone to give a complex structure that is decomposed by diluted chlorhydric acid to give cyclohexanone (109).

Pervaporation Membranes

Membranes prepared by copolymerization of VBC and N-phenylmaleimide are more permeable to

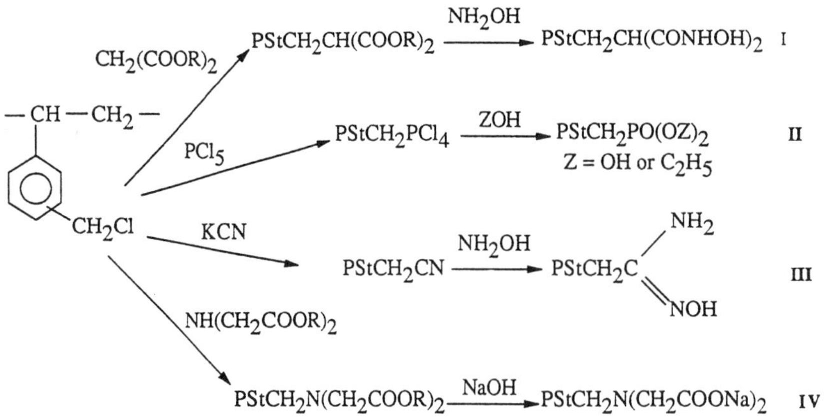

Scheme IX. Preparations of chelating polymers from P(VBC).

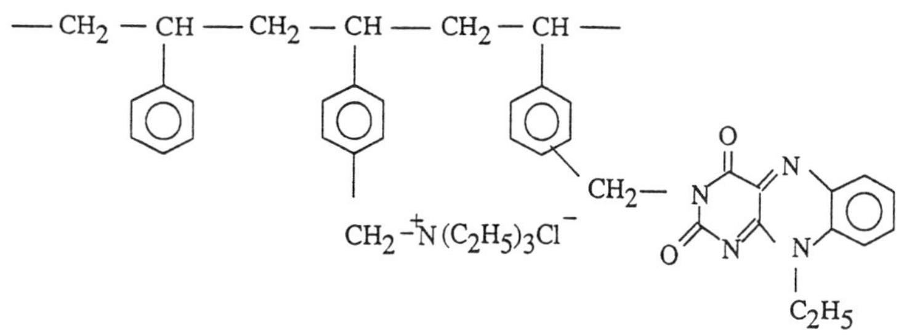

Poly I

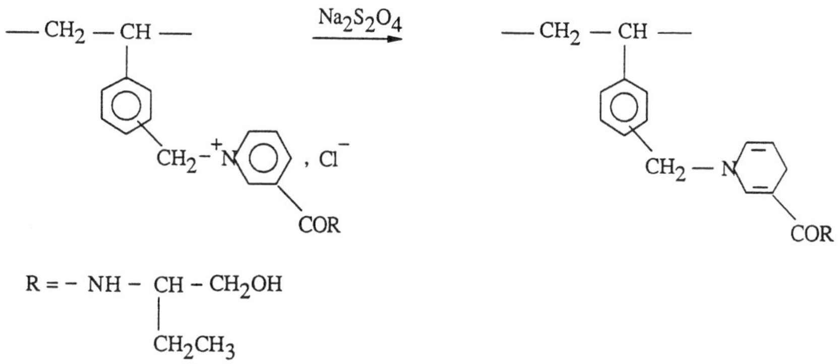

$R = -NH-CH-CH_2OH$
CH_2CH_3

Poly II

Scheme X. Preparations of polymeric reagents from P(VBC).

water than to ethanol (*110*). Membranes of P(VBC) grafted with sulfene complexed by Co(II) salts enable a partial separation of oxygen–nitrogen mixture (*111*).

Other membranes from VBC have been prepared either from ionic derivatives of P(VBC) (applications of P(VBC) as polyelectrolytes) or by polymerization of a VBC derivative carrying a carbon–silicon, carbon–tin, or carbon–germanium bond. These substituted styrenes are prepared by reaction of the Grignard compound of VBC with chlorotrimethylsilane, chloromethyltin, or chlorotrimethylgermanium, and the resulting $StCH_2X(CH_3)_3$ (X = Si, Sn, or Ge) is polymerized with a radical initiator. The corresponding polymers have generally good oxygen-permeation behavior (*34–112*).

Elastomers

Copolymers of VBC, styrene, and isoprene are treated with sodium acetylacetonate then with a palladium(II) salt. The chelation reaction provides a copolymer with

high elastic properties that has been characterized by means of ^1H NMR spectroscopy (*115*).

Optical Lenses

Materials for optical lenses must have a low T_g, they must swell in water to form hydrogels and be insoluble in all solvents. Such properties can be found in copolymers of VBC with alkenes or dienes. Three procedures have been used: VBC reacts with potassium hydroxide potassium sulfide, or with a difunctional nucleophile to give a symmetric diene $StCH_2X–CH_2St$ (X = O, S, OCH_2CH_2O, SC_6H_4S, etc.) (*113, 114*). The reaction of methacrylic derivatives (acid or sodium salt) with VBC monomer gives an unsymmetric diene (*115*). Trialkoxysilanes can also be added to the double bond of VBC, and the silylated derivative is substituted with a methacrylic derivative to give $R_3SiStCH_2OCOC-(CH_3)=CH_2$ (R=$(CH_3O)_3C$) (*116*). These functionalized monomers can be polymerized or copolymerized, and the resulting products have high refractive indexes and good oxygen permeability (*116*).

Optoelectronic Polymers

Homopolymers and copolymers of modified VBC have been widely used for optical fibers and negative photoresists that are photocross-linkable. These polymers are synthesized by nucleophilic substitution of the chlorine atom of VBC by a fluorocarbon radical $(C_nF_{2n+1}CH_2)$ followed by polymerization (117). Optical disk substrate with 89% transmittance is prepared by copolymerization of styrene with a diene VBC derivative, which was synthesized from VBC and bisphenol A (118).

Nonlinear optical (NLO) dyes linked to a double bond and an aromatic nucleus with an electron-withdrawing or -donating group can be polymerized or copolymerized to obtain the thin films that exhibit NLO response. NLO derivatives of VBC were prepared recently by nucleophilic substitution of the chlorine atom with substituted amines, phenols, or benzoic acid followed by polymerization. The resulting products show high second harmonic generation after corona poling (119). A full discussion on NLO polymers is provided in another chapter.

Biomedical Polymers

Polymer latexes prepared from VBC or VBC derivatives have received much attention as carriers for antibodies and antigens for immunoassays. New functionalized styrenes have been prepared by reaction of VBC with glutaric or succinic anhydride or 2-mercaptoethanol and copolymerized with styrene. These new copolymers have been patented as sorbents of proteins (120). The antibacterial activity of copolymers of trialkyvinyl-benzylammonium (or phosphonium) chloride has been well demonstrated (28–33), and a recent article (121) described the use of such a polymer as percutaneous drug absorption enhancer. Modification of VBC with pentahalogenated phenol or pyridine 2-thiol-1-oxide followed by copolymerization with styrene provides products having mildew resistance (122).

Conclusion and Future Prospects

At present there are several thousand articles and patents dealing with the VBC monomer, its polymer and numerous copolymers, as well as the corresponding materials obtained by chloromethylation of polystyrene. This interest relates largely to the importance of chloromethylated polymer supports for solid-phase synthesis, but VBC polymers and copolymers also have been examined for use in the field of microelectronics, nonlinear optics, membranes, and biomedicals products. None of these so far has led to any major product development, but interest in VBC polymers is likely to continue unabated.

References

1. The Dow Chemical Company *Vinylbenzylchloride (VBC)*; Developmental Products, Technical Data, Dow Chemical Company: Midland, MI, 1975.
2. Camps, M.; Chatzopoulos, M.; Monthéard, J. P. *J. Macromol. Sci. Macromol. Rev.* **1982**, C22(3), 343.
3. Ionics Inc. U.S. Patent 2780604, 1957; Clarke, J. T.; Hammerschag, A. H. *Chem. Abstr.* **1957**, 51, 12144.
4. Camps, M.; Chatzopoulos, M.; Camps, J.; Monthéard, J. P. *J. Macromol. Sci. Macromol. Rev.* **1988**, C27, 505.
5. Pinell, R. P.; Khune, G. D.; Khatri, N. A.; Manatt, S. L. *Tetrahedron Lett.* **1984**, 25, 3511.
6. *Synthesis and Separation Using Functional Polymers*; Sherrington, D. C.; Hodge, P., Eds.; John Wiley & Sons: Chichester, Great Britain, 1988.
7. *Functionalized Polymers and Their Applications*; Akelah, A.; Moet, A., Eds.; Chapman and Hall: London, 1990.
8. Gelbard, G. *L'actualité Chim.* **1984**, 7.
9. *Ion Exchangers*; Dorfner, K., Ed.; Walter de Gruyter: Berlin, Germany, 1991.
10. Wichterle, O.; Cerny, J. *Chem. Listy.* **1955**, 49, 1038.
11. Dow Chemical Br. Patent 792859, 1958; Mc Master, E. L.; Stowe, S. C. *Chem. Abstr.* **1959**, 53, 2156.
12. Dow Chemical Br. Patent 792860, 1958; Mc Master, E. L.; Stowe, S. C. *Chem. Abstr.* **1959**, 53, 2156.
13. Hoffenberg, D. S. *Ind. Eng. Chem. Prod. Res. Develop.* **1964**, 3, 113.
14. Mitsubishi Petrochemical Jpn. Patent 62 13 8442, 1987; Inaba, M.; Kataoka, M. *Chem. Abstr.* **1988**, 108, 55639.
15. Monthéard, J. P.; Camps, M.; Chatzopoulos, M.; Pham, Q. T. *Makromol. Chem.* **1985**, 186, 2513.
16. Hatada, K.; Nakanishi, H.; Ute, K.; Kitayama, T. *Polym. J.* **1986**, 18, 581.
17. Nyquist, R. A. *Appl. Spectros.* **1986**, 40, 196.
18. Bobenrieth, M. J.; Delolme, F.; Fraisse, D.; Monthéard, J. P.; Camps, M.; Boinon, B. *Org. Mass. Spectros.* **1989**, 24, 963.
19. Höpken, J.; Möller, M. *Macromolecules* **1992**, 25, 1461.
20. Chao, D.; Itsumo, S.; Ito, K. *Polym. J.* **1991**, 23, 1045.
21. Manecke, G.; Kraemer, A. *Makromol. Chem.* **1981**, 182(11), 3017.
22. Kamogawa, H.; Suguyama, K.; Hanawa, H.; Nanasawa, M. *J. Polym. Sci. Polym. Chem. Ed.* **1976**, 14, 511.
23. Motoi, M.; Nagahara, S.; Akiyama, H.; Horuishi, M.; Kanoh, S. *Polym. J.* **1989**, 21, 1001.
24. Mutiah, J.; Mathias, L. J. *J. Polym. Sci., Part A, Polym. Chem.* **1991**, 30, 509.
25. Amadei, H.; Monthéard, J. P.; Boiteux, G.; Lucas, J. M.; Seytre, G. *Makromol. Chem.* **1985**, 186, 1321.
26. Pugh, C.; Percec, V. *Polym. Bull.* **1985**, 14, 109.
27. Sumi, K.; Anson, F. C. *J. Phys. Chem.* **1986**, 90, 3845.
28. (a) Ikeda, T.; Tazuke, S.; Suzuki, Y. *Makromol. Chem.* **1984**, 185, 869; (b) Senuma, M.; Iwakura, M.; Ebihara, S.; Shimura, Y.; Kaeriyama, K. *Angew. Makromol. Chem.* **1993**, 204, 119.
29. Ogura, K.; Itoh, K.; Isogai, S.; Kondo, S.; Tsuda, K. *J. Macromol. Sci. Chem.* **1982**, A17(9), 1371.
30. Otsu, T.; Yamashita, K.; Tsuda, K. *Macromolecules* **1986**, 19, 287.

31. Endo, T.; Meruoka, S.; Yokozawa, T. *J. Polym. Sci., Part A, Polym. Chem.* **1987**, *25*, 2925.

32. Tomoi, M.; Ogawa, E.; Hosokawa, Y.; Kakiuchi, H. *J. Polym. Sci., Polym. Chem. Ed.* **1982**, *20*, 3015.

33. Kanazawa, A.; Ikeda, T.; Endo, T. *J. Polym. Sci., Part A, Polym. Chem.* **1993**, *31*,1467.

34. Kawakami, Y.; Karasawa, H.; Aoki, T.; Yamamura, Y.; Hisada, H.; Yamashita, Y. *Polym. J.* **1985**, *17*, 1159.

35. Arshady, R.; Ledwith, A. *React. Polym.* **1983**, *1*, 159.

36. Margel, S.; Nou, E.; Fisher, I. *J. Polym. Sci., Part A, Chem. Ed.* **1991**, *29*, 347.

37. Monthéard, J. P.; Camps, M.; Boinon, B.; Ainad-Tabet, D.; Pham, Q. T. *Makromol. Chem.* **1987**, *188*, 2417.

38. Mohanraj, S.; Ford, W. T. *Macromolecules* **1985**, *18*, 351.

39. Boinon, B.; Ainad-Tabet, D.; Monthéard, J. P. (a) *Polym. Degrad. Stab.* **1990**, *28*, 197; (b) *J. Anal. Appl. Pyrol.* **1988**, *13*, 171.

40. Monthéard, J. P.; Camps, M.; Kawaye, S.; Pham, Q. T.; Seytre, G. *Makromol. Chem.* **1982**, *183*, 1191.

41. Turner, S. R.; Anderson, C. C.; Kolterman, K. M. *J. Polym. Sci., Part C, Polym. Lett.* **1989**, *27*, 253.

42. Yeh, H. C.; Eichinger, B. E.; Andersen, N. H. *J. Polym. Sci., Polym. Chem. Ed.* **1982**, *20*, 2575.

43. Etoh, T.; Miyazaki, M.; Iwatake, M.; Harada, K.; Nakayama, N.; Sugii, A. *J. Appl. Polym. Sci.* **1992**, *46*, 517.

44. Matsui, S.; Mori, K.; Saigo, K.; Shiokawa, T.; Toyoda, K.; Namba, S. *J. Vac. Sci. Technol.* **1986**, *B 4*, 845.

45. Iwamoto, N.; Eichinger, B. E.; Andersen, N. H. *Rubber Chem. Technol.* **1984**, *57*, 944.

46. Tunigaki, K.; Suzuki, M.; Ohnishi, Y. *J. Electrochem. Soc.* **1986**, *133*, 979.

47. Saigo, K. *J. Polym. Sci., Part A, Polym. Chem.* **1989**, *27*, 2203.

48. DuPont, J. G.; Allen, C. W. *Macromolecules* **1979**, *12*, 169.

49. Mixon, D. A.; Novembre, A. E. *J. Vac. Sci. Technol.* **1989**, *B7*, 1723.

50. Fazio, R. C.; Taylor, L. D. *Polym. Bull.* **1989**, *22*, 449.

51. Bao, Y. T.; Loeppky, R. N. *Chem. Res. Toxicol.* **1991**, *4*, 382.

52. Affrosman, S.; Bakshaee, M.; Branley, D.; Chow, F.; Dix, C.; Hendy, P.; Jones, M.; Ledwith, A.; Mills, M.; Tate, P. M.; Pethric, R. A. *Polym. Int.* **1992**, *28*, 209.

53. Kondo, S.; Ohtsuka, T.; Ogura, K.; Tsuda, T. *J. Macromol. Sci. Chem.* **1979**, *A13*, 767.

54. He, B. L.; Fang, W. K. *Chin. Sci. Bull.* **1989**, *34*, 925, *Chem. Abstr.* **1990**, *112*, 199–169.

55. Takeda, K.; Akiyama, M.; Yamamizu, T. *React. Polym.* **1985**, *4*, 11.

56. Guyot, A.; Bartholin, M. *Prog. Polym. Sci.* **1982**, *8*, 277.

57. Arshady, R. *Adv. Mater.* **1991**, *3*, 182.

58. Revillon, A.; Guyot, A.; Yuan, Q.; da Prato, P. *React. Polym.* **1989**, *10*, 11.

59. Balakrishnan, T.; Lee, J.; Ford, W.T. In *Macromolecular Syntheses*; Stille, J. K., Ed.; Robert E. Krieger: Malabar, Florida, 1990; Vol 10, p 19.

60. Barrett, E. P.; Joyner, L. G.; Halenda, P. H. *J. Am. Chem. Soc.* **1951**, *73*, 373.

61. Brun, M.; Quinso, J. F.; Spitz, R.; Bartholin, M. *Makromol. Chem.* **1982**, *183*, 62.

62. Ruckenstein, E.; Hong, L. *Chem. Mater.* **1992**, *4*, 122.

63. Arai, K.; Ogiwara, Y. *J. Appl. Polym. Sci.* **1988**, *36*, 1651.

64. Taxi, M.; Tardi, M.; Polton, A.; Sigwalt, P. *Br. Polym. J.* **1987**, *19*, 369

65. Nakagawa, H.; Tsuge, S.; Mohanraj, S.; Ford, W. T. *Macromolecules* **1988**, *21*, 930.

66. Monthéard, J. P.; Chatzopoulos, M.; Camps, M. *J. Macromol. Sci., Macromol. Rev.* **1988**, *C 28*, 503.

67. Nishikubo, T.; Kondo, T.; Inomata, K. *Macromolecules* **1989**, *22*, 3827.

68. Jones, R. G.; Matsubayashi, Y. *Eur. Polym. J.* **1989**, *25*, 701.

69. Charlesby, A. *Proc. Roy. Soc. A.* **1954**, *222*, 542.

70. Ledwith, A.; Mills, M.; Hendy, P.; Brown, A.; Clements, S.; Moody, R. *J. Vac. Sci. Technol.* **1985**, *B 3*, 339.

71. Jurek, M. J.; Novembre, A. E.; Heyward, I. P.; Gooden, R.; Reichmanis, E. *Polym. Prepr. (Am. Chem. Soc. Div. Polym. Chem.)* **1988**, *29*, 546.

72. Mora, H.; Fabre, J. M.; Giral, L.; Montginoul, C.; Sagnes, R.; Schué, F. (a) *Makromol. Chem.* **1992**, *193*, 1337. (b) *Angew. Macromol. Chem.* **1992**, *197*, 89.

73. Imato, T.; Ogawa, H.; Morooka, S.; Kato, Y. *J. Membr. Sci.* **1983**, 53.

74. Nakanishi, T.; Komiyama, J.; Ijima, T. *J. Membr. Sci.* **1986**, *26*, 263.

75. Kawahara, T.; Ihara, H.; Mizutani, Y. *J. Appl. Polym. Sci.* **1987**, *33*, 1343.

76. Reiner, A.; Ledjeff, K. *J. Membr. Sci.* **1988**, *36*, 535.

77. Mitsubishi Kasei Corp Eur. Patent 0444643 A2 (1990) invs.; Tomoi, M.; Kiyokama, A. *Chem. Abstr.* **1992**, *116*, 42785.

78. Ford, W. T.; Lee, J.; Tomoi, M. *Am. Polym. Prepr. (Chem. Soc. Div. Polym. Chem.)* **1982**, *23*, 183.

79. Bernard, M.; Ford, W. T.; Taylor, T. W. *Macromolecules* **1984**, *17*, 1812.

80. Tomoi, M.; Ogawa, E.; Hosokama, Y.; Kakiuchi, H. *J. Polym. Sci. Polym. Chem. Ed.* **1982**, *20*, 3421.

81. Maeda, Y.; Taniguchi, H.; Uchida, M. *Chem. Express* **1988**, *3*, 691.

82. Wang, M. L.; Wu, H. S. *Ind. Eng. Chem. Res.* **1992**, *31*, 490.

83. Montgomery, D. D.; Shigehara, K.; Tsuchida, E.; Anson, F. *J. Am. Chem. Soc.* **1984**, *106*, 7991.

84. Montgomery, D. D.; Anson, F. C. *J. Am. Chem. Soc.* **1985**, *107*, 3431.

85. Bolts, J. M.; Mariella Jr., R. P. *J. Appl. Polym. Sci.* **1990**, *39*, 785.

86. Ageishi, K.; Endo, T.; Okawara, M. *Macromolecules* **1983**, *16*, 884.

87. Olmsted, J.; Mc Clanahan, S. F.; Danielson, E.; Younathan, J. N.; Meyer, T. J. *J. Am. Chem. Soc.* **1987**, *109*, 3297.

88. Younathan, J. N.; Mc Clanahan, S. F.; Meyer, T. J. *Macromolecules* **1989**, *22*, 1048.

89. Tsou, Y. M.; Liu, H. Y.; Bard, A. J. *J. Electrochem. Soc.* **1988**, *135*, 1669.

90. (a) Kamogawa, H.; Kikushima, K.; Nanasawa, M. *J. Polym. Sci., Part A, Polym. Chem.* **1989**, *27*, 393. (b) Kitamura, N.; Nambu, Y; Endo, T. *J. Polym. Sci. Part. A, Polym. Chem.* **1988**, *26*, 993.

91. Hamaide, T.; Germanaud, L.; Le Perchec, P. *Makromol. Chem.* **1985**, *187*, 93.

92. Hassanein, M.; Akelah, A.; Selim, A.; El Amshary, H. *Indian J. Chem. Sect.* **1990**, *29 B*, 763.

93. Srinivasan, S.; Ford, W. T. *Polym. Mater. Sci. Eng.* **1991**, *64*, 355.

94. Yang, Y. B.; Nevejans, F.; Verzele, M. *Chromatographia* **1985**, *20*, 735.

95. Tomoi, M.; Yanai, N.; Shiiki, S.; Kakiuchi, H. *J. Polym. Sci., Polym. Chem. Ed.* **1984**, *22*, 911.

96. Shan, Y.; Kang, R. H.; Li, W. *Ind. Eng. Chem. Res.* **1989**, *28*, 1289

97. Yaacoub, E.; Le Perchec, P. *React. Polym.* **1988**, *8*, 285.

98. Hirotsu, T.; Katoh, S.; Sugasaka, K.; Ichimura, S. K.; Sudo, Y.; Fujishima, M; Abe, Y; Misonoo, T. *J. Polym. Sci. Polym. Chem. Ed* **1986**, *24*, 1953.

99. Alexandratos, S. D.; Quillon, D. R. *React. Polym.* **1990**, *13*, 255.

100. Egawa, H.; Nonaka, T.; Tsukamoto, H. *Polym. J.* **1991**, *23*, 1037.

101. CNRS Eur. Patent 0287 436 A1, 1988, invs.; Handel, H.; Chaumeil, H.; *Chem. Abstr.* **1989**, *111*, 116–254.

102. Tomoi, M.; Gioto, M.; Kakiuchi, H. *J. Polym. Sci. Part A, Polym. Chem.* **1987**, *25*, 77.

103. Kameda, K.; Takemoto, T.; Imanaka, T. *Chem. Lett.* **1989**, 1759.

104. Paczkowski, J.; Neckers, D. C. *Macromolecules* **1985**, *18*, 1245.

105. Hargreaves, J. S.; Webber, S. E. *Can. J. Chem.* **1985**, *63*, 1320.

106. Wöhrle, D.; Paliuras, M. *Makromol. Chem.* **1991**, *192*, 819

107. Bootsma, J. P. C.; Rupert, L. A. M.; Ghalla, G. *J. Polym. Sci., Polym. Chem. Ed.* **1984**, *22*, 2169.

108. Losset, D.; Dupras, G.; Bourguignon, J.; Queguinier, G. *Polym. Bull.* **1989**, *21*, 649.

109. Yokoyama, M.; Ikeda, K.; Sugiki, T.; Kojima, K. *Sulfur Lett.* **1989**, *9*, 219.

110. Yoshikawa, M.; Yokio, H.; Sanui, K.; Ogata, N. *Polym. J.* **1985**, *17*, 363.

111. Delaney, M. S.; Reddy, D.; Wessling, R. A. *J. Membr. Sci.* **1990**, *49*, 15.

112. Kawakami, Y.; Hisada, H.; Yamashita, Y. *J. Polym. Sci., Part A, Polym. Chem.* **1988**, *26*, 1307.

113. Seiko Epson Jpn. Patent, 1988, invs.; Murata, T.; Koinyma, V.; Sano, V.; Mogami, T. *Chem. Abstr.* **1988**, *109*, 56205.

114. Mitsubishi Gas Chemical Co. Jpn. Patent, 1989, Invs.; Kawaki, T.; Kobayashi, M., Aoki, O.; Iwao, S.; Iwai, T., Shintani, N.; Yamazaki, Y. *Chem. Abstr.* **1990**, *112*, 200–208

115. Menicon Co. Ltd Eur. Patent 0381005 A2, 1990, invs.; Yanagawa, H.; Kamiya, N. *Chem. Abstr.* **1991**, *115*, 57227.

116. Alcon. Lab. U.S. Patent, 1987, invs.; Park, J.; Falcetta, J. T.; *Chem. Abstr.* **1987**, *106*, 201796

117. Sagami Chemical Research Jpn. Patent 01009207 A2, 1989, invs.; Matsui, K.; Ishihara, K.: Kogine, R. *Chem. Abstr.* **1989**, *111*, 31353.

118. Tokuyama Soda Co. Jpn. Patent 01042447 A2, 1989, invs.; Ueda, M; Kanehira, N; Matsumoto, Y. *Chem. Abstr.* **1989**, *111*, 215426.

119. Hayashi, A.; Goto, Y.; Nakayama, M.; Kalysynski, K.; Sato, H.; Kondo, K.; Wanatabe, T.; Miyata, S. *Chem. Mater.* **1992**, *4*, 555.

120. Eastman Kodak Eur. Patent 0466220 A2., 1992, invs.; Sutton, R. C. *Chem. Abstr.* **1992**, *116*, 195089.

121. Aoyagi, T.; Terashima, O.; Suzuki, N.; Matsui, K.; Nagase, Y. *J. Controlled Release* **1990**, *13*, 63.

122. Toyo Boseki Kabushiki Kaisha U.S. Patent, 1990, invs.; Mitamura, H.; Arimutsu, Y. *Chem. Abstr.* **1991**, *114*, 103038.

Functional Polymers via Anionic Polymerization

Akira Hirao

This chapter presents a general review of synthesis of functional polymers by anionic polymerization of the respective functional monomers, and especially on living systems, which represent the most common form of anionic polymerization. Polymerization systems discussed include those of styrene, 1,3-butadiene, isoprene, vinylpyridines, alkyl (meth)acrylates, (meth)acrylonitrile, acrylamides, and the functional derivatives of all of these compounds. Recent advances in the study of living anionic polymerization, initiator systems, and new functional monomers are outlined, and the importance of protecting groups in synthesis of functional polymers by living anionic polymerization is emphasized.

Vinyl monomers undergoing anionic polymerization are substituted with electron-withdrawing groups. Moreover, conjugation between the vinyl group and the substituent is also an important structural factor for vinyl monomer in anionic polymerization. Typical examples of these monomers are 1,3-butadiene, isoprene, styrene, vinylpyridines, alkyl acrylates and methacrylates, vinyl ketones, acrylonitrile, vinylidene cyanide, nitroethylene, and cyano- and nitro-acrylates. The Q (conjugation) and polarization (e) values of these monomers are listed in Table I (1). Ethylene can be polymerized anionically to produce oligomers with molecular weights of 10^3 daltons (2). Many heterocyclic monomers may also be polymerized under the conditions of anionic polymerization. They include oxiranes, cyclic sulfides, lactones, lactams, cyclosiloxanes, cyclic silanes, and cyclic phosphanes (3).

The choice of anionic initiator is also important and often crucial in polymerization. One must always consider suitable matching of the nucleophilicity of the initiator and the electrophilicity of the monomer to realize the anionic polymerization. Scheme I shows typical matching among initiators and representative vinyl monomers. The initiators can be divided roughly

into four classes by their nucleophilicities (4). Similarly, the monomers are also divided into four classes by their electrophilicities, and their e values are useful for this division (Table I). Good matchings for these classes (both initiator and monomer) are shown by arrows.

The most reactive initiator class consists of alkali metals, their organometallics, and metal–arene complexes. They can in principle polymerize all classes of monomers. The second class of initiators are less nucleophilic than the first one and are too weak to initiate the polymerization of 1,3-butadiene, isoprene, and styrene listed in the first class of monomers. They can initiate all the monomers in the second through fourth classes. The third class of initiators involves the salts of organic alkali metals derived from alcohols, cyclopentadiene, diethyl malonate, and some related compounds (their pK_a values are in the range of 10–20). The monomers in the third and fourth classes are polymerized with these anionic species, but no polymerizations of the first and second monomer classes occur with these initiators. The fourth (the least reactive) initiator class is only effective in the polymerization of the most electrophilic monomers in the fourth class.

Table I. *Q* and *e* Values of Monomers of Interest for Anionic Polymerization

Monomer	Q	e
Isoprene	1.99	−0.55
1,3-Butadiene	1.70	−0.50
Styrene	1.00	−0.80
4-Vinylpyridine	2.47	0.84
Methyl methacrylate	0.78	0.40
Methyl acrylate	0.45	0.64
Acrylonitrile	0.48	1.23
Methyl vinyl ketone	0.66	1.05
α-Cyano acrylate	4.91	0.91
Vinylidene cyanide	14.22	1.92

Source: Data compiled from reference 1.

In practice, the matching of anionic monomers and initiators involves a number of complication and limitations. Side reactions often occur to some extent during the polymerizations of the monomers in the second, third, and fourth classes with use of the most reactive initiators. For example, methyl methacrylate (MMA) undergoes anionic polymerization with organolithium compounds (the first class of initiator) while competing with ester carbonyl attack (*5*). More seriously, polymerizations of the monomers in the third and forth classes with organolithium compounds are generally difficult due to extensive side reactions before initiation and at early stages of the polymerization (*6, 7*). Accordingly, combinations between parallel classes of initiators and monomers, (e.g., first vs. first, second vs. second, etc.) are more

suitable in practice. The matching shown in Scheme I may also be influenced to some extent by temperature, solvents, or additives. Thus, careful matching and control of reaction conditions are required to produce an ideal system where the polymerization proceeds satisfactorily without termination (*8*).

The growing chain ends remain active after consumption of the monomer under the indicated conditions. Such a system was first discovered by Szwarc (*9*) in the anionic polymerization of styrene and was called *living polymerization*. Many advances and developments in the field of anionic polymerization undoubtedly stem from the discovery of living polymers. The nonterminating nature of this system facilitates the controlled synthesis of polymers of desired molecular weights; narrow molecular-weight distributions; block copolymers; specially shaped polymers such as star, comb, and cyclic structures; and end-functionalized polymers (*10*) (Scheme II). Thus, living anionic polymerization is the method of choice to construct a wide variety of polymers with molecular architectures and to design functional polymers.

Many efforts have been made in the search for other anionic living polymerization systems since the first discovery by Szwarc, and a number of monomer types suitable for anionic living systems have been found so far (*10*). The most suitable monomers are conjugated hydrocarbons such as styrene, 1,3-butadiene, and isoprene. Living polymers can also be obtained from a few styrene derivatives substituted with alkyl, aryl, and alkoxy functions (*11–13*). Under careful conditions, 2-vinylpyridine (*14*) and some alkyl methacry-

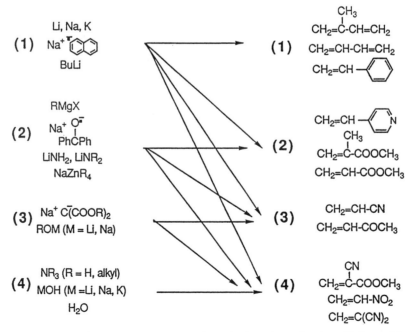

Scheme I. Matching among initiators and monomers for anionic polymerization. (Adapted from reference 4.)

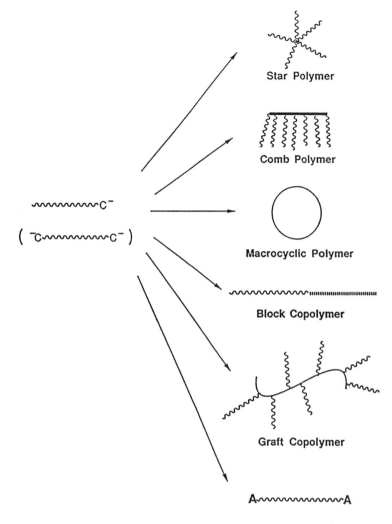

Scheme II. Specially shaped polymers derived from anionic living polymers. (Reproduced with permission from reference 10. Copyright 1983 Academic Press.)

lates (*15*) underwent living anionic polymerization. Although living polymerizations of acrylonitrile and vinyl ketone were reported previously, full details of these systems were not well established.

More recently, a remarkable development was made (*16, 17*) in the living anionic polymerization of styrene derivatives with a wide variety of useful functional groups. New types of living polymers were also produced from some functional 1,3-butadienes and methacrylic esters (*17*). Thus, the range of anionic living polymers available is now broadened to a considerable extent. Since the early 1980s, new living-polymerization systems have been developed for other polymerization processes, although these systems are still regarded as specialties (*18–21*). This chapter presents a general review of anionic polymerization of functional vinyl monomers and focuses on living systems, which are the most characteristic systems of anionic polymerization.

Styrene and 1,3-Butadiene Derivatives

Living Polymerization of Styrene, 1,3-Butadiene, and Isoprene

Styrene, 1,3-butadiene, and isoprene are classified as the least-reactive monomers under the condition of anionic polymerization. Highly reactive initiators (Group 1 in Scheme I) are, therefore, required to polymerize these monomers. These initiators include alkali metals, their organometallics, and complexes of alkali metal and polynuclear aromatic compounds. Grignard reagents can also initiate the polymerization of styrene and 1,3-butadiene, but only in the presence of N,N,N',N',N'',N''-hexamethylphosphoric triamide (HMPT) (*22*). Barium, strontium, and calcium naphthalene anion radicals are also possible initiators for the polymerization of styrene (*23*). Furthermore, styrene can undergo anionic polymerization with less-basic alkali metal amides and the disodium salt of benzophenone; however, low initiator efficiencies and low polymer yields are observed.

The advantage of anionic polymerization of these monomers is to produce living polymers. Although other systems capable of producing living polymers were developed recently, the most well-established living polymerization system from a preparative point of view is currently that of styrene and 1,3-dienic monomers (10). In this method, molecular weight of the polymer can be controlled over the wide range of 10^2 to 10^6 daltons, and the resulting polymers have very narrow molecular-weight distributions. Ratios of weight-average (M_w) to number-average (M_n) molecular weights of less than 1.05 are typically obtained. Moreover, this method offers an advantage by preparing a variety of well-defined macromolecules with molecular architectures that involve star-, comb-, and cyclic-shaped polymers; block and graft copolymers; and end-functionalized polymers including macromonomers (Scheme II).

Homopolymerization of styrene, 1,3-butadiene, and isoprene has not been fully investigated, and new developments are still being reported. The regiospecific polymerization of 1,3-butadiene to afford the polymer with 1,2-vinyl structure (>99 %) was realized by Halasa et al. (24). Fetters et al. (25, 26) demonstrated that quantitative hydrogenation of polydienes obtained by means of the anionic living polymerization can lead to amorphous polyolefins that retain the desirable characteristics of the parent polydienes. The resulting polyolefins with well-controlled chain lengths are used as model polymers to investigate physical properties in bulk and in solutions.

A large number of block and graft copolymers have been synthesized by the anionic method (3, 8, 10, 27) (Table II). Because these copolymers are generally microphase-separated and show mesomorphic phases, they exhibit interesting properties. Block copolymers of styrene with 1,3-butadiene and isoprene are anionically synthesized and are used as thermoplastic elastomers on an industrial scale (28). Recently, an interesting hyperbranched graft-like polymer was synthesized (29) by the repeated reaction of living polystyrene with the chloromethylated polystyrene as shown in Scheme III.

Star- and comb-shaped polymers were synthesized (30) by the reaction of anionic living polymers with multifunctional electrophiles to create macromolecules with 3–18 branches. The use of a living polymer as an initiator for the polymerization of a small amount of divinyl monomer is also interesting for the

Table II. Block Copolymers Prepared by Sequential Addition of Two Different Monomers

Monomer A	Monomer B	Block Copolymer
Styrene	1,3-Butadiene	A–B–A and A–B
Styrene	Isoprene	A–B–A and A–B
1,3-Butadiene	Styrene	A–B–A and A–B
Isoprene	Styrene	A–B–A and A–B
Styrene	2-Vinylpyridine	A–B
Styrene	MMA	A–B
1,3-Butadiene	MMA	A–B
2-Vinylpyridine	MMA	A–B
Styrene	Ethylene oxide	A–B

Source: Reproduced with permission from references 3, 8, and 10. Copyrights 1988, 1967, and 1983, respectively.

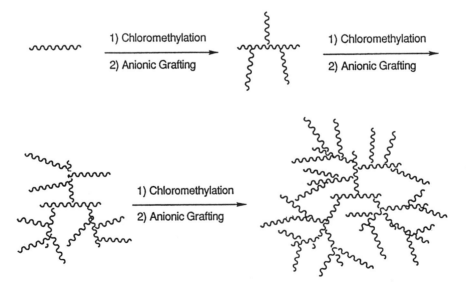

Scheme III. Hyperbranched polymers derived from living polystyrene by repeated chloromethylation and anion grafting. (Reproduced from reference 29. Copyright 1991 American Chemical Society.)

synthesis of star macromolecules, but the number of branches cannot be controlled precisely by this method (*31*).

Cyclic macromolecules can be synthesized by the reaction between a difunctional living polymer and a difunctional electrophile at high dilutions (Scheme II). Yields of cyclic polymers are 20–60%, because the end-to-end cyclization always competes with the chain extension reaction of the original living polymer (*32–34*). The hydrodynamic volumes of cyclic polymers in solutions theoretically are predicted to be much smaller than those of linear polymers as demonstrated by viscosity and gel permeation chromatographic measurements.

The stable but highly reactive carbanionic nature of living polymers allows controlled termination reactions with a wide variety of electrophiles to produce end-functionalized polymers (so-called telechelic polymers and macromonomers) (Scheme II). Telechelic polymers with terminal OH, Cl, or COOH groups are useful industrial materials for synthesis of segmented polyurethanes, polyamides, and polyesters, as well as for chain extension and cross-linking agents (*35*). Macromonomers provide easy access to graft copolymers with other monomers by free-radical initiation. Living polymerization is in principle the best process to regulate the polymer chain length and distribution and to produce polymers with desired functionalities on either one or both chain ends.

Although many of the previous end-functionalization reactions to anionic polymers were not fully characterized, some excellent examples have recently been reported. Polystyrenes, polybutadiene, and polyisoprenes with functional termini such as COOH, OH, SO_3H, NH_2, naphthalene, and pyrene fluorescent groups and vinyl derivatives were synthesized successfully (*36–39*). Thus, anionic living polymerization of styrene and 1,3-dienes provides a powerful tool to produce a variety of designed functional polymers with precise molecular architectures.

Living Polymerization of Functional Styrenes

A major problem of anionic living systems of styrene and 1,3-dienes is the incompatibility of highly reactive carbanions at their growing chain ends with most useful functional groups. Therefore, the production of functional living polymers from styrene and 1,3-dienes by anionic polymerization is difficult. Nakahama and I (*16*) recently proposed to overcome this difficulty by protecting the functional group during anionic polymerization, as outlined in Scheme IV.

The key factor in this method is to find a suitable protecting group that is completely stable under the conditions of living anionic polymerization and is quantitatively removed under mild conditions without polymer degradation or cross-linking. If these two conditions can be realized, a general synthetic strategy would be created for producing new functional polymers from functional monomers by anionic polymerization, and a wide range of living anionic polymers would become accessible.

So far by using this strategy, new polystyrenes with a variety of functional groups (OH, NH_2, SH, CHO, $COCH_3$, COOH, PO_3H, and C≡C) were synthesized successfully (Scheme V) (*40–43*). The resulting polymers have predictable molecular weights of 10^3 to 10^5 daltons and narrow molecular-weight distributions, M_w/M_n values, usually ≤1.1. Some representative results are summarized in Table III. Block copolymers with these polymer segments were also prepared with classical monomers such as styrene, isoprene, and MMA (*16*). The current strategy applied to functional styrene derivatives should also be extended to 1,3-butadiene, 2-vinylpyridine, and alkyl methacrylates.

Vinyl Monomers with Electron-Withdrawing Substituents

Vinylpyridines

Polyvinylpyridines and related polymers with pyridine rings are attractive functional polymers and have a variety of applications (*44*). They are also easily quaternized with alkyl halides and organic and inorganic acids to yield water-soluble polyelectrolytes.

From their *e* values (*see* Table I) and cross-over reactions with other monomers, vinylpyridines are, in anionic polymerization, more reactive than 1,3-dienes and styrene but less reactive than alkyl (meth)acrylates. Soon after the discovery of anionic living polymerizations of styrene (*9*), the living anionic polymerizations of 1,3-butadiene, isoprene, and 2- and

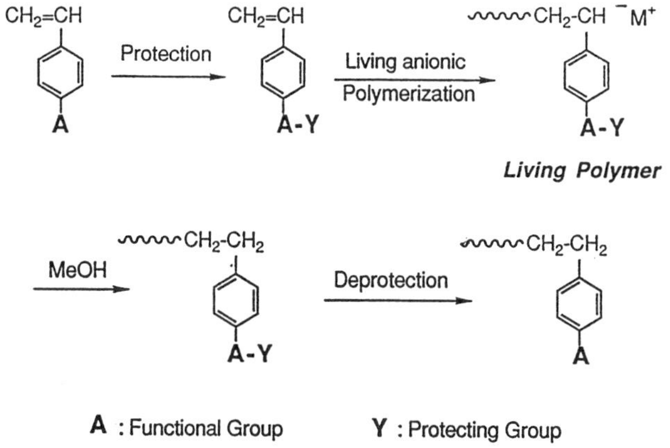

A : Functional Group   **Y** : Protecting Group

Scheme IV. Protection and living anionic polymerization of functional monomer. (Reproduced with permission from reference 16. Copyright 1990 Elsevier.)

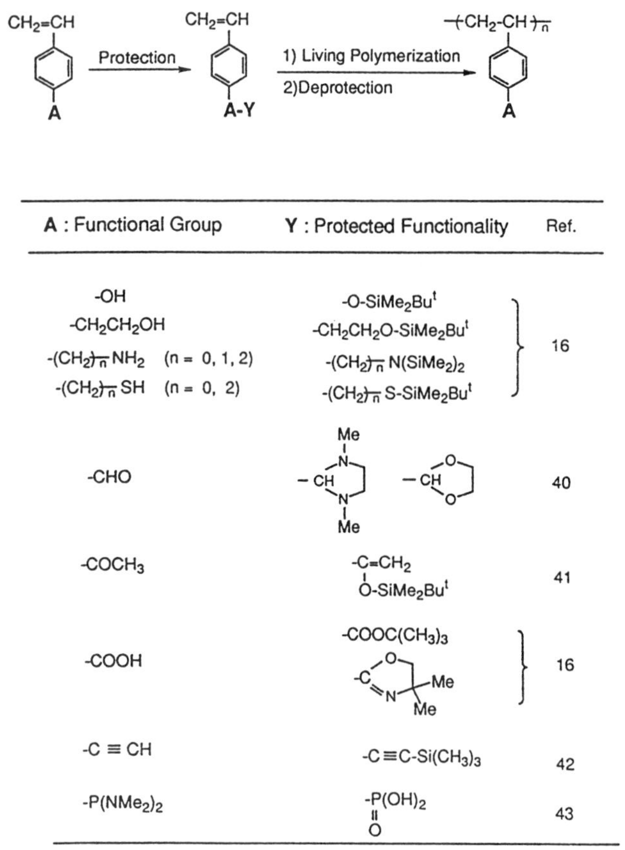

Scheme V. Synthesis of functional polystyrenes via anionic polymerization. (Reproduced with permission from reference 16. Copyright 1990 Elsevier.)

4-vinylpyridines were suggested (10). However, recent detailed investigations indicate that anionic polymerization of 2-vinylpyridine proceeds well in a living manner, but the polymerization of the 4-isomer is not straightforward. In fact, poly-2-vinylpyridines with narrow molecular-weight distributions were obtained (14, 45, 46). Well-defined block copolymers including poly-2-vinylpyridine segments were also synthesized.

Similarly, 2-isopropenylpyridine recently was reported to undergo living anionic polymerization (47).

Interestingly, the polymerization of 2-vinylpyridine with organo-Mg compounds in toluene or benzene was stereospecific and gave rise to essentially isotactic macromolecules (14). Several excellent studies on the conformation of these polymers were undertaken by Hogen-Esch and co-workers (48) to understand various factors controlling stereoregulation in these systems.

By contrast, considerable complications arise in anionic polymerization of 4-vinylpyridine due to both side reactions and polymer insolubility. Very recently, Varshney et al. (49) suggested living anionic polymerization of 4-vinylpyridine in a mixed solvent, either (THF–HMPT) or THF–dimethylformamide. They also demonstrated the successful synthesis of diblock copolymers with 4-vinylpyridine and *tert*-butyl methacrylate (1) Scheme VI. However, to my knowledge, no studies on anionic polymerization of 3-vinylpyridine have been reported.

Acrylic and Methacrylic Esters

Living Anionic Polymerization

Because the ester functionality of methacrylic esters is incompatible with carbanionic species, anionic polymerization of these monomers is complicated. In fact, in early studies of anionic polymerization of MMA with RLi in hydrocarbon solvents, the ester attack by the initiator and anion chain ends was observed (5, 50). For example, the reaction of the ester group of MMA with *n*-butyl lithium (BuLi) produces isopropenyl ketone and lithium methoxide (Scheme VII). The resulting isopropenyl ketone, a more reactive monomer, rapidly reacts with the growing chain end to produce a relatively unreactive end group, which terminates the growing chain. Similarly, the ester function of MMA [or poly(MMA) (PMMA)] may react with the chain-end anion. The formation of a cyclic trimer via cyclization (Scheme VII) also contributes to the termination (51).

Table III. Anionic Polymerization of Protected Styrene Derivatives

| Protected Functionality | Functional Group | $M_n \times 10^{-3}$ (dalton) | | M_w/M_n |
		Calculated	Observed	
–O–SiMe$_2$-*t*-Bu	–OH	48	53	1.04
–CH$_2$CH$_2$OSiR$_3$	(–CH$_2$CH$_2$OH)	55	54	1.11
–N(SiMe$_3$)$_2$	(–NH$_2$)	30	30	1.03
–CH$_2$CH$_2$SSiR$_3$	(–CH$_2$CH$_2$SH)	53	67	1.03
1,3-Dimethylimidazole	(–CHO)	16	14	1.08
–C(CH$_2$)OSiMe$_2$-*t*-Bu	(–COCH$_3$)	38	39	1.08
–C≡C–SiMe$_3$	(–C≡CH)	42	40	1.08

Note: Polymerization in THF at –78 °C.

Source: Compiled from references 16, 40–42.

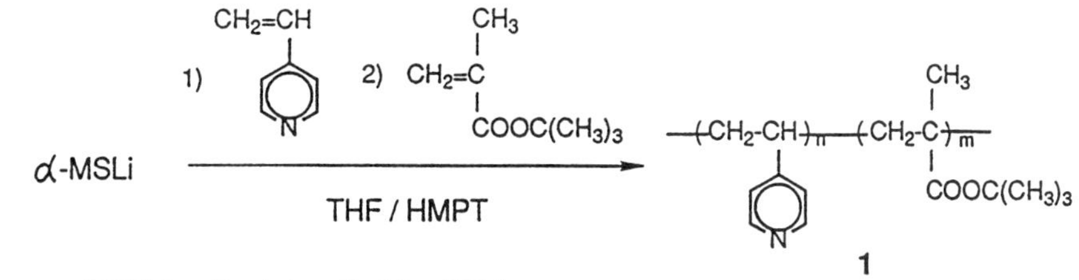

CH₂=CH ... CH₃ ... CH₃

Scheme VI. Diblock copolymer of 4-vinylpyridine and *tert*-butyl methacrylate produced by anionic polymerization. (Reproduced from reference 49. Copyright 1993 American Chemical Society.)

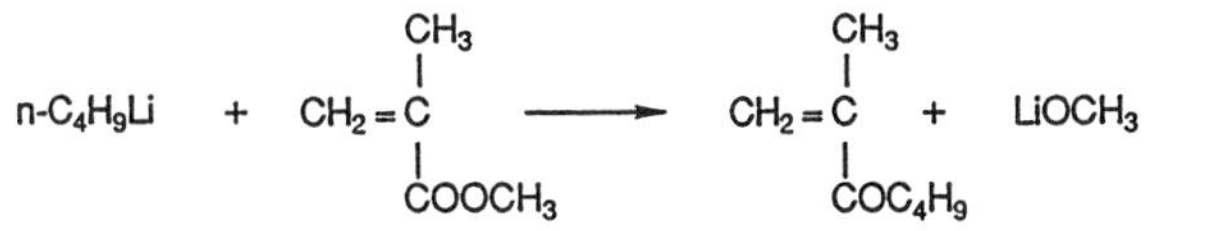

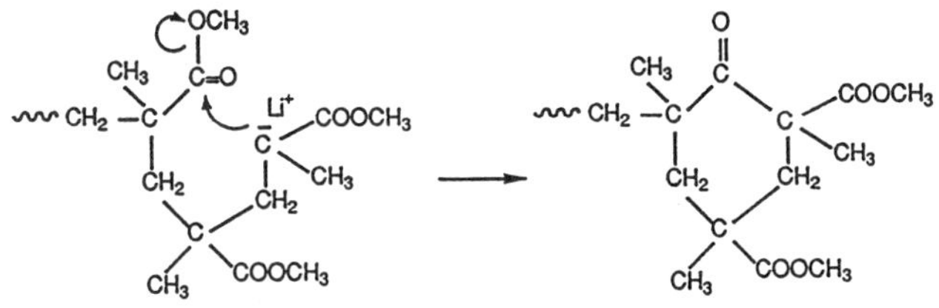

Scheme VII. Possible termination reactions via quenching of the initiator or the macroanion by the ester group in methyl methacrylate and poly(methyl methacrylate). (Reproduced with permission from reference 50. Copyright 1988 Elsevier.)

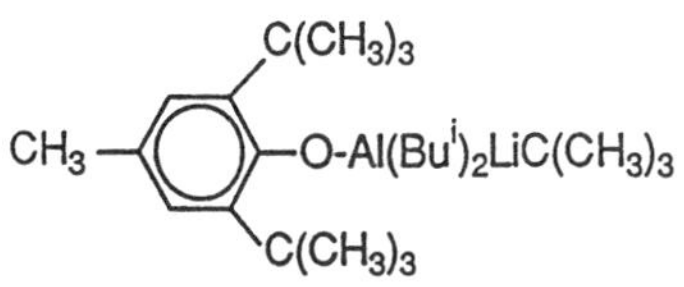

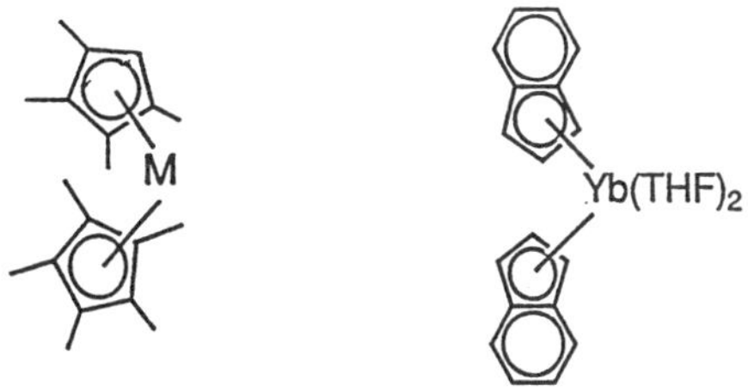

3a, M = Yb(THF)₂
3b, M = Sm(THF)₂
3c, M = Eu(OEt)₂(THF)

3d

Chart I. LiAlR(R')₂(OR") and organolanthanide complexes as initiators for living polymerization of MMA (**2**) and alkyl methacrylates (**3**).

However, by using bulky and less nucleophilic initiators in polar solvents, successful living polymerizations of MMA and related methacrylates were reported. Schulz and co-workers (*52*) showed that an almost unperturbed living polymerization of MMA was observed at temperatures below –50 °C with either benzylic sodium or cesium in THF or 1,2-dimethoxyethane (*52*). DuPont researchers (*53*) also found that the living character of polymerization of MMA was maintained at temperatures below –20 °C with the adduct of BuLi and pyridine in a mixture of pyridine and toluene (*53*). They also demonstrated that by using 1,1-diphenylhexyllithium (from BuLi and 1,1-diphenylethylene), polymerization of MMA at –70 °C produced a living polymer. Recently, Hatada and co-workers (*54, 55*) prepared living polymers of MMA and related monomers in toluene with either *t*-BuMgBr or with *t*-BuLi–Bu₃Al.

More recently, Ballard et al. (*56*) successfully demonstrated the living anionic polymerization of MMA with bulky lithium aluminum alkyls (**2**)(Chart I). The important contribution of this group is to show that the polymerization can be carried out at elevated temperatures around 0 °C in hydrocarbon solvents like toluene.

Yasuda et al. (*57, 58*) also developed a new initiator system based on organolanthanide complexes (**3**, Chart I) by which the polymerizations of MMA and other methacrylates proceed well without any chain termination. By this system, PMMAs with molecular weights of $\leq 500 \times 10^3$ and M_w/M_n values of <1.05 could be obtained even at temperatures of ≤ 40 °C (Table IV). Such well-controlled PMMAs with high molecular weights have been difficult to obtain by conventional living anionic polymerization, and no successful example has been reported.

In contrast to the successful living polymerization of alkyl methacrylates, polymerization of alkyl acrylates generally is complicated under similar conditions. In addition to the ester function, the labile α-hydrogen in acrylates presents further problems during the polymerization. Thus, polymerization of alkyl acrylates has so far resulted in low yields and broad molecular-weight distributions (*59*).

Table IV. Polymerization of MMA with Various Organolanthanide Complexes as Initiators

Initiator	Temperature (°C)	$M_n \times 10^{-3}$ (dalton) Calculated	Observed	M_w/M_n
$[SmH(C_5Me_5)_2]_2$	0	50	58	1.02
$[SmH(C_5Me_5)_2]_2$	40	50	55	1.03
$[SmH(C_5Me_5)_2]_2$	0	150	215	1.03
$[SmH(C_5Me_5)_2]_2$	0	300	563	1.04
$YbMe(C_5Me_5)_2(THF)_2$	0	50	48	1.04
$LuMe(C_5Me_5)_2(THF)_2$	0	50	61	1.04
$[YMe(C_5H_5)_2]_2$	0	50	82	1.22

Note: Polymerization were carried out in toluene.

Source: Compiled from reference 58.

Teyssié and co-workers (*59*) recently found an interesting additive effect of LiCl on the anionic polymerization of *tert*-butyl acrylate. Polymerization could be controlled to proceed quantitatively without side reactions. The resulting poly(*t*-butyl acrylate)s possessed predictable molecular weights and narrow molecular-weight distributions. Furthermore, Teyssié and co-workers (*60*) synthesized a novel block copolymer of styrene and *t*-butyl acrylate with a high degree of block homogeneity.

Reetz (*61*) developed interesting initiator systems based on stable tetrabutylammonium salts of $-CR-(CO_2R')_2$, $-C(CH_3)(CN)_2$, and $-SR$. These metal-free anions could effectively initiate the polymerization of *n*-butyl acrylate to afford the corresponding polymers with relatively narrow molecular-weight distributions ($M_w/M_n = 1.2$–1.5). Thus, the situation for anionic polymerization of simple acrylic esters appears very promising, although the methods are not yet adequate for the living polymerization of more reactive acrylic esters.

A recent topic in the polymerization of acrylates and methacrylates is the discovery of group transfer polymerization (GTP) by Webster et al. (*62*). GTP is an

extension to a polymerization reaction of the C–C bond formed between silyl ketene acetals with Michael-type electrophiles. This mechanism for MMA is a characteristic one, in which the silyl group is always transferred in each propagating step (Scheme VIII). Alkyl methacrylates, and even acrylates, can be polymerized by GTP in a way akin to a living mechanism. GTP also works effectively for a variety of electrophilic monomers such as polyunsaturated esters, lactones, vinyl methyl ketones, *N,N*-dimethyl methacrylamide, and acrylonitrile. Under the carefully controlled conditions, some of these monomers also undergo nearly termination-free polymerization, resulting in quantitative yields of the corresponding polymers with relatively controlled structures (*63*).

New Developments in Anionic Polymerizations of Methacrylic Esters with Functional Groups

Okamoto and Yashima (*64*) reported an interesting asymmetric anionic polymerization of triphenylmethyl methacrylate (TrMA) (*64*). The optically active polymer could be obtained by the polymerization of TrMA with either a chiral organolithium compound or

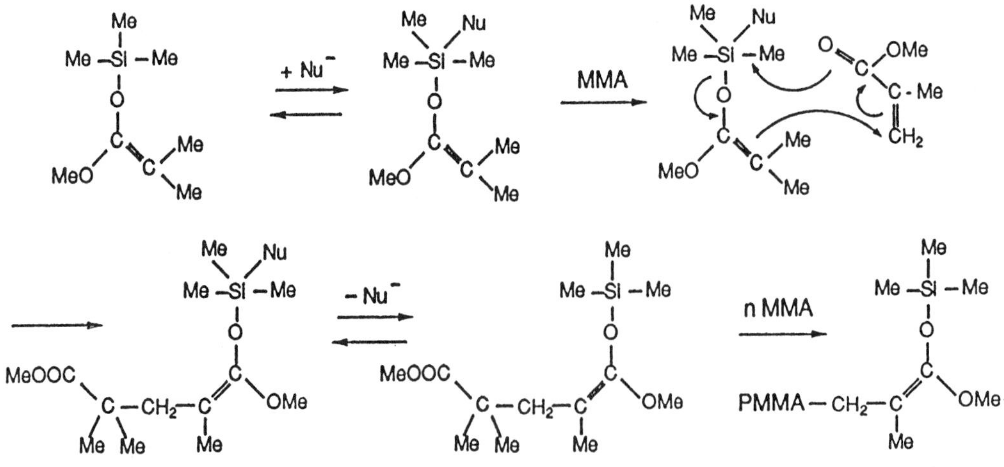

Scheme VIII. Mechanism of group transfer polymerization of MMA. (Reproduced from Reference 62. Copyright 1983 American Chemical Society.)

BuLi in the presence of a chiral ligand such as sparteine in toluene. The optical activity of this polymer with large optical rotation is attributed to the chirality of the helix conformation of a single screw sense stabilized by the bulky triphenylmethyl group. A variety of related optically active polymers with helical structures and bulky groups such as diphenyl-2-pyridylmethyl and cyclohexyldiphenylmethyl groups similarly have been synthesized (65). Under careful experimental conditions, some of these methacrylates are suggested to undergo living polymerization.

The helical polymethacrylates thus produced often undergo helix–helix transition. For example, changes from one-handed helix to the mixture of right- and left-handed helices and from positive to negative rotation and reversible transition are observed. The resulting chiral polymers are very effective for optical resolution of a variety of racemic compounds and are used commercially as chromatographic packing for this purpose (66, 67).

Co-workers and I (68) successfully synthesized poly(2-hydroxyethyl methacrylate)s (PHEMA) with controlled polymer chain lengths by living anionic polymerization of trimethylsilyl-protected HEMA followed by deprotection (Scheme IX). Well-defined block copolymers of styrene and HEMA were also synthesized in a similar manner (68). These hydrophilic block copolymers show excellent nonthrombogenic activity compared with commercially available segmented polyurethane (Biomer) and poly(ethylene oxide)-grafted Biomer (69).

Poly(acrylic acid) and poly(methacrylic acid) with predictable molecular weights and narrow distributions were prepared by means of living anionic polymerization of *t*-butyl acrylate and methacrylate followed by deprotection (Scheme IX) (70). Thus, these well-defined polyelectrolytes are now accessible.

Teyssié et al. (71) reported that anionic polymerization of glycidyl methacrylate (**4** in Chart II) proceeded well in THF at –78 °C without any chain termination. A quantitative yield of the polymer and a relatively narrow molecular-weight distribution were obtained. Very recently, Finkelmann and Bohnert (72) demonstrated the synthesis of side-chain liquid-crystalline polymers by living anionic polymerization of 6-[4-(methoxyphenoxycarbonyl)phenoxy]hexyl methacrylate (**5** in Chart II). These polymers with well-regulated chain lengths may be useful in elucidation of the relationship between molecular weights and phase behavior, and this relationship is still an unresolved question in liquid-crystalline polymers.

A detailed study of anionic polymerization of methyl α-trifluoromethylacrylate (**6**) was reported by Ito et al. (73). As expected, typical anionic initiators of MMA were marginally effective, whereas various metal salts with 18-crown-6 and pyridine readily poly-

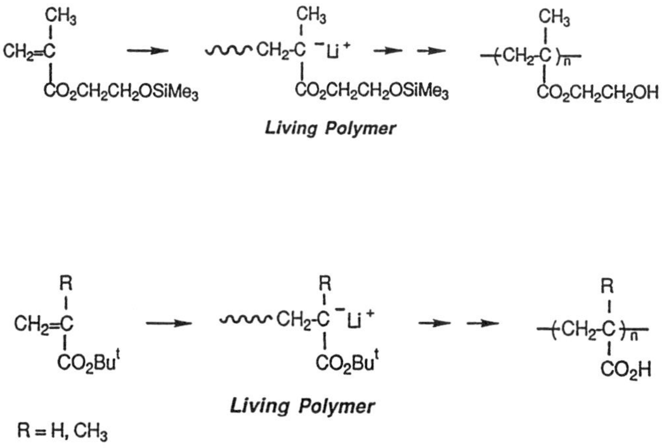

Scheme IX. Synthesis of PHEMA and poly(methacrylic acid) by living anionic polymerization of the protected monomers followed by deprotection. (Reproduced from references 68 and 70. Copyright 1986 American Chemical Society.)

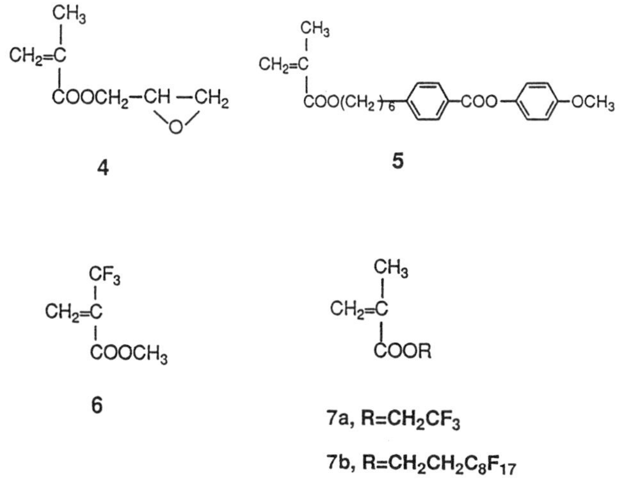

Chart II. Examples of functional methacrylates polymerized by anionic polymerization: glycidyl methacrylate (**4**), mesogenic alkyl methacrylate (**5**), methyl α-trifluoromethyl acrylate (**6**), and perfluoroalkyl methacrylates (**7**).

merized this monomer. The strong electron-withdrawing character of the CF_3 group may enhance the anionic polymerizability of the monomer, but the CF_3 group can also be the source of complicated side reactions when used with more reactive initiators.

Narita et al. (74, 75) reported the anionic polymerizations of a series of perfluoroalkyl methacrylates. These monomers were generally less reactive than MMA, and the polymers were produced in low to moderate yields under typical anionic conditions applied to MMA. However, Nakahama and I recently found that methacrylates **7a** and **7b** (Chart II) undergo anionic polymerization smoothly to quantitatively afford the corresponding polymers in THF at –78 °C. Block copolymers of styrene and these monomers were also synthesized, and the polymerizations are suggested to proceed by a living mechanism (76).

Acrylamides and Methacrylamides

Structures of polyacrylamides obtained by anionic polymerization are very characteristic (77). Acrylamides can be polymerized anionically by proton-transfer reaction to the double bond and result in the formation of polyamides, whereas free-radical polymerization results in predominantly vinyl-type polymers. Similarly, the base-catalyzed polymerization of 4-vinylbenzamide proceeds mainly via proton-transfer to give a linear polyamide with a small amount of vinyl polymer. Interestingly, the presence of LiCl changes the polymerization mode completely to vinyl polymerization (78).

Recently, Jung and Heitz (79) demonstrated that polymerization of 4-vinylbenzamide with cesium *tert*-butoxide in HMPT proceeds only by proton-transfer reaction to exclusively afford a polyamide. The resulting polyamide with one terminal vinyl bond can also be used as a macromer (**8** in Chart III). N-Alkyl-substituted acrylamides also undergo proton-transfer polymerization with typical anionic initiators such as BuLi, NaH, and metal *tert*-butoxide (80). On the other hand, the corresponding methacrylamides are generally less reactive under similar conditions and produce low yields of low molecular-weight oligomers (80).

Anionic polymerization of N,N-dialkyl acrylamides gave crystalline polymers insoluble in most organic solvents (81). However, Hogen-Esch and Xie (82) recently reported that anionic polymerization of N,N-dimethyl acrylamide with diphenylmethylcesium proceeds well to give soluble polymers with relatively narrow molecular-weight distributions.

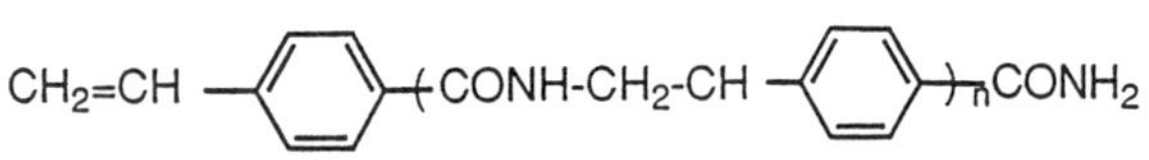

Surprisingly, N,N-dialkyl methacrylamides were observed not to undergo anionic polymerization with initiators such as organolithiums, organomagnesiums, metal hydrides, or metal alkoxides, although the reason is not clear yet. N-Aziridine-substituted methacrylamide (**9**) is the only monomer so far anionically polymerized (83). The polymer obtained with BuLi in toluene seemed to be 100% isotactic. Okamoto et al. (84) succeeded in the asymmetric polymerization of N,N-diphenylacrylamide and its derivatives with organolithium initiators in the presence of chiral ligands. The resulting polymers were optically active and had one-handed helical conformation. N-Ethyl and N-phenyl maleimides (**10a** and **10b**, Chart III) were reported by Narita, Hagiwara, and co-workers (85) to undergo anionic polymerization with alkali metal *tert*-butoxides, Grignard reagents, alkylzinc, and -ate complexes of R$_2$Zn and R'$_3$Al with R"Li (85).

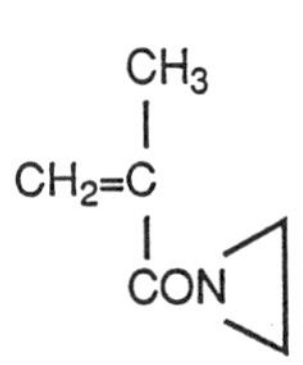

Other Monomers

Acrylonitrile and methacrylonitrile have been known for a long time to anionically polymerize with a wide variety of initiators, including alkyllithiums (6), Grignard reagents (86), metal alkoxides and hydroxides (87), and tertiary phosphines (88). Unfortunately, intermolecular cyclization, attack of the nitrile group with anionic species, and proton abstraction often occur during the polymerizations of these monomers. The resulting polymers generally have broad molecular-weight distributions and are yellow colored due to cyclization structures.

The growing chain ends of polyacrylonitrile and polymethacrylonitrile have been investigated by IR and UV spectroscopies and other methods based on molecular-orbital calculation. In addition to the carbanion species of these monomers, several other species produced by nitrile addition to the anionic centers exist in the polymer (89). Suzuki et al. (90) attempted to synthesize block copolymers of acrylonitrile and methacrylonitrile with ethylene oxide by anionic polymerization using the disodium salt of poly(ethylene oxide). The resulting block copolymer of methacrylonitrile with ethylene oxide was obtained in a reasonable yield in the presence of dicyclohexyl-18-crown-6. However, extensive metalation took place in the case of acrylonitrile and led to low yields and branched structures.

Ito et al. (91) observed that α-(trifluoromethyl)-acrylonitrile readily polymerized with BuLi-18-crown-6 or pyridine. This observation is interesting because this monomer is resistant to free-radical polymerization. The polymer is highly sensitive to deep-UV and electron-beam radiations and is of interest for lithography. A new conjugated polymer was obtained by anionic polymerization of trifluoromethylacetylene (**11** in Scheme X) with BuLi, although the yields were

$CH_2=CH$ —⟨benzene⟩—[$CONH-CH_2-CH$—⟨benzene⟩—]$_n$$CONH_2$

8

CH$_3$
|
CH$_2$=C
|
CON⟨aziridine⟩

9

O=⟨maleimide ring⟩=O
|
N
|
R

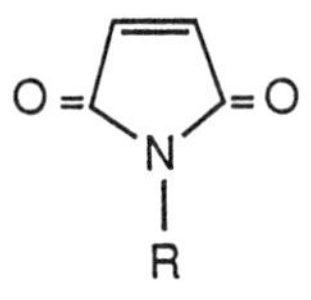

10a, R = Et

10b, R = Ph

Chart III. Amide-containing functional monomers: polyamide with styryl terminal group (**8**), N-methacroylaziridine (**9**), and N-ethyl- and phenylmaleimides (**10**).

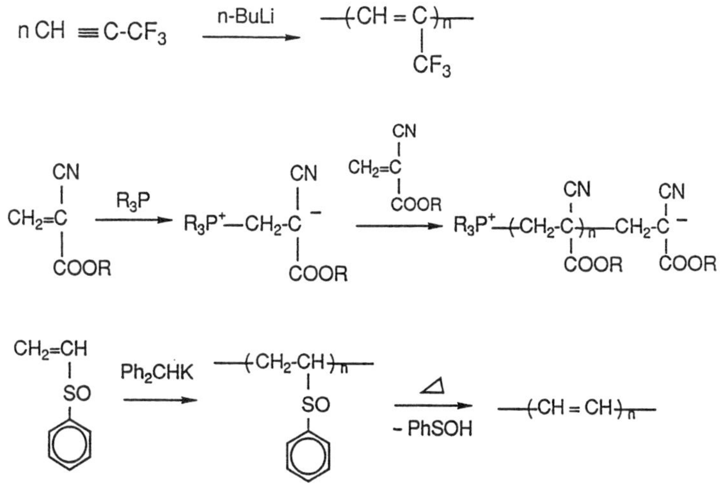

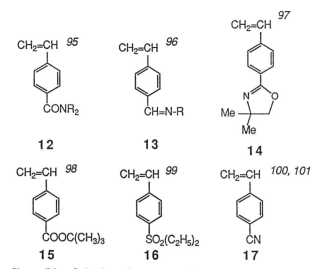

Scheme X. Miscellaneous functional polymers produced by anionic polymerization.

relatively low (*92*). The effects of n- and p-type doping on the electrical conductivity of this polymer were examined.

α,α'-Disubstituted ethylenes with electron-withdrawing groups are known to be the most anionically reactive class of monomers on the basis of their *e* values (Table I), and some are used as adhesives. Johnston and Pepper (*93*) systematically studied the anionic polymerization of α-cyanoacrylates with phosphines and amines via a zwitterionic process (Scheme X) and obtained the corresponding polymers with very high molecular weights ($>10^6$ daltons). These polymers are used widely for the manufacture of biomedical and drug-deliveries devices.

Hogen-Esch and co-workers (*94*) recently reported the living anionic polymerization of vinyl phenyl sulfoxide. The A–B and B–A–B block copolymers of styrene (A) and vinyl phenyl sulfoxide (B) were synthesized successfully. Moreover, the resulting polymer was transformed into polyacetylene by thermal elimination of phenylsulfenic acid (Scheme X). Thus, nearly monodispersed polyacetylenes were obtained via anionic polymerization, although conductivity measurements of the iodine-doped sample indicated a relatively short conjugation length.

Ishizone, co-workers, and I (*95–101*) recently succeeded in the new living anionic polymerization of several styrene derivatives substituted with electron-withdrawing groups (**12–17**, Chart IV). A number of new well-defined functional polystyrenes and their block copolymers were synthesized. The results of homopolymerization are summarized in Table V. The success of living polymerization in this case may be attributed to the carbanion stabilization by the electron-withdrawing effect of the functional groups and by the extension of the conjugation from the carbanions to these groups. These styrenes were observed to be anionically more reactive than styrene itself and

Chart IV. *p*-Substituted styrenes with electron-withdrawing functional groups used for the synthesis of functional polymers by living anionic polymerization.

to have reactivities similar to 2-vinylpyridine and MMA.

Tsuruta and Nagasaki recently reported the reactions of divinylbenzenes with a variety of amines catalyzed by lithium alkylamides. Interestingly, a variety of 1:1 adducts, oligomers, and alternating copolymers were prepared selectively by choosing appropriate reaction conditions (*see* Chapter 1.12).

Monomers with Reactive Silyl Groups

Polymers containing silicon atoms in their main chains are widely used (*102*). The best examples are poly-(dimethylsiloxane)s, which are used as specialty elastomer, oils, greases, and membranes. Polycarbosilanes (*103*) and polysilazanes (*104*) are also used as industrial precursors of silicon carbide and silicon nitride fibers. Recently, polysilanes have been a subject of sci-

Table V. Anionic Polymerization of *p*-Substituted Styrenes with Electron-Withdrawing Functional Groups

Monomer	Functional Group	$M_n \times 10^{-3}$ (dalton)		M_w/M_n
		Calculated	Observed	
12	(CONEt$_2$)	48	48	1.08
13	(CH=N-C$_6$H$_{11}$)	22	24	1.06
14	(Oxazoline)	67	75	1.07
15	(COO-*t*-Bu)	63	59	1.18
16	(SO$_2$NEt$_2$)	33	38	1.09
17	(CN)	30	33	1.04

Note: Polymerization in THF at –78 °C. *See* Chart IV for monomers.

Source: Compiled from references 95–101.

entific and technological interest because of their characteristic optical and electronic properties (*105*).

The introduction of a silicon atom on the polymer side chains may greatly influence the physical and chemical properties of the polymer. Creation of new functional polymers by modification of the silicon atom is also possible. Effectiveness of such polymers will be discussed in more detail in Chapter 1.12. However, little attention had been paid to the synthesis and application of these polymers until recently, and only a few syntheses, mainly by radical polymerization, have been reported (*106*).

Herz and co-workers (*107*) and Yamazaki et al. (*108*) independently reported the first successful anionic living polymerization of 4-trimethylsilyl-styrene in the early 1980s. Subsequently, co-workers and I (*17, 109–111*) demonstrated the anionic living polymerization of several monomers containing reactive silyl groups, including 4-substituted styrenes with Si–OR (**18**), Si–O–Si (**19**), Si–H (**20**), Si–N (**21**), Si–Si (**22**), and Si–Si–Si bonds (**23**); 2-(trialkoxysilyl)-1,3-butadienes (**24**) (*112, 113*); and methacrylates with (RO)$_3$Si groups (**25**) (*114*) (Chart V).

The success of living polymerization of these monomers clearly indicates that silyl groups are stable in the presence of the carbanions and highly reactive anionic initiators. This result is quite surprising, because most of these silyl groups are not compatible with reactive carbanionic species like BuLi. As expected, all the resulting polymers showed predictable molecular weights and narrow molecular-weight distributions. Well-defined block copolymers with styrene, 1,3-butadiene, isoprene, and MMA were synthesized (*17*). The results are summarized in Tables VI and VII.

The reactive silyl groups on the polymers can participate in a wide variety of useful reactions based on the silicon atom, including reaction with inorganic compounds, hydrosilylation, photolysis, and insertion reactions. Polymers with such excellent structural merits are expected to have many potential applications, especially where molecular design with precise architecture is required.

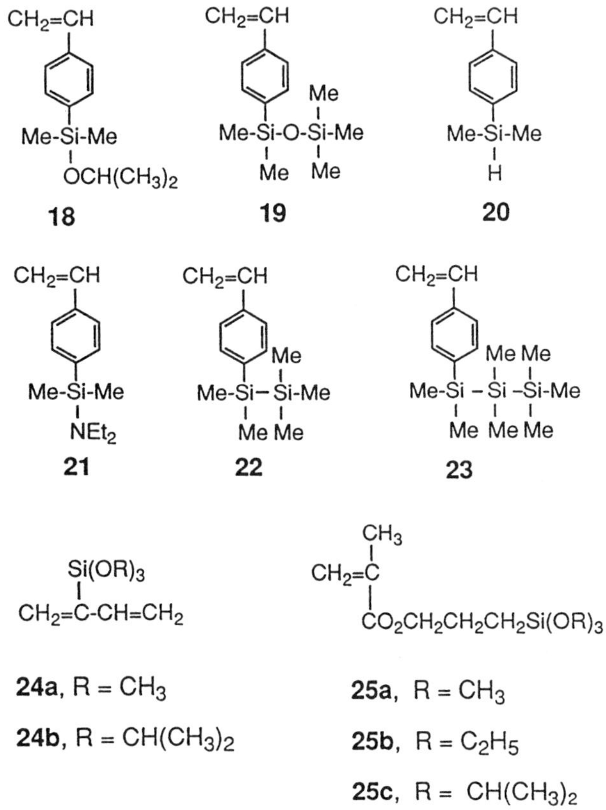

Chart V. Monomers with reactive silyl groups.

Table VI. Anionic Polymerization of Monomers with Reactive Silyl Groups

Monomer	$M_n \times 10^{-3}$ (dalton)		M_w/M_n
	Calculated	Observed	
18	35	33	1.06
19	14	13	1.06
20	49	45	1.10
21	19	22	1.05
22	31	33	1.03
23	36	39	1.04
24a	14	14	1.10
24b	36	31	1.11
25c	50	58	1.02

Note: Polymerization in THF at –78 °C. *See* Chart V for monomers.

Source: Compiled from references 17, 109–114.

Table VII. Block Copolymers by Sequential Anionic Polymerization

Type	Monomer A	Monomer B	$M_n \times 10^{-3}$ *(dalton)*		[A]/[B]	
			Calculated	*Observed*	*Calculated*	*Observed*
A–B–A	**5**	Isoprene	62	65	51/49	51/49
A–B–A	**20**	Isoprene	51	52	50/50	40/60
B–A–B	**20**	Styrene	25	24	38/62	39/61
A–B–A	**21**	Styrene	45	45	36/64	39/61
A–B	**22**	Styrene	15	15	30/70	30/70
B–A	**24b**	Isoprene	30	30	10/90	10/90
A–B	**25c**	Styrene	18	17	24/76	24/76
B–A	**25c**	MMA	17	19	14/86	11/89

Note: Polymerization in THF at –78 °C. *See* charts for monomers. Values of M_w/M_n were 1.03–1.13.

Source: Compiled from reference 17.

Poly[2-(triisopropoxysilyl)-1,3-butadiene] obtained by anionic polymerization had a perfect (*E*)-1,4 structure (cis configuration with respect to the main chain) (*112, 113*). To my knowledge, this report is the first demonstration of both stereo- and regiospecific polymerizations in the field of living anionic polymerization. As expected, this polymer is a viscous liquid and has a low glass-transition temperature of about –37 °C. It contains an alkoxysilyl group not only cross-linkable but also reactive toward inorganic compounds, and it is expected to find use as a new elastomer.

Co-workers and I (*115*) recently demonstrated that a nearly monodispersed copolymer of MMA and 3-(triethoxysilyl)propyl methacrylate obtained by anionic polymerization has 10 times greater resistance toward a CBrF$_3$ plasma than does PMMA homopolymer. Niobium thin-film microbridges showing superconductivity could be fabricated at the nanometer level from this new copolymer by using electron-beam lithography.

Nakahama, Lee, and I (*116, 117*) developed a new method to prepare microporous membranes from well-defined block copolymers with reactive silyl groups in microphase-separated microdomains as illustrated in Scheme XI. The method involved casting a block copolymer film with a microphase-separated lamella structure, fixation of the microdomain of the poly(4-[(2-propoxy)dimethylsilyl]styrene) block by hydrolysis, and cross-linking and oxidative etching of the polyisoprene microdomain with ozone to form micropores. Microporous membranes with uniform pores of about 10 nm were thus prepared from a series of block copolymers with different compositions (*116, 117*).

Thus, the lammela structure of the original block copolymer film was directly reflected in the shape and size of the micropores produced by this process. In other words, the microstructure of a produced porous membrane can be controlled by the block lengths of the starting block copolymer. Nakahama et al. (*118*) also recently demonstrated that a new glucose sensor with rapid response can be made by immobilizing glucose oxidase inside the microporous membrane prepared on a platinum electrode by a similar route, as shown in Scheme XII.

Anionic polymerization of vinyltrimethylsilane was studied in 1960s by Nametkin et al. (*119*), and

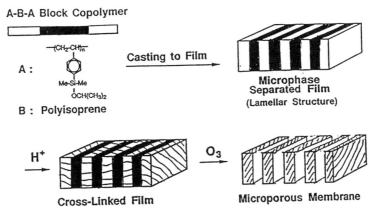

Scheme XI. Preparation of microporous membrane from well-defined block copolymer. (Reproduced from references 116 and 117. Copyrights 1988 and 1989, respectively, American Chemical Society.)

they suggested that it proceeded via a living mechanism. However, recent work by Oku et al. (*120, 121*) indicated that, in the presence of *N,N,N',N'*-tetramethylethylenediamine, an intramolecular rearrangement occurred exclusively from the propagating carbanion to the methyl group of the trimethylsilyl side chain (Scheme XIII). In comparison with the polymer obtained by free-radical polymerization, a new polycarbosilane was thus obtained quantitatively by anionic polymerization.

Concluding Remarks and Future Prospects

A large number of functional monomers can be anionically polymerized, although this number is still limited in comparison with radical polymerization. In general, careful selection of solvent, initiator, and temperature is important for successful anionic polymerization. Sensitivity of anionic species toward oxygen and moisture remains an inconvenient factor inherent in anionic polymerization. On the other hand, the production of living polymers is the most advantageous aspect of anionic polymerization over free-radical poly-

merization. In particular, substantial advance has been made in producing living polymers with a wide variety of functional groups, and it is now possible to produce new functional polymers with well-defined structures. To this end, many novel functional block copolymers previously unknown can be prepared by living anionic polymerization, providing access to a variety of specialty polymers and molecular devices in the future. However, the problem of controlled anionic polymerization of highly reactive monomers has not been solved so far, and this is a challenging target in the field of anionic polymerization.

References

1. Greenley, R. Z. *Polymer Handbook*; Brandrup, J.; Immergut, E. H., Eds.; Wiley-Interscience: New York, 1989; Vol. II, pp 267–274.
2. Hay, J. N.; Harris, D. S.; Wiles, M. *Polymer* **1976**, *17*, 613–617.
3. Rempp, R.; Franta, E.; Herz, J. E. *Adv. Polym. Sci.* **1988**, *86*, 147–173.
4. Tsuruta, T. *Anionic Polymerization*; Kagakudoujin: Tokyo, Japan, 1973, pp 7–24.

Scheme XII. Glucose sensor based on microporous enzyme membrane prepared from well-defined block copolymer on a Pt electrode. (Reproduced with permission from reference 118. Copyright 1991 Elsevier.)

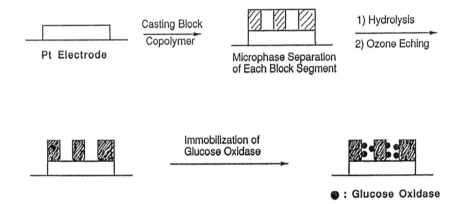

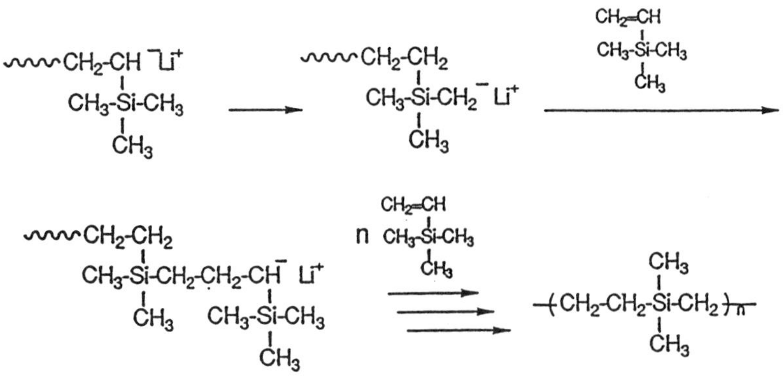

Scheme XIII. Anionic polymerization of trimethylvinylsilane.

5. Kawabata, N.; Tsuruta, T. *Makromol. Chem.* **1965**, *86*, 231–252.

6. Kawabata, N.; Tsuruta, T. *Makromol. Chem.* **1966**, *98*, 262–274.

7. Lyons, A. R.; Catterall, E. *Eur. Polym. J.* **1971**, *7*, 839–848.

8. Morton, M.; Fetters, L. J. *Macromol. Rev.* **1967**, *2*, 71–113.

9. Szwarc, M. *Nature (London)* **1956**, *178*, 1168–1169.

10. Morton, M. *Anionic Polymerization: Principles and Practices*; Academic: Orlando, FL, 1983; pp 13–85, 179–239.

11. Chaumont, P.; Beinert, G.; Herz, J.; Rempp, P. *Makromol. Chem.* **1979**, *180*, 2061–2071.

12. O'Malley, J. J.; Yanus, J. F.; Pearson, J. M. *Macromolecules* **1972**, *5*, 158–161.

13. Engel. D.; Schulz, R. C. *Eur. Polym. J.* **1983**, *19*, 967–972.

14. Fontanille, M.; Sigwalt, P. *Bull. Soc. Chim. Fr.* 1967, 4083–4087, 4087–4095, 4095–4100.

15. Lohr, G.; Schulz, G. V. *Makromol. Chem.* **1973**, *172*, 137–149.

16. Nakahama, S.; Hirao, A. *Prog. Polym. Sci.* **1990**, *15*, 299–335.

17. Hirao, A.; Nakahama, S. *Prog. Polym. Sci.* **1992**, *17*, 283–317.

18. Doi, Y.; Keii, T. *Adv. Polym. Sci.* **1986**, *73/74*, 201–248.

19. Sawamoto, M.; Aoshima, S.; Higashimura, T. *Makromol. Chem. Macromol. Symp.* **1988**, *13/14*, 513–526.

20. Grubbs, R. H.; Tumas, W. *Science (Washington, D.C.)* **1989**, *243*, 907–915.

21. Aida, T.; Kuroki, M.; Sugimoto, H.; Watanabe, T.; Adachi, T.; Kawamura, C.; Inoue, S. *Makromol. Chem., Macromol. Symp.* **1993**, *67*, 125–135.

22. Tomoi, M.; Kakiuchi, H. *Kogyo Kagaku Zasshi.* **1970**, *73*, 2367–2371.

23. Nuffer, R.; Francois, B. *Eur. Polym. J.* **1984**, *20*, 389–393.

24. Halasa, A. F.; Lohr, D. F.; Hall, J. E. *J. Polym. Sci., Polym. Chem. Ed.* **1981**, *19*, 1357–1360.

25. Xu, Z.; Hadjichristidis, N.; Carella, J. M.; Fetters, L. J. *Macromolecules* **1983**, *16*, 925–929.

26. Mays, J.; Hadjichristidis, N.; Fetters, L. J. *Macromolecules* **1984**, *17*, 2723–2728.

27. McGrath, J. E. *J. Chem. Educ.* **1981**, *58*, 914–921.

28. Holden, G.; Bishop, E. T.; Legge, N. R. *J. Polym. Symp., Part C.* **1969**, *26*, 37–57.

29. Gauthier, M.; Moller, M. *Macromolecules* **1991**, *24*, 4548–4553.

30. Hadjichristidis, M.; Fetters, L. J. *Macromolecules* **1980**, *13*, 191–193.

31. Zilliox, J. G.; Rempp, P.; Parrod, J. *J. Polym. Symp., Part C.* **1968**, *22*, 145–156.

32. Geiser, D.; Hocker, H. *Macromolecules* **1980**, *13*, 653–656.

33. Roovers, J.; Toporowski, M. *Macromolecules* **1983**, *16*, 843–849.

34. Hild, G.; Strazielle, C.; Rempp, P. *Eur. Polym. J.* **1983**, *19*, 721–727.

35. French, D. M. *Rubber Chem. Tech.,* **1969**, *42*, 71–109.

36. Quirk, R. P.; Yin, J.; Guo, S-H.; Hu, X-W.; Summers, G.; Kim, J.; Zhu, L-F.; Schock, L. E. *Makromol. Chem., Macromol. Symp.* **1990**, *32*, 47–59.

37. Quirk, R. P.; Yin, J.; Guo, S-H.; Hu, X-W.; Summers, G.; Kim, J.; Zhu, L-F.; J. J. Ma.; Takizawa, T.; Lynch, T. *Rubber Chem. Technol.* **1991**, *64*, 648–660.

38. Hirao, A.; Ueda, K.; Nakahama, S. *Macromolecules* **1990**, *23*, 939–945.

39. Hirao, A.; Nagahama, H.; Ishizone, T.; Nakahama, S. *Macromolecules* **1993**, *26*, 2145–2150.

40. Ishizone, T.; Kato, R.; Ishino, Y.; Hirao, A.; Nakahama, S. *Macromolecules* **1991**, *24*, 1449–1454.

41. Hirao, A.; Kato, K.; Nakahama, S. *Macromolecules* **1992**, *25*, 535–540.

42. Ishizone, T.; Hirao, A.; Nakahama, S.; Kakuchi, T.; Yokota, K.; Tsuda, K. *Macromolecules* **1991**, *24*, 5230–5231.

43. Kase, T.; Imahori, M.; Kazawa, T.; Isono, Y.; Fujimoto, T. *Macromolecules* **1991**, *24*, 1714–1719.

44. Takemoto, K. *J. Macromol. Sci., Rev. Macromol. Chem.* **1970**, *C5*, 29–102.

45. Soom, A. H.; Fontanille, M. *Makromol. Chem.* **1980**, *181*, 799–808.

46. Takaki, M.; Asami, R.; Tanaka, S.; Hayashi, H.; Hogen-Esch, T. E. *Macromolecules* **1986**, *19*, 2900–2903.

47. Soum, A. H.; Tien, C-F.; Hogen-Esch, T. E.; D'Accorso, N. B.; Fontanille, M. *Makromol. Chem., Rapid Commun.* **1986**, *7*, 671–678.

48. Hogen-Esch, T. E. *Makromol. Chem., Macromol. Symp.* **1993**, *67*, 43–66.

49. Varshney, S. K.; Zhong, X. F.; Eisenberg, A. *Macromolecules* **1993**, *26*, 701–706.

50. Hatada, K.; Kitayama, T.; Ute, K. *Prog. Polym. Sci.* **1988**, *13*, 189–276.

51. Glusker, D. L.; Lysloff, I.; Yoncoskie, B. *J. Polym. Sci.* **1961**, *49*, 297–313 and 315–334.

52. Warzelhan, V.; Hocker, H.; Schulz, G. V. *Makromol. Chem.* **1980**, *181*, 149–163.

53. Anderson, B. C.; Andrews, G. D.; Arthur, P., Jr.; Jacobson, H. W.; Playtis, A. J.; Sharkey, W. H. *Macromolecules* **1981**, *14*, 1599–1601.

54. Hatada, K.; Ute, K.; Tanaka, K.; Okamoto, Y.; Kitayama, T. *Polym. J.* **1986**, *18*, 1037–1047.

55. Kitayama, T.; Shinozaki, T.; Sakamoto, T.; Yamamoto, M.; Hatada, K. *Makromol. Chem., Suppl.* **1988**, *15*, 167–185.

56. Ballard, D. G. H.; Bowles, R. J.; Haddleton, D. M.; Richards, S. N.; Sellers, R.; Twose, D. L. *Macromolecules* **1992**, *25*, 5907–5913.

57. Yasuda, H.; Yamamoto, H.; Yokota, K.; Miyake, S.; Nakamura, A. *J. Am. Chem. Soc.* **1992**, *114*, 4908–4910.

58. Yasuda, H.; Yamamoto, M.; Yamashita, K.; Yokota, K.; Nakamura, A.; Miyake, S.; Kai, Y.; Kanehisa. N. *Macromolecules* **1993**, *26*, 7134–7143.

59. Fayt, R.; Forte, R.; Jacobs, C.; Jerome, R.; Ouhadi, T.; Teyssie, Ph.; Varshney, S. V. *Macromolecules* **1987**, *20*, 1442–1444.

60. Hautekeer, J. P.; Varshney, S. K.; Fayt, R.; Jacobs, C.; Jerome, R.; Teyssie, P. H. *Macromolecules* **1990**, *23*, 3893–3898.

61. Reetz, M. T. *Angew. Chem. Int. Ed. Engl.* **1988**, *27*, 994–998.

62. Webster, O. W.; Hertler, W. R.; Sogah, D. Y.; Farnham, W. B.; RajanBabu, T. V. *J. Am. Chem. Soc.* **1983**, *103*, 5706–5708.

63. Hertler, W. R.; RajanBabu, T. V.; Ovenall, D. W.; Reddy, G. S.; Sogah, D. Y. *J. Am. Chem. Soc.* **1988**, *110*, 5841–5853.

64. Okamoto, Y.; Yashima, E. *Prog. Polym. Sci.* **1990**, *15*, 263–298.

65. Okamoto, Y.; Mohri, H.; Nakano, T.; Hatada, K. *J. Am. Chem. Soc.* **1989**, *111*, 5952–5954.

66. Okamoto, Y.; Okamoto, I.; Yuki, H.; Murata, S.; Noyori, R.; Takaya, H. *J. Am. Chem. Soc.* **1981**, *103*, 6971–6973.

67. Chance, J. M.; Geiger, J. H.; Okamoto, Y.; Aburatani, R; Mislow, K. *J. Am. Chem. Soc.* **1990**, *112*, 3540–3547.

68. Hirao, A.; Kato, H.; Yamaguchi, K.; Nakahama, S. *Macromolecules* **1986**, *19*, 1294–1299.

69. Nojiri, C.; Okano, T.; Koyanagi, H.; Nakahama, S.; Park, K. D.; Kim. S. W. *J. Biomater. Sci., Polym. Ed.* **1992**, *4*, 75–88.

70. Bugner, D. E. *Am. Chem Soc., Polym. Prepr.* **1986**, *27(2)*, 57–58.

71. Leemans, L.; Fayt, R.; Teyssie, Ph. *J. Polym. Sci., Polym. Chem. Ed.* **1990**, *28*, 2187–2193.

72. Bohnert, R.; Finkelmann, H. *Makromol. Chem., Rapid Commun.* **1993**, *14*, 139–146.

73. Ito, H.; Miller, D. C.; Willson, C. G. *Macromolecules.* **1982**, *15*, 915–920.

74. Narita, T.; Hagiwara, T.; Hamana, H. *Makromol. Chem., Rapid. Commun.* **1985**, *6*, 175–178.

75. Narita, T; Hagiwara, T.; Hamana, H.; Miyasaka, T.; Wakayama, A.; Hatta, T. *Makromol. Chem.* **1987**, *188*, 273–279.

76. Hirao, A.; Nakahama, S., Unpublished results.

77. Breslow, D. S.; Hulse, G. E.; Matlack, A. S. *J. Am. Chem. Soc.* **1957**, *79*, 3760–3763.

78. Asahara, T.; Yoda, N. *J. Polym. Sci., Polym. Lett.* **1966**, *4*, 921–925.

79. Jung, H.; Heitz, W. *Makromol. Chem., Rapid Commun.* **1988**, *9*, 373–379.

80. Kennedy, J. P.; Otsu, T. *J. Macromol. Sci., Rev. Macromol. Chem.* **1972**, *C6*, 237–298.

81. Butler, K.; Thomas, P. R.; Tyler, G. H. *J. Polym. Sci.* **1960**, *48*, 357–366.

82. Xie, X.; Hogen-Esch, T. E. *Am. Chem. Soc., Polym. Prepr.* **1993**, *34(1)*, 118–119.

83. Okamoto, Y.; Yuki, H. *J. Polym. Sci., Polym. Chem. Ed.* **1981**, *19*, 2647–2650

84. Okamoto, Y.; Hayashida, H.; Hatada, K. *Polym. J.* **1989**, *21*, 543–549.

85. Hagiwara, T.; Sato, J.; Hamana, H.; Narita, T. *Makromol. Chem.* **1987**, *188*, 1825–1831.

86. Tsvetanov, Ch. B. *Eur. Polym. J.* **1979**, *15*, 503–507.

87. Zilkha, A.; Feit, B-A.; Farankel, M. *J. Polym. Sci.* **1961**, *49*, 231–240.

88. Eisenbach, Jaack, V.; Schnecko, H.; Kern, W. *Makromol. Chem.* **1974**, *175*, 1329–1357.

89. Tsvetanov, Ch. B.; Panayotov, I. M. *Eur. Polym. J.* **1975**, *11*, 209–214.

90. Suzuki, T.; Murakami, Y.; Takegami, Y. *Polym. J.* **1982**, *14*, 431–439.

91. Ito, H.; Renaldo, A. F.; Ueda, M. *Macromolecules* **1989**, *22*, 45–51.

92. Rubner, M.; Deits, W. *J. Polym. Sci., Polym. Chem. Ed.* **1982**, *20*, 2043–2051.

93. Johnston, D. S.; Pepper, D. C. *Makromol. Chem.* **1981**, *182*, 393–435.

94. Kanga, R. S.; Hogen–Esch, T. E.; Randrianalimanana, E.; Soum, A.; Fontanille, M. *Macromolecules* **1990**, *23*, 4241–4246.

95. Ishizone, T.; Wakabayashi, S.; Hirao, A.; Nakahama, S. *Macromolecules* **1991**, *24*, 5015–5022.

96. Hirao, A.; Nakahama, S. *Macromolecules.* **1987**, *20*, 2968–2972.

97. Hirao, A.; Ishino, Y.; Nakahama, S. *Macromolecules* **1986**, *19*, 2307–2309 and 1988, 21, 561–565.

98. Ishizone, T.; Hirao, A.; Nakahama, S. *Macromolecules* **1989**, *22*, 2895–2901.

99. Ishizone, T.; Tsuchiya, J.; Hirao, A.; Nakahama, S. *Macromolecules* **1992**, *25*, 4840–4847.

100. Ishizone, T.; Hirao, A.; Nakahama, S. *Macromolecules* **1991**, *24*, 625–626.

101. Ishizone, T.; Sugiyama, K.; Hirao, A.; Nakahama, S. *Macromolecules* **1993**, *26*, 3009–3018.

102. Yilgor, I.; McGrath, J. E. *Adv. Polym. Sci.* **1988**, *86*, 1–86.

103. Yajima, S.; Hayashi, J.; Omori, M. *Nature (London).* **1976**, *261*, 683–685.

104. Seyferth, D.; Wiseman, G. H.; Prud'homme, C. *J. Am. Ceram. Soc.* **1983**, *66*, C13–C14.

105. Miller, R. D.; Michl, J. *Chem. Rev.* **1989**, *89*, 1359–1410.

106. Greber, G.; Reese, H. *Makromol. Chem.* **1964**, *77*, 7–12 and 13–25.

107. Chaumont, P.; Beinert, G.; Herz, J.; Rempp. P. *Makromol. Chem.* **1982**, *183*, 1181–1190.

108. Yamazaki, N.; Nakahama, S.; Hirao, A.; Shiraishi, Y.; Phung, H. M. *Contem. Top. Polym. Sci.* **1984**, *4*, 379–385.

109. Hirao, A.; Nagawa, T.; Hatayama, T.; Yamaguchi, K.; Nakahama, S. *Macromolecules* **1985**, *18*, 2101–2105 and **1987**, *20*, 242–247.

110. Hirao, A.; Hatayama, T.; Nakahama, S. *Macromolecules* **1987**, *20*, 1505–1509.

111. Taki, K.; Hirao, A.; Nakahama, S. *Macromolecules* **1991**, *24*, 1455–1458.

112. Takenaka, K.; Hattori, T.; Hirao, A.; Nakahama, S. *Macromolecules* **1989**, *22*, 1563–1567.

113. Takenaka, K.; Hirao, A.; Hattori, T.; Nakahama, S. *Macromolecules* **1987**, *20*, 2034–2035 and **1992**, *25*, 96–101.

114. Ozaki, H.; Hirao, A.; Nakahama, S. *Macromolecules* **1992**, *25*, 1391–1395.

115. Uzawa, Y.; Hirose, N.; Harada, Y.; Sano, M.; Sekine, M.; Yamaguchi, K.; Ozaki, H.; Hirao, A.; Yoshimori, S.; Kawamura, M. *IEICE Trans.* **1991**, *E74*, 2015–2019.

116. Lee, J-S.; Hirao, A.; Nakahama, S. *Macromolecules* **1988**, *21*, 274–276.

117. Lee, J-S.; Hirao, A.; Nakahama, S. *Macromolecules* **1989**, *22*, 2602–2606.

118. Lee, J-S.; Hirao, A.; Nakahama, S. *Sens. Actuators* **1991**, *B3*, 215–219.

119. Nametkin, N. S.; Topchiev, A. V.; Durgav'yen *J. Polym. Sci.* **1963**, *C4*, 1053–1059.

120. Asami, R.; Oku, J.; Takeuchi, M.; Nakamura, K.; Takaki, M. *Polym. J.* **1988**, *20*, 699–702.

121. Oku, J.; Hasegawa, T.; Takeuchi, M.; Nakamura, K.; Takaki, M.; Asami, R. *Polym. J.* **1991**, *23*, 195–199.

Functional Polyolefins via Ziegler–Natta Polymerization: The Borane Approach

T. C. Chung

This chapter provides a brief overview of Ziegler–Natta polymerization and discusses a new method for the preparation of functionalized polyolefins. This method involves borane-containing intermediates (monomers and polymers) that can be easily converted to the desired functional polymers under mild conditions. Both direct and postpolymerization processes can be used, and most trialkylboranes examined are stable to Ziegler–Natta catalysts. These borane-containing α-olefins can be homo- or co-polymerized with α-olefins to the corresponding borane-containing polyolefins with high molecular weights. In the postpolymerization, the hydroboration of unsaturated polyolefins is also effective. The good solubility of borane derivatives in hydrocarbon solvents provides the conditions for the formation of high molecular weight borane-containing polymers, as well as for the modification of unsaturated polyolefins. A broad range of functionalized polyolefins have been prepared with interesting chemical and physical properties. Potential applications are interfacial compatilizers and supported catalysts.

Polyolefins, especially polyethylene, polypropylene, poly(1-butene), and their copolymers, are used in a wide range of applications. These polyolefins incorporate an excellent combination of mechanical, chemical, and electronic properties; processibility; recyclable capabilities; and low cost (1). Nevertheless, the lack of reactive groups in the polymer structure may limit some of their end uses, particularly where adhesion, dyeability, capability for painting, printability, or compatibility with other functional polymers is paramount. Accordingly, the chemical modification of polyolefins has been an area of increasing interest as a route to higher value products, and various methods of functionalization (2–5) have been employed to alter their chemical and physical properties.

Ziegler–Natta (Z–N) polymerization is the most important method for preparing polyolefins (6), but direct polymerization of functional monomers by this method is normally very difficult because of catalyst poisoning and side reactions (7). The Lewis acid components (Ti, V, Zr, and Al) of this catalyst tend to complex with nonbonded electron pairs on N, O, and halogens of functional monomers, in preference to complexation with the π-electrons of the double bonds. The net result is the deactivation of the active polymerization sites by formation of stable complexes between the catalyst and functional groups, thus inhibiting polymerization. Also, postpolymerization processes modifying the preformed polyolefins suffer from other problems such as degradation of polymer backbone (8). A fundamental need exists to develop new chemistry that can address the challenge of preparing functionalized polyolefins with controllable molecular weight and functional-group concentration. This chapter provides an outline of recent developments based on the use of borane intermediates for solving this problem. But first, an overview of Z–N polymerization is in order.

Z–N Polymerization

Olefins are only polymerized to linear high molecular weight polymers by transition metal catalysts, usually Z–N catalysts (6). In addition, Z–N catalysis is one of the few methods capable of stereoregularly polymerizing α-olefins. Isotactic polyolefins with repeating stereoregular structure (Structure 1) lead to helical chains that can efficiently pack together in crystals. The catalyst-controlled isotactic structure is responsible for the excellent thermal and mechanical properties of stereoregular polyolefins.

Catalyst Development

The discovery of Z–N catalysts started with Karl Ziegler at the Max Planck Institute in Germany. While studying the reactions and oligomerization of ethylene by aluminum alkyls, known as "Aufbau" reaction (6), he found by accident that contamination from the nickel reaction vessel altered the reaction. With this clue, Ziegler began investigating reactions of a series of transition metal salts with $AlEt_3$ and ethylene. In 1953 the first actual Ziegler polymerization, as we know it, used zirconium acetylacetonate and $AlEt_3$ to form high molecular weight linear polyethylene. The process was soon improved by the use of $TiCl_4$ and $AlEt_3$. Giulio Natta, a consultant with Montecatini Company under an agreement with Ziegler, soon undertook further studies of these catalysts. In 1954, the second major breakthrough took place. Natta realized that the Ziegler catalysts could polymerize propylene to a crystalline, stereoregular high polymer. Immediately after these initial experiments, larger research groups were set up in the United States, Italy, and Germany to study these compounds and their resulting polymers. Ziegler and Natta were jointly awarded the Nobel Prize for chemistry in 1963 for their roles in the start of the "Golden Age" of polymer science (6).

In the broadest definition Z–N catalysts are a "mixture of a metal alkyl of base metal alkyls of group I to III and a transition metal salt of metals of group IV to VIII" (6). The base metal component, most often an aluminum alkyl, serves as an alkylating agent for the transition metal salt. The most common one, $TiCl_3$, is a crystal of alternating layers of Ti^{+3} and Cl^{-1} ions. At the edges, cracks, or defects the alkylation takes place on titanium atoms with an unsaturated coordination sphere. As shown in Structure 2, two different active species are proposed: the monometallic and bimetallic active centers. There is evidence that both species might be present in the heterogeneous catalysts.

Structure 1. Molecular structure of isotactic polyolefin.

Mono-metallic   Bi-metallic

Structure 2. Two active species in Ziegler–Natta catalysts.

Both homogeneous and heterogeneous catalysts are useful in Z–N processes. In the past, the heterogeneous systems offered better stereospecific addition for the preparation of isotactic structure. The soluble systems are generally used for polyethylene and ethylene propylene (EP) copolymers where tacticity is irrelevant. However, a new generation of soluble metallocene catalysts with high isospecific control will be discussed later.

Polymerization Mechanism

To this day there is still controversy over the exact mechanism of Z–N polymerization. Even the propagating active center of the reaction is disputed. The most reasonable theory states that the polymer growth takes place at a transition metal–carbon bond. "Cossie was the first to propose a mechanism in which the polymer chain grows by successive olefin insertion" (9) as shown in Scheme I. Cossie's mechanism was later proven by deuterium labeling experiments done by Grubbs (10).

The first step of the reaction is the coordination of the C=C bond's π electrons to the vacant d orbitals of the octahedral Ti (a). The olefin complex then inserts into the Ti–C bond as shown (b and c). The empty orbital and the alkyl group can then switch octahedral positions (d) to generate the correct stereochemistry for the next insertion. These reactions (a–d) are repeated thousands of times to successively add to the growing polymer chain. As long as the empty orbital migration step takes place faster than the next insertion an isotactic structure is produced.

Recent Advances in Catalysis

The discovery of Z–N catalysts sparked a wave of research in organometallic catalysis and polymer sci-

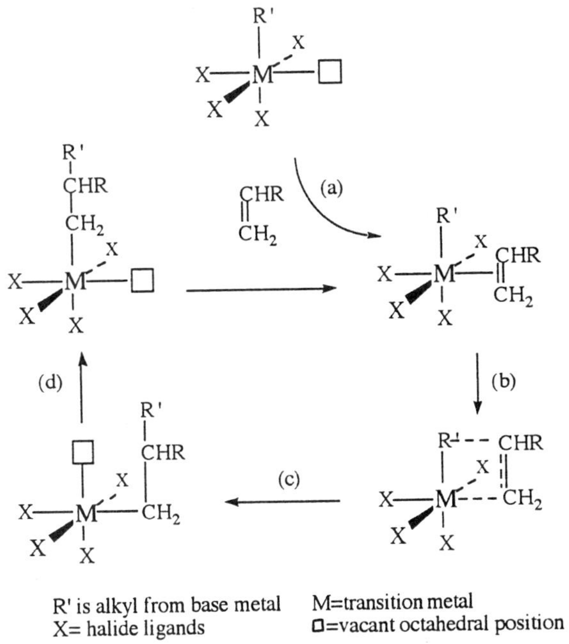

R' is alkyl from base metal   M=transition metal
X= halide ligands   □=vacant octahedral position

Scheme I. Cossie's mechanism of Ziegler–Natta polymerization.

ence that is still growing. The most important advances in Z–N catalysis are in the area of soluble metallocene systems. One example is 1,1'-ethylenedi-η^5-indeylzirconium dichloride in which rotation of the η^5-indenyl about the metal center is prevented by the presence of a bridge (ethylene or silane) between the complexed aromatic rings (11–13). The racemic mixture with the presence of methaluminoxane yields polypropylene (PP) with an isotactic index of 98% and activities as high as 43,000 kg of PP per gram of Zr, an order of magnitude greater than heterogeneous catalysts (14). Usually, the soluble catalysts have a single active site unlike the numerous different active sites found on the heterogeneous transition metal catalysts. The resulting polymers have much narrower molecular weight distributions. Often reactivity differences between comonomers are lowered and more control of comonomer composition is allowed.

Borane-Containing Monomers in Z–N Polymerization

Despite the advances in catalysis and the favorable chemical and physical properties, polyolefins also have certain shortcomings that must be addressed, especially involving other materials such as pigments, paints, glass fibers, and carbon black. Polyolefins exhibit inadequate compatibility with other synthetic polymers and virtually no adhesion to metal or glass. Most of these difficulties would be resolved by the introduction of chemical functionalities or by grafting with a suitable polar monomer, which could enhance

interactions between polyolefin and other materials in polymer blends and composites. Unfortunately, most of the functionalization chemistries reported, including both direct and postpolymerization processes, have not adequately provided the answers to the problem.

In the past few years, we have been investigating a new approach for functionalizing polyolefins by using borane intermediates (15–19). Borane-containing polyolefins were prepared by both direct and post-polymerizations, and the polyolefins were then converted to various functional polymers under mild reaction conditions. The success of this chemistry relies on several important properties of borane intermediates (20):

- stability of borane moiety to transition metal catalysts
- solubility of borane compounds in hydrocarbon solvents (hexane and toluene) used in transition metal polymerizations
- ease of hydroboration reactions
- versatility of the borane groups, which can be transformed to a wide variety of functionalities

Many new functionalized polyolefins with various molecular architectures have been obtained based on this chemistry, and most of these polyolefins would be very difficult to obtain by other methods.

Stability of Boranes to Z–N Catalysts

The feasibility of polymerizing alkenylboranes by Z–N catalysts can be deduced from the stability of such catalysts in organoborane solutions. Several reactions were used to examine this stability. One such example was the homopolymerization of 1-octene in the absence or presence of an equimolar amount of Et_3B. Gel permeation chromatography results in Figure 1 show that Et_3B has no retarding effect on the molecular weight of the polymer.

In addition, Et_3B was recovered by distillation after the polymerization. The same ^{11}B NMR chemical shift (~88 ppm vs. BF_3) was observed in both the original and recovered materials. All results imply that no significant reaction takes place in mixing organoborane and the Z–N catalyst and suggest that borane-substituted α-olefins may be polymerized by Z–N catalysts.

Homopolymerization of Borane Monomers

On the basis of stability tests between trialkylborane and Z–N catalysts, the polymerization capabilities of a series of borane monomers (α-olefin containing an ω-borane group) (16) was examined (Scheme II). Various Z–N catalysts, including both homogeneous and het-

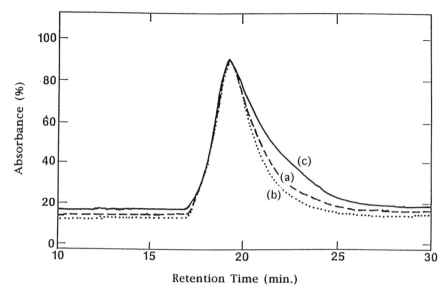

Figure 1. Effect of Et₃B on GPC profiles of poly(1-octene): (a), polymerization in the absence of Et₃B (56% conversion); (b) and (c), Et₃B/1-octene mole ratio, 1:1 (23% and 83% conversions, respectively).

R is alkyl groups, except H and CH₃.

Scheme II. Polymerization of borane monomers by using Ziegler–Natta catalyst.

erogeneous catalysts, have been used for this polymerization. In most cases, high molecular weight polymers were obtained with high yields during relatively short reaction times.

Figure 2 shows the ^{11}B NMR spectrum of a borane-containing polyolefin, which displays a single chemical shift at 88.2 ppm, the same as that of the monomer. The coexistence of the same trialkylborane in both monomer and polymer is crucial and strongly supports the important criteria of stability and polymerization capability of an alkenylborane with Z–N catalysts. Moreover, the polyborane was as easily converted to many functional polymers as the corresponding small borane molecule (*20*).

Direct Copolymerization of α-Olefins and Borane Monomers

Scheme III shows the copolymerization of borane monomers and α-olefins with Z–N catalysts and the conversion of the resulting polymer to typical functional polymer derivatives (*21–23*). The copolymerizations between borane monomer 5-hexenyl-9-borabicyclo[3.3.1]nonane (5-hexenyl-BBN) and various α-olefins such as ethylene, propylene, 1-butene, and 1-octene were carried out in an inert gas atmosphere

by using both heterogeneous catalysts, such as TiCl₃–Et₂AlCl, and homogeneous metallocene catalysts (*13*), such as Cp₂ZrCl₂ and Et(Ind)₂ZrCl₂ with aluminoxane (MAO). In general, the homogeneous catalysts were very effective for polyethylene copolymer syntheses, and isospecific heterogeneous catalysts were used for propylene polymerization to obtain PP copolymers with high melting points. The state of the polymerization mixture was very dependent on the nature of the α-olefin. In the 1-octene case, a homogeneous solution was observed throughout the whole copolymerization reaction. For propylene, on the other hand, almost immediately a white precipitate could be seen in the deep purple solution, and this precipitate is due to the PP crystalline structure with a high isotactic content. The copolymerization was terminated after a certain reaction time by addition of isopropanol to destroy the active metal species. Excess isopropanol was used to ensure the complete coagulation of polymer from solution. The resulting borane-containing copolymers were isolated from solution by filtration and then washed repeatedly with isopropanol.

The resulting borane-containing copolymers poly(ethylene-*co*-5-hexenyl-BBN) (PE-B), poly(propylene-*co*-5-hexenyl-BBN) (PP-B), and poly(1-butene-*co*-5-hexenyl-BBN) (PB-B) containing low concentration of borane monomers (<10 mol%) are insoluble in common organic solvents at room temperature but are soluble at higher temperatures. On the other hand, the copolymers of poly(1-octene-*co*-5-hexenyl-BBN) (PO-B) in all compositions are soluble in most hydrocarbon solvents at room temperature. The borane concentration can be measured by solution ^{11}B NMR spectroscopy. By using a known amount of triethylborate

(a)
(b)
200
150
100
50
0
ppm
BF3·OEt2

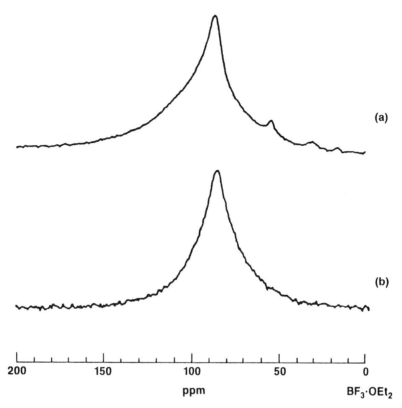

Figure 2. ^{11}B NMR spectra of (a) poly(propylene-*co*-hexenyl-BBN) with 3.5 mol% borane concentration and (b) 5-hexenyl-BBN.

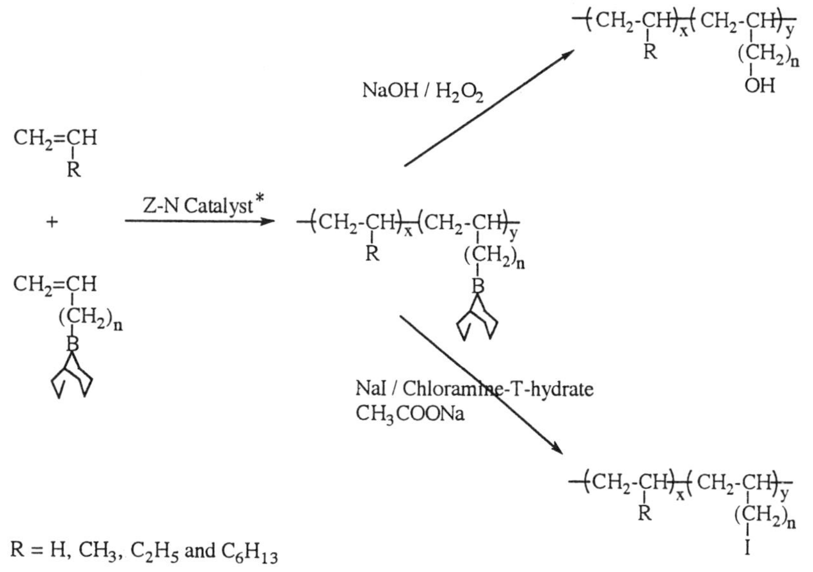

NaOH / H2O2
Z-N Catalyst*
NaI / Chloramine-T-hydrate
CH3COONa
R = H, CH3, C2H5 and C6H13
* Z-N Cat. = TiCl3/EtAlCl2, Cp2ZrCl2/MAO and C2H4[Ind]2ZrCl2/MAO

Scheme III. Copolymerization of borane monomer and α-olefin in Ziegler–Natta process.

(chemical shift at 19 ppm) as a reference and comparing the integrated peak areas, the borane content in the copolymer can be determined quantitatively.

The borane-containing copolymers are stable for long periods of time (6 months in dry-box) or at elevated temperatures (90 °C during NMR measurement) as long as O_2 is excluded. By exposing a copolymer to air, it becomes insoluble at any temperature. The cross-linking reaction is due to free radical couplings, which take place during the oxidation of borane groups by oxygen.

For controlled oxidation reactions, polymers were treated with NaOH–H_2O_2 at 40 °C for 3 h. The borane groups were completely converted to the corresponding hydroxy groups even in the PE-B, PP-B, and PB-B heterogeneous cases. The effective reaction is apparently due to the semicrystalline microstructure of the copolymers. Figure 3a shows ^{1}H NMR spectrum of the

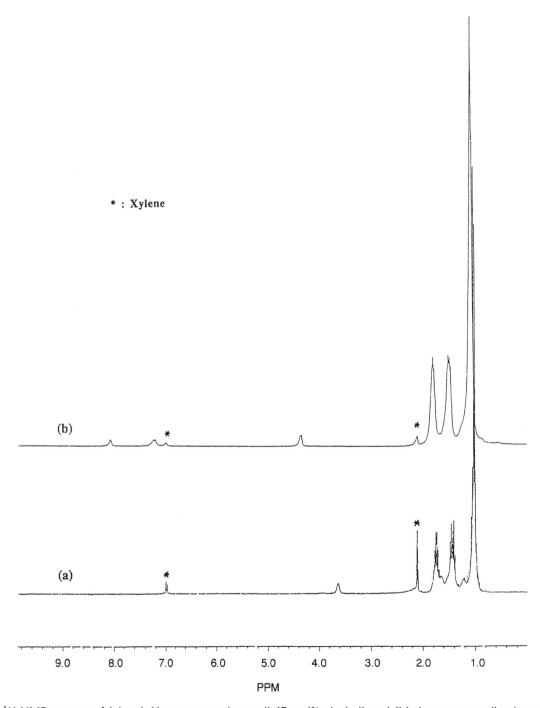

Figure 3. ^{1}H NMR spectra of (a) poly(1-propene-co-hexenol) (5 mol% alcohol) and (b) the corresponding benzoate ester.

hydroxylated polypropylene (PP–OH) containing 5 mol% of hydroxy groups. The peak at 3.5 ppm corresponds to the protons (–CH$_2$–OH) adjacent to the primary alcohol. The PP–OH polymer can be further modified to its ester form by benzoyl chloride. The esterification was complete and showed the disappearance of the primary alcohol peak at 3.5 ppm and the appearance of CH$_2$–O–C=O at 4.0 ppm and two aromatic bands around 7.2 and 8.1 ppm (Figure 3b).

To determine the copolymer compositions, the peak at 3.5 ppm in Figure 3a was integrated in reference to the three (major) overlapping peaks between 1.95 ppm and 0.72 ppm to compare the number of protons adjacent to the alcohol (–CH$_2$–OH) to the rest of the aliphatic protons. Likewise, the complete oxidation reactions were shown in other polyolefin copolymers. Figure 4 shows ^{1}H NMR spectra of hydroxylated poly(1-butene) (PB–OH) and poly(1-octene) (PO–OH) containing 7 and 20 mol% hydroxy groups, respectively.

Borane-containing polymers were also converted to the corresponding iodine-containing polymers. Conversion of borane to iodine groups was accomplished by using NaI and chloramine-T-hydrate (*N*-chloro-*p*-toluenesulfonamide, sodium salt) under basic conditions at room temperature. The mild oxidizing agent chloramine-T and NaI generate I$^+$ in situ, which reacts with the borane base complex by an S$_E$2 mechanism. The facile conversion is similar to those of small borane molecules (*20*) and borane-terminated telechelic polymers (*18*). The resulting iodine-containing polymers were soluble in hydrocarbon solvents. However, the iodide PP was initially soluble in xylene at 130 °C and soon became yellowish and increasingly less soluble. The reason for the resulting cross-linking is unknown.

Reactivity Ratios

The best way to describe a copolymerization is to measure the reactivity ratios (*24*) of the comonomers. Thus, a series of experiments were carried out by varying monomer-feed ratios and comparing the polymer composition at low conversion (Table I). The polymer samples PP-B and PB-B were obtained from the copolymerization of 5-hexenyl-BBN with propylene and 5-hexenyl-BBN with 1-butene, respectively. The difference in reactivities of propylene and 5-hexenyl-BBN makes it difficult to determine the amount of borane groups in the copolymers at low 5-hexenyl-BBN concentration. On the other hand, the reactivity of 1-butene is much closer to 5-hexenyl-BBN, and hence their reactivity ratios can be worked out more readily.

Reactivity ratios of propylene or 1-butene (M_1) (r_1 = k_{11}/k_{12}) and 5-hexenyl-BBN (M_2) ($r_2 = k_{22}/k_{21}$) are

estimated by the Kelen–Tudos method (*25*). The calculation is based on the following equation.

$$\eta = r_1 \xi - r_2/\alpha(1 - \xi)$$

$$\eta = G/\alpha + F \text{ and } \xi = F/\alpha + F$$

where η is intrinsic viscosity, variable $x = [M_1]/[M_2]$ in feed and variable $y = d[M_1]/d[M_2]$ in copolymer; $G = x(y - 1)/y$, $F = x^2/y$; and $\alpha = (F_m \times F_M)^{1/2}$. Variables F_m and F_M are the lowest and highest values of F, respectively. Figure 5 shows the plot η versus ξ and the least-squares best-fit line.

The extrapolation to $\xi = 0$ and $\xi = 1$ gives $-r_2/\alpha$ and r_1. We obtain $r_1 = 70.476$. $r_2 = 0.028$, and $r_1 r_2 = 1.973$ for propylene–5-hexenyl-BBN and $r_1 = 7.13$, $r_2 = 0.41$, and $r_1 r_2 = 2.92$ for 1-butene–5-hexenyl-BBN, respectively. Obviously, both copolymerization reactions are not ideal cases. The values of $r_1 r_2$ are far from unity, and the reaction is favorable for α-olefin incorporation, especially in the copolymerization of propylene and 5-hexenyl-BBN. In the batch reaction with the fixed monomer ratio of propylene–5-hexenyl-BBN, either a narrow compositional distribution was obtained at low conversion (at extremely low yield) or a broad distribution of copolymer composition was obtained for high conversions.

Continuous Polymerization

More uniform composition of copolymer feasibly can be obtained by an engineering approach, such as the control of monomer feed ratio during the copolymerization. In the preliminary experiments, the more reactive α-olefin monomer was added gradually to keep its concentration constant relative to the borane monomer. The α-olefin was added in decreasing amounts to account for the consumption of borane monomer in the feed (*see* Experimental section). This approach can produce copolymers with narrower compositional distributions and at a higher yield (Table II).

Because of the susceptibility of borane groups to air oxidation, the borane-containing polymers were oxidized to the corresponding hydroxylated polymers for characterization. The molecular weights of the polymers were determined by viscosity measurement in a cone/plate viscometer at 135 °C in decalin solution. To enhance the solubility of functionalized polymers, the hydroxylated polymers were esterified completely with benzoyl chloride. The viscosity-average molecular weights (M_v) were calculated using the Mark–Houwink equation, $[\eta]_0 = K(M_v)^a$, where $K = 11.0 \times 10^{-3}$ (mL/g) (*26*) and $a = 0.80$. As shown in Table II, M_v values are high (about 200,000 g/mol) for all samples. The absence of any significant change in the

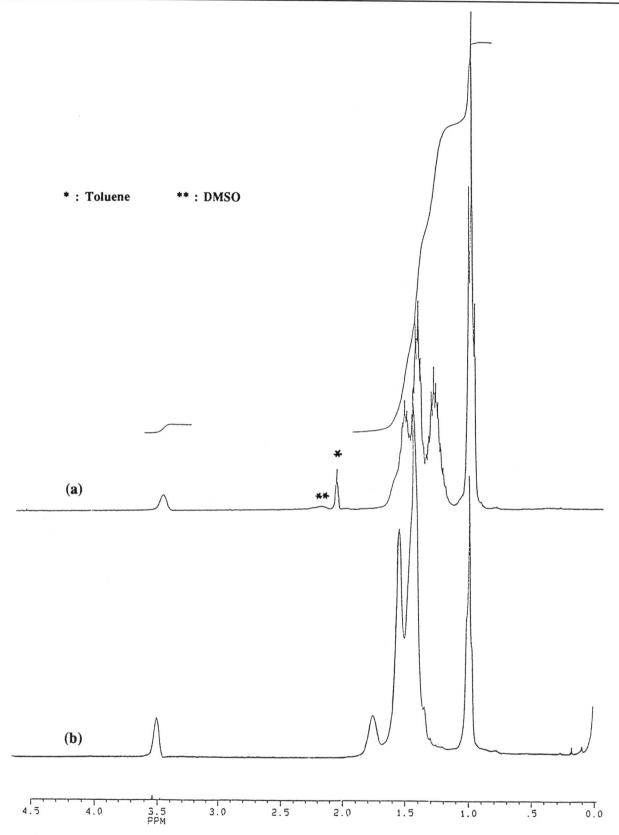

Figure 4. ¹H NMR spectra of (a) poly(1-butene-*co*-hexenol) (7 mol% of alcohol) and (b) poly(1-octene-*co*-hexenol) (20 mol% of alcohol).

Table I. Summary of Copolymerization Between Borane Monomer (M_2) and α-Olefins (M_1) at Low Concentration

M_1	Mol% M_2 in Feed	Mol% M_2 in Copolymer [a]	Yield (%)
Propylene	33.33	1.2	15.2
Propylene	50.00	1.7	11.0
Propylene	66.66	2.7	8.2
1-Butene	14.29	2.4	9.9
1-Butene	25.00	3.8	8.9
1-Butene	33.33	7.0	9.4
1-Butene	50.00	15.5	6.3

[a]Values determined by ^{11}B NMR spectroscopy.

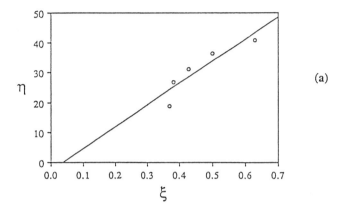

(a)

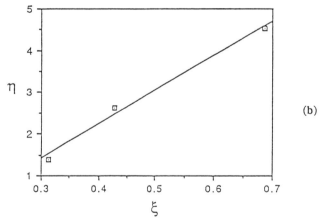

(b)

Figure 5. Kelen–Tudos plot of η verse ξ for (a) propylene and (b) 1-butene with hexenyl-BBN.

molecular weight due to the addition of the borane monomer is quiet interesting, especially in the heterogeneous reaction condition. Obviously, no catalyst poisoning by borane is indicated. In addition, the solubility of the borane monomer appears to be similar to those of the α-olefins. Table III shows the fractionation results of PP–OH copolymers obtained by both continuous and batch processes using 10/1 and 1/1 monomer feed ratios, respectively, and near complete monomer conversion.

Both samples from continuous and batch processes were subjected to fractionation by sequential Soxhlet extractions by using methanol, methyl ethyl ketone (MEK), heptane, and xylene all under N_2. The solvents were chosen so as to separate by polarity (OH content) and crystallinity (isotacticity and PP sequence length). The PP–OH copolymers with >60% alcohol content are soluble in MeOH. The MEK fraction was a rubbery, tacky material indicative of low isotacticity. Because of the low boiling point of heptane, its fraction represents polymer with intermediate tacticity or with more hexenol units that reduce crystal formation. Xylene should dissolve all the remaining highly isotactic polymer. The continuous reaction offers much more narrow composition distribution and high crystallinity.

Postpolymerization Functionalization

Another approach (27) to prepare borane-containing PP involves hydroboration reaction of unsaturated PP and subsequent oxidation to convert borane groups to hydroxy groups (Scheme IV). The hydroboration reactions were carried out under heterogeneous reaction conditions by suspending the powder form of poly-(propylene-*co*-1,4-hexadiene) (PP-1,4-HD) in tetrahydrofuran (THF). Both diborane (BH_3) and dialkyborane (BBN) were used for the hydroboration reaction. One BH_3 molecule can react with up to three double bonds. This process can cause cross-linking reactions between polymer chains. However, cross-linking did not cause any notable difference in the heterogeneous state of the reaction mixture. To ensure the complete reaction of the internal double bonds, the reaction was run for 5 h at 65 °C in THF.

Table II. Copolymerization of α-Olefin (M_1) with 5-Hexyl-BBN (M_2) by Continuous Reaction

Polymer	Mol% M_2 in Feed	Mol% OH in Polymer[a]	Reaction Time (h)	Yield (%)	Intrinsic Viscosity (η)	M_v (g/mol)
PP	0	0	2	93	2.07	230,000
PP–OH	10	3	3	62	1.78	183,000
PP–OH	13	5	5	35	1.71	174,000
PB–OH	5	2.5	2	70	—	—
PB–OH	10	6.5	2	66	—	—

[a]Values determined by ^{1}H NMR spectra.

Table III. Fractionation of PP–OH Obtained by Batch and Continuous Methods

	Wt% of Polymer Extracted				
Method	Methanol	MEK	Heptane	Xylene	Insoluble
Continuous [a]	None	8.5	14.2	77.7	None
Batch [b]	42.6	18.9	7.6	23.1	7.7

[a]Propylene–5-hexyl-9-BBN mole ratio, 10/1.

[b]Propylene–5-hexyl-9-BBN mole ratio, 1/1.

The borane groups in the resulting polymers were oxidized by using NaOH–H_2O_2 reagents at 40 °C for 3 h. Remarkably, a complete conversion was achieved in a heterogeneous mixture by involving a crystalline–hydrophobic polyolefin with an aqueous reagent under mild reaction conditions. Figure 6 compares the ^{1}H NMR spectra of poly(propylene-co-1,4-hexadiene) before and after hydroboration and oxidation reactions.

The concentration of the unsaturated monomer units, corresponding to the chemical shift at 5.5 ppm, decreases to the limit of NMR sensitivity and shows the secondary alcohol peak appearing at 3.5 ppm. Apparently, both hydroboration and oxidation reactions were not inhibited by the insolubility of PP. While the PP segments are crystallized, the side chains containing the double bonds are expelled out to the amorphous phase, which is capable of swelling by the solvent used for the reaction. In addition, the high reactivities of hydroboration and oxidation reactions certainly enhance the effectiveness of functionalization.

The concentration of functional groups can be controlled by the percentage of double bonds in polymer as well as the quantity of borane reagents. The functional group is secondary due to the internal double bonds in PP. Unfortunately, only internal double bonds can be introduced into polyolefins by Ziegler–Natta polymerization. In the case of direct copolymerization using borane monomer, the primary functional

Table IV. Main Properties of PP-1,4-HD and Its Hydroxy Derivative

Polymer	Intrinsic Viscosity	Mol Wt (g/mol)	Melting Point (°C)	Heat of Fusion (J/g)
PP–1,4-HD	1.373	131,900	153	50
PP–OH	1.425	138,300	151	37

group located at the end of the side chain is obtained in the functionalized PP.

The molecular weights of polymers were determined by intrinsic viscosity measurement in a cone–plate viscometer at 135 °C in decalin solution. The hydroxylated polymer (PP–OH–C) was esterified with benzoyl chloride to enhance its solubility. Table IV lists the main properties of PP copolymers and its hydroxy derivative. There is no appreciable change in intrinsic viscosity and molecular weight after functionalization.

Physical Properties of Hydroxylated PP

Thermal properties of copolymers were studied by both thermogravimetric analysis (TGA) and differential scanning calorimetry (DSC). Figure 7 compares the TGA results between PP–OH copolymer with 5 mol% hydroxy groups (b) obtained from direct copolymerization and two homopolymers, isotactic PP (a) and polyhexenol (c), under argon (top) and air (bottom) atmospheres.

The thermal stability of PP–OH is higher than that of pure isotactic PP. The decomposition temperature of PP–OH is above 280 °C in argon and about 205 °C in air. A slightly better resistance in decomposition may be due to the relatively high thermal stability of the primary alcohol and the fact that the functional group is pendant on a side chain. In fact, the thermal decomposition temperature of poly(vinyl alcohol) is below 150 °C. Primary alcohols are more thermally stable than secondary alcohols. In addition, the decomposi-

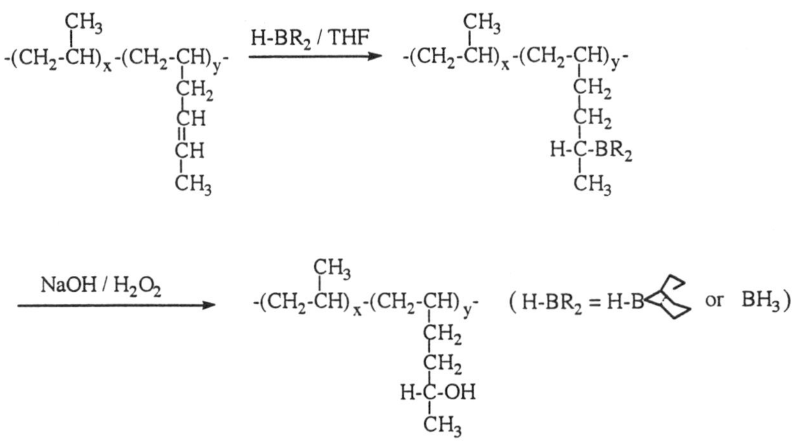

Scheme IV. Hydroboration and oxidation of poly(propylene-co-1,4-hexadiene).

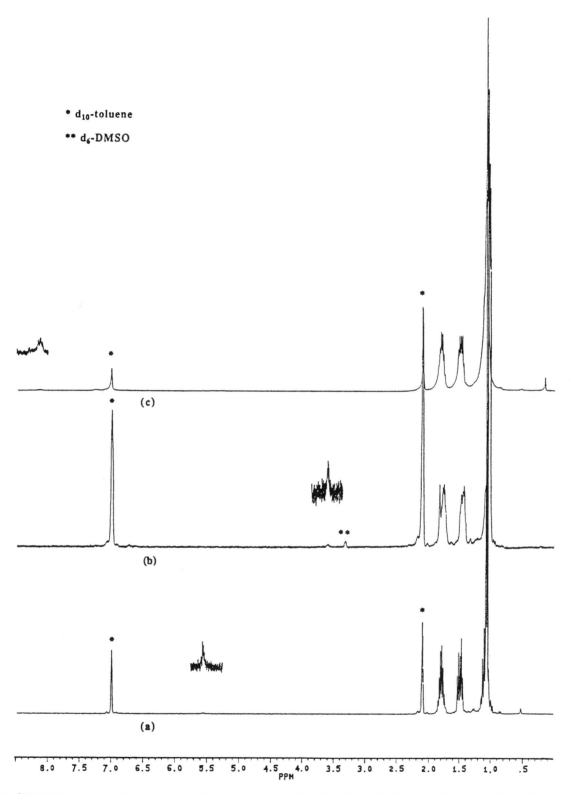

* d$_{10}$-toluene

** d$_6$-DMSO

Figure 6. ^{1}H NMR spectra of (a) poly(propylene-*co*-1,4-hexadiene) (1.7 mol% 1,4-hexadiene) and its (b) hydroxylated and (c) benzoated derivatives.

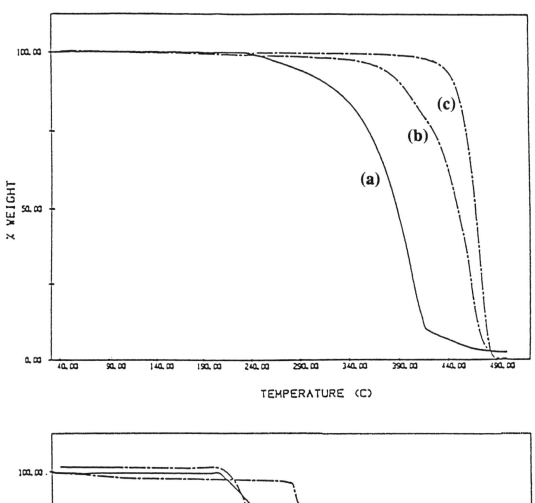

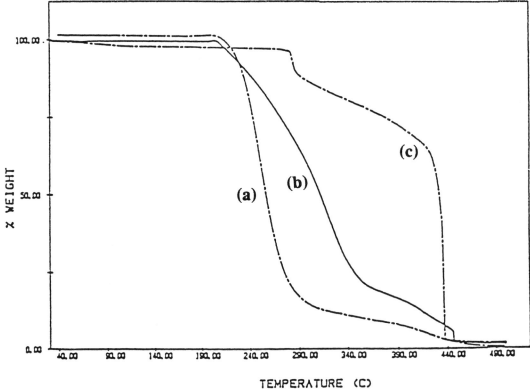

Figure 7. TGA of (a) isotactic PP, (b) PP–OH copolymer obtained by direct copolymerization, and (c) polyhexenol under argon (top) and air (bottom) atmospheres.

tion of functional groups in the pendent side chains will not affect the polymer backbone; therefore, the copolymer will maintain its high molecular weight and mechanical properties.

DSC was used to determine the effect of copolymer composition on the crystallinity. Figure 8 shows DSC curves of the two PP–OH copolymers, from direct copolymerization (5 mol% OH) and by modification of unsaturated PP (1.7 mol% OH), and isotactic PP. All samples have quite similar melting points (T_m ~ 160 °C); the actual depression in melting point caused by functional groups is <10 °C. Higher thermostability and slightly lower melting point in PP–OH copolymers ensure their processibility under conditions similar to those for isotactic PP.

The crystalline structure of functionalized PP was also studied by a polarized optical microscope. Figure 9 shows the micrographs for the three polymers in Figure 8.

All samples were treated in parallel under similar thermal conditions. The spherulites are crystalline, whereas the nonspherulitic regions are amorphous. It is not surprising to see bigger spherulites in pure isotactic PP and smaller ones in PP–OH. Overall, the trend is consistent with DSC results. Both PP–OH copolymers are highly crystalline with smaller spherulites and slightly lower T_m values compared with isotactic PP. Effective preservation of crystallinity in

PP–OH copolymers is apparently due to the sufficient length of PP sequences for forming crystalline structures. Most of the hydroxy groups in the PP–OH copolymers are located in the amorphous phases (Figure 10). The hydroxy groups are very mobile and react with $EtAlCl_2$ almost quantitatively even under heterogeneous reaction conditions.

Co-Crystallization Between PP and PP–OH

The co-crystallization of PP and PP–OH was studied by DSC analysis. A drawn PP film and PP–OH copolymer were melted together, and the DSC endotherm was observed over the normal melting range of these two components. Upon cooling, crystallization occurred at approximately 110–130 °C, and peak shape and location seemed to be the net result of individual PP and PP–OH crystallization. However, when the sample was reheated after cooling, co-crystallization of PP and PP–OH was shown by the presence of a single melting peak intermediate below the melting region of PP (Figure 11b).

A laminated sample containing PP and PP–OH layers was also studied by DSC (Figure 11c). The melting endotherm of the PP–PP–OH laminate sample consists of three peaks. The peak in the middle is at approximately the same position as the previously identified co-crystallization peak in Figure 11b,

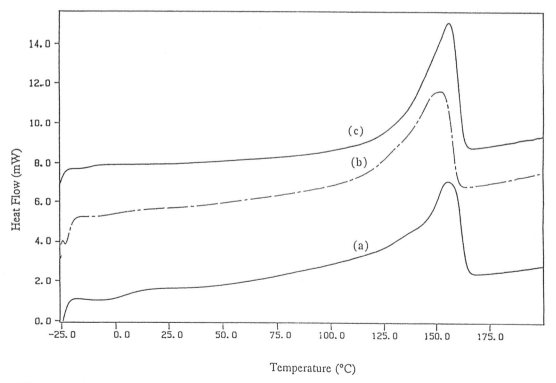

Figure 8. DSC curves of PP–OH copolymers from (a) direct copolymerization and (b) functionalization and (c) that of isotactic PP.

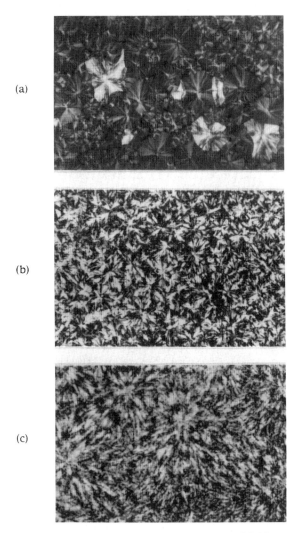

(a)

(b)

(c)

Figure 9. The polarized optical micrographs of (a) isotactic PP, (b) poly(propylene-*co*-hexe-6-ol) with 5 mol% OH, and (c) poly(propylene-*co*-hexe-4-ol) with 1.7 mol% OH.

whereas the other two peaks are indicative of pure PP and PP–OH. Accordingly, co-crystallization occurs at the PP–PP–OH interface in the laminate.

Adhesion Studies of PP–Al and PP–Glass Laminates

The study of adhesion between PP and glass or aluminium (*28, 29*) has both scientific and commercial interests. The hydroxylated PP was used as a surface modifier to improve PP adhesion. The flexibility of hydroxy groups located at the ends of side chains in PP–OH may enhance the interaction of PP–OH to substrates. On the other hand, the high crystallinity PP–OH (similar T_m as PP) may co-crystallized with PP, which provides strong adhesion. Both drawn and undrawn PP films (commercial products) were laminated with PP–OH treated substrates. As shown in Table V, the peel test results are compared with those

obtained from standard acid-etched samples, which involve surface modification of both PP and Al by a dichromate–sulfuric solution and then the same lamination procedure. PP–Al laminates bonded by PP–OH exhibited a remarkable fivefold increase in peel strength over acid-etched samples.

Contact angles of water drops on peeled Al and PP surfaces revealed typical hydrophobic surfaces with angles of 130–140°. The same results were revealed in scanning electron microscopy (SEM) studies. As shown in Figure 12, both peeled Al and PP surfaces show similar morphologies.

Cohesive failure (not adhesive failure) occurs and gives rise to the high peel strengths observed for these PP–Al laminates. When peeling Al from well-bonded PP, the failure path appeared to propagate within the PP matrix. This failure path also helps to explain why the peel strength of drawn PP–Al laminates are greater than those of undrawn PP sample.

The same results were observed in the PP–glass laminates. Table VI shows the 90° peel test results of several PP–glass laminates, including both undrawn and drawn PP samples with various glass surfaces. Most of the specimens have strong peel strength. The drawn PP–acid-etched E-glass (a calcium-boroaluminosilicate glass) laminates, produced with PP–OH interfacial agent and hot-press process, have considerably high peel strength (about 1100 to 1300 N/m). For comparison, a control experiment by laminating acid-etched PP film with acid-etched E-glass, without the use of PP–OH, shows very low adhesion (too low to get any significant number in the 90° peel test). This result indicated the important role of the PP–OH layer as the interfacial agent between glass and PP film. The results of SEM and the contact angles of water drops on the peeled surfaces all indicate cohesive failures in PP–PP–OH matrix.

The strong adhesion between PP–OH and glass surface is primarily due to chemical bonding. The reflection IR studies provide the experimental evidence of the chemical reaction between free Si–OH groups on the glass surface and hydroxy groups in PP–OH. Figure 13 shows the IR spectra of two samples: silica glass surface and coated PP–OH–glass after thermal treatment.

The IR absorption of the glass surface in Figure 13a shows two absorptions: the broad band between 3000 to 3700 cm⁻¹ due to H_2O and vicinal Si–OH absorption modes and the small peak at 3745 cm⁻¹ corresponding to the free Si–OH groups on the surface. After PP–OH coating with thickness about 1000 Å, several additional absorption peaks between 3000 and 2800 cm⁻¹, corresponding to saturated hydrocarbon, appear in Figure 13b. At this stage, the PP–OH thin film does not interact chemically with the glass surface and can be redissolved in xylene at elevated tempera-

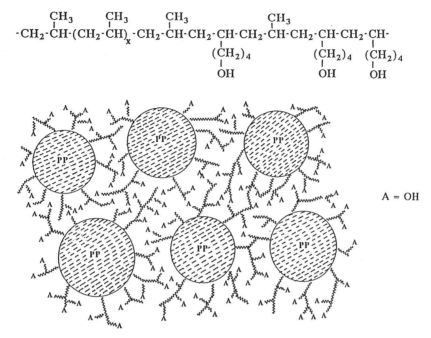

Figure 10. Pictorial illustration of PP–OH morphology.

DSC Curves

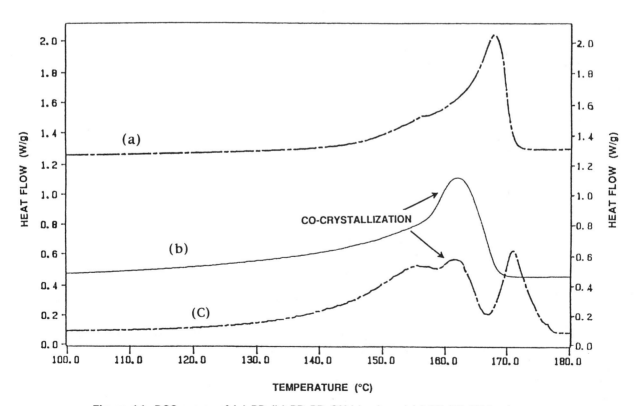

Figure 11. DSC curves of (a) PP, (b) PP–PP–OH blend, and (c) PP–PP–OH laminate.

Table V. Peel Strength of PP–Al Laminates with PP–OH

Sample	Peel Strength (N/m)
Acid-etched	
Undrawn PP–Al	126 ± 26
Drawn PP–Al	130 ± 34
PP–OH solution cast	
Undrawn PP–Al	675 ± 44
Drawn PP–Al	1155 ± 52

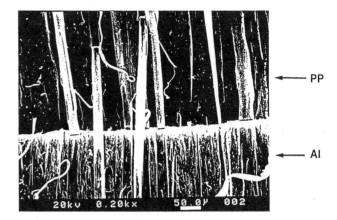

Figure 12. Scanning electron micrographs of the peeled surfaces of drawn PP–Al laminate.

Table VI. Peel Strength of PP–Glass Laminates with PP–OH

	Undrawn PP		Drawn PP	
Glass Surfaces	a	b	a	b
E-glass				
Cleaned by acetone and hydrated	223 ± 26	327 ± 23	297 ± 28	415 ± 61
Acid-etched	416 ± 77	1092 ± 141	520 ± 83	1373 ± 18
Quartz				
Cleaned by acetone and hydrated	362 ± 30	529 ± 78	488 ± 42	672 ± 86

Note: Peel strength values are in N/m. Method a was solution casting with 20 μm PP–OH thickness. Method b was hot-press process with 25 μm PP–OH thickness.

ture. After thermal treatment of PP–OH–glass at 140 °C for 2 h, the peak at 3745 cm^{-1} is almost diminished (Figure 13c). In addition, the PP–OH thin film becomes completely insoluble in xylene even up to 150 °C. These results strongly suggest effective chemical reaction of free silanol groups and primary hydroxy groups to form –Si–O–C– bonds, which may be responsible for the high adhesion at the PP–OH–glass interface.

Immobilization of Lewis Acids on Functionalized Polyolefins

Another application of functionalized polyolefins is in the chemical immobilization of soluble catalysts, so that catalysts can be recovered and reused for many reaction cycles. This process presents several important advantages in chemical production processes. These advantages include waste reduction, lower cost, and simpler purification and all save energy and are environmentally friendly. These considerations become even more important when the reactions require a large quantity of catalyst, such as in the oligomerization of olefins.

The semicrystalline functionalized polyolefins, such as the hydroxylated PP or poly(1-butene), in various forms (film, fiber, and particle) were reacted with Lewis acids, such as EtAlCl$_2$ and BF$_3$ (Scheme V). The immobilization reaction (30, 31) was usually carried out at room temperature by stirring the hydroxylated polyolefin solid with excess EtAlCl$_2$ solution for a few hours. The unreacted reagent was removed by washing the resulting supported catalyst with pure solvent several times. Solid-state ^{27}Al NMR spectroscopy was used to analyze the catalytic species in the supported catalyst. The spectra of three supported catalysts were compared with their corresponding soluble catalysts. Figure 14a shows the ^{27}Al spectrum of the catalyst sample prepared by reacting hydroxylated PP with EtAlCl$_2$ at room temperature. Only a single peak at 89 ppm, corresponding to –OAlCl$_2$, was observed. No peak was found at 170 ppm to correspond to EtAlCl$_2$.

The same chemical reaction was also observed in a reference sample by using 1-pentanol instead of the hydroxylated PP in a reaction with a stoichiometric amount of EtAlCl$_2$. Figure 14b is the solution ^{27}Al NMR spectrum of the resulting C$_5$H$_{11}$-OAlCl$_2$ that indicates the same chemical shift corresponding to a single –OAlCl$_2$ species.

In the case of BF$_3$, the hydroxy groups in PP–OH copolymer were converted to lithium alkoxide groups by simply mixing polymer particles with *n*-butyl lithium solution. After washing out excess *n*-butyl lithium, the polymer particles were mixed with a

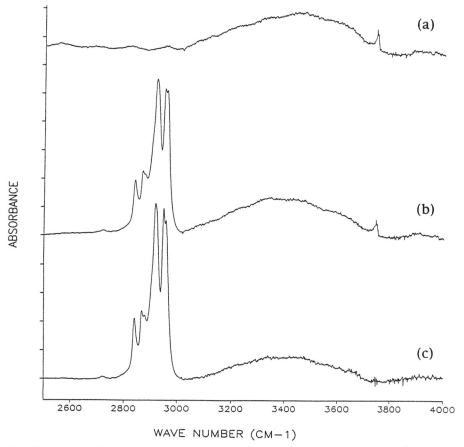

Figure 13. Reflection IR spectra of (a) glass surface, (b) PP–OH (thickness about 1000 Å) coated on glass, and (c) PP–OH–glass after thermal treatment.

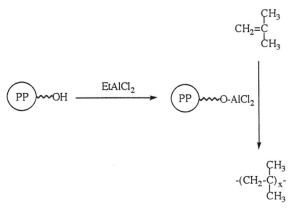

Scheme V. Polymerization of isobutylene by an immobilized catalyst prepared by the reaction of PP–OH and EtAlCl$_2$.

BF_3–CH_2Cl_2 solution. Figure 15 compares the ^{11}B NMR spectra of the immobilized catalyst and the corresponding soluble catalyst prepared from 1-pentanol, and both spectra are almost identical with a peak at 0 ppm corresponding to –OBF_2 group. This result was also confirmed by elemental analysis, which showed a mole ratio of 1:2 between B and F in the PP–OBF_2 sample.

The polymer-supported catalyst (*30, 31*) was used as the Lewis acid catalyst in the carbocationic polymerization of isobutylene (Scheme V). After the reaction, polyisobutylene (PIB) was obtained by filtering out the supported catalyst then removing solvent and unreacted monomers under vacuum. The recovered catalyst was mixed with another isobutylene–hexane solution, followed by the same separation and recovery processes. This reaction cycle was repeated a number of times. Table VII summarizes the results of PIB obtained by using the fine powder form (particle size, <1 mm) of PP-supported –$OAlCl_2$ catalyst. Typically, the monomer (isobutylene) to catalyst (–$OAlCl_2$) ratio was about 500 to 1.

In most reaction cycles, the conversion from monomer to polymer was complete within 15 min. The same catalyst activity was maintained for several reaction cycles. This unusually high and stable catalyst activity in the polyolefin-supported system must be due to a high and stable surface area, which could be a consequence of small particle size, the crystallinity of polyolefin, and the flexibility of the side chain between the catalyst and polymer backbone.

On the other hand, the catalyst PP–O–BF_2 produced relatively low molecular weight PIB (Table VIII).

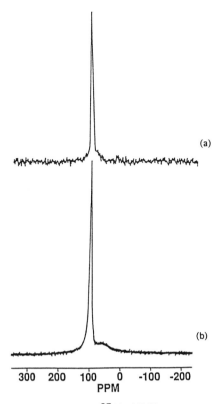

Figure 14. Solid-state ^{27}Al NMR spectrum of (a) PP–O–AlCl$_2$ and ^{27}Al NMR spectrum of (b) C$_5$H$_{11}$-OAlCl$_2$ in toluene.

Table VII. Preparation of PIB in the Presence of PP–OH–AlCl$_2$ Powder in Hexane

Run	Temp. (°C)	Time (min)	M_n (g/mol)	PDI
1	25	90	1050	2.0
2	25	60	1150	1.6
3	25	20	1150	1.8
4	25	15	1140	1.6
8	25	15	1180	1.5
10	0	15	4540	2.6
C$_5$–O–AlCl$_2$	25	15	1180	2.3
C$_5$–O–AlCl$_2$	0	15	5450	2.6

Note: The isobutylene monomer to OAlCl$_2$ catalyst ratio was about 500 to 1. M_n is number-average molecular weight. PDI is polydispersity index. Yields for all runs were 100%.

Table VIII. Preparation of PIB by PB–O–BF$_2$ Powder in Hexane

Run	Temp. (°C)	Time (min)	M_w (g/mol)	PDI
1	25	15	400	1.1
2	25	15	450	1.2
3	25	15	450	1.2
4	25	15	420	1.4
5	0	15	580	1.2
8	0	15	640	1.5
C$_5$–O–BF$_2$	25	5	500	1.9
C$_5$–O–BF$_2$	0	5	1080	2.0

Note: The isobutylene monomer to –OBF$_2$ catalyst ratio was about 500 to 1. M_w is weight-average molecular weight. PDI is polydispersity index. Yields for all runs were 100%.

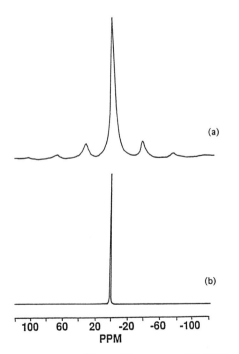

Figure 15. Solid-state ^{11}B NMR spectra of (a) PP–O–BF$_2$ and (b) C$_5$H$_{11}$-OBF$_2$ in toluene.

The monomer (isobutylene) to catalyst (–OBF$_2$) ratio was about 500 to 1. The molecular structures, especially the unsaturation aspects, in PIB polymers prepared by the immobilized catalysts. The PIB with terminal double bonds is a very desirable material, called "reactive PIB", which can be further functionalized under mild reaction conditions. Figure 16 shows the ^1H NMR spectra of PIB polymers prepared by PP–O–AlCl$_2$ and PP–O–BF$_2$ catalysts at room temperature.

Figure 16a is very similar to that of PIB prepared by soluble Al catalysts such as AlCl$_3$, EtAlCl$_2$, Et$_2$AlCl, and the controlling C$_5$H$_{11}$–O–AlCl$_2$ catalyst. Two major peaks, at 0.95 and 1.09 ppm, are due to CH$_3$ and CH$_2$ protons of PIB. Some weak peaks in the olefinic region occurred between 4.5 and 6.0 ppm. This unsaturated double bond in the polymer chain is evidence that a proton chain-transfer reaction (β-proton elimination) occurs in the polymerization process. There are two quartets at 5.4 and 5.2 ppm and two singlets at 4.9 and 4.6 ppm. The singlets at 4.9 and 4.6 ppm are indicative of two types of nonequivalent vinylidene hydrogens that may be located at the end of polymer chain. The quartets at 5.4 and 5.2 ppm are the olefinic hydrogens

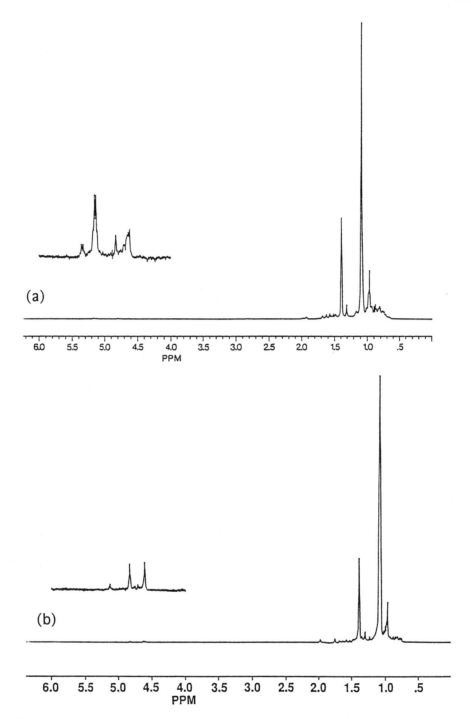

Figure 16. ^{1}H NMR spectra of PIB prepared by (a) PP–O–AlCl$_2$ and (b) PB–O–BF$_2$.

coupled to the methyl group and are due to the internal double bonds. A significantly high amount of internal double bonds with various structures are present, and this result indicates that a carbocationic isomerization is taking place by the Lewis acid catalyst following the polymerization. A very different ^{1}H NMR spectrum of PIB is observed for supported catalyst PP–O–BF$_2$ (Figure 16b). The chemical shifts in the double bond region consist of two major singlets at 4.9 and 4.6 ppm that correspond to terminal double bond, and a small peak at 5.15 ppm corresponds to an internal double bond. By comparing the integrated intensities between olefinic peaks, the PIB has more than 85% of terminal double bonds. The reason for such a high percentage of terminal double bonds in PP–O–BF$_2$ may be due to the absence of isomerization.

Experimental Details

Instruments and Measurements

The 90° peel test (ASTM D 773–47) was performed on an Instron 4201 Tensile Tester at a rate of 1 cm/min by peeling PP backing materials from PP–OH coated matrices. Peel strengths were measured on 1 × 3 in. specimens that were clamped to the supports. The peel strength was calculated from peel force–width of test specimen. Five samples were measured to obtain the average peel strength. The contact angle was measured directly on the sessile drop of the distilled water on each sample with the help of the Rame Hart telescope goniometer. The sessile drop has a diameter of at least 4 mm in order to eliminate the size influence on the contact angle. The contact angles were obtained from the average of 10 points on each sample surface. The ISI 40× scanning electron microscope was used to examine the features of the specimens obtained from the peel strength tests. Specimens were lightly coated with gold by sputtering. DSC was performed on a Perkin Elmer DSC-7, TAC-7 instrument controller, from –30 °C to 200 °C with a heating rate of 20 °C/min.

Reflectance IR spectra of the adsorbates were recorded on a Digilab model FTS 60 spectrometer. The sample was prepared by coating PP–OH (0.25% by weight in xylene) solution onto the silica glass surface. For high reflection sensitivity, a thin layer (60 Å) of silica glass was deposited on the gold surface by a sol–gel process (tetraethylorthosilicate–aluminum butoxide solution spin coating on a vapor-deposited gold surface). After the thermal treatment of coated PP–OH film at 140 °C for 2 h, the excess PP–OH was removed by rinsing with pure solvent. A uniform PP–OH film with thickness about 1000 Å was coated on the silica surface. Both thicknesses of PP–OH and silica films were determined by an ellipsometer operating at a wavelength of 632.8 nm and an incident angle of 70°.

Batch Polymerization

In a typical copolymerization, 15.5 mL (10.09 g, 0.2401 mol) of dried propylene liquid at –78 °C was transferred into a 500-mL schlenk flask containing 200 mL of degassed frozen toluene. This reaction flask was then warmed to room temperature before bringing into the dry-box, where 49.03 g (0.2401 mol) of hexenyl-BBN was added via the side-arm bulb. The residual monomer was washed in with 10 mL toluene.

Meanwhile 4.093 g (56 mmol) AlEt$_2$Cl in 15 mL toluene was added dropwise to the burgundy-colored slurry of 15 mL toluene and 0.901 g (5.97 mmol) of solid TiCl$_3$AA (AA is reduced and activated aluminum). This catalyst was premixed for 0.5 h before adding to the reactor via the side-arm bulb. After 2 h of polymerizing at room temperature, cold isopropanol (20 mL) was added to terminate the polymerization as shown by the color change from deep burgundy to clear–pale brown. The flask contents were then poured into a bottle containing 300 mL of isopropanol. The bottle was sealed and placed in the dry-box freezer overnight to facilitate the polymer precipitation or coagulation (depending on the borane content). The polymer was isolated by filtration, washed with more isopropanol, and squeeze dried, all in the dry-box. A small amount of white rubbery polymer was vacuum dried, dissolved in d_{10}-o-xylene, and analyzed by ^{11}B-NMR at 120 °C by using triethylborate as a standard.

Continuous Polymerization

In an argon-filled dry box, 15.477 g 5-hexenyl-BBN and 200 mL of hexane were placed in a Parr 450-mL stirred-pressure reactor and sealed. Outside the box, 4.6 g of propylene was added under N$_2$ pressure. A slurry of 1.027 g of TiCl$_3$•AA and 4.705 g of AlEt$_2$Cl in 80 mL of toluene was then added under N$_2$ pressure to initiate the polymerization. Additional portions of propylene (4.10, 3.60, 3.20, 2.80, and 2.00 g) were added at 30-min intervals. One hour after the last monomer charge, the reaction was terminated by injection of 100 mL of isopropanol. The mixture was stirred for 0.5 h before venting the excess pressure and then taken into the dry box for further purification with isopropanol.

Hydroboration of Poly(propylene-co-1,4-hexadiene)

In an argon-filled dry box, 0.550 g of inhibitor-free poly(propylene-co-1,4-hexadiene) containing 1.7% 1,4-hexadiene units was placed in a suspension of 25 mL of dry–degassed toluene. A solution of 0.069 g of BBN in 15 mL of dry–degassed THF was added to the polymer suspension. The suspension was heated to 65 °C in a flask equipped with a condenser. After stirring for 5 h, the polymer was precipitated into 150 mL of dry–degassed isopropanol and isolated by filtration in the dry box. The borane-containing polymer was then placed in a suspension of THF for oxidation reaction.

For the hydroboration reaction with BH$_3$, 0.451 g of inhibitor-free poly(propylene-co-1,4-hexadiene) containing 1.7% 1,4-hexadiene units was placed in a suspension of 15 mL of dry, O$_2$-free THF in an argon-filled dry box. The polymer was hydroborated by the addition of 0.953 g of 1 M BH$_3$/THF solution (d = 0.898 g/mL). The polymer slurry was heated to 65 °C for 5 h in a flask equipped with a condenser.

Oxidation and Iodization Reactions

In a typical example, the borane-containing polymer (PP-B) was placed in a 2000-mL round-bottom flask equipped with septum and stirrer in 700 mL of THF to form a slurry. A solution of 19 g of NaOH in 60 mL of water, 30 mL of THF, and 20 mL of MeOH was degassed and added dropwise into the reactor. The flask was then cooled to 0 °C before slowly adding a degassed solution of 87.6 g of 30% aqueous H_2O_2 in 75 mL of THF via a double-tipped needle. The reaction was allowed to slowly warm to room temperature before heating up to 40 °C for 6 h. The hydroxylated polymer was then precipitated in water, squeeze dried, and placed in a slurry of 500 mL of methanol. After 3 h of vigorous stirring, approximately 75 mL of MeOH was distilled under N_2 to remove the boric acid–methanol azeotrope. Again the polymer was precipitated in water, squeeze dried, washed with acetone, and dried under high vacuum at 45 °C.

For the iodization reaction, 0.231 g of white, powdery PP-B was placed in a suspension of THF in a 150-mL round-bottom flask equipped with a magnetic stir bar and a rubber septum. A degassed solution of 0.012 g of sodium acetate in 10 mL of methanol was added via a syringe under N_2. This procedure was followed by the dropwise addition of 0.020 g NaI in 5 mL of degassed water. The cloudy white polymer suspension turned pale yellow, which faded after adding 0.016 g chloramine-T-hydrate in 10 mL of degassed MeOH. After 2 h of stirring at room temperature the reaction was terminated by addition of aqueous sodium thiosulfate, and 300 mL of dilute HCl was added to precipitate the polymer. The polymer was repeatedly washed with water and acetone. The solid was placed in a refluxing MeOH slurry and precipitated in water two times before drying under vacuum. The IR spectrum was recorded by using a KBr pellet of the white polymer.

Lamination Procedures

The PP–OH layer was coated on a glass surface by either solution casting or a hot press. In solution casting, a hot PP–OH solution (1 wt% in xylene) at 140 °C was spread on glass (plain E-glass, acid-etched E-glass, and quartz slides) to form a complete coating layer. The solvent was then allowed to evaporate at room temperature and humidity. After the designed evaporation time, films that formed were placed in a vacuum oven at 50 °C for 24 h to remove any residual solvent in the film before bonding with PP backing materials. Usually, the PP–OH layer is approximately 20 mm thick measured by SEM.

In the hot-press process, a PP–OH powder was spread on a glass surface and melted between two Teflon-coated aluminum plates at 180 °C. After at least 20 min to ensure complete melting, the sample was pressed at minimum applied pressure for 25 min. During this procedure, hydroxy groups of PP–OH reacted with silanol groups on the glass surface. The molten polymer was then cooled to induce solidification of the PP–OH. The resultant layer was approximately 25-mm thick estimated by SEM.

PP–glass laminates were prepared in a Carber hot press. Laminate with a PP film on top of PP–OH coated glass was held at 180 °C for 25 min under minimum applied pressure and then slowly cooled under pressure for 5 h. The resulting laminate retained most of the original dimensions of PP films based on SEM micrograph. Usually, both undrawn PP and drawn PP films after lamination are 170–180 mm thick.

Immobilization of EtAlCl₂ and BF₃

In a typical example, hydroxylated PP (150 mg, 5 mol% OH) suspended in toluene (15 mL) was mixed with excess aluminum compounds (~10 mmol). After 3 h, polyolefin was filtered and washed with hexane repeatedly to remove the remaining aluminum compound. In the case of reaction with BF_3, the hydroxylated polyolefin (150 mg) was treated with 0.1 mL (1 mmol) n-BuLi (10 M) in 7 mL toluene for 1 h. Polyolefin was filtered and washed with hexane to remove excess lithium compounds. Traces of solvent were removed from the polyolefin powder under vacuum. To this polymer a saturated BF_3 solution in dichloromethane was added. The mixture was stirred at room temperature for 3 h. Dichloromethane and excess BF_3 were removed on a vacuum line.

Polymerization of Isobutylene

Polymerization was carried out in a high vacuum apparatus that consists of two 100 mL flasks (A and B) and one stopcock (a) was used to separate the flasks. The other stopcock (b) was used to control the vacuum conditions and nitrogen flow. In the dry box, a supported Lewis acid catalyst (100 mg) was charged to the flask A, the valve (b) was then closed. The whole apparatus was moved to a vacuum line and was pumped to high vacuum before closing the valve (b). Isobutylene (4 mL, 50 mmol) was condensed in flask B and dissolved in about 20 mL of hexane, which was vacuum-distilled into flask B. Isobutylene solution was warmed up to the required temperature and transferred to the catalyst in flask A. After stirring the reaction mixture for the required time, PIB solution was separated from the supported catalyst by filtration in the dry box. PIB was obtained by evaporating the solvent under vacuum. The supported catalyst was then recharged to flask A and the entire process was repeated.

Conclusion

In this chapter, I have introduced a new method, the borane approach, to prepare functionalized polyolefins by both direct polymerization and post-polymerization functionalization. The success of this chemistry relies on the stability of the borane moiety to transition-metal catalysts, the solubility of borane compounds in hydrocarbon solvents (hexane and toluene) used in transition-metal polymerizations, the high efficiency of hydroboration, and the versatility of borane groups that can be transformed to a wide variety of functionalities. Many new functionalized polyolefins with various molecular architectures have been obtained based on this chemistry. Most of these polyolefins would be very difficult to obtain by other methods. The preservation of thermal stability and crystallinity in functionalized PP offer the good processibility and unique morphology, in which the functional groups are located on the surface of the crystalline phases. This type of copolymer has been demonstrated as a very effective interfacial modifier to improve adhesion between polyolefin and substrates. The copolymer is also a very interesting support material for immobilized catalysts such as EtAlCl$_2$ and BF$_3$.

Acknowledgment

Financial support from the Polymer Program of the National Science Foundation is gratefully appreciated.

References

1. Baijal, M. D. *Plastics Polymer Science and Technology;* John Wiley & Sons: New York, 1982.
2. Carraher, E. C., Jr.; Moore, J. A. *Modification of Polymers;* Plenum: Oxford, 1982.
3. Pinazzi, C.; Guillaume, P.; Reyx, D. *J. Eur. Polym.* **1977,** *13,* 711.
4. Chung, T. C., Raate, M., Berluche, E.; Schulz, D. N. *Macromolecules* **1988,** *21,* 1903.
5. Chung, T. C. *J. Polym. Sci., Polym. Chem. Ed.* **1989,** *27.* 3251.
6. Boor, Jr., J. *Ziegler–Natta Catalysts and Polymerizations;* Academic: Orlando, FL, 1979.
7. Purgett, M. D. Ph.D. Thesis, University of Massachusetts, 1984.
8. Ruggeri, G.; Aglietto, M.; Petragnani, A.; Ciardelli, F. *Eur. Polym. J.* **1983,** *19,* 863.
9. Crabtree, R. *The Organometallic Chemistry of the Transition Metals;* John Wiley and Sons: New York, 1988.
10. Grubbs, H. *J. Am. Chem. Soc.* **1985,** *107,* 3377.
11. Cheng, H.; Ewen, J. *Makrol. Chem.* **1989,** *190,* 1931.
12. Zambelli, L.; Grassi, A.; Resconi, L.; Albizzati, E.; Mazzocchi, R.; *Macromolecules* **1988,** *21,* 617.
13. Kaminsky, W.; Bark, A.; Spiehl, R. *Isotactic Polymerization of Olefins with Catalyst for Olefin Polymerization;* Kaminsky, W; Sinn, H., Eds.; Springer-Verlag: 1988.
14. Odian, G. *Principals of Polymerization,* 3rd ed.; Wiley-Interscience: New York, 1991.
15. Chung, T. C. U.S. Patents 4 734 472 and 4 751 276, 1988.
16. Chung, T. C. *Macromolecules* **1988,** *21,* 865.
17. Chung, T. C.; Ramakrishnan, S.; Kim, M. W. *Macromolecules* **1991,** *24,* 2675.
18. Chung, T. C.; Chasmawala, M. *Macromolecules* **1991,** *24,* 3718.
19. Ramakrishnan, S.; Chung, T. C. *Macromolecules* **1990,** *23,* 4519.
20. Brown, H. C. *Organic Synthesis via Boranes;* Wiley-Interscience: New York, 1975.
21. Chung, T. C.; Rhubright, D. *Macromolecules* **1991,** *24,* 970.
22. Chung, T. C. *CHEMTECH* **1991,** *21,* 496.
23. Ramakrishnan, S.; Berluche, E.; Chung, T. C. *Macromolecules* **1990,** *23,* 378.
24. Chung, T. C.; Rhubright, D. *Macromolecules* **1993,** *26,* 3019.
25. Kelen, T.; Tudos, F. *React. Kinet. Catal. Lett.* **1974,** *1,* 487.
26. Kinsinger, J. B.; Higher, R. E. *J. Phys. Chem.* **1959,** *63,* 2002.
27. Chung, T. C.; Rhubright, D. *J. Polym. Sci., Polym. Chem. Ed.* **1993,** *31,* 2759.
28. Chinsirikul, W; Chung, T. C.; Harrison, I. *J. Thermoplast. Compos. Mater.* **1993,** *6,* 18.
29. Lee, S. H.; Li, C. L.; Chung, T. C. *Polymer* **1994,** *35,* 2979.
30. Chung, T. C.; Kumar, A. *Polym. Bull.* **1992,** *28,* 123.
31. Chung, T. C.; Rhubright, D.; Kumar, A. *Polym. Bull.* **1993,** *30,* 385.

Living Cationic Polymerization: Synthesis of End-Functionalized Polyisobutylenes

Munmaya K. Mishra and Joseph P. Kennedy

This chapter presents an overview of living cationic polymerization and its application to the synthesis of end-functional polyisobutylenes (PIBs). Recent developments in the synthesis of tertiary-chlorine end-functional PIBs and other end-functional PIBs useful for the synthesis of specialty telechelics are outlined. Derivatization of end-functional PIBs leading to new materials such as block copolymers, amphiphilic blocks and networks, polyepoxides, ionomers, and polyurethanes is also discussed.

Since the discovery of living anionic polymerization and of the synthetic usefulness of these processes, great efforts have been made to develop a living carbocationic polymerization. Finally, the discovery of a living carbocationic polymerization (1, 2) of isobutyl vinyl ether and shortly thereafter the isobutylene polymerization were disclosed. The most effective route to design high molecular weight polymers is by living or quasiliving carbocationic polymerization. These techniques offer the following useful, important, synthetic incentives:

- The degree of polymerization (*DP*) can be controlled by controlling monomer (*M*) and initiator (*I*) concentration: $DP_n = [M_o]/[I_o]$.
- Molecular weight distributions can be controlled, provided the rate of initiation (R_i) is greater than the rate of propagation (R_p).
- Various end groups can be designed.
- Block copolymers, thermoplastic elastomers, etc., can be designed.

During the last decade cationic polymerization underwent dramatic developments (1). Intensive research efforts on understanding and controlling elementary events (e.g., initiation, propagation, termination, and transfer) of carbocationic polymerization led to the discovery of living carbocationic polymerization systems in the mid-1980s. The key requirements for living carbocationic polymerization are conditions under which chain transfer is absent and termination is absent or reversible.

Carbocations are unstable, and the absence of chain transfer is difficult to achieve; however, chain transfer can be strongly suppressed or virtually eliminated by relatively stable (i.e., less-ionized) propagating species. According to current experimental evidence (1), living carbocationic polymerization systems are quasiliving in nature; that is, these polymerizations involve reversible termination. This result signifies that there is an equilibrium between an inactive (dormant) species and active (living) propagating species. The equilibria between these species are faster than the rate of propagation. The equilibria can be represented by the following equation:

$$[D] \rightleftharpoons [L] \rightleftharpoons [C^+]$$

3469–8/97/0057$15.00/0 © 1997 American Chemical Society

where $[D]$, $[L]$, and $[C^+]$ denote dormant, living, and conventional ionic species, respectively. In living polymerization, propagation mainly occurs by $[L]$.

Living polymerization is essentially dependent on controlled propagation. Controlled propagation proceeds uninterruptedly in the absence of chain-breaking (i.e., via chain transfer or termination). Controlled propagation may also occur in the presence of rapidly reversible chain-breaking events: when the rates of reversible chain transfer or termination are higher than that of propagation. On the basis of this scenario two kinds of living polymerizations occur:

- ideal living polymerization, in which chain transfer (tr) and termination (t) are absent and their rates (R) are zero $(R_{tr} = R_t = 0)$.
- quasiliving polymerizations, in which rapidly reversible chain transfer or termination is present, and the rate of these processes is faster than that of propagation.

From the practical point of view both ideal and quasiliving polymerizations are equally important and valuable. A necessary but insufficient criterion for living polymerization is that $R_t = 0$; a rigorous definition is that both R_t and R_{tr} are zero.

The diagnostic proof for the existence of living polymerization (1, 2) is the rectilinear M_n (number-average molecular weight) versus W_p (weight of polymer formed, conversion) plot passing through the origin when molecular weights are plotted versus conversion. Necessarily, the plot of the number of moles of polymer formed (N) versus conversion must be horizontal (N is independent of time), and the plot must start at the height of the number of moles of the initiator, I_o (i.e., $N = I_o$). Such data can be obtained by intermittent sampling of a polymerization and by analyzing the samples. Because the carbocationic polymerization of isobutylene (IB) is often rapid (100% conversion may be reached within seconds), intermittent sampling for analysis may be impossible. The kinetic results were obtained by using the incremental-monomer-addition (IMA) technique. The IMA technique was developed to follow the M_n versus conversion (gram of polymer formed) profile, where the withdrawing of samples at representative low or intermediate conversion would have been difficult.

In contrast to the diagnostic IMA technique, the conventional all-monomer-in (AMI) procedure is carried out by adding the coinitiator to a charge containing initiator solvent and all the monomer. For an AMI series, the polymerizations are carried out by varying the monomer concentration at a constant initiator amount. Just as with the IMA series, the AMI experiment concludes by determining M_n and plotting M_n

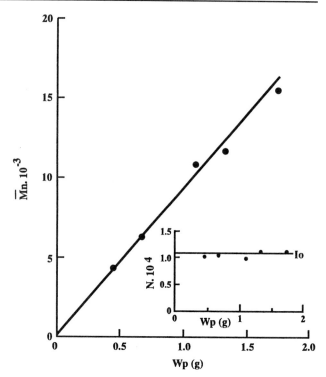

Figure 1. Plot of M_n and N versus W_p.

versus W_p, and N versus W_p (Figure 1). These kinetic results are critical for the diagnosis of the living nature of a polymerization. A comprehensive view of the living carbocationic polymerization and mechanistic details were published recently (1).

Cationic Polymerization of Isobutylene

Acid-induced polymerization of IB was mentioned by Russian authors in 1873 (3). The next milestone was a series of discoveries by Thomas and co-workers (4, 5) during the mid-1930s on the polymerization of IB to very high molecular weight products and, more significantly, the copolymerization of IB with small quantities of isoprene and other conjugated dienes. Subsequently, the kinetics and mechanism of carbocationic polymerizations were studied during the late 1940s (6, 7). In the late 1970s and early 1980s the field was further advanced with the discovery of the inifer (*initiator and transfer*) technique (8), which for the first time led to well-defined novel telechelics and block copolymers.

The era of designed synthesis of polyisobutylene (PIB)-based products began by the discoveries of living carbocationic polymerization of IB by Lewis acids complexed with esters (9, 10), ethers (11, 12), alcohols (13–15), and peroxides (16, 17). Living polymerization of IB provided a new beginning in controlled polymer homogeneity, functionalization, and block copolymerization. After the discovery of initiating systems leading to living IB polymerizations, research has continued in many directions and has expanded the scope of living carbocationic polymerization. Many publications have described variations of the original initiating systems (for a recent summary of the field, *see* reference 1). Living cationic polymerization and the inifer systems invariably produce tertiary-alkyl-chloride (t-Cl) end-group PIBs by end quenching or termination (*1*). The t-Cl end group is the starting site for most of the other end-functional products described in the next section.

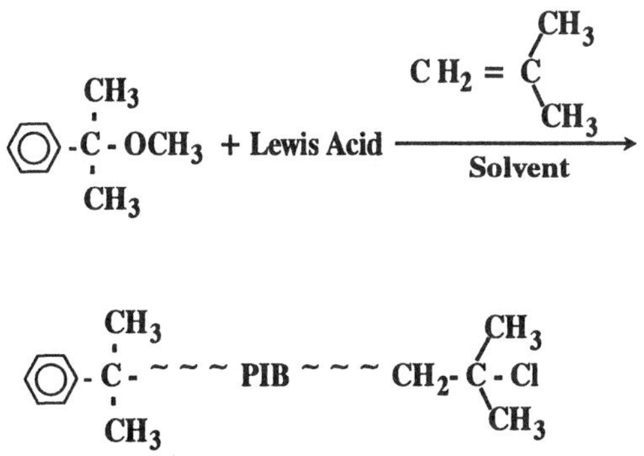

Design of End-Functional Polyisobutylenes

Despite the large number of well-known synthetic polymers representing a wide array of structures and properties, a strong demand exists for new polymers for specific applications of high performance materials. Different synthetic routes can lead to new materials. One important method is designed functionalization of PIB by end quenching (*1*).

Macromolecules capped by a reactive end group are termed telechelic polymers (or telechelics). The word *telechelic*, derived from the Greek (*tele* distant and *chelos* claw), was proposed by Uraneck et al. (*18*) to describe linear polymers possessing two reactive terminal functions. In recent years the terminology has been extended (*19, 20*) to accommodate polymers carrying three or more reactive end groups as well as star-shaped polymers.

End-functional telechelics are of great interest. Industrial interest in telechelics is mainly due to their ease of processibility, because they are usually low or moderate molecular weight liquids. Quantitative end functionality is mandatory for efficient use in applications such as chain extension and network formation. The synthesis of end-functional polymers from telechelics can be achieved by end quenching or by end functionalization. In end quenching, the quenching agent terminates the polymerization and converts the chain end to the desired functional end group. This process is normally a one-pot method. In contrast, end functionalization is a postpolymerization process that involves quantitative conversion of a well-defined polymer end group into another well-defined end-functional group. Functional PIBs can be prepared by either of these methods.

Synthesis of Polyisobutylenes with Tertiary Chlorine End Groups

The end group of PIB obtained by end quenching is invariably t-Cl (*1*). Other functional groups have been

Scheme I. General reaction scheme for IB polymerization.

synthesized from the t-Cl end of PIB by various quantitative end-functionalization methods. The polymerization scheme of IB is shown in Scheme I.

Schemes II–VI summarize initiating systems for IB polymerization that lead to t-Cl end-group PIBs and include inifers (*21*), alcohols (*13–15*), esters (*9, 10*), ethers (*2, 11, 12*), and peroxides (*16, 17*). The initiating system comprises a Lewis acid (BCl_3 or $TiCl_4$) and an organic compound such as tertiary alcohol, esters, and ethers. The reaction between the organic compound and Lewis acid produces a carbocation that initiates the IB polymerization. The molecular weight of the product can be controlled by the ratio of monomer concentration over initiator concentration (i.e., alcohols, ethers, and esters). Quenching of the polymerization with nucleophiles such as methanol and pyridine or thermal quenching (by heating above decomposition temperature of the initiating complex) yields tertiary chlorine end groups quantitatively independent of the initiating system. End-group analysis of PIBs has been carried out mainly by NMR spectroscopy (*38–41*).

A detailed discussion of the initiating systems and other conversion methods (with variation of the original system) is available elsewhere (*1*). However, a few examples for the synthesis of mono-, di-, and triarm PIBs with t-Cl end groups by using tertiary ethers are provided in Tables I–III.

Experimental details of IB polymerization were described elsewhere (*2, 12, 30*). The polymerization experiments were carried out in a series of pressure tubes by charging a specific amount each of a diluent, initiator, and monomer at a specific temperature (between –10 °C and –70 °C). Methyl chloride and IB gases were purified according to the standard procedures (*42*) and condensed and charged into the pressure tubes. The sequence of addition was solvent (methyl chloride or methylene chloride), then tertiary ether, followed by IB. The polymerization was initiated by the addition of a required amount of BCl_3. The polymerization was observed as being extremely

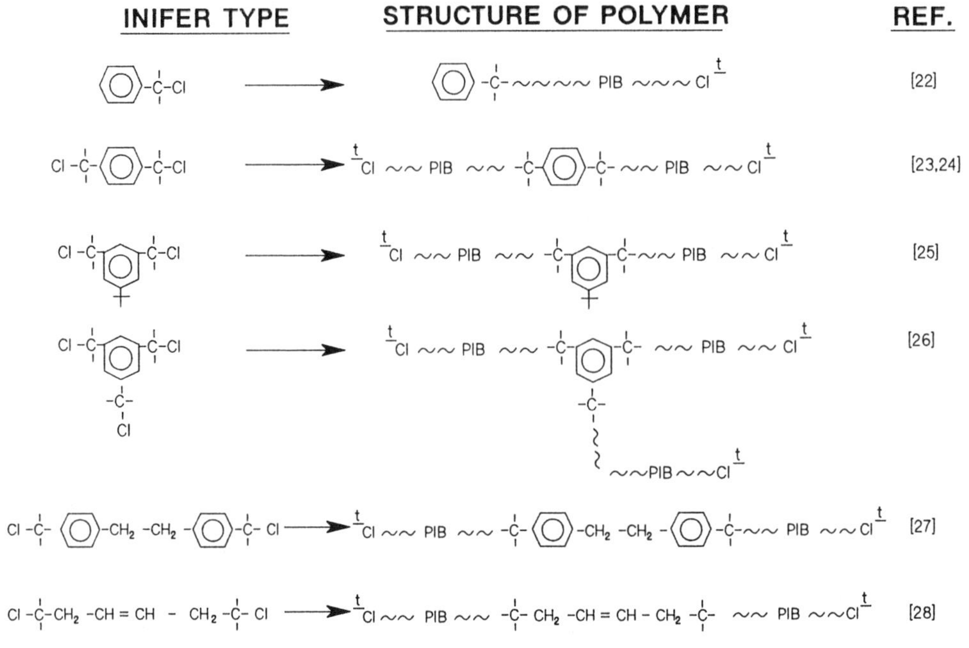

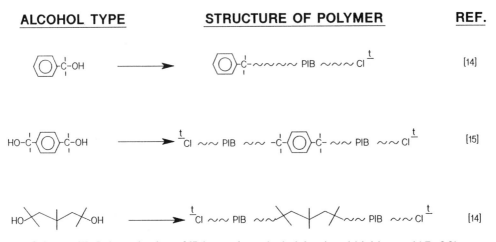

Scheme II. Polymerization of IB by infer–Lewis acid initiators (*21*).

Scheme III. Polymerization of IB by tertiary alcohol–Lewis acid initiators (*13, 29*).

rapid; the reaction appeared to be over in a few minutes. To ensure that the polymerization was complete, the pressure tubes were left undisturbed for about 30 min, after which the reaction was quenched by the addition of a small amount of methanol. The polymers were recovered by evaporation of volatiles, dissolution in *n*-hexane, filtration of inorganic residues such as Lewis acid residues, and removal of hexane by evaporation. The resultant polymers were characterized by spectroscopy and gel permeation chromatographic techniques.

The polymerization results are presented in Tables I–III. The results indicate that tertiary ethers are suitable initiators for the preparation of terminally functionalized PIBs. The experimental molecular

weight (M_e) of the PIBs is given by the following equation:

$$M_e = (M_o/I_o) \times IB_M$$

where M_o is initial monomer concentration, I_o is initial concentration of tertiary ether, and IB_M is molecular weight of IB.

Polyisobutylenes with Other End-Functional Groups

By exploiting the high selectivity and ease of derivatization of t-Cl end groups, a wide variety of macromonomers, telechelics, and block and graft copoly-

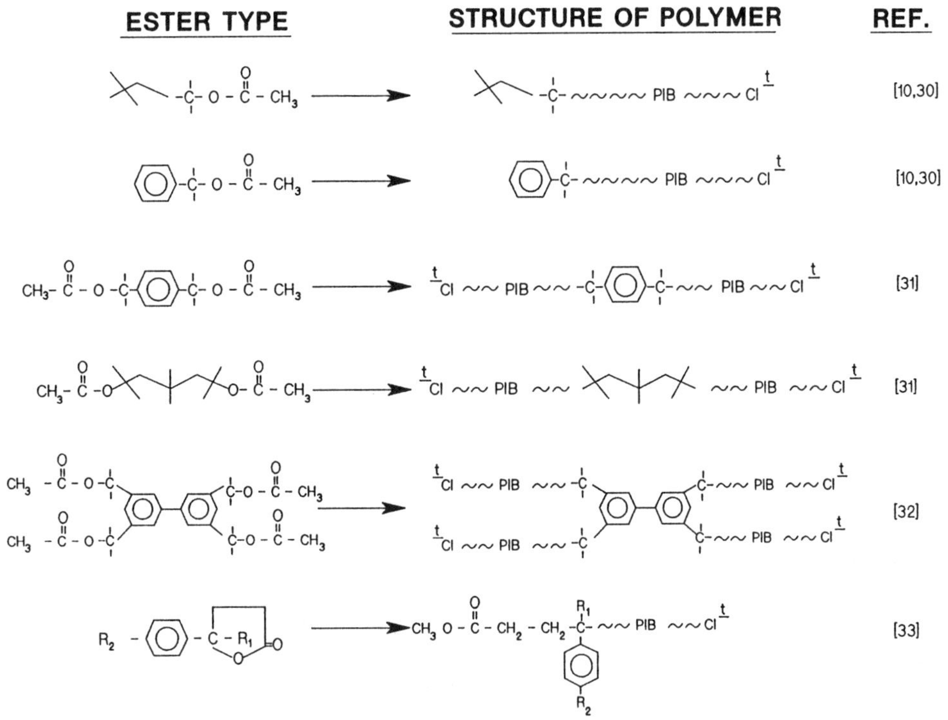

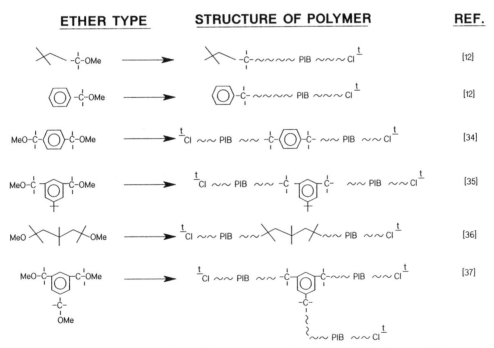

Scheme IV. Polymerization of IB by tertiary esters–Lewis acid initiators (*9*).

Scheme V. Polymerization of IB by tertiary ethers–Lewis acid initiators (*11*).

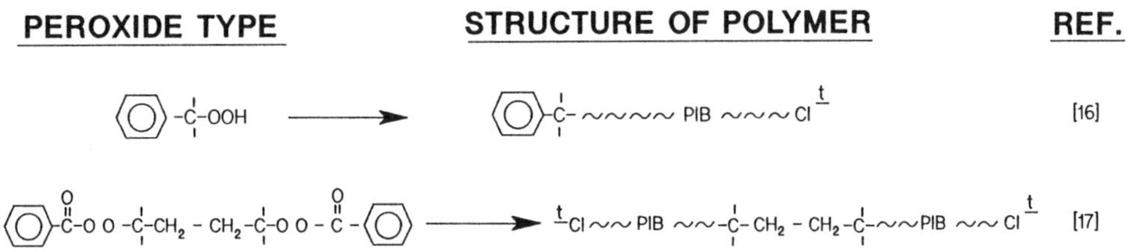

Scheme VI. Polymerization of IB by tertiary peroxides–Lewis acid initiators (*16, 17*).

Table I. Experimental Data for Synthesis of Monofunctional PIBs with Tertiary Chlorine End Groups by Using Cumyl Methyl Ether (CumOMe)–BCl_3 Initiator

[CumOMe] (mol/L)	Solvent	Temp. (°C)	[IB] (mol/L)	[BCl_3] (mol/L)	M_n (g/mol)	I_{eff} (%)
7.7×10^{-2}	CH_2Cl_2	−30	1.17	0.37	1500	—
2.3×10^{-2}	CH_2Cl_2	−30	1.17	0.37	4650	—
2.3×10^{-2}	CH_2Cl_2	−10	1.17	0.37	5000	—
7.4×10^{-3}	CH_2Cl_2	−30	1.17	0.37	13,700[a]	—
7.4×10^{-3}	CH_2Cl_2	−10	1.17	0.37	12,000[a]	—
7.7×10^{-2}	CH_3Cl	−30	1.17	0.37	960	—
2.3×10^{-2}	CH_3Cl	−30	1.17	0.37	3600	—
7.4×10^{-3}	CH_3Cl	−30	1.17	0.37	147,000[a]	—
5.7×10^{-3}	CH_2Cl_2	−10	0.282	0.155	3080	94
5.7×10^{-3}	CH_2Cl_2	−10	0.564	0.155	7100	79
5.7×10^{-3}	CH_2Cl_2	−10	0.864	0.155	11,900	70
5.7×10^{-3}	CH_2Cl_2	−10	1.228	0.155	13,300	84
5.7×10^{-3}	CH_2Cl_2	−10	1.58	0.155	14,900	93

Note: Time was 30 min. Reaction was quenched with precooled methanol. Conversion was ~100%. I_{eff} is initiator efficiency.

[a]Conversion was 100% after 1 min.

Source: Data adapted from reference 12.

Table II. Experimental Data for Synthesis of Difunctional PIBs with Tertiary Chlorine End Groups by Using Initiator of Dicumyl Methyl Ether (DiCumOMe)–BCl_3

[DiCumOMe] (mol/L)	Solvent	Temp. (°C)	[IB] (mol/L)	[BCl_3] (mol/L)	M_n (g/mol)	I_{eff} (%)
5.35×10^{-2}	CH_3Cl	−70	0.935	0.25	1300	97
1.38×10^{-2}	CH_3Cl	−70	0.935	0.25	4050	100
3.8×10^{-3}	CH_3Cl	−70	0.935	0.25	13,650	92
1.66×10^{-2}	CH_3Cl	−70	1.12	0.10	4800	85

Note: Time was 30 min. Reaction was quenched with precooled methanol. Conversion was 100%. I_{eff} is initiator efficiency.

Source: Data adapted from reference 34.

Table III. Experimental Data for Synthesis of Trifunctional PIBs with Tertiary Chlorine End Groups by Using Initiator of TriCumyl Methyl Ether (TriCumOMe)–BCl_3

[TriCumOMe] (mol/L)	Solvent	Temp. (°C)	[IB] (mol/L)	[BCl_3] (mol/L)	M_n (g/mol)	I_{eff} (%)
4.35×10^{-3}	CH_2Cl_2	−10	0.266	0.19	3950	87
4.35×10^{-3}	CH_2Cl_2	−10	0.543	0.19	8200	85
4.35×10^{-3}	CH_2Cl_2	−10	0.694	0.19	9650	93
4.35×10^{-3}	CH_2Cl_2	−10	0.840	0.19	10,900	99
4.35×10^{-3}	CH_2Cl_2	−10	1.20	0.19	15,250	100
5.25×10^{-2}	CH_2Cl_2	−30	0.94	0.51	1300	98
5.25×10^{-2}	CH_3Cl	−30	0.94	0.51	1200	100
5.59×10^{-2}	CH_3Cl	−30	1.0	0.51	1300	—
1.86×10^{-2}	CH_3Cl	−30	1.0	0.51	4800	—
6.2×10^{-3}	CH_3Cl	−30	1.0	0.51	14,700	—
5.59×10^{-2}	CH_2Cl_2	−30	1.0	0.51	1350	—
1.86×10^{-2}	CH_2Cl_2	−30	1.0	0.51	5500	—
6.2×10^{-3}	CH_2Cl_2	−30	1.0	0.51	13,700	—

Note: Time was 30 min. Reaction was quenched with precooled methanol. Conversion was 100%. I_{eff} is initiator efficiency.

Source: Data adapted from reference 37.

mers were prepared from Cl-ended PIBs. Various other end-functional PIBs also were prepared by controlled initiation by using monofunctional initiators such as cumyl chloride, cumyl alcohol, and cumyl methyl ether. The monofunctional initiator becomes the head-group of the new chain and gives rise to the end functionality. A wide range of end-functional PIBs obtained by different methods are tabulated in Appendixes I and II.

Application of End-Functional Polyisobutylenes

Synthesis of Block Copolymers

Understanding the elementary steps of carbocationic polymerization led to the design of unique block and graft copolymers obtained by using functional PIBs. With the ready availability of functional polymers by the new polymerization techniques, efforts were made to prepare block copolymers by multimode polymerization (transformation reaction) (43). Such transformation can be achieved, for example, by preparation of a t-Cl-terminated PIB, dehydrohalogenation, and then reaction with butyllithium. The resulting allylic carbanion can then be used to induce anionic polymerization of butadiene (BD) (44) (Scheme VII). The PIB-*b*-PBD block copolymer contains no detectable homo-PBD at low BD concentrations. However, at higher BD concentrations the homo-PBD content becomes noticeable. Poly(methyl methacrylate-*b*-isobutylene-*b*-methyl methacrylate) (PMMA-*b*-PIB-*b*-PMMA) triblock copolymer was prepared (45) by lithiation of α,ω-*p*-tolyl PIB followed by

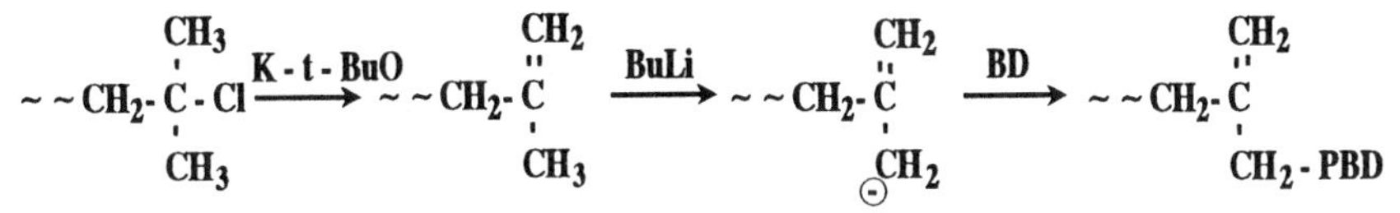

$$\sim\sim CH_2\text{-}\underset{\underset{CH_3}{|}}{\overset{\overset{CH_3}{|}}{C}}\text{-}Cl \xrightarrow{K\text{-}t\text{-}BuO} \sim\sim CH_2\text{-}\underset{\underset{CH_3}{|}}{\overset{\overset{CH_2}{||}}{C}} \xrightarrow{BuLi} \sim\sim CH_2\text{-}\underset{\underset{\underset{\ominus}{CH_2}}{|}}{\overset{\overset{CH_2}{||}}{C}} \xrightarrow{BD} \sim\sim CH_2\text{-}\underset{\underset{CH_2\text{-}PBD}{|}}{\overset{\overset{CH_2}{||}}{C}}$$

Scheme VII. Synthesis of block copolymers of IB and BD by anionic polymerization of BD by using functional PIB.

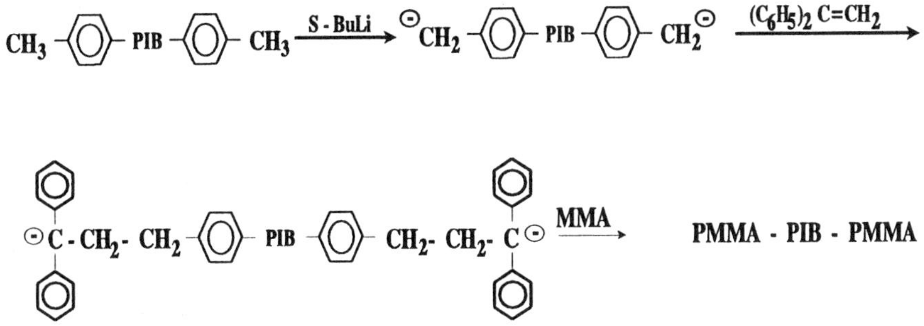

CH$_3$ —⬡— PIB —⬡— CH$_3$ $\xrightarrow{S\text{-}BuLi}$ $^{\ominus}$CH$_2$ —⬡— PIB —⬡— CH$_2$$^{\ominus}$ $\xrightarrow{(C_6H_5)_2\,C=CH_2}$

$^{\ominus}$C-CH$_2$-CH$_2$—⬡— PIB —⬡— CH$_2$-CH$_2$-C$^{\ominus}$ $\xrightarrow{MMA}$ **PMMA - PIB - PMMA**

Scheme VIII. Synthesis of triblock copolymer of IB and MMA by anionic polymerization of MMA by using end-functional PIB.

$$HOCH_2\!\sim\sim\sim PIB \sim\sim\sim CH_2\,OH$$
$$\downarrow \quad n\text{-}BuLi$$
$$LiOCH_2\!\sim\sim\sim PIB \sim\sim\sim CH_2\,OLi$$
$$\downarrow \quad \textbf{D}_3 \textbf{ (Hexamethyl Trisiloxane)}$$

$$\underset{\underset{Me}{|}}{\overset{\overset{Me}{|}}{LiOSi}}\text{-}(O\text{-}\underset{\underset{Me}{|}}{\overset{\overset{Me}{|}}{Si}}\text{-})\,mOCH_2\!\sim PIB \sim CH_2O(\text{-}\underset{\underset{Me}{|}}{\overset{\overset{Me}{|}}{Si}}\text{-}O)\,m\text{-}\underset{\underset{Me}{|}}{\overset{\overset{Me}{|}}{Si}}OLi \xrightarrow{\hspace{1cm}} \overset{(CH_3)_2\,SiCl_2}{}$$

Excess (CH$_3$)$_3$SiCl

$$(CH_3)_3\,SiO\underset{\underset{Me}{|}}{\overset{\overset{Me}{|}}{Si}}\text{-}(O\text{-}\underset{\underset{Me}{|}}{\overset{\overset{Me}{|}}{Si}}\text{-})\,mOCH_2\!\sim\sim\sim PIB\sim\sim\sim CH_2O(\text{-}\underset{\underset{Me}{|}}{\overset{\overset{Me}{|}}{Si}}\text{-}O)\,m\text{-}\underset{\underset{Me}{|}}{\overset{\overset{Me}{|}}{Si}}OSi\,(CH_3)_3$$

PDMS - b - PIB - b - PDMS

$$[\,\text{-}O\underset{\underset{Me}{|}}{\overset{\overset{Me}{|}}{Si}}\text{-}(\text{-}O\text{-}\underset{\underset{Me}{|}}{\overset{\overset{Me}{|}}{Si}}\text{-})\,mOCH_2\!\sim PIB \sim CH_2O(\text{-}\underset{\underset{Me}{|}}{\overset{\overset{Me}{|}}{Si}}\text{-}O)\,m\underset{\underset{Me}{|}}{\overset{\overset{Me}{|}}{Si}}\text{-}O\text{-}\underset{\underset{Me}{|}}{\overset{\overset{Me}{|}}{Si}}\text{-}\,]\,n$$

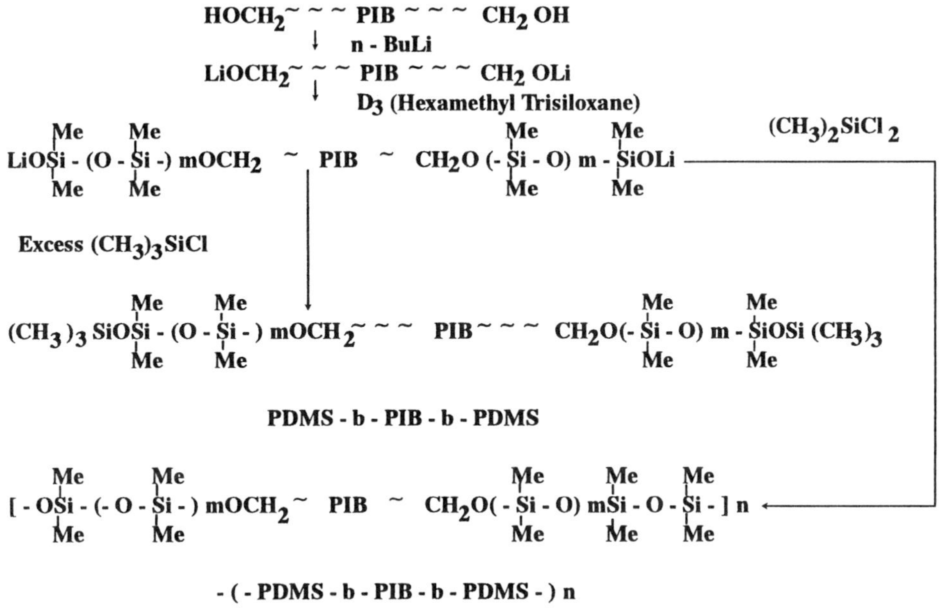

- (- PDMS - b - PIB - b - PDMS -) n

Scheme IX. Synthesis of block copolymers of IB and hexamethyltrisiloxane by anionic ring-opening polymerization of hexamethyltrisiloxane by using functional PIB.

the addition of 1,1-diphenylethylene to the PIB dianions (Scheme VIII).

PIB–siloxane di- and multiblock copolymers were prepared (*46*) by using a macroinitiator. The macroinitiator was prepared by converting hydroxyl telechelic PIB to the alkoxide with butyllithium followed by the anionic ring-opening polymerization of hexamethylcyclotrisiloxane (D$_3$) (Scheme IX). Deactivation of living D$_3$-polymer with Me$_3$SiCl yielded the target triblock, whereas stoichiometric amounts of Me$_3$SiCl$_2$ gave multiblock copolymers. Initiation of D$_3$ polymerization by PIB with an Li phenolate end was also studied (Table IV). However, the phenolate exhibited a much lower initiating efficiency for D$_3$ polymerization

than the alcoholate under similar conditions. The resulting triblock copolymer exhibited surprisingly high hydrolytic resistance even in concentrated hydrochloric acid (*46*).

Hatada and co-workers (*47*) also prepared triblock copolymers comprising PMMA end blocks flanking a PIB center block by using α,ω-dihydroxyl PIBs as starting materials. The synthesis involved the preparation of a α,ω-dilithiated PIB and its use for the low temperature polymerization of MMA (Scheme X).

Nylon 6–PIB diblock, triblock, and triarm radial block copolymers were synthesized (*48*) by anionic polymerization of ε-caprolactam by using functional PIBs as initiator. PIBs having terminal isocyanate

Table IV. Synthesis of Triblock (Me_3Si-PDMS-b-PIB-b-PDMS-SiMe$_3$) and Multiblock -(PDMS-b-PIB-b-PDMS-SiMe$_2$-)$_n$ by Anionic Polymerization in THF at 26 °C

LiO-PIB-OLi (mmol/L)	D_3 (mol/L)	Conver. (%)	Triblock M_n Theor.	Exp.	Multiblock Exp. M_n
10	0.99	30	10,900	9800	30,800
10	0.99	71	19,700	18,500	—
5	0.52	20	8900	8100	25,400
5	0.52	47	15,400	13,800	47,200
5	0.52	80	22,800	19,600	—
5	0.22	65	10,600	9600	44,800

Note: THF is tetrahydrofuran. For LiO-PIB-OLi, M_n is 4200. Conver. is conversion; Theor. is theoretical; Exp. is experimental.

groups were prepared by the reaction of hydroxy telechelic functional PIB with hexamethylene diisocyanate, toluene diisocyanate, or *N*-chlorocarbonyl diisocyanate (Scheme XI). Derivatization with *N*-chlorocarbonyl diisocyanate and subsequent anionic caprolactam polymerization gave the highest yields and blocking efficiencies (40–80%).

The first polystyrene–polyisobutylene (PSt–PIB) diblock was prepared by Peyrot (*49*). The synthesis involved three steps:

1. polymerization of IB by $C_6H_5CH_2Cl$–Et_2AlCl initiator leading to benzyl-headed PIB (C_6H_5-CH_2–PIB)
2. chloromethylation yielding $ClCH_2C_6H_4CH_2$–PIB
3. polymerization of styrene by the PIB–p-$CH_2C_6H_4$-CH_2Cl–Et_2AlCl combination

The products, however, were not fully characterized.

Terminally chlorinated PIB obtained by direct polymerization was also used for the polymerization of styrene. The t-Cl-ended PIB (*50*) prepared by the H_2O–BCl_3 initiating system induced the polymerization of styrene in the presence of Et_2AlCl to produce PIB-b-PSt diblocks. However, the block copolymers were contaminated with homopolymers.

Asymmetric telechelic PIB (*51, 52*) carrying an olefinic head group and a t-Cl end group was prepared by initiating the polymerization of IB with the $(CH_3)_2C=CHCH_2Cl$–BCl_3 combination. The use of this polymer with Et_2AlCl for the polymerization of styrene (*53*) or α-methylstyrene (*51, 52*) yielded PIB-b-PSt or PIB-b-PαMeSt. Chain transfer was evident during the polymerization of styrene, but no chain transfer (or very little) was evident in the case of α-methylstyrene. In this manner relatively clean PIB-b-PαMeSt block copolymers with $(CH_3)_2C=CH$–CH_2– head groups were prepared. Asymmetric α,ω-dichloro telechelic PIB (PIB with a neopentyl-chlorine head group and a t-Cl end group) was used with Et_2AlCl to initiate the polymerization of α-methylstyrene to produce the PIB-b-PαMeSt block copolymer (Scheme XII) (*54*).

α,ω-Di-*tert*-chloro–PIB (t-Cl-PIB-t-Cl) with Et_2AlCl induced the polymerization of α-methylstyrene at both ends of this telechelic prepolymer (*53*). Extraction of the crude product with a series of hydrocarbon solvents revealed a broad compositional distribution. The linear di- and triarm star block copolymers of IB and

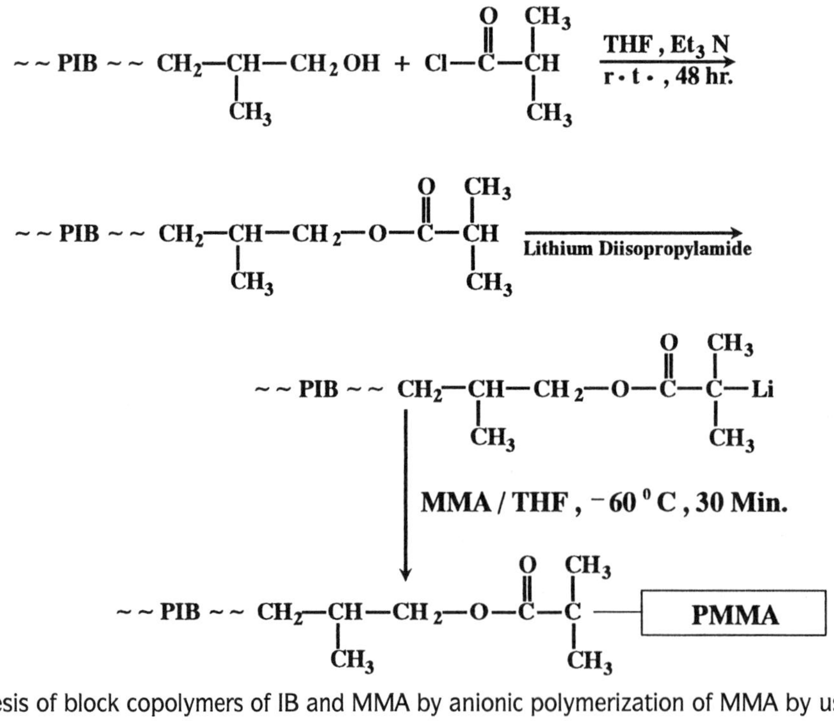

Scheme X. Synthesis of block copolymers of IB and MMA by anionic polymerization of MMA by using functional PIB.

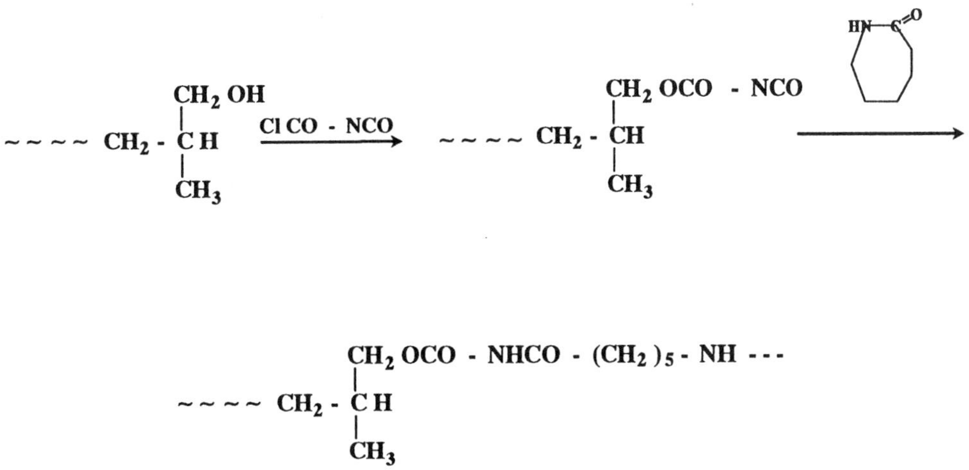

Scheme XI. Synthesis of block copolymers of IB and nylon-6 by anionic ring-opening polymerization of ε-caprolactam by using functional PIB.

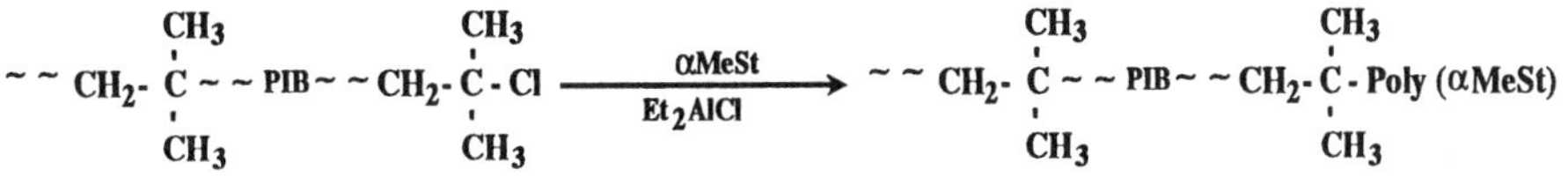

Scheme XII. Synthesis of block copolymers of IB and αMeSt by anionic polymerization of αMeSt by using asymmetric dichloro telechelic PIB.

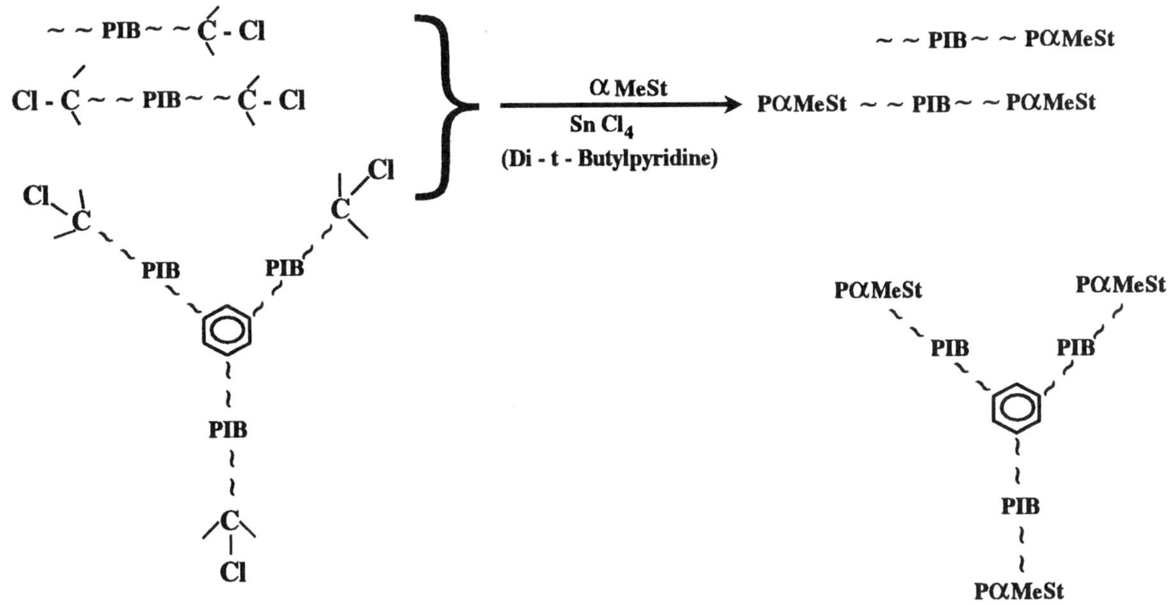

Scheme XIII. Synthesis of block copolymers of IB and αMeSt by anionic polymerization of αMeSt by using PIB with t-Cl end groups (3-arm star PIB).

αMeSt were synthesized in the presence of a proton trap, 2,6-di-*tert*-butylpyridine (DtBP) (*55*). Polymerization of α-methylstyrene by linear or three-arm star PIB having one, two, or three t-Cl termini occurred with SnCl₄ coinitiator in the presence of DtBP (Scheme XIII). The block copolymers were synthesized with 70–95% efficiency.

Amphiphilic Block Copolymers

Amphiphilic block copolymers consist of hydrophilic and hydrophobic polymer segments, and these copolymers dissolve in both water and organic solvents. Because of their favorable solubility and mechanical properties, they find applications in the manufacture

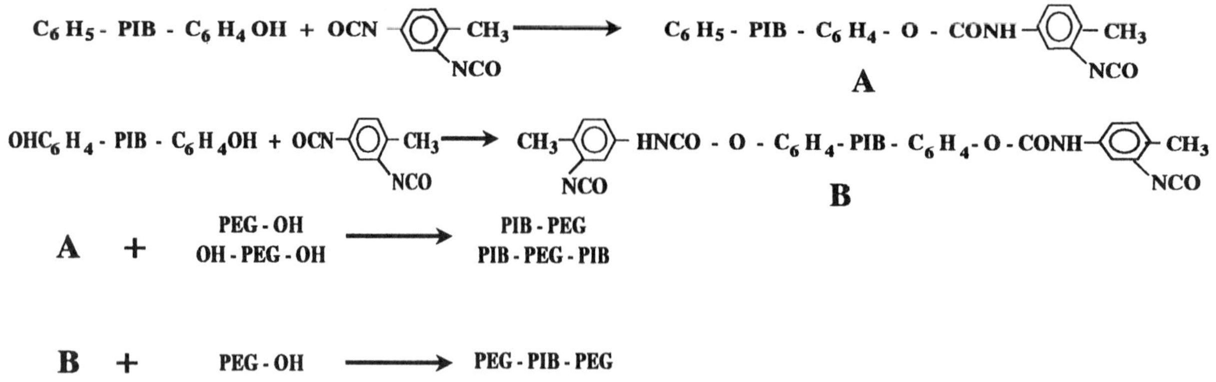

$C_6H_5 - PIB - C_6H_4OH + OCN - \langle O \rangle - CH_3 \longrightarrow C_6H_5 - PIB - C_6H_4 - O - CONH - \langle O \rangle - CH_3$
                            NCO   **A**   NCO

$OHC_6H_4 - PIB - C_6H_4OH + OCN - \langle O \rangle - CH_3 \longrightarrow CH_3 - \langle O \rangle - HNCO - O - C_6H_4 - PIB - C_6H_4 - O - CONH - \langle O \rangle - CH_3$
                                  NCO   NCO   **B**   NCO

$$\textbf{A} + \begin{array}{c} \text{PEG - OH} \\ \text{OH - PEG - OH} \end{array} \longrightarrow \begin{array}{c} \text{PIB - PEG} \\ \text{PIB - PEG - PIB} \end{array}$$

$$\textbf{B} + \text{PEG - OH} \longrightarrow \text{PEG - PIB - PEG}$$

Scheme XIV. Synthesis of di- and triblock copolymers of PIB and PEG by using functional PIB.

$$\text{OCN - OCO - PIB - -OCO - NCO} + \text{HO - PEG - OH} \longrightarrow$$

$$\text{-} (\text{CONH - OCO - PIB - OCO - NHCO - O PEG - O})_n$$

$$\sim\sim PIB \sim\sim - \langle O \rangle - CH_3 \xrightarrow{\text{NBS}} \sim\sim PIB \sim\sim - \langle O \rangle - CH_2Br$$

$$\Big\downarrow \text{KO - PEG - OK}$$

$$\text{PIB - PEG - PIB}$$

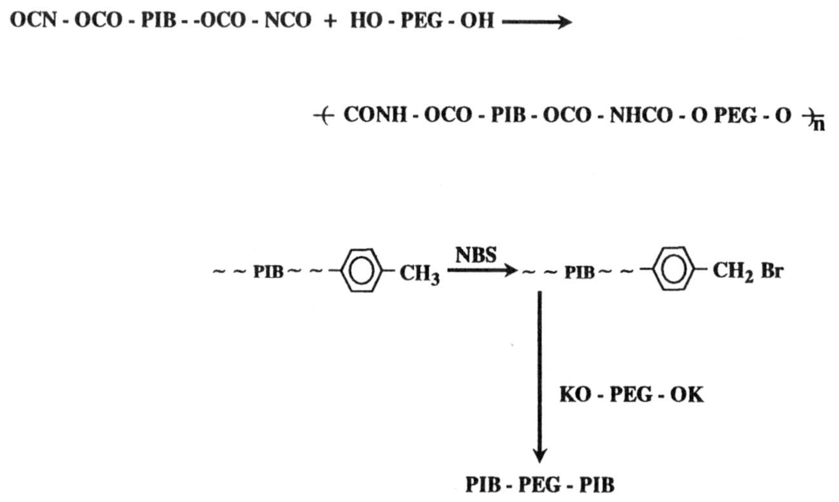

Scheme XV. Synthesis of triblocks of PIB and PEG by using functional PIB.

of detergents, surfactants, pharmaceuticals, cosmetics, and polymer-blends and in oil recovery, coatings, and agricultural products.

Amphiphilic PIB–polyethylene glycol (PEG) di- and triblock copolymers were prepared (56) by coupling mono- and dihydroxyl-terminated (HO–PEG–OH) with isocyanate-ended PIB (Scheme XIV). Relatively low blocking efficiencies (75–85%) were obtained in the presence of stannous octoate catalyst. Amphiphilic multiblock copolymers of PIB and PEG also were prepared (57) by coupling α,ω-oxycarbonyl isocyanate PIB with HO–PEG–OH in solution and in bulk. Higher degrees of extension ($n = 6–7$) were obtained in bulk than in solution.

Very recently another approach for the synthesis of amphiphilic triblock copolymers of PIB and PEG was reported (58). The triblocks were prepared by coupling the potassium salt of dihydroxy-terminated PEG (KO–PEG–OK) with bromotolyl-ended PIB (blocking efficiencies were ca. 85%), prepared by brominating (49) tolyl-ended PIB with N-bromosuccinimide (NBS) (Scheme XV). These triblock copoly-

mers are useful as polymeric surfactants in emulsion polymerization of vinyl monomers.

Amphiphilic block copolymers of PIB and poly(N-acetylethyleneimine) or polyethyleneimine were synthesized by using linear and three-arm star tosyl-telechelic PIB macroinitiators for the ring-opening polymerization of 2-methyl-2-oxazoline (59). The blocking efficiencies were about 75%.

Amphiphilic Networks

Amphiphilic networks are two-component random networks comprised of hydrophilic and hydrophobic chains, in which one type of chain may function as the cross-linking moiety for the other. Telechelic macromonomers are suitable precursors for the preparation of amphiphilic networks by copolymerization with conventional monomers that give water-soluble or swellable segments during polymerization. Thus, N-vinyl-2-pyrrolidone or vinyl acetate were copolymerized with di- or tri-styryl telechelic PIBs (linear and three-arm star PIBs with p-styryl end groups) by a free-

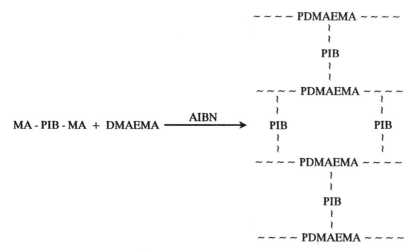

Scheme XVI. Synthesis of an amphiphilic network by using PIB with methyl acrylate (MA–PIB–MA) and dimethylaminoethyl methacrylate (DMAEMA) end groups.

radical initiator (*60, 61*). Also, the free-radical copolymerization of methacrylate-ended ditelechelic PIB macromonomers (MA–PIB–MA) with 2-(dimethylamino)ethyl methacrylate (DMAEMA) yielded amphiphilic networks (*62*) (Scheme XVI). Variation of the molecular weight of MA–PIB–MA and its concentration in the network dramatically affects the network swellability (from 170% to 20% in water and from 40% to 170% in heptane) and other mechanically properties. A network containing 53–58% of MA–PIB–MA implanted dorsally or abdominally in rats showed excellent biocompatibility and biostability (*63*). These amphiphilic networks were studied as potential implantable, controlled drug-release systems (*63*).

Copolymerization of MA–PIB–MA with *N,N*-dimethylacrylamide and 2-trimethylsilyl-oxyethyl methacrylate yielded the corresponding poly(*N,N*-dimethyl acrylamide)-l-polyisobutylene and poly(2-hydroxyethyl methacrylate)-l-polyisobutylene (*64*). The networks were clear and transparent and had satisfactory mechanical properties for implantation. Both networks exhibited interesting sustained drug-release characteristics.

Polyepoxides

Epoxy resins are versatile engineering materials for applications in construction, transportation, composites, adhesives, and coatings. These resins conventionally are prepared by curing epoxy-terminated prepolymers with multifunctional curing agents such as amines. Novel epoxy telechelic PIBs that are curable by conventional epoxy technologies could become important intermediates for such applications.

The rubbery PIBs with glycidyl ether ends are promising as additives to improve toughness and moisture, temperature, and oxidative resistance of conventional epoxy resins. Curing of these new PIB-based epoxides with triethylenetetramine gave (*65, 66*) tough, transparent, flexible materials that exhibited outstanding hydrolytic stability.

The curing of α,α'-disubstituted epoxy telechelic PIBs (*67*) obtained by epoxidation of the corresponding isopropenyl-terminated PIBs was slow, probably due to steric hindrance. However, satisfactory cure rates were observed with epoxy telechelic PIBs prepared from allyl-terminated prepolymers (*68, 69*). Cross-linking of this epoxy-functional PIB with triethylenetetramine yielded rubbers with ca. 300% elongation. In another experiment, dianiline–PIBs and $H_2NAr–PIB–ArNH_2$ (*70*) were used to cross-link bisphenol-A diglycidyl ether. The cured samples exhibited outstanding hydrolytic and heat stability and good mechanical properties.

Ionomers

Ionomers are polymers containing typically less than 15% ionic moieties. Their mechanical and rheological properties are significantly affected by the small quantities of ionic groups attached to the polymer chains because of coulombic interactions between the ionic groups. The polymers form ionic cross-linking or clusters through ionic association. The segregated ionic domains can be disrupted by heat so that ionomers are processable at moderately elevated temperatures and are useful as hot-melt adhesives, coatings, and films.

PIB-based ionomers with terminal $-SO_3^-M^+$ (M = Na^+, K^+, Zn^{2+}, Ca^{2+}, Cr^{3+}, La^{3+}, etc.) were prepared by sulfonation of terminal $-CH_2-C(CH_3)=CH_2$ groups of linear and three-arm star telechelic PIBs, followed by neutralization with a base (*71, 72*). The effects of the nature of counterions, amount of neutralizing agent, PIB molecular weight, and molecular architecture on mechanical properties were studied extensively (72, 73–78). Three-arm star PIB-based ionomers of M_n 10,000–35,000 exhibited high extension at break (as high as ca. 1000%), and these ionomers displayed low permanent set and low hysteresis (73). The melt viscosity of these ionomers increased dramatically with the use of 100% excess of KOH neutralizing agent (74). Zinc-neutralized ionomers exhibited a lower viscosity than potassium- or calcium-neutralized products (75). Linear and three-arm star PIB ionomers exhibited unique behavior in nonpolar solvents such as *n*-hexane. The overall mechanical properties of PIB-based ionomers having no dangling chain ends were comparable (and in some cases superior) to ionomers based on ethylene-propylene diene monomers and carrying randomly distributed $-SO_3^-M^+$ groups (72).

Polyurethanes

Polyurethanes, $[(RNHCOO)_x R'_y]_n$, are multiblock copolymers consisting of –RNHCOO– hard segments, (R is usually an aromatic residue) and flexible, rubbery, soft segments (R'). The soft segments are conventionally polyethers or polyesters, or for specialty applications, polybutadienes or polysiloxanes.

Polyurethanes prepared from hydroxy telechelic PIBs exhibited outstanding hot-air and hydrolytic stability, low gas permeability, and negligible water absorption (67, 79–82). Table V summarizes characteristic properties of such materials. The high hydrolytic and oxidative stabilities are due to the lack of moisture permeability and water absorption by the nonpolar PIB domains and saturated structures, respectively. Because of this advantageous combination of properties, PIB-based polyurethanes are of interest as sealants, gaskets, rocket-propellant binders, and various medical applications.

Hydroxy telechelic PIBs obtained from isopropenyl- (83, 84) or allyl-terminated precursors (68, 69) were used to prepare polyurethanes containing PIB soft segments (68, 69, 85). Chain extension of PIB-diols or -triols with di- or triisocyanates gave model networks (82, 86–90). The tensile strength of PIB model networks increased with decreasing molecular weight (in grams per mole) between two cross-links (M_c) up to a limiting value, whereas elongation decreased monotonously with decreasing M_c (82). However, mechanical properties of PIB-based polyurethanes were inferior compared with conventional polyurethanes (91–93). Polyisobutylene-based urethane foams prepared from telechelic PIB-triols and tolylene diisocyanate exhibited

Table V. Physical Characteristics of Hydroxy Telechelic PIB-Based Polyurethane Networks

Parameter	Value
Hydrolytic stability (4 days at 85 °C; ASTM D3137)	~100%
Heat aging (3 days at 128 °C; ASTM D573)	~100%
Water absorption (3 days at 76 °C; ASTM D570)	~0%
O_2 permeability (1 day at 30 °C; ASTM D1434)	—[a]

[a] Equivalent to or better than butyl rubber.

excellent thermal, environmental, barrier, and mechanical properties (79).

Other Applications

Functional PIBs, particularly those derived from succinic anhydride, have many applications, such as adhesives, coatings, dispersants, emulsifiers, detergents, membranes, metal complexing resins, paper modifiers, printing inks, sealants–caulks, and flocculating agents for oil-well drilling (94). One of the earliest patents (95) on the succinic anhydride functional PIB disclosed the use of the product as a corrosion inhibitor. Some of the earlier patents (96–99) also disclosed the use of succinic anhydride functional PIBs as dispersants in fuels and lubricants. After these patents, many patents have appeared along these lines. Much of the patent literature up to 1980 was cited in a book on maleic anhydride (94).

Concluding Remarks and Future Prospects

One of the most significant recent developments is the emergence of living carbocationic polymerization leading to specialty telechelics. The most important long-range significance of living carbocationic polymerization is probably the production of block, particularly di- and triblock, copolymers by simple sequential monomer-addition process. However, certain specific functional groups have been synthesized by direct end capping by using living carbocationic polymerization of IB. In this review we described the design and application of various functional PIBs. PIB-based macromonomers are another class of new materials that are rich in possibilities. Copolymerization of these macromonomers with specific monomers holds promise for biomedical applications in the form of amphiphilic networks for drug-delivery systems, prostheses, or artificial organs. We foresee sustained activity in various specialty fields, including biomedical applications.

Appendixes

Appendix I. Various End-Functional PIBs Prepared by Different Methods

Functional Group	Synthesis Method	Ref.
From t-Cl end group (–CH$_2$–C(CH$_3$)$_2$–Cl)		
–CH$_2$–C(CH$_3$)=CH$_2$	Exo-selective dehydrochlorination by base	100–102
–CH$_2$–C(CH$_3$)$_2$–CH$_2$–CH=CH$_2$	Allylation by allyltrimethylsilane in presence of Friedel–Crafts catalyst	103, 104
–CH$_2$–C(CH$_3$)$_2$–p-C$_6$H$_4$–R (R = H, CH$_3$, OH, OCH$_3$)	Friedel–Crafts alkylation of C$_6$H$_5$–R	70, 105, 106
–CH$_2$–C(CH$_3$)$_2$–p-C$_6$H$_4$–CH(R)–CH$_2$–Br (R = H, CH$_3$)	By Friedel–Crafts alkylation	107
–CH$_2$–C(CH$_2^-$)=CH$_2$	One-pot dehydrochlorination-metallation by alkyllithium	108, 109
From isopropylidine end group {–CH$_2$–C(CH$_3$)=CH$_2$}		
–CH$_2$–C(CH$_3$)–CH$_2$–Si(CH$_3$)$_2$–Cl	Hydrosilylation reaction	110–112
–CH$_2$–C(CH$_3$)–CH$_2$OH	Hydroboration and oxidation	83, 84, 113
–CH$_2$–C(CH$_3$)$_2$–p-C$_6$H$_4$–OR (R = H, CH$_3$)	By Friedel–Crafts alkylation	106, 107
2-Methylepoxypropane	Peroxidation with m-chloroperbenzoic acid	114
–CH$_2$–C(=CH$_2$)–CH$_2$SO$_3$H	Sulfonation with acetyl sulfate	71, 72, 115
–CH$_2$–C(CH$_2^-$)=CH$_2$	Metallation by alkyl lithium	116
–CH$_2$–C(CH$_3$)–CH$_2$–S–(CH$_2$)$_n$–COOH	UV-induced chain extension by a dithiol	117
Succinic anhydride		94
From primary hydroxy end group (–CH$_2$–C(CH$_3$)–CH$_2$OH)		
–N=C=O	Reaction with excess toluene diisocyanate	118
–CH$_2$–C(CH$_3$)–CH$_2$O–CO–N=C=O	Reaction with N-chlorocarbonyl isocyanate	57
–CH$_2$–C(CH$_3$)–CH$_2$OOCRCOOH (R = p-C$_6$H$_4$ or (CH$_2$)$_4$)	Reaction with excess dicarboxylic acid chlorides followed by hydrolysis	119
–CH$_2$–C(CH$_3$)–CH$_2$OOC–C$_6$H$_2$–(COOH)$_3$	Esterified with excess pyromellitic dianhydride in the presence of 4-N,N'-dimethylaminopyridine	120
–CH$_2$–C(CH$_3$)–CH$_2$–O–CH$_2$–CH=CH$_2$	Reaction with allylbromide and alkali in the presence of tetrabutylammonium bisulfate (TBAB, phase-transfer catalyst)	121
–CH$_2$–C(CH$_3$)–CH$_2$–O–CH$_2$–C– CH	Reaction with propargyl bromide	122
–CH$_2$–C(CH$_3$)–CH$_2$–O–CH$_2$–CH$_2$–CN	Reaction with acrylonitrile using different catalysts	123
–CH$_2$–C(CH$_3$)–CH$_2$OOC–p-C$_6$H$_4$–CN	Reaction with p-cyanobenzoyl chloride, or p-cyanobenzoic acid using different catalysts	123
–CH$_2$–C(CH$_3$)–CH$_2$CN	Methanesulfonation with methanesulfonyl chloride, followed by reaction with NaCN in the presence of phase transfer catalyst	124
–CH$_2$–C(CH$_3$)–CH$_2$OTS (Tosyl-ended PIB)	Tosylation in the presence of catalyst	59
–CH$_2$–C(CH$_3$)–CH$_2$OCO–CR=CH$_2$ (R = H, CH$_3$, CN)	Reaction with acryloyl or methacryloyl chloride	22, 125, 126
From other end groups		
–CH$_2$–C(CH$_3$)$_2$–CH$_2$–CH=CH$_2$	Allylation of living polymer chains	68, 69
4,4-dimethyl-1,2-epoxypentane	Epoxidation of allyl-caped macromonomer (i.e., above)	68, 69
1,2-Epoxy-3-(4-t-butyl)phenoxypropane	Reacting phenol-ended PIB with epichlorohydrin	65
–CH$_2$–C(CH$_3$)$_2$–C$_6$H$_4$–C(R)=CH$_2$ (R= H, CH$_3$, etc.)	Dehydrohalogenation of 2-bromoethylbenzene-ended PIB	108
–CH$_2$–C(CH$_3$)–CH$_2$–O–CH$_2$–CH$_2$–NH$_2$	Reduction of cyano-ended PIB with LiAlH$_4$	124

Continued on next page

Appendix I. Various End-Functional PIBs Prepared by Different Methods—*continued*

Functional Group	Synthesis Method	Ref.
$-CH_2-C(CH_3)_2-C_6H_4-O-CH_3-CH_2-O-CH=CH_2$	Reaction of 2-chloroethyl vinyl ether with phenol-ended PIB or metallated PIB	127
$-CH_2-C(CH_3)_2-C_6H_4-A$ (A = $COCH_3$, SO_2Cl, NO_2)	Acetylation, chlorosulfonation, or nitration of the aromatic ring of the phenyl-ended PIB	128
$-CH_2-C(CH_3)_2-C_6H_4-NH_2$	Reduction of nitrophenyl-ended PIB	128
$-CH_2-C(=CH_2)-CH_2-B$ (B = CH_2CH_2OH, COOH)	Oxyethylation or carbonation of the metallated PIB	110
$-CH_2-CH(CH_3)-CH=O$	Isomerization of epoxy-ended PIB by $ZnBr_2$	129
$-CH_2-CH(CH_3)-CH_2-SiH(CH_3)_2$	Reduction of chlorosilyl-ended PIB by $LiAlH_4$	111
$-CH_2-CH(CH_3)-CH_2-Si(CH_3)_2-C_5H_4$	Reaction of Si–Cl-terminated PIB with sodium cyclopentadienyl	112

Appendix II. Other Groups Synthesized by Different Initiating Systems

Head Group	Initiator System	Ref.
$(CH_3)_2C=CH-CH_2-$	$(CH_3)_2C=CH-CH_2Cl/BCl_3$	51, 52
$CH_2=CH-CH(CH_3)-$	$CH_2=CH-CH(CH_3)Cl/(CH_2=CH)_3Al$	130
$CH_2=CH-C_6H_4-CH_2-$	$CH_2=CH-C_6H_4-CH_2Cl/Et_3Al$	131, 132
$Br-CH_2-CH_2-C_6H_4-CH_2-$	$BrCH_2-CH_2-C_6H_4-CH_2Cl/Me_3Al$	133
$OH-C_6H_4-C(CH_3)_2-$	$p-HO-C_6H_4-C(CH_3)_2-Cl/BCl_3$	134
1-Methyl-4-(methyl)ethyl-1-cyclohexene	$p-MeC_6H_8-C(CH_3)_2-Cl/BCl_3$	135
$CH_2=C(CH_3)-OCO-$	1-Chlorodicyclopentadiene (DCp-Cl)/Et_2AlCl	136
$Cl-C_6H_4-C(CH_3)_2-$	2-Pinanylchloride formed in situ from *cis*-2-pinanol/BCl_3	137

References

1. Kennedy, J. P.; Ivan, B. *Designed Polymer by Carbocationic Macromolecular Engineering: Theory & Practice;* Hanser: Munich, Germany, 1991.
2. Mishra, M. K. In *Frontiers in Macromolecular Science and Applications;* Sivaram, S.; Mishra, M. K., Ed.; *Indian J. Technol.* (special volume) **1993**, 31, 197.
3. Butlerov, A.; Goryanov, V. *Ber. Dtsch. Chem. Ges. Ber.* **1873**, 6, 561.
4. Thomas, R. M.; Lightbown, I. E.; Sparks, W. J.; Frolich, K. P.; Murphree, E. V. *Ind. Eng. Chem.* **1940**, 32, 1283.
5. Thomas, R. M.; Sparks, W. J. U.S. Patent 2 356 128, 1944.
6. Evans, A. G.; Meadows, G. W.; Polanyi, M. *Nature (London)* **1946**, 158, 94.
7. Plesch, P. H.; Polanyi, M.; Skinner, H. A. *J. Chem. Soc.* **1947**, 257.
8. Kennedy, J. P.; Marichal, E. *Carbocationic Polymerization;* Wiley Interscience: New York, 1982; and references therein.
9. Kennedy, J. P.; Faust, R. U.S. Patent 4 910 321, 1990
10. Faust, R.; Kennedy, J. P. *Polym. Bull.* **1986**, 15, 317.
11. Kennedy, J. P.; Mishra, M. K. U.S. Patent 4 929 683, 1990.
12. Mishra, M. K.; Kennedy, J. P. *J. Macromol. Sci. Chem.* **1987**, A24, 933.
13. Nguyen, H. A.; Kennedy, J. P. *Polym. Bull.* **1981**, 6, 47, 55.
14. Mishra, M. K.; Chen, C. C.; Kennedy, J. P. *Polym. Bull.* **1989**, 22, 455.
15. Chen, C. C.; Kaszas, G.; Puskas, S. E.; Kennedy, J. P. *Polym. Bull.* **1989**, 22, 463.
16. Mishra, M. K.; Osman, A. U.S. Patent 4 908 421, 1990.
17. Mishra, M. K.; Osman, A. U.S. Patent 4 943 421, 1990.
18. Uraneck, C. A.; Hsieh, H. L.; Buck, O. G. *J. Polym. Sci.* **1962**, 46, 148.
19. Kennedy, J. P. *J. Appl. Polym. Sci.* **1984**, 39, 21.
20. *Telechelic Polymers;* Goethals, E. J., Ed.; CRC: Boca Raton, FL, 1989.
21. Kennedy, J. P.; Smith, R. A.; Ross, L. R. U.S. Patent 4 276 394, 1981.
22. Kennedy, J. P.; Hiza, M. *J. Polym. Sci., Polym. Chem. Ed.* **1983**, 21, 1033.
23. Kennedy, J. P.; Smith, R. A. *J. Polym. Sci., Polym. Chem. Ed.* **1980**, 18, 1523.
24. Chang, V. S. C.; Kennedy, J. P.; Ivan, B. *Polym. Bull.* **1980**, 3, 339.
25. Santos, R.; Fehervari, A.; Kennedy, J. P. *J. Polym. Sci., Polym. Chem. Ed.* **1984**, 22, 2685.
26. Kennedy, J. P.; Ross, L. R.; Lackey, J. E.; Nuyken, O. *Polym. Bull.* **1981**, 4, 67.
27. Akatsu, S.; Watnabe, R. Jpn. Patent 61/73701 A2, 1986.
28. Mah, S.; Faust, R.; Zsuga, M.; Kennedy, J. P. *Polym. Bull.* **1988**, 18, 433.
29. Nguyen, H. A.; Kennedy, J. P. *Polym. Bull.* **1983**, 9, 507.

30. Faust, R.; Kennedy, J. P. *J. Polym. Sci., Polym. Chem. Ed.* **1987**, *25*, 1847.
31. Faust, R.; Nagy, A.; Kennedy; J. P. *J. Macromol. Sci. Chem.* **1987**, *A24*, 595.
32. Huang, K. J.; Zsuga, M.; Kennedy, J. P. *Polym. Bull.* **1988**, *19*, 43.
33. Fehervari, A. F.; Faust. R.; Kennedy, J. P. *J. Macromol. Sci. Chem.* **1990**, *A27*, 1571.
34. Mishra, M. K.; Kennedy, J. P. *Polym. Bull.* **1987**, *17*, 7.
35. Wang, B.; Mishra, M. K.; Kennedy, J. P. *Polym. Bull.* **1987**, *17*, 213.
36. Wang, B.; Mishra, M. K.; Kennedy, J. P. *Polym. Bull.* **1987**, *17*, 205.
37. Mishra, M. K.; Wang, B.; Kennedy, J. P. *Polym. Bull.* **1987** *17*, 307.
38. Mishra, M. K.; Santee, E.; Kennedy, J. P. *Polym. Bull.* **1986**, *15*, 469.
39. Tessier, M.; Marechal, E. *Polym. Bull.* **1983**, *10*, 152.
40. Tessier, M.; Nguyen, H. A.; Marechal, E. *Polym. Bull.* **1981**, *4*, 111.
41. Fehervari, A. F.; Kennedy, J. P.; Tudos, F. *J Macromol. Sci. Chem.* **1980**, *15*, 215.
42. Mishra, M. K.; Sar-Mishra, B.; Kennedy, J. P. *Macromol. Synth.* **1990**, *10*, 53.
43. Yagci, Y.; Mishra, M. In *Macromolecular Design: Concept and Practice*; Mishra, M. K., Ed.; Polymer Frontiers International., Inc.: New York, 1994; Chapter 10.
44. Nemes, S.; Kennedy, J. P. *J. Macromol. Sci. Chem.* **1991**, *A28*, 311.
45. Kennedy, J. P.; Price, S. *Polym. Mater. Sci. Eng.* **1991**, *64*, 40.
46. Wilczek, L.; Mishra, M. K.; Kennedy, J. P. *J. Macromol. Sci. Chem.* **1987**, *A24(9)*, 1033.
47. Kitayama, T.; Nishiwa, T.; Hatada, K. *Polym. Bull.* **1991**, *26*, 513.
48. Wondraczek, R. H.; Kennedy, J. P. *J. Polym. Sci. Polym. Chem. Ed.* **1982**, *20*, 173.
49. Peyrot, J. Ger. Patent 2 161 859, 1972.
50. Kennedy, J. P.; Feinberg, S. C.; Huang, S. Y. *J. Polym. Sci., Polym. Chem. Ed.* **1978**, *16*, 243.
51. Kennedy, J. P.; Hunag, S. Y.; Smith, R. A. *Polym. Bull.* **1979**, *1*, 371.
52. Kennedy, J. P.; Huang, S. Y.; Smith, R. A. *J. Macromol. Sci. Chem.* **1980**, *A14*, 1085.
53. Kennedy, J. P.; Smith, R. A. *J. Polym. Sci., Polym. Chem. Ed.* **1980**, *18*, 1539.
54. Kennedy, J. P.; Chen, F. J.-Y. *Polym. Prepr. (Am. Chem. Soc. Div. Polym. Symp.)* **1979**, *20(2)*, 310.
55. Kennedy, J. P.; Guhaniyogi, S. C.; Ross, L. R. *J. Macromol. Sci., Chem.* **1982**, *A18*, 119.
56. Kennedy, J. P.; Hongu, Y. *Polym. Bull.* **1985**, *13*, 115.
57. Wondraczek, R. H.; Kennedy, J. P. *Polym. Bull.* **1981**, *4*, 445.
58. Sar, B. Ph.D. Dissertation, University of Akron, Akron, OH, 1993.
59. Percec, V.; Guhaniyogi, S. C.; Kennedy, J. P.; Ivan, B. *Polym. Bull.* **1982**, *8*, 25.
60. Keszler, B.; Kennedy, J. P. *J. Macromol. Sci. Chem.* **1984**, *A21*, 319.
61. Kennedy, J. P. U.S. Patent 4 486 572, 1984.
62. Chen, D.; Kennedy, J. P.; Allen, A. S. *J. Macromol. Sci. Chem.* **1988**, *A25*, 389.
63. Kennedy, J. P. *CHEMTECH* **1994**, *February*, 24.
64. Ivan, B.; Kennedy, J. P.; Mackey, P. W. *Polym. Prepr. (Am. Chem. Soc. Div. Polym. Symp.)* **1990**, *31(2)*, 215, 217.
65. Kennedy, J. P.; Guhaniyogi, S. C.; Percec, V. *Polym. Bull.* **1982**, *8*, 571.
66. Kennedy, J. P.; Guhaniyogi, S. C. U.S. Patent 4 429 099, 1984.
67. Hartman, K. O.; Bohrer, E. K. *Polym. Bull.* **1987**, *18*, 403
68. Ivan, B.; Kennedy, J. P. *Polym. Mater. Sci. Eng.* **1988**, *58*, 869.
69. Ivan, B.; Kennedy, J. P. *J. Polym. Sci., Polym. Chem., Ed.* **1989**, *28*, 89.
70. Kennedy, J. P.; Hiza, M. *J. Polym. Sci., Polym. Chem. Ed.* **1983**, *21*, 3573.
71. Kennedy, J. P.; Storey, R. F. *Org. Coat. Appl. Polym. Sci.* **1982**, *46*, 182.
72. Mohajer, V.; Tyagi, D.; Wilkes, G. L.; Storey, R. F.; Kennedy, J. P. *Polym. Bull.* **1982**, *8*, 47.
73. Bagrodia, S.; Mohajer, Y.; Wilkes, G. L.; Storey, R. F.; Kennedy, J. P. *Polym. Bull.* **1982**, *8*, 281; *Polym. Bull.* **1983**, *9*, 174; *Polym. Prepr. (Am. Chem. Soc. Div. Polym. Symp.)* **1983**, *24(2)*, 88.
74. Bagrodia, S.; Pisipati, R.; Wilkes, G. L.; Stoney, R. F.; Kennedy, J. P. *J. Appl. Polym. Sci.* **1984**, *29*, 3065.
75. Bagrodia, S. R.; Wilkes, G. L.; Kennedy, J. P. *J. Appl. Polym. Sci.* **1985**, *30*, 2179; *Polym. Eng. Sci.* **1986**, *26*, 662.
76. Mohajer, Y.; Bagodia, S.; Wilkes, G. L.; Storey, R. F.; Kennedy, J. P. *J. Appl. Polym. Sci.* **1984**, *29*, 1943.
77. Tant, M. R.; Wilkes, G. L.; Storey, R. F.; Kennedy, J. P. *Polym. Prepr. (Am. Chem. Soc. Div. Polym. Symp.)* **1984**, *25(2)*, 118; *Polym. Bull.* **1985**, *13*, 541.
78. Tant, M. R.; Wilkes, G. L.; Read, M. D.; Kennedy, J. P. *J. Polym. Sci., Polym. Lett. Ed.* **1986**, *24*, 619.
79. Ako, M.; Kennedy, J. P. *J. Appl. Polym. Sci.* **1989**, *37*, 1351.
80. Chang, V. S. C.; Kennedy, J. P. *Polym. Bull.* **1982**, *8*, 69.
81. Kennedy, J. P. *CHEMTECH* **1986**, 694.
82. Kennedy, J. P.; Lackey, J. E. *J. Appl. Polym. Sci.* **1987**, *33*, 2449.
83. Ivan, B.; Kennedy, J. P.; Chang, V. S. C. *J. Polym. Sci., Polym. Chem. Ed.* **1980**, *18*, 3177.
84. Kennedy, J. P. U.S. Patent 4,276,394, 1982.
85. Kennedy, J. P.; Ivan, B.; Chang, V. S. C. In *Urethane Chemistry and Applications*; Edwards, K. N., Ed.; ACS Symposium Series 172; American Chemical Society: Washington, DC, 1981; p 383.
86. Jiang, C.-Y.; Mark, J. E.; Chang, V. S. C.; Kennedy, J. P. *Polym. Bull.* **1984**, *11*, 319.
87. Kennedy, J. P.; Lackey, J. E. *Polym. Mater. Sci. Eng.* **1983**, *49*, 69.
88. Lackey, J. E.; Chang, V. S. C.; Kennedy, J. P.; Zhang, Z.-M.; Sung, P-H; Mark, J. E. *Polym. Bull.* **1984**, *11*, 19.
89. Miyabayashi, T.; Kennedy, J. P. *J. Appl. Polym. Sci.* **1986**, *31*, 2, 523.
90. Sung, P.-H.; Pan, S.-J.; Mark, J. E.; Chang, V. S. C.; Lackey, J. E.; Kennedy, J. P. *Polym. Bull.* **1983**, *9*, 375.
91. Speckhard, T. A.; Gibson, P. E.; Cooper, S. L.; Chang, V. S. C.; Kennedy, J. P. *Polymer* **1985**, *26*, 55.
92. Speckhard, T. A.; Hwang, K. K. S.; Cooper, S. C.; Chang, V. S. C.; Kennedy, J. P. *Polymer* **1985**, *26*, 70.

93. Speckhard, T. A.; Cooper, S. L. *Rubber Chem. Technol.* **1986**, *59*, 405.
94. Trivedi, B. C.; Culbertson, B. M. *Maleic Anhydride;* Plenum: New York, 1982, and references therein.
95. Landis, P. S.; Norris, H. D.; White, R. V. U.S. Patent 2 568 876, 1951.
96. Anderson, R. G.; Drummond, A. Y.; Stuart, F. A. U.S. Patent 3 024 237, 1962.
97. Le Suer, W. M. U.S. Patent 3 087 936, 1963.
98. Osuch, C. U.S. Patent 3 288 714, 1963.
99. Meinhardt, N. A.; Davis, K. E. U.S. Patent 4 234 435, 1980.
100. Mishra, M. K.; Sar, B.; Kennedy, J. P. *Polym. Bull.* **1985**, *13*, 435.
101. Mishra, M. K.; Sar, B.; Kennedy, J. P. *Macromol. Synth.* **1990**, *10*, 49.
102. Kennedy, J. P.; Chang, V. S. C.; Smith, R. A.; Ivan, B. *Polym. Bull.* **1979**, *1*, 575.
103. Wilczek, L.; Kennedy, J. P. *J. Polym. Sci., Polym. Chem. Ed.* **1987**, *25*, 3255.
104. Fujisawa, H.; Noda, K.; Yonezawa, K. Jpn. Patent 04154815, A2, 1992
105. Kennedy, J. P.; Guhaniyogi, S. C.; Percec, V. *Polym. Bull.* **1982**, *8*, 563.
106. Mishra, M. K.; Sar, B.; Kennedy, J. P. *Polym. Bull.* **1986**, *16*, 47.
107. Keszler, B.; Chang, V. S. C.; Kennedy, J. P. *J. Macromol. Sci Chem.* **1984**, *A21*, 307.
108. Kennedy, J. P.; Peng, K. L.; Wilczek, L. *Polym. Bull.* **1988**, *19*, 441.
109. Nemes, S.; Peng, K. L.; Wilczek, L.; Kennedy, J. P. *Polym. Bull.* **1990**, *24*, 187.
110. Chang, V.S.C.; Kennedy, J. P. *Polym. Bull.* **1981**, *5*, 379.
111. Fang, T. R.; Kennedy, J. P. *Polym. Bull.* **1983**, *10*, 82.
112. Kennedy, J. P.; Carlson, G. M. *J. Polym. Sci., Polym. Chem. Ed.* **1983**, *21*, 2973.
113. Sar, B.; Mishra, M. K.; Kennedy, J. P. *Macromol. Synth.* **1990**, *10*, 45.
114. Kennedy, J. P.; Chang, V. S. C.; Francik, W. P. *J. Polym. Sci., Polym. Chem. Ed.* **1982**, *20*, 2809.
115. Storey, R. F.; Lee, Y. *J. Polym. Sci., Polym. Chem. Ed.* **1991**, *29*, 317.
116. Peng, K. L.; Kennedy, J. P.; Wilczek, L. *J. Polym. Sci., Polym. Chem. Ed.* **1988**, *26*, 2235.
117. Nuyken, O.; Chang, V. S. C.; Kennedy, J. P. *Polym. Bull.* **1981**, *4*, 61.
118. Wondraczek, R. H.; Kennedy, J. P. *Polym. Bull.* **1980**, *2*, 675.
119. Liao, T.-P.; Kennedy, J. P. *Polym. Bull.* **1981**, *5*, 11.
120. Percec, V.; Guhaniyogi, S. C.; Kennedy, J. P. *Polym. Bull.* **1982**, *8*, 319
121. Percec, V.; Guhaniyogi, S. C.; Kennedy, J. P. *Polym. Bull.* **1982**, *8*, 551
122. Percec, V.; Guhaniyogi, S. C.; Kennedy, J. P. *Polym. Bull.* **1983**, *9*, 570.
123. Percec, V.; Kennedy, J. P. *Polym. Bull.* **1983**, *10*, 31.
124. Percec, V.; Guhaniyogi, S. C.; Kennedy, J. P. *Polym. Bull.* **1983**, *9*, 27.
125. Liao, T.-P.; Kennedy, J. P. *Polym. Bull.* **1981**, *6*, 135.
126. Kennedy, J. P.; Midha, S.; Gadkari, A. *J. Macromol. Sci. Chem.* **1991**, *A28*, 209.
127. Nemes, S.; Pernecker, T.; Kennedy, J. P. *Polym. Bull.* **1991**, *25*, 633.
128. Kennedy, J. P.; Hiza, M. *J. Polym. Sci.; Polym. Chem. Ed.* **1983**, *21*, 3573.
129. Kennedy, J. P.; Chang, V. S. C.; Francik, W. P. *J. Polym. Sci., Polym. Chem. Ed.* **1982**, *20*, 2809.
130. Mandal, B. M.; Kennedy, J. P. *J. Polym. Sci., Polym. Chem. Ed.* **1978**, *16*, 833.
131. Kennedy, J. P.; Frisch, K. C. *Prepr. Macro Florence* **1980**, *2*, 162.
132. Kennedy, J. P.; Frisch, K. C. U.S. Patent 4,327,201, 1982.
133. Kennedy, J. P.; Lo, C. Y. *Polym. Prepr. (Am. Chem. Soc. Div. Polum. Symp.)* **1982**, *23(2)*, 99.
134. Kennedy, J. P.; Carter, S. D. *Polym. Prepr. (Am. Chem. Soc. Div. Polum. Symp.)* **1986**, *27(2)*, 29.
135. Kennedy, J. P.; Hiza, M. *Polym. Bull.* **1982**, *8*, 557.
136. Kennedy, J. P.; Carlson, G. M.; Ribel, K. *Polym. Bull.* **1983**, *9*, 268.
137. Nemes, S.; Kennedy, J. P. *Polym. Bull.* **1989**, *21*, 293.

Perfect Diels–Alder Ladder Polymers: Precursors for Extended π-Conjugation

Arnulf-Dieter Schlüter, Matthias Löffler, Adelheid Godt, and Karsten Blatter

After an excursion into the strategy of the synthesis of structurally perfect ladder polymers, a whole new family of double-stranded Diels–Alder (DA) polymers is presented. Insight into how rigorously their structures were proved as well as data on the achievable molecular weights is provided. Molecular structure has considerable influence on the shape of these polymers, which can even form two-dimensional coils. The second part of this chapter discusses the potential of the new generation of ladder polymers as precursors for their fully unsaturated counterparts, which may be envisaged as either "graphitic ribbons" or "fullerene peels". The materials aspect of these target structures is briefly discussed, and the chapter culminates in two experiments where a polymer-analogous transformation of DA precursor ladders could actually be brought about.

Ladder (ribbon) polymers consist of cyclic subunits that are connected to each other by two links that are attached to different sites of the respective subunits. Thus, ladder polymers have two independent strands of regularly tied bonds that do not merge into a single or double bond or cross each other like in a spiro connection (Figure 1) (1). In the initial phase of the history of ladder polymers, this unique structural feature was believed to make the polymers ideal candidates for applications requiring materials with high thermal, mechanical, and chemical stability. This belief was rationalized because the molecular weight of ladder polymers obviously remains constant even if one of the two strands break. However, due to their poor solubility and infusibility, processing of these polymers was almost impossible. Consequently, useful materials with the expected properties could not be obtained, and the first generation of ladder polymers did not gain industrial importance (1, 2).

Since the mid-1980s, much interest has existed in the nonlinear optical and electrical properties of rigid-rod polymer films. High optical nonlinearities derived from π-conjugation, high laser-damage thresholds, and capability to form electrically conductive materials after doping suggested conjugated ladder polymers were potentially very interesting candidates for such applications. These characteristics led to a revival of interest in this class of polymers (3–7).

However, progress in the synthesis of ladder polymers has not kept pace with the rapidly growing technological importance of the materials. The strategies used today still stem from the early days of ladder-polymer synthesis, some 30 years ago. Almost no conceptual development toward the synthesis of well-defined and fully characterizable ladder polymers has taken place. The situation is best described by the statement of Yu and Dalton (8), "In fact, no one to date has ever made and unambiguously characterized a complete (classical) ladder polymer". Undoubtedly, the synthesis of a truly double-stranded polymer is a real synthetic challenge, considering that even the synthesis of well-defined single-stranded polymers is sometimes difficult to achieve. In the case of ladder polymers, two links have to be tied *n*-times between every two subunits without generating defects!

This chapter intends to show that there are previously disregarded tools available to the chemist that allow the synthesis of a whole new family of struc-

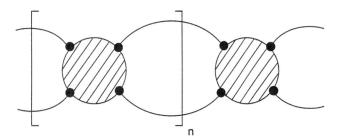

Figure 1. General representation of the structure of a ladder polymer with cyclic subunits and two independent strands of bonds.

turally perfect ladder polymers (9). We demonstrate that these polymers can be designed so that they may serve as precursors for either fully unsaturated "graphitic ribbons" or "fullerene peels". If a precursor strategy is to be applied reasonably, the precursors must be fully characterized before their transformation into the final target structures. Therefore, we describe how the precursor ladder polymers are obtained and into what depth their structures and shapes are investigated before we present attempts to convert them into their unsaturated counterparts.

Some of these attempts have been rather successful; the feasibility of others still awaits to be proven.

Classical Routes to Ladder Polymers: A Critical Perspective

Most of the known syntheses of ladder polymers fall into two categories. In the first, two tetrafunctional monomers are made to react with each other, whereby—assuming the reaction proceeds in the desired way—the double-stranded structure grows from the very beginning (Figure 2a) (9, 10). In the second category, single-stranded polymers are synthesized to carry the required functionalities at defined, regular distances along the chain. These functionalities are then used to generate the second strand in a series of polymer-analogous reactions (Figure 2b) (9, 11–13). Both strategies have serious disadvantages.

In the first case, it is difficult to see which factors, under homogenous reaction conditions, would force the monomers to react exclusively in the desired way. One wrong linkage, as indicated in Figure 2a, inevitably leads to inter-ribbon cross-linking: one of the reasons for the insolubility of most of the known

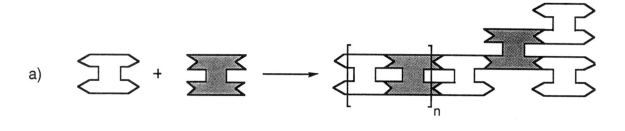

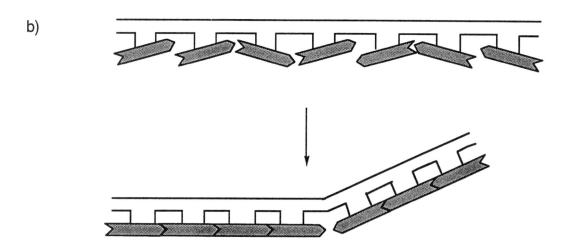

Figure 2. The two classical strategies to ladder polymers: Polycondensation of tetrafunctional monomers (a) or polymer-analogous cyclizations of appropriately substituted, single-stranded prepolymers (b). For both cases, structural irregularities likely to occur are shown.

ladder polymers. In addition to this topological issue, problems also arise from the condensation reactions that are most often used and are difficult to drive to completion. (Because of the rigidity of the ladder polymers, the reaction mixtures become very viscous and polymerization stops at relatively low conversions.) This deficiency results not only in the formation of low molecular-weight ribbons but also in ribbons with incompletely cyclized repeat units.

The second strategy (Figure 2b) looks very elegant on paper but could only be successful if the polymer-analogous reactions (cyclizations) did not proceed randomly but rather started at one terminus of the single-stranded prepolymer and proceeded, one after the other along the chain like a zipper. However, such a reaction course is clearly against statistics and, therefore, very unlikely to take place.

In addition to these strategy-specific drawbacks, there is another serious problem inherently associated with ladder polymers: namely, their poor solubility. Because of their rigid backbones, they show strong intermolecular interactions. These interactions are not only reflected in the desirable mechanical properties like high tensile strength but also in the undesirable insolubility that makes all attempts at structure elucidation very difficult. Therefore, if progress is to be made in the synthesis of well-defined and characterizable ladder polymers, two main issues have to be addressed: suppression of side reactions and achievement of increased solubility.

Diels–Alder Route to Ladder Polymers

In the course of a DA cyclization, each two reaction partners pass through a transition state that is stabilized by electron delocalization, or as chemists put it, by the partially aromatic character. This stabilization is the key to perfect structures of the products of this reaction, because it forces the reactants into the ideal relative geometry for a clean (regiospecific) reaction to take place. Thus, side reactions are suppressed and yields are high (*14*). For these reasons, the DA methodology was selected as the main tool for structure control in a new synthetic strategy to ladder polymers, a general representation of which is shown in Scheme I. [Bailey was one of the first to realize the usefulness of the DA reaction for the synthesis of ladder polymers, but, unfortunately, met with limited success because of the lack of suitable monomers (e.g., *see* reference 15).]

The second issue, the insolubility problem, is accounted for in Scheme I by the attachment of flexible alkyl chains, rings, or loops. This method significantly increases the solubility of rigid molecules in both polymer and low molecular weight chemistry. A strategy based on DA polyaddition necessarily yields ribbons whose molecular structures consist of a rather complex (generally statistical) sequence of exo–endo stereoisomers. This feature, which is associated with kinks in the backbone, may be considered as aesthetically unpleasing, but it certainly helps keep the polymer in solution and, therefore, is an additional plus of the strategy. Also, the DA strategy inevitably yields ladders that contain saturated (sp^3-hybridized) carbon atoms in the backbone. Depending on the respective monomers, the backbone may contain heteroatom bridges or (adjacent) hydrogen atoms. If the targeted polymer-analogous conversion of these precursor ladders is to be brought about, these sites of saturation ought to be removable by some very mild and controllable chemistry.

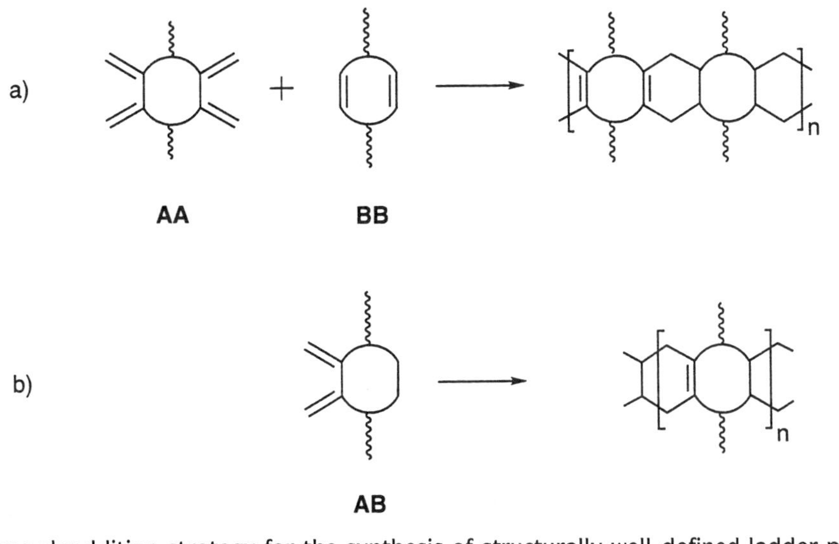

Scheme I. Diels–Alder polyaddition strategy for the synthesis of structurally well-defined ladder polymers by using (a) bifunctional dienes (AA-type monomers) and bifunctional dienophiles (BB-type monomers) and (b) bifunctional DA components with both diene and dienophile functionalities (AB-type monomer). The solubility enhancing alkyl chains are indicated as wavy lines (typically hexyl).

Suitable bifunctional DA monomers for the strategy outlined in Scheme I were selected in strict accordance with the following criteria:

- good accessibility on the gram or larger scale
- existence of structural features that help increase the solubility of the polymer
- irreversibility of each single DA cyclization

The sets of synthesized monomers are shown in Chart I. The monomers consist of bisdienes, bisdienophiles, and dienedienophiles, which are classified as AA-, BB-, and AB-type monomers, respectively. The monomers were designed so that the equilibria of the addition reactions between them lie far on the product side. Quite a few bifunctional DA compounds are known in the literature (16–20), but unfortunately, none of them meets all the criteria. Therefore, their use in polymer synthesis did not seem advisable. The DA strategy for the synthesis of double-stranded polymers also was used by others (21).

Monomers 2–5, 6, and 9 (22–28) were prepared and used on the 1–10-g scale; monomers 1 (29), 7 (30), and 8 (27) were generated from the precursor molecules and used in situ, typically on the 5-g scale. The formulae of 1, 7, and 8 do not represent proven molecular structures but serve as a rationalization for the observed DA reactivity of the intermediates. Monomers 1, 3, and 9 differ from known compounds because they are substituted with alkyl chains or loops only. Monomers 5 and 9 are bridged over by flexible alkyl loops (ansa compounds). These loops are effective solubilizers for entropic reasons (like straight alkyl chains) and also for enthalpic reasons because they disturb the packing of the corresponding polymers.

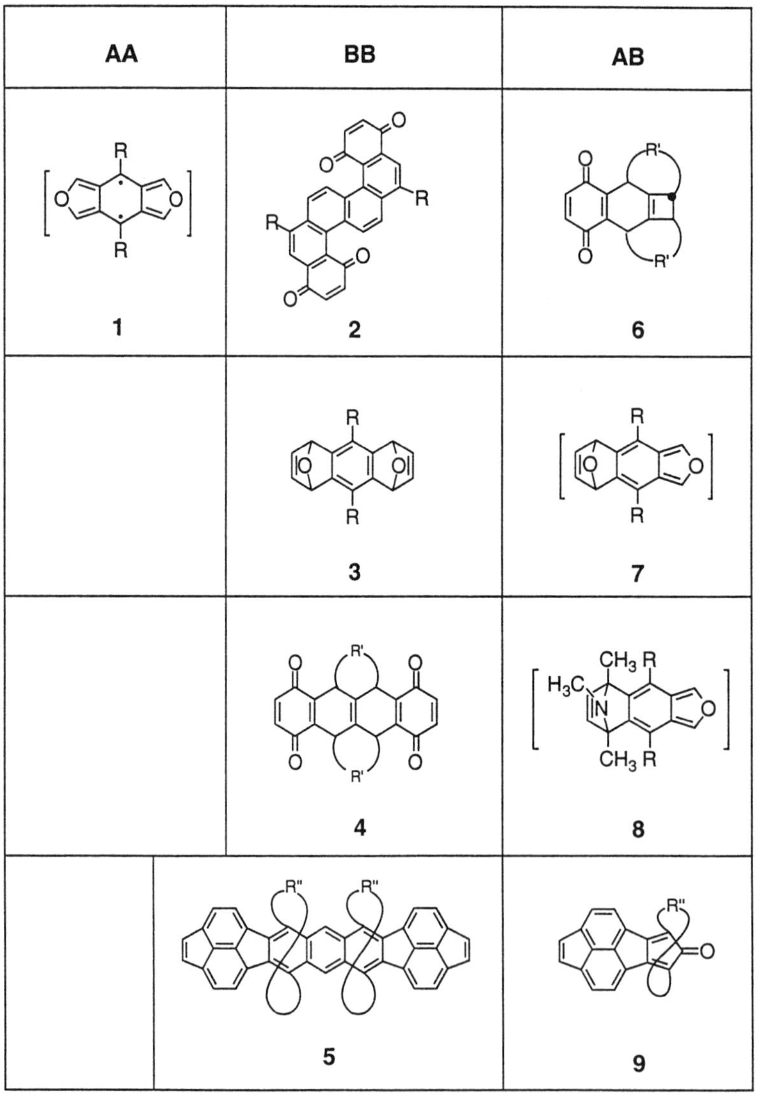

Chart I. Construction sets of AA-, BB-, and AB-type Diels–Alder building blocks for the synthesis of linearly and angularly annulated ladder polymers: R, straight alkyl chains, typically C_6 to C_{12}; R', hexa-1,6-diyl; R'', dodeca-1,12-diyl.

Polymers **10–13** (Chart II) were obtained by combining these monomers properly like building blocks in a construction set. Thus, the AA-monomer **1** was reacted with exactly the same stoichiometric amount of any of the BB monomers **2–5** to give polymers **14** (from **2**) (*30*), **10** (from **3**) (*23*), **12** (from **3**) (*24*), **15** (from **4**) (*31*), and **17** (from **5**) (*32*). On the other hand, simple thermal treatment of the AB-building blocks **6–9** gave polymers **13** (*25*), **10** (*23*), **11** (*27*), and **16** (*33*), respectively (Chart II). Yields were almost quantitative. All ladder polymers obtained turned out to be soluble in common organic solvents like chloroform, tetrahydrofuran, and benzene at room temperature, a result that nicely underlines the effectiveness of the alkyl substitution. The high solubility enabled careful structure elucidation of the polymers.

Chart II divides polymers **10–17** into three groups according to the structural features of their backbones. This division will gain some importance in the last part of this chapter in the discussion of the polymer-analogous transformation of these polymers into fully unsaturated ladders. Group 1 polymers **10–13** consist of a linear sequence of six-membered rings. Group 2 polymers **14** and **15** are also made up of six-membered rings, but they are annulated angularly. Finally, group 3 polymers **16** and **17** are again linearly annulated, but they contain six-membered as well as five-membered rings.

Molecular Structure and Overall Shape

The molecular structures of the repeat units of all new ladder polymers were investigated and proven by

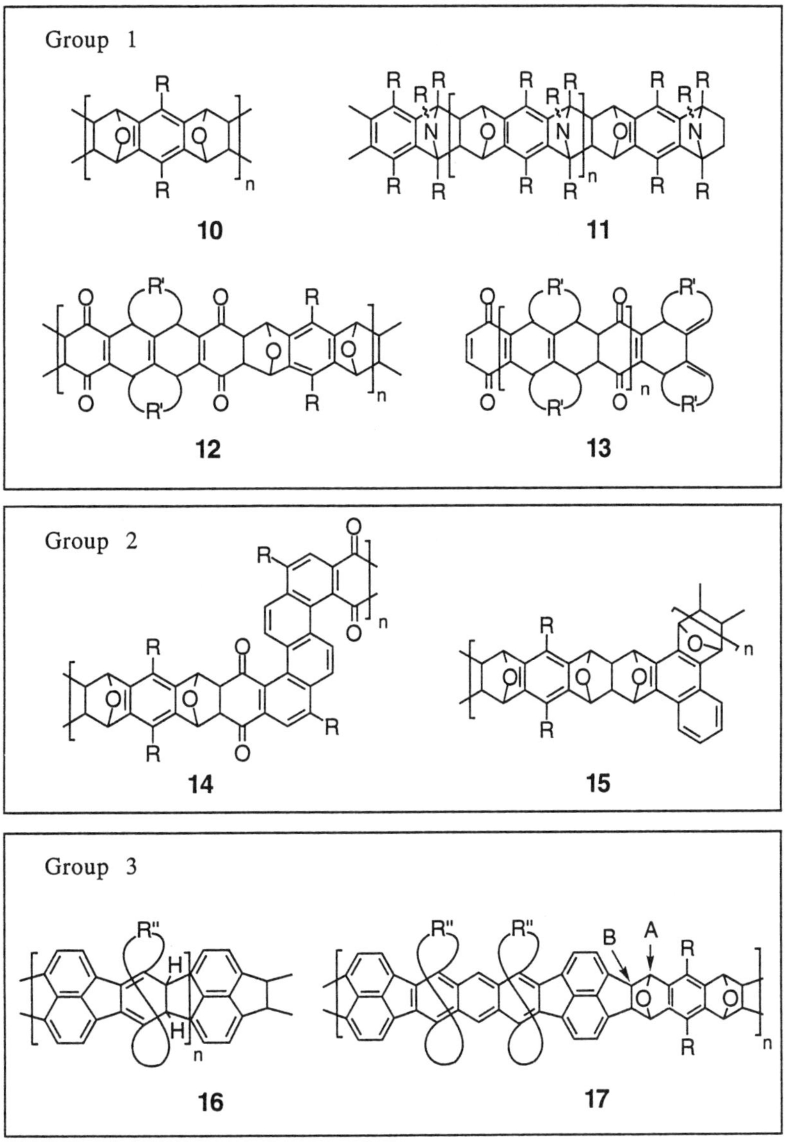

Chart II. Prepared Diels–Alder ladder polymers: R, straight alkyl chains, typically C6 to C12; R′, hexa-1,6-diyl; R″, dodeca-1,12-diyl.

high-resolution NMR spectroscopy. In most cases, the assignment of signals rests on detailed analyses of the NMR spectra of corresponding model compounds. Particularly informative is the structure elucidation of ribbon **14** (Chart II). For comparison, the 15-ring system **18** (Figure 3) was prepared (as a mixture of stereo isomers) (*29*). Figure 3 shows the high resolution NMR spectrum of this model (a) and of ribbon polymer **14** (c). The match of both spectra is excellent except for the marked signals of **18** at about δ = 190 ppm, which do not have a counterpart in the polymer spectrum. However, these signals were caused by the quinoid end groups and should not appear in the polymer spectrum. The observation of these signals establishes the proposed structure of **14**.

Figure 3 contains another interesting piece of information related to the solubility question. Polymer **14** was also prepared with the shorter hexyl chains, but in this case the solubility of the material turned out to be too low for recording a high resolution ^{13}C NMR spectrum. Thus, even a relatively dense substitution with hexyl chains is not sufficient in all cases. The structure of polymer **14** (R = hexyl) was confirmed by recording a solid-state cross-polarization magic angle spinning (CP MAS) ^{13}C NMR spectrum (Figure 3b). Despite the greater line widths, the match with the spectra (a) and (c) is convincing.

A second example may illustrate the great depth in which the structures of DA ladder polymers were investigated. Figure 4 shows the ^{1}H NMR spectrum of **13**. The main signals are obviously in full agreement with its structure. Attention was focused on the signals marked α–ε, which have very low intensities. What is the origin of these signals? This question requires explanation if the molecular structure of the ladder polymer is to be rigorously proven. Fortunately, it was possible to show that the signals α–ε are due to end groups and therefore are not caused by structural defects of any kind (*34*).

Stereochemical aspects of the new ribbons were investigated. As indicated previously, the backbones of all the ladder polymers synthesized using DA polyaddition contain complex sequences of repeating units with different stereochemistry. (In the case of the angular annulated ladders the situation is even more complex. New monomers may react with the growing chain in a *cisoid* or *transoid* fashion relative to the terminal repeat unit.) The ^{13}C NMR shifts of the atoms of a repeat unit generally depend on its stereochemistry and differ by a few parts per million. The shifts may also depend on the stereochemistry of the two neighboring repeat units. As a result, the spectra are quite complex, and a reliable correlation of their stereogeometries or sequences with certain signals is difficult (if not impossible) to achieve. To obtain at least some quantitative data on how many of which stereoisomers are present

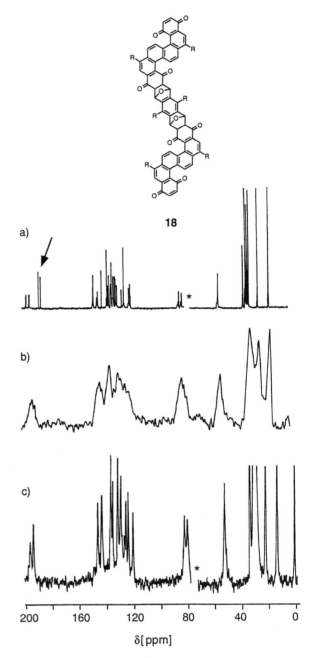

Figure 3. High resolution ^{13}C NMR spectra of model compound **18** (a), solid-state CPMAS ^{13}C NMR spectrum of polymer **14** (R = hexyl) (b), and ladder polymer **14** (R = dodecyl) (c). The signals of the quinoid end group of **18** are marked (arrow). Solvent signals (chloroform) in spectra (a) and (c) are erased for clarity (*).

in the backbone, and to get as much information about the microstructure of the polymers as possible, the following procedure was pursued for polymer **13**.

1. Pieces of the polymer backbone containing the relevant stereoisomeric forms were synthesized.
2. Single crystals of these pieces were grown and their stereochemistry was proven by X-ray diffraction.

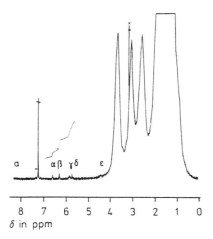

Figure 4. ^{1}H NMR spectrum (CDCl$_3$, 20 °C) of polymer **13**. The signals due to end groups are marked with α–ε (+, benzene; x, *tert*-butyl methyl ether).

3. The NMR spectra of the dissolved single crystals was recorded, observed shifts were correlated with those in the polymer spectrum, and the desired data were extracted.

This procedure not only has the function of answering the stereochemical questions, but the results can be thought of as a starting point for future attempts to achieve the ultimate goal of a controlled stereospecific DA polymerization.

Polymer **13** contains repeat units in two stereoisomeric forms, endo and exo. Among others, the model compounds (Chart III) *endo*-**19**, *exo*-**19**, and **20** (containing both exo and endo repeat units) were prepared, and their structures were determined using X-ray diffraction (*35*). On the basis of the analysis of the NMR spectra of these compounds it was possible to assign groups of signals in the ^{13}C NMR spectrum of **13** (Figure 5) to determine whether they are associated with exo or endo repeat units. Quantitative analysis of the intensities of relevant signals yielded an exo:endo ratio of approximately 1:1. Furthermore, other information on the sequence was obtained. For example, the signal at δ = 132 ppm was assigned to all carbons at the 4a(8a) positions in exo repeating units centered in endo–exo–endo or endo–exo–exo triads (*24*).

The Shape of DA Ladders

After having elucidated the molecular structures and some aspects of their stereochemical nature, the logical next step was to investigate the secondary structure of the new ribbons. This step is clearly a very complex matter for which no experimental answers are available to date. However, to gain insight into the secondary structures, the X-ray data of the models for polymer **13** were used as a data set for a computer

program that allows the assembly of molecule fragments to form larger ones. [The structure was generated using the INSIGHT program on a Silicon Graphics work station (IRIS 4070 GT).] All available stereochemical information was considered in this process. By this means a computer model of polymer **13** was generated whose three-dimensional views should give a realistic picture of its overall shape (Plate 1). According to this picture, polymer **13** has a coiled structure similar to the single-stranded polymers. From an inspection of the two possible stereoisomeric repeat units (*see* structures of exo- and endo-**19** in Chart III) it is evident that the endo unit is not responsible for the three-dimensionally coiled backbone. This unit has a mirror plane that is symmetrical to the one in which chain propagation occurs.

The exo unit, however, has a kink in its structure that already undergoes a flipping process near room temperature. This characteristic was proven by dynamic NMR studies using exo-**19** as a model, and it reflects some tortional flexibility of the polymer backbone at each exo repeat unit (*24*). Polymer **13** and presumably polymer **12** behave more or less like normal polymers in that they take on randomly coiled shapes, whereas polymers **10**, **11**, **16**, and **17** may behave differently. These structures do not contain a cyclohexene ring but do contain a linear sequence of conformationally rigid fragments (like 7-oxa(aza)norbornenes, benzenes, naphthalenes, and flat five-membered rings) and have the proper symmetry elements for the polymer to show plane symmetry. If this picture holds, these polymers should have the unique shape of a two-dimensional coil or a disc (Figure 6).

Compound **21**, a model for **10**, was selected to demonstrate this point. The X-ray structures of two diastereomers of **21** (Figure 7) show that both isomers have a mirror plane in which all oxygen atoms lie. However, these structures and the vague description of the fragments of **10** as being rigid are not sufficient to lead to the conclusion that **10** attains the shape of a (snake-like) two-dimensional coil. This result would depend on the degree of anisotropy of the flexibility of **10** in the plane and out of the plane. We applied all-atom molecular dynamics (MD) to determine the shape of **10** in solution (*35*). Polymer **22**, which consists of seven repeat units of fragments of **10** carrying no alkyl chains and terminated by benzene rings, was selected as a model. The MD studies were carried out with two different starting conformations of **22** (A and B, *see* Plate 2) to avoid a dependence of the result on the initial structure and in the presence of solvent molecules (benzene) because polymer/solvent effects have considerable influence on the conformation of a polymer chain. The starting conformation A of model **22** was constructed by using bond lengths and angles

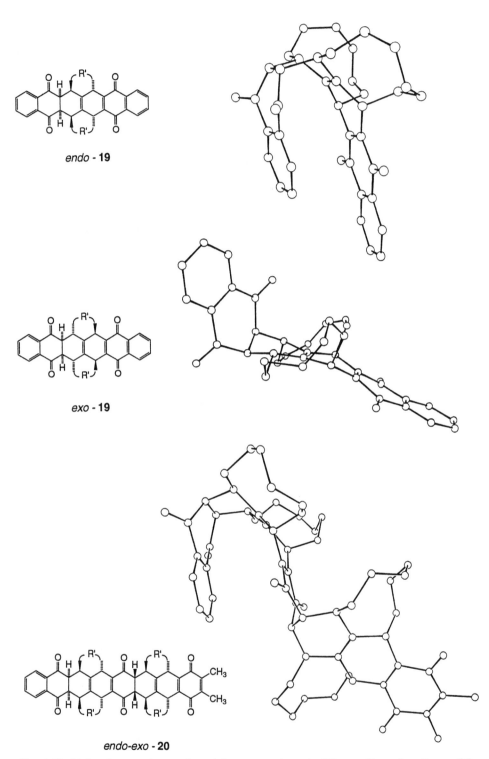

endo - 19

exo - 19

endo-exo - 20

Chart III. Molecular structures of model compounds *endo*-**19**, *exo*-**9**, and *endo-exo*-**20**.

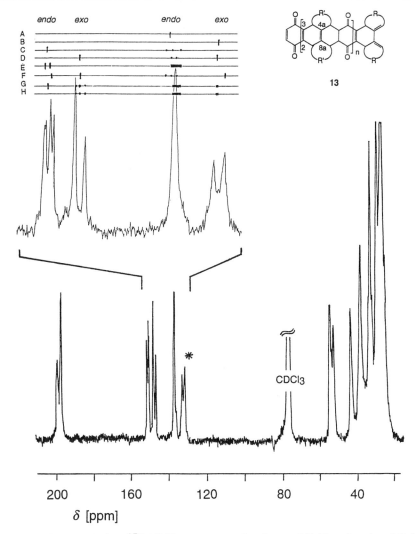

Figure 5. Fully reproducible high-resolution ^{13}C NMR spectrum of polymer **13**. The signals of C-2(3) and C-4a(8a) are grouped around δ = 135 and δ = 150 ppm, respectively. The assignment is based on eight different model compounds (A–H) as is indicated in the inserted enlargement of the olefinic region of the spectrum. The structures of A–H are not shown. The signal at δ = 132 ppm is marked (*).

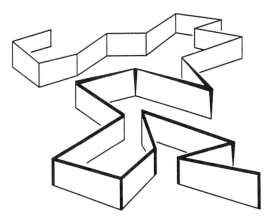

Figure 6. Schematic representation of the expected secondary structure of polymers **10**, **11**, **16**, and **17**, which is a two-dimensional coil.

from the X-ray structures of **21** and the AMBER data bank (*36*), which show plane symmetry.

Conformation B is an equilibrated structure (after 30 ps) that was created from A by performing an MD calculation in vacuo. During the calculation (without solvent) the initially two-dimensional coil starts to deviate significantly from planarity, which is initiated by higher fluctuations of the termini.

Both starting structures were placed in a box of constant volume filled with benzene molecules of realistic density at 300 K, and the molecular dynamics were simulated under periodic boundary conditions. Conformation A keeps its two-dimensional shape over the entire period of simulation (130 ps). Visualization of the molecular dynamics in a movie showed that the structure, on a time average, retains the symmetry σ_{xy}.

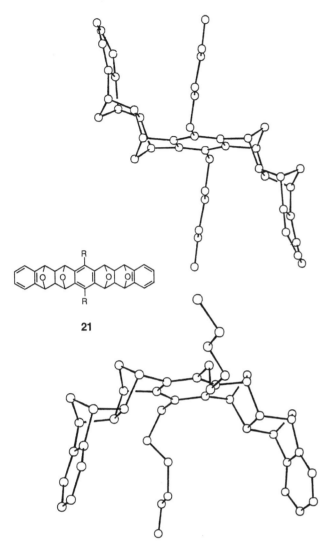

21

Figure 7. Structures in the crystal of two diastereomers of model compound **21** (ORTEP).

For starting structure B, the simulation in benzene at 300 K did not lead to any significant change of the gross conformation even at extended simulation time (200 ps). Therefore, the system was heated to 700 K at 100 K intervals. After each run had equilibrated, the system was cooled to 300 K. Finally, at 700 K for this conformation, a change in a plane-symmetrical geometry took place. As in the case of conformation A, the visualization of the molecular motion revealed that conformation B was not spherical anymore, but two dimensional and disclike (on a time average).

Thus, regardless of the starting conformation, structure **22** in benzene solution equilibrates to a two-dimensional conformation, which indicates a unique structural feature of **10** and other ladder polymers of comparable rigidity and symmetry. However, this calculation does not account for other effects that might play a role during synthesis of the polymer. As the

polymer chain grows, larger fragments of the same polymer chain might irreversibly overlap each other, or intermolecular effects (for example, formation of entanglements) might prevent the polymer from relaxing into a two-dimensional geometry. Small-angle X-ray scattering is an appropriate method to investigate this problem.

Molecular Weights

The new generation of ladder polymers are soluble in common solvents at reasonable concentrations. This characteristic is an enormous advantage in that all typical polymer analytical techniques can be applied. Thus, it is reasonable to assume that the molecular weight data obtained are more accurate than those for classical ladder polymers. However, sufficient solubility alone does not solve all the problems. For instance, the important method, gel permeation chromatography (GPC), is of limited value because appropriate standards for ladder polymers with little flexibility are not available. Nevertheless, GPC measurements were carried out but not interpreted quantitatively. In most cases, relatively broad and monomodal molecular-weight distributions were obtained and indicated clean polymerizations. Occasionally, the elution curves showed formation of some low molecular-weight material that could easily be removed. In one case it was shown that a double-stranded cycle (a [6]beltene derivative) had been formed (25). Table I contains a selection of representative GPC data.

The molecular weights of the ribbon polymers were determined by using vapor and membrane osmometry. To rule out possible aggregation phenomena, only dilute solutions were measured and each molecular weight was reproduced over a range of different concentrations. According to these measurements, representative samples have about 50 repeat units on average corresponding to number-average

Table I. Selected Molecular Weight Data of DA Ladder Polymers Obtained from Unfractionated Material Using Gel Permeation Chromatography (Versus Polystyrene Standard)

Polymer	M_n	P_n
10	14,000–19,400	40–55
11	35,200–51,500	90–131
13	12,700–15,900	36–45
16[a]	3500–7900	10–25
17[b]	12,000–27,000	10–25

NOTE: P_n is number average degree of polymerization. For polymer structures, *see* Chart II.

[a]R" is $(CH_2)_{12}$.

[b]R is C_6H_{13} and R" is $(CH_2)_{12}$.

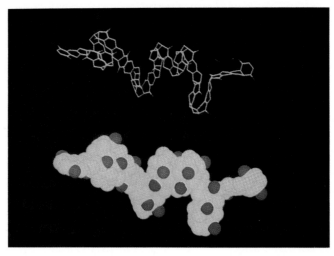

Plate 1. Computer generated three-dimensional model of polymer **13** without hydrogen atoms and flexible alkyl rings (top) and the corresponding van der Waals plot (below). Carbon is in green; oxygen is in red.

22

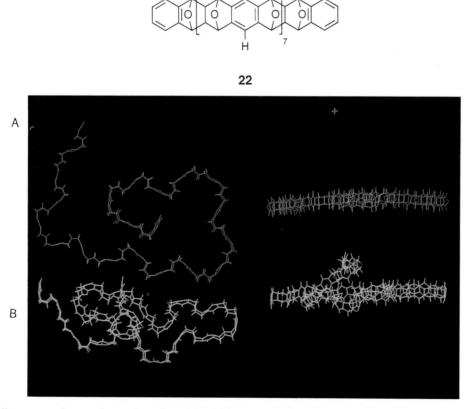

Plate 2. Two different starting conformations A and B of **22**. Top and side view of A (red) and B (green).

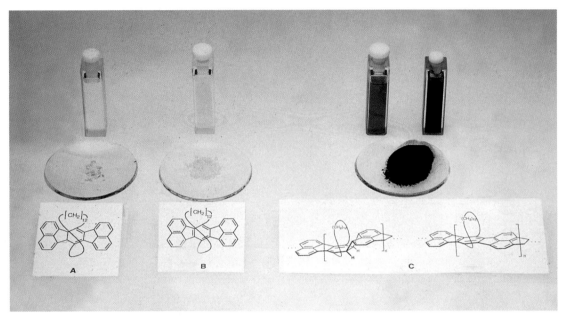

Plate 3. Representation of solutions and solids of the model compounds **27** (A) and **28** (B) and the polymer **26** (C).

molecular weights (M_n) of roughly 20,000 g/mol. Samples with higher molecular weights were easily obtained through fractionating techniques. Thus, for polymer **13**, gram fractions of material (still soluble) were prepared that had an M_n of ca. 100,000 g/mol. This result translates into a ribbon of 1000 linearly annulated six-membered rings and clearly demonstrates the power of DA polyaddition for the synthesis of double-stranded polymers. In accordance with their relatively high molecular weight, DA ladder polymers can be cast into flexible films from solution. Group 3 polymers have lower molecular weights, as was already indicated by the GPC data (Table I). We believe that this phenomenon is not inherent, but that it reflects polymerization procedures that are not yet optimized for these recently developed monomers.

During thermal treatment, conventional ladder polymers undergo nonconcerted bond cleavage processes at sites randomly distributed over the two strands of bonds. The molecular weights remain constant until two opposing bonds are coincidentally cleaved. Because DA cyclization–recyclization is an equilibrium process, the previously described new generation of ladder polymers may exhibit an additional mode of decomposition: retro-DA cleavage. By this reaction, two bonds of the same six-membered ring are cleaved in a concerted manner. Such a process would have a detrimental effect on the properties of DA ladder polymers in that each single retro-DA step breaks the backbone apart into two independent fragments and thus decreases the molecular weight. This issue was investigated in some depth by using polymer **10** as a representative. Fortunately, no retro-reaction occurred

within a reasonable temperature range (*37*). Thermal treatment of **10** with a monofunctional DA component in large excess led neither to a decrease of molecular weight nor to a broadening of the distribution.

Unsaturated, Double-Stranded Hydrocarbon Polymers

The chemist who is dealing with the synthesis of graphitic ribbons or fullerene peels has to, in effect, "do the splits". On the one hand, the unsaturated structures that the chemist is trying to prepare should have low band gaps, for example. On the other hand, these structures should be stable enough so that work can be done with them. Low band gap is a very important feature for many potential applications and is automatically associated with high reactivity. Parent polypyrrole, for example, cannot be obtained, because it is instantaneously oxidized after exposure to air. Other such polymers will simply react with themselves. The chemist, therefore, must move on a very narrow path between uninteresting but accessible and interesting but inaccessible materials.

As mentioned earlier, groups 1 and 2 of DA polymers may serve as suitable precursor polymers for graphitic ribbons (Figure 8), whereas group 3 representatives serve as precursors for high-molecular weight, planarized analogs of fragments of fullerenes (Figure 9). According to nomenclature introduced by Balaban (*38*), graphitic ribbons are called *polycatafusenes*. Polycatafusenes should be very exciting materials, because their properties ought to lie somewhere in between those of graphite and single-stranded

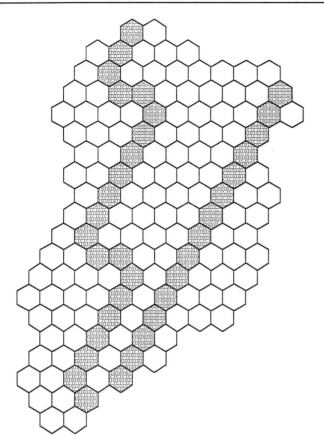

Figure 8. Sections of the graphite lattice (a) the left and (b) the right shadowed structure shown in the graphitic lattice for which the polymers **14**, **15**, and **10–13**, respectively, are potential precursors.

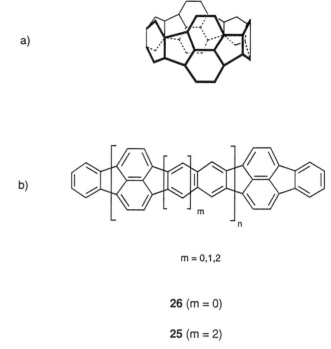

a)

b)

$m = 0,1,2$

26 $(m = 0)$

25 $(m = 2)$

Figure 9. Fragments of fullerenes: (a) the hypothetical equator region of C_{60} and (b) linear polymeric analogs (and homologs) of this equator region (fullerene peel).

polyacetylene. Theoretical calculations predict low band gaps, high electrical conductivities, and high optical nonlinearities for extended sections of a graphite lattice such as polycatafusenes (*39–46*). The fullerene-derived structures obtained from group 3 polymers would also be of great interest, not only for comparison with certain fullerene properties, but also because they might form semiconducting materials that are persistent in the undoped state and that have a high charge-carrier mobility. Potential applications would, therefore, be in the areas of photoconductivity and photovoltaics (*47–49*). Theoreticians might also be interested in these polymers in light of recent interest in molecules with nonalternant structures (*50, 51*).

At this point, it seems appropriate to make another general comment on the envisaged polymer-analogous transformations. Fully unsaturated, hydrocarbon ladder polymers have, on the time average, a planar board-like structure. Undoubtedly, the method successfully applied to increase solubility of the DA-ladders by the attachment of flexible alkyl substituents will not be powerful enough to keep flat polymers in solution. This reason is why the precursor concept was used throughout the whole enterprise. Its applica-

tion allows the complete characterization of the DA-ladders *before* they are transformed—by some defined chemistry—into the insoluble final structures. The solubility of the precursor polymers also allows the polymers to be processed into thin films in which the conversion may then be induced. Precursor concepts have proven to be very useful in the synthesis of insoluble yet highly important polymers. The most prominent examples are the Durham polyacetylene (*52*) and the Amherst poly(phenylene vinylene) (*53, 54*).

Toward Graphitic Ribbons

Except for polymer **13**, all groups 1 and 2 ribbon polymers contain π-conjugation extending over different areas (1–4 aromatic rings), which are interrupted regularly for inherent synthetic reasons. Clar showed (*55*) that the reactivity of π-systems (e.g., toward oxygen) increases with their increasing linear extension. Once generated, the polycatafusenes are expected to be quite reactive species. In light of this expectation, the interruption of the π-conjugation may have facilitated the isolation and characterization of the new ladder polymers.

Despite this assumed reactivity, optimism concerning the feasibility of generating polycatafusenes from the precursor ribbons is justified for two reasons. As mentioned, the reactivity of the linear acenes, which are perfect models for the described polymers, increases with the number of the fused aromatic rings (55). This result is reflected by their ionization potentials (IPs) which from naphthalene to hexacene decrease from 8.15 to 6.36 eV (56). These IPs, however, seem to converge at a potential not much lower than that for hexacene. This phenomenon means that the maximum reactivity of such compounds is almost reached by hexacene; therefore, generating and processing polycatafusenes under inert conditions may not be too difficult. This argument is supported by Chapman and Fang's recent success (57) in the synthesis of octacene, a compound that they could still work with and one that is only moderately more reactive than its lower homolog heptacene (57). Furthermore, they presented sound evidence for the existence of nonacene, where they unfortunately reached the point of zero solubility in the series of acenes. Some interesting approaches in this direction were recently undertaken in Müllen's laboratory (58). Additionally, substituents may stabilize polycatafusenes sterically or electronically in regard to self-DA reaction.

The second argument, in short, is stabilization through crystallization. Molecules (not only polyacetylene) that are very sensitive in solution may form quite stable materials in the solid state. In the case of acenes, Chapman and Fang proved (57) that solid heptacene is significantly more persistent than the same compound in solution. On the basis of these observations, a process in which the growing polycatafusene crystallizes might be considered worthwhile.

Table II contains a selection of band-gap calculations of a variety of fully unsaturated ladder polymers. Depending on the method applied, the values for the gap of a certain polymer vary quite significantly. In principle, reliable knowledge about the band gap would be invaluable for the chemist when deciding the target structure.

Because a polymer-analogous aromatization of groups 1 and 2 precursors is a rather complex problem, it was approached from different directions. All approaches involved the removal of the bridges that serve as protecting groups for the fully unsaturated π-system. Present studies used models and polymers with either oxygen (ether) or oxygen and nitrogen (imino) bridges, which are generally accompanied by other sites of saturation (hydrogen-bearing carbons). Attempts to remove oxygen bridges from **10** were reported (9). Only partial success was realized, however, presumably because wet-chemical procedures were required. Treatment of a film of **10** with a solution of the oxophilic reagent trimethylsilyl iodide, for example, led to an almost complete removal of the oxygen bridges (CP MAS ^{13}C NMR). However, the films became so brittle that further investigation was impossible.

As a consequence, polymer **11** was targeted, which opens up the possibility of a removal of the

Table II. Calculated Band Gaps for Selected, Fully Unsaturated Ladder Polymers

Polymer	Band Gap (eV)	Ref.
Poly([2,3-8,9]benzanthracene)	2.86 (VEH)	I
Poly(acene)	0.09 (PMO)	II
	0.0 (VEH)	III
	0.002 (EHMO)	IV
Poly(peri-naphthaline)	0.54 (PMO)	II
	0.73 (PPP)	II
	0.44 (VEH)	V
	1.2	VI
Poly(peri-anthracene)	2.26 (VEH)	V
Poly(biphenylene) (linear)	0.1 (VEH)	VII
Poly(biphenylene) (angular)	3.0 (VEH)	VII

NOTE: References are as follows: I: Toussaint, J. M.; Brédas, J. L. *Synth. Met.* **1992**, *46*, 325; II: Pomerantz, M.; Cardona, R.; Rooney, P. *Macromolecules* **1989**, *22*, 304; III: Brédas, J. L.; Chance, R. R.; Baughman, R. H.; Silbey, R. J. *J. Chem. Phys.* **1982**, *76*, 3673; IV: Woodward, R. B.; Hoffmann, R. *Proc. R. Soc. London* **1979**, *366*, 23; V: Brédas, J. L.; Baughman, R. H. *J. Chem. Phys.* **1983**, *83*, 1316; VI: Tyutyulkov, N., Tadjer, A.; Mintcheva, I. *Synth. Met.* **1990**, *38*, 313; VII: Brédas, J. L.; Baughman, R. H. *J. Polym. Sci. Polym. Lett. Ed.* **1983**, *21*, 475.

bridges by thermal or photochemical treatment. Now all that is required is to remove the oxygen bridges by classical procedures, cast the resulting polymer into films, and generate the linear polycatafusene by heating the film. Experiments using low molecular weight model compounds look promising. Dehydration as well as the complete thermally induced elimination of the nitrogen bridges are feasible. Furthermore, the synthesis of a very high molecular weight polymer 11 (R = CH₃) was successful (27). The nitrogen bridge in this polymer, however, carries CH₃ groups instead of N(CH₃)₂ groups, as would normally be required for a thermally induced (cheletropic) elimination. The future will show if this last obstacle can be overcome.

Not very much can be said about the angularly annulated (group 2) precursors, except that they are presently being investigated in our laboratory. According to Clar's concept (55), angular polycatafusenes should be more persistent than linear ones. Some of them, in fact, will be as stable as a rock and do not have particularly interesting electronic properties.

Fragments of Fullerenes

The generation of defined, unsaturated, double-stranded polymers has borne more fruit with the

1

23

− 2 H₂O

24

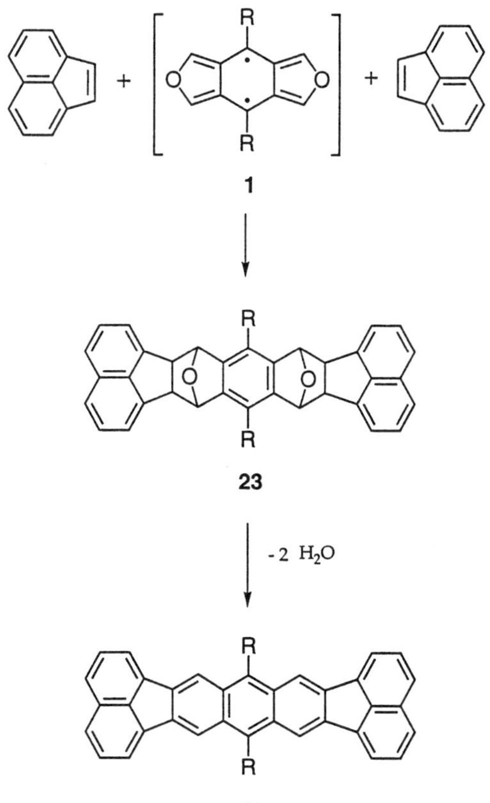

Scheme II. Synthesis of model compound **23** and its dehydration to the benzodifluoranthene **24**.

group 3 precursor polymers 16 and 17 than with the others. Compound 23 (Scheme II: a model for polymer 17) can be dehydrated, regardless of its actual stereochemistry, with *p*-toluene sulfonic acid (*p*-TsOH) to virtually 100%. In very thoroughly conducted experiments, the benzodifluoranthene 24 was obtained in isolated yields of 97.8–98.5% (59). To our knowledge, this attempt is the only one of numerous attempts where the dehydration of the precursor molecules into polycyclic aromatic compounds proceeded reproducibly and absolutely cleanly. This dehydration does not lead to any kind of side reaction, which is of crucial importance for the targeted polymer-analogous application. Another characteristic of this dehydration is that it is does not seem to proceed catalytically. An important consequence is that the number of water molecules removed, and thus the degrees of both planarization and solubility, can be controlled by stoichiometry.

For the polymer-analogous dehydration studies, a solution of 17 in toluene was refluxed with 1.4 equivalents of *p*-TsOH per repeat unit. After a few hours the color of the initially bright yellow solution turned red but stayed homogenous. Normal work up yielded a material whose UV spectrum showed a considerable bathochromic shift in comparison with the starting material (Figure 10) (60). Additionally, the intensities of the NMR signals of carbons A and B (structure 17 in Chart II) are greatly reduced to the advantage of the aromatic carbons. If the same experiment was carried out with an excess of *p*-TsOH, the dehydration could even be driven to completion. Polymer 25 was recovered as an insoluble material. Its UV and CP MAS ¹³C NMR spectra are depicted in Figures 10 and 11, respectively. Sparingly soluble, low molecular weight material was used for the UV spectrum, which shows a significant increase in the intensity of the long wavelength absorptions but no further bathochromic shift of λ_{max}. The NMR spectrum gives no indication of "residual water" in the structure: The oxygen- and hydrogen-carrying carbon atoms of 17 in question (A and B) typically absorb in the ranges of δ = 75–85 ppm and δ = 50–60 ppm.

A success was also accomplished using polymer 16, even though its aromatization required a polymer-analogous dehydrogenation rather than a dehydration (Scheme III). The idea was to try to simply "titrate" the precursor polymer 16 with 2,3-dichloro-5,6-decyano-benzoquinone (DDQ), which is a powerful dehydrogenation agent. Depending on the amount of DDQ used, the degree of aromatization ought to be controlled and adjustable to the respective requirements. Ultimately, stoichiometric amounts should furnish the fully unsaturated polymer 26.

Very recent experiments have proven (61) that polymer-analogous dehydration can actually be

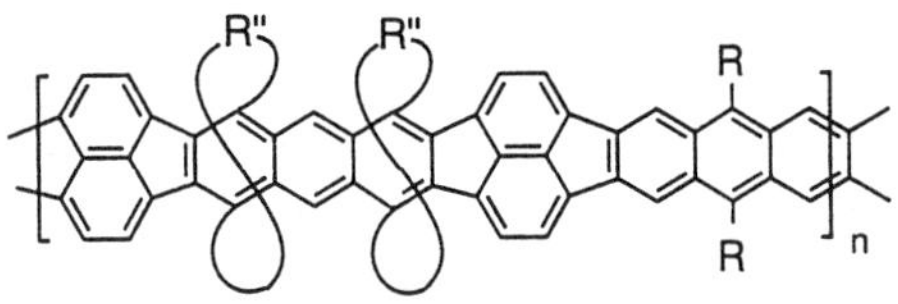

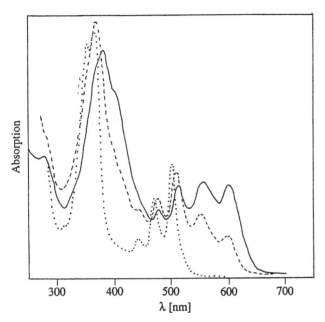

25

Figure 10. UV spectra (chloroform) of precursor polymer **17** (.....), the 80–90% dehydrated polymer **17** (-----), and the fully dehydrated, unsaturated polymer **25** (_____).

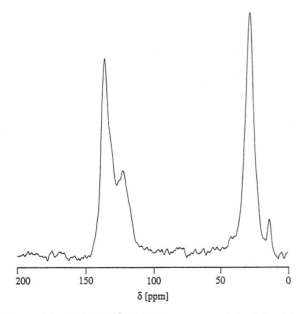

Figure 11. CP MAS ^{13}C NMR spectrum of the fully dehydrated unsaturated polymer **25**.

brought about. Figure 12 shows the ^{13}C NMR spectra of two model compounds, **27** and **28**, and a CP MAS ^{13}C NMR spectrum of fully unsaturated polymer **26**. The signal of **27** at approximately δ = 50 ppm does not appear in the polymer spectrum. The UV evidence is also convincing. Figure 13 compares the spectra of **27**, **28**, and **26**. Finally, Plate 3 shows the appearance of **27**, **28**, and **26** in solution and neat.

Outlook

"So what? All this effort for just a few unambiguously characterized double-stranded polymers with no immediate applications", a material scientist may ask provocatively. This criticism is correct and yet it is not. True, the new generation of ladder polymers are still far from any application. Suggestions have been made on which research direction looks most promising. Also, the first fully unsaturated ladders with defined structures are now available for physicist to investi-

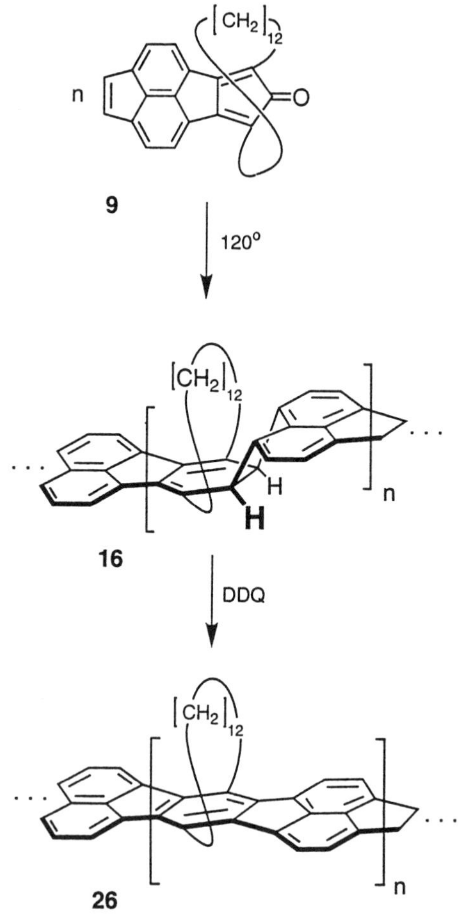

Scheme III. Polymerization of the AB-type Diels–Alder monomer **9** and dehydrogenation of the resultant prepolymer **16** to give the fully unsaturated ladder polymer **26**.

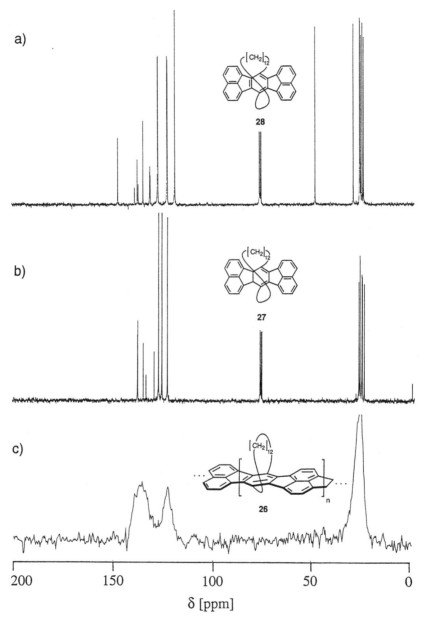

Figure 12. ^{13}C NMR spectra of the model compounds **27** and **28** and CP MAS ^{13}C NMR spectrum of the fully unsaturated polymer **26**. The signal in spectrum (a) at approximately δ = 50 ppm does not appear in spectra (b) and (c).

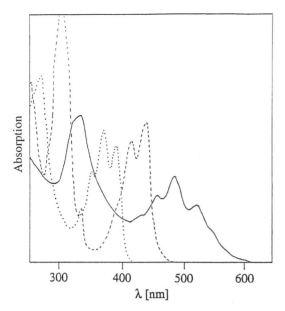

Figure 13. The UV spectra (solvent) of the model compounds **27** (.....), **28** (-----), and the polymer **26** (____).

gate. However, criticism from the material scientist is not justified, because the synthesis of a well-defined polymer is necessarily the first and one of the most important steps toward good materials. Within the next few years one will know whether perfect ladder polymers with extended π-conjugation can be converted into perfect materials with useful applications.

Acknowledgments

We thank the following people who contributed to this work in one or another way: T. Vogel, R. Packe, A. Dahms, B. Schürmann, and V. Enkelmann, G. Wegner (Mainz) and M. Schmidt (Bayreuth), and H. Schirmer. Financial support of this work was generously provided by the Bundesministerium für Forschung und Technologie, the Max Planck Society, Fonds der Chemischen Industrie, and Deutsche Forschungsgemeinschaft. For help with the English we are very thankful to P. Winchester.

References

1. Overberger, C. G.; Moore, J. A. *Adv. Polym. Sci.* **1970**, *7*, 113.
2. Yu, L. P.; Chen, M.; Dalton, L. R. *Chem. Mater.* **1990**, *2*, 649.
3. Yu, L. P.; Dalton, L. R. *Synth. Met.* **1989**, *29*, E463.
4. Dalton, L. R.; McLean, M.; Polis, D.; Yu, L. P.; Young, C. *ACS, Polym. Mater. Sci. Eng.* **1989**, *60*, 410.
5. Dalton, L. R.; *Nonlinear Optical Effects in Organic Polymers;* Messier, J. et al., Eds.; Kluwer Academic: Dordrecht, Netherlands, 1989; p 123.
6. Belaish, I.; Davidov, D.; Selig, H.; McLean, M. R.; Dalton, L. R. *Angew. Chem., Adv. Mater.* **1989**, *101*, 1601.
7. Dahm, J.; Davidov, D.; Macho, V.; Spiess, H. W.; McLean, M. R.; Dalton, L. R. *Polym. Adv. Technol.* **1990**, *1*, 247.
8. Yu, L. R.; Dalton, L. R. *Macromolecules* **1990**, *23*, 3439.
9. Schlüter, A.-D. *Adv. Mater.* **1991**, *3*, 282.
10. Bi, X.-T.; Litt, M. H. *Polymer* **1987**, *28*, 2346.
11. Scherf, U.; Müllen, K. *Synthesis* **1992**, 23.
12. Chmil, K.; Scherf, U. *Makromol. Chem., Rapid Commun.* **1993**, *14*, 217.
13. Tour, J. M.; Lamba, J. J. S. *Polym. Prepr. (Am. Chem. Soc. Div. Polym. Symp.)* **1993**, *34*, 203.
14. Sauer, J. *Angew. Chem. Int. Ed. Engl.* **1967**, *6*, 16.
15. Bailey, W. J.; Feinberg, B. D. *Polym. Prepr. (Am. Chem. Soc. Div. Polym. Symp.)* **1967**, *8*, 165.
16. Christophel, W. C.; Miller, L. L. *J. Org. Chem.* **1986**, *51*, 4169.
17. Kohnke, F. H.; Slawin, A. M. Z.; Stoddart, J. F.; Williams, D. J. *Angew. Chem.* **1987**, *99*, 941.
18. Luo, J.; Hart, H. *J. Org. Chem.* **1988**, *53*, 1343.
19. Hart, H.; Raju, N.; Meador, M. A.; Ward, D. L. *J. Org. Chem.* **1983**, *48*, 4357.
20. Le Houllier, C. S.; Gribble, G. W. *J. Org. Chem.* **1983**, *48*, 1682.
21. Wegener, S.; Müllen, K. *Macromolecules* **1993**, *26*, 3037.
22. Blatter, K.; Schlüter, A.-D.; Wegner, G. *J. Org. Chem.* **1989**, *54*, 2396.
23. Vogel, T.; Blatter, K.; Schlüter, A.-D. *Makromol. Chem., Rapid Commun.* **1989**, *10*, 427.
24. Godt, A. Ph.D. Thesis, University of Mainz, Mainz, Germany, **1991**.
25. Godt., A.; Enkelmann. V.; Schlüter, A.-D. *Angew. Chem. Int. Ed. Engl.* **1989**, *28*, 1680.
26. Godt, A.; Schlüter, A.-D. *Chem. Ber.* **1991**, *124*, 149.
27. Löffler, M.; Enkelmann, V.; Schlüter, A.-D. *Acta Polym.* **1993**, *44*, 50.
28. Löffler, M.; Schlüter, A.-D. *Synlett* **1994**, 75.
29. Blatter, K.; Schlüter, A.-D. *Chem. Ber.* **1989**, *122*, 1351.
30. Blatter, K.; Schlüter, A.-D. *Macromolecules* **1989**, *22*, 3506.
31. Packe, R.; Enkelmann, V.; Schlüter, A.-D. *Makromol. Chem.* **1992**, *193*, 2829.
32. Löffler, M.; Schlüter, A.-D., unpublished results.
33. Löffler, M.; Schlüter, A.-D. *Polym. Prepr. (Am. Chem. Soc. Div. Polym. Symp.)* **1993**, *34*, 199.
34. Godt, A.; Schlüter, A.-D. *Makromol. Chem.* **1992**, *193*, 501.
35. Schürmann, B.; Löffler, M.; Enkelmann, V.; Schlüter, A.-D. *Angew. Chem. Int. Ed. Engl.* **1993**, *32*, 123.
36. Weiner, S. J.; Kollman, P. A.; Case, D. A.; Singh, U. C.; Ghio, C.;Alagona, G.; Profeta, P.; Weiner, P. *J. Am. Chem. Soc.* **1984**, *106*, 765.
37. Löffler, M.; Packe, R.; Schlüter, A.-D. *Macromolecules* **1992**, *25*, 4213.
38. Balaban, A. T. *Pure Appl. Chem.* **1982**, *54*, 1075.
39. Kivelson, S.; Chapman, O. L. *Phys. Rev.* **1983**, *B 28*, 7236.
40. Brédas, J. L.; Baughman, R. H. *J. Chem. Phys.* **1985**, *83*, 1316.
41. Kertesz, M. *Int. Rev. Phys. Chem.* **1985**, *4*, 125.
42. Lowe, J. P.; Kafafi, S. A.; La Temina, J. P. *J. Phys. Chem.* **1986**, *90*, 6602.
43. Bakhshi, A. K., Ladik, J. *Synth. Met.* **1985**, *30*, 115.
44. Tyutyulkov, N.; Tadjer, A.; Mintcheva, I. *Synth. Met.* **1990**, *38*, 313.

45. Hosoya, H.; Kumazaki, H.; Chida, K.; Ohuchi, M.; Gao, Y.-D. *Pure Appl. Chem.* **1990**, *62*, 445.

46. Toussiant, J. M.; Brédas, J. L. *Synth. Met.* **1992**, *46*, 325.

47. Suzuki, H.; Meyer, H.; Simmerer, J.; Yang, J.; Haarer, D. *Adv. Mater.* **1993**, *5*, 743 and refs. cited therein.

48. Dyreklev, P.; Inganäs, O. *J. Appl. Phys.* **1992**, *71*, 2816.

49. Grem, G.; Leditzky, G.; Ullrich, B.; Leising, G. *Adv. Mater.* **1992**, *4*, 36.

50. Tyutyulkov, N.; Dietz, F.; Müllen, K.; Baumgarten, M.; Karabunarliev, S. *Theor. Chim. Acta* **1993**, *86*, 353.

51. Murata, I. *Pure Appl. Chem.* **1993**, *65*, 97.

52. Feast, W. J.; Edwards, J. H. *Polymer* **1980**, *21*, 595.

53. Wessling, R. A. *J. Polym. Sci., Polym. Symp.* **1985**, *72*, 55.

54. Lenz, R. W.; Han, C.-C.; Stenger-Smith, J.; Karasz, F. E. *J. Polym. Sci., Polym. Chem.* **1988**, *26*, 3241.

55. Clar, E. *The Aromatic Sextet*; Wiley: London, 1972.

56. Biermann, D.; Schmidt, W. *J. Am. Chem. Soc.* **1980**, *102*, 3163.

57. Fang, T. Ph.D. Thesis, University of California, Los Angeles, CA, 1986.

58. Wegener, S. Ph.D. Thesis, University of Mainz, Mainz, Germany, 1992.

59. Schirmer, H.; Schlüter, A.-D.; Enkelmann, V. *Chem. Ber.* **1993**, *126*, 2543

60. Löffler, M.; Schlüter. A.-D.; Gessler, K.; Saenger, W.; Toussaint, J.-M.; Brédas, J.-L. *Angew. Chem. Int. Ed. Engl.* **1994**, *33*, 2209.

61. Schlüter, A.-D.; Löffler, M.; Enkelmann, V. *Nature (London)* **1994**, *368*, 831.

Functional Polymers via Ring-Opening–Closing Alternating Copolymerization

Shiro Kobayashi, Hiroshi Uyama, and Jun-ichi Kadokawa

This review describes the general aspects of ring-opening and ring-closing polymerizations and then a new concept, namely, ring-opening-closing alternating copolymerization. In this new case, the reaction of a cyclic monomer (A) and a noncyclic monomer (B) takes place without any initiator to give an alternating copolymer structure with a ring-opened unit from monomer A and a ring-closed unit from monomer B. The cyclic monomers are cyclic phosphonites and 2-oxazoline. For monomer B, the use of muconic acid, dialdehydes, di(meth)acrylic anhydrides, and N-methyldiacrylamide is described. A mechanism involving zwitterion intermediates is considered for these copolymerizations.

Since Staudinger discovered that ring-opening polymerization of ethylene oxide produces linear poly(oxyethylene) in 1929, ring-opening polymerization has been studied extensively (Scheme I) (*1–3*). On the other hand, Butler found (*4*) in 1949 that bifunctional monomers such as nonconjugated dienes underwent polymerization to give soluble polymers with ring structures formed via intramolecular cyclization at each propagation step (cyclopolymerization or ring-forming polymerization) (Scheme I). The present chapter provides a review of the preparation of functional polymers based on a new type of copolymerization that combines the two modes of ring-opening polymerization and ring-closing polymerization. The general polymerization scheme is shown in Scheme II. The combination of a cyclic monomer (A) and a noncyclic monomer (B) affords a copolymer whose structural unit has a ring-opened part from monomer A and a ring-closed part from monomer B in an alternating manner [*ring-opening-closing alternating copolymerization* (ROCAC)] (*5–7*).

Ring-opening polymerization chemistry has received much interest in both commercial development and academic research (*1–3*). Several polymers, typically polyepoxides, polyacetals, and silicones, are produced industrially by using ring-opening polymerizations. A lot of papers discuss ring-opening polymerizations of a variety of monomers with different ring structures and sizes. Cyclic phosphonites (**1**) and 2-methyl-2-oxazoline (**3**) were employed as cyclic monomers in ROCAC. Electrophilic (cationic) ring-opening polymerization (*8*) of cyclic phosphorus(III) compounds (*9*) has long been known. Generally, a cyclic phosphonite undergoes cationic ring-opening polymerization via Arbuzov-type reaction to yield poly(phosphonite) (**2**) (Scheme III). During the polymerization, the phosphorus atom of the monomer is oxidized from the oxidation state +3 to +5, whereas the OCH$_2$ carbon in the monomer is reduced from –1 to –3 during conversion to the PCH$_2$. Therefore, an intramolecular oxidation–reduction takes place when an Arbuzov-type reaction is involved.

2-Methyl-2-oxazoline (**3**) is cationically polymerized to produce poly(N-acetylethylenimine) (**4**) (Scheme III) (*10*). During the polymerization, the imino ether group in **3** is isomerized to the amide group. The resulting polymer is regarded as a polymeric analog of N,N-dimethylacetamide (DMAc).

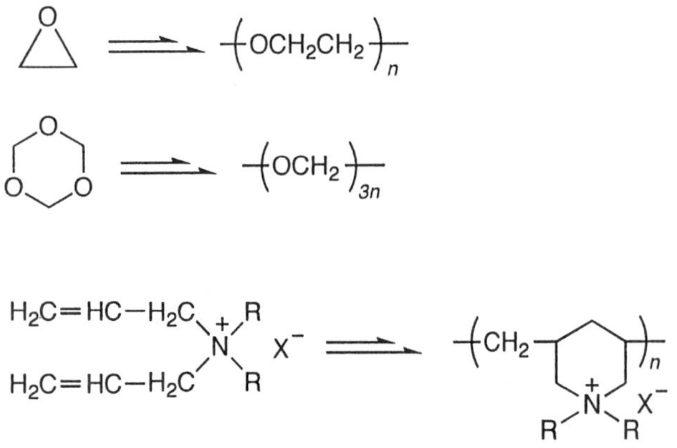

Scheme I. Typical examples of ring-opening polymerization and cyclopolymerization (ring-closing polymerization).

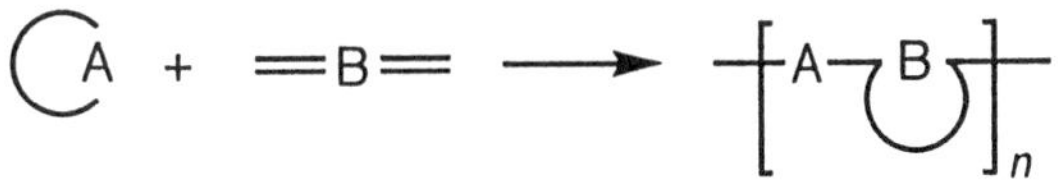

Scheme II. General scheme of ROCAC.

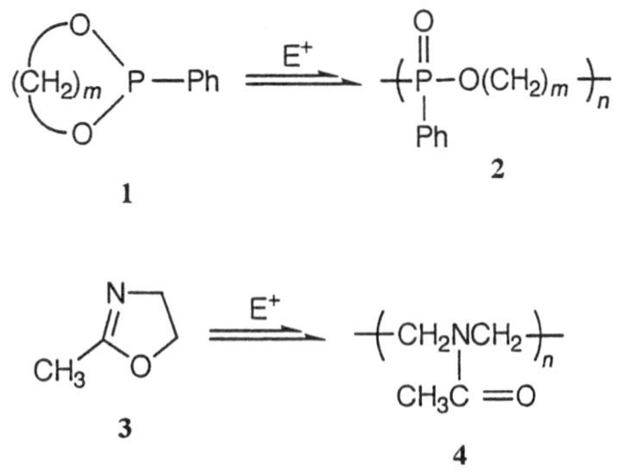

Scheme III. Ring-opening polymerizations of cyclic phosphonites (1) and 2-methyl-2-oxazoline (3).

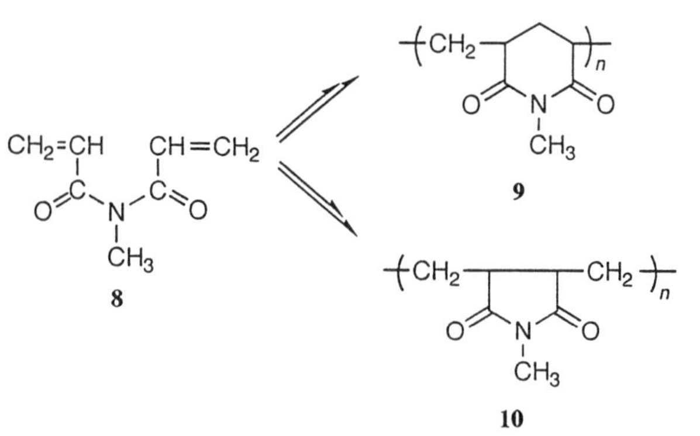

Scheme IV. Cyclopolymerization (ring-closing polymerization) of N-methlyldiacrylamide (8).

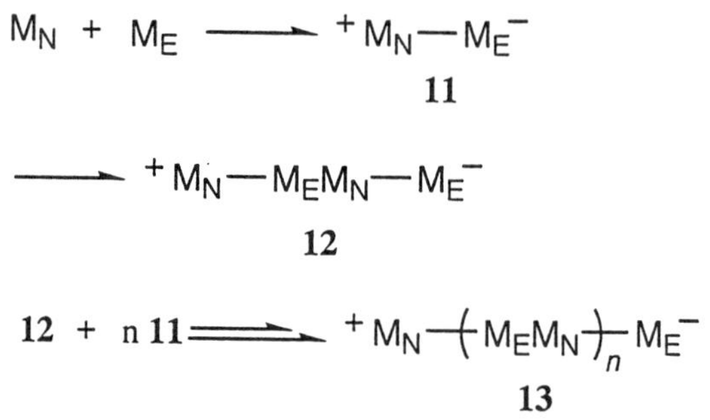

$$M_N + M_E \longrightarrow {}^+M_N\!-\!M_E^-$$
11

$$\longrightarrow {}^+M_N\!-\!M_E M_N\!-\!M_E^-$$
12

$$12 + n\,11 \Longrightarrow {}^+M_N\!-\!(M_E M_N)_n\!-\!M_E^-$$
13

Scheme V. General scheme of zwitterionic alternating copolymerization without using initiator.

DMAc is an aprotic polar solvent and solubilizes various polar organic polymers. The characteristic properties of DMAc are observed in its polymer analog (4). Recently, various functional polymers based on 4 have been developed (10–12).

Cyclopolymerizations provide a cyclic structure in the main chain of the polymer. Until now, a variety of multiply bonded structures such as alkenes, alkynes, carbonyls, isocyanates, and nitriles have been reported to undergo cyclopolymerization (4). Ring-closing monomers used in ROCAC were muconic acids (5), dialdehlydes (6), di(meth)acrylic anhydrides (7), and N-methyldiacrylamide (8). Glutaraldehyde has been polymerized with no initiator or cationically polymerized to give a polymer mainly having a tetrahydropyran ring structure (4). The resulting polymer is thermally unstable and depolymerization to monomer occurs. Succinaldehyde is also polymerized with no initiator or cationically. Cyclopolymerization of o-phthalaldehyde has been extensively explored. The

monomer readily homopolymerizes by using γ-ray irradiation, cationic initiators, anionic initiators, and coordinated anionic initiators.

Di(meth)acrylic anhydrides (7) have been reported to be radically cyclopolymerized. The polymer from dimethacrylic anhydride consists of a six-membered anhydride unit. The polymerization of diacrylic anhydride produces a polymer with a five- or six-membered structure, depending on the polymerization condition.

N-Substituted dimethacrylamides have been polymerized using radical and anionic initiators to yield soluble polymers (4). Two different cyclic structures in the resulting polymer are possible: a six- or five-membered ring, depending on the orientation of the second vinyl group at the intramolecular propagation step, head-to tail (9), or head to head (10), respectively (Scheme IV). In anionic polymerization of 8, only the five-membered cyclic unit was obtained.

ROCAC proceeds via a zwitterion intermediate without any initiator. A large number of zwitterionic copolymerizations have been reported (13), and ROCAC is one of the examples of this type of copolymerization. Two kinds of monomers, one nucleophilic (M_N) and one electrophilic (M_E), are used in the alternating copolymerization without using an initiator. The general scheme is shown in Scheme V.

When the monomers combine, they form a zwitterion intermediate **11**, which is a so-called "genetic zwitterion" and is responsible for both initiation and propagation. Two molecules of **11** react with each other to produce the first propagating species **12** ("macrozwitterion") that grows by successive reactions with **11**. As the reaction proceeds, the concentration of macrozwitterion of various sizes **12** and **13** increases, and reaction of the macrozwitterions takes place.

On the basis of this concept, many combinations of M_N and M_E have been investigated. A typical example is the copolymerization between 2-oxazoline (**14**) and β-propiolactone (**15**) (Scheme VI). An equimolar reaction of both monomers takes place without added initiator to quantitatively give the linear polymer **16**. For this copolymerization, a mechanism involving zwitterion intermediates has been proposed, in which **14** behaves as M_N and **15** as M_E, and these reactions give a genetic zwitterion **17**. Two molecules of **17** afford a dimeric zwitterion **18**; the carboxylate group of **17** attacks the oxazolinium site of another **17** to open the ring. Successive reactions of **18** with **17** give rise to the alternating copolymer **16** (*14*).

ROCAC of Cyclic Phosphonites with Muconic Acid

The first example of this novel copolymerization was the reaction of cyclic phosphonites (**1**) with muconic acid (**5**) (Scheme VII) (*15, 16*). The monomers used were five-, six-, and seven-membered cyclic phosphonites (**1a**, **1b**, and **1c**), and *cis,trans-* and *cis,cis*-muconic acids (**5a** and **5b**). In all runs, the reaction of the 1:1 monomer feed proceeded without any added initiator to produce alternating copolymer **19** having a repeating unit of ring-opened structure of **1** and ring-closed

structure of **5** (Table I). The structure of the copolymer was confirmed by ^1H NMR, ^{13}C NMR, ^{31}P NMR, and IR spectroscopies.

The time–conversion curves for the copolymerization of **1a** with **5a** and the copolymer composition are shown in Figure 1. The monomer unit ratio in copolymer **19a** was about 0.5 at every stage of the copolymerization and confirmed the alternating structure of **19a**.

Figure 2 is a copolymer composition curve with varying monomer feed ratios in the copolymerization of **1a** with **5a**. In the region where **1a** is more than 50 mol% in the feed, alternating propagation took place. In the area where **1a** is less than 50 mol%, however, both the alternating as well as homo-propagation of **5a** occurred. The apparent monomer reactivity ratios were determined as $r_{1a} = 0.06$ and $r_{5a} = 0.00$.

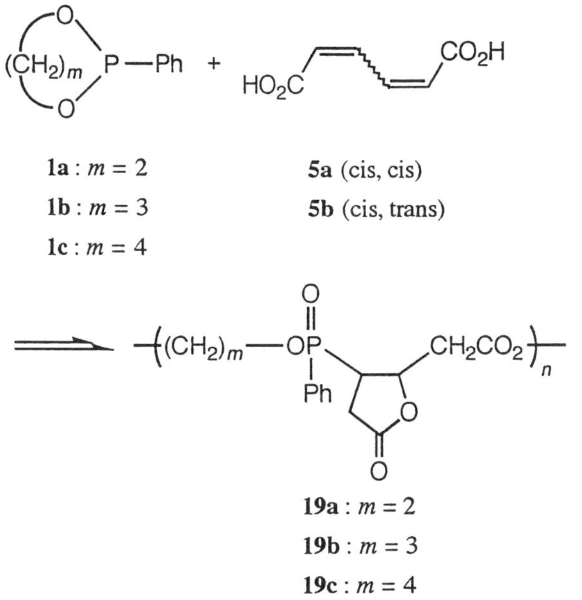

1a : m = 2   **5a** (cis, cis)
1b : m = 3   **5b** (cis, trans)
1c : m = 4

19a : m = 2
19b : m = 3
19c : m = 4

Scheme VII. ROCAC of cyclic phosphonites (1) with muconic acids (5).

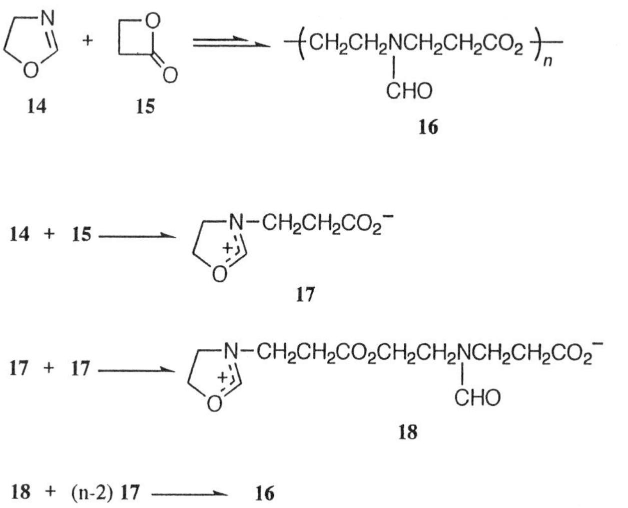

16

17

18

18 + (n-2) **17** ⟶ **16**

Scheme VI. Alternating copolymerization of 2-oxazoline (14) with β-propiolactone (15).

Table I. ROCAC of Cyclic Phosphonites and Muconic Acids

| | Copolymerization[a] | | | | Copolymer | |
Entry	Monomer 1	5	Temp. (°C)	Time (h)	Yield (%)[b]	Structure	M_w (daltons)[c]
1	1a	5a	35	786	68	19a	1200
2	1a	5a	100	17	57	19a	3500
3	1b	5a	100	6	54	19b	1200
4	1c	5a	100	70	51	19c	1200
5	1a	5b	100	9	80	19a	2300
6	1b	5b	100	6	85	19b	2100

[a]A mixture of 1.0 mmol of each monomer in 1.5 mL of DMF.
[b]Isolated yield of the diethyl ether insoluble part.
[c]Values determined by gel permeation chromatography.
Source: Data adapted from reference 16.

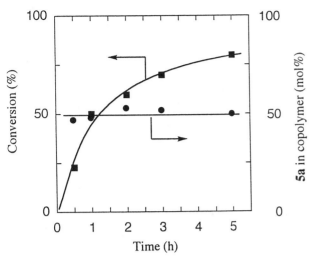

Figure 1. Time-conversion curve (■) and copolymer composition (●) in the copolymerization of cyclic phosphonite (1a) with muconic acid (5a) at 100 °C in N,N-dimethylformamide (DMF): [1a] = [5a] = 0.1 mmol in 0.15 mL of DMF.

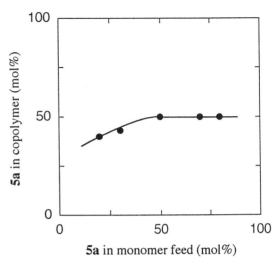

Figure 2. Copolymerization of cyclic phosphonite (1a) with muconic acid (5a) at 100 °C in N,N-dimethylformamide (DMF) for 0.5 h. Polymer compositions at various comonomer feeds: [1a] + [5a] = 0.2 mmol in 0.15 mL of DMF.

To confirm the structure of **19**, the following model reaction was carried out at 100 °C in dimethylformamide (DMF) for 24 h (Scheme VIII). Expected product **21** was obtained quantitatively, and this result confirmed the unit structure in **19**.

On the basis of the above data, the following zwitterionic mechanism is suggested to explain the course of the copolymerization (Scheme IX). The first step of the reaction is an intermolecular Michael-type addition of **1** to **5** to form a phosphonium–carbanion intermediate (**22**), followed by a hydrogen transfer to give a transient zwitterion (**23**). Then, **23** is converted into a key species of genetic zwitterion **24** via a ring-closing process involving an intramolecular Michael-type addition and a hydrogen-transfer process. In the next step, an Arbuzov-type reaction occurs between two molecules of **24**. The propagation proceeds via the successive attack of genetic zwitterions **24** onto dimeric zwitterions to lead to the alternating copolymer (**19**) of a macrozwitterion structure.

The ^{31}P NMR spectral data of the copolymerization mixture (in situ at 60 °C) showed two small signals at +0.7 and −14.9 ppm that were ascribable to structures **24a** and the spiro pentacovalent phosphorus compound (**25**), respectively, together with the main signal due to the copolymer. Therefore, **24** is in equilibrium with **25**.

ROCAC of Cyclic Phosphonites with Dialdehydes

A second example of ROCAC is shown by the reaction of cyclic phosphonites and dialdehydes (Scheme X)

(*17, 18*). In this case, ring-opening monomers **1a–c** and salicylyl phenyl phosphonite (**1d**) and ring-closing monomers succinaldehyde (**6a**), glutaraldehyde (**6b**), adipinaldehyde (**6c**), and phthalaldehyde (**6d**) were used.

Copolymerization of **1a–c** with **6a** or **6b** in equimolar feed ratio proceeded without any initiator to give the alternating copolymer **26**, whose repeating unit consists of one part formed by monomer **1** and the other part (cyclic ether) by monomer **6**. The copolymerization results are summarized in Table II. In all runs, both the ring-opening and ring-closing steps took place quantitatively.

A novel type of cyclic phosphonite **1d** was also copolymerized with **6b**. This polymer involves the ring-opening of **1d** and the complete cyclization of **6b** to produce **26g**. The copolymerization took place at 0 °C (entry 7 in Table II). ROCAC using adipinaldehyde (**6c**) produced cross-linking materials (entry 8), probably owing to the unfavorable formation of a seven-membered ring during the cyclization of **6c**.

An aromatic dialdehyde can also be a candidate as a ring-closing monomer. Phthalaldehyde **6d** was

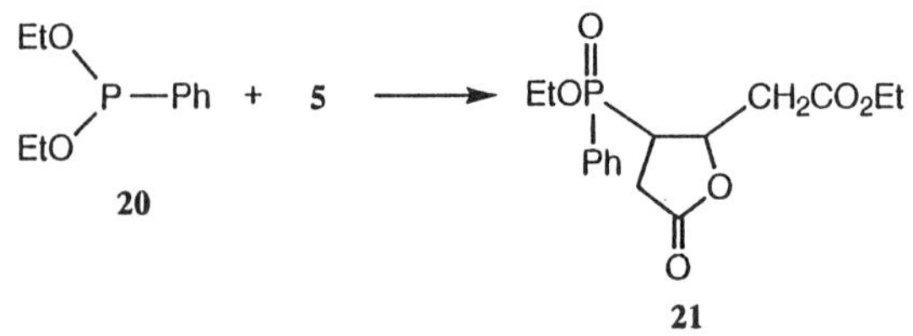

Scheme VIII. Reaction of diethyl phenylphosphonite (20) with muconic acid (5).

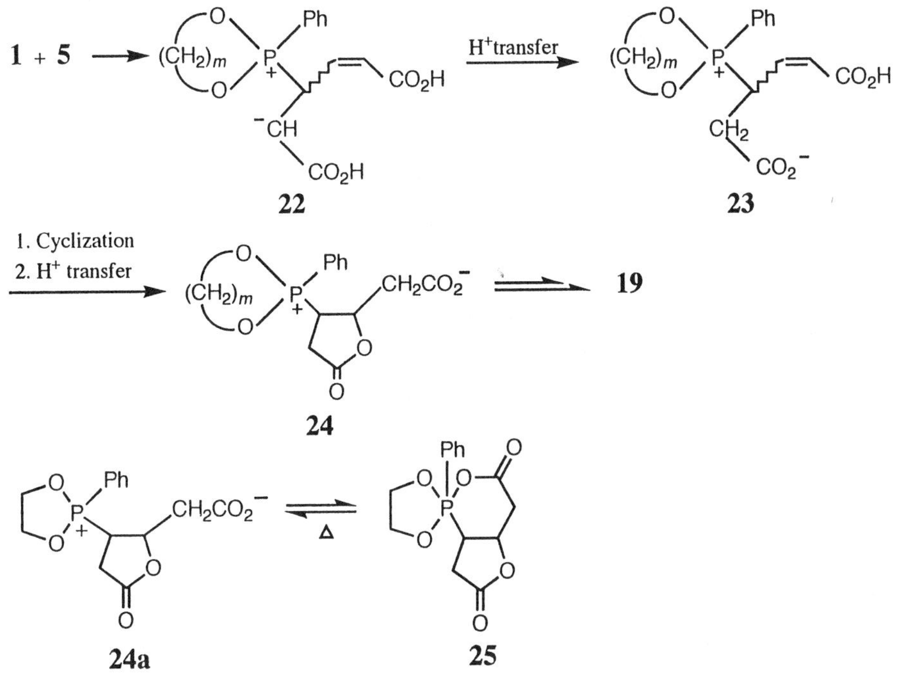

Scheme IX. Reaction mechanism of ROCAC of cyclic phosphonite (1) with muconic acid (5).

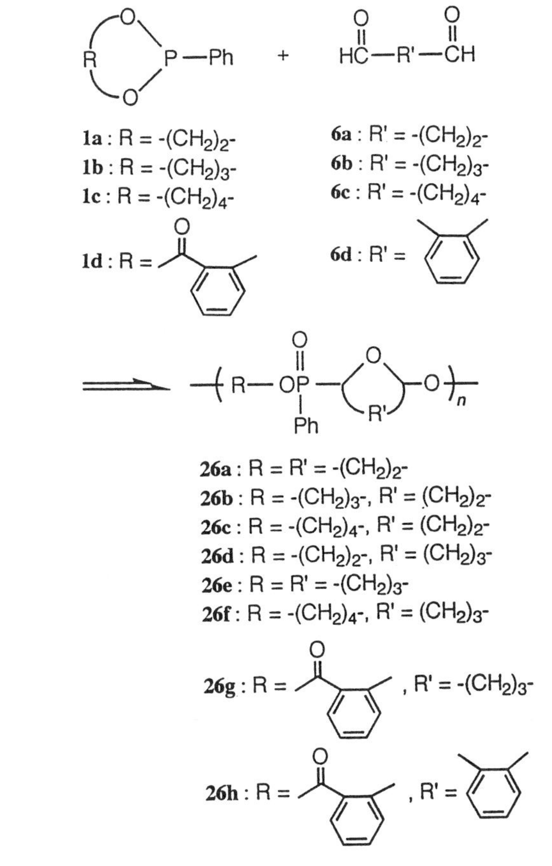

Scheme X. ROCAC of cyclic phosphonites (1) with dialdehydes (6).

employed for copolymerization with cyclic phosphonites. In copolymerization of **6d** with **1a–c**, the phosphorus atom in the cyclic phosphonite attacked not only the carbonyl carbon of the aldehyde but also the oxygen atom of the aldehyde group (Scheme XI). The nucleophilic attack of phosphorus in **1** on the carbon on **6** affords a phosphinate-type product, whereas attack on oxygen produces a phosphonate unit. In the case of copolymerization at 100 °C, the ratios of the carbon attack for **1a–c** were 30, 10, and 21%, respectively. In the reaction of **1d** with **6d**, on the other hand, the phosphorus atom exclusively attacked on the carbon. The copolymerization proceeded much faster than that with an aliphatic dialdehyde. It was exothermic even at room temperature and produced a powdery polymer with high molecular weight (entry 9).

To confirm the copolymer structure, the following model reaction was performed. The reaction of diethyl phenylphosphonite (**20**) with **6b** produced a product **27**, indicating that the ring-closing step is definitely involved in the copolymerization (Scheme XI).

On the basis of the above data, copolymerization of **1** with **6** can be explained by the mechanism shown in Scheme XII. The first step of the reaction is the nucleophilic attack of the phosphorus atom of **1** on the aldehyde carbon of **6** to form the intermediate **28**. The second step is the fast intramolecular cyclization (ring-closing) of **28** that gives rise to a genetic zwitterion **29**. The copolymer **26** is then produced by the reaction of the anionic oxygen of **29** and/or macrozwitterions with the methylene carbon adjacent to the OP⁺ of **29**

Table II. ROCAC of Cyclic Phosphonites with Dialdehydes

| | Copolymerization[a] | | | | | Copolymer | | |
| | Monomer 1 | Monomer 6 | Solvent | Temp. (°C) | Time (h) | Yield (%)[b] | Structure | M_w (daltons)[c] |
Entry								
1	**1a**	**6a**	CHCl$_3$	80	20	72	**26a**	3200
2	**1b**	**6a**	Toluene	100	4	42	**26b**	1500
3	**1c**	**6a**	Toluene	100	4	30	**26c**	1600
4	**1a**	**6b**	Toluene	100	4	79	**26d**	1200
5	**1b**	**6b**	Toluene	100	4	57	**26e**	2000
6	**1c**	**6b**	Toluene	100	4	59	**26f**	1300
7	**1d**	**6b**	CHCl$_3$	0	9	48	**26g**	1200
8	**1a**	**6c**	CHCl$_3$	50	8		cross-linking	
9	**1d**	**6d**	CDCl$_3$	rt	48	63	**26h**	4000

[a]A mixture of 1.0 mmol of each monomer in 0.3 mL of solvent.
[b]Isolated yield of the diethyl ether insoluble part.
[c]Values determined by gel permeation chromatography.

Source: Data adapted from reference 18.

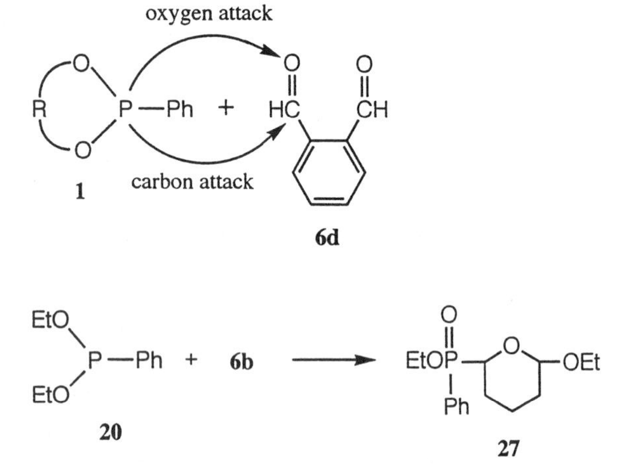

Scheme XI. Reaction pathway of cyclic phosphonite (1) with phthalaldehyde (6d) and reaction scheme of diethyl phenylphosphonite (20) with glutalaldehyde (6b).

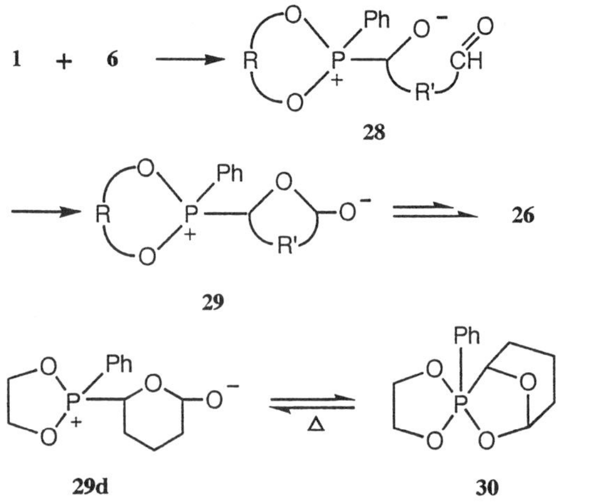

Scheme XII. Reaction mechanism of ROCAC of cyclic phosphonite (1) with dialdehyde (6).

and/or macrozwitterions via an Arbuzov type reaction with ring-opening. Thus, the repeating unit of **26** possesses one part from the ring-opening of monomer **1** and the other part from the ring-closing of **6**.

The copolymerization of **1a** and **6b** was monitored in situ at low temperature (25 °C) by ^{31}P NMR spectroscopy. After mixing the monomers, two peaks appeared at −24.3 and +41.4 ppm in addition to the monomer peak. The peak at −24.3 ppm was assigned to the phosphorus atom of a pentacovalent spirophosphorane (**30**), and the peak at +41.4 ppm was assigned to that of the phosphinate unit of the copolymer. The peak for the spiro-compound was observed from −20 °C to +25 °C. In the case of polymerization at a higher temperature (80 °C), the spiro compound was not detected. These results indicate that at lower temperatures the zwitterion intermediate **29** is in equilibrium with the spiro compound **30**, which at elevated tem-

peratures undergoes a thermally induced ring-opening polymerization to form copolymer **26**.

ROCAC of Cyclic Phosphonites with (Meth)acrylic Anhydrides

Copolymerization of cyclic phosphonite (**1a–c**) with methacrylic and acrylic anhydrides (**7a** and **7b**) as a novel ring-closing monomer was investigated (*19*). The resulting copolymers (**31**) have phosphinate–cyclic acid anhydride structures as ring-opened and ring-closed units, respectively (Scheme XIII). In case of the copolymerization of **1a** and **7a**, the ring opening of **1a** and the quantitative ring closing of **7a** took place to give the alternating copolymer **31a**. In the copolymerizations of other monomer combinations, however, ^{1}H NMR spectra of **31** showed peaks due to (meth)acrylic ester group. These data indicate that **7** was not quantitatively ring-closed. In this case, the structure of the product can be assigned as a random copolymer having two alternating units, **31** and **32**.

Copolymerization results are shown in Table III. The molecular weight of the copolymer obtained from **1c** and **7c** was relatively higher than that using **1a** and **7b** as cyclic monomer. The highest molecular weight (8900 daltons) was obtained in the copolymerzation of **1b** and **7b** in acetonitrile at 100 °C for 18 h (entry 5). A further increase in temperature up to 150 °C afforded a cross-linked material (entries 6 and 8), probably owing to a thermal or radical polymerization of the pendant (meth)acrylate group.

The copolymerization of **1** with **7** can be explained reasonably by zwitterion propagation intermediates (Scheme XIV). The first step is an intermolecular addition of **1** to **7** to form **33**, followed by an intramolecular cyclization (via Michael addition) to give a transient zwitterion **34**. In the next step, an Arbuzov-type reaction occurs between two molecules of **34** to give a dimeric zwitterion **35**. The subsequent propagations are the reactions between a phosphonium site and an oxygen anion site of zwitterion **34** and **35** and/or macrozwitterions. Depending on the polymerization conditions, the Arbuzov-type propagation of zwitterion **33** takes place to form the alternating unit **32**. The cyclization of **33** to **34** and the propagation of **33** to **32** are competitive and produce a random copolymer of units **31** and **32**.

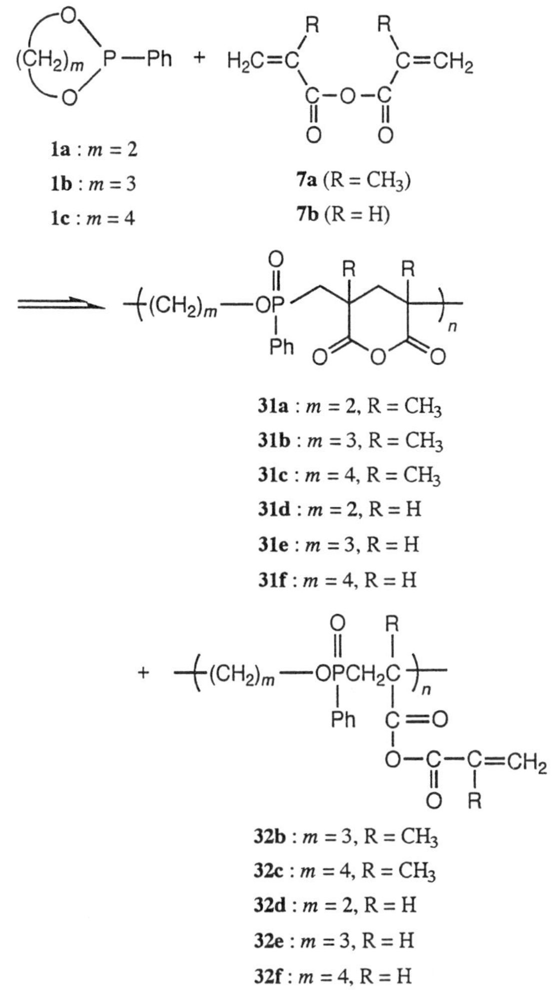

ROCAC of 2-Methyl-2-oxazoline with *N*-Methyldiacrylamide

Recently we have found that a combination of 2-methyl-2-oxazoline (**3**, a cyclic monomer) and *N*-methyldiacrylamide (**8**, a noncyclic monomer) underwent the ROCAC without any initiator (Scheme XV)

Scheme XIII. ROCAC of cyclic phosphonites (**1**) with di(meth)acrylic anhydrides (**7**).

Table III. ROCAC of Cyclic Phosphonites with Di(meth)acrylic Anhydrides

	Copolymerization[a]					Copolymer		
Entry	Monomer 1	Monomer 7	Solvent	Temp. (°C)	Time (h)	Yield (%)	Structure	Cyclization (%)[b]	M_w (daltons)[c]
1	**1a**	**7a**	CHCl$_3$	100	20	43	**31a**	98	2400
2	**1b**	**7a**	CHCl$_3$	100	20	35	**31b + 32b**	91	1500
3	**1c**	**7a**	CH$_3$CN	100	20	46	**31c + 32c**	84	6300
4	**1a**	**7b**	CHCl$_3$	100	18	92	**31d + 32d**	94	2100
5	**1b**	**7b**	CH$_3$CN	100	18	72	**31e + 32e**	89	8900
6	**1b**	**7b**	CHCl$_3$	150	24		cross-linking		
7	**1b**	**7c**	CHCl$_3$	100	18	100	**31f + 32f**	83	5800
8	**1b**	**7c**	CHCl$_3$	150	24		cross-linking		

[a]A mixture of 1.0 mmol of each monomer in 1.0 mL of solvent.
[b]Extent of cyclization values determined by ^{1}H NMR spectroscopy.
[c]Values determined by gel permeation chromatography.

Source: Data adapted from reference 19.

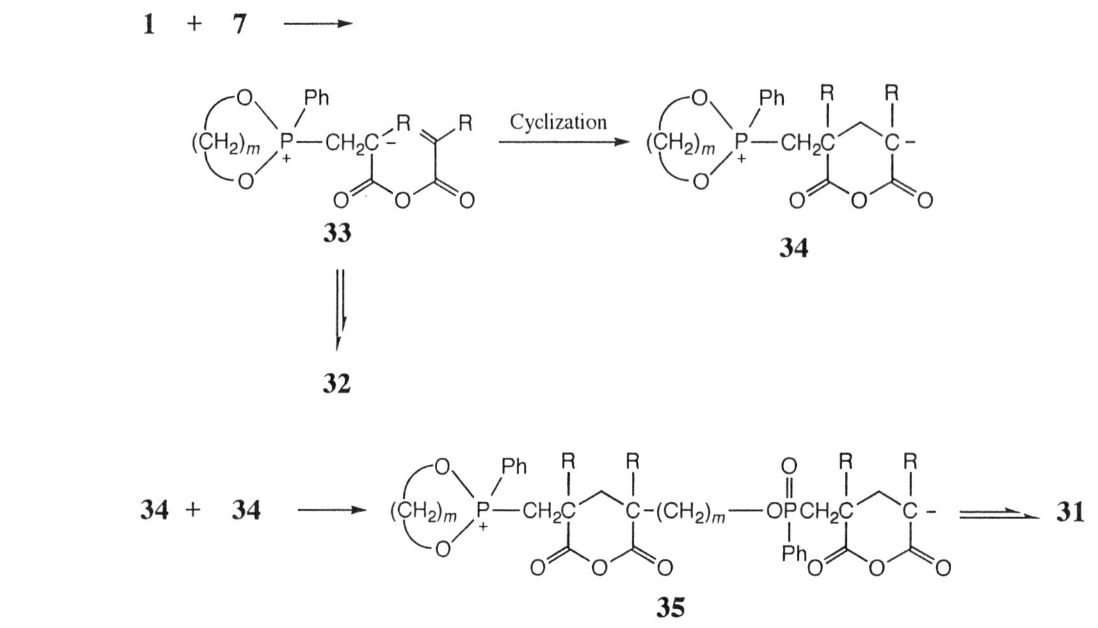

1 + 7 ⟶

Scheme XIV. Reaction mechanism of ROCAC of cyclic phosphonite (1) with di(meth)acrylic anhydride (7).

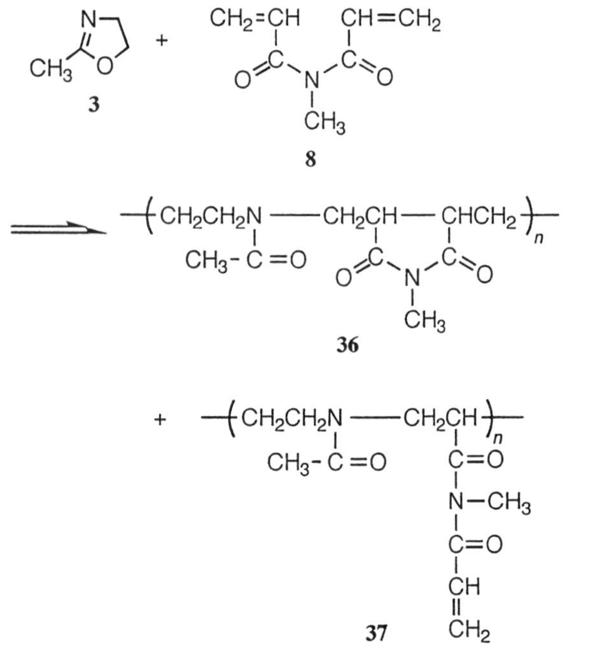

Scheme XV. ROCAC of 2-methyl-2-oxazoline (3) with N-methyldiacrylamide (8).

(20). The copolymerzation in chloroform at 100 °C for 24 h produced the copolymer with a molecular weight of 4100 daltons in 60% yield. In the ^{1}H NMR spectrum, peaks due to vinyl protons of the acrylamide group were observed, and this result indicated that cyclization of **8** was not quantitative; the resulting copolymer had two random units of **36** and **37** (ca. 65% of the cyclized unit). When the copolymerization was continued longer (48 h), an insoluble polymer was obtained, probably by cross-linking reaction of the pendant double bonds of **37**. From NMR and IR analyses, the copolymer possessed the five-membered structure of an N-methylimide.

Summary

We have demonstrated the syntheses of a series of novel 1:1 alternating copolymers with one part of the repeating unit coming from the ring-opening of one monomer, and the other part from the ring-closing of another monomer. In these studies, the cyclic monomers were cyclic phosphonites and 2-methyl-2-oxazoline; and the noncyclic monomers were muconic acid, dialdehyde, (meth)acrylic anhydride, and N-methyl-diacrylamide. The copolymerizations of the cyclic and noncyclic monomers in the equimolar feed ratio proceeded without any initiator to produce alternating copolymers. These reactions represent a new type of copolymerization, termed *ring-opening-closing alternating copolymerization* (ROCAC).

The resulting copolymers have cyclic structures of lactone, cyclic ether, cyclic acid anhydride, and cyclic imide in the main chain. The polymers are thus reactive and represent a highly versatile class of functional copolymers. The phosphinate and amide group in the copolymer may be used as model compounds of natural polymers and as new biodegradable polymers.

References

1. *Ring-Opening Polymerization;* Frisch, K. C.; Reegen, S. L., Eds.; Marcel Dekker: New York, 1969.
2. *Ring-Opening Polymerization;* Ivin, K. J.; Saegusa, T., Eds.; Elsevier Applied Science: London, 1984.

3. *Ring-Opening Polymerization: Mechanisms, Catalysis Structure, Utility;* Brunelle, D. J., Ed.; Hanser: Munich, Germany, 1993.

4. Butler, G. In *Encyclopedia of Polymer Science and Engineering*, 2nd ed.; John Wiley & Sons: New York, 1986; Vol. 4, pp 543–598.

5. Kobayashi, S. *Makromol. Chem., Macromol. Symp.* **1991,** *42/43,* 93.

6. Lundmark, S.; Kobayashi, S. *Makromol. Chem., Macromol. Symp.* **1992,** *54/55,* 107.

7. Kobayashi, S.; Lundmark, S.; Kadokawa, J.; Uyama, H.; Shoda, S. *Makromol. Chem., Macromol. Symp.* **1993,** *73,* 137.

8. Kobayashi, S. In *Ring-Opening Polymerization: Kinetics, Mechanisms, and Synthesis;* McGrath, J. E., Ed.; ACS Symposium Series No. 286; American Chemical Society: Washington, DC, 1985; pp 291–312.

9. Lapienis, G.; Penczek, S. In *Ring-Opening Polymerization;* Ivin, K. J.; Saegusa, T., Eds.; Elsevier Applied Science: London, 1984; pp 919–1053.

10. Kobayashi, S. *Prog. Polym. Sci.* **1990,** *15,* 751.

11. Kobayashi, S.; Uyama, H. *Polym. News* **1991,** *16,* 70.

12. Chujo, Y.; Saegusa, T. In *Ring-Opening Polymerization;* Brunelle, D. J., Ed.; Hanser: Munich, Germany, 1993; pp 239–262.

13. Kobayashi, S.; Saegusa, T. In *Alternating Copolymers;* Cowie, T. M. G., Ed.; Plenum: New York, 1985; pp 189–238.

14. Saegusa, T.; Kobayashi, S.; Kimura, Y. *Macromolecules* **1974,** *7,* 1.

15. Kobayashi, S.; Kadokawa, J.; Uyama, H.; Shoda, S.; Lundmark, S. *Macromolecules* **1990,** *23,* 3541.

16. Kobayashi, S.; Kadokawa, J.; Uyama, H.; Shoda, S.; Lundmark, S. *Macromolecules* **1992,** *25,* 5861.

17. Kobayashi, S.; Lundmark, S.; Kadokawa, J.; Albertsson, A. C. *Macromolecules* **1991,** *24,* 2129.

18. Kobayashi, S.; Lundmark, S.; Kadokawa, J.; Albertsson, A. C. *Macromolecules* **1992,** *25,* 5867.

19. Lundmark, S.; Kadokawa, J.; Kobayashi, S. *Macromolecules* **1992,** *25,* 5873.

20. Kadokawa, J.; Matsumura, Y.; Kobayashi S. *Macromol. Chem. Phys.* **1994,** *195,* 3689.

Functionalized Polymers
via Enzymatic Synthesis

Helmut Ritter

The first part of this chapter deals with the use of multienzyme systems of living cells to produce linear polyesters from bacteria. Recent progress on the controlled production of special polypeptides from genetically modified microorganisms is also briefly discussed. In the second part of the chapter, the use of isolated enzymes for the synthesis of new monomers and for polycondensates is reviewed. For example, the enantioselective synthesis of new (meth)acryloyl monomers with the help of enzymes, the lipase-catalyzed synthesis of macromonomers, and the preparation of crystalline cellulose in the presence of a glucosidase are outlined. Enzymatic modification of polymers for degradation or substrate attachment is also covered.

The existence of life is based on biochemical transformations of a wide variety of substrates catalyzed by enzymes under physiological conditions. This knowledge has long inspired many attempts to use enzymes as catalysts for the development of low molecular weight organic molecules. In recent years, the use of enzymes for the synthesis of macromolecular compounds also has attracted considerable interest.

This chapter provides an overview of enzymatic catalysis for the synthesis of functionalized macromolecules in vivo and in vitro. The first part of the chapter deals with the application of multienzyme systems of whole cells to produce new polyesters and polypeptides. In the second part of the chapter, a review of recent progress on the use of isolated enzymes in polymer chemistry is presented.

From the viewpoint of polymer chemistry, even nucleic acids (RNA and DNA), which bear the genetic code, are functionalized polyesters formed from phosphoric acid and a pentose (ribose in RNA and deoxyribose in DNA). RNA and DNA contain a purine or pyrimidine base in the N-β-glucosidal position of the sugar component. The nature of the N-heterocyclic

components (adenine, thymine, cytosine, and guanine) is the basis of the genetic code and determines the structure of proteins. The genetic information is normally transmitted via a messenger RNA (mRNA) in the cell to control the biosynthesis of enzymes. Some of these enzymes are themselves responsible for the synthesis of other biopolymers, which may be divided into nucleic acids, proteins, polysaccharides, lignin, polyisoprens, and polyesters.

The first section of this chapter focuses on progress in the area of bacterially produced polyesters (*1*). Some of these polyesters, such as poly(2-hydroxybutyrates) (P-2-HB), have reached the commercial stage. In addition, special polypeptides from living organisms are described that are mainly characterized by an alternating amino acid composition. The latter polymers are present, for example, in natural silk; however, recently they were also produced in genetically modified bacteria (*2*).

In the second section of this chapter, advances in the area of enzymes in polymer chemistry are reviewed. Progress in technical production of enzymes made it possible to apply these biocatalysts for the synthesis of many types of low molecular weight compounds in vitro (3). For example, the stereochemistry of certain condensation reactions can be controlled under mild physiological conditions. Also, isolated enzymes have attracted increasing interest in polymer chemistry. Initially enzymes were used to synthesize new types of monomers. Later, they were applied successfully to produce macromolecules from certain monomers. Under carefully controlled conditions, they can also modify certain polymers in the main chain or in the side chains (4).

Synthesis of New Polymers in Living Cells

Polyesters from Bacteria

Many bacteria have the enzymatic capacity to produce poly(hydroxyalkanoates) (PHA, **1**) as intracellular energy and carbon storage material (1). Initially, PHA was isolated from *Bacillus megaterium* in 1926. The structure of this material was characterized to be P-3-HB (**2**), and it had a high optical activity (5–7). PHAs are generally linear hydrophobic polyesters with thermoplastic properties.

$$-(-O-*CHR-(CH_2)_x-CO-)_n-$$

1

$$-(-O-*CH(CH_3)-CH_2-CO-)_n-$$

2

where the degree of polycondensation (n) = 100–30,000, x = 1–4, and R = alkyl group.

Isolated PHB tends to crystallize in a right-handed 2_1-helix and to have a melting point of about 180 °C. The value of n depends on various factors and may reach 30,000 (8).

Agricultural and horticultural applications of PHAs have been suggested. The biodegradable polymer may be used, for example, to prepare plant growth regulators or pesticides formulations for sustained release. Additionally, PHAs may be useful for medical implants, pharmaceutical drug release, and as plastic films for packaging (1).

The biosynthesis and degradation of the PHAs in living cells occur via cyclic metabolic processes (1, 8). PHB deposits in the cytoplasmic fluid as granules that have diameters of about 0.3–1.0 μm. As mentioned previously, many types of bacteria are able to accumulate PHB in amounts of 20–80% of their dry cellular weight. Under appropriate nutritional conditions, *Alcanigenes eutrophus* can accumulate PHB up to 96% of its dry weight.

In some microorganisms, such as *Escherichia coli*, the PHB is located not in the cytoplasm but in the cytoplasmic membrane. In this case, a complex mixture is present consisting of PHB, calcium ions, and polyphosphate. The PHB chains consist of 120–200 subunits, whereas the polyphosphate chain contains 130–170 repeating units (9–12).

Much interest has focused on the behavior of two organisms, namely *A. eutrophus* and *Pseudomonas oleovorans*. These bacteria were used to produce various types of new PHAs. *A. eutrophus* produces PHAs with relatively short side chains, whereas *P. oleovorans* is characterized by the production of PHAs with longer side chains (1). *P. oleovorans* accepts fatty acids with more than 5 carbon atoms as a single carbon source for successful fermentation (13–15).

Recently, it was observed that *P. oleovorans* can produce unsaturated and lightly branched polyesters (16, 17), polyesters bearing halogen atoms (18, 19), nitrile groups (20), and phenyl (21) or phenoxy groups (22). In any case, a carbon feed has to be applied that contains the corresponding functional group as a substituent on the fatty acid. For example, the incorporation of phenoxyalcanoic acid into a copolyester is shown in Scheme I.

Polypeptides from Genetically Modified Bacteria

The statistical nature of classical polymerization reactions means that synthetic polymers are necessarily heterogeneous. Thus, conventional synthetic polymers consist of mixtures of polymer chains characterized by a broad chain-length distribution. In contrast, most biologically produced polycondensates are well defined in

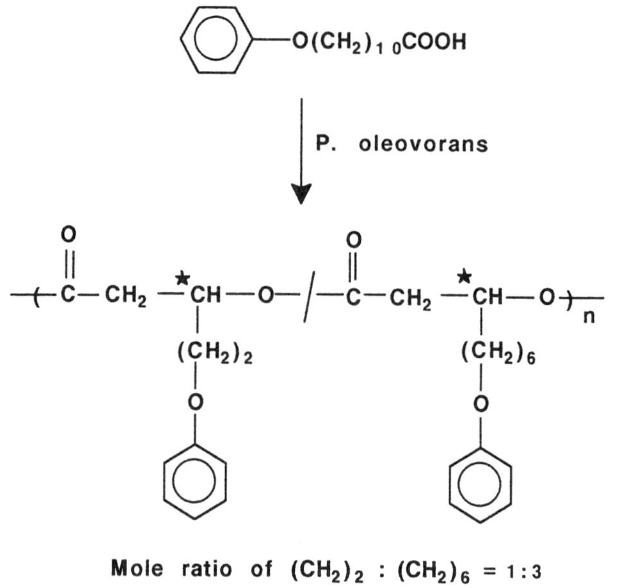

Scheme I. Bacterial production of a phenoxyalkyl substituted polyester.

relation to their molecular composition and chain length. For example, all enzymes (i.e., proteins) have precise primary, secondary, and tertiary structures.

The advent of recombinant DNA technology has provided a basis for the controlled synthesis of new types of peptides and proteins via genetically modified bacteria (24). Recently, genetically directed synthesis of new proteins with well-defined amino acid sequences and molecular weights (e.g., 3) were realized (23):

$$-(AlaGly)_3ProGluGly-$$

3

To this end, the primary amino acid sequence of the desired polypeptide is first encoded into a complementary sequence of DNA, which may be constructed via solid-phase synthesis and enzymatic ligation (25). Insertion of the synthetic gene into an expression vector is then followed by a transformation of a bacterial host organism, in which protein expression is anticipated. The process is completed by fermentation and product isolation. Production of peptide materials in multigram quantities is possible by this route in laboratory-scale fermentation equipment. The genetical control of amino acid sequence in the products opens the possibility to design polypeptides that may form lamellar crystals with definite chain-folded arrangement. The analysis of such a macromolecular structure was accomplished with the help of matrix-assisted laser-desorption mass spectrometry. This accurate mass determination allowed the discovery of two previously undetected mutations in the DNA code for proteins (26).

Synthesis and Modification of Macromolecules by Isolated Enzymes

Enzymes act as biological catalysts and enhance the rate of chemical reactions normally taking place within cells. The reactants in enzyme-catalyzed reactions are termed "substrates", and each enzyme is more or less specific in acting on those functional groups of the substrates that can interact effectively with its active center. Enzymes are, of course, extensively applied as selective catalysts in low-molecular-weight organic chemistry (27). The realization that enzymes can act not only in aqueous media but also in dry organic solvents has opened a major field of enzymatic catalysis (28–30). This realization means that even high molecular weight compounds that are only soluble in organic solvents can also become potential substrates for enzymes.

Enzymatic Synthesis of Functional (Meth)Acrylates

Optically active polymers have found applications as catalysts and reagents for asymmetric synthesis and as chiral absorbents for separation of racemic mixtures. Most of the current optically active synthetic polymers have been prepared by chemical means. Recently, the chemoenzymic synthesis of optically active (meth)-acryloyl derivatives has been achieved by enantioselective hydrolysis of racemic N-methacryloyl alanine methyl ester in the presence of an esterase (hydrolase) (Scheme II) (31).

The enzymatically controlled transesterification between an oxime acrylate and racemic 2-ethylhexanol in the presence of porcine pancreas lipase produced chiral acrylates in different organic solvents (tetrahydrofuran, benzene, hexane, or $CHCl_3$) (32). Lipase-catalyzed transesterifications are often slow, depending on the structure of the substrates. Conventional acrylate esters (e.g., ethyl or butyl esters) react extremely slowly under usual conditions. The introduction of cyclohexyl oxime as an electron-withdrawing residue leads to the formation of activated acrylate esters. Oximes act as irreversible acyl-transfer agents analogous to enol esters or phenyl esters, and the leaving groups do not participate in the back reaction (Scheme III).

A similar transesterification was applied to produce optically active polymers that had enantiomeric excess (ee) values from 76 to 98%. From trifluoroethyl (meth)acrylate or vinylacrylate, racemic alcohols and amines were transesterified in the presence of lipase and subtilisin (Table I). These optically active mono-

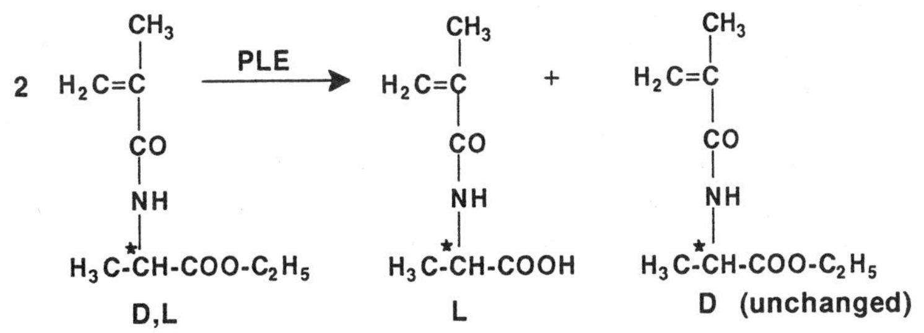

Scheme II. Enantioselective hydrolysis of a racemic monomer mixture with an esterase from porcine liver (PLE).

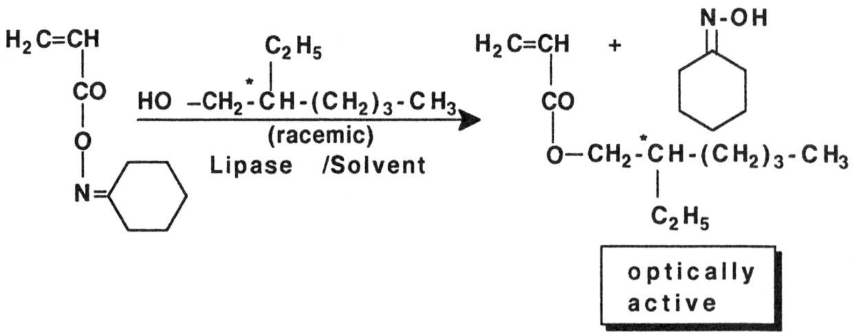

Scheme III. Transesterification of cyclohexyl oxime acrylate with a racemic alcohol in the presence of lipase to produce optically active acrylates.

Table I. Chemoenzymatic Synthesis of Optically Active Poly(meth)acrylates and Polymethacrylamides from Racemic Nucleophiles

Substrates	Enzyme	Chiral Monomer	ee (%)	ee of Polymer Side Chains (%)
(R,S)-1-Indanol trifluoroethyl acrylate	Lipase	(R)-Acrylate	98	98
(R,S)-2-Benzylpropanol trifluoroethyl methacrylate	Lipase	(S)-Methacrylate	90	90
(R,S)-1,2,3,4-Tetrahydro-1-naphthol vinyl acrylate	Lipase	(R)-Acrylate	99	98
(R,S)-1-(1-Naphthyl)ethylamine trifluoroethyl methacrylate	Subtilisin	(S)-Methacrylamide	>98	96
(R,S)-Phenylalaninamide trifluoroethyl methacrylate	Subtilisin	(S)-Methacrylamide	77	76

Source: Reproduced from reference 33. Copyright 1991 American Chemical Society.

mers were polymerized by a free-radical mechanism yielding the corresponding homopolymers with chiral side groups (33). In the presence of immobilized lipase from *Candida cylindracea* as catalyst, the transesterification of an activated vinyl acrylate with various alcohols in isooctane was also successfully performed.

The ease of transesterification in this work was observed to depend on the structure of the alcohol substrates and was as follows: *n*-hexyl > *n*-octyl > *n*-butyl alcohol with aliphatic alcohols and β-phenethyl >> benzyl alcohol for alcohol containing a phenyl group. The yield of the resulting acrylic esters after 50 h at 37 °C was about 40–66% for *n*-hexyl and β-phenethyl alcohol. As mentioned previously, acetaldehyde as a leaving group does not allow back reactions, and therefore the equilibrium is shifted toward the products. However, the released acetaldehyde may decrease the lipase activity. As an example, the chemoenzymatically synthesized β-phenethyl acrylate was polymerized in suspension yielding a high molecular weight polymer (34). In this work, the chemoenzymatic synthesis of chiral (meth)acryl derivatives incorporates the enzymatic resolution of racemic monomeric substrates into high molecular weight polymers.

The chemoenzymatic synthesis of functionalized polymers with comb-like structures has also been performed by lipase-catalyzed esterification of 11-methacryloylaminoundecanoic acid with a series of alcohols (isobutyl alcohol, cyclohexanol, menthol, cholesterol, and testosterone) and subsequent radical polymerization of the obtained esters (Scheme IV).

The kinetics of esterification strongly depends on the structure of the alcohol, especially in the case of hindered structures. For example, isobutyl alcohol has the lowest steric hindrance among the substrates examined, and it is therefore the most reactive (about 90% yield after 1 day). Esterification of cyclohexanol proceeds at a slightly lower rate and has a yield of about 80% after 24 h. The least-reactive substrate was testosterone (OH on C-17). This result can be explained by the neighborhood of a sterical hindering axial C-18 methyl group of the steroid. Only 20% yield of testosterone ester was obtained after 8 days. Surprisingly, the reactivity of the OH-group of the steroid cholesterol (C-3 position) was in the same order as cyclohexanol (35).

The lipase-catalyzed condensation of 11-methacryloylaminoundecanoic acid with 12-hydroxylauric acid yields a methacrylamide macromonomer with a free-carboxylic end group. From fast atom bombardment mass spectrometry, the existence of different degrees of esterification in the mixture was established (Scheme V). The enzymatically prepared macromonomer and the corresponding homopolymer can crystallize. The melting point and the melting enthalpy of the long side chains of the polymethacrylamide are slightly lower compared with the values of the unpolymerized macromonomer. The values obtained from differential scanning calorimetric measurements for melting point and melting enthalpy decreased from 77.1 °C and 150 J/g for the macromonomer to 72.0 °C and 110 J/g for the homopolymer, respectively.

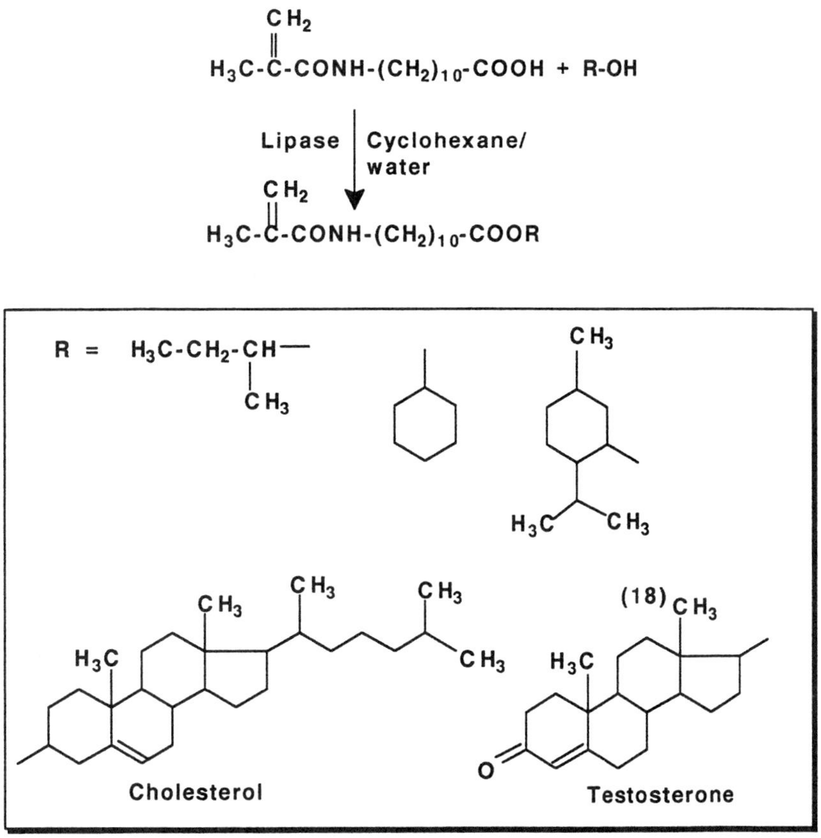

$$\text{H}_3\text{C}-\overset{\overset{\displaystyle\text{CH}_2}{\|}}{\text{C}}-\text{CONH-(CH}_2)_{10}\text{-COOH} + \text{R-OH}$$

Lipase | Cyclohexane/ water

$$\text{H}_3\text{C}-\overset{\overset{\displaystyle\text{CH}_2}{\|}}{\text{C}}-\text{CONH-(CH}_2)_{10}\text{-COOR}$$

R = $\text{H}_3\text{C-CH}_2\text{-CH}-$ with CH_3

Cholesterol   (18)   Testosterone

Scheme IV. Lipase catalyzed esterification of *N*-methacryloylaminoundecanoic acid.

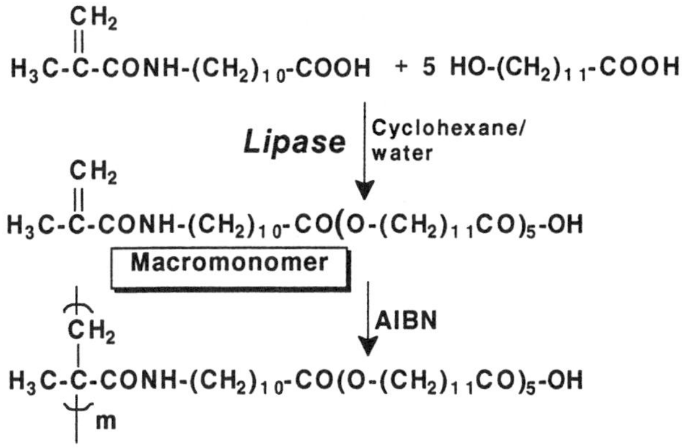

$$\text{H}_3\text{C}-\overset{\overset{\displaystyle\text{CH}_2}{\|}}{\text{C}}-\text{CONH-(CH}_2)_{10}\text{-COOH} + 5\ \text{HO-(CH}_2)_{11}\text{-COOH}$$

Lipase | Cyclohexane/ water

$$\text{H}_3\text{C}-\overset{\overset{\displaystyle\text{CH}_2}{\|}}{\text{C}}-\text{CONH-(CH}_2)_{10}\text{-CO}\big(\text{O-(CH}_2)_{11}\text{CO}\big)_5\text{-OH}$$

Macromonomer

AIBN

$$\text{H}_3\text{C}-\overset{\overset{\displaystyle\text{CH}_2}{\|}}{\text{C}}-\text{CONH-(CH}_2)_{10}\text{-CO}\big(\text{O-(CH}_2)_{11}\text{CO}\big)_5\text{-OH}\Big]_m$$

Scheme V. Lipase catalyzed synthesis of a polymerizable oligoester.

The free carboxylic end groups of the methacrylamide macromonomer and the corresponding homopolymer can be enzymatically esterified with a suitable monoalcohol (e.g., isopropanol or fluorenyl methanol). During this reaction, a partial transesterification with some of the ester groups of the oligoester chain was observed by NMR spectroscopy as a side reaction (*36*).

The incorporation of sugar residues into polymers may extend the application of these materials into different areas such as chiral separation for chromatography and as medical devices. However, the direct and selective synthesis of mono(meth)acryl of polyhydroxy compounds is complicated. The usual way to avoid multiple linkages between the sugar and the polymer is to apply specific protecting groups such as trityl groups and then to induce deprotection. The use of enzymes as highly selective catalysts for regioselective sugar acylation has been successfully demonstrated for this purpose. For example, the regioselective enzymatic acylation of sucrose with vinyl acrylate was conducted to give a 60% yield of the mono acrylate derivative after 5 days at 45 °C (Scheme VI) (*37*).

A similar chemoenzymatic approach has been developed to prepare monosaccharide-substituted polyacrylates. Lipase from *Pseudomonas cepacia* was used to catalyze the transesterification in pyridine of a variety of monosaccharides with vinyl acrylate to give mainly 6-acryloyl esters. These acrylates were polymerized radically to give the corresponding polyacrylates. This method led to the synthesis of a highly water-soluble poly(methyl 6-acryloyl-β-galactoside) that had a number-average molecular weight (M_n) of 58,000 daltons. The copolymerization of the sugar-containing acrylates with the cross-linking agent ethyleneglycol dimethacrylate resulted in materials that swell in aqueous solution. For example, a cross-linked

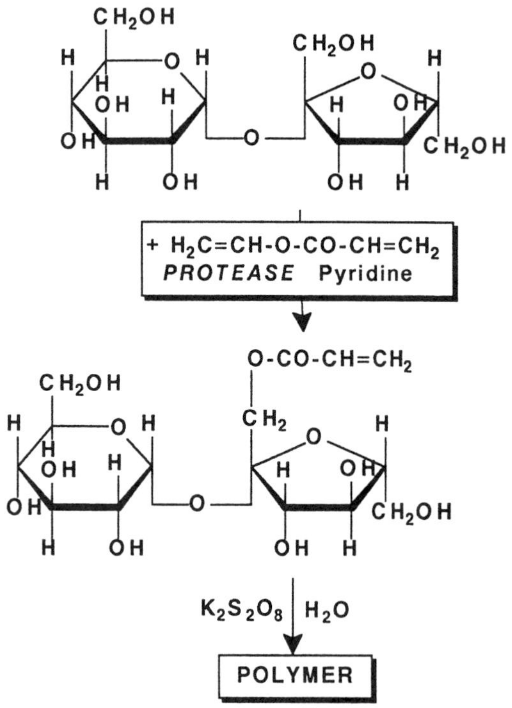

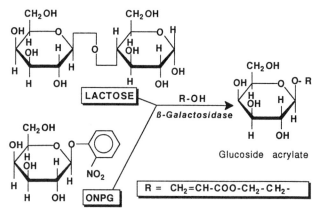

Scheme VI. Chemoenzymatic synthesis of hydrophilic polyacrylate containing disaccharides in the side chains.

Scheme VII. Synthesis of a glycoside acrylmonomer from ONPG and lactose in the presence β-galactosidase.

copolymer from methyl-6-acryloyl-β-galactoside and 0.3% dimethacrylate was shown to hold nearly 50-fold its weight in water. Such materials may be useful as biocompatible hydrogels for biomedical and membrane applications (*38*). Additionally, sugar groups play an important role in cell recognition phenomena. Therefore, the creation of polyacrylates with well-defined sugar side chains is of great interest. For example, the synthesis of vinyl monomers with sugar groups by classical chemical methods normally leads to mixtures of α- and β-anomers. The application of a glucosidase for the synthesis of sugar acrylates was successfully performed. 2-Acryloyloxyethyl β-D-galactopyranoside can be directly prepared by trans-glycosylation reaction of lactose or *o*-nitrophenyl β-D-galactopyranoside (ONPG) with 2-hydroxyethyl acrylate in the presence of β-galactosidase in a phosphate buffer (Scheme VII). The yield and the rate of reaction of ONPG were superior to those values of lactose. The yields of the acryloyl derivatives were >90% after 30 min for ONPG but about 50% after 2.5 h for lactose. The resulting monomers polymerized readily with a free-radical initiator (*39*).

Enzymatic Polycondensation

Many attempts have been made to produce high molecular weight polyesters from suitable alcohols and carboxylic acids and from hydroxyacids via enzymatic catalysis at moderate temperatures. However, most of these efforts led to the formation of oligomers. These observations may be explained because enzymes not only catalyze the condensation step but also the depolymerization reaction. Additionally, the condensation reaction leads to the formation of oligomers that are less soluble than the monomers. Therefore, the reactivity of the oligomer end groups probably decreases and gradually diminishes the rate of poly-condensation. For example, condensation of dicarboxylic acids containing 6 to 14 carbon atoms with ethylene glycol or 1,3-propanediol was performed in the presence of a lipase from *Aspergillus niger* and produced oligomers with degree of polymerization values of 5–7 (*40*).

The enzymatic polycondensation of 12-hydroxy-octadecanoic acid, 12-hydroxy-*cis*-9-octadecenoic acid, 16-hydroxyhexadecanoic acid, or 12-hydroxydodecanoic acid was carried out in water–isooctane by using lipase from *Candida rugosa* or *Chromobacterium viscosum*. The resulting oligomers from the acids bearing primary hydroxyl groups show relatively narrow molecular weight distributions, but those oligomers from acids with secondary hydroxyl groups show broader molecular weight distributions. The M_n values of the resulting products were ≤1,300 daltons (*41*).

Enzymatic ring-opening polymerization of lactones was achieved by using lipase as a catalyst. The polymerization of ε-caprolactone by *Pseudomonas fluorescens* lipase at 60 °C in bulk afforded a polyester with M_n values of 7000 daltons. The relatively low molecular weight was verified by NMR spectroscopy of the end group (*42*).

The enzymatic polyesterification of linear ω-hydroxyesters (**4**) was carried out in dry organic media in the presence of porcine pancreatic lipase. Surprisingly, the molecular weight of the resulting polyesters (**5**) increased significantly at higher temperatures. The highest degree of polycondensation (105) was achieved with methyl 6-hydroxyhexanoate in *n*-hexane at the boiling temperature (69 °C). The resulting polyester has film forming properties (43).

$$HO(CH_2)_mCOOCH_3$$

4

$$H-\{O(CH_2)_mCO\}_n-OCH_3$$

5

$$4 + lipase \rightarrow 5$$

where $n = 4$ and $m = 5$.

The chiral selection of substrates by enzymes is a fascinating tool for the synthesis of optically active polymers. In this connection, the enantioselective oligocondensation of bis(2,2,2-trichloroethyethyl) *trans*-3,4-epoxyadipate with 1,4-butanediol was carried out successfully in the presence of lipase in anhydrous ethyl ether as the solvent (44) (Scheme VIII).

As shown in Scheme VIII, only one of the enantiomers is incorporated into the oligoester, and the other is not a substrate for the lipase. Therefore, the resulting oligoester is optically active. The unchanged enantiomer of the diester was shown to have an enantiomeric purity of >95%. Here again, only a relatively low molecular weight (ca. 5300 daltons) could be achieved.

The oligocondensation of the prochiral hydroxyester, dimethyl-β-hydroxyglutarate, with various hydrolytic enzymes led to the formation of optically active dimers and trimers (Scheme IX). The resulting chirality is due to the newly created asymmetric center in the product. The ee values were 30–37% depending on the kind of enzyme. An important feature of the

resulting chiral oligomers is their functional groups, which may be used for cross-linking or other residues. This approach opens the possibility to produce chiral polymers from achiral monomers (45).

The incorporation of saccharides into chemoenzymatically prepared polycondensates was described for sucrose polymers. As shown in Scheme X, the disaccharide can be condensed at C-6 with bis(2,2,2-trifluoroethyl)adipate in organic solvents to form activated diesters in 20% yield. The subsequent polycondensation of the purified diester with ethylenediamine was carried out under mild conditions. The resulting polyesteramide was a semicrystalline solid that was soluble in a variety of polar organic solvents (37).

Recently, the synthesis of crystalline cellulose from β-D-cellobiosyl fluoride (activated monomer) in the presence of cellulase was achieved. The condensation has been performed in a mixture of acetonitrile and acetate buffer (5:1), and a 54% yield of a water-insoluble part was obtained. This water-insoluble fraction of the product has been shown by solid-state ^{13}C-NMR, IR, and X-ray spectroscopies to be synthetic cellulose with a high crystallinity (46). This product has been converted to the soluble triacetate whose molecular weight was at least 6,300 daltons, a value indicating a degree of polymerization >22 (Scheme XI).

In a similar way, maltose oligomers with degrees of polycondensation as great as those for heptamers have been prepared from α-D-maltosylfluoride in the presence of an α-amylase as catalyst in a mixed solvent of methanol–phosphate buffer at pH 7. The most appropriate reaction conditions for the production of higher oligomers were to use methanol–buffer (2:1) as solvent and an amylase enzyme from *Aspergillus oryzae* as catalyst (47).

Oxidative oligocondensations in the presence of peroxidases also were performed. For example, a solution of *o*-phenylene diamine in 1,4-dioxane and phosphate buffer (pH 7.0) was treated with hydrogen peroxide and horseradish peroxidase at room temperature,

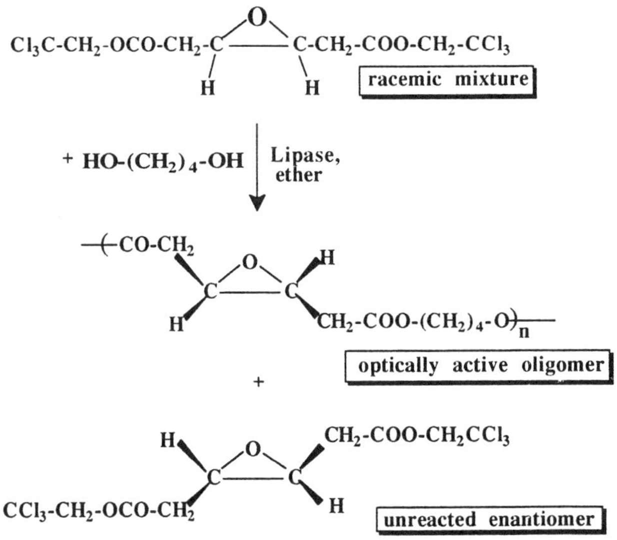

Scheme VIII. Enantioselective oligocondensation.

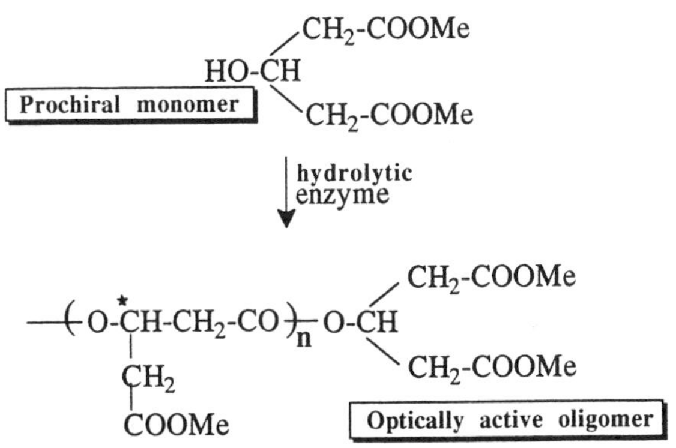

Scheme IX. Enzymatic synthesis of an optically active oligoester.

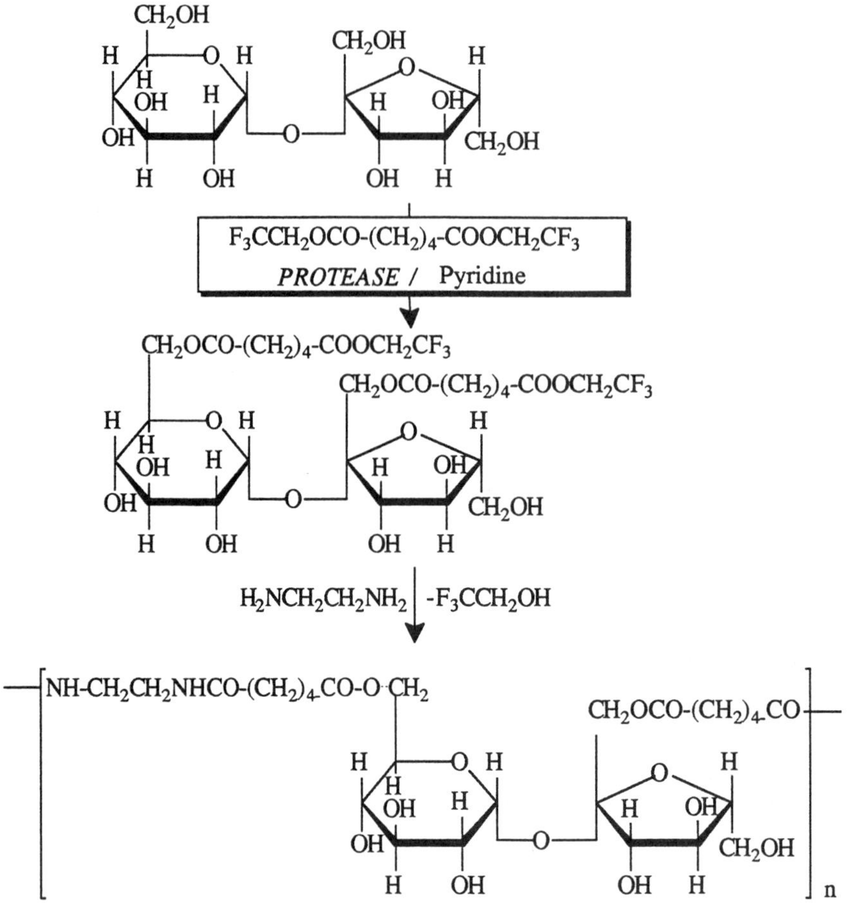

Scheme X. Chemoenzymatic synthesis of a polyesteramide containing sugar residues in the main chain.

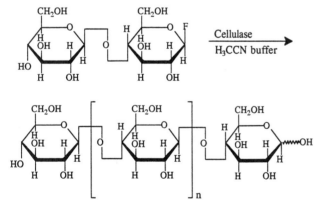

Scheme XI. Cellulase catalyzed synthesis of crystalline cellulose starting from β-D-cellobiosyl fluoride.

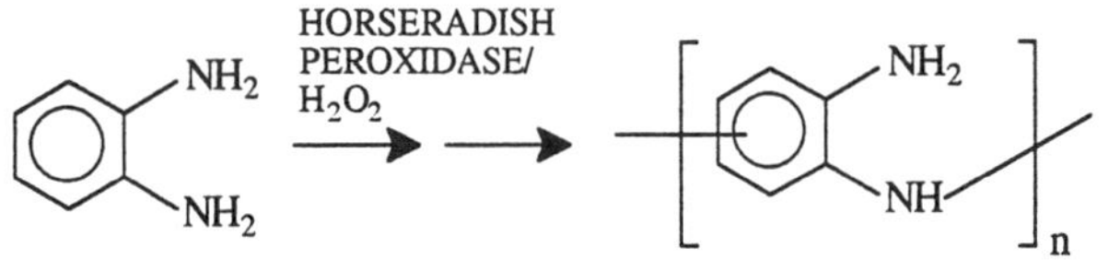

Scheme XII. Oxidative polycondensation in the presence of a peroxidase.

and polymers with average molecular weights of 20,000 daltons were obtained (Scheme XII). The polymerization of the diamine without the enzyme gave only oligomers (*48*).

Horseradish peroxidase also catalyzes the copolymerization of *p*-cresol with lignin in aqueous organic solvent mixtures in the presence of H_2O_2. Almost all of the lignin and more than one-third of the phenol are consumed and yield a highly insoluble material, probably the product of lignin cross-linking with the phenol. In the absence of cresol, a dimethylformamide-soluble lignin material is obtained (*49*).

Free-radical polymerization of acrylamide has been achieved in the presence of glucose, $FeCl_2$, and glucose oxidase as biocatalysts under mild conditions. In these reactions, the enzymatically produced H_2O_2 is decomposed by the reducing agent to form starting OH-radicals (*52*) (Scheme XIII). Enzymatically produced hydrogen peroxide can also be used to epoxydize methacryl monomers containing chalkone moieties (Scheme XIV) under mildly basic conditions in high yields (*53*).

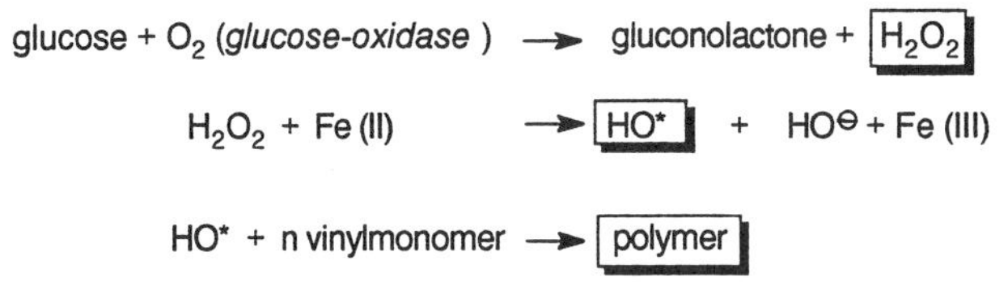

$$glucose + O_2 \; (\textit{glucose-oxidase}) \longrightarrow gluconolactone + \boxed{H_2O_2}$$

$$H_2O_2 + Fe\,(II) \longrightarrow \boxed{HO^*} + HO^\ominus + Fe\,(III)$$

$$HO^* + n\,vinylmonomer \longrightarrow \boxed{polymer}$$

Scheme XIII. Chemoenzymic initiation of free-radical polymerization of water-soluble vinylmonomers (e.g., acrylamide).

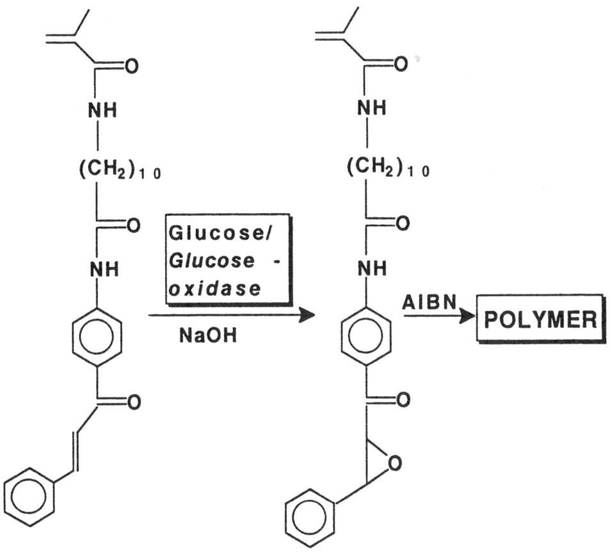

Scheme XIV. Epoxidation of a methacryl monomer by means of glucose oxidase–glucose.

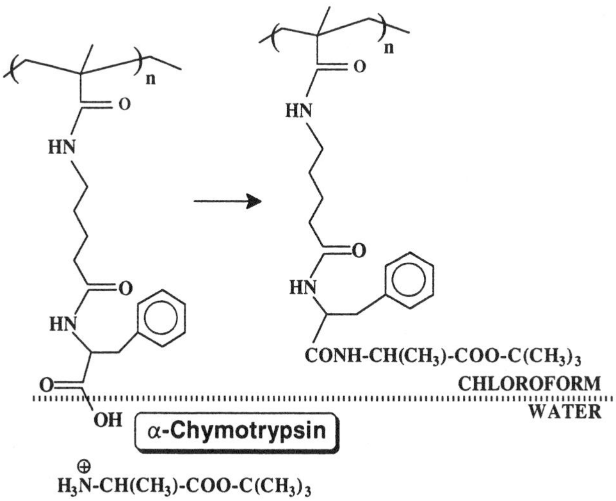

Scheme XV. Enzymatic anchoring of an amino acid at a soluble polymer in water–chloroform.

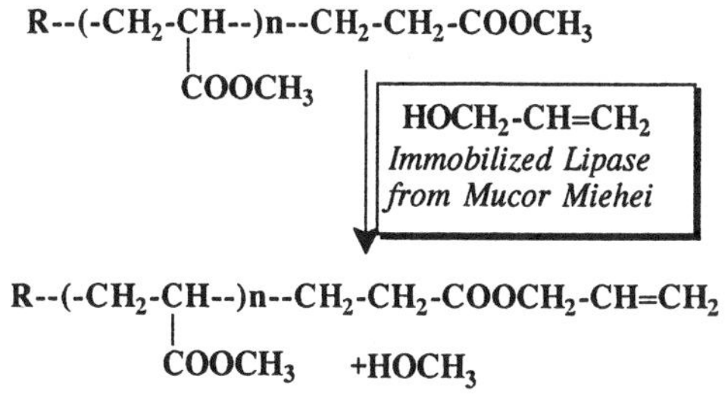

Scheme XVI. Lipozyme-catalyzed transesterification of oligo-(methacrylates) with allylalcohol.

Enzymatic Modification and Degradation of Synthetic Polymers

The enzymatic degradation of polymers such as polyesters, proteins, and polysaccharides and the cleavage of side chains from synthetic polymers are well known (*54*). In contrast, only a few papers have appeared describing the enzymatic anchoring of low molecular weight compounds to polymers. For example, it was recently demonstrated that phenylalanine residues at the end of polymer side chains can be coupled with alanine *t*-butyl ester in the presence of α-chymotrypsin in water–chloroform (Scheme XV). The condensation reaction is highly favored because the more apolar peptide component migrates into the organic phase. The highest conversion of the free carboxylic groups of the *N*-acylated phenylalanine component in the side chains with racemic alanin *t*-butylester was 35% after 7 days at room temperature (*55*).

Lipase-catalyzed esterification of allyl alcohol with oligo-methylacrylates that contain methyl ester chain ends occurs regioselectively only at the chain ends (*56*). The sterically more hindered methyl ester side groups are not effected under these conditions (Scheme XVI).

Lipase-catalyzed acetylation of side chain racemic OH-groups in high molecular weight poly methacrylates was achieved with a maximum conversion of 40%. A low enantioselectivity of this esterification was established by measuring the optical rotations of the modified polymers (up to –1.20 at room temperature) (*57*).

Concluding Remarks

Enzymes have been long used in research and industry for the synthesis and chemical transformation of low molecular weight compounds. In the area of synthetic polymer chemistry, the enormous potential of enzymes has only recently been realized. Enzymatic synthesis of monomers and polycondensates as well as modification of polymers with regio- and enantioselective control present a highly promising area of exploration in macromolecular chemistry.

Up until now, mainly enzymatically catalyzed hydrolysis and condensation reactions have been described in polymer chemistry. However, enantioselective reduction of carbonyl groups with reduced nicotinamide adenine dinucleotide–alcohol-dehydrogenase, hydroxylation of alicyclic compounds with cytochrome P450, and the formation of C–C bonds have not yet been applied for the creation of new macromolecules.

In the long term, the genetic creation of special catalytically active antibodies in addition to usual enzymes should further enhance the use of biocatalysts in polymer chemistry.

References

1. Doi, Y. *Microbial Polyesters;* VCH Publishers (UK): Cambridge, United Kingdom, 1990.
2. Tirrel, D. A.; Fournier, J.; Mason, T. L. *Mater. Res. Soc. Bull.* **1991**, *16*, 23–28.
3. Davis H. G.; Green R. H.; Kelley D. R.; Roberts S. M. *Biotransformations in Preparative Organic Chemistry: Best Synthetic Methods;* Academic: London, 1989.
4. Ritter, H. *Trends in Polym. Sci.* **1993**, *1*, 171–173.
5. Lemoigne, M. *Bull. Soc. Chim. Biol.* **1926**, *8*, 770–782.
6. Lemoigne, M.; Grelet, N.; Crosan, U.; Le Treis, M. *Bull. Soc. Chim. Biol. (Paris)* **1945**, *27*, 90–95.
7. Lemoigne, M. *Helv. Chim. Acta* **1946**, *29*, 1303–1306.
8. Lafferty, R. M.; Kosatko, W. In *Biotechnology;* Rehm H. J.; Reed, G., Eds.; 6B; VCH: Weinheim, Germany, 1988; pp 135–176.
9. Steinbüchel, A. In *Biomaterials;* Byrom, D., Ed.; Macmillan: London, 1991, 123–213.
10. Reusch, R. N.; Sadoff, H. L. *Proc. Natl. Acad. Sci. U.S.A* **1988**, *85*, 4176.
11. Reusch, R. N.; Sadoff, H. L. *J. Bacteriol.* **1983**, *156*, 778–788.
12. Reusch, R. N. *Proc. Soc. Exp. Biol. Med.* **1989**, *154*, 870–878.
13. DeSmet, M. J.; Eggink, G.; Witholt, B.; Kingma, J.; Wynberg, H. *J. Bacteriol.* **1983**, *154*, 870–878.
14. Lageveen, R. G.; Huisman, G. W.; Pseusting, H.; Ketella, P.; Egging, G.; Witholt, B. *Appl. Environ. Microbiol.;* **1988**, *54*, 2924.
15. Ballistreri, A.; Montaudo, G.; Impallomeni, G.; Lenz, R. W.; Kim, Y. B.; Fuller, R. C.; *Macromolecules* **1990**, *23*, 5059–5064.
16. Fritzsche, K.; Lenz, R. W.; Fuller, R. C. *Int. J. Biol. Macromol.* **1990**, *23*, 5059–5064.
17. Fritzsche, K.; Lenz, R. W.; Fuller, R. C. *Int. J. Biol. Macromol.* **1990**, *12*, 92–101.
18. Abe, C.; Taima, Y.; Nakumura, Y.; Doi, Y. *Polym. Commun.***1990**, *31*, 404–406.
19. Doi, Y.; Abe, C. In *Microbial Polyesters;* VCH: New York, **1990**; p 57.
20. Kim, Y. B.; Lenz, R. W.; Fuller, R. C. *J. Bioact. Compat. Polym.* **1990**, *6*, 382–392.
21. Fritzsche, K.; Lenz, R. W. *Makromol. Chem.* **1990**, *191*, 1957–1965.
22. Ritter, H.; Gräfin v. Spee, A. *Makromol. Chem. Phys.* **1994**, *195*, 1665–1672.
23. Ritter, H. *Angew. Chem. Int. Ed. Engl.* **1991**, *30*, 677– 678.
24. McGrath, K. P.; Tirrell, D. A.; Kawai, M.; Mason, T. L.; Fournier, M. J. *Biotechnol. Prog.* **1990**, *6*, 188–192.
25. Typical procedures are represented in Sambrook, J.; Frisch, E. F.; Maniatis, T. *Molecular Cloning, a Laboratory Manual,* 2nd ed.; Cold Spring Harbor Laboratory: New York, 1989.
26. Beavis, R. C.; Chait, B. T.; Creel, H. S.; Fournier, M. J.; Mason, T. L.; Tirrell, D. A. *J. Am. Chem. Soc.* **1992**, *114*,7584–7585.
27. Wong, C. H.; Whitesides, G. M. *Enzymes in Synthetic Organic Chemistry;* Tetrahedron Organic Chemistry Series; Series Editor: Balwin, J. E.; Elsevier Advanced Technology: Oxford, England, 1994; Vol. 12.
28. Riva, S.; Chopineau, J.; Kieboom, A. P. G.; Klibanov, A. M. *J. Am. Chem. Soc.* **1988**, *110*, 584–589.
29. Therisod, M.; Klibanov, A. M. *J. Am. Chem. Soc.* **1986**, *108*, 5683–5640.
30. Kirchner, G.; Scollar, M.P.; Klibanov, A.M. *J. Am. Chem. Soc.* **1985**, *107*, 7072–7076.
31. Ritter, H.; Siebel, C. *Makromol. Chem.; Rapid Commun.* **1985**, *6*, 521–525.
32. Ghorare, A.; Kumar, G. S. *J. Chem. Soc., Chem. Commun.* **1990**, 134–135.
33. Margolin, A. L.; Fitzpatrick, P. A.; Dubin, P. L. *J. Am. Chem. Soc.* **1991**, *113*, 4693–4694.
34. Ikeda, I.; Tunaka, J.; Suzuki, K. *Tetrahedron Lett.* **1991**, *32*, 6865–6866.
35. Pavel, K.; Ritter, H. *Makromol. Chem.* **1991**, *192* 1941–1949.
36. Pavel, K.; Ritter, H. *Makromol. Chem.* **1993**, *194*, 3369–3376.
37. Patil, D. R.; Dordick, J. S.; Rethwisch, D. G. *Macromolecules* **1991**, *24*, 3462–3463.
38. Martin, B. D.; Ampofo, S. A.; Linhardt, R. J.; Dordick, J. S. *Macromolecules* **1992**, *25*, 7081–7085.
39. Matsumura, S.; Kubokawa, H.; Toshima, K. *Makromol. Chem. Rapid Commun.* **1993**, *14*, 55–58.
40. Okumura, S.; Iwai, M.; Tominaga, Y. *Agric. Biol. Chem.* **1984**, *48*, 2805–2808.
41. Matsumura, S.; Takahashi, J. *Macromol. Chem., Rapid Commun.,* **1986**, *7*, 369–373.
42. Uyama, H.; Kobayashi, S. *Chem. Lett., Chem. Soc. Jpn.* **1993**, 1149–1150
43. Knani, D.; Gutman, A. L.; Kohn, D. H. *J. Polym. Sci., Part A: Polym. Chem.* **1993**, *31*, 1221–1232.
44. Wallace J. S.; Morrow, C. J. *J. Polym. Sci., Part A; Polym. Chem.* **1989**, *27*, 2553–2567.
45. Gutman, A. L.; Bravdo, T. *J. Org. Chem.* **1989**, *54*, 5645–5646.
46. Kobayashi, S.; Kashiwa, K.; Kawasaki, K.; Shoda, S.; *J. Am. Chem. Soc.* **1991**, *113*, 3079–3084.
47. Kobayashi, S.; Shimada, J.; Kashiwa, K.; Shoda, S. *Macromolecules* **1992**, *25*,3237–3241.
48. Kobayashi, S.; Kaneko, I.; Uyama, H. *Chem. Lett., Chem. Soc. Jpn.* **1992**, 393–394.
49. Blinkovsky, A. M.; Dordick, S. J. *J. Polym. Sci., Part A: Polym. Chem.* **1993**, *31*, 1839–1846.
50. Hale, P. D; Boguslavsky, L. I.; Inagaki, T.; Karan, H. I.; Lee, H. S.; Skotheim, T. A.; Okamoto, Y. *Anal. Chem.* **1991**, *63*, 677–682.
51. Lee, H. S.; Liu, L.-F.; Hale, P. D.; Okamoto, Y. *Heteroat. Chem.* **1992**, *3*, 303–309.
52. Ritter, H. West German Patent 3,743,198 A1, 1990, CA 112(2):8037u.
53. Goretzki, Ch.; Ritter, H. *Macromol. Rep.* **1995**, *A32(182)*, 237–245.

54. *Agricultural and Synthetic Polymers: Biodegradability and Utilization;* Glass, J. E.; Swift, G., Eds.; ACS Symposium Series 433; American Chemical Society: Washington, DC, 1990.

55. Rehse, H.; Ritter, H. *Makromol. Chem.* **1988**, *189*, 529-539.
56. Lailot, T.; Brigotiot, M.; Marechal, E. *Polym. Bull. Berlin* **1991**, *26*, 55–62
57. Pavel, K.; Ritter, H. *Makromol. Chem.* **1992**, *193*, 323–328.

Polymers Containing Phosphonic and Bis(phosphonic acid) Groups

Mats J. Sundell and Jan H. Näsman

During recent years polymers containing phosphonic and bis(phosphonic acid) groups have found a number of interesting applications. These applications include the use of these polymers for separation processes, as adhesives, in medical applications, and as a ligand for catalytically active metals. In this chapter, the preparation of polymers containing phosphonic and especially bis(phosphonic acid) groups is reviewed. The polymers are synthesized either by polymerization of a functional monomer or by chemical modification of a preformed polymer. Some interesting applications using low molecular weight bis(phosphonic acid) compounds are also discussed.

Polymers containing acidic groups have found widespread use throughout the field of chemical technology. Traditionally, these resins have been based on sulfonic acid functionality. The idea of using sulfonated polystyrene as an ion-exchange resin goes back as far as 1930 and has led to the development of a number of separation processes. In the 1950s these polymers also found extensive use as polymer-bound acidic catalysts. Catalytically active metals also have been loaded onto polymers containing sulfonic acid and used as polymer-supported catalysts. Because the ease of separation has been a main goal in all these applications, it is very often preferable to use cross-linked polymers. However, the lack of selectivity of the sulfonic acid resin for specific ions can limit its practical use.

Several workers have shown that the ion-exchange reaction alone does not have a wide enough range in the free energy of reaction to allow selectivity. For example, the change in enthalpy of reaction for the sulfonic acid functionality with alkali metal ions is only 1.5 kcal/mol (1). Additionally, a small value for the free energy of reaction implies that the metal can be lost to the environment in catalytic applications because the binding would be relatively weak. For both separation and catalytic applications, a higher selectivity must be achieved for the metal ion of interest. In our laboratory, we have been looking at the possibility of using phosphorus-containing acids for this purpose. We have been concentrating on polymers containing phosphonic acid, and especially bis(phosphonic acid). Very little data on these polymers are available in the literature.

Bis(phosphonic acid) derivatives are unusual in their ability to form strong complexes with a number of heavy metals even in highly acidic solutions. Some bis(phosphonic acid) derivatives also have a strong affinity for a number of radioactive metal ions. Additionally, a number of interesting studies have been done using low molecular weight analogs. These studies include the use of bis(phosphonic acid) derivatives in solvent extraction processes (2), metal oxide dissolution (3), ion flotation (4), medical applications (5). For example, hydroxyethyl-1,1-diphosphonic acid and vinylidene-1,1-diphosphonic acid are highly effective stripping agents for Am, Pu, and U (6). These extractants also are thermally unstable and can readily be oxidized by using heat- or metal-catalyzed H_2O_2 oxidation (7). Hydroxyethyl-1,1-diphosphonic acid also was used recently to inhibit bioprosthetic tissue calcification in bioprosthetic heart valves (8).

Phosphonic acids belong to the class of the lower oxyacids of phosphorus and can be deemed to have been derived from orthophosphoric acid. The nomenclature for this family of acids is given in Chart I. Phosphorus-containing polymers tend to be flame retardant (9, 10). This effect is also observed for polymers containing phosphonate (11) and bis(phosphonate) (12) groups. Other interesting features of phosphorus polymers are adhesion to metals, metal ion binding characteristics, and increased polarity. Polymers based principally on phosphorous-containing monomers are expensive and therefore have not attained much commercial interest. However, polymers containing phosphonic acid are used industrially for ion exchange, adhesion, and scale inhibition. Some of these polymers form water-soluble coatings used as primers in metal protection, photolithographic plates, dental applications, and in control of bone growths. Polymers containing phosphonic acid have a hydrolytically stable P–C bond, where phosphorus is a part of a pendant side chain, as in polyvinyl(phosphonic acid) (1) (Chart I).

Phosphonic acid derivatives are generally alkyl or aryl compounds giving rise to the organic acids and esters, of which a very large number have been made and characterized, including unsaturated monomers (13, 14). Phosphonic acids are normally crystalline solids with melting points in the 100–200 °C range. The aliphatic esters are oils, but aromatic esters are crystalline solids with lower melting points than the corresponding acids.

In this review a short introduction to the synthesis of polymers containing phosphonic acid and bis(phosphonic acid) is given. Special attention is focused on materials suitable for applications based on the ability to bind species either through an ion-exchange or a coordination mechanism. Two pathways to this class of polymers are presented: polymerization of monomers containing phosphonic acid and methods based on chemical modification of a preformed polymeric network.

Synthetic Approaches to Polymers Containing Acidic Groups

Several factors are important in selecting the type of polymer to be used to support a given functional group. These factors include the ease of preparation of potentially suitable polymers with the appropriate functional groups and the ease with which the polymer can be obtained in a good physical form. Depending on the mode of operation chosen, one particular polymer structure and morphology might be more effective than another. Also, apart from those groups that are intended to react, the polymer backbone should be chemically inert under the conditions of use. This characteristic is especially important if the intention is to recycle the reactive polymer, because after many reaction cycles even quite small and otherwise unimportant side reactions can seriously impair the properties of the polymer.

Therefore, it is not surprising that polystyrene cross-linked with divinylbenzene has found the most widespread use. This polymer offers several advantages:

1. A number of well-documented functionalization routes are available.
2. The degree of cross-linking can be controlled, so polymers having a high swelling or ones that do not substantially swell can be obtained.
3. Polystyrene is compatible with a number of organic solvents used in organic synthesis.
4. The polymer is not degradable by most chemical reagents under routine conditions.

A very large number of polymers containing acidic groups have been synthesized (Chart II). Polymers containing carboxylic (2), sulfonic (3, 4), phosphonic (5, 6) and phosphonic acid (7) groups are among the most important of these. The acidic strength of some of these groups on polystyrene are summarized in Table I (15). The pK values for the bis(phosphonic acid) group are given as obtained for ethenylidenebis(phosphonic acid) (15) (16).

The chemical modification route is particularly attractive with commercially available polystyrene-based resins, because their aromatic rings can be modified readily by electrophilic aromatic substitution or

Phosphorus acid	$(HO)_3P$
Phosphonous acid	$HP(OH)_2$
Phosphinous acid	$H_2P(OH)$
Phosphoric acid	$(HO)_3PO$
Phosphonic acid	$HPO(OH)_2$
Phosphinic acid	$H_2PO(OH)$
Methylene bis(phosphonic acid)	$H_2C(PO(OH)_2)_2$
Poly(vinylphosphonic acid)	

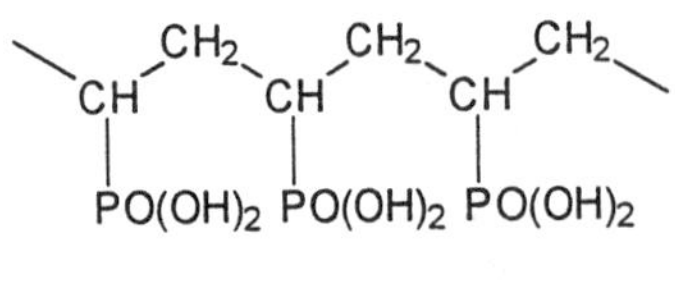

1

Chart I. Chemical structures and nomenclature of lower oxyacids of phosphorus and polyvinyl(phosphonic acid).

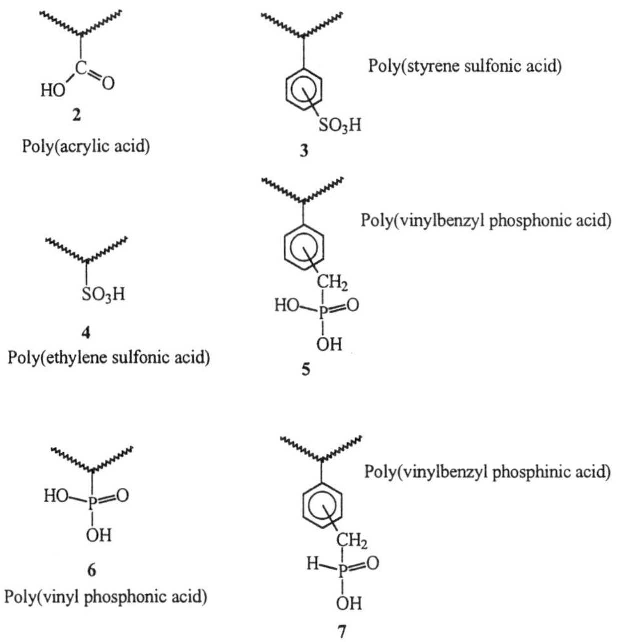

Chart II. Structures of typical polymers containing acidic groups.

Table I. pK_a Values for Acidic Groups Attached to Polystyrene (*see* Chart II)

Hydrogen Form	pK_a1 Range	pK_a2 Range
–SO₃H	<1	—
–PO₃H₂	2–3	7–8
–CH–COOH	5.7–7.1	—
–C(CH₃)–COOH	6.6–7.95	—
–CH₂C(PO(OH)₂)₂	1.4	2.1

through other simple reactions. Although this approach has been very successful in its application for the preparation of resins having a great variety of structures, it is a route that should be considered with great care. Sulfonation is an excellent example of introducing acidic groups into a polymer through chemical modification and is today the most common method of preparing polymers containing sulfonic acid groups (Scheme I, top). Sulfonation reagents include sulfur trioxide or its complexes, sulfuric or chlorosulfonic acid and acetyl sulfate. Aromatic polymers may be sulfonated by treating the polymer with 100% H₂SO₄ in the presence of Ag₂SO₄ catalyst or by a number of other methods (*17*). The degree of sulfonation is close to unity, and yields are above 90% (*18*).

Poly(benzylsulfonic acid) (**9**), which is less sensitive to desulfonation, can be made under mild conditions by treating poly(vinylbenzyl chloride) (**8**) with NaSO₃ (Scheme I, bottom) (*19, 20*).

The preparation of useful functional polymers from functional monomers has received relatively less sustained attention than the chemical modification

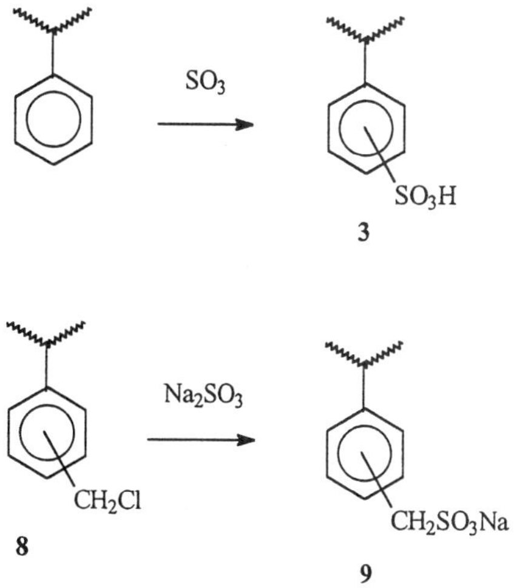

Scheme I. Sulfonation of polystyrene (top) and poly(VBC) (bottom).

approach. A large number of functional monomers are available. However, many researchers are reluctant to venture into the preparation of polymers with suitable physical and chemical properties. The preparation of 1–2% cross-linked gel-type and porous polymers by suspension polymerization is a relatively easy task. Variables such as the nature and amount of pore-forming diluent, the stirring rate, reactor design, and the relative proportions and properties of the monomers can be adjusted to control characteristics of the final resins. Sulfonated polymers may be obtained by polymerization of ethylenesulfonic acid (*21*). However, most of the research in this field is directed toward the polymerization of styrenesulfonic acid (*22*).

Synthesis of Monomers Containing Phosphonic or Bis(phosphonic acid)

Monomers Containing Phosphonic Acid

Although very few monomers containing phosphonic acid are readily available, a large number of vinyl and diene phosphonate monomers have been described, including the commercially available monomers bis(2-chloroethyl)vinyl phosphonate, diethylvinyl phosphonate, and vinylphosphonic acid (*23*). A number of phosphonate-containing monomers are also available through laboratory chemicals suppliers.

Phosphonic acids can be prepared by oxidation of the corresponding phosphine or phosphorous acid (*24*). However, such procedures are not in widespread use. Most synthetic routs include the Arbuzov or the Arbuzov–Michaelis reaction (*25*). The synthesis of diethylvinylbenzyl phosphonate (**11**) from vinylbenzyl chloride (VBC) (**10**) and triethyl phosphite is mentioned in a patent (*26*). However, the reaction is not as simple as inferred from the patent literature. For example, full conversion of chloromethylated polystyrene required reflux of the polymer solution at ca. 150 °C for 8 h (27). These conditions could not be applied to VBC without polymerization. Yu and co-workers found (28) that 6-*tert*-butyl-2,4-dimethylphenol is a very effective polymerization inhibitor for the conversion of VBC to vinylbenzyl phosphonate at 93 °C (Scheme II).

Thus, reacting VBC (0.1 mol) with triethyl phosphite (0.5 mol) for 72 h gave 1 70% yield of diethylvinylbenzyl phosphonate (11). Acrylate or methacrylate phosphonates (i.e., acroyloxy-diethyl-methyl phosphonate (13)) have been claimed in patents but have not been commercialized (Scheme II) (29, 30).

Monomers Containing Bis(phosphonic acid)

Ethenylidenebis(phosphonic acid) (15) is prepared via thermal dehydration of tetrasodium (1-hydroxyethylidene)bis(phosphonate) (14) at high temperatures (Scheme III) (31, 32). Disadvantages of this procedure include the need for a precise purification process. Tetraalkyl esters of ethenylidenebis(phosphonic acid) can be prepared by the reaction of free acid (15) with trialkyl(*ortho*-formate)s (33). Thermogravimetric and differential thermal analyses were used to establish the optimum conditions for the intramolecular dehydration of the salts (14) (34). The tested salts underwent water elimination at different temperatures. Consequently, they are of different values for the synthesis of 15. The dibarium salt was found to outperform the sodium salt.

More recently, Degenhardt and Burdshall reported (35) a convenient two-step, single-pot method for

preparing tetraalkyl ethenylidenebis(phosphonates). This method is a modification of a procedure originally developed by Pudovic and co-workers to prepare α-phosphonoacrylate esters (17) from the corresponding phosphonoacetates (16) (36, 37). Reaction of the phosphonoacetate in methanol with paraformaldehyde and a catalytic amount of piperidine followed by an acid-catalyzed elimination step afforded the desired esters (17) in good overall yield (Scheme IIIb).

Applying this method for the synthesis of ethenylidenebis(phosphonate) (20) presented some practical problems due to slow and incomplete reactions and difficult product workup. Reaction of 18 with paraformaldehyde could be conducted efficiently by using diethylamine instead of piperidine in a 1:1:5 molar ratio of 18:diethylamine:paraformaldehyde (the initial concentrations of 18 were about 0.35 M). Under these conditions a methosylated intermediate (19) is formed, which on refluxing in toluene for 14 h with a catalytic amount of *p*-toluenesulfonic acid affords the target ester 20, (Scheme IIIc). This ester can be purified with bromotrimethylsilane followed by hydrolysis (38, 39). A considerable difference in the rates of reaction of the starting esters 18 with paraformaldehyde was observed depending on the steric bulk of the alkyl group, as can be seen from Table II.

Another interesting route to tetraisopropyl ethenylidenebisphosphonate (20c) was reported by a Russian team (40). The reaction of tetraisopropyl methylene bisphosphonate with $(Et_2N)_2CH_2$ and a small amount of hydroquinone at 170–180 °C gave 58% of 20c.

We reported (41) a convenient synthetic route to 1-(vinylphenyl)propane-2,2-bis(phosphonic acid) and its alkyl esters from VBC (10) and tetraalkyl methyl-

a)

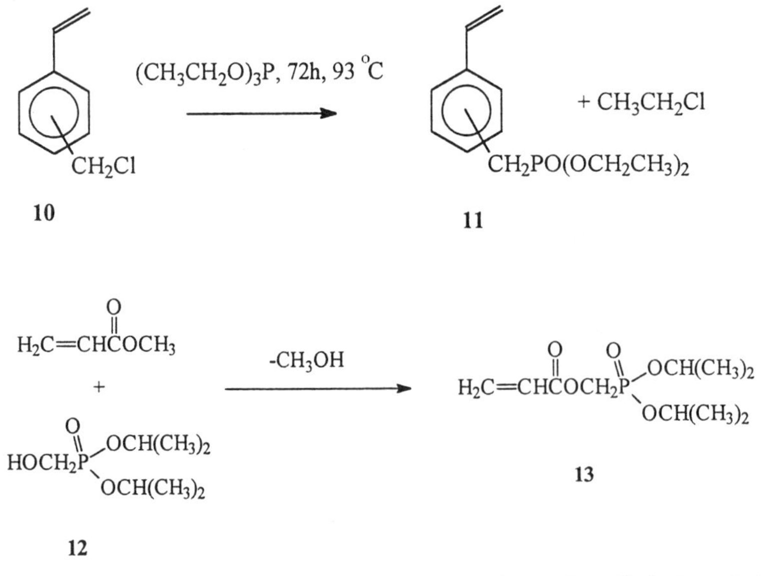

Scheme II. Synthesis of vinylbenzylphosphonate diethyl ester (top) and acroyloxydiethyl-methylphosphonate (bottom).

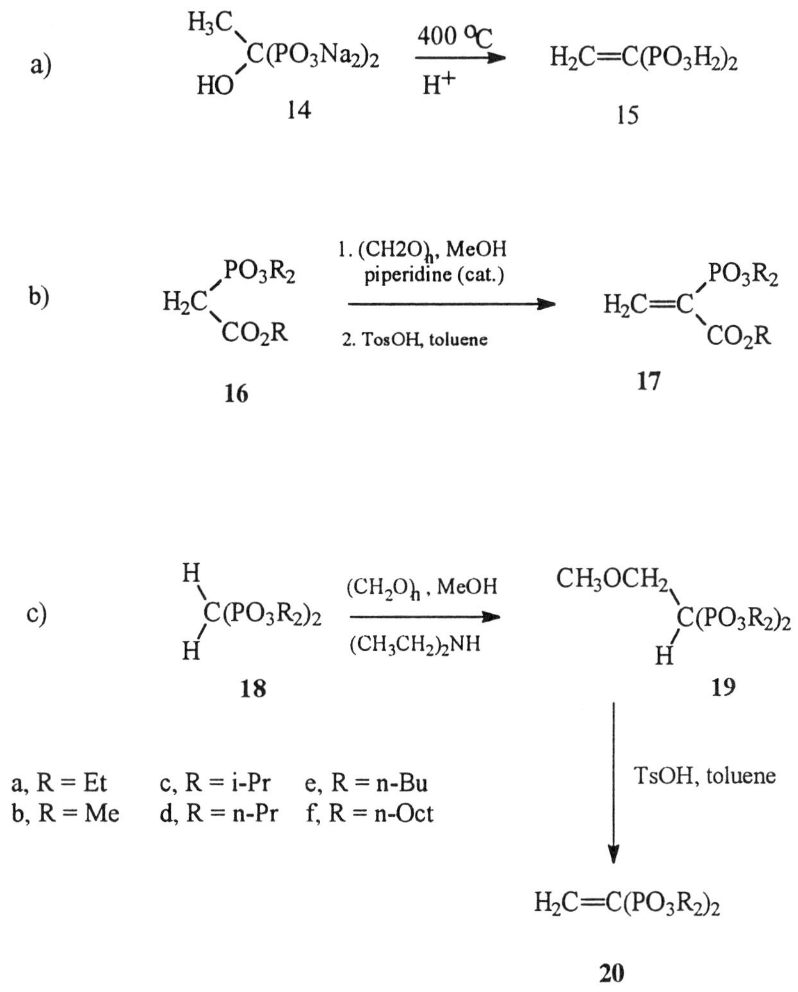

a)

$$H_3C \atop HO \!\!\diagdown\!\! C(PO_3Na_2)_2 \quad \xrightarrow[\text{H}^+]{400\ ^\circ C} \quad H_2C{=}C(PO_3H_2)_2$$

14   15

b)

$$H_2C \atop \diagdown \!\! C \!\!\diagup\!\! {PO_3R_2 \atop CO_2R} \quad \xrightarrow[\text{2. TosOH, toluene}]{\text{1. (CH2O)}_n,\ \text{MeOH} \atop \text{piperidine (cat.)}} \quad H_2C{=}C \!\!\diagup\!\! {PO_3R_2 \atop CO_2R}$$

16   17

c)

$$H \atop H \!\!\diagdown\!\! C(PO_3R_2)_2 \quad \xrightarrow[\text{(CH}_3\text{CH}_2)_2\text{NH}]{\text{(CH}_2\text{O)}_n,\ \text{MeOH}} \quad CH_3OCH_2 \atop H \!\!\diagdown\!\! C(PO_3R_2)_2$$

18   19

a, R = Et   c, R = i-Pr   e, R = n-Bu
b, R = Me   d, R = n-Pr   f, R = n-Oct

|
TsOH, toluene
↓

$$H_2C{=}C(PO_3R_2)_2$$

20

Scheme III. Synthesis of monomers containing bis(phosphonic acid) (a); thermal dehydration of tetrasodium (1-hydroxy-ethylidene)bis(phosphonate) (b); preparation of phosphonoacrylate esters (c); synthesis of tetraalkyl ethenylidenebis-(phosphonates) (**35**).

Table II. Synthesis of Some Tetraalkyl Ethenylidenebisphosphonates

Bisphosphonate Ester	No.	Reaction Time (h)	Yield (%)
Tetramethyl ethenylidenebisphosphonate	**18b**	2	72
Tetraethyl ethenylidenebisphosphonate	**18a**	24	79
Tetraisopropyl ethenylidenebisphosphonate	**18c**	117	55
Tetra-*n*-propyl ethenylidenebisphosphonate	**18d**	46	72
Tetra-*n*-butyl ethenylidenebisphosphonate	**18e**	46	66
Tetra-*n*-octyl ethenylidenebisphosphonate	**18f**	50	64

enebis(phosphonate) (Scheme IV). Tetraethyl methyl-enebis(phosphonate) (**18a**) was prepared in 50% yield from sodium diethylphosphite and dichloromethane in a one-pot synthesis as reported previously (*42*). By using **18a** as starting material, a small amount of divinyl bis(phosphonate) (**24**) is also formed. Interestingly, this monomer can be used as a convenient cross-linker in the monomer for the subsequent polymerization if a cross-linked polymer is to be synthesized. However, we found that the formation of this side

product can be avoided by using tetraethyl ethane-1,1-bis(phosphonate) for the nucleophilic substitution reaction with VBC. The methyl group blocks the formation of disubstituted bisphosphonate. The reaction was performed by adding **21** (0.1 mol) to sodium hydride (0.11 mol) in dry toluene under nitrogen. When the release of hydrogen ceased, the stirring was continued for an additional hour at room temperature. Then, VBC (0.12 mols; 10:30, meta:para) was added, and the resulting mixture was stirred until the

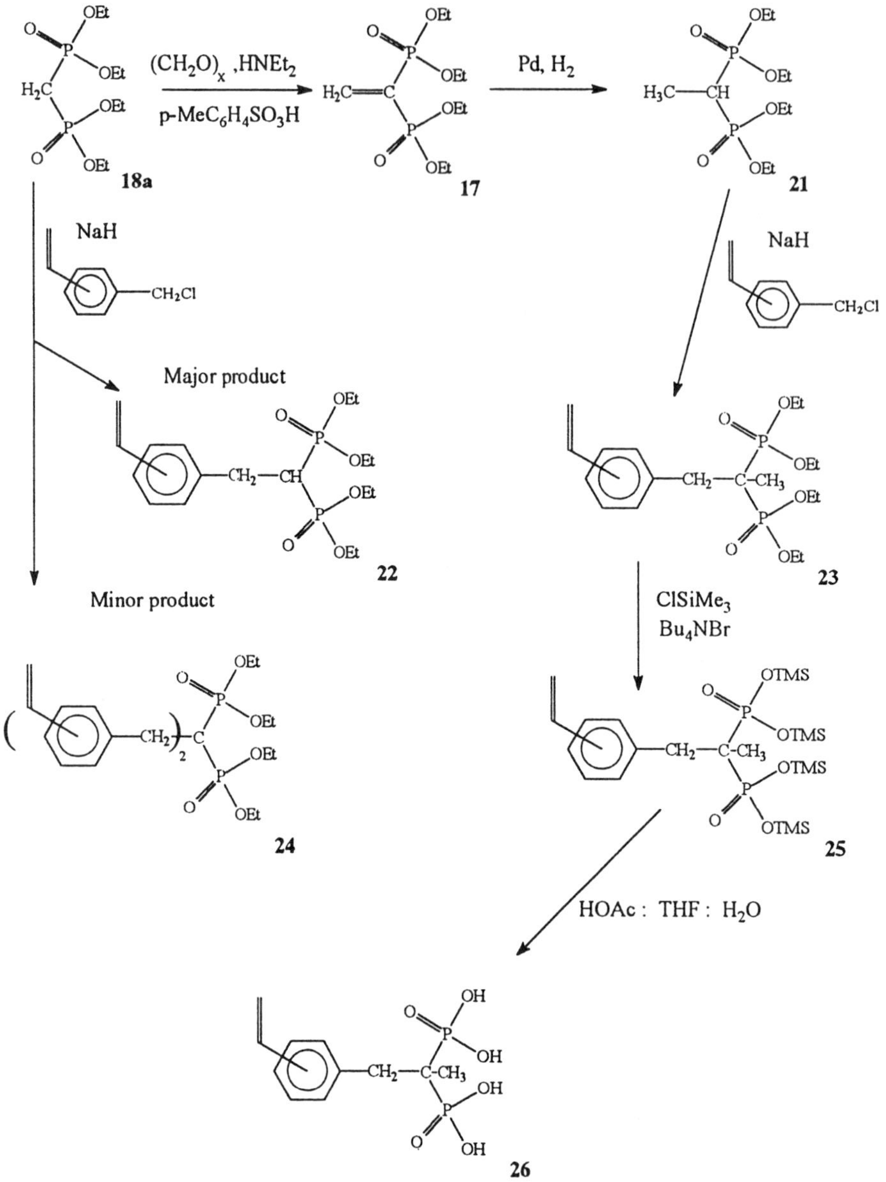

Scheme IV. Synthesis of some 1-(vinylphenyl)ethane-2,2-bis (phosphonic acid) derivatives.

reaction was complete (2 weeks at ambient temperature). The reaction mixture was concentrated under reduced pressure, and the residue was dissolved in diethyl ether, washed three times with water, dried over sodium sulfate, and concentrated again to give an 81% yield of the crude product tetraethyl 1-(vinylphenyl)propane-2,2-bisphosphonate (23). The bisphosphonate monomer may be converted to free acidic by reaction with bromotrimethylsilane, hydrolysis, and precipitation in ethyl acetate.

A U.S. patent by Omura et al. describes (43) the synthesis of N-methacryloyl-12-aminododecanoyl ethanebis(phosphonic acid) (28), which exhibited very good adhesion toward the hard tissues of the living body and also toward metal surfaces. The monomer is synthesized by a multistep synthesis from methacryl-

oyl chloride, aminododecanoic acid, and 1-hydroxy-ethane-1,1-diphosphonic acid (27) (Scheme V). The synthesis of methacryloyloxyethane-1,1-diphosphonic acid (29) from 1-hydroxyethane-1,1-diphosphonic acid (27) is described in a Japanese patent for use in dental adhesives (Scheme V) (44).

Preparation of Polymers Containing Phosphonic Acid or Bis(phosphonic acid)

Polymerization of Monomers Containing Phosphonic Acid

The free-radical homopolymerization of vinyl phosphonic acid was patented over 30 years ago by Hoechst (45). Since then, various papers and patents

a $\quad H_2C{=}\overset{\underset{\displaystyle CH_3}{|}}{C}{-}COCl \;+\; H_2N{-}(CH_2)_{11}{-}COOH \;\longrightarrow\; H_2C{=}\overset{\underset{\displaystyle CH_3}{|}}{C}{-}CONH{-}(CH_2)_{11}{-}COOH$

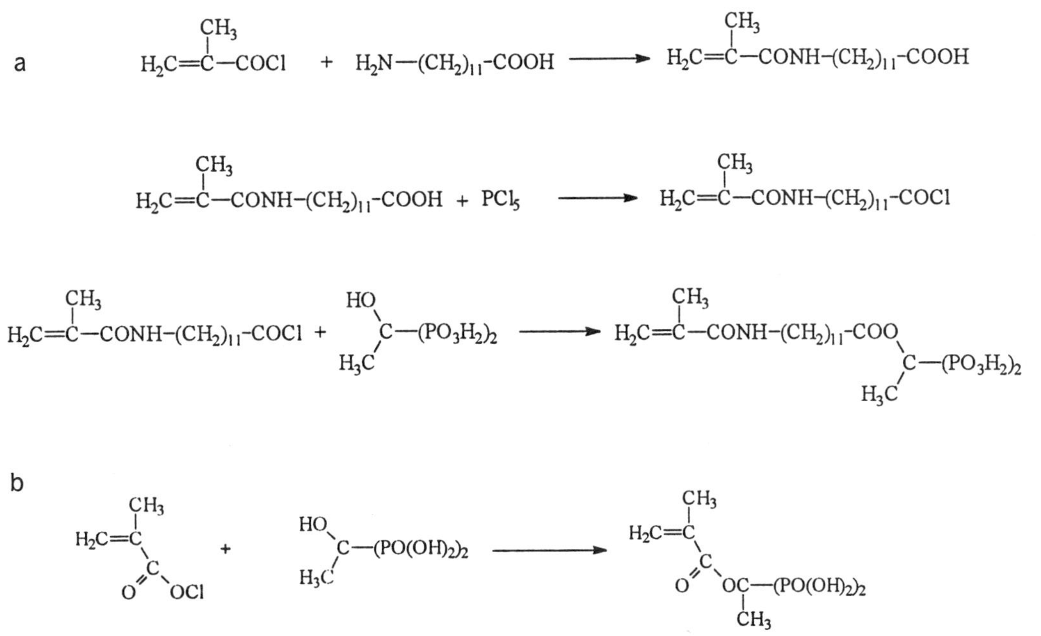

$H_2C{=}\overset{\underset{\displaystyle CH_3}{|}}{C}{-}CONH{-}(CH_2)_{11}{-}COOH \;+\; PCl_5 \;\longrightarrow\; H_2C{=}\overset{\underset{\displaystyle CH_3}{|}}{C}{-}CONH{-}(CH_2)_{11}{-}COCl$

$H_2C{=}\overset{\underset{\displaystyle CH_3}{|}}{C}{-}CONH{-}(CH_2)_{11}{-}COCl \;+\; \underset{\displaystyle H_3C}{\overset{\displaystyle HO}{\diagdown}}C{-}(PO_3H_2)_2 \;\longrightarrow\; H_2C{=}\overset{\underset{\displaystyle CH_3}{|}}{C}{-}CONH{-}(CH_2)_{11}{-}COO\underset{\displaystyle H_3C}{\diagdown}C{-}(PO_3H_2)_2$

b $H_2C{=}\underset{\displaystyle O^{\diagdown}C{\diagdown}OCl}{\overset{\displaystyle CH_3}{|}}C \;+\; \underset{\displaystyle H_3C}{\overset{\displaystyle HO}{\diagdown}}C{-}(PO(OH)_2)_2 \;\longrightarrow\; H_2C{=}\underset{\displaystyle O^{\diagdown}C{\diagdown}OC{-}(PO(OH)_2)_2 \;\;\; CH_3}{\overset{\displaystyle CH_3}{|}}C$

Scheme V. Synthesis of (a) *N*-methacryloyl-12-aminododecanoyl ethanebisphosphonic acid and (b) methacryloyloxyethane-1,1-diphosphonic acid.

have been published on the homo- and copolymerization of vinylphosphonic acid and related monomers (*46*). The radical homopolymerization of vinylphosphonic acid in water is slow. Concentrated systems, high amounts of initiator added in portions throughout the reaction, and long reaction times are needed to obtain high yields (*47*). Vinylphosphonates homopolymerize slowly and also copolymerize poorly with styrene. They are prone to chain transfer through the alkyl groups of the monomer and the polymer (*48*). On the other hand, they copolymerize well with vinyl halides, vinyl acetate, and acrylonitrile and fairly well with acrylates and methacrylates (*49*). Typically, azo and peroxide initiators are used, and the polymers tend to be of quite low molecular weight.

The Alfrey–Price copolymerization parameters Q and e for bis(2-chloroethyl)vinylphosphonate are 0.23 and 1.73, respectively, reflecting an electron-poor double bond and low resonance stabilization of the free radical (*50*). Dialkyl vinylphosphonates are polymerized to high molecular weight polymers by sodium naphthalide (*51, 52*) or trialkylaluminium (*53*). Ziegler–Natta catalysts are claimed to promote copolymerization of vinylphosphonates with propylene, although relatively large amounts of catalyst are required (*54*).

Polymerization of Monomers Containing Bis(phosphonic acid)

Tetraalkyl ethenylidenebis(phosphonate) can be polymerized using azo or peroxy initiators. However, the polymerization is slow, and generally only low molec-

ular weight products are obtained. 1-Vinylphenylpropane-2,2-bis(phosphonic acid) (**26**) is insoluble in aromatic solvents but can easily be copolymerized with styrene in solvents such as lower alcohols or tetrahydrofuran (THF). However, copolymerization of **26** with styrene and divinylbenzene (DVB) to obtain a cross-linked resin is complicated by the insolubility of the bis(phosphonic acid) monomer in the styrene monomers. The monomer **26** could be dispersed in a water–styrene–DVB emulsion. Sodium bis(2-ethylhexyl)sulfosuccinate (AOT), an anionic surfactant with two branched hydrocarbon tails, is especially suitable for the preparation of this type of emulsions. When AOT is dissolved in hydrocarbon solvents, it is capable of emulsifying up to 50 molecules of water per surfactant molecule. The hydrophilic–lipophilic properties of AOT are well balanced, and it forms an emulsion as a function of ionic strength and temperature without the necessity of blending with other surfactants (*55*).

Solid porous materials can be obtained by polymerization of an AOT-stabilized water-in-oil (W/O) emulsion of water in styrene–DVB (or another suitable divinyl monomer). A scanning electron micrograph of the polymer obtained is shown in Figure 1. In this system, it is possible to disperse up to 15% 1-vinylphenylpropane-2,2-bis(phosphonic acid) based on total monomer content in a 5% water, 15% AOT emulsion. These values are remarkably high compared with pure water, which can solubilize only up to 10% of the monomer. A solution of 15% AOT in styrene has no emulsifying effect at all. This behavior is due to the ability of the bis(phosphonic acid) monomer to act as a

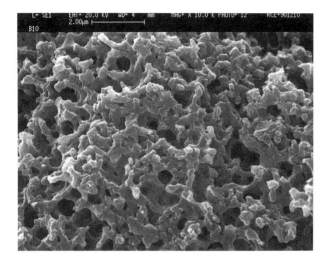

Figure 1. Scanning electron micrograph of a polymer obtained by polymerization of a water/(AOT + styrene) emulsion. The emulsion consisted of 86.2% styrene monomers (containing 20 mol% bis(vinylphenylethane)), 9.6% AOT, and 4.2% water.

co-surfactant in the emulsion system. On increasing the content of 1-vinylphenylpropane-2,2-bis(phosphonic acid), the W/O micelles in the system transform into somewhat larger, but still narrow sized, aggregates of a Gaussian distribution. Polymerization of **26** in the water/(AOT + styrene + DVB) system can be effected by azo-isobutyronitril (AIBN) and heating. After polymerization, porous solids are obtained. The W/O micelles in the system coalesce to some degree during polymerization to form a loosely interconnected structure to which the functional co-surfactant is anchored (Figure 2).

The amount of accessible bis(phosphonic acid) groups on the polymer was determined by binding Cu^{2+} ions from an aqueous solution. Because the functional groups within the heavily cross-linked polystyrene matrix are not accessible to the water phase, only the groups that reside on the pore surface will be able to bind Cu^{2+} ions. The ground (0.5–0.125 mm) polymer was placed in a column, treated with 1 M KOH, and washed with water to pH 7. After this process, a 10-fold excess $CuCl_2$ solution (200 mL) was circulated through the column overnight. The polymer was washed thoroughly with water, after which Cu^{2+} was eluted from the polymer with 1 M HCl. Direct current plasma analysis of Cu^{2+} in the acidic eluents, together with elemental analysis of the polymer, confirmed that a very high degree of the acid groups are present on surfaces available to the aqueous Cu^{2+} solution. The composition of the emulsions employed and the degree of accessible bis(phosphonic acid) groups in the polymers are presented in Table III. According to

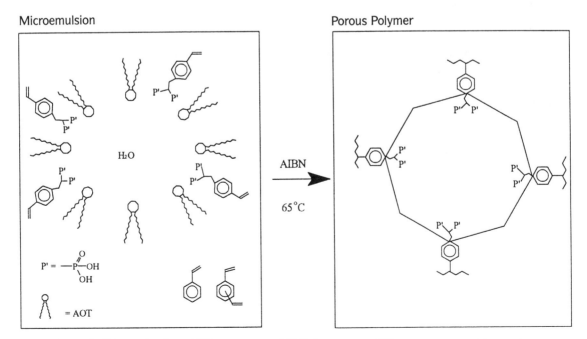

Figure 2. Polymerization of **26**, styrene, and DVB in a water–AOT–monomer emulsion system.

Table III. Experimental Data for Preparation of Porous Cross-Linked Polystyrene Possessing Surface-Supported Bis(phosphonic acid) Groups by Complete Emulsion Polymerization System

	Composition of Emulsion				*Phosphonic Acid in Polymer*		
Emulsion	*AOT (%)*	*Water (%)*	*Styrene/20% DVB (%)*	*(26)[a] (%)*	*Theory (mmol/g)*	*P Analysis (mmol/g)*	*$Cu_{complexed}$ (mmol/g)*
A1	10.5	4.1	70.2	15.2	0.58	0.48	0.36
A2	11.4	4.4	76.4	7.7	0.3	0.29	0.17
A3	10.2	7.8	67.5	14.6	0.58	0.48	0.3
A4	11	8.4	73.1	7.4	0.3	0.27	0.18

[a]1-Vinylphenylpropane-2,2-bis(phosphonic acid).

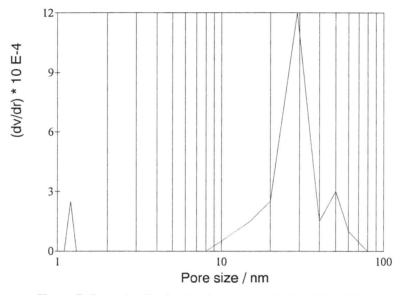

Figure 3. Pore-size distribution for polymer A1 (*see* Table III).

Brunauer–Emmett–Teller (BET) adsorption analysis, polymer A1 had a surface of 16.7 m²/g. The pore volume distribution calculated from the adsorption–desorption isotherm shows quite a narrow distribution of approximately 250-Å pores (which is approximately the same size as the water pools) (Figure 3).

The formation of porous resins also can be achieved by another procedure. 1-Vinylphenyl-propane-2,2-bis(phosphonic acid) (26) is insoluble in styrene but soluble in most short chain alcohols. Many such alcohols may be used as pore-forming diluents in the synthesis of porous polystyrene used, for example, as separation media. A suitable diluent should be a solvent for all the monomers but a nonsolvent for the formed polymer. The polymerization of styrene and cross-linker in the presence of a pore-forming diluent proceeds through a rapid phase separation at the point where the diluent no longer can solubilize the formed polymer. If the bis(phosphonic acid) monomer is added to this diluent we would expect to see a concentration of bis(phosphonic acid) groups to the alcohol-rich pores formed during the phase separation

(Figure 4). This result is confirmed by quantitative assessment of the metal complexation (by elemental analysis of the polymer and analysis of the Cu²⁺ solutions) presented in Table IV.

Although this second method gave lower accessibilities (43–65%) than the emulsion polymerization, it is likely that the bis(phosphonic acid) groups indeed are attracted toward the polymer–alcohol interface formed during polymerization. Higher amounts of diluent make it possible to solubilize more bis(phosphonic acid) monomer in the system, but unfortunately this increase also lowers the copolymer yield considerably. This result is due to the formation of soluble 1-vinylphenylpropane-2,2-bis(phosphonic acid) homopolymer or oligomer in the alcohol-filled pores after phase separation.

Alexandratos and co-workers prepared (56) styrene-based resin containing bis(phosphonic acid), trademarked Diphonix, by copolymerization of tetra-alkyl ethenylidenebis(phosphonate) with styrene, divinylbenzene, and acrylate comonomers. Hydrolysis of the ester to the diphosphonic acid yields a ligand

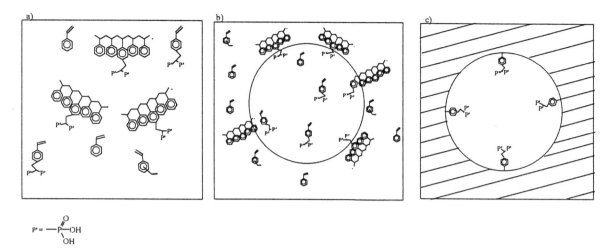

Figure 4. Schematic illustration of the polymerization of styrene, DVB, and **26** in the presence of a pore-forming diluent: a, early stage of polymerization; b, functional monomer and bis(phosphonic acid) rich segments concentrated in the alcohol-filled pores after phase separation has occurred; c, formation of a porous structure with the acidic groups anchored to the pore surface.

Table IV. Experimental Data for Preparation of Porous Cross-Linked Polystyrene Possessing Surface-Supported Bis(phosphonic acid) Groups via Polymerization in Presence of Diluent

Nr	Diluent	Styrene/20% DVB (%)	P_{theory} (%)	$P_{analysis}$ (mmol/g)	Access (%)
B1	Propanol, 40%	0.41	0.19	0.59	43
B2	Butanol, 40%	0.41	0.19	0.55	40
B3	Hexanol, 40%	0.41	0.19	0.51	34
B4	Butanol, 50%	0.25	0.25	0.33	65

Notes: P_{theory} is 2[1-vinylphenylpropane-2,2-bis(phosphonic acid) (mmol)]/[monomers (g)]. $P_{analysis}$ is amount of phosphorus from elemental analysis. Access is $Cu_{complexed}/P_{analysis}$.

that is highly selective toward actinide and heavy metal ions. The rate of complexation was greatly enhanced when the polystyrene part of the polymer was sulfonated. Unfortunately, the radical copolymerization of tetraalkyl ethenylidenebis(phosphonate) is slow and is prone to chain transfer to the alkyl groups. Thus, high amounts of cross-linker and initiator are needed to obtain reasonable yields: 30 mol% divinylbenzene and at least 5% benzoyl peroxide gave a 92% incorporation of the bis(phosphonic acid) monomer and a 70% copolymer yield.

Chemical Modification of Polymers

An alternative method to polymers containing phosphonic acid is to introduce phosphonic acid groups into a preformed polymer. Several routes are available. Phosphonic acid groups reportedly were introduced into polyolefins by oxidative phosphonylation followed by hydrolysis or alcoholysis to yield the acid or

ester (Scheme VI) (*57*). The Arbuzov reaction can be used to introduce phosphonic acid groups on halogenated polymers. The Cl atom in poly(vinyl chloride) cannot be replaced (i.e., producing polymers with phosphorus attached directly to the carbon chain) due to rapid dehydrochlorination of the polymer.

A European patent in 1988 describes the introduction of bis(phosphonate) groups into vinylphosphonate polymers by a Michael addition (Scheme VII). The reaction is performed by treating the polymer with a strong base at a low temperature and reacting the obtained anion with ethenylidenebis(phosphonate) (*58*). A typical example involves the reaction of 0.3 M *n*-butyllithium with 96.8 g (0.59 mol) poly-(diethyl vinylphosphonate) in 2000 mL THF at –78 °C for 1 h. Nest, 0.3 M of **20a**, was added and the temperature was maintained at –78 °C for 1 h. The solution was allowed to warm to room temperature, 75 mL of water was added, and the solvents were evaporated. The residue was dissolved in 1500 mL HCl concentrated HCl, refluxed for 4 h, and concentrated to provide the hydrolyzed product. The crude product was chromatographed on Sephadex G-25 (5000 molecular weight cutoff) to provide 41 g of a polyvinylphosphonic acid polymer containing bisphosphonic acid. The molecular weight of the polymer was 9500. P[31] NMR analysis gave a phosphonic acid to bis(phosphonic acid) ratio of approximately 2.5.

Ethenylidenebis(phosphonate)s and their acids are known to undergo facile Michael-type addition reactions with N, P, or S nucleophiles (but not with O or C nucleophiles) to give C-substituted methylenebis-(phosphonates). These types of reactions have been performed with low molecular weight nucleophiles (*59, 60*) but not yet with polymers.

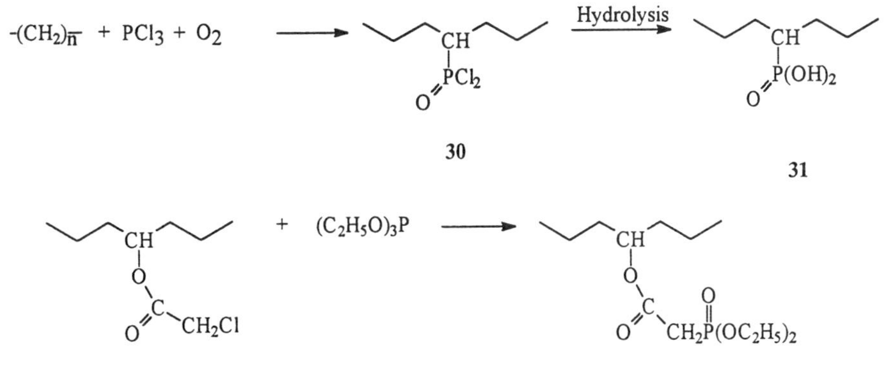

Scheme VI. Phosphonylation of polyethylene and halogenated polymers.

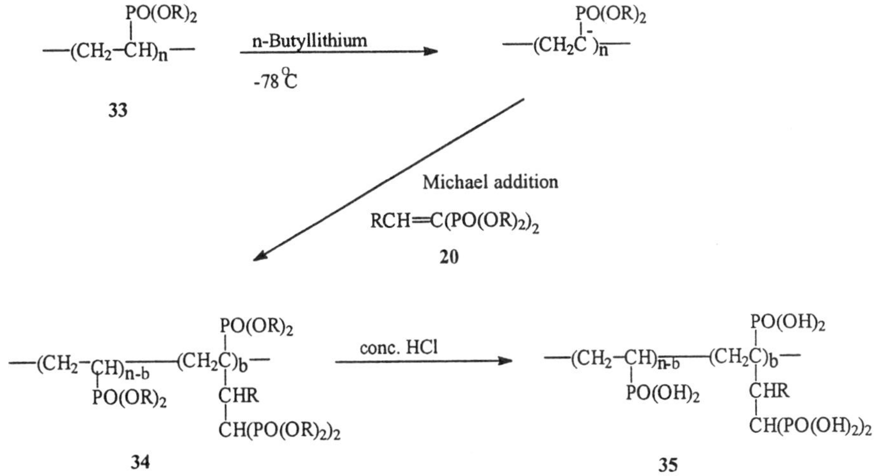

Scheme VII. Introduction of bisphosphonates into vinylphosphonate polymers using Michael addition.

Acrylic polymers containing bis(phosphonic acid) can be synthesized by the reaction of phosphorous acid or a precursor to phosphorous acid such as PCl$_3$ with a carboxylic polymer in a polar organic solvent, such as THF or sulfolane (Scheme VIII) (61). The carboxylic polymer can be poly(acrylic acid) (2), a copolymer with methylester units, or derivatives of maleic anhydride copolymers. A useful carboxylic copolymer may contain alkyl acrylate residues to render the polymer soluble in the solvent used in the reaction. A typical example involves reaction of 125 g polyacrylic acid (2) (molecular weight, 2100), dissolved in 25.9 g water and 300 g sulfolane, with 1.44 M PCl$_3$ at 45 °C for 1 h. The solution was heated to 100 °C for 2 h before allowing the reaction mixture to cool. The polymer was precipitated in CHCl$_3$ and collected by vacuum filtration. After this procedure, the polymer was redissolved in water, and the solution was refluxed for 18 h. Precipitation gave 72 g of the final product. ^{31}P NMR analysis indicated that 43 mol% of the phosphorus in the product was present as hydroxybisphosphonic acid. The total phosphorus content was 12.28%. This result gave a carboxylic to bisphosphonic acid molar ratio of about 4.

A process for the synthesis of a bis(phosphonic acid) polymer from a nitrile-containing polymer is described in a U.S. patent by Chai et al. (62). Polystyrene is one of the polymers most widely used as a functional polymer precursor (63, 64). Phosphonic acid and phosphonate ester resin can be synthesized by treating the chloromethylated polymer (8) with PCl$_3$ in the presence of AlCl$_3$. Hydrolysis or alcoholysis yields the free acidic form (38a), the ester form (38b), or a mixture thereof (Scheme VIIIb) (65). If the pure diester resin is to be prepared, mild reaction conditions and anhydrous ethanol are used. Alexandratos et al. also reported (65) that a rapid addition of ethanol leads to the buildup of heat from the highly exothermic reaction, which allows some of the liberated HCl to react with the ethoxy groups to form ethyl chloride and monoprotic or diprotic phosphorus ligands. Controlled heating to 80 °C leads to the formation of a resin that is almost entirely the monoethyl ester (38c) of poly(vinylbenzyl phosphonic acid).

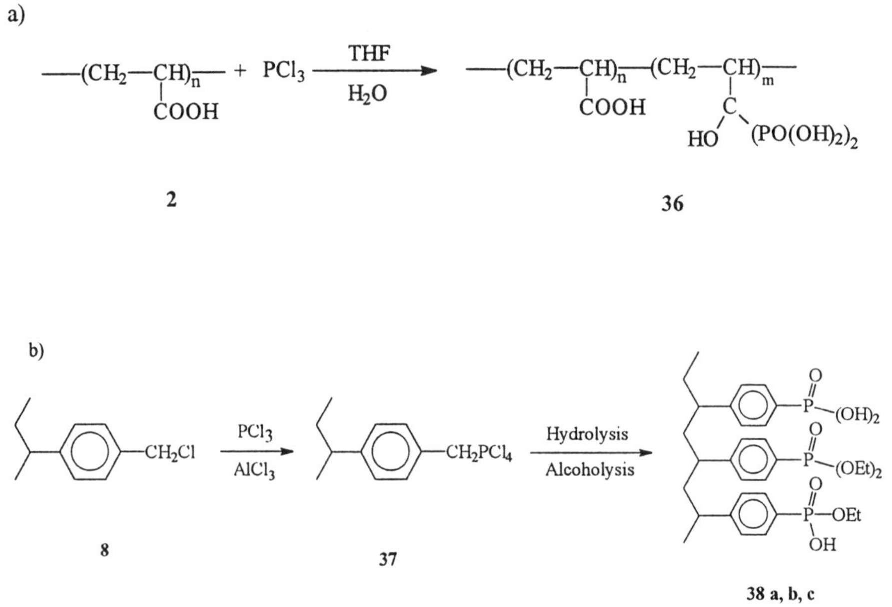

a)

$$—(CH_2—CH)_n— + PCl_3 \xrightarrow[\text{H}_2\text{O}]{\text{THF}} —(CH_2—CH)_n—(CH_2—CH)_m—$$

2   36

b)

8   37

38 a, b, c

Scheme VIII. Introduction of bis(phosphonic acid) groups into acrylic acid polymers (a), and synthesis of a mixed phosphonic acid, phosphonate polystyrene resin (b).

Grafting on Preformed Polymers

Another useful method for preparing functional polymers is grafting. Especially, radiation grafting offers promising new opportunities (66). This method involves taking a polymer, with appropriate morphology and physical properties, and generating free radicals on the polymer chain by irradiation. The free radicals can either combine to give cross-links (e.g., polyethylene) or cause chain scission (e.g., polypropylene). In the presence of vinyl monomers, on the other hand, the free radicals can initiate graft copolymerization, producing modified polymers that may exhibit many useful properties. Attributes of such graft copolymers include the following.

1. Side chains are not cross-linked.
2. Chemically inert and mechanically rigid preformed polymers may be chosen.
3. High extents of grafting are possible.
4. High accessibility may be achieved if the grafted polymer is compatible with the solvent in use.
5. Different polymeric materials, such as films, fibers, membranes, or beads may be grafted offering new materials suitable for, among other things, flow processes (67).

In our laboratory, we recently studied the synthesis of poly(ethylene-g-VBC) (39) copolymer by radiation-induced grafting. In contrast to most earlier work, which focused on the use of low dose rate τ-rays from ^{60}C sources, we have preferred to use high energy electrons from accelerators with high dose rates. These accelerators are suitable for on-line applications, which makes radiation-based chemical processes more attractive commercially. Thus, 200-μm thick polyethylene films were irradiated with 100 kGy, and the films were immediately immersed in a solution of VBC, which was continuously purged with nitrogen to remove oxygen (Figure 5). The degree of grafting, determined by gravimetric and elemental analyses of chlorine, was followed as a function of reaction time (Figure 5). The degree of grafting reached in 30 h by using pure VBC was 270% (based on the weight of the initial polyethylene). This result corresponds to a capacity of 4.1 mmol Cl/g, compared with 6.5 mmol Cl/g polymer for poly(VBC) (8). The poly(ethylene-g-VBC) (39) film swollen in toluene was used as a starting material for the nucleophilic substitution reaction

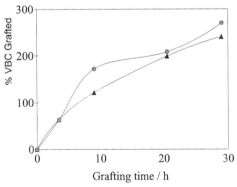

Figure 5. Degree of grafting of VBC on polyethylene as a function of time: radiation dose, 100 kGy; ●, elemental analysis of chlorine; ▲, weight increase based on starting polyethylene.

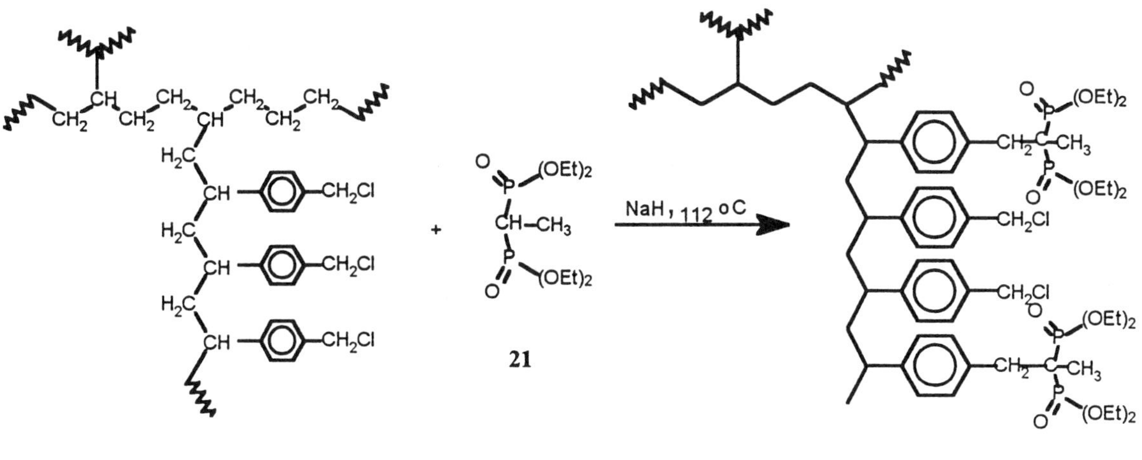

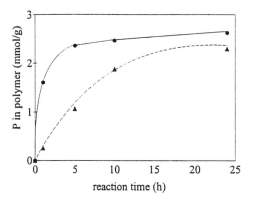

Scheme IX. Anchoring bisphosphonates onto a poly(ethylene-*g*-VBC) copolymer.

with bis(phosphonate) esters (Scheme IX) (*68*). Two different bisphosphonate derivatives, tetraisopropyl methane-bisphosphonate sodium salt and tetraethyl ethane-bisphosphonate sodium salt, were used for this reaction.

The functionalization was followed by elemental analysis of phosphorus in the film. The reaction with tetraethyl ethanebis(phosphonate) (**21**) was faster than that of tetraisopropylmethanebis(phosphonate) (**18**) and produced higher yields (Figure 6).

The reactions were also followed by studying the transversal distribution of chlorine and phosphorus in the films as a function of reaction time using scanning energy dispersive X-ray analysis. Figure 7 shows relative phosphorus and chlorine distributions over a cross-section of a 200-μm thick film (160% VBC grafted) after 1- and 10-h functionalization with tetraethyl ethanebis(phosphonate) in toluene at 112 °C. The film

was also scanned for sodium to ensure that the chlorine present in the film was in the form of unreacted poly(VBC) and not in the form of sodium chloride. As can be seen from Figure 7, mass transfer into the film was rapid, and the reaction rate was constant throughout the film. Obviously, the film did not contain any chlorine that could not be reached by the reaction solution. The obtained polyethylene-*g*-tetraethyl-1-(vinylphenyl)propane-2,2-bis(phosphonate) (**40**) could be converted to the free acid by hydrolysis with a strong mineral acid.

Applications of Polymers Containing Phosphonic Acid or Bis(phosphonic acid)

Medical Uses

Aqueous solutions of poly(vinylphosphonic acid) form polyelectrolyte cements with a wide range of cation-releasing metal oxides or ion-leachable glasses. Similar materials, polycarboxylate cements, are currently in widespread use in dentistry. Thus, the use of polyphosphonates as either a bulk restorative or adhesive substrate was reported (*69*). Poly(vinylphosphonic acid) is a much stronger acid than poly(acrylic acid) and has a greatly increased reactivity toward metal oxides used in tooth restorative materials (*70*). When used as a bulk restorative material, the faster acting of the metal oxide polyphosphonate cements seems to produce a cement that is less susceptible to hydrolysis. Furthermore, the cements had a very rapid rate of set and the potential to yield translucent materials. Unfortunately, it seems to be difficult to obtain poly(vinylphosphonic acid) cements with sufficient mechanical strength. Bulk restorative materials containing phosphonic acid groups also have been prepared by incor-

Figure 6. Phosphorus content of the grafted film as a function of time: reaction with (●) tetraethylethane-bis-phosphonate and (▲) tetraisopropyl ethane-bisphosphonate.

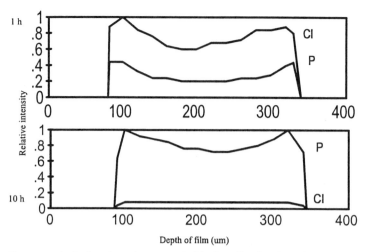

Figure 7. Energy dispersive X-ray analysis for phosphorus and chlorine distributions in a cross-section of a poly(ethylene-*g*-vinylbenzyl chloride) film (160% grafted) after bisphosphonylation for 1 h and 10 h.

porating monomers such as vinylphosphonic acid in commercial composite resin systems, and increased adhesion was observed (*71*). These cements show a number of important differences in performance in comparison with poly(acrylic acid)-based systems such as increased hydrolytic stability (*72*).

Adhesion to the hard tissues of the body is of major importance in a number of applications. In particular, because they are used in the body, the adhesion must be satisfactory under wet conditions at physiological pH. The adhesive bond between a phosphonate-containing polymer and the tooth is expected to be very strong due to the exchange of phosphonate (RPO_3^{2-}) for phosphate (PO_4^{3-}) anions on the surface of hydroxyapatite. This view is supported by studies on the adsorption of phosphonated polymers onto hydroxyapatite that show that a monolayer of practically irreversibly adsorbed phosphonated polymer is formed on the tooth enamel (*73*, *74*).

The use of a monomer containing bis(phosphonic acid) (**28**) as the active component in adhesive compositions used for coating the tooth cavity before filling is described in a U.S. patent by Omura et al. (*43*). The bis(phosphonic acid) end of the active monomer binds strongly to the tooth dentine, whereas the double bond may be cured with the filling material by using a room temperature curing agent. Because of the strong interaction between the bis(phosphonic acid) and the Ca^{2+} in the dentine, this type of adhesive exhibits higher strength than adhesives based on phosphate or phosphonic acid derivatives. The same monomer is also suggested for use as an adhesive in orthopedic applications. The compound may be employed as a bone cement for bonding and fixing a ceramic or metal artificial joint or splint to the bone. Other features of phosphonate polymers are their pronounced anticarciogenic (*75*) and antibacterial (*76*) activity.

Organic polyphosphonates also are nontoxic, and therefore they have been suggested as additives in dentifrice and mouthwash (*77, 78*).

The effect of poly(vinyl phosphonate) and 1-hydroxyethane-1,1-bis(phosphonate) on the mineralization of ectopic bone also was investigated. The results corroborate the assumption that the effects of poly(vinyl phosphonate) on hard tissues are confined to the extracellular space, whereas the effects of bis(phosphonate) derivatives on bone development are due to intracellular activities (*79*).

Ion Separation

Charged species are capable of undergoing many reactions such as ion exchange, oxidation/reduction, coordination, chelation, and precipitation (*80*). The selective binding of a targeted metal ion by a given ligand is an important objective for many applications. The pK_a values typically range from 2 to 3 and 7 to 8 for phosphonic acid polymers, compared with <1 for sulfonic acid polymers (*see* Table I); therefore, we would expect an increased ion selectivity for the phosphonic acid polymers in ion-exchange reactions. Increased selectivity is also induced by the Lewis basicity of the phosphoryl oxygen. Thus, many polymers based on phosphonic acids have the ability to bind species both through ion exchange and coordination.

Alexandratos and co-workers investigated (*81, 82*) binding properties of polystyrene containing phosphonic acid or a phosphonic acid and phosphonate ester. Affinity series were determined from highly acidic solutions held at constant ionic strength. The dominant absorption mechanism under these conditions is suggested to be coordination through the phosphoryl oxygen, with competition for ligand sites between the metal salt of interest and the $NaNO_3$

background. The selectivity series displayed by the phosphonic resin has a pattern of Fe > Hg > Ag > Mn (concentration, 10^{-5} to 10^{-1} M). In another study by the same group, the phosphonate monoester was more effective than a phosphonic acid resin for the extraction of americium in the presence of excess sodium (Figure 8). This result is probably a consequence of the increased Lewis basicity of the phosphoryl oxygen due to the electron donating nature of the ester group.

Styrene copolymers containing bis(phosphonic acid) (i.e., the previously mentioned Diphonix resin) exhibit high selectivity toward actinide and heavy metal ions. Diphonix has strong affinity for Th and U even in 10 M nitric, hydrochloric, or sulfuric acid, and it is able to uncomplex and sorb a number of transition metal ions from solutions containing chelating agents such as citric acid and EDTA at pH 2–3 or higher (*83*). Diphonix also possesses high selectivity for Cr, Mn, Fe, Co, Ni, Cu, Zn, Cd, Hg, Sn, Pb, and Bi over Ca, Mg, and Na in a wide pH range. Polymers with these characteristics have the potential to aid in reduction of the volume of radioactive wastes that must be disposed of in a deep geological repository, treatment of industrial waste streams, and purification of potable water. Relative selectivities of Diphonix toward a number of transition metal ions at pH 5 also were repeated recently (*84*).

Other Uses

Metal surfaces treated with poly(vinylphosphonic acid) exhibit corrosion resistance and improved adhesion to coatings (*85–87*). Another feature of adhesives based on poly(vinylphosphonic acid) is excellent hydrophilicity even when stained with oil (*88*). This type of coating finds use for treatment of aluminum fins in heat exchangers. Poly(vinylphosphonic acid)-based coatings can also be deposited onto metal surfaces through an electrochemical process. The insoluble metal oxide–organic polymer complex formed is composed of anodic oxide combined with the polyacid, which forms a protective layer on the metal (*89*).

The strong interaction between bis(phosphonic acids) and heavy metals made us interested in using a polymer containing bis(phosphonic acid) as a ligand to support a catalytically active metal. This porous bis(phosphonic acid) polymer (A1 in Table III) was treated with Pd^{2+} ions, followed by reduction to palladium metal, and the polymer-supported catalyst used for the hydrogenation of octene (*41*). X-ray photoelectron spectroscopy (XPS) of the polymer-bound catalyst before and after activation showed complete reduction of the Pd^{2+} ions to Pd^{0}. The XPS spectrum shows the palladium 3d bands are narrower and shifted 2.3 eV toward lower binding energy (Figure 9). Hydrogenation of 1-octene was carried out in the presence of

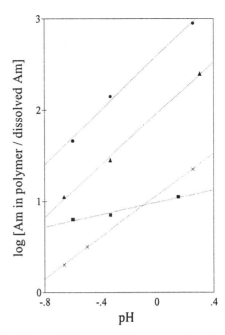

Figure 8. Effect of pH on americium extraction in the presence of excess sodium by (▲) phosphinic acid, (●) phosphonate monoester, (×) phosphonic acid, and (■) sulfonic acid.

this polymer-supported catalyst under an atmospheric hydrogen pressure. The half lives and rates calculated from the hydrogen consumption are presented in Table V.

After each reaction cycle, the catalyst was separated, washed, and reused. No decrease in activity could be detected during 10 hydrogenation cycles. Neither could any palladium be detected with plasma analysis of the reaction solutions after the supported catalyst was removed by filtration. The hydrogenation rate using the polymer-supported catalyst (S_{BET} area = 16.7 m^2/g) was comparable with the rate found for palladium on activated carbon in an identical experiment.

Concluding Remarks

Polymers containing bisphosphonic acid groups are useful for water and waste stream treatment, adhesives, medical applications, scale inhibition, and catalysis. Adhesion to the hard tissues of the body is one field where polymers containing bis(phosphonic acid) have a great potential. Because compounds containing bis(phosphonic acid) have excellent adhesion to metal materials and ceramics, they may also be useful as adhesives for machines, electrical applications, cans, and ceramics. They may also be useful as paint primer and coating agents. In the field of separations, these polymers are of interest for separation of a number of actinides. Most of the information concerning applica-

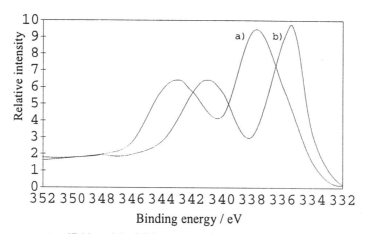

Figure 9. X-ray photoelectron spectra (3d bands) of Pd supported on a porous phosphonic acid polymer (A1 in Table III): a) fresh catalyst and b) activated catalyst.

Table V. Half-Life ($t_{1/2}$) and Rate of Hydrogenation of 1-Octene with Polymer-Supported Pd Catalyst for 10 Cycles

Parameter	1	2	3	4	5	6	7	8	9	10
Catalyst remained (mg)	131	130	126	126	115	110	109	108	107	10
1-Octene $t_{1/2}$ (min)	12	10.5	12	13.5	13.5	14	16.5	14	15	14
H_2 consumption (mL/min g)[a]	0.7	0.83	0.65	0.65	0.71	0.69	0.7	0.7	0.71	0.75

[a]H_2 consumed in mL/(min g) polymer catalyst.

tions employing polymers containing bis(phosphonic acid) is found in patents, and little information is available in the open literature. To obtain a broader knowledge of potential applications of polymers containing bisphosphonic acid, understanding of available synthetic routes to this class of materials is important. After creating new polymers containing bis(phosphonic acid), great care should be given to the physical properties of the polymer. This concern is especially true for polymers to be used in separation processes. These materials should be mechanically strong, have high capacity, and allow a high accessibility of active sites. In particular, we found it useful to use graft copolymers for this purpose. The matrix polymer can be polyethylene, polypropylene, or poly-(vinylidenefluoride) in the form of a film, fiber, fabric, or membrane. These materials are of potential interest for a wide range of applicators including, in particular, catalytic and separator processes.

References

1. Boyd, G. E.; Vaslow, F.; Lindenbaum, S. J. *Phys. Chem.* **1967**, *71*, 2214.
2. Horwitz, E. P.; Gatrone, R. C.; Nash, K. L. WO 9005115, 1990.
3. Horwitz, E. P.; Chiarizia, R. WO 9117124, 1991.
4. Yamashita, H.; Omori, K.; Nozaki, T. *Nippon Kagaku Kaishi* **1991**, *2*, 168.
5. Francis, M. D.; Centner, R. L. *J. Chem. Educ.* **1978**, *55(12)*, 760.
6. Horwitz, E. P.; et al. *Process Metall.* **1992**, *7A*, 357.
7. Nash, K. L.; Rickert, P. G. *Sep. Sci Technol.* **1993**, *28(1–3)*, 25.
8. Johnston, T. P.; Webb, C. L.; Schoen, F. J.; Levy, R. J. *J. Controlled Release* **1993**, *25*, 227.
9. *Fire Retardancy of Polymeric Materials;* Kuryla, W. C.; Papa, A. J., Eds.; Marcel Dekker: New York, 1973, Vols. 1 and 2; 1975, Vol. 3; 1978, Vol. 4; 1979, Vol. 5.
10. Wail, E. D. In *Kirk–Othmer Encyclopedia of Chemical Technology;* Grayson, M., Ed.; Wiley Interscience: New York, 1980; Vol. 10.
11. Axelrod, R. J. Eur. Patent 0 147 724 A2, 1984.
12. Carrol, R, L; Crutchfield, M. M. U.S. Patent 3,544,509, 1970.
13. Gefter, E. L. *Organophosphorus Monomers and Polymers;* Kosolapoff, G. M., Transl; Jacolev, L., Ed.; Associated Technical Services, Inc.: Glen Ridge, NJ, 1962.
14. Sander, M.; Steininger, E. *J. Macromol. Sci. Rev.* **1967**, *CI*, 1.
15. In *Encyplopedia of Polymer Science and Engineering;* Mark, H. F.; Bikales, N. M.; Menges, G.; Kroschwitz, J. I., Eds.; John Wiley: New York, 1985.
16. *Dictionary of Organic Compounds;* Chapman & Hall: New York, 1982; Vol. 5.
17. Goethals E. J. *Top. Sulfur Chem.* **1977**, *3*, 1.
18. Wink, H. *Makromol. Chem.* **1981**, *182*, 279.
19. Jones, G. D. U.S. Patent 2,909,508, 1959.
20. Dösher, F.; Klein, J.; Pohl, F.; Widdecke, H. *Makromol. Chem.* **1983**, *184*, 1585.
21. Breslow, D. S.; Hulse, G. E. *J. Am. Chem. Soc.* **1954**, *76*, 6399.

22. Kangas, D. A. In *Functional Monomers*; Yocum, R. H.; Nyqvist, E. B., Eds.; Marcel Dekker: New York, 1973; Vol. 1, p 489.
23. Weil, E. D. In *Enciclopedia of Polymer Science and Engineering*; Mark, H. F.; Bikales, N. M.; Menges, G.; Kroschwitz, J. I., Eds.; John Wiley: New York, 1985.
24. Van Wazer, J. R. *Phosphorus and Its Compounds*; Interscience: New York, 1958; Vol. 1.
25. Sandler, S. R.; Karo, W. In *Polymer Synthesis*; Academic: Orlando, FL, 1992; Vol. 1.
26. McMaster, E. L.; Glesner, W. K. U.S. Patent 2,980,721 1961.
27. Sun J.; Cabasso I. *J. Polym. Sci., Polym. Chem. Ed.* **1989**, *27*, 3985.
28. Yu, Z.; Zhu, W.; Cabasso, I. *J. Polym. Sci., Polym. Chem. Ed.* **1990**, *28*, 227.
29. O'Brien, J. L.; Lane, C. A. U.S. Patent 2,934,555, 1960.
30. Goldborn, P. U.S. Patent 3,780,146, 1973.
31. Carrol, R. L. U.S. Patent 3 686 290, 1972.
32. Carrol, R. L.; Crutchfield, M. M. Can. Patent 811 736, 1969.
33. Nicholson, D. A.; Cilley, W. A.; Quimby, O. T. *J. Org. Chem.* **1970**, *35*, 3149.
34. Surowiec, J. *J. Therm. Anal.* **1987**, *33(4)*, 1091.
35. Degenhardt, C. R.; Burdshall, D. C. *J. Org. Chem.* **1986**, *51*, 3488.
36. Pudovik, A. N.; Nikitina, V. I.; Kurguzova, A. M. *J. Gen. Chem. USSR (Engl. Transl.)* **1970**, *40*, 261.
37. Semmelhack, M. F.; Tomesch, J. C.; Czarny, M.; Boettger, S. *J. Org. Chem.* **1978**, *43*, 1259.
38. Rabinowitz, R. *J. Org. Chem.* **1963**, *28*, 2975.
39. Blackburn, M.; Ingelson, D. *J. Chem. Soc., Perkin Trans.* **1980**, *1*, 1150.
40. Prishcenko, A. A.; et al. *Zh. Obshch. Khim.* **1991**, *61(4)*, 1018.
41. Sundell, M. J.; Pajunen, E. O.; Hormi, O. E. O.; Näsman, J. H. *Chem. Mater.* **1993**, *5*, 372.
42. Hormi, O. E. O.; et al. *Synth. Commun.* **1990**, *20*, 1865.
43. Omura; et al. U.S. Patent 4,499,251, 1985.
44. Jpn. Patent 49557, 1982.
45. Herbst, W.; Rochlitz, F.; Vielsek, H.; Wagner, E. Ger. Patent 1,106,963, 1961.
46. Thomson, R. A. M. In *Developments in Ionic Polymers*; Wilson, A. D.; Prosser, H. J., Eds.; Elsevier: London, 1986; Vol. 2.
47. Duersch, W.; Herwig, W.; Engelhardt, F. *Ger. Offen.* 3248491, 1984.
48. Muray, B. J. *J. Polym. Sci. Part C: Polym. Symp.* **1969**, *16*, 1869.
49. Mikofalvy, B. K. U.S. Patent 3,489,706, 1970.
50. Gallagher, R.; Hwa, J. C. H. *J. Polym. Sci. Part C: Polym. Symp.* **1978**, *64*, 329.
51. Tsuda, T.; Yamashita, Y. *Kogyo Kagaku Zasshi* **1962**, *65*, 811.
52. *Chem. Abstr.* **1962**, *57*, 15344.
53. Coover, H. W.; Mc Call, M. A. U.S. Patent 3,043,841, 1962.
54. Marktschaffel, F. B.; Turbak, A. F. U.S. Patent 3,282,904, 1966.
55. Kunieda, H.; Shinoda, K. *J. Colloid Interface Sci.* **1980**, *75*, 601.
56. Alexandratos, S. D. *Abstracts of Papers*, 205th National Meeting of the American Chemical Society, Minneapolis, MN; American Chemical Society: Washington, DC, 1993, Part 1.
57. Scroeder, J. P.; Sopchak, W. P. *J. Polym. Sci.* **1960**, *47*, 417.
58. Degenhardt, C. R.; Kozikovski, B. A. Eur. Patent 0321233A1, 1989.
59. Hutchinson, D. W.; Thornton, D. M. *J. Organomet. Chem.* **1988**, *346(3)*, 341.
60. Alfer'ev, I. S.; Mikhalin, N. V.; Kotlyarevskii, I. L. *Izv. Akad. Nauk SSSR, Ser. Khim.* **1984**, *5*, 1122.
61. Masler U.S. Patent 4,207,405, 1980.
62. Chai, B. J; et al. U.S. Patent 4,239,695 1980.
63. Farral, M. J.; Frechet, J. M. J. *J. Org. Chem.* **1976**, *41*, 3877.
64. Pepper, K. W.; et al. *J. Chem. Soc.* **1953**, 4097.
65. Alexandratos, S. D.; Quillen, D. R.; Bates, M. E. *Macromolecules* **1987**, *20*, 1191.
66. Guyot, A.; et al. *React. Polym.* **1991**, *16*, 233.
67. Tundo, P. *Continuous Flow Methods in Organic Synthesis*; Ellis Horwood: New York, 1991.
68. Sundell, M. J.; et al. *React. Polym.* in press.
69. Ellis, J. *Materials Based on Polyelectrolytes*; Thames Polytechnic: London, 1989.
70. Ellis, J.; Wilson A. D. *Dent. Mater.* **1992**, *8(2)*, 79.
71. Anbar, M.; Farley, E. P. *J. Dent. Res.* **1974**, *53*, 879.
72. Ellis, J.; Anstice, M.; Wilson, A. D. *Clin. Mater.* **1991**, *7(4)*, 341.
73. Anbar, M.; St. John, G. A. *J. Dent. Res.* **1971**, *50*, 778.
74. Bartels, T.; Arends, J. *J. Caries Res.* **1979**, *13*, 218.
75. Anbar, M.; St. John, G. A.; Scott, A. C. *J. Dent. Res.* **1974**, *53(4)*, 867, 1240.
76. Gaffer, A.; Nabi, N.; Afflitto, J. Br. Patent. 2 235 201, 1991.
77. Gaffer, A. Br. Patent 2,090,265, 1982.
78. Gaffer, A. Eur. Patent 492998, 1992.
79. Rath, N. C.; Dimitrijevich, S.; Anbar, M. *Chem. Biol. Interact.* **1984**, *48(3)*, 339.
80. Dorfner, K. *Ion Exchangers*; Ann Arbor Science Publishers Inc.: Ann Arbor, MI, 1990.
81. Alexandratos, S. D.; Crick, D. W.; Darrell, W.; Quillen, D. R. *Ind. Eng. Chem. Res.* **1991**, *30(4)*, 772,
82. Alexandratos, S. D.; Quillen, D. R. *React. Polym.* **1990**, *13(3)*, 255.
83. Horwitz, E. P.; Chiarizia, R.; Diamond, H.; Gatrone, R.C.; Alexandratos, S.D.; Trochimczuk, A. Q.; Crick, D. W. *Solvent Extr. Ion Exch.* **1993**, *11(5)*, 943.
84. Chiarizia, R.; Horwitz, E. P.; Gatrone, R. C.; Alexandratos, S. D.; Trochimczuk, A. Q.; Crick, D. W. *Solvent Extr. Ion Exch.* **1993**, *11(5)*, 967.
85. Herbst, W.; et al. Ger. Patent 1,184,588, 1964.
86. Herbst, W.; et al. Ger. Patent 1,188,411, 1965.
87. Herbst, W.; et al. Ger. Patent 1,207,760, 1965.
88. Toyose, K.; Hatanaka, K.; Fukui, M. Jpn. Patent 61264040, 1986.
89. Platzer, S. J. W. U.S. Patent 4578156A, 1986.

Amine-Containing Polymers: Amine-Functionalized Polyolefins

Hai-Qi Xie, Warren E. Baker, and Reza Arshady

This chapter presents a brief overview of amine-containing polymers in general, and then focuses on the preparation, characterization, and applications of amine-functionalized polyolefins. Modification of preformed polyolefins in homogeneous solution and in multiphase media by ammonia plasma and melt grafting of aminoalkyl methacrylates onto polyethylene and polypropylene is discussed. Detailed experimental results for melt grafting of 2-(N,N-dimethylamino)ethyl and 2-(N-t-butylamino)ethyl methacrylates are summarized. Ziegler–Natta copolymerization of masked amine-containing olefins for the preparation of primary amine functionalized polyolefins is also reviewed. Applications of amine-functionalized polyolefins as polymer blending compatibilizers are discussed, and comments on future prospect of amine-functionalized polyolefins are also provided.

Polymeric materials containing amino groups are among the most widely studied and technically useful functional polymers. This feature is undoubtedly due to the chemical and structural versatility of the amino function. The chemical reactions of amines include those of bases and nucleophiles, two of the fundamental affinities in chemistry. Structurally, polymer-bound amines may be in the form of primary, secondary, tertiary, aliphatic, linear, alicyclic, or aromatic; and the amino function may be located in the polymer main chain or in the side chains. Chemical structures of typical amine-containing (amine-bound) polymers are shown in Chart I.

This structural and functional versatility of amines leads to an exceedingly diverse variety of applications ranging from polymer precursors for other functional polymers to ion-exchange resins, analytical and preparative chromatography, solid-phase synthesis and catalysis, drug delivery, and other biomedical uses to metal ion extraction (hydrometallurgy), light harvesting, and chemical sensors.

Another line of interest in polymer-bound amine functionality is the improvement of physical–mechanical properties of the polymer by enhancing its adhesion and miscibility in polymer blends and composites. This type of property enhancement is particularly useful in polyolefins because unmodified polyolefins lack any other strongly interacting residues needed for miscibility or adhesion.

A number of amine-containing polymers and their applications are alluded to throughout this book (especially in Part Four), and a new type of amino polymer with amine and silicon residues in the main chain is discussed in Chapter 1.12. This chapter provides a brief overview of amine-containing polymers in general, followed by a detailed discussion of amine-modified polyolefins with a view to their enhanced physical–mechanical properties.

Overview of Amine-Containing Polymers

Polymeric Amines Obtained by Conventional Methods

In common with other functional polymers, amine-containing polymers can be produced by both direct

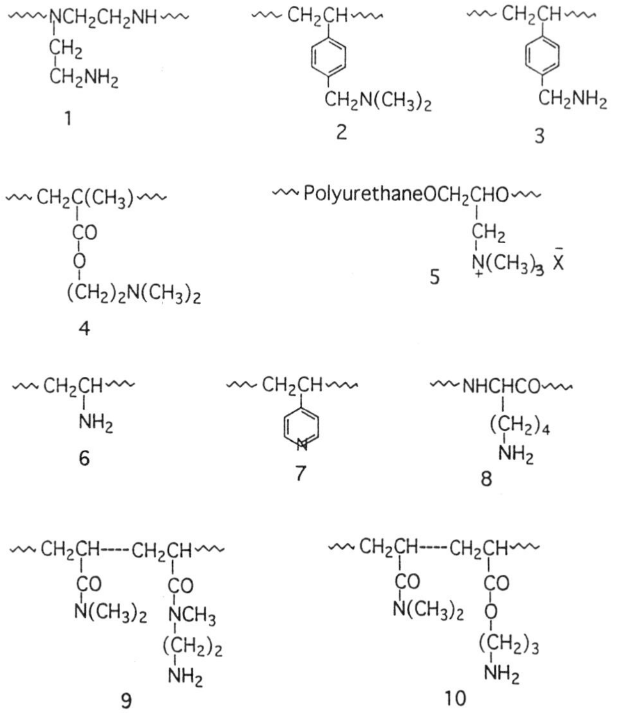

Chart I. Chemical structures of various types of amino-containing polymers.

polymerization and postpolymerization functionalization. In practice, however, each type of amino polymer is likely to be more readily accessible by one or the other method. For example, commercially available polyethyleneimine (PEI, **1**, Chart I) is obtained by ring-opening polymerization of ethyleneimine as a highly complex branched polymer containing primary (NH$_2$, 25%), secondary (–NH–, 25%), and tertiary (–N–, 50%) amino groups. Linear PEI containing only secondary amino groups can be obtained by polymerization of 2-oxazoline, followed by hydrolysis (*1*).

Commercially available PEI is used widely for the preparation of ion-exchange resins (**2**) and for a wide range of other applications where a polymeric amine (or base) is needed without specific requirements for polymer structure. PEI derivatives have been extensively studied as catalysts (the so-called enzyme models, enzyme-like polymers, or enzyme mimics) for ester hydrolysis (*3, 4*). In particular, polymeric imidazoles based on PEI or other polymers also have long attracted interest as enzyme-like polymers for hydrolytic transformations (*3*). More recent studies, however, involve more highly sophisticated PEI derivatives, an example of which is shown in Scheme I (*5*).

Among amino derivatives of polystyrene (PS), ion exchangers are obtained by treatment of chloromethylated polystyrene resins with dimethyl- or trimethylamine. These ion exchangers are probably the most widely used amino polymers. Syntheses of these resins (**2**) are basically similar to those of soluble polychloromethylstyrene described in Chapter 1.1. Crosslinked polystyrenes with methylamino groups (**3**) are obtained by various routes and are used as polymer supports for solid-phase peptide synthesis (Merrifield method). Another class of amine-based ion exchangers (**4**) is obtained by copolymerization of *N,N*-dimethylaminoethyl acrylate with ethylene dimethacrylate. Polyurethane ionomers (**5**) carrying tertiary amine or quaternary ammonium salts also were described (*6*). Polyvinylamine (**6**) is a potentially very useful material, but its preparation is rather difficult. The Hoffman degradation of the amide groups on acrylamide-containing polymers (**7**) is relatively more practicable than other methods, even though not a satisfactory synthetic route.

Soluble (**8**) and cross-linked (**9**) polymers and copolymers of 2- and 4-vinylpyridines (**7**) are obtained by direct polymerization, and were studied widely for chemical and physicochemical applications, examples of which can be found in Part Four of this book. Another interesting amino polymer is polylysine (**8**), which is obtained by polymerization of the γ-pro-

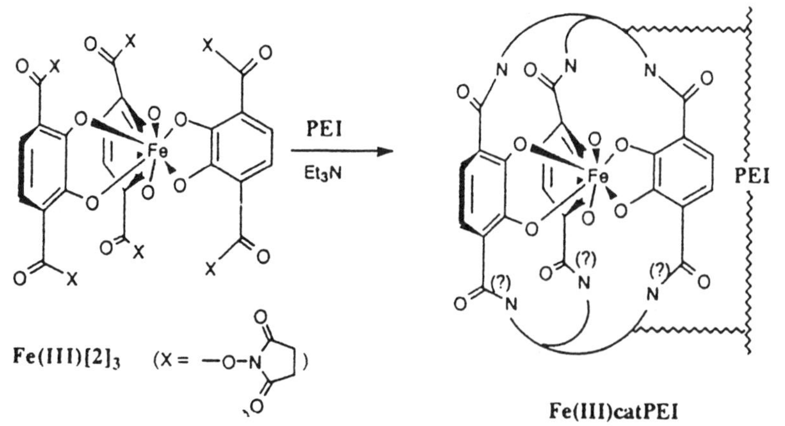

Scheme I. Synthesis of a metaloenzyme model polymer prepared by derivatization of polyethylenimine.

tected lysine-*N*-carboxyanhydride (the NCA method) (*10*). Polylysine is of considerable interest as a precursor for polymeric drug carriers and other biologically active materials (*11*).

A series of hydrophilic resins (**9**, **10**) carrying primary amino groups at the end of a spacer arm have been developed for solid-phase peptide synthesis. These polymers are obtained by water-in-oil suspension copolymerization of dimethylacrylamide with a diacrylamide cross-linker and an N-protected aminoalkylacrylamide, followed by deprotection (*12*). A related series of hydrophilic resins based on γ-formamidoalkyl acrylate (CH_2=CHCOO–R–NHCHO, R = alkyl) as the functional monomer was developed similarly (*13*). In this case, the formyl groups can be removed by aqueous HCl to produce the corresponding amino polymers, or they can be treated with toluenesulfonyl chloride in pyridine to obtain the corresponding isocyano derivatives (–NC).

Polymeric Amines Obtained by Active Ester Synthesis

As outlined previously, a wide variety of amino-containing polymers are prepared by as many different synthetic routes. For research purposes a desired amine functionality can often be attached to practically any type of polymer backbone. For most technological applications, however, the choice of the synthetic route and the type of polymer backbone for the development of amino-containing polymers are strongly influenced by the following three criteria.

- labor and cost of producing the desired polymer
- effect of the chemical structure of the polymer backbone on the functional performance of the polymer
- need for a spacer arm between the polymer backbone and the amino functionality

Among the amino-containing polymers discussed previously, PEI (**1**), polystyrene-based ion exchangers (**2**), polyvinylpyridines (**7**), and polylysine (**8**) are commercially available and are relatively inexpensive. Most other amino polymers are obtained by multistep monomer synthesis or polymer modification and are usually very costly to produce. The problem of the chemical structure is, in most cases, related to the hydrophobic/hydrophilic nature of the polymer backbone (compare structures in Chart I and Scheme II), and often a balance of both hydrophobicity and hydrophilicity is required for optimum functional performance. It has also been established that for most chemical and physicochemical applications of functional polymers a spacer arm between the functional residue and the polymer backbone enhances the functional performance of the polymer.

To meet these three criteria, a new synthetic approach recently was introduced on the basis of the chemistry of activated esters, as illustrated in Scheme II. This synthesis scheme, termed active ester synthesis (*14, 15*), is practically simple, generally applicable for the introduction of different amino (and other functional) groups, and usually provides a spacer arm between the amino group and the polymer backbone. Typical examples of amine-containing polymers available by this new method are indicated in Scheme II.

Amine-Functionalized Polyolefins: Introduction

The development of modern polyolefin synthesis technology has resulted in a great variety of inexpensive and very useful polymeric materials (*16, 17*). A plethora of commodity polyolefins such as polyethylene and polypropylene is widely available. For example, the range of density for commercially available polyethylene grades (*18*) varies from less than 0.86 g/cm^3 for very low density polyethylene to more than 0.95 g/cm^3 for high density polyethylene, each having its distinctive properties and uses. Polyethylenes with short-chain or long-chain branching, and ultra-high molecular weights, are also available (*17*). New grades of metallocene polyolefins have even better controlled microstructure, molecular weight distribution, and mechanical and chemical properties. Polyolefins are thermoplastics with various elasticity, ductility, brittleness, and other unique properties ranging from rubbery ethylene–propylene rubber and ethylene–propylene diene modified rubber to brittle polypropylene. In general, polyolefins are light, chemically inert, and hydrophobic. They can be used to replace many tradi-

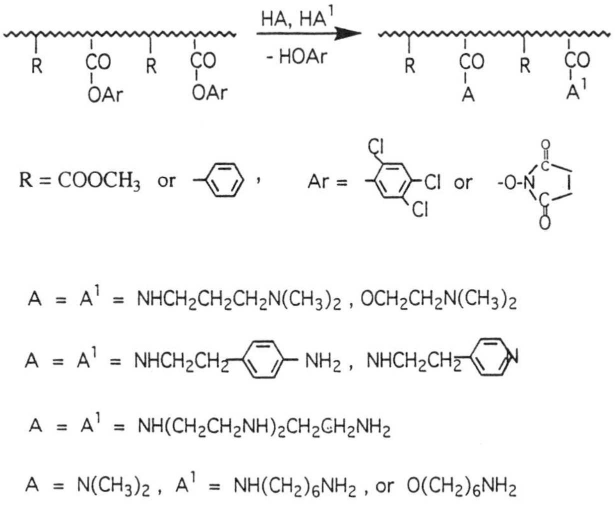

Scheme II. Synthesis of amine-containing polymers by active ester synthesis.

tional materials such as metals, cellulosics, and ceramics in applications where high strength–mass ratio, corrosion resistance, and moisture barrier are required (*18*).

These advantages of polyolefins are mainly because polyolefins are inert nonpolar hydrocarbon materials. These characteristics, however, also limit their usefulness in other applications where polarity, hydrogen-bonding, or reactivity are required. For example, the nonpolar character of polyolefins results in their immiscibility with other polar materials such as polyamides, polyesters, polycarbonates, glasses, and ceramics. Hence blends and composites of polyolefins with polar materials usually display inferior physical properties to those predicted from their compositions. Useful blends and composites of polyolefins with other substances cannot be obtained without careful consideration of compatibilization issues (*19–21*).

To overcome these disadvantages of polyolefins, much effort has been devoted to modify polyolefins into more polar and reactive polymers while preserving their desired properties. For example, polyethylene (PE) does not stick to metal substrates, but when PE is grafted with maleic anhydride, it shows such a strong adhesion to metal that it can be used in metal-coating applications (*22*). In another example, ethylene-propylene rubber (EPR) grafted with maleic anhydride is used in the manufacture of toughened polyamides used as engineering resins.

In the past decades, a great number of studies have been dedicated to the functionalization of polyolefins with acidic and electrophilic groups such as carboxylic acid and anhydride (*23*), leading to the commercialization of a variety of acid- and anhydride-functionalized polyolefins. They are produced from either copolymerization of functional monomers (*24*) or post-synthesis modifications of polyolefins. Well-known products include DuPont Canada's Fusabond series resins (polyolefins grafted with maleic anhydride), Arco's Dylark poly(styrene-*co*-maleic anhydride) resins, Dow Chemical's Primacor, and DuPont's Nucrel (copolymers of ethylene and (meth)acrylic acids). However, base- and nucleophile-functionalized polyolefins have attracted less attention.

The importance of base- and nucleophile-functionalized polyolefins can be seen from two aspects. First, nonpolar media greatly enhance the reactivity of bases and nucleophiles through the mechanism called *desolvation* relative to polar media such as water and alcohol (*25*). Such desolvation can be explained by a ground-state destabilization that is larger than that of the transition state (Figure 1). The root cause of the difference in destabilization is the fact that the ground state (reactants) has a higher charge density than the

transition state. This difference results in increased reactivity for bases and nucleophiles in nonpolar media. Polyolefins in solid or molten state can be regarded as nonpolar media. Therefore, polyolefins containing bases and nucleophiles could be used as very reactive immobilized bases and nucleophiles—a novel type of specialty chemical. Second, polyolefins carrying basic and nucleophilic groups can have strong interaction or form stable covalent bonds with acids and anhydrides. New categories of useful blends and composites of polyolefins with other polymers and inorganic materials may thus be obtained, which in turn may spark new applications for acid- and anhydride-functionalized polyolefins.

For the purpose of this discussion, amine-functionalized polyolefins should carry limited amounts of amino groups chemically attached to the polyolefin backbones or side chains (Chart II). The amount of amine functionality should be so low that the end products retain most of their bulk properties such as hydrophobility, corrosion and chemical resistance, and useful mechanical properties. Homopolymers of amine-containing monomers have properties too much deviated from those of polyolefins. For example, PEI and its derivatives are freely soluble in water and react strongly with acids, whereas polyolefins are hydrophobic and resistant to strong acids.

Amines are weak bases and nucleophiles in polar solvents such as water and alcohol, and they are strong bases and nucleophiles in nonpolar polyolefin due to the desolvation mechanism discussed previously. The basicity of amines can result in strong ionic or polar interactions with carboxylic acids and anhydrides (Scheme III). Primary and unhindered secondary amines are strong nucleophiles. Amidation and imidation of these amines with carboxylic acids and anhydrides result in the formation of stable amide and imide covalent bonds. On the other hand, tertiary and hindered secondary amines act as bases and form ionic bonds with acids. Covalent bonding and ionic and polar interactions are the main thermodynamic driving forces for reactive compatibilization of polymer blends and interfacial adhesion in polymer composite preparation.

Preparation of Amine-Functionalized Polyolefins

Amines are introduced to polyolefin by either copolymerization of amine-containing monomers with olefins or functionalization of preformed polyolefins. These two approaches are outlined subsequently.

Modification of commercially available and inexpensive polyolefins is now recognized as a viable way of making new polymeric materials. Its main advantage over copolymerization is that modified polymers

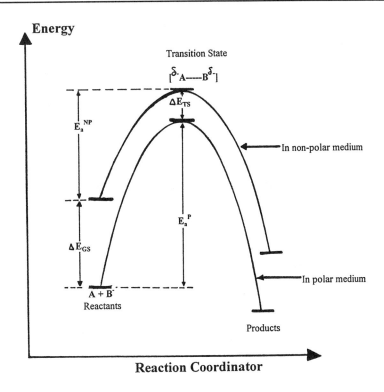

ΔE_{GS} : ground state desolvation energy; ΔE_{TS}: transition state desolvation energy;

E_a^P: activation energy in polar medium; E_a^{NP}: activation energy in non-polar medium

Figure 1. Mechanism of desolvation in polar media: ΔE_{GS} is ground-state desolvation energy, ΔE_{TS} is transition-state desolvation energy, E_a^P is activation energy in polar medium, and E_a^{NP} is activation energy in nonpolar medium.

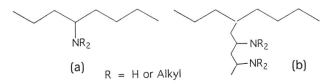

Chart II. Different types of amine-functionalized polyolefins: amine bound to polyolefin (a) backbones or (b) side chain.

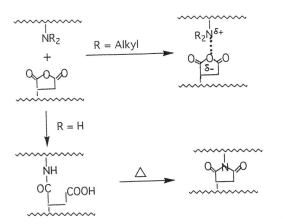

Scheme III. Ionic, polar, and covalent interactions of amines with carboxylic acids and anhydrides.

with predictable properties can be obtained without designing new macromolecules.

Free-radical grafting is a preferred way to add functionality to polyolefins. Polyolefins are inert nonpolar hydrocarbon macromolecules. An effective way to increase their reactivity is to create macroradicals on their backbones or side chains through free-radical reactions. The abundance of C–H bonds in polyolefins allows the creation of macroradicals through hydrogen abstraction by free radicals generated from the decomposition of initiators such as organic peroxides. Homopyrolysis of C–C bonds induced by heating, high energy irradiation, and mechanical shearing can also generate polyolefin macroradicals for free-radical grafting. Polyolefin macroradicals can add to the double bonds of vinyl monomers carrying amino groups. The newly formed macroradicals may add to another vinyl monomer (graft chain propagation) or terminate the grafting by macroradical combination, transfer, or disproportionation. In addition to the desired grafting reactions, polyolefin cross-linking, degradation, or monomer homopolymerization may also occur. The net result of grafting reactions is that the vinyl monomers together with their amino groups are chemically bound to polyolefins. Amine functionalization of polyolefins through free-radical mechanisms can be realized in solution (solution grafting), at the interface (multiphase grafting), or in polymer melt (melt grafting).

Solution Grafting

Because of difficulties to transmit high energy irradiation and to generate intensive shearing in solution, free-radical initiators are often the sources for primary radicals. A common solvent system is usually found for polyolefins, amine-containing monomers, and free-radical initiators. Because hydrogen abstraction is characterized by high activation energies, relatively high reaction temperatures, and hence high boiling point, solvents are often used. High temperature also facilitates the dissolution of the polymers. Aromatic hydrocarbon solvents such as benzene, chlorobenzene, toluene, and xylene are preferred over saturated hydrocarbons (mineral oils). This fact is because in aromatic solvents, free radicals generated from initiator decomposition can selectively attack the polymer chains, whereas in saturated hydrocarbon solvents hydrogen abstraction can occur on solvent molecules as well as the polyolefin. In addition, highly concentrated polyolefin solutions are hard to prepare and handle. Concentrations in the range of 4–10 wt% polymer are usually used. As a result, solvent molecules are usually more abundant than polyolefin macromolecules. This characteristic makes the grafting to solvent more probable than to polyolefins. Handling of a large quantity of solvents and control of environment pollution are very expensive and challenging for solution-grafting processes. These processes are not widely used in commercial production. The major advantage of solution grafting is that the grafting is realized in a homogeneous system, and reaction conditions can be accurately controlled. Accordingly, solution grafting is mostly used in model studies for particular grafting processes.

Multiphase Grafting

In this method, the polyolefin, vinyl monomers to be grafted, and radical sources are distributed in more than one phase. Free-radical initiators and high energy irradiation are possible primary radical sources. The grafting is realized at the interface or in the boundary area of two phases. For example, polymer surface modification can be achieved by passing a polymer film with its surface coated with free-radical initiator through a hot monomer solution. The initiator is not soluble in the solvent in which the monomer is dissolved. The grafting occurs on the surface of the polymer film. Multiphase grafting has the advantage of minimizing monomer homopolymerization. It is

mostly suitable for polymer surface modification while the composition, structure, and morphology of polymer beneath the surface area are to be preserved.

Melt Grafting

Melt grafting is more proactive than solution and multiphase grafting. It consists of mixing rapidly the small molecules (monomers, initiators, and additives) with the polymer melt at elevated temperature and simultaneously generating polyolefin macroradicals. The easy realization of elevated temperature and intensive shearing used in polymer melt processing is the main advantage of melt grafting. This realization allows thermal and mechanical generation of macroradicals directly from polymer backbones. Melt-grafting temperature used is often close to or higher than the homopolymer ceiling point of the monomers to be grafted. This temperature will reduce homopolymer formation from amine-containing monomers. Because no solvent is involved, the possibility of initiator radicals attacking solvent molecules is eliminated. This result increases initiator grafting efficiency. More importantly, the absence of large quantities of solvents minimizes environmental problems and reduces manufacturing costs. The elimination of solvent also results in high process compactness (the reactor space per unit quantity of product). Melt grafting can be realized in a continuous way on conventional polymer processing equipment, such as single- or twin-screw extruders. Commercialization of melt-grafting processes can be realized without major capital investment in new production facilities.

In recent years, progress has been made in our laboratory on melt grafting of amine-containing vinyl monomers onto LLDPE. The research aims at finding a viable way to make amine-functionalized polyolefins. Progress on melt grafting of amine-containing monomers onto LLDPE will be detailed in later sections.

Copolymerization of Amine-Containing Monomers

Amine-functionalized polyolefins can also be prepared by Ziegler–Natta and free-radical copolymerization of olefins with small amount of vinyl monomers containing amino groups. For Ziegler–Natta processes, interference of active hydrogen atoms of primary and secondary amino groups with metal catalysts must be considered. Monomer reactivity ratios are the key factor in free-radical copolymerization.

Ziegler–Natta Copolymerization

Because hydrogen atoms of amines can deactivate metal catalysts, Ziegler–Natta polymerization cannot be directly applied to copolymerization of olefins with vinyl monomers containing primary or secondary amino groups. Recently, a method was developed to mask primary and secondary amines by metal chelating. In this way, primary and secondary amine-containing monomers can be copolymerized with olefin monomers. After polymerization, the amino groups are released by alcoholysis. Tertiary amino groups have no active hydrogen atom and do not interfere with catalysts. Tertiary amine-containing monomers can be directly copolymerized with olefins by Ziegler–Natta processes (*26*).

Free-Radical Copolymerization

Amine-functionalized polyolefins also can be prepared by free-radical copolymerization of amine-containing vinyl monomers with olefins. Amine content in the final products is determined by the reactivity ratios of different monomers.

Characterization of Amine-Functionalized Polyolefins

Bulk properties such as melt flow index and glass transition temperature of amine-functionalized polyolefins can be determined using standard testing procedures. When measuring molecular weight and molecular weight distribution by gel permeation chromatography, care must be taken to choose the appropriate separation columns. Neutral or basic materials should be used as column-filling materials. Acidic gel-filled columns are not suitable because of the interactions of amines with acids.

Identifying the amine functionality is one of the major tasks in characterizing amine-functionalized polyolefins. Specific chemical and physical methods as well as methods stemming from applications of other amino polymers are used for both qualitative and quantitative analyses of amine-functionalized polyolefins, as outlined subsequently.

Titration

Amine-functionalized polyolefins can be characterized using titration with standard acid solutions, but these polymers have limited solubility in polar solvents like water and alcohol where titration is usually carried out. To overcome this obstacle, mixtures of alcohol with toluene, xylene, or chlorobenzenes are used for titration. For example, the amount of grafted tertiary amines on EPR was determined by potentiometric titration with 0.10 M HCl in a mixture of *o*-dichlorobenzene–ethanol 9/1 (v/v) (*27*).

Nuclear Magnetic Resonance Spectroscopy

High resolution NMR spectroscopy is a particularly powerful method in the characterization of amine-functionalized polyolefins. Hydrogen and carbon atoms adjacent to amino groups have characteristic chemical shifts. This fact can be used to identify, as well as quantitatively analyze, amino groups in amine-functionalized polyethylene (*28–32*) and EPR (*33, 34*). Proton NMR spectroscopy was used to determine concentrations as low as 0.2 wt% of DMAEMA and TBAEMA grafted on LLDPE (*30, 31*).

Infrared Spectroscopy

IR spectroscopy is frequently used for the characterization of amine-functionalized polyolefins (*27–36*). Primary aliphatic amines have three typical N–H absorption bands at 3450–3250, 1645–1585, and 910–770 cm^{-1} (*33, 34*). N–H bands of secondary amines can be found typically at 3450–3250 and 740–690 cm^{-1} (*37*). These IR absorption bands can be used to identify amino groups in amine-functionalized polyolefins. Also, IR spectroscopy can be used to quantitatively analyze amine content in polyolefins if reliable calibration curves can be obtained by other primary analytical methods such as titration, NMR spectroscopy, and elemental analysis (*28, 29, 33–36*). IR spectra of reaction products of amine-functionalized polyolefins with carboxylic acid and anhydride can also be used as indirect evidence for the existence of amino functionality in polyolefins. For example, only ammonium salt was identified in reactive blends of tertiary amine-functionalized polyethylene with styrene–maleic anhydride copolymer (*38*). Both ammonium salt and amide were characterized in blends of secondary amine-grafted polyethylene with carboxylic acid functionalized polyethylene and styrene–maleic anhydride copolymer (*29, 39*). IR spectra were also used in the identification of imide formation from reactive compatibilization of amine-functionalized polyethylene with poly(methyl methacrylate) containing carboxylic acid (*40*).

Nitrogen Analysis

Nitrogen analysis provides quantitative information on nitrogen content in amine-functionalized polyolefins.

Other Methods

Roberts (*41*) studied acid–base interactions in the adhesion of rubber surface by using rolling adhesion measurements and wettability experiments. In the rolling adhesion experiments, cylindrical samples of smooth-surfaced vulcanized rubber were timed for their rolling descent down an inclined glass track that was acid treated. The wettability experiments were conducted by measuring the contact angles of a range of organic and inorganic liquids on rubber surfaces that allowed the determination of acidity and basicity of the rubber surface. Such tests are simple and useful in practice (*41*).

Examples of Amine-Functionalized Polyolefins and Their Applications

This section describes several examples of amine-functionalized polyolefins and their applications. The discussion is organized according to the type of amine incorporated into polyolefins: namely tertiary, secondary, and primary.

Tertiary Amine-Functionalized Polyolefins

Functionalization of polyolefins with tertiary amines is relatively easy. Monomers such as acrylates and methacrylates containing tertiary amines are commercially available, and they can be copolymerized with olefins or grafted onto polyolefins by solution-, multiphase-, and melt-grafting processes. In addition, tertiary amine functionalized polyolefins can also be prepared via derivatization of carboxylic acid and anhydride-functionalized polyolefins with primary–tertiary diamines (Scheme IV).

Bonvicini et al. (*42*) described a multiphase-grafting process to modify polypropylene surface with DEAEMA and its analogs. The process consisted of hydroperoxidizing polypropylene surface by dipping polypropylene articles in oxygen under 1–10 atmosphere at about 50 °C, and up to 10 wt% oxygen was introduced to polypropylene articles. The surface hydroperoxidized articles were then placed in

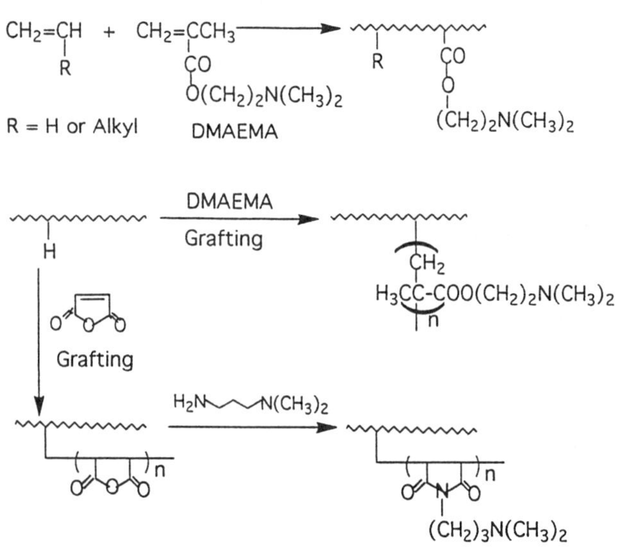

Scheme IV. Preparation of tertiary amine functionalized polyolefins.

DEAEMA solution at 50–100 °C. The multiphase grafting of DEAEMA onto polypropylene surface was initiated by free radicals generated from the decomposition of surface hydrogen peroxide, leading to substantial DEAEMA grafting. Surfaces of such modified polypropylene articles had greatly enhanced dyeability toward acid dyes, improved wettability, and increased resistance to accumulating electrostatic charge.

Immerzi et al. (*27*) reported two different approaches to prepare tertiary amine functionalized EPR by solution grafting. The first approach used maleic anhydride functionalized polyolefins as starting materials and derivatized them with DMAPA (Scheme IV). The primary amino group of DMAPA reacted with anhydride group to first form an amidic acid linkage at low temperature, then the amino group cyclized under heating to form a more stable imide linkage. The tertiary amine does not participate in the reaction and remains as the new functionality grafted on EPR. The amidic acid intermediate was identified by IR spectroscopy when the reaction was carried out at room temperature. When the amidic acid intermediate was brought to 90 °C, cyclization occurred rapidly transforming amidic acid to a 5-membered imide ring structure that links the tertiary amino group to polymer backbone through the propyl group. With amine–anhydride molar ratio of at least 4/1, the imidation reaction was complete.

The second approach explored by Immerzi et al. was the solution grafting of DMAEMA onto EPR by using dicumyl peroxide initiator. Relatively high concentrations of the monomer and peroxide were used. For example, in a typical grafting reaction, 18.66 g DMAEMA and 2.34 g peroxide were added to 5.0 g polymer dissolved in 100 mL chlorobenzene. The degree of grafting was 5.3 wt% as determined by potentiometric titration of tertiary amino group. Optimal grafting conditions were determined as 60 min at 132 °C with 1.3 wt% peroxide initiator.

Tertiary amine functionalized EPR reacted with anhydride- and acid-functionalized EPR resulting in gelation (*27*) and elastomeric network formation (*43*). IR spectroscopy indicated strong polar interaction between anhydride and tertiary amine, ammonium-carboxylate ionic interaction, as well as hydrogen bonding in resulting gel products (*27*).

Simmons and Baker (*28*) first reported the more proactive melt grafting of DMAEMA onto LLDPE by using an organic peroxide initiator on a Banbury type batch mixer. The organic peroxide initiator was 2,5-dimethyl-2,5-di(*t*-butylperoxy)hexyne-3 (Lupersol 130). Three processing systems were studied:

- LLDPE alone
- LLDPE and peroxide initiator
- LLDPE, peroxide, and monomer

Comparison of the torque–time curves for the three systems indicated that the aminomethacrylate monomer is capable of inhibiting cross-linking of polyethylene by peroxide radicals (Figure 2). This evidence was the first for a melt-grafting reaction: the quench of polyethylene cross-linking by aminomethacrylate showed signs of macroradical addition to aminomethacrylate monomer. Amine graft was subsequently identified on purified grafted products by IR and NMR spectroscopies, respectively (Figures 3 and 4). Products containing up to 3 wt% aminomethacrylate graft were prepared. Polymer cross-linking, as indicated by product melt flow index, was very sensitive to initiator concentration (Figure 5). The choice of optimum feed compositions and processing conditions

allowed the production of a material containing the maximum amount of graft while minimizing polymer cross-linking. DSC studies showed that the product had an onset degradation temperature 20 °C higher than that of the starting polyethylene.

Song and Baker (35) reported a systematic investigation on the same melt-grafting system. The melt grafting was also carried out on a Banbury type batch mixer. Monomer conversion, homopolymer formation, graft level, and grafting efficiency were measured under a wide range of experimental conditions. High graft level was obtained at 140 °C. Product melt flow index was reduced linearly with reaction temperature (Figure 6). Surprisingly, the plot of graft level versus reaction time showed a maximum value at 4 min (Figure 7). The high graft level obtained at short reaction time suggested that the melt grafting could be efficiently carried out in single- or twin-screw extruders where short residence times are usually encountered.

Subsequently, Song and Baker (36) studied the melt grafting of DMAEMA onto LLDPE in an intermeshing twin-screw extruder. Graft levels of up to 2.5 wt% were obtained without excessive polyethylene cross-linking by using a high monomer feed rate with low initiator concentration. Optimal processing temperature was determined to be 130–160 °C. Too high a processing temperature resulted in low graft level and low grafting efficiency. The optimal residence time was about 5 min with 23 wt% monomer load and 0.56 wt% peroxide initiator. Measurement of product melt flow index indicated that the aminoalkyl methacrylate monomer significantly suppressed cross-linking of polyethylene chains. The ability of the monomer to compete for initiator and polymer radicals and to reduce polymer cross-linking was further demon-

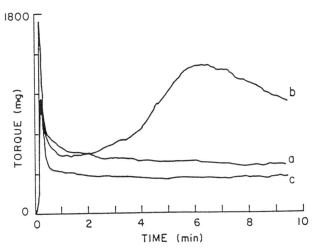

Figure 2. Torque versus time for a, LLDPE; b, LLDPE + Lupersol 130; and c, LLDPE + Lupersol 130 + DMAEMA processed at 160 °C and 100 rpm. (Reproduced with permission from reference 28. Copyright 1988.)

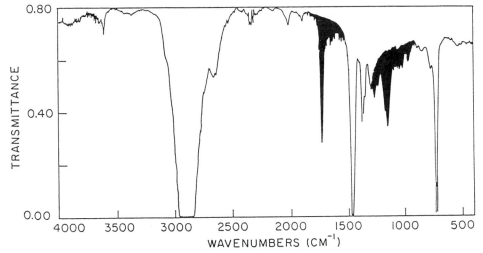

Figure 3. IR spectra of LLDPE and LLDPE-*g*-DMAEMA. Shaded areas show new absorbances from purified grafted polymer. (Reproduced with permission from reference 28. Copyright 1988.)

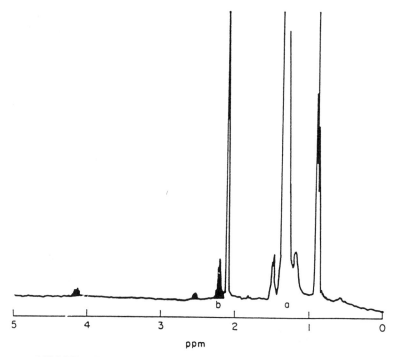

Figure 4. ^{1}H NMR spectrum of LLDPE-*g*-DMAEMA. Shaded areas are not observed for pure LLDPE. (Reproduced with permission from reference 28. Copyright 1988.)

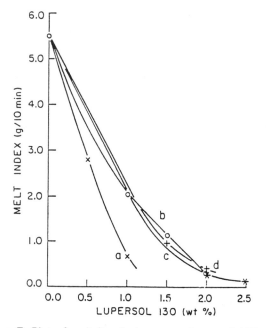

Figure 5. Plot of melt flow index versus Lupersol 130 concentration after 15 min at 160 °C and 100 rpm for a, 10 wt% DMAEMA; b, 20 wt% DMAEMA; c, 30 wt% DMAEMA; and d, 40 wt% DMAEMA. (Reproduced with permission from reference 28. Copyright 1988.)

low graft level and less cross-linking. The percentage fill in these studies was not well controlled and probably lead to phase separation in the reaction zone. With a high degree of fill in the reaction zone, the grafting efficiency was improved significantly (*44*).

Because of the elevated temperature experienced in the melt-grafting processes, oxidation of amines by peroxide initiators was observed. A direct consequence of the amine oxidation was the severe discoloration of melt-grafted products. This discoloration detracts from the major merits of melt-grafting processes, because discolored materials usually have much lower commercial value than the colorless ones. In addition, oxidation reduces the amine content of the polymer. To overcome this problem, some conventional nonoxidizing azo initiators like 2,2'-azobisisobutyronitrile were used in melt grafting of DMAEMA to LLDPE, but without success (*28*). Recently, a phenylazo initiator, 2-phenylazo-2,4,-dimethyl-4- methoxyvaleronitrile (trade name V-19) was very useful to overcome this obstacle (*30, 31*). In the absence of monomers, V-19 causes polyethylene cross-linking and polypropylene degradation. Experiments on melt grafting of DMAEMA onto LLDPE demonstrated that this phenylazo initiator is capable of initiating melt grafting with lower discoloration of products. An additional advantage of V-19 for melt grafting is its high thermal stability ($T_{t1/2=10h}$ = 120 °C), which is suitable for melt-grafting processes.

Felmine et al. (*32*) investigated free-radical grafting of DMAEMA onto squalane, a C-30 saturated

strated by the study of sequential addition of monomer and initiator along the extruder. Injecting the initiator into the extruder before the monomer yielded high graft level but brought about extensive polymer cross-linking. The reversed injection mode resulted in

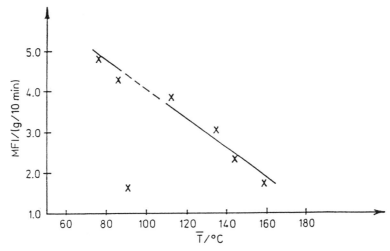

Figure 6. Variation of melt flow index (MFI) with average reaction temperature: reaction time, 6 min; LLDPE, 40 g; DMAEMA, 12 g; Lupersol 130, 0.6 g. (Reproduced with permission from reference 35. Copyright 1990 Huthig & Verlag, Basel, Switzerland.)

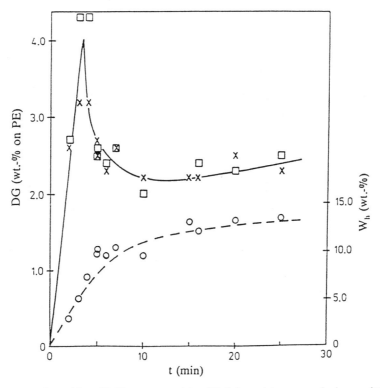

Figure 7. Variation of degree of grafting (DG) measured by IR (×) and by mass balance (□) and weight percent of homopolymer (○) with reaction time at 160 °C: LLDPE = 40 g, DMAEMA = 12 g, Lupersol 130 = 0.6 g. (Reproduced with permission from reference 35. Copyright 1990 Huthig & Verlag, Basel, Switzerland.)

hydrocarbon with six methyl substituents. This work served as a model study for melt grafting in a twin-screw extruder, and the resulting DMAEMA graft and homopolymer were separated and characterized. When the grafting was carried out at 150 °C, the graft chain length was estimated to be 1.4–7 monomer units and homopolymer degree of polymerization was 30–50. Kinetic studies indicate that the grafting reaction is first order with respect to monomer.

Tertiary amine functionalized polyethylene prepared by melt grafting was used in reactive compatibilization of polyethylene and polystyrene blends (*38, 39*), in which polystyrene contained anhydride groups. Improved morphology and mechanical properties were observed for reactive blends compared to nonreactive blends (Figures 8 and 9). These improvements were attributed to strong polar, ionic, and hydrogen-bonding interactions, as evidenced by IR spectroscopy (*38, 39*).

Secondary Amine Functionalized Polyolefins

Acrylates or methacrylates carrying unhindered secondary amino groups are not stable, because they undergo intra- and intermolecular alkylation by the double bond and amidation of the ester groups at high temperatures. However, acrylates and methacrylates containing hindered secondary amines are more stable, and they have stronger basicity than the corresponding tertiary amines. For example, TBAEMA has an intrinsic dissociation constant (pK_0) of 9.12 (*45, 46*), as compared with a pK_0 of 7.94–8.14 for DMAEMA (*45–47*). Therefore, TBAEMA-grafted polyolefins have stronger basicity than DMAEMA-grafted polymers.

Radiation-induced grafting of TBAEMA onto low density polyethylene was reported by Odian et al. (*48, 49*). *n*-Hexane can accelerate grafting of TBAEMA onto polyethylene film (multiphase grafting). Increases in grafting rates were observed with *n*-hexane content ranging from 10–70 wt%, the maximum rate being at TBAEMA:*n*-hexane ratio of 70:30. This effect was attributed to the insolubility of growing graft chains in the solvent and the greater accessibility of monomer to the growing graft chains caused by the swelling effect of the monomer. This rate acceleration phenomenon is similar to the Trommsdorff effect observed in radiation-induced grafting of styrene onto polyolefin in methanol (*50*).

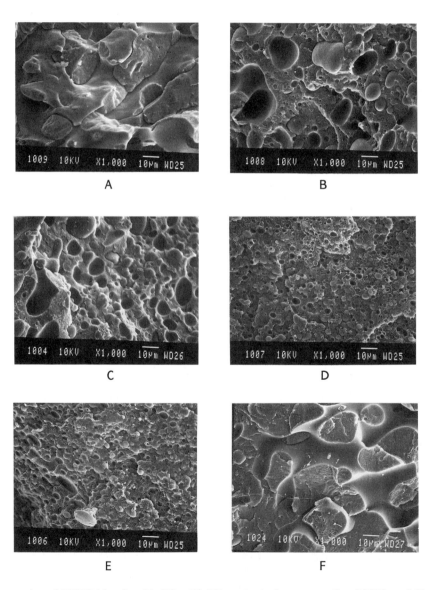

Figure 8. SEM micrographs of PE/PS blends with 30 wt% PE content: A, nonreactive PE/PS-*co*-MAn; B, PE-*g*-DMAEMA (purified)/PS-*co*-MAn; C, PE-*g*-DMAEMA (crude)/PS-*co*-MAn; D, PE-*g*-TBAEMA (purified)/PS-*co*-MAn; E, PE-*g*-TBAEMA (crude)/PS-*co*-MAn; F, PE/PE-*g*-DMAEMA (crude)/PS-*co*-MAn 22.5/7.5/70. (Reproduced with permission from reference 39. Copyright 1992 John Wiley & Sons, Inc.)

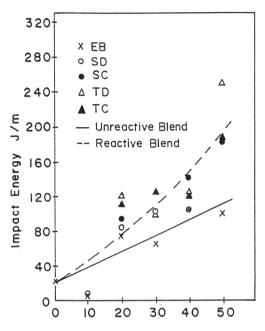

Figure 9. Variation of impact strength with PE content for blends of PE/PS-*co*-MAn: EB, LLDPE; SD, LLDPE-*g*-TBAEMA (purified); SC, LLDPE-*g*-TBAEMA (crude); TD, LLDPE-*g*-DMAEMA (purified); TC, LLDPE-*g*-DMAEMA.

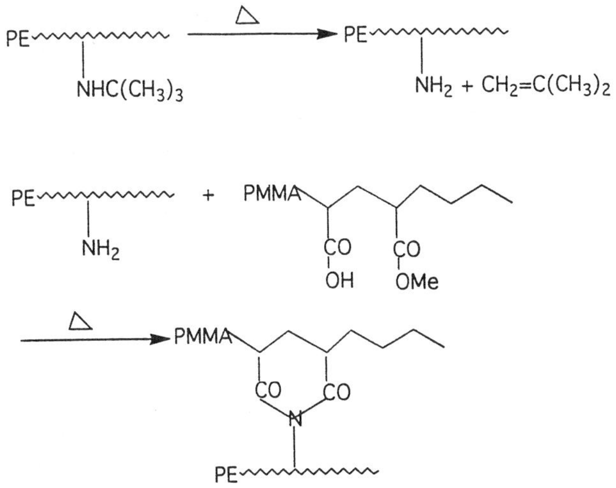

Scheme V. Proposed mechanism of imide formation in PE–PMMA reactive blends. (Reproduced with permission from reference 40. Copyright 1993.)

Song and Baker studied melt grafting of TBAEMA onto LLDPE by using peroxide initiator Lupersol 130 in a Banbury batch mixer (*29*). TBAEMA graft was characterized using IR and ¹H, and ¹³C NMR spectroscopies. The effects of feed composition and reaction time were investigated. By using 1.14 wt% Lupersol 130, 13 wt% monomer, and relatively short reaction time (2 min), a mixture containing 3 wt% grafted TBAEMA, 5 wt% TBAEMA homopolymer with minor cross-linking was obtained. The high graft level and grafting efficiency achieved in short reaction time were very attractive for continuous processes such as single- and twin-screw extrusion. DSC showed higher stability of secondary amine grafted polyethylene than the starting resin.

Trials were also conducted to increase the grafting and to minimize the homopolymerization by adding various reagents such as polymerization catalysts, chain-transfer agents, and chain terminators. *p*-Benzoquinone suppressed TBAEMA homopolymerization and significantly increased grafting efficiency, but other reagents had no effect on grafting efficiency or brought about adverse effects on grafting.

Reactions of secondary amine functionalized polyethylene with succinic acid, carboxylic acid modified polyethyelene, and copoly(styrene-maleic anhydride) were studied by IR (*29*). Both ionic interaction and covalent bonding (amide formation) were detected. Secondary amine grafted polyethylene was then used

in a subsequent study (*39*) for the compatibilization of PE/PS blends. Significant increase in impact energy and much finer morphology were observed for reactive blends relative to nonreactive blends. The use of TBAEMA functionalized PE in PE–poly(methyl methacrylate) blends resulted in enhanced polymer compatibility (morphology observed by SEM) and improved impact and tensile strength for reactive blends relative to nonreactive blends. The reaction sequence shown in Scheme V was proposed to explain the in-situ reactive compatibilization (*40*).

Recently, Hsu et al. found that polyethylene grafted with secondary amine can be used as Ziegler–Natta catalyst support material (*see also* Chapter 1.3). TBAEMA-grafted polyethylene was purified by dissolving the crude product in refluxing toluene and precipitating the polymer in methanol (*51*). In this way, TBAEMA homopolymer was eliminated from mixture. The secondary amine functionalized polyethylene was treated with BuMgCl in toluene and then with TiCl₄ to obtain a supported catalyst containing 1–2 wt% Ti. The catalyst showed a reactivity of 15–20 kg PE/g Ti per hour for ethylene polymerization under 30 psi pressure at 80 °C in heptane. However, no catalytic activity was observed for catalyst prepared using the same procedure but with the corresponding tertiary amine (DMAEMA) functionalized polyethylene. This result suggests the possible interaction of magnesium with the secondary amine as shown in Scheme VI, but no direct evidence was obtained for this reaction mechanism.

Primary Amine Functionalized Polyolefins

Primary amines are strong bases and nucleophiles. Their high reactivity makes primary amine functionalized polyolefins very attractive in polymer reactive compatibilization and other applications. However, monomers containing primary amines are less readily available than the secondary and tertiary ones, and they are rarely used to make primary amine functionalized polyolefins. Nevertheless, nonconventional free-radical chemistries were used successfully to prepare primary amine functionalized polyolefins, as exemplified subsequently.

PE Functionalization with Ammonia

Chapell and Brown (52) reported the use of low-pressure ammonia plasma in surface modification of extended chain polyethylene fibers, leading to the incorporation of primary amino groups to polyethylene fiber surface. The treated fiber surface was analyzed by X-ray photoelectron spectroscopy and spectrophotometry. Composites of ammonia plasma treated extended chain polyethylene fiber with epoxy and polyester resins showed a marked increase in interlaminar shear strength over composites made from untreated fiber. This observation could be explained by the reaction of the amino groups on PE with epoxy and ester functions on epoxy resins and polyester.

Copolymerization of Olefins Carrying Masked Amine

As mentioned previously, olefins containing primary and secondary amines cannot be directly polymerized by Ziegler–Natta catalysts, because their active hydrogen atoms deactivate the metal catalysts. This obstacle has been overcome by complexing the amines with metal atoms before polymerization (53). The complexing metal can be part of the catalyst formulation or specifically added to amino monomer. Such complexed amines are harmless to Ziegler–Natta catalysts. This characteristic allows the copolymerization of amine-containing olefins with ethylene, propylene, and other olefins. The primary amines were subsequently regenerated by alcoholysis during the deashing operation that is an integral part of Ziegler–Natta polymerization to obtain primary and secondary amine functionalized EPR and EPDM (Scheme VII) (33, 34).

IR and NMR spectroscopies were used for the characterization of the resulting amine-functionalized polyolefins. IR absorbance at 1619 cm^{-1} due to N–H groups was used to determine the concentration of

amine groups in EPR and EPDM. Gel permeation chromatography and solvent fractionation experiments showed uniform distribution of amino groups across the molecular weight distribution. The distribution of amino groups was believed to be random along the polymer backbone, but the reactivity ratios were not studied.

Primary amine functionalized polyolefins were used in the preparation of compatibilizers for polypropylene and EPR blends (54). The compatibilizers were prepared by reacting primary amine functionalized EPR with maleic anhydride functionalized polypropylene to form diblock graft copolymer EPR-g-PP (Scheme VIII). The formation of diblock graft copoly-

Masking (Protecting):

$$\text{Olefin-NHR} \xrightarrow[{-R^1\text{-H}}]{{+R^1\text{-Y}}} \text{Olefin-NR-Y}$$

Polymerization:

Ethylene + Propylene + Olefin-NR-Y ⟶ EPR∿∿∿
                                             |
                                           NR-Y

Demasking:

$$\text{EPR}\diagdown\hspace{-2pt}\underset{\text{NR-Y}}{|} + R^2\text{-OH} \xrightarrow[{-Y\text{-OR}^2}]{{+H^+}} \text{EPR}\diagdown\hspace{-2pt}\underset{\text{NHR}}{|}$$

R, R^1, R^2 = H or Alkyl   Y = Masking Group

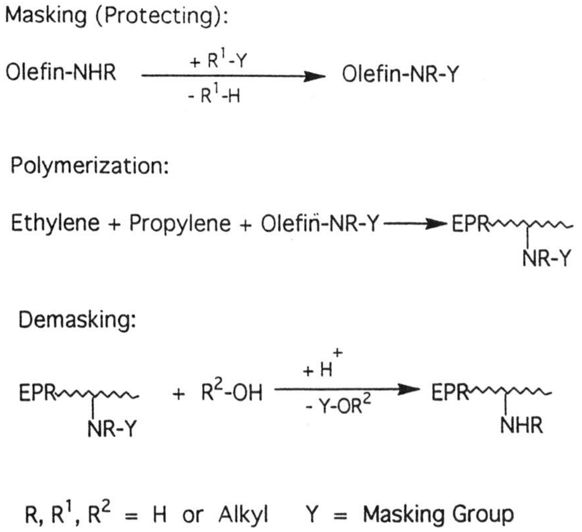

Scheme VII. Ziegler–Natta copolymerization of olefins with masked amines. (Adapted from references 33 and 34.)

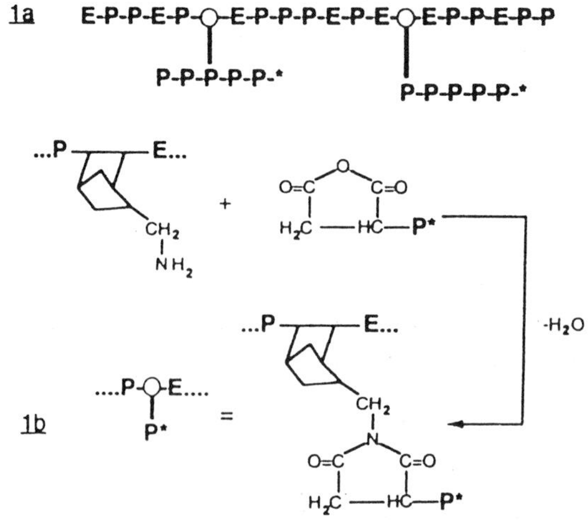

Structure (1a) and imidization reaction (1b) at the linking point for the EP-g-iPP polymer. P* is the polymeric fragment of the isotactic polypropene, and E and P are the ethene and propene residues of the ethene–propene copolymer.

Scheme VIII. Formation of diblock graft copolymer EPR-g-PP. (Reproduced from reference 53. Copyright 1992 American Chemical Society.)

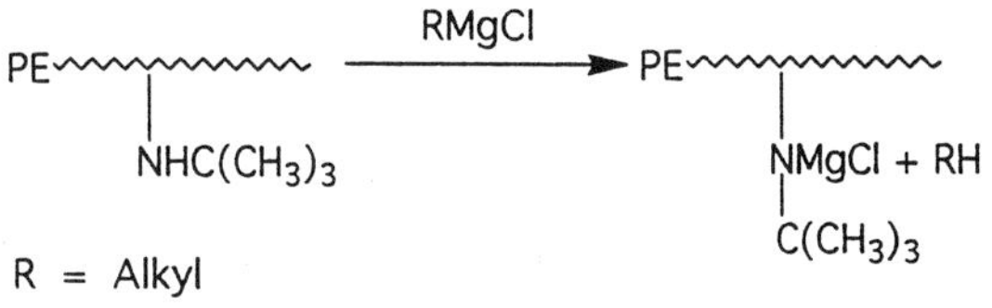

Scheme VI. Proposed reaction of PE-g-TBAEMA with BuMgCl.

mer EPR-*g*-PP was confirmed by IR spectroscopy, polymer fractionation, and DSC.

Addition of a small amount of the diblock compatibilizer to EPR/PP blends has a large effect on blend morphology and mechanical properties: EPR domain size was reduced by a factor of 4 (Figure 10), ductile to brittle transition temperature was dropped by 20 °C (Figure 11), and mechanical properties were significantly improved (*54*). Primary amine functionalized EPR was also used in the manufacturing of metal sludge and varnish dispersing agents in lubricating oil formulations (*55–57*).

Concluding Remarks

Amine-containing polymers are available in an extensive range of polymer type and functionality (primary, secondary, and tertiary), and they are among the most widely studied and technologically useful functional polymers. Major applications of amino polymers include ion exchange technology, solid-phase synthesis and catalysis, hydrometallurgy, and drug delivery. As a focus of this chapter, amine-functionalized polyolefins are emerging as a class of very useful specialty polymers. Polyolefins functionalized with tertiary and hindered secondary amino groups can be prepared by direct copolymerization or from preformed polyolefins by solution-, multiphase-, and melt-grafting processes. Polyolefins with primary amino groups can also be obtained by direct polymerization of masked amino-olefins, as well as by modification of polyolefins with ammonia plasma. Plasma grafting of ammonia to polyolefins is particularly useful because it uses inexpensive ammonia as grafting monomer

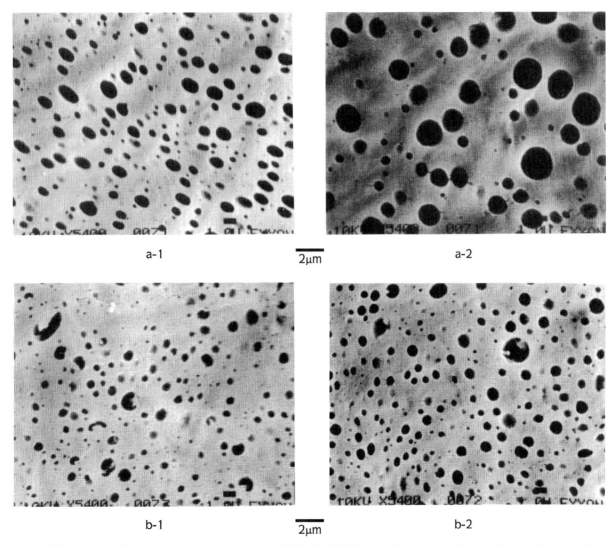

a-1 $\overline{2\mu m}$ a-2

b-1 $\overline{2\mu m}$ b-2

Figure 10. SEM graphs of hexane-extracted fractions of EPR-PP 20/80 blend before (a-1) and after (a-2) thermal annealing. Similar micrographs for EPR–PP 20/80 blend with 5% amine functionalized EPR before (b-1) and after (b-2) thermal annealing. (Reproduced from reference 54. Copyright 1993 American Chemical Society.)

and produces highly reactive primary amine functionalized polyolefins. The specific application of amine-functionalized polyolefins is to enhance the compatibility of these polymers in blends and composites. Thus, the development of amine-functionalized polyolefins is expected to facilitate the formation of polyolefin blends and composites.

Abbreviations

DEAEMA 2-(*N,N*-Diethylamino)ethyl methacrylate
DMAEMA 2-(*N,N*-Dimethylamino)ethyl methacrylate
DMAPA 3-(*N,N*-Dimethylamino)propylamine
DSC Differential scanning calorimetry
EPDM Ethylene propylene diene modified rubber
EPR Ethylene propylene rubber
LLDPE Linear low density polyethylene
MAn Maleic anhydride
PE Polyethylene
PP Polypropylene
iPP Isopolypropylene
SEM Scanning electron microscopy
TBAEMA 2-(*N-t*-Butylamino)ethyl methacrylate

References

1. Saegusa, T.; Ikeda, H.; Fuji., H. *Macromolecules* **1972**, *5*, 108. This communication also contains reference to ring opening polymerization of ethyleneimine.
2. Seko, M.; Myaka, T.; Imamura K. Jpn. Kokai 76,73087, 1976.
3. Inaki, Y. *The Synthesis of Enzyme-Like Polymers, in Functional Monomers and Polymers*; Takemoto, K.; Inaki, Y.; Ottenbrite, R. M., Eds.; Marcel Dekker: New York, 1987; pp 507–539.
4. Suh, J. *Acc. Chem. Res.* **1992**, *25*, 273–279
5. Suh, J.; Lee, K. S.; Paik, H-J. *Inorg. Chem.* **1994**, *33*, 3–4.
6. George, M. H.; Arshady, R.; Kakati, D. K.; Al-Shahi, W.; Gosain, R. *Polymeric Materials Encyclopedia*; Salmone, J. C., Ed.; in press.
7. Tanaka, H. *J. Polym. Chem. Polym. Lett. Ed.* **1978**, *16*, 87–89.
8. Ferere, Y.; Gramain, P. *Macromolecules* **1992**, *25*, 3184–3189
9. Maillard-terrier, M. C.; Caze, C. *Eur. Polym. J.* **1984**, *20* 113–117.
10. Fasman, G. D.; Idelson, M.; Blout, E. E. *J. Am. Chem. Soc.* **1961**, *83*, 709–712.
11. Arnold L. J., Jr. *Meth. Enzymol.* **1985**, *112A*, 270–285.
12. Arshady, R. *Colloid Polym. Sci.* **1990**, *268*, 948–958.
13. Arshady, R. *Polymer* **1982**, *23*, 1099–1100.
14. Arshady, R. *Adv. Polym. Sci.* **1993**, *111*, 1–41
15. Arshady, R. *Trends Polym. Sci.* **1993**, *1*, 86–91
16. *World Polyethylene Industry 1982–1983*; Chem Systems Inc.: Tarrytown, NY, 1983.
17. *Crystalline Olefin Polymers*; Raffe, R.; Doak, K. W., Eds.; John Wiley & Sons: New York, 1965.
18. Doak, K. W. *Encycl. Polym. Sci. Eng.* **1986**, *6*, 383.

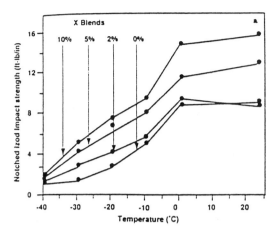
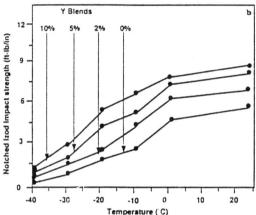
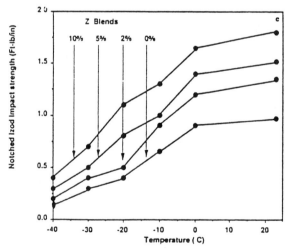

Figure 11. Notched Izod impact strength as a function of temperature and the amount of graft diblock EPR-*g*-PP copolymer compatibilizer for blends of EPR–PP (a) 20/80, (b) 15/85, and (c) 20/80 (high molecular weight PP). (Reproduced from reference 54. Copyright 1993 American Chemical Society.)

19. *Polymer Blends and Composites in Multiphase Systems;* Han, C. D., Ed.; Advances in Chemistry 206; American Chemical Society: Washington DC, 1984.
20. *Polymer Blends;* Paul, D. R.; Newman, S., Eds.; Academic: Orlando, FL, 1978.
21. *Frontiers in Materials Technology;* Myers, M. A., Ed.; Elsevier: Amsterdam, Netherland, 1985.
22. Wong, C. S. U.S. Patent 5,091,260, assigned to DuPont Canada Inc.; February 25, 1992.
23. Liu, N. C.; Baker, W. E. *Adv. Polym. Tech.* **1992,** *11,* 249.
24. Degorre, F.; Kiffer, D.; Laloi, M.; Terrier, F. *Soc. Chim. Fr.* **1988,** 415.
25. *Functional Monomers;* Yocum, R. H.; Nyquist, E. B., Eds.; Marcel Dekker, Inc.: New York, 1974; Vol. 2.
26. Langer, A. W., Jr. US Patent 4,094,818, assigned to Exxon, June 13, 1978.
27. Immerzi, B.; Lanzetta, N.; Laurienzo, P.; Maglio, G.; Malinconico, M.; Martuscelli, E.; Palumbo, R. *Makromol. Chem.* **1987,** *188,* 951.
28. Simmons, A.; Baker, W. E. *Polym. Eng. Sci.* **1988,** *29,* 1117.
29. Song, Z.; Baker, W. E. *Polymer* **1992,** *33,* 3266.
30. Xie, H. Q.; Seay, M.; Oliphant, K.; Baker, W. E. *Polym. Prepr. (Am. Chem. Soc. Div. Polym. Chem.)* **1992,** *67,* 110.
31. Xie, H. Q.; Seay, M.; Oliphant, K.; Baker, W. E. *J. Appl. Polym. Sci.* **1993,** *48,* 1199.
32. Felmine, J. B.; Baker, W. E.; Russell, K. E. *Polym. Prepr. (Am. Chem. Soc. Div. Polym. Chem.)* **1991,** *32,* 181.
33. Datta, S. In *High Value Polymers;* Fawcett, A., Ed.; Royal Society of Chemistry: London, 1990; pp 33–57.
34. Datta, S.; Kresge, E. N. U.S. Patent 4,987,200, January 22, 1991.
35. Song, Z.; Baker, W. E. *Angew. Makromol. Chem.* **1990,** *181,* 1
36. Song, Z; Baker, W. E. *J. Appl. Polym. Sci.* **1990,** *41,* 1299.
37. Pouchert, C. J. *The Aldrich Library of Infrared Spectra,* 2nd ed.; Aldrich Chemical Company: Milwaukee, WI, 1975; p 147.
38. Simmons, A.; Baker, W. E. *Polym. Commun.* **1990,** *31,* 20.
39. Song, Z.; Baker, W. E. *J. Appl. Polym. Sci.* **1992,** *44,* 2167.
40. Liu, T.; Xie, H. Q.; O'Callaghan, K. J.; Rudin, A.; Baker, W. E. *J. Polym. Sci., B. Polym. Phys.* **1993,** *31,* 1347.
41. Roberts, A. D. *Langmuir* **1992,** *8,* 1479.
42. Bonvicini, A.; Monaci; Cappuccio, V. U.S. Patent 3,131,990, May 5, 1964.
43. Greco, P.; Laurienzo, P.; Maglio, G.; Malinconico, M.; Martuscelli, E. *Makromol. Chem.* **1986,** *188,* 961.
44. Oliphant, K.; Baker, W. E. *Proceeding of the 52nd Annual Technical Conference of the Society of Plastics Engineers;* Society of Plastics Engineers: Brookfield, CT, 1994.
45. Dokolina, G. S.; Tur'yan, Ya. I; Korshunov, M. A. *Zh. Obshch. Khim.* **1969,** *39,* 1203; *Chem. Abstr. 71,* 80554n.
46. Tur'yan, Ya. I.; Dokolina, G. S.; Korshunov, M. A. *Zh. Obshch. Khim.* **1970,** *40,* 1894; *Chem. Abstr. 74,* 87203t.
47. Pradny, M.; Sevcik, S. *Makromol. Chem.* **1985,** *186,* 111.
48. Odian, G.; Acker, T.; Sobel, M. *J. Appl. Polym. Sci.* **1963,** *7,* 245.
49. Odian, G.; Acker, T. *U.S. At. Energy Comm.* **1962,** *TID-7643,* 233.
50. Odian, G. G.; Sobel, M.; Rossi, A.; Klein, R. *J. Polym. Sci.* **1961,** *55,* 663.
51. Mteza, S.; Bacon, D. W.; Hsu, C. C. Department of Chemical Engineering, Queen's University, Kingston, Ontario, K7L 3N6, Canada; unpublished results.
52. Chapell, P. J. C.; Brown, J. R. *Surf. Interface Anal.* **1991,** *17,* 143.
53. Datta, S.; Morrar, F. T. *Macromolecules* **1992,** *25,* 6430.
54. Datta, S.; Lohse, D. J. *Macromolecules* **1993,** *26,* 2064.
55. Patil, A. O.; Datta, S.; Gardiner, J. B.; Lundeberg, R. D. U.S. Patent 5,030,370, assigned to Exxon Chemical Patents, Inc., July 9, 1991.
56. Patil, A. O.; Datta, S.; Lundeberg, R. D. Eur. Patent Application EP 432939, assigned to Exxon Chemical Patents, Inc, June 19, 1991.
57. Patil, A. O.; Datta, S.; Lundeberg, R. D. PCT International Application WO 91 13954, assigned to Exxon Chemical Patents, Inc., March 8, 1990.

Metal-Containing Polymers and Interpenetrating Networks

Reza Arshady, Benedetto Corain, and Marco Zecca

Metal-containing polymers (MCPs) represent a very broad range of macromolecular compounds and polymeric materials that may carry the metal atoms in the main chain or side chains. MCPs are produced by direct polymerization, polycondensation, and polycoordination, as well as by the introduction of metallic species into preformed polymers. In most cases, the metal is attached to the polymer via chelation or coordination bonds, but many MCPs are also based on ionic and covalent polymer–metal links. MCPs are of considerable interest as new materials for emerging technologies in electronics, sensors, specialty coatings, and catalysis. This chapter provides a general introduction to the subject of MCPs and discusses their synthesis and properties. Recent work in our laboratories on the synthesis and characterization of metal-containing interpenetrating polymer networks (MC-IPNs) is also discussed, and potential use of MC-IPNs as polymer-supported transition metal catalysts is illustrated.

Metal-containing organic polymers (or organometallic polymers) represent a very broad spectrum of materials, ranging from vinyl polymers with metal-containing side chains to condensation and coordination polymers carrying metal atoms in the main chains. An extensive range of metal-containing polymers with various degrees of complexity and sophistication have been described during the past three decades (1, 2). Chemical structures of typical examples of these polymers are shown in Figure 1.

The terms metallo-organic polymer (MOP), organometallic polymer (OMP), and metal-containing polymer (MCP) are used by different authors synonymously or for describing different classes of polymers shown in Figure 1. For the present discussion, we adopt MCP as a general term, MOP for polymers containing metal ions, and OMP for all other metal-containing polymers. This terminology is not completely satisfactory, because it leaves at least three groups of structurally different polymers indistinguishable from each other. However, the adoption of a more elaborate terminology is beyond the scope of the present review.

MCPs are obtained by a variety of synthetic routes as may be expected from the types of structures shown in Figure 1. Different synthetic routes may involve vinyl polymerization, polycondensation, polycoordination, or the chemical modification of preformed polymers. The creation of sophisticated MCP structures per se presents a fascinating field of activity for organometallic and synthetic polymer chemists. However, interest in the synthesis and study of MCPs is largely related to their applications in, among others, electronics, membrane processes, coatings, and catalysis.

We have in recent years reported the synthesis of a number of soluble MCPs of potential interest for electrode modification (3–5) and beaded MCPs as polymer-supported (or hybrid) transition metal catalysts (6–9). We have also developed a novel synthetic approach for the synthesis of a new class of MCPs in the form of beaded interpenetrating networks (10–12). These new materials are produced by polymerization of metal-containing monomers within the matrixes of preformed beaded polymers, and they provide ready

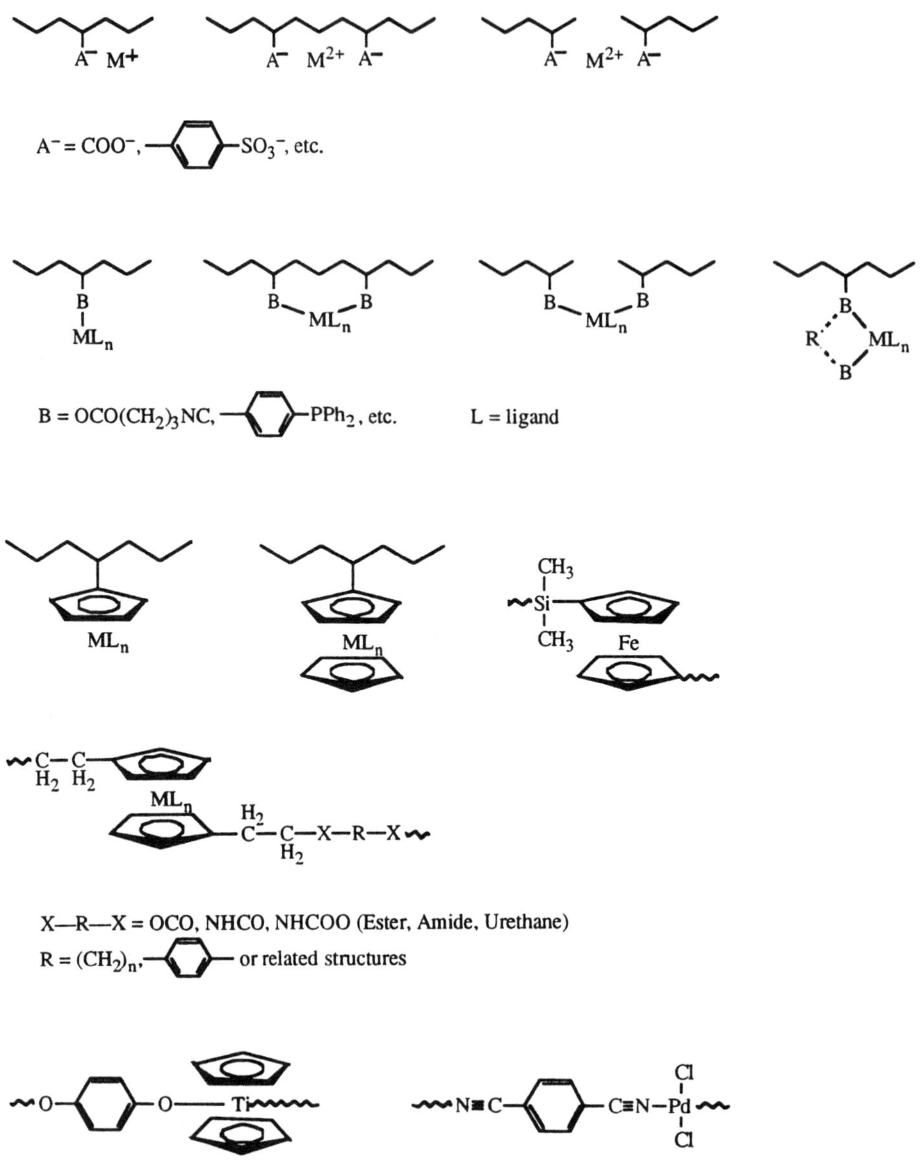
A⁻ M+
A⁻ M2+ A⁻
A⁻ M2+ A⁻
A⁻ = COO⁻, —⟨⟩—SO3⁻, etc.
B
MLn
B = OCO(CH2)3NC, —⟨⟩—PPh2, etc.
L = ligand
R
MLn
CH3
Si
CH3
Fe
C—C
H2 H2
C—C—X—R—X
X—R—X = OCO, NHCO, NHCOO (Ester, Amide, Urethane)
R = (CH2)n, —⟨⟩— or related structures
O
Ti
N≡C
C≡N—Pd
Cl

Figure 1. Typical examples of metal-containing polymers.

access to beaded resins containing highly dispersed organometallic or metallic species.

This chapter presents a brief overview of metal-containing monomers and polymers, followed by a review of our own work on polymeric transition metal complexes and metal-containing interpenetrating networks (MC-IPNs). Characterization of the new materials by electron microscopy, X-ray microprobe analysis, Mössbauer spectroscopy, swelling behavior, and gel permeation are discussed. The potential of MC-IPNs as polymer-supported transition metal catalysts is also outlined. Two specific examples of organometallic polymers, namely polymers of stannylene and germylene, are fully covered in another chapter, and the use of a number of MCPs for energy transfer studies and sensor applications is discussed in other chapters in this volume. Literature on MCPs is extensive, including a number of recent books and reviews with key word titles of metal-containing monomers (*2, 13*), organometallic polymers (*14–17*), metal-containing polymers (*13, 18, 19*), inorganic and metal-containing polymers (*1, 20, 21*) and applications of organometallic polymers (*22*). For a general text on organometallic chemistry, and the related topic of metal–polymer composites, *see* references 23 and 24, respectively.

Metal-Containing Monomers

Typical structures of metal-containing monomers can be gleaned from Figure 1, except that some of the structures may not be stable in monomer form. The simplest metal-containing monomers are those carrying metal ions, and they include alkali salts of (meth)-acrylic acid, (meth)acryloxysulfonic acid, 4-vinylben-zenesulfonic acid (23), and propanediolsulfonates and related ionic diols used for the synthesis of polyure-thane ionomers (25). A whole series of metal acrylates, methacrylates, sulfonates, and related compounds are commercially available from specialist monomer sup-pliers. However, the synthesis of organometallic monomers is somewhat more complicated. Metal-con-taining monomers that have a truly organometallic structure (i.e., M–C bond) can be obtained from the corresponding Grignard reagents by displacing Mg–X with the desired metal or metal derivative (26, 27). But this synthetic route is of limited interest for the pur-pose of the present discussion.

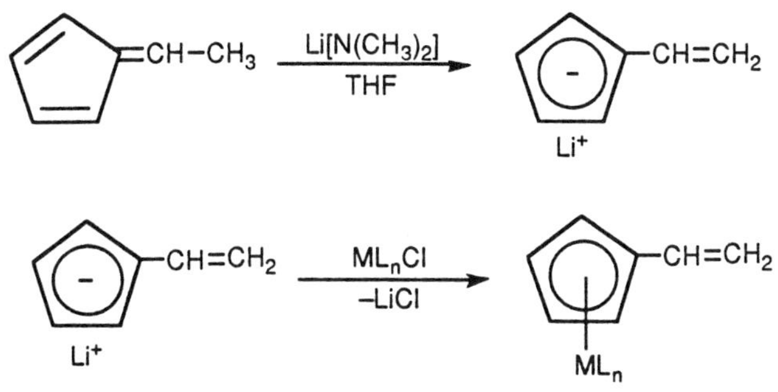

M = Cu, Rh, Ir, Ti, Mo, W

Figure 2. General route for the synthesis of η^5-(cyclopen-tadienyl)-metal monomers. (Adapted from reference 9.)

Metal-containing monomers in which the metal is attached to aromatic structures (π-complex) have, on the other hand, attracted considerable attention. One particularly interesting example is the general route reported by Macomber et al. (28, 29) for the synthesis of (η^5-vinyl cyclopentadienyl)-metal type monomers (Figure 2).

The synthesis of a wide range of titanium mono- and diacrylates and methacrylates and allyloxy deriva-tives of dicyclopentadienyl titanium (Cp-Ti) from the corresponding $(Cp)_2TiCl_2$ has been reported by Russ-ian workers. This group also described an extensive variety of tetraalkoxy derivatives of titanium $(C_4H_9O)_3$-TiOR), in which the R group contains a double and/or triple bond with different structures. These derivatives are obtained by trans-esterification of the correspond-ing tetrabutoxy derivative with the desired HO–R (2).

Another class of MCPs are those in which the metal center is complexed (or chelated) by a uni- or multidentate ligand. From an organometallic point of view, the synthesis of this class of monomers is straightforward. It involves the synthesis of functional monomers carrying the desired ligand, followed by a ligand exchange reaction. The alternative route (i.e., the generation of a polymerizable vinyl residue on the metal complex) is also applicable if the metal–ligand bonds are sufficiently stable. This topic has been reviewed recently (13), and structures of typical monomers carrying metal chelate are shown in Figure 3. Note that the last structure in Figure 3 represents an interesting chelating species suitable for both polycon-

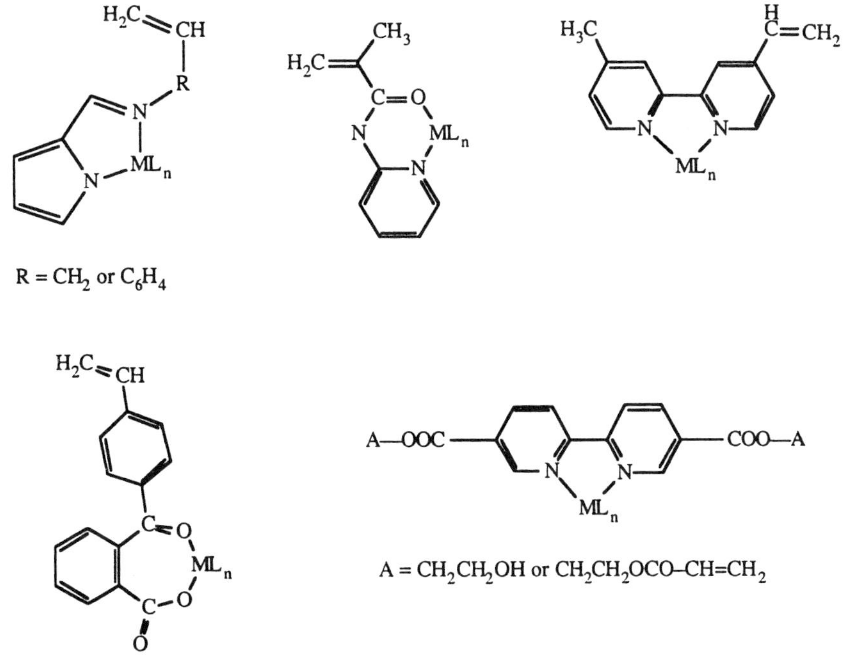

Figure 3. Structures of typical polymerizable chelating ligands (metal-chelating monomers). The chelated M is generally a transition metal such as Cu, Fe, Co, Cr, Pd, or Ru in their medium-high oxidation state.

densation (A = CH$_2$CH$_2$OH) and vinyl polymerization (A = CH$_2$CH$_2$O–OCCH=CH$_2$).

An interesting class of metal-containing monomers based on a monodentate ligand is the cis-square planar isocyano derivative indicated in Figure 4 (30, 31). A wide range of isocyanoalkyl acrylates and their formamido precursors are readily accessible (32–34) from the corresponding amino alcohols. Monomers used for the synthesis of MCPs via polycondensation and polycoordination are usually complementary monomer pairs: that is, one monomer is the metal center and the other is a difunctional (dichelating) compound (see below).

Metal-Containing Polymers

The variety of MCPs indicated in Figure 1 is obtained by four different routes, depending on the availability, stability, and polymerizability of the starting monomers:

1. polycondensation of metal-containing monomers (with complementary difunctional monomers)
2. polycoordination (polychelation) of transition metal centers with bi-bidentate ligands
3. addition polymerization or copolymerization of metal-containing monomers
4. attachment of metal centers to preformed polymers

Methods 1 and 2 are, from a synthetic point of view, relatively simple. However, these two methods are less generally employed in practice, and hence their chemistry is less fully developed than those of methods 3 and 4. In general terms, the chemistries of polycondensation and polycoordination are rather similar. Therefore, these two method are discussed briefly together, followed by the discussion of vinyl polymerization and polymer modification in somewhat more detail.

Condensation and Coordination Polymers

A variety of metal-containing polycondensates can be obtained by the reaction of metal dihalides with dinucleophiles. In these reactions the metal dihalide behaves rather similar to a diacid chloride (Figure 5) (35–39). These polycondensation reactions are best carried out in nonaqueous media or under interfacial conditions and in the presence of a base (tertiary amine > HO$^-$). The reaction usually proceeds to completion very rapidly (within 2 min) with or without a catalyst. Metal-containing polycondensates can also be obtained by difunctional monomers carrying metal centers as part of their structure, rather than their reactive group. An interesting example is the Knoevenagel polycondensation represented by the last reaction shown in Figure 5 (40–42).

Another recent example was described by Dembek et al. (43), which involves the polycondensation of (p-phenylenediamine)Cr(CO)$_3$ with terephthaloyl chloride to produce the corresponding organometallic liquid crystalline polymer. The presence of the metal center in the polymer is reported to enhance polymer solubility, and hence to facilitate the processing of the intractable aramide-type polymer. Following the processing (e.g., film formation), the metal can be removed from the polymer by oxidative decomposition.

Molecular weight of metal-containing polycondensates may vary widely depending on the metal, two complementary monomers, and most importantly on the exact details of stoichiometry and reaction conditions in common with other polycondensations in general. For titanocene ether, for example, molecular weights of up to 100,000 were reported (37). Polymer properties also depend on the metal and the chemical structures of the ligands and organic units. Metal-containing polycondensates have been examined for a wide range of applications, as will be highlighted in the final section.

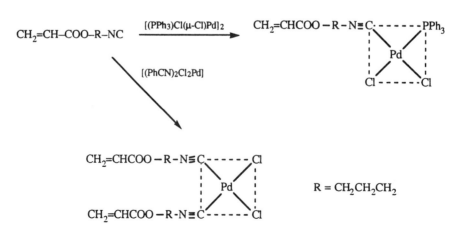

Figure 4. Synthesis of cis-square planar complexes of Pd with 3-isocyanopropyl acrylate.

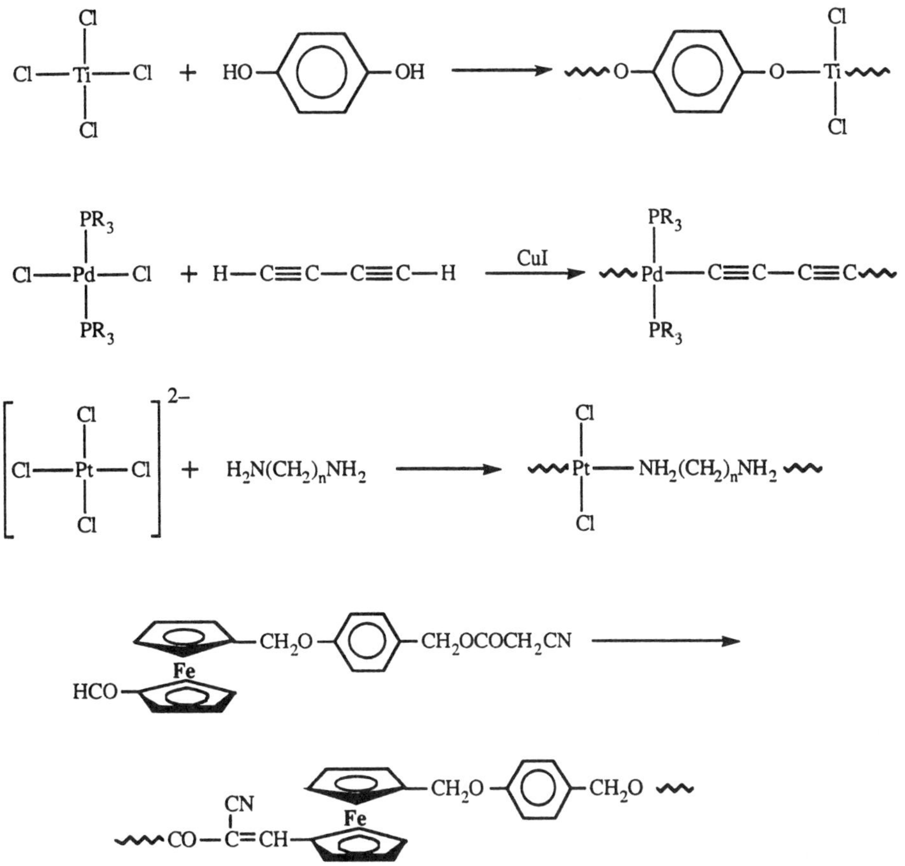

Figure 5. Examples of metal-containing polymers available via polycondensation reactions.

Metal-containing coordination polymers (i.e., coordination MCPs) are essentially similar to condensation MCPs, except that the metal–organic linkages are often (but not always) bidentate coordination bonds with various degree of ionic character. Preparation of MCPs by polycoordination (polychelation) follows the usual procedures of ligand exchange in organometallic chemistry (23), and a number of general guidelines were expressed by Bailor (44). A wide variety of structural configurations (including linear, sheet-like, tridimensional, and rigid rods) are obtained, depending on the metal center and the coordinating or chelating ligand (18, 45, 46). The majority of reported coordination MCPs involve bis-bidentate (chelating) ligands. However, many bis-monodentate ligands also produce stable coordination polymers with various transition metal centers. Figure 6 shows different structures proposed for coordination polymers of Pd, Pt, and Rh with diisocyanides as an example of bis(monodentate) ligands.

Coordination polymers of transition metals with quinine and quinoids have attracted considerable interest as redox active materials, in which polymer conductivity can be controlled by altering the structures of both bridging ligand (the quinine) and the metal center (i.e., the metal and its other ligands) (47,

48). A recent example is the orthoquinone polymer $[Ru_2C_6H_4O_2(CO)_4]_n$. This new polymer is suggested to represent the first example of a two-dimensional structure, which is produced by the contribution of both the bidentate *ortho*-oxygen ligand and the dienyl sphere (47).

A highly intriguing class of MCPs with repeat units of *p*-phenylene-2-5-(cobaltocyclopentadiene) recently was produced by polymerization of 1,4-phenyldiacetylene with cobaltocyclopentadiene diphenylphosphine [CpCo(PPh$_3$)$_2$]. This new method of MCP formation, termed metal cyclizing polymerization (49), probably involves a concerted reaction pathway comprising the polyaddition of the triple bonds, simultaneous with (or followed by) polycondensation–polycoordination of the resulting species with [CpCo(PPh$_3$)$_2$].

Addition Polymers

Polymerization and copolymerization of metal-containing vinyl monomers can, in principle, be accomplished by free radical, anionic, and cationic mechanisms. In practice, however, the success of each of these mechanisms depends mainly on two factors: the distance (i.e., the number of bonds) between the metal and the vinyl group, and the inertness or interference

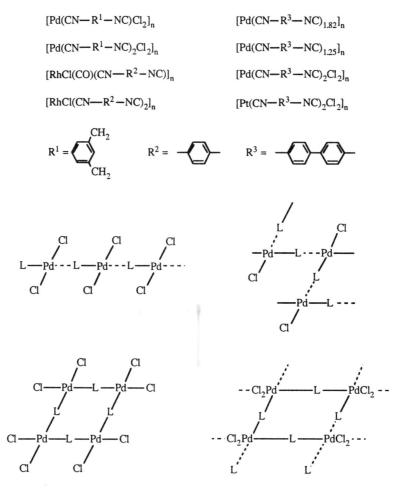

$[Pd(CN-R^1-NC)Cl_2]_n$

$[Pd(CN-R^1-NC)_2Cl_2]_n$

$[RhCl(CO)(CN-R^2-NC)]_n$

$[RhCl(CN-R^2-NC)_2]_n$

$[Pd(CN-R^3-NC)_{1.82}]_n$

$[Pd(CN-R^3-NC)_{1.25}]_n$

$[Pd(CN-R^3-NC)_2Cl_2]_n$

$[Pt(CN-R^3-NC)_2Cl_2]_n$

Figure 6. Structures proposed for coordination polymers of Pd, Pt, and Rh with diisocyanides (*see* references 9, 10, 45).

of the metal center toward the active center of propagating species.

Polymerization of monomers in which the metal is directly attached to the vinyl group (e.g., CH_2=CH–MLn) apparently has not been studied. The metallic side chain in most such monomers would probably be too bulky for homopolymerization, although their copolymerization with non-metallic monomers may be practicable. An additional difficulty may be the interference of the metal atom toward the directly bound active center of the polymerization reaction (i.e., $-CH_2-C^*H(MLn)$, where the asterisk represents a free radical, anionic, or cationic species). Another class of metal-containing vinyl monomers that deserves special mention is metal salts of (meth)acrylates and vinylbenzene sulfonates. These monomers can be polymerized routinely by free-radical initiators (usually in aqueous solutions) but not by ionic mechanisms.

Between the two classes of monomers outlined above, exists a broad range of other metal-containing monomers in which the metal center is attached to the monomer side chain via covalent or coordination bonds (*see* Figures 3 and 4), and it is the (co)polymerization of these monomers that has attracted the most

attention. A great deal of literature in this area involves the study of vinylmetallocenes, and several reviews of the subject are available (*1, 2, 13–22*).

The general observation is that most vinyl monomers containing a covalent or coordinated metal center polymerize by free-radical initiators to produce high molecular weight polymers. Copolymerization of a wide range of metal-containing monomers with non-metallic monomers has also been reported, and the corresponding copolymerization data are available (*2, 14, 18*). In most cases, however, branching, crosslinking, and other side reactions may take place because of chain transfer with the organic moieties directly attached to the metal. Azo initiators are usually preferred for these polymerizations, because peroxy initiators may lead to oxidation of the metal. Cationic polymerization using Lewis acids is also feasible for many monomers, but anionic polymerization is practicable only when the metal does not destabilize or scavenge the anionic active center.

The polymerization of metal-containing monomers in which metallocenes (*50, 51*) or pentadienetricarbonyl iron (*52*) are separated from the double bond has also been studied. Interesting examples of metal-containing vinyl monomers unrelated to metallocenes are the two Pd-containing monomers shown in Figure 4 (*30, 31*). Most other metal-containing vinyl monomers, whose structures can be gleaned from Figures 1–4 also can be produced by known synthetic methods. In practice, however, vinyl type MCPs are more readily accessible by metal incorporation into preformed polymers, as outlined below.

Metal Incorporation into Preformed Polymers

From a practical point of view, incorporation of metal centers into side chains of vinylic polymers presents a relatively convenient route to metal-containing polymers. This route is usually the method of choice for the synthesis of polymer-supported transition metal catalysts in beaded form (i.e., polymer supports). In common with other functional monomers, the genera-

tion of the desired complexing or chelating ligands on the polymer may be accomplished by direct copolymerization or polymer functionalization (*8, 9, 53–55*).

In our work on transition metal complexes of beaded isocyano polymer supports, we adopted a compromise approach in which immediate precursors of the isocyano polymers are obtained in the form of their formamido derivatives by direct suspension copolymerization, followed by dehydration to produce the desired isocyano ligands on the polymer (Figure 7) (*7–9*).

Incorporation of the desired metal into the polymer (i.e., the formation of polymeric metal complex or chelate) follows the principle of ligand exchange reactions (*23*). However, the multisite nature of polymeric ligands, and particularly that of cross-linked polymers, creates a special situation where the number of polymer-bound ligands coordinated to each metal center is

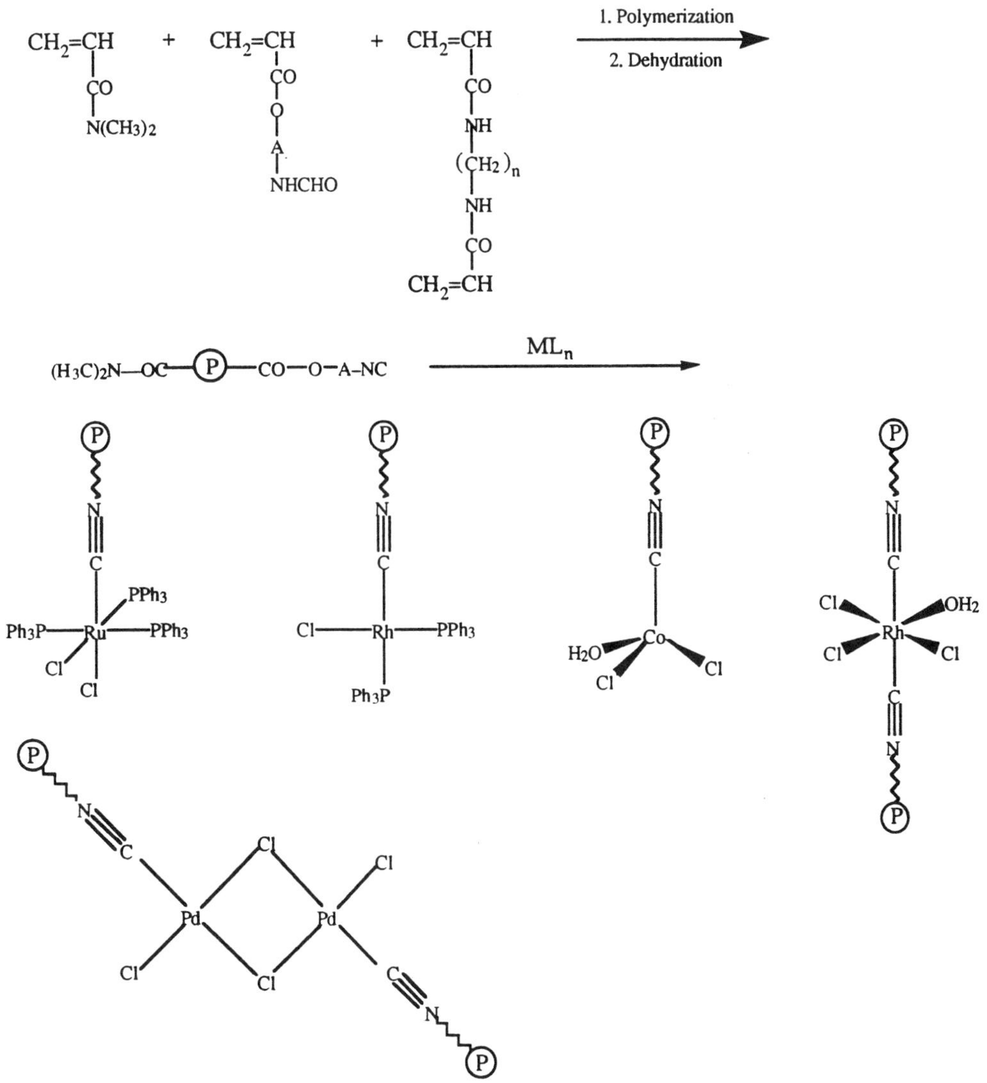

Figure 7. Example of polymer-supported transition metal complexes produced by incorporation of metal into preformed polymers carrying isocyano ligand.

determined by all of the experimental variables, rather than simply the metal-to-ligand ratio. This means that, in practice, often a mixture of complex species is formed on the polymer, and that highly coordinated metal centers on the polymer may form as a result of slow diffusion of the metal substrate, whatever the metal-to-ligand ratio (9).

We have also employed a similar compromise method for the synthesis of three classes of soluble metal-containing polymers based on polystyrene (3), maleic anhydride copolymers (4), and copolymers of styrene with acrylates and acrylamides (5). All of these polymers are obtained by copolymerization of functional monomers, followed by conversion of the functional groups to the desired ligands on the polymers, and subsequent introduction of metal centers by lig-

and exchange. Structures of typical examples of these metal-containing polymers that were produced as potentially useful materials for electrode modification (electrocatalysis) are shown in Figure 8. These structures are in good agreement with the analytical data for the polymers containing about 3–25% metal (3–5). Our synthetic method for the preparation of styrene-based metal-containing polymers also was adapted for the synthesis of related polymers for electron transfer and photochemical studies (56, 57).

Metal-Containing Interpenetrating Networks

Interpenetrating polymer networks (IPNs) (58–62) are a special class of polymer composites (63–65) in which

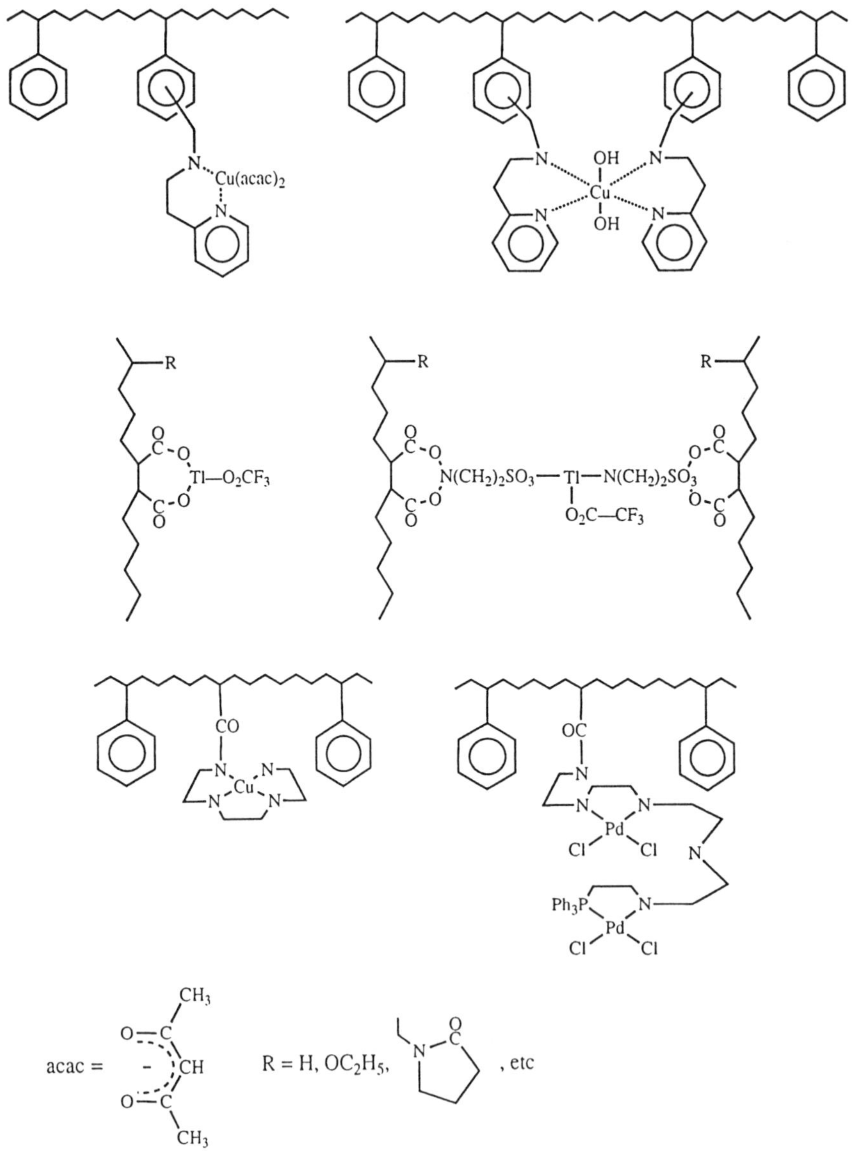

Figure 8. Typical examples of metal-containing polymers based on polystyrene, copolymers of maleic anhydride with vinyl monomers, or styrene with acrylamides. (Adapted from references 3–5.)

two tridimensional polymer networks (or matrixes) are intermeshed (or interwoven) together. In this section we present an outline of our recent work and related literature reports on the synthesis and characterization of IPNs in which one of the polymer networks contains metal centers in the form of complex or ionic species. For the purpose of the present discussion we use the term IPN for systems in which both networks are covalently cross-linked. To our knowledge, there is only one report on the synthesis of metal-containing semi-IPNs (i.e., systems in which only one of the networks is covalently cross-linked) (66). For more elaborate terminology and definitions of different types of IPNs, *see* references 61–65.

In the field of materials engineering, polymer IPNs (and polymer composites in general) are developed to combine the mechanical properties of two or more individual polymers. For functional polymers, composite formation may facilitate synthetic and manufacturing processes, and/or enhance mechanical and handling properties of the resulting functional polymers. The aim of our initial approach in this field (10) was to produce beaded functional polymers (including metal-containing polymers) from monomers that could not be converted into beaded polymers by conventional polymerization methods. In the case of MCPs for catalytic applications, an additionally important bonus of composite formation is the generation of highly dispersed metal complexes or metallic powders within preformed beaded polymeric materials.

Our synthetic scheme is illustrated in Figure 8. It involves the impregnation of a suitably chosen beaded cross-linked polymer with a solution of the desired monomer (or monomer mixture) containing the initiator, followed by polymerization. This evidently leads to the formation of a second network (matrix) within the existing network (matrix), and hence the formation of two "interpenetrating" networks. When desired, we adjust experimental conditions in such a way so that the impregnating monomer solution is completely absorbed by the polymer beads, and hence polymerization takes place virtually inside the beads.

As long as the preformed polymer network (or matrix) is chemically inert and does not affect the course of the polymerization, the formation of the second network may be accomplished by either of the initiator systems known to be suitable for the polymerization of the desired metal-containing monomer (*see* above). Our initial studies on the formation of beaded IPNs (10) involved the use of azo-bis-isobutyronitrile (AIBN) as free radical initiator. However, all of our MC-IPNs discussed here were produced by γ-ray-induced free-radical polymerization (11).

In our first attempt (10), we also formed coordinated poly(isocyano-palladium) species of the type shown in Figure 6 inside cross-linked polydimethylacrylamide (PDMA) beads, but the study of this system was not pursued. Subsequently, a covalently cross-linked polymer of vinylferrocene and dimethylacrylamide was chosen as a model organometallic network, and the preformed nonmetallic network was the above mentioned PDMA. Both the non-metallic PDMA and the organometallic IPNs were formed by γ-radiation-induced polymerization (11). Figure 10 shows the separate structures of the two interpenetrating networks and an X-ray microprobe of the IPN product. As clearly evident from the X-ray microprobe picture, the Fe-containing polymer is homogeneously distributed throughout the IPN.

To ascertain the integrity of the ferrocenyl residues during the polymerization, the Fe-containing

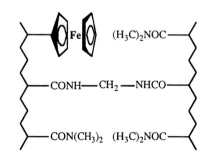

Organic Matrix

Organometallic Guest

Figure 10. Chemical structures of the two separate networks (matrixes) of beaded IPNs. The organometallic matrix is formed by radiation-induced polymerization inside the organic matrix (preformed polydimethylacrylamide beads).

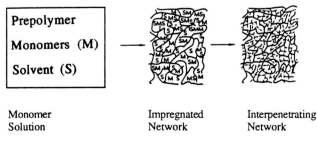

Prepolymer		
Monomers (M)		
Solvent (S)		

Monomer	Impregnated	Interpenetrating
Solution	Network	Network

Figure 9. General scheme for the synthesis of beaded IPNs.

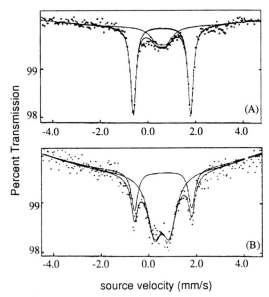

Figure 11. Mössbauer spectra of Fe-containing IPN (*see* Figure 10) A, before and B, after oxidation with hydrogen peroxide. (Adapted from reference 11.)

IPN was examined by Mössbauer spectroscopy, and the spectrum was indistinguishable from that of ferrocene (Figure 11A). This result indicates that the organometallic structure is not affected under the chosen polymerization conditions (*11*). Mössbauer spectroscopy was also useful in the evaluation of the chemical accessibility of the metal centers inside the IPN. Figure 11 shows the spectra of the IPN before (A) and after (B) oxidation with hydrogen peroxide. The oxidation reaction was also studied quantitatively by using one equivalent of aqueous $AgNO_3$ and was found to proceed to completion within 22 h (30 min, 63%; 3 h, 87%) (*67*).

Following the synthesis and characterization of the ferrocene-containing IPN, we focused our attention on palladium-containing IPNs for their potentials as polymer-supported catalysts. The first example of a Pd-containing IPN recently was prepared by copolymerization of Pd-(3-isocyano-propyl acrylate) (*see* Figure 4) with DMA and methylene-bis(acrylamide) within a preformed DMA matrix. The polymerization was carried out by γ-ray-induced polymerization in essentially the same way as described for Fe-containing IPNs (*11*). Figure 12 shows the X-ray microprobe images of chlorine and palladium distribution in a cross-section of the composite. Here too, it is evident that the Pd-monomer is homogeneously distributed throughout the bulk of the IPN. For comparison, the Pd-containing matrix was also prepared without the preformed DMA matrix, and in both cases the isocyano-Pd complex was found by IR spectroscopy to be unaffected by the employed polymerization conditions. Details of compositions of three different

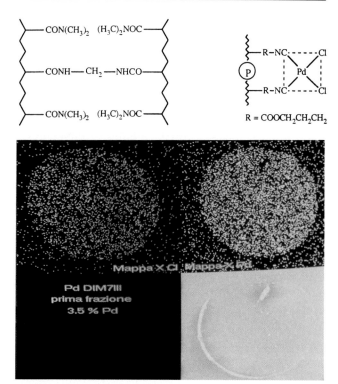

Figure 12. Chemical structure and X-ray microprobe images of a cross-section of a Pd-IPN bead showing Cl (left) and Pd (right) images of the particle. The sample was produced by γ-ray-induced polymerization of Pd-coordinated 3-isocyanopropyl acrylate within a preformed DMA matrix (*see* Table I).

organic matrixes and Fe- and Pd-containing IPNs described above are given in Table I.

The generally lower experimental values obtained for the elemental composition of the polymers is probably due to (at least in part) the difficulty of complete burning of amide-containing cross-linked polymers. This problem can often be minimized by adding catalytic quantities of WO_2 to the sample. In addition, all of the polymers listed in Table I are strongly hygroscopic, a fact that also contributes to the lower experimental values recorded in the table.

The only other reported MC-IPN we are aware of is a Cr-containing semi-IPN produced by the copolymerization of acrylonitrile (AN) and divinylbenzene (DVB) in a solution of chromium acrylate (AIBN, 60 °C, 1 h) (*66*). The reported semi-IPNs contain about 1–6% of the Cr-containing polymer (ca. 0.5–3% Cr), which acts as a thin non-cross-linked matrix for the cross-linked network.

Table II provides a summary of the compositions and properties of a series of Cr-containing semi-IPNs reported in reference 66. It is interesting that as the proportion of poly(chromium acrylate) (PCA) in the IPN decreases, product yield, the extractable polymer, and the swelling of IPN decrease. This appears to indicate that PCA hinders the cross-linking of the copoly(AN–

Table I. Experimental Data for Synthesis of DMA Matrixes and MCP-IPNs

DMA Matrix or MCP-IPN	Monomers Used (wt%)[a]			Composition Found (%) (Expected)			
	DMA	BIS	MCM	C	H	N	M
Matrix 1	94.3	5.7	—	50.07 (60.3)	8.08 (9.0)	11.48 (14.4)	— —
Matrix 2	94.3	5.7	—	56.5 (60.3)	8.9 (9.0)	12.1 (14.4)	— —
Matrix 2	94.3	5.7	—	52.1 (60.3)	8.3 (9.0)	11.9 (14.4)	— —
Fe-IPN-1	32.3	3.2	17.2	59.50 (55.8)	8.63 (7.6)	10.66 (12.3)	3.85 (4.0)
Fe-IPN-2	37.0	3.0	13.7	55.1 (59.1)	8.3 (8.4)	11.3 (11.3)	3.2 (4.1)
Pd-IPN	30.9	3.0	17.2	51.3 (56.2)	7.9 (8.1)	11.5 (13.0)	3.5 (4.0)

Note: DMA is dimethylacrylamide; BIS is methylene-bis(acrylamide); MCM is metal-containing monomer. Dashes indicate that data is not present.

[a]Values include the weight of the DMA matrix used.

Table II. Compositions and Properties of Semi-IPNs Based on Copoly(acrylonitrile–divinylbenzene)–poly(chromium acrylate) (AN–DVB–PCA)

IPN Sample	Molar Conc. for Network Formation			Yield (%)	Percent Extracted in DMF	Swelling		M_c (g/mol)[a]
	PCA ($\times 10^4$)	AN ($\times 10^2$)	DVB ($\times 10^4$)			In DMF	In Dioxan	
IPN-3	4.5	1.5	0.56	29.3	24	82	44	326
IPN-4	3.0	1.5	0.56	28.0	30	54	37	276
IPN-5	1.5	1.5	0.56	27.2	19	37	34	118
IPN-6	0.75	1.5	0.56	18.8	10	28	25	97
IPN-10	4.5	1.5	0.28	28.5	44	94	45	431
IPN-11	4.5	1.5	0.14	25.7	56	308	131	2432
IPN-12	4.5	1.5	0.84	43.0	—[b]	—[b]	—[b]	—[b]
IPN-13	4.5	3.0	0.56	12.5	26	99	43	358
IPN-14	4.5	2.25	0.56	24.7	26	90	40	348
IPN-15	4.5	0.75	0.56	33.7	26	66	35	236

[a]Average molecular weight between the cross-links.
[b]Data not given.

Source: Compiled from reference 67.

DVB) network, but the actual reason is probably the inhibition of polymerization by PCA. The effect of acrylonitrile concentration on the polymer yield reflects the lower rate of polymerization (smaller reactivity ratio) of this monomer as compared with that of DVB.

In addition to the data in Table II, it is also reported that by increasing the molecular weight of PC, the swelling of the IPN decreases (*66*). The Cr-containing IPNs have also been examined by scanning electron microscopy (SEM) to establish their phase-separation behavior.

Matrix and Morphology of Fe-Containing IPNs

Organic matrixes and metal-containing IPNs listed in Table I were all produced in the form of small particles

(micrometer to millimeter size range). The particulate form was obtained from heterogenized monomer mixtures (matrix 1) or after crumbling and crushing of the fibrous products during recovery and washing (matrix 2). This particle morphology is considered desirable for the intended handling of the materials as models for polymer-supported catalysts. Characterization of the material was also undertaken in this respect and included microanalysis, X-ray microprobe analysis, SEM, swelling and permeation behavior, and electron spin resonance (ESR) spectroscopy.

SEM micrographs of the DMA matrix-1 at three different magnifications (1000, 10,000, and 20,000) are shown in Figure 13, and the corresponding organometallic interpenetrating networks are shown in Figure 14. The micrographs in Figure 13 show a typically low porosity resin matrix with relatively large individ-

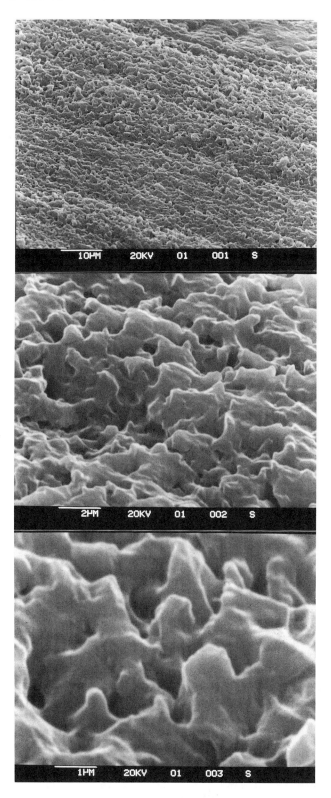

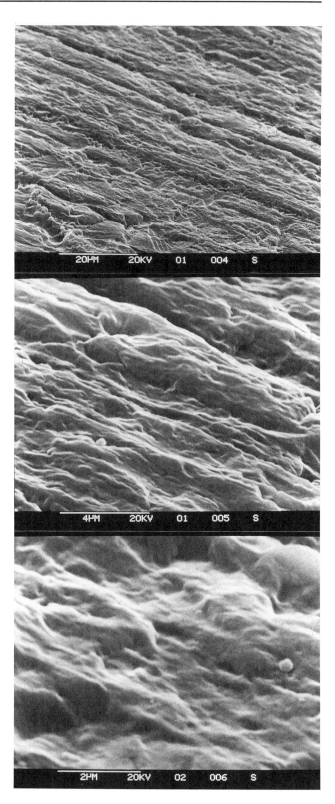

Figure 13. Scanning electron micrographs of surface of DMA matrix-1 at magnification of 700×, 7000×, and 14,000×.

Figure 14. Scanning electron micrographs of surface of IPN-1 at magnification of 700×, 7000×, and 14,000×.

ual pores (channels) of up to about 1 μm. This structure is consistent with what is expected on the basis of the experimental conditions employed (Figure 9). Under these conditions, a low porosity matrix is formed by coagulation of a cross-linked polymer in a good solvent, as compared with more highly porous matrixes obtained by precipitation–aggregation of the polymer in a precipitant–poor solvent.

In the context of the present discussion, however, it is remarkable that the IPN structures shown in Figure 14 appear virtually nonporous up to a magnification of 20,000 presently examined. In other words, the formation of the second matrix (network) within the first porous matrix leads to virtually complete disappearance of all of the pores (channels). It should not be assumed, however, that the second matrix is formed simply inside the individual pores. The DMA matrix swells homogeneously in the monomer mixture used for the formation of the second matrix (see Figure 9), and hence reverts to a homogeneous matrix with far more numerous pores. Under these conditions, the second matrix is formed homogeneously throughout the bulk of the first matrix, and hence produces a nearly ideal example of a molecular organometallic composite. Further discussion on the nature of pores in the DMA matrix-1 and the corresponding MC-IPN is continued in the light of the gel permeation experiments.

An important characteristic of polymer-supported catalysts is their swelling behavior in different solvents. The degree of swelling (or solvent uptake) can be measured and expressed in terms of weight increase (see Table II), bulk expanded volume (V_B), or swollen volume (V_S). V_S is obtained by subtracting the interparticle spaces (excluded volume, V_E) from the bulk expanded volume:

$$V_S = V_B - V_E$$

Table III shows the swollen volumes of matrixes 1 and 2 and their Fe-containing composites in water, dichloromethane (DCM), and tetrahydrofuran (THF). The extent of swelling is generally controlled by the concentration of the cross-linking units in each case. The relative swollen volumes of different materials

confirm several aspects of the materials expected from their chemical structure and the polymerization conditions employed for their formation. Thus, the largely dimethylacrylamide structure of the materials is known to be compatible with water and DCM, but not with THF (68). Similarly, the relatively more porous matrix 1 formed in the presence of water (ice nuclei) shows relatively higher solvent uptake in all three solvents examined.

Interestingly, both IPNs show smaller swollen volumes relative to the corresponding PDMA matrixes. This result reflects an intrinsic property of two interwoven networks, where the two covalently cross-linked networks mutually limit the expansion of each other. However, both of the networks with the indicated degrees of cross-linking (see Table I) remain relatively flexible in good solvents such as water and DCM.

Another practically important criterion of polymer-supported catalysts is particle porosity (see reference 69). However, in swellable materials the state of the swollen structure (the gel structure) is more informative than particle porosity in the dry state. Chemically useful information in the swollen (or gel) state includes the mobility of the chain segments, size of the substrates that can readily permeate the gel, viscosity effects, and accessibility of the reactive sites.

One interesting method for gaining insight into the pore structure and substrate permeability in the swollen state is gel permeation (or size exclusion) chromatography (GPC or SEC). Experimentally, this method is exactly the same as GPC (SEC) procedures employed for the determination of polymer molecular weights on porous supports. However, by using substrates of known molecular weights (and shapes), the permeation process can be worked out to provide information about the matrix structure of the support material. Gel permeation for the characterization of polymer matrixes has been studied by, among others, Halàsz (70) and Jeràbek (71, 72). Jeràbek also employed the method (73) for the evaluation of the PDMA matrixes and Fe-containing IPNs described here.

Table IV shows a summary of the gel permeation studies presented in the form of chain density distribution (i.e., the volume fraction of different chain densities within the water-swollen matrix or IPN). The data are obtained by eluting a series of molecular weight standards (D_2O, sugars, and dextrans; weight-average molecular weights of 20–40,000, diameter = 3–70 Å) through a packed column of the material, followed by numerical analysis of the elution volumes (71–73). Although the present data are not complete and do not show the whole chain density picture of the samples, they illustrate the potential of the method for studying the matrix–pore structures of

Table III. Swollen Volume of DMA Matrixes 1 and 2 and Corresponding Fe-IPNs 1 and 2

Matrix or Composite	Swollen Volume (mL/g)^a		
	In Water	In DCM	In THF
Matrix 1	3.77	3.38	2.13
Fe-IPN 1	2.14	2.44	1.59
Matrix 2	2.54	2.81	1.56
Fe-IPN 2	1.91	2.22	1.39

^{a}Swollen volume = bulk expanded volume − excluded volume.

Table IV. Chain Density Distribution of Water-Swollen DMA Matrix 1 And Fe-IPN 1 Estimated by Solute Permeation

Chain Density $(nm/nm^3)^a$	Volume Fraction (mL/g Dry Material)^b	
	DMA Matrix 1	Fe-IPN 1
0.1	0.000	0.074
0.2	0.930	0.108
0.4	0.212	0.068
2.0	1.340	1.603
All	3.77	2.14

^{a}Calculated by integration of the entire range of volume fractions, which is essentially the same as the data in Table III.

^{b}Solvated volume of chains with indicated density.

swollen PDMA matrixes and the corresponding MC-IPNs.

Matrix 1 shows two distinct regions of low density (0.2 nm/nm^3) and high density (2.0 nm/nm^3) chains. These two regions may correspond, at least qualitatively, to the porous and nonporous regions of the particles as observed by electron microscopy in the dry state (*see* Figures 13 and 14). But in the IPN, the low density regions largely disappear, and the higher density regions are the dominant feature. However, the Fe-containing matrix cannot be assumed to be formed only in the low density regions of matrix 1. Indeed our expectation is that, because matrix 1 swells appreciably under the polymerization conditions, the Fe-containing matrix is formed throughout the lower and higher density regions. Also, matrix 1 and IPN-1 are composed of flexible polymer chains, and the chain densities presented in Table IV must be regarded as dynamic, rather than static, features.

Spin labelling (i.e., electron spin resonance (ESR) spectroscopy) provides a strong tool for the study of molecular and segmental mobility in polymers (74) and is especially suitable for the study of multicomponent polymer systems (75). Figure 15 shows the ESR spectra of the nitroxide spin probe 2,2,6,6-tetra-methylpiperidinyl-1-oxyl (TEMPONE) in water, and in water-swollen matrix 1 and IPN-1. All three spectra show only a *fast* spin motion (76), a result indicating an amorphous and dynamically mobile environment (77, 78). This observation (i.e., no static porosity) is consistent with the expected gel behavior of the polymers (68) on the basis of their flexible matrix structure. The reduced signal intensity of the probe in the water-swollen matrix and IPN reflects the lower mobility (increased viscosity) of these two environments, as compared with that in water.

A series of further ESR data for matrix 1 and IPN-1, swollen in either water, DCM, or THF, is given in Table V (67). These data are presented in terms of rotational correlation time (τ): the time required by the probe molecule for sweeping one radian during its rotational diffusion (79). Qualitatively, these data reflect the swelling behavior of the polymers in the three solvents (*see* Table III). However, the quantitative picture is made rather complicated by the widely differing viscosities of the solvents themselves and by the substantially higher viscosity of the water-swollen systems (H-bonding) as compared with those of DCM and THF. Accordingly, the ESR data reflect the reduced mobility of the samples in terms of the combined effects of relative mobility of the swollen matrix and IPN per se, and the effects of polymer–solvent interactions (i.e., H-bonding in the case of water). Also, the hydrophobic probe is expected to have little (if any) interaction with matrix-1 but a more strong interaction with IPN-1, which contains the relatively hydrophobic ferrocene units.

The effect of temperature on the rotational correlation time (τ) of TEMPONE in the water-swollen matrix 1 and IPN-1 is presented in Figure 16. The τ plots for the water-swollen matrix 1 and IPN-1 above room temperature are essentially similar to that of water, and hence represent a viscosity effect only. However, the IPN-1 curve below room temperature shows a dramatic decrease in mobility. This interesting

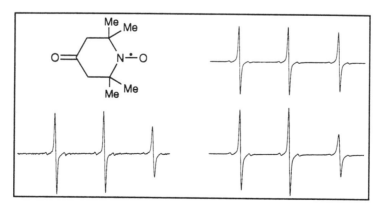

Figure 15. Electron spin resonance (ESR) spectra of the nitroxide probe, 2,2,6,6-tetramethylpiperidinyl-1-oxyl (TEMPONE), in water, water-swollen DMA matrix-1, and the corresponding Fe-containing IPN-1.

Table V. Rotational Correlation Times of TEMPONE in Solvent-Swollen DMA Matrix 1 and Fe-IPN 1

Matrix or IPN	Rotational Correlation Time (t, ps) in Solvent		
	Water	DCM	THF
None	54	23	16
DMA Matrix 1	220	76	330
Fe-IPN 1	520	136	331

Source: Compiled from reference 79.

behavior is thought to represent the increased level of hydrophobic interactions between the probe and the ferrocenyl residues in the polymer, as mentioned above.

Current and Potential Applications

MCPs are useful for a wide range of applications, including analysis and catalysis (chemical, electro, and photo), optical and electronic devices, colorants and coatings, ceramic precursors, structural composites, ceramics, controlled-released medicaments, and biocides. General reviews of these and other applications have been presented by Sheats (*18*), Pittman and Carraher (*22*), and others (*1*). The use of polymer-supported catalysts has been reviewed extensively (*1*, *9*, *53–55*), and other chapters in this book cover photocatalytic and photoanalytical application of polymeric metal complexes. Here, we provide a brief account of our most recent results on the use of a Pd-containing DMA matrix and the corresponding Pd-containing IPN (Figure 12) as hydrogenation catalysts.

Table VI. Hydrogenation of 4-Nitrotoluene in Presence of Pd-Containing DMA Matrix or PD-IPN (*see* Figure 2)

Catalyst	Conversion (%)		Turnover
	22 h	44 h	
Pd-Matrix	2.0	10	101
Pd-Matrix (Recycled)	38	57	570
Pd-IPN	52	100	1000
Pd-IPN (Recycled)	43	58	577

Note: Reaction conditions: solvent, methanol; substrate, 1 M; Pd, 0.001 M; temperature, 20 °C; H_2 pressure, 105 KPa.

Hydrogenation of 4-nitrotoluene to 4-aminotoluene was performed at room temperature in methanol, saturated with H_2 at atmospheric pressure, in the presence of the required quantity of the supported catalyst to contain 0.1 mole% palladium. The results are summarized in Table VI. In the first cycle, both the Pd-containing matrix and the Pd-containing IPN show substantially different activities, but the reason for this difference is not clear at this stage. In the second cycle, however, both the matrix and the IPN show more or less comparable activities, and their activities are maintained for at least three cycles presently studied. It is, however, known (*8, 9*) that the isocyano function is reduced to the corresponding methylamine derivative under the indicated experimental conditions. It appears that Pd(II) is also reduced to metallic Pd, and hence the generation of highly dispersed metallic palladium in the IPN, as indicated in Figure 17. Furthermore, we observed that about 85% of the chlorine content of the polymer is also removed during the reduction process.

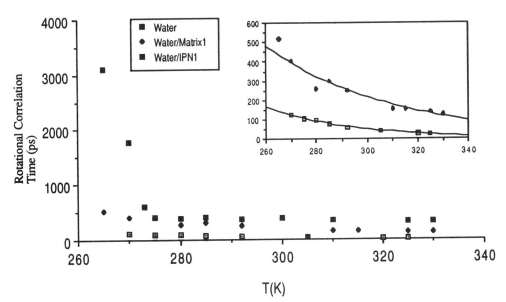

Figure 16. Effect of temperature on the rotational correlation time, τ, of TEMPONE in water, water-swollen DMA matrix-1, and water-swollen Fe-containing IPN-1). Magnified τ scale for water and matrix-1 is shown in the insert. (Adapted from reference 79.)

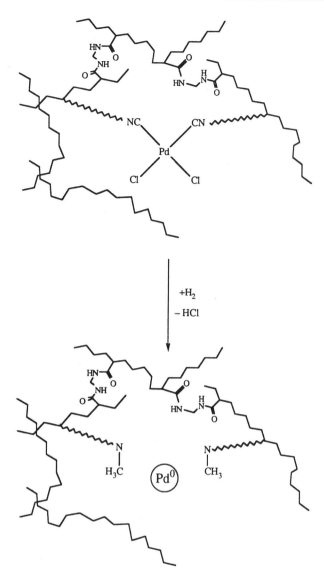

Figure 17. Schematic presentation of the formation of dispersed metallic palladium within the IPN by hydrogenation of the corresponding IPN containing isocyano coordinated Pd^{II}.

Conclusions and Future Prospects

A wide variety of metal containing polymers (MCPs) are available via polymerization, polycondensation, and polycoordination of suitably chosen monomeric units, as well as via the introduction of metal centers into preformed polymers. The metal atom may reside in the main polymer chain by covalent links or coordination, or it may be attached to the side chains in the form of cation, covalent, or complexed species. MCPs are of considerable attraction as new materials for emerging technologies in, among other fields, electronics, sensors, and catalysis. Metal containing interpenetrating polymer (MC-IPNs) represent a relatively new form of MCPs, which may provide synthetic and han-

dling advantages for such applications. In most cases, however, technological applications of MCPs are in their infancy, and hence substantial grounds remain to be established before viable directions can be defined.

Acknowledgment

This chapter is dedicated to Professor Ivar Ugi on the occasion of his 65th birthday.

Part of this work was supported by Progetto Finalizzato Chimica Fine II, Italian Research Council (Rome).

References

1. *Inorganic and Metal-Containing Polymeric Materials;* Sheats, C. E.; Pittman, C. U; Zeldin, M.; Currell, B., Eds.; Plenum: New York 1990.
2. Pomogailo, A. D.; Savostyanov, V. S. *JMS–Rev. Macromol. Chem. Phys.* **1985**, *C25(3)*, 375–479.
3. Reddy, B. S. R.; Arshady, R.; George, M. H. *Polymer* **1984**, *25*, 115–120.
4. Arshady, R.; Reddy, B. S. R.; George, M. H. *Polymer* **1984**, *25*, 716–721.
5. Arshady, R.; Reddy, B. S. R.; George, M. H. *Polymer* **1986**, *27*, 769–775.
6. Arshady, R. *Makromol. Chem. Rapid Commun.* **1983**, *4*, 237–241.
7. Arshady, R.; Corain, B. *Trans. Met. Chem.* **1983**, *8*, 182–184.
8. Arshady, R.; Basato, M.; Corain, B.; Della Giustina, L.; Lora, S.; Palma, G.; Roncato, M.; Zecca, M. *J. Mol. Catal.* **1989**, *53*, 111–128.
9. Arshady, R.; Zecca, M; Corain, B. *React. Polym.* **1993**, *20*, 147–173.
10. Arshady, R. *Polymer* **1982**, *23*, 1870–1872.
11. Arshady, R.; Basato, M.; Corain, B.; Lora, S.; Zecca, M. *Adv. Mater.* **1990**, *2*, 412–414.
12. Corain, B.; Corvaja, C.; Jerábek, K.; Lora, S.; Palma, G.; Zecca, M. In *Advanced Syntheses and Methodologies in Inorganic Chemistry: New Compounds and Materials;* Daolio, S.; Fabrizio, M; Guerriero, P.; Tondello, E.; Vigato, P. A., Eds.; Photograph: Padova, Italy, 1992; Vol. 1, pp 151–154.
13. Uflyand, E; Pomogailo, A. D. *Russ. Chem. Rev.* **1991**, *60*, 1532–1552.
14. Pittman, C. U., Jr.; Rausch, M. D. *Pure Appl. Chem.* **1986**, *4*, 617–622.
15. *Organometallic Polymers;* Carraher, C. E., Jr.; Sheats, J. E.; Pittman, C. U., Jr., Eds.; Academic: Orlando, FL, 1978; pp 1–353.
16. *Metallocene Polymers;* Neuse, E. W.; Rosenberg, H., Eds.; Marcel Dekker: New York, 1970.
17. Pittman, C. U., Jr.; Carraher, C. E., Jr.; Reynolds, J. R. In *Encyclopedia of Polymer Science and Engineering;* Mark, H.; Bikales, N.; Overberger, C.; Menges, J., Eds.; John Wiley and Sons: New York, 1987; Vol. 10, pp 541–594.
18. Sheats, J. E. *Kirk–Othmer Encyclopedia of Chemical Technology;* John Wiley and Sons: New York, 1981; Vol. 15, pp 184–220.

19. *Metal-Containing Polymer Systems;* Sheats, J. E.; Carraher, C. E., Jr.; Pittman, C. U., Jr., Eds.; Plenum: New York, 1985.

20. *Advances in Organometallic and Inorganic Polymer Science;* Carraher, C. E., Jr.; Sheats, J. E.; Pittman, C. U., Jr., Eds.; Marcel Dekker: New York, 1982.

21. *Inorganic and Organometallic Polymers: Macromolecules Containing Silicon, Phosphorus, and Other Inorganic Elements;* Zeldin, M.; Wynne, K. J.; Allcock, H. R., Eds.; ACS Symposium Series 360; American Chemical Society: Washington, DC, 1988.

22. Pittman, C. U., Jr.; Carraher, C. E., Jr. *Appl. Polym. [Proc. Am. Chem. Soc. Symp. O. A. Battista, Appl. Polym. Sci.]* **1988**, 113–124.

23. *Comprehensive Organometallic Chemistry;* Wilkinson, G.; Abel, E. W.; Stone, F. G., Eds.; Pergamon: Oxford, England, 1982.

24. *Metal–Polymer Composites;* Delmonte J., Ed.; Van Nostrand: New York, 1990.

25. Kakati, D. K.; Gosain, R.; George, M. H. *Polymer* **1984**, *35*, 398–402.

26. Fujita, N.; Sonogashira, K. *J. Polym. Sci., Polym. Chem. Ed.* **1974**, *12*, 2845–2856.

27. Cardin, C. J.; Cardin, D. J.; Lappert, M. F. *J. Chem. Soc., Dalton Trans.* **1977**, 767–779.

28. Macomber, D. W.; Hart, W. P.; Rausch, M. D.; Priester, R. D.; Pittman, C. U., Jr. *J. Am. Chem. Soc.* **1982**, *104*, 884–887.

29. Macomber, D. W.; Hart, W. P.; Rausch, M. D. *J. Organometal. Chem.* **1983**, *250*, 311–318.

30. Casellato, U.; Corain, B.; Zecca, M.; Michelin, R. A.; Mozzon, M.; Graziani, R. *Inorg. Chim. Acta* **1989**, *156*, 165–167.

31. Corain, B.; Sam, F. O.; Zecca, M.; Lora, S.; Palma, G. *Angew. Chem.* **1990**, *29(4)*, 384–385.

32. Arshady, R. *Polymer* **1982**, *23*, 1099–1100.

33. Arshady, R.; Ugi, I. *Polymer* **1990**, *31*, 1164–1169.

34. Corain, B.; Sam, F. O.; Zecca, M.; Veronese, A. C.; Palma, G.; Lora, S. *Makromol. Chem. Rapid Commun.* **1989**, *10*, 697–704.

35. Carraher, C. E., Jr. In *Interfacial Synthesis: Polymer Applications and Technology;* Millich, F.; Carraher, C. E., Jr., Eds.; Marcel Dekker: New York, 1977; Vol. 2, pp 367–416.

36. Carraher, C. E., Jr. In *Organometallic Polymers;* Carraher, C. E., Jr.; Sheats, J. E.; Pittman, C. U., Eds.; Academic: Orlando, FL, 1978, pp. 79-86.

37. Carraher, C. E., Jr.; Molloy, H. M.; Taylor, M. L.; Tiernan, T. O.; Yelton, R. O.; Schroeder, J. A.; Bogdan, M. R. *Org. Coat. Plast. Chem.* **1979**, *41(2)*, 197–202

38. Takahashi, S.; Morimoto, H.; Murata, E.; Katsoka, S.; Sonogashira, K.; Nagihara, N. *J. Polym. Sci. Polym. Chem. Ed.* **1982**, *20*, 565–773.

39. Sonogashira, K.; Fujikura, Y.; Yatake, T.; Toyoshima, N.; Takahashi, S.; Hagihara, N. *J. Organometal. Chem.* **1978**, *145*, 101–108.

40. Wright M. E.; Mullick, S. *Macromolecules* **1992**, *25*, 6045–6049.

41. Wright M. E.; Toplika, E. G. *Macromolecules* **1992**, *25*, 6050–6054.

42. Wright M. E.; Sigman, M. S. *Macromolecules* **1992**, *25*, 6055–6058.

43. Dembek, A. A.; Burch, R. R.; Feiring, A. E. *Polym. Prepr. (Am. Chem. Soc. Div. Polym. Chem. Symp.)* **1993**, *34(1)*, 172–173.

44. Bailor, J. C., Jr. In *Preparative Inorganic Reactions;* Jolly, W. L., Ed.; Interscience Publishers: New York, 1964; Vol. 1, pp 1–27.

45. Pittman, C. U., Jr.; Carraher, C. E, Jr.; Sheats, J. E.; Timken, M. D.; Zeldin, M. In *Inorganic and Metal-Containing Polymeric Materials;* Carraher, C. E., Jr.; Pittman, C. U, Jr.; Zeldin, M.; Currell, B., Eds.; Plenum: New York, 1990; pp 1–27.

46. Efraty, A.; Feinstein, I.; Frolow F.; Wackerle, L. *J. Am. Chem. Soc.* **1980**, *102*, 6341.

47. Bohle, D. S.; Goodson, P. A. *Polym. Prepr. (Am. Chem. Soc. Div. Polym. Chem. Symp.)* **1993**, *34 1*, 358–359.

48. Burch, R. R. *Chem. Mater.* **1990**, *2*, 633–635.

49. Nishihara, H.; Shimura, T.; Ohkubo, A.; Matsuda, N.; Aramaki, K. *Adv. Mater.* **1993**, *5*, 752–754.

50. Lai, J. C.; Rounsefell, T. D.; Pittman, C. U., Jr. *Macromolecules*, **1971**, *4*, 155–161.

51. Pittman, C. U.; Marlin, G. V. *J. Polym. Sci. Polym. Chem. Ed.* **1973**, *11*, 2753–2765.

52. Pittman, C. U., Jr.; Ayers, O. E.; McManus, S. P. *J. Macromol. Sci. Chem.* **1973**, *A7(8)*, 1563–1579.

53. Carrou, P. E.; Gates, B. C. In *Polymer-Bound Transition Metal Complex Catalysts;* Sherrington, D. C.; Hodges, P., Eds.; John Wiley and Sons: Chichester, England, 1988; pp 123–146.

54. Hartley, F. R. *Supported Metal Complexes;* D. Reidel Publishers: Dordrecht, Netherlands, 1985.

55. Chauvin, Y.; Commereuce, D.; Dawans, F. *Prog. Polym. Sci.* **1977**, *5*, 95–226.

56. Baxter, S. M.; Jones, W. E., Jr.; Danielson, E.; Worl, L.; Strouse, G.; Younathan, J.; Meyer, T. J. *Coord. Chem. Rev.* **1991**, *111*, 47–71.

57. Younathan, J.; Jones, W. E., Jr.; Meyer, T. J. *J. Phys. Chem.* **1991**, *95*, 488–492.

58. Sperling, L. H. *CHEMTECH* **1988**, February, 104–109.

59. Sperling, L. H.; *Interpenetrating Polymer Networks and Related Materials;* Plenum: New York, 1981.

60. Frisch, H. L; Frisch, K. C.; Klempner, D. *CHEMTECH* **1977**, 188–191.

61. Lipatov, Y. S.; Sergeeva, L. M. *Russ. Chem. Rev.* **1976**, *45(1)*, 63–74.

62. Klempner, D. J. *Angew. Chem. Int. Ed. Engl.* **1978**, *17*, 97–106.

63. *Multicomponent Polymer Materials;* Paul, D. R.; Sperling, L. H., Eds.; Advances in Chemistry 211; American Chemical Society, Washington DC, 1986.

64. Ilschner, B.; Lees, J. K.; Dhingre, A. K.; McCullough, R. L. *Ullman's Encyclopedia of Industrial Chemistry;* VCH: Weinheim, Germany, 1986; Vol. A7, pp 396–410.

65. Ashbey, M. F.; Jones, D. R. H. *Engineering Materials 2— An Introduction to Microstructure, Processing and Design;* International Series on Materials Science and Technology 39; Pergamon: Oxford, England, 1986; Chapter 25, pp 241–254.

66. Gupta, N.; Srivastava, A. K. *High Perform. Polym.* **1992**, *4*, 225–235.

67. Corain, B.; Zecca, M.; Corvaja, C.; Palma, G.; Lora, S.; Jerábek, K. *Adv. Mater.* **1993**, *5*, 367

68. Arshady, R. *Colloid Polym. Sci.* **1990**, *268*, 948–958.
69. Guyot, A. In *Syntheses and Separations Using Functional Polymers*; Sherrington, D. C.; Hodge, P., Eds.; John Wiley and Sons: London, 1988; pp 1–42.
70. Halász, I.; Martin, K. *Angew. Chem. Int. Ed. Engl.* **1978**, *17*, 901–908.
71. Jerábek, K. *J. Anal. Chem.* **1985**, 1595–1598.
72. Jerábek, K. *J. Anal. Chem.* **1985**, 1598–1602.
73. Corain, B.; Jerábek, K.; Lora, S.; Palma, G.; Zecca, M. *Adv. Mater.* **1992**, *4*, 97–99.
74. Törmälä, P.; Weber, G.; Lindberg, J. J. *Spin Label and Probe Studies of Relaxations and Phase Transitions in Polymeric Solids and Melts;* MMI Press Symposium Series, 1 (Molecule Motion Polymers, ESR); MMI, 1980; pp 115–134.
75. Müller, G.; Stadler, R.; Schlick, S. *Makromol. Chem. Rapid Commun.* **1992**, *13*, 117–124.
76. Schlick, S.; Harvey, R. D.; Alonso-Amigo, M. G.; Klempner, D. *Macromolecules* **1989**, *22*, 822–830.
77. Rex, G. C.; Schlick, S. *Polymer* **1987**, *28*, 2134–2138.
78. Rex, G. C.; Schlick, S. *J. Phys. Chem.* **1985**, *89*, 3598–3601.
79. *Spin Labeling: Theory and Application;* Berliner, L. J., Ed.; Academic: Orlando, FL, 1976.

Germylene and Stannylene Polymers

Shiro Kobayashi, Shin-ichiro Shoda, and Satoru Iwata

This chapter provides a brief overview of germanium- and tin- containing polymers, followed by details of a new class of polymers synthesized by using germylene and stannylene. These polymerizations involve a combination of an oxidant monomer (M_{ox}) and a reductant monomer (M_{red}), in which the germylene or stannylene behaves as M_{red}. The M_{ox} groups employed are p-benzoquinones, N-phenyl-p-quinoneimines, cyclic α,β-unsaturated ketones, and substituted acetylenes. The oxidative addition of a poly(disulfide) into a germylene has also been demonstrated. In addition, a "ligand substitution polymerization" has been developed and affords a polygermane of higher molecular weight. Most polymers prepared according to this new concept of polymerization have alternative structures.

The creation of a novel class of functional polymer having an unexpected physical property requires the development of new methodologies in the field of polymer synthesis. Conventional syntheses can produce polymers of restricted structures from which only well-known chemical or physical properties are predicted. Organometallic polymers are very interesting substances from the viewpoint of both basic organic chemistry and applied materials science (1). Recently, much attention has been paid to polymers having a group 14 element (2–4) in the main chain, especially the syntheses and applications of silicon-containing polymers (5–10). However, few studies on polymers having germanium or tin atom have been reported (11–29). Organogermane polymers show photoactivity and bleaching behavior (11), thermochromism (12), and semiconductivity (13); tin-containing polymers show biological activity such as fungicidal properties (14, 15).

Among the divalent species of group 14 elements, carbenes and silylenes are normally quite reactive and very unstable. Germylenes (16, 17) and stannylenes (18) are more stable and can be isolated even by distil-lation, provided that an appropriate ligand is attached to the germanium or tin atom. These chemical species that possess a strong reducing ability are potentially useful for construction of various polymer structures because they can form two new chemical bonds via α,α-addition on the germanium or tin atom by changing their valence state from two to four (Scheme I). In this chapter, we provide an overview of Ge- and Sn-containing polymers, followed by a review of the copolymerization of divalent species of Ge and Sn as comonomers based on their reducing properties.

Overview of Ge- and Sn-Containing Polymers

The conventional method for preparation of polymers having a germanium or tin atom in the main chain is based on polycondensation (Scheme II). This process normally involves the elimination of a small molecule X-Y as a result of acid–base or dehydration reactions. For example, in the case of polycondensation between a metal chloride and a diol, two hydrogen chloride molecules are generated in the course of the polycon-

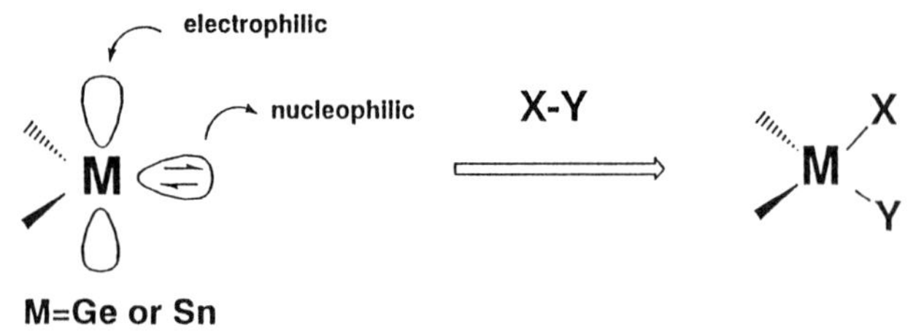

M=Ge or Sn

Scheme I. Formation of two new chemical bonds via α,α-addition on germanium or tin atom by using lone pair (nucleophilic center) and vacant orbital (electrophilic center) of germylene or stannylene.

densation. Consequently, it is necessary to select an appropriate base to scavenge the eliminated acid.

The syntheses of metal-containing polyesters by the interfacial technique has been achieved where M is silicon, germanium, or tin (19, 20) (Scheme II). In general organometallic dihalides in an organic solvent are added to stirred solutions of diacid salts in water. The Ge-containing polyesters exhibit a broad absorption band in the region of 3100–3500 cm^{-1} attributed to the presence of Ge–OH end groups. The products have a low number-average molecular weight (M_n) in the range of 2,000–3,000 daltons, which represents an average degree of polymerization in the range of 7–12. Polyesters having a tin atom in the main chain are thermally stable and possess antibacterial and antifungal properties. Tin polyesters are soluble only in 2-chloroethanol and trifluoroacetic acid, whereas germanium polyesters are soluble in N,N-dimethylformamide, acetone, and dimethylsulfoxide.

Interfacial condensation was used to synthesize oligomeric silicon-, germanium-, and tin-containing ferrocene polyesters (21) (Scheme II). The yield of the polycondensation decreases in the order of Sn > Ge > Si and I > Br > Cl.

The polymer reaction of natural products with a reactive tin compound also produces biologically active materials. Hydroxy-containing natural products, including sucrose, xylan, dextran, and cellulose have been condensed with alkyl and aryl tin chlorides to produce tin-containing materials (22). Many of these materials inhibit selected microorganisms when the materials are used in bulk and as additives. Inhibition is generally in the order of R = methyl > ethyl > propyl > butyl. Tin-containing lignin was prepared by reacting kraft lignin products such as Indulin AT or Indulin C with tributyltin chloride (23). Good activity was found for the tributyltin-modified product against *Candida albicans*, which is responsible for yeast infections in women, but the product showed no activity against *Staphylococcus aureus* or *Escherichia coli*.

Dihydroxy(phthalocyanino)silicon and dihydroxy(hemiporphyrazino)germanium react with bivalent alcohols to give polymers with silyl ether and germyl ether structure (13) (Scheme III). Dichloro(phthalocyanino)silicon and dichloro(hemiporphyrazino)germanium also react with phenol and hydroquinone to create polymeric phenoxy derivatives. The reaction of these silicon or germanium compounds with mono- and dibasic carboxylic acids leads to the corresponding esters (13). The resulting polymers

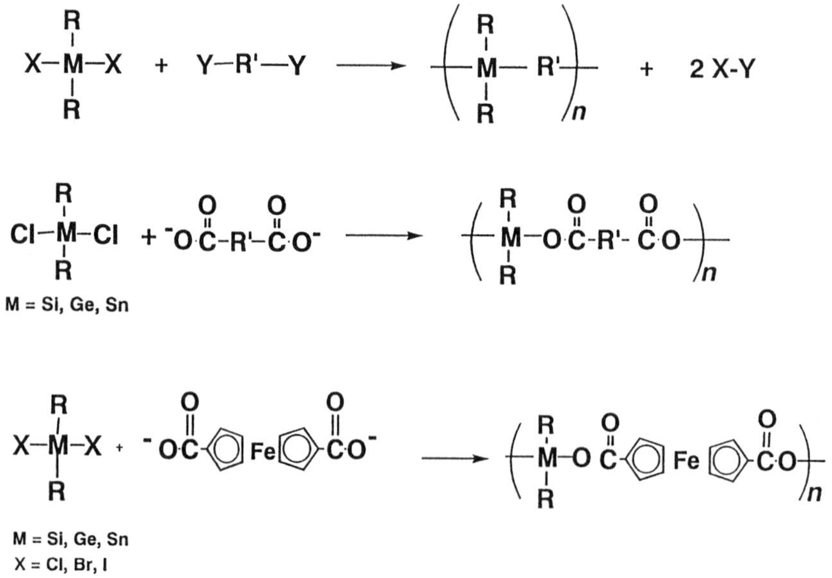

Scheme II. Formation of organometallic polymers via polycondensation (top); general scheme for metal-containing polyesters starting from a tetravalent silicon, germanium, or tin chloride (bottom); and the formation of silicon-, germanium-, and tin-containing ferrocene polyesters by interfacial polycondensation (bottom).

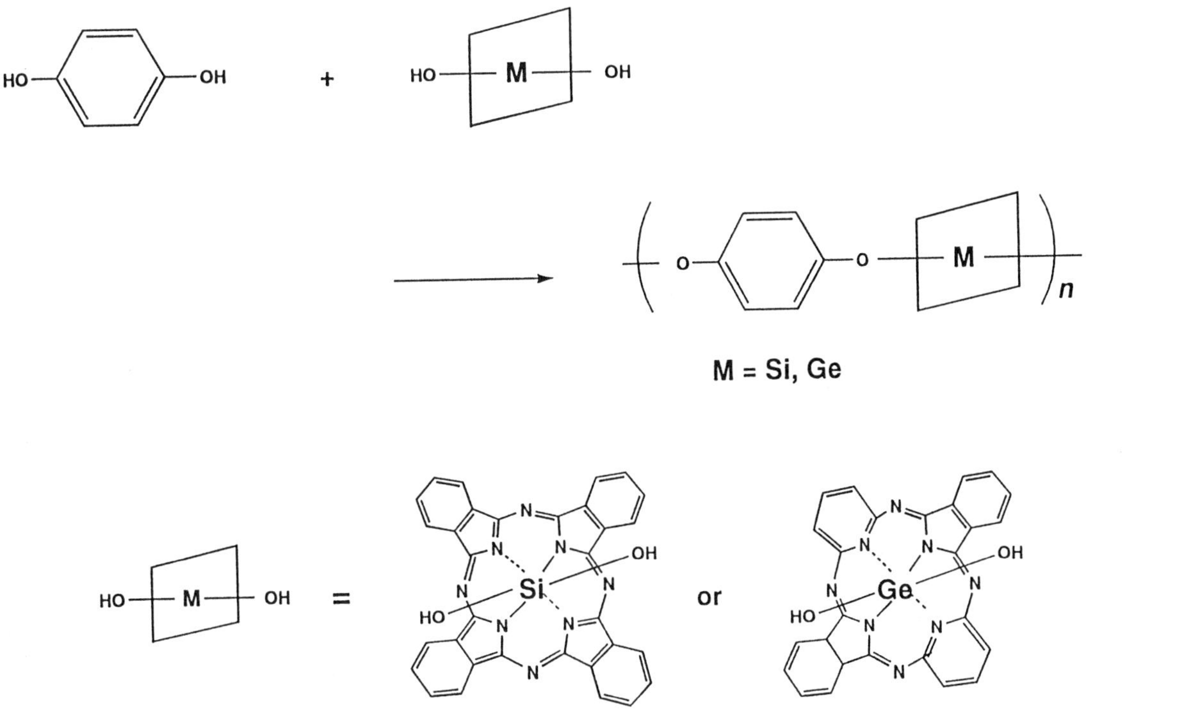

Scheme III. Synthesis of silicon- and germanium-containing oligomers having a phthalocyano and hemiporphyrazino group as ligand.

show semiconductivity and have a conductance (κ) at 298 K of 10^{-7} to 10^{-16} S cm^{-1}).

Polymers having a tin–carbon bond in the main chain have been synthesized by a hydrostannylation reaction between an organotin dihydride and an α,ω-diolefin (Scheme IV). Phenyltinhydrides add to the olefin more easily than alkyltinhydrides. The weight-average molecular weights of the resulting copolymers were $M_w = 1 \times 10^4$ to 2×10^4 daltons when aluminum hydride was added as a catalyst (24). Poly[3,3'-(1,3-dioxo-1,3-digermoxanediyl)bispropanoic acid] has been prepared via hydrogermylation of acrylic acid and subsequent hydrolysis (25, 26).

Ring-opening polymerization of a tetravalent cyclic germanium compound also affords a polymer having a tetravalent germanium unit in the main chain. The thermal ring-opening polymerization of ferroceno-phane containing a germanium atom in the bridge between the cyclopentadienyl ligands was reported (27). The resulting polymers were stable to atmospheric moisture both in the solid state and in solution. A ger-macyclopentane derivative fused by a furan or a thio-phene ring was polymerized by *n*-butyllithium and hexamethylphosphoramide in tetrahydrofuran (THF) (28, 29). The molecular weight of the resulting polymer having a furan moiety was $M_n = 8100$ daltons, and the polymer was thermally stable to 370 °C.

New functional polymers having germanium and tin atoms in the main chain were synthesized by the combination of a reductant monomer (M_{red}) and oxi-

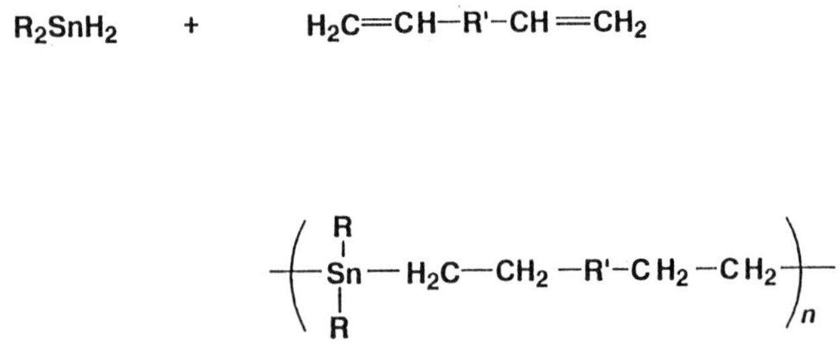

Scheme IV. Synthesis of organotin polymers by hydrostan-nylation of organotin dihydride and diolefin.

dant monomer (M_{ox}), in which the divalent species germylene and stannylene behave as M_{red} (Scheme V). The reminder of this chapter is devoted to this topic.

Copolymerization of Germylene with *p*-Benzoquinone Derivatives

Oxidation–Reduction Alternating Copolymerization

Germylene 1 and 2 (30, 31) copolymerized with vari-ous *p*-benzoquinone derivatives 3 to give copolymers 4 and 5, respectively, having an alternating germa-nium(IV) unit and a *p*-hydroquinone unit in the main chain (Scheme VI). The resulting copolymers are of relatively high molecular weight and are soluble in organic solvents. During the copolymerization, ger-mylene (M_{red}) is oxidized and the *p*-benzoquinone

$$M_{red} + M_{ox} \longrightarrow \left(M_{red}\ M_{ox} \right)_n$$

Reductant   Oxidant   Alternating
Monomer   Monomer   Copolymer

Scheme V. Oxidation–reduction alternating copolymerization by the combination of reductant monomer (M_{red}) and oxidant monomer (M_{ox}).

derivative (M_{ox}) is reduced (*oxidation–reduction copolymerization*) (32). The copolymerization took place smoothly at –78 °C in toluene without any catalyst and was complete within 1 h. The sterically hindered *p*-benzoquinone derivatives were also usable as oxidant monomers, and this use indicated the high nucleophilicity of germylene toward compounds having a carbon–oxygen double bond.

The resulting copolymers are white fine powders and their solubility in *n*-hexane, benzene, and chloroform is generally very high. The copolymers are insoluble in acetone and acetonitrile, and this insolubility enables their purification by precipitation from acetonitrile. The copolymers have relatively high molecular weights ($M_w > 2.9 \times 10^4$ daltons). A copolymer of especially high molecular weight ($M_w = 1.4 \times 10^6$ daltons) was obtained with 1,4-naphthoquinone as the oxidant monomer. Copolymer structures have been determined by ^{1}H NMR and ^{13}C NMR spectroscopies as well as elemental analysis. The ^{1}H NMR spectrum of the copolymer obtained from germylene and *p*-benzoquinone in toluene exhibited signals at δ 0.25 ppm ascribable to the methyl protons of the trimethylsilyl group, and at δ 6.92 ppm due to the aromatic protons of the *p*-hydroquinone units. These results support the alternating structure (Figure 1). The result of copolymerization of **1** with various *p*-benzoquinone derivatives (**3a–3i**) are summarized in Table I.

Stability of Copolymers

Copolymers are thermally stable and melt without decomposition. The glass transition and melt transition temperatures of **4a** and **4g** determined by differential scanning calorimetry were $T_g = 75.6$ °C, $T_m = 234.7$ °C and $T_g = 46.1$ °C, $T_m = 234.9$ °C, respectively. The stability of the resulting copolymers toward moisture was examined by the following hydrolysis experiment.

An aqueous THF solution of copolymer **4b** or **4g** was stirred at room temperature for one day, and molecular weight change was monitored by gel permeation chromatography. The analysis showed no decrease of molecular weight for **4b** (Figure 2). The Si–N, Ge–N, and Ge–O bonds are easily cleaved by the attack of nucleophiles such as alcohol and water. The steric hindrance of the amide substituent on the germanium atom may be the reason for observed enhancement of stability. Copolymer **4g** was gradually hydrolyzed and its molecular weight dispersity increased; therefore, the naphthalene ring is not as bulky as a *t*-butyl group (Figure 2).

The degradability of the polymers under γ-ray irradiation was also investigated. The polymers were irradiated with a ^{60}Co γ-ray source at 7,000 Ci in vacuum and in air. The gauss values for the scission (G_s) determined by comparing the M_n values before and after irradiation indicated that the scission mainly takes place in the main chain and not in the side chain (33). This selective degradation behavior by irradiation is in contrast to that of poly(siloxane), which causes gellation due to cross-linking.

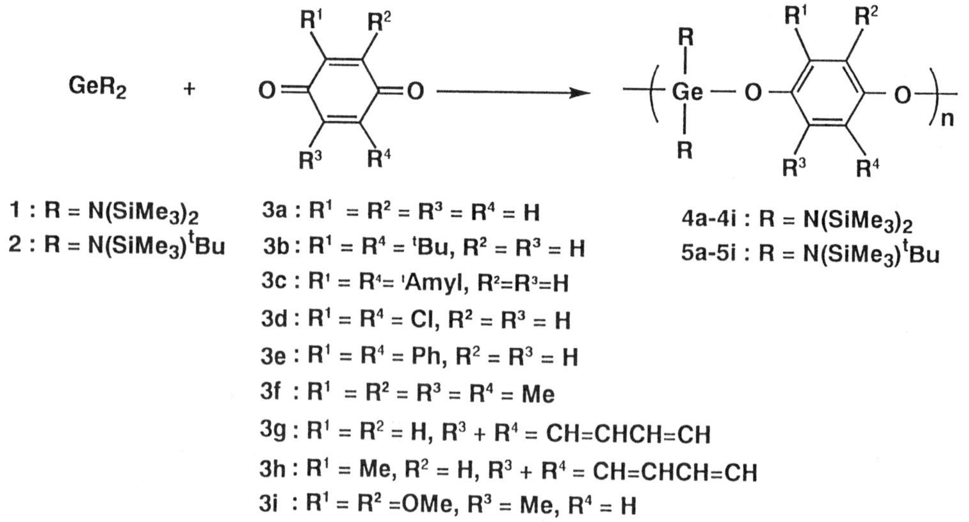

GeR$_2$ + [*p*-benzoquinone derivative structure] ⟶ [copolymer structure]

1 : R = N(SiMe$_3$)$_2$
2 : R = N(SiMe$_3$)^{t}Bu

3a : R^1 = R^2 = R^3 = R^4 = H
3b : R^1 = R^4 = ^{t}Bu, R^2 = R^3 = H
3c : R^1 = R^4 = ^{t}Amyl, R^2 = R^3 = H
3d : R^1 = R^4 = Cl, R^2 = R^3 = H
3e : R^1 = R^4 = Ph, R^2 = R^3 = H
3f : R^1 = R^2 = R^3 = R^4 = Me
3g : R^1 = R^2 = H, R^3 + R^4 = CH=CHCH=CH
3h : R^1 = Me, R^2 = H, R^3 + R^4 = CH=CHCH=CH
3i : R^1 = R^2 = OMe, R^3 = Me, R^4 = H

4a-4i : R = N(SiMe$_3$)$_2$
5a-5i : R = N(SiMe$_3$)^{t}Bu

Scheme VI. Copolymerization of germylene as reductant monomer (M_{red}) with *p*-benzoquinone derivative as oxidant monomer (M_{ox}).

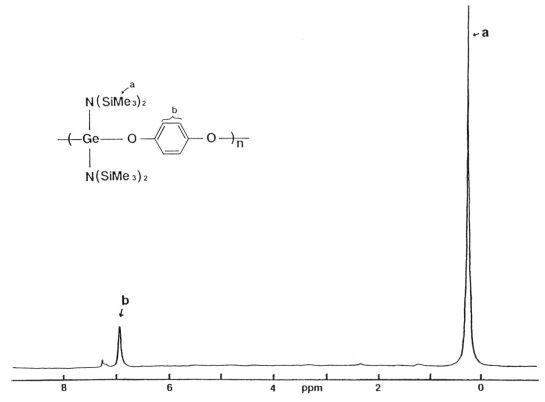

Figure 1. ¹H NMR spectrum of copolymer from germylene and *p*-benzoquinone (in CDCl₃).

Table I. Oxidation–Reduction Copolymerization of Germylene 1 with *p*-Benzoquinone Derivatives

Quinone	Yield (%)	M_w	M_w/M_n
3a	85	1.3×10^5	2.36
3a	quant.[a]	1.4×10^5	2.32
3b	94	5.8×10^4	2.51
3c	92	2.9×10^4	1.98
3d	quant.	2.9×10^4	1.98
3e	89	8.6×10^4	2.72
3f	92	—[b]	—[b]
3g	96	1.4×10^6	2.44
3h	95	3.9×10^4	2.75
3i	90	2.6×10^5	3.57

Note: All reactions were performed in toluene at –78 °C except where noted. For quinone derivatives, *see* Scheme VI. All molecular weight values are in daltons and were determined by gel permeation chromatography.

[a]Reaction performed at 0 °C.

[b]Polymer was insoluble in organic solvents.

Source: Data adapted from reference 32.

Copolymerization Mechanism

The reaction of germylene with 2,6-di-*t*-butyl-*p*-benzoquinone was taken as a model for the copolymerization (Scheme VII) (*34*). The electron spin resonance (ESR) spectrum of the reaction mixture showed a triplet signal ascribable to a semiquinone biradical species **6** whose structure was determined by converting it to a seven-membered cyclic peroxide **7** by air oxidation. The structure of **7** was confirmed by X-ray crystallographic analysis. Also, the biradical **6** is in equilibrium with compound **8** under argon. The formation of **8** is explained by a coupling reaction of the semiquinone biradical derivative **6** at the aromatic carbon atom giving rise to a five-membered cyclic structure. All these results clearly indicate that semiquinone biradical species can be generated via a one-electron transfer process from germylene into one of the oxygens on the *p*-benzoquinone moiety.

In the polymerization of germylenes with 2,5-di-*t*-butyl-*p*-benzoquinone, a germanium biradical species **9** was detected by ESR spectroscopy of the copolymerization mixture by using excess germylene. The spectrum shows a quintet with a ratio of 1:2:3:2:1 ascribable to a germyl radical located at both polymer chain ends (Figure 3). The observed coupling (10.06 G) is due to two nitrogen atoms spin quantum number (l = 1) on the germanium. Ten satellite peaks due to ⁷³Ge (I = 9/2) were also observed. The smaller g value (g = 1.9986) compared with Me₃Ge· (*g* = 2.010) and Ph₃Ge· (*g* = 2.017) (*35*) clearly indicates that the geometry of **9** is different from that of the germyl radicals bound to three carbon atoms; hence, its structure may have

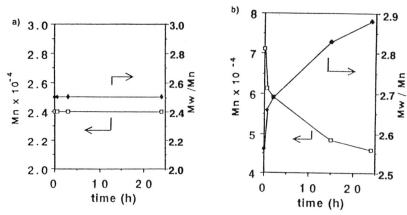

Figure 2. Change of molecular weight (M_n) and its distribution (M_w/M_n) during hydrolysis of (a) copolymer from germylene and 2,5-di-*t*-butyl-*p*-benzoquinone and (b) copolymer from germylene and naphthoquinone. Key: ♦, change of molecular weight distribution; □, change of molecular weight.

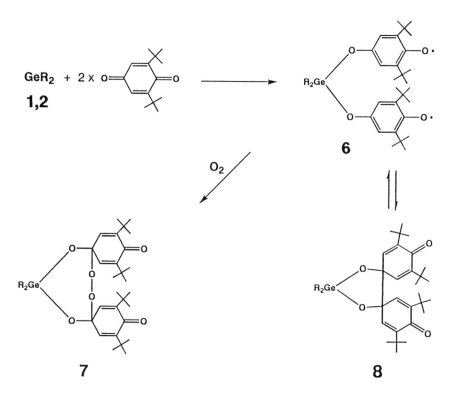

Scheme VII. Formation of semiquinone biradical and its conversion to a seven-membered cyclic peroxide.

pyramidal shape due to the electronegative oxygen and nitrogen atoms bound to the germanium atom.

To confirm the germyl biradical structure, the biradical species (**9**) was quenched with dibenzyl disulfide. When the polymerization was carried out in the presence of the disulfide (Scheme VIII), the oligomer **10** having a benzylmercapto group at both chain ends was obtained. The biradical species was also treated with tetramethyl piperidinyloxy (TEMPO) radical to quantitatively afford the TEMPO group-terminated oligomer **11** (Scheme VIII). The presence of two

TEMPO groups in **11** was proved by molecular weight determination by means of vapor pressure osmometry (VPO), ^{1}H NMR spectroscopy, as well as elemental analyses.

To determine the concentration of the radical species, the following spin trapping experiments were performed. Germylene and 2,5-disubstituted *p*-benzoquinone were reacted in *n*-hexane at −78 °C for 5 min to produce an alternating copolymer (**12**), which is expected to have germyl radicals at both ends (Scheme IX). Then, **12** was quenched at room temper-

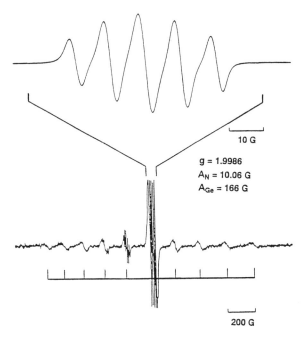

Figure 3. ESR spectrum of germanium biradical in the copolymerization mixture by using excess germylene.

ature for 1 h by 2,6-di-*t*-butyl-*p*-benzoquinone to give **13**. Transformation of germyl radicals to semiquinone radicals was supported by a color change from brown to red. The ESR measurement showed the complete disappearance of signals from the germyl radical.

Only the signal due to the terminal oxyradical of the product was observed. Thus, the concentration of the radical was determined.

Copolymers were isolated and M_n was determined by VPO. From these data, polymer functionality (number of the radical per polymer chain) was calculated as 1.60 for $R^1 = t$-butyl group and 2.15 for $R^1 = t$-amyl group. These values agree considerably well with the expected value of 2.00 and confirm the view that the copolymerization proceeds via biradical propagating species. Similar results also were obtained by quenching radical species with a nitroso compound such as 2,4,6-tri-*t*-butylnitrosobenzene.

On the basis of these findings, a copolymerization mechanism was proposed (Scheme X). The first step is the formation of a biradical, a short-lived intermediate whose electrons are delocalized on the germanium atom and semiquinone biradical. The resulting semiquinone biradical further reacts with the germylene to afford a germyl biradical. By repeating these processes, the elongation of the polymer chain can be achieved. The homopropagation of the Ge–Ge bond formation is sterically impossible. The benzoquinone homo-unit formation is also unlikely because of the unfavorable O–O bond formation, thus allowing the alternating arrangement. During the polymerization, radical coupling reactions between oxygen radicals and germyl radicals may participate in the polymer chain extension.

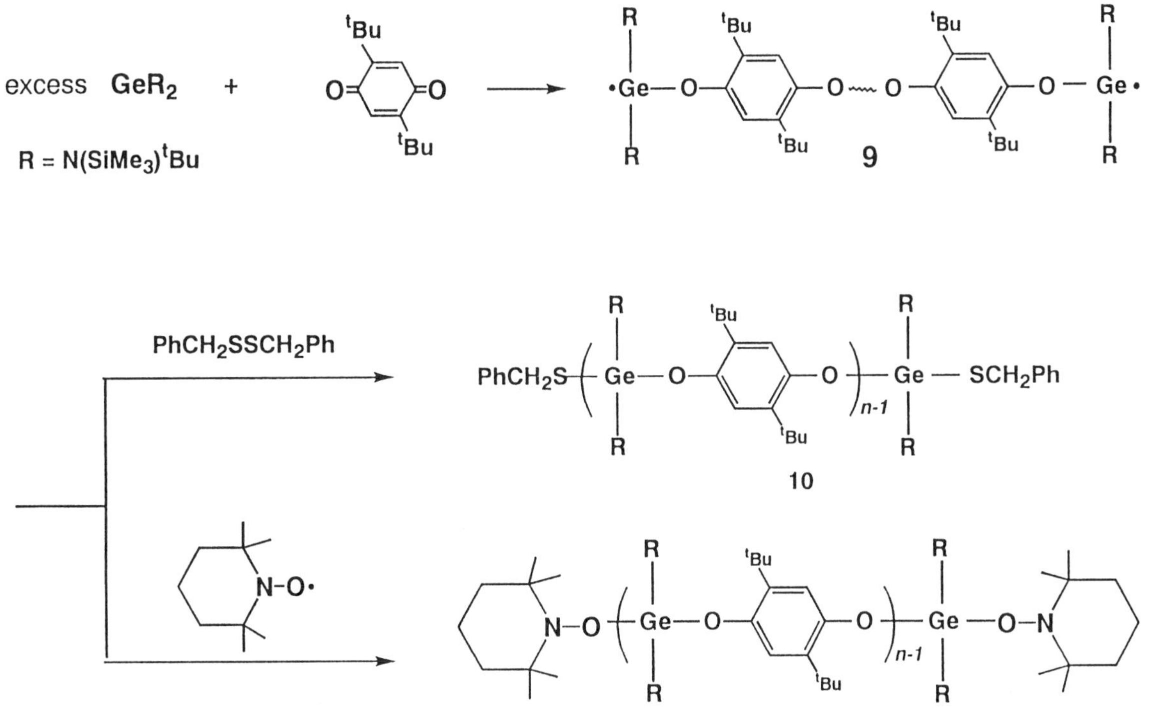

Scheme VIII. Formation of oligomers by the termination of germanium biradical by dibenzyldisulfide and TEMPO radical.

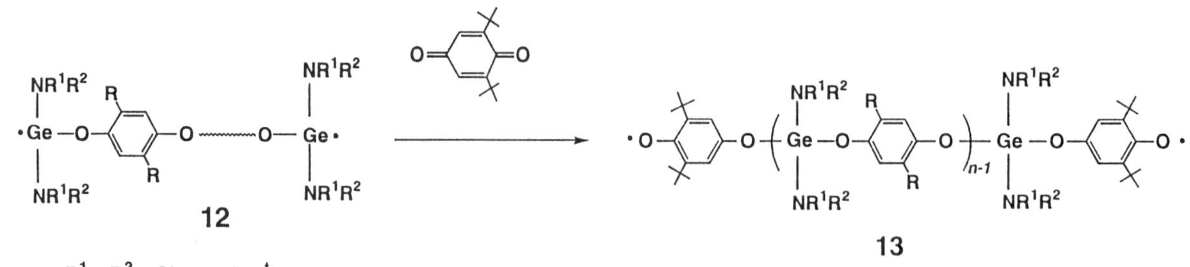

$R^1 = R^2 = SiMe_3$, $R = {}^tBu$
$R^1 = R^2 = SiMe_3$, $R = {}^tAmyl$
$R^1 = SiMe_3$, $R^2 = {}^tBu$, $R = {}^tBu$
$R^1 = SiMe_3$, $R^2 = {}^tBu$, $R = {}^tAmyl$

Scheme IX. Spin trapping of germanium biradical by 2,6-di-*t*-butyl-*p*-benzoquinone in copolymerization mixture.

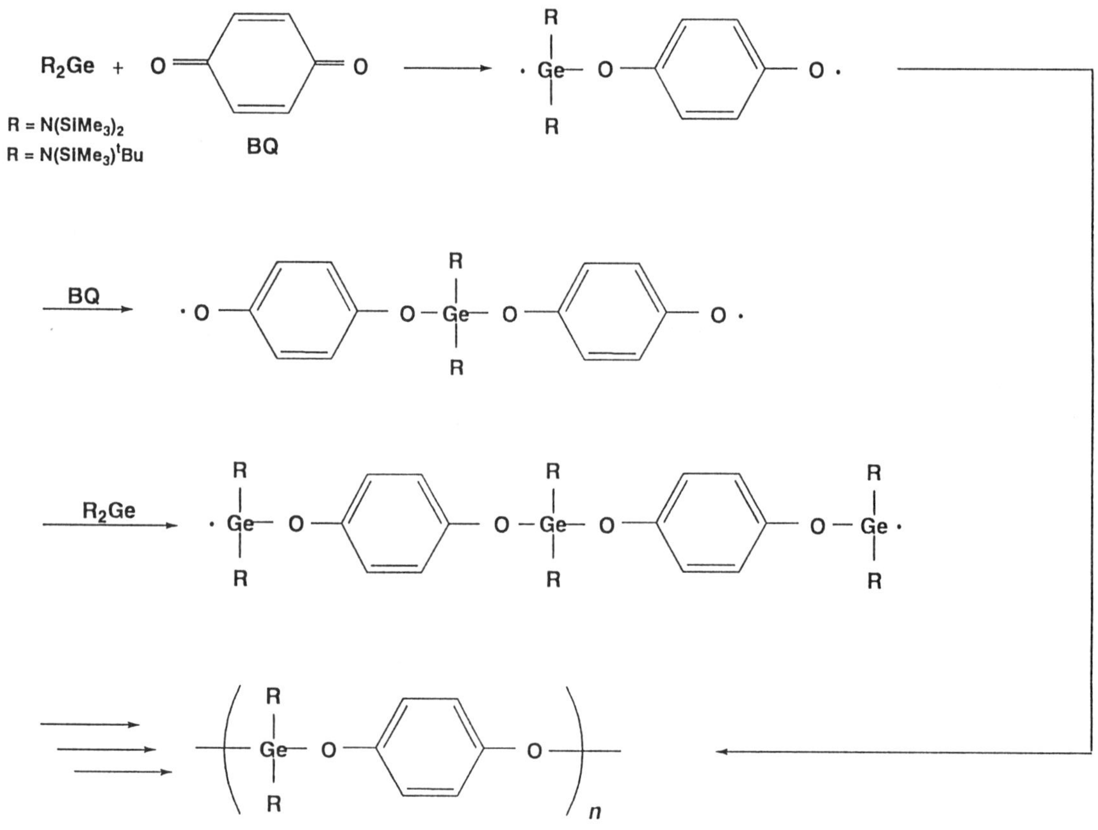

$R = N(SiMe_3)_2$
$R = N(SiMe_3)^tBu$

Scheme X. Copolymerization mechanism of germylene and *p*-benzoquinone involving a germyl biradical.

Copolymerization of Germylene with p-Quinoneimine Derivatives

As an extension of the alternating copolymerization of the germylene and *p*-benzoquinone derivative, a *p*-quinoneimine derivative was used as M_{ox}. The reaction of germylene with N-phenyl-*p*-quinoneimine took place at −78 °C in toluene and gave rise to an alternating copolymer **14** at 63% yield (Scheme XI) (*36*). The structure of the resulting copolymer was confirmed by ^1H and ^{13}C NMR spectra. The molecular weight of **14** was $M_w = 1.5 \times 10^4$ daltons, and the

value of M_w/M_n was 1.94. The resulting copolymers are soluble in chloroform, diethyl ether, and *n*-hexane.

Copolymerization of Stannylene with *p*-Benzoquinone Derivatives

p-Benzoquinone derivatives copolymerized with various tin(II) compounds to afford the corresponding alternating copolymers, most of which are insoluble in organic solvents. By the hydrolysis of the resulting copolymers, *p*-hydroquinone could be obtained in over 90% yields, results indicating that the oxida-

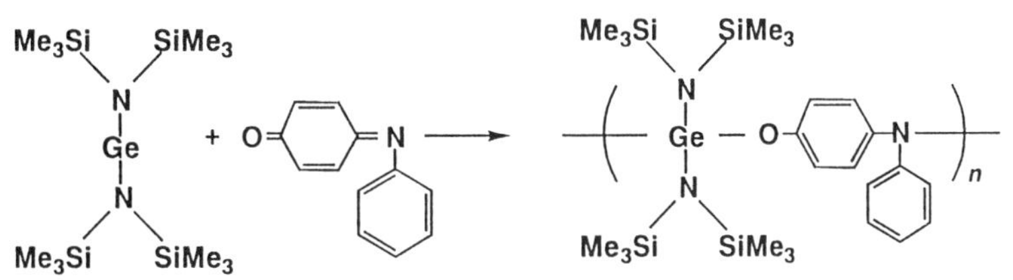

Scheme XI. Copolymerization of germylene with *p*-quinoneimine.

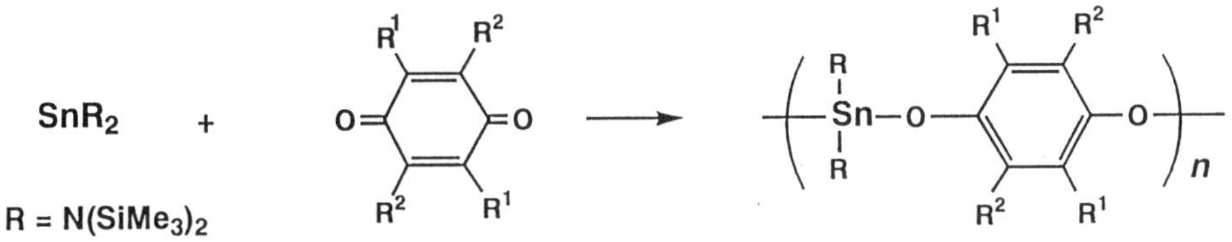

$R^1, R^2 = H$, alkyl

Scheme XII. Copolymerization of stannylene with *p*-benzoquinone derivatives.

tion–reduction proceeded effectively. Among tin(II) compounds screened, bis[bis(trimethylsilyl)methyl]-tin(II) showed the highest reactivity. For example, this compound reacted with sterically hindered *p*-benzoquinone derivatives such as di-*t*-butyl-*p*-benzoquinone or duroquinone that could not be reduced by tin acetate(II), tin chloride(II), tin alkoxide(II), or dimethylstannocene (Sn(C$_5$H$_4$Me)$_2$). The ESR spectra of the reaction mixture suggested a mechanism involving radical species.

The reaction of bis[bis(trimethylsilyl)amide]tin(II) with various *p*-benzoquinone derivatives proceeded smoothly at –42 °C in toluene to give the corresponding copolymers **15** having a tetravalent tin unit and a hydroquinone unit alternating in the main chain (Scheme XII) (*37*). When 2,5-di-*t*-butyl-*p*-benzo-quinone or 2,5-di-*t*-amyl-*p*-benzoquinone was used as the oxidant monomer, the resulting copolymers were soluble in organic solvents and were stable in air.

Poly(germanium enolate)s

Formation of Cyclic Germanium Enolates

In the course of our investigation for the development of new polymerizations by the combined use of M$_{ox}$ and M$_{red}$, we found that the germylene undergoes rapid addition reactions toward a variety of α,β-unsaturated carbonyl compounds with *s*-cis-conformation to give the corresponding cyclic germanium enolates **16** (Scheme XIII) (*38*). These enolates have a germanium atom regioselectively bound to an oxygen atom. All the addition reactions of the germylene to the α,β-unsaturated carbonyl compounds occurred instantaneously at room temperature. By using this procedure, various cyclic germanium enolates derived from

Scheme XIII. Formation of cyclic germanium enolate by reaction of germylene and α,β-unsaturated carbonyl compounds having *s*-cis conformation.

α,β-unsaturated esters, ketones, and aldehydes were obtained in good yields under mild conditions.

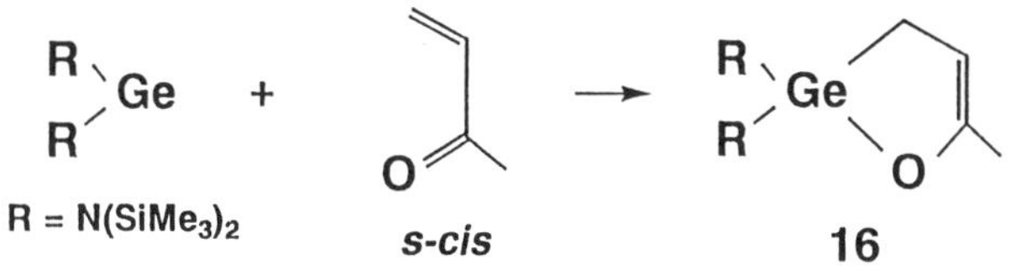

The reaction was carried out in benzene-*d*^{6}, so the formation of the enolates could be directly observed by ^{1}H NMR spectroscopy. The structures of the resulting enolates were determined by comparison of ^{1}H NMR spectra with those of the corresponding 1-oxa-stannacyclopent-4-ene derivatives reported by Neumann and Hillner (*39*). These enolates showed a nucleophilic reactivity that is characteristic of a metal–enolate compound. For example, when the enolate derived from methyl acrylate was treated with chloral as an electrophile at room temperature, the aldol-type reaction occurred smoothly, and the corresponding 1-oxa-2-germacyclopentane derivative was obtained in good yield. The treatment of the enolate with excess acetaldehyde also afforded a cyclic product.

Alternating Copolymerization

We found that the reaction of a germylene and cyclic α,β-unsaturated ketones having *s*-trans structure gives alternating copolymers **17** having a germanium enolate structure in the main chain (Scheme XIV) (*40, 41*). The copolymerization took place smoothly at 0 °C in THF in the presence of catalytic amounts of lithium com-

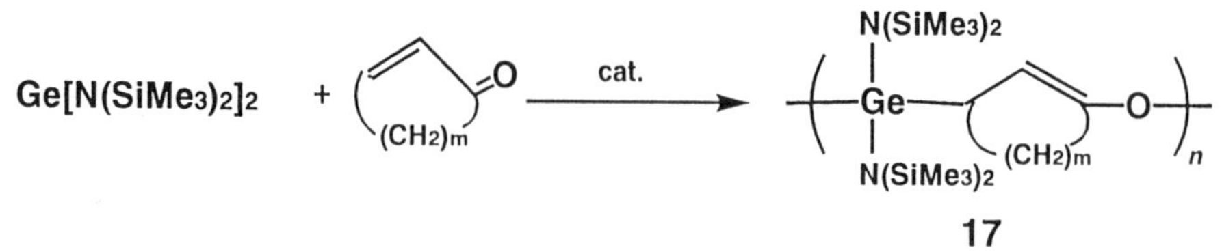

$$Ge[N(SiMe_3)_2]_2 \quad + \quad \text{(cyclic unsaturated ketone)} \quad \xrightarrow{\text{cat.}} \quad \text{(polymer 17)}$$

17

Scheme XIV. Alternating copolymerization of germylene with cyclic α,β-unsaturated ketones having *s*-trans conformation.

pounds such as lithium bis(trimethylsilyl)amide and *n*-butyllithium (Table II). A neutral salt of lithium chloride also acted as a catalyst, which enabled us to carry out the copolymerization under essentially neutral conditions. Other metal salts such as lithium fluoride, sodium chloride, potassium chloride, and potassium chloride–18-crown-6 did not produce copolymers efficiently. When the reaction was carried out at a lower temperature (–42 °C), no copolymerization took place.

The ^{1}H NMR spectrum of the copolymer obtained from germylene and 2-cyclopenten-1-one exhibited a signal at δ = 4.70 ppm due to the olefinic proton of the enolate structure. This result indicated that the resulting copolymer has an *O*-germylated enolate structure in the main chain. The ^{13}C NMR spectrum of the copolymer showed six peaks at δ 6.3, 25.5, 34.6, 39.7, 103.0, and 157.1 ppm ascribable to carbon atoms denoted as a–f in Figure 4. Elemental analysis of the

Table II. Copolymerization of Germylene 1 with 2-Cyclopent-1-one

Catalyst	Catalyst (mol%)	Yield (%)	M_w	M_w/M_n
None	0.0	0	—	—
LiN(SiMe$_3$)$_2$	0.9	90	48.2	4.74
n-BuLi	1.0	89	5.24	2.65
EtOLi	8.8	99	4.47	2.56
LiF	1.0	0	—	—
LiCl	1.0	85	10.7	2.63
LiBr	1.0	83	5.38	1.31
NaCl	—	0	—	—
KCl	—	0	—	—
KCl–18-crown-6	—	0	—	—

Note: Copolymerization was performed in THF at 0 °C for 3 h. Yields are isolated yields. All molecular weight values are in daltons (× 10^{-4}) and were determined by gel permeation chromatography.

Source: Data adapted from reference 40.

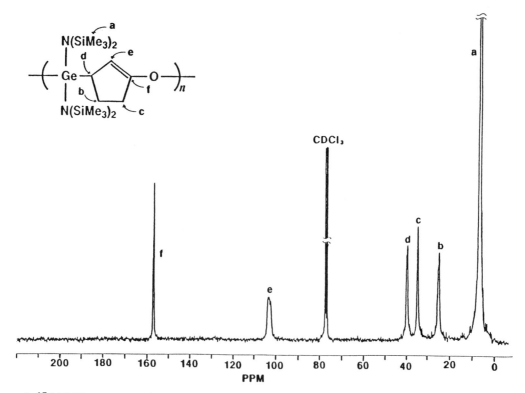

Figure 4. ^{13}C NMR spectrum of poly(germanium enolate) prepared from germylene and 2-cyclopent-1-one.

product also confirmed the structure of the 1:1 alternating copolymer. These results clearly indicate that a polymer of regioselectively O-germylated enolate type was formed; a repeating unit derived from the isomeric C-germylated enolate structure was not detected.

The resulting copolymers are fine white powders and are soluble in common organic solvents such as *n*-hexane and chloroform. The copolymers are reasonably stable against moisture. An aqueous THF (95%) solution of the copolymer was stirred for 24 h at room temperature. The gel permeation chromatography analysis of the recovered copolymer indicated little change in molecular weights. The copolymers have high molecular weights ($M_w > 10^5$ daltons) and give transparent films by casting. Differential scanning calorimetry analysis of the copolymer prepared from germylene and 2-cyclopenten-1-one indicated a T_g of 40.2 °C and a melting point of 220. 8 °C.

The copolymerization mechanism may be explained as follows (Scheme XV). The lithium catalyst reacts with germylene to produce a germyl anion species (18) that adds the cyclic ketone via a Michael-type addition and produces an enolate anion (19). The germyl anion is regenerated by the reaction of the enolate with the germylene. The steric hindrance of the bulky silyl amide group of germylene and the very poor homopolymerizability of the cyclic ketone prevent the homo unit formation and hence allow the alternating propagation.

Oxidative Addition of Poly(disulfide) into Germylene

New sulfur-containing polymers (20) having a germanium(IV) unit and dithiol unit alternating in the main chain were synthesized by the oxidative addition of a poly(disulfide) into the germylene (Scheme XVI) (36). This reaction proceeds almost quantitatively. When the germylene was treated with an equimolar amount of poly(disulfide), the ratio of the germanium(IV) units and dithiol units in the resulting copolymer was 1:1, which was confirmed by ^{1}H NMR spectroscopy. The copolymers were soluble in *n*-hexane, diethyl ether, and chloroform. The degree of polymerization of the resulting copolymers was 18 units smaller than that of the starting poly(disulfide) (42).

Copolymerization of Germylene with Acetylene Derivatives

Germylene was copolymerized for the first time with acetylene monomers in the presence of a rhodium catalyst giving rise to copolymers (21) with a structure having acetylene units in excess (Scheme XVII) (42). Among the catalysts used, a rhodium norbornadiene complex was most effective for the copolymerization.

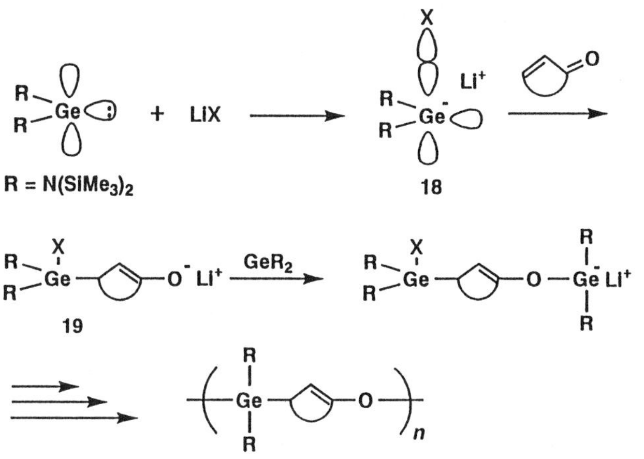

Scheme XV. Copolymerization mechanism for poly(germanium enolate) involving germyl anion.

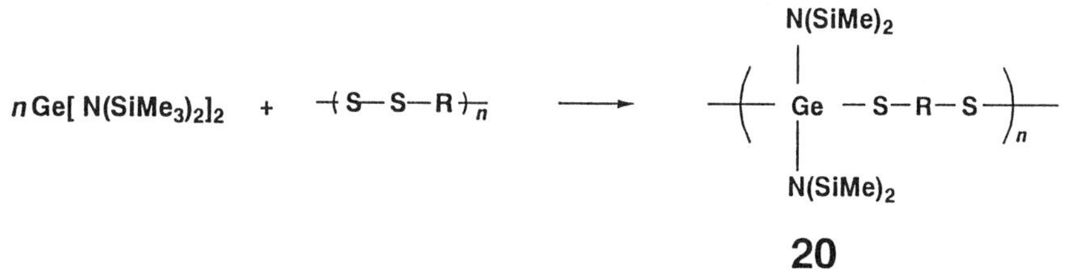

20

Scheme XVI. Oxidative addition of polydisulfide into germylene.

21

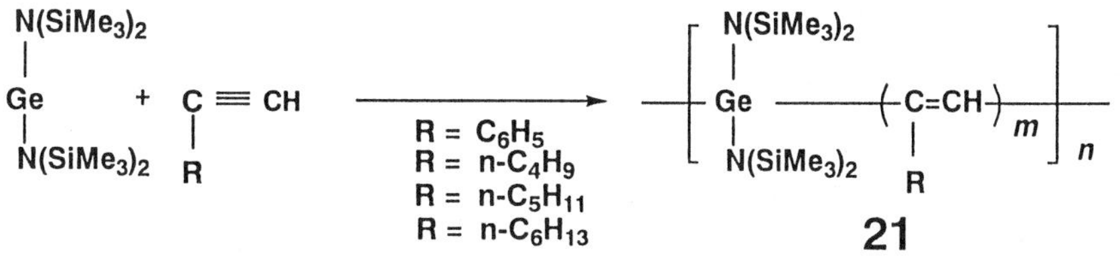

Scheme XVII. Copolymerization of germylene with acetylene derivatives catalyzed by rhodium compounds.

$$\text{GeCl}_2 \cdot \text{C}_4\text{H}_8\text{O}_2 \ + \ \text{R-Li} \longrightarrow \ \cdot(\text{GeR}_2)_n$$

R = CH$_3$
R = n-C$_4$H$_9$ **22**
R = C$_6$H$_5$

Scheme XVIII. Formation of polygermane by ligand substitution polymerization using germanium dichloride-1,4-dioxane complex and alkyl lithium.

Ligand-Substitution Polymerization of Germanium Dichloride

We developed a new polymerization reaction for the synthesis of polygermane (**22**) via a ligand-substitution polymerization of germanium dichloride with organolithium compounds (Scheme XVIII) (43). A lower reaction temperature was favorable for the formation of higher molecular weight polymers. Only low molecular weight oligomers were formed from both alkyl and phenyl lithium compounds when the reaction was carried out at 0 °C, whereas higher molecular weight poly-alkylgermanes ($M_w > 1.7 \times 10^4$ daltons) were formed at –78 °C with alkyl lithium compounds.

Summary

Novel copolymerizations for the synthesis of germanium- and tin-containing polymers by using divalent germanium and tin species (germylene and stannylene) have been developed. The copolymerization involves a redox process between a germylene or a stannylene (M_{red}) and an oxidant monomer (M_{ox}). The mechanism for the copolymerization of germylenes and p-benzoquinone derivatives has been definitely established as a biradical polymerization system involving a germyl radical and a semiquinone radical. This system provides the first clear evidence for a biradical mechanism in polymer chemistry. Polymer synthesis using germylene and stannylene species is thus well documented. In addition, a ligand-substitution polymerization is a new concept in polymerization chemistry and affords a polygermane of high molecular weight. Notably, these novel organometallic polymers are impossible to prepare according to the conventional methodologies of polycondensation.

References

1. *Inorganic and Organometallic Polymers*; Zeldin, M.; Wynne, K. J.; Allcock, H. R., Eds.; ACS Symposium Series 360; American Chemical Society: Washington, DC, 1988.
2. Stark, F. O.; Falender, J. R.; Wright, A. P. In *Comprehensive Organometallic Chemistry*; Wilkison, G.; Stone, F. G. A.; Abel, E. W., Eds.; Pergamon: Oxford, England, 1982; Vol. 2, pp 305–363.
3. Rivière, P.; Rivière-Baudet, M.; Satgé, J. In *Comprehensive Organometallic Chemistry*; Wilkison, G.; Stone, F. G. A.; Abel, E. W., Eds.; Pergamon: Oxford, England, 1982; Vol. 2, pp 399–518.
4. Davies, A. G. In *Comprehensive Organometallic Chemistry*; Wilkison, G.; Stone, F. G. A.; Abel, E. W., Eds.; Pergamon: Oxford, England, 1982; Vol. 2, pp 519–627.
5. West, R. In *Comprehensive Organometallic Chemistry*; Wilkison, G.; Stone, F. G. A.; Abel, E. W., Eds.; Pergamon: Oxford, England, 1982; Vol. 2, pp 365–397.
6. Burkhard, C. A. *J. Am. Chem. Soc.* **1949**, *71*, 963.
7. Yajima, S.; Okumura; Hayashi, J.; Omori, M. *J. Am. Ceram. Soc.* **1976**, *59*, 324.
8. West, R.; David, L. D.; Djurovich, P. I.; Stearley, K. L.; Srinivasan, K. S. V.; Yu, H. *J. Am. Chem. Soc.* **1981**, *103*, 7352.
9. West, R. *J. Organomet. Chem.* **1986**, *300*, 327.
10. Wolff, A. R.; West, R. *Appl. Organomet. Chem.* **1987**, *1*, 7.
11. Trefonas, P.; West, R. *J. Polym. Sci., Polym. Chem. Ed.* **1985**, *23*, 2099.
12. Miller, R. D.; Sooriyakumaran, R. *J. Polym. Sci., Polym. Chem. Ed.* **1987**, *25*, 111.
13. Meyer, G.; Woehrle, D. *Makromol. Chem.* **1974**, *175*, 714.
14. Carraher, C. E., Jr.; Giron, D. J.; Cerutis, D. R.; Burt, W. R.; Venkatachalem, R. S.; Gehrke, T. J.; Tsuji, S.; Blaxall, H. S. In *Biological Activities of Polymers*; Carraher, C. E., Jr.; Gebelein, C. G., Eds.; ACS Symposium Series 186; American Chemical Society: Washington, DC, 1982; pp 13–25.
15. Andersen, D. M.; Mendoza, J. A.; Garg, B. K.; Subramanian, R. V. In *Biological Activities of Polymers*; Carraher, C. E., Jr.; Gebelein, C. G., Eds.; ACS Symposium Series 186; American Chemical Society: Washington, DC, 1982; pp 27–33.
16. Satgé, J. *Pure Appl. Chem.* **1984**, *56*, 137.
17. Neumann, W. P. *Chem. Rev.* **1991**, *91*, 311.
18. Neumann, W. P.; Grugel, C.; Schriewer, M. *Angew. Chem. Int. Ed. Engl.* **1979**, *18*, 543.
19. Carraher, C. E., Jr.; Dammeier, R. L. *Makromol. Chem.* **1971**, *141*, 245.
20. Carraher, C. E., Jr.; Dammeier, R. L. *J. Polym. Sci., Part A-1* **1972**, *10*, 413.
21. Carraher, C. E., Jr.; Jorgensen, S.; Lessek, P. J. *J. Appl. Polym. Sci.* **1976**, *20*, 2255.
22. Carraher, C. E., Jr.; Butler, C.; Naoshima, Y.; Foster V. R.; Giron, D.; Mykytiuk, P. D. *Polym. Sci. Technol.* **1988**, *38*, 175.
23. Carraher, C. E., Jr.; Sterling, D. C.; Butler, C.; Ridgway, T. H.; Louda, J. W. *Polym. Mater. Sci. Eng.* **1990**, *62*, 241.
24. Neumann, W. P.; Schneider, B. *Ann. Chem.* **1967**, *707*, 20.
25. Tsutsui, M.; Kakimoto, N.; Axtell, D. D.; Asai, K. *J. Am. Chem. Soc.* **1976**, *98*, 8287.
26. Kakimoto, N.; Akiba, M.; Takada, T. *Synthesis* **1985**, 272.
27. Honeyman, C.; Foucher, D. A.; Mourad, O.; Rulkens, R.; Manners, I. *Polym. Prepr. (Am. Chem. Soc. Div. Polym. Chem.)* **1994**, 330.
28. Zhou, S. Q.; Weber, W. P.; Mazerolles, P.; Laurent, C. *Polym. Bull.* **1990**, *23*, 583.
29. Liao, X.; Weber, W. P.; Mazerolles, P.; Lauent, C.; Faucher, A. *Polym. Bull.* **1991**, *26*, 499.

30. Harris, D. H.; Lappert, M. F. *J. Chem. Soc., Chem. Commun.* **1974**, 895.

31. Gynane, M. J. S.; Harris, D. H.; Lappert, M. F.; Power, P. P.; Rivière, P.; Rivière-Baudet, M. *J. Chem. Soc., Dalton Trans.* **1977**, 2004.

32. Kobayashi, S.; Iwata, S.; Abe, M.; Shoda, S. *J. Am. Chem. Soc.* **1990**, *112*, 1625.

33. Kobayashi, S.; Iwata, S.; Shoda, S.; Yamaoka, H., unpublished result.

34. Kobayashi, S.; Iwata, S.; Abe, M.; Shoda, S. *J. Am. Chem. Soc.* **1995**, *117*, 2187.

35. Bennett, S. W.; Eaborn, C.; Hudson, A.; Hussain, H. A.; Jackson, R. A. *J. Organomet. Chem.* **1969**, *16*, 36.

36. Kobayashi, S.; Iwata, S.; Abe, M.; Shoda, S. *Polym. Prepr. Jpn.* **1990**, *39*, 1781.

37. Kobayashi, S.; Shoda, S.; Iwata, S. *Polym. Prepr. Jpn.* **1989**, *38*, 252.

38. Kobayahsi, S.; Shoda, S.; Iwata, S. *Chem. Express* **1989**, *4*, 41.

39. Hillner, K.; Neumann, W. P. *Tetrahedron Lett.* **1986**, *27*, 5347.

40. Kobayashi, S.; Iwata, S.; Yajima, K.; Yagi, K.; Shoda, S. *J. Am. Chem. Soc.* **1992**, *114*, 4929.

41. Kobayashi, S.; Shoda, S. *Adv. Mater.* **1993**, *5*, 57.

42. Kobayashi, S.; Cao, S. *Chem. Lett.* **1993**, 25.

43. Kobayashi, S.; Cao, S. *Chem. Lett.* **1993**, 1385.

Silicon-Containing Monomers and Polymers

Yukio Nagasaki

This chapter presents a review of silicon-containing organic macromonomers and polymers and their potential applications. One class of Si-containing polymers is based on polystyrene and is obtained by free-radical or anionic polymerization. The other class of interesting Si-containing polymers discussed is the polysilamine oligomers (heterotelechelics) obtained by anionic polyaddition of dimethyldivinylsilane with N,N'-diethylethylenediamine. Various aspects of the synthesis and characterization of these polymers are reviewed. Potential applications of styrene-based Si-containing polymers for gas separation, electron-beam resists, and biomaterials are outlined. Polysilamines containing both silicon and amine functionality in their main chains show very interesting transition properties in aqueous media as a result of protonation/deprotonation. These polymers are also of potential interest for developing graft copolymers, surface modifiers, stimuli-sensitive gels, and other materials.

Organosilicon-containing polymers are of great interest as candidates for new functional materials such as gas separation membranes (1), resists (2), semiconducting polymers (3), and surface modification agents (4). The strategy for synthesis of organosilicon-containing polymers having new structures is to create novel synthetic routes based on reaction design and methodology newly developed by co-workers and me (5). In 1986, co-workers and I found that 4-methylstyrene (1) in the presence of lithium diisopropylamide (2) is metalated at the methyl group to form 4-vinylbenzyllithium (3) without any side reaction (Scheme I). By using 3 as an intermediate, several kinds of Si-containing monomers, oligomers, and polymers having new structure were synthesized. Hence, the first part of this review deals with the synthesis of Si-containing styrene monomers and their polymerization reactivities. The second part discusses synthesis of Si-containing polymers with reactive vinyl groups by using anionic polymerization initiated by lithium alkylamide. Lithium diisopropylamide (2) can initiate anionic copolymerization of 1,4-divinylbenzene with 4-trimethylsilylstyrene to form an Si-containing polymer with pendant vinyl groups without cross-linking. This technique can be applied to block copolymer synthesis, and poly(divinylbenzene-*b*-silicone-*b*-divinylbenzene) could be synthesized easily.

The third topic of this chapter is related to end-reactive oligomers (macromonomers and telechelic oligomers) containing alternate organosilyl and amino groups in the main chain obtained via anionic polyaddition of lithiated diamines with divinylsilane derivatives. The resulting oligomer shows unique characteristics such as pH and temperature dependencies of solubility. Current and potential applications of the Si-containing polymers including gas separation membranes and electron-beam resists are also discussed in the final section.

Silicon-Containing Monomers and Polymers

Lithium alkylamide has been widely used as a strong base to lithiate active hydrogen compounds (6). When lithium alkylamide is mixed with conjugated olefins such as isoprene and styrene, however, an addition reaction takes place instead of metalation (7). In this way, lithium alkylamides can be used as both metalation and amination agents. Lithium diisopropylamide (2), however, shows a unique reactivity toward conjugated olefins owing to its bulky substituents. In 1986, a co-worker and I found (5a) that 2 induces lithiation of 4-methylstyrene (1) (instead of addition to the double bond) to form 4-vinylbenzyllithium (3; Scheme I). Notably, toluene is not lithiated by 2 under the same reaction conditions. The result may be explained by stabilization of the carbanion owing to 10π-conjugation in 3. Compound 3 can be regarded as a very unique carbanion having a reactive vinyl group. By using 3 as an intermediate, co-workers and I were able to synthesize a variety of novel monomers and oligomers.

Silicon-Containing Styrene Monomers

Syntheses of organosilyl-containing styrene monomers were carried out by adding corresponding organosilyl chloride slowly to 3 in tetrahydrofuran (THF) (Figure 1) (5b). The gas chromatogram of the reaction mixture is also shown in Figure 1. Two new peaks appear at about 10 and 15 min. On the basis of gel permeation chromatographic (GPC) measurements, neither polymerization nor oligomerization took place during the reaction with trimethylsilyl chloride. In gas chromatographic–mass spectrometric analysis, parent ion signals of m/z 190 and 262 were observed, which correspond to mono- and doubly trimethylsilylated 1. From ^{1}H and ^{13}C NMR analyses, these two products were identified to be 4-(trimethylsilylmethyl)styrene (5) and 4-[bis(trimethylsilyl)methyl]styrene (7), respectively. On the basis of these results, the silylation reaction of 4-methylstyrene was concluded to proceed as shown in Figure 1.

The yield of the silylated monomers was dependent on the initial mole ratio of the reactants 1 to 2. When the reaction was carried out with $[1]_0/[2]_0 = 2$, the yield of 5 reached 82%, and the yield was more than 60% for 7 with $[1]_0/[2]_0 = 0.5$.

The equilibrium constant for the lithiation reaction of 1 by 2 (Scheme I) was 0.54 in THF at 20 °C, and the mole ratio of 3 to 2 was calculated as 26:74 from the equilibrium constant. However, the yield of the products (5 or 7) was much higher than the value expected from the equilibrium constant. This result

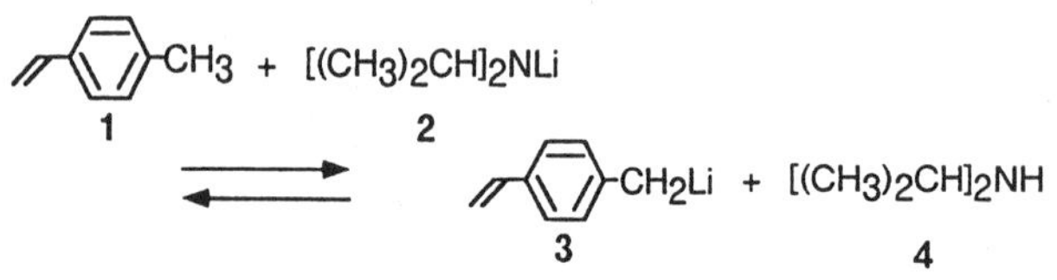

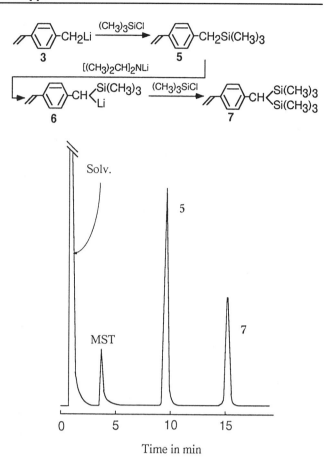

Figure 1. Gas chromatogram of products from the silylation of 4-methylstyrene (MST) in the presence of lithium diisopropylamide. (Reproduced with permission from reference 5b. Copyright 1986 Hütig & Wepf Verlag Basel.)

may be explained by the low reactivity of 2 with trimethylsilyl chloride owing to the steric hindrance. Thus, trimethylsilyl chloride reacts preferentially with 3 as compared to 2, even though the concentration of 2 is much higher than that of 3. Actually, no coupling product between 2 and trimethylsilyl chloride could be detected by gas chromatography after the reaction is completed. High yields of silylated styrenes could be achieved in this way.

Several types of alkylmetal-containing 4-methylstyrene derivatives were also obtained by adding trialkylmetal chlorides to the equilibrium mixture prepared from 1 and 2 (5f). Yields of the products were strongly dependent on the structure of alkyl substituents adjacent to the metal atom. Structures and maximum yields of several organometal-containing 4-methylstyrenes are summarized in Table I. Because compound 5 was metalated by 2 to form 6 (Figure 1), styrene derivatives having trimethylsilyl and other trialkylmetal groups can also be synthesized (Table I).

Metalation of 4-methylstyrenes by 2 can be applied to synthesis of other types of new monomers having a siloxyalkyl group by using an oxirane as a

Scheme I. Lithiation of 4-methylstyrene (1) by lithium diisopropylamide (2).

Table I. Synthesis of Si-Containing Styrenes

Monomers	Yield (%)	Monomers	Yield (%)
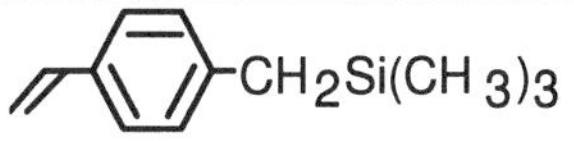 $CH_2Si(CH_3)_3$	82	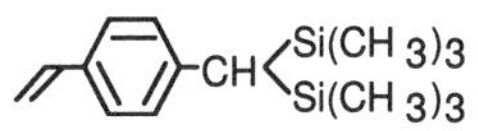 $CH\langle^{Si(CH_3)_3}_{Si(CH_3)_3}$	65
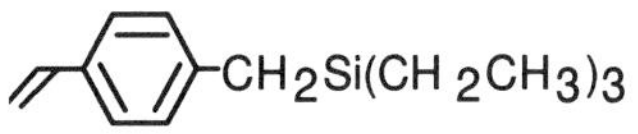 $CH_2Si(CH_2CH_3)_3$	79	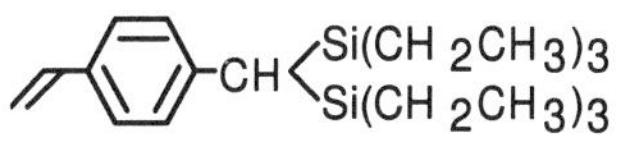 $CH\langle^{Si(CH_2CH_3)_3}_{Si(CH_2CH_3)_3}$	41
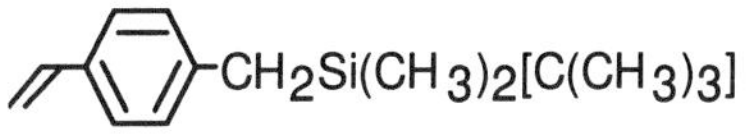 $CH_2Si(CH_3)_2[C(CH_3)_3]$	49	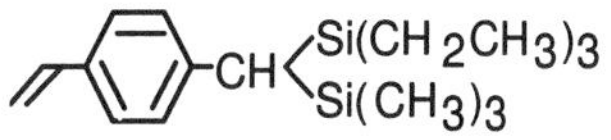 $CH\langle^{Si(CH_2CH_3)_3}_{Si(CH_3)_3}$	69
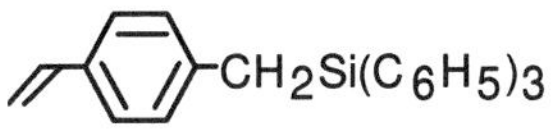 $CH_2Si(C_6H_5)_3$	67	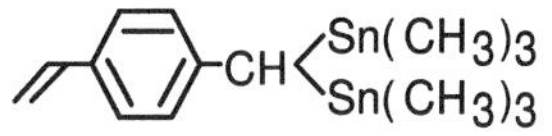 $CH\langle^{Sn(CH_3)_3}_{Sn(CH_3)_3}$	50
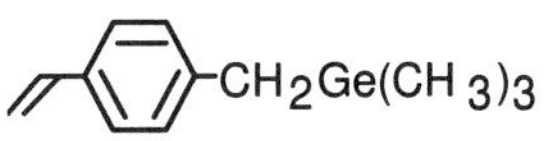 $CH_2Ge(CH_3)_3$	34	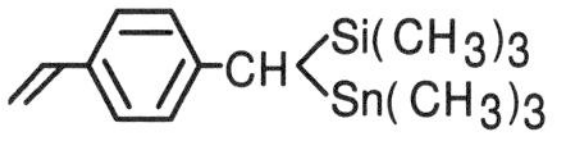 $CH\langle^{Si(CH_3)_3}_{Sn(CH_3)_3}$	33
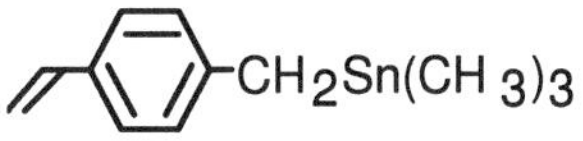 $CH_2Sn(CH_3)_3$	63		

reactant (*5d, 8*). Thus, when an oxirane derivative was added to the equilibrium mixture containing **3** or **6**, anionic ring opening of oxirane by carbanion proceeded to form styrenes containing alkoxy groups (Scheme II). In this reaction, oxirane reacted preferentially with carbanion rather than with the amide anion. Structures and maximum yields of the resulting monomers are summarized in Table II.

Styrene monomers carrying residues with three or four Si atoms on the 4-position can be obtained in just one step. Goronwicz and West (*9*) reported that *tert*-butyllithium induced a metalation of trimethylsilyl chloride to form α-silylmethyl carbanion (**9**) at low temperature. When trimethylsilyl chloride was treated with **2** in THF at –40 °C, a similar reaction took place to form a chlorosilane having three Si atoms. Thus, because the coupling reaction between **2** and trimethylsilyl chloride is slow, trimethylsilyl chloride was metalated by **2** to form **9**. The carbanion **9** then reacted with trimethylsilyl chloride to form **10**. Subsequently, metalation of **10** and repeated coupling with trimethylsilyl chloride formed a silyl chloride carrying three Si atoms (**11**) (Scheme III). By adding **1** to the reaction mixture after the formation of **11** and by increasing the temperature to 20 °C, metalation of **1** took place by excess **2** to form **3**; then, coupling took place between **3** and **11** to form styrenes with three Si atoms. When **5** was used as a reactant instead of **1**, styrenes carrying four Si atoms could be synthesized (*10*).

This reaction was extended to other monomers having active hydrogen, such as 4-allylstyrene (*11*) and 4-isopropenyltoluene (*12*). When 4-allylstyrene was treated with **2**, metalation proceeded at –40 °C in

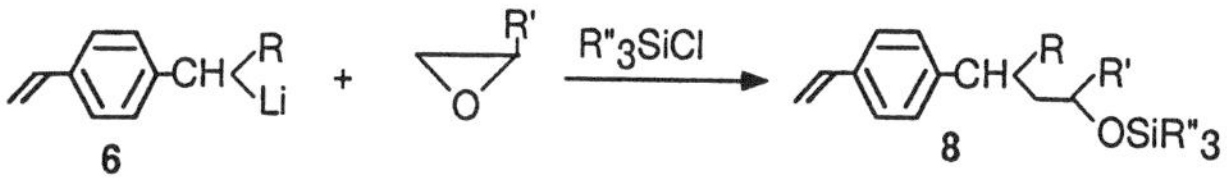

$$CH\langle^{R}_{Li} + \overset{R'}{\underset{O}{\triangle}} \xrightarrow{R''_3SiCl} CH\langle^{R\quad R'}_{OSiR''_3}$$
$$\mathbf{6} \qquad\qquad\qquad\qquad \mathbf{8}$$

Scheme II. Synthesis of styrene monomers with trialkylsiloxy group.

THF to form carbanion **14**. When trimethylsilyl chloride was added to the reaction mixture, mono- and bis(trimethylsilylated) allylstyrene was obtained (Scheme IV).

Organosilyl-containing α-methylstyrenes can also be synthesized in the same way. When 4-isopropenyltoluene was used instead of 4-methylstyrene, lithiation of the methyl group directly on the ring took place to form 4-isopropenylbenzyllithium selectively; then, silylation proceeded smoothly to form mono- and di(trimethylsilylated) monomers without any side reaction when trimethylsilyl chloride was used (Scheme IV).

Silicon-Containing Oligomers with Reactive Vinyl Groups

As stated previously, the silylation reaction of 4-methylstyrene (**1**) by trimethylsilyl chloride in the presence of lithium diisopropylamide (**2**) produced the doubly trimethylsilylated product **7**. In this way, the methyl group of **1** behaved as a difunctional group in the **2**-induced silylation reaction. If coupling between vinylbenzyllithium (**3**) and silyl chloride proceeds preferentially even when a difunctional silyl chloride is used, the formation of a novel organosili-

Table II. Synthesis of Si-Containing Styrenes

Monomers	Yield (%)	Monomers	Yield (%)
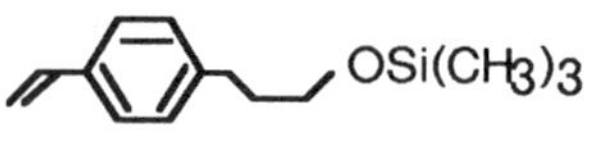	29	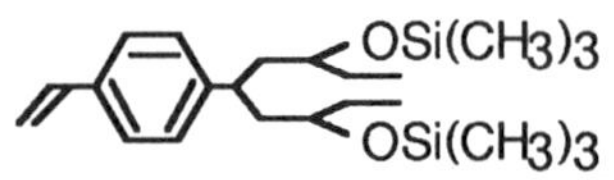	10
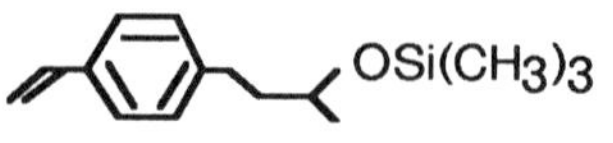	47	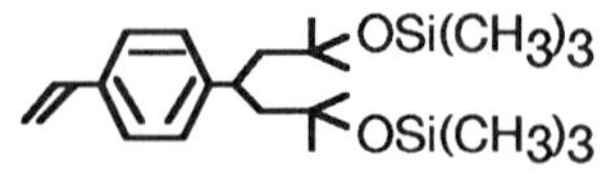	6
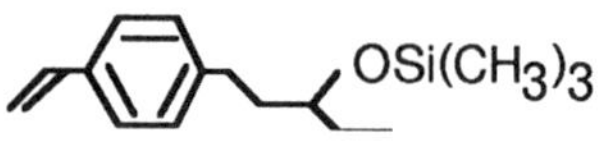	56	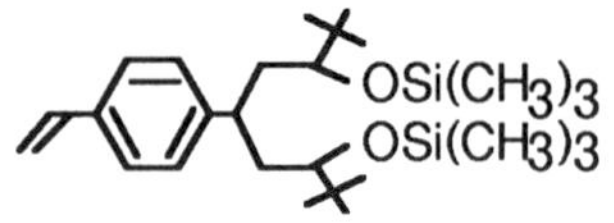	4
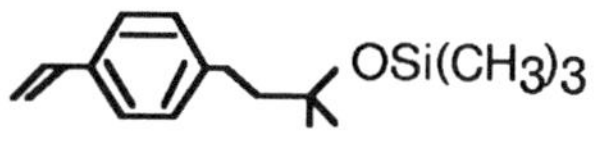	61	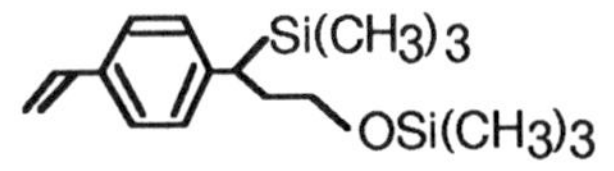	14
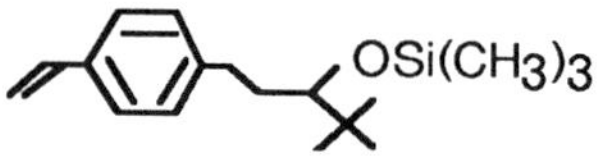	75	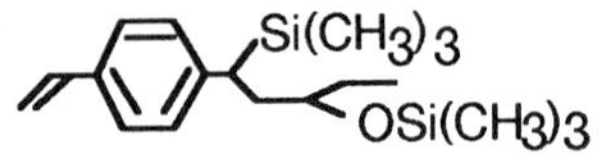	72
		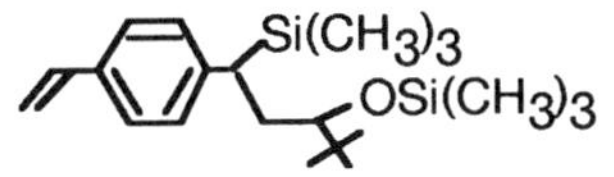	40

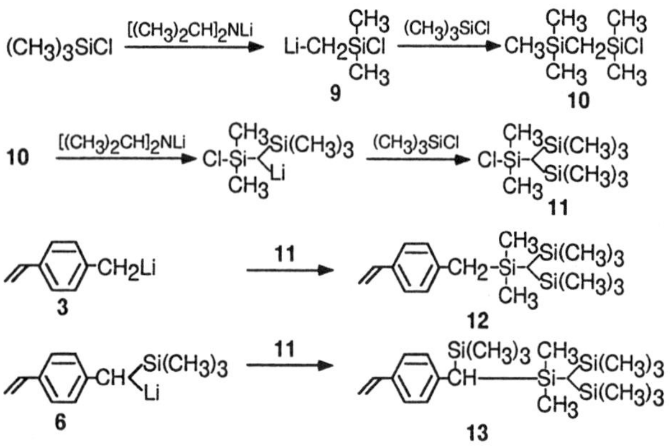

Scheme III. Synthesis of styrene monomers with multi-organosilyl groups starting from 4-methylstyrene (**1**).

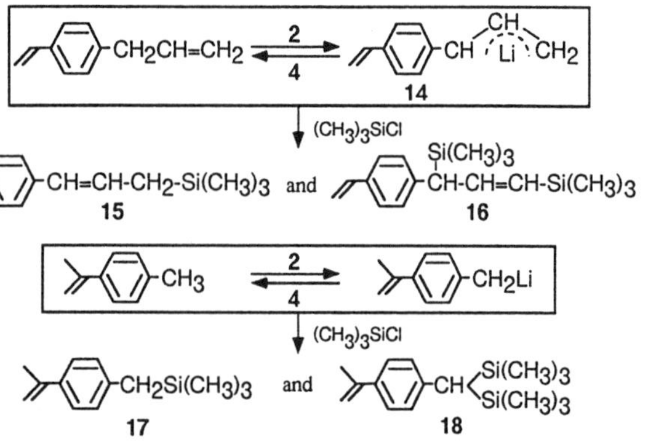

Scheme IV. Synthesis of styrene monomers with multi-organosilyl groups starting from 4-allylstyrene and iso-propenyltoluene.

con polymer is expected through a polycondensation reaction. By using 1,2-bis(chlorodimethylsilyl)ethane (**19**) as a difunctional silyl chloride, possibilities for the polycondensation reactions were examined.

When a THF solution of **19** was added slowly into the equilibrium mixture containing **3** and **2**, a non-volatile material was formed that was soluble in a variety of organic solvents (*5g*). The molecular weight of the product was in the range of 200 to 2000. In addition, the product had UV absorption at 296 nm (indicative of the vinyl phenyl group), and these results suggest that an anionic polycondensation reaction took place.

From 500-MHz ^{1}H NMR spectrum of the product (also shown in Figure 2), further information about anionic polycondensation was obtained. By using compounds **7** and **19** as references, the product was confirmed to contain the structural units of both **1** and **19**. From quantitative analysis using vinyl β-protons and phenyl protons, the vinyl group of **1** remained intact. The resulting oligomer, **21**, has a unique structure containing silyl groups in the main chain and reactive vinyl groups in the side chain, and it is a candidate for novel functional materials. When dichloro-dimethylsilane was used as the difunctional counterpart, a similar anionic polycondensation reaction took place to form the corresponding oligomers.

This anionic polycondensation reaction was also applicable to other monomers having two active hydrogens. When trimethylallylsilane was used instead of **1**, an Si-containing oligomer with a double bond in the main chain (**22**) could be obtained. When a mixture of **1** and trimethylallylsilane was used, co-oligomers having both components (**23**) were obtained (Scheme V) (*13*).

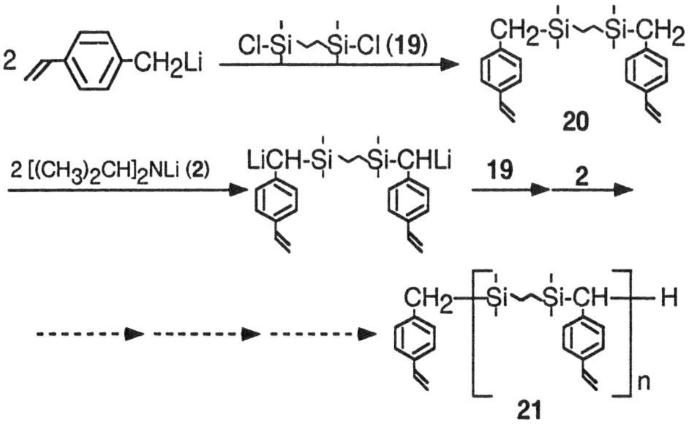

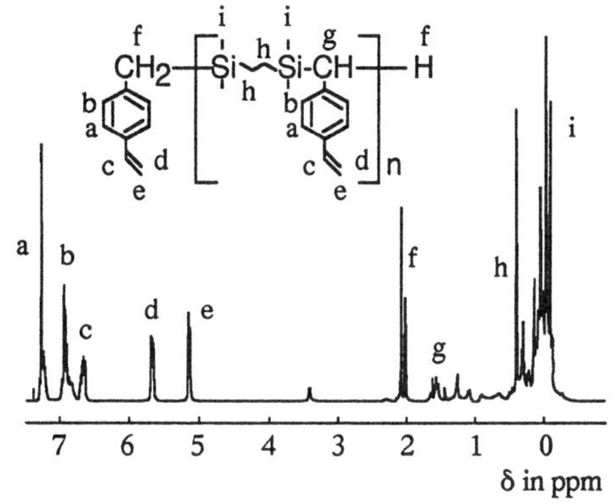

Silicon-Containing Polystyrene Macromonomers

The carbanion structures such as **3** or **6** are stabilized enough to exist for at least 30 min without any side reaction under the reaction conditions (20 °C in THF). In addition, the reactivity of 4-(trimethylsilylmethyl)-styrene (**5**) toward nucleophilic reagents is much lower than that of styrene itself owing to σ–π hyperconjugation (*14*). For monomer **5** to undergo an anionic polymerization in the equilibrium mixture containing **2** and **6** (Figure 3), co-workers and I carried out the reaction by using a relatively high concentration of **5**. Our idea was to synthesize novel macromonomers from **5**. Under suitable concentration conditions, and at 0 °C for 60 h, the reaction gave a nonvolatile material, and the molecular weight was 200–3000 (*5c*). The oligomer thus obtained had UV absorption at 296 nm (the vinyl benzyl group).

The ^{1}H NMR spectrum of the product is also shown in Figure 3. By using **5**, dimer of **5** and polystyrene macromonomers as reference compounds, the assignments of proton signals of the product were carried out. Even though there were two anionic species in the equilibrium system, the resulting oligomers exhibit no methyl proton signal around δ = 1.0 originating from a diisopropylamino group. This result suggests that the anionic polymerization of **5** was induced selectively with **6** as initiator. Actually, the average molecular weight of the oligomers determined by GPC (1040) agreed well with that from ^{1}H NMR measurement (1100) by assuming one vinyl group per each oligomer molecule.

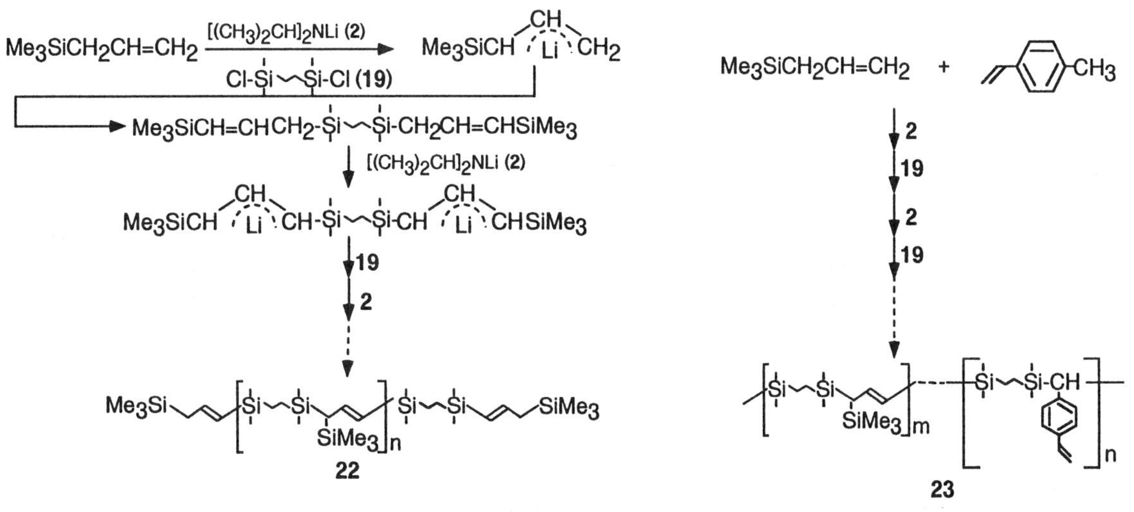

Figure 2. ^{1}H NMR spectrum of the oligomer, **21**, obtained by anionic polycondensation. (Reproduced with permission from reference 5g. Copyright 1990 Hütig & Wepf Verlag Basel.)

Scheme V. Synthesis of Si-containing oligomers with reactive double bonds in the main chain (**22**) and in the main chain and side chain (**23**) through anionic polycondensations.

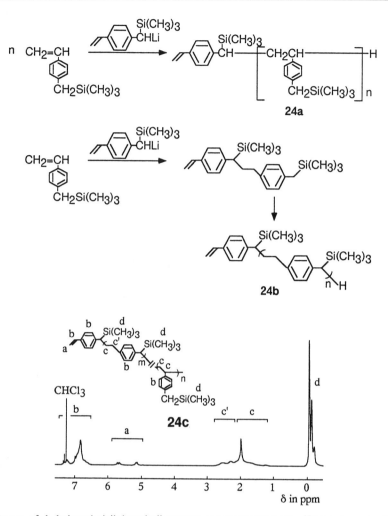

Figure 3. ^{1}H NMR spectrum of 4-(trimethylsilylmethyl)styrene macromonomers. (Reproduced with permission from reference 5c. Copyright 1989 Hütig & Wepf Verlag Basel.)

Another key consideration in the anionic polymerization of **5** is the mode of enchainment of the repeating units. Two possible modes may be envisaged. The first one is the classical mode in which the vinyl polymerization of **5** is initiated by **6** to form a vinyl-type macromonomer (**24a**). The second mode is a novel one in which a one-to-one addition takes place between **5** and **6**, followed by a transmetalation reaction by **4**. After repeating this addition cycle, a polyaddition macromonomer (**24b**) is obtained.

A quantitative analysis of the ^{1}H NMR spectrum in Figure 3 showed that the two modes of polymerization had taken place concurrently to form a copolymer (**24c**). For instance, the contents of $-CH_2-Si(CH_3)_3$ and $>CH-Si(CH_3)_3$ enchainments for the copolymer (**24c**) shown in Figure 3 are 52.5% and 47.5%, respectively. By increasing the concentration of the transmetalation agent (**4**), the content of $>CH-Si(CH_3)_3$ linkage in the macromonomers increased. Actually, when the reaction was run under the condition $[5]_0/[2]_0/[4]_0 =$

$0.5/1/2.9$ at 50 °C, the mode of polyaddition reached more than 92%.

Silicon-Containing Polystyrenes by Radical Polymerizations

To estimate the reactivity of the monomers thus obtained and to obtain Si-containing (co)polymers with new structures, radical (co)polymerizations of these monomers were carried out. The reactivity of organosilicon compounds is known to be greatly influenced by their substituents (*15*). For example, R-Si groups in phenyl-substituted silicon compounds show an electron-withdrawing effect (*16*). On the other hand, an electron-releasing effect is observed for benzyl-type silicon compounds (*17*). Because silicon-containing monomers and macromonomers such as **5**, **7**, and **24** have a benzylsilane-type structure, these monomers should exhibit interesting reactivities resulting from the electron-releasing effect of the organosilyl substituent.

Radical copolymerizations of styrene with 4-(trimethylsilylmethyl)styrene (5) and with 4-[bis(trimethylsilyl)methyl]styrene (7) were carried out. The mole fraction of 5 in the copolymer tended to be lower than that in the feed. A similar tendency was observed in copolymerization of 7 with styrene. In the copolymerization of 5 or 7 with methyl methacrylate, on the other hand, the alternating tendency of the copolymer obtained is higher than that of the copolymerization of styrene with methyl methacrylate.

By using the Kelen–Tüdós plot, monomer reactivity ratios were determined. Values for Q and e, which are defined as conjugation and polarization factors, respectively (18), were also determined from the monomer reactivity ratios (1.85 and −1.74 for 5; 1.83 and −1.81 for 7, respectively). The larger negative e values indicate that the vinyl groups of these monomers are more electron-rich than styrene, and this difference is due to the electron-releasing character of the silylmethyl group. Excess net charges, relative to styrene, at the β-carbons of the vinyl group of 4-substituted styrenes were calculated by the AM1 method. The order of excess charge densities is parallel with that of the observed electron-donating effect. The effect is attributable to σ-π hyperconjugation between C–Si bond and the vinylphenyl conjugated system, in which the polar C–Si bond acts as a π-electron source.

From ^1H NMR, UV, and semi-empirical molecular orbital calculations, other Si-containing styrenes mentioned previously have almost similar reactivities to those of 5 and 7. Actually, several Si-containing styrenes thus obtained easily gave homopolymers with high molecular weight by radical polymerization. Even allylstyrenes, 15 and 16, gave homopolymers by bulk polymerization. However, polymerization of 4-allylstyrene itself gives only a gel under similar conditions due to cross-linking by the allyl group.

Silicon-Containing Polystyrenes by Anionic Polymerizations

As stated previously, it is easy to synthesize Si-containing polystyrenes by radical (co)polymerization. For several applications, it is important to synthesize well-defined polymers in terms of molecular weights and end groups. To fulfill these demands, anionic polymerization of 4-[bis(trimethylsilyl)methyl]styrene (7) was examined. Figure 4 shows GPC diagrams of the resulting polymer obtained in THF by using butyllithium as an initiator (2a).

The polymers had relatively narrow molecular weight distribution (weight-average to number-average molecular weight ratio, M_w/M_n = 1.06–1.07) even at 20 °C, whereas the molecular weight distribution of polystyrene obtained under similar conditions was

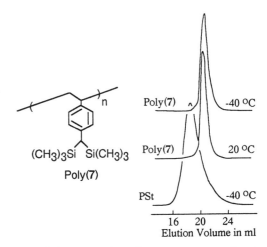

Figure 4. Gel permeation chromatograms of polymers obtained by anionic polymerizations. (Reproduced with permission from reference 2a. Copyright 1990 John Wiley & Sons, Ltd.)

relatively broader (M_n = 6200; M_w/M_n = 1.80) even at −40 °C. This difference may be explained by lower propagation rate of 7 owing to its electron-donating character.

When styrene was added into the reaction mixture after 7 was consumed completely, anionic polymerization proceeded to form a block copolymer consisting of poly(7) and polystyrene segments. From GPC measurements, the peak from the prepolymer(7) almost diminished and indicated the living nature of chain ends in the anionic polymerization of 7.

Trimethylsilyl group is a well-known stabilizer of carbanions if in the α-position (19). Actually, bis(trimethylsilyl)methane is easily metalated by *tert*-butyllithium to form the corresponding carbanion [bis-(trimethylsilyl)methyllithium] (20). In the anionic polymerization of 7, however, the methine group adjacent to two trimethylsilyl groups and a phenyl group was not metalated by the growing carbanion even at 20 °C, evidently due to the steric hindrance of these three substituents.

Living anionic polymerization of 7 can also be applicable for the synthesis of other block copolymers. Nakahama and co-workers (21) reported that 2-(trimethylsiloxy)ethyl methacrylate (25) can be polymerized anionically under suitable conditions. When 25 was added to the anionic polymerization mixture after the monomer was consumed completely, the color of the carbanion diminished immediately and polymerization of 25 proceeded at −74 °C. Polymerization of 25 proceeded rapidly (within 5 min), and the poly(7-b-HEMA) (26) could be obtained after cleavage of trimethylsilyl protect groups by HCl. In this way, Si-containing hydrophilic–hydrophobic block copoly-

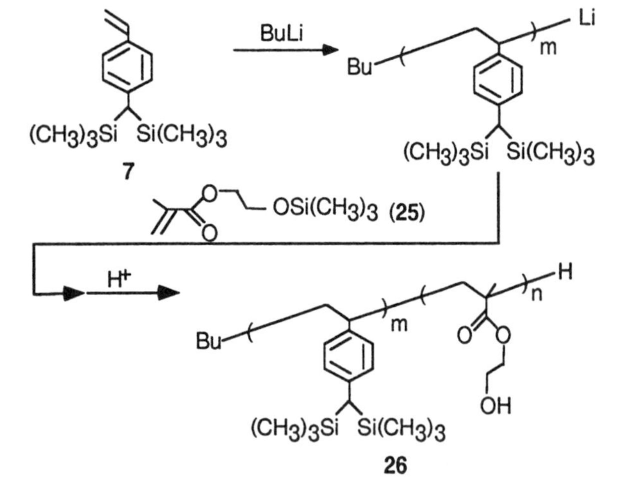

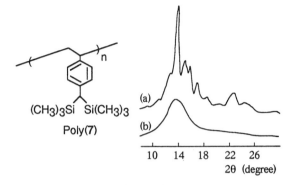

Scheme VI. Poly{4-[bis(trimethylsilyl)methyl]styrene-*b*-2-hydroxyethyl methacrylate} obtained by living anionic polymerization.

Figure 5. Wide-angle X-ray diffractograms of poly{4-[bis-(trimethylsilyl)methyl]styrene} obtained by anionic polymerization in hexane (a) and radical polymerization (b). (Reproduced with permission from reference 24. Copyright 1993 Hütig & Wepf Verlag Basel.)

mers can be synthesized easily by anionic polymerization (Scheme VI).

Another attractive feature of anionic polymerizations is to control microstructures of the polymers and to investigate structure–property relationships (22); microstructure–crystalline relations are especially attractive. Highly tactic polymers can be obtained in coordination polymerizations of styrenes (23), and some of the polymers show high crystallinity. In anionic polymerization with alkali metal counter ions, however, there are very few reports on crystalline polystyrenes. In most cases, the crystallinity was not high even after separation of the fraction insoluble in methyl ethyl ketone (24). For example, Cazzaniga and Cohen (20d) reported that butyllithium–lithium *tert*-butoxide complex in hexane at –40 °C gave atactic-rich polystyrene, in which the fraction (20–30%) insoluble in methyl ethyl ketone showed isotacticity and crystallinity (degree of crystallinity was between 10 and 25%).

Because the poly(7) prepared by radical polymerization was soluble in hexane owing to organosilyl groups, co-workers and I investigated (25) solvent effect on microtacticities of the polymer formed by anionic polymerization. Butyllithium (BuLi) itself did not induce any reaction toward **7** in hexane at –40 °C (26). By adding *N,N,N′,N′*-tetramethylethylenediamine into the reaction mixture, the color of the mixture turned brownish red immediately. After a 15-min reaction, the reaction mixture became heterogeneous and kept the color of the carbanion. Contrary to our anticipation, however, the polymer obtained (85% after 4 h) was insoluble in any organic solvent at ambient temperature. At elevated temperature, the polymer dissolved in benzene, chloroform, and THF.

From the X-ray diffraction patterns of poly(7) shown in Figure 5, poly(7) synthesized by radical

polymerization showed one broad peak, indicating the polymer has only amorphous structure. However, the anionically synthesized poly(7) showed sharp diffraction patterns, indicating a high crystallinity. The degree of crystallinity of poly(7) was determined to be 52% by assuming the 100% crystalline polymer showed the same integration intensity in diffractograms as the 100% amorphous sample. From ^{13}C NMR analysis, poly(7) obtained by anionic polymerization was found to have a high stereoregularity (25).

Anionic polymerization of α-methylstyrene is known to be an equilibrium reaction (27). Over the ceiling temperature, depolymerization proceeds predominantly and no polymer is obtained. Because one Si-containing monomer (18) has an α-methylstyrene skeleton, the same phenomena can be anticipated; hence, anionic polymerizability of this monomer was examined (12). When 1,1-diphenylhexyllithium (DPL) was used as an initiator, no polymerization of 18 took place, a result indicating that the carbanion of DPL is more stable than the propagating carbanion, even at –50 °C. On the other hand, both *n*- and *tert*- BuLi gave polymer almost quantitatively at –50 °C in THF. Initiator efficiencies of *n*- and *tert*-BuLi for the polymerization, however, were extremely low (8.9 and 13.0%, respectively), even though the molecular weight distributions of the polymers were not high (M_w/M_n = 1.20 and 1.23, respectively). These results strongly suggest that aggregation of BuLi affects initiator efficiency in the present anionic polymerization system. Actually, the polymerization proceeded quantitatively with almost 100% of initiator efficiency when living oligo-(α-methylstyryl) lithium was used as initiator, which is known not to form strong aggregates. When temperature of the reaction mixture was raised from –50 °C to ambient temperature after the polymerization attained in equilibrium, depolymerization proceeded and almost no polymer was retained around the same

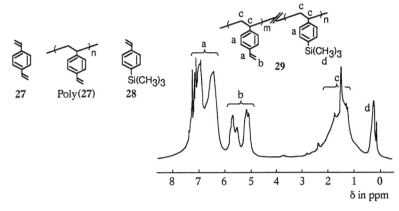

Figure 6. ^1H NMR spectrum of poly(1,4-divinylbenzene-*co*-4-trimethylsilylstyrene). (Reproduced with permission from reference 31. Copyright 1992 Hütig & Wepf Verlag Basel.)

molecular weight region. *sec*-BuLi also gave the poly(**18**) with expected molecular weight and M_w/M_n =1.2. In this way, the anionic polymerization of **18** with certain anionic initiators proceeded without any termination (living system), and the polymer obtained had controlled molecular weight and molecular weight distribution.

Silicon-Containing Polymers with Reactive Vinyl Groups

Nitadori and Tsuruta (*28*) reported that lithium diisopropylamide (**2**) in the presence of diisopropylamine (**4**) initiated an anionic polymerization of 1,4-(or 1,3)-divinylbenzene (**27**) to form a soluble polymer (**27**) (Figure 6). The reason for the formation of the soluble poly(**27**) by the **2**-induced polymerization was discussed elsewhere (*29*).

By using this reaction, co-workers and I carried out the synthesis of several vinyl-containing soluble polymers including poly(**27**), its crown ether end-capped analog (*30*), and its copolymers with other common monomers (*31*). In the anionic copolymerization of **27** with styrene, the styrene units in the copolymer were only 10% with an equimolar monomer feed. In the copolymerization of **27** with 2-vinylpyridine, the copolymer contained 94% 2-vinylpyridine units.

As stated previously, the trimethylsilyl group directly attached to the phenyl ring shows electron-withdrawing character owing to the empty *d*-orbital at the Si atom. Actually, 4-trimethylsilylstyrene (**28**) shows higher reactivity in nucleophilic addition reactions compared with styrene (*32*). This result can be explained by low electron density at the vinyl β-carbon due to electron withdrawing by trimethylsilyl group through phenyl π-conjugation. In anionic addition reactions with lithium diethylamide, the reactivity of **28** is very close to that of **27** [$k_{28} = 34.4 \times 10^{-4}$ dm^3 mol^{-1} s^{-1}); $k_{27} = 43.5 \times 10^{-4}$ dm^3 mol^{-1} s^{-1}].

Anionic copolymerization of **27** with **28** was expected to provide an opportunity to synthesize soluble silicon-containing polymers with a controlled amount of reactive vinyl groups.

Anionic copolymerizations of **27** with **28** had a polymer yield of more than 95%. The copolymer was soluble in common organic solvents such as THF, chloroform, benzene, toluene, dimethylformamide, and dioxane. The copolymers had molecular weights of up to 10^5 (values were determined by GPC).

Figure 6 shows ^1H NMR spectrum of the polymer (**29**) in $CDCl_3$. The signals appearing around 5–6 ppm were assigned to vinyl β-proton originated from **27**, and the signals appearing around 0 ppm were from silylmethyl proton of **28**. However, **2** did not initiate the anionic homopolymerization of **28**.

Figure 7 shows a copolymerization diagram of **27** and **28**. The concentration of **27** in the copolymer (**29**) can be controlled in the range 0–95%.

Two types of Si-containing polymers with reactive vinyl groups in the side chain can be synthesized. One type is obtained by anionic polycondensation between

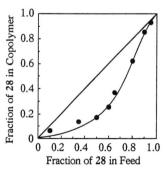

Figure 7. Copolymer composition curve for the lithium diisopropylamide induced anionic copolymerization of 1,4-divinylbenzene with 4-trimethylsilylstyrene. (Reproduced with permission from reference 31. Copyright 1992 Hütig & Wepf Verlag Basel.)

4-vinylbenzyllithium and disilyl chloride, and the other type is obtained by anionic copolymerizations of 1,4-divinylbenzene with Si-containing styrene. A third synthetic method for reactive Si-containing polymers is described. In this method, an end-functional poly(27) is treated with end-functional silicon-containing oligomers.

In the lithium-amide-induced anionic polymerization, the initiator **2** adds to the double bond of **27**, followed by polymerization. The resulting polymer has one diisopropylamino group at the end of the polymer chain. Thus, if a lithium amide having another functional group can initiate the polymerization of **27**, end-reactive poly(27) can be synthesized. Co-workers and I investigated (33) functional lithium amides such as lithium bis(trimethylsilyl)amide (**30**) and lithiated *N,N'*-bis(trimethylsilyl)ethylenediamine (**31**) for this purpose.

When lithium bis(trimethylsilyl)amide (**30**) was used as an initiator for **27** in THF, no polymerization took place, a result indicating the reactivity of **30** to be much lower than **2**. Lower nucleophilic reactivity of **30** may be attributable to the electron-withdrawing effect of trimethylsilyl groups. Actually, the acidity of bis(trimethylsilyl)amine ($pK_a = 25.8$) (34) is much larger than that of diisopropylamine ($pK_a = 35.7$).

Nucleophilic reactivity of lithium alkylamides was strongly dependent on the structure of alkylamino group. For example, the reactivities of lithiated *N,N'*-diethylethylenediamine toward styrene were much higher than that of lithium diethylamide. Because lithiated *N,N'*-bis(trimethylsilyl)ethylenediamine (**31**) has a diamine structural unit, it was anticipated that **31** should show relatively higher reactivity toward **27**. Actually, **31** initiated the polymerization of **27** in good yields (40–80%). The polymer obtained was soluble in solvents such as THF, benzene, toluene, chloroform, and DMF. Figure 8A shows the ^{1}H NMR spectrum of the polymer. From quantitative measurements using vinyl β-protons and phenyl protons, the polymer was confirmed to have almost one vinyl group per monomeric unit. Signals appearing around 0 ppm are assignable to trimethylsilyl protons and indicate *N,N'*-bis(trimethylsilyl)ethylenediamine moiety at the chain ends.

To synthesize soluble poly(27) with NH_2 chain ends, a cleavage reaction of N–Si bond at the chain end was carried out by using tetrabutylammonium fluoride (Bu_4NF). As shown in the spectrum of polymer after being treated with Bu_4NF (Figure 8B), signals assignable to trimethylsilyl protons disappeared completely, a result indicating the cleavage reaction of N–Si bond to have proceeded completely. From GPC, UV, and nitrogen titrations, the polymer was found to have exactly one ethylenediamine unit for each chain.

End-reactive poly(27) can provide an opportunity for multicomponent materials. By using the poly(27) with the primary amine end, co-workers and I examined a synthesis of A–B–A triblock copolymer. For this synthesis, the amino polymer (M_n, 7.7×10^3) and α,ω-bis(3-aminopropyl)oligo(dimethylsiloxane) (telechelic silicone; M_n, 4.6×10^3) were condensed by using 4,4'-diphenylmethanediisocyanate as a coupling agent (Figure 9). The reaction proceeded homogeneously and no gel was formed. The polymer obtained was a white powder soluble in several solvents. Figure 9 shows the GPC chromatograms of the amino polymer before (A) and after (B) the coupling reaction with telechelic-silicone. As shown in the figure, the polymer obtained after the coupling reaction shifted to the higher molecular weight side, indicating that the amino-ended polymer reacted with isocyanate-ended silicone chemically but not their mixture. From quantitative GPC measurements, M_n was 2.0×10^4, which was in good accordance with the value calculated as A–B–A type tri-blockcopolymer (Figure 9).

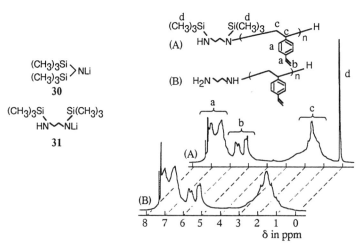

Figure 8. ^{1}H NMR spectra of poly(1,4-divinylbenzene) with silylamine (A) and primary amine (B). (Reproduced with permission from reference 32. Copyright 1990 Comité van Beheer Van Het Bulletin v.z.w.)

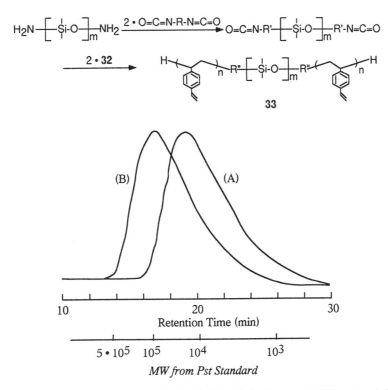

Figure 9. Gel permeation chromatograms of amino-ended poly(1,4-divinylbenzene) (A) and poly(1,4-divinylbenzene-*b*-silicone-1,4-divinylbenzene) (B). (Reproduced with permission from reference 32. Copyright 1990 Comité van Beheer Van Het Bulletin v.z.w.)

New Functional Oligomers with Alternating Organosilicon and Amino Groups in Main Chain

Kataoka and co-workers (*35*) reported that polymers having alternating structures consisting of ethylenediamine and *p*-diethylenephenylene units in the main chain showed unique pH sensitivity, because the elasticity of the polymers changes drastically by the protonation of the amino groups. Recently, co-workers and I investigated (*36*) reactivities of vinylsilane compounds toward nucleophilic reagents. When lithium dialkylamide was used as a nucleophile, an addition reaction toward the double bond of vinylsilanes took place selectively to form one-to-one adducts. The reactivity of the double bond of vinylsilanes toward lithium diethylamide was of the same order as that of para-substituted styrenes. For example, the rate constant of lithium diethylamide addition to dimethyldivinylsilane (**34**) [k_{34} = 31.2 dm^3 (mol^{-1} s^{-1})] was of the same order as the rate constant for lithium diethylamide addition to 1,4-divinylbenzene (**27**) [k_{27} = 30.4 dm^3 (mol^{-1} s^{-1})]. Actually, the reaction between **34** and lithium diethylamide gave a monoadduct, 2-(*N*,*N*-diethylamino)ethyldimethylvinylsilane, and a diadduct, bis[2-(*N*,*N*-diethylamino)ethyldimethylsilane. When a lithiated diamine (e.g., *N*-lithio-*N*,*N*'-diethylethylenediamine) was used, anionic polyaddition

between **34** and the diamine produced a new type of reactive oligomer having alternating repeating units of diamine and organosilyl groups (*37*).

To obtain polymers having desired molecular mass through polyaddition reactions, co-workers and I used the monoadduct (6-ethyl-3,3-dimethyl-6,9-diaza-3-sila-1-undecene, **35**) as starting material for the polyaddition reaction (Figure 10). Anionic polyaddition of **35** was carried out in the presence of lithium diisopropylamide as metalating agent. The reaction proceeded smoothly, and a colorless nonvolatile material remained after evaporation of low boiling compounds. The oligomer thus obtained (**36**) was soluble in a variety of solvents such as benzene, toluene, hexane, THF, dioxane, ether, chloroform, carbon tetrachloride, methanol, ethanol, acetone, and acidic water. From GPC analysis, the oligomer had a molecular weight ranging from a few hundred to 5000.

Figure 10 shows ^{1}H NMR spectra of **35** and the resulting oligomer (**36**). By using trimethylvinylsilane, 2-(*N*,*N*-diethylamino)ethyldimethylvinylsilane, diethylamine, and triethylamine as reference compounds, the NMR signals were assigned as indicated in Figure 10.

Because the intensity of α-end protons of the oligomers, a' (6.0) and f (3.0), agreed stoichiometrically with those of ω-end protons, c' (3.0) and e' (4.0), the oligomers were confirmed to have one polymeriz-

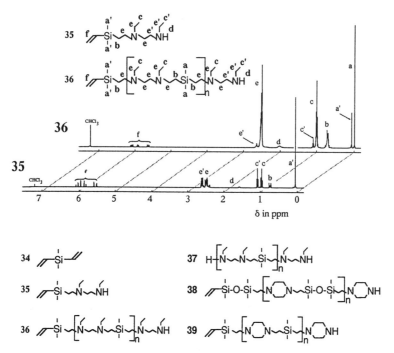

Figure 10. ^{1}H NMR spectra of 6-ethyl-3,3-dimethyl-6,9-diaza-3-sila-1-undecene (**35**) and poly(silamine) macromonomer (**36**). Peak intensities appearing for **36** were as follows: a, 41.0; a', 6.0; b, 30.5; c, 46.8; c', 3.0; e, 94.7; e', 4.0; and f, 3.0. (Reproduced with permission from reference 36. Copyright 1992 Marcel Dekker, Inc.)

able vinyl group at one end of each polymer chain and one *sec*-amino group at another. Thus, the oligomer can be regarded as a heterotelechelic oligomer. Under appropriate reaction conditions, the molecular weight of **36** was controllable in the range between a few hundred to 5000. Telechelic oligomers with *sec*-amino groups at both chain ends (**37**) can be obtained very easily by adding N,N'-diethylethylene-diamine to the mixture after the polyaddition reaction is complete (*38*).

This anionic polyaddition also can be applied to other diamines and divinylsilanes such as N-lithio-piperazine and 1,1,3,3-tetramethyldivinyldisiloxane. The molecular weights of these oligomers can be controlled between a few hundred and a few thousand. The oligomers obtained in this way (Figure 10) can be used for the synthesis of graft copolymers by radical copolymerization.

Applications

One of our interests is to determine structure–characteristic relations between Si-containing polymers and certain properties. Co-workers and I have been synthesizing several kinds of new Si-containing monomers, oligomers, and polymers. In this section, I describe applications of Si-containing polymers as gas separation membranes and electron-beam resists. Potential applications for Si-containing reactive polymers and polyamines are also described.

Gas Separation Membrane from Silicon-Containing Polymers

Gas separation through membrane is one of the most attractive industrial processes because of low energy consumption and ease of operation (*39*). Polydimethyl-siloxane (silicone) is known to have one of the highest permeation coefficients for several gases. However, there are several problems for oxygen permselective membranes, such as very low mechanical strength and low permselectivity for oxygen against nitrogen. Several approaches can create new gas-separation membranes with high permeability coefficients, permselectivities, and mechanical strength. For this purpose, silicon-containing polymers such as poly(1-trimethyl-silylpropyne) (*40*), polyfumalates (*41*), and poly(tri-methylvinylsilane) (*42*) are very promising.

Because our Si-containing monomers and oligomers have a styrene skeleton, macromolecular structure such as molecular weights, membrane formability, and surface morphology are easy to control through homo- and co-polymerizations. In this section, gas permeation properties of Si-containing polystyrenes are described.

The poly(7) obtained from radical polymerization shows unique properties such as a very high glass transition temperature (T_g) (150 °C), low density (0.923 g/dm^3), and high silicon content (21.4 wt%). Because the organosilyl group is known to show affinity toward oxygen, polymers having such a high Si con-

Table III. Gas Permeation Properties in Poly(**7**) Membrane

Gas	P
N_2	12
O_2	45
CH_4	45
CO_2	200
H_2	480

Note: Thickness of membrane, 120 μm; P is permeation coefficient in 10^{-10} cm^3 (STP) × cm cm^{-2} s^{-1} (cm Hg^{-1}).

Source: Reprinted with permission from reference 1a. Copyright 1989.

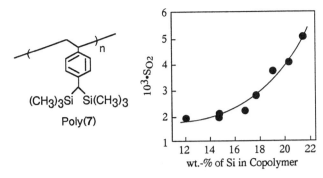

Figure 11. Plots of oxygen solubility coefficients (S_{O2}) versus Si content in the copolymer of **5** and **7** with common monomers. (Reproduced with permission from reference 1b. Copyright 1990 Hütig & Wepf Verlag Basel.)

tent should have favorable gas permeation properties toward oxygen. Table III shows several gas permeation properties of poly(**7**). As anticipated, the membrane showed fairly high permeation coefficients (P) toward small molecules. Especially, the permeation coefficient for oxygen (P_{O2}) was much higher than that of other vinyl polymers. Also, a high permselectivity for oxygen against nitrogen ($P_{O2}/P_{N2} = 3.8$) was observed, a result conforming to the high permeation coefficient for oxygen [$P_{O2} = 4.5 \times 10^{-9}$ cm^3 (STP) cm cm^{-2} s^{-1} (cm Hg)$^{-1}$] (*1b*).

To obtain further information about silicon-containing polymer membranes, gas permeation behavior, especially for oxygen and nitrogen, of several homo- and copolymers of **7** was investigated. Figure 11 shows plots of oxygen-solubility coefficients (S_{O2}) versus Si content in the copolymers (*1a*). Increasing Si content in the copolymers significantly increases S_{O2}. Thus, oxygen permeability in the Si-containing polymers was considered to be dominated by the solubility factor. Permselectivity of the membrane for oxygen against nitrogen was governed by polymer constitu-

tion rather than by silicon content. Polymer membranes having fluoroalkyl or siloxy groups showed high permeation coefficients for both oxygen and nitrogen and consequently low oxygen permselectivity. High permselectivity was obtained in the case of copolymer of **7** with butyl acrylate (**40**). Figure 12 shows plots of P_{O2} and P_{O2}/P_{N2} against copolymer composition. Oxygen permselectivity of the membrane increased up to 4.0–4.6 with increasing composition of **40** in the copolymer.

Gas permeation properties for several other silicon-containing polymers are summarized in Table IV. As seen from Table IV, gas permeation is not consistent with Si content. Actually, gas permeability of poly(**12**) (Run 9 in Table IV) was one order of magnitude lower than that of poly(**7**) even though Si-content of poly(**12**) is much higher than that of poly(**7**). In general, membranes with lower T_g show relatively higher gas per-

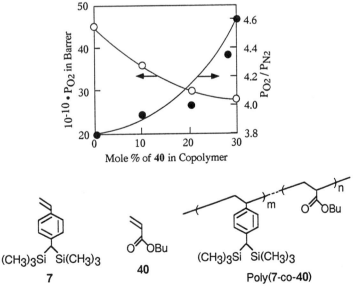

Figure 12. Plots of oxygen permeation coefficient (P_{O2}) and oxygen permselectivity (P_{O2}/P_{N2}) versus butyl acrylate content in poly(**7**-*co*-**40**). (Reproduced with permission from reference 1b. Copyright 1990 Hütig & Wepf Verlag Basel.)

Table IV. Permeation Behavior of Si-Containing Polystyrenes

Run	X in	P_{O2}	P_{N2}	P_{O2}/P_{N2}	T_g (°C)	Si (%)
1	CH_2SiMe_3 (1d)	14	3.6	3.9	88	14.7
2	CH_2SiEt_3	22	6.6	3.3	ND	12.1
3	$CH[SiMe_3]_2$	45	12	3.8	148	21.4
4	$CH[SiMe_3][SiEt_3]$	18	4.0	4.5	124	18.4
5	$CH[SiMe_3]CH_2CH(Et)OSiMe_3$	22	6.2	3.6	ND	16.8
6	$CH[SiMe_3]$(oligo5)	14	3.5	4.1	ND	14.7
7	$CH=CHCH_2SiMe_3$	17	5	3.4	78	13.0
8	$CH[SiMe_3]CH=CHSiMe_3$	26	8	3.3	102	19.4
9	$CH_2SiMe_2CH[SiMe_3]_2$	4	1.2	3.4	82	25.1
10	$CH[SiMe_3]SiMe_2CH[SiMe_3]_2$	18	5	3.5	86	27.6

Note: Values for P are in 10^{-10} cm^{-3} (STP) × cm cm^{-2} s^{-1}. ND is not determined.

meability (43). However, permeation behavior of poly(**12**) does not follow this pattern, because T_g of poly(**12**) was much lower than that of poly(**7**) (82 °C vs. 150 °C). This deviation may be explained by molecular motion of the side chains, especially because spinning of trimethylsilyl groups (44) is very high and the main chain cannot follow their movements; hence, a high T_g results. Thus, poly(**7**) at ambient temperature shows a state between T_g and second-order transition (45). Such molecular motions of the side chain repulse each other to provide relatively larger free space in the membrane. Actually, density of poly(**7**) was much lower than that of polystyrene itself ($d = 0.923$ g/dm^3 vs. 1.04–1.065 g/dm^3 (46)). Such a polymer with large free space in the solid state allows oxygen molecules to pass easily through the membrane. On the other hand, polymers with flexible side chains such as poly(**12**) decrease the free space due to the packing by the side chains. Consequently, density of poly(**12**) ($d = 0.935$ g/dm^3) is higher than that of poly(**7**).

Electron Beam Resists from Silicon-Containing Polymers

The desire to develop smaller patterns of electronic devices demands increasingly more high performance from resists for submicrometer lithography. Electron beam (EB) resist is one of the most promising technologies for nanometer dimensions. To get high resolution patterns, very thin resist films must be used. For this purpose, however, organic polymers do not have enough resistivity against oxygen reactive ion etching. There have been many attempts (47) to use organometallic polymers as EB resists, especially Si-containing polymers such as Si-containing polystyrenes (48), polymethacrylates (49), and silicones (50). Not many reports, however, have been published on the relationship between polymer structure and resist properties.

Poly(**7**) has several characteristics such as high silicon content (24.1 wt.%) and fairly high T_g (140–160 °C) and ease of molecular weight control through living

anionic polymerization. When poly(**7**) was exposed to an electron beam, the film was completely insolubilized; therefore, the polymer is a negative-type resists (2a). Typical sensitivity curve for poly(**7**) is shown in Figure 13 (51). A lithographically useful sensitivity is represented by the dose required to produced half relative thickness ($D_g^{0.5}$). The sensitivity of negative resists is well known to improve in proportion to M_w. Values of $D_g^i \times M_w$ (52) were employed to compare sensitivity of several negative resists: D_g^i is defined as a minimum EB dose for detectable gel formation. A contrast parameter, γ, was defined according to the literature (53) as follows: $\gamma = 0.5(|\log(D_g^{0.5}/D_g^i)|)^{-1}$. Lithographic parameters determined from the curves for several Si-containing polymers are summarized in Table V. Sensitivities of these homopolymers [poly(**1**), poly(**5**), and poly(**7**)] were independent of Si content in the polymer, a result indicating no remarkable reaction taking place on the C–Si bond under the indicated condition. Miller et al. reported (54) that crosslinking of polystyrene by EB exposure was due to the formation of methine radical and subsequent coupling reaction in the main chain. Similar mechanisms for trimethylsilylated polymers [poly(**5**) and poly(**7**)] can be considered.

Good correlation exists between T_g and EB sensitivity in Si-containing polymers, and this correlation indicates polymer mobility plays an important role in intermolecular coupling reactions. Introduction of glycidyl methacrylate (41) into the polymer induced a significant increase in sensitivity. Therefore, incorporation of **41** into the polymer brought about a lower T_g and reactive groups. This incorporation may be the primary reason for different patterns of correlation between homopolymers and copolymers.

The contrast parameter (γ) of Si-containing homologs of polystyrene (poly(**1**)) showed a different pattern from sensitivity characteristics: increasing Si-content in the polymers increased the γ value. For example, poly(**1**) obtained by radical polymerization had a γ value of 1.3 (55). For the Si-containing poly-

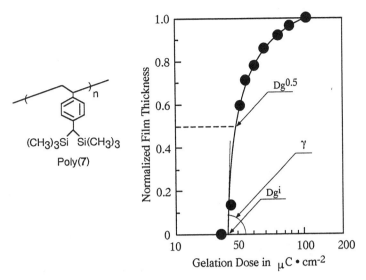

Figure 13. Sensitivity curve for poly(**7**) toward EB-exposure: D_g^i is a minimum EB dose for detectable gel formation; $D_g^{0.5}$ is a dose required to produced half relative thickness; γ is a contrast parameter and was defined according to the literature (*1*) as follows: $\gamma = 0.5(|\log(D_g^{0.5}/D_g^i)|)^{-1}$.

Table V. Properties of Si-Containing Polymers as EB Resists

Polymer	M_w (10^{-4})	M_w/M_n	*Si Content* *(wt. %)*	T_g (°C)	D_g^i $(\mu C/cm^2)$	γ
Poly(**1**) (**2e**)	22.0	1.75	0	93	9.7	1.3
Poly(**5**)	6.9	1.60	14.7	82	28	2.4
Poly(**7**)	13.8	1.78	21.4	154	20	3.8
Poly(**7**)	10.6	1.17	21.4	156	26	8.1
Poly(**7**-*co*-**41**) (86.4)[a]	7.8	1.53	19.7	140	22	2.5
Poly(**7**-*co*-**41**) (70.2)[a]	4.6	1.55	17.4	106	12	1.9
Poly(**24**)	15.4	1.17	53.5[b]	140	7.7	4.9

[a]**7** content in mol%.

[b]Sn content.

mers, poly(**5**) and poly(**7**) formed by radical polymerization, γ values were 2.4 and 3.8, respectively. These values cannot be explained by only polymer T_g values, which are 85 °C for poly(**5**), 93 °C for poly(**1**), and 150 °C for poly(**7**). The organosilicon moiety in polystyrene may play an important role in increasing the γ value. Poly(**7**) had a relatively narrow molecular weight distribution (M_w/M_n = 1.17) and showed an extremely high γ value (8.1) owing to its favorable characters. In the case of copolymer of **7** with **41**, γ values decreased with increasing **41** content in the copolymer.

From a lithographic point of view, Si-containing polymers are useful as a base resin of a top layer for a bilayer system. Even though poly(**7**) has a very high resolution parameter (γ value = 8.1), it is not sensitive enough ($D_g^{0.5}$ = 30 $\mu C/cm^2$). To improve sensitivity against EB exposure and to keep the γ value high, we examined an organostannyl-containing polymer, (poly-{4-[bis(trimethylstannyl)methyl]styrene}) poly(**42**), which has the same skeleton as poly(**7**) (*5f*). Exposure

of poly(**42**) to an electron beam induced cross-linking; therefore, the polymer works as a negative resist. As anticipated, sensitivity of poly(**42**) (D_g^i = 7.7 $\mu C/cm^2$) was three times higher than that of poly(**7**) owing to high sensitivity of the Sn atom to EB exposure (*56*). However, γ of poly(**42**) (M_w/M_n = 1.18; γ = 4.9) was lower than that of poly(**7**) (M_w/M_n = 1.17; γ = 8.1). This difference may be attributable to the lower T_g of poly(**42**) (T_g = 140 °C) compared with that of poly(**7**) (T_g = 154 °C). Polymers based on 4-[bis(trimethylsilyl)-methyl]isopropenylbenzene (**18**) also appear promising as resist materials due to the high T_g of the polymer (T_g = 224 °C) and their depolymerization tendency (*21*).

Silicon-Containing Polymers with Reactive Vinyl Groups

Polymers having vinyl groups at chain end, in the main chain or side chains, can be regarded as one class of "functional" or "reactive" polymers. We synthesized several Si-containing polymers with vinyl groups. Several opportunities exist to use these reactive groups for applications such as functionalization of the vinyl groups, cross-linking for resist materials, and cross-linking to improve mechanical strength and membrane formability. Some of these examples are described in this section.

Soluble poly(divinylbenzene) [poly(27)] obtained through anionic polymerization induced by lithium amide has a vinylphenyl group in each monomeric unit. When diethylamine was added to the reaction mixture after the polymerization was almost completed, transmetalation took place between diethylamine and 2 to form lithium diethylamide, which was in turn added to the pendant vinyl groups in the polymer. Finally, polymers with pendant tertiary amino groups were obtained. Degree of amination of vinyl groups could be controlled from 0 to 65% (28b). Notably, 2 itself does not react with pendant vinyl groups. This reaction can also be applied to the other primary or secondary amines such as N,N'-diethylethylenediamine and N-methylethylenediamine to form reactive polymers with secondary or primary amino groups in the side chain. These reactions proceed smoothly by one-pot synthesis without any gel formation.

Oxidation and cross-linking of vinyl groups on the polymer also proceed easily, even in membranes, when the sample is heated up to its T_g. Figure 14 shows IR spectra of poly(27) membrane before and after heating up to 200 °C. In the spectrum after heating, absorptions of OH and C=O groups appeared around 3300 cm^{-1} and 1650 cm^{-1}, respectively, and indicate that the polymer has hydroxy and carbonyl groups (CO and COO) owing to oxidation. The resulting membrane cannot dissolve in any common solvent.

Si-containing polymers with vinyl pendant groups including poly(27-co-28), 21, and 23 can be functionalized as mentioned previously. In this way, these polymers may be useful for negative-resist materials with

high sensitivity. Cross-linked membranes also may be developed for gas separation. Müller et al. (57) reported that copolymerization of methacryloyl chloride with common monomers proceeds easily to form soluble copolymers. We examined copolymerizations of 4-[bis(trimethylsilyl)methyl]styrene (7) with methacryloyl chloride to form a corresponding soluble copolymer. The resulting polymer may be useful as a precursor of an ultra thin film prepared through interfacial polymerization with diamines for gas separation.

Silicon-Containing Polyamines

Polyamines are of interest for their pH dependency in aqueous media. Figure 15 shows acid–base and turbidity changes of an aqueous solution of poly(silamine) macromonomer (36) against pH change (37). The two-stage titration curve of 36 is attributable to the two-stage deprotonation of ethylenediamine structure (58). When the second-stage deprotonation of protonated 36 was reached (at about pH = 8.5), turbidity increased drastically.

The degree of protonation of basic compounds in water is well known to be a function of temperature (59). Actually, with increasing temperature of the solution of 36 at pH 7.2, the solution became turbid at 60 °C owing to deprotonation and decreased hydration of the polymer. The oligomer (36) shows low critical solution temperature (LCST) (60): By increasing the pH, LCST decreased to 50 °C at pH 7.4 and to 40 °C at pH 7.8. Thus, LCST of the oligomers can be controlled by the degree of protonation (61).

Because 36 has a vinyl group at one chain end and a sec-amino group at the other end, its chemical modification is easy. Actually, when the oligomer was mixed with 3-mercaptopropyltrimethoxysilane in the presence of radical generator (AIBN), an SH group

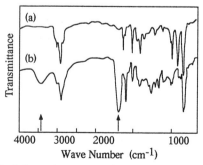

Figure 14. Infrared spectra of poly(1,4-divinylbenzene) before (a) and after (b) heating to 200 °C. (Reproduced with permission from reference 28b. Copyright 1989 Marcel Dekker, Inc.)

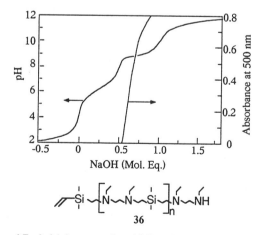

Figure 15. Acid–base and turbidity titration of protonated poly(silamine) (36). (Reproduced with permission from reference 36. Copyright 1992 Marcel Dekker, Inc.)

was added to the double bond to form poly(silamine) oligomers with the silane moiety (62). The poly(silamine)s thus obtained are of potential interest for developing functional graft copolymers for stimuli-sensitive gel and surface modifiers.

Concluding Remarks and Future Prospects

Co-workers and I used a metalation reaction of 4-methylstyrene to form 4-vinylbenzyllithium without any side reactions. By using 4-vinylbenzyllithium as an intermediate, we were able to synthesize several kinds of Si-containing monomers and oligomers. The Si-containing polymers show interesting characteristics as gas separation membranes and EB resists. By using anionic polymerization of 1,4-divinylbenzene induced by lithium alkylamide, several Si-containing copolymers with reactive pendant groups could be synthesized, and these copolymers can be used as functional materials after modification of pendant double bonds.

A selective-addition reaction of lithium alkylamides to vinylsilanes opened a new synthetic route to polymers having both organosilicon and amino groups in the main chain. Such polymers have unique pH dependency in aqueous solution. The new organosilicon polymers have numerous possibilities for materials development. Thinking about "what is going on in the flask" is of primary interests for a polymer synthetic chemist. If polymeric materials with new structures are created in our flask, and the polymers have certain novel functions, then there is no pleasure superior to that. I close this chapter hoping the polymers described here are also good enough to play a part in the society at large.

Acknowledgments

I am grateful to Professor Teiji Tsuruta, formerly with the Research Institute for Biosciences, Science University of Tokyo, for his cooperation with the synthetic work. I thank Professors Masao Kato and Kazunori Kataoka, Department of Materials Science and Technology, Science University of Tokyo, for their cooperation with the application studies. I would like to express my sincere appreciation to Norikiyo Kato and many students for their continuous efforts to promote this research project.

Part of this work was supported financially by a Grant-in-Aid for Scientific Research on Priority Areas (Synthetic Process and Control of Functionality Materials, Reaction Design for Synthesis of Functionality Materials, No. 04205123), The Ministry of Education, Science and Culture, Japan.

References

1. (a) Nagasaki, Y.; Suda, M.; Tsuruta, T.; Ishihara, K.; Nagase, Y. *Makromol. Chem., Rapid Commun.* **1989**, *10*, 255. (b) Nagasaki, Y.; Suda, M.; Tsuruta, T.; Ishihara, K.; Nagase, Y. *Makromol. Chem.* **1990**, *191*, 2297. (c) Kawakami, Y.; Karasawa, H.; Aoki, T.; Yamamura, Y.; Yamashita, Y. *Polym. J.* **1985**, *17*, 1159. (d) Barrer, R. M.; Chio, H. T. *J. Polym. Sci., Part C: Polym. Symp.* **1966**, *10*, 111.

2. (a) Kato, N.; Nagasaki, Y.; Kato, M. *Polym. Adv. Technol.* **1990**, *1*, 341. (b) Taylor, G. N.; Wolf, T. M.; Moran, J. M. *J. Vac. Sci. Technol.* **1981**, *19*, 872. (c) Hayase, S.; Horiguchi, R.; Ohnishi, Y.; Ushiroguchi, T. *Macromolecules* **1990**, *22*, 2933. (d) Saigo, K.; Watanabe, F. *J. Polym. Sci., Part A: Polym. Chem.* **1989**, *27*, 2611. (e) Hartney, M. A.; Tarascon, R. G.; Novembre, A. E. *J. Vac. Sci. Technol.* **1985**, *B3*, 360.

3. (a) van der Laan, G. P.; de Haas, H. P.; Hummel, A.; Frey, H.; Sheiko, S.; Hoeller, M. *Macromolecules* **1994**, *27*, 1897. (b) Ohshita, T.; Kanaya, D.; Ishikawa, M. *Appl. Organomet. Chem.* **1993**, *7*, 269. (c) Allred, A. L.; van Beek, D. A., Jr. *Polyhedron* **1991**, *10*, 1227.

4. Rocow, E. G. *Silicon and Silicone*; Springer-Verlag: Berlin, Germany, 1987.

5. (a) Nagasaki, Y.; Tsuruta, T. *Makromol. Chem.* **1986**, *187*, 1583. (b) Nagasaki, Y.; Tsuruta, T. *Makromol. Chem., Rapid Commun.* **1986**, *7*, 437. (c) Nagasaki, Y.; Tsuruta, T. *Makromol. Chem., Rapid Commun.* **1989**, *10*, 403. (d) Nagasaki, Y.; Takahashi, S.; Tsuruta, T. *Makromol. Chem.* **1990**, *191*, 2297. (e) Nagasaki, Y.; Tsuruta, T. *New Polym. Mater.* **1991**, *2*, 357. (f) Nagasaki, Y.; Kurosawa, K.; Tsuruta, T. *Bull. Chem. Soc. Jpn.* **1990**, *63*, 3036. (g) Nagasaki, Y.; Kato, N.; Kato, M.; Tsuruta, T. *Makromol. Chem., Rapid Commun.* **1990**, *11*, 651.

6. For reviews, see (a) Wakefield, B. J. *Organolithium Methods*; Academic: London, **1988**. (b) *New Applications of Organometallic Reagents in Organic Synthesis*; Seyferth, D., Ed.; Elsevier: Amsterdam, Netherlands, 1976.

7. (a) Imai, N.; Narita, T.; Tsuruta, T. *Tetrahedron Lett.* **1971**, *38*, 3517. (b) Narita, T.; Nitadori, Y.; Tsuruta, T. *Polym. J.* **1977**, *9*, 191. (c) Nagasaki, Y.; Higuchi, A.; Goan, H.; Yoshino, N.; Tsuruta, T. *Makromol. Chem.* **1989**, *190*, 53.

8. Nagasaki, Y.; Shimidzu, H.; Tsuruta, T. *Makromol. Chem., Rapid Commun.* **1988**, *9*, 381.

9. Goronwicz, G. A., West, R. *J. Am. Chem. Soc.* **1968**, *90*, 4478.

10. Nagasaki, Y.; Hashimoto, Y.; Kato, M.; Kimijima, T. *Macromol. Chem., Rapid Commun.* **1994**, *15*, 619.

11. Nagasaki, Y.; Hashimoto, Y.; Kato, M.; Kimijima, T. *Polym. J.* **1994**, *26*, 745.

12. Nagasaki, Y.; Yamazaki, N.; Kato, N.; Kato, M. *Macromolecules* **1994**, *27*, 3702.

13. Kato, N.; Kimura, H.; Nagasaki, Y.; Kato, M.; Kimijima, T. *Polymer* **1994**, *35*, 4228.

14. Nagasaki, Y.; Tsuruta, T.; Hirano, T. *Makromol. Chem.* **1989**, *190*, 1855.

15. For reviews, see (a) *Comprehensive Organometallic Chemistry*; Wilkinson, G., Ed.; Pergamon: Oxford, England, 1982; Vol. 2. (b) Cunico, R. F. *J. Organomet. Chem.* **1986**, *109*, 1.

16. (a) Taft, R. W.; Price, E.; Fox, E. R.; Lewis, I. C.; Andersen, K. K.; Davis, G. T. *J. Am. Chem. Soc.* **1963**, *85*, 3146. (b) Brown, J. F.; Prescott, P. I. *J. Am. Chem. Soc.* **1964**, *86*, 1402.

17. Baker, B. G. *Electronic Transitions in Organometalloids;* Academic: Orlando, FL, 1969.

18. Alfrey, T.; Bohrer, J. J., Jr. *Copolymerization;* Interscience: New York, 1952; p 64.

19. Magnus, P. D.; Sarkar, T.; Djuric, S. *Organosilicon Compounds in Organic Synthesis;* Wilkinson, S. G., Ed.; Comprehensive Organometallic Chemistry; Pergamon: Oxford, England, 1982.

20. Gröbel, B. T.; Seebach, D. *Chem. Ber.* **1977**, *110*, 852.

21. Hirao, A.; Kato, H.; Yamaguchi, K.; Nakahama, S. *Macromolecules* **1986**, *19*, 1294.

22. (a) *Anionic Polymerization; Kinetics, Mechanisms, and Synthesis;* McGrath, J. E., Ed.; ACS Symposium Series 166; American Chemical Society: Washington, DC, **1981**. (b) *Recent Advances in Anionic Polymerization;* Hogen-Esch, T.; Smid, J., Eds.; Elsevier: New York, 1987.

23. Natta, G.; Pino, P.; Corradini, P.; Danusso, F.; Mantica, E.; Mazzanti, G.; Moraglio, G. *J. Am. Chem. Soc.* **1955**, *77*, 1708. (b) Ishihara, N.; Seimiya, T.; Kuramoto M.; Uoi, M. *Macromolecules* **1986**, *19*, 2464.

24. (a) Williams, J. L.; VenDenBerghe, J.; Dunham, K. R.; Dulmage, W. J. *J. Am. Chem. Soc.* **1957**, *79*, 1716. (b) Kern, W.; Braun, D.; Herner, M. *Makromol. Chem.* **1958**, *28*, 1. (c) Dainton, F. S.; Wiles, D. M.; Wright A. N. *J. Polym. Sci.* **1960**, *45*, 111. (d) Cazzaniga, L.; Cohen, R. E. *Macromolecules* **1989**, *22*, 4125.

25. Kato, N.; Nagasaki, Y.; Kato, M. *Makromol. Chem., Rapid Commun.* **1993**, *14*, 569.

26. Young, R. N.; Quirk, R. P.; Fetters, L. J. *Anionic Polymerizations of Nonpolar Monomers Involving Lithium;* Advances in Polymer Science 56; Springer-Verlag: Berlin, Germany, 1984.

27. McCormick, H. W. *J. Polym. Sci. Lett.* **1957**, *25*, 488.

28. Nitadori, Y.; Tsuruta, T. *Makromol. Chem.* **1978**, *179*, 2069.

29. (a) Nagasaki, Y.; Ito, H.; Tsuruta, T. *Makromol. Chem.* **1986**, *187*, 23. (b) Nagasaki, Y.; Tsuruta, T. *J. Macromol. Sci. Chem.* **1989**, *A26(8)*, 1043.

30. (a) Akashi, R.; Nagasaki, Y.; Tsuruta, T. *Makromol. Chem.* **1987**, *188*, 719. (b) Nagasaki, Y.; Tamura, Y.; Tsuruta, T. *Makromol. Chem., Rapid Commun.* **1988**, *9*, 31.

31. Nagasaki, Y.; Taniuchi, M.; Tsuruta, T. *Makromol. Chem.* **1988**, *189*, 723.

32. Nagasaki, Y.; Han, S.-B.; Kato, M.; Tsuruta, T. *Makromol. Chem.* **1992**, *193*, 1663.

33. Nagasaki, Y.; Kato, M.; Kato, N.; Mitsuhata, Y.; Nishizuka, H.; Tsuruta, T. *Bull. Soc. Chim. Bel.* **1990**, *99*, 957.

34. Fraser, R. R.; Mansour, T. S.; Savard, S. *Can. J. Chem.* **1985**, *63*, 3505.

35. (a) Koyo, H.; Tsuruta, T.; Kataoka, K. *Polym. J.* **1993**, *25*, 141. (b) Kataoka, K.; Koyo, H.; Hirai, H.; Tsuruta, T. *Macromolecules* **1995**, *28*, 3336.

36. Nagasaki, Y.; Morishita, S.; Kato, M.; Tsuruta, T. *Bull. Chem. Soc. Jpn.* **1992**, *65*, 949.

37. Nagasaki, Y.; Honzawa, E.; Kato, M.; Kihara, Y.; Tsuruta, T. *J. Macromol. Sci. Pure Appl. Chem.* **1992**, *A29*, 457

38. Nagasaki, Y.; Honzawa, E.; Kato, M.; Kataoka, T.; Tsuruta, T. *Polym. Prepr.* **1993**, *34*, 304.

39. Vansant, E. F.; Dewolfs, R. *Gas Separation Technology;* Elsevier: Amsterdam, Netherlands, 1990.

40. Masuda, T.; Isobe, E.; Higashimura, T.; Takada, K. *J. Am. Chem. Soc.* **1983**, *105*, 7473.

41. Choi, S.-B.; Takahara, A.; Amaya, N.; Murata, Y.; Kajiyama, T. *Polym. J.* **1989**, *21*, 433.

42. For reviews, see: Yampol'skii, Y. P.; Volkov, V. V. *J. Membr. Sci.* **1991**, *64*, 191.

43. Kawakami, Y.; Sugisaka, T. *J. Membr. Sci.* **1990**, *50*, 189.

44. Rochow, E. G. *Silicon and Silicones;* Springer-Verlag: Berlin, Germany, 1990.

45. Van Kreevelen, D. W. *Properties of Polymers,* 3rd ed.; Elsevier: Amsterdam, Netherlands, 1990.

46. *Styrene, Its Polymers, Copolymers and Derivatives;* Boundy, R. H.; Boyer, R. F., Eds.; Reinhold: New York, 1952.

47. For reviews *see* (a) Sugita, K.; Ueno, N. *Prog. Polym. Sci.* **1992**, *17*, 319. (b) *Polymers in Microlithography;* Reichmanis, E.; MacDonald, S. A.; Iwayanagi, T., Eds.; ACS Symposium Series 412; American Chemical Society: Washington, DC, 1989.

48. (a) MacDonald, S. A.; Ito, H.; Willson, C. G. *Microelectron. Eng.* **1983**, *1*, 269. (b) Saigo, K.; Watanabe, F. *J. Polym. Sci., Part A.: Polym. Chem. Ed.* **1989**, *7*, 2611.

49. (a) Novembre, A. E.; Reichmanis, E.; Davis, M. *Proc. SPIE* **1986**, *631*, 14. (b) Jones, R. G.; Cragg, R. H.; Davies, R. D. P.; Brambley, D. R. *J. Mater. Chem.* **1992**, *2*, 371.

50. Noguchi, T.; Nito, K.; Seto, J.; Hata, I.; Sato, H.; Tsumori, T. *Proc. SPIE* **1988**, *920*, 198.

51. Kato, N.; Takeda, K.; Nagasaki, Y.; Kato, M. *Ind. Eng. Chem. Res.,* **1994**, *33*, 417.

52. Imamura, S.; Tamamura, T.; Harada, K. *J. Appl. Polym. Sci.* **1982**, *27*, 937.

53. Atoda, N.; Kawakatsu, H. *J. Electrochem. Soc.* **1976**, *123*, 1519.

54. Miller, A. A.; Lawton, E. J.; Balwit, J. S. *J. Polym. Sci.* **1954**, *14*, 503.

55. Harteny, M. A.; Tarascon, R. G.; Novembre, A. E. *J. Vac. Sci. Technol.* **1985**, *B3*, 360.

56. Kato, N.; Yamazaki, N; Nagasaki, Y.; Kato, M. *Polym. Bull.* **1994**, *32*, 55.

57. Müller, H.; Nuyken, O.; Strohriegl, P. *Makromol Chem., Rapid Commun.* **1992**, *13*, 125.

58. Kataoka, Y. *Cell Separation Science and Technology;* Kompala, D. S.; Todd, P., Eds.; ACS Symposium Series 464; American Chemical Society, Washington, DC, 1991; Chapter 11, pp 159–174.

59. Perrin, D. D. *Austral. J. Chem.* **1964**, *17*, 484.

60. *Water-Soluble Polymers: Synthesis, Solution Properties, and Applications;* Shalaby, S. W.; McCormick, C. L.; Butler, G. B., Eds.; ACS Symposium Series 467; American Chemical Society: Washington, D. C., 1991.

61. Nagasaki, Y.; Honzawa, E.; Kato, M.; Kataoka, K.; Tsuruta, T. *Macromolecules* **1994**, *27*, 4848.

62. Nagasaki, Y.; Tsujimoto, H; Honzawa, E.; Kato, M.; Kataoka, K.; Tsuruta, T. *Polym. Prepr. (Am. Chem. Soc. Div. Polym. Chem.)* **1995**, *36*, 59.

Functional Polymer Surfaces Produced from Cold Plasma

J. C. Brosse, G. Legeay, and F. Poncin-Epaillard

This chapter presents a general review of functional polymer surfaces obtained by plasma and a brief description of various applications of the resulting modified polymers. A cold plasma is defined as being the state of ionized gas (e.g., oxygen, nitrogen, halogen, or organic monomer) composed of positively charged species, electrons, free radical, and UV or visible radiation. Cold plasma affects the polymer surface in several ways. It may create functional groups on the outer surface or produce cross-linking and degradation. Cold plasma can also be generated with an organic molecule, which can be either a gas or a liquid, coated on the polymer surface (plasma-induced polymerization) and leading to the deposition of a cross-linked functional polymer layer on the polymer substrate. These plasma processes and various applications of plasma-modified polymers in microelectronic, adhesion, and biomedical applications are discussed.

Typical plasma polymerization and modification of polymer materials by plasma are carried out under low pressure (10^{-1}–10^{-2} mbar), and gas-phase reactions (ionization of molecules by electron bombardment and ion–molecule reactions) control the course of surface modification or surface polymerization. The term *plasma* describes the state of an ionized gas consisting mainly of atoms and electrons and positively charged molecules.

Definitions

Cold Plasma and Plasma Chemistry on Polymers

The classical definition of plasma limits the term to an appreciably ionized gas or vapor that conducts electricity and is electrically neutral, fluid, 'hot', and viscous. The modern definition is less restrictive and denotes plasma as an approximately ionized gas (1). Plasma created by an electric glow discharge is often called low temperature plasma to distinguish it from hot plasma. There are many types of electric discharges, all characterized by the presence of free electrons or an electric field. Glow discharge is the most frequently used in polymer treatment. Plasma physics is a major field of basic research and good reviews of the subject are available (2).

Polymer Surfaces

Classical surface chemistry assumes that solid surfaces, such as metals, are rigid, immobile, and at equilibrium in dry or wet atmospheres. These assumptions allow one to probe adsorption and wetting processes purely from the point of view of the liquid phase, because one assumes that the solid phase does not at all reorient or otherwise change in contact with different liquids or atmospheric environments. Although such assumptions may be partially correct for truly rigid solids, they are wholly inappropriate for polymers. Polymer structures and properties are, in general, time and temperature dependent. Because of the relatively large size and high molecular weight of polymers, it is unlikely that most polymeric solids ever achieve a true equilibrium. Solid polymers are, therefore, inherently nonequilibrium structures and as such exhibit a range of relaxation times and properties that have been largely neglected or ignored in polymer surface chemistry and physics. Considerable evidence exists to show that the surface properties of polymers are time, temperature, and environment dependent (3). In addition to polymer dynamics and motion at its surface, surface composition and structure show an almost inevitable difference with the bulk composition and structure, and interface phenomenon must be taken into account (4). For example, oxidation degree is more important at the surface then in bulk, because oxidation proceeds through atmospheric oxygen at the surface exposed to radiation.

Polymer Surface Treatment

Polymer surfaces are treated because their properties are inadequate. These treatments must lead to the generation of reactive groups and hence to interactions between different components in a product without changing physical and bulk properties of the individual components. Techniques for surface treatment are divided into three categories: chemical, mechanical, and physical (5, 6).

A typical chemical reagent is chromic acid solution. Surface oxidation of polypropylene by chromic acid, as compared with oxygen plasma, seems to be selective in preferentially etching the amorphous zones (7). Other proposed oxidizing agents include potassium permanganate, nitric acid, and hydrogen peroxide (8). Etch process is also possible by treating fluorinated polymers with sodium alloys to lead to more adhesive surfaces (9). Halogenation (fluorination or chlorination) of polymer surfaces is achieved under mild conditions by using diluted molecular fluorine or chlorine (10). Adhesion primer treatments also offer an increase of adhesion for different polymers (mostly fluorinated polymers) through a surface modification involving the creation of C=O, C=C, CH_3, CH_2, and NH_2 groups (6).

Mechanical and abrasive treatments (abrasion, sand, and hot air projections) increase the contact area of the polymer surface. Thus, roughness is increased and oxidation could be noticed. In some cases, these treatments need a cleaning stage with an agent that is not a solvent for the polymer. Mixture of air and oxygen (10%) gives a stable and oxidative flame leading to a surface combustion, which increases the surface energy. This well-controlled treatment is an efficient and lasting procedure that confers adhesion to polymers, mostly polyolefins (11).

More recent procedures such as radiation (UV), electron or ion bombardments, and plasma or Corona discharges have been developed successfully. Far-UV radiation from a 193 nm pulsed-excimer laser is highly effective in modifying polymer surfaces because of the short penetration depth (<300 nm) and high quantum yield for bond breaking (12). Electron-beam interaction with polymer surface seems to be less degradative than KrF laser (13). Plasma treatment, under reduced or atmospheric pressure (Corona discharge), will be fully described in the following section. Figure 1 shows a schematic presentation of a plasma-treatment experimentation.

Knowledge of Polymer Surface

Analytical techniques for polymer surfaces are less well developed than those for polymer bulk. Surface analysis is mostly conducted by sophisticated techniques that may not even produce quantitative data. We consider three aspects of surface analysis.

Visual Aspects

The very first analysis that should be performed on a polymer surface is to note its aspect. The eyes, with the aid of a small magnifying lens, can see surface features to less than 100 μm. Using staining techniques and back-scattered electron detection in scanning electron microscopy, one can easily obtain resolutions of about 100 nm.

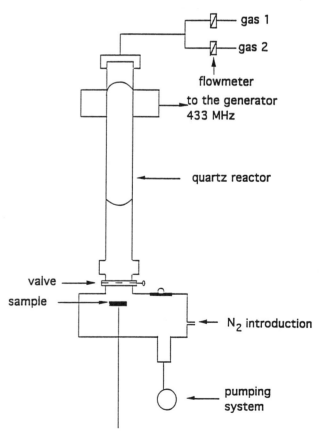

Figure 1. Schematic presentation of a typical experimental setup for plasma treatment of polymers.

Physicochemical Aspects

Surface tension, capillary, and contact angle measurements give information on the outermost few angstroms of the surface based on solid–liquid–vapor or solid–liquid–liquid contact angles. Required equipment is relatively simple and inexpensive. Although interpretation of the results is dependent on a number of assumptions, a first-order interpretation is possible and has proved very useful in practically all areas of surface science and engineering. In this respect, polymers have been classified into four types:

- those with weak surface energy (E_s = 10–35 mJ/m^2) such as fluorinated polymers and polyethylenes that are hydrophobic and have poor wettability
- those having a high surface energy (E_s > 50 mJ/m^2) such as polymers that are very hydrophilic and have perfect wettability (adhesion is possible)
- those with intermediate surface energy (E_s = 35–50 mJ/m^2) such as those polymers that present a hydrophilic–hydrophobic character and have good adhesiveness
- Polyurethanes that are composed of hydrophobic and hydrophilic segments with different lengths and present adhesion properties of a hydrophilic or hydrophobic material

Wettability is also successfully employed for acid-base titration on modified polyethylenes, and it is more sensitive than X-ray photoelectron spectroscopy (XPS) to the rearrangement process (*14*).

Chemical Aspects

Secondary ion mass spectroscopy (SIMS) is a surface analysis technique where an ion beam is focused over a sample surface. The ion beam interacts with the surface destructively by breaking the interatomic and intermolecular bonds and by producing fragments that are detected by a mass spectrometer. Because SIMS is a destructive technique, there have been considerable questions about its use. Clearly, polymer molecular weights are being changed by chain scission and degradation or cross-linking during such an analysis. As these effects become better understood, applications to practical polymer analysis will expand. But no quantitative analysis is possible currently.

Ion scattering spectroscopy (ISS) differs from SIMS in that the ion beam that is used to interrogate the surface is the same ion beam analyzed. The primary ion beam samples only the outermost one or two atomic layers of the surface. Applications of ISS to polymers have been rather limited, partly because of the charge-neutralization problem and partly because of the unavailability of reliable ISS instrumentation.

For large surface areas, XPS is an acceptable technique. The information derived from XPS is as follows (*4, 15*):

- elements present, except H and He
- approximate surface concentration of elements ±10%
- bonding state or oxidation level
- information on aromatic or unsaturated structures
- information on surface electrical properties from charging studies
- nondestructive depth profile and surface heterogeneity obtained by using angular-dependent XPS studies
- destructive profile obtained by using argon etching (but precaution in interpretation for some polymers application is needed)

In practice, a combination approach is often needed. Different poly(methylacrylate)s are analyzed with both XPS and SIMS (*16*). XPS does not provide the information necessary to distinguish between polymers that contain similar short-range chemistry. However, SIMS and valence-band XPS spectra may often contain peak structures that are sufficiently unique to identify specific molecular structures. More recently, imaging XPS analysis is also available.

Polymer surface structures are also accessible to IR or Raman spectroscopic investigations (*17*). The principles of these methods are based on the phenomenon of total internal reflection. The depth of penetration (dp) increases with increasing wavelength and increases as refractive indexes approach each other. The phenomenon of surface-enhanced Raman spectroscopy induced by a rough surface of particles of certain metals, primarily silver, offers interesting new possibilities of polymer surface analysis (depth = 20 nm) (*18*).

Chemically specific reactions in the gas or liquid phase have been used to characterize surface functional groups, mostly for XPS detection (*19, 20*). However, a chemical titration also can be useful and more quantitative, but the choice of reagent is limited because side reactions must be avoided (*21*).

Polymer Surface Modification with Cold Plasma

Search for Polar and Apolar Surfaces

In industrial surroundings, plastic materials are not always used alone, but they are integrated in sophisticated components and composite materials. Thus, a need for good adhesion between the different components necessitates the enhancement of surface properties. For a polyolefin, surface adhesion properties are poor, and different preactivation methods that lead to better surface properties have been developed. Two opposite effects are usually required:

1. Low surface energy, weak wettability, and apolar surfaces are usually applied for biological and industrial applications (e.g., tribology and antiadhesives).
2. High surface energy, strong wettability, and polar surfaces are used for adhesion with other macromolecules, metals, adhesives, paints, and varnishes.

Cold-plasma treatment is the only versatile method for producing polymer surfaces with low or high surface energy, depending mostly on the choice of treatment conditions. Tables I and II summarize the different possibilities. Depending on the conditions of plasma treatment, the contact angle with distilled water may vary from 0° to more than 100°. Therefore, plasma treatment induces a large scale of wettability. Nitrogen and nitrogen-derivative (especially nitrogen dioxide) plasmas lead to the lowest wettability, and carbon dioxide produces a higher wettability (Table I). Apolar surfaces with low surface tension are formed when the polymer is treated by fluorinated plasma. SF_6 seems to be less efficient than CF_4 plasma (Table II). For proper control of different surfaces, wettability by plasma treatment, a comprehension of plasma chemistry, and comprehension of the interfaces of plasma-polymer chemistry are essential.

Table I. Effects of Plasma Treatments on Hydrophilicity of Different Polymers Measured by Contact Angle in Distilled Water (θ H_2O)

Polymer	Untreated Sample, θ H_2O (°)	Hydrophilic Character		Ref.
		Plasma	θ H_2O (°)	
Polyethylene	102	NO_2	52	22
		N_2	48	
		O_2	46	
		NO	28	
		CO	16	
		CO_2	8	
Ultrahigh strength polyethylene	92	N_2	34	23
		Ar	32	
		CO_2	32	
	103	N_2	65	24
Isotactic polypropylene	94	CO_2	45	25
	94	CO_2	49	25
	94	N_2	50	26
Polypropylene	95.9	N_2	34.6	27
	95	O_2	23	28
Poly(tetrafluoroethylene)	112	O_2	91	29
		Air	86	
		Ar	79	
		H_2O	74	
PFA	118	Air	70	30
PMMA	70.5	O_2/H_2O	45.5	31
		H_2O	29.7	
Polycarbonate	80	O_2	48	32
Polystyrene	90	O_2	47	32
	90	O_2	11	33
Polyimide	74.7	N_2	11.6	34
		Ar	11.4	
		CO	10.5	
		O_2	9.6	
		CO_2	8.4	
		NO_2	8.0	
		NO	7.6	

Note: PFA is poly(tetrafluoroethylene-*co*-perfluoroalkoxyvinylether).

Table II. Effects of Plasma Treatments on Hydrophobicity of Different Polymers Measured by Contact Angle in Distilled Water (θ H$_2$O)

Polymer	Untreated Sample, θH_2O (°)	Hydrophobic Character		
		Plasma	θH_2O (°)	Ref.
Polyethylene	95	CF$_4$	125	35
	95	SF$_6$	111	35
	103	CF$_4$	121	36
	103	SF$_6$	117	36
Polypropylene	104	CF$_4$	123	36
		SF$_6$	117	
PMMA	70	CF$_4$	108	37
Polyamide 6	0	CF$_4$	141	38
	63.5	CF$_4$	108.5	4
Poly(ethylene terephthalate)	121	CF$_4$	136	38
	69	CF$_4$	98	4
Polystyrene	88	CF$_4$	111	36
		SF$_6$	108	
Cotton	0	CF$_4$	129.5	38
Wool	129	CF$_4$	134	38
Silk	0	CF$_4$	136	38

Mechanism of Surface Modification with Cold Plasma

Plasma is composed of different reactive species whose concentrations depend on operating conditions. The species present are mostly radicals, ions, photons, excited molecules, and their fragments. All these species are known to induce different chemical reactions on the polymer surface such as cross-linking, degradation, functionalization, and activation, (i.e., formation of radicals able to postreact with any foreign molecule). Thus, during plasma–polymer interactions, all of these reactions take place, and their courses depend on both nature of the plasma and the polymer.

Cross-Linking

Polytetrafluoroethylene has been treated in different plasmas of air, oxygen, argon, and others (*29*). The main observations are fluorine abstraction and cross-linking on the polymer surface, whereas only a low level of oxygen-containing groups are generated on the surface. The same effect is also noticed for polyolefin treatment (*39*). The cross-linked surface extends to a depth of about 30 nm, apparently independent of exposure time, and enhances cohesive strength and improves adhesive bonding. The reactions shown in Scheme I are suggested for plasma treatment of polyethylene.

This mechanism is described as cross-linking by activated species of inert gas, where activated species such as He* initiate branching. Experiments on the

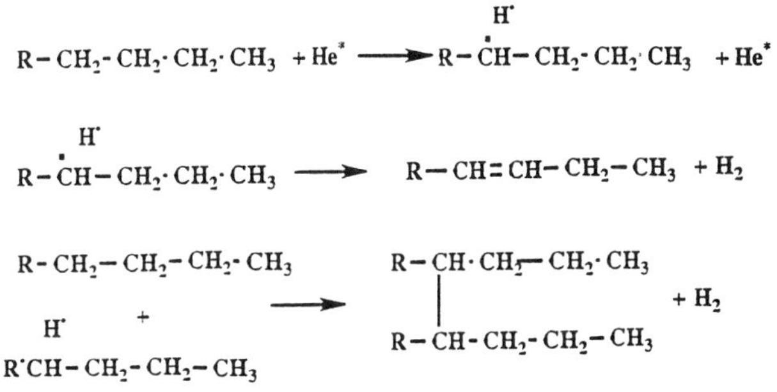

Scheme I. Reactions suggested for the plasma treatment of polyethylene.

treatment of ethylene–tetrafluoroethylene copolymer by an argon plasma (*40*) or a high density polyethylene in a hydrogen plasma (*41*) demonstrate that cross-linking is due to indirect energy transfer from excited species (M*) at least in the outermost few monolayers. At greater depth, UV radiation may become a more important agent (Figure 2). Physical contact between the plasma and the polymer, that is treatment in the discharge, is not required (*41*). Clark and Dilks (*40*) proposed that the reaction is a first-order reaction for each component. Mathematical expressions are as follows:

- at the surface (*S*)

$$d(X_S)/dt = k_M(X_S)(M^*) + k_L(X_S)I_0(1 - e^{-kd})$$

- in the bulk (*b*)

$$d(X_b)/dt = k_L(X_b)I_0 e^{-kd}$$

After integration, the two rate constants are as follows:

$$k_S = k_M(M^*) + k_L I_0 (1 - e^{-kd})$$

$$k_b = k_L I_0 e^{-kd}$$

where X is concentration of structural feature, I_0 is incident radiation, k_L is rate constant for UV reaction, k_M is the rate constant for reaction with reactive species, and d is the thickness of the cross-linked layer under M* bombardment.

Mathematical treatment and comparison with XPS results determined that k_S is one order of magnitude higher than k_L; that is, surface reactions are mostly induced by ions and metastable species (40). On the other hand, UV radiation induced mostly cross-linking. Because gelation measurements are focused on the bulk material, these results are not contradictory to the results of Hudis and Prescott (41, 42), who compared a gelation curve to an exponentially attenuated light theory applied to a spectrum of light (120–190 nm) and a diffusion theory.

Poly(vinyl chloride) films irradiated with hydrogen plasma are also characterized as surface cross-linked materials, and some cross-linking in the bulk is also observed (43). Studies of solubility, intrinsic viscosity, and swelling data show that free radicals are formed at the beginning of irradiation at the surface and migrate into the interior of the sample and induce more cross-linking. Plasticizers enhanced the action of plasma, whereas stabilizers and fillers reduced the action (Figure 3).

Degradation

Degradation here is synonymous with product ablation, especially of volatile products. In microelectronics, it is usually described as an etching process. However, degradation is more generally appropriate because it reflects the morphological and physicochemical changes of polymers. Examples of different degradation effects are given subsequently. The nature of the polymer greatly influences the pattern of degradation (Table III). The major conclusions are (38, 45) as follows:

1. Presence of oxygen in the polymer structure renders the polymer susceptible to plasma, but nitrogen seems to have the opposite effect.
2. The most susceptible structure is aliphatic polyether with –O– in the backbone.
3. Polyethers with cyclic rings such as polysaccharides are also susceptible.
4. Effect of oxygen is somewhat offset when the structure contains aromatic nitrogen adjacent to the oxygen atom and oxygen in the pendent moiety.
5. The absence of oxygen or the presence of nitrogen make the polymer structure less susceptible to plasma (for examples, polyimides).

In many cases, a simple ablation of the surface is observed, except for polyethylene and polypropylene, where a highly oxidized layer is created (44). The rapid reactions that occur at the polymer-film–oxygen-radical interface are essentially unaffected by the presence of phenolic antioxidants over a wide range of concentrations (44). Polyacrylonitrile degradation in oxygen plasma, analyzed with coupled mass spectrometry, induces the formation of C, H_2, N_2, H_2O,

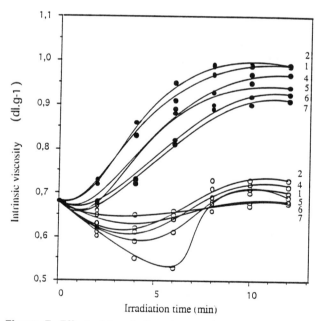

Figure 3. Effect of irradiation time on intrinsic viscosity of PVC (plasma energy, 7.5 eV): stored period 7 days (●) or 10 min (○); 1, pure PVC; 2, PVC + 20% dibutylphthalate (DBPhth); 4, PVC + 20% tricresyl phosphate (TCPhos); 5, PVC + 20% TCPhos + 10% CaCO₃; 6, PVC + 20% TCPhos + 10% CaCO₃ + 5% dibasic lead phosphate (DBLPh); 7, PVC + 20% TCPhos + 10% CaCO₃ + 10% DBLPh. (Reproduced with permission from reference 43. Copyright 1979.)

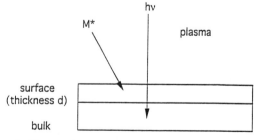

Figure 2. Theoretical model of (in)direct energy absorption (M* is reactive species; d is thickness of the cross-linked layer). (Reproduced with permission from reference 40. Copyright 1977.)

Table III. Rate of Weight Loss of Polymers Irradiated with Different Plasmas

Polymer	O_2 Plasma	He Plasma
LDPE	5.1	1.2 (45), 0.7 (38)
Irradiated LDPE (10 Mrad)	7.0	
Irradiated LDPE (105 Mrad)	8.5	
HD ethylene butene copolymer	6.7	
Polypropylene	7.1	0.8 (45), 0.3 (38)
Chlorinated HDPE	1.03	
Natural rubber	7.0	
Natural rubber + sulfur raw stock	2.5	
Natural rubber + sulfur vulcanizate	0.3	
Natural rubber + peroxide raw stock	6.2	
Polystyrene	2.6	
Poly(vinyl fluoride)	5.2	
PVC		0.3 (38)
PTFE	1.3	
PMMA	4.4	
Poly(acrylic acid)		16.2 (45)
Poly(methacrylate acid)		15.4 (45)
PAN		0.1 (38)
Polyimide	2.4	
Polycarbonate	5.3	
Poly(ethylene terephthalate)	3.7	1.7 (45)
Polyamide 6	5.7	1.1 (45), 1.0 (38)
Poly(vinyl pyrrolidone)		11.9 (45)

Note: LDPE is low-density polyethylene; HDPE is high-density polyethylene; PVC is poly(vinyl chloride); PTFE is poly(tetrafluoroethylene); PAN is poly(acrylonitrile). Values for O_2 plasma [× 10^3 g/(cm² min)] were adapted from reference 44. Values for He plasma [g/(cm² min)] were adapted from the reference in parentheses.

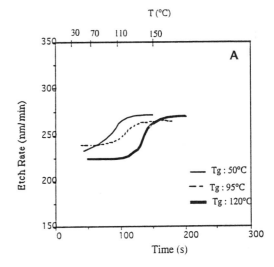

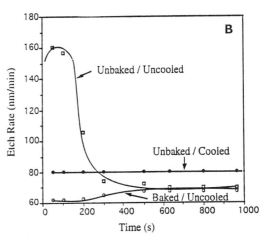

Figure 4. (A) Etch rate of novolac resins with different glass transition temperatures in O_2 plasma as a function of time and temperature. (Reproduced with permission from reference 50. Copyright 1991.) (B) Apparent etch rate in an electron cyclotron resonance reactor versus time of novolac polymer for different bake conditions. (Reproduced with permission from reference 52. Copyright 1990.)

CO, and CO_2 species. No traces of monomer are formed (unlike in thermal degradation) (46).

The degradation mechanism also was studied in relation to polymer crystallinity (47, 48). Ablation with air plasma occurs mainly in the amorphous region of polypropylene, and with prolonged treatment the polymer surface becomes rough enough to have a microdomain structure (47). Polymers like poly(ether-urethane)s (PEU) that contain hard and soft segments show a surface mostly composed of hard segments (urethane) when submitted to a plasma etching, and these modifications seem to be temporary over a period of 6 months (48).

The rate of plasma etching depends also on viscoelastic properties of polymers. Studies on etching of novolac in an oxygen microwave multipolar plasma show that above the glass transition temperature, polymer mobility allows the extension of surface degradation to the entire thickness by a self-diffusion mechanism (49–51) (Figure 4A). Free volume of the polymer seems to control the etch rate, because during oxygen-plasma treatment, the polymer layers become

hot and may undergo volume relaxation. Therefore, an unbaked and uncooled novolac resin will degrade in a few seconds of treatment (Figure 4B) (52).

An etching process needs a degrading plasma. Most etching gases are fluorinated gases, such as CF_4, SF_6, or oxidized gases such as oxygen. However, this fact does not mean that other plasmas do not induce degradation. They do, but the degradation rate is slow. Plasma parameters such as duration, gas flow, or pressure (concentration of etching species and discharge power) may control the degradation rate (Figures 5 and 6) (38). The relationship between discharge power and degradation rate also emphasizes the notion of W/FM (where W is electrical discharge power, F is monomer flow, and M is monomer molecular weight;

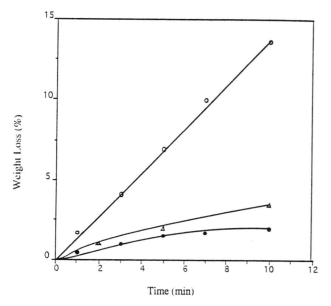

Figure 5. Weight loss of nylon 6 fiber during plasma treatment by (o) air plasma, (●) N_2 plasma (power, 50 W; flow rate, 100cm³/min¹), and (△) CF_4 plasma (power, 150 W; flow rate, 100 cm³/min¹). (Reproduced with permission from reference 38. Copyright 1982.)

see **Plasma Polymerization**). Weight loss increases linearly with *W/FM* ratio until a maximum, and then decreases. If *W/FM* is low, the degradation rate is mostly dependent on the discharge power; if *W/FM* is high, degradation rate depends on the gas flow (*38*).

Ion bombardment, elevates temperatures, and possibly photon absorption can enhance etch rates (*49–51, 53*).

In such fluorinated or oxidized plasma, the reagents are mostly oxygen and fluorine atoms (*53*). Atomic oxygen can add to unsaturated groups or saturated hydrocarbons to form various radicals and peroxides leading to chain scission, aromatic ring opening, and functionalization in polyimide treated in O_2 plasma (*54*). Addition of fluorine-containing gases is also known to increase the O atom concentration relative to that obtained in pure O_2 plasmas. In CF_4-rich plasma, CF_2 is produced rapidly and induces mostly functionalization of the polymer surface (*55*). However, in O_2-rich regimes, these radicals are quenched by the oxygen atom. The increase in O atom production after addition of CF_4 to O_2 plasmas is probably due to changes in electron density and energy distribution (*56*). Etching-rate dependence of O_2–CF_4 gas feed (Figure 7) displays an increase after CF_4 addition and a characteristic maximum in O_2 rich plasmas (*55*). A dependence on chemical structure of the polymer is also observed and leads to the reaction of fluorine atom with the polymer. This reaction may be due to hydrogen substitution or fluorine addition to the double bond. D'Agostino proposed (*56*) a series of possible reaction pathways leading to volatile etching products for both saturated and unsaturated groups in O_2–CF_4 plasmas (Scheme II). Fluorine-atom reaction on unsaturated polymer leads to addition reactions and free-radical formation in the α position, which can be oxi-

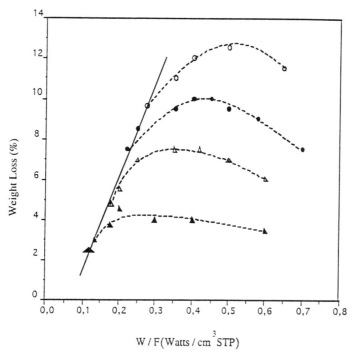

W / F(Watts / cm³ STP)

Figure 6. Relationship between values of *W/FM* and weight loss of nylon 6 treated with air plasma for 5 min: (o) 100 W, (●) 70 W, (△) 50 W, (▲) 30 W. (Reproduced with permission from reference 38. Copyright 1982.)

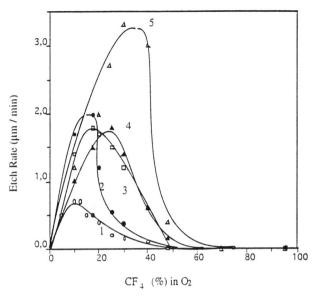

Figure 7. Etch rate behavior for different polymers downstream of a plasma containing O_2–CF_4 mixtures containing 5% Ar: 1, polyisoprene; 2, polystyrene; 3, polybutadiene; 4, polyethylene; 5, poly(vinyl alcohol). (Reproduced with permission from reference 55. Copyright 1990.)

dized or peroxidized. In saturated macromolecules, the predominant reaction with fluorine atom is hydrogen abstraction, leading to a radical transformation to peroxide, followed by chain scission.

Functionalization

The nature of surface functionalization depends on the structure of plasma gas, and functionalization reactions can be classified into three main types:

- nitration through incorporation of nitrogen and its derivatives
- oxidation through attachment of oxygen and its derivatives
- halogenation, mostly fluorination, through interactions of fluorinated plasma

Nitration

The most common plasma gases used are N_2, NH_3, H_2/N_2, NO, NO_2, and their mixtures. Interactions between nitrogen, ammonia, and the mixture of hydrogen and nitrogen lead to attachment of different nitrogen functions (amino groups, amide, nitrile, and imides) (*57, 58*). However, their surface characterization does not produce exact information on the nature of functional groups and the α-carbon and the surface structure changes with time.

One attraction of such modifications is to provide surfaces with high levels of hydrophilicity and with attached sites having polar and basic characters that may be involved in adhesion through Lewis acid–base interactions (*58, 59*). Good examples of increased adhesion strength with nitrogen and ammonia plasmas are those of treated polyaramid fibers and polyaramid-reinforced resin-matrix (*60*). Poly(*p*-phenylene terephthalamide) can be functionalized with N_2, NH_3, or methylamine plasma.

Experiments on poly(methyl methacrylate) (PMMA) and polypropylene reveal that during a short treatment with NH_3 plasma, chain scission takes place but nitrogen is not incorporated (*61, 62*). NH_3 plasma also was used as pretreatment to form active free radical and basic sites on polyurethane substrates (*63*).

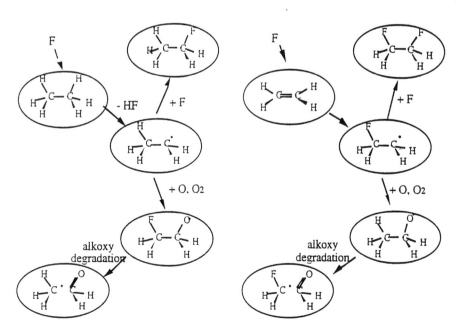

Scheme II. Possible reaction pathways for the formation of volatile etching products from saturated and unsaturated groups in plasmas containing O and F atoms. (Reproduced with permission from reference 56. Copyright 1990.)

Effects of nitrogen and nitrogen oxides plasmas on polyolefins and polyimides also were investigated and compared to those of oxygen and oxygen-derivative plasmas treatments (*22, 23, 34, 64*). In NO plasma, the main components are fragmented species due to NO molecules, and the minor components are due to O_2 and N_2 molecules, oxygen, and nitrogen atoms. For NO_2 plasma, NO molecules and oxygen atoms are the main components, and NO_2, N_2, and O_2 are minor molecules. NO plasma treatment is the preeminent process. Some of the imide groups in the Kapton film are cleaved by plasma treatment, and as a result secondary amide groups are formed.

Oxidation

A variety of oxygen-derivative gases are employed, but these treatments are more drastic because oxidation is usually followed by degradation. An oxidized surface, obtained from plasma treatment, presents all kinds of oxygen-derivative groups or functions characterized in XPS analysis (*20, 31, 47, 65–69*): –$\underline{C}$–OR, –$\underline{C}$OOR, and –$\underline{C}$OOH.

Also, polyolefins (low-density polyethylene and polypropylene) treated in air plasma (*47*) or in CO_2 plasma (*25*) lead to the formation of trans C=C bonds. Relatively small amounts of aromatic groups in the polymer chain convey substantially increased resistance to plasma oxidation (*68*). Some general features can be pointed out (*70*): $\underline{C}$–O linkages originate mostly from the aliphatic component of a polymer, whereas phenyl rings yield predominantly $\underline{C}$=O and O–$\underline{C}$=O species; carbonyl groups produce mainly O–$\underline{C}$=O.

As for different nitrogen-derivative plasmas, CO and CO_2 plasmas were compared for polyethylene surface modification (*67*). With CO_2 plasma, treated surfaces possess 7–10 times more oxygen groups than CO plasma. Introducing polar groups by using oxidative plasma on polymer surfaces, surface interactions, and thus adhesion to polymer and metals can be improved (*65, 71*). This treatment is applied to fluoropolymers due to their lack of adhesive properties. Functionalization is run with H_2O/O_2 or H_2O/H_2 plasma treatment (*71*). By using hydrogen gas as a fluorine-atom scavenger, oxygen incorporation on the surface matrix could be well controlled.

In oxidation processes, the proposed mechanism involves reactive species such as vacuum UV (VUV) radiation, ions, radicals, and oxygen atoms (*70, 72–74*). Surface modification of a polymer by an oxygen plasma can be modeled by VUV radiation. However, competing etching–ablation processes also occur, which continuously expose unreacted polymer; this phenomenon does not happen during UV irradiation. Therefore, a greater depth of oxidation is found with plasma irradiation (*70*). Ions are important in the oxidation only if there is a negative self-bias in the poly-

mer surface (*75*). Feed of oxygen plasma with iodine, known as a radical scavenger, inhibits the oxidation of the polymer surface (*72*). Downstream discharge treatment allows a selection of different reactive species in O_2 plasmas. This process was studied for an understanding of the basic mechanism (*76, 77*) and a study of polymer behavior in a low-earth orbit environment (*73, 74*). The $O(^3P)$ oxygen atom is the most reactive species for oxidation.

Fluorination

Fluorination is most commonly conducted to obtain a polymer surface having a strong hydrophobic character and thus a low friction coefficient (*78*), a weak adhesion, and at least an anti-adhesion for biological applications (*79*). The prevalent group is CF_2, but some CF and CF_3 groups are also detected (*35, 36, 80–83*).

Different fluorinated gases can be used: CF_4 and its derivatives, CF_3H, CF_3Cl, and CF_3Br. The CF_4 plasma is the only treatment whereby fluorinated polypropylene surfaces are obtained directly without the deposition of a thin plasma-polymerized film (*81*). SF_6 decomposes to liberate fluorine atoms and SF_x radicals. Because sulfur-containing groups apparently are not fixed to polymer surfaces, SF_6 plasmas simply act as a convenient source of active F atoms that can be exchanged for hydrogen in the polymer surface. Thus, SF_6 plasma is considered more for an etching process than for surface fluorination (*36, 80, 82*). Fluorination of different crystalline polymers was investigated (*84*). Plasma treatment did not change the film crystallinity (Figure 8).

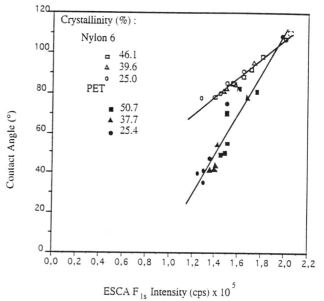

Figure 8. Correlation between contact angle in water and XPS F_{1s} intensity of nylon 6 and poly(ethylene terephthalate) films treated with CF_4 plasma. (Reproduced with permission from reference 84. Copyright 1988.)

A mechanism for fluorination of polyimide and polyethylene by CF_4 plasma treatment and subsequent removal of volatile molecules was proposed (*85*). In fluorination, the fluorinated layer thickness increases with time and the outer surface reaches a constant composition. In defluorination, pit formation is observed. When pitting begins, the composition changes and the fluorine level decreases. Some models of the polyimide fluorination mechanisms were proposed (*86*). CF_4 plasma is considered as the only source of fluorine atoms. Abstraction of hydrogen from saturated carbon is almost as easy as radical–radical reactions at the plasma–polymer interface.

The important role of the photon component of fluorinated plasma was documented (*87*). Photochemical reactions were performed with a plasma as the only source of radiation, and polymer specimens were isolated from all other components. Results establish that radiation in the VUV range enhances the reactivity of the polymer surface toward fluorine gas, and these findings are consistent with what is known about photochemistry of the reactants.

Activation

Activation means creation of reactive species on the polymer surface, mostly radicals that are able to react through one of three previously described effects or to postreact with a foreign reagent like a monomer. This postreaction leads to a grafted surface. Treatment of polyethylene surfaces by Corona discharge followed by reaction with a free-radical compound, diphenyl-picrylhydrazyl (DPPH, which enables the detection of 10^{13} radicals/cm²), indicate that Corona treatment produces free radicals, probably in the form of peroxides (*88*). An electron spin resonance study (*35, 89*) of radicals produced by residual-air glow discharge and oxygen plasma showed two common types of radicals and nitroxide radicals. Alkyl radicals were unstable at room temperature and reacted rapidly with oxygen to form peroxide (*35, 89*). Nitroxide radical was also detected in the air glow-discharge treatment.

In the case of air, oxygen, hydrogen, or inert gas plasmas, the generated radicals, with a few exceptions, appear to be similar to those obtained by γ-irradiation or high energy electron beams (*90*). In treatment of polypropylene with nitrogen plasma, dependence of radical generation on plasma parameters was studied (*21*). The longer the plasma treatment, the higher the concentration of free radical.

Fluoropolymers treated in argon plasma produced three kinds of radicals: mid-chain, end-chain, and dangling bond sites (*91*). The corresponding peroxy radicals formed on subsequent air exposure (*30, 35, 92*) are assigned to two types of oxygen remarkably stable at room temperature, one bonded to carbons in the main chains and the other to carbons arising from chain scission. This radical formation is probably due to the absence of hydrogen in fluoropolymers or the formation of a stabilized structure.

Aging

The aging process cannot be classified as one of the main effects of plasma treatment of polymers. Because cross-linking, degradation, and functionalization proceed through a radical mechanism and because chain mobility is not negligible, aging could appear and have an effect on the polymer. Two aspects of aging are distinguished: chemical and physicochemical.

The chemical aspects are controlled by the reactivity of the attached groups or functions on the polymer surface after plasma treatment. Examples have been already given with peroxidation of radicals (*30, 35, 92*). Hydrolysis of certain imino functions in nitrogen plasma also was noted (*93*).

The physicochemical aspects have a lifetime dependence on the short range (rotational) and long range (diffusional) mobility of polymer segments near the surface, and these aspects are influenced by surface treatment and storage. Poly(ether ether ketone) treated in oxygen plasma (Figure 9) (*94, 95*) is an example. The hydrophilicity and oxygen-atom concentration decreased, a result clearly indicating the dependence of aging on storage conditions. Annealing of the substrate may induce some surface reorganization (*33*).

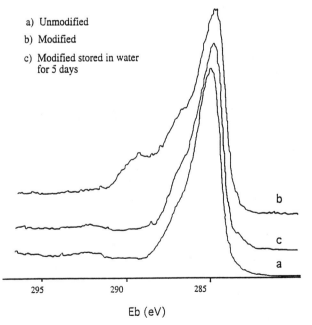

a) Unmodified

b) Modified

c) Modified stored in water for 5 days

Figure 9. XPS spectra of the C_{1s} region of poly(ether ether ketone). (Reproduced with permission from reference 94. Copyright 1987.)

Postplasma Treatment: Grafting

Through different radiation techniques such as plasma irradiation, surface graft polymerization can be achieved and can offer an effective means of introducing desirable properties into the polymer surface without affecting the architecture of the polymer backbone. In this grafting process, a radical mechanism is predominant, but an ionic mechanism is also involved. Different grafting techniques, like other radiations (γ, electron beam, and UV), are classified in three groups of processes: simultaneous, pre-irradiation, and peroxide.

Simultaneous Processes

This technique is a one-step synthesis of a graft copolymer on the surface. Monomer and the polymer surface are mutually irradiated by a cold plasma. Plasma phase can be constituted from the monomer vapor to be grafted, and the plasma polymerization leads to a grafted layer on the polymer surface (*1, 96*). However, the deposited layer is adherent but not really chemically grafted; the polymer surface does not contain reactive sites able to form covalent bonds with the monomer or plasma–polymer fragments. Even if the deposited layer is grafted, the product will never fit with the definition of a graft copolymer. In the same manner, plasma-induced polymerization should lead to a grafted layer on the polymer surface (*97*). A preadsorbed monomer or polymer layer can be treated in cold plasma. However, grafting is not the only reaction; cross-linking and degradation also take place (*98*).

Pre-Irradiation Processes

This method of graft copolymer synthesis involves a two-step process. The first step is an activation of the polymer substrate. As described before, most reported plasmas are inert gases, nitrogen, or residual atmosphere plasmas, all of which generate free radicals. These radicals (always kept under vacuum conditions) are used for grafting through plasma polymerization (*1*), plasma-induced polymerization (*97*), monomer-vapor adsorption (*99–103*), or dipping into a monomer solution (*104–107*).

Surface grafting through monomer-vapor adsorption mostly involves surface activation with argon plasma (*99–103*). The monomer vapor (glycidyl methacrylate (*99*), 2-chloroethyl vinyl ether, vinylsulfonic acid (*100*), acrylic acid (*101*), vinylpyridine (*102*), or *N,N*-dimethylacrylamide (*103*)) is adsorbed on the surface and then reacts with dangling radicals of the activated surface to give a graft layer.

When grafting is run in a monomer solution such as acrylic acid (*104, 107*), 2-hydroxyethyl methacrylate,

acrylamide (*104*), *N,N*-dimethylaminoethyl methacrylate (*105*), or methyl acrylate (*106*), the plasma gas is usually residual air (*104, 107, 108*) or argon (*106*). In such a case, grafting yield could reach levels up to 670% of substrate weight. Such a surface grafting involves diffusion phenomena (*108*) and is of little interest. Grafting yield depends on activation conditions: that is, duration of nitrogen plasma treatment (*104, 106*) and discharge power (*106*). When translating these physical parameters into chemical parameters such as concentrations of radical and functional (amino) groups (*109*), the yield of acrylic acid grafting is linearly dependent on the functional-group concentrations. The hydrophobic polymer (e.g., polypropylene) to be grafted is first made compatible with a hydrophilic solution (aqueous monomer solution) by introduction of amino groups on the surface. Thus, monomer molecules are attracted by the functional groups, and grafting yield increases as amine concentration on the polypropylene surface increases. Grafting kinetics are also dependent on grafting conditions: monomer concentration, bath temperature, and postreaction time (*105*).

Peroxides Processes

Peroxide formation in cold-plasma treatment can be achieved by sequential and simultaneous processes.

Sequential Processes

The polymer substrate is treated in an inert gas plasma, mostly argon, and then the activated surface is subjected to an air or oxygen atmosphere (*110–112*). Alkyl radicals formed during plasma treatment are peroxidized in contact with air or oxygen, and high yields are obtained (*110*). However, radical recombination is a competitive reaction and leads to surface cross-linking rather than peroxidation (*111*). Another consequence is a decrease in monomer diffusion (*110*). Grafting kinetics require special attention because this process involves a surface thickness of >0.1 μm. The surface analyses used are only semiquantitative, and grafting kinetics are compared with those of homopolymerizations. Cooper and co-workers proposed (*113*) that graft chains should have the same length as the corresponding homopolymers. When applied to grafting acrylic acid onto polypropylene in a nitrogen plasma, with or without peroxidation phase (*109*), the following conclusions were drawn for peroxide initiation:

1. Initiation rate depends mainly on monomer diffusion.
2. Propagation rate increases linearly with time.
3. Chain transfer to monomer and solvent are important.

4. For long grafting times (>20 h), the concentration of the growing sites and chain molecular weight increase with time.

The following conclusions were drawn for alkyl radical initiation:

1. Long chains are obtained within short periods.
2. Monomer diffusion is not important.
3. Transfer reactions exist and do not allow an increase of molecular weight after 5 h of grafting.

Simultaneous Processes

This method involves a direct surface irradiation with an air or oxygen plasma (*112, 114, 115*). If peroxide concentration is too high, decomposition reactions are followed by recombination and the formation of inactive oxide products (*114*), and these conditions lead to lower grafting yields. Adsorbed oxygen does not favor grafting, except if the reaction is run in the presence of riboflavin during UV radiation (*115*). Peroxide-induced grafting seems to lead to higher yields than with other initiations, and crystalline areas are also covered by grafting (*110*).

Plasma and Plasma-Induced Polymerization

Plasma Polymerization

Plasma polymerization is used to describe the deposition of a polymer layer from an organic vapor fragmented in electrical discharge. Different organic fragments recombine and polymerize to form a tri-dimensional network (*1, 56*).

Nature of Plasma Polymer

When the discharge power is turned on, vapor molecules are fragmented by collisions. Their rearrangement leads to the deposited layer whose chemical structure reflects more the nature of the fragments than the molecules. An ethylene plasma leads to a plasma-polyethylene composed of methylene groups, residual double bonds, and branching and aromatic residues (*116*) (Chart I). The chemical structure of the deposited layers depends on the monomer structure and on plasma parameters. The deposited layer can be a powder, a film, or an oil (*117, 118*).

Three kinds of monomers can be distinguished on the basis of electrical discharge power (W), monomer flow (F), and monomer molecular weight (M) as defined by the W/FM ratio (*119*) (Figure 10).

1. Group I monomers are compounds that contain a triple bond or aromatic or heteroaromatic struc-

tures and have a low $(W/FM)_c$ (critical wattage for a stable plasma) (Figure 10A).
2. Group II monomers contain a double bond or cyclic structure with a low $(W/FM)_c$ and are dependent on vapor flow (Figure 10B).
3. Group III monomers are saturated hydrocarbons with a high $(W/FM)_c$ and a strong dependence on vapor flow (Figure 10B).

Yasuda and Hirotsu provided (*119*) evidence that the slopes of the straight lines giving the dependence of $(W/FM)_c$ on the monomer flow in Figure 10B are proportional to the concentration of abstracted hydrogen.

Despite orders of magnitude in differences among deposition rates of monomers at various discharge conditions, the data plotted in Figure 11 show relatively small differences between different monomers. These results indicate that all monomers polymerize largely in a similar fashion (*1, 120*).

Initiator Species

Ionization of vapor molecules is a necessary step for plasma generation. However, before reaching the ionization energy level, other molecular scissions can take place (Table IV); the dissociation reactions will induce free-radical formation.

The major reactive species present in plasma are free radicals, whose concentrations in the vapor phase are 3–5 orders of magnitude higher than those of ions. Radicals are also found in the solid phase and are partially produced by UV radiation (*1, 121*). Group I compounds (*see* Nature of Plasma Polymer) present the highest radical concentration, and group III present the lowest (Figure 10). Ions seem to act only as etching species (*122*), and their dependence on electron-density variation is not quite clear (*123*).

Mechanism of Layer Growth

In 1976, a monoradical mechanism was proposed (*124*) involving the generation of free radicals and growing chain ends in the gas phase, followed by an adsorption of the monomer and macroradicals on the reactor walls and a propagation step on the surface. Yasuda proposed (*1*) the rapid-step growth polymerization, which involves a bicyclic formation process (Scheme III). Cycle I in Scheme III corresponds to the plasma polymerization of group III monomers, and cycle II corresponds to group I. An ionic mechanism also was proposed for plasma polymerization of methane and styrene (*125, 126*).

Plasma chemistry of fluorinated gases indicates that, depending on the chemical nature of the gas, decomposition of the gas during electrical discharge can lead to fluorine atoms and also CF_x radicals (*127, 128*) (Scheme IV). Fluorine atoms most commonly

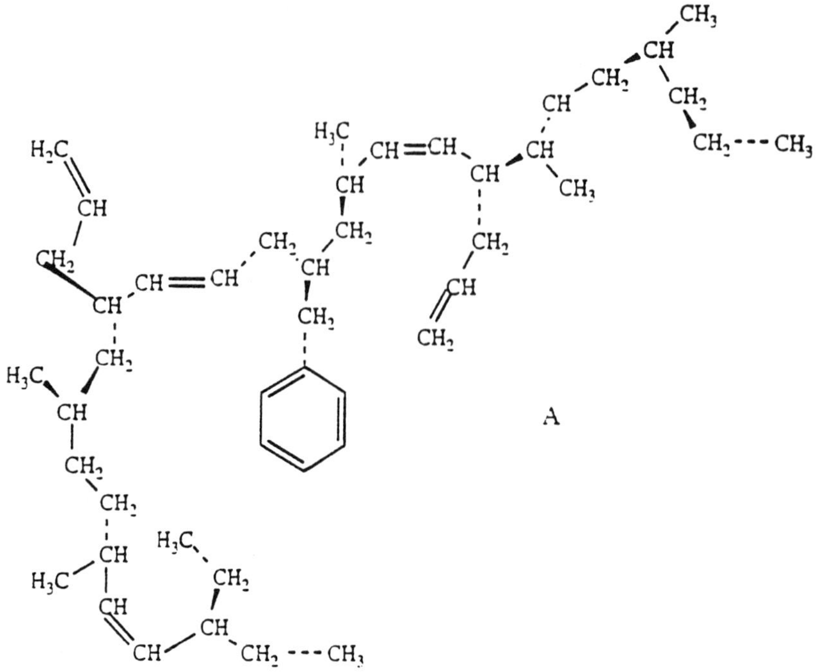

Chart I. Model structure of plasma polyethylene (A) oil and (B) film. (Reproduced with permission from reference 116. Copyright 1976.)

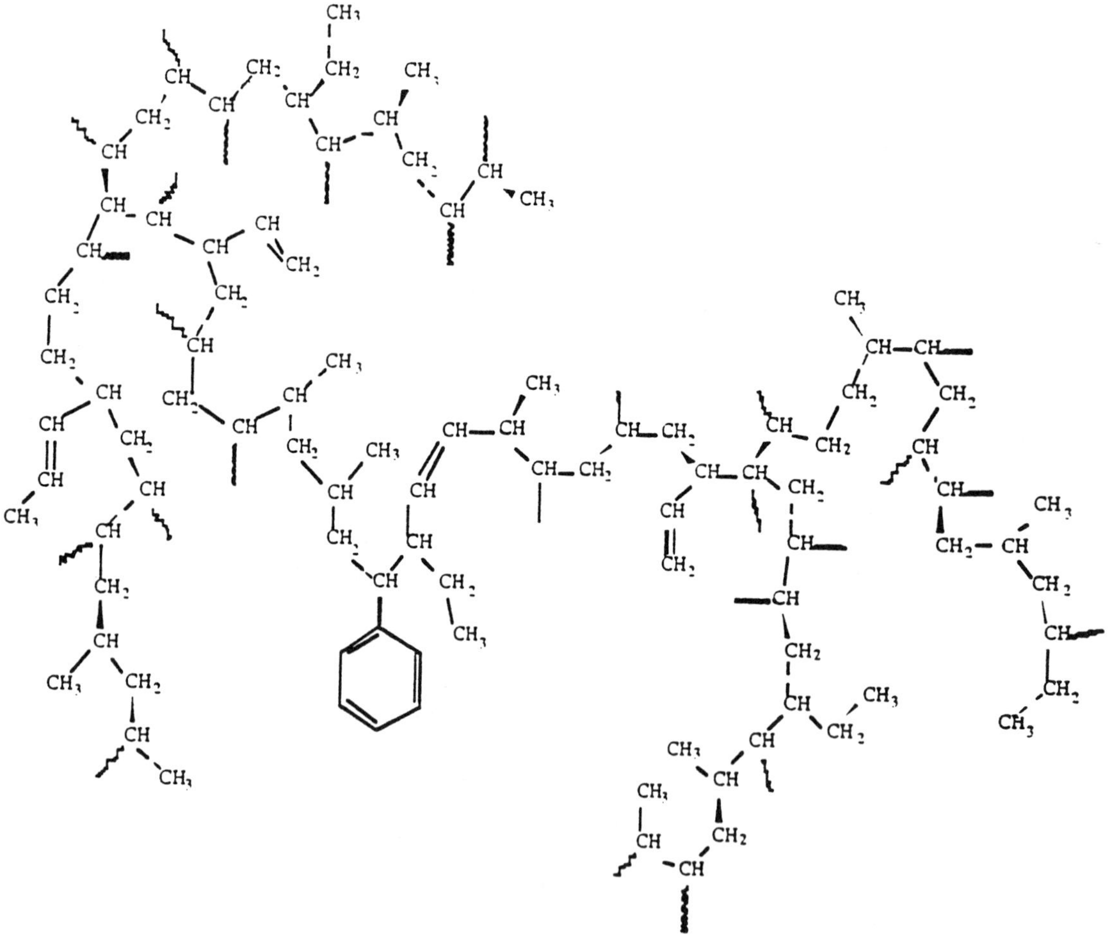

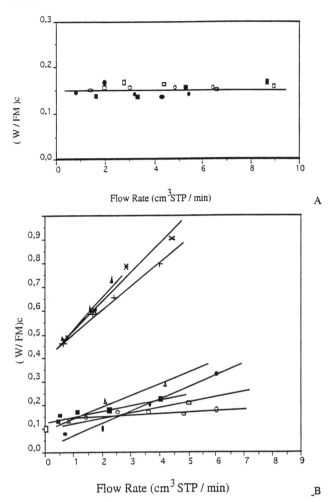

Table IVa. Dissociation Energy of Different Bonds of Interest in Plasma Polymerization

Bond	Dissociation Energy (eV)
C–C	3.61
C=C	6.35 (σ bond)
C=C	2.74 (π bond)
C–H	4.30
C–N	3.17
C=N	9.26
C–O	3.74
C=O	7.78
C–F	5.35
C–Cl	3.52
N–H	4.04
O–H	4.83
O–O	1.52

Table IVb. Dissociation, Metastable, and Ionization Energies of Gases of Interest in Plasma Polymerization

Gas	Dissociation Energy (eV)	Metastable Energy (eV)	Ionization Energy (eV)
He	—	19.8	24.6
Ne	—	16.6	21.6
Ar	—	11.5	15.8
Kr	—	9.9	14.0
Xe	—	8.32	12.1
H_2	4.5	—	15.6
N_2	9.8	—	15.5
O_2	5.1	—	12.5

Figure 10. Plots of $(W/FM)_c$ against the flow rate of different monomers: A: CH_2=CH–CN (○), C_6H_6 (●), HC=CN (■), H_2C=CH–C_6H_5 (□). B: CH_3–$(CH_2)_4$–CH_3 (▲), CH_3–CH_3 (x), CH_4 (+), CH_3–CH=CH_2 (●), CH_2=CH_2 (○), $H(CH_3)_2$–Si–O–Si$(CH_3)_2$H (■), C_6H_{12} (△), C_6H_{10}. (Reproduced with permission from reference 119. Copyright 1978.)

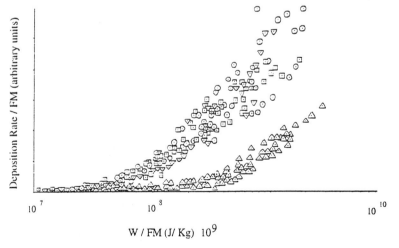

Figure 11. Effect of *W/FM* on deposition rate for various compounds: (○) organosulfurs, (□) organosilicones, (▽) fluorocarbons, (△) hydrocarbons. (Reproduced with permission from reference 1. Copyright 1985.)

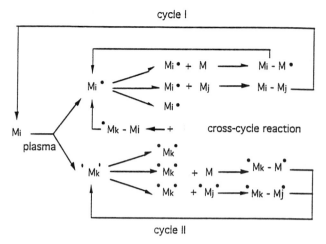

Scheme III. Schematic representation of step-growth polymerization (Mi, Mj, and Mk represent different oligomers). (Reproduced with permission from reference 1. Copyright 1985.)

induce degradation, whereas CF_x radicals favor the deposition of the plasma polymer. The plasma polymer presents a deficiency of fluorine as compared with that of the monomer vapor. Addition of hydrogen favors polymer deposition, and oxygen has the opposite effect on the polymer (*129, 130*). A general scheme of competitive ablation polymerization was proposed (*1*) for all kinds of monomers (Scheme V).

Comparison with Other Gas-Phase Polymerizations

Because evidence of free-radical formation and UV radiation was already given, plasma polymerization can be compared to polymerization of parylene in vacuum at high temperature (or during UV radiation) (Figure 12). For this polymerization, the following aspects may be pointed out.

1. The polymerization mechanism is based on reactions between diradicals rather than the chain-growth addition mechanism that is based on reactions between reactive species and monomer molecules.
2. The ceiling temperature of polymer formation is the limiting thermodynamic factor for polymerization in vacuum.
3. Polymer formation in a fixed location within a vacuum reactor can be enhanced by the addition of an inert gas (a kinetic effect).

All these aspects are common with plasma polymerization. However, plasma polymerization involves more reactive species than parylene polymerization.

Plasma polymerization also was compared with UV-vacuum deposition (UV-VD radiation) (*131*). Monomers that polymerize by UV-VD polymerization also do so in plasma polymerization. Radical concen-

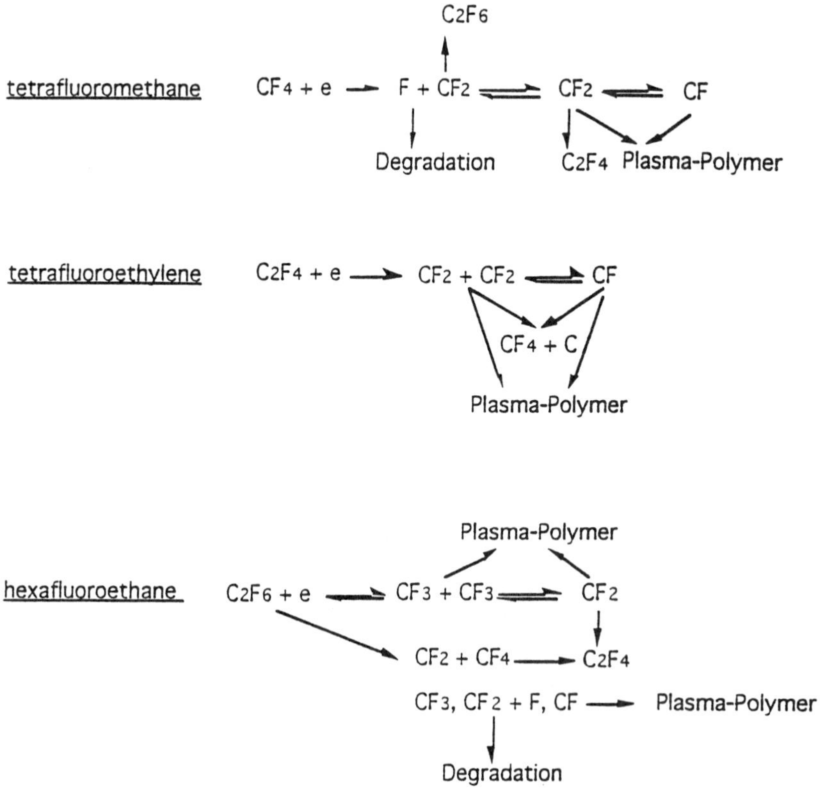

Scheme IV. Decomposition of fluorinated gases in electrical discharge.

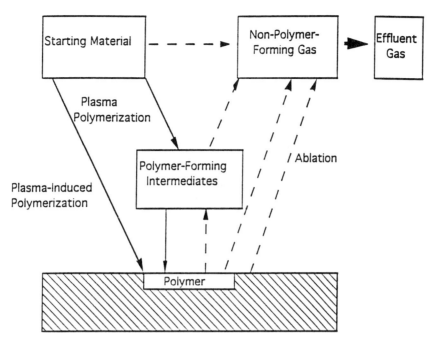

Scheme V. Competitive ablation and polymerization scheme. (Reproduced with permission from reference 1. Copyright 1985.)

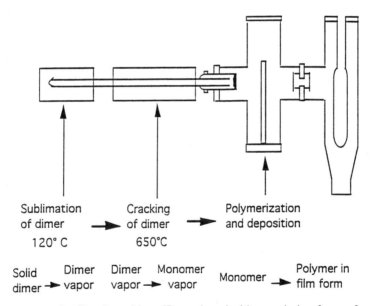

Figure 12. Parylene reactor system for film deposition. (Reproduced with permission from reference 1. Copyright 1985.)

tration in the solid film is high in both cases. Therefore, free radicals act as a plasma polymerization initiator.

Plasma-Induced Polymerization

In plasma-induced polymerization, two phases (liquid and gas) or two steps can be distinguished. The plasma phase is not necessarily the polymerization phase, where the monomer can be a liquid or a solid. As for postgrafting reactions, the initiation step can be completely different from the propagation and termination steps.

Static plasma-induced polymerization involves a plasma initiation at the top of a sealed tube, which contains the frozen monomer in bulk or in solution in a partial vacuum. As a result of the electrical discharge, an oily product appears on the tube walls and initiates postpolymerization of the monomer. Experimental data for static plasma-induced copolymerization of some acrylic monomers are summarized in Table V (*132–138*). These yields depend mostly on the nature of

Table V. Experimental Data for Plasma-Induced Polymerization of Acrylic Monomers

Monomer	Polymerization Medium	Plasma Conditions	Postpolymerization Conditions	Yield (%)	Ref.
Methyl methacrylate	Bulk	60 s	100 h, 25°	40	132
	Bulk	60 s, 100 W	144 h, 25°	7	133
	Bulk	30–60 s, 30–50 W	13 h, 65°	8.6	134
	Bulk	30 s, 20 W	96 h, 4°	4.5	135
	Bulk	30 s, 20 W	24 h, 50°	24.5	135
	Chloroform	30 s, 20 W	144 h, 4°	7	136
	Chloroform	30 s, 20 W	144 h, 70°	100	136
Ethyl methacrylate	Bulk	60 s	168 h, 56°	1	132
	Bulk	30–60 s, 30–50 W	61 h, 65°	Trace	134
Ethyl acrylate	Bulk	60 s, 100 W	48 h, 25°	0	133
	H$_2$O	60 s, 100 W	24 h, 25°	39	133
n-Butyl methacrylate	Bulk	30–60 s, 30–50 W	72 h, 65°	3.1	134
	MI	30–60 s, 30–50 W	3 h, 65°	8.3	134
	Bulk	60 s, 100 W	72 h, 25°	8	133
	H$_2$O	60 s, 100 W	72 h, 25°	100	133
	Bulk	30 s	168 h, 5°	1–2	132
n-Butyl acrylate	Bulk	20 s	168 h, 5°	0	132
	Bulk	60 s, 100 W	100 h, 25°	0.5	133
	H$_2$O	60 s, 100 W	24 h, 25°	35	133
Methyl acrylate	Bulk	180 s	200 h, 20°	0	132
	Bulk	900 s	168 h, 5°	0	132
Methacrylic acid	Bulk	30 s	168 h, 5°	3	132
	H$_2$O	15 s	90 h, 5°	80	132
	H$_2$O	30–60 s, 30–50 W	2 h, 65°	0.7	137
	MI	30–60 s, 30–50 W	17 h, 65°	10.1	137
	Benzene	30–60 s, 30–50 W	17 h, 65°	Trace	137
Acrylic acid	Bulk	30 s	168 h, 5°	3	132
	H$_2$O	15 s	90 h, 5°	50	132
	H$_2$O	30–60 s, 30–50 W	1 h, 65°	95	137
	MI	30–60 s, 30–50 W	1 h, 65°	20.8	137
	Benzene	30–60 s, 30–50 W	2 h, 65°	Trace	137
Acrylamide	Bulk	40 s	45 h, 20°	Trace	132
	H$_2$O	15 s	45 h, 20°	60	132
Methacrylamide	Bulk	120 s	45 h, 20°	Trace	132
	H$_2$O	12 s	45 h, 20°	80	132
	H$_2$O	60 s, 100 W	48 h, 25°	100	138

Note: MI is methyl isobutyrate.

solvent and temperature. Solvent effect and monomer selectivity are strong. These monomers are also known to thermally polymerize methylmethacrylate and methacrylic acid (AM). Addition of a comonomer just after the plasma-induced polymerization allows block copolymer synthesis and results in high molecular weight (*139*). This kind of polymerization is sometimes called living radical (co)polymerization.

The mechanism of plasma-induced polymerization is thus completely different from that of plasma polymerization. Most of the results suggest a free-radical polymerization. An important feature for PMMA obtained by plasma-induced polymerization is stereoregularity close to that of PMMA obtained by UV radiation (Table VI) (*135, 149*) or by any radical process.

Reactivity ratios of MMA and styrene comonomers are similar to those obtained under UV-induced polymerization, with or without inhibition by a radical scavenger (DPPH, no inhibition by water) (*149*).

Despite the evidence of radical polymerization, there are a few peculiar phenomena such as ultrahigh molecular weight polymers (up to 18×10^6), strong solvent effect, and monomer selectivity (Tables V and VI) (*139–153*). For example, styrene polymerizes by cationic, anionic, and radical mechanisms and does not form a polymer by plasma pre-irradiation (*132*).

Plasma-induced polymerization can also be compared with thermal polymerization, because in both cases, ultrahigh molecular weight polymer is obtained. A diradical initiator structure was proposed (*154*) for

Table VI. Molecular Weights and Stereoregularity of Polyacrylates Obtained by Plasma-Induced Polymerization

Polymer	M_n	M_w	Stereoregularity (%)	Ref.
PMMA	3.0–14.6	0.8–18.7	68 (syndio)	133, 135
			31 (hetero)	136, 140
			<1 (iso)	141
Poly(MMA-*co*-styrene)	11.3	18.3		142, 143
Poly(isobutyl methacrylate)		10–34		144, 145
Poly(acrylonitrile)		1–3		146
Poly(methacrylonitrile-*co*-styrene)		1.0–1.16		147
PSSHMA		6.9		148

Note: Number-average (M_n) and weight-average (M_w) molecular weight values are 10^{-6} g/mol. PSSHMA is poly(sodium sulfohexylmethacrylate).

thermal polymerization. Diradical initiator formation was tested with pentafluorostyrene (*154*). Monomers that form diradicals thermally or spontaneously can be polymerized by plasma-induced polymerization. Because of diradical polymerization, the resulting polymers have high molecular weights.

Dynamic plasma-induced polymerization is quite different from the static mode, because the plasma and the monomer are two different entities and because the polymerization stops if the plasma phase is not maintained in contact with the monomer. Monomer crystals, such as trioxane and tetraoxane, polymerize in residual air plasma, and the corresponding polymer is highly crystalline and possesses interesting morphological features (*155*).

Monomers with high molecular weights, such as heavy acrylates (mono- or multifunctional), were cho-

sen for minimizing evaporation in vacuum. The plasma-induced polymerization of multifunctional acrylates is controlled by several parameters (*156*). Its rate decreases with increasing monomer functionality and increases with discharge power and reaches a plateau. The kinetics are affected by the gas nature. Oxygen and its derivative plasmas inhibit the polymerization.

Figure 13 describes the initiation rate of the ethyleneglycol diacrylate polymerization for different plasmas or UV radiations coming from these plasmas, and dynamic plasma-induced polymerization is described in terms of direct or indirect energy transfer from the plasma to the monomer. The indirect transfer is related to the absorption of UV-vis emissions by the monomer, and the direct transfer corresponds to the collision of reactive species with the monomer molecule. The

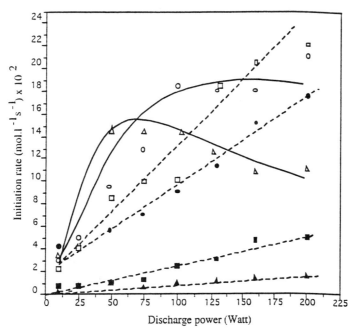

Figure 13. Effect of discharge power on initiation rate of polyethyleneglycoldiacrylate polymerization for different plasmas: (□) N_2, (■) UV-N_2, (○) He, (●) UV-He, (△) CF_4, (▲) UV-CF_4 plasmas. (Reproduced with permission from reference 156. Copyright 1989.)

direct transfer leads to competitive reactions (addition, termination, degradation, and surface functionalization). The indirect energy transfer induces mostly polymerization. A radical mechanism is the most probable mechanism. The mechanism shows similarities with UV-radiation polymerization and initiator decomposition, and efficiency is related to plasma emission. This plasma initiator has a specific efficiency depending on plasma parameters, and the initiator increases the cross-linked thickness. Tertiary amines used as coinitiator increase, in some cases, efficiency of initiation.

Applications of Plasma Technology in Polymer Chemistry

Plasma technology can be employed to obtain materials with new surface properties useful in microelectronics, biomedical technology, and applied research. More conventional technologies, such as polymer deposition and coating, also produce new materials that have good properties and have industrial applications.

Microelectronic Applications

The microelectronics industry employs plasma technology in photoresist etching processes such as phenolic-resin (novolac) degradation in O_2 or O_2/CF_4 plasma. Publications and reviews focusing on the study of etching kinetics and on the most efficient industrial process are numerous and too specialized to be described in this chapter.

Applied Research and Technology

Most research on plasma treatment of polymers is devoted to the study of the increase or decrease of surface energy and modification of the hydrophilic, tribologic, and adhesive characteristics of polymer surfaces. Some of these studies are concerned with basic research, but other are useful for material transport through plasma polymers, barrier and protective films, electrical applications, abrasion-resistant coating, and optical applications.

Material Transport Through Plasma Polymers

The presence of a fluorinated surface layer reduces the migration of additive to the surface (*157*) and reduces the initial solvent-permeation rates (*158*). For example, gas chromatographic silica supports are modified successfully by benzene plasma–polymer deposits, and their selectivity is enhanced (*159*).

Barrier and Protective Films

Plasma treatment is also characterized by an ability to deposit highly adherent, highly cross-linked, and pinhole-free polymers. These polymers are sometimes employed as barrier layers to permeation of different solvents (*160*).

Electrical Applications

Plasma polymerization of phthalocyanine compounds produces films with good rectification and photovoltaic and electrochrome effects (*161*).

Abrasion-Resistant Coating

The abrasion-resistant property depends on the chemical structure of the plasma–polymer. The diamond-like structure presents a real abrasion-resistant property, but any structural default leads to a decrease in abrasion-resistant property.

Optical Application

Polymer films prepared by glow-discharge polymerization from methane, tetramethylsilane, and tetramethyltin present optical properties, including protection of the surface of active photodevices against exposure to high humidity and oxidant gases (*162*).

Adhesion Properties

Plasma modification of polymer surfaces is studied with the goal of increasing the adhesion and the compatibility of dissimilar phases used in composites. Noble-gas plasma etching of fluoropolymers produces a partially defluorinated surface that improves adhesion to epoxy resins (*163*). Interfacial properties between fibers and the matrix contribute to overall properties in high performance composites. The O_2 and N_2/H_2 plasma treatments are most effective for introducing oxygen-containing functionalities at the fiber surface and for improving the interfacial shear strength of carbon fiber–epoxy composites (*164*). A relationship between the oxygen concentration at the fiber surface and the interfacial shear strength is demonstrated in Figure 14. A possible interpretation of this effect is that oxygen- and nitrogen-containing groups at the surface lead to a decisive improvement in the interfacial shear strength by enhancing adhesion and by preventing slippage in the fiber–matrix interface during tension.

Except for pure fluorinated plasmas, plasma treatment leads mostly to higher adhesion (Table VII). Plasma treatment of polymers can give 3–20 times the improvement in bond strength. However, no amount of surface treatment can give strong bonds on a weak surface. Studies of plasma-treatment conditions show that, in some cases, the highest bond strength does not occur at conditions of the best wetting (*165*).

Anti-Adhesion Properties

The presence of fluorinated groups corresponding to covalent carbon–fluorine bonds with a high strength

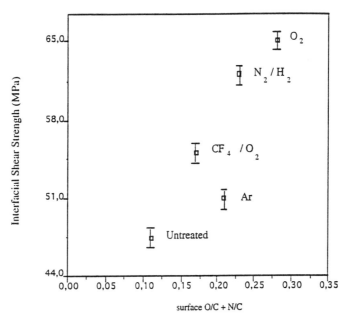

Figure 14. Interfacial shear strength as a function of the surface O/C + N/C (measured by XPS; substrate, carbon-fiber adhesive epoxy composite).

Table VII. Examples of Lap Shear Bonding Improvement by Plasma Treatment

Polymer	Control (psi)	After Plasma (psi)
Polyimide–graphite	420	2600
Polyphenylene sulfide	290	1360
Polyethersulfone	130	3140
PE–PTFE	<10	3200
HDPE	315	3125
LDPE	370	1450
Polypropylene	370	3080
Polycarbonate	410	928
Nylon	850	4000
Polystyrene	570	4000
Mylar A	530	1660
PVDF	280	1300
PTFE	75	750

Note: PE is polyethylene; PTFE is poly(tetrafluoroethylene); HD is high density; LD is low density; PVDF is poly(vinylidene fluoride).

Source: Data reproduced from reference 165.

energy (E_{C-F}) of 108–116 Kcal/mol lead to an increase of hydrophobic and anti-adhesion properties. Applications of treatments are focused on biomaterials and polymers used in biological surroundings (*166, 167*). Fluorine can be provided either by chemical modifications or by fluorinated polymer deposition. Fluorination by a cold plasma is well adapted to biomaterial modification, and patents have been issued in cardiovascular (*168*) and ophthalmic (*37, 169, 170*) domains.

Three commercial fluorine gases are most commonly used: SF_6, NF_3, and CF_4. Carbon tetrafluoride is nontoxic and is usually preferred. Its efficiency is high: fluorine-atom concentration calculated from XPS

analysis on PMMA is higher than carbon-atom concentration (*169*). High resolution XPS spectra give information on chemical structure and repartition of fluorine–carbon bond. The CF_3 substituents are present in a concentration of ca. 5%, the other substituents are as follows: CF_2 (0–15%), CF (10%), CO–F (10%), CO–O (15–20%), C–O (20%), and C–C and C–H (25–30%). Hydrophobic character is pronounced; contact angle with distilled water increases from 59° to 104°.

Hemocompatibility tests on fluorinated polyethylene indicate that platelet aggregation is minimum when fluorine concentration reaches ca. 10% (*168*). Higher or lower values do not lead to improved effects, and sometimes even they show negative effects. Results of biological tests lead to optimization of the plasma treatment. For the treatment of cuprophan dialysis membranes, medium values of plasma-treatment conditions lead to the most attractive biological effects (*171*). Aging is also considered, especially when treated biomaterial is applied in wet surroundings (hydrophilic), which can modify the structural properties of the biomaterial (*172*).

Plasma polymerization of fluorinated monomers such as C_2F_4, C_2F_6, C_3F_6, and C_3F_8 and characterization of the corresponding deposited layers have been long established (*173*). These layers have a very low surface energy (8.1 mJ/m² for perfluoropropane plasma polymer (*170*)), and hence they have reduced adhesion. Typical uses include the reduced adhesion of corneal endothelial cells (on PMMA implant intraocular lenses), adhesion of fibrinogen on silicon overlapped with fluorinated polymers, and adhesion of platelets on polyesters used as cardiovascular prosthesis (*174*).

Conclusion

Simple surface modification (functionalization) of polymers can be achieved by treatment with different plasmas known as oxidation, nitration, and halogenation plasmas. Thicker plasma-generated layers also can be deposited on the polymer surface by using a monomer plasma (i.e., plasma-induced graft polymerization). Adhesion between the polymer substrate and the plasma-generated polymer is normally weak, but this problem can be overcome by pretreatment of the surface. Thus, cold-plasma technology provides access to a wide variety of tailored polymer surfaces, and potential applications exist in all areas where surface interactions or diffusion play a part. Examples of these applications include adhesion, especially coatings, in electronic and biomedical technologies. Side reactions during plasma treatment and the difficulty in controlling the course of plasma-induced polymerization represent the immediate challenges of plasma polymer chemistry. Steady progress in these areas is expected to enhance the scope of cold-plasma technology for the creation of speciality polymer surfaces with increasing sophistication and specificity.

References

1. *Plasma Polymerization*; Yasuda, H., Ed.; Academic: Orlando, FL, 1985.
2. *Processes in Discharge Plasmas*; Proud, J. M.; Luessen, L. H., Eds.; NATO ASI Series B; Plenum: New York, 1986.
3. *Surface and Interfacial Aspects of Biomedical Polymers*; Andrade, J., Ed.; Plenum: New York, 1985; Vol. 1.
4. Ratner, B. D. *Polymers in Medecine*; Migliaresi, C., Ed.; Elsevier: Amsterdam, Netherlands, 1988; Vol. 3, p 87.
5. Waddell, W. H.; Evans, L. R.; Gillick, J. G.; Shuttleworth, D. *Rubber Chem. Technol.* **1992**, *65(3)*, 687.
6. Legeay, G.; Brosse, J. C.; Epaillard, F. *Double Liaison Chim. Peint.* **1986**, *367-368*, 111.
7. Mullera, J.; Feijoo, J. L.; Villamizar, C. A.; Vasquez, P. E. *Metal Finishing* **1986**, *84(4)*, 57.
8. Silverstein, M. S.; Breuer, O. *Polym. Mater. Sci. Eng.* **1992**, *67*, 26.
9. Tung, B.; Polinski, P. T.; Terney, S. T. *SAMPE Q.* **1988**, *19(3)*, 36.
10. Cross, E. M.; McCarthy, T. J. *Macromolecules* **1992**, *25*, 2603.
11. Martz, P.; Hami-Eddine, M.; Wu, D.; Papirer, E.; Vallat, M. F.; Schultz, J. *Proceedings of the 2nd Congress of ADHECOM*; Bordeaux, France, 1989; p 119.
12. Srinivasan, R.; Lazare, S. *Polymer* **1985**, *26*, 1297.
13. Kawanishi, S.; Shimizu, Y.; Sugimoto, S.; Suzuki, N. *Polymer*, **1991**, *32(6)*, 979.
14. Wamser, C. C.; Gilbert, M. I. *Langmuir* **1992**, *8(6)* 1608.
15. Laguës, M. *Actual. Chim.* **1990**, *Janvier-Fevrier*, 17.
16. Stickle, W. F.; Moulder, J. F. *J. Vac. Sci. Technol.* **1991**, *9(3)*, 1441.
17. Zhao, Q.; Long, F. J., Liu, X. S. *J. Appl. Polym. Sci.* **1991**, *43(1)*, 19.
18. Allara, D. L.; Murray, C. A.; Bodoff, S. In *Physicochemical Aspects of Polymer Surfaces*; Mittal, K. L., Ed.; Plenum: New York, 1985; Vol. 1, pp 33–43.
19. Ohmichi, T.; Tamaki, H.; Kawasaki, H. In *Physicochemical Aspects of Polymer Surfaces*; Mittal, K. L., Ed.; Plenum: New York, 1985; Vol. 2, pp 793–800.
20. Gerenser, L. J.; Elman, J. F.; Mason, M. G.; Pochan, J. M. *Polymer* **1985**, *26*, 1162.
21. Poncin-Epaillard, F.; Chevet, B.; Brosse, J. C. *J. Adhes. Sci. Technol.* **1994**, *8(4)*, 455.
22. Inagaki, N.; Tasaka, S.; Kawai, H.; Kimura, Y. *J. Adhes. Sci. Technol.* **1990**, *4(2)*, 99.
23. Hild, D. N.; Schwartz, P. S. *J. Adhes. Sci. Technol.* **1992**, *6(8)*, 879.
24. Steinhauser, H.; Ellinghorst, G. *Angew. Makromol. Chem.* **1984**, 120.
25. Poncin-Epaillard, F.; Chevet, B.; Brosse, J. C. *Eur. Polym. J.* **1990**, *26(3)*, 333.
26. Poncin-Epaillard, F.; Chevet, B.; Brosse, J. C. *Makromol. Chem.* **1991**, *192*, 1589.
27. Marchant, R. E.; Chou, C. J.; Khoo, C. *J. Appl. Polym. Sci., Appl. Polym. Symp.* **1988**, *42*, 125.
28. Novis, Y.; Chaib, M.; Caudano, R. *Br. Polym. J.* **1989**, *21*, 171.
29. Youxian, D.; Griesser, H. J.; Mau, A. W.; Schmidt, R.; Liesegang J. *Polymer* **1991**, *32(6)*, 1126.
30. Momose, Y.; Tamura, Y.; Ogino, M.; Okazaki, S.; Hirayama, M. *J. Vac. Sci. Technol. A* **1992**, *10(1)*, 229.
31. Vargo, T. G.; Gardella, J. A. *J. Polym. Sci., Polym. Chem. Ed.* **1989**, *27*, 1267.
32. Morra, M.; Ochiello, E.; Garbassi, F. *Angew. Makromol. Chem.* **1991**, *189*, 125.
33. Garbassi, F.; Morra, M.; Ochiello, E.; Pozzi, L. *Proceedings of the 9th International Symposium of Plasma Chemistry*; D'Agostino, Ed.; Bari, Italy, 1989; p 1595.
34. Inagaki, N.; Tasaka, S.; Hibi, K. *J. Polym. Sci., Polym. Chem.* **1992**, *30(7)*, 1425.
35. Legeay, G.; Rousseau, J. J.; Brosse, J. C. *Eur. Polym. J.* **1985**, *21(1)*, 1.
36. Strobel, M.; Lyons, C. S. *Polym. Mater. Sci. Eng.* **1987**, *56*, 232.
37. Eloy, R.; Parrat, D.; Tran Minh, Duc; Legeay, G.; Bechetoille, A. *J. Cataract Refractive Surg.* **1993**, *19*, 364.
38. Yasuda, T. *Org. Coat. Appl. Polym. Sci. Proc.* **1982**, *47*, 313.
39. Coopes, I. H.; Gifkins, K. J. *J. Macromol. Sci. Chem.* **1982**, *17(2)*, 217.
40. Clark, D. T.; Dilks, A. *J. Polym. Sci., Polym. Chem. Ed.* **1977**, *15*, 2321.
41. Hudis, M.; Prescott, L. E. *J. Polym. Sci., Polym. Lett. Ed.* **1972**, *10*, 179.
42. Hudis, M. *J. Appl. Polym. Sci.* **1972**, *16*, 2397.
43. Zahran, A. H.; Nofal, E.; El-Azmirly, M. A.; Elsabee, M. Z. *J. Appl. Polym. Sci.* **1979**, *24*, 1875.
44. Hansen, R. H.; Pascale, J. V.; De Benedictis, T.; Rentzepis, P. M. *J. Polym. Sci.* **1965**, *A(3)*, 2205.
45. Yasuda, H.; Lamaze, C. E.; Sakaoku, K. *J. Appl. Polym. Sci.* **1973**, *17*, 137.
46. Ivanov, S. I. *Eur. Polym. J.* **1984**, *20(4)*, 415.
47. Ogita, T.; Ponomarev, A. N.; Nishimoto, S.; Kagiya, T. *J. Macromol. Sci. Chem.* **1985**, *A22(8)*, 1135.

48. Vargo, T. G.; Hook, D. G.; Gardella, J. A.; Eberhardt, M. A.; Meyer, A. E.; Baier, R. E. *J. Polym. Sci., Polym. Chem. Ed.* **1991**, *29*, 535.

49. Paniez, P.; Joubert, O.; Pons, M.; Oberlin, C.; Vachette, T.; Weill, A.; Pelletier, J.; Fiori, C. *Microelectron. Eng.* **1991**, *13(1-4)*, 57.

50. Joubert, O.; Paniez, P.; Pelletier, J.; Pons, M. *Appl. Phys. Lett.* **1991**, *58(9)*, 959.

51. Joubert, O.; Fiori, C.; Oberlin, C.; Paniez, P.; Pelletier, J.; Pons, M.; Vachette, T.; Weill, A. *J. Appl. Phys.* **1991**, *69(3)*, 1697.

52. Paniez, P.; Pons, M.; Joubert, O. *Microelectron. Eng.* **1990**, *11*, 469.

53. Joubert, O.; Pelletier, J.; Arnal, Y. *J. Appl. Phys.* **1989**, *65(12)*, 5096.

54. Chou, N. J.; Parazsczak, J.; Babich, E.; Chaug, Y. S.; Goldblatt, R. *Microelectron. Eng.* **1986**, *5*, 375.

55. Egitto, F. D. *Pure Appl. Chem.* **1990**, *62(9)*, 1699.

56. *Plasma Deposition, Treatment, and Etching of Polymers*; D'Agostino, R., Ed.; Academic: Orlando, FL, 1990.

57. Foerch, R.; Beamson, G.; Briggs, D. *Surf. Interface Anal.* **1991**, *17(12)*, 842.

58. Andre, V.; Arefi, F.; Amouroux, J.; De Puydt, Y.; Bertrand, P.; Lorang, G. *Vide, Couches Minces* **1989**, *246*, 227.

59. Benoist, P.; Legeay, G.; Morello, R.; Poncin-Epaillard, F. *Eur. Polym. J.* **1992**, *28(11)*, 1383.

60. Brown, J. R.; Chappell, P. J.; Mathys, Z. *J. Mater. Sci.* **1991**, *26(15)*, 4172.

61. Lub, J.; Van Vroonhoven, F. C.; Benninghoven, A. *J. Polym. Sci., Polym. Chem. Ed.* **1989**, *27*, 4035.

62. Arefi, F.; Andre, V.; Montazerrahmati, P.; Amouroux, J. *Pure Appl. Chem.* **1992**, *64(5)*, 715.

63. Giroux, T. A.; Cooper; S. L. *J. Appl. Polym. Sci.* **1991**, *43(1)*, 145.

64. Inagaki, N.; Tasaka, S.; Ohkubo, J.; Kawai, H. *J. Appl. Polym. Sci., Appl. Polym. Symp.* **1990**, *46*, 399.

65. Carlsson, C. M.; Johanson, K. *Polym. Mater. Sci. Eng.* **1992**, *67*, 21.

66. Occhiello, E.; Morra, M.; Morini, G.; Garbassi, F.; Humphrey, P. *J. Appl. Polym. Sci.* **1991**, *42*, 551.

67. Inagaki, N.; Tasaka, S.; Hibi, K. *Polym. Prepr.* **1990**, *31(2)*, 380.

68. Moss, S. J.; Jolly, A. M.; Tighe, B. J. *Plasma Chem. Plasma Process* **1986**, *6(4)*, 401.

69. Clark, D. T.; Wilson, R. *J. Polym. Sci., Polym. Chem. Ed.* **1983**, *21*, 837.

70. Shard, A. G.; Badyal, J. P. *Macromolecules* **1992**, *25(7)*, 2053.

71. Vargo, T. G.; Gardella, J. A.; Meyer, A. E.; Baier, R. E. *J. Polym. Sci., Polym. Chem. Ed.* **1991**, *29*, 555.

72. Munro, H. S.; Beer, H. *Polym. Comm.* **1986**, *27*, 79.

73. Golub, M. A.; Wydeven, T. *Polym. Degrad. Stab.* **1988**, *22*, 325.

74. Golub, M. A. *Makromol. Chem., Macromol. Symp.* **1992**, *53*, 379.

75. Liston, E. M. *Proceedings of the 9th International Symposium of Plasma Chemistry*; D'Agostino, Ed.; Bari, Italy, 1989; p L7.

76. Normand, F.; Marec, J.; Leprince, P.; Clouet, F. *Vide Couches Minces* **1991**, *Suppl. 256*, 256.

77. Shi, K.; Clouet, F. *J. Appl. Polym.* **1992**, *46*, 2063.

78. Chasset, R.; Legeay, G.; Touraine, J. C.; Arzur, B. *Eur. Polym. J.* **1988**, *24(11)*, 1049.

79. Legeay, G.; Poncin-Epaillard, F.; Brosse, J. C. *Proceedings of the 2nd Congress on Adhecom*; 1989; p 59.

80. Strobel, M.; Thomas, P. A.; Lyons, C. S. *J. Polym. Sci., Polym. Chem. Ed.* **1987**, *25*, 3343.

81. Strobel, M.; Corn, S.; Lyons, C. S.; Korba, G. A. *J. Polym. Sci., Polym. Chem. Ed.* **1985**, *23*, 1125.

82. Poncin-Epaillard, F.; Legeay, G.; Brosse, J. C. *J. Appl. Polym. Sci.* **1992**, *44*, 1513.

83. Clark, D. T.; Feast, D. T.; Musgrave, W. K.; Ritchie, I. *J. Polym. Sci., Polym. Chem. Ed.* **1975**, *13*, 857.

84. Yasuda, T.; Okuno, T.; Yoshida, K.; Yasuda, H. *J. Polym. Sci., Polym. Phys. Ed.* **1988**, *26*, 1781.

85. Matienzo, L. J.; Emmi, F.; Egitto, F. D.; Van Hart, D. C. *J. Vac. Sci. Technol.* **1988**, *A6(3)*, 950.

86. Scott, P. M.; Matienzo, L. J.; Babu, S.V. *J. Vac. Sci. Technol.* **1990**, *A8(3)*, 2382.

87. Corbin, G. A.; Cohen, G. A.; Baddour, R. F. *Macromolecules* **1985**, *18*, 98.

88. Carley, J. F.; Kitze, P. T. *Polym. Eng. Sci.* **1980**, *20(5)*, 330.

89. Dandurand, S. P.; Wertheimer, M. R.; Yelon, A. *Polymer* **1983**, *24*, 1581.

90. Kuzuya, M.; Noguchi, A.; Ito, H.; Kondo, S.; Noda, N. *J. Polym. Sci., Polym. Chem. Ed.* **1991**, *29*, 1.

91. Kuzuya, M.; Ito, H.; Kondo, S.; Noda, N.; Noguchi, A. *Macromolecules* **1991**, *24*, 6612.

92. Momose, Y.; Ohaku, T.; Chuma, H.; Okazaki, S.; Saruta, T.; Masui, M.; Takesuchi, M. *J. Appl. Polym. Sci., Appl. Polym. Symp.* **1992**, *46*, 153.

93. Gerenser, L. J. *J. Adhes. Sci. Tech.* **1987**, *1(4)*, 303.

94. Munro, H. S.; McBriar, D. I. *Polym. Mater. Sci. Eng.* **1987**, *56*, 337.

95. Brennan, W. J.; Feast, W. J.; Munro, H. S.; Walker, S. A. *Polymer* **1991**, *32(8)*, 1527.

96. Schram, D. C.; Kroesen, G. M.; Beulens, J. J. *Polym. Mater. Sci. Eng.* **1990**, *62*, 25.

97. Epaillard, F.; Brosse, J. C.; Legeay, G. *Makromol. Chem.* **1988**, *189*, 2293.

98. Terlingen, J. G.; Hoofman, A. S.; Teijen, J. *Proceedings of the 10th International Symposium of Plasma Chemistry*; Ehlemann, U.; Leigon, H. G.; Wiegemann, K., Eds.; Bochum, Germany, 1991; Vol. 2, No. 5, p 3p1.

99. Inagaki, N.; Tasaka, S.; Horikawa, Y. *Polym. Bull.* **1991**, *26(3)*, 283.

100. Grunwald, H.; Geckeler, K. E.; Munro, H. S. *Proceedings of the 10th International Symposium of Plasma Chemistry*; Ehlemann, U.; Leigon, H. G.; Wiegemann, K., Eds.; Bochum, Germany, 1991; Vol. 2, No. 5, pp 15-l6.

101. Vigo, F.; Uliana, C.; Traverso, M. *Eur. Polym. J.* **1991**, *27(8)*, 779.

102. Sherry, B. G.; Yasuda, H.; El-Nokaly, M.; Friberg, E. *J. Dispersion Sci. Technol.* **1983**, *4(3)*, 275.

103. Onishi, M.; Shimura, K.; Seita, Y.; Yamashita, S.; Takahashi, A.; Masuoka, T. *Radiat. Phys. Chem.* **1992**, *39(6)*, 569.

104. Osada, Y.; Iriyama, Y. *Thin Solid Films* **1984**, *118*, 197.

105. Hirotsu, T.; Arita, A. *J. Appl. Polym. Sci.* **1991**, *42*, 3255.

106. Yamaguchi, T.; Nakao, S.; Kimura, S. *Macromolecules* **1991**, *24(20)*, 5522.

107. Masuda, T.; Kotoura, M.; Tsushihara, K.; Higashimura, T. *J. Appl. Polym. Sci.* **1991**, *43*, 423.

108. Hirotsu, T.; Isayama, M. *J. Membr. Sci.* **1989**, *45(1-2)*, 137.

109. Poncin-Epaillard, F.; Chevet B.; Brosse, J. C. *J. Appl. Polym. Sci.* **1994**, *53*, 1291.

110. Matsuoka, T.; Hirasa, O.; Suda, Y.; Ohnishi, M. *Radiat. Phys. Chem.* **1989**, *33(5)*, 421.

111. Suzuki, M.; Kishida, M.; Iwata, H.; Ikada, Y. *Macromolecules* **1986**, *19*, 1804.

112. Fujimoto, K.; Tadokoro, H.; Minato, M.; Ikada, Y. *Polym. Mater. Sci. Eng.* **1990**, *62*, 736.

113. Cooper, W.; Vaughan, G.; Madden, R. W. *J. Appl. Polym. Sci.* **1959**, *1(3)*, 329.

114. Ito, Y.; Kotera, S.; Inaba, M.; Kono, K.; Imanishi, Y. *Polymer* **1990**, *31*, 2157.

115. Ichijima, H.; Okada, T.; Uyama, Y.; Ikada, Y. *Makromol. Chem.* **1991**, *192*, 1213.

116. Tibbitt, J. M.; Shen, M.; Bell, A. T. *J. Macromol. Sci. Chem.* **1976**, *A10(9)*, 1623.

117. Kobayashi, H.; Shen, M.; Bell, A. T. *J. Macromol. Sci. Chem.* **1974**, *A8(2)*, 373.

118. Ohno, M.; Ohno, K.; Sohma, J. *J. Polym. Sci., Polym. Chem.* **1987**, *25*, 1273.

119. Yasuda, H.; Hirotsu, T. *J. Polym. Sci., Polym. Chem. Ed.* **1978**, *16*, 743.

120. Yasuda, T.; Gazicki, M.; Yasuda, H. *J. Appl. Polym. Sci., Appl. Polym. Symp.* **1984**, *38*, 201.

121. Morosoff, N.; Crist, B.; Burmgarner, B.; Hzu, B.; Yasuda, H. *J. Macromol. Sci. Chem.* **1976**, *A(10)*, 451.

122. D'Agostino, R.; Fracassi, F.; Illuzi, F. *Polym. Mater. Sci. Eng.* **1990**, *62*, 157.

123. Moisan, M.; Barbeau, C.; Claude, R.; Margot-Chaker, J.; Sauve, G. *Vide Couches Minces* **1989**, *Suppl. 246*, 148.

124. *Plasma Chemistry of Polymers*; Bell, A. T., Ed.; Marcel Dekker: New York, 1976.

125. Smolinsky, G.; Vasile, M. J. *J. Macromol. Sci.* **1976**, *A10(3)*, 473.

126. Thompson, L. F.; Mayhan, K. G. *J. Appl. Polym. Sci.* **1972**, *16*, 2317.

127. Kay, E. *Proceedings of the International Ion Engineering Congress*; ISSAT'83 and IPAT'83, Kyoko, Japan, 1983; p 1657.

128. D'Agostino, R. *Polym. Mater. Sci. Eng.* **1987**, *56*, 221.

129. McLaughlin, K. L.; Butler, S. L.; Edgar, T. F.; Trachtenberg, I. *J. Electrochem. Soc. Technol.* **1991**, *138(3)*, 789.

130. Lamontagne, B.; Wertheimer, M. R. *Proceedings of the 2nd Annual International Conference on Plasma Chemistry and Technology*; Berniy, N. V., Ed.; San Diego, CA, 1989, p 1045.

131. Yeh, Y. S.; Yasuda, H. *Vide Couches Minces* **1989**, *Suppl. 246*, 216.

132. Osada, Y.; Bell, A. T.; Shen, M. *Plasma Polymerization*; Shen, M.; Bell, A. T., Eds.; ACS Symposium Series 108; American Chemical Society: Washington, DC, 1979; pp 253–261.

133. Osada, Y.; Takase, Y. *J. Polym. Sci., Polym. Lett. Ed.* **1983**, *21*, 643.

134. Kuzuya, M.; Kawagushi, T.; Daiko, T.; Okuda, T. *J. Polym. Sci., Polym. Lett. Ed.* **1983**, *21*, 510.

135. Akovali, G.; Demirel, G.; Ozkar, S. *J. Macromol. Sci. Chem.* **1983**, *A20(8)*, 887.

136. Demirel, G.; Akovali, G. *J. Polym. Sci., Polym. Chem. Ed.* **1985**, *23*, 2377.

137. Kuzuya, M.; Kawagushi, T.; Yanagihara, Y.; Nakai, S.; Okuda T. *J. Polym. Sci., Polym. Chem. Ed.* **1986**, *24*, 707.

138. Osada, Y.; Takase, M.; Iriyama, Y. *Polym. J.* **1983**, *15(1)*, 81.

139. Simionescu, C. I.; Chelaru, C. *Polym. Bull.* **1994**, *32*, 611.

140. Osada, Y.; Bell, A. T.; Shen, M. *J. Polym. Sci., Polym. Lett. Ed.* **1978**, *16*, 309.

141. Simionescu, C. I.; Simionescu, B. C.; Ioan, S. *Makromol. Chem., Rapid Commun.* **1985**, *4*, 549.

142. Simionescu, C. I.; Simionescu, B. C. *Makromol. Chem.* **1985**, *184*, 829.

143. Ioan, S.; Simionescu, B. C.; Simionescu, C. I. *Polym. Bull.* **1982**, *6*, 421.

144. Simionescu, B. C.; Ioan, S.; Flondor, A.; Simionescu, C. I. *Angew. Makromol. Chem.* **1987**, *152*, 121.

145. Simionescu, C. I.; Ioan, S.; Flondor, A. *Makromol. Chem.* **1988**, *189*, 2331.

146. Simionescu, B. C.; Ioan, S.; Bercea, M.; Simionescu, C. I. *Eur. Polym. J.* **1991**, *27(7)*, 589.

147. Simionescu, B. C.; Ioan, S.; Simionescu, C.I. *Polym. Bull.* **1982**, *6*, 409.

148. Chen, W. Y.; Sun, D. H.; Chang, G. L.; Feng, X. D.; Chang, G. B. *Polym. Sci., Polym. Lett. Ed.* **1983**, *21*, 335.

149. Johnson, D. R.; Osada, Y.; Bell, A. T.; Shen, M. *Macromolecules* **1981**, *14*, 118.

150. Kuzuya, M.; Kamiya, K.; Kawagushi, T.; Okada, T. *J. Polym. Sci., Polym. Lett. Ed.* **1983**, *21*, 509.

151. Simionescu, C. I.; Petrovan, S.; Simionescu, B. C. *Eur. Polym. J.* **1983**, *19(10/11)*, 1005.

152. Paul, C. W.; Bell, A. T.; Soong, D. S. *Macromolecules* **1986**, *19*, 1436.

153. Yang, M. L.; Ma, Y. G.; Shen, J. C. *Polymer* **1992**, *33(13)*, 2757.

154. Iriyama, Y.; Yasuda, H. *Polym. Mater. Sci. Eng.* **1990**, *62*, 162.

155. Osada, Y.; Mizumotoa *Macromolecules* **1985**, *18*, 302.

156. Epaillard, F.; Brosse, J. C.; Legeay, G. *J. Appl. Polym. Sci.* **1989**, *38*, 887.

157. In-Houng, L.; Cohen, E.; Baddour, R. F. *J. Appl. Polym. Sci.* **1986**, *31*, 901.

158. Corbin, G. A.; Cohen, R. E.; Baddour, R. F. *J. Appl. Polym. Sci.* **1985**, *30*, 1407.

159. Gavrilova, T. B.; Nitikin, T. B.; Vlasenko, E. V.; Topalova, I.; Petsev, N.; Chanev, C. *J. Chromatogr.* **1991**, *552*, 179.

160. De Mendez, M.; Boeda, J. C.; Legeay, G. "Amélioration des Propriétés de Surface des Matériaux Plastiques Isolants par Greffage Plasma Basse Temperature;" Final Report of MRT Contract Nos. 82–A 0775 and 82 –A 0776; Ministère de la Recherche et de la Technologie: Paris, 1984.

161. Osada, Y; Mizumoto, A.; Tsuruta, H. *J. Macromol. Sci. Chem.* **1987**, *24(3/4)*, 403.

162. Akovali, G. *J. Appl. Polym. Sci.* **1986**, *32*, 4027.

163. Kelber, J. A. *Polym. Mater. Sci. Eng.* **1988**, *119*, 255.

164. Bian, X. S.; Ambrosio, L.; Kenny, J. M.; Nicolais, L.; Occhiello, E.; Morra, M.; Garbassi, F.; Debenedetto, A. T. *J. Adhes. Sci. Technol.* **1991**, *5(5)*, 377.

165. Liston, E. M. *Polym. Mater. Sci. Eng.* **1990**, *62*, 423.

166. Hoffman, A. S. *J. Appl. Polym. Sci., Appl. Polym. Symp.* **1988**, *42*, 251.
167. Legeay, G.; Poncin-Epaillard, F. *Vide Couches Minces* **1993**, *Suppl. 266*, 251.
168. Legeay, G.; Gardette, J.; Brosse, J. C.; De La Faye, D.; Legendre, J. M. Eur. Patent 88 402 380, 1988.
169. Bechetoille, A.; Legeay, G.; Legeais, V.; Gravagna, P.; Mercier, L. Eur. Patent 0 487 418, 1992.
170. Ratner, B. D.; Mateo, N. U.S. Patent 5 091 204, 1992.
171. Man, N. K.; Legeay, G.; Jehenne, G.; Tiberghien, D.; De La Faye, D. *Artif. Organs* **1990**, *14* (2), 44.
172. Andrade, J. D.; Lee, M. S.; Gregonis, D. E. In *Surface and Interfacial Aspects of Biomedical Polymers*; Andrade J. D., Ed.; Plenum: New York, 1985; Vol. 1, pp 249–292.
173. *Plasma Science and Technology*; Boënig, H. V., Ed.; C.H. Verlag: Munich, Germany, 1982.
174. Kiaei, D.; Hoffman, A. S.; Ratner, B. D.; Herbett, T. A. *J. Appl. Polym. Sci., Appl. Polym. Symp.* **1988**, *42*, 269.

Functional Cellulose Derivatives

Kurt E. Geckeler and Manfred Anders

In recent years substantial research efforts have been directed toward the development of new methods for chemical derivatization of cellulose to meet increasingly specific technological demands. We review general aspects of the chemical modification of cellulose and discuss a series of important cellulose derivatives. We also outline technological and industrial applications of functional cellulose derivatives reported in recent years.

Cellulose is the most abundant polymer in nature. In addition to its biological role, cellulose is also among the most industrially useful materials. New uses of cellulose are reported on the basis of its wide-ranging biocompatibility and other specific properties. In contrast to synthetic polymers derived from petroleum, natural gas, and coal, which have uncertain future availability, cellulose is environmentally friendly, made from air and water, and biodegraded into air and water.

Cellulose is a linear polymer made up of β-D-glucose units linked by 1,4-bonds, and the polymer shows polydispersity like other polysaccharides (1–4). Depending on its source and pretreatment, cellulose is characterized by degrees of polymerization of between 200 and 5000. Because of the β-1,4-glucosidic linkage, the anhydroglucose units are rotated around the main axis of the cellulose molecule (180°, alternating). Consequently, the anhydrocellobiose, consisting of two monosaccharide units, represents the basic unit of the cellulose chain (Figure 1).

Because of the versatile properties of cellulose derivatives, they have been applied in many fields including catalysis, chromatography, and biomedicine (3–6). Several specific properties of cellulose derivatives have recently found new applications. For example, the stereoregular structure is useful for chromatographic separation of optically active compounds and for stereoselective synthesis. The molecular stiffness of the cellulose molecules can be used for the preparation of liquid crystalline polymers.

Derivatization of Cellulose

Cellulose can be transformed to alcoholates, ethers, esters, or oxidized derivatives. The derivatization can occur at the primary and at the secondary hydroxyl groups (Figure 1).

The hydroxyl groups also can be exchanged by nucleophilic species. Thus, the following six major reaction types of chemical modification of cellulose can be discerned:

- etherification of the hydroxyl groups to O-alkyl products
- esterification of the OH groups to esters
- nucelophilic substitution to deoxy derivatives
- oxidation
- graft polymerization
- elimination reactions to generate double bonds

Because of the glycosidic linkages in its backbone, cellulose can be hydrolyzed easily by acids or degraded by enzymes (2). Nucleophilic substitution is the main reaction type for producing other functional derivatives of cellulose, and some of these derivatives

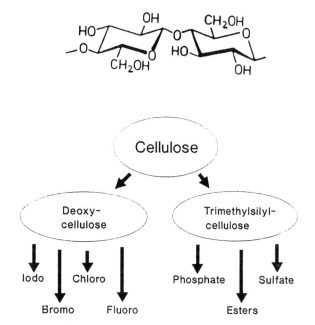

Figure 1. General scheme for the preparation of functional cellulose derivatives.

can serve as intermediates. Among the fundamental questions discussed in this review are the preparation of uniformly substituted cellulose derivatives by specific procedures and the advantages attained using such procedures.

The existence of numerous polar hydroxyl groups along the polymer chain gives access to a great variety of derivatization reactions (*1, 4, 7, 8*). The strong intermolecular interactions between the hydroxyl groups and the high degree of molecular orientation (partial crystallinity) in cellulose make this polymer relatively rigid and intractable. Thus, a major problem for the chemical modification of cellulose is the intra- and intermolecular hydrogen bonds between the neighboring hydroxlyl groups that limit their direct accessibility. To be able to chemically modify cellulosic hydroxyl groups, a preceding activation step is usually necessary to remove the hydrogen bonds. This process can be achieved by using strongly swelling (H-bond breaking) polar media. However, only a partial access to the hydroxyl groups is attained.

Homogeneous Phase Modification Procedures

Modification of cellulose under homogeneous conditions can be achieved in three ways:

- dissolution of cellulose in special nonderivatizing solvent systems such as dimethylacetamide–lithium chloride or *N*-methylmorpholine-*N*-oxide, followed by derivatization

- derivatizing solvent systems (e.g., N_2O_4–dimethylformamide (DMF) or dimethyl sulfoxide (DMSO)–paraformaldehyde) to bring the cellulose into solution with the formation of soluble, unstable derivatives (e.g., nitrites)

- modification of soluble cellulose derivatives, which can be isolated and used as reactive intermediates because of their high reactivity (good leaving groups) for nucleophilic substitutions

Chlorodeoxycellulose, tosyl cellulose, and trimethylsilylcellulose are particularly useful in the modification of soluble cellulose derivatives. Trimethylsilylcellulose has the additional advantage that the polarity of the soluble polymer can be tailored according to the degree of substitution, enabling the use of a great variety of conventional solvents, reagents, and synthetic techniques (Figure 2).

Halogenated Deoxycellulose

Halogenated deoxycellulose derivatives can be formally regarded as cellulose esters of hydrohalic acids (i.e., cellulose-X, where X is F, Cl, Br, I, etc.). Halogenation of cellulose alters its properties and decreases its flammability but increases rot-resistance and water-resistance. In some cases the materials even become water repellant (*9–13*). Properties of halogenated deoxycelluloses depend on the type of halogen involved and also on the degree of substitution. Despite the interesting properties mentioned, halodeoxycelluloses are not of economical importance, but they play an important role as reactive intermediates. Chlorodeoxycellulose is especially valuable for this purpose.

Chlorodeoxycellulose

Heterogeneous chlorination of cellulose has been studied extensively (*7, 14, 15*). The conventional method for synthesizing halodeoxycellulose is the nucleophilic displacement of good leaving groups by halides. As leaving groups, the tosylate and mesylate are most frequently used (*16, 17*). Other leaving groups, such as nitrate and DMF from a formiminium salt intermediate, are also used (*18*). Three reactive cellulose derivatives, namely tosylate, nitrate, and sulfate, were compared to study the effect of aqueous and nonaqueous solvents consisting of LiCl, KBr, and KI (*16*). Interestingly, cellulose sulfate with a degree of substitution (DS) of 0.6 did not react, and cellulose nitrate was less reactive than the toslyate in aqueous medium. A DS of 0.34 was obtained for the preparation of iododeoxycellulose by reaction of aqueous NaI with cellulose tosylate for 35 h. A faster reaction was observed in organic solvents (e.g., cyclohexanone and *n*-butanol).

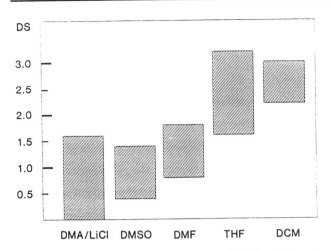

Figure 2. Comparison of the solubility of TMS cellulose in different solvents: DMA is dimethylacetamide; THF is tetrahydrofuran; DCM is dichloromethane.

A high DS of 2.4 was reached by McCormick and Callais (19), who reported the conversions of cellulose to the tosylate in a homogeneous system of lithium chloride and *N,N*-dimethylacetamide. Subsequent heating of a solution of this product in DMF–LiCl at 100 °C for 3 h yielded chlorodeoxycellulose with a DS of 2.3. In some cases the chlorination of cellulose can already take place during the tosylation reaction, because the homogeneous synthesis of tosyl cellulose in *N,N*-dimethylacetamide–LiCl is accompanied by the solvolysis of tosyl chloride. This reaction of the Vilsmeier–Haack type leads to a competitive formation of tosyl cellulose and chlorodeoxycellulose (20).

Because the chloro derivatives of cellulose are themselves mainly of importance as synthetic interme-diates, it is preferable to convert the reactive tosylates directly to the desired product. Also, a variety of other reagents can be used for chlorination in DMF, which is the solvent most frequently used. DMF is an essential factor in the development of chlorocellulose, because it is an excellent acid scavenger and it forms an iminium salt that facilitates the chlorination. In Table I, the experimental conditions of different preparation meth-ods for chlorodeoxycellulose are summarized.

Degradation of cellulose during deoxychlorination in DMF is inhibited by the formation of chlorodi-methylformiminium chloride [Cl–CH=N(CH$_3$)$_2$Cl$^-$]. This iminium salt is produced from DMF and thionyl chloride (*12, 18, 21, 22*), methanesulfonyl chloride (mesyl chloride) (*23–26*), and phosphoryl chloride (*27–29*). Chlorodimethyl formiminium chloride reacts with cellulose as an intermediate during the chlorina-tion reaction (Scheme I) (*18, 24, 27, 30*). The identifica-tion of 3,6-dichloro-3,6-dideoxyallose units in the chloro derivative of cellulose suggests that the replace-ment of (CH$_3$)$_2$N$^+$=CHO groups by chlorine follows the S$_N$2 mechanism (*22, 25*).

Reaction conditions employed for these conver-sions strongly influence the nature of the final product. For example, treating native cotton with POCl$_3$ in DMF at 75 °C for 1 h yielded fibrous cellulose (DS 0.55) with good retention of strength. At high POCl$_3$ concentra-tions chlorination was predominant, whereas at low POCl$_3$ concentrations phosphorylation was favored (*27*). Selective substitution of the primary hydroxyl groups by chlorine yielding 6-chloro-6-deoxycellulose was described by Horton et al. (*26*). They treated differ-ent cellulose materials with methanesulfonyl chloride in DMF and obtained products with DS values of

Table I. Experimental Conditions for the Preparation of Chlorodeoxycellulose

Starting Cellulose	Reagents	Temp. (°C)	Time (h)	Degree of Substitution	Ref.
Heterogeneous					
Cell–OH	SOCl$_2$–DMF	—	—	–1	21
Cell–OH	SOCl$_2$–Pyridine	26	4	0.29	36
Cell–OH	SOCl$_2$–Pyridine	69	1	1.04	36
Cell–OH	SOCl$_2$–Pyridine	50	4	0.67	36
Cell–OH	SOCl$_2$–DMF	100	1	0.35	18
Cell–ONa	CH$_3$SO$_2$Cl–DMF	—	—	0.35	24
Cell–OH	POCl$_3$–DMF	75	1	0.55	27
Cell–OH	P–Cl$_2$–DMF	—	—	≈0.5	63
Cell–OH	SO$_2$Cl$_2$–Pyridine–CHCl$_3$	–78	—	1.7	31
Cell–OH	P(C$_6$H$_5$)$_3$–*N*-Chlorosuccinimide–DMF	—		0.6–0.8	55
Homogeneous					
Cell–Tos	DMF–LiCl	100	3	2.3	19
Cell–OH (DMA–LiCl)	P(C$_6$H$_5$)$_3$–*N*-Chlorosuccinimide	50	6	0.2–1.86	32
Cell–OH (CH$_3$SO$_2$Cl)	Chloral–DMF	75	1	0.8	25
Cell–OH (CH$_3$SO$_2$Cl)	Chloral–DMF	75	2–3	1.3	25

Note: Dashes indicate that no data was given in the reference.

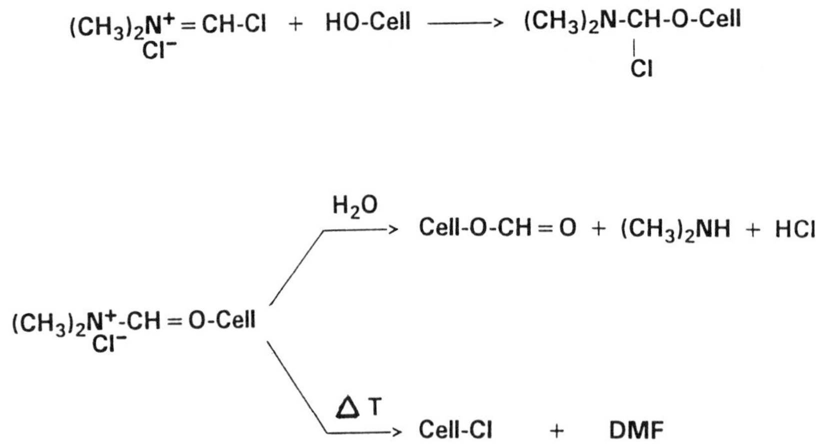

$$(CH_3)_2\overset{+}{\underset{Cl^-}{N}} = CH\text{-}Cl \; + \; HO\text{-}Cell \longrightarrow (CH_3)_2N\text{-}CH\text{-}O\text{-}Cell$$

$$(CH_3)_2\overset{+}{\underset{Cl^-}{N}}\text{-}CH = O\text{-}Cell$$

$$\overset{H_2O}{\longrightarrow} \; Cell\text{-}O\text{-}CH = O \; + \; (CH_3)_2NH \; + \; HCl$$

$$\overset{\Delta T}{\longrightarrow} \; Cell\text{-}Cl \; + \; DMF$$

Scheme I. Chlorination of cellulose by chlorodimethylformiminium chloride (formed from DMF with thionyl chloride or other chlorinating agents): Cell is cellulose backbone.

0.1–0.8. No substitution of secondary cellulosic hydroxyl groups was observed.

Another method to prepare chloro derivatives of cellulose includes the use of sulfuryl chloride. The reaction in pyridine at ambient temperature yielded a highly substituted chlorocellulose (DS 1.7), but reaction at −78 °C produced chlorosulfates. In the chlorination reaction, the substitution took place first at the C-6 hydroxyl and then at the secondary C-3 hydroxyl group, but no reaction at the C-2 position occurred (31). The detection of 3,6-dichloro-3,6-dideoxyallose moieties indicated the replacement of the chlorosulfate group by the chloride anion in accordance with an S_N2 mechanism.

Furuhata et al. used (32) N-chlorosuccinimide triphenylphosphine with LiCl–DMF in the homogenous phase for the chlorination of cellulose. They confirmed the results of a single replacement of the 6-hydroxyl in the early stage. However, the C-3 hydroxyl group was replaced at a lower rate with Walden inversion. Thus, a maximum DS of 1.86 was reached in this work. The halogenation of cellulose with this system is regarded to proceed through the formation of cellulose phosphonium salts and the subsequent attack of halide ions on the phosphonium ester bonds (Scheme II) (33, 34).

Polyhalodeoxycelluloses were prepared from 5,6-unsaturated cellulose by using free-radical initiators and compounds such as CCl_4. For example, 5-chloro-6-trichloromethyldeoxycellulose was synthesized in this way (35). Several processes, such as mercerization (12, 21, 36), regeneration (23), or preswelling (18), enhanced the accessibility on cellulose OH groups to the chlorinating agent. The best results, however, were obtained when working under homogeneous reaction conditions (20, 32). Table II provides a survey of the use of chlorocellulose as an intermediate for the preparation of a variety of cellulose derivatives.

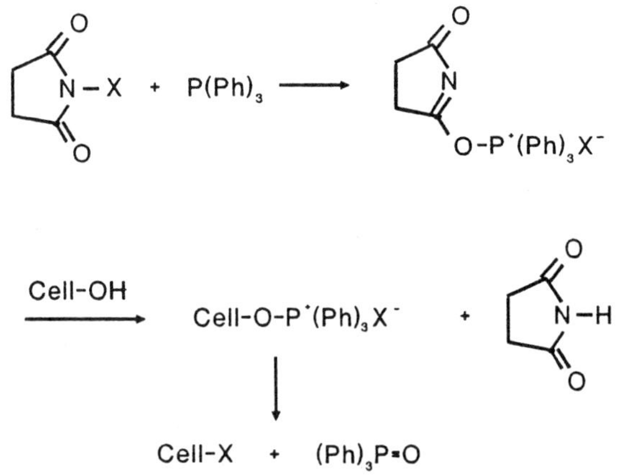

Scheme II. Chlorination of cellulose by N-chlorosuccinimide and triphenylphosphine.

Other Halogenated Derivatives of Cellulose

Halogenated celluloses are accessible by reaction of the mesyl cellulose with fluorides, chlorides, bromides, or iodides (9, 57). Tosylcellulose can be prepared similarly (16, 58). Other reagents used to prepare halodeoxycellulose include N-halosuccinimides. For example, 2,3-diacetylcellulose reacts with N-bromo-, N-chloro-, or N-iodosuccinimide in the presence of triphenylphosphine in DMF via hydrolysis of the acetyl groups, followed by the formation of the corresponding 6-halodeoxycellulose (DS 0.6–0.8) (59).

Fluorocellulose

Halide displacement reactions are less suitable for the preparation of fluorodeoxycellulose. These reactions yield only low degrees of substitution because the fluoride ion is a poor nucleophile. Better results are obtained by direct fluorination using sulfur tetrafluo-

Table II. Preparation of Cellulose Derivatives Using Chlorodeoxycellulose as an Intermediate

Reagent	Substituent Introduced	Ref.
KI–DMF	I	37
NaI–2,5-hexanedione	I	38
NH_3–C_2H_5OH	NH_2	39, 40
[benzene ring with NH_2 and COOH]	NH–[benzene ring with COOH]	14
H_2N–CH_2R–n-BuOH	$NHCH_2R$ (R = alkyl)	12
$H_2N(CH_2)_nNH_2$	$NH(CH_2)_nNH_2$ or cross-linking	41–43
$HN(CH_2CH_2NH_2)_2$	$N(CH_2CH_2NH_2)_2$	44
$S=C(NH_2)_2$	NH–$C(=S)$–NH_2	45
H_2N–NH_2	NH–NH_2	46
$H_2N(CH_2)_2NH_2 \cdot H_2S_3$	$(S)_nCH_2$–Cell	47
$N(CH_2COOH)_3$	$N^+(CH_2COOH)_3 \, Cl^-$	45
NaN_3	N_3	23, 48, 49
KCN	CN	50
NaSCN	SCN	51
HS–$(CH_2)_n$–COOH	S–$(CH_2)_n$–COOH	12
1. $Na_2S_2O_3$– 2. HS–CH_2COOH	SH	52
t-BuOK–DMSO	5,6-Cellulosene	53
$HN(CH_2$–COOH)	$N(CH_2COOH)_2$	54, 56

ride at ambient temperature, and DS values of up to 1.5 are reached. However, the use of acid scavengers, like pyridine or sodium fluoride, is necessary to prevent strong degradation (60).

Access to fluorodeoxycellulose by directly replacing the cellulosic hydroxyl with fluoride was investigated by using three different methods by Frazier and Glasser (61). They applied tosylation–mesylation, triflation, and dialkylamino sulfurtrifluoridation for the derivatization of cellulose and found that the fluorine derivatives were obtained in very low yields by all three methods. Other reagents, such as dialkylamino sulfurtrifluorides (R_2NSF_3), were also used for direct fluorination of cellulose (62).

Bromocellulose

Bromination of cellulose in a homogeneous phase reaction using *N*-bromosuccinimide-triphenylphosphine, in LiBr–dimethylacetamide, was reported by Furuhata et al. to give (63) DS values of 0.9 for bromination. Interestingly, even for the highest degree of substitution, only the hydroxyl groups in the 6-position were replaced. *N*-Bromosuccinimide and triphenylphosphine were also employed for the heterogeneous bromination of cotton, but the bromine content attained was only 2.7% (64). The use of molecular bromine and red phosphorus as brominating agents leads to products with bromine contents of up to 16%. However, the homogeneity of the resulting products is questionable, because the products showed phosphorus and carbonyl absorptions when examined by IR spectroscopy (65).

Iodocellulose

A simple way to prepare iododeoxycellulose starting from chlorodeoxycellulose is the nucleophilic exchange of chlorine by sodium iodide (37, 38). Because of the strong nucleophilic character of the iodide ion, the chlorine atoms are substituted almost completely by NaI in 2,5-hexanedione to produce the corresponding iodocellulose. C-2 and C-3 substituted cellulose derivatives with DS >1.0 were also prepared by using different sulfonates (66). Similarly, the 6-O-trityl-2(3)-4-nitrobenzene sulfonate of cellulose was treated with sodium iodide in DMF to produce the corresponding iodo derivative.

Applications

By using halogenated cellulose as activated intermediates, amino acids can be covalently bound to cellulose (40). Also, by introducing iminodiacetic acid (45, 54, 56), thiourea (45), or amino groups (41–44), materials can be prepared that act as complexing ligands and are useful materials suitable for heavy metal enrichment. Similarly, thioethers of cellulose containing carboxyl groups can be prepared (12). Thiodeoxycellulose was obtained by reaction with sodium thiosulfate and subsequent treatment with mercaptoacetic acid (52). Hydrazo-deoxycellulose and cellulose derivatives carrying primary amino groups (*see* subsequent discussion) are also used as ion-exchangers and matrices for immobilization of biologically active compounds.

Silicon-Functionalized Cellulose

Silicon-functionalized cellulose derivatives recently attracted great attention because of their interesting properties. A salient feature of the trimethylsilyl group is its hydrophobic nature. Therefore, the introduction of this group to cellulose is used to modify cellulose fibers and tissues. However, the low hydrolytic stability of Si–O–C bonds restricts the practical application of such materials. This drawback can be overcome partially by the introduction of relatively long-chain organo-siloxane substituents.

Cellulose ethers containing silicon moieties of high hydrolytic stability were prepared by treating cellulose with α-chloro-ω-trimethylsilyl poly(DMSO) in pyridine (Scheme IIIA) (67). Cellulose ethers obtained in this work had a DS range of 0.2 for $n = 16$ to 0.52 for $n = 4$. By increasing length of the substituent, the

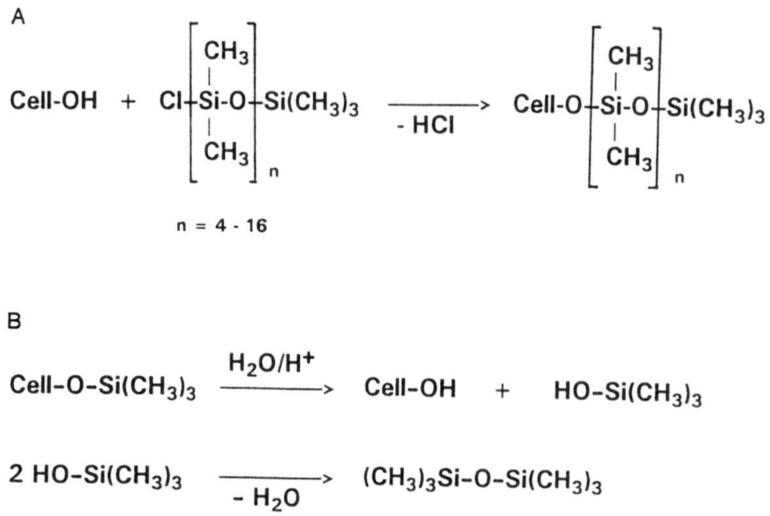

A

Cell-OH + Cl$-$Si$-$O$+$Si(CH$_3$)$_3$ $\xrightarrow{\text{- HCl}}$ Cell-O$+$Si$-$O$+$Si(CH$_3$)$_3$

n = 4 - 16

B

Cell-O-Si(CH$_3$)$_3$ $\xrightarrow{\text{H}_2\text{O/H}^+}$ Cell-OH + HO-Si(CH$_3$)$_3$

2 HO-Si(CH$_3$)$_3$ $\xrightarrow{\text{- H}_2\text{O}}$ (CH$_3$)$_3$Si-O-Si(CH$_3$)$_3$

Scheme III. Synthesis of Cell-poly(DMSO) and its acidic hydrolysis.

hydrolytic stability of the Si–O–C bond was enhanced, which is attributed to the shielding effect of the siloxane residue. For example, 90% of the Si–O–C bonds was hydrolyzed when trimethylsilylcellulose ($n = 0$) was boiled in water for 10 h, compared with only 24% for the derivative containing a long-chain substituent ($n = 16$) (14).

Trimethylsilylcellulose ethers are stable in cold water and in the absence of acid or base, but they easily hydrolyze in treatment with dilute aqueous acids (68, 69). The trimethylsilanol thus formed condenses to hexamethyldisiloxane (Scheme IIIB).

Trimethylsilylcellulose as an Intermediate

The solubility of trimethylsilylcellulose in all common solvents can be controlled to obtain products with low DS values soluble in strong polar solvents (e.g., DMSO) or with high DS values soluble in nonpolar solvents (e.g., n-hexane). The resulting trimethylsilyl products are stable, easy to handle, and are useful as intermediates for the preparation of regioselective esters under homogeneous conditions (72–78). Silylation in the presence of chlorotrimethylsilane–pyridine leads to soluble products with a DS > 1.9. In contrast, when using methylamine–DMSO or ammonia–DMF, the products with a DS range of 0.4–3.0 were completely soluble. Accordingly, trimethylsilylcellulose offers the possibility for using common solvents and reagents for the synthesis of cellulose derivatives in homogenous systems.

A survey of reaction conditions reported for synthesis of soluble trialkylsilylcelluloses is provided in Table III. Notably, the trimethylsilyl groups can be used as blocking or protecting groups as well as an activating group for attaining regioselectivity.

Applications

Because of the easy cleavage of the trimethylsilyl groups by dilute acids, trimethylsilylation can be used to produce regenerated cellulose in the form of spherical particles, layers, or membranes. Another feature of trimethylsilylcellulose is the formation of a lyotropic mesophase at a concentration of 27 wt% in dichloromethane. Thus, cellulosic materials with defined DS and distribution of substituents can be prepared for chromatography, polymeric supports for drugs, bioactive carriers, liquid-crystalline polymers, photoactive layers, and membranes. However, so far trimethylsilylcellulose has been used mainly as a soluble intermediate for the homogeneous esterification of cellulose and for the preparation of water-soluble sulfates and phosphates.

Preparation of Regenerated Cellulose Fibers

The sensitivity of silicon-containing cellulose toward diluted aqueous acids, leading to rapid hydrolysis of the Si–OEt bonds, can be used for the production of regenerated fibers (68–70) (trimethylsilyl procedure). Thus, a new route to the production of regenerated cellulose fibers is offered that is environmentally more friendly and economical. The relatively expensive reagent trimethylchlorosilane, as well as other reagents and solvents, can be recycled. The conditions for heterogeneous phase desilylation of trimethylsilylcellulose are given in Table IV.

Carboxylic Esters of Cellulose

For the esterification of cellulose, carboxylic acid chlorides in the presence of tertiary amines or pyridine conventionally are used. At relatively high tempera-

Table III. Experimental Conditions for the Synthesis of Trimethylsilylcellulose

Silylating Agent	Reaction Medium	Temp. (°C)	Time (h)	DS	Cell–TMS Soluble in	Ref.
TMS–Cl	Pyridine	110	3	2.5	Insoluble	81
TMS–Cl	Pyridine–xylene	110	4	2.46	Benzene	86
TMS–Cl	Pyridine–toluene	110	1	2.82	Toluene	85
TMS–Cl	Pyridine–petrolether	15	3	3	Hexane	83
TMS–Cl	Pyridine	20	3	2.65	THF	82
TMS–Cl	Pyridine	20	3	1.37	DMF	82
TMS–Cl	NH$_3$	–70	12	1.46	DMF	89
TMS–Cl	NH$_3$	–70	12	1.50	DMF	89
(TMS)$_2$NH–HCl	DMF	100	1–3	2.2–2.6	CH$_2$Cl$_2$	71
(TMS)$_2$NH–HCl$^-$	DMF	110	8	2.19	—	97
(TMS)$_2$NH	DMAc–LiCl	80	1	2.7–2.9	THF	95
(TMS)$_2$NH	Formamide	70	2	3.0	CHCl$_3$	93
(TMS)$_2$O	DMAc–LiCl	80	1.5	1.55	CH$_2$Cl$_2$	98
TMS–NH–Ac	Melt	170–180	6	2.95	Hexane	91

TMS–N=C–CH$_3$
     |
   TMS–O
	N-Methyl-pyrrolidone–xylene	150–160	5	2.73	Xylene	99

TMS–O–CO–NH–TMS	DMF	110	8	2.2	THF	70
TMS–O–CO–NH–TMS	THF	20	14	2.1	THF	70

Table IV. Reaction Conditions for the Heterogeneous Desilylation of TMS Cellulose

Initial DS	Solvent	Hydrolysis Medium	Time (h)	Final DS
2.92	THF	Water	24	1.38
2.92	THF	Water	8	2.80
1.55	DMF	Water	1	1.55
			4	1.46
0.88	DMF	Water	1	0.80
			4	0.35
2.92	THF	2% NaHCO$_3$	8	2.90
1.55	DMF	2% NaHCO$_3$	8	1.50
0.88	DMF	2% NaHCO$_3$	8	0.76
2.92	THF	0.1 N NaOH	8	2.25
1.55	DMF	0.1 N NaOH	8	0.58
0.88	DMF	0.1 N NaOH	8	0.15
2.92	THF	0.1 N HCl	10 min	0.02
1.55	DMF	0.1 N HCl	4 min	0.02
2.92	THF	0.1 N Pyridine/HCl	15 min	0.02
2.43	THF	0.1 N Pyridine/HCl	15 min	0.02

Note: DS is degree of substitution.

Source: Data are taken from reference 70.

tures, all remaining hydroxyl groups of the trimethylsilylethers are esterified (Scheme IV). 4-Dimethylaminopyridine acts as a catalyst for the synthesis of cellulose esters via trimethylsilylcellulose (78). The acylation of low molecular weight alcohols can be accelerated significantly by catalytic amounts of 4-dimethylaminopyridine (79). However, by applying this catalyst to the esterification of cellulose in the heterogeneous phase, only a slight improvement was observed (80). In contrast, by performing the reaction via the soluble trimethylsilylcellulose in the homogeneous phase, a considerable catalytic effect was observed (78). Table V lists the experimental data for acylation of trimethylsilylcellulose (70).

Cellulose Sulfate

In recent years, the preparation of water-soluble cellulose sulfates under homogenous conditions (in the presence of N$_2$O$_4$–DMF) was investigated. However, reactions with N$_2$O$_4$, either alone or with DMF, create toxicological and ecological problems. Therefore, sulfation of cellulose via trimethylsilylcellulose is an interesting alternative (72, 73, 76). A rapid conversion of trimethylsilylcellulose (DS, 1.55) to cellulose sulfate was observed when trimethylsilylcellulose was allowed to react with SO$_3$·MF (SO$_3$ content, 1.5–3.0 wt%) at room temperature in DMF (72). Under these conditions, the maximum DS of the sulfate halfester was attained after 30 min (Scheme V). The mechanism of sulfation is characterized by a selective attack on the trimethylsilyl group. The direct attack of SO$_3$ on the Si–O bond is an insert reaction leading to a cellulose trimethylsilyl sulfate, which then can be hydrolyzed with NaOH to the corresponding cellulose sulfuric acid halfester (Scheme V).

Cellulose Phosphate

Cellulose phosphates can be synthesized by reacting trimethylsilylcellulose (DS, 1.5–2.3) with phosphoryl

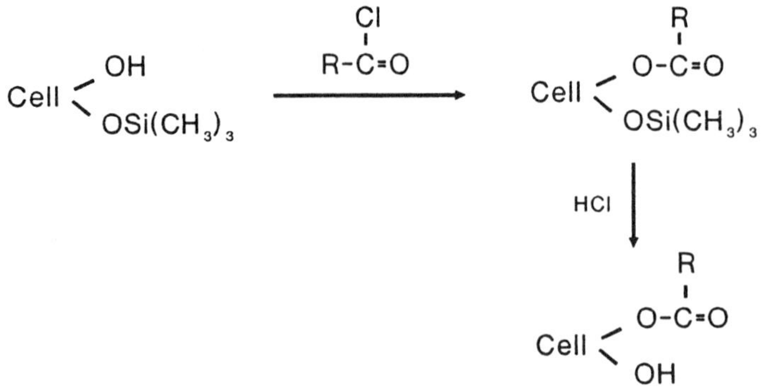

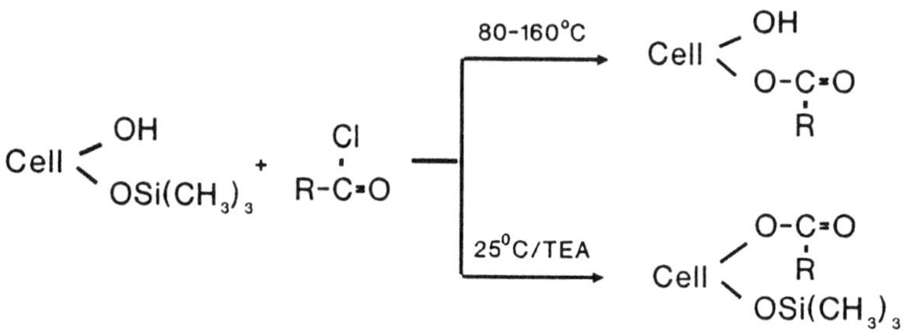

Scheme IV. Preparation of carboxylic esters of cellulose by acylation of Cell-trimethylsilane.

Table V. Experimental Data for Several Acylations of Cell–TMS by Carboxylic Acid Chlorides (R–COCl)

R in R–COCl	DS in Cell–TMS	Reaction Medium	Product DS (Si)	DS (Ester)
C_6H_5–CH=CH–	1.55	DMF–$N(C_2H_5)_3$	1.31	0.98
CH_3–$(CH_2)_{14}$–	1.55	DMF–$N(C_2H_5)_3$	1.27	1.05
Cl–CH_2–	1.55	DMF–$N(C_2H_5)_3$	1.28	1.55
4–NO_2–C_6H_4–	1.99	C_6H_6–$N(C_2H_5)_3$– DMAP	1.95	0.46
4–NO_2–C_6H_4–	1.99	C_6H_6–$N(C_2H_5)_3$– DMAP	0.02	0.43
4–NO_2–C_6H_4–	1.99	C_6H_6–$N(C_2H_5)_3$– DMAP	0.02	0.39

Note: DS is degree of substitution.

Source: Data are taken from reference 70.

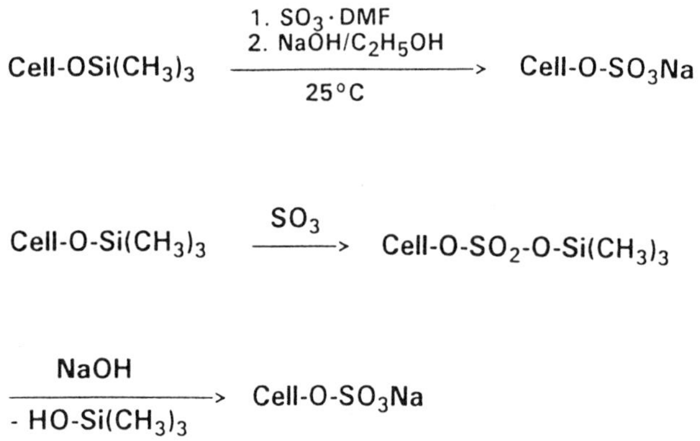

chloride in the presence of equimolar amounts of triethylamine at room temperature followed by hydrolysis (Scheme VI) (72). The resulting cellulose phosphates were noncross-linked and water soluble and had a DS range of 0.3–0.6.

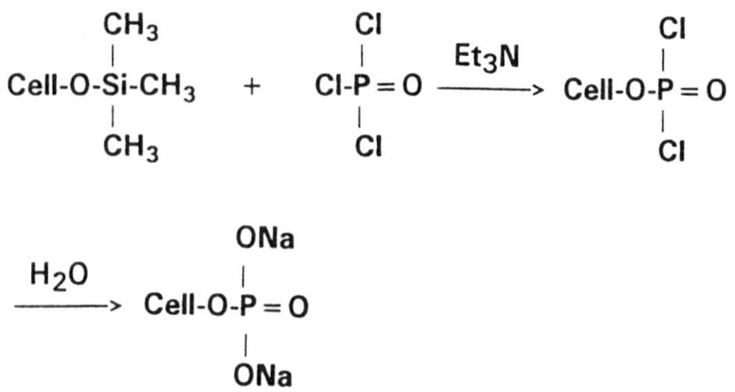

Silylation of Cellulose

The most frequently prepared and extensively studied silylethers of cellulose are the trimethylsilylcelluloses. They are produced by using chlorotrimethylsilane, N-trimethylsilylacetamide, hexamethyldisilazane, and

Scheme V. Preparation of Cell-sulfates via Cell-trimethylsilane by SO₃–DMF complex.

Scheme VI. Preparation of Cell-phosphate via Cell-trimethylsilane by phosphoryl chloride.

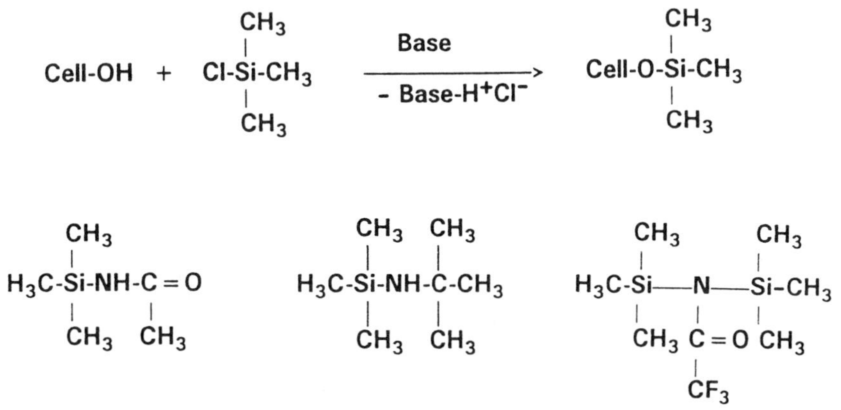

Scheme VII. Silylation of cellulose by chlorotrimethylsilane. Structures of other silylating agents for this reaction are also shown.

N,N-bis(trimethylsilyl)trifluoroacetamide (Scheme VII). Chlorotrimethylsilane is relatively inexpensive. It was allowed to react with cellulose in the presence of tertiary amines, usually pyridines (*81–86*) or ammonia (*87–90*). *N*-Trimethylsilylacetamide was used for the preparation of 2,3,6-tristrimethylsilylcellulose in the form of a melt. The reaction was carried out at 170–180 °C for 6 h (*91*). Hexamethyldisilazane was used to obtain trimethylsilylcellulose with DS values of 2.2–2.6. This reaction was carried out in DMF at 100 °C in the presence of catalytic amounts of chlorotrimethylsilane and pyridine for 3 h (*92*).

A completely trimethylsilylated cellulose derivative was obtained from hexamethyldisilazane heated for 2 h at 70 °C (*93*). The effect of hexamethylphosphoric acid triamide on the reaction of cellulose with this reagent was also studied (*94*). It is still questionable whether the derivatization of cellulose with chlorotrimethylsilane in liquid ammonia also proceeds via the intermediate hexamethyldisilazane (*87, 88*). On the other hand, hexamethyldisilazane is regarded as a low reactive silylating agent that needs to be activated by addition of catalytic amounts of chlorotrimethylsilane (*94*).

Homogeneous Phase Systems

For the preparation of highly substituted trimethylsilylated celluloses, Schempp et al. reported (*95*) the use of cellulose solution in dimethylacetamide–LiCl. With this solution, the reaction proceeds without chain degradation or side reactions under relatively mild reaction conditions (80 °C, 1 h). Interestingly, even celluloses with a degree of polymerization >> 2000 could be transformed into soluble products. Therefore, this method is suitable for the study of the molecular mass distribution of celluloses by gel permeation chromatography (*95*).

N,N-Bis(trimethylsilyl)trifluoroacetamide also allows access to homogeneous phase preparation of silylated cellulose under mild conditions. For example, cellulose dissolved in DMSO–paraformaldehyde was transformed with this reagent to silylated cellulose of DS 2.4 within 15 min at room temperature (*96*).

Concluding Remarks

Applications of functional cellulose derivatives have expanded considerably from their early use. Two dominant areas for further development are use of novel functional derivatives of cellulose for specific applications and access to new cellulose derivatives through derivatization reactions in the homogenous phase. The number of cellulose derivatives produced in solution (or in a homogenous phase) can be assumed to increase significantly because of their inherent advantages. Limitations in this direction could stem primarily from environmental aspects with regard to the use of organic solvents. From the ecological viewpoint, functional cellulose derivatives are important as renewable resource materials. The intelligent design of such derivatives for specific needs is a challenge for interdisciplinary research. In addition, expanded biomedical applications can be expected.

References

1. *Modified Celluloses;* Rowell, R. M.; Young, R. A., Eds.; Academic: Orlando, FL, 1978.
2. *Cellulose and Other Natural Polymer Systems: Biogenesis, Structure, and Degradation;* Brown, M. R., Jr., Ed.; Plenum: New York, 1982.
3. Daniel, J. R. In *Encyclopedia of Polymer Science Engineering,* 2nd Ed.; Kroschwitz, J. I., Ed.; Wiley: New York, 1985; Vol. 3, pp 90–123.
4. Doelker, E. *Adv. Polym. Sci.* **1993,** *107,* 199–265.
5. Geckeler, K. E.; Pillai, R. N.; Mutter, M *Adv. Polym. Sci.* **1981,** *39,* 65.
6. Bayer, E.; Schumann, W.; Geckeler, K. E. In *Renewable-Resource Materials;* Carraher, C. E., Jr.; Sperling, L. H., Eds.; Plenum: New York, 1986; p 115.

7. Krylova, R. G. *Russ. Chem. Rev.* **1987**, *56*, 175–189.
8. Anders, M.; Geckeler, K. E. *Abstracts of Papers*, International Conference on Advanced Polymer Materials, Dresden, Germany, 1993; p 169.
9. Schwenker, R. F., Jr.; Pascu, E. *Ind. Eng. Chem.* **1958**, *50*, 91–96
10. Ishii, T.; Ishizu, A.; Nakano, J. *Sen' i Gakkaishi* **1978**, *34*, T505–T509.
11. Vigo, T. L.; Daigle, D. J.; Welch, C. M. *Text. Res. J.* **1973**, *43*, 715–718.
12. Vigo, T. L.; Welch, C. M. *Text. Res. J.* **1970**, *40*, 109–115.
13. Ishizu, A. In *Wood and Cellulosic Chemistry*; Hon, D. N.-S; Shiraishi, N., Eds.; Marcel Dekker, New York, 1990; Chapter 16, pp 757–800.
14. Gal'braikh, L. S.; Rogovin, Z. A. In *High Polymers: Cellulose and Cellulose Derivates*; Ott, E.; Spurlin, H. M.; Grafflin, M. W.,Eds.; Wiley-Interscience: New York, 1971; Vol. 5, Part. 5.
15. Vigo, T. L. In *Encyclopedia of Polymer Science Engineering*, 2nd ed.; Kroschwitz, J. I., Ed.; Wiley: New York, 1985; Vol. 3, pp 124–139.
16. Sletkina, L. S.; Polyakov, A. I.; Rogovin, Z. A. *Vysokomol. Soedin.* **1965**, *7*, 199–204.
17. Klein, E.; Snowden, J. E. *Ind. Eng. Chem.* **1958**, *50*, 80–82.
18. Vigo, T. L.; Daigle, D. J.; Welch, C. M. *J. Polym. Sci. Part B: Polym. Lett.* **1972**, *10*, 397–406.
19. McCormick, C. L.; Callais, P. A. *Polymer* **1987**, *28*, 2317–2323.
20. McCormick, C. L.; Dawsey, T. R.; Newman, J. K. *Carbohydr. Res.* **1990**, *208*, 183–191.
21. Polyakov, A. I.; Rogovin, Z. A. *Vysokomol. Soedin.* **1963**, *5*, 11–17.
22. Ishizu, A.; Ishii, T.; Itoh, T.; Nakano, J. *Sen' i Gakkaishi* **1977**, *33*, T-91–T-94.
23. Croon, I.; Manley, R. St. J. In *Methods in Carbohydrate Chemistry*; Whistler, R. L.; BeMiller, J. N., Eds.; Academic: Orlando, FL, 1963; Vol. 3, pp 274–277.
24. Srivastava, H. C.; Harshe, S. N.; Gharia, M. M. *Text. Res. J.* **1972**, *42*, 150–154.
25. Ishii, T.; Ishizu, A.; Nakano, J. *Carbohydr. Res.* **1977**, *59*, 155–163.
26. Horton, D.; Luetzow, A. E.; Theander, O. *Carbohydr. Res.* **1973**, *26*, 1–19.
27. Vigo, T. L.; Welch, C. M. *Carbohydr. Res.* **1974**, *32*, 331–338.
28. Smits, J.; Van Grieken, R. *Angew. Makromol. Chem.* **1978**, *72*, 105–113.
29. Gennaro, M. C.; Sarzanini, C.; Mentast, E.; Baiocchi, C. *Talanta* **1985**, *32*, 961–966.
30. Vigo, T. L.; Daigle, D. J. *Carbohydr. Res.* **1972**, *21*, 369–377.
31. Krylova, R. G.; Usov, A. I.; Shashkov, A. S. *Bioorg. Khim.* **1981**, *7*, 1586–1593.
32. Furuhata, K.; Chang, H.; Aoki, N.; Sakamoto, M. *Carbohydr. Res.* **1992**, *230*, 151–164.
33. Krylova, R. G. *Russ. Chem. Rev.* **1987**, *56*, 175–189.
34. Bose, A. K.; Lal, B. *Tetrahedron Lett.* **1973**, 3937–3940.
35. Dimitrov, D. G.; Gal'braikh, L. S.; Rogovin, Z. A. *Cellul. Chem. Technol.* **1986**, *2*, 375–389.
36. Boehm, R. L. *J. Org. Chem.* **1958**, *23*, 1716–1720.
37. Nishioka, N.; Matsumoto, K.; Kosai, K. *Polym. J.* **1983**, *15*, 153–158.
38. Ishii, T. *Carbohydr. Res.* **1986**, *154*, 63–70.
39. Polyakov, A.; Rogovin, Z. A. *Vysokomol. Soedin.* **1963**, *5*, 147–149.
40. Simionescu, C. I.; Dumitriu, S.; Popa, M.; Dumitriu, M. *Polym. Bull.* **1986**, *16*, 319–325.
41. Nakamura, S.; Amano, M.; Saegusa, Y. In *Cellulose Structural and Functional Aspects*; Kennedy, J. F.; Phillips, G. O.; Williams, P. A., Eds.; John Wiley: New York, 1989; p 225.
42. Nakamura, S.; Amano, M.; Saegusa, Y.; Sato, T. *J. Appl. Polym. Sci.* **1992**, *45*, 265–271.
43. Priksane, A.; Prikulis, A. *Latv. PSR Zinat. Akad. Vestis, Kim. Ser.* **1990**, 489–493.
44. Smits, J.; Van Grieken, R. E. *Angew. Makromol. Chem.* **1978**, *72*, 105–113.
45. Mentasti, E.; Sarzanini, C.; Gennaro, M. C.; Porta, V. *Polyhedron* **1987**, *6*, 1197–1202.
46. Machida, S.; Sueyoshi, Y. *Angew. Makromol. Chem.* **1976**, *49*, 171–176.
47. Vigo, T. L. *Text. Res. J.* **1976**, *46*, 261–264.
48. Petrus, L.; Petrusova, M. *Chem. Pap.* **1986**, *40*, 519–22.
49. Furubeppu, S.; Kondo, T.; Ishizu, A. *Sen'i Gakkaishi* **1991**, *47*, 592–597.
50. Polyakov, A. I.; Rogovin, Z. A. *Vysokomol. Soedin.* **1963**, *5*, 11–17.
51. Vigo, T. L.; Danna, G. F.; Welch, C. M. *Carbohydr. Res.* **1975**, *44*, 45–52.
52. Gemeiner, P.; Zemek, J. *Collect. Czech. Chem. Commun.* **1981**, *46*, 1693–1700.
53. Srivastava, H. C.; Harshe, S. N.; Gharia, M. M. *Text. Res. J.* **1972**, *40*, 150–154.
54. Priksane, A.; Prikulis, A.; Isaeva, D. A. *Latv. PSR Zinat. Akad. Vestis, Kim. Ser.* **1990**, 494–497.
55. Usov, A.I.; Krylova, R. G.; Suleimanova, F. R. *Izv. Akad. Nauk. SSSR Ser. Khim.* **1977**, 2158–2160.
56. Chan, W. H.; Lam-Leung, S. Y.; Fong, W. S.; Kwan, F. W. *J. Appl. Polym. Sci.* **1992**, *46*, 921–930.
57. Malm C. J.; Tanghe, L. J.; Laird, B. C. *J. Am. Chem. Soc.* **1948**, *70*, 2740–2747.
58. Sletkina, L. S.; Rogovin, Z. A. *Vysokomol. Soedin. Ser. B* **1967**, *9*, 37–39.
59. Usov, A. I.; Krylova, R. G.; Suleimanova, F. R. *Izv. Akad. Nauk. SSSR Ser. Khim.* **1975**, 2122–2123; *Izv. Akad. Nauk. SSSR Ser. Khim.* **1977**, 2158–2160.
60. Gorbunov, B. N.; Nazarov, A. A.; Protopopov, P. A.; Khardin, A. P. *Vysokomol. Soedin. Ser. A* **1972**, *14*, 2527–2530.
61. Frazier, Ch. E.; Glasser, W. G. *Polym. Prepr. (Am. Chem. Soc., Div. Polym. Chem.)* **1990**, *31*, 634.
62. Middleton, W. J. US. Patent 3,976,691, 1976.
63. Furuhata, K.; Kogonei, K.; Chang, H-S.; Aoki, N.; Sakamoto, M. *Carbohydr. Res.* **1992**, *230*, 165–177.
64. Ziderman, I. I. *Text. Res. J.* **1981**, *51*, 777–781.
65. Rath, H.; Rau, J. H.; Brink, G. *Melliand Textilber.* **1966**, *47*, 909–912.
66. Nikologorskaya, L. G.; Gal'braikh, L. S.; Kozlova; Y. S.; Rogovin, Z. A. *Vysokomol. Soedin.* **1970**, *12*, 2762–2767.
67. Ivanov, N. V.; Rogovin, Z. A.; Andrivanov, K. A.; *Zellulose und ihre Derivate*; Verlag der Akademie der Wissenschaften der UDSSR: Moskau, Russia, 1963; pp 44–47.
68. Greber, G.; Paschinger, O. *Proceedings of the 5th International Dissolving Pulps Conference*; Technical Association

of the Pulp and Paper Industry: Atlanta, GA, 1980; pp 94–99.

69. Greber, G.; Paschinger, O. *Papier* **1981**, *35*, 547–554.
70. Stein, A. Ph. D. Thesis, University of Jena, 1991.
71. Cooper, G. K.; Sandberg, K. R.; Hinck, J. F. *J. Appl. Polym. Sci.* **1981**, *26*, 3827–3836.
72. Klemm, D.; Schnabelrauch, M.; Stein, A.; Philipp, B.; Wagenknecht, W.; Nehls, I. *Papier* **1990**, *44*, 624–632.
73. Klemm, D.; Schnabelrauch, M.; Stein, A.; Heinze, T.; Erler, U.; Vogt, S. *Papier* **1991**, *45*, 773–778.
74. Stein, A.; Klemm, D. *Makromol. Chem., Rapid Commun.* **1988**, *9*, 569–573.
75. Klemm, D.; Schnabelrauch, M.; Stein, A. *Makromol. Chem.* **1990**, *191*, 2985–2991.
76. Wagenknecht, W.; Nehls, I.; Stein, A.; Klemmm, D.; Philipp, B. *Acta Polym.* **1992**, *43*, 266–269.
77. Horton, D.; Lehmann, J. *Carbohydr. Res.* **1978**, *61*, 553–556.
78. Klemm, D.; Schumann, P.; Hartmann, M. *Z. Chem.* **1984**, *24*, 62.
79. Höfle, G.; Steglich, W.; Vorbrüggen, H. *Angew. Chem.* **1978**, *90*, 602–615.
80. Philipp, B.; Fanter, C.; Wagenknecht, W.; Hartmann, M; Klemm, D.; Geschwend, G.; Schuhmann, P. *Cellul. Chem. Technol.* **1983**, *17*, 341–353.
81. Schuyten, H. A; Weaver, H. A.; Reid, I. D.; Jürgens, J. F. *J. Am. Chem. Soc.* **1948**, *70*, 1919–1920.
82. Greber, G.; Paschinger, O. Fr. Patent 2,477,158, 1981.
83. Keilich, G.; Tihlarik, K.; Husemann, E. *Makromol. Chem.* **1968**, *120*, 87–95.

84. Klebe, J. F.; Finkbeiner, H. L. *J. Polym. Sci. Part.A1* **1969**, *7*, 1947–1958.
85. Klebe, J. F. U.S. Patent 3,418,312, 1968.
86. Klebe, J. F.; U.S. Patent 3,418,313, 1968.
87. Greber, G.; Paschinger, O. *Proceedings of the 5th International Dissolving Pulps Conference*; Technical Association of the Pulp and Paper Industry: Atlanta, GA, 1980; pp 94–99.
88. Greber, G.; Paschinger, O. *Papier* **1981**, *35*, 547.
89. Greber, G.; Paschinger, O. Fr. Patent 2,477,157, 1981.
90. Stein, A.; Wagenknecht, W.; Klemm, D.; Philipp B.; Ger. (East) Patent 298,644, 1992.
91. Bredereck, K.; Strunck, K.; Menrad, H. *Makromol. Chem.* **1969**, *126*, 139–146.
92. Cooper, G. K.; Sandberg, K. R.; Hinck, J. F. *J. Appl. Polym. Sci.* **1981**, *26*, 3827–3836.
93. Harmon, R. E.; De, K. K.; Gupta, S. K. *Carbohydr. Res.* **1973**, *31*, 407–409.
94. Nagy, J.; Borbély-Kuszmann, A.; Becker-Pálossy, K.; Zimonyi-Hegedüs, E. *Makromol. Chem.* **1973**, *165*, 335–338.
95. Schempp, W.; Krause, T.; Seifried, U.; Koura, A. *Das Papier* **1984**, *38*, 607–610.
96. Shiraishi, N.; Miyagi Y.; Yamashita, S.; Yokota, T.; Hayashi, Y. *Sen'i Gakkaishi* **1979**, *35*, T466–T478.
97. Green, J. G. U.S. Patent 4 390 692, 1983.
98. Pawlowski, W. P.; Sankar, S. S.; Gilbert, R. D. *J. Polym. Sci. Part A* **1987**, *25*, 3355.
99. Finkbeiner, H. L.; Klebe, J. F. U.S. Patent 3 432 488, 1969.

Chitin and Chitosan Derivatives

Keisuke Kurita

Recent progress of chitin chemistry and applications as well as some basic aspects of chitin as a functional material are reviewed. Among many kinds of polysaccharides, chitin and the derived chitosan are attracting much attention due to their useful properties and their abundance in nature. Although structurally similar to cellulose, chitin and chitosan are amino polysaccharides and thus have much higher potential for the development of sophisticated functional polymeric materials. To this end, special attention has been focused on the development of modification reactions and on the structure–property relationship of the resulting modified products. As a result, significant achievements have been made in both basic research and application of chitin and chitosan in the past decade.

Natural polymers are classified into four categories: proteins, nucleic acids, polysaccharides, and rubbers. Of these compounds, nucleic acids and certain proteins participate in various biologically important activities. In contrast, polysaccharides have been considered to be less significant in terms of biological functions, and they are regarded primarily as structural materials and suppliers of water and energy. However, they are attracting increasingly more attention from various viewpoints as their inherent properties and biological functions are better understood.

Many kinds of polysaccharides are produced in large quantities in nature, but cellulose boasts of the largest production. In fact, it is the most abundant organic compound on earth, and chitin is the next. Cellulose is synthesized mainly in plants, whereas chitin is synthesized mainly in lower animals. Alginic acid, agar, and carrageenan are also important biological resources produced in marine algae. There are many other polysaccharides such as amylose, dextran, pullulan, curdlan, xanthan, hyaluronic acid, and polygalactosamine. A great deal of interest has been aroused in these polysaccharides as specialty polymeric materials having specific properties such as biodegradability, biocompatibility, and bioactivity.

They have high potential for developing sophisticated functional polymers quite different from those of synthetic polymers.

Among polysaccharides, cellulose and chitin are the most important biopolymers. Although cellulose has been studied extensively, only limited attention has been paid to chitin. Despite its huge annual production of about 1×10^{10} to 1×10^{11} tons, chitin remains an almost unused biomass resource primarily due to its intractable bulk structure.

Chitin is structurally similar to cellulose, but it is an amino polysaccharide having acetamide groups at the C-2 positions in place of hydroxyl groups (Chart I). Deacetylation transforms chitin into chitosan, which has free amino groups. Chitin is thus expected to have higher potential than cellulose, and chitin is becoming increasingly important as a natural resource and a novel functional polymeric material.

Special emphasis has been put on the chemical modifications of chitin to explore its full potential. Modifications of chitin are, however, generally difficult due to lack of solubility. Therefore, many efforts have focused on exploration of facile modification reactions that may open a way to use of chitin. As a result, recent progress of basic chitin chemistry is note-

3469–8/97/0239$15.25/0 © 1997 American Chemical Society

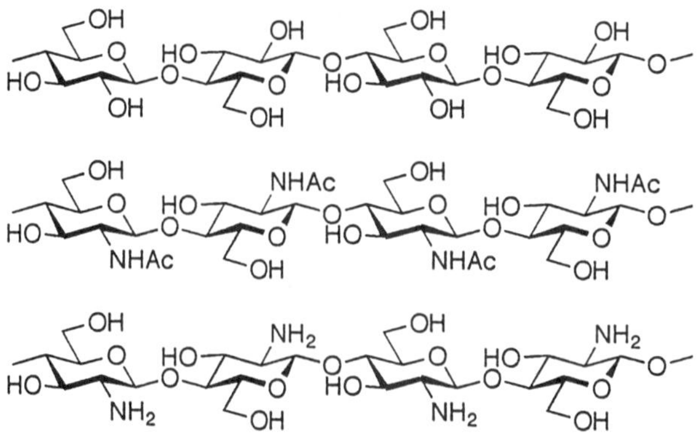

Cellulose

Chitin

Chitosan

Chart I. Molecular structures of cellulose, chitin, and chitosan.

worthy. Studies on its applications are also quite encouraging, and several products have been put on the market recently. On the basis of the rapidly growing interests in chitin, two books were published (*1, 2*).

This chapter deals with the fundamental aspects of chitin and recent progress of modifications leading to the development of novel advanced functions. The resulting characteristic properties will be discussed in relation to the perspectives for practical applications.

Sources of Chitin and Chitosan

Although chitin is found widely in nature (from fungi and algae to lower animals), arthropod shells (exoskeltons) are the most easily accessible sources of chitin. These shells contain 20–50% chitin on a dry weight basis, and proteins and calcium carbonate are two other major components. From a practical viewpoint, shells of crustaceans such as crabs and shrimps conveniently are available as waste from seafood processing industry and are used for commercial production of chitin.

Squid pens also contain a form of chitin that is classified as β-chitin in comparison with the ordinary α-chitin in crustacean shells. β-Chitin is quite attractive as another form of chitin having peculiar characteristics considerably different from those of α-chitin, and the chemistry is rapidly advancing. However, β-chitin is not yet produced commercially. Other possible sources for practical chitin production include krill, clams, oysters, fungi, and insects. Cell walls of some fungi (Zygomycetes) contain chitosan as well as chitin and may be used as sources of chitosan.

Preparation

Chitin

The most common sources of chitin are crab and shrimp shells. These shells also contain other substances such as proteins and pigments and calcium

carbonate necessary for hardening the shells. Furthermore, chitin is assumed to have polypeptide side chains linked directly to some of the C-2 amino groups through amide bonds.

Chitin is not soluble in ordinary solvents and hence cannot be isolated by extraction. Therefore, it is obtained as the residue remaining after decomposing other shell components with acid and alkali. Shells are first treated with dilute hydrochloric acid to remove calcium carbonate. The decalcified shells are then pulverized and heated in 1–2 M sodium hydroxide to decompose proteins. To ensure removal of proteins and pigments, the alkaline treatment is repeated a few more times. If necessary, the resulting product is again treated with dilute hydrochloric acid and alkali and washed with deionized water until neutral. On drying, chitin is obtained as an almost colorless material. The yield is 30–35% based on dried shrimp shells (*3*).

A simpler procedure is used for the isolation of β-chitin from squid pens, because the molecular arrangement of β-chitin is less tight. Moreover, pens are almost exclusively composed of chitin and proteins with only trace amounts of metal salts (*4*). During the isolation procedure, some acetyl groups are removed, and the resulting chitin has a degree of deacetylation of 0.08–0.15.

Because of the insoluble nature of chitin, colloidal chitin is often used as a substrate to assay chitin-degrading enzymes and as a carbon and nitrogen source for chitin-digesting microorganisms. It is prepared by dissolving chitin in hot concentrated hydrochloric acid followed by pouring the solution into water. It is obtained as a fine powder and can be dispersed in water. However, the molecular weight is lowered as a result of degradation by hydrochloric acid.

Chitosan

Chitosan itself occurs in nature in some fungi and can be isolated from their cell walls. Further studies, however, seem to be necessary for practical production of chitosan, although the feasibility has been demonstrated (5). Chitosan is considered to be formed by the action of chitin deacetylase on chitin.

Chitosan is usually prepared by deacetylating α-chitin with 40–50% aqueous alkali at 120–160 °C under heterogeneous conditions. These conditions lead to a degree of deacetylation of 70–95%. Complete deacetylation is possible by repeating alkaline treatment (6, 7). To prepare chitosan from β-chitin isolated from squid pens, milder conditions (40% sodium hydroxide at 80 °C) are applied to suppress discoloration (4).

Water-Soluble Chitin

When chitin is steeped in concentrated aqueous sodium hydroxide and treated with crushed ice, an alkaline chitin solution results. Under these conditions, deacetylation proceeds efficiently. An alkali chitin solution in 10% sodium hydroxide left for 70 h at room temperature gives a product with about 50% deacetylation that is soluble in neutral water (3, 8). Higher or lower deacetylations fail to lead to complete solubilization, and the degree of deacetylation should be 0.45–0.55 for water-solubility (*see* subsequent discussion). Alternatively, random N-acetylation of chitosan to a degree of acetylation of around 0.5 gives rise to a water-soluble product (9).

Chitin Oligomers

Chitin and chitosan backbones are hydrolyzed with hot hydrochloric acid to afford D-glucosamine. Degradation under milder conditions affords mixtures of oligomers consisting of N-acetyl-D-glucosamine and D-glucosamine, respectively (Scheme I). The resulting oligomers are separated by column chromatography. Recent size exclusion chromatography enabled the separation of 15 oligomers from a chitosan hydrolyzate (10). D-Glucosamine oligomers can be converted into N-acetyl-D-glucosamine oligomers by selective N-acetylation. These two series of oligomers up to hexamers or heptamers are available commercially.

Enzymatic degradation of chitin is another way to prepare oligomers. Certain chitin-digesting bacteria, *Vibrio anguillarum* E-383a and *Bacillus licheniformis* X-7u, degrade colloidal chitin to give di-*N*-acetylchitobiose ($(GlcNAc)_2$) at 40–50% yield (11). Degradation of chitosan with a chitosanase from *Bacillus* sp. No. 7-M gives the corresponding dimer to pentamer without producing the monomer (12). A cellulase of *Trichoderma viride* hydrolyzes chitosan and leads to the formation of the hexamer to octamer, which can be isolated in crystalline forms (13).

Some chitinolytic enzymes such as chitinases and lysozyme have high transglycosylation activity and thus catalyze the formation of oligosaccharides from smaller oligosaccharides. With a chitinase, for example, $(GlcNAc)_4$ gives $(GlcNAc)_6$ and $(GlcNAc)_2$ in 40 and 56% yields, respectively, and with lysozyme, $(GlcNAc)_2$ gives $(GlcNAc)_6$ and $(GlcNAc)_7$ (14).

Acetolysis of chitin affords oligomers in the form of peracetates. The peracetate of $(GlcNAc)_2$ is prepared from α-chitin at generally <10% yield, but the yield is improved to 16% with powdered colloidal chitin (15). In contrast, β-chitin gives the dimer at 17% yield without any special pretreatment because of the higher solubility of this chitin variety (16).

Properties

Crystalline Structure

Chitin is a linear polysaccharide consisting of β-1,4-linked *N*-acetyl-D-glucosamine. On the basis of its crystalline structures, chitin is classified into three forms: α-, β-, and γ-chitins. The most abundant form is α-chitin, where the molecules are aligned in an antiparallel fashion (Chart II) (17). This molecular arrangement is favorable for the formation of strong intermolecular hydrogen bonding, and α-chitin is the most stable form of the three.

In β-chitin, the molecules are packed in a parallel way, resulting in a weaker intermolecular force. β-Chitin is therefore less stable than α-chitin and can be transformed into α-chitin. γ-Chitin is considered to be

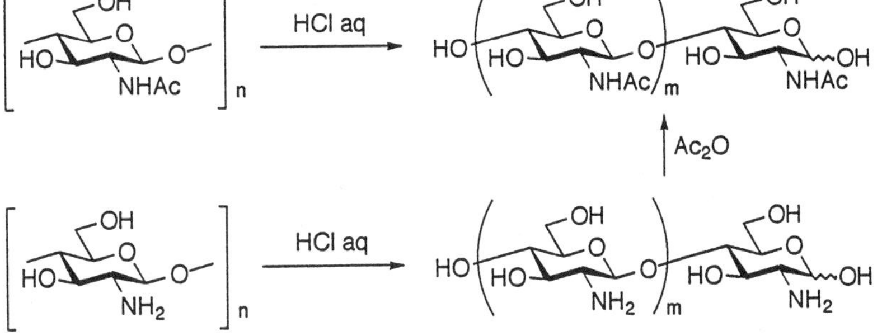

Scheme I. Preparation of chitin and chitosan oligomers.

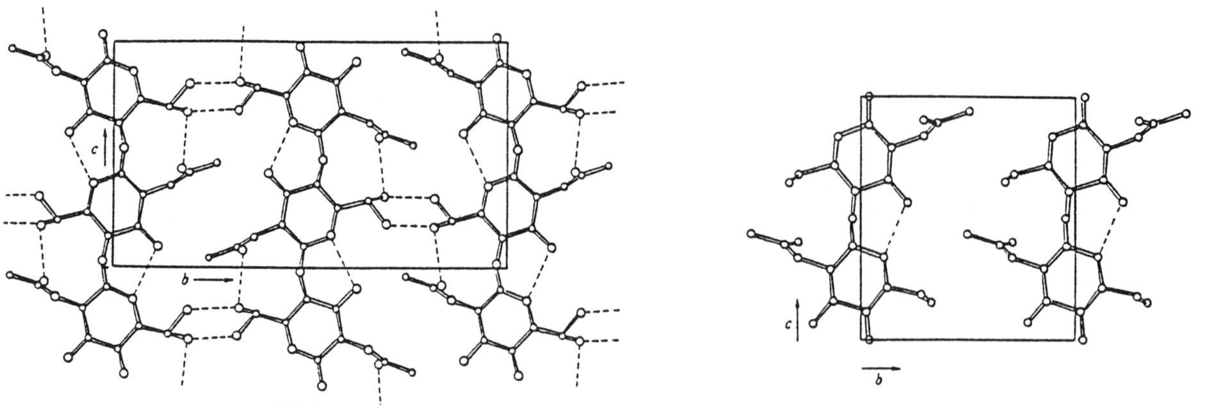

α-Chitin   β-Chitin

Chart II. Molecular arrangement of α- and β-chitins (adapted from reference 17).

a mixture of α- and β-forms and has both parallel and antiparallel arrangements.

Degree of N-Acetylation

Not all the C-2 amino groups of chitin are considered to be acetylated in nature, and free amino groups are present to some extent. Moreover, deacetylation takes place during isolation. Therefore, chitin samples may have widely different degrees of deacetylation depending on their origin and mode of isolation. To prepare chitin with a uniform structure (i.e., poly(N-acetyl-D-glucosamine)), selective N-acetylation of the water-soluble chitin (18) or β-chitin (19) is necessary.

The degree of deacetylation is thus one of the most important factors for specifying chitin. Although many methods were reported for assessing the degree of deacetylation, each method has some merits and demerits (2); acid–base and colloid titration (3), conductometric titration (20), and IR spectroscopy are relatively easy and reliable. For IR spectroscopic determination, the absorbance ratio of A_{1550}/A_{2878} was first suggested (21), but several other ratios [A_{1655}/A_{3450} (22) and A_{1655}/A_{2867} (23)] were reported.

Solubility

The ordinary chitin (α-chitin) is not soluble and does not swell appreciably in common organic solvents. It is soluble only in special solvents such as N,N-dimethylacetamide (DMAc) containing lithium chloride, although the extent of solubility is dependent on the sources of chitin (24). A mixture of DMAc and N-methyl-2-pyrrolidone (NMP) containing 5–8% lithium chloride frequently is employed to cast films (25). Hexafluoroacetone and hexafluoro-2-propanol are also good solvents for chitin.

In sharp contrast to the intractable nature of α-chitin, β-chitin swells highly in water because of weak intermolecular hydrogen bonding. β-Chitin is even

soluble in formic acid. Although the water-soluble chitin (see previous discussion) is soluble in neutral water, chitosan is soluble only in aqueous acidic solutions such as acetic and hydrochloric acids. The extent of solubility depends on the acid, and some acids such as phosphoric and sulfuric acids are not suitable for solubilization.

Hydrophilicity

Like other polysaccharides, α-chitin is hygroscopic. β-Chitin shows much higher hygroscopicity because of the loose arrangement of its molecules and the cotton-like fluffy state compared with the more solid state of α-chitin. Chitosan derived from α-chitin shows higher hydrophilicity than α-chitin, but the reverse is true of the chitosan derived from β-chitin (Figure 1) (4). The water-soluble chitin is highly hygroscopic and almost like hyaluronic acid.

Biodegradability

Both chitin and chitosan are degraded in nature by many varieties of microorganisms. Most of the chitinases in microorganisms hydrolyze N-acetyl-β-1,4-glucosaminide linkages randomly. Chitinases also are present in higher plants, even though plants do not have chitin as a structural component. This characteristic may be related to the self-defense activity of plants against pathogenic microbes and insects that have chitin. Chitinases have been studied extensively, but not much is known about chitosanases.

The degree of deacetylation of chitin affects its biodegradability, and the sample with a degree of deacetylation of 0.7 is most susceptible to lysozyme (26). Introduction of mercapto groups also enhances susceptibility to lysozyme (27). These results are interpreted in terms of the destruction of crystalline structures. O-Carboxymethyl-chitin, which has the substituents mainly at C-6, is degraded by lysozyme more

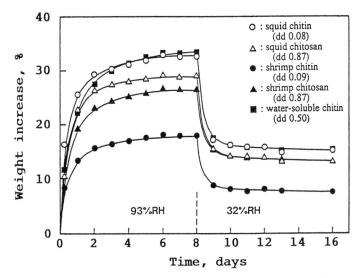

Figure 1. Hygroscopic characteristics of chitins and chitosans (adapted from reference 4).

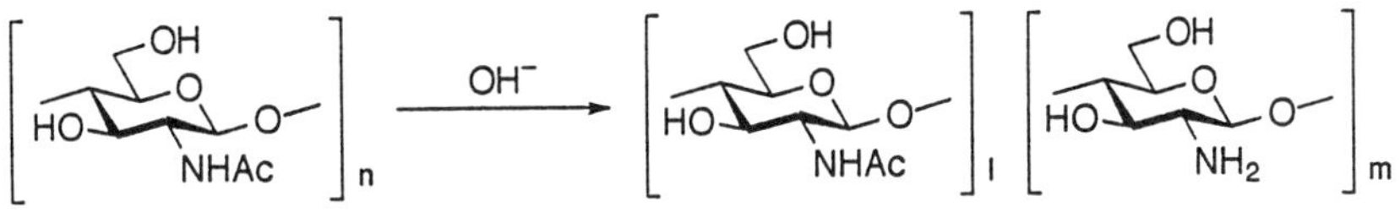

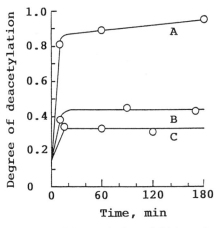

Scheme II. Deacetylation of chitin.

readily than the original chitin, but the degradation may be slower if the carboxymethyl group is at C-3 (*28*). Contradictory results were reported about the susceptibility of *N*-acylchitosans to chitinase, but certain derivatives having small substituents appear to be hydrolyzed more readily than chitin itself (*29*).

Modifications

Because chitin and chitosan have strong intermolecular forces and highly crystalline structures, they are insoluble in most organic solvents and much less accessible to potential reactants than cellulose. Therefore, modifications of chitin and chitosan generally have been performed under heterogeneous conditions, with a few exceptions (*30*). However, more efficient modification reactions are being developed that use chitosan, β-chitin, the water-soluble chitin, and organosoluble derivatives; reported modifications include acylation, tosylation, and tritylation.

Deacetylation

Deacetylation of chitin is effected with strong aqueous bases such as sodium and potassium hydroxides (Scheme II). The reaction is carried out by heating a suspension of chitin flakes or powders in the alkali at 120–160 °C to give chitosan. When deacetylation is conducted with dilute alkali (20 or 30% sodium hydroxide) at gentle reflux, the degree of deacetyla-

Figure 2. Progress of deacetylation of chitin under heterogeneous conditions: (A) 40% NaOH at 130 °C; (B) 30% NaOH at 112 °C; (C) 20% NaOH at 105 °C (adapted from reference 31).

tion levels off early (Figure 2) (*31*). This result suggests that deacetylation under these conditions proceeds preferentially in amorphous regions of chitin, and the products are supposed to be block-type copolymers composed of *N*-acetyl-D-glucosamine and D-glucosamine. As shown in Figure 2, deacetylation with 40% sodium hydroxide at 130 °C is quite satisfactory, and the deacetylation degree increases with time without leveling off. During the deacetylation process, however, degradation of the main chain occurs as evident by considerable decrease in the molecular weight listed in Table I (7, 32).

β-Chitin shows much higher reactivity than α-chitin in deacetylation (Figure 3) because of relatively weak intermolecular hydrogen bonding, but β-chitin tends to discolor during the reaction. Deacetylation at 80 °C produces almost colorless chitosan (4).

Deacetylation proceeds more efficiently in homogeneous alkali chitin solution (Figure 4), and the kinetics were examined (33). Under these homogeneous conditions, deacetylation was considered to take place randomly along the main chain, and the samples with about 50% deacetylation were water-soluble (3, 8). Partially deacetylated chitins with degrees of deacetylation <0.45 or >0.55 showed lower solubility as indicated by regions other than C in Figure 4 (31). The water solubility of the samples with a degree of deacetylation about 0.5 is ascribable to the greatly enhanced hydrophilicity due to the random distribution of acetyl groups at half of the amino groups. Interestingly, samples with the same degree of deacetylation prepared by conventional heterogeneous hydrolysis are not soluble owing to the blockwise distribution of the acetyl groups. The difference in distribution of acetyl groups was confirmed by physical data and lysozyme hydrolysis (34) and by NMR spectroscopy (35). The water-soluble chitin can also be prepared by random N-acetylation of chitosan to an extent of about 0.5 (9). The water-soluble chitin is useful as a reactive precursor for further modification, because the reactions can be performed in homogeneous aqueous solution or in a highly swollen state in organic solvents.

Acylation

Because α-chitin is not soluble in most solvents suitable for acylation, extensive acetylation is attained only under rather severe conditions (e.g., with acetic anhydride and hydrogen chloride). Full acetylation is possible in methanesulfonic acid with acetic anhydride (Scheme III) (36). The reaction mixture is heterogeneous initially, but the acetylated derivative goes into solution as the degree of acetylation increases. A mixture of trichloroacetic acid and 1,2-dichloroethane dissolves chitin, and so acetylation proceeds in solution (37). These solvent systems, however, are strongly acidic and cause degradation of the backbone. DMAc containing lithium chloride also allows reactions in solution, and acylated or carbamoylated chitins were prepared (38) as models for controlled-release herbicides.

Although β-chitin is also not soluble in common solvents, it swells considerably in many solvents including methanol. Free amino groups present in β-chitin can thus be fully acetylated directly in methanol with acetic anhydride to give structurally uniform chitin, poly(N-acetyl-D-glucosamine) (Scheme III) (19). α-Chitin, however, does not undergo acetylation under similar conditions. When β-chitin is treated with acetic anhydride in pyridine in the presence of 4-dimethylaminopyridine, fully acetylated chitin is prepared quite readily, and this result indicates much

Table I. Molecular Weight Change During Alkaline Treatment of Chitin

Time (h)	M_w ($\times 10^3$)	M_n ($\times 10^3$)	M_w/M_n
0.5	1487 ± 98	322 ± 27	4.63 ± 0.18
1.0	1142 ± 94	232 ± 21	4.93 ± 0.25
2.0	925 ± 63	186 ± 8	4.98 ± 0.19
3.0	846 ± 24	177 ± 7	4.79 ± 0.15
4.0	775 ± 61	161 ± 12	4.81 ± 0.09
5.0	667 ± 26	139 ± 8	4.80 ± 0.16

Note: Treatment performed in 50% NaOH at 100 °C. M_w and M_n are weight- and number-average molecular weights, respectively.

Source: Reproduced with permission from reference 32. Copyright 1978.

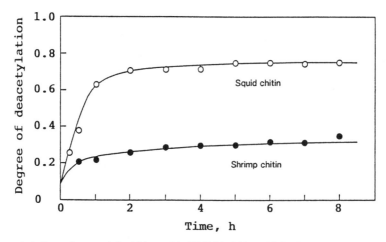

Figure 3. Deacetylation of α- and β-chitins with 30% NaOH at 100 °C (adapted from reference 4).

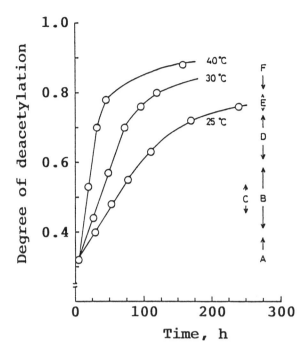

Figure 4. Deacetylation of chitin under homogeneous conditions: A and E, precipitation on dilution; B, no precipitation on dilution or neutralization; C, isolated sample is soluble in water; D, precipitation on neutralization; F, gelation (adapted from reference 31).

higher potential of β-chitin as a versatile starting material for chemical modifications (*19*).

Chitosan is soluble in aqueous acetic acid and does not precipitate on dilution with an equal amount of methanol; therefore, N-acylation of chitosan is facile. With carboxylic acid anhydrides or chlorides, acylation proceeds smoothly at the free amino groups and then more slowly at hydroxyl groups (*39*), but gelation is usually observed during reaction as a result of reduced solubility (*40, 41*). Chitosan is also acylated with long chain aliphatic acid chlorides in pyridine–chloroform, and a soluble derivative with a degree of acylation of 4 was obtained (*42*).

The water-soluble chitin (*see* previous discussion) is used conveniently as a starting material for modifi-

cation, because this chitin is soluble in neutral water and also forms a highly swollen gel on pouring the aqueous solution into organic solvents suitable for the reactions. The water-soluble chitin is thus selectively N-acetylated in homogeneous solution with acetic acid and dicyclohexylcarbodiimide to give fully N-acetylated chitin (*18*). With a highly swollen water-soluble chitin gel in pyridine or DMAc, many modification reactions can be conducted in almost homogeneous solutions under mild conditions. An example is acylation with aromatic anhydrides such as phthalic, trimellitic, and pyromellitic anhydrides. In these cases, acylation followed by dehydration gives rise to the formation of the imide derivatives (Scheme IV). The phthaloyl derivative was soluble in dimethyl sulfoxide (DMSO). The trimellitic and pyromellitic derivatives were reactive toward nucleophiles due to the presence of carboxyl or acid anhydride groups (*43*).

To further improve solubility in organic solvents, N-phthaloylation of 100% deacetylated chitosan has been examined in detail. The reaction is accomplished by treating a chitosan suspension in *N,N*-dimethylformamide (DMF) with excess phthalic anhydride at 130 °C. The resulting fully N-phthaloylated chitosan proved to be readily soluble in polar solvents such as pyridine, DMF, DMAc, and DMSO (*44*). Also, N-phthaloylation proceeds more readily with chitosan derived from β-chitin than that from α-chitin, a result indicating that the difference in crystalline structures of chitins affects their reactivity even after deacetylation (*45*).

With *N*-phthaloyl-chitosan, various modification reactions proceed quantitatively and regioselectively in homogeneous solutions in organic solvents. Tritylation, for example, quantitatively takes place at C-6 hydroxyl groups in pyridine. Several other reactions also were confirmed to be facile and quantitative compared with the conventional sluggish reactions on chitin (*46*). These procedures based on *N*-phthaloyl-chitosan are applicable for the preparation of various derivatives with well-defined structures. Some typical examples are shown in Scheme V.

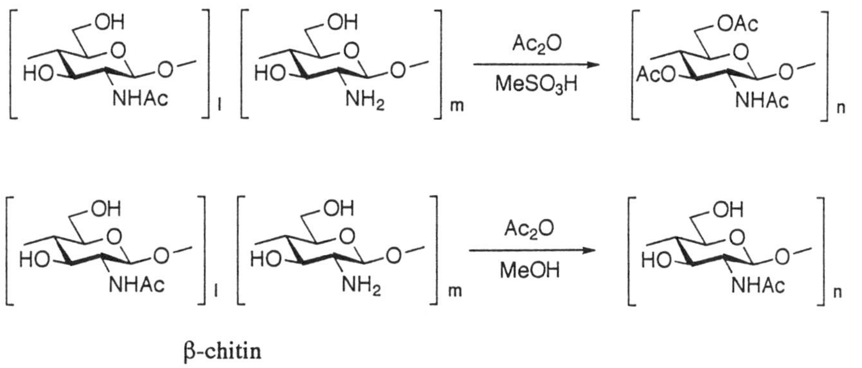

β-chitin

Scheme III. Peracetylation and N-acetylation of partially deacetylated chitins.

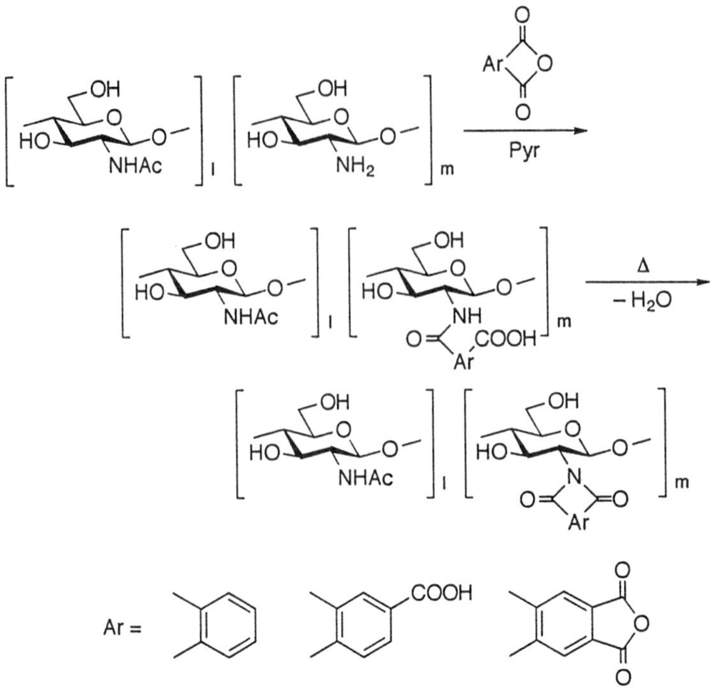

Scheme IV. Acylation of the water-soluble chitin with cyclic acid anhydrides.

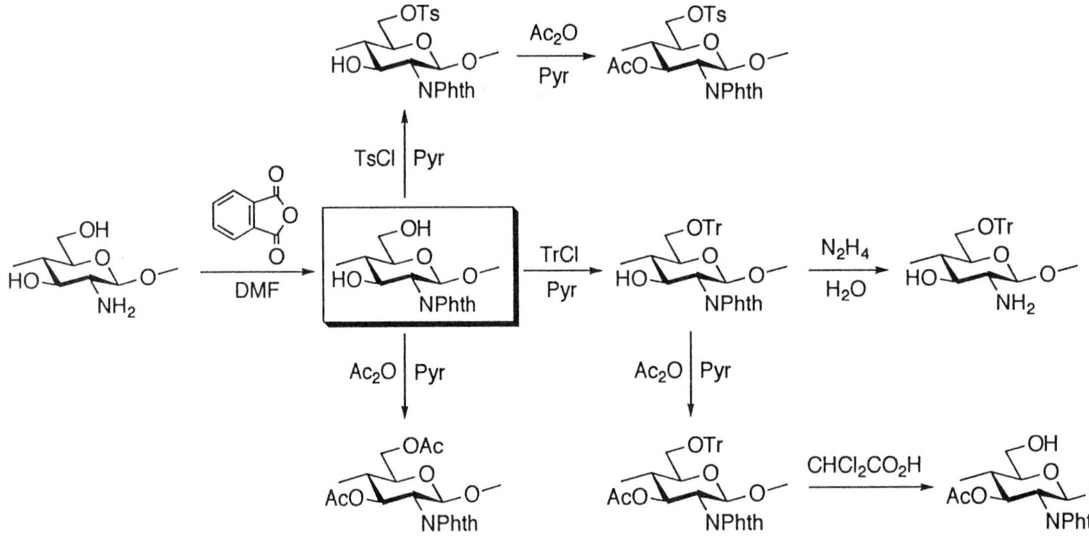

Scheme V. Preparation of *N*-phthaloyl-chitosan and its modifications.

The phthaloyl group can be removed with hydrazine to generate the free amino group. Dephthaloylation of *N*-phthaloyl-6-trityl-chitosan, for example, gives 6-trityl-chitosan. 6-Trityl-chitosan can be transformed into amphiphilic derivatives by regioselective introduction of palmitoyl groups at the C-2 amino and C-3 hydroxyl groups, followed by detritylation and then sulfation at the C-6 hydroxyl group. The resulting derivative shows characteristic amphiphilic properties and forms Langmuir monolayers with high collapse pressures (*47*).

A highly swollen chitosan gel in pyridine is suitable for the preparation of the water-soluble chitin (Scheme VI) (*9*). This procedure appears to be superior to deacetylation of chitin in homogeneous alkali solution because product isolation is easier. Thus, the water-soluble chitins with desired molecular weights may be prepared starting from chitosans. Partial acetylation is also possible in homogeneous solution in aqueous acetic acid–methanol (*48*).

Acylation of chitosan can also be accomplished with cyclic esters (lactones). When chitosan is treated

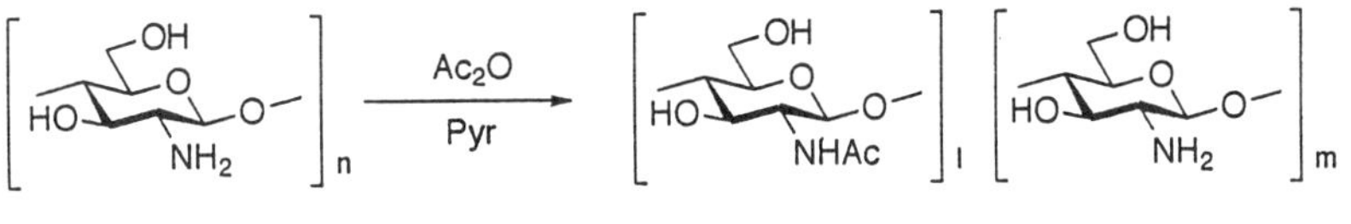

Scheme VI. Preparation of the water-soluble chitin from chitosan.

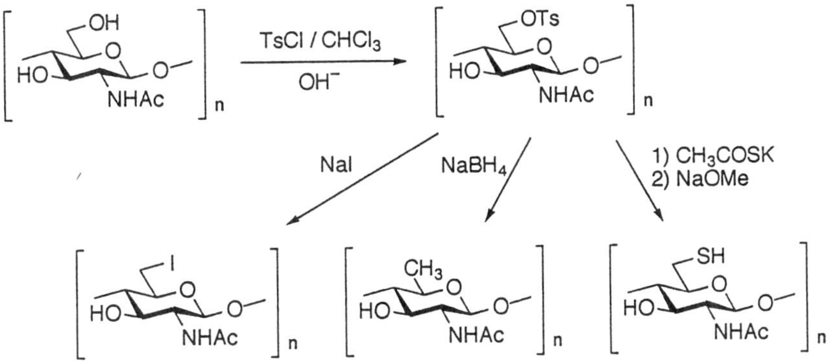

Scheme VII. Preparation of tosyl-chitin and its modifications.

with γ-butyrolactone in DMSO, the amino groups are acylated (*49*), whereas with β-propiolactone the conflicting results of acylation (*50*) and alkylation (*51*) were reported.

Tosylation

Incorporation of *p*-toluenesulfonyl (tosyl) groups into chitin is useful for preparing soluble precursors that enable controlled modifications under mild conditions, because the tosyl group enhances solubilization and tosylates are highly reactive. Although tosylation of chitin is quite sluggish in pyridine even at 100 °C, it proceeds smoothly in a two-phase mixture of an aqueous alkali chitin solution and tosyl chloride in chloroform (Scheme VII) (*52*). The substitution degree reaches 1.0 with excess tosyl chloride and then levels off. This result suggests that tosyl groups had been introduced preferentially at the less hindered C-6 positions. The resulting tosyl-chitins with substitution degrees >0.5 are soluble in polar organic solvents such as DMSO.

Tosyl-chitins are very convenient reactive intermediates for further modifications in homogeneous solution. They can be converted into iodo-chitins with sodium iodide and deoxy-chitins with sodium borohydride (*53*). When tosyl-chitins are treated with potassium thioacetate and then with methoxide to remove *S*-acetyl groups, mercapto chitin derivatives are obtained (Scheme VII) (*27*).

Alkylation

Hydroxyethylation (or glycolation) of chitin is effected by treating alkali chitin with ethylene oxide to give hydroxyethyl-chitins or glycol-chitins. However, the reaction is carried out under strongly alkaline conditions, and N-deacetylation takes place at the same time (Scheme VIII). Moreover, ethylene oxide may polymerize and result in the introduction of oligoethylene glycol side chains, as observed in the hydroxyethylation of cellulose under similar conditions. The derivatives are readily soluble in water and are used as substrates for assaying chitinolytic enzymes. In place of ethylene oxide, 2-chloroethanol can be used. In a similar manner, propylene oxide gives hydroxypropyl-chitin.

Hydroxypropylation of chitosan preferentially takes place at hydroxyl groups under alkaline conditions, but N-alkylation occurs almost exclusively under acidic conditions (*54*). Hydroxyalkylation is also possible with glycidol or 3-chloropropane-1,2-diol (glycerol α-monochlorohydrin). The resulting chitosan derivatives are soluble in water and attractive as components of toiletries.

On treatment with *N,N*-diethylaminoethyl chloride, alkali chitin gives diethylaminoethyl-chitins (*20*). The resulting derivatives with substitution degrees >0.5 are soluble in water and swell highly even in nonpolar solvents such as benzene. They are useful as adsorbents for column chromatography (*55*).

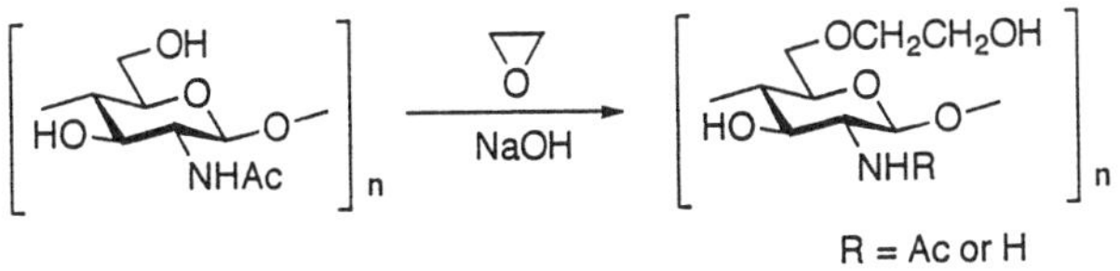

Scheme VIII. Preparation of hydroxyethyl-chitin.

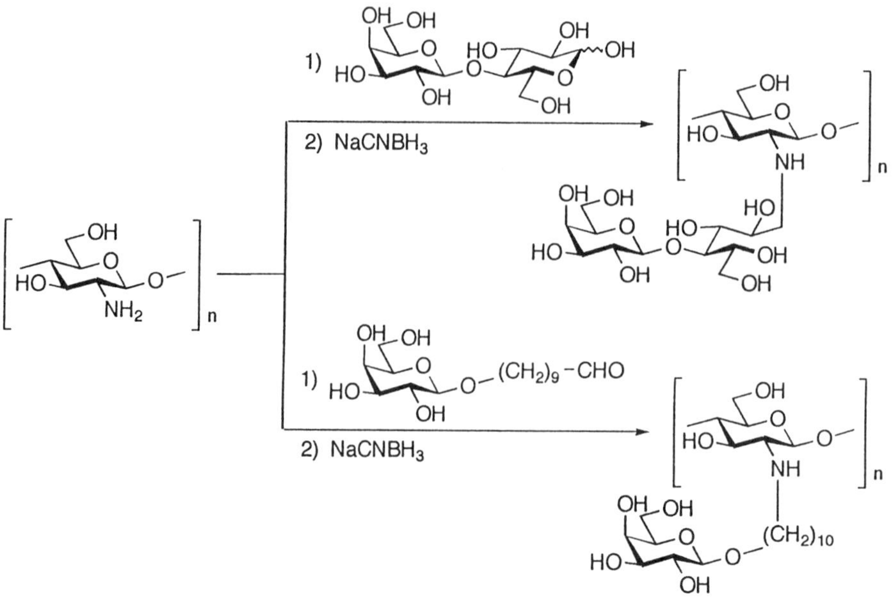

Scheme IX. Reductive alkylation of chitosan with sugars.

Reductive alkylation is another interesting procedure for N-alkylation. Chitosan is first treated with an aldehyde to give an imine (Schiff base), which is converted into N-alkyl derivative by reduction with sodium borohydride or sodium cyanoborohydride. With reducing sugars including glucose, galactose, lactose, and cellobiose in the presence of sodium cyanoborohydride, sugar groups are introduced into chitosan through reductive alkylation (Scheme IX) (56, 57). The degrees of substitution are generally high (Table II), and the derivatives are soluble in water or dilute acids. The derivatives are interesting as water-soluble polysaccharides with peculiar viscosity behavior. An aqueous solution of the chitosan-lactose showed unusual non-Newtonian features such as low-shear Newtonian behavior, a medium shear viscosity increase (dilatancy), and a high-shear viscosity drop

(pseudoplasticity) (58). Among the three chitosan derivatives prepared from glucose, maltose, and maltotriose, only the one from maltotriose showed a rapid decrease in viscosity with an increase in the shear rate. These derivatives were also confirmed to be good adsorbents for metal cations.

Sugar compounds having an aldehyde group in the aglycones are incorporated similarly without opening the sugar ring. Aqueous acetic acid solutions of these chitosan derivatives carrying glucose or galactose through a C_{10} spacer (Scheme IX) form gels during heating to 50 °C and revert to solution on cooling (59).

When chitosan is treated with aliphatic aldehydes and then with a reducing agent, N-alkylation takes place. In this manner alkyl groups such as methyl, ethyl, and propyl are introduced. The resulting derivatives are soluble in dilute acetic acid and have film-forming properties. They adsorb various metal cations (60).

Carboxylation

As in the preparation of carboxymethyl-cellulose, carboxymethylation of chitin proceeds with alkali chitin and monochloroacetic acid. Use of appropriate organic solvents facilitates the reaction (61). Because of the strongly alkaline conditions, N-deacetylation also takes place and leads to the formation of amphoteric polymers having both carboxyl and amino groups (Scheme X). The degree of deacetylation may reach >0.5. Carboxymethylation is supposed to proceed preferentially at C-6 as implied from the results of backbone hydrolysis (62). Carboxymethylated chitin is soluble in water. When chitosan is treated with mono-

Table II. Reductive *N*-Alkylation of Chitosan with Carbohydrates

Carbohydrate	mol/ Glucosamine	Time (h)	DS	Solubility[a]
Glucose	3.00	8	0.9	2
Glucosamine	7.8	48	0.67	2
Galactose	2.22	4	0.97	2
Lactose	4.0	96	0.95	1, 2
Cellobiose	2.92	48	0.4	1, 2
Maltose	1.7	30	0.6	1
Maltotriose	4.16	144	0.54	1

Note: Procedure performed with 1% AcOH–MeOH at room temperature. DS is degree of substitution determined by elemental analysis.

[a] Solubility is 1 in neutral water (pH 7) and 2 in dilute acidic solution (pH 5–6).

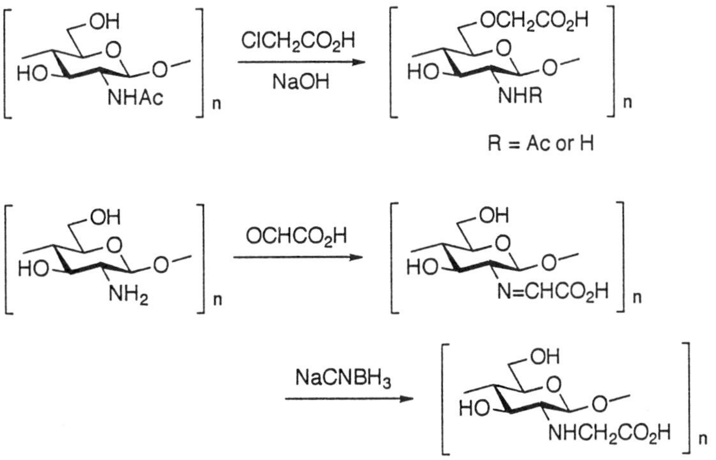

Scheme X. O- and N-Carboxymethylation of chitin.

chloroacetic acid, N-,O-dicarboxymethylated chitosan is produced.

Chitosan and glyoxylic acid form a Schiff base that is converted into N-carboxymethyl-chitosan on reduction (Scheme X) (*63*). Several other N-carboxyalkylated chitosans were prepared in a similar manner from carboxylic acids having an aldehyde or keto group including pyruvic, β-hydroxypyruvic, and 2-ketoglutaric acids. These carboxyalkylated derivatives are generally soluble in water and are attracting much attention for collection of metal cations, sustained release of drugs, and biomedical uses. Cyclic carboxylic anhydrides such as phthalic, succinic, and maleic anhydrides react with the free amino groups of chitosan to give amic acid derivatives.

Sulfation

Sulfation of chitin and chitosan was studied extensively, primarily for preparing anticoagulants, because of the structural similarities to heparin. Heparin is a partially sulfated and N-acetylated polysaccharide composed of D-glucosamine, D-glucuronic acid, and L-iduronic acid. The use of various sulfating agents (including concentrated or fuming sulfuric acid and sulfur trioxide with pyridine, trimethylamine, DMF, or sulfur dioxide) was reported, but chlorosulfonic acid is most widely employed. In chitin, hydroxyl groups are sulfated, whereas with chitosan, sulfation occurs at both hydroxyl and amino groups. The sulfur trioxide–pyridine complex is selective for sulfation of chitosan amino groups (*64*). Regioselective introduction of sulfate groups at C-6 is possible by sulfation of chitosan whose C-2 amino and C-3 hydroxyl groups are protected by copper ions (*65*). N-Carboxymethyl-chitosan (*66*) and O-carboxymethyl-chitin (*67*) were also sulfated to develop anticoagulants.

Schiff Base Formation

Free amino groups of chitosan react with an aldehyde to give the corresponding Schiff base. The resulting imine linkage is fairly stable in neutral and alkaline solutions, but under acidic conditions, it is readily hydrolyzed. Therefore, the amino groups can be protected with aldehydes to allow hydroxyl groups to be modified. On the basis of this technique, full acylation of chitosan is attained, which is difficult by direct acylation. For example, fully acetylated chitin is prepared in this way by protection of the amino groups, O-acetylation, deprotection, and finally N-acetylation (*68*).

Reactions of chitosan with aldehydes are carried out conveniently in a mixture of aqueous acetic acid and methanol to give derivatives with a degree of substitution ≤1.0. During the reaction, the originally homogeneous solution gels due to poor solubility of the Schiff base (*69*). Thermal characteristics of chitosan derivatives having long alkylidene groups were examined, and the glass transition temperature of the decanal derivative was 126 °C (*70*).

Quaternary Salt Formation

Because of the presence of free amino groups, chitosan forms salts with organic and inorganic acids. This characteristic accounts for the solubility of chitosan in aqueous acid solutions. It is readily soluble in hydrochloric or nitric acid (≤1%), less soluble in phosphoric acid, and not at all soluble in sulfuric acid. Many aqueous organic acids dissolve chitosan; and formic, acetic, lactic, and pyruvic acids are frequently used to make solutions.

N-Trimethylation of chitosan has been effected with excess methyl iodide and sodium hydroxide to give polycation derivatives soluble in water in a wide pH range. Trimethylation behavior was studied in detail recently (71), and the addition of sodium iodide led to higher degrees of substitution. The degree of trimethylation increased up to 0.64 with increasing reaction time. Similarly, N-triethylation is possible with ethyl iodide. N-Dimethylated chitosan prepared by reductive alkylation using formaldehyde also gives the N-trimethylated derivative on treatment with methyl iodide (72).

When chitosan is treated with a quaternized epoxide such as glycidyltrimethylammonium chloride, quaternization is achieved at the incorporated side chains to give water-soluble derivatives (Scheme XI) (73). These derivatives are analogous to cationized cellulose prepared from glycol-cellulose and glycidyltrimethylammonium chloride, which is a common polycation component of shampoos.

Chitosan and some derivatives form polyion complexes with polyanions. On mixing a chitosan solution with a polyanion solution, a polyion complex precipitates. The compositions and properties vary widely depending on concentration, molar ratio, pH, and order of mixing (74). Various polyanions, including dextran sulfate, carboxymethyl-dextran, carboxymethyl-cellulose, carboxymethyl-chitosan, alginic acid, poly(aspartic acid), keratin derivatives, and poly(acrylic acid), have been used for this purpose. The resulting complexes are useful as anticoagulant biomedical materials, microcapsule walls, membranes, and carriers for drug-delivery systems.

Miscellaneous Reactions

Chitin forms esters with inorganic acids such as nitric and phosphoric acids. Unlike the preparation of cellulose nitrate, chitin degrades to considerable extents with a mixture of nitric and sulfuric acids. Concentrated nitric or fuming nitric acid gives chitin nitrates. Similarly, chitosan nitrates are produced with a mixture of concentrated nitric acid, acetic anhydride, and acetic acid. However, the chitin or chitosan backbone

is considered to be cleaved considerably under these strongly acidic conditions.

On heating chitin with phosphoric acid and urea in DMF, phosphorylated chitins were prepared (75). The reaction is more facile with phosphorus pentaoxide in methanesulfonic acid and gives highly phosphorylated products (76). These derivatives show high affinity for metal cations.

When chitin is treated with carbon disulfide in strongly alkaline media, chitin xanthate is formed (like cellulose), which can be used to prepare fibers and films. When treated with carbon disulfide in concentrated ammonia, chitosan gives an ammonium salt of dithiocarbamate derivative, which is useful for recovery of metal cations (77). The reaction of partially deacetylated chitin with carbon disulfide reaches an equilibrium during 3 h at room temperature, and about 30% of the amino groups are modified under these conditions (78).

Graft copolymerization onto chitin and chitosan is another interesting field, but it has not been explored much. Vinyl monomers are graft copolymerized onto chitin with cerium(IV) (79) or irradiation (80). The resulting chitin derivatives having poly(acrylic acid) or polyacrylamide grafts show high hydrophilicity. Iodochitin is also interesting as a trunk polymer having initiating groups for graft copolymerization of vinyl monomers. It actually undergoes graft copolymerization of styrene by a cationic mechanism in the presence of a Lewis acid such as tin(IV) chloride, as well as by a radical mechanism on UV irradiation (81). Graft copolymerization on chitosan is performed with 2,2'-azobisisobutyronitrile (AIBN), cerium(IV), or a redox system.

The free amino groups of chitosan or the water-soluble chitin are expected to work as initiators for ring-opening polymerization of N-carboxy-α-amino acid anhydrides (NCAs). NCAs are very susceptible to hydrolysis, and no graft copolymerization was observed with chitosan. However, copolymerization of

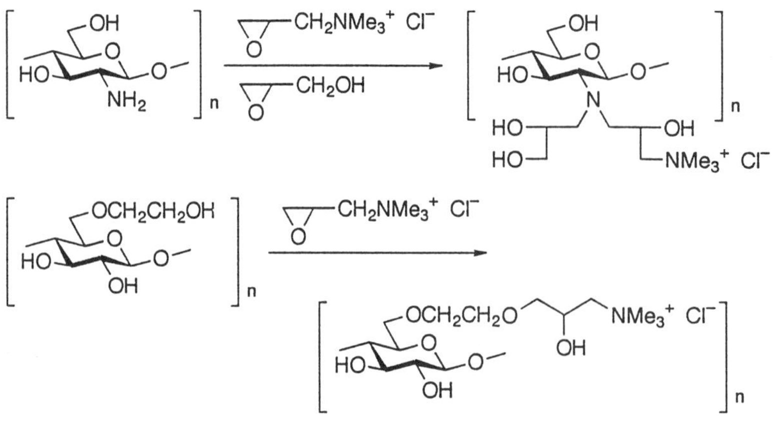

Scheme XI. Introduction of quaternary ammonium groups into chitosan and cellulose.

γ-methyl L-glutamate NCA on the water-soluble chitin proceeds efficiently in a two-phase system of water and ethyl acetate at 0 °C (Scheme XII) (82). Grafting efficiencies are very high, and most of the NCA is consumed for graft copolymerization (Table III). The resulting graft copolymers show varying degrees of solubility depending on graft length. Ester groups of the poly(γ-methyl L-glutamate) grafts can be hydrolyzed to give chitin-*graft*-poly(L-glutamate)s that are readily soluble in water (83).

Applications

Wastewater Treatment

Chitosan is soluble in dilute organic acids to give a polycation solution useful as a coagulant for wastewater treatment. It is highly effective in coagulation and dewatering of activated sludge. Chitosan is less toxic than synthetic polycations such as copolymers of acrylamide and methacrylates (84), and it is also biodegradable. Therefore, it is a favorable coagulating agent for environmental protection.

Chitosans with higher molecular weights are generally better coagulants (85). Coagulation of wastewater from food processing plants with chitosan may become of practical importance for recovering proteins and thereby producing livestock feed. Chitosan is an effective flocculant in acidic media but not in alkaline conditions because of precipitation. Poly-(acrylic acid)-grafted chitosan demonstrated high floc-

culation ability over a wide pH range, and the efficiency was greatest at pH 10 (86).

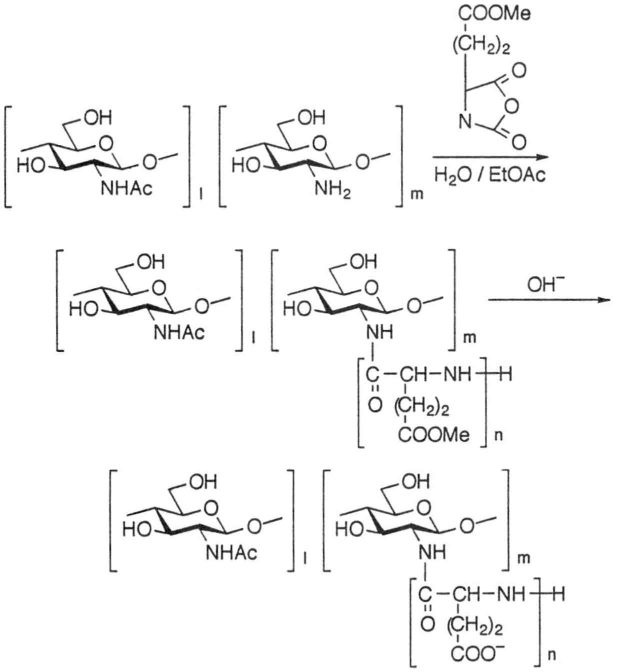

Toiletries

Water-soluble chitosan and chitin derivatives having polycationic and film-forming properties are highly desirable components for skin lotions, creams, hair set lotions, hair sprays, and shampoos. Skin and hair surfaces are negatively charged when wet; therefore, such polycations are essential for coating and protection. Chitosan salts of organic acids are film-forming materials, but the hygroscopicity depends greatly on the acid components (Table IV) (87).

Quaternized chitin derivatives have found applications in toiletries. Derivatives prepared by treating chitosan with glycidyltrimethylammonium chloride and glycidol are useful as skin and hair-care products (73). The water-soluble chitin is characterized by high moisture retention, an attractive property for cosmetics, and it is added to skin lotions and other toiletry products. Hydroxypropyl-chitin and O-carboxymethyl-chitin are also commercially available as water-soluble toiletry components.

Medicine

Various biological, physiological, and pharmacological activities have been known for chitin and chitosan,

Scheme XII. Graft copolymerization of γ-methyl L-glutamate NCA onto the water-soluble chitin.

Table III. Graft Copolymerization of γ-Methyl L-Glutamate NCA onto the Water-Soluble Chitin

NCA/ Glucosamine	Degree of Polymerization		Grafting Efficiency (%)	
	Weight	IR	Weight	IR
2.3	2.1	1.8	91	78
8.4	6.2	6.0	73	71
13.6	10.4	10.9	76	80
16.1	13.1	12.3	81	76
26.6	22.0	20.8	83	78

NOTE: Copolymerization performed in water–ethyl acetate at 0 °C for 2 h. Average degree of polymerization of side chains calculated from weight increases or IR spectroscopy. Grafting efficiency calculated by the following equation: (amount of NCA graft copolymerized/amount of NCA added) × 100.

Table IV. Hygroscopicity of Chitosan Salt Films

Acid Component	Hygroscopicity (%)
L-Glutamic acid	17.9
D,L-Lactic acid	17.6
Formic acid	13.8
Benzoic acid	6.5
Salicylic acid	2.6
p-Aminobenzoic acid	0.7

Note: Values are weight increases of the films when relative humidity was changed from 35% to 70%.

including antitumor activity, immune enhancement, hypolipidemic activity, hemostatic activity, acceleration of wound healing, promotion of growth of *L. bifidus,* and virucidal and fungicidal effect. Chitosan selectively aggregates L1210 leukemia cells in the presence of normal murine erythrocytes and bone marrow cells, and it exhibits a pronounced effect in increasing the ratio of normal leukocytes to L1210 cells in venous blood (Table V). Chronic treatment with low doses of chitosan extended lifespans of mice bearing P388 or Ehrich ascites tumors (*88*).

Macrophage activation by partially deacetylated chitins is noteworthy. It is dependent on the degree of deacetylation, and maximum activation occurs at 0.7. This result indicates that the material is a potential immuno adjuvant (*89*). Oligomers of chitin and chitosan are also interesting because of the immune-enhancing activity. They are soluble in water and hence are easily injected intravenously. Of the oligomers, hexamers of both D-glucosamine and *N*-acetyl-D-glucosamine have considerable antitumor and antimicrobial activities (*90*). A chitosan derivative prepared by oxidation of the hydroxymethyl group to carboxyl group also inhibits growth of L1210 cells (*91*).

As dietary fibers, chitin and chitosan exhibit hypolipidemic activity. They reduce cholesterol and triglyceride levels in blood plasma (serum) and liver effectively (Table VI) (*92*). The effect is much more prominent with chitosan than chitin (*93*). Molecular weight of chitosan does not affect the activity much, but chitosan oligomers are not effective at all (*94*).

Nonwoven fabrics made of chitin fibers are produced commercially as artificial skins to cover burns and injuries (*95*). Cotton-like chitin and chitosan are useful as wound dressings or filling agents for animals (*96*). These fabrics and cottons promote wound healing and enable high quality of cosmetic restoration. *N*-Carboxybutyl-chitosan prepared by reductive alkylation of chitosan with levulinic acid shows high antimicrobial activity and is promising as a wound-dressing material (*97*).

Because chitin and chitosan consist of *N*-acetyl-D-glucosamine and D-glucosamine, sulfated chitin and chitosan are probable substitutes for heparin. Sulfated derivatives have been prepared and shown to have anticoagulant activity. However, considerable scatter exists in reported activities, probably because of the difference in the fine structures, and thus it is still difficult to thoroughly discuss structure–activity relationship. Chitosan sulfates show higher activity in general than chitin sulfates, and the activity may reach that of heparin. Sulfamino groups are considered to play an important role, but selectively N-sulfated chitosan was inactive. Subsequent O-sulfation led to high activity (*64*). The activity generally increases with the sulfation degree, but toxicity increases also. It is necessary to prepare well-defined sulfated derivatives to develop active heparinoids with low toxicity.

As implied by the structure of heparin, the presence of carboxyl groups may be significant, and sulfated *N*-carboxymethyl-chitosan was highly active (*66*). *O*-Carboxymethyl-chitins were also sulfated, and the influence of the degree of substitution was discussed (*67*). Although all the sulfated *O*-carboxy-methyl-chitins were active, the derivatives with degrees of carboxymethylation of 0.5–0.6 showed high activities similar to that of heparin. Because the sulfated *O*-carboxymethyl-chitins have higher antithrombogenic activity than the corresponding sulfated chitins, the importance of the carboxyl group has been suggested. Sulfated *N*-carboxymethyl-chitosan is also interesting as an antiviral agent that inhibits the propagation of the human immunodeficiency virus type 1and Rauscher murine leukemia virus (*98*).

Chitin derivatives may be used as biodegradable carriers for drug-delivery systems. Drugs such as methamphetamine and neocarzinostatin were conjugated with *O*-carboxymethyl-chitin through covalent bonding or entrapment. Subcutaneous injection of the resulting polymeric prodrugs resulted in sustained release of the drugs into blood (*28*). Drug conjugates prepared from *N*-succinyl-chitosan or *O*-carboxy-methyl-chitin and mitomycin C show favorable effects

Table V. Effects of Chitosan Treatment on Dissemination of Intraperitoneally Implanted L1210 Cells in Peripheral Blood of Mice

Daily Dose of Chitosan (mg/kg)	*Total Leukocyte Count (Day 8) ÷ Total Leukocyte Count (Day 0)*	*Normal Leukocytes[a] ÷ L1210 Cells*
80	7.9 ± 2.7	22.5 ± 2.5
40	10.3 ± 4.0	25.0 ± 4.0
20	3.4 ± 0.5	67.0 ± 27
10	8.5 ± 1.5	3.2 ± 0.2
0	12.9 ± 3.0	0.9 ± 0.3

[a] Values determined in venous blood.

Table VI. Effects of Fiber Additives on Cholesterol and Triglyceride Levels of Rats

Additive	*Cholesterol Level*		*Triglyceride Level*	
	Serum (mg/dL)	*Liver (mg/g)*	*Serum (mg/dL)*	*Liver (mg/g)*
Cellulose	244 ± 35	44 ± 11	563 ± 72	172 ± 33
Pectin	175 ± 13	42 ± 7	404 ± 78	156 ± 17
Pectin-Al salt	164 ± 10	36 ± 6	382 ± 58	138 ± 12
Chitosan	137 ± 9	15 ± 4	303 ± 33	87 ± 20

Note: Rats were fed a diet containing 4% nondigestible fiber for 4 weeks.

at high doses compared to mitomycin C alone (99). Incorporation of 5-fluorouracil to chitosan reduces its toxicity and results in extended lifespans of tumor-bearing mice (100).

Higher dissolution rates of drugs are observed when mixed with powdery chitin or chitosan than when mixed with microcrystalline cellulose, and bioavailability is enhanced (101). Therefore, chitin and chitosan may improve dissolution behavior of drugs and be useful for tablet preparation.

Fungus Control

Chitosan has been identified as a potential control agent for microorganisms (102). It suppresses the growth of various bacteria including *E. coli*, *P. aeruginosa*, *B. subtilis*, and *S. aureus*, probably as a result of binding on the cell surface (103, 104). The results of growth inhibition of *E. coli* revealed that chitosan with a lower molecular weight exhibited higher antibacterial activity. Chitosan also shows antifungal activity. It inhibits the growth of *Fusarium solani*, a typical plant pathogenic fungus. Chitosan with a higher deacetylation degree is more efficient. Chitosan oligomers with degrees of polymerization above 5 also show antibacterial and antifungal activities. *N*-Carboxymethyl-chitosan suppresses both growth and aflatoxin production by *Aspergillus flavus* and *A. parasiticus* (105).

Chitin improves soil condition, probably as a result of favorable control of microorganisms in soil. The effects of addition of chitin and chitosan to soil or spraying chitosan solutions on vegetables were described in patents, and several forms of powders or solutions are now commercially available. Chitosan has virucidal and fungicidal activities and sometimes prevents plant diseases and improves crops. Recent experiments revealed that plant seeds coated with chitosan showed high germination ratios and rapid growth leading to increased crops (106).

Food Processing

Possible applications of chitin and chitosan as ingredients for many sorts of food were discussed in patents. They may be beneficial as thickening agents and functional additives with biological effects including lowering cholesterol level and stimulating growth of *L. bifidus*. The chitosan esters of aliphatic acids were claimed as emulsifiers. Chitin and chitosan are potentially useful for food processing in many ways (107). Chitosan is effective for clarifying beverages such as fruit juices (108) and coffee as well as extracting acidic components. Antimicrobial activity of chitosan and its oligomers may find uses in food preservation as claimed in patents, and some kinds of food were shown to be preserved for prolonged periods.

Analysis and Separation

Chitosan and its N-trimethylated derivatives form a polyion complex with a polyanion in water, and the reaction is used for determining the amount of polymeric ions (109). The method, referred to as colloid titration, is now frequently used to titrate polymeric ions. Glycol-chitosan and N-trimethylated glycol-chitosan are soluble in water in a wide pH range, even in alkaline solutions. These compounds have become standard polycation reagents for titration and are commercially available. These polycations bind to bacterial surfaces and show bactericidal activity. They are also used for titration of bacteria suspensions (110).

Chitin and chitosan powders are used as adsorbents for thin-layer chromatography to separate various organic compounds including amino acids (111). They are also useful in column chromatography, and some chitin gels for high-performance liquid chromatography (HPLC) are on the market. Separation of some dyes by a column containing pulverized diethylaminoethyl-chitins is dependent on the degree of substitution, and a high substitution degree is favored to separate methylene blue and sudan red (55).

The chirality of chitin can be used for asymmetric recognition and optical resolution of racemic mixtures. Macroporous silica gels coated with fully carbamated chitosan prepared from chitosan and excess phenyl isocyanate are advantageous as packing materials for HPLC to separate certain racemic mixtures (Table VII) (112). Separation efficiency depends on the racemic compounds. Resolution is also possible with phenyl carbamate derivatives of *N*-arylidene-chitosans (113). *N*-Acylated chitosan gels separate D,L-amino acids by LC (114).

Chitin potentially is useful for the affinity chromatography separation of some biopolymers. Lysozyme and wheat germ agglutinin are purified by affinity chromatography on chitin (115). Introduction of some active groups into chitosan enables the development of specific affinities for various bioactive species. For instance, chitosan having immobilized trypsin inhibitor is suitable for affinity precipitation of trypsin for purification (116). Chitosan derivatives

Table VII. Chromatographic Separation of Racemic ATFE and TB on Phenylcarbamates of Polysaccharides

Carbamates	ATFE			TB		
	k_1'	α	R_s	k_1'	α	R_s
Cellulose	1.42(−)	1.47	1.38	1.02(+)	1.39	1.73
Amylose	0.61	1.00		0.77(+)	1.28	1.10
Chitosan	0.55(−)	2.25	2.97	1.19	1.00	

Note: ATFE is 1-(9-anthryl)-2,2,2-trifluoroethanol; TB is Troeger base; k_1' is capacity factor for less retained enantiomer; α is separation factor; R_s is resolution factor.

bearing pendant 1-thio-β-D-glucopyranosyl groups are used for affinity chromatography of β-D-glucosidases (117).

Membranes

Transparent and tough chitosan films are prepared by solution casting from aqueous acetic acid solution followed by treatment with alkali to remove the acid. Cross-linking of chitosan films is effected with glutaraldehyde. The membrane can be used to transfer chloride ions from an acid solution (0.1 M HCl) to an alkaline solution (0.1 M NaCl, pH 13). This transference is accomplished as a result of salt formation at the acidic interface and release at the alkaline interface. This phenomenon is an interesting example of active transport (118).

Recent interest in membrane separation has focused on pervaporation to separate ethanol–water mixtures. Chitosan membranes selectively permeated water for concentration of ethanol (119). Similarly, aqueous DMSO is concentrated through chitosan membranes.

Compared with chitosan, chitin is more difficult to fabricate into films. However, chitin is soluble in a mixture of DMAc, NMP, and LiCl, and the preparation of chitin films with considerable strength is enabled. Either homogeneous dense membranes or asymmetric porous membranes can be prepared depending on the preparative procedures. These chitin membranes may be useful for dialysis, ultrafiltration, and reverse osmosis (120).

Immobilization of Bioactive Species

As supports for immobilization of biologically active species, chitin and chitosan have been rated high because they are nontoxic and generally superior to other supports in the amounts and activities of immo-

bilized species (121). Chitosan may be employed more conveniently than chitin because of the presence of free amino groups. Various enzymes have been immobilized on chitin and chitosan through several immobilization procedures, but the glutaraldehyde method is the most convenient. For example, lactase immobilized on chitin with glutaraldehyde showed about 60% of the original activity (122). Enzymes immobilized similarly on chitosan usually retain 30–80% of activities of soluble enzymes, and considerable activities are retained after prolonged storage or repeated uses. Porous chitosan beads are suited for immobilization of proteases (123). Without glutaraldehyde, enzymes can be immobilized on chitosan by ionic interactions, and α-chymotrypsin thus immobilized showed higher activity than that immobilized with glutaraldehyde (124).

Immobilization of microbial cells is sometimes more advantageous than that of enzymes, and chitin and chitosan are effective supports for cell immobilization. Chitosan beads were used for cell immobilization, and the immobilized yeast cells showed 1.5 times higher activity in ethanol production than those on calcium alginate (125). Polyion complexes are often used for cell immobilization, and those prepared from chitosan and alginate allow the concurrent immobilization and permeability of plant cells while maintaining some cell viability (126).

To develop polymeric, asymmetric, reducing agents, the active site of coenzyme reduced nicotinamide adenine dinucleotide (reduced form) was generated on the water-soluble chitin and chitosan. The water-soluble chitin and chitosan were transformed into derivatives having 1,4-dihydronicotinamide groups by N-nicotinoylation followed by quaternization with benzyl chloride and reduction (Scheme XIII). The resulting derivatives show high asymmetric recognition in the reduction of ethyl benzoylformate to ethyl mandelate, although chemical yield is low due

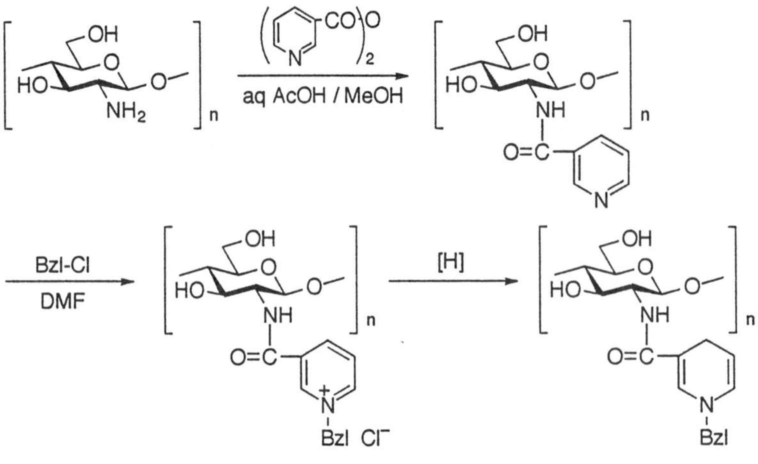

Scheme XIII. Introduction of the 1,4-dihydronicotinamide group into chitosan.

to steric hindrance (*127*). Incorporation of oligoalanine spacer arms between the chitin backbone and the active site highly improves the chemical yield, but the asymmetric selectivity is decreased somewhat (*128*).

Adsorption of Metal Cations

Chitin and chitosan show high affinity for metal cations, and the adsorption behavior was studied extensively. Chitosan, however, has much higher capacity than chitin for collecting heavy metal cations, and it is one of the best adsorbents occurring in nature. It adsorbs a wide variety of metal cations such as copper, mercury, cadmium, iron, nickel, zinc, lead, and silver (*129*). It also collects highly toxic organomercury compounds (*130*) and is expected to remove toxic heavy metals from industrial wastewater.

The metal adsorption ability is influenced greatly by the degree of deacetylation, and collections of both mercury and copper increase with increasing deacetylation degree of chitins prepared by the conventional heterogeneous deacetylation method (*131*). The influence is especially evident in the low deacetylation region (Figure 5). The samples prepared by deacetylation in homogeneous solution, however, show maxima at around 50% deacetylation (Figure 5), and these maxima correspond to the deacetylation of the water-soluble chitin. This result indicates that adsorption capacity is not determined only by the amount of the free amino groups but also by enhanced hydrophilicity resulting from destruction of the crystalline structure (*131*).

Cross-linking with glutaraldehyde (*132*) and introduction of nonanoyl groups (*133, 134*) at the amino groups show similar influence on adsorption capacity. The capacity increases markedly in both cases with low degrees of substitution, and then it decreases with increased substitution (Figure 6). This phenomenon is also interpreted in terms of increased hydrophilicity

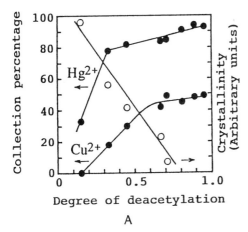 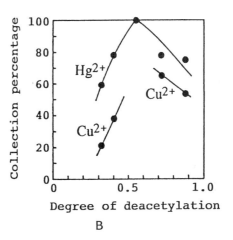

Figure 5. Collection of mercury (II) and copper (II) ions (A) by partially deacetylated chitins prepared under heterogeneous conditions, and (B) by partially deacetylated chitins prepared under homogeneous conditions (adapted from reference *131*).

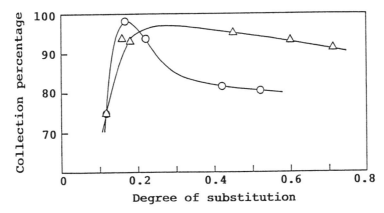

Figure 6. Effects of the degree of substitution of acylated chitosans on the collection of copper(II) ion; acylated chitosan, 50 mg; copper(II) solution, 5×10^{-4} mol/L, 25 mL; △, acetylated chitosans; ○, nonanoylated chitosans prepared from chitosan with a degree of acetylation of 0.12 (adapted from reference *134*).

Table VIII. Selectivity in Adsorption of Metal Cations by Chitin Phosphate and Chitosan Phosphate

Adsorbent	Mg^{2+}	Ca^{2+}	Mn^{2+}	Co^{2+}	Ni^{2+}	Cu^{2+}	Zn^{2+}	Cd^{2+}	UO_2^{2+}
P-Chitin	11.1	0.0	16.7	10.9	9.7	28.3	14.7	17.2	68.4
P-Chitosan	15.3	12.3	28.2	19.7	17.9	43.0	29.8	37.2	76.1

Note: Values are in percent. Adsorbent was 50 mg; metal ion solution was 50 mL; metal ion concentration was 4×10^{-4} mol/L each.

caused by destruction of the crystalline structure by low degrees of cross-linking or acylation.

Phosphorylated chitins and chitosans are also high performance adsorbents for metal cations. Adsorption selectivity was studied by using a mixture of nine different metal cations, and both phosphorylated chitin and chitosan have particularly high affinity for uranyl ions (Table VIII) (75). Chitosan derivatives prepared by reductive alkylation with dehydroascorbic acid (135) or α-ketoglutaric acid (136) efficiently adsorb metal cations including uranyl ions. These results suggest possibilities of collecting uranium from seawater.

Fibers, Textiles, and Composites

Chitin and chitosan fibers are made by wet spinning, but chitosan fibers are attracting more attention because of the antimicrobial characteristics. Fibers obtained by spinning a mixture of polynosic cellulose xanthate solution and finely powdered chitosan show considerable antibacterial and antifungal activities, and chitosan has no effect on the original mechanical strength and feeling of polynosic fibers. Therefore, they are beneficial as antimicrobial and deodorant fibers for underwears and socks (137). Chitosan-treated porous polyurethane films also have desirable properties as inner layers of sports wears due to the antimicrobial activity as well as hygroscopic nature. These products are now on the market, and the need for these functional fibers, textiles, and coatings will grow further.

Composite films were prepared by mixing a chitosan solution in dilute acetic acid with a suspension of homogenized cellulose, followed by evaporation. The resulting films have high oxygen-gas barrier capacity and are degraded to fine fragments by a cellulase or a chitosan-degrading bacterium (138).

Concluding Remarks and Future Prospects

Despite the remarkable abundance, chitin began attracting attention as a functional polymeric material only recently. Progress of the chemistry is, however, quite rapid as evident by a sharp increase in the number of papers and patents, particularly in the past decade.

The unique structure of chitin should enable sophisticated molecular design through controlled modification reactions. Development of various modes of regioselective modifications may overcome the difficulty in tailoring chitin derivatives with well-defined structures. These modifications should in turn help to establish the structure–property relationship necessary to accumulate information on how to develop specifically desirable functions.

Besides conventional application in water treatment, chitin, chitosan, and their derivatives such as the water-soluble chitin, carboxymethyl-chitin (chitosan), glycol-chitin (chitosan), hydroxypropyl-chitin (chitosan), and quaternized chitins (chitosans) recently have found practical applications as novel functional polymers mainly for toiletries, immobilization of bioactive species, and biomedical materials.

Extensive studies are being conducted in different fields to develop toiletries; water and moisture retention agents; drugs; virucidal, bactericidal, and fungicidal agents; biocompatible and bioactive materials; supports for enzyme and cell immobilization; polymeric reagents; specialty fabricated materials such as films, fibers, and papers; membranes; metal cation adsorbents; and agrochemical agents for seed coating, improved growth, and soil amendment. Some of the most promising and immediate applications may be for toiletries, biomedical materials, drugs, and food; and other areas may follow in the near future.

Although a great variety of potential applications have been suggested in patents, sometimes only expectations or presumptions are offered without detailed supporting data based on fundamental understanding of the complicated characteristics of this amino polysaccharide. Therefore, basic research in chitin chemistry is more and more necessary to fill in this knowledge gap.

References

1. Muzzarelli, R. A. A. *Chitin*; Pergamon: Oxford, 1977.
2. Roberts, G. A. F. *Chitin Chemistry*; Macmillan: London, 1992.
3. Sannan, T.; Kurita, K.; Iwakura, Y. *Makromol. Chem.* **1976**, *177*, 3589.
4. Kurita, K.; Tomita, K.; Tada, T.; Ishii, S.; Nishimura, S.; Shimoda, K. *J. Polym. Sci., Part A: Polym. Chem.* **1993**, *31*, 485.

5. McGahren, W. J.; Perkinson, G. A.; Growich, J. A.; Leese, R. A.; Ellestad, G. A. *Process Biochem.* **1984**, *19*, 88.
6. Mima, S.; Miya, M.; Iwamoto, R.; Yoshikawa, S. *J. Appl. Polym. Sci.* **1983**, *28*, 1909.
7. Domard, A.; Rinaudo, M. *Int. J. Biol. Macromol.* **1983**, *5*, 49.
8. Sannan, T.; Kurita, K.; Iwakura, Y. *Makromol. Chem.* **1975**, *176*, 1191.
9. Kurita, K.; Kamiya, M.; Nishimura, S. *Carbohydr. Polym.* **1991**, *16*, 83.
10. Domard, A.; Cartier, N. *Int. J. Biol. Macromol.* **1989**, *11*, 297.
11. Takiguchi, Y.; Shimahara, K. *Lett. Appl. Microbiol.* **1988**, *6*, 129; *Agric. Biol. Chem.* **1989**, *53*, 1537.
12. Izume, M.; Ohtakara, A. *Agric. Biol. Chem.* **1987**, *51*, 1189.
13. Muraki, E.; Yaku, F.; Kojima, H. *Carbohydr. Res.* **1993**, *239*, 227.
14. Usui, T.; Matsui, H.; Isobe, K. *Carbohydr. Res.* **1990**, *203*, 65.
15. Nishimura, S.; Kuzuhara, H.; Takiguchi, Y.; Shimahara, K. *Carbohydr. Res.* **1989**, *194*, 223.
16. Kurita, K.; Tomita, K.; Ishii, S.; Nishimura, S.; Shimoda, K. *J. Polym. Sci., Part A: Polym. Chem.* **1993**, *31*, 2393.
17. Blackwell, J.; Gardner, K. H.; Kolpak, F. J.; Minke, R.; Classey, W. B. *Fiber Diffraction Methods*; French, A. D.; Gardner, K. H., Eds.; ACS Symposium Series 141; American Chemical Society: Washington, DC, 1980; pp 315–334.
18. Kurita, K.; Sannan, T.; Iwakura, Y. *Makromol. Chem.* **1977**, *178*, 2595.
19. Kurita, K.; Ishii, S.; Tomita, K.; Nishimura, S.; Shimoda, K. *J. Polym. Sci., Part A: Polym. Chem.* **1994**, *32*, 1027.
20. Kurita, K.; Inoue, S.; Koyama, Y. *Polym. Bull.* **1989**, *21*, 13.
21. Sannan, T.; Kurita, K.; Ogura, K.; Iwakura, Y. *Polymer* **1978**, *19*, 458.
22. Moore, G. K.; Roberts, G. A. F. *Int. J. Biol. Macromol.* **1980**, *2*, 115.
23. Miya, M.; Iwamoto, R.; Yoshikawa, S.; Mima, S. *Int. J. Biol. Macromol.* **1980**, *2*, 323.
24. Rutherford, F. A.; Austin, P. R. *Proceedings of the 1st International Conference on Chitin Chitosan*; MIT Sea Grant Report MITSG 78–7; Massachusetts Institute of Technology: Cambridge, MA, 1978; p 182.
25. Uragami, T.; Ohsumi, Y.; Sugihara, M. *Polymer* **1981**, *22*, 1155.
26. Sashiwa, H.; Saimoto, H.; Shigemasa, Y.; Ogawa, R.; Tokura, S. *Int. J. Biol. Macromol.* **1990**, *12*, 295.
27. Kurita, K.; Yoshino, H.; Nishimura, S.; Ishii, S. *Carbohydr. Polym.* **1993**, *20*, 239.
28. Tokura, S.; Miura, Y.; Kaneda, Y.; Uraki, Y. In *Polymeric Delivery Systems: Properties and Applications*; El-Nokaly, M. A.; Piatt, D. M.; Charpentier, B. A., Eds.; ACS Symposium Series 520; American Chemical Society: Washington, DC, 1993; pp 351–361.
29. Hirano, S.; Matsumura, T. *Carbohydr. Res.* **1987**, *165*, 120.
30. Kurita, K. In *Chitin in Nature and Technology*; Muzzarelli, R.; Jeuniaux, C.; Gooday, G. W., Eds.; Plenum; New York, 1986; p 287.
31. Kurita, K.; Sannan, T.; Iwakura, Y. *Makromol. Chem.* **1977**, *178*, 3197.
32. Wu, A. C. M.; Bough, W. A. *Proceedings of the 1st International Conference on Chitin/Chitosan*; MIT Sea Grant Report MITSG 78–7; Massachusetts Institute of Technology: Cambridge, MA, 1978; p 88.
33. Sannan, T.; Kurita, K.; Iwakura, Y. *Polym. J.* **1977**, *9*, 649.
34. Aiba, S. *Int. J. Biol. Macromol.* **1991**, *13*, 40; **1992**, *14*, 225.
35. Vårum, K. M.; Anthonsen, M. W.; Grasdalen, H.; Smidsrød, O. *Carbohydr. Res.* **1991**, *211*, 17; **1991**, *217*, 19.
36. Nishi, N.; Noguchi, J.; Tokura, S.; Shiota, H. *Polym. J.* **1979**, *11*, 27.
37. Ando, T.; Kataoka, S. *Kobunshi Ronbunshu* **1980**, *37*, 1; *Chem. Abstr.* **1980**, *92*, 131066y.
38. McCormick, C. L.; Lichatowich, D. K. *J. Polym. Sci., Polym. Lett. Ed.* **1979**, *17*, 479.
39. Kurita, K.; Chikaoka, S.; Kamiya, M.; Koyama, Y. *Bull. Chem. Soc. Jpn.* **1988**, *61*, 927.
40. Hirano, S.; Ohe, Y. *Agric. Biol. Chem.* **1975**, *39*, 1337.
41. Moore, G. K.; Roberts, G. A. F. *Int. J. Biol. Macromol.* **1980**, *2*, 73.
42. Fujii, S.; Kumagai, H.; Noda, M. *Carbohydr. Res.* **1980**, *83*, 389.
43. Kurita, K.; Ichikawa, H.; Ishizeki, S.; Fujisaki, H.; Iwakura, Y. *Makromol. Chem.* **1982**, *183*, 1161.
44. Nishimura, S.; Kohgo, O.; Kurita, K.; Vittavatvong, C.; Kuzuhara, H. *Chem. Lett.* **1990**, 243.
45. Kurita, K.; Tomita, K.; Tada, T.; Nishimura, S.; Ishii, S. *Polym. Bull.* **1993**, *30*, 429.
46. Nishimura, S.; Kohgo, O.; Kurita, K. *Macromolecules* **1991**, *24*, 4745.
47. Nishimura, S.; Miura, Y.; Ren, L.; Sato, M.; Yamagishi, A.; Nishi, N.; Tokura, S.; Kurita, K.; Ishii, S. *Chem. Lett.* **1993**, 1623.
48. Aiba, S. *Int. J. Biol. Macromol.* **1989**, *11*, 249.
49. Kurita, K.; Nishimura, S.; Takeda, T. *Polym. J.* **1990**, *22*, 429.
50. Loubaki, E.; Sicsic, S.; Le Goffic, F. *Eur. Polym. J.* **1989**, *25*, 379.
51. Otuski, S.; Iwamoto, R. *Kobunshi Ronbunshu* **1982**, *44*, 115; *Chem. Abstr.* **1987**, *107*, 176358q.
52. Kurita, K.; Inoue, S.; Nishimura, S. *J. Polym. Sci., Part A: Polym. Chem.* **1991**, *29*, 937.
53. Kurita, K.; Yoshino, H.; Yokota, K.; Ando, M.; Inoue, S.; Ishii, S.; Nishimura, S. *Macromolecules* **1992**, *25*, 3786.
54. Maresch, G.; Clausen, T.; Lang, G. In *Chitin and Chitosan*; Skjak-Bræk, G.; Anthonsen, T.; Sandford, P., Ed.; Elsevier: Essex, England, 1989; p 389.
55. Kurita, K.; Inoue, S.; Koyama, Y.; Nishimura, S. *Macromolecules* **1990**, *23*, 2865.
56. Hall, L. D.; Yalpani, M. *J. Chem. Soc., Chem. Commun.* **1980**, 1153.
57. Yalpani, M.; Hall, L. D. *Macromolecules* **1984**, *17*, 272.
58. Yalpani, M.; Hall, L. D.; Tung, M. A.; Brooks, D. E. *Nature (London)* **1983**, *302*, 812.
59. Holme, K. R.; Hall, L. D. *Macromolecules* **1991**, *24*, 3828.
60. Muzzarelli, R. A. A.; Tanfani, F.; Enamuelli, M.; Mariotti, S. *J. Membr. Sci.* **1983**, *16*, 295.
61. Trujillo, R. *Carbohydr. Res.* **1968**, *7*, 483.
62. Miyazaki, T.; Matsushima, Y. *Bull. Chem. Soc. Jpn.* **1968**, *41*, 2723.
63. Muzzarelli, R. A. A.; Tanfani, F.; Emanuelli, M.; Mariotti, S. *Carbohydr. Res.* **1982**, *107*, 199.
64. Warner, D. T.; Coleman, L. L. *J. Org. Chem.* **1958**, *23*, 1133.
65. Terbojevich, M.; Carraro, C.; Cosani, A. *Makromol. Chem.* **1989**, *190*, 2847.

66. Muzzarelli, R. A. A.; Tanfani, F.; Emanuelli, M.; Pace, D. P.; Chiurazzi, E.; Piani, M. *Carbohydr. Res.* **1984**, *126*, 225.
67. Nishimura, S.; Nishi, N.; Tokura, S. *Carbohydr. Res.* **1986**, *156*, 286.
68. Moore, G. K.; Roberts, G. A. F. *Int. J. Biol. Macromol.* **1982**, *4*, 246.
69. Hirano, S. *Agric. Biol. Chem.* **1978**, *42*, 1939.
70. Kurita, K.; Ishiguro, M.; Kitajima, T. *Int. J. Biol. Macromol.* **1988**, *10*, 124.
71. Domard, A.; Rinaudo, M.; Terrassin, C. *Int. J. Biol. Macromol.* **1986**, *8*, 105.
72. Muzzarelli, R. A. A.; Tanfani, F. *Carbohydr. Polym.* **1985**, *5*, 297.
73. Lang, G.; Clausen, T. In *Chitin and Chitosan*; Skjak-Bræk, G.; Anthonsen, T.; Sandford, P., Eds.; Elsevier: Essex, England, 1989; p 139.
74. Kikuchi, Y.; Takebayashi, T. *Bull. Chem. Soc. Jpn.* **1982**, *55*, 2307.
75. Sakaguchi, T.; Horikoshi, T.; Nakajima, A. *Agric. Biol. Chem.* **1981**, *45*, 2191.
76. Nishi, N.; Ebina, A.; Nishimura, S.; Tsutsumi, A.; Hasegawa, O.; Tokura, S. *Int. J. Biol. Macromol.* **1986**, *8*, 311.
77. Muzzarelli, R. A. A.; Tanfani, F.; Mariotti, S.; Emanuelli, M. *Carbohydr. Res.* **1982**, *104*, 235.
78. Synowiecki, J.; Sikorska-Siondalska, A.; El-Bedaway, A. E. *Biotechnol. Bioeng.* **1987**, *29*, 352.
79. Kurita, K.; Kawata, M.; Koyama, Y.; Nishimura, S. *J. Appl. Polym. Sci.* **1991**, *42*, 2885.
80. Takahashi, A.; Sugahara, Y.; Hirano, Y. *J. Polym. Sci., Part A: Polym. Chem.* **1989**, *27*, 3817.
81. Kurita, K.; Inoue, S.; Yamamura, K.; Yoshino, H.; Ishii, S.; Nishimura, S. *Macromolecules* **1992**, *25*, 3791.
82. Kurita, K.; Kanari, M.; Koyama, Y. *Polym. Bull.* **1985**, *14*, 511.
83. Kurita, K.; Yoshida, A.; Koyama, Y. *Macromolecules* **1988**, *21*, 1579.
84. Arai, K.; Kinumaki, T.; Fujita, T. *Bull. Tokai Reg. Fish. Res. Lab.* **1968**, *56*, 89.
85. Bough, W. A.; Wu, A. C. M.; Campbell, T. E.; Holmes, M. R.; Perkins, B. E. *Biotechnol. Bioeng.* **1978**, *20*, 1945.
86. Kim, Y. B.; Jung, B. O.; Kang, Y. S.; Kim, K. S.; Kim, J. I.; Kim, K. H. *Pollimo* **1989**, *13*, 126; *Chem. Abstr.* **1989**, *111*, 59884e.
87. Gross, P.; Konrad, E.; Mager, H. *Parfuem. Kosmet.* **1983**, *64*, 367.
88. Sirica, A. E.; Woodman, R. J. *J. Natl. Cancer Inst.* **1971**, *47*, 377.
89. Nishimura, K.; Nishimura, S.; Seo, H.; Nishi, N.; Tokura, S.; Azuma, I. *Vaccine* **1987**, *5*, 136.
90. Tokoro, A.; Tatewaki, N.; Suzuki, K.; Mikami, T.; Suzuki, S.; Suzuki, M. *Chem. Pharm. Bull.* **1988**, *36*, 784.
91. Horton, D.; Just, E. K. *Carbohydr. Res.* **1973**, *29*, 173.
92. Nagyvary, J. J.; Falk, J. D.; Hill, M. L.; Schmidt, M. L.; Wilkins, A. K.; Bradbury, E. L. *Nutr. Rep. Inst.* **1979**, *20*, 677.
93. Sugano, M.; Fujikawa, T.; Hiratsuji, Y.; Nakashima, K.; Fukuda, N.; Hasegawa, Y. *Am. J. Clin. Nutr.* **1980**, *33*, 787.
94. Sugano, M.; Watanabe, S.; Kishi, A.; Izume, M.; Ohtakara, A. *Lipids* **1988**, *23*, 187.
95. Kifune, K. In *Advances in Chitin and Chitosan*; Brine, C. J.; Sandford, P. A.; Zikakis, J. P., Ed.; Elsevier: Essex, England, 1992; p 9.
96. Minami, S.; Okamoto, Y.; Matsuhashi, A.; Sashiwa, H.; Saimoto, H.; Shigemasa, Y.; Tanigawa, T.; Tanaka, Y.; Tokura, S. In *Advances in Chitin and Chitosan*; Brine, C. J.; Sandford, P. A.; Zikakis, J. P., Ed.; Elsevier: Essex, England, 1992; p 61.
97. Muzzarelli, R.; Tarsi, R.; Fillipini, O.; Giovanetti, E.; Biangini, G.; Varaldo, P. E. *Antimicrob. Agents Chemother.* **1990**, *34*, 2019.
98. Sosa, M. A. G.; Fazely, F.; Koch, J. A.; Vercellotti, S. V.; Ruprecht, R. M. *Biochem. Biophys. Res. Commun.* **1991**, *174*, 489.
99. Song, Y.; Onishi, H.; Nagai, T. *Chem. Pharm. Bull.* **1992**, *40*, 2822; *Biol. Pharm. Bull.* **1993**, *16*, 48.
100. Ouchi, T.; Banba, T.; Fujimoto, M.; Hamamoto, S. *Makromol. Chem.* **1989**, *190*, 1817.
101. Sawayanagi, Y.; Nambu, N.; Nagai, T. *Chem. Pharm. Bull.* **1983**, *31*, 2064.
102. Allan, C. R.; Hadwiger, L. E. *Exp. Mycol.* **1979**, *3*, 285.
103. Hadwiger, L. A.; Fristensky, B.; Rigglemen, R. C. In *Chitin, Chitosan, and Related Enzymes*; Zikakis, J. P., Ed.; Academic: Orlando, FL, 1984; p 291.
104. Uchida, Y.; Izume, M.; Ohtakara, A. In *Chitin and Chitosan*; Skjak-Bræk, G.; Anthonsen, T.; Sandford, P., Ed.; Elsevier: Essex, England, 1989; p 373.
105. Cuero, R. G.; Osuji, G.; Washington, A. *Biotechnol. Lett.* **1991**, *13*, 441.
106. Hirano, S.; Yamamoto, T.; Hayashi, M.; Nishida, T.; Inui, H. *Agric. Biol. Chem.* **1990**, *54*, 2719.
107. Knorr, D. *Food Technol.* **1984**, 85; In *Biotechnology of Marine Polysaccharides*; Colwell, R. R.; Pariser, E. R.; Sinskey, A. J., Eds.; Hemisphere: New York, 1985; p 313.
108. Soto-Peralta, N. V.; Müller, H.; Knorr, D. *J. Food Sci.* **1989**, *54*, 495.
109. Terayama, H. *J. Polym. Sci.* **1952**, *8*, 243.
110. Terayama, H. *Arch. Biochem. Biophys.* **1954**, *50*, 55.
111. Nagasawa, K.; Watanabe, H.; Ogano, A. *J. Chromatogr.* **1970**, *47*, 408.
112. Okamoto, Y.; Kawashima, M.; Hatada, K. *J. Am. Chem. Soc.* **1984**, *106*, 5357.
113. Ohga, K.; Oyama, H.; Muta, Y. *Anal. Sci.* **1991**, *7*, 653.
114. Seo, T.; Gan, Y. A.; Kanbara, T.; Iijima, T. *J. Appl. Polym. Sci.* **1989**, *38*, 997.
115. Bloch, R.; Burger, M. M. *Biochem. Biophys. Res. Commun.* **1974**, *58*, 13.
116. Senstad, C.; Mattiasson, B. *Biotechnol. Bioeng.* **1989**, *33*, 216.
117. Holme, K. R.; Hall, L. D.; Armstrong, C. R.; Withers, S. G. *Carbohydr. Res.* **1988**, *173*, 285.
118. Uragami, T.; Yoshida, F.; Sugihara, M. *Makromol. Chem., Rapid Commun.* **1983**, *4*, 99.
119. Uragami, T.; Takigawa, K. *Polymer* **1990**, *31*, 668.
120. Uragami, T.; Ohsumi, Y.; Sugihara, M. *Polymer* **1981**, *22*, 1155.
121. Muzzarelli, R. A. A. *Enzyme Microb. Technol.* **1980**, *2*, 177.
122. Stanley, W. L.; Watters, G. G.; Chan, B.; Mercer, J. M. *Biotechnol. Bioeng.* **1975**, *17*, 315.
123. Hayashi, T.; Ikeda, Y. *J. Appl. Polym. Sci.* **1991**, *42*, 86.

124. Muzzarelli, R. A. A.; Barontini, G.; Rocchetti, R. *Biotechnol. Bioeng.* **1976**, *18*, 1445.
125. Shinonaga, M.; Kawamura, Y.; Yamane, T. *J. Ferment. Bioeng.* **1992**, *74*, 90.
126. Knorr, D.; Daly, M. M. *Process Biochem.* **1988**, *48*, 48.
127. Kurita, K.; Koyama, Y.; Murakami, K.; Yoshida, S.; Chau, N. *Polym. J.* **1986**, *18*, 673.
128. Kurita, K.; Iwawaki, S.; Ishii, S.; Nishimura, S. *J. Polym. Sci., Part A: Polym. Chem.* **1992**, *30*, 685.
129. Muzzarelli, R. A. A.; Rocchetti, R. *Talanta* **1974**, *21*, 1137.
130. Masri, M. S.; Friedman, M. *Environ. Sci. Technol.* **1972**, *6*, 745.
131. Kurita, K.; Sannan, T.; Iwakura, Y. *J. Appl. Polym. Sci.* **1979**, *23*, 511.
132. Kurita, K.; Koyama, Y.; Taniguchi, A. *J. Appl. Polym. Sci.* **1986**, *31*, 1169; **1986**, *31*, 1951.
133. Kurita, K.; Chikaoka, S.; Koyama, Y. *Chem. Lett.* **1988**, 9.
134. Kurita, K.; Koyama, Y.; Chikaoka, S. *Polym. J.* **1988**, *20*, 1083.
135. Muzzarelli, R. A. A. *Carbohydr. Polym.* **1985**, *5*, 85.
136. Muzzarelli, R. A. A.; Zattoni, A. *Int. J. Biol. Macromol.* **1986**, *8*, 137.
137. Seo, H.; Mitsuhashi, K.; Tanibe, H. In *Advances in Chitin and Chitosan*; Brine, C. J.; Sandford, P. A.; Zikakis, J. P., Ed.; Elsevier: Essex, England, 1992; p 32.
138. Hosokawa, J.; Nishiyama, M.; Yoshihara, K.; Kubo, T. *Ind. Eng. Chem. Res.* **1990**, *29*, 800.

Development of Novel Materials from Proteins

Semih Erhan

Proteins are the most underrated and underused polymers. This characteristic is due to their moisture sensitivity, which leads to plasticization and enzymatic degradation. There are, however, some proteins such as barnacle cement and mussel byssus that not only set in the presence of water but resist enzymatic degradation and adhere to metals. Other proteins have resilience as good or better than any synthetic rubber, such as resilin, which is found in insect flight muscles, and abductin, which is found in bivalve adductor muscles. These observations suggest that by the application of appropriate chemical treatments, proteins could be made to perform up to the highest technical requirements. The purpose of this chapter is to provide evidence, albeit preliminary, that it is possible to pick a protein and convert it to a high-performance material such as rubber for specialty applications.

Proteins are biopolymers formed by 20 amino acids held together by peptide bonds. These amino acids have acidic, basic, hydroxylated, hydrophobic, and sulfur-containing side chains (Figure 1). As soon as proteins are formed they acquire secondary structures consisting of helices, pleated β-sheets, or random coils. Formation of secondary structures is followed by the formation of tertiary and in certain cases quaternary structures as a result of secondary-force interactions among different functional groups on the side chains of amino acids. Because amino acid composition of proteins differs widely, so do their secondary and tertiary structures, which depend solely on the sequence of their amino acids. These acquired conformations confer to the proteins their biological and structural functions.

Most proteins have no covalent bonds that crosslink their molecules together. Those few that have intermolecular cross-links (e.g., collagen, resilin, and abductin) have specific functions: collagen is a structural protein found in connective tissues; resilin and abductin are protein rubbers that function as energy-absorbing molecules in insect flight muscles and bivalve adductor muscles, respectively.

Because of the hinge-like polypeptide backbone, which enables a protein molecule to have intimate contact with most substrates, and some detergency, which allows wetting of those substrates, proteins, especially animal glue, have been used throughout history as excellent adhesives for materials with porous and uneven surfaces. Their traditional use as adhesives in certain applications such as gummed tape has declined, and their use in cork and paper has just about disappeared because of the superior qualities of synthetic adhesives and the new developments in vegetable adhesives (1).

On the other hand, the use of proteins in the manufacture of caulks, plasters, and paints has grown. Gelatin and its enzymatic-degradation products are used predominantly in cosmetic, pharmaceutical, and photographic industries (2). Several attempts were made (3, 4) to modify proteins by grafting various monomers onto insoluble fibrous proteins such as silk, cotton, and wool to improve such qualities as dyeabil-

3469–8/97/0261$15.00/0 © 1997 American Chemical Society

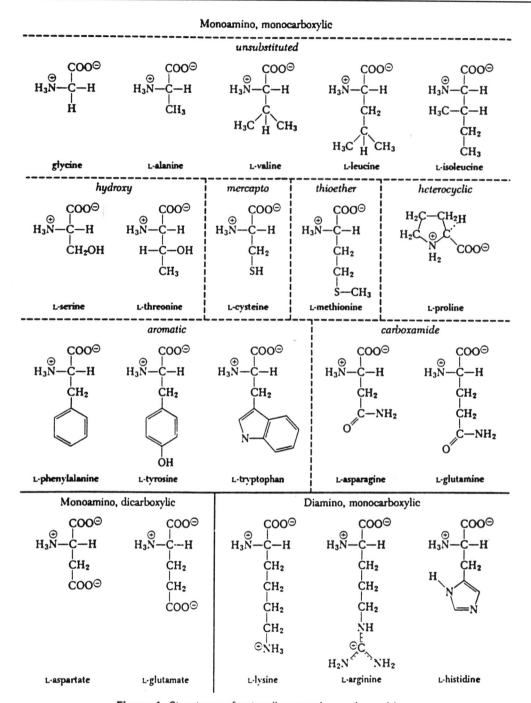

Figure 1. Structures of naturally occurring amino acids.

ity and moisture resistance. Also, the use of microorganisms to produce biological and biodegradable polymers (5) and synthetic polypeptides to study self assembly (6) are growing trends.

Interestingly, proteins have never been considered fit for specialty adhesive applications because they are sensitive to moisture, which plasticizes them, and in the moist state they are attacked by extracellular enzymes produced by microorganisms. Also, they do not inherently adhere to metals. However, some proteins not only set in the presence of water but also adhere to all kinds of surfaces, including metals and Teflon, and resist enzymatic degradation.

The ability to resist enzymatic degradation suggests very high cross-linking density, and the ability to bind to metals is generally attributed to the presence of diphenolic groups on the protein molecule. Consequently, proteins that possess a high degree of these three-dimensional cross-links can overcome sensitivity toward moisture and resist enzymatic degradation. Hence, appropriate chemical manipulation of proteins to achieve tight three-dimensional cross-linking should confer them the ability to resist the ravages of water and good mechanical characteristics. Ability to bond to metals would come from the introduction of diphenolic moieties to the protein backbone.

Adhesive Development

As mentioned, amino acid composition of proteins varies greatly, and some amino acids are absent altogether in certain proteins. To attain the desired characteristics, the proteins need to be subjected to certain modifications. These modifications will alter the chemical nature of the amino acid side chains. For decades biochemists and molecular biologists have developed many reactions for the modification of amino acids (7) because they wanted to find out which amino acids are found at the active site of an enzyme. Because even a temporary inactivation of the enzymatic activity was indicative of the presence of the amino acid being modified at the active site, these scientists were not interested in developing other reactions that yielded a stable bond.

When the aim is to produce new structural materials, such as adhesives and coatings, the modification reaction has to satisfy the following requirements:

- formation of irreversible covalent bonds
- use of a reagent that has selectivity toward preferably one amino acid
- presence of another reactive or potentially reactive moiety on the reagent, so that further chemical reactions can be affected after the modification

Very few reactions listed in the chemical literature (8) satisfy all of these requirements simultaneously. The sole exception is the reductive alkylation applicable to lysine residues. However, appreciable concentrations of lysine are found only in biologically active molecules, such as enzymes and hormones. Isolation of large enough quantities of such proteins in a pure state is extremely expensive and economically unfeasible.

Thus, one has to contend with proteins like whey protein and soy protein isolate, which have adequate contents of lysine and tyrosine. However, at least during the development of the procedure, use of a protein that has been thoroughly studied and whose amino acid sequence is known is as important as the higher content of these amino acids. This requirement is necessary to monitor each step of the reactions and to follow the progression of the process.

Industrial Adhesives

Gelatin–Epoxy Grafts

Although the amino acid composition of gelatin is not ideal, commercially available 300 Bloom gelatin has been chosen for this study, for the reasons given previously. Furthermore, gelatin has extremely good adhesive properties and is the main constituent of carpenters' glue. This commercially available gelatin consists predominantly of protein chains with over 1000 amino acids and smaller peptides formed during the preparation of gelatin. It has 36–38 lysines and 5 tyrosines distributed unevenly along its backbone.

Modification of Lysine Residues

Lysine residues were modified by reductive alkylation by using phenolic aldehydes such as 2,3-, 2,4-, and 3,4-dihydroxybenzaldehyde in the presence of sodium cyanoborohydride (Scheme I) (9). Reaction of an amine with an aldehyde yields a Schiff base, which is rather unstable, especially with aliphatic aldehydes. Although aromatic aldehydes yield more stable products, the reaction is never complete even when a selective reducing agent such as sodium cyanoborohydride is used. With gelatin, a conversion of only 60–65% of the available lysines was possible under the most favorable conditions.

Because the lysine content of gelatin is very low (36–38 lysines), considerable effort was expended to study the reaction with model compounds by using free lysine. Replacement of an electron-donating hydroxyl group with electron-withdrawing nitro groups in the aldehyde was only partially successful; with *p*-nitrobenzaldehyde, a conversion rate of 70–71% was achieved (Table I). Removal of the reaction water with azeotropic distillation or performing the reaction in the presence of a molecular sieve did not improve the reaction yield appreciably (10).

Because gelatin is a denatured protein, accessibility of aldehydes to lysines cannot be the reason for the low conversion. Furthermore, the nonreacting lysines react with epichlorohydrin and dinitrofluorobenzene nearly quantitatively. Hence, the incomplete conversion has to be attributed to the reversibility of the Schiff base (11) (Scheme II). The arylamine derivative that is formed after the reduction of the double bound, however, is stable and resistant to acid degradation.

Modification of Arginine Residues

There are about 50 arginine residues in the gelatin backbone. The guanidinium group of arginine is very difficult to modify due to its very high alkalinity. Such modifications have been achieved in the past by using glyoxal, malondialdehyde, or benzyl, and some of the reactions were reported to be reversible (12, 13). 4-Phenoxybenzil, on the other hand, reacts with both free arginine and arginine residues on gelatin (Scheme III).

The reaction yields one major and a minor product (14, 15). The model compound (a) is stable in boil-

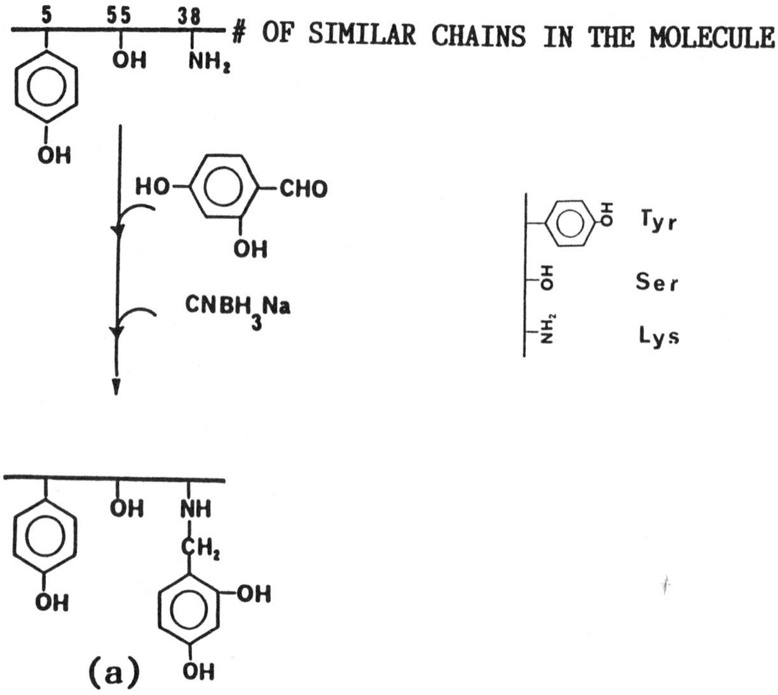

Scheme I. Modification of gelatin.

Table I. Extent of Modification of Free Lysine by Different Substituted Benzaldehydes

Benzaldehyde Substituent	Alkylation (%)
2,4-Dihydroxy	60.0
2,4-Dimethoxy	50.0
o-Nitro	59.7
m-Nitro	61.3
p-Nitro	71.8

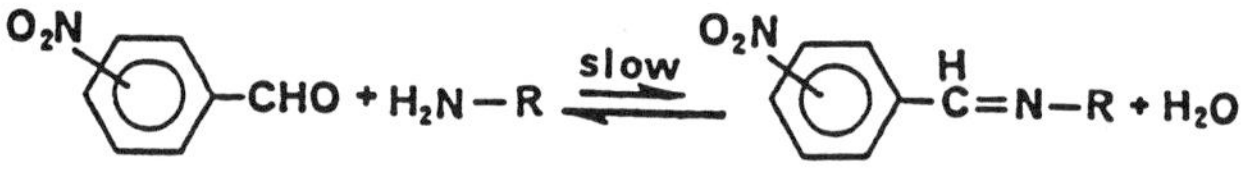

Scheme II. Mechanism of reductive alkylation.

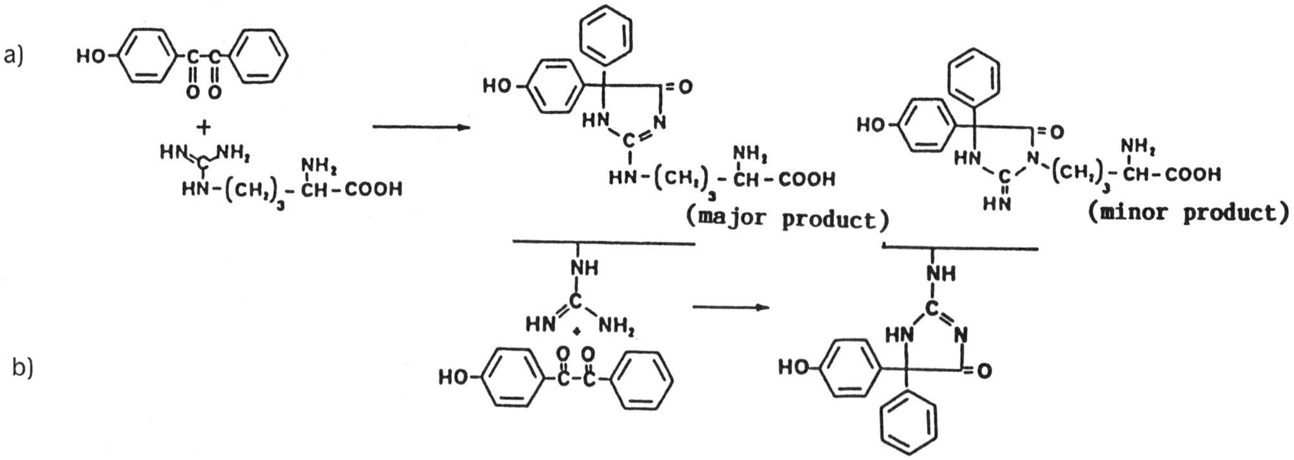

Scheme III. Modification of arginine.

ing with acids and alkali. Because of the high cost of 4-phenoxybenzil, this reaction is not likely to be useful for modifying arginine for industrial applications. Nevertheless, other chemicals that can react with the guanidinium group or ways to economically convert this group to ornithine can overcome this problem.

Modified gelatin was reacted in 2 M sodium hydroxide solution with 50 M excess epichlorohydrin in dimethyl sulfoxide (DMSO) at 60 °C for 5 h. The reaction mixture was filtered from any precipitated inorganic salts through packed glass wool, and the filtrate was poured into excess acetone. The precipitate was filtered, washed extensively with acetone, and dried in vacuo overnight (Scheme IV).

Epoxidized gelatin was dissolved in DMSO and heated to 50 °C. A 5 M excess of bisphenol A or any other polyphenol, such as resorcinol sulfide, was added and stirred until dissolved. An equimolar amount of sodium hydroxide (40% solution) was added gradually. After 6 h the reaction mixture was filtered to remove precipitated sodium hydroxide, and the product was obtained by pouring the filtrate into excess acetone. After extensive washing the product was dried in vacuo (Scheme V).

The product obtained after grafting with a polyphenol was once again reacted with epichlorohydrin, as described in a previous section. The resulting material was soluble in water, DMSO, ethylene glycol, and *N*-methylpyrrolidone (Scheme VI). Curing can be achieved by using any standard curing agent used for epoxy resins, such as aliphatic or aromatic amines, anhydrides, or Lewis acids (Scheme VII).

A typical curing mixture containing a graft prepared with resorcinol sulfide was slurred in XAMA-2 with pyromellitic-anhydride-bonded aluminum coupons after being heated 4 h at 150 °C. (XAMA-2 is a polyaziridine that can react with free carboxyl groups.) The coupons were cleaned with a chromic acid–sulfuric acid mixture after degreasing. The product had a tensile strength of 4000 psi. With sandblasted steel coupons, only 1500 psi was achieved.

Gelatin-Phenolic Resin Grafts

By using a similar stepwise procedure, a gelatin–resorcinol–formaldehyde graft was prepared. Modified gelatin was dissolved in DMSO and reacted with 40–80 M excess of formaldehyde at pH 12 and stirring

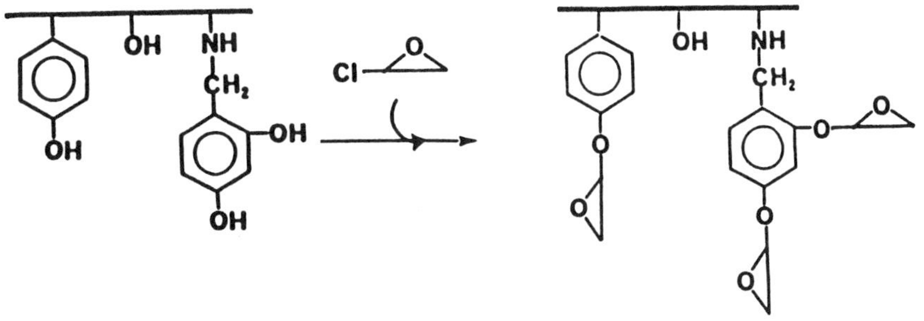

Scheme IV. Epoxidation of modified gelatin.

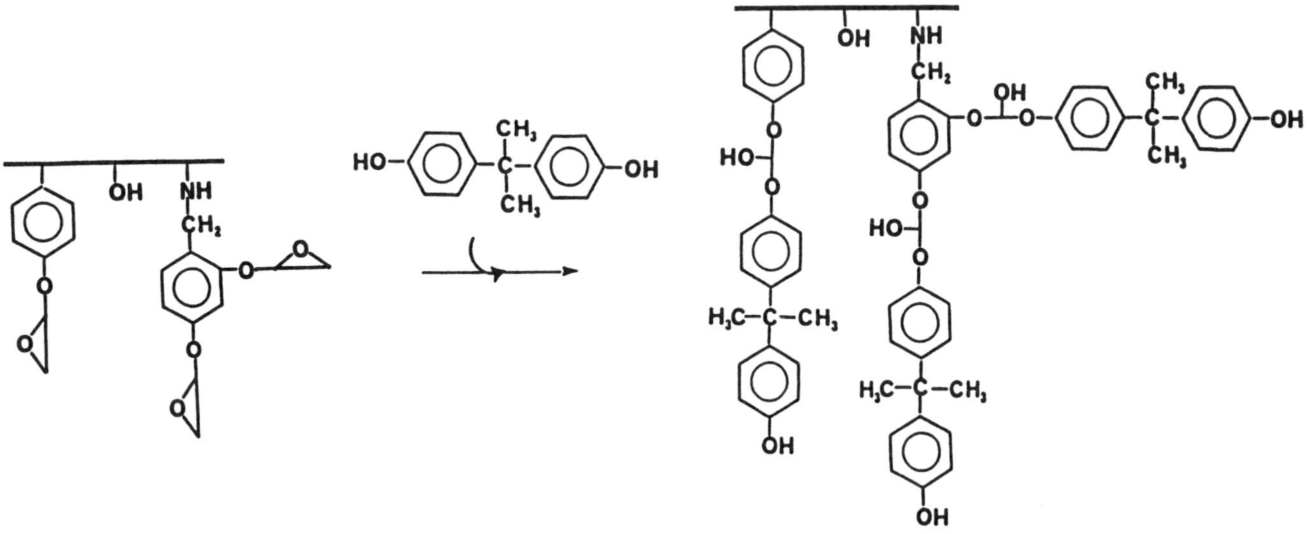

Scheme V. Grafting of polyphenol.

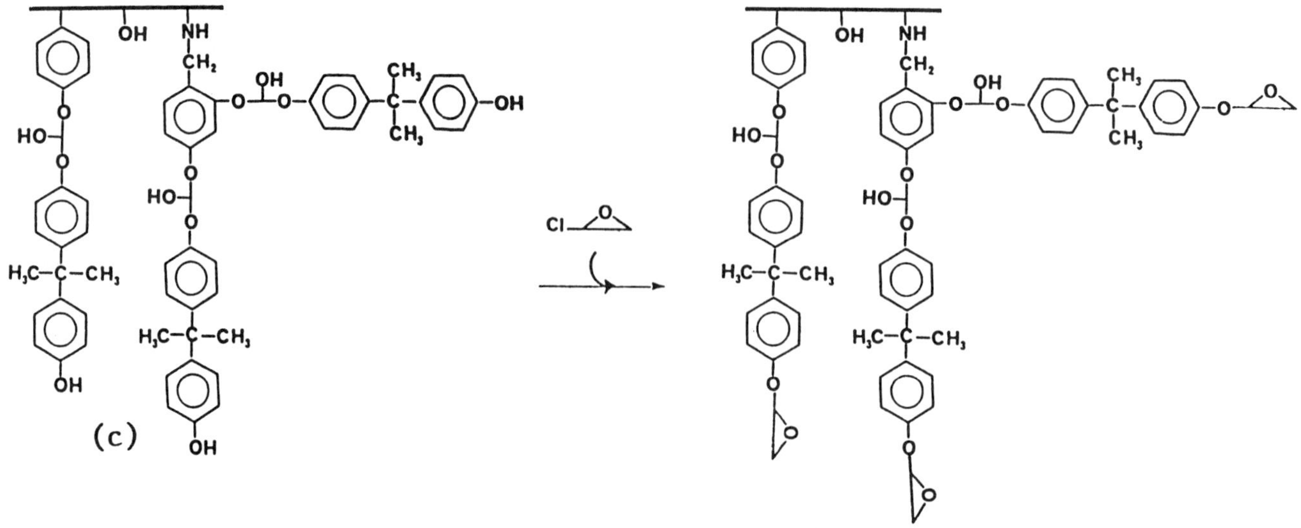

Scheme VI. Epoxidation of gelatin–polyphenol graft.

GELATIN–POLYPHENOL GRAFT $\xrightarrow{\text{DIAMINE, ANHYDRIDE ETC.}}$ CURED ADHESIVE

Scheme VII. Curing of graft.

at 90 °C for 5 h. After cooling, the reaction mixture was filtered through glass wool, and the product was obtained by pouring the filtrate into excess acetone. The precipitate was filtered, washed with acetone, and dried in vacuo at room temperature. The product was soluble in water. A nearly quantitative conversion was achieved.

The formylated product was dissolved in DMSO, a 50 M excess resorcinol was added with aqueous alkali to attain a pH of 12, and the mixture was heated at 70 °C for 4 h. The product was obtained by pouring into excess acetone. The precipitate was washed with methanol and dried in vacuo overnight.

The product is a thermoset material that can be cured with formaldehyde, epoxy resins, and other standard curing agents at appropriate temperatures. The graft can be mixed with fillers and formed with heat and pressure into three-dimensional objects by using suitable molds. Interestingly, the formulated gelatin can also be heat cured without any additional curing agent.

Bioadhesives

For many years barnacle cement has been considered an ideal bioadhesive because of the following properties:

1. It sets in the presence of water in a few minutes.
2. It adheres to teeth, bones, metals, and even to Teflon.
3. It does not shrink when dry.

4. It resists degradation by proteolytic enzymes.

However, the mechanism of action of barnacle cement had to be gleaned from indirect observations including histochemistry, histoenzymology, model chemical reactions, and amino acid composition studies because no inhibitor could be found to prevent its setting. The presence of phenolic residues (tyrosines) and the enzyme polyphenoloxidase (PPO) in the secreted cement as well as the cement gland suggests Scheme VII may explain this bonding.

According to this scheme, PPO first hydroxylates the tyrosine residues to L-dihydroxyphenylalanine (DOPA) and then oxidizes the residues to dopaquinone analog. The quinone then spontaneously reacts with the free-amino group of lysines residues. Model studies using different amines, quinones, and hydroquinones has supported this mechanism (*16*). Amino acid composition of barnacle cement is uncertain because the quinone-amine reaction introduces an amino group to the aromatic ring, which is resistant to acid and base hydrolysis. Thus, the very low tyrosine and lysine content seen in many amino acid composition studies on barnacle cement represents only those that have not been involved in the setting reaction!

Additional support for this view is provided by mussel byssus, which works with a very similar mechanism. The byssus contains several proteins. One of these, which is believed to be involved in the adhesion process, contains many repeats (70–80 times) of a unique decapeptide, ala-lys-pro-ser-tyr-hyp-hyp-thr-DOPA-lys, which contains L-DOPA in its sequence (*17*).

The lysine and tyrosine content of this decapeptide is 20%, an observation that seems to support the idea that amino acid analyses of barnacle cement provide low values for these two amino acids. This fact is important because the cross-linking density of this adhesive depends on the presence of a high content of phenolic groups, and the resistance of barnacle cement to proteolytic degradation indicates very high cross-linking density.

The significance of this problem was realized only after studies to produce a barnacle-cement mimic failed to reach a product similar to the native adhesive. For example, when the amine content of gelatin was increased by adding ca. 6% poly-L-lysine, the reaction mixture did not set for at least 20 min, and the product did not have the adhesive characteristics of barnacle cement.

Because the modification involved the introduction of phenolic groups, reductive alkylation with phe-

nolic aldehydes was used for the preparation of barnacle–cement mimic. As the first step, model reactions were carried out with different phenols and PPO. The reactions were followed by UV spectroscopy. As expected PPO accepted tyrosine (Figure 2), phenol, and catechol as substrates but not resorcinol, which cannot form a quinone. In the presence of an amine, new absorption peaks appear that are characteristic of mono- and disubstituted quinones. (3,4-Dihydroxybenzaldehyde-modified gelatin is similarly affected by PPO Figure 1.) Even in very dilute solutions a black precipitate forms after 20–30 min (Figures 2 and 3).

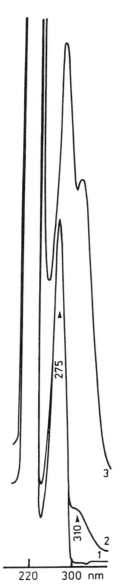

Figure 2. UV spectra of the reaction between tyrosine and PPO: 1, tyrosine alone; 2, tyrosine and PPO after 2 min and after 10 min, 275-nm peak is characteristic of catechol.

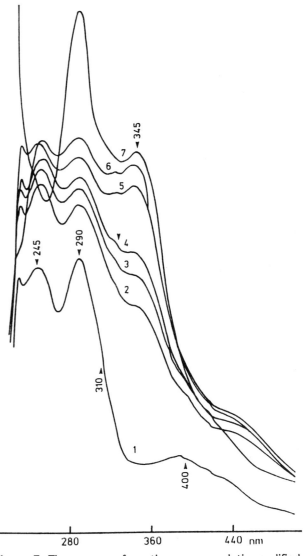

Figure 3. Time course of reaction among gelatin modified with 3,4-dihydroxybenzaldehyde alone and in the presence of spermine, and spermidine with methyl catechol used as a model for tyrosine: 1–6, spectra of reaction mixture containing modified gelatin and spermine at 3, 45, 84, 120, 300, and 600 min, respectively; 7, reaction mixture containing methyl catechol PPO and spermidine after 150 min, 245-nm peak is due to benzaldehyde and 345-nm peak is due to disubstituted benzoquinone.

Even though enzymatic reactions are very simple to perform, making an adhesive by this process requires further considerations. When 60% modified gelatin is used for these studies, about 15–16 out of the 36–38 available lysines will retain their free amino groups; therefore, there are not enough amino groups for efficient cross-linking. To compensate for this deficiency, either an equivalent amount of unmodified gelatin or other polyamines such as polylysine or spermine have to be added to the modified gelatin to achieve adequate cross-linking.

When a mixture consisting of PPO, 60% modified gelatin, and unmodified gelatin (60% of the quantity of the modified gelatin) is mixed as 2–2.5% solutions in phosphate buffer (pH 6.5) and placed between slices of horse femur and clamped for 20 min, bonding is achieved. Similar reactions can be carried out by using polylysine, spermine, or spermidine with modified gelatin. Obviously, the adhesion occurs through mechanical interlocking of the adhesive into the pores of the bone (18).

The major problem using gelatin for this purpose, in addition to its unsatisfactory amino acid composition, is the shrinkage that the adhesive undergoes during curing and drying. Because of the rod-like structure of the adhesive, which leads to solutions of very high viscosity, gelatin solutions can only be prepared at 2–2.5% concentrations at ambient temperature. Higher concentrations of up to 10–12% are possible at higher temperatures. However, such high temperatures destroy the catalytic activity of the PPO. Selection of other proteins with higher lysine and tyrosine content and other polyamine could overcome these problems.

Coating Development

The ability of barnacle cement and mussel byssus to adhere to metals is probably due to the presence of diphenolic groups on their protein backbones. In the biological realm, diphenols can only be formed from tyrosines by PPO. The presence of L-DOPA with many repeats in a byssus protein demonstrates that L-DOPA does indeed occur in bioadhesives, and that it could be responsible for the adhesion of these biological materials to metals.

A completely unrelated set of observations made on quinone and its derivatives also support the idea that binding to metals is through the presence of diphenolic moieties (19). By using cyclic voltammetry, such compounds were demonstrated to bind even to gold and platinum, displace water, and cover the metal surface completely. A volatile compound, p-benzoquinone, when coating the surfaces of noble metals, resisted 10^{-9} mm vacuum. The mechanism was suggested to be chemisorption providing very strong bonding.

Polymers based on the reaction between diamines and quinones, where each repeat unit contains two quinone oxygens and two imino nitrogens, also bind to metals with extraordinary affinity. They displace water from wet, rusty steel surfaces and have excellent anticorrosive activity (20–22). For these reasons, it was decided to find out whether the introduction of diphenolic groups onto a protein backbone could enable a water-soluble protein to coat metals and become insoluble in water. The protein chosen for this purpose was whey protein, which is a water-soluble by-product of the dairy industry. It contains 3–4% tyrosine and 7–9% lysine. Whey protein was subjected to reductive alkylation with 2,4- and 3,4-dihydroxy-benzaldehyde. The modified protein was completely soluble in water and DMSO. A 20–30% solution of the modified protein in DMSO, applied to a metal panel and allowed to dry, formed a coating that remained attached to the metal after 165 h in water even after being scored. This result demonstrates that the presence of diphenolic moieties allows a water-soluble protein to adhere to metals and become water insoluble. The coatings produced in this way seem to have some anticorrosive activity.

Attempted Preparation of Protein Rubbers

Among the most remarkable products of nature are the protein rubbers resilin and abductin. Both of these materials conform closely to the classical rubber-elasticity theory. Both act as mechanical-energy stores. Both are able to recover completely after applied deformation. They also represent random coiled proteins cross-linked by tyrosine–tyrosine bonds. Resilin has a resilience of 96–97%, which is as good or better than the very best synthetic rubbers (23).

Resilin occurs in insect wings and legs. During the flight, resilin in the wing hinge is deflected at extremes of the wing strokes, slows the wing down, and stores the kinetic energy of the wings as "strain energy". The stored energy is then delivered back to the wing for the start of the next stroke, as kinetic energy, and it enables the wing to accelerate more quickly into the next stroke. Resilin is also used to store energy received at a relatively slow rate by muscular contraction, then it releases the energy at far faster rate and enables the jump of the locust.

After finding out that protein rubbers do exist, the question that has to be addressed is: Is it possible to pick a protein and to treat it chemically or enzymatically so that it can "mimic" the characteristics of resilin? The answer is "yes".

Because proteins are brittle solids in the dry state, any viscoelastic material produced has to be plasticized. This process can be achieved by moistening the

Table II. Tensile Strength of Flexibilized Gelatin Samples

Thickness (cm)	Width (cm)	Area ($m^2 \times 10$)	Elongation (cm)	Elang. (%)	Load (kg)	Tensile at Break (kg/m^2)	Initial Modulus ($kg/m^2 \times 10^7$)
Gelatin 537-3A[a]							
0.0498	0.2633	13	0.42	21.00	1.260	969,230	1.723
0.0492	0.2671	13	0.48	24.00	1.370	1,053,846	1.750
0.450XXX	0.2537	11	0.65	32.50	1.284	1,167,273	1.672
0.0454	0.2432	11	0.59	29.50	1.119	1,017,273	1.745
0.0474	0.2568	12	0.54	26.75	1.258	1,051,905	1.723
Gelatin 537-3B[b]							
0.0435	0.2470	11	0.91	45.50	0.580	539,811	0.5091
0.0436	0.2578	11	0.88	44.00	0.655	582,736	0.7455
0.0434	0.2529	11	0.83	41.50	0.582	546,255	6.182
0.0431	0.2599	11	0.87	43.50	0.612	546,346	0.6000
0.0434	0.2544	11	0.87	43.63	0.607	549,787	0.6182

[a]3A is 5 g epoxidized decaglycerol and 5 g gelatin/300 mL water.

[b]3B is 5 g epoxidized decaglycerol and 4 g gelatin/300 mL water.

protein powder and kneading. This process, however, will yield temporary plasticity. Glycerol, because of its hygroscopic nature, was used successfully for this purpose (24). By using a polyglycerol (a thermally produced condensation product of glycerol) instead of glycerol, one can have prolonged plasticization of gelatin. This approach using an artificial skin model was recently described (25). Thus, by mixing gelatin with epoxidized polyglycerol, one can obtain a permanently plasticized material with impressive mechanical characteristics (Table II).

By starting with these observations, it was possible to produce a ball that bounced as well as any rubber ball and a disk that had memory (i.e., when stretched to twice its length, it receded to its original length; unpublished observations).

Earlier studies (26) elucidated the structural characteristics of resilin as random coils cross-linked by tyrosine–tyrosine and tyrosine–tyrosine–tyrosine bonds (26). When tyrosine is treated with peroxidase, one obtains di- and tritysoines (unpublished observations). Also, when starting with skin fibroin, a resilin-like rubbery material was synthesized by incubation with peroxidase and hydrogen peroxide (27).

Consequently, one could start with other proteins or with synthetic polypeptides to enzymatically prepare rubber-like materials. How these materials can be used remains to be seen.

Conclusions and Future Prospects

Proteins are certainly unique as functional polymers in their own biological functions, but not as high-performance polymeric materials. However, in this chapter the possibility of producing high-performance polymeric materials from rather amorphous proteins such as gelatin, soya, and whey protein by-products has been demonstrated for the first time. Of course, all of the products described in this chapter require much more intensive studies before optimum characteristics and uses can be defined. This research will depend on whether those that support research can, for once, learn to bet on unconventional approaches. If successful, these new developments will help reduce the problem of disposal of a large portions of by-products of various industries and certainly reduce dependence on petroleum-based raw materials. Furthermore, protein-based polymeric products are likely to be biodegradable.

References

1. *Kirk–Othmer Concise Encyclopedia of Chemical Technology*; Wiley-Interscience: New York, 1985; pp 554, 565.
2. Ward, A. G.; Courts, A. *The Science and Technology of Gelatin*; Academic: Orlando, FL, 1977.
3. Nayale, P. L.; Lenka, S.; Pati, N. C. *Angew. Makromol. Chem.* **1979**, *75*, 29.
4. Varma, D. S.; Sadhir, R. K. *Angew. Makromol. Chem.* **1979**, *81*, 179.
5. *Plastics from Microbes: Microbial Syntheses of Polymers and Polymer Precursors*; Mobley, D. P., Ed; Hanse Publishers: Munich, Germany, 1994.
6. *Hierarchically Structured Materials*; Aksay, I. A.; Baer, E.; Sarikaya, M.; Tirrell, Eds; Materials Research Society: Pittsburgh, PA, 1992.
7. Means, G. E.; Feeney, R. E. *Chemical Modification of Proteins*; Holden-Day: San Francisco, CA, 1971.
8. Means, G. E. *J. Prot. Chem.* **1984**, *3*, 121.
9. Kaleem, K.; Chertok, F.; Erhan, S. *Nature (London)* **1987**, *325*, 328–329.
10. Kaleem, K.; Chertok, F.; Nithianandam, V. S.; Erhan, S. *Int. J. Peptide Protein Res.* **1990**, *35*, 542–544.
11. Borsch, R. F.; Berstein, M. D.; Durst, H. D. *J. Am. Chem. Soc.* **1971**, *93*, 2897–2902.
12. Toi, K.; Bynum, E.; Norris, E.; Itano, H. A. *J. Biol. Chem.* **1967**, *242*, 1036–1043.

13. Itano, H. A.; Yamada, S. *Anal. Biochem.* **1972**, *48*, 483–490.
14. Kaleem, K.; Chertok, F.; Erhan, S. *J. Biol. Phys.* **1987**, *15*, 63–68.
15. Kaleem, K.; Chertok, F.; Erhan, S. *J. Biol. Phys.* **1987**, *15*, 71–74.
16. Lindner, E. *Proceedings of the 1973 Pacific Conference on Chemistry and Spectroscopy*; San Diego, CA, 1973.
17. Waite, J. H. *J. Biol. Chem.* **1983**, *258*, 2911–2915.
18. Kaleem, K.; Chertok, F.; Erhan, S. *Angew. Makromol. Chem.* **1987**, *155*, 31–43.
19. Soriaga, M. P.; Hubbard, A. T. *J. Am. Chem. Soc.* **1982**, *104*, 2735–2742.
20. Kaleem, K.; Chertok, F.; Erhan, S. *Prog. Org. Coatings* **1987**, *15*, 63–71.
21. Nithianandam, V. S.; Kaleem, K.; Chertok, F.; Erhan, S. *J. Appl. Polym. Sci.* **1991**, *42*, 2893–2897.
22. Nithianandam, V. S.; Chertok, F.; Erhan, S. *J. Coatings Technol.* **1991**, *63*, 47–50.
23. Vincent, J. F. V. *Protein Rubbers; Structural Biomaterials*; John Wiley: New York, 1982; pp 59–61.
24. Yannas, I. C.; Tobolsky, A. V. *J. Appl. Polym. Sci.* **1968**, *12*, 1–8.
25. Shinde, G. S.; Nithianandam, V. S.; Kaleem, K.; Erhan, S. *Biomed. Mater. Eng.* **1992**, *2*, 123–126.
26. Andersen, S. O. *Biochim. Biophys. Acta.* **1964**, *93*, 213–215.
27. Andersen, S. O. *Acta Physiol. Scand.* **1966**, *66*, 8–81.

Part Two

Radiation Effects and Applications

Polymer Photochemistry and Photo-Cross-Linking

Hans-Joachim Timpe

This chapter presents a general overview of photo-cross-linking of polymers. It discusses the basic phenomena of light absorption by polymers and subsequent processes leading to polymer cross-linking. The actual cross-linking between photosensitive groups (chromophores) on the polymer chains (or chain segments) may be induced either directly by light or by the use of photoinitiators such as ketones, nitrenes (generating free radicals), or protonic or Lewis acids (generating cations). Mechanistic aspects of these processes are discussed, and advantages and limitations of different methods are outlined. Suggestions are also made as to the future development of photo-cross-linkable polymers, and the need for systematic studies of photo-cross-linking of polymers is emphasized.

The oldest reported example of photo-cross-linking in a polymer-based system dates back to pre-Biblical times. For the preparation of Egyptian mummies linen cloths were dipped into a solution of lavender oil containing Syrian asphalt. The asphalt contains several unsaturated aromatic and heterocyclic compounds, which harden under the influence of light. In the last century, a similar photohardening of asphalt was used by Niepce in the first imaging material for permanent photographs (1822). Later in the nineteenth century, the immobilization of gelatin by exposure to sunlight in the presence of potassium dichromate was discovered. This type of photo-cross-linking of polymers formed the basis of the first photoresist reaching the market, and this resist is still used in some countries for the preparation of particular colored linens.

In recent years, photo-cross-linking of polymers has attracted considerable attention as a rapid, energy efficient, and in some cases nearly pollution-free method for the application of polymers for emerging technologies. Nevertheless, compositions based on the photo-cross-linking of polymers represent only a rela-

tively small fraction of the whole market in the area of the so-called photopolymers, because they are mostly derived from photoinduced polymer formation processes starting from monomeric or oligomeric compounds. This situation signifies current limitations for the use of photo-cross-linking of polymers, as well as the potential for further developments.

Despite numerous current and potential applications of photo-cross-linking, available information on the fundamentals of photo-cross-linking of polymers is relatively limited. This deficiency is largely because special experimental techniques necessary for monitoring cross-linking processes, complicated syntheses of specifically functionalized polymers, and the influence of the polymeric environment on photo-cross-linking make reproducible results difficult to obtain. Furthermore, research on this area requires experimental techniques related to several fields of science and technology.

Thus, in this chapter attention will be directed mainly to fundamentals of polymer photochemistry and photo-cross-linking. Further chemical and practical aspects of this topic are more fully detailed in subsequent chapters. Within this framework, I introduce the field from a personal point of view, and also discuss the potentials and limitations of photo-cross-linking of polymers for technological applications.

Photo-Induced Processes in Polymeric Media

Excitation and Deactivation Processes

Before considering details of the pathways in which polymers may be photochemically cross-linked, it is necessary to provide a general framework for the interaction between the incident light (photons) and macromolecules. Because most applications of photo-cross-linking of polymers are concerned with processes that occur in thin films (*1–5*), the discussion here is devoted mainly to amorphous polymer systems (for further information, *see* references *6–8*).

Photochemical reactions of macromolecules are initiated by the absorption of light in the UV or visible range, and hence they are fundamentally different from thermal (dark) reactions. The absorption of light leads to an excited state (an electronic isomer of the ground state) due to changes of energy and spin state of some of the electrons. This excited state exhibits physical and chemical properties different from those of the ground state, which in many cases provides reaction pathways not possible by thermal activation.

Photochemical processes can be divided into three groups, and they are usually represented by a Jablonski diagram (Figure 1).

- absorption step (A) leading to the excited molecule

- photophysical processes including vibrational relaxation (SR), internal conversion (IC), intersystem crossing (ISC), fluorescence (F), and phosphorescence (P)
- photochemical processes (CH) giving rise to chemical conversion and the formation of new compounds

The main photophysical processes include the following:

- fluorescence: radiative transition between singlet states (rate constant, $k_f = 10^5$ to 10^9 s^{-1})
- phosphorescence: radiative transition between triplet state and singlet ground state (rate constant, $k_p = 10^{-1}$ to 10^6 s^{-1})
- internal conversion: nonradiative transition between states of the same spin multiplicity via release of heat (rate constant, $k_{ic} = 10^8$ to 10^{12} s^{-1})
- intersystem crossing: nonradiative transition between states of different spin multiplicity (rate constant, k_{isc}^S or k_{isc}^T, 10^7 to 10^{10} s^{-1})
- vibrational relaxation: transition between different vibrational excited states (rate constant, $k_r = 10^{10}$ to 10^{12} s^{-1})

As follows from Figure 1, a photochemical process (CH) of the macromolecule either in singlet state (rate constant, k_r^S) or triplet state (rate constant, k_r^T) must compete with all of the photophysical processes. Only when the photochemical pathways are faster than the photophysical processes is the yield of the chemical reaction sufficiently high to lead to the formation of new products [i.e., $k_r^S \approx \Sigma(k_f + k_{ic} + k_{isc}^S)$ and $k_r^T \approx \Sigma(k_p + k_{isc}^T)$]. As a consequence of the typical rate constants

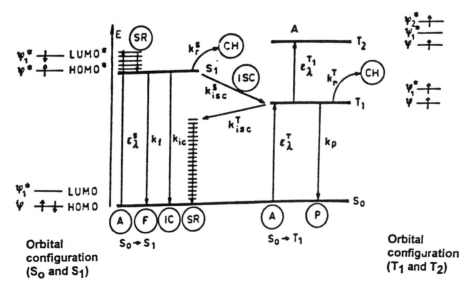

Figure 1. Jablonski diagram of photoinduced processes and orbital situation of frontier electrons. For explanation of the symbols, *see* text. The process indicated with ελ (extinction coefficient) is the absorption of photons.

given, singlet states are pronounced only for mono-molecular photoreactions (e.g., intramolecular rearrangement and homolytic bond cleavage). Because of its longer lifetime, the triplet state can be postulated to be more reactive in bimolecular photoreactions.

It can, therefore, be noted that:

1. Photoinduced processes proceed very rapidly as compared with thermal reactions (i.e., those starting from the ground state).
2. Rate constants for photophysical processes of the triplet excited state are several order of magnitude lower than those of the singlet state.
3. Rate constants and efficiencies of ISC processes may be high compared with those of IC processes (typical examples are ketones).
4. Rate constants of ISC process are higher if heavy atoms are present in the molecule or if solvents containing heavy atoms are used as reaction media.

For many kinetic discussions of photoinduced processes it is useful to introduce the lifetime (τ) of the excited state as defined in eq 1. To express the effectivity of photoinduced processes the overall quantum yield (ϕ) must be defined. For the chemical deactivation process of excited states, this is an experimental value indicating the ratio of converted molecules (n_a) to absorbed photons (n_q), as shown in eq 2. Similar kinetic expressions can be developed for all photophysical processes. As a typical example, and with the help of the process steps outlined in Figure 1, the fluorescence quantum yield (ϕ_f) is obtained by eq 3.

$$\tau = 1/\Sigma k_i \tag{1}$$

$$\phi = n_a/n_q \tag{2}$$

$$\phi_f = k_f\tau_s = k_f/(k_f + k_{ic} + k_{isc}) \tag{3}$$

Light Absorption in Macromolecules

The fundamental law of photochemistry states that only light absorbed by the molecule (i.e., also by a macromolecule) can initiate a photochemical process (Grotthus–Draw law). Two conditions are required for light absorption by a molecule.

1. The wavelength (λ) (or the frequency, ν) of the light must correspond to the energy of a possible electronic transition in the molecule, as expressed by eq 4, where h is Planck constant, and E_a and E_g are energies of the excited and ground state, respectively.

$$h\nu = E_a - E_g \tag{4}$$

2. The interaction between electrical components of the incident light and the polymer molecule must lead to a change of either charge distribution or dipole moment of the molecule (i.e., the transition moment integral must be non-zero).

Both of the above conditions can be used to explain several fundamental aspects of photochemical and photophysical processes.

The bonding (frontier) electrons of an organic molecule oscillate with a frequency of 10^{15} to 10^{16} s^{-1}. Therefore, it can easily be explained why these electrons interact with photons of incident light having wavelenghts of λ = 200–700 nm ($\nu = c/\lambda$, where c is the velocity of photons). These electrons can be contained only in particular molecular orbitals (MOs), the formation of which occurs within the framework of the Hückel theory by combination of two atomic orbitals (LCAO model; linear combination of atomic orbitals). Different MOs give rise to discrete energy states (*see* Figure 1), and transition between these states is possible either by absorption or emission of energy.

Molecules have the minimum electronic energy in their ground state, which normally is symbolized as S_o. Then, in all molecules, two electrons with antiparallel spin are occupied in the highest bonding orbital (highest occupied molecular orbital; HOMO), in accord with the Pauli exclusion principle. The lowest energetic transition occurs by interaction of the molecule with the incident light, when one of these electrons is promoted to the lowest unoccupied molecular orbital (LUMO) exhibiting antibonding character. The state thus produced is the first excited state, symbolized either as S_1 (both frontier electrons exhibit antiparallel spin) or T_1 (parallel spin of both electrons) depending on the spin configuration of this state.

Most light absorption processes obey the Frank–Condon principle, and in consequence, the first excited state formed is initially both electronically and vibrationally excited. As mentioned previously, in the amorphous solid environment of the polymeric molecule, excess vibrational energy is dissipated very rapidly by SR processes, leading to the lowest vibrational level of the appropriate electronically excited state.

If incident light is used having appropriate energy, then electronic transitions to energetically higher lying orbitals are also possible (e.g., S_2, S_3 or T_2, T_3). Normally, these upper excited states deactivate very fast by IC processes to the lowest state of the same spin configuration, and their existence can be largely ignored in the present discussion. In all cases, singlet states possess greater energies than the corresponding triplet states, because of the greater repulsion energy in the former.

By using the singlet–triplet classification, three electronic transitions can be distinguished (Figure 1):

- S_o, S_n transitions
- T_n, T_{n+1} transitions
- S_o, T_n transitions

Only the first two transitions are characterized by an effective interaction between the light and the electrons of the molecule (allowed transitions). The S_o, T_n transition type of electronic excitation is strongly restricted (spin forbidden).

Another classification of the electronic transitions uses the different orbitals existing in a large organic molecule. These examples include the following transitions: σ, σ^*; n, σ^*; n, π^*; and π, π^* (Figure 2).

The indicated orbitals are associated with typical structural groups contained in a molecule. Within the photochemical framework, such units are called chromophores or chromophore groups. As follows from Figure 2, the transitions of these chromophores are characterized by different transition energies, and therefore, by different spectral ranges in which they occur. Typical examples for spectral parameters of chromophores incorporated in polymers are given in Table I and can serve as a guideline for excitation energies of a particular macromolecule.

Because of the various structural elements in most macromolecules, all of the previously mentioned transitions can occur simultaneously in these compounds. However, only n, π^* and π, π^* transitions are important for photo-cross-linking processes of polymers discussed here. If different types of orbitals are present in a given macromolecule, then several types of transitions may occur when energetic conditions corresponding to eq 4 are obeyed. As a first approximation, different electronic transitions mentioned previously are characterized by specific chemical reactions (see later discussion).

For quantitative description of the interaction between light and the macromolecule being irradiated, two further aspects have to be considered. The first aspect is that when the light penetrates the polymer layer, it will be partly reflected at the phase boundaries (e.g., air–polymer or polymer–substrate), absorbed within the polymer layer, or transmitted through it. The reflected and the transmitted light are both lost as far as the photochemical conversion of the macromolecule is considered. The ratio between the transmitted and the absorbed light is closely related to the chemical structure of the macromolecule. The absorption of light (A) by a macromolecule layer can be evaluated by the Beer–Lambert law according to eq 5, where I_o and I are intensities of the incident and transmitted light, respectively, d is layer thickness (cm), c_m is concentration of the macromolecule within the layer, and $\varepsilon\lambda$ is the decadic extinction coefficient (L cm^{-1} mol^{-1}). Although this law is strictly valid only for highly dilute solutions of low molecular weight compounds, in many cases it also gives well-approximated A values for solid polymer layers.

$$A = \log(I_o/I) = \varepsilon_\lambda c_m d \qquad (5)$$

The second aspect is that the value of the extinction coefficient is, in a first approximation, associated

Table I. Typical Chromophores Present in Macromolecules and Their Spectral Parameters

Chromophore	Structural Formula	λ_{max} (nm)	ε_{max} (l mol^{-1}cm^{-1})	Transition Type
Alkane	$R_3C–H$	125	1,000	σ, σ^*
	$R_3C–CR_3$	135	1,000	σ, σ^*
Diene	$R_2C=CR^1CR^2C=CR_2$	220	20,000	π, π^*
Ether	$R_3C–O–CR_3$	185	250	n, σ^*
Alkyl halide	$R_3C–Cl$	175	400	n, σ^*
	$R_3C–Br$	210	600	n, σ^*
Alkene	$R_2C=CR_2$	200	10,000	$\pi, \pi*$
Aromatic	$R^1–C_6H_5$	255	200	π, π^*
	$R_2C=CR^1C_6H_5$	285	1,000	π, π^*
Ketone or	$R^1–CO–R^2$	190	1,000	π, π^*
aldehyde		280	20	n, π^*
	$Ar–CO–R^1$	310	30	n, π^*
	$R_2C=CR^1COR^2$	330	25	n, π^*
Nitrile	$R_3C–CN$	220	150	π, π^*
Ester	$R^1–COOR^2$	210	50	n, π^*
Amide	$R^1–CONR_2$	220	150	n, π^*
Nitro	$R_3C–NO_2$	300	100	π, π^*
		600	20	n, π^*

Note: R, R^1, R^2 is H or alkyl, Ar is aryl.

Source: Data are taken from reference 3.

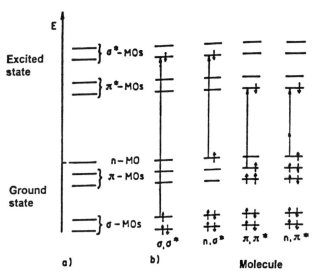

Figure 2. Hypothetical energetic scheme for orbitals (a) and different electronic transitions (b) in a molecule. Electronic configuration represents the singlet states.

with the type of electronic transition occurring after the absorption of light. In general, the $\varepsilon\lambda$ values of π,π^* transitions are much higher compared to those of n,π^* transitions. Furthermore, forbidden transitions (for example S_0, T_1 transitions in Figure 1) exhibit very small $\varepsilon\lambda$ (i.e., $\varepsilon\lambda)\lambda < 1$ cm^{-1} mol^{-1}). Typical values of $\varepsilon\lambda$ for most chromophore groups in macromolecules are summarized in Table I. The data show a variation of several orders of magnitude for $\varepsilon\lambda$ indicating the wide range of excitations possible by different electrons. Thus, for a constant layer thickness, macromolecules with different chromophore groups lead to different degrees of absorbance.

For example, calculation using eq 5 and 10^{-3} M of a chromophore with $\varepsilon\lambda = 10^4$ L cm^{-1} mol^{-1} shows that even in a layer thickness of $d = 0.2$ cm, 99% of the incident light will be absorbed. Accordingly, transformations take place only near the surface of this layer, and all macromolecules outside of this thin layer remain unaffected. In contrast, if the extinction coefficient is small (e.g., 10 L cm^{-1} mol^{-1}), most of the light will be transmitted through the layer. The exposure time must be prolonged to excite many macromolecules and to achieve high photochemical reaction yields. These calculations show the importance of the extinction coefficient for an effective use of the incident light.

Energy Transfer in Excited Molecules

Following irradiation, the excitation energy flows between different states of the same molecule. However, energy can also be transferred to other molecules with different chemical structures in their ground state levels (9). This energy transfer process can be described by eq 6, where D and A represent donor and acceptor, respectively. The fundamental energetic prerequisite for such energy transfer is given by eq 7. In most cases, the reaction based on eq 6 proceeds as a downhill process only.

$$D^* + A \xrightarrow{k_{et}} D + A^* \tag{6}$$

$$E_{D^*} \geq E_{A^*} \tag{7}$$

At the moment of the transfer, donor and acceptor molecules are coupled and form a new entity. The energy transfer in this entity can be described by two different models: namely coulombic or dipolar interaction (Förster mechanism) (10, 11) or electron exchange (Dexter mechanism) (12, 13).

The simplified form of the Förster mechanism indicates that light absorption by a molecule of D produces D*, followed by fluorescence of D*, and reabsorption of the emitted light by a molecule of A leads to A*. Therefore, the rate of energy transfer depends on the fluorescence intensity (e.g., oscillator strength

of the fluorescence light) and on the fluorescence lifetime (τ_s^D) of the donor, as well as on the spectral overlap between the fluorescence spectrum of D and the absorption spectrum of A. In general, the Förster model can successfully described the energy transfer between singlet states.

In many cases it has been observed that a coulombic interaction between two molecules is possible only over a distance of up to 50 Å. The general relationship between the rate constant of the energy transfer (k_{et}) and the molecular separation (R) can be written as eq 8, where R_0 is the critical distance at which the rate of energy transfer and the rate of the spontaneous deactivation of D* are equal. The sixth-power dependence on separation of the two molecules is a most important feature of this transfer mechanism; the transfer rate decreases sharply as the distance between D and A increases. However, a diffusional encounter between the two molecules is not required, which makes the energy transfer possible in viscous media.

$$k_{et} = 1/\tau_s^D (R_0/R)^6 \tag{8}$$

The exchange mechanism (Dexter mechanism) is based on the orbital overlap between the donor and the acceptor molecules as it is given in Figure 3. This form of electronic interaction works only when D* and A are within a collisional distance (in the order of 10 Å). If the HOMO-LUMO energy gaps of donor and acceptor molecules are similar, then electron return can occur leading to low energy transfer efficiency.

As follows from Figure 3, spectral overlap between the two molecules is necessary again, and therefore, $E_{D^*} - E_{A^*} < 0$. However, exchange transfer is rather nonspecific and is also less sensitive to changes in multiplicity of the electronic states between which the energy transfer takes place. Thus, the theoretical principles of the exchange mechanism can be used to

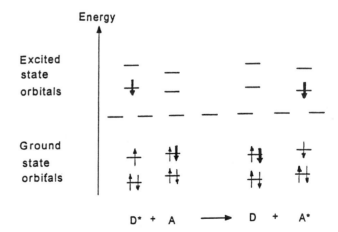

Figure 3. Singlet–singlet energy transfer described by electron exchange mechanism represented in a Hückel orbital diagram. The signed electrons have been exchanged.

interpret energy transfer processes between singlet states, triplet states, and between singlet and triplet states. Irrespective of the mechanism of a particular energy transfer process, the rate constants for these processes are $k_{et} = 10^9$–10^{10} L mol^{-1} s^{-1}.

Energy transfer processes play an important role in the photo-cross-linking of polymers because they end up either with the deactivation of the excited state responsible for cross-linking or with the formation of that excited state. The deactivation case is a quenching reaction, and a compound functioning in this way is a quencher. This reaction leads to a loss of excitation energy and decreases the efficiency of cross-linking. Either the Stern–Volmer expression (eq 9) or the Perrin formula (eq 10) can be used to describe that process quantitatively. In eqs 9 and 10, τ is the lifetime of the corresponding excited state in the presence of the quencher concentration [A], τ_o is lifetime in the absence of the quencher, V is volume of the quenching sphere, and N is Avogadro number. Equation 9 is useful for dilute solutions of D and A, whereas eq 10 is better suited for energy transfer in solid systems.

$$\tau/\tau_o = 1 - k_{et}\tau_o[A] \qquad (9)$$

$$\tau/\tau_o = \exp(NV[A]) \qquad (10)$$

By using the previously mentioned values of lifetimes and eq 9, it follows that at small quencher concentrations triplet states are quenched more efficiently compared to singlet states. Because of the high k_{et} values, the quencher must be present in the system in concentrations of 10^{-3} to 10^{-4} L/mol for triplet states and above 10^{-2} L/mol for singlet states to give high quenching rates. Typical examples important for the subject discussed in this chapter are the quenching of triplet excited ketones by C=C double bonds bearing functional groups (14) or by oxygen (15).

However, using energy transfer processes for photo-cross-linking of polymers is also desirable. As follows from eq 6, energy transfer makes it possible to populate an excited state of a particular molecule, A, without light absorption by this molecule. In cases where the spectral mismatch between the light emitter and the polymer of A does not allow an efficient light absorption by this compound, such a process extends the action spectrum of the system to longer wavelengths. This process is termed spectral sensitization, and the component D, which transfers the energy, is called the sensitizer.

Sensitization is commonly used for the generation of triplet states. The operation of this form of excited state population is illustrated in Figure 4, from which the following conditions can be deduced.

1. The S_o,S_1 transition of the sensitizer should occur at longer wavelengths than that of A, and the

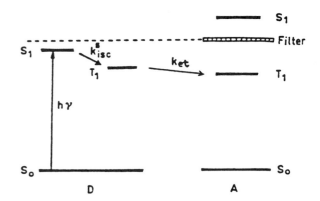

Figure 4. Scheme of a triplet–triplet energy transfer. In consequence of the filter used (i.e., glass filter or colored solutions), a direct S_o,S_1 transition of the energy acceptor (A) is impossible.

response of D should be high in the spectral region where A does not absorb.
2. Both the rate constant of ISC process and the triplet lifetime of the sensitizer must be high to increase the probability of energy transfer between D* and A.
3. The triplet energy (E_T) of the sensitizer must lie above the E_T value of the acceptor to provide an exothermic energy transfer.
4. Chemical deactivation pathways of the sensitizer (especially such processes with the acceptor) should be very depressed.

Ketones, polyaromatic hydrocarbons, heteroaromatics, and dyes are usually suitable for sensitized generation of triplet states (16).

Electron Transfer in Excited Molecules

Because of their half-filled HOMOs and LUMOs, many excited molecules are strongly polarized. Therefore, an interaction with molecules polarizable in their ground state can take place leading to generation of complexes (17, 18). Analogous to complex formation in ground states [charge transfer (ct) complexes (19)], the interaction between both partners requires a suitable geometric arrangement. In solid layers of macromolecules, some of the molecules may possess the correct geometry and high concentrations of functional groups needed for an efficient complex formation during irradiation. These species exist only in excited states and exhibit no analogous entities in the ground state. As with other molecules, both singlet and triplet excited states are possible. However, the bonding energy of these complexes, resulting from the cooperative orbital overlap between both reaction partners, is small.

Two types of such complexes, namely excimers (eq 11) and exciplexes (eq 12), can be distinguished.

$$D\!:^* + D\!: \rightarrow (D...D)^* \leftrightarrow (D^+\!:^-D) \leftrightarrow (D^-\!:D^+) \quad (11)$$

$$D\!:^* + A \rightarrow (D...A)^* \leftrightarrow (D^+A^-) \quad (12)$$

An energetic scheme for the generation of these excited complexes is depicted in Figure 5. Excimers and exciplexes are able to deactivate into the ground states of the components D: and A nonradiatively (k_d) and radiatively (k_{em}) analogous to other excited states. The emission spectra of excimers and exciplexes are red-shifted compared with those of individual compounds (*see* the energetic situation in Figure 5). Furthermore, no vibrational structure is present in these spectra.

However, several (but not all) exciplexes can also deactivate by pathways desirable for chemical conversions. Because of the different electronic properties of D: and A, an asymmetrical separation of the bonding electrons takes place within the exciplex (*see* the distribution of electronic density in eqs 11 and 12). Such processes can end up in a single electron transfer (SET) to the acceptor A, leading to ion-radicals if both partners are neutral species (*20, 21*). As a result, one partner of the reaction couple (D:) is oxidized, and the other (A) is reduced. The incident light can be absorbed either by D: or A without influencing the overall result of this process. The resulting ion-radicals then react by dark reactions to produce the final products. The chemical conversion of the nonabsorbing compound is sensitized by the other compound if its spectral response is red-shifted. This type of sensitization is sometimes termed chemical sensitization.

The free energy, ΔG_{el}, for a photoinduced SET may be calculated by an expression derived by Weller (*22*) (eq 13). Here, $E(D\!:/D^+)$ is the oxidation potential of the donor (D:), $E(A^-/A)$ is reduction potential of the acceptor (A), ΔE_{00}^* is excitation energy of the partner that absorbs the light, and E_{coul} is coulombic energy between the two partners. For many compounds, the physical parameters given in eq 13 are known. Therefore, evaluating the thermodynamic conditions for SET of a particular reaction couple is possible.

$$\Delta G_{el} = E(D\!:/D^+) - E(A^-/A) - \Delta E_{00}^* - E_{coul} \quad (13)$$

The rate constants of SET (k_{el})processes depend on the ΔG_{el} values and also on the distance between the electron donor and acceptor molecules. The optimal distance between the two partners is ca. 0.7 nm. However, if the free energy is negative, then high rate constants in the range of 10^8–10^6 s^{-1} are obtained even at distances of 1.5–1.7 nm. These high values are due to the through-bond interactions, similar to those in the case of σ-bonds (*23*). Thus, by assuming appropriate concentrations of the reaction partners, SET can successfully compete with physical deactivation pathways of excited states. This result follows from eq 9 using the rate constant of SET instead of k_{et}.

Photoinduced SET processes can function as a starting point for cross-linking reactions of polymers. The efficiency of this step depends on the type of excited state involved. As shown in Figure 5, electron return can occur within the radical pair initially formed. In this way, all of the excitation energy is lost owing to the generation of ground state of molecules D: and A. Because of spin restrictions, this return process is depressed for triplet radical pairs, generated from triplet excited state of either D: or A. In this case, the efficiency of SET is much higher, as compared with pathways involving only the singlet states.

However, the yields of the SET depend also on a number of other factors. The physical deactivation pathways of excimers and exciplexes are associated with the loss of energy absorbed either by D: or A molecules. Therefore, for efficient generation of ion radicals (the end products of the SET), these pathways should be very depressed, which normally is difficult to achieve. Furthermore, the excimer formation followed by deactivation is the mechanism of the so-called self-quenching of excited molecules. This reaction plays an important role when the concentration of D: in the reaction medium is high, a condition usually met in many polymer systems.

Energy Migration in Macromolecules

Both energy and electron transfer processes were discussed as processes occurring between two separate molecules. In practice, macromolecules contain a great number of pendent functional groups capable of energy transfer or a tandem of energy and electron transfer. Similar conditions may also occur in blends of

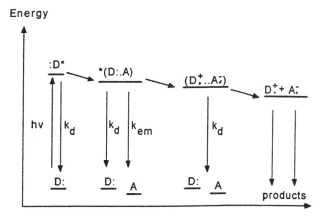

Figure 5. Deactivation of excited states via exciplex formation. D* represents either the singlet or the triplet state of D:; k_d summarizes all physical deactivation processes of the species.

different polymers. When these units are arranged in a desirable manner, energy or electron transfer may take place either between functionalities of one macromolecule or between those contained in different macromolecules. These transfer processes are the basis of energy migration in polymers (24, 25).

A number of simplified models for possible reaction pathways based primarily on energy transfer processes are given in Scheme I. In examples a and b, the energy transfer occurs intramolecularly: the structural and geometric conditions exist in one macromolecule (intrachain reactions). The long-range energy transfers shown in Scheme I were confirmed by experimental results involving the emission properties of polymers and copolymers (26–30). For example, in copolymers of styrene and 1-vinylnaphthalene (99:1 mass ratio), naphthalene phosphorescence was observed almost exclusively. In contrast, in blends of poly(1-vinylnaphthalene) with polystyrene (99:1 mass ratio), mainly the phosphorescence of polystyrene is observed (as in neat polystyrene) (31). In the case of copolymers of styrene and 1-vinylnaphthalene, the light absorbed by the phenyl units (due to their higher c_m value; see eq 5) is transferred intramolecularly to the naphthalene moieties, which function because of their lower lying triplet states as energy traps. Thus, although the naphthalene functionalities have not directly absorbed the incident light photons, they are excited via energy migration leading to their normal photophysical and photochemical processes. This travelling of energy through the pendent groups is termed down-chain migration.

A polymer chain may act as an antenna directing the excitation energy to reactive sites in a fashion similar to antennas for radio or TV receivers (antenna effect). In this respect, some polymers capable of photo-cross-linking exhibit properties of true solid-state systems (1).

The average number of individual transfer steps (n_{et}) possible in one macromolecule can be calculated by eq 14. For these calculations, the lifetime (τ) of the state involved and the time for one transfer step t_h (hopping time) must be known. To achieve a high number of transfer steps, the excitation produced at the first unit must be freely mobile in the polymer system to reach the functionality suitable for energy or electron trap. It can be predicted that these processes are more likely for triplet states rather than singlet excited states.

$$n_{et} = (\tau/t_h)^{1/2} \qquad (14)$$

The average distance, d_{et}, of energy migration is given by eq 15, where Λ is diffusion constant (m²/s), and a is average structural distance between the chromophores.

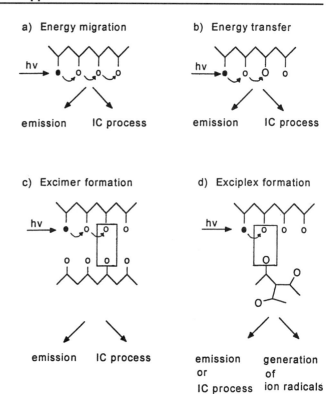

Scheme I. General schemes for energy migration and transfer in macromolecules (i.e., without initiators).

$$d_{et} = (2\Lambda\tau)^{1/2} = a(\tau/t_h)^{1/2} \qquad (15)$$

The energy migration can be described with the general models of energy transfer discussed previously. Thus, as a consequence of eq 8, the rate constants of energy migration also depend strongly on the distance between the chromophores. For evaluations, the quantity of a in eq 15 can be used instead of R in eq 8. Values for R_o of several chromophores are known (32). The number of chromophores and their distance relate to the structure of the polymer, including polymer composition, tacticity, and comonomer distribution in copolymers.

The solvent used for the dissolution of polymers can also influence the results due to different geometric entities formed in solutions. Furthermore, in dilute polymer solutions, the energy is restricted to an intramolecular path along the polymer backbone, whereas in solid polymers the energy may migrate from one polymer chain to another.

When large numbers of chromophores are present in the polymer chains, simultaneous excitations of several groups in one chain are also possible. In solid polymer systems, excited chromophores of different molecules may be close to each other, leading to another type of interaction between excited states as shown in eq 16.

$$T_1^* + T_1^* \rightarrow S_1^* + S_o \qquad (16)$$

Thus, when two triplet excited molecules approach each other, disproportionation occurs and leads to the formation of two molecules: one in its singlet excited state and the other in singlet ground state. This process is termed triplet–triplet annihilation (*T,T* annihilation). As a result, the S can emit fluorescence light measurable by normal emission equipment, but its appearance is delayed compared with the usual prompt fluorescence light (delayed fluorescence). This reaction pathway destroys the excitation energy and lowers the efficiency of photoinduced polymer cross-linking, but all of these excitation energy annihilation processes are the basis for a high photostabilization of polymers (2).

Because of the complexity of the possible reaction pathways after excitation, it is difficult to predict exactly the results for a particular cross-linking system. Typical results for compounds (1–4) (Chart I) are discussed here to illustrate the principles involved (33–36). The triplet lifetime (τ_t) of isopropyl benzophenone (1) is 5 μs, whereas polyvinylbenzophenone (2) exhibits a value of 0.7 μs only. The values for copolymers (3) are between these two data, depending on the m:n ratio (*see* Chart I). Both *T,T* annihilation and self-quenching are assumed to take place in the polymers, although self-quenching is seldom observed for monomeric carbonyl compounds (37, 38). The *T,T* annihilation, indicated via a delayed naphthalene fluorescence, also took place in the copolymers (4). In this case the following reaction cascade is thought to occur:

$$\text{naph.}(S_1) \rightarrow \text{PhCO}(S_1) \rightarrow \text{PhCO}(T_1) \rightarrow \text{naph.}(T_1) \rightarrow$$
$$\text{naph.}(S_1) + \text{naph.}(S_o)$$

The effectiveness of energy migration (η) is expressed by eq 17, where n_t is the number of transferred photons, and n_q is the number of photons absorbed by the donor. Typical results of energy transfer from naphthalene chromophore to anthracene and phenyl ketone substituents in polymers are summarized in Table II. These data underline the dependence of η on different structural and solvent parameters discussed previously.

$$\eta = n_t/n_q \qquad (17)$$

If functional groups capable of energy or electron transfer are part of the polymer backbone, but they are not adjacent to each other and the chain is not sufficiently flexible, then intramolecular down-chain energy migration is restricted. In this case, only intermolecular interactions between excited groups of one chain with ground state units in another chain are possible. Examples of this situation are copolymers (5) (39) and (6) (30) (Chart II). In dilute solutions of these polymers, predominant fluorescence of the chromophores (*p*-phenylene diacrylate in 5 and anthracene in 6) is observed, together with a certain amount of excimer fluorescence. The excimer fluorescence indicates that the polymer chains can fold upon themselves to a limited extent, bringing a few chromophores in a contact for producing excimer fluorescence. However, when solid films of these polymers are irradiated, no monomer fluorescence is seen and only strong interchain excimer emission is observed.

Efficient exciplex formation is observed with copolymers incorporating both donor and acceptor units. Typical examples are terephthalates 7 having pendant

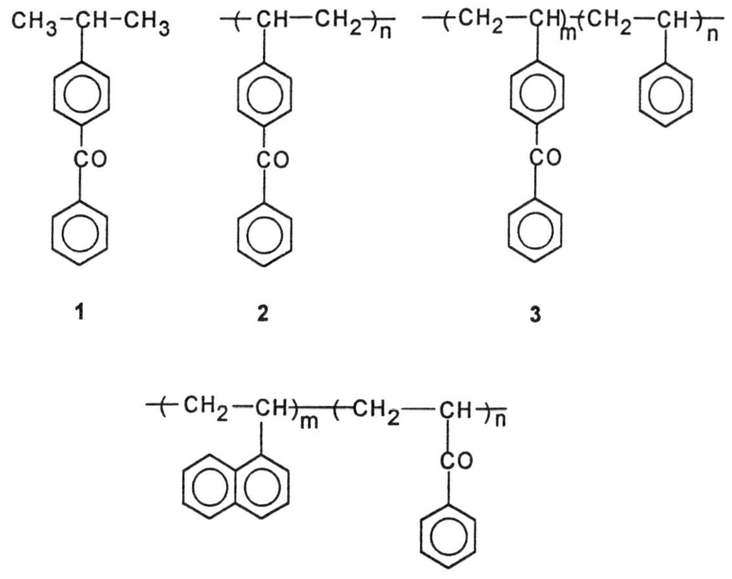

Chart I. Examples of polymers suitable for energy transfer processes in excited states.

Table II. Efficiency (η) of Energy Transfer from Naphthalene Chromophore to Anthryl and Phenyl Ketone Units in Polymers

Comonomer Units	Molar Ratio	Solvent[a]	η (%)
1-Methylnaphthalene acrylate and	1:0,174	THF	9,9
9-methylanthracene methacrylate	1:0,665	THF	30
	1:2,06	THF	84
2-Vinylnaphthalene and	1:0,17	MTHF	12
phenyl vinyl ketone	1:0,995	MTHF	49
	1:3,8	MTHF	71
	1:12	MTHF	94
1-Methylnaphthalene methacrylate		THF	10
and 9-anthracene end groups		dioxane	11
		dioxane–MeOH (60/40; v/v)	21
		dioxane–MeOH (20/80; v/v)	66
		poly(methyl methacrylate)	43

[a]THF is tetrahydrofurane; MTHF is 2-methyl tetrahydrofurane; and MeOH is methanol.

Source: Data are taken from reference 3.

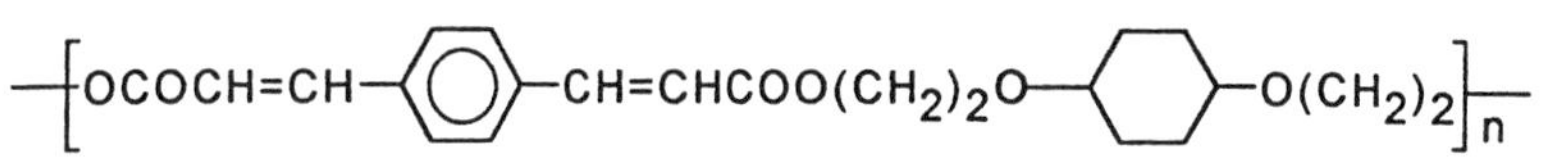

5

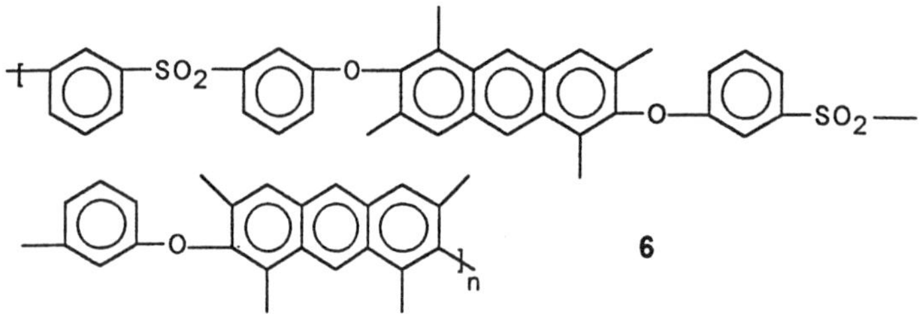

6

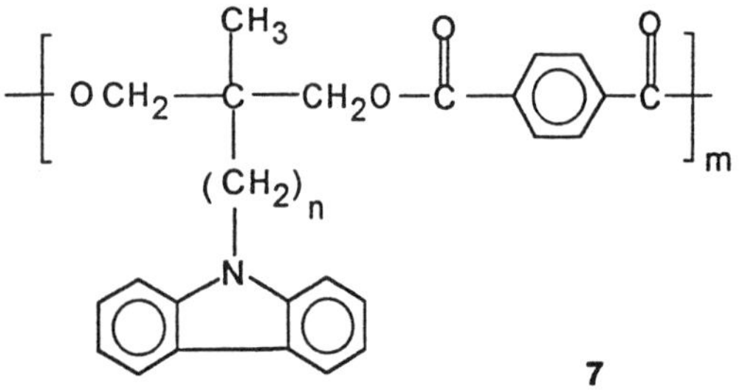

7

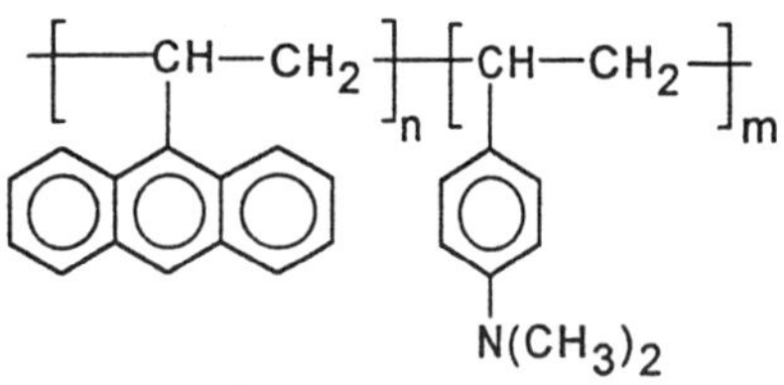

8

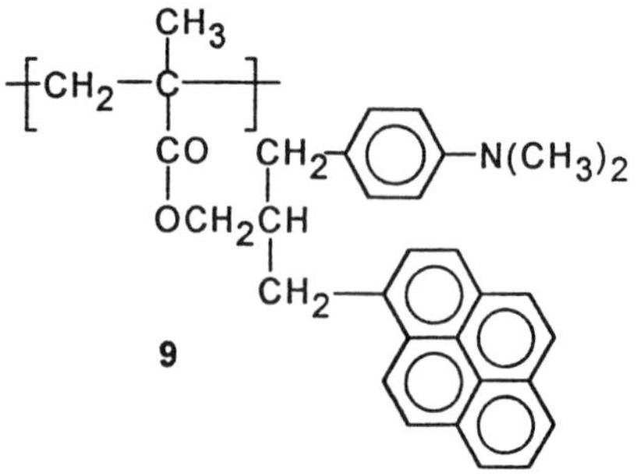

9

Chart II. Examples of typical polymers exhibiting interchain excimer fluorescence.

ω-carbazolylalkyl groups (40), copolymers 8 of 9-vinylphenanthrene and *p-N,N*-dimethylaminostyrene 8 (41), and polymethacrylates 9 having 2-[(1-pyrenyl)-methyl]-2-[4-(dimethylamino)benzyl]ethyl side chains (42). In some cases, ct complex formation in the ground state also takes place (43). Because of their high spatial concentration, more than one donor and acceptor unit can be involved in the complex generation (ter-molecular complexes). Sometimes these ct complexes absorb with a slight bathochromic shift compared to the parent compounds, however most with highly decreased extinction coefficients. At the first approximation, the irradiation of these complexes leads to the same products as those obtained via the exciplex formation (44).

Reactivity Distribution in Photo-Cross-Linking of Amorphous Polymers

A high inherent photosensitivity of the cross-linking groups is usually necessary to achieve an efficient photo-cross-linking of the polymer chains. In many cases, however, this prerequisite is not a sufficient condition for a high cross-linking efficiency. This characteristic is especially true for photo-cross-linking in amorphous solid layers. Even the same photosensitive groups may behave differently, depending on the nature of the polymeric matrix. The following parameters also influence the photosensitivity of the polymer: concentration of the functional groups, their spatial distribution and mobility in the matrix, mobility of the excitation energy, and the presence or absence of energy or electron traps. In either case, each functional group has a situation different from that of the other. Thus, a broad reactivity distribution must exist, which was indeed found for many reactions in solid polymers (45–47).

The influence of these parameters is particularly important for photo-cross-linking, because the process is a bimolecular reaction, usually with second-order kinetics requiring sufficient mobility of the reacting groups. Furthermore, these groups must contact each other during the short lifetime (micro- or nanoseconds time scale) of the excited state of the active group.

For a better understanding of the effects of structural parameters on photo-cross-linking, models of reactant sites can be studied (1, 48). A reactant site is simply defined as a functional group on the polymer together with its immediate surrounding. Thus, each reactant site possesses its individual reactivity as conditioned by the previously named parameters. Accordingly, the reactivity of a particular photo-cross-linking system observed macroscopically represents the overall effect of an ensemble of different site reactivities. Quantitative data of a photo-cross-linking reaction in solid polymer films must be dependent on the exposure time owing to the heterogeneity of the site

reactivity. At the beginning of the process, only the most reactive sites are consumed and give rise to higher reactivity parameters (cross-linking rate and quantum yield) as compared with lower values at the end of the reaction, when the least reactive sites take part in the reaction.

Typical examples for such a situation are the photo-cross-linking of polyesters of *p*-phenylenediacrylic acid 5 (Chart II) (49, 50), poly(vinyl cinnamate)s (10), and diphenylcyclopropene derivatives of poly-(vinyl alcohol) (11) (Chart III). In the case of 5 and 10, the original quantum yield of the photoinduced [2 + 2] cycloaddition of these compounds is approximately $\phi = 0.25$, but it decreases with increased conversion. The reaction stops ($\phi = 0$) well before all double bonds are converted (80% for 5 and 50% for 10). This lack of reactivity is attributed to functional groups located in unreactive reactant sites. The nonreactivity of these sites is caused by the absence of reaction partners in their vicinity. Similar conditions also were met in free-radical cross-linking reactions (51, 52).

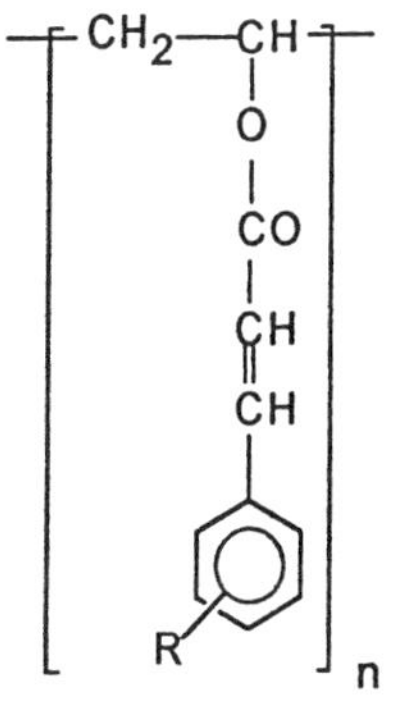

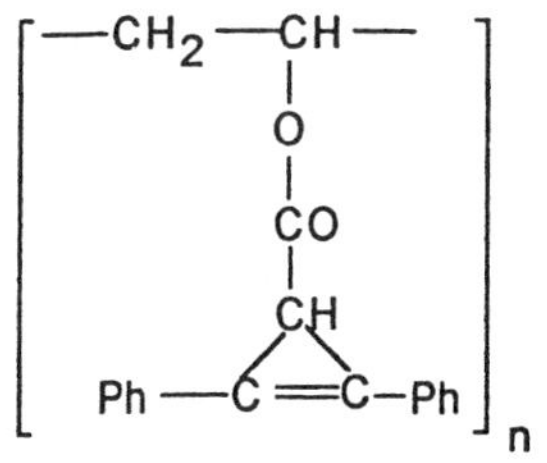

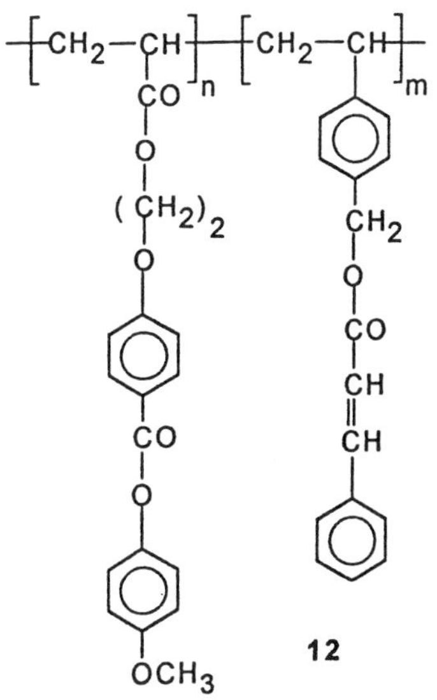

Chart III. Examples of polymers suitable for photoinduced cycloaddition.

For all cross-linking processes, the character of the initial ensemble of reactant sites is changed during the irradiation time, which can be described quantitatively by iterative procedures (50). This reactivity analysis also provides insights into the mechanism of the photo-cross-linking reaction.

Within the context of this discussion, it is not surprising that chromophore concentration (in the backbone or as pendant) is of secondary importance in photo-cross-linking of solid polymer films, contrary to solution kinetics (53). Often, by changing the concentration of the reactive groups, segment mobility of the polymer chain may be adversely affected. It seems, however, that an increase in molecular polymer weight has a positive effect on photo-cross-linking (53): the necessary degree of cross-linking for insolubilization decreases with increasing molecular weight.

Measurements of temperature effects in many photo-cross-linkable polymers show, as expected, clear correlation between glass transition temperature (T_g) and the reaction rate or quantum yield of photo-cross-linking (54–56). Above T_g, the rate of cross-linking is usually much higher compared with that in the glassy state. This difference is thought to be due to restrictive parameters such as segment mobility and encapsulation of functional sites. Below T_g, the process often follows the same kinetics as observed for solutions. Under these conditions, physical and chemical parameters for each reactive group equally determine the collective reactivity of the functional sites.

To overcome the negative influence of the randomness of reactive sites in most photo-cross-linkable polymers, preorganization of reactive groups was proposed for enhanced reactivity (30, 57–60). The preordering principle so far studied is based on electron-acceptor interactions. Typical examples of such preorganization effects are observed in the copolymer **6** (Chart II) and blends of poly(vinyl cinnamate)s (**10**) (Chart III). In **6**, donor and acceptor moieties are incorporated in the same macromolecule, whereas in **10** the electron donor (R = CH_3O or CH_3) and electron acceptor (R = Cl or CH_3CO) come from two different polymer chains. In the systems for **10**, the electronic interaction between the reactant groups in their ground state is sufficiently high to give a suitable spatial arrangement between the C=C double bonds of two polymer chains (stabilization energies are 3–8 kcal/mol (59)). In approximately equimolar mixtures of donor–acceptor-substituted poly(vinyl cinnamate)s, a significant enhancement of intermolecular photoreactivity was observed. When the functional groups are attached to the backbone via a spacer unit, the probability of pair formation needed for an intermolecular cross-linking is reduced, and therefore, a decreased photosensitivity of such blends is found.

A second principle to use preorganization effects for an effective photo-cross-linking is based on polymeric liquid crystals with photosensitive groups incorporated either in the polymer main chain or as side group. Recently, several such polymers were described containing the cinnamate chromophore (61–66) (typical example, **12** in Chart III) or bis(benzylidene)cycloalkanone groups (67). In the mesophase, these liquid crystal polymers have a higher photo-cross-linking ability than in the isotropic phase. In the isotropic phase, the random orientation of the chromophores again drastically curtails the cross-linking rate.

On the basis of these results, only a few guidelines can be given for the photoreactivity of a particular photo-cross-linkable polymer in bulk.

Photo-Cross-Linking Processes

Cross-linking of polymers by light can be accomplished by two alternative methods:

- excitation of the macromolecules by direct light absorption or sensitization
- excitation of an initiator molecule, followed by reaction of the resulting photolytic products with the macromolecules

In the first method, the macromolecule must contain chromophore groups able to absorb the light emitted by the light source. In the second, the macromolecule must contain functionalities capable of dark reactions with free radicals or ionic species formed photolytically.

Direct Photo-Cross-Linking Processes

Polymer cross-linking via direct excitation of macromolecules can be broadly presented in terms of steps 1–4 shown in Scheme II. In the first step, the chromophore group (Y) absorbs the incident light and is converted into its excited state (either singlet or triplet state). Cross-linking then follows by interaction with a ground state chromophore located in another polymer chain, leading to the formation of the cross-link Z, which also functions as a branching center. This reaction cycle repeats itself many times.

In the early stages, the cross-linking leads to an increase in the average molecular weight of the polymer, and the molecular weight distribution is also dramatically effected. The great number of chromophore groups (Y) in macro-molecules with the same probability of excitation leads to the formation of a three-dimensional network (Scheme III). As a further consequence, the larger molecules in the system will be cross-linked faster as compared to smaller molecules.

(1) P_n-Y $\xrightarrow{h\nu}$ $(P_n$-Y$)^*$

(2) $(P_n$-Y$)^* + P_m$-Y $\longrightarrow$ P_n-Z-P_m

(3) P_n-Y $\xrightarrow{h\nu}$ $(P_n$-Y$)^*$

(4) $(P_n$-Y$)^* + P_n$-Z-P_m $\longrightarrow$ P_n-Z-P_m-Z-P_n

P_n; P_m = polymer chains; Y = chromophore

Scheme II. General reaction scheme of direct photo-cross-linking of macromolecules: P_m and P_n are polymer chains; Y is a chromophore.

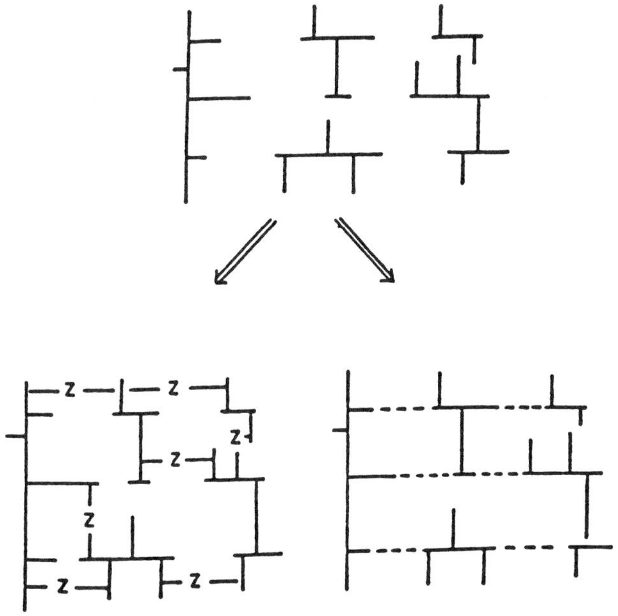

Scheme III. General scheme of network formation: Increase of branching points (a); increase of numbers of linear chains (b).

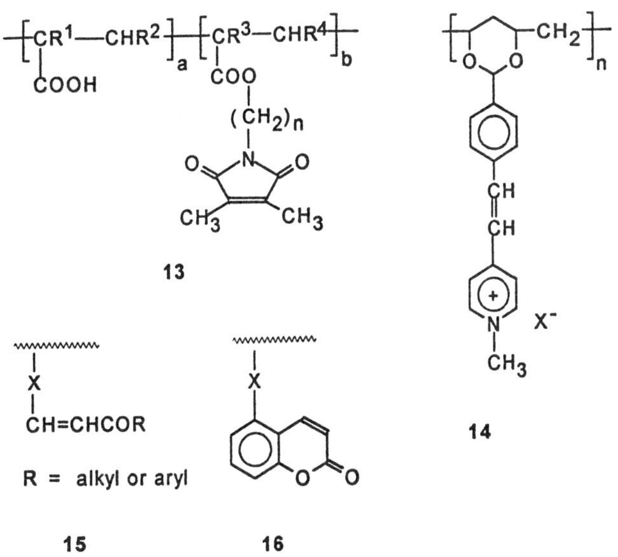

13

R = alkyl or aryl

15   **16**

14

Chart IV. Typical polymer structures suitable for photoinduced [2 + 2] cycloadditions.

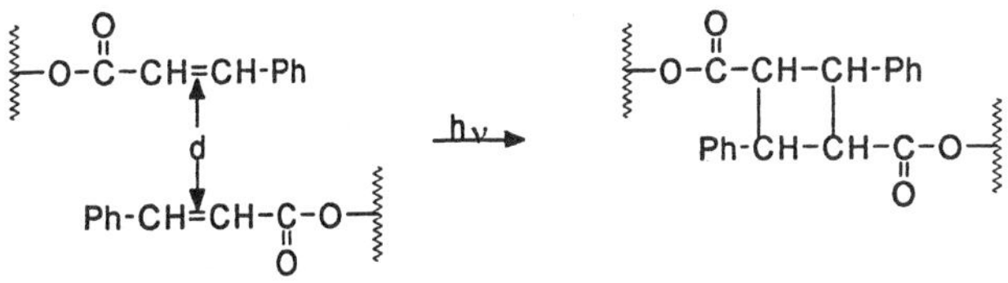

Scheme IV. Photo-cross-linking of poly(vinyl cinnamate)s **10**. The best distance (*d*) for this type of cross-linking is 0.35–0.41 nm.

However, the highly branched supermolecule will hinder the efficient reaction of these chromophores (*see* the discussion about reactant sites).

As follows from Scheme II, for an efficient network formation, the stepwise absorption of many photons is required. Therefore, the overall cross-linking quantum yields are below unity due to the loss of excitation energy by some photophysical processes. The direct excitation of macromolecules with commercially available light sources needs chromophore groups, which absorb above 320 nm and have moderate to high extinction coefficients (7). However, for other polymers special equipment based on quartz materials and low pressure mercury arc lamps can be used.

The data in Table I show that systems with conjugated C=C bonds or with α,β-unsaturated carbonyl groups absorb near the range suitable for practical applications. In these respects, poly(vinyl cinnamate)s (**10**) (*49, 57–60, 68*) and polymers having dimethylmaleimide side chains (**13**) (Chart IV) (*69–71*), styrylpyridinium side chains (**14**) (*72–74*), chalcone side chains (**15**) (*67, 75*), cumarine side chains (**16**) (*75*), or anthrylene side chains (**6** in Chart II) (*76, 77*) absorb near or above 300 nm.

The photo-cross-linking of the polymers **10** and **13–16** proceeds via [2 + 2] cycloaddition to form cyclobutanes (*78*) (Scheme IV). In the ground state, this process is forbidden by orbital symmetry (Hoffmann–Woodward rules), but it is allowed if the symmetry is changed because one of the partners reacts in its excited state. Cross-linking takes place when the singlet excited C=C bond in one molecule interacts with the ground state of another one in a second macromolecule. The best distance (*d*) of the two double bonds for such a reaction is between 0.35 and 0.41 nm. The light sensitivity of these polymers depends on parameters such as T_g, chromophore concentration, distribution (random, alternate, or pre-organized), and polarity of the polymer. To overcome the difficulties for the mismatch between the absorption spectrum of some polymers and the spectral emission of the light sources used commercially, different long-wavelength energy absorbing or electron-transfer sensitizers can be used (e.g., Michler's ketone, thioxanthone (79)). However, the electron-transfer sensitization of the cycloaddition shown in Scheme IV leads to a change of the cross-linking mechanism.

Anthryl-substituted polymers (mainly at the 9-position of the anthracene ring) also react via a [4 + 4] cycloaddition involving the S_1 excited state of one of the reactants. For polymers **6** (Chart II) exhibiting such structural moiety, excimers are intermediates in their photo-cross-linking (30). However, by a random distribution of the pendant anthracene groups, extensive association of chromophores occurs on individual chains leading to a high extent of intrachain photodimerization also in dilute solutions (28, 80, 81). In the initial stage of this process, cross-links are formed between nearest-neighbor side chains.

The increase of the average molecular weight due to the irradiation of the previously named polymers, even at partial conversion, leads to substantial reduction of their solubility. This fact is an important prerequisite for the application of these polymers for imaging materials (82) or photoresists (1, 2).

Another mechanism for direct photo-cross-linking is the photodissociation of bonds as a result of macromolecular excitation. However, the individual processes involved in this route are less specific than photocycloaddition. Cross-linking via photodissociation is illustrated in Scheme V. When a bond in a functional group (X) of the main chain is broken, the polymer is mainly degraded into small fragments (83). If side chain groups (X-Y) are cleaved, and the resulting radi-

cals are stabilized by electronic effects, cross-linked polymers are the end products of the overall process. More generally, however, degradation and cross-linking compete with each other. For example, in polystyrene and its derivatives, photoirradiation at 254 nm induces both decomposition and cross-linking, depending on the stability of the intermediary radicals (84). In the presence of vinylic monomers or oligomers the extent of cross-linking can be increased.

Because of the non-selectivity of the radical processes following the homolytic bond cleavage, only a few examples are given in the literature, including those of poly(4-bromoacetylstyrene) (85), polyepisulfides (86, 87), poly(vinyl pyridine-N-oxide) (88), and polysilanes (89).

Photo-Cross-Linking via Photoinitiators

Azide Initiators

Azides are among the most typical photoinitiators used for photo-cross-linking of polymers. At electronic excitation, azides lose nitrogen to form nitrenes with quantum yields of about 0.3–1.0 (90–92). The primary processes of azide photolysis and the secondary processes possible with important functional groups contained in macromolecules are represented in Scheme VI.

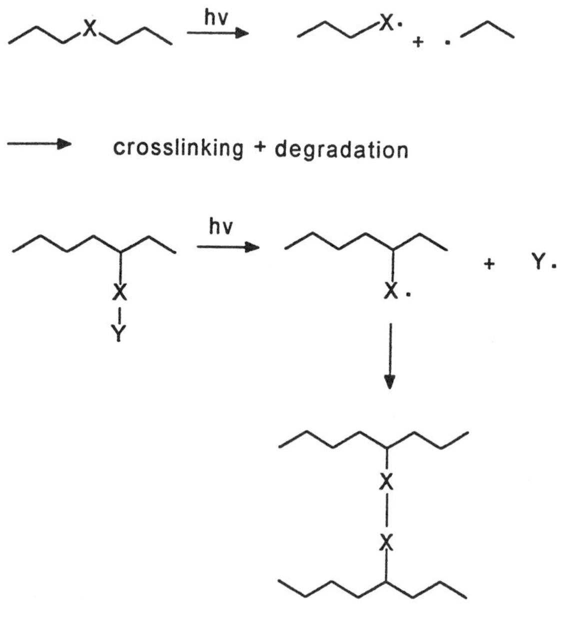

Scheme V. General reaction scheme of polymer photo-cross-linking via homolytic bond scission.

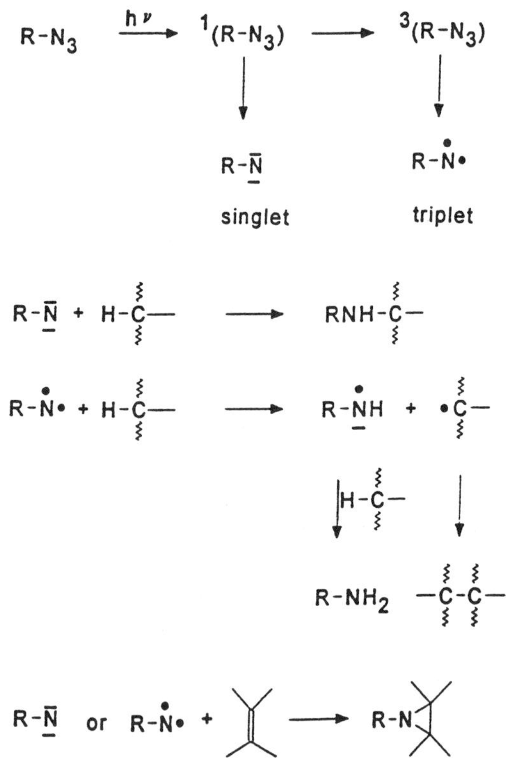

Scheme VI. Photolysis of azides and reactions of nitrenes for cross-linking of polymers.

Nitrenes, which exist either as singlet species (two coupled electron pairs at nitrogen) or triplet species (two unpaired electrons and one electron pair), are highly reactive intermediates for reactions with different compounds. Both the singlet nitrene (electrophilic character) and triplet nitrene (biradical type) exhibit specific reaction behavior. Thus, singlet nitrene prefers to react via insertion reaction with polymeric C–H bonds to give polymeric secondary amines. In the case of nitrenes derived from low molecular weight azides, this reaction has little influence on the average molecular weight of the polymer. In contrast, triplet nitrenes react by hydrogen abstraction and form an amino radical and a carbon-centered macromolecular radical simultaneously. After spin conversion, the combination of the polymeric radicals leads to cross-linking. The amino radical may also abstract hydrogen from the macromolecule or combine with the carbon radical. The abstraction process enhances thermal cross-linking and thus increases cross-linking efficiency. The reaction with unsaturated reactants (e.g., aromatics and ethylenic double bonds) can be initiated with both types of nitrenes.

By application of bifunctional azides (bis-azides) in a nonphotoreactive substrate, reactions also enable cross-linking or network formation with singlet nitrenes due to the stepwise photolysis of the two functionalities (Scheme VII). The R residue between the two azido groups functions as a spacer between the two polymer chains (P_n–H). However, even in this case the triplet pathway has a higher photosensitivity. Thus, for increased cross-linking, the chemical structure of bisazides should favor ISC processes in the

excited state or the initially formed nitrenes. For this reason, azides in practical systems contain conjugated carbonyl groups (*93–95*), which are known to increase the probability of ISC. Additionally, these compounds absorb at approximately 340 nm, a region highly suited for practical applications.

The azido group may also be attached directly to the polymer (*96, 97*). The irradiation of these polymers leads to very efficient cross-linking. A drawback of the application of nitrenes for cross-linking is their sensitivity toward oxygen. This problem can be avoided by performing the cross-linking process in the absence of oxygen.

Free-Radical Initiators

Polymer cross-linking by free radicals formed photochemically is a relatively simple process and can be employed in two different ways.

- reactions of free radicals with ethylenic groups in the polymer backbone as end groups or side chains
- hydrogen abstraction from macromolecule by excited molecules, followed by coupling of the resulting carbon radicals

Network formation via the first pathway is shown in Scheme III. Many photoinitiators normally used for free-radical polymerization can also serve as a source of radicals for photo-cross-linking (*98–100*). Very recently, macromolecular radical photoinitiators were developed for this purpose (*101*). The mechanism of this type of photo-cross-linking is similar to that of free-radical polymerization. However, the termination step is caused mainly by second-order processes between polymer and primary initiating radicals (*102*).

Typical examples of the photoinduced hydrogen abstraction are the interaction between macromolecules and excited triplet ketones or quinones having n,π^* electronic transitions (*103*). A simplified reaction scheme using benzophenone for this purpose is outlined in Scheme VIII. Following excitation and ISC, the

$$N_3\text{-R-}N_3 \xrightarrow{\ h\nu\ } \text{:N-R-}N_3 + N_2$$

$$P_n\text{-H} + \text{:N-R-}N_3 \longrightarrow P_n\text{-NH-R-}N_3$$

$$P_n\text{-NH-R-}N_3 \xrightarrow{\ h\nu\ } P_n\text{-NH-R-N: } + N_2$$

$$P_n\text{-H} + P_n\text{-NH-R-N:} \longrightarrow P_n\text{-NH-R-NH-}P_n$$

P_n = polymer chain

Typical example for R:

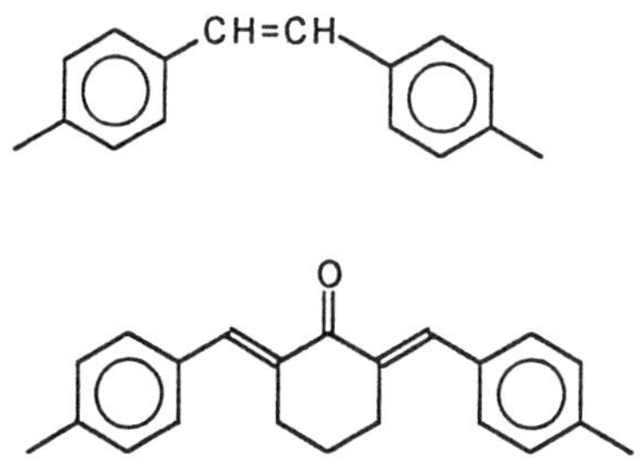

Scheme VII. Photo-cross-linking of polymers with bis(azides) as light-absorbing components. The nitrene intermediates are of singlet type.

$$Ph_2C{=}O \xrightarrow{\ h\nu\ } Ph_2\overset{\bullet}{C}\text{-O}\cdot\ (T_1)$$

$$Ph_2\overset{\bullet}{C}\text{-O}\cdot + P_n\text{-H} \longrightarrow Ph_2\overset{\bullet}{C}\text{-OH} + P_n\cdot$$

$$P_n\cdot + P_n\cdot \longrightarrow P_n\text{-}P_n$$

$$P_n\cdot + Ph_2\overset{\bullet}{C}\text{-OH} \longrightarrow P_n\text{-C(OH)}Ph_2$$

$$Ph_2\overset{\bullet}{C}\text{-OH} + Ph_2\overset{\bullet}{C}\text{-OH} \longrightarrow Ph_2C(OH)\text{-}C(OH)Ph_2$$

P_n = polymer chain

Scheme VIII. Photo-cross-linking of polymers with benzophenone as light-absorbing component.

triplet state of the ketone is formed and is capable of hydrogen abstraction due to its biradical character. The coupling of two polymer radicals leads to cross-links. However, cross-linking and degradation processes for these radicals compete with each other.

The type of photo-cross-linking shown in Scheme VIII is well suited for polyolefines (104–108) and polyimides (109). Hydrogen abstraction, and hence cross-linking, take place at tertiary and secondary carbon atoms along the chains. However, free radical photo-cross-linking is nonspecific. To achieve cross-linking several bimolecular reaction steps are necessary, and these steps are difficult to guarantee in amorphous polymers. Therefore, the efficiency of most of these processes is relatively low (quantum yields < 0.1).

Cationic Initiators

In contrast to free-radical systems, only a small number of polymer-bound functionalities are susceptible to cationic cross-linking (110). Such reactions are found in compounds having C=C bonds with electron-releasing substituents (e.g., $ROCR=CR_2$) and in strained cyclic ethers such as epoxides (111) or glycidyl ethers (112, 113). Of these, only the C=C functionalities can be contained in the polymer backbone, the others are pendant groups attached to the polymer chain (e.g., **17** and **18** in Chart V (111)).

So far, only epoxides and glycidyl ethers have found application in cationic cross-linkable systems (111, 113–115). However, compared with the widespread radical systems, the importance of such photopolymers is relatively limited due to the lack of availability of appropriate macromolecules and suitable photoinitiators. For recent reviews on photoinitiators for cationic processes, *see* references 116–118.

In general, the photoinduced cationic ring-opening cross-linking of epoxide-containing macromolecules can be represented as outlined in Scheme IX.

Protonic (e.g., Brönsted) acids as well as organic or inorganic Lewis acids are capable of initiating the cationic cross-linking process. The driving force of this reaction is the high strain energy (about 114 kJ/mol) of the three-membered epoxide ring (Scheme IX, indicated as Y). The actual cross-linking reaction takes place in step (3). The chain transfer reactions (4) and (5) are not particularly harmful in the case of polymeric multifunctional epoxies. The resulting cation (in most cases an oxonium ion) or protonic acid are able to reinitiate the cross-linking process. However, when stronger basic compounds (e.g., amines) are present in the system, the propagation chain will be interrupted, and the efficiency of the process will be lowered. The same situation holds with nucleophilic counterions (A^-) when onium salts (e.g., iodonium or sulfonium salts) are used as photoinitiators. In this case, the reaction ends up in a simple addition product of the acid at the epoxide ring (step (2)).

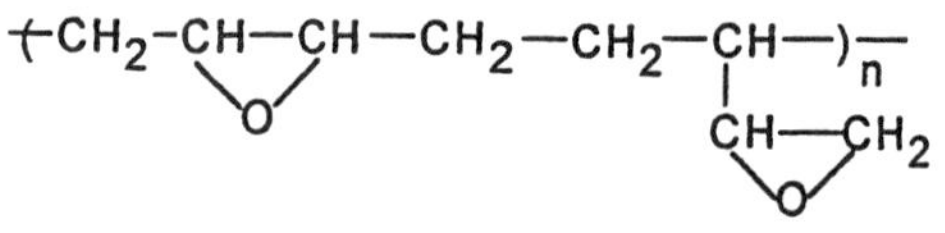

17

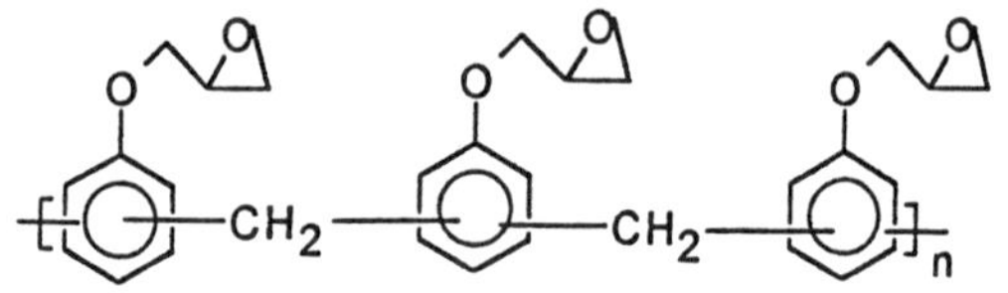

18

Chart V. Typical polymers with epoxide and glycidyl ether groups suitable for cationic photo-cross-linking.

		hν	
(1)	Cationic initiator	⟶	$H^+ A^-$ (Bronsted acid) or
			MtX_n (Lewis acid)
(2)	$P\text{-}Y + H^+ A^-$	⟶	$P\text{-}Y\text{-}H^+ A^-$ $P\text{-}CH(OH)\text{-}CH_2\text{-}A$
			($P^+ A^-$)
(3)	$P\text{-}Y + P^+ A^-$	⟶	$P\text{-}CH(OH)\text{-}CH_2\text{-}Y^+\text{-}P\ A^-$
			($P_n^+ A^-$)
(4)	$P_n^+ A^- + P\text{-}Z$	⟶	$P\text{-}Z^+\text{-}P\text{-}Y\ A^-$
(5)	$P^+ A^-$ (or $P_n^+ A^-$) $+ H_2O$	⟶	$P\text{-}OH + H^+ A^-$

Scheme IX. Schematic mechanism of photoinduced cationic ring-opening cross-linking of epoxide-containing macromolecules: P represents a polymer chain; Y is an epoxide group; Mt is a metal atom; X is a halogen atom typically for Lewis acids; P^+A^- and $P_n^+A^-$ are macromolecular oxonium ions; Z is a nucleophilic group in the macromolecule. When Lewis acids are produced by step (1), in all formulas H is exchanged against MtX_{n-1} and A^- against X^-.

Cationic photo-cross-linking has certain inherent advantages. The process is unaffected by the presence of air or oxygen, which is one of the main drawbacks of free-radical cross-linking. The species formed photochemically from the initiators (step (1) in Scheme IX) have relatively long lifetimes, and due to this extended lifetime, reactions continue after the light source has been removed. Therefore, systems that are UV- as well as heat-curable are easy to develop. The long lifetime of the initiating species means that it is possible for numerous thermal reactions to take place between the functional groups capable of cationic cross-linking. This phenomenon can result in a high cure dose response. However, at room temperature the overall cross-linking rates of cationic systems are low compared with those of free-radical systems. Despite this strong drawback, there is presently considerable interest in cationic photosensitive polymers for emerging technologies.

Concluding Remarks

Photo-cross-linking of polymers is a technology enjoying increasing interest due to its potentials for the study of polymer reactions and for a wide variety of applications. The rapid growth of this technology is reflected in recent progress in the development of both suitable macromolecules and related technical equipment. Various applications of photo-cross-linkable polymers in the emerging industries are outlined in subsequent chapters. The application of laser-induced processes to photoimaging, microelectronics, three-dimensional machining, holographic optical elements, and materials for information recording and storage will stimulate the future research on this field. However, industrial use of lasers requires macromolecules sensitive at particular wavelengths emitted by commercial lasers (e.g., in visible spectral or near-IR ranges). The synthesis of such products is not well developed so far. Another way to go is the application of sensitizers for the cross-linking, which absorb at the wavelengths of the laser sources. Water-borne UV-curable formulations, curing of encapsulated liquid crystals, emulsion processes, and surface polymerization of polymeric materials are also likely to attract increasing interest. A number of other basic topics of photo-cross-linking of polymers have not hitherto been investigated. I believe more quantitative research should be devoted to the correlation between photochemical and thermal reaction steps, to polymer properties and cross-linking efficiency, to change of polymer properties versus cross-linking density, and to the characterization of the end products. These studies should enable us, among other things, to develop more highly reactive macromolecules and more suitable photoinitiators for their manipulation by more readily accessible light sources.

References

1. Reiser, A. *Photoreactive Polymers;* John Wiley & Sons: New York, 1989.
2. Böttcher, H.; Bendig, J.; Fox, M. A.; Hopf, G.; Timpe, H.-J. *Technical Applications of Photochemistry;* Deutscher Verlag für Grundstoffindustrie: Leipzig, Germany, 1991.
3. Timpe, H.-J.; Baumann, H. *Photopolymere—Prinzipien und Anwendungen;* Deutscher Verlag für Grundstoffindustrie: Leipzig, Germany, 1988.
4. Roffey, C. G. *Photopolymerization of Surface Coatings;* John Wiley & Sons: Chichester, England, 1982.
5. *Radiation Curing—Science and Technology;* Pappas, S. P., Ed.; Topics in Applied Chemistry; Plenum: New York, 1992.
6. Turro, N. J. *Modern Molecular Photochemistry;* Benjamin-Cummings: Menlo Park, CA, 1978.
7. *Einführung in die Photochemie,* 3rd ed.; Becker, H. G. O., Ed.; Deutscher Verlag der Wissenschaften: Berlin, Germany, 1991.
8. Wayne, R. P. *Principles and Applications of Photochemistry;* Oxford Science: Oxford, England, 1988.
9. Turro, N. J. *Pure Appl. Chem.* **1977,** *49,* 405.
10. Förster, T. *Die Fluoreszenz Organischer Verbindungen;* Vanderhoek & Ruprecht: Göttingen, Germany, 1951.
11. Lamola, A. A. *Energy Transfer and Organic Photochemistry;* Interscience: New York, 1969.
12. Dexter, D. L. *J. Chem. Phys.* **1948,** *21,* 836.
13. Balzani, V.; Bolleta, F.; Scandola, F. *J. Am. Chem. Soc.* **1980,** *102,* 2152.
14. Timpe, H.-J.; Kronfeld, K.-P. *J. Photochem. Photobiol. A: Chem.* **1989,** *46* 253.
15. *Singlet Oxygen;* Wassermann, H. H.; Murray, R. W., Eds.; Academic: Orlando, FL, 1979.
16. Engel, P. S.; Monroe, B. M. *Adv. Photochem.* **1972,** *8,* 246.
17. *The Exciplex;* Gordon, M.; Ware, W. R., Eds.; Academic: Orlando, FL, 1975.
18. Lim, E. C. *Acc. Chem. Res.* **1987,** *20,* 8.
19. Briegleb, G. *Elektronen-Donator-Acceptor-Komplexe;* Springer Verlag: Berlin, Göttingen, Heidelberg, Germany, 1961.
20. Kavarnos, G. J. *Top. Curr. Chem.* **1990,** *156,* 23.
21. *Fundamentals of Photoinduced Electron Transfer;* Kavarnos, G. J., Ed.; VCH Verlag: Weinheim, Germany, 1993.
22. *Photoinduced Electron Transfer;* Fox, M. A.; Channon, M., Eds.; Elsevier: Amsterdam, Netherlands, 1988.
23. Hurst, J. R.; Mc Donald, J. D.; Schuster, G. B. *J. Am. Chem. Soc.* **1982,** *104,* 2065.
24. Guillet, J. E. *Polymer Photophysics and Photochemistry;* Cambridge University: Cambridge, England, 1984.
25. Webber, S. E. In *New Trends in the Photochemistry of Polymers;* Allen, N. S.; Rabek, J. F., Eds.; Elsevier Applied: London, 1985; p 19.
26. Klöpffer, W. *J. Chem. Phys.* **1969,** *50,* 2337.
27. David, C.; Demarteau, W.; Geuskens, G. *Eur. Polym. J.* **1972,** *6,* 1397.
28. Osada, Y.; Koike, M.; Katsumura, E. *Chem. Lett.* **1981,** 809.
29. Guillet, J. E.; Rendall, W. A. *Macromolecules* **1986,** *19,* 224.

30. Bruss, J. M.; Sahyun, M. R. V.; Schmidt, E.; Sharma, D. K. *J. Polym. Sci. Part A: Polym. Chem.* **1993**, *31*, 987.
31. Cozzens, R. F.; Fox, R. B. *J. Chem. Phys.* **1969**, *50*, 1532.
32. Berlman, I. B. *Energy Transfer Parameters of Aromatic Molecules*; Academic: Orlando, FL, 1973.
33. Kiwi, W.; Schnabel, W. *Macromolecules* **1975**, *8*, 430.
34. Hayashi, K.; Irie, M.; Kiwi, W. *Polym. J.* **1977**, *9*, 41.
35. Schnabel, W. *Makromol. Chem.* **1979**, *180*, 1487.
36. Das, P. K.; Scaiano, J. C. *Macromolecules* **1981**, *14*, 683.
37. Wagner, P. J. *Acc. Chem. Res.* **1989**, *22*, 306.
38. Formosinho, S. J.; Arnaut, L. G. *Adv. Photochem.* **1991**, *16*, 67.
39. Graley, M.; Reiser, A.; Roberts, A. J.; Phillips, D. *Macromolecules* **1981**, *14*, 1752.
40. Tazuke, T.; Matsuyma, Y. *Macromolecules* **1977**, *10*, 215.
41. Iwai, K.; Itoh, Y.; Furue, M.; Nazukura, S.-I. *J. Polym. Sci.; Polym. Chem. Ed.* **1983**, *21*, 2439.
42. Iwaya, Y.; Tazuke, S. *Macromolecules* **1982**, *15*, 396.
43. Tazuke, S.; Yuan, H. L.; Matsumaru, T.; Yamaguchi, Y. *Chem. Phys. Lett.* **1982**, *92*, 81.
44. Timpe, H.-J. *Top. Curr. Chem.* **1990**, *156*, 168.
45. Smets, G. *Pure Appl. Chem.* **1972**, *30*, 1.
46. Doba, T.; Ingold, K. U.; Siebrand, W.; Wildmann, T. A. *Faraday Discuss. Chem. Soc.* **1984**, *78*, 175.
47. Timpe, H.-J.; Strehmel, B.; Schiller, K.; Stevens, S. *Makromol. Chem., Rapid Commun.* **1988**, *9*, 749.
48. Lebedew, J. S. *Kinet. Katal.* **1978**, *19*, 1367.
49. Reiser, A.; Egerton, P. L. *Photogr. Sci. Eng.* **1979**, *23*, 144.
50. Pitts, A.; Reiser, A. *J. Am. Chem. Soc.* **1983**, *105*, 5540.
51. Kloosterboer, J. G. *Adv. Polym. Sci.* **1988**, *84*, 3.
52. Strehmel, B.; Anwand, D.; Timpe, H.-J. *Prog. Colloid Polym. Sci.* **1992**, *90*, 70.
53. Tazuke, S. In *Developments in Polymer Photochemistry*; Allen, N. S., Ed.; Applied Science: Barking, England, 1982; Vol. 3, p 53.
54. Tazuke, S.; Tanabe, T. *Macromolecules* **1979**, *12*, 853.
55. Kawai, W. *Eur. Polym. J.* **1977**, *13*, 413.
56. Smets, G. *Adv. Polym. Sci.* **1983**, *18*, 50.
57. Watanabe, S.; Ichimura, K. *J. Polym. Sci., Polym. Chem. Ed.* **1982**, *20*, 3261.
58. Lin, A. A.; Reiser, A. *Macromolecules* **1990**, *22*, 3898.
59. Lin, A. A.; Chu, C.-F.; Reiser, A. *Macromolecules* **1990**, *23*, 3611.
60. Huang, W.-Y.; Lin, A. A.; Reiser, A. *Macromolecules* **1991**, *24*, 4600.
61. Krigbaum, W. R.; Ishikawa, T.; Watanabe, J.; Toriumi, H.; Kubota, K. *J. Polym. Sci., Polym. Chem. Ed.* **1983**, *21*, 1851.
62. Gulielminetti, J. M.; Decobert, G.; Dubois, J. C. *Polym. Bull.* **1986**, *16*, 411.
63. Ikeda, T.; Itakura, H.; Lee, C.; Winnik, F. M.; Tazuke, C. *Macromolecules* **1988**, *21*, 3536.
64. Koch, T.; Ritter, H.; Bochholz, N.; Knochel, F. *Makromol. Chem.* **1989**, *190*, 1369.
65. Whitcombe, M. J.; Gilbert, A.; Mitchell, G. R. *J. Polym. Sci., Polym. Chem. Ed.* **1992**, *30*, 1681.
66. Yamashita, K.; Kyo, S.; Miyagawa, T.; Nango, M.; Tsuda, K. *J. Appl. Polym. Sci.* **1994**, *52*, 577.
67. Gangadhara; Kishor, K. *Macromolecules* **1993**, *26*, 2995.
68. Egerton, P. L.; Pitts, E.; Reiser, A. *Macromolecules* **1981**, *14*, 95.
69. Zweifel, H. *Photogr. Sci. Engin.* **1982**, *27*, 114.
70. Finter, J.; Widmer, E.; Zweifel, H. *Angew. Makromol. Chem.* **1984**, *128*, 71.
71. Finter, J.; Haniotis, Z.; Lohse, F.; Meier, K.; Zweifel, H. *Angew. Makromol. Chem.* **1985**, *133*, 147.
72. Ichimura, K.; Watanabe, S. *J. Polym. Sci., Polym. Chem. Ed.* **1982**, *20*, 1419.
73. Ichimura, K. *J. Polym. Sci., Polym. Chem. Ed.* **1982**, *18*, 1411.
74. Ichimura, K.; Oohara, N. *J. Polym. Sci., Polym. Chem. Ed.* **1987**, *25*, 3063.
75. Williams, J. L. R.; Farid, S. Y.; Doty, J. C.; Daly, R. C.; Specht, D. P.; Searle, R.; Borden, D. G.; Chang, H. J.; Martic, P. A. *Pure Appl. Chem.* **1977**, *49*, 32.
76. Niume, K.; Toyofuku, K.; Toda, F.; Uno, K.; Hasegawa, M.; Iwakura, Y. *J. Polym. Sci., Polym. Chem. Ed.* **1982**, *20*, 665.
77. Hargreaves, J. S. *J. Polym. Sci. Part A: Polym. Chem.* **1989**, *27*, 203.
78. Dilling, W. I. *Chem. Rev.* **1983**, *83*, 1.
79. Meier, K.; Zweifel, H. *J. Photochem.* **1986**, *35*, 353.
80. Tazuke, S.; Hayashi, N. *J. Polym. Sci. Part A: Polym. Chem. Ed.* **1978**, *16*, 2729.
81. Torii, T.; Ushiki, H.; Horie, K. *Polym. J. (Tokyo)* **1992**, *24*, 1057.
82. Cohen, A. B.; Walker, P. In *Imaging Processes and Materials*; Sturge, J.; Walworth, V.; Shepp, A., Eds.; Van Nostrand Reinhold: New York, 1989; p 226.
83. Schnabel, W. *Polymer Degradation*; Akademie-Verlag: Berlin, Germany, 1981.
84. Shevlyakov, A. S.; Samokhvalova, L. M. *Plast. Massy* **1977**, 49.
85. Tsunooka, M.; Sasaki, H.; Tanaka, M. *J. Polym. Sci., Polym. Lett. Ed.* **1980**, *18*, 407.
86. Egawa, H.; Ishikawa, M.; Tsunooka, M.; Ueda, T.; Tanaka, M. *J. Polym. Sci., Polym. Chem Ed.* **1983**, *21*, 479, 1233.
87. Tsunooka, M.; Tanaka, S.; Tanaka, M.; Egawa, H.; Nonaka, T. *J. Polym. Sci., Polym. Chem. Ed.* **1985**, *23*, 2495.
88. Decout, J. L.; Lablache-Combier, A.; Louchex, C. *J. Polym. Sci., Polym. Chem. Ed.* **1980**, *18*, 2371, 2391.
89. Rosilio, C.; Rosilio, A.; Serre, B. *Microelectron. Eng.* **1988**, *8*, 55.
90. Scriven, E. F. V. *Azides and Nitrenes*; Academic: Orlando, FL, 1984.
91. Budyka, M. F.; Kantor, M. M.; Alfimov, M. V. *Usp. Khim.* **1992**, *61*, 48.
92. Gritsan, N. P.; Pritchina, E. A. *Usp. Khim.* **1992**, *61*, 910.
93. Uchino, S.; Tanaka, T.; Ueno, T.; Iwayanagi, T.; Hayashi, N. *J. Photopolym. Sci. Technol.* **1992**, *5*, 93.
94. Reinisch, G.; Wigant, L.; Hiller, W.; Schmolke, R. *Angew. Makromol. Chem.* **1992**, *197*, 149.
95. Treushnikov, V. M.; Telepneva, T. V.; Oleinik, A. V.; Sorin, E. L., Korshak, V. V.; Krongauz, E. S.; Belomoina, N. M. *Vysokomol. Soedin., Ser. A* **1986**, *28*, 2129.
96. Koseki, K.; Shibata, T.; Yamaoka, T. *Kobunshi Ronbunshu* **1987**, *44*, 173.
97. Nishibuko, T.; Iizawa, T.; Saito, Y. *J. Polym. Sci., Polym. Chem. Ed.* **1983**, *21*, 2291.
98. Gruber, H. F. *Prog. Polym. Sci.* **1992**, *17*, 953.
99. Davidson, R. S. *J. Photochem. Photobiol. A: Chem.* **1993**, *73*, 81.

100. Baumann, H.; Timpe, H.-J. *Kunststoffe* **1989**, *79*, 696.
101. Davidson, R. S. *J. Photochem. Photobiol. A: Chem.* **1993**, *69*, 263.
102. Müller, U.; Jockusch, S.; Timpe, H.-J. *J. Polym. Sci. Part A: Polym. Chem.* **1992**, *30*, 2755.
103. Wagner, P.; Park, B. S. *Org. Photochem.* **1991**, *11*, 227.
104. Zamotaev, P. V. *Makromol. Chem., Makromol. Symp.* **1988**, *28*, 287.
105. Qu, B.; Xu, Y.; Shi, W.; Rangby, B. *Macromolecules* **1992**, *25*, 5215.
106. Zamotaev, P. V.; Esaulenko, G. B.; Sergienko, S. A.; Bezruk, L. I. *Vysokomol. Soedin., Ser. A* **1992**, *34*, 45.
107. Zamotaev, P. V.; Luzgarev, I. M. *Angew. Makromol. Chem.* **1989**, *173*, 47.
108. Cha, Y.; Tsunooka, M.; Tanaka, M. *J. Polym. Sci., Polym. Chem. Ed.* **1984**, *22*, 2973.
109. Lin, A. A.; Sastri, V. R.; Tesoro, G.; Reiser, A.; Eachus, R. *Macromolecules* **1988**, *21*, 1165.
110. Goethals, E. J. *Cationic Polymerization and Related Processes*; Academic: Orlando, FL, 1984.
111. Lohse, F.; Zweifel, H. *Adv. Polym. Sci.* **1986**, *78*, 61.
112. May, C. B.; Tanaka, Y. *Epoxy Resins—Chemistry and Technology*; Marcel Dekker: New York, 1975.
113. Franzke, M.; Lorkowski, H.-J.; Timpe, H.-J.; Wigant, L. *Acta Polym.* **1988**, *29*, 277.
114. Crivello, J. V. *Adv. Polym. Sci.* **1984**, *62*, 3.
115. Timpe, H.-J. *Eur. Coating J.* **1993**, *7/8*, 498.
116. Sahyun, M. R. V.; De Voe, R. J.; Olofson, P. M. In *Radiation Curing in Polymer Science and Technology*; Fouassier, J. P.; Rabek, J. F., Eds.; Elsevier Science: Barking, England, 1993; Vol. II, p 505.
117. Timpe, H.-J. In *Radiation Curing in Polymer Science and Technology*; Fouassier, J. P.; Rabek, J. F., Eds.; Elsevier Science: Barking, England, 1993; Vol. II, p 529.
118. Dietlicker, K. K. In *Chemistry & Technology of UV&EB Formulations for Coatings, Inks & Paints*; Oldring, P. K. T., Ed.; SITA Technology: London, 1991; Vol. III.

Photosensitive Polymers

Sung Il Hong, So Young Joo, and Doo Whan Kang

Recent progress in polymer photochemistry has led to the development of photosensitive polymers as a new source of photofunctional as well as photosensitive materials. Photosensitive polymers can be applied in areas such as printing, electronics, paints, biomaterials, information recording, and UV curing inks. In this chapter, the current state of the art and potential applications in microlithography and high technology materials are reviewed. Structural characteristics of photosensitive polymers with photosensitive groups in the main or side chains, those with photoactive compounds, silicon-containing photosensitive polymers, and photodegradable polymers are reviewed.

Photochemical and photophysical reactions in polymers occur following the absorption of light photons in the wavelength range of 100–700 nm. Light absorption leads to excitation, and the excited molecules can lose their excitation energy by photochemical and photophysical reactions. Photochemical reactions that lead to the formation of new chemical products include free-radical formation, cyclization, elimination, and intramolecular rearrangement. Photophysical processes include conversion to thermal energy, internal conversion, intersystem crossing, and radiative dissipation.

Most molecules that absorb a photon in the ground singlet electronic state (S_0) are activated to the excited singlet state (S_1, S_2, S_3, etc.). The lowest excited singlet state (S_1) is formed by the activation of the ground singlet state or by the internal conversion of higher excited singlet states (S_2, S_3, etc.) due to vibrational relaxation. Molecules in the lowest excited singlet state can lead to chemical reaction, excited complex formation, intersystem crossing, radiationless internal conversion, and radiative deactivation. The lowest excited triplet state (T_1), where the spin of the excited electron is parallel, may be formed only by intersystem crossing because of spin-forbidden transitions. Chemical reaction, phosphorescence, and internal conversion take place according to the deactivation processes of the lowest excited triplet state. This T_1 state may transfer to the higher triplet states (T_2, T_3, etc.) by absorbing a new photon. Further discussion and illustration of the underlying Jablonski diagram are given in reference 1.

Polymers that have sufficient energy following the absorption of a photon may be homolytically cleaved to give free radicals. These polymers may then undergo secondary reactions such as cyclization, hydrogen abstraction, or fragmentation. Polymers capable of these photoreactions are called *photosensitive polymers*. Various chemical reactions of photosensitive polymers such as polarity change, cross-linking, photografting, and photodegradation were investigated and applied to photofabrication. Photofabrication can be classified into graphic arts, chemical milling, and microfabrication. Conventional uses of graphic arts include planography, intaglio, gravure, and screen printing. Chemical milling, which is a fabrication process of large objects with the help of resist patterns and etchants, is applied for large decorative patterns. Microfabrication, which was recently developed, includes circuit boards, printed circuits, integrated devices, integrated optics, and thin film devices (2). Microfabrication of photosensitive polymers has contributed significantly to the modern semiconductor technology, and this process plays a crucial role in microlithography (3–5).

Among photoreactions, photopolymerization has several important nonimage relief uses for UV curing of varnishes and printing inks, radiation curing in the copying of the videodisks, and replication of lenses (6). Trans–cis isomerization, ring opening-closing, tautomerism, and valence isomerization reactions may take place in photochromic reactions (7). Photochromic polymers have covalently bound photochromic groups such as stilbene and spiropyran groups in the side chains or main chain. In general, polymers are either insulating or poorly conductive in the dark; however, photoconductive polymers become more conductive by irradiation. These photoconductivities of polymers are currently used only in electrophotography and are being explored in battery electrodes, solar cells, semiconductors, and molecular-sized electron devices (8).

Structural characteristics of photosensitive polymers are reviewed in this chapter. Their potential applications in microlithography and in other high technology materials are outlined. The review is organized on the basis of the chemical structure of the polymers and their photosensitive groups. First, polymers carrying various photosensitive groups as part of their structure are discussed, followed by photosensitive systems containing a reactive polymer and a low molecular weight photoactive compound, and then silicon-containing photosensitive polymers. A brief discussion about photodegradation of polymers is also included.

Polymers with Photosensitive Groups

A wide variety of polymers carrying photosensitivity have been reported during the past 70 years. The classical example of photosensitive polymers are based on the cinnamoyl group. We discuss other polymer-bound photosensitive groups, including diazonaphthoquinone, diazonium, azide, o-nitrobenzyl, and phenylester group.

Table I. Absorption Behaviors of Unsensitized and Sensitized Poly(vinyl cinnamate)

Sensitizer	Absorption Maximum (nm)	Sensitivity Maximum (nm)
None	276	270
p-Nitroaniline	370	370
5-Nitroacenaphthene	374	385
Picramide	418	420

Source: Reproduced with permission from reference 9. Copyright 1965 Chemical Society of Japan.

Polymers with Cinnamoyl Groups

Cinnamic acid is directly photodimerized by UV irradiation to form cyclobutane derivatives. The classic negative resist, poly(vinyl cinnamate), is cross-linked by photoaddition between an excited cinnamoyl group of one polymer chain and a ground-state cinnamoyl group of another. In this case, carbonyl and phenyl groups provide favorable polarization to the double bond. Thus, the photoreaction of cinnamoyl groups takes place easily to form photo-cross-linking, as discussed in Chapter 2.1. Chalcone, coumarin, styryl-pyridium, and diphenylcyclopropene (Chart I) are similar to compounds used as a negative resist.

Various polymers containing cinnamoyl groups are prepared for use as negative resists; however, they are particularly ineffective for cross-linking because of the low quantum yield (ca. 0.1). Two methods are generally known to improve the quantum yield. The first is to use cinnamic acid derivatives containing extended conjugate groups, and the second is to use sensitizers. Table I shows maximum absorption and sensitivity maximum of unsensitized and sensitized poly(vinyl cinnamate) (9, 10). Sensitivity is defined as the input incident energy required to attain a certain degree of chemical response in the resist.

Polyimides with cinnamoyl derivatives drew much attention because of the direct imaging poten-

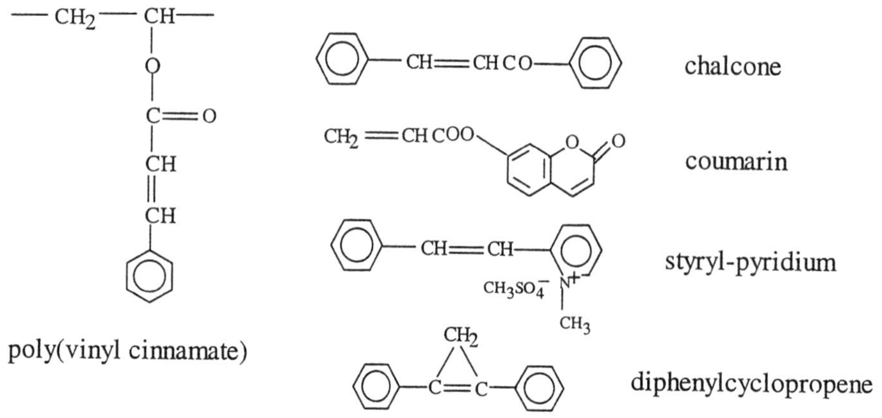

Chart I. Structures of compounds photodimerized by UV irradiation to cross-links.

tials (*11, 12*). Cinnamoyl derivatives are attached to the carboxylic acid group of polyamic acids via an ester linkage (Scheme I). The solubility of the polyamic acids with cinnamoyl derivatives is reduced to cross-linking by irradiation, and thermal imidization may eliminate all photosensitive cinnamoyl groups. This negative photosensitive polyimide exhibits high photospeed, contrast, and resolution. However, its use is limited by low quantum yield, excessive volume contraction, and short shelf-life problems (*13, 14*).

Polymers with Diazonaphthoquinone Groups

The primary photochemical reaction of diazonaphthoquinone group in various polymers is the elimination of nitrogen. Thus, diazonaphthoquinone yields a reactive ketene intermediate, which rearranges to carboxylic acid by Wölff rearrangement (*15*). Various polymers with diazonaphthoquinone as pendant groups are used as positive resists. Yamashita et al. (*16*) applied a cresol novolac resin substituted with a diazonaphthoquinone group as a negative resist. By exposing the polymer to deep-UV irradia-

tion, they could prepare cross-linked polymers by a radical mechanism. Organosilicon polymers with this substituent in the side chains were used as a negative or positive photoresist, because they have good thermal stability, good resistance to etching in oxygen plasma, and excellent coating properties due to the very low surface energy. Babich and co-workers (*17*) demonstrated the possibility of using diazonaphthoquinone-substituted polysiloxane in lithography. These polysiloxanes were prepared from the condensation of poly(dimethyl-*co*- aminopropylmethylsiloxane) and naphthoquinone-1,2-diazide-5-sulfonoylchloride (Scheme II).

Kang et al. (*18*) prepared polycarbosilane with different Si–H group contents by controlling the thermal decomposition condition of polymethylsilane and introduced diazonaphthoquinone in the side chain. Sensitivities of the resulting diazonaphthoquinone-substituted aminopropyl polycarbosilane were compared with those of cinnamoyl-substituted polysiloxane. The sensitivity and contrast of diazonaphthoquinone-substituted polycarbosilane and cinnamoyl-substituted polysiloxane are summarized in Table II.

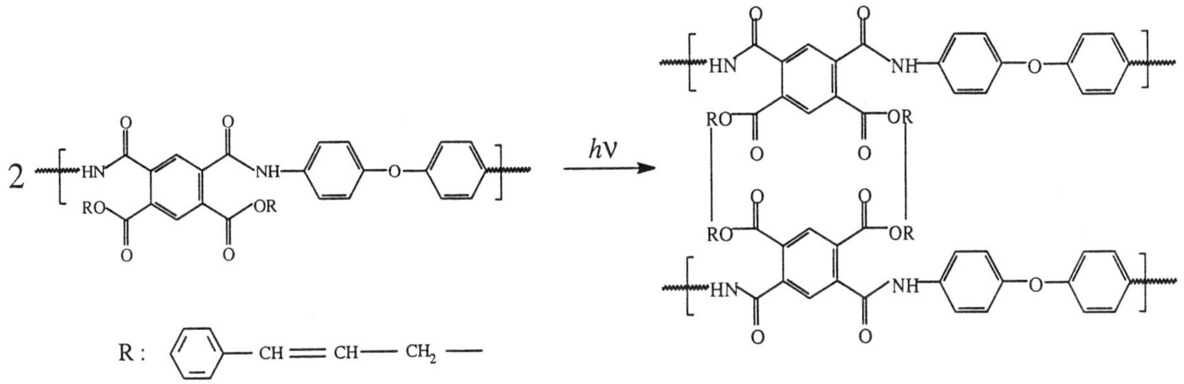

Scheme I. Cross-linking of polyamic acid substituted with cinnamoyl derivatives.

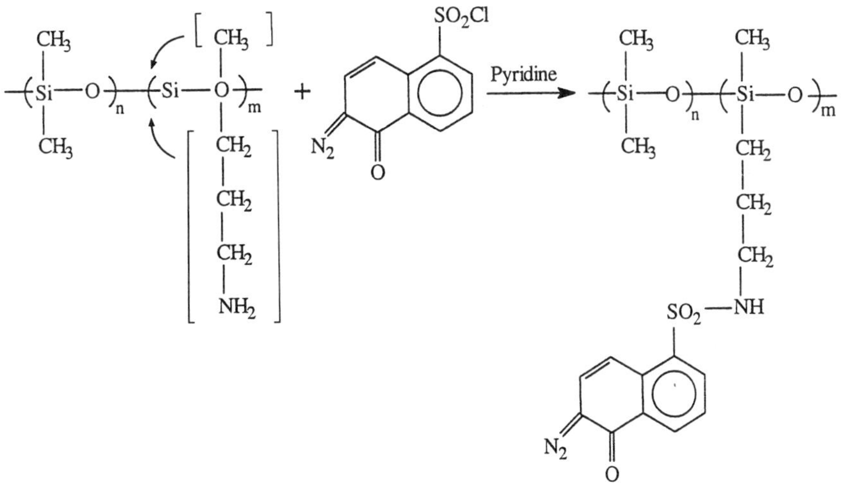

Scheme II. Preparation of diazonaphthoquinone-substituted polysiloxane.

Table II. Sensitivity and Contrasts Behaviors of Diazonaphthoquinone-Substituted Polycarbosilane (NDSP) and Cinnamoyl-Substituted Polysiloxane (CPS) with Different Sensitizers

Polymer	Sensitizer	Sensitivity (mJ/cm^2)	Contrast (γ)
NDSP	NONE	350	1.48
	Picramide	240	2.07
	2,6-Dichloro-4-nitroaniline	280	1.63
CPS	NONE	194	1.24
	5-Nitro-acenaphthene	126	1.68
	Benzophenone	158	1.33

Source: Reproduced with permission from reference 18. Copyright 1992 Polymer Society of Korea.

Polymers with Diazonium Groups

Because the nitrogen in the diazonium group can be removed easily, diazonium ion becomes carbonium ion by UV irradiation and then reacts with chloride ion to form the nonionic chloride derivative (Scheme III).

Polymers with diazonium ions are soluble in water, but they become insoluble in the aqueous developer by decomposition of diazonium group by UV irradiation. These polymers are used in the printing industry and have the advantage of being soluble in aqueous developer. Also, the activated diazonium ion reacts with secondary amine to produce a quaternary ammonium ion (*19*). Thus, a cross-linking reaction can take place (Scheme III).

Polymers with Azide Groups

An azide pendant group in a polymer is activated to nitrene by UV irradiation, and various reactions can take place according to the activated states of nitrene (*20*). Aromatic azides excited by light absorption lose nitrogen and transform to extremely reactive singlet-state nitrenes. Singlet-state nitrene is an electronic deficient species and can readily react with C–H, O–H, and N–H groups or add to double bonds. It can also decay to the ground triplet state. The triplet-state nitrene can abstract hydrogen and produce amino and carbon radicals. The amino radical can abstract hydrogen from hydrocarbon to produce primary amine, but

it cannot cross-link by recombination with the carbon radical because of spin direction. The triplet-state nitrene may also add to double bonds. Nitrene–nitrene coupling reaction to form azo dyes can occur in both singlet and triplet nitrenes. For example, a nitrene of activated poly(vinyl-*p*-azidobenzoate) may react with another nitrene to form an azo bond, combine with the polymer, or cross-link by hydrogen abstraction (Scheme IV) (*21*).

These reactions can take place in the polymers having phenylazo, aromatic sulfonylazo, and carbonylazo groups. Phenylazides are commonly used in photo-cross-linking.

Polymers with o-Nitrobenzyl Groups

Intramolecular photooxidation of *o*-nitrobenzaldehyde gives *o*-nitrobenzoic acid (Scheme Va) (*22*). This photoreaction was applied to positive-type photoresists by using carboxyl-protecting behaviors of the *o*-nitrobenzyl group. Reichmanis et al. (*23*) compared the photogeneration efficiency of carboxylic acid from various group-substituted 2-nitrobenzyltrimethylacetates in solution and in solid polymer matrix. *o*-Nitrobenzyl-substituted polymers with an additional *o*-nitro or ester substituent significantly increased the quantum yield. Meanwhile, photoreaction in a polymer matrix was retarded compared with photoreaction in acetonitrile solution because of the more hindered motion in the solid state. Thus, the appropriate selection of the substituent on the nitrobenzyl derivatives determined the photoresist sensitivity.

Wilkins et al. (*24*) reported a two-component system of methyl acrylate–methacrylic acid copolymer and *o*-nitrobenzyl cholate as a positive resist material. This system has high sensitivity and contrast due to its transparency to deep-UV irradiation. However, it has less resistance in plasma-etching environments than aromatic polymers with *o*-nitrobenzyl pendant groups.

In one-component systems such as poly(*o*-nitrobenzyl acrylate), a solubility difference results from carboxylic acid produced due to the photochemical reaction of *o*-nitrobenzyl group (Scheme Vb) (*25*). Also, polymers such as polyethers and poly(nitroanilides) containing *o*-nitrobenzyl groups have effective sensitivity, because the solubility difference results from

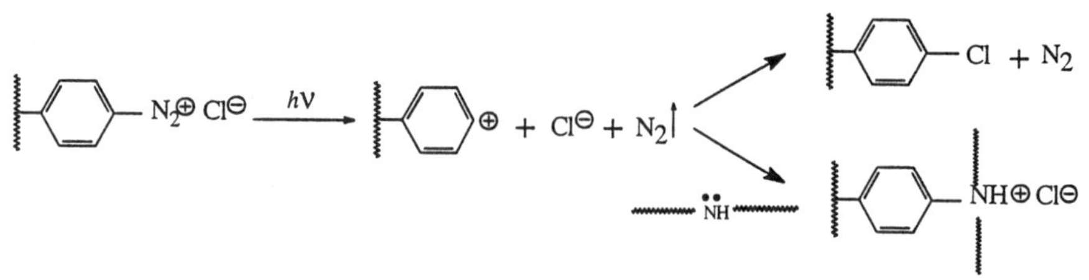

Scheme III. Photodecomposition and cross-linking reactions of polymer containing a diazonium ion.

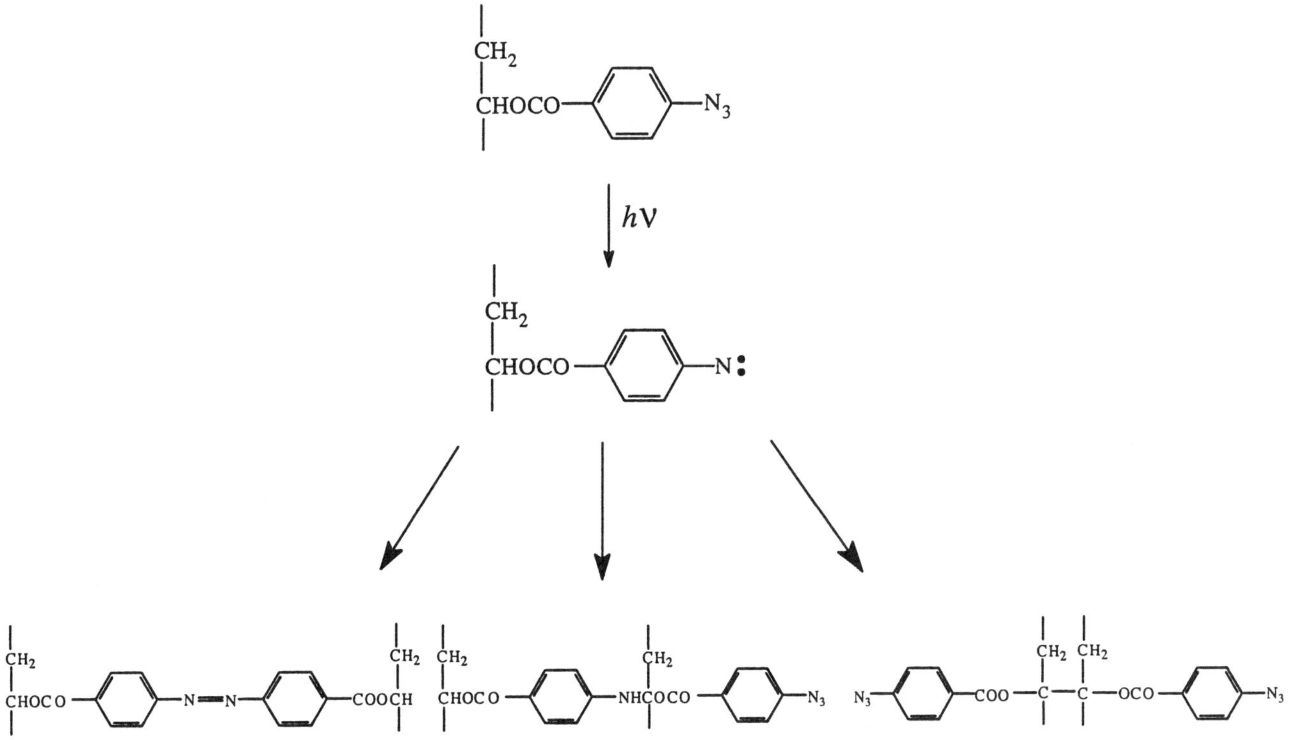

Scheme IV. Cross-linking reactions of poly(vinyl-*p*-azidobenzoate)s.

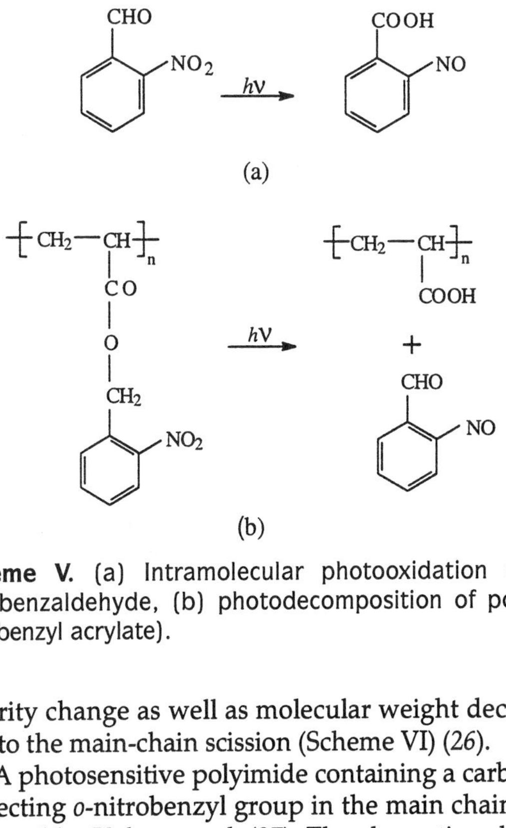

Scheme V. (a) Intramolecular photooxidation of *o*-nitrobenzaldehyde, (b) photodecomposition of poly(*o*-nitrobenzyl acrylate).

polarity change as well as molecular weight decrease due to the main-chain scission (Scheme VI) (*26*).

A photosensitive polyimide containing a carboxyl-protecting *o*-nitrobenzyl group in the main chain was reported by Kubota et al. (*27*). The absorption due to the nitrobenzyl group occurs in the vicinity of 250 nm and disappears after UV irradiation. This result indicates that nitrobenzyl groups undergo a photochemi-

cal reaction to produce the corresponding polymer and the free carboxylic acid soluble in alkaline aqueous solution. Recently, we employed (*28*) the same route for the synthesis of a modified *o*-nitrobenzyl polyamic ester containing *m*-phenylene groups (PAPAA) and benzophenone groups (PBPAA) to improve the sensitivity of this type of polymer (Scheme VII) (*28*). UV spectrum changes of PAPAA and PBPAA by irradiation take place at 238 nm and 276 nm, respectively. For PBPAA containing the benzophenone structure, the red-shift effect is quite significant and indicates that structural modifications can be accomplished for developing near-UV resists.

Polymers with Phenyl Ester Groups

Kobsa (*29*) and Kalmus and Hercules (*30*) found that organic molecules with phenyl esters, phenyl carbonates, or phenyl ethers undergo photo-Fries rearrangement by UV irradiation. Thus, the polymers containing phenyl esters, phenyl carbonates, or phenyl ethers can be used as resist materials. For example, phenyl acetate in ethanol produces *o*- and *p*-hydroxyacetophenones and phenol by photo-Fries rearrangement (Scheme VIII). By UV irradiation, a homolytic cleavage of phenyl acetate with the lowest excited singlet state occurs at the C–O bond in ArO–COR and produces two types of radicals. These radicals combine to form *o*- and *p*-hydroxyacetophenone or to abstract hydrogen to form phenol.

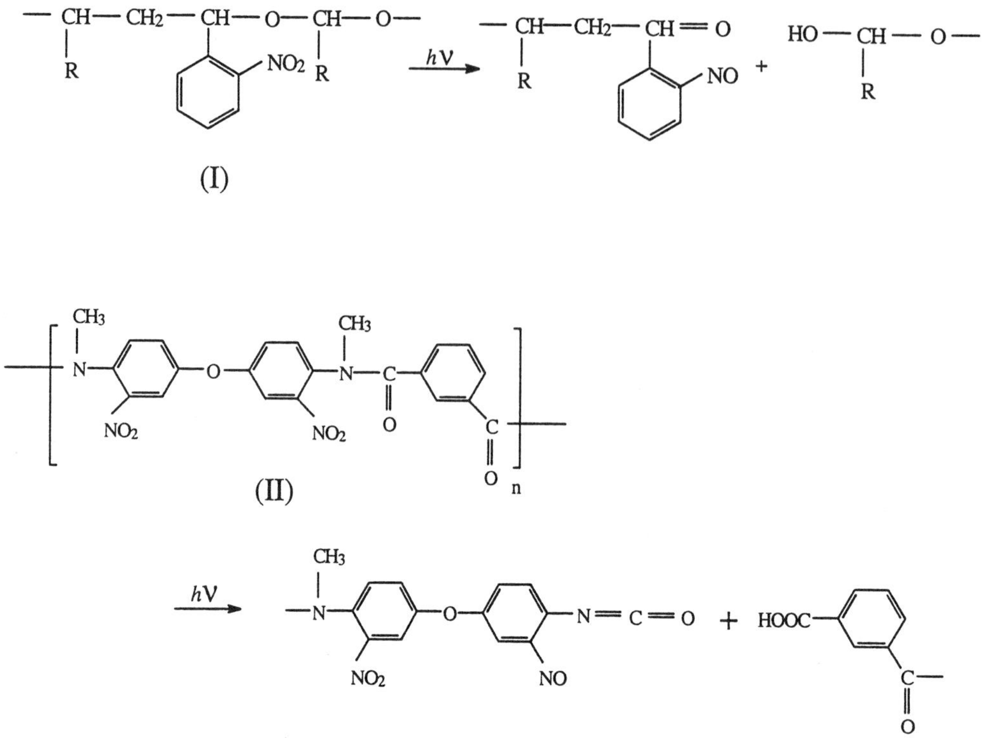

(I)

(II)

Scheme VI. Photodecomposition of *o*-nitrobenzyl-substituted polyether (I) and poly(*m*-nitroanilide) (II).

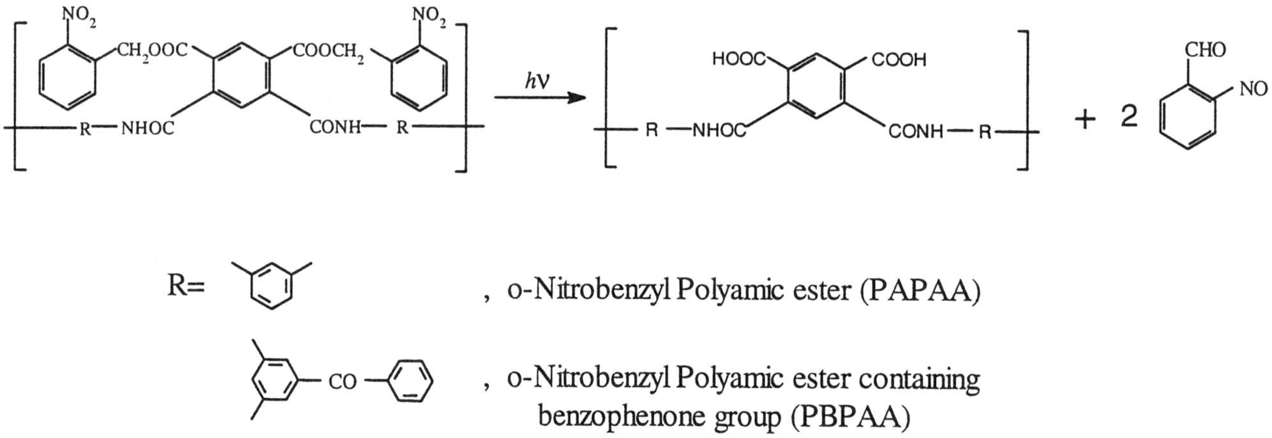

R=   , o-Nitrobenzyl Polyamic ester (PAPAA)

    , o-Nitrobenzyl Polyamic ester containing
benzophenone group (PBPAA)

Scheme VII. Photosensitive polyimide precursors containing *o*-nitrobenzyl groups and their photoconversion to alkali soluble polymers.

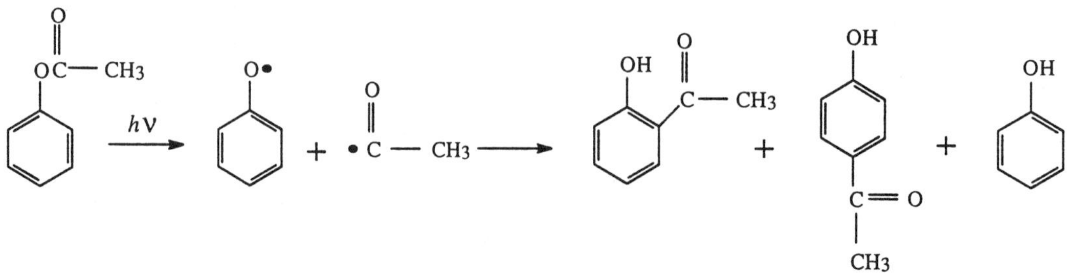

Scheme VIII. Photo-Fries rearrangement of phenyl acetate.

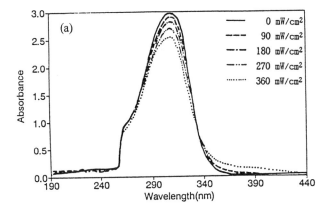

Scheme IX. Thermal and photo-Fries conversion of model diacid (I) to the corresponding diimide (II) and diamic acid (III).

The diacid model compound and diimide model compound (I and II in Scheme IX) were investigated (*31*). UV spectra (Figure 1) indicate that the diacid model compound (I) undergoes photo-Fries rearrangement at 320 nm to give OH-containing diamic acid (III) (Scheme IX). The diimide type compound (II) has no UV absorption changes, because rearrangement requires too much bending energy. Thus, to explain the efficiency of the photo-Fries rearrangement, we need an exact chain dimension and chain flexibility of chromophores.

A typical example of the photo-Fries reaction is the rearrangement of poly(phenyl acrylate) to yield *o*- and *p*-hydroxyphenone pendant groups (Scheme Xa) (*32*). Analogous polymers containing photosensitive groups in the pendant chain are poly(phenyl methacrylate), poly(*p*-acetoxystyrene), poly(*p*-formyloxystyrene), and poly(*p*-acetamido-styrene). All of these polymers can have selective solubility in a polar or nonpolar solvent (*33*, *34*). Under UV irradiation, these polymers are converted to hydroxy-containing polymers that can dissolve in a polar solvent, but they become insoluble in a nonpolar solvent. Thus, it is expected that these polymers can be applied to dual tone resists. Similarly, poly(aryl esters) and poly(aryl carbonates) undergo photo-Fries reaction to produce photostabilizing effects (Scheme Xb) (*35*).

Photosensitive polyimides are attractive as direct imaging materials. However, the majority of photosensitive polyimide research is concentrated on negative-type resists, which have swelling and low selectivity of dissolution in the developer due to photocross-linking (*36*). More recently, positive-type photosensitive polyimides had a limited usage due to the volume shrinkage by the elimination of photosensitive groups. Joo (*31*) developed new types of photosensitive polyimides that can be used as positive-type resists. Polyamic acids containing phenyl esters (Scheme Xc) undergo photo-Fries rearrangement by UV irradiation to yield the corresponding poly(amic acid)s containing hydroxy groups, which are soluble in alkaline aqueous solution. The poly(amic acid)s are imidized by heating. Hydroxyl groups produced by photo-Fries rearrangement enhance dissolution in aqueous alkaline solution. The ketone groups produced in exposed regions decrease chain flexibility compared with the ester group. Thus, the postexposure baking can induce the imidization selectivity. The curing kinetics analyzed by IR spectros-

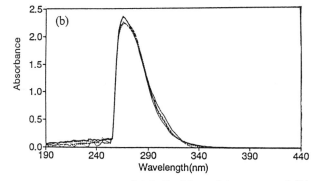

Figure 1. UV spectra of (a) diacid model compound (I in Scheme IX), and (b) diimide model compound (II in Scheme IX) by exposure.

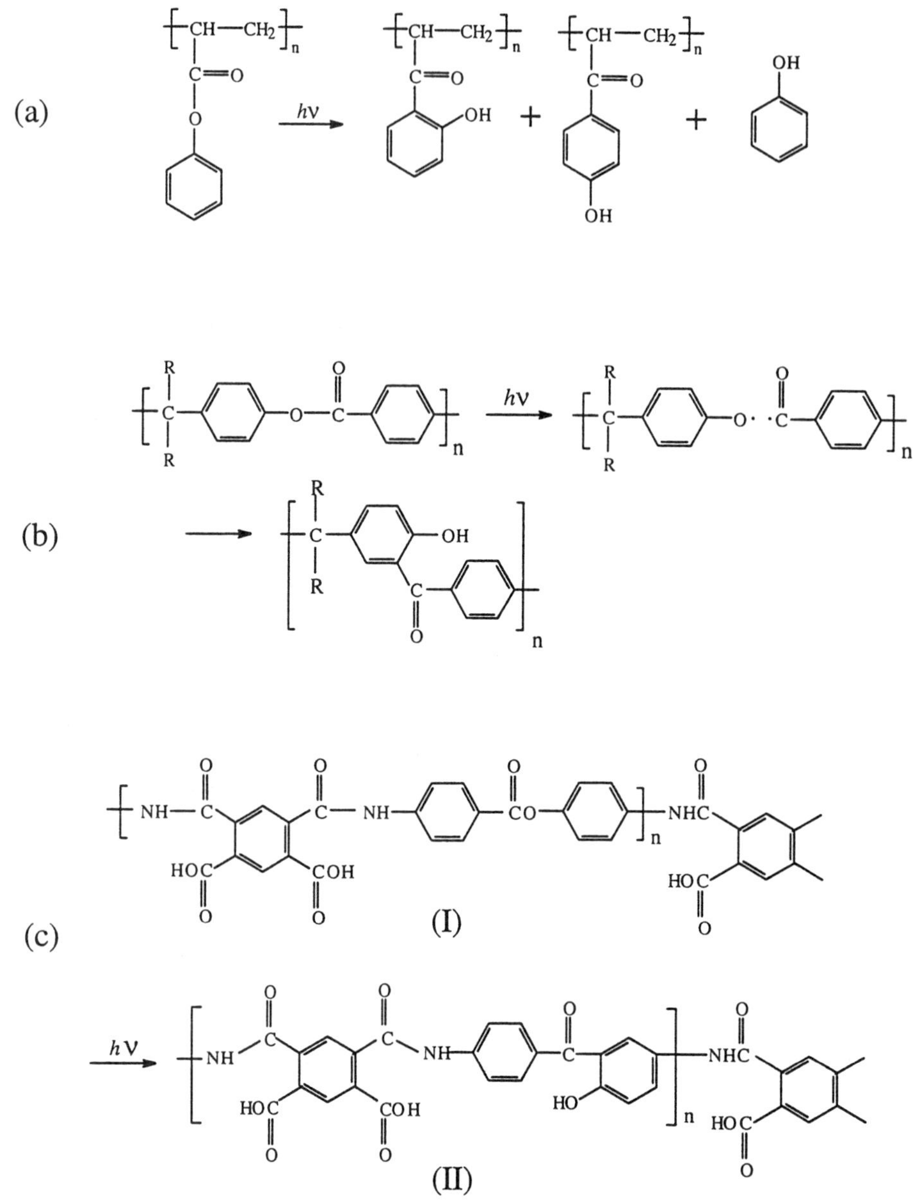

Scheme X. (a) Photo-Fries rearrangement of poly(phenyl acrylate)s, (b) photostabilizing effect by the photo-Fries rearrangement, and (c) photo-Fries rearrangement of poly(amic acid)s.

copy show that the imidization rate of the unexposed region is faster than that of exposed film (Figure 2). Because these aromatic polyimides with ester groups have good thermal stability (T_d > 450 °C) and high glass transition temperature (T_g > 300 °C), they can be applied to dielectric and passivation layers that can be direct-patterned.

Photosensitive Polymers with Photoactive Compounds

High performance resists with high resolution of less than 0.5 μm and good sensitivity will be required in advanced microelectronic devices. However, the effi-

ciency of the photochemical reaction is limited by the quantum yield for the process, expressed as molecules transformed per each photon absorbed. For example, quantum yield of diazonaphthoquinone is between 0.2 and 0.3. This fact means a diazonaphthoquinone must absorb 3–5 photons to induce a photochemical reaction, and therefore diazonaphthoquinone has limited sensitivity. The concept of chemical amplification (*37, 38*), whereby a single photoevent initiates numerous subsequent reactions, is based on the use of photoactive low molecular weight compounds (or photochemical catalysts). Photoactive compounds substantially increase the quantum yield of the photoreactions for polymer transformations such as cross-

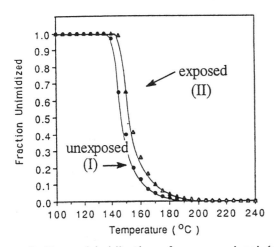

Figure 2. Thermal imidization of unexposed poly(amic acid) (I in Scheme X) and exposed poly(amic acid) (II in Scheme X). (Reproduced from reference 31.)

$$Ph_3S^+X^- \xrightarrow{h\nu} Ph_2S^{+\cdot}X^- + Ph\cdot$$

$$Ph_2S^{+\cdot}X^- + RH \longrightarrow Ph_2S^+\!-\!HX^- + R\cdot$$

$$Ph_2S^+\!-\!HX^- \longrightarrow Ph_2S + H^+X^-$$

Scheme XI. Photodecomposition mechanism of triphenyl-sulfonium salts.

linking (*39–41*), depolymerization (*40–43*), or deprotection of acid-labile polymers (*44–46*).

In this section, we present an outline of the mechanism of the most widely used photoactive compounds, the so-called photochemical acid generators, as well as diazides and azides. A comprehensive survey of the polymer chemistry and resist technology of chemically amplified imaging systems is provided in another chapter in this book.

Compounds producing strong acid by photoreaction are used in resist design with new imaging mechanisms. Various onium salts (diaryliodonium and triarylsulfonium salts) (*47, 48*), nitrobenzyl esters (*49, 50*), and imino sulfonates (*51*) were widely used as photochemical acid generators. For example, in the photodecomposition of triphenyl sulfonium salts the major photoproduct obtained from UV irradiation is a Brönsted acid (Scheme XI).

The first photoreaction shown in Scheme XI involves the homolytic cleavage of the Ph–S bond to yield the cation radical (Ph$_2$S•) and phenyl radical (Ph•). The cation radical may abstract a hydrogen atom from an alkyl group and produce an acid. Triphenyl sulfonium salts have high thermal stability (T_d > 200 °C) and have been used extensively for photocuring of polymer films and photoresists. However, they have limited usage because their radical and radical cation can lead to undesirable reactions and result in low solubility of polymers. To overcome the undesirable properties of phenyl sulfonium salts, acid generators without ion or cation radicals were widely investigated (*49–51*). As a typical example, *o*-nitrobenzyl esters of sulfonic acids generate the corresponding sulfonic acid by intramolecular rearrangement (Scheme XIIa). Because *o*-nitrobenzyl esters have poor thermal stability, modification of their structures such as introduction of an electron-withdrawing

group or an additional nitrobenzyl substituent is considered to improve their physical properties (*52*). Iminosulfonates prepared from a sulfonyl chloride and an oxime have high sensitivity. Iminosulfonates are decomposed by UV irradiation by the cleavage of the N–O bond, and then they abstract a hydrogen from an alkyl group (a polymer) to produce a sulfonic acid (Scheme XIIb).

Most chemically amplified resist materials are based on the deprotection of *tert*-butoxycarbonyl (*t*-BOC) or tetrahydropyranyl groups of polymer side chains (*53–54*) and the depolymerization of acid-labile polymers such as poly(phthalaldehyde). These resists usually exhibit high performance in the fabrication of very large scale integrated circuits. Recently, cross-linking based on electrophilic aromatic substitution was used in the design of acid-catalyzed negative resists (*39–41*). An electrophilic group caused by a photoacid generator is combined with the activated phenolic ring (Scheme XIII).

Kim recently prepared (*55*) an interesting thermally stable acid-catalyzed resist by formulating *N*-tosyloxyphthalimide as a photochemical acid generator and *t*-BOC-protected aromatic polyamides. The UV absorption intensity for *N*-tosyloxyphthalimide at 255–280 nm decreased as the irradiation time increased (Figure 3). *N*-Tosyloxyphthalimide generates *p*-toluenesulfonic acid by photolysis, and the resulting compound can participate in the deprotection reaction of the protected polymer (Scheme XIV).

Baking the acid-catalyzed resist with a photogenerated acid compound to an appropriate temperature after the exposure leads to acid-catalyzed thermolysis in the exposed regions but not in the unexposed regions. Postexposure baking time and temperature are very important to obtain a clear pattern. If the postexposure baking conditions are insufficient, the exposed regions that must be removed are not removed completely, because the *t*-BOC protecting group is removed by the thermal catalytic reaction (*56, 57*).

Photosensitive polymer blends based on diazonaphthoquinone and phenolic resin are some of the most important positive resists used in the semiconductor and printing industries. Diazonaphthoquinone derivatives are insoluble and also reduce the solubility of phenolic resin (a low molecular weight novolac) in

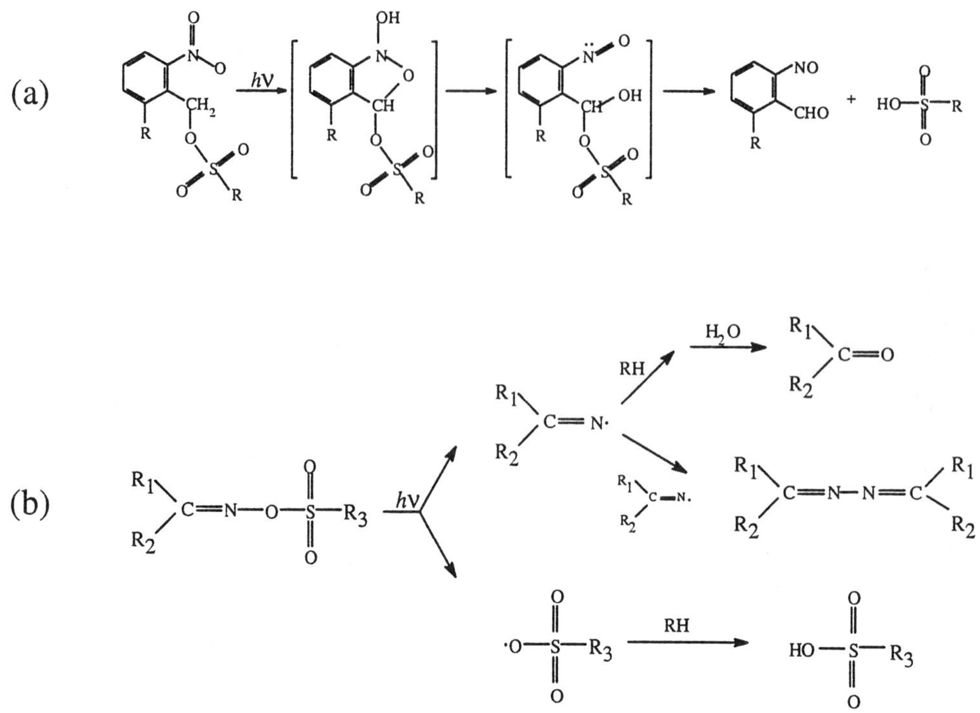

(a)

(b)

Scheme XII. Photodecomposition mechanism of *o*-nitrobenzyl ester of sulfonic acids (a) and imino sulfonates (b).

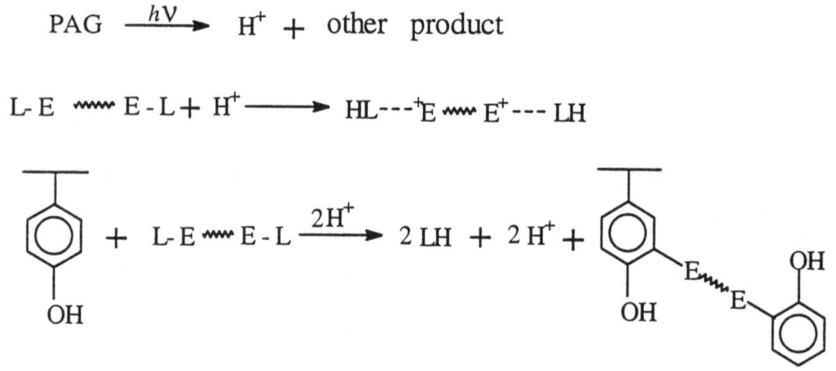

$$PAG \xrightarrow{h\nu} H^+ + \text{other product}$$

$$L\text{-}E \,\text{\small www}\, E\text{-}L + H^+ \longrightarrow HL\text{---}^+E \,\text{\small www}\, E^+ \text{---} LH$$

PAG : Photoacid generator

L- E www E - L : Acid-activated crosslinking agent

Scheme XIII. Cross-linking of a phenolic resin by a photochemical acid generator.

alkaline aqueous solution. However, these derivatives, when rearranged by UV irradiation, become soluble in the alkaline aqueous solution. The azo-coupling reaction of novolac and diazonaphthoquinone photoactive compounds in unexposed regions also was used (*58–62*). The most popular photoactive compound is obtained by esterification of diazonaphthoquinone sulfonyl chloride and polyhydroxybenzophenone (*63*). A blend of bisarylazide and partially cyclized rubber is widely used in the microelectronics industry as negative photoresists (*64–66*). However, the system has limited resolution due to swelling of rubber during the developing process and is being replaced,

gradually, by positive high resolution photoresists such as phenolic resins (*67*). Thus, the phenolic resins such as cresol novolac and poly(*p*-hydroxy styrene) are used to prevent swelling during the developing process.

Photosensitive Polysilanes

Organosilicon polymers with a silicon–silicon bond in the main chain and with two organic substituents at each silicon atom have very interesting physical and chemical properties as a new class of photosensitive polymers (*68, 69*). Although polysilane derivatives

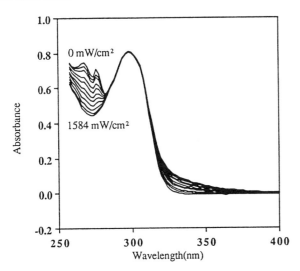

Figure 3. Spectrum changes of 2×10^{-5} mol *N*-tosyloxyphthalimide by UV irradiation in acetonitrile/H_2O: 1/1.

sodium at above 100 °C in toluene. Similarly, copolymers were synthesized by the reaction of two different diorganodichlorosilanes (Scheme XV) (70).

These polysilanes can be applied to new areas such as photoresists in microelectronics, photoinitiators in radical polymerization, and photoconductors. Photosensitive characteristics in polysilanes are attributed to their unusual electronic structure. The cumulated silicon–silicon bond in the backbone has a strong electron delocalization. This phenomenon leads to special photoconductivity and photosensitivity. The σ-delocalization of the silicon–silicon bond in polysilanes closely resembles conjugation in unsaturated polymers such as polyacetylene (Chart II) (71). For electrons in the Si–Si σ-bond of polysilanes, the ionization energies are much lower than covalent bonds of C–C, C–H, etc. For example, the ionization potentials are $CH_3CH_2CH_2CH_3$, 10.6 eV; $H_2C{=}CH{-}CH{=}CH_2$, 9.1 eV; and $Me(SiMe_2)_4Me$, 8.0 eV (72). Because the σ-electrons in polysilane are loosely bound, a delocalized model is considered.

High molecular weight polysilanes strongly absorb UV irradiation between 300 and 400 nm. The molar absorptivity per silicon–silicon bond increases rapidly with increasing molecular weight and approaches a

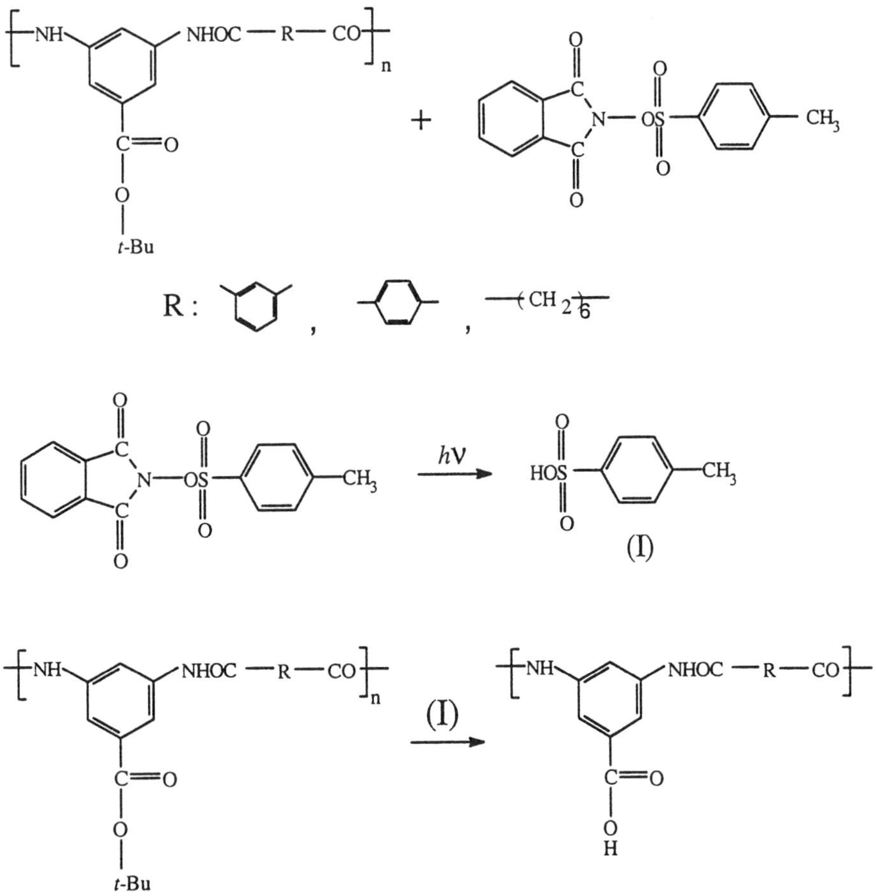

were prepared over 60 years ago by Kipping, these materials attracted little attention because of their intractable and insoluble properties. Recently, soluble high molecular weight polysilanes were prepared from dialkyl-, diaryl-, and alkylaryldichlorosilane with

Scheme XIV. Chemical reactions in the chemically amplified resist based on aromatic polyamides and *N*-tosyloxyphthalimide.

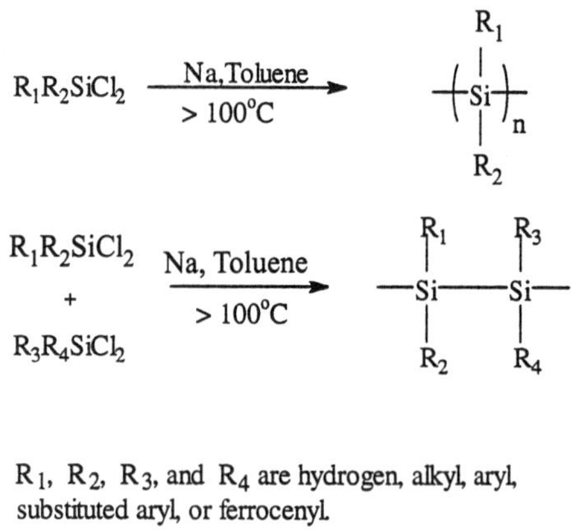

$$R_1R_2SiCl_2 \xrightarrow[> 100°C]{Na, Toluene} \left(\underset{R_2}{\overset{R_1}{Si}} \right)_n$$

$$R_1R_2SiCl_2 \underset{R_3R_4SiCl_2}{\overset{+}{}} \xrightarrow[> 100°C]{Na, Toluene} -\underset{R_2}{\overset{R_1}{Si}} - \underset{R_4}{\overset{R_3}{Si}} -$$

R₁, R₂, R₃, and R₄ are hydrogen, alkyl, aryl, substituted aryl, or ferrocenyl.

Scheme XV. Preparation of soluble polysilanes and silane copolymers.

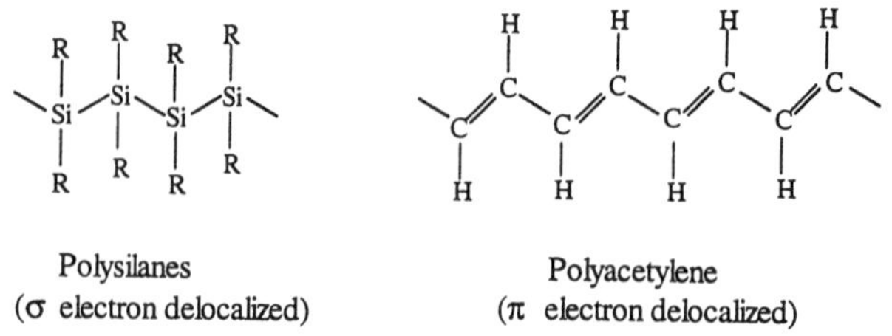

Polysilanes
(σ electron delocalized)

Polyacetylene
(π electron delocalized)

Chart II. Comparison between electronic structures of polysilane and polyacetylene.

limiting value at the degree of polymerization of about 40–50 (73). These absorptions are attributed to the transition from silicon–silicon σ-orbital to anti-bonding σ-orbital in alkyl-substituted polysilanes (74). The position of this absorption depends on the configuration of the backbone. Absorption wavelength (λ_{max}) increases as the number of trans silicon–silicon configurations

increases. Also, UV absorption in polysilanes with linear alkyl substituents depends on the nature of the substituent on the silicon atom (75).

During UV irradiation, polysilanes undergo chain scissions. The photolysis mechanism of polysilanes includes homolysis to silyl radicals, elimination of silylene, and these two processes simultaneously (Scheme XVI). However, when UV light with a wavelength longer than 300 nm is used, elimination of silylene does not take place, and the homolytic scission of silicon–silicon bond (Scheme XVI(A)) becomes the major process (70). Photochemical cross-linking of these polysilanes by the reaction of silyl radical with unsatu-

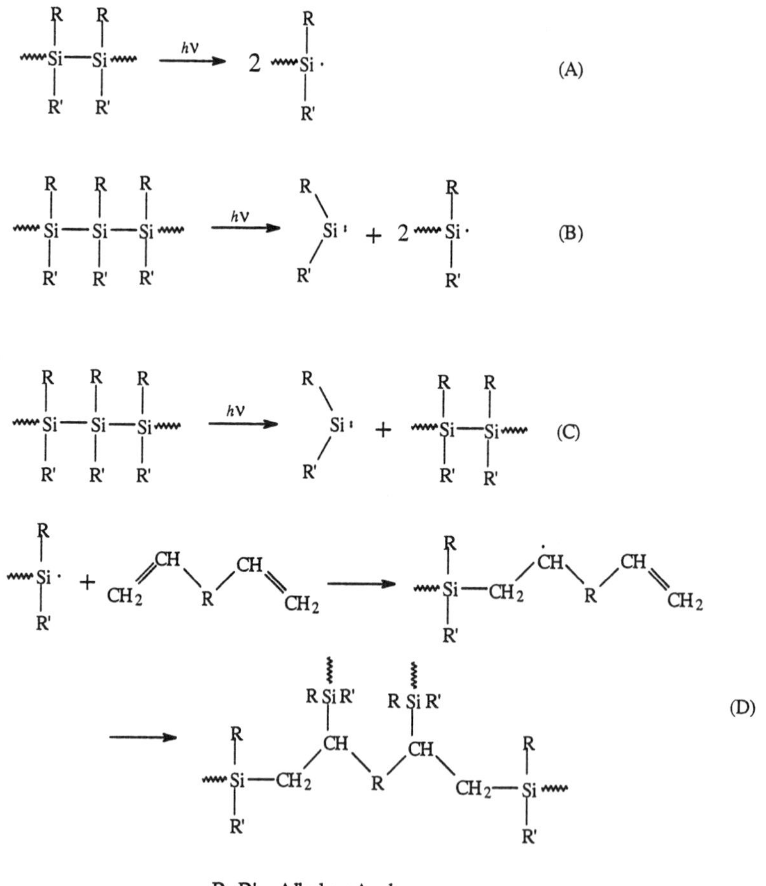

R, R'= Alkyl or Aryl

Scheme XVI. Photolysis of polysilanes (A, B, and C) and photochemical cross-linking of polysilanes radical with unsaturated hydrocarbons (D).

rated hydrocarbons takes place readily to form cross-linked structures (Scheme XVI(D)). In aryl-substituted polysilanes, scission of silicon–silicon bonds by UV irradiation produces silicon=carbon double bonds and Si–H compounds by hydrogen transfer from α-carbon atoms of silyl radicals (Scheme XVII) (70).

Ishikawa et al. prepared (76) poly[p-(1,2-dimethyldiphenyldisilanylene)phenylene] and examined photochemical behaviors in solution and solid film. During UV radiation, polymer I (Scheme XVII) degraded to silyl radicals by homolytic scission of silicon–silicon bonds, followed by hydrogen abstraction from methyl of silyl radical intermediates to produce silene.

Padmanaban et al. synthesized (77) photosensitive polydisilane derivatives with high thermal stability by introducing photoactive disilane units to aromatic polyamide or polyimide (Scheme XVIII). These compounds are easily degraded by UV irradiation to reactive intermediates such as radicals, anions, or cations. These intermediates undergo further reactions with unsaturated hydrocarbons or alkenylpolysilanes. Also, silicon–silicon bonds are homolytically cleaved at elevated temperature or by shear force.

Polysilanes undergo photoscission with light to produced silyl radicals. They are used as photoinitiators for free-radical polymerization of vinyl monomers.

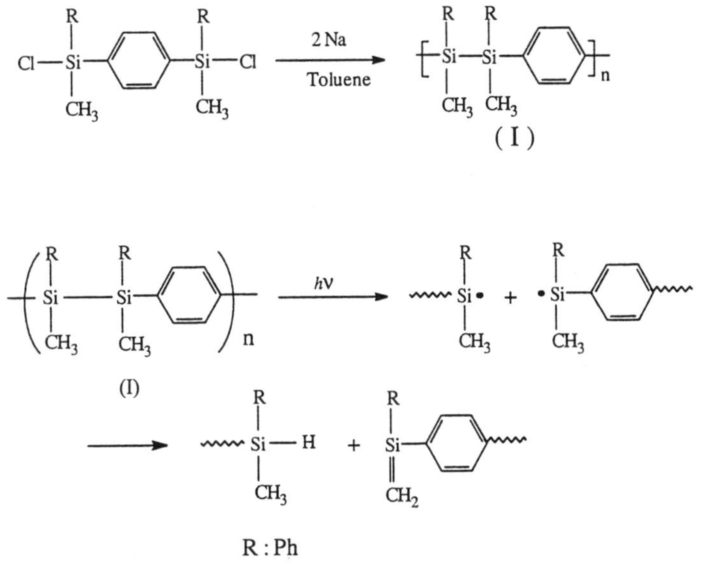

Scheme XVII. Synthesis and photolysis of poly[p-(1,2-dimethyldisilanylene) phenylene].

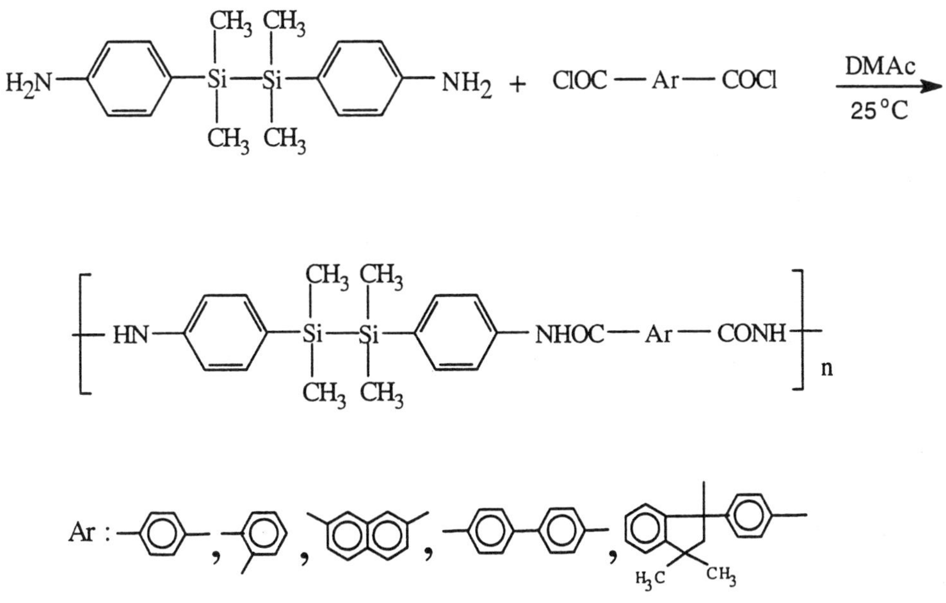

Scheme XVIII. Preparation of aromatic polyamides with disilane units.

However, photoinitiator efficiency is low when compared with conventional photoinitiators. In the case of photopolymerization of styrene by using a phenyl-methylpolysilane initiator, initiation efficiency was about 10 (78). This efficiency is due to the formed silyl radicals partially changing to a silane and silene by the hydrogen transfer from silyl radicals (Scheme XVII). Because the silicon–carbon double bond in produced silene is known as a very effective scavenger of oxygen, free-radical polymerization in an oxygen atmosphere is possible by using polysilanes as initiators.

The use of polysilanes as electrical conductors is due to the electron delocalization in the silicon–silicon bonds of polysilanes. Polysilanes become semiconductors when doped with strong electron acceptors such as SbF_5 or AsF_5. Although polysilanes are almost insulators and have a conductivity of 10^{-11} $\Omega^{-1} \cdot cm^{-1}$, the conductivity of cross-linked phenylmethylpolysilane (mol. wt, 300,000; $PhMeSi/Me_2Si = 5$) exposed to 1.33 Pa of AsF_5 was ~0.5 $\Omega^{-1} \cdot cm^{-1}$ (79).

Another interesting class of silicon compounds is acylsilanes photoisomerizing to carbenes, which are very effective cross-linking intermediates like nitrenes. Carbenes (or activated acylsilanes) react with functional groups such as OH or NH groups on polymers (80). Thus, bifunctional acylsilanes can be used as reagents for photo-cross-linking of polymers such as poly(vinylalcohol) (Scheme XIX) and can be applied as negative resists.

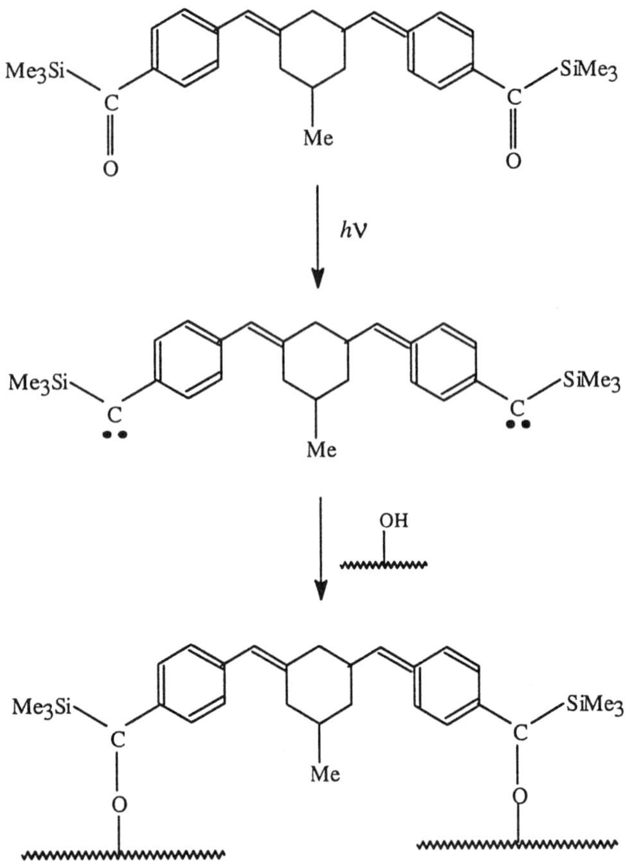

Photodegradable Polymers

Most polymers may undergo photodegradation because the quantum energies associated with light or UV irradiation can break the chemical bonds in polymers, and the photostability of polymers determines their end use. Early investigations of photodegradability of polymers were aimed at better understanding and hence reducing or preventing the degradation of commodity polymers such as plastics (81). However, with increasing accumulation of photostable polymers in the environment, the study of polymer photodegradation is being directed in the opposite direction: that is, to enhance polymer degradation and reduce environmental pollution (82). In addition, enhancement of polymer photodegradation is also of considerable interest in the development of X-ray and electron-beam resists (83, 84).

Mechanism of Photodegradation

Photodegradation of polymers involves random degradation, depolymerization, and photooxidative degradation, all mediated by polymer free radicals (Scheme XX). These free radicals, such as polymer ($\sim C\bullet$), polymer alkyloxy ($\sim O\bullet$), polymer alkylperoxy

Scheme XIX. Photo-cross-linking of OH-containing polymers by difunctional acylsilanes.

($\sim OO\bullet$), and hydroxyl ($HO\bullet$), result in chain scission, cross-linking, and secondary oxidation. The initiation step generally results from different photophysical processes such as the formation of electronically excited species, energy transfer, or direct photodissociation of chemical bonds. A few examples of polymer photodegradation are discussed briefly.

Polyketones

Polyketones containing a pendant carbonyl group are photodegraded by Norrish type I and II reactions (Scheme XXIa) or are photoreduced, or they form cyclobutanol (85). Norrish type II reactions may induce main-chain scission. However, radicals produced by Norrish type I reactions participate in repolymerization with the double bond of Norrish type II products. Photooxidative degradation may also occur simultaneously to produce alkylperoxyradicals, which abstract hydrogen to form hydroxyperoxy groups leading to main-chain scission through β-scission. Polymers containing carbonyl groups in the main chain also undergo photodegradation by Norrish type I and II reactions (Scheme XXIb) (86).

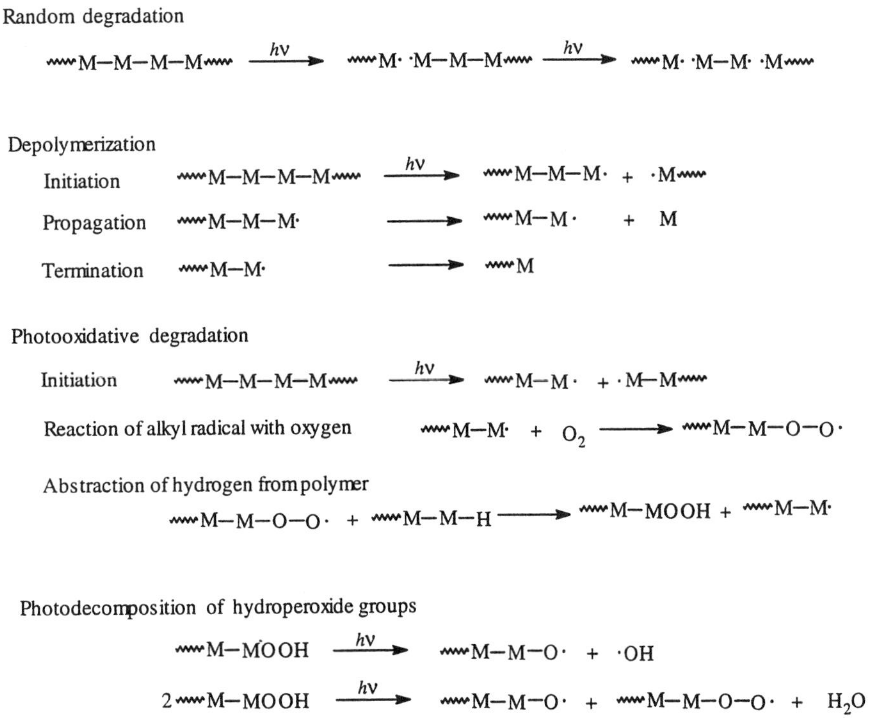

Random degradation

$$\text{wwwM—M—M—M www} \xrightarrow{h\nu} \text{wwwM· ·M—M—M www} \xrightarrow{h\nu} \text{wwwM· ·M—M· ·M www}$$

Depolymerization

Initiation $\qquad$ wwwM—M—M—M www $\xrightarrow{h\nu}$ wwwM—M—M· + ·M www

Propagation $\qquad$ wwwM—M—M· $\longrightarrow$ wwwM—M· + M

Termination $\qquad$ wwwM—M· $\longrightarrow$ wwwM

Photooxidative degradation

Initiation $\qquad$ wwwM—M—M—M www $\xrightarrow{h\nu}$ wwwM—M· + ·M—M www

Reaction of alkyl radical with oxygen $\qquad$ wwwM—M· + O_2 $\longrightarrow$ wwwM—M—O—O·

Abstraction of hydrogen from polymer

wwwM—M—O—O· + wwwM—M—H $\longrightarrow$ wwwM—MOOH + wwwM—M·

Photodecomposition of hydroperoxide groups

wwwM—MOOH $\xrightarrow{h\nu}$ wwwM—M—O· + ·OH

2 wwwM—MOOH $\xrightarrow{h\nu}$ wwwM—M—O· + wwwM—M—O—O· + H_2O

Scheme XX. Schematic presentation of various routes for photodegradation of polymers.

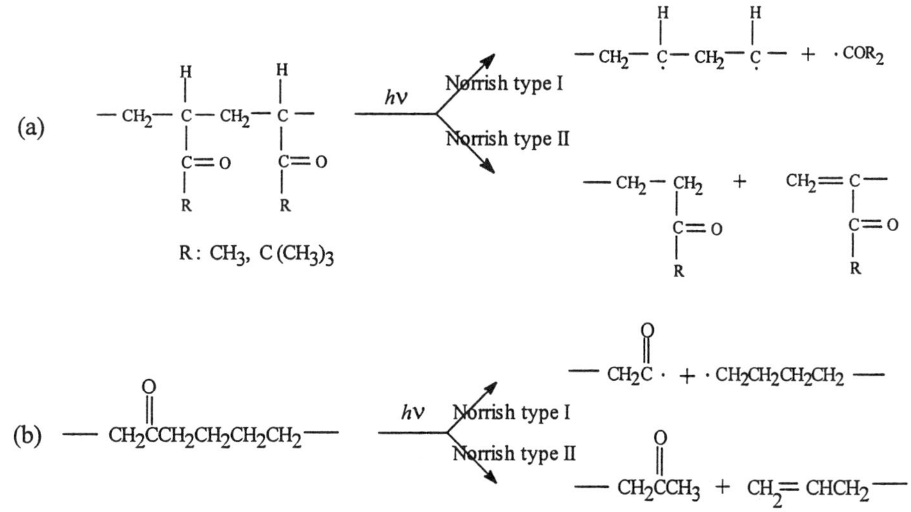

Scheme XXI. Norrish type photodegradation of side chain polyketones (a) and main-chain polyketones (b).

Polyacrylates and Polymethacrylates

Polymers containing ester groups also undergo extensive degradation and chain scission by Norrish type I and II reactions (Scheme XXII) (*87*). Introduction of bulky groups and substitution with polar electronegative groups, which reduce the polymer main chain, enhance the sensitivity of photodegradation (*88*). Other photodegradations of these polymers include side-chain photodecomposition and photooxidative degradation.

Polymers with ester groups such as poly(methyl methacrylate) (PMMA) are a classical example of electron beam resists. PMMA has very high resolution and is used in mask making and direct writing. Mitsuoka et al. investigated (*89*) the effect of wavelength on the degradation of PMMA and found that direct side-chain scission due to irradiation of wavelength shorter than 280 nm initiated the main-chain scission. Various methacrylate esters and substitutions at the α-position of acrylate have been widely applied to the advanced resists (*90*).

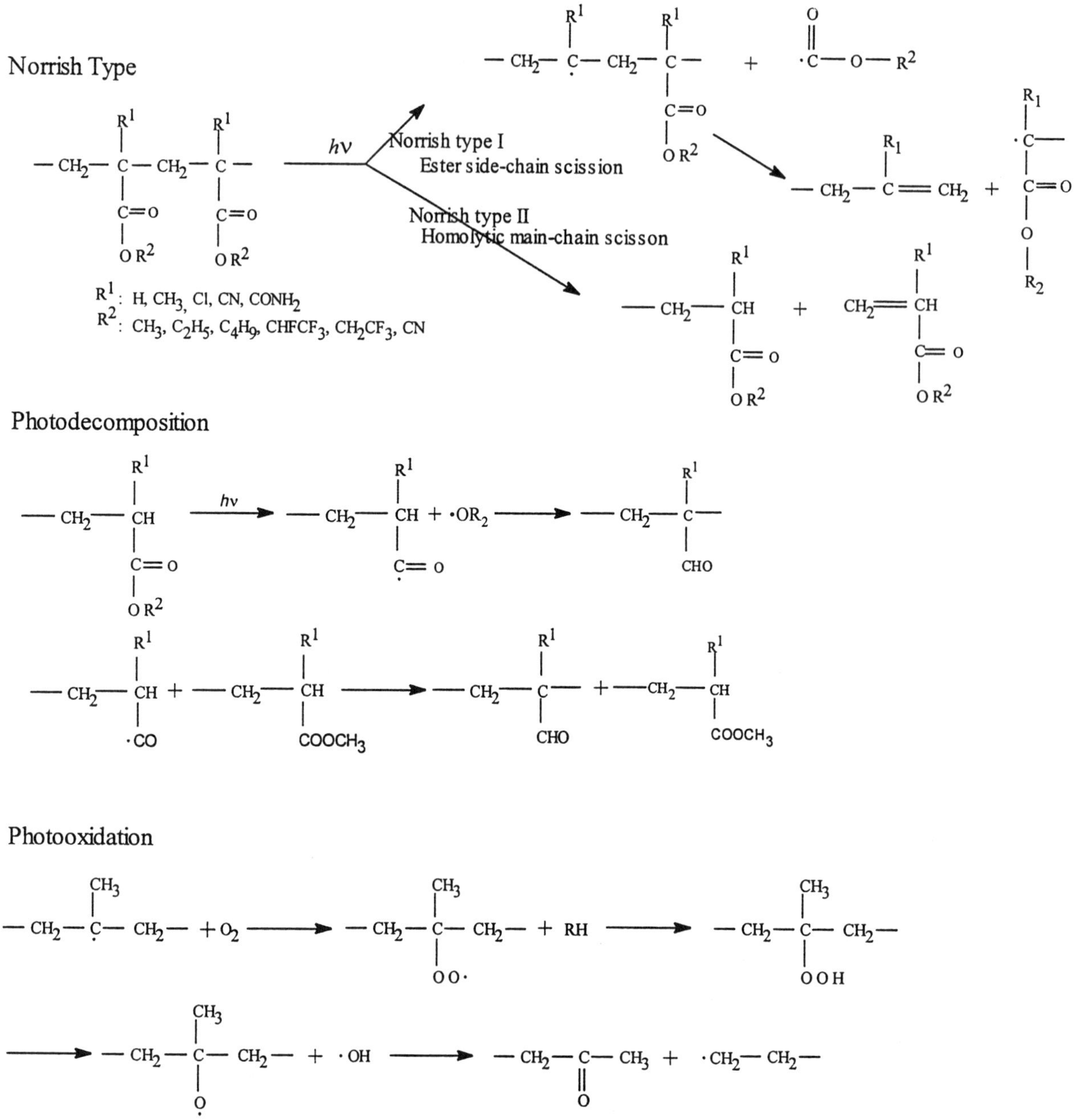

Norrish Type

R^1: H, CH$_3$, Cl, CN, CONH$_2$
R^2: CH$_3$, C$_2$H$_5$, C$_4$H$_9$, CHFCF$_3$, CH$_2$CF$_3$, CN

Photodecomposition

Photooxidation

Scheme XXII. Photodegradation of polyacrylates and polymethacrylates by Norrish type degradation, photodecomposition, and photooxidation of polyacrylates.

Polysulfones

Polysulfones are easily photodegraded by 300–350-nm wavelength irradiation, because the C–S bond energy is only 60 Kcal/mol compared to 80 Kcal/mole of the C–C bond. Scheme XXIII illustrates a proposed degradation pathway for polysulfones (91).

The electron loss from a polyolefinsulfone chain generates a radical cation and leads to the elimination of SO$_2$. As a result, the polymer undergoes significant main-chain degradation. Poly(butene sulfone) has been used as a mask in the fabrication of chrome photomasks (92). However, the low T_g and the unusually

high sensitivity of this polymer can be problematic in these applications.

Concluding Remarks

This chapter discusses photosensitive polymers, combination of polymers with photoactive compounds, photoactive polysilanes, and photodegradable polymers. A major application of these polymers is in resist development to obtain sub-micrometer resolution. Resist science and technology is an expanding field, and there is increasing demand for more sensitive photopolymers and exposure equipment. Thus, new pho-

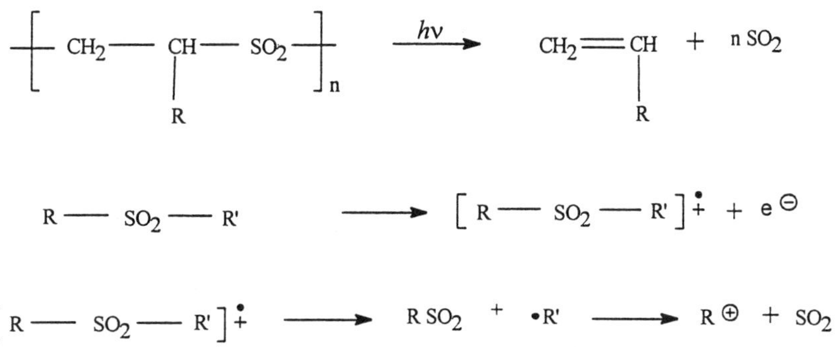

Scheme XXIII. A proposed photodegradation pathway for polysulfones.

tosensitive polymers will continue to make significant contributions to electronics and photonics through improvements of material and related technologies, especially for deep-UV, electron beam, and X-ray lithography and chemical amplification resist systems.

References

1. Cowan, D. C. *Elements of Organic Photochemistry;* Plenum: New York, 1976; pp 3–17.
2. *Practical Technology of New Photosensitive Polymers;* Akamatsu, K., Ed.; CMC: Tokyo, Japan, 1987 (in Japanese).
3. Reiser, A. *Photoreactive Polymers;* John Wiley & Sons: New York, 1989.
4. Willson, C. G.; Bowden M. J. In *Electronic and Photonic Applications of Polymers;* Bowden, M. J.; Turner, S. R., Eds.; Advances in Chemistry Series 218; American Chemical Society: Washington, DC, 1988, pp 78–108.
5. *Polymers for Microelectronics;* Tabata, Y.; Mita, I.; Nonogaki, S.; Horie, K.; Tagawa, S., Eds.; Kodansha: Tokyo, Japan, 1990.
6. *UV Curing: Science and Technology;* Pappas, S. P., Ed.; Technology Marketing Corp.; Norwalk, CT, 1978.
7. Krongauz, V. In *Polymers for Microelectronics;* Tabata, Y.; Mita, I.; Nonogaki, S.; Horie, K.; Tagawa, S., Eds.; Kodansha: Tokyo, Japan, 1990; pp 529–540.
8. Law, K. Y. *Chem. Rev.* 1993, 93, 449–486.
9. Shim, J. S.; Kato, J.; Yoshinaga, T.; Kikuchi, S. *Kogyo Kagaku Zasshi* 1965, 68, 262–268.
10. Hong, S. I.; Son, T. H. *J. Soc. Text. Eng. Chem.* 1981, 18, 31–37.
11. Yoda, N.; Hiramoto, H. *J. Macromol. SCL-CHEM.* 1984, A21, 1641–1663.
12. Rohde, O.; Riediker, M.; Schaffner, A. *SPIE Adv. Resist Technol. Process. II* 1985, 539, 175–180.
13. Pottiger, M. T. *Solid State Technol.* 1989, S1–S4.
14. Li, Z.; Zhu, P.; Wang, L. *J. Appl. Polym. Sci.* 1992, 44, 1365–1370.
15. Pacansky, J.; Lyerla, J. R. *IBM J. Res. Develop.* 1979, 23, 42–55.
16. Yamashita, Y.; Kawazu, R.; Kawamura, K.; Ohno, S.; Asuno, T.; Kobayashi, K.; Magamatsu, G. *J. Vac. Sci. Technol.* 1985, B3, 314–318.
17. Hatzakis, M.; Shaw, J.; Babich, E.; Paraszczak, J. *J. Vac. Sci. Technol.* 1988, B6(6), 2224–2228.
18. Kang, D. W.; Chung, N. J.; Baeh, J. H. *Korea Polym. J.* 1994, RC17, 8–15.
19. *Practical Technology of New Photosensitive Polymers;* Akamatsu, K., Ed.; CMC: Tokyo, Japan, 1987; pp 68–74 (in Japanese).
20. Swenton, J. S.; Ikeler, T. J.; Williams, B. H. *J. Am. Chem. Soc.* 1970, 92(10), 3103–3109.
21. Oh, K. N.; Hong, S. I. *J. Kor. Soc. Text. Eng.* 1983, 20, 21–27.
22. Reichmanis, E.; Gooden, R.; Wilkins, C. W.; Schonhorn, H. *J. Polym. Sci.* 1983, 21, 1075–1083.
23. Reichmanis, E.; Smith, B. C.; Gooden, R. *J. Polym. Sci.: Polym. Chem. Ed.* 1985, 23, 1–8.
24. Wilkins, C. W.; Reichmanis, E.; Chandross, E. A. *J. Electrochem. Soc. Solid-State Sci. Technol.* 1982, 129, 2552–2555.
25. Reiser, A. *Photoreactive Polymers: The Science and Technology of Resists;* John Wiley & Sons: New York, 1989; pp 264–267.
26. Iizawa, T.; Kudou, H.; Nishikubo, T. *J. Polym. Sci.: Part A: Polym. Chem.* 1991, 29, 1875–1882.
27. Kubota, S.; Moriwaki, T.; Ando, T.; Fukami, A. *J. Appl. Polym. Sci.* 1987, 33, 1763–1775.
28. Moon, S. J.; Joo, S. Y.; Chun, B. C.; Hong, S. I. *J. Korean Fiber Soc.* 1992, 29, 63–70.
29. Kobsa, H. *J. Org. Chem.* 1962, 27, 2293–2298.
30. Kalmus, C. E.; Hercules, D. M. *J. Am. Chem. Soc.* 1973, 96(2), 449–456.
31. Joo, S. Y. Ph. D. Dissertation, Seoul National University, Seoul, Korea, 1993.
32. Li, S.-K.; Guillet, J. E. *Macromolecules* 1977, 10, 840–844.
33. Merie-Aubry, L.; Holden, D. A.; Merie, Y.; Guillet, J. E. *Macromolecules* 1980, 13, 1138–1143.
34. Tessier, T. G.; Fréchet, J. M. J.; Willson, C. G.; Ito, H. In *Materials for Microlithography:Radiation-Sensitivity Polymers;* Thompson, L. F.; Willson, C. G.; Fréchet, J. M. J.; Eds.; ACS Symposium Series 266; American Chemical Society: Washington, DC, 1984; 269–292.
35. Cohen, S. M.; Young, R. H.; Markhart, A. H. *J. Polym. Sci.: Part A-1* 1971, 9, 3263–3299.
36. Nishizawa, H.; Sato, K.; Kojima, M.; Satou, H. *Polym. Eng. Sci.* 1992, 32, 1610–1612.
37. Willson, G. C.; Ito, H.; Fréchet, J. M. J.; Tessier, T. G.; Houlihan, F. M. *J. Electrochem. Soc.: Solid-State Sci. Technol.* 1986, 133, 181–187.
38. Reichmanis, E.; Houlihan, F. H.; Nalamansu, O.; Neenan, T. X. *Am. Chem. Soc.* 1991, 3, 394–407.
39. Fréchet, J. M. J.; Matszczak, S.; Reck, B.; Stover, H. D. H.; Willson, C. G. *Macromolecules* 1991, 24, 1746–1754.
40. Fréchet, J. M. J.; Matuszczak, S.; Lee, S. M.; Fahey, J. *Polym. Eng. Sci.* 1992, 32, 1471–1475.
41. Schaedeli, U. *Polym. Eng. Sci.* 1992, 32, 1523–1529.

42. Ito, H.; Willson, C. G. *Polym. Eng. Sci.* **1983**, *23*, 1012–1018.
43. Fréchet, J. M. J.; Bouchard, F.; Eichler, E. *Polym. J.* **1987**, *19*, 31–49.
44. Przybilla, K.-J.; Dammel, R.; Pawlowski, G.; Roschhert, H. *Polym. Eng. Sci.* **1992**, *32*, 1516–1522.
45. Ito, H.; Ueda, M.; England, W. P. *Macromolecules* **1990**, *23*, 2589–2598.
46. Allen, R. D.; Wallraff, G. M.; Hinsberg, W. D. *J. Vac. Sci. Technol. B* **1991**, *9*, 3357–3361.
47. Crivello, J. V.; Lam, J. H. W. *J. Polym. Sci.: Polym. Chem. Ed.* **1980**, *18*, 2677–2695.
48. Ueno, T.; Schlegel, L.; Hayashi, N.; Shirashi, H.; Iwayanagi, T. *Polym. Eng. Sci.* **1992**, *32*, 1511–1515.
49. Houlihan, F. M.; Neenan, T. X.; Reichmanis, E.; Kometani, J. M.; Thompson, L. F.; Chin, T.; Kanga, R. S. *J. Vac. Sci. Technol. B* **1990**, *8*, 1461–1465.
50. Houlihan, F. M.; Neenan, T. X.; Chin, E.; Reichmanis, E.; Kometani, J. M. *Polym. Eng. Sci.* **1992**, *32*, 1509–1510.
51. Shirai, M.; Kinoshita, H.; Tsunooka, M. *Eur. Polym. J.* **1992**, *28*, 379–385.
52. Neenan, T. X.; Houlihan, F. M.; Reichmanis, E.; Kometani, J. M.; Bachman, B. J.; Thompson, L. F. *Macromolecules* **1990**, *23*, 145–150.
53. Taylor, G. N.; Stillwagon, L. E.; Houlihan, F. M. *J. Vac. Sci. Technol. B* **1991**, *9*, 3348–3356.
54. Hesp, S. A. M.; Hayashi, N.; Ueno, T. *J. Appl. Polym. Sci.* **1991**, *42*, 877–883.
55. Kim, J. S. Master Thesis, Seoul National University, Seoul, Korea, 1993.
56. Long, T.; Obendorf, K. *Polym. Eng. Sci.* **1992**, *32*, 1589–1593.
57. Nalamasu, O.; Reichmanis, E.; Hanson, J. E.; Kanga, R. S.; Heimbrook, L. A.; Emerson, A. B.; Baiocchi, F. A.; Vaidya, S. *Polym. Eng. Sci.* **1992**, *32*, 1565–1570.
58. Hiraoka, H.; Gutierrez, A. R. *J. Electrochem. Soc. Solid-State Sci. Technol.* **1979**, *126*, 860–865.
59. Meyerhofer, D. *IEEE Trans. Electron Devices* **1980**, *ED-27*, 921–926.
60. Bowden, M. J.; Thompson, L. F.; Fahrenholtz, S. R.; Doerries, E. M. *J. Electrochem. Soc.: Solid-State Sci. Technol.* **1981**, *128*, 1304–1312.
61. Hanabata, M.; Oi, F.; Furuta, A. *Polym. Eng. Sci.* **1992**, *32*, 1494–1499.
62. Kishimura, S.; Yamaguchi, A.; Yamada, Y.; Nagata, H. *Polym. Eng. Sci.* **1992**, *32*, 1550–1555.
63. Joo, S. Y.; Hong. S. I. *J. Korean Ind. Eng. Chem.* **1990**, *1*, 116–123.
64. Willson, C. G. In *Introduction to Microlithography*; Thompson, L. F.; Willson, C. G.; Bowden, M. J., Eds.; ACS Symposium Series 219, American Chemical Society: Washington, DC, 1983; pp 88–159.
65. Kim, J. H. Master Thesis, Seoul National University, Seoul, Korea, 1989.
66. Uchhino, S.-I.; Tanaka, T.; Ueno, T.; Iwayanagi, T.; Hayashi, N. *J. Vac. Sci. Technol. B* **1991**, *9*, 3162–3165.
67. Nonogaki, S.; Toriumi, M. *Macromol. Chem., Macromol. Symp.* **1990**, *33*, 233–241.
68. Trujillo, R. E. *J. Organomet. Chem.* **1980**, *198*, C27–C28.
69. Miller, R. D.; Michl, J. *Chem. Rev.* **1989**, *89*, 1359–1410.
70. West. R.; Maxka, J. In *Inorganic and Organometallic Polymers*; Zeldin, M.; Wynne, K. J.; Allcock, J. R., Eds.; ACS Symposium Series 360; American Chemical Society: Washington, DC, 1988; pp 6–20.
71. Mark, J. E.; Allcock, J. R.; West. R. In *Inorganic Polymers*; Prentice-Hall: Englewood Cliffs, NJ, 1992; pp 206–210.
72. Bock, H.; Ensslin, W. *Angew. Chem. Int. Ed.* **1971**, *10*, 404–405.
73. Trefnas, P., III; West, R.; Miller, R. D. *J. Am. Chem. Soc.* **1985**, *107*, 2737–2742.
74. Takeda, K.; Teremae, H.; Matsumoto, N. *J. Am. Chem. Soc.* **1986**, *108*, 8186–8190.
75. Miller, R. D.; Rabolt, J. F.; Sooriyakumaran, R.; Fleming, W.; Fickes, G. N.; Farmer, B. L.; Kuzmany, H. In *Inorganic and Organometallic Polymers*; Zeldin, M.; Wynne, K. J.; Allcock, H. R., Eds.; ACS Symposium Series 360; American Chemical Society: Washington, DC, 1988; pp 43–60.
76. Ishikawa, M.; Nate, K. In *Inorganic and Organometallic Polymers*; Zeldin, M.; Wynne, K. J.; Allcock, H. R., Eds.; ACS Symposium Series 360; American Chemical Society: Washington, DC, 1988; pp 209–223.
77. Padmanaban, M.; Toriumi, M.; Kakimoto, M.; Imai, Y. *Polym. Prepr.* **1990**, *31(2)*, 282–283.
78. West, R. *J. Organomet. Chem.* **1986**, *300*, 327–346.
79. West, R.; David, L. D.; Djurovich, P. I.; Stearley, K. L.; Srinivasan, K. S. V.; Yu, H. *J. Amer. Chem. Soc.* **1981**, *103*, 7352–7354.
80. Reiser, A. *Photoreactive Polymers*; John Wiley & Sons: New York, 1989; pp 39–42.
81. Rabek, J. F. *Photostabilization of Polymers—Principles and Applications*; Elsivier Applied Science: London, 1990; pp 1–41.
82. Aclean, J. E. *Polm. Degrad. Stab.* **1990**, *29*, 65–71.
83. Lingnau, J.; Dammel, R.; Theis, J. *Solid State Technol.* **1989**, *September*, 105–111.
84. Lingnau, J.; Dammel, R.; Lindley, C. R.; Pawlowski, G.; Scheunemann, U.; Theis, J. In *Polymers for Microelectronics*; Tabata, Y.; Mita, I.; Nonogaki, S.; Hories, K.; Tagawa, S., Eds.; Kodansha: Tokyo, Japan, 1990; pp 445–462.
85. Bargon, J. *J. Polym. Sci.* **1978**, *A1*, 2747–2758.
86. Rabek, J. F. *Mechanisms of Photophysical Processes and Photochemical Reactions in Polymers*; John Wiley & Sons: Chichester, England, 1987.
87. Reiser, A. *Photoreactive Polymers: The Science and Technology of Resists*; John Wiley & Sons: New York, 1989; pp 269–272.
88. Kuzuya, M.; Sawada, K.; Takai, T.; Noguchi, A. *Polym. J.* **1993**, *25*, 75–81.
89. Mitsuoka, T.; Torikai, A.; Fueki, K. *J. Appl. Polym. Sci.* **1993**, *47*, 1027–1032.
90. Bowden, M. J. In *Materials for Microlithography: Radiation-Sensitive Polymers*; Thompson, L. F.; Willson, C. G.; Fréchet, J. M. J., Eds.; ACS Symposium Series 266: American Chemical Society: Washington, DC, 1984; pp 66–75.
91. Reiser, A. *Photoreactive Polymers: The Science and Technology of Resists*; John Wiley & Sons: New York, 1989; pp 323–324.
92. Thompson, L. F.; Bowden, M. J. *J. Electrochem. Sci. Solid-State Sci. Technol.* **1973**, *120*, 1722–1725.

Functional Polymers for Microlithography: Nonamplified Imaging Systems

Hiroshi Ito

This chapter reviews functional polymers employed as resist materials in microlithography for semi-conductor-device fabrication and describes conventional resist systems based on nonamplified imaging mechanisms. After a brief introduction to microlithographic technology, various resist systems are described on the basis of their imaging chemistries. The imaging processes described include single-component and two-component systems and bilayer and silylation techniques. A comprehensive survey of the various polymers used or studied for these processes is also provided.

The manufacture of modern integrated circuit devices is a complex process involving the integration of insulator, conductor, and semiconductor structures with high-purity single crystals of semiconductor materials such as silicon. The microelectronic technology has shown astonishing progress during the past decade. This progress includes the improvements of integrated circuit devices by increasing the number of components per chip by continually reducing the minimum feature sizes on the chip. Sixteen-megabit *dynamic random access memory* (DRAM) devices are currently in production and have features as small as 0.5 μm (ten times smaller than a typical bacterium and 100 times smaller than the diameter of a human hair). The fabrication of 64-Mbit (0.35-μm minimum feature size) and 256-Mbit (0.25-μm minimum feature size) DRAM is in a development stage and is moving toward manufacturing.

These complex devices are fabricated by a technology known as microlithography, in which radiation-sensitive polymeric materials called resists are used in generating minute stencils for subsequent processes (Figure 1). The resist material is applied as a thin coating, typically by spin casting over substrate (wafer),

followed by heating (postapply bake, preexposure bake, or prebake) to remove the casting solvent. The resist film is subsequently exposed in an imagewise fashion through a mask (in photo- and X-ray lithography) or directly with finely focused electron beams. The exposed resist film is then developed, typically by immersion in a developer solvent to generate three-dimensional relief images. The exposure may render the resist film more soluble in the developer and produce a *positive-tone* image of the mask; conversely, the film may become less soluble after exposure and result in generation of a *negative-tone* image. The resist film remaining after the development functions as a protective mask during transfer of the resist image onto the substrate by etching and related processes. The resist film must "resist" the etchant and protect the underlying substrate while the bared areas are being etched. The remaining resist film is finally stripped, and an image of the desired circuit is left on the substrate. The process is repeated many times to fabricate complex semiconductor devices.

For a resist polymer to be useful, it must be capable of spin-casting from solution into a thin and uniform film that adheres to various substrates such as

3469–8/97/0311$17.25/0 © 1997 American Chemical Society

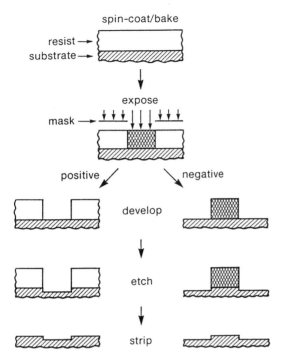

Figure 1. Schematic sequence of lithographic processes.

metals, semiconductors, and insulators. Furthermore, the resist must possess high radiation sensitivity and high resolution capability. The resist must also be able to withstand extremely harsh environments (high temperatures, strong corrosive acids, and plasmas) to be viable for practical use.

Resist systems can be classified on the basis of their design, namely one-component systems and two- or multicomponent systems. One-component resist systems consist of pure polymers that must combine all the necessary attributes such as substrate protection, radiation sensitivity, and film-forming characteristics. One-component resists are currently used only in photomask fabrication with electron-beam (e-beam) exposure, but they were heavily studied more than a decade ago.

The most popular resist designs in modern lithography are based on two-component systems in which resist functions are provided by two separate components. For example, the polymer in the two-component systems is in many cases inert to radiation but functions only as a binder. However, there are cases where the polymer, though radiation insensitive, participates in a secondary reaction induced by radiolysis of the radiation-sensitive component. Typical examples are the so-called "chemical-amplification" resist systems, in which generation of an acid by radiation induces a cascade of subsequent chemical transformations and provides a gain mechanism. This chapter provides a review of conventional resist systems. Chemically amplified resist systems are described separately in Chapter 2.4, because they constitute an

entire family of emerging technologies for use in device manufacturing with the minimum feature sizes below 0.5 μm.

Resist systems can also be divided into three groups on the basis of the radiation source: UV or photoresists, e-beam resists, and X-ray resists. Photolithography using UV light has been the predominant technology in semiconductor manufacture and continues to be so for the foreseeable future. X-ray lithography is capable of producing high resolution and high aspect ratio (height–width) images and is considered to be the future technology, whereas e-beam lithography is used in mask fabrication. Photolithography can be further subdivided into near-UV (350–450 nm), mid-UV (300–350 nm), and deep-UV (<300 nm) technologies (Figure 2), depending on the wavelength of the exposure. Because the resolution is proportional to the exposing wavelength and inversely proportional to the numerical aperture (NA) of the lens, shifting from G line (436 nm) to I line (365 nm) lithography with a high-NA lens is the dominant technology in the manufacture of 16-Mbit DRAM (0.5-μm) devices. Krypton fluoride (KrF, 248 nm) and argon fluoride (ArF, 193 nm) excimer laser technologies are emerging as the minimum feature size continues to shrink below 0.5 μm. These new technologies necessitate development of new resist materials and present an ever-increasing challenge and opportunity to polymer chemists (1–8).

Resist performance is generally presented in the form of sensitivity (or contrast) curves (Figure 3). The thickness of the exposed resist film (typically normal-

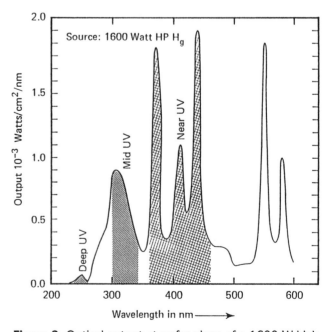

Figure 2. Optical output at wafer plane of a 1600-W high-pressure mercury lamp.

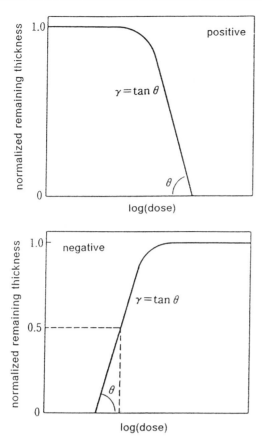

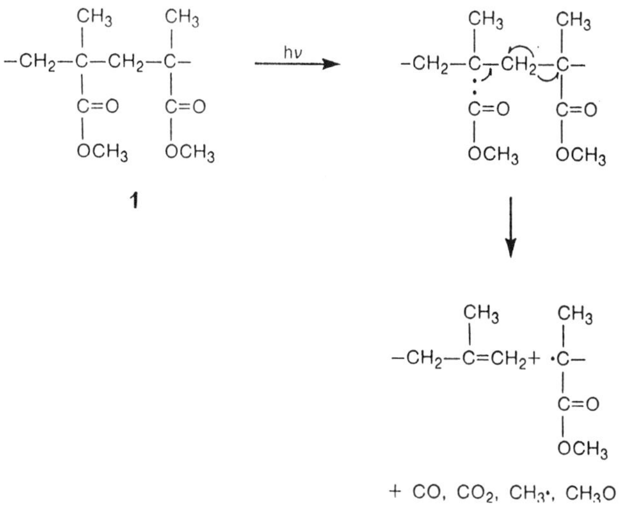

Scheme I. Mechanism of radiation-induced chain scission in PMMA.

must satisfy all the requirements, which may sometimes contradict each other.

Positive Resist System

Degradable Polymers

A one-component positive resist comprises a polymer that undergoes main-chain scission by irradiation. A classical example is poly(methyl methacrylate) (PMMA, **1**, Scheme I), which was the very first e-beam (*9*) and deep-UV resist (*10*) and has been known as a high resolution resist. Photolysis of PMMA results in chain scission through homolysis of the side-chain followed by decarbonylation to form a stable tertiary radical on the main chain. This radical then undergoes a β-scission to generate an acyl-stabilized tertiary radical (Scheme I) (*11, 12*). This process results in molecular-weight reduction and enhanced dissolution of the exposed region. High energy radiations such as e-beam, X-ray, and proton-beam are about 10 times more efficient than deep-UV in main-chain scission and result in removal of fewer ester groups (*13*). The intrinsic scission susceptibility of polymeric materials is expressed by G_s, defined as the number of scissions per 100 eV of absorbed dose. Values of G_s can be calculated from the Charlesby equations (*14*). PMMA has a G_s value of 1.3.

In addition to molecular weight reduction, solvent mobility in the PMMA film strongly affects polymer dissolution. Increased free volume (or decreased density due to gaseous product formation) increases the rate at which the developer solvent diffuses into the film in the exposed regions (*15*). The effects of e-beam irradiation and molecular weight on the dissolution rate, S (Å/min), of PMMA in amyl acetate are shown in Figure 4. The rate of increase in S with decreasing mol-

---

Figure 3. Sensitivity (contrast) curves for positive and negative resists.

ized to the initial thickness) remaining after development is plotted as a function of the logarithm of exposure dose (typically, mJ/cm^2 for UV and X-ray exposures and μC/cm^2 for e-beam irradiation). In positive resists, the sensitivity is defined as the dose at which the exposed regions are cleanly developed to the substrate and minimum thickness loss occurs in the unexposed areas. The sensitivity of negative resists is typically defined as the dose at which 50% of the thickness is retained in the exposed regions. Another widely used performance parameter is the contrast (γ), which is determined from the slope of the linear portions of the curves. Higher contrasts generally lead to higher resolutions.

Single-Component Resists

Radiation- and photosensitive polymers that degrade or cross-link by irradiation could be used as single-component positive or negative resists, respectively. In addition, photochemical transformation of nonpolar groups to polar groups can be used to design dual-tone resists that provide either positive or negative images depending on the polarity of the developer. In this one-component design, the functional polymer

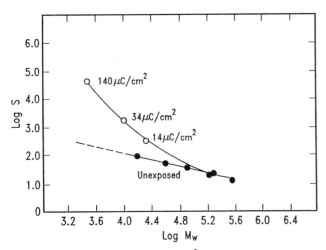

Figure 4. Dissolution rate, S (Å/min), of 1-μm-thick PMMA film in amyl acetate at 24.5 °C vs. molecular weight: (●) unexposed and (○) exposed. (Reproduced with permission from reference 15. Copyright 1978 The Society of Plastics Engineers, Inc.)

ecular weight is much faster in the exposed (○) than in the unexposed (●) PMMA.

As a resist material, PMMA has many desirable properties such as availability, ease of handling, excellent film-forming characteristics, wide process latitude, and high resolution. PMMA is more conventionally produced by radical polymerization, but living anionic polymerization (*16*) and group-transfer polymerization (*17*) provide polymers with narrow molecular weight distribution. The glass transition temperature (T_g) of PMMA is ca. 100 °C, and the onset of thermal decomposition is observed at ca. 250 °C. Degradation behavior depends on the end group (*18*).

However, PMMA lacks sensitivity and dry-etch resistance for pattern transfer. Many PMMA analogs and copolymers were prepared and evaluated for increasing the sensitivity primarily for e-beam applications. Introduction of electron-withdrawing groups such as CN, halogen, or haloalkane increased the radiation sensitivity of these polymers (*19*). However, such substituents may alter the electron density and the bulkiness of the monomer and may change their polymerization behavior (*20*). Among polymethacrylates, FBM, a poly(fluorobutyl methacrylate)-based resist, seems to be the most sensitive to deep-UV radiation (94 mJ/cm²) and to electron beams (*21*). The increased sensitivity of FBM has been ascribed to improved dissolution characteristics rather than to an increase in G_s.

Poly(glycidyl methacrylate) (PGMA) is a well-known negative e-beam resist (*22*), but it also functions as a positive resist when exposed to deep-UV exposure (*23*). The epoxide functionality responsible for cross-linking in e-beam exposure does not absorb in the deep-UV region; therefore, the response of PGMA to deep-UV irradiation is due to the n-π* tran-

sition of the carbonyl chromophore leading to main-chain scission like PMMA.

Methacrylate copolymers were also reported as deep-UV resists. Incorporation of 3-oximino-2-butanone methacrylate into the PMMA structure by radical copolymerization (**2** in Chart I) improves the absorption characteristics of the polymer in the 220–260 nm range and provides an alternative path for scission with enhanced sensitivity (*24*). Similarly, copolymer of indenone with MMA (**3**) has an increased UV absorption in the 230–330 nm region that results in enhanced sensitivity (*25*). The higher sensitivity is explained in terms of the steric strain in the cyclopentanone moiety.

However, PMMA itself is highly transparent below 300 nm, and hence it is very attractive for ArF excimer laser (193 nm) lithography. In fact, exposure of PMMA to an ArF excimer laser was reported more than 10 years ago (*26*). Aromatic polymers cannot be used as a single-layer ArF resist because of their excessive absorption at 193 nm. On the other hand, aliphatic polymers such as polymethacrylates are transparent but lack dry-etch resistance. However, polymethacrylates of alicyclic alcohols show good plasma resistance (*27*) and open the possibility of ArF excimer laser lithography as an emerging technology for device manufacturing. The majority of ArF resists are chemically amplified and will be described in Chapter 2.4.

A terpolymer of methyl methacrylate, methacrylic acid, and methacrylic anhydride (**4**) was developed as a thermally stable e-beam resist that is more sensitive than PMMA (*28*). The terpolymer was synthesized by heating a copolymer of methyl methacrylate with methacrylic acid (anhydride ring formation through elimination of water and methanol). Other methacrylic anhydride copolymers were also studied as positive e-beam resist materials. Electron-rich monomers such as α-methylstyrene, methyl vinyl ether, and isopropenyl acetate were chosen as comonomers in radical copolymerization with methacrylic anhydride, and these reactions tend to provide alternating copolymers because both comonomers are reluctant to undergo radical homopolymerization (*29*).

Poly[(meth)acrylic acid] can be converted to polyglutarimide by treating the film at 200 °C with a vapor of ammonia or amine, and the resulting imide film functions as a positive resist (*30*). Poly(dimethyl glutarimide) (PMGI, **5**) is sensitive to deep-UV irradiation below 280 nm, soluble in aqueous base, resistant to common organic solvents, and thermally stable to ca. 180 °C (*31*).

Another approach to sensitivity enhancement of methacrylic polymers involves cross-linking through anhydride formation from methacrylic acid or methacryloyl chloride during prebake near 200 °C (*32*). The pre-cross-linking renders the unexposed resist essentially insoluble in a developer solvent and raises the T_g

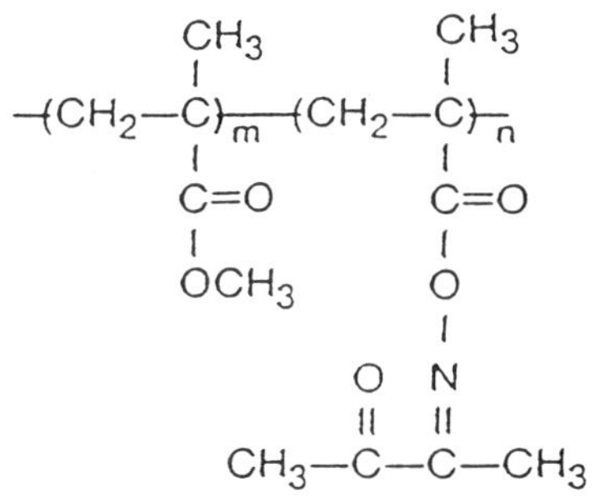

2

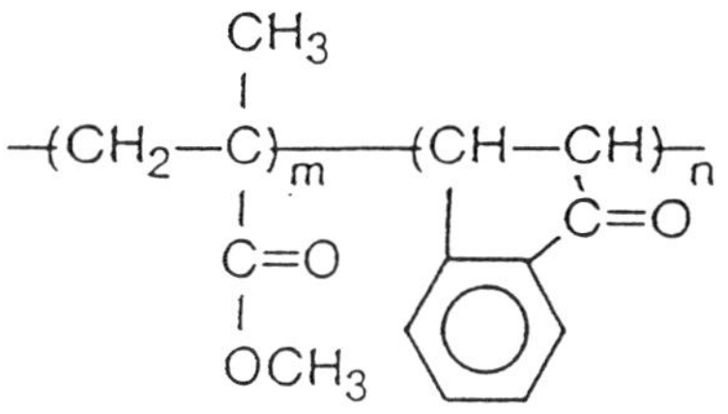

3

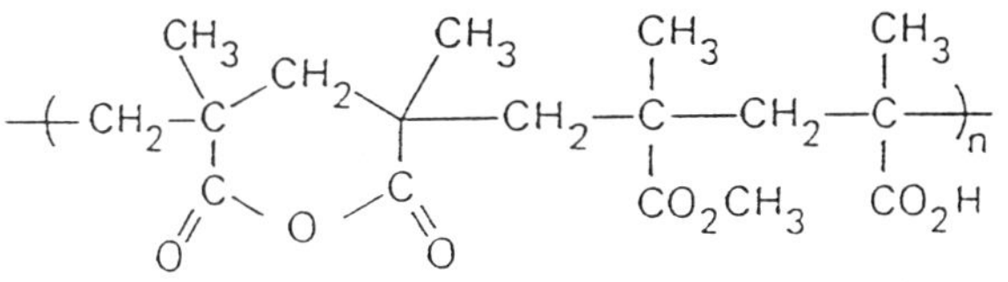

4

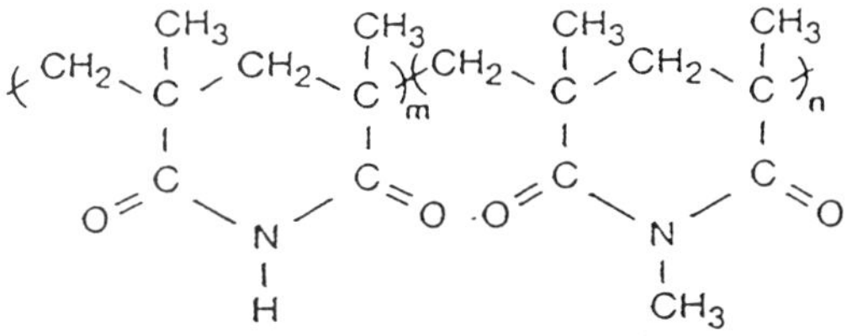

5

Chart I. Examples of photo- and radiation-degradable polymers.

to 130–150 °C, whereas the exposed areas become soluble by chain scission (Figure 5).

These polymers that undergo radiation-induced degradation are typically aliphatic and also degrade rapidly in plasmas, and these characteristics preclude their use in device fabrication. Therefore, several attempts were made to combine the high susceptibility of chain scission with the high dry-etch resistance of aromatic polymers. For example, poly(phenyl methacrylate) withstands reactive sputter etching twice as much as PMMA, but it is much less sensitive than PMMA to e-beam irradiation (due to a sponge effect) (*33*). Copolymers of phenyl methacrylate with methacrylic acid, pre-cross-linked during prebake, offer a sensitivity of ca. 20 μC/cm^2 without losing the high dry-etch durability of poly(phenyl methacrylate) (*34*).

Another important class of degrading deep-UV positive resists is based on isopropenyl ketone polymers. Poly(methyl isopropenyl ketone) (PMIPK, **6** in Scheme II) exhibits a weak absorption at 285 nm due to the carbonyl chromophore, and it is about seven times more sensitive than PMMA as a positive deep-UV resist (*35*). Vinyl and isopropenyl ketones are reac-

tive monomers prepared by the Mannich reaction or Pd-catalyzed tin-coupling reaction (*36*), and these monomers undergo radical or anionic polymerization quite readily. To increase the quantum efficiency for the chain scission in the solid state, the methyl group of **6** was replaced with a *t*-butyl group (*37*).

In contrast to the high polymerizability of other vinyl ketones, isopropenyl *t*-butyl ketone (IPTBK) does not undergo radical homopolymerization but polymerizes anionically at cryogenic temperatures (*38, 39*). Because of its rigidity, poly(IPTBK) (**7**) has a high T_g of 135 °C. In this polymer, the carbonyl group is located between two quaternary centers (Scheme II), and the Norrish type I α-cleavage on either side of the carbonyl carbon generates a stable tertiary radical. Conversely, α-cleavage of **6** yields either a *t*-butyl or methyl radical. Because the stability of these species is significantly different, cleavage only generates the tertiary radical. Thus, the replacement of the methyl group in **6** with a *t*-butyl group increases the propensity for the Norrish type I photochemical degradation, which is a predominant pathway in the solid state. The quantum yield of chain scission at 313 nm in films of **7** is 12 times higher than that of **6**.

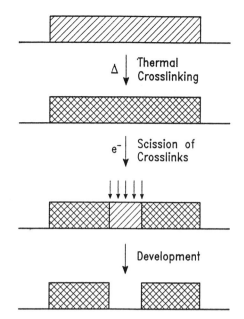

Figure 5. Precross-linked positive-resist process.

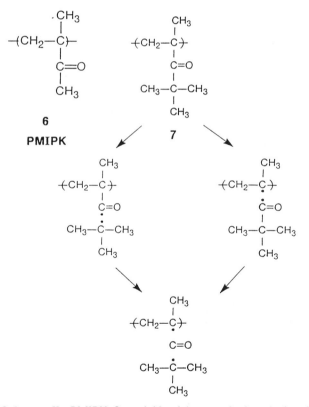

Scheme II. PMIPK **6** and Norrish type I photolysis of PIPTBK **7**.

Aromatic groups provide a convenient chromophore for UV absorption. Copolymers of isopropenyl phenyl ketone with MMA show spectral sensitivity up to the mid-UV range (*40*).

Poly(olefin sulfones) (usually alternating) are prepared by radical copolymerization of olefins with sul-

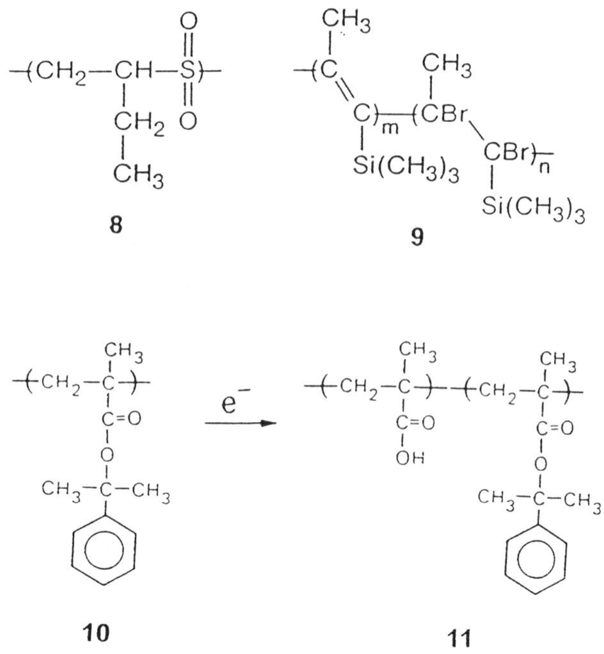

Scheme III. Poly(olefin sulfone), brominated polysilylacetylene, and radiation-induced deesterification.

fur dioxide, typically at cryogenic temperatures because of their low ceiling temperatures (T_c). Poly-(butene-1 sulfone) (**8**, Scheme III) was reported in 1970 to degrade by γ-irradiation and to have a G_s value of about 10; these characteristics make it one of the most radiation-sensitive polymers known (*41*). Therefore, a great deal of research was prompted for developing sensitive, positive e-beam resists based on this class of polymer. Polymer **8** undergoes efficient main-chain scission upon exposure to an electron beam to produce sulfur dioxide and butene. Irradiation results in scission of the main-chain C–S bond, followed by depolymerization to release volatile monomeric species. This polymer is also sensitive to 185-nm radiation (*42*) but does not absorb above 200 nm. Poly(2-methyl-1-pentene sulfone) undergoes "vapor development" due to depolymerization upon e-beam exposure at 90 °C (*43*). Incorporation of pendant aromatic rings into the polysulfone structure extends its photosensitivity to deep-UV and mid-UV regions (*44*). Carbonyl-containing poly(olefin sulfones) such as poly(5-hexen-2-one sulfone) are sensitive to UV-induced degradation (*45*). Silicon-containing polysulfones are described in a subsequent section.

As mentioned previously, incorporation of aromatic groups increases dry-etch resistance. Positive-resist materials that undergo efficient main-chain scission upon irradiation generally lack dry-etch durability. However, incorporation of ca. 50 mol% styrenic comonomer provides the dry-etch resistance comparable with aromatic homopolymers (*46–48*).

Polyacetylenes are known to degrade upon γ-irradiation in the presence of oxygen (*49*). Photolysis of substituted polyacetylenes was also reported (*50*). Substituted polyacetylenes are prepared by polymerization with catalysts such as WCl_6 or $MoCl_5$ and a cocatalyst such as tetraphenyltin (*51*). Brominated poly(1-trimethylsilylpropyne) (**9** in Scheme III) was described as a positive deep-UV resist (*52*).

Polysilanes (*53*) and poly(disilanylenephenylene) (*54*) contain Si atoms in the backbone. These polymers undergo molecular weight reduction upon irradiation, and hence they function as positive resists. Polysilanes are obtained by the Wurtz coupling reaction of dichlorosilane with metallic sodium, and usually they have a broad molecular weight distribution (bimodal in many cases) (*55*). Ring-opening polymerization of cyclosilanes and anionic polymerization of protected disilenes provide polysilanes with improved structural control (*56*). Such organometallic resists will be discussed in a later section.

Polyphthalaldehyde "self-develops" due to depolymerization upon excimer laser irradiation without solvent development (*57*). Copolymers of aliphatic aldehydes were also reported to undergo depolymerization upon e-beam exposure and to function as self-developing e-beam resists (*58*). Addition of a photochemical acid generator to the polyphthalaldehyde derivatives provides highly sensitive self- or "thermal development", which will be described in detail in Chapter 2.4.

Certain polysilanes undergo "photochemical ablation" upon irradiation of an excimer laser, and this phenomenon allows all dry development (*59, 60*). The phenomenon of photochemical ablation was first accomplished by 193-nm ArF excimer laser radiation (*61*). The ablation reaction is attributed to the high absorption of almost all the incident light in a thin upper layer of the polymer film, followed by bond cleavage and ejection of small molecules. Table I summarizes lithographic sensitivities of representative positive resist systems based on main-chain degradation.

Positive Resist Based on Polarity Change

Poly(α,α-dimethylbenzyl methacrylate) (**10**) de-esterifies to some extent upon e-beam exposure and functions as a sensitive positive resist when developed with an alcoholic solution of alkoxide (Scheme III) (*62*).

Negative Resists

Cross-Linkable Polymers

The first useful negative e-beam resist materials were designed on the basis of radiation-induced cross-linking involving oxirane or epoxy groups. Copolymers of glycidyl methacrylate and ethyl acrylate (**12**, Scheme IV) are widely used in photomask fabrication (*63*). The measured G_x value (the number of cross-linking events per 100 eV of absorbed dose) of epoxy polymers is ca. 10. The high radiation sensitivity of these polymers was ascribed to the chain-reaction mechanism for cross-linking (*64*) and also to efficient dissociative electron capture to form a radical anion through β-opening (*65*). However, these negative-resist systems tend to suffer from the so-called "dark reaction" or "post-polymerization", which occurs after the exposure and induces line-width shift from standing (especially in vacuum) after exposure (*66*). Because of resolution limitations due to swelling, low T_g, and poor resistance to plasma, the glycidyl methacrylate negative-resist systems are losing favor.

On the other hand, negative resists based on cross-linking of aromatic polymers were developed

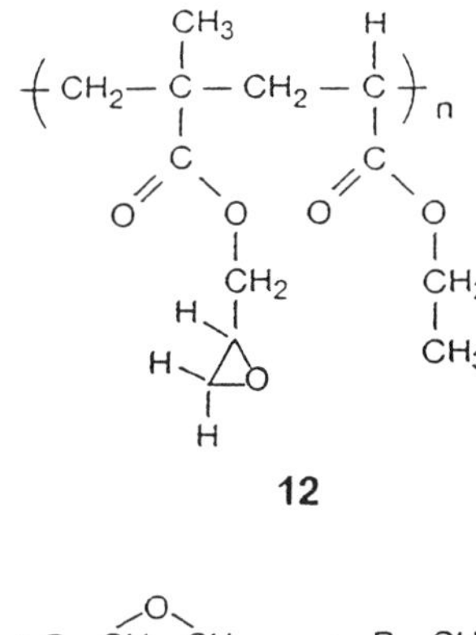

Table I. Lithographic Sensitivities of Representative Positive Resists Based on Main-Chain Degradation

Polymer	UV (mJ/cm²)	Electron Beam (µC/cm² at 20 kV)	X-ray (mJ/cm²)
1	700 (200–240 nm)	80–100	500–1000 (A1)
2	130 (240–270 nm)	30–50	
3	60 (230–300 nm)		
4		6.5	
5	1500 (200–280 nm)		
6	140 (230–320 nm)		
8	5 (185 nm)	0.2–1.0	94 (Pd)
9	20–25		
FBM	94 (200–260 nm)	0.4	100 (A1), 52 (Mo)

12

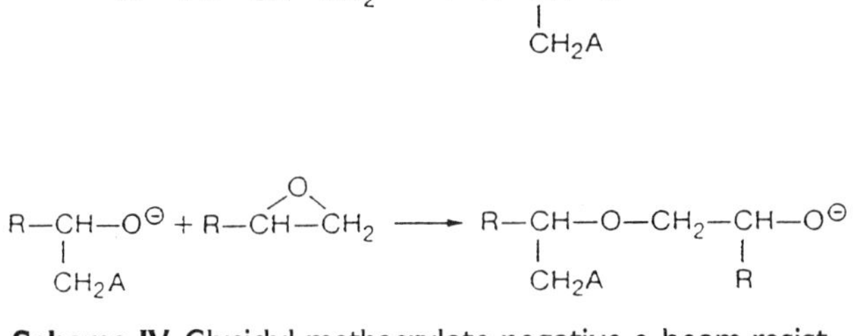

Scheme IV. Glycidyl methacrylate negative e-beam resist.

primarily for dry-etching applications in e-beam lithography. Aromatic rings in polymers also provide absorption of deep-UV and mid-UV radiation due to π-π* transitions of the aromatic compounds. Polystyrene is a negative-working resist, but it is much less sensitive than PMMA. Chloromethylation or chlorination of polystyrene has produced a class of sensitive negative-resist systems (13, Chart II), which provide high sensitivity and high contrast without dark reaction (Figure 6) (67–69). The imaging mechanism involves dissociative electron capture and cleavage of the carbon–halogen bond to generate a radical, which leads to cross-linking. Figure 6 shows the thickness remaining after development as a function of storage time after exposure for an epoxy resist 12 and poly(chloromethylstyrene) 13. Polymer 13 suffers from

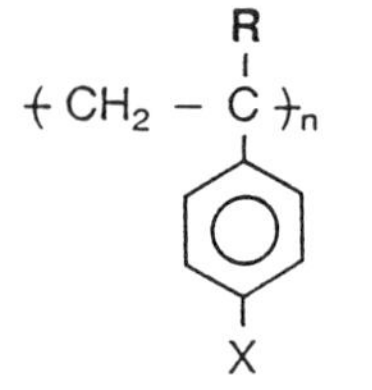

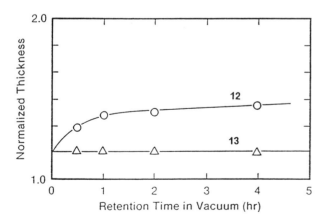

13

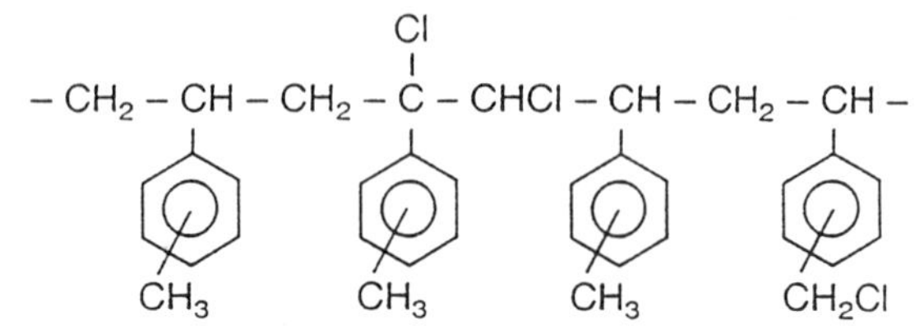

14

Chart II. Chlorinated and chloromethylated polystyrene negative resists.

2.0

Normalized Thickness

12

13

1.0

Retention Time in Vacuum (hr)

Figure 6. Change in thickness remaining after development vs. storage time after exposure for 12 (○) and 13 (△). (Reproduced with permission from reference 67. Copyright 1979 Electrochemical Society.)

no dark reaction in contrast to the thickness increase observed in the epoxy resist after exposure.

Because styrenic polymers can be readily prepared by radical, cationic, and anionic polymerization, the effect of molecular weight and dispersity on the sensitivity and contrast has been studied. Poly(4-chlorostyrene) is synthesized by radical polymerization of 4-chlorostyrene, and a fractionated sample with a narrow molecular weight distribution offers a high contrast (68). Decreased dispersity at constant number-average molecular weight (M_n) results in increased contrast. Monodispersed poly(chloromethylstyrene) can be prepared by chloromethylation of monodispersed polystyrene to provide a high resist contrast (67). Increased molecular weight at constant dispersity tends to result in increased sensitivity (Figure 7). Chloromethylation was also carried out (70) on nearly monodispersed poly(α-methylstyrene) to provide a high resolution negative resist with a high T_g of 170 °C. Another approach to incorporate chlorine into styrene-based polymers was chlorination of poly(4-methylstyrene), which produces the chlorinated polymer 14 (Chart II) (71). Two different types of radiation-sensitive groups were also combined by copolymerization of glycidyl methacrylate with 3-chlorostyrene or with vinylbenzylchloride (chloromethylstyrene) (72). Table II shows lithographic sensitivities of representative cross-linking negative-resist systems.

Negative Resists Based on Polarity Change

Pyridinium ylide groups (15, Scheme V) that undergo photolysis to diazepines have been attached to polymer chains (73) The change from the polar ylide to the nonpolar diazepine (Scheme V) allows negative imaging after development with water. This resist was developed for mid-UV exposure, and it absorbs intensely, bleaches significantly, and requires a high exposure dose. In principle, the polarity change should also provide positive imaging, but concomitant cross-linking precludes positive imaging in this case.

Dual-Tone Resist Systems

The photo-Fries rearrangement was incorporated in resist design as a suitable mechanism for achieving the desired change in the side-chain polarity. This system is based on poly(4-formyloxystyrene) (16) that undergoes Fries-like homolysis and decomposition upon exposure to deep-UV radiation to produce polyhydroxystyrene (17, Scheme VI) and carbon monoxide in the exposed regions (74). The phenolic photoproduct is soluble in polar solvents such as alcohol or aqueous base, whereas the formyl ester precursor in the unexposed areas is completely insoluble in these solvents. Consequently, development with polar solvents generates positive-tone images of the mask. Conversely,

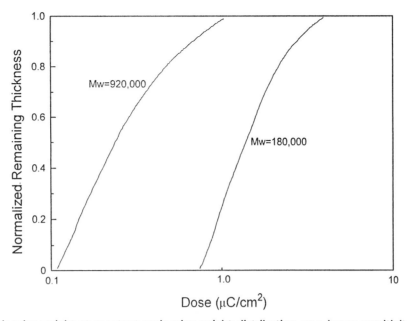

Figure 7. Effect of molecular weight at constant molecular weight distribution on e-beam sensitivity of poly(chloromethylstyrene) **13**. (Reproduced with permission from reference 67. Copyright 1979 Electrochemical Society.)

Table II. Lithographic Sensitivities of Representative Negative Resists Based on Cross-Linking

Polymer	UV (mJ/cm²)	Electron Beam (μC/cm² at 20 kV)	X-ray (mJ/cm²)
12		0.6	15–20 (A1), 175 (Pd)
13	50 (245 nm)	0.4–39	8–29 (Mo), 25 (Pd)
14		0.4–1.6	29 (Mo)

the formyl ester polymer is soluble in nonpolar solvents such as chlorobenzene or anisole, in which the phenolic photoproduct is completely insoluble. Therefore, development in these nonpolar solvents results in negative imaging.

Two-Component Resist Systems

As the minimum feature size continues to shrink, the demands placed on resist materials have become so stringent that one-component systems no longer satisfy all the desired criteria in resist development. As a consequence, modern lithographic technologies are almost exclusively based on two- or multicomponent resists, in which each component performs one or more specific functions. This design provides greater flexibility and therefore higher performance. Some of the radiation-sensitive polymers described in previous sections allow addition of a second component as a sensitizer (sensitivity enhancement), but these arrangements are not regarded as multicomponent systems. The polymers employed in the formulation of two- or multicomponent resists are typically inert to radiation and primarily function as a binder (matrix resin). Thus, these polymers may not be considered as functional polymers, but they have profound effects on the lithographic performance.

Negative-Resist Systems

Rubber–Azide Systems

Negative-resist systems consisting of photoactive aromatic azides and polymer binders have been well known since the 1930s (75), and negative photoresists based on an azide and a cyclized cis-1,4-polyisoprene have been widely used in the microelectronics industry. The matrix resin is a synthetic rubber of poly(cis-isoprene) cyclized to increase its T_g and toughness. The cyclized rubber material is highly soluble in nonpolar solvents such as toluene, xylene, or halogenated alkanes. The bisarylazide most commonly used for conventional near-UV lithography is 2,6-bis(4'-azidobenza)-4-methylcyclohexanone (Figure 8). Cyclized polyisoprene sensitized with bisazide generates, upon irradiation, an insoluble cross-linked network. The arylazide is decomposed in the excited state into a reactive nitrene intermediate that can undergo a variety of reactions (Scheme VII) including nitrene–nitrene coupling (forming azo dyes), insertion into carbon–hydrogen bonds (forming secondary amines), abstraction of hydrogen from the rubber backbone (forming an imino radical and a carbon radical with subsequent coupling reactions), and insertion into the double bond of polyisoprene (forming three-membered azilidine linkages). Polybutadiene was also used instead of polyisoprene.

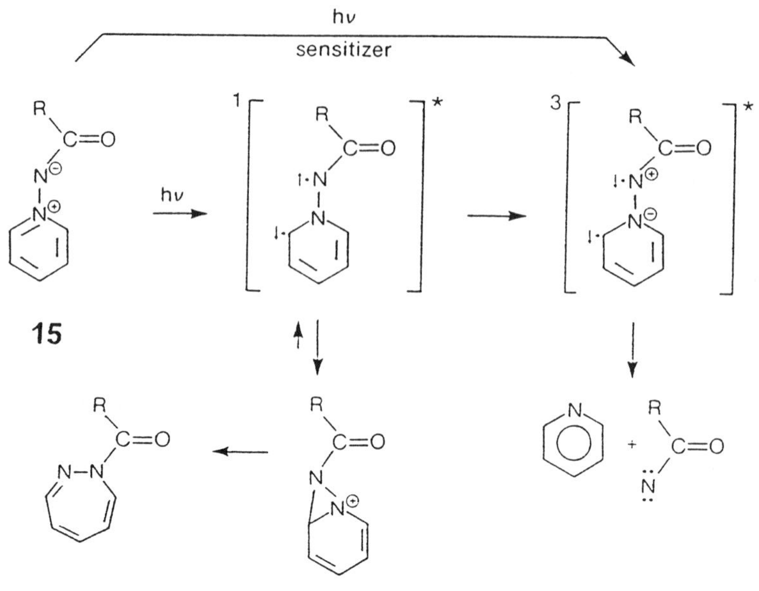

Scheme V. Photolysis of polar pyridinium ylide to nonpolar diazepine.

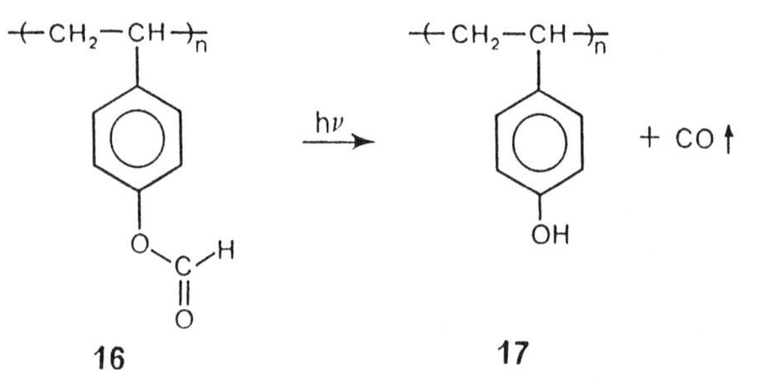

Scheme VI. Photo-Fries decomposition for polarity change.

The cyclized rubber–bisazide formulations offer high sensitivity, ease of handling, and wide process latitude. Several bisazides with absorption maxima between 240 and 290 nm were evaluated as deep-UV sensitizers for cyclized polyisoprene (76). However, the resolution of these systems was limited by relatively low contrast and distortion of resist patterns (bridging or snaking) as a result of polymer swelling during development.

Phenolic Resin–Azide Negative Resists

Negative resists that develop without swelling were formulated with polyhydroxystyrene (**17**) as the matrix resin and bisarylazides as the photoactive compound. The first example of such resists was designed for deep-UV lithography by using commercially available **17** and 3,3'-bisazidophenyl sulfone (77). The base-soluble phenolic resin is manufactured by radical polymerization of 4-vinylphenol and has a T_g of ca. 160 °C and an M_w of 1500–7000 (broad molecular weight distribution). Dissolution inhibition occurs upon exposure, and the unexposed areas of the resist film are dissolved in aqueous base in an etching-type

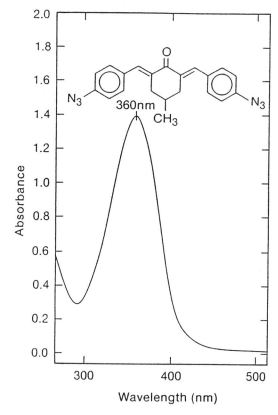

Figure 8. UV spectrum of commonly used bisarylazide in solution.

dissolution without swelling. Hydrogen abstraction from the polymer by nitrene and subsequent polymer–radical recombination result in increased molecular weight of the polymer, and the exposed areas are rendered less soluble in the aqueous base. The differential solubility is not the result of cross-linking at nor-

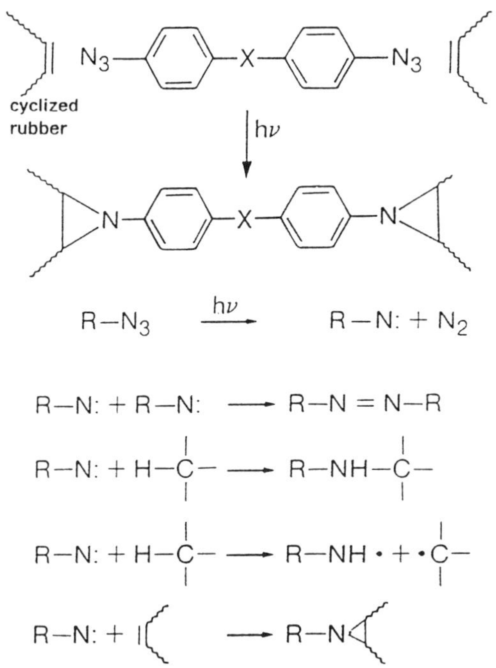

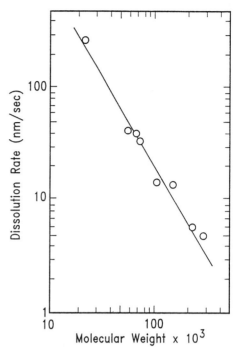

Scheme VII. Cross-linking mechanism of rubber/azide negative resist.

Figure 9. Dissolution rate of **17** in aqueous base as a function of molecular weight. (Reproduced with permission from reference 78. Copyright 1978 International Society for Optical Engineering.)

mal exposure doses. The exposed areas remain soluble but dissolve much more slowly because the polymer shows a very steep change in dissolution rate as a function of molecular weight (*78*) (Figure 9). Because of the intense absorption of the commercial samples of **17** used as the matrix resin in the deep-UV region, the photochemical reaction leading to the decreased solubility occurs mainly in the upper regions of the film and results in undercut profiles. Preparation of more transparent **17** will be described in a subsequent section. This resist system was modified by varying the structure of the bis-azide for deep-UV, mid-UV (*79*), and I line (*80*) sensitivity.

Other Negative-Resist Systems Based on Phenolic Resins

The phenolic resin continues to be important in modern lithographic technologies, because it allows swelling-free aqueous-base development and offers high dry-etch resistance. New sensitizers for phenolic resins were reported recently. Pyridine *N*-oxides are soluble in aqueous base. They are reduced upon UV exposure to pyridines, which are insoluble in aqueous developers and function as dissolution inhibitors (*81*). Thus, the two-component system provides negative images after development with aqueous base. The molecular weight of the phenolic polymer did not change after irradiation in the presence of pyridine *N*-oxide according to gel permeation chromatographic analysis.

Masking of the base-solubilizing OH functionality of phenolic resins through photochemical esterifica-

tion or protection was used recently in the design of deep-UV negative-resist systems. 1,3-Dioxin-4-ones undergo cycloreversion to acylketenes, which react with dipolar dienophiles to produce heterocyclic compounds upon 254-nm irradiation. Ketoacid ester and a carbonyl compound were detected after UV irradiation of 1,3-dioxin-4-ones in the presence of *m*-cresol (Scheme VIII) (*82*).

Negative-resist systems consisting of **17** and organic polyhalides such as trichloromethylated 1,3,5-triazines and trichloromethylated benzenes were evaluated (*83*) for excimer laser lithography. Postexposure baking is required for this system (120 °C, 2 min), as is the case with the pyridine *N*-oxide system (150 °C, 2 min) and with the 1,3-dioxin-4-one system (120 °C).

Molecular weights of **17** and polystyrene increase when irradiated in the presence of tris(trichloromethyl)triazine, but 2,4-dichlorobenzotrichloride does not affect the molecular weight of **17** when UV-irradiated. When a film of **17** containing the *s*-triazine is exposed and postbaked, the IR intensity of the phenolic hydroxyl group decreases with an increasing C=O intensity at 1740 cm, which is assignable to the phenyl ester of the aromatic carboxylic acid. Thus, it appears that the insolubilization of the two-component resists is the result of photochemically induced esterification of the phenolic matrix resin (Scheme IX). The use of aromatic compounds substituted with two or more CCl$_3$ groups results in radiation-induced molecular weight

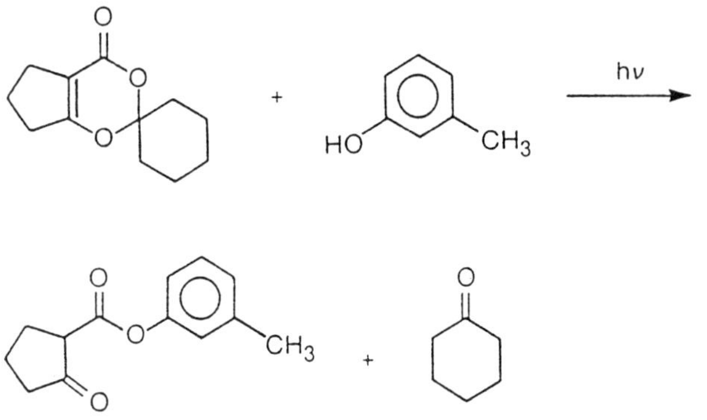

Scheme VIII. Esterification of phenol by photolysis of 1,3-dioxin-4-one.

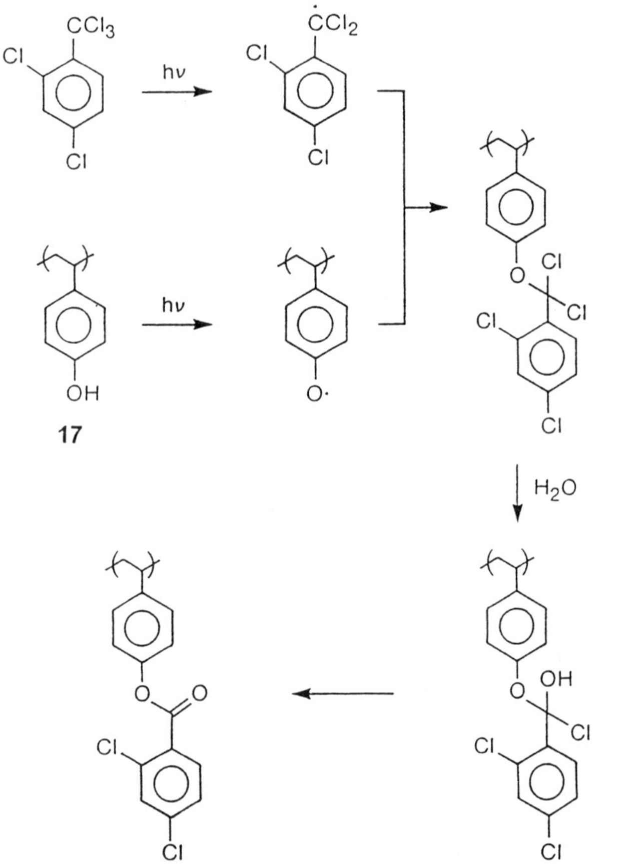

Scheme IX. Esterification of phenol by photolysis of polyhalide.

increase or cross-linking, which participates in the negative imaging. However, the mono-CCl$_3$-substituted benzene provides good resist performance, even without molecular weight increase, and this result indicates that the photochemically induced esterification of the phenolic matrix resin is the primary mechanism for the negative imaging. Another phenol protection scheme involves the photolysis of diphenyltetrazole to diphenylnitrilimine, which is highly reactive toward phenolic OH groups (Scheme X) (84).

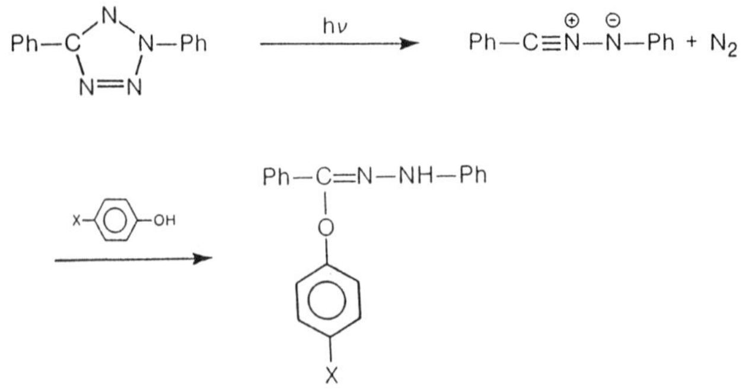

Scheme X. Protection of phenol by photolysis of diphenyltetrazole.

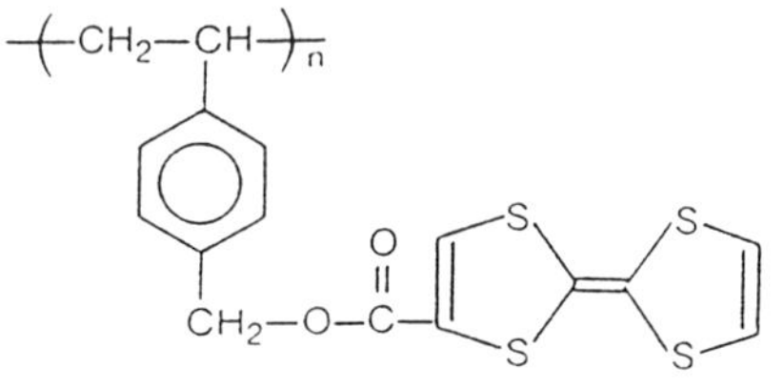

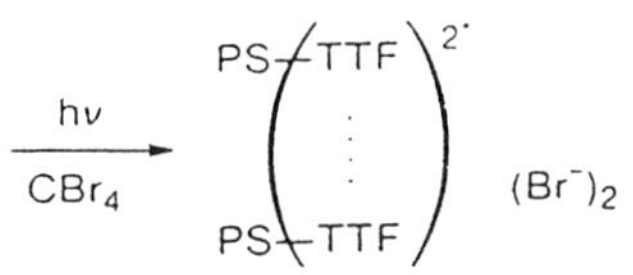

Scheme XI. Radiation-induced change from nonpolar to polar state for negative imaging.

Negative Resists Based on Polarity Change

Polystyrene–tetrathiofulvalene (**18**, Scheme XI) was reported as a nonswelling high-contrast resist (85). The differential dissolution is achieved through a change in the polarity of pendant groups rather than through generation of a cross-linked network. Exposure of **18** sensitized with a perhaloalkane such as carbon tetrabromide results in the generation of dimeric tetrathiofulvalene bromide salts that are polar in nature and hence insoluble in nonpolar organic solvents. Consequently, development with a nonpolar solvent selectively dissolves the unexposed nonpolar polymer and provides high resolution negative patterns with no evidence of swelling-induced distortion.

Positive-Resist Systems

Diazonaphthoquinone–Novolac Resists

Positive-resist systems consisting of a cresol–formaldehyde novolac resin (**19**) and diazonaphthoquinone as a photosensitive dissolution inhibitor have been the workhorses of microelectronics for their high resolution, high thermal stability, and resistance to dry-etching conditions. Novolac resins have low molecular weight and broad molecular weight distribution (M_n

= 300–1000) and T_g of 90–120 °C, and they become cross-linked by extended heating above 130 °C. These resins are soluble in common organic solvents and aqueous alkali and can be cast from solution to form isotropic glassy films of high quality.

Diazonaphthoquinone is soluble in common organic solvents but not in aqueous base. It functions as a dissolution inhibitor for aqueous-base development of the novolac-based resist. The photo-excited naphthoquinone chromophore releases nitrogen to form a highly reactive carbene intermediate with a quantum yield of 0.2–0.3 mol/einstein. The carbene intermediate undergoes the Wolff rearrangement to ketene, which rapidly reacts with water (absorbed in the novolac resin film) to produce an indenecarboxylic acid (Scheme XII) (*86*). Thus, the dissolution-inhibiting diazonaphthoquinone is photochemically transformed to a base-soluble indenecarboxylic acid, and this transformation allows the exposed area of the resist film to dissolve much faster than the unexposed region in an aqueous-base developer. This photochemically induced dissolution-rate differentiation is the basis of generation of positive relief images. Figure 10 shows the dissolution rate of the unexposed and exposed films of a novolac resist as a function of the diazonaphthoquinone concentration and illustrates the dissolution-inhibition effect of the photoactive compound and the enhanced dissolution rate of the exposed film in aqueous base.

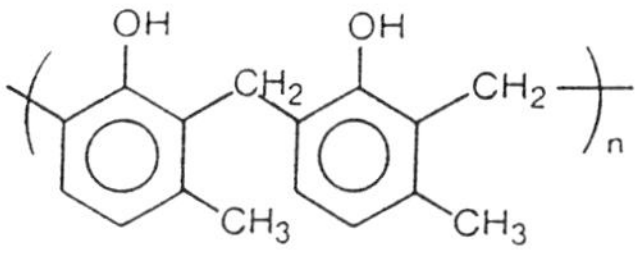

Interaction of diazonaphthoquinone with a phenolic matrix resin to retard the dissolution rate is not well understood, but it may involve hydrogen-bonding effects or chemical reactions between diazoquinone and the novolac resin catalyzed by the base developer. The role of the novolac matrix resin as a base-soluble binder is not as minor as it may seem. For example, the ratio of *o*-cresol to *m*-cresol in the novolac, the ratio of ortho to para backbone linkages, the molecular weight, and molecular weight distribution all affect the dissolution kinetics and the thermal flow resistance (T_g) of the resist (Table III) (*87–92*).

The diazonaphthoquinone isomers that are used in the commercial photoresist formulations are 5- and 4-arylsulfonates. The 5-arylsulfonates are almost exclusively used and characterized by an absorbance maximum at ca. 400 nm and a slightly stronger second maximum at 340 nm. The 4-sulfonate isomers have a

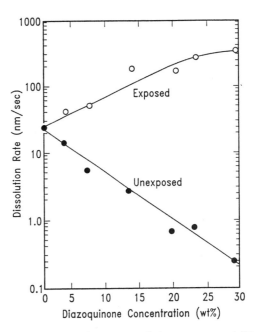

Figure 10. Dissolution rates of the unexposed (●) and exposed (○) films of a novolac/diazoquinone resist in aqueous base.

Table III. Effects of Novolac Resin Properties on Lithographic Parameters

Parameter	M_w	o,o-*Sub.*	*m/p*	M_w/M_n
Sensitivity	↓	↓	↑	
Dissolution inhibition	↑	↑	↓	↓
Dissolution promotion	↓		↑	↑
Contrast		↑	↓	
Exposure latitude		↑	↑	
Thermal stability	↑	↓	↓	

Note: M_w is weight-average molecular weight; Sub. is substitution; m/p is *m*- to *p*-cresol ratio; M_n is number-average molecular weight; ↑ is increase; ↓ is decrease.

19

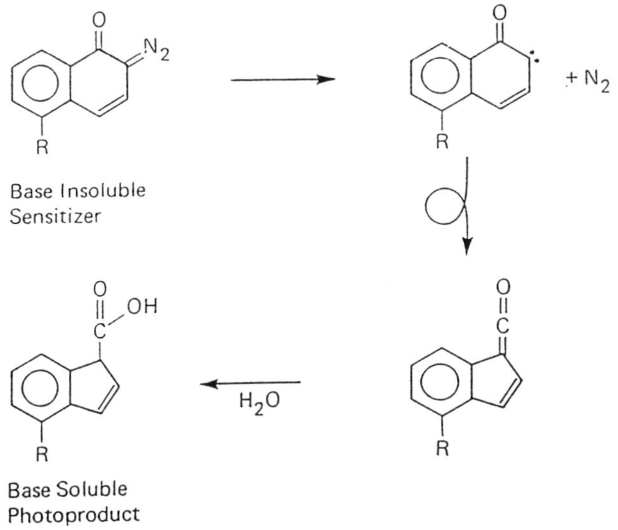

Scheme XII. Novolac resin and photolysis of diazonaphthoquinone.

Base Insoluble Sensitizer

Base Soluble Photoproduct

single absorbance maximum in the near-UV region centered at ca. 380 nm, and they have little or no absorbance at 436 nm (G line). The 6-arylsulfonates are more red-shifted than the 5-isomer and provide higher sensitivities at the G line and at 488 nm (Ar ion laser) (93).

An increase in resolution (R) can be accomplished by reduction of the exposing wavelength (λ), by increasing the lens NA, or by improving the process-dependent constant (k); R = kλ/NA. The diazo-quinone–novolac resist systems developed for broad-band and G line exposure consist of 1-oxo-2-diazo-5-sulfonates of polyhydroxybenzophenone. Although the drive toward higher resolution has necessitated the shift from G line to I line (365 nm), the classical G line photoresist could not be used in I line lithography, because the 5-sulfonates used in the G line resist yield photoproducts that absorb at 365 nm, although they bleach cleanly at 436 nm. Newer formulations for I line exposure typically employ photoactive chromophores directly bonded to a phenolic resin such as novolac. Excellent new I line resists were developed by optimizing transmission and bleaching characteristics at 365 nm and by maximizing the contrast and process latitude to support 0.50–0.35-μm lithography for 16- and 64-Mbit DRAM manufacture. Figure 11 presents a scanning electron micrograph of 0.4-μm line–space patterns printed in a diazonaphtho-quinone–novolac resist on an I line stepper.

A mid-UV (300–350 nm) resist was successfully formulated (94) by using a novolac resin more transparent at 313 nm and a diazonaphthoquinone sulfonate of an aliphatic alcohol that absorbs significantly and bleaches cleanly at 313 nm. However, extension of the diazonaphthoquinone–novolac resists to deep-UV (200–300 nm) lithography has been severely hampered, primarily due to high absorption of the novolac resin and nonbleachable absorption of diazoquinone in the deep-UV region.

Deep-UV Dissolution-Inhibition Resists

In the deep-UV imaging, the diazonaphthoquinone–novolac systems suffer from the absorption of both the photoproduct and novolac resin. However, significant progress has been made recently in application of the conventional design to deep-UV lithography. For example, a high resolution of 0.3 μm was demonstrated in a 0.5-μm thick film at 62 mJ/cm^2 on a KrF excimer laser stepper (NA = 0.42) by employing a 4-sulfonate of 2,3,4,4'-tetrahydroxybenzophenone and by optimizing the developer concentration to induce surface induction without sacrificing the photospeed (95). Although the resist film was significantly more transparent at 248 nm than the conventional photoresists [optical density (OD) = 1.25/μm vs. 1.7/μm before exposure and 1.05/μm vs. 1.65/μm after exposure], the 0.3-μm line–space patterns had sloped sidewalls at the thickness of 0.5 μm.

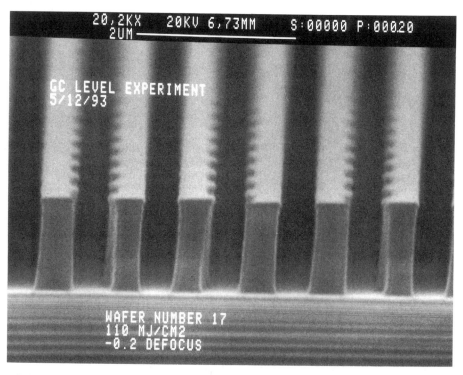

Figure 11. Positive 0.4-μm line–space patterns delineated in a novolac–diazoquinone resist by projection printing at 365 nm and aqueous-base development (courtesy of M. Platt, IBM Microelectronics Division).

For deep-UV applications, several photoactive dissolution inhibitors were reported. 5-Diazo-Meldrum's acid exhibits an intense bleachable absorbance at 254 nm with a sharp cutoff above 300 nm and is converted to volatile photoproducts (bleaching) (*96*). Photolysis results in a Wolff rearrangement to produce a ketene intermediate that decomposes further into carbon monoxide and acetone (Scheme XIII). The resists formulated with these photoactive compounds and a novolac resin exhibited a reasonable sensitivity but suffered from volatility and solubility problems of the photoactive compound and the high absorbance of the novolac resin. This aliphatic diazoketone family has been expanded to include cyclic 1,3-diacyl-2-diazo compounds containing heteroatoms (A, Scheme XIII) (*97*). To increase the solubility and to decrease the volatility of the diazo-Meldrum's acid, 2-diazo-1,3-diketocyclohexane-5-carboxylic esters (*98*) and acyclic analogs such as 1,7-bis(4-chlorosulfonylphenyl)-4-diazo-3,5-heptanedione (B and C, Scheme XIII) were evaluated, and the acyclic analogs blended with deep-UV-transparent poly(styrene-*co*-maleic acid half ester) (*99*).

α-Diazoacetoacetates of steroids and polyfunctional alcohols undergo photolysis to give carboxylic acids and exhibit excellent bleaching effects (Chart III) (*100*). These dissolution inhibitors were mixed with base-soluble poly(4-hydroxybenzylsilsesquioxane) (**20**), and the two-component resists were evaluated on a KrF excimer laser stepper (*100*).

The classical *o*-nitrobenzyl photochemistry was also used in the design of dissolution inhibitors. *o*-Nitrobenzyl esters are photolyzed to carboxylic acids and *o*-nitrosobenzaldehyde, resulting in the transformation of the dissolution-inhibiting lipophilic ester to a dissolution-promoting hydrophilic acid. *o*-Nitrobenzyl cholate is insoluble in aqueous base and inhibits the dissolution of a base-soluble copolymer of methyl acrylate and methacrylic acid, which is transparent in the deep-UV region. Photolysis of the cholate leads to the formation of the corresponding carboxylic acid and *o*-nitrosobenzaldehyde, and conversion of the dissolution-inhibiting ester into a dissolution-promoting acid results (Scheme XIV) (*101*).

Certain onium salts (cationic photoinitiators) are also a new class of dissolution inhibitors (*102*). Diphenyliodonium and triphenylsulfonium salts generate strong acids upon irradiation (Scheme XV) and can inhibit efficiently the dissolution of a novolac resin in aqueous base to provide a two-component deep-UV resist with a sensitivity of 25 mJ/cm². Figure 12 shows dissolution kinetics of unexposed and exposed (20 mJ/cm² at 254 nm) films of **19** containing 4.8 wt% of triphenylsulfonium hexafluoroantimonate in aqueous base. Whereas the unexposed film dissolves very slowly, UV exposure renders the film highly soluble.

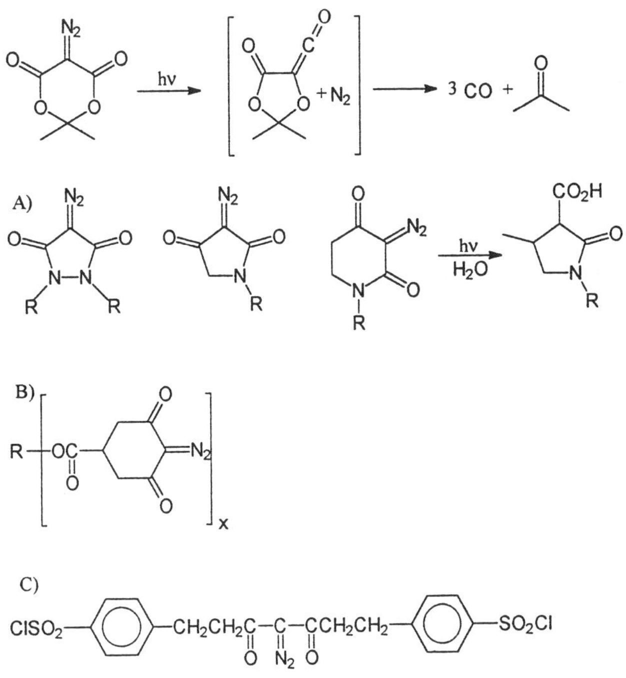

Scheme XIII. Deep-UV dissolution inhibitors.

This dissolution-inhibition effect of certain photochemical acid generators must be taken into consideration in the design of chemical-amplification resists, as discussed in Chapter 2.4. Addition of triphenylsulfonium hexafluoroarsenate to positive resists consisting of acyclic diazoketone and poly(styrene-*co*-maleic acid half ester) or phenolic polymers (**17**) significantly enhances the performance (sensitivity and contrast) of the KrF excimer laser resist through improved dissolution kinetics (*103*).

Poly(Olefin Sulfone)–Novolac Resists

The novolac (**19**)–diazonaphthoquinone resist, which is resistant to dry etching, is insensitive to e-beam or X-ray radiation, whereas poly(olefin sulfone) resists, which are sensitive to e-beam radiation, suffer from poor dry-etch durability. Accordingly, the high radiation sensitivity of poly(olefin sulfone)s was combined with the high dry-etch resistance of the novolac-based resists for e-beam direct writing (*104, 105*). In this two-component formulation, the lipophilic poly(olefin sulfone) functions as a polymeric dissolution inhibitor of the novolac matrix (**19**) in aqueous base. However, novolac and poly(olefin sulfone) are not miscible in general. The use of a special casting solvent allowed the formation of homogeneous films from the mixture of a *m*-cresol novolac resin and poly(2-methyl-1-pentene sulfone) (*106*). To enhance the miscibility of the two polymers, either novolac (*104*) or polysulfone (*105*) was modified. Figure 13 shows the dissolution-inhibition effect of a modified poly(olefin sulfone) and the effect of e-beam irradiation (3, 5, and 10 μC/cm² at

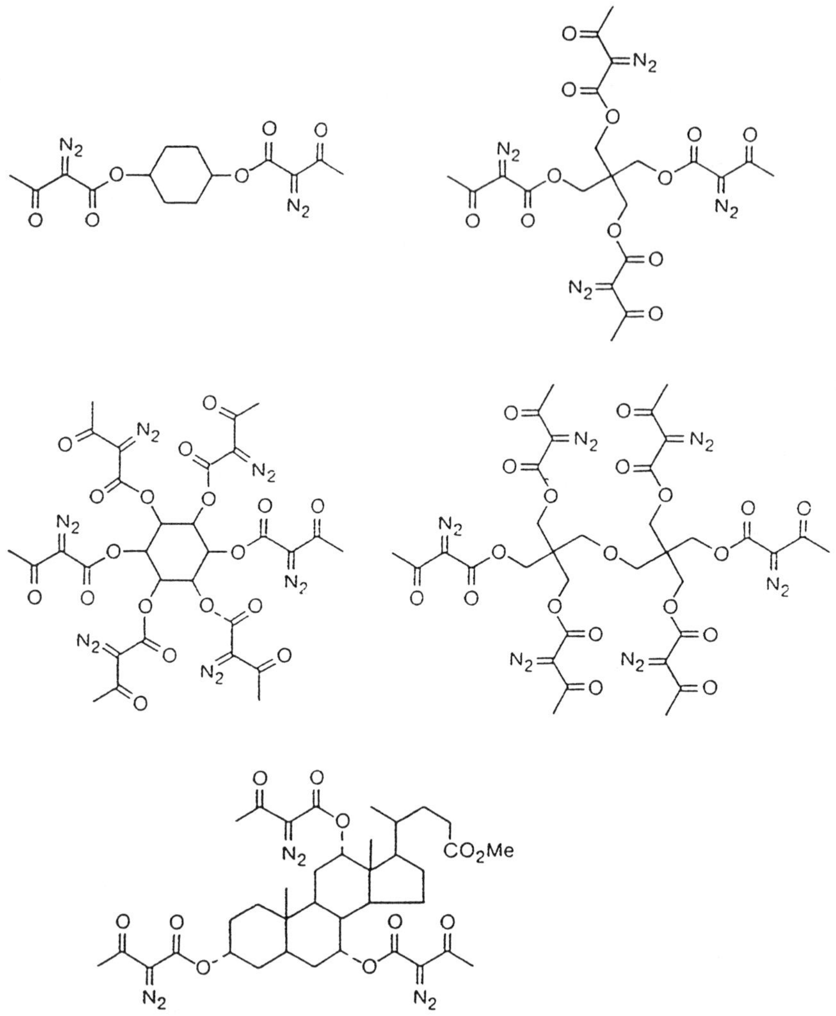

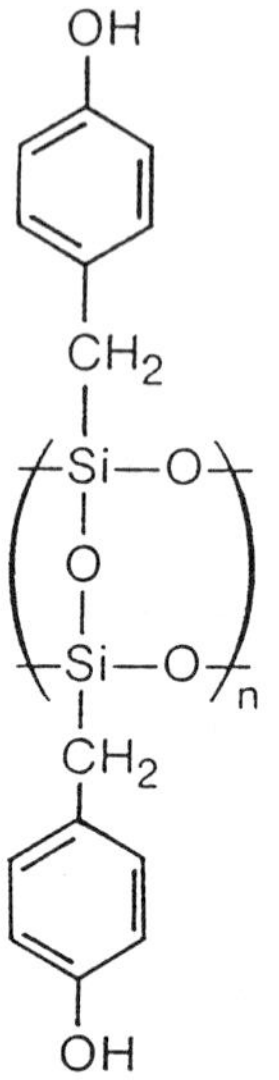

Chart III. α-Diazoacetoacetates and base-soluble polysilsesquioxane.

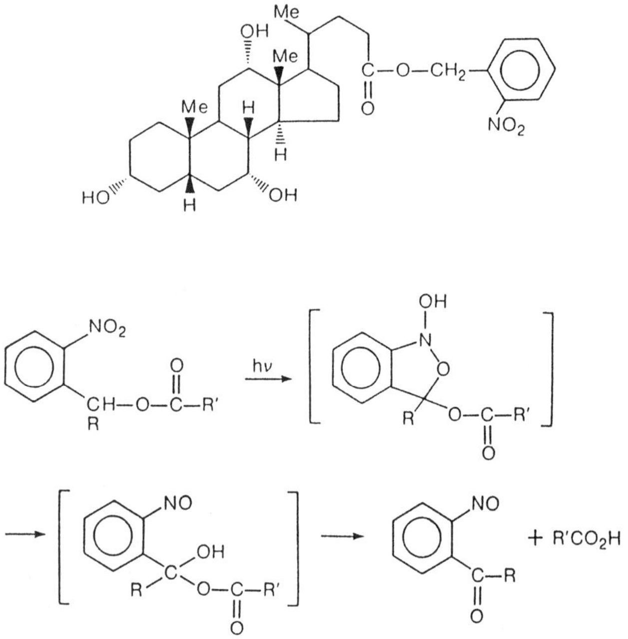

Scheme XIV. *o*-Nitrobenzyl cholate dissolution inhibitor and its photolysis.

$$ArN_2^+ \; MX_n^- \xrightarrow{h\nu} ArX \; + \; N_2 \; + \; \underline{MX_{n-1}}$$

$$Ar_2I^+ \; MX_n^- \xrightarrow{h\nu} ArI \; + \; \underline{HMX_n} \; + \; others$$

$$Ar_3S^+ \; MX_n^- \xrightarrow{h\nu} Ar_2S \; + \; \underline{HMX_n} \; + \; others$$

$$MX_n \; = \; BF_4, \; PF_6, \; AsF_6, \; SbF_6, \; etc.$$

Scheme XV. Onium salt acid generators.

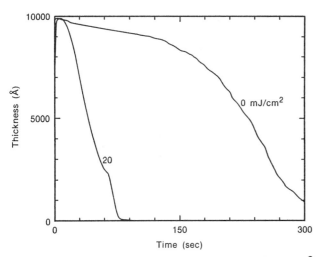

Figure 12. Dissolution kinetics of unexposed (0 mJ/cm²) and exposed (20 mJ/cm²) films of **19** containing 4.8 wt% of triphenylsulfonium hexafluoroantimonate in aqueous base. (Reproduced with permission from reference 102. Copyright 1988 Electrochemical Society.)

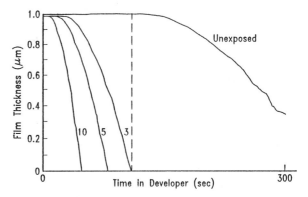

Figure 13. Dissolution curves of novolac–sulfone resist exposed to 0, 3, 5, and 10 μC/cm² of 20-kV electron beam. (Reproduced with permission from reference 105. Copyright 1988 Electrochemical Society.)

20 kV) on the dissolution kinetics of the blend film in aqueous base. The film exposed to 3 μC/cm² completely dissolves to the substrate, whereas the unexposed film is essentially insoluble for at least 2 min.

Multilayer and Dry-Resist Systems

Linewidth control over steps and wafer topography becomes extremely challenging as the feature size shrinks. Reflection from the topographic features produces linewidth variation. Another important issue in optical lithography is depth of focus (DOF). The resolution can be increased by reducing the exposing wavelength or by increasing the lens NA as mentioned earlier. However, DOF is directly proportional to the wavelength and inversely proportional to NA^2 (DOF = $k\lambda/NA^2$). Thus, the higher resolution is achieved in optical lithography at the expense of DOF. The multilayer scheme provides solutions to these problems encountered in optical lithography (*107*). Furthermore, the use of thin imaging layers in the multilayer scheme could improve resolution and linewidth control. The vertical dimensions have remained essentially constant (ca. 1 μm), whereas the lateral dimensions have shrunken rapidly over several generations of semiconductor devices, which demand high aspect-ratio (height–width) imaging.

The dry-etching process employing plasmas has replaced the wet technique in fabrication of the substrate to achieve a tighter control of the image-transfer process, and resist chemists have been motivated to design resist systems that can be developed with plasmas without use of developer solvents. Two general schemes are available in the use of plasmas, especially

oxygen reactive ion etching (RIE) in the resist development processes, which typically employ organometallic polymers that are resistant to oxygen plasma. Organosilicon polymers are of particular interest. RIE can produce vertical wall profiles because of the highly anisotropic nature of the etching process. Silicon-containing polymers are converted to a thin layer (ca. 50 Å) of silicon oxide in oxygen RIE, and the silicon oxide is impervious to further etching.

The first approach involves coating of the substrate with a thick (1–2 μm) planarizing layer of an organic (aromatic) polymer, which is then coated with a rather thin (ca. 0.5 μm) organosilicon resist (a bilayer system). Silicon is selectively removed from either the exposed or the unexposed regions of the organosilicon resist by development, which provides a protective stencil for image transfer by oxygen RIE and the planarizing organic polymer layer.

The second approach to oxygen RIE imaging is selective incorporation of silicon in organic resist films after exposure, and this approach allows the design of single-layer all-dry processes. Gas-phase functionalization is particularly useful when employed in the top surface imaging (TSI) scheme (explained later).

Silicon-Containing Resist Systems

The bilayer lithographic scheme using an organometallic resist has evolved from trilayer lithography, which employs an organic imaging layer, a thin film of silicon oxide as an oxygen RIE barrier, and a thick organic polymer layer for planarization (Figure 14). In the trilayer scheme, the top resist film is imaged with a developer in a conventional manner and is used as a mask to etch the oxide layer with a fluorocarbon plasma. The imaged oxide layer now protects the underlying polymer film during oxygen RIE pattern transfer.

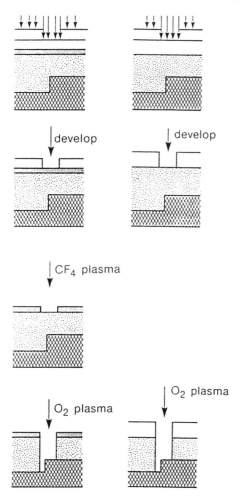

Figure 14. Trilayer and bilayer lithographic imaging processes.

In the bilayer scheme, the functions of the intermediate silicon oxide layer and the top imaging layer are combined to simplify the time-consuming trilayer lithography. Various organometallic resists have been developed for this purpose. In general, all bilayer resist systems contain significant percentages of an element that forms a refractory oxide upon oxygen plasma treatment. Hard-baked (cross-linked) novolac resists or polyimides are commonly used as the planarizing layer.

Semi-Dry Bilayer Lithography (Wet Development–Dry Pattern Transfer)

The first organometallic polymers used in the bilayer scheme were commercial polysiloxanes (**21, 22**; Chart IV), which function as cross-linking negative resists (*108*). The etch-rate ratio of a novolac resist to polysiloxane in an oxygen plasma is reportedly 50:1, a result indicating that only a 40-nm-thick layer of polysiloxane is sufficient to protect a 2-μm-thick novolac resist. A chloromethyl group was incorporated into the benzene ring of polydiphenylsiloxane (**23**) to

improve its deep-UV sensitivity and to produce a highly sensitive e-beam resist (*109*). The chloromethylated polydiphenylsiloxane has a $T_g > 25$ °C, whereas the parent polymer is a grease at room temperature. The chloromethylated polydiphenylsiloxane was further treated with potassium methacrylate, and the resulting polymer was mixed with a photochemical radical initiator to formulate a sensitive near-UV negative resist (*110*).

An aqueous-base soluble polysiloxane with a phenolic group directly attached to Si (**24**) was evaluated as a matrix resin in a two-component system containing a diazonaphthoquinone dissolution inhibitor (*111*). Silsesquioxane oligomers (**25**) were converted to alkaline-soluble resins by Friedel–Crafts acetylation with acetyl chloride–AlCl₃ and subsequent hydrolysis. The base-soluble silsesquioxane oligomer was used in conjunction with diazonaphthoquinone for near-UV, deep-UV, e-beam, and X-ray exposures [poly(2-methyl-1-pentene sulfone) was used as a polymeric dissolution inhibitor for e-beam imaging] as well as with azides for cross-linking negative-resist systems (*110, 112*).

In addition to the linear and ladder siloxane polymers, three-dimensional polysilphenylenesiloxane (**26**), which is soluble in methyl isobutyl ketone, was developed as a negative KrF excimer laser resist for bilayer lithography (Chart IV) (*113*). This polymer consists of a rigid three-dimensional mesh of a silphenylenesiloxane core that is surrounded by functional groups. The rigid core contributes to low swelling, high thermal stability, high oxygen RIE stability, and high deep-UV absorption. In a similar fashion, a micro-resinoid siloxane resin bearing chloromethylbenzene moieties as cross-linking groups was reported (*114*) as a sensitive e-beam and KrF excimer laser bilayer resist.

Polysilanes (or polysilylenes, **27**, Chart V) have attracted much attention because of their unique and interesting properties, including their potential as a lithographic material (*53, 115*). These materials function as positive-tone deep-UV and mid-UV resists and also as e-beam resists (*116, 117*). Backbone scission of polysilanes induced by irradiation leads to a decrease in chain length, which results in a blue shift of absorption maximum and a reduction in the extinction coefficient. Aliphatic polysilanes such as poly-(cyclohexylmethylsilane) were successfully used as deep-UV imaging layers for oxygen RIE pattern transfer, but the required exposure doses (125–150 mJ/cm²) were higher than desirable for commercial deep-UV exposure systems.

Two approaches have been taken to increase the exposure sensitivity of these polymers: namely, the synthesis of new polysilanes that are intrinsically more sensitive to photodegradation in the solid state,

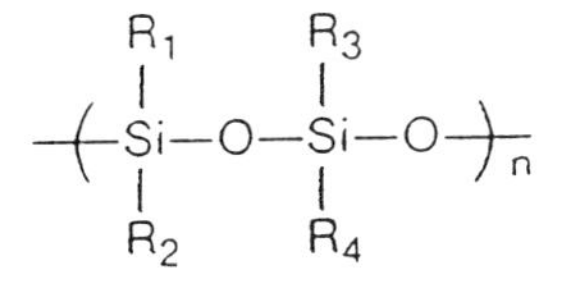

Chart IV. Polysiloxanes reported for bilayer lithography.

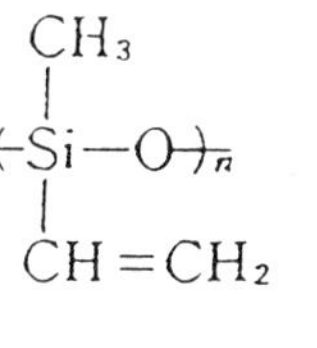

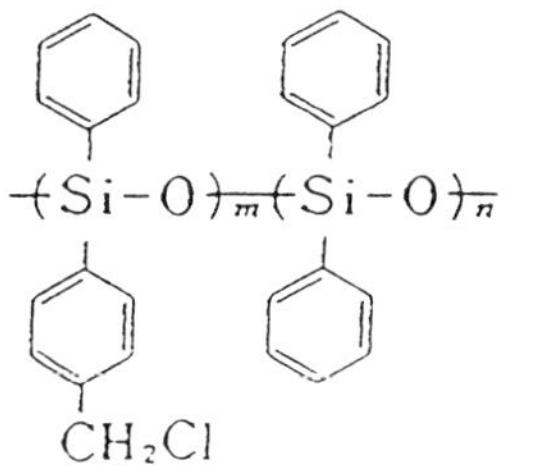

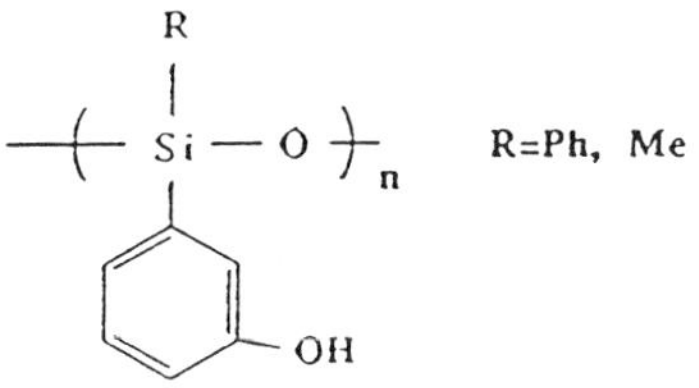

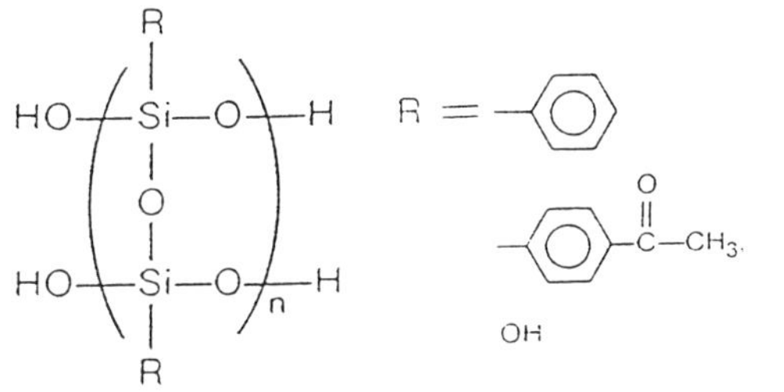

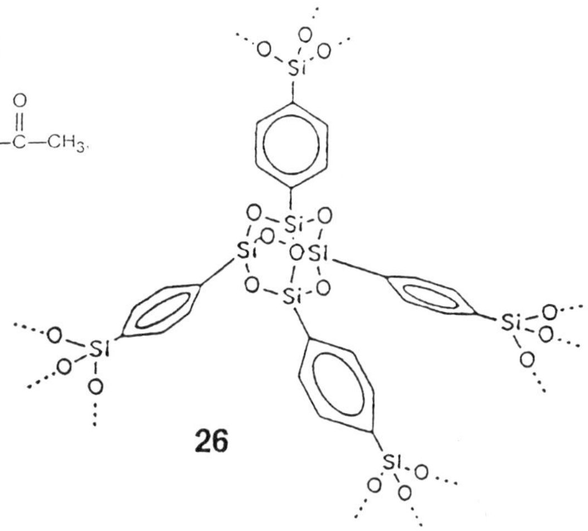

and doping with sensitizing additives (*118*). Aryl-substituted silane homopolymers that are significantly more photolabile than the standard poly(methylphenylsilane) were prepared and tested. In addition, a number of additives that quantitatively quench the polymer fluorescence in the solid state were identified. The combination of the new polysilane materials such as poly[(4-*t*-butylphenyl)ethylsilane] with sensitizing additives has allowed submicrometer deep-UV imaging at exposure doses as low as 20 mJ/cm^2 or less in a bilayer configuration.

Poly(*p*-disilanylenephenylene) (**28**, Chart V) has a Si–Si linkage in the backbone interrupted by a benzene ring as a spacer and is more transparent in the deep-UV region than polysilanes (*54*). Aqueous-base soluble polysilanes were prepared recently (*111*) by introducing pendant phenolic or carboxylic acid functionalities (**29**), and these polysilanes were evaluated as single-component KrF excimer laser resists (>1.2 J/cm^2).

Polysilynes (network polysilanes, **30**) contain only one pendant alkyl group per silicon atom and exist as

Si–Si σ-conjugated networks (no structural similarity to polyacetylenes) (*119*). These network materials represent structural and stoichiometric intermediates between linear polysilanes and amorphous silicon, and they also show intermediate optical and electronic properties. The σ-σ* electronic transition in the Si backbone results in strong UV absorbance. Irradiation in air leads to photooxidation of the Si–Si network and the formation of a rigid insoluble siloxane (*120*).

The UV spectra of alkyl-substituted polysilynes irradiated with 190-nm radiation show significant bleaching down to 190 nm, whereas poly(phenylsilyne) remains absorptive at wavelengths less than 230 nm (*120*). The bleaching in UV absorption is due to the replacement of the Si–Si network with the O–Si–O network. The resist can be wet-developed with toluene in a negative mode by cross-linking via intermolecular Si–O–Si bond formation or dry-developed by using HBr RIE by photooxidation to induce etch selectivity (*120*). The sensitivity to 193-nm radiation in either case is 20–200 mJ/cm^2, depending on the resist

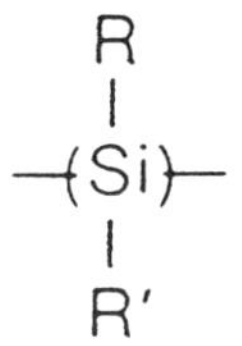

R
│
─(Si)─
│
R′

27

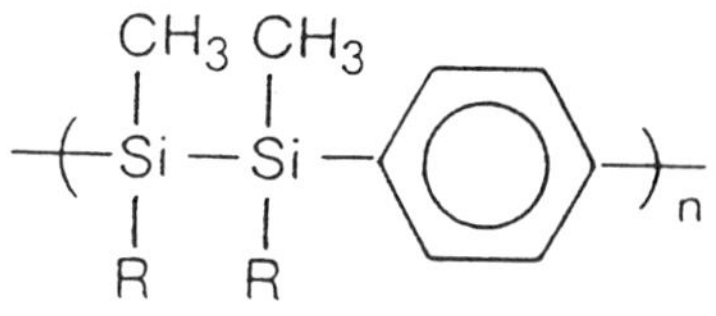

CH₃ CH₃
│   │
─(Si ─ Si ─⟨benzene⟩─)ₙ
│   │
R   R

28

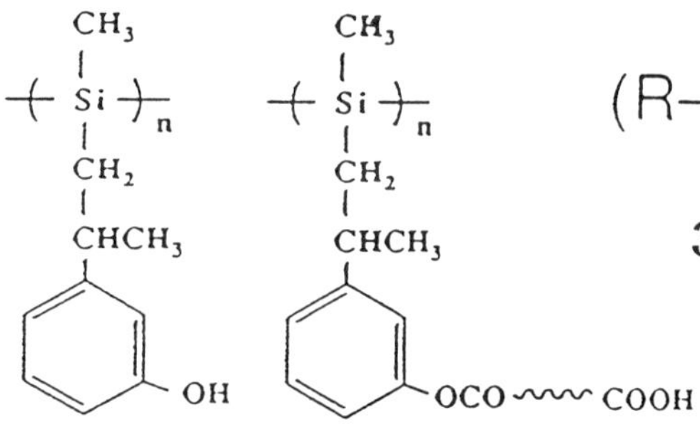

CH₃
│
─(Si)ₙ─
│
CH₂
│
CHCH₃
│
⟨benzene⟩
OH

CH₃
│
─(Si)ₙ─
│
CH₂
│
CHCH₃
│
⟨benzene⟩
OCO ∿ COOH

29

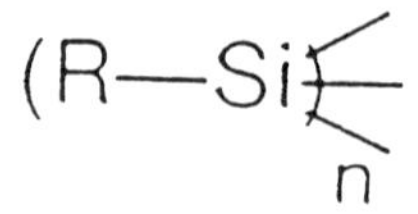

(R─Si⟨⟩ₙ

30

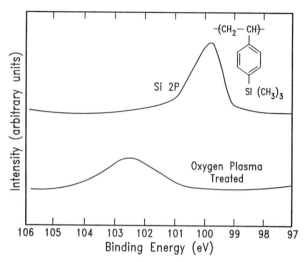

Figure 15. ESCA spectra of poly(4-trimethylsilylstyrene) before and after oxygen plasma treatment. (Reproduced with permission from reference 121. Copyright 1983 North-Holland.)

Chart V. Polysilanes and analogs reported for bilayer lithography.

formulation. The best resist compositions are obtained with polysilynes that contain predominantly small aliphatic pendant groups (*n*-butyl). By using a 0.33 NA catadioptric lens with a phase shift mask at 193 nm, equal line-and-space features as small as 0.15 μm were obtained (*120*) and transferred through 1.0 μm of a planarizing layer by using oxygen RIE.

Metallic species can also be placed in polymer side chains to enhance oxygen plasma resistance. A representative example is poly(trimethylsilylstyrene) (*121, 122*). Electron spectroscopy for chemical analysis spectra presented in Figure 15 clearly show that oxygen plasma treatment converts the organosilicon polymer

to silicon oxide. The high oxygen plasma resistance of trimethylsilyl- and trimethylstannylstyrene polymers was combined with the high deep-UV sensitivity of chlorinated or chloromethylated polystyrene by radical copolymerization (**31**, Chart VI) (*121, 122*). A small amount of Si (3 wt%) in an organic polymer has a substantial effect on the oxygen plasma etch rate, and incorporation of ca. 10 wt% Si is sufficient to produce an effective etch barrier.

Terpolymers of Si-containing methacrylate, MMA, and methacrylic acid (**32**, Chart VI) were evaluated (*123*) as positive deep-UV resists for a bilayer application and exhibited a sensitivity of 0.5–3.6 J/cm² in development with organic solvents. Trimethylsilyl-methyl methacrylate was radically copolymerized (*124*) with chloromethylstyrene for the design of nega-

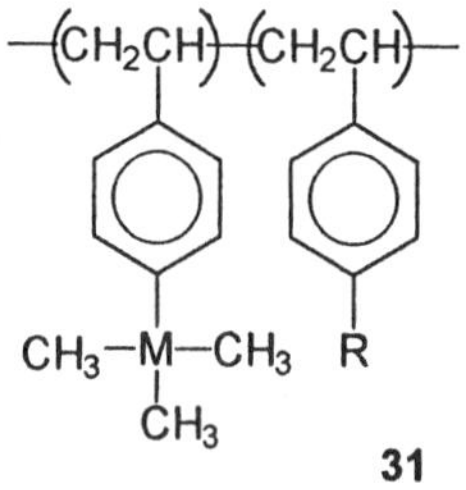

31

M= Si or Sn
R= Cl or CH₂Cl

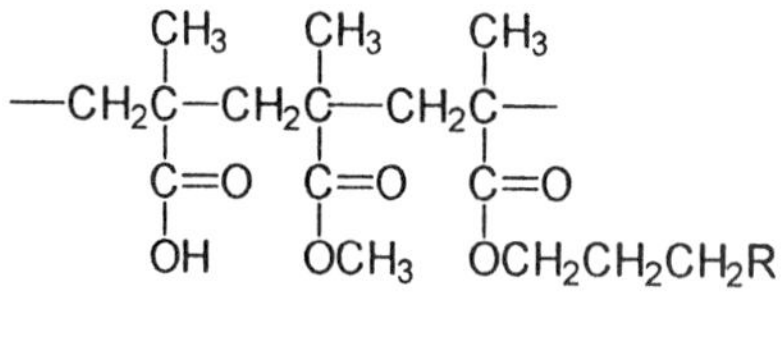

32

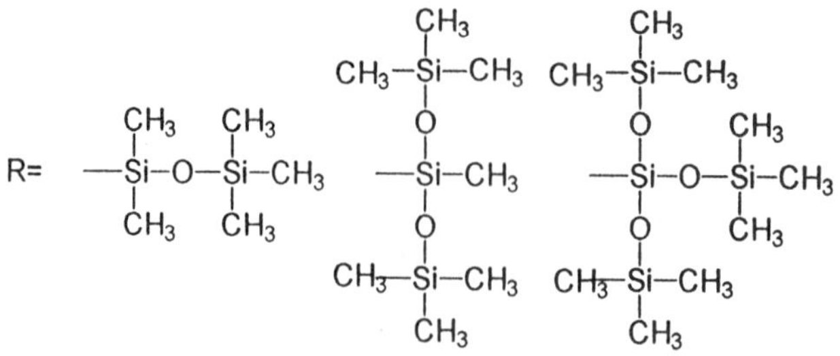

Chart VI. Polymers containing Si (or Sn) in side chain.

tive-resist systems. A copolymer containing at least 90 mol% of the Si-bearing units is required in this case for acceptable oxygen RIE resistance. As mentioned earlier, brominated poly(1-trimethylsilylpropyne) with the degree of bromination of 0.1–0.2 per monomer unit functions as a positive deep-UV resist and has a sensitivity of 20–25 mJ/cm^2 and γ of 3.5–4.0 when postbaked at 140 °C for 1 h (*52*).

To render the highly radiation-sensitive poly(olefin sulfone)s (**8**) resistant to oxygen plasma for an e-beam imaging layer in bilayer lithography, copolymers of silicon-containing olefins with sulfur dioxide (**33**, Chart VII) were produced (*125*). Introduction of silicon into poly(olefin sulfone) side chains causes marked changes in solubility and chemical and thermal stability. In an effort to improve the processing characteristics of these alkenylsilane derivatives, a series of polysulfones made from the linear and cyclic alkenyl(di)silane derivatives were recently examined. The resistance of these linear and poly(ω-alkenyl(di)silane sulfone)s to oxygen plasma shows a marked dependence on the structure of the copolymer as well as development and plasma-processing conditions (*126*).

Poly(silylstyrene sulfone)s (**34**) exhibit excellent stability, film-forming properties, and excellent oxygen RIE resistance without the passivation treatment (*127*). Poly(4-trimethylsilylstyrene sulfone) has a T_g of 188 °C and an e-beam sensitivity of 5–8 μC/cm^2. To further

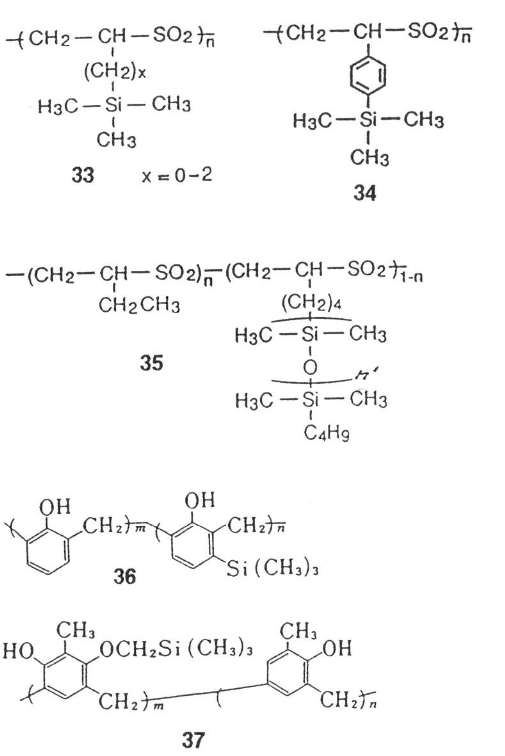

extend this concept, linear oligosiloxane was incorporated into poly(olefin sulfone) structures (**35**) by terpolymerization of 5-hexenyl-functionalized poly(dimethylsiloxane) macromonomers with 1-butene and sulfur dioxide (*128*). The comb-like architecture limits the size of the phase-separated domains to the order of 10 Å, and this architecture has no deleterious effect on the quality of the lithographic pattern. Above a certain composition, polysulfone forms a continuous phase and its properties dominate the physical properties of cast films. Although the polysiloxane chain exists largely as phase-separated domains, the domain concentration is such that films exhibit excellent stability in reactive oxygen plasmas.

In attempts to use the commonly available diazoquinone–novolac photoresists as an imaging layer in the oxygen RIE bilayer scheme, silicon-containing novolac resins were synthesized (**36, 37**; Chart VII) (*129–131*). Representative organosilicon polymers for use in the bilayer lithography are tabulated in Table IV.

All-Dry Processes

Generation of primary resist images with use of a plasma was first reported in 1979. In this plasma-developable photoresist (PDP) process (*132*), the resist is coated in the usual fashion and exposed. The exposed film is then heated to produce a relief image of a negative tone, which is then transferred into the underlying substrate by RIE (Figure 16). The PDP concept was extended to the design of an X-ray resist system, which is based on poly(2,3-dichloropropyl acrylate) (**38**) as a host polymer and a silicon-containing monomer (Scheme XVI) (*133–135*). X-ray exposure of the resist film initiates radical cross-linking of the polymer, and hence it incorporates the silicon-containing monomer into the network (fixing). The exposed film is then heated to remove the unreacted organometallic monomer from the unexposed areas. Development is accomplished by treating the baked film with an oxygen plasma.

Chart VII. Si-containing poly(olefin sulfone)s and novolac resins.

Table IV. Lithographic Sensitivities of Representative Si-Containing Resists for Bilayer Lithography

Resist	Sensitivity
22	1.5 μC/cm^2 (20 kV)
23	20 mJ/cm^2 (deep UV)
	2–5 μC/cm^2 (20 kV)
	50 mJ/cm^2 (Mo X-ray)
27	100 mJ/cm^2 (mid UV)
31	5–20 mJ/cm^2 (254 nm)
	1–4 μC/cm^2 (20 kV)
32	500–3600 mJ/cm^2 (260 nm)
33	1.0–1.5 μC/cm^2 (20 kV)
9	20–25 mJ/cm^2 (160 nm)
34	5–8 μC/cm^2 (20 kV)

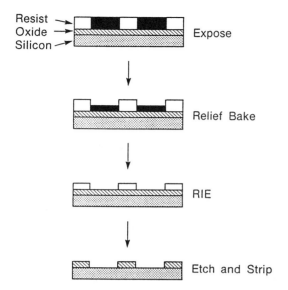

Figure 16. Plasma-developable photoresist process.

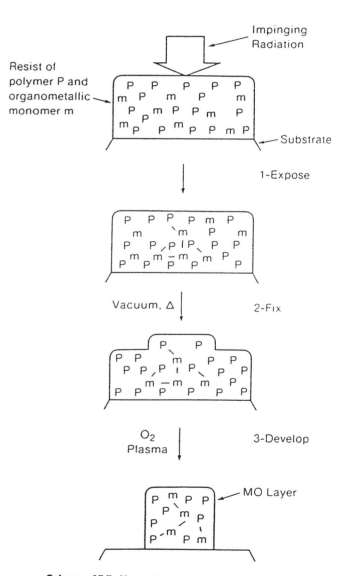

Scheme XVI. X-ray dry-development process.

The PDP concept is based on radiation-induced attachment of silicon-containing molecules to a host polymer followed by removal of unreacted organosilicon compounds by heating; therefore, the process provides negative plasma development. If organosilicon moieties are covalently bonded to the resist polymer in such a way that they can be scissioned selectively by radiation, then positive-tone plasma development is also possible. An acrylic copolymer containing silicon side groups that can be cleaved from the polymer by exposure to UV light is such an example of dry-developable deep-UV resist (*see* Scheme XVII) (*136*). The photochemically generated, volatile, silicon-containing species are removed from the exposed regions by subsequent heating, and the oxygen plasma etch rate in the exposed areas is higher than the unexposed areas that contain the polymer-bound silicon species.

Gas-Phase Functionalization and Silylation

Another approach for designing plasma-developable resists or oxygen RIE pattern transfer is to introduce metallic species into selected (exposed or unexposed) areas of an organic resist film. A bisazide–polyisoprene resist was treated with an inorganic halide such as $SiCl_4$ after UV exposure. Incorporation of silicon resulted predominantly in the unexposed areas and produced a positive-tone image when developed in an oxygen plasma (*137*). The scheme has been extended to other polymer systems in conjunction with the use of $TiCl_4$, in which the reactivity is markedly dependent on humidity (*138*). Layers of water selectively absorbed in photooxidized areas of polymer films were reacted with $TiCl_4$ to generate an oxygen RIE barrier layer (*139*).

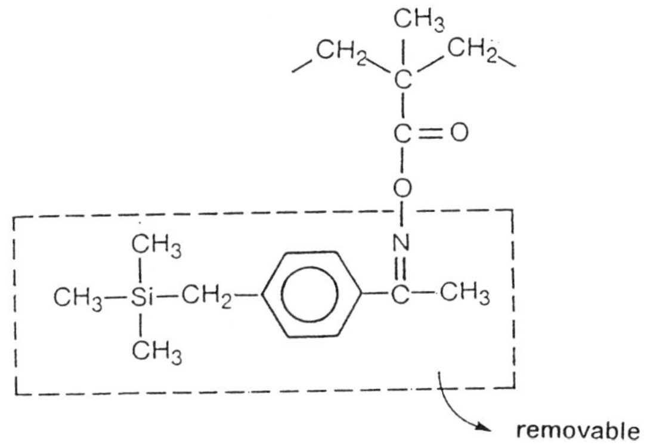

Scheme XVII. Positive-tone dry-development resist.

Silicon oxide formation at the near surface of UV irradiated polymers was applied to surface imaging (*140*). For example, the polymer precursor **39** generates polymer-bound sulfonic acid residues upon irradiation (Scheme XVIII). Water is absorbed from the atmosphere into the exposed regions of a resist film. When the exposed film is treated with the vapor of alkoxysilanes, silicon oxide is formed selectively only at the irradiated surface, which functions as an oxygen RIE mask was provided for negative dry development (*140*).

The commonly available novolac–diazonaphthoquinone photoresists can also be rendered dry-developable with oxygen plasma by the diffusion-enhanced silylating resist (DESIRE) process, which is based on selective silylation of a phenolic resin controlled by differential diffusion of a silylating reagent (Figure 17) (*141*). The novolac resist undergoes thermal cross-linking in the unexposed areas after postbake, which prevents diffusion of the silyl compound; whereas the gaseous silylating agent diffuses at a much faster rate into the exposed regions and reacts with the surrounding phenolic groups. Negative imaging results after oxygen RIE. Because the thermally cross-linked phenolic resin in the unexposed areas reacts with the silylating reagent to a minor degree, preburning with fluorocarbon plasma before oxygen RIE is employed to remove a thin top layer of organosilicon polymer (*142*).

The DESIRE process (originally developed for near-UV exposures) was evaluated extensively for device fabrication (*143*) and also was applied to deep-UV imaging as a top surface imaging (TSI) because of the high opacity of the photoresist below 300 nm (*144*). For deep-UV applications, certain adjustments to the resist formulation are necessary to prevent cross-linking by exposure. Cross-linking of the resist

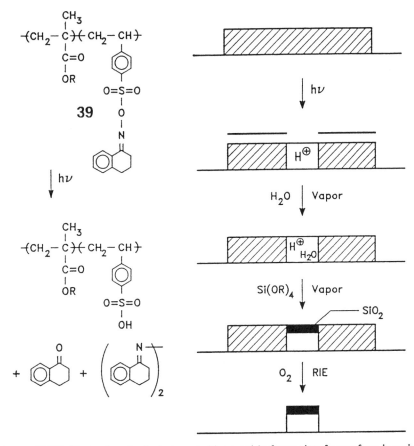

Scheme XVIII. Photochemically induced silicon oxide formation for surface imaging.

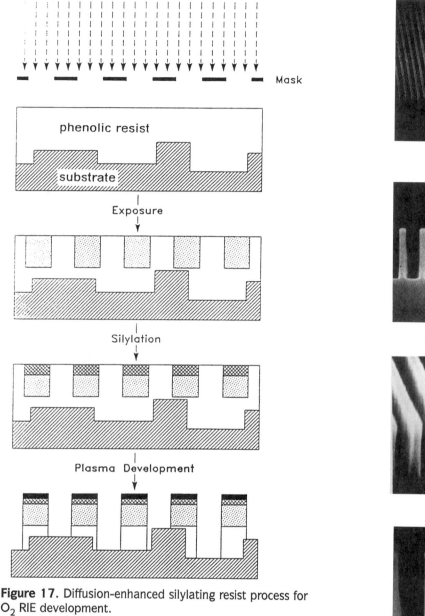

Figure 17. Diffusion-enhanced silylating resist process for O_2 RIE development.

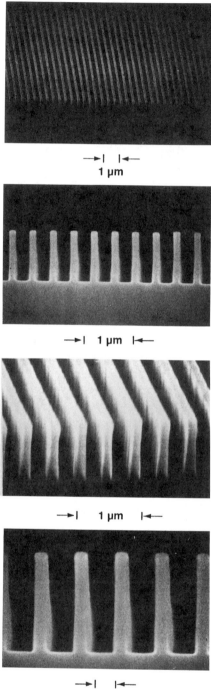

Figure 18. Positive 0.2-μm line–space patterns delineated in a phenolic polymer film by ArF excimer laser exposure followed by silylation and O_2 RIE (courtesy of M. Hartney, MIT Lincoln Laboratory).

by imagewise deep-UV exposure followed by near-UV flood exposure reverses the tone of silylation and is called "positive re-imaging by dry etching" (*145*).

Novolac-based resists were evaluated as 193-nm ArF excimer laser positive-TSI resists. Pure phenolic polymers (**17**) and *m*-cresol novolac resin undergo efficient cross-linking when irradiated with an ArF 193-nm excimer laser, even in the absence of cross-linkers, and thus can be selectively silylated in the unexposed regions for positive-tone TSI by oxygen RIE development (*146*). TSI using oxygen RIE is a very logical approach to 193-nm lithography because almost all organics (especially aromatics) absorb excessively at this ArF excimer laser wavelength. Figure 18 shows positive 0.2-μm line–space images delineated in a phe-

nolic polymer (**17**) film by ArF excimer laser exposure followed by gas-phase silylation and oxygen RIE development. The silylation process can be carried out in solution (*147*) or in the gas phase.

Phenolic resists can also be silylated after wet development (*148, 149*). In the "silicon-added bilayer resist" process (Figure 19), a diazonaphthoquinone–novolac resist is coated on top of a planarizing layer, followed by imaging in the conventional fashion by using an aqueous-base developer. The phenolic polymer remaining in the unexposed areas after development is silylated in a gas phase (*148*) or in solution (*149*) to provide oxygen RIE resistance for dry etching of the underlying layer.

Oxygen RIE is sometimes accompanied by linewidth loss due to insufficient etch resistance of the top resist and nonvertical side walls of the top resist patterns. A bilayer process called "chemical amplification of resist lines" (Si-CARL) was proposed to minimize this problem. Si-CARL involves biasing the top resist images by aqueous silylation of anhydride-containing resists (*150–152*). The Si-CARL bilayer process employs maleic anhydride copolymers that are developed with aqueous base and subsequently reacted with amine-ended silicons (Scheme XIX). The silicon-diamine diffuses into the solid polymer film during the aqueous silylation at room temperature, a strongly cross-linked network with a high silicon content is formed by reaction of amine with the anhydride groups, and oxygen RIE transfers the top resist image to the planarizing layer. Although copolymers of maleic anhydride exhibit good performance as a top resist, poor solubility (slow hydrolysis of the anhydride groups) in alkaline developers results in low sensitivity with use of a diazonaphthoquinone dissolution inhibitor. To improve the base solubility for the Si-CARL process, the maleic anhydride comonomer was partially replaced with maleimide.

The effects of silylating reagents and resin structure on the silylation kinetics and resist performance have been studied quite extensively for gas-phase silylation (*148, 153–156*).

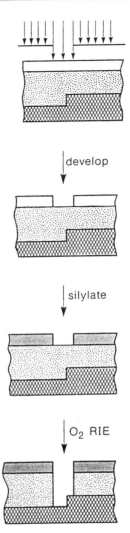

Figure 19. Silylation after wet development of phenolic resist and O_2 RIE pattern transfer.

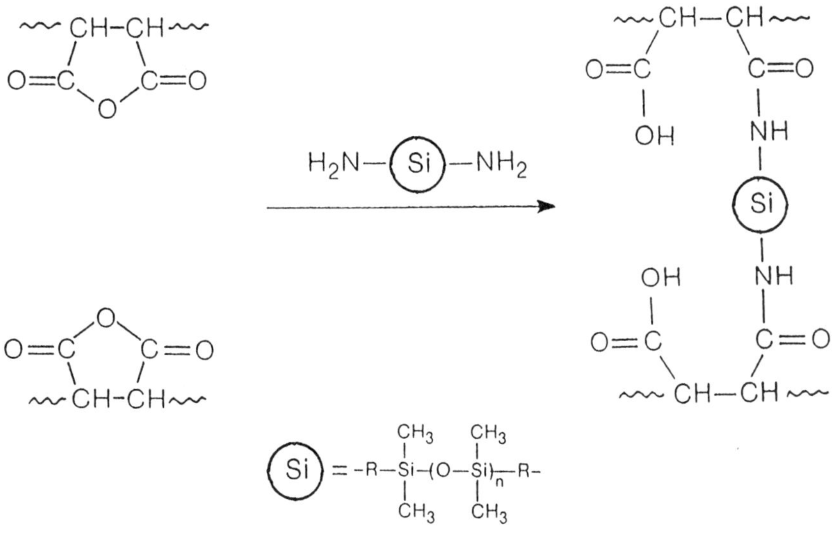

Scheme XIX. Wet development and silylation–cross-linking of maleic anhydride resist for bilayer O_2 RIE pattern transfer.

Conclusions and Future Prospects

Sensitivities in the conventional resist systems are dictated primarily by the efficiency of radiation events (quantum yields and *G* values). Some of these systems are very old but still are used in the semiconductor industry. PMMA is known to provide a very high resolution and has been used in fabrication of experimental devices. The diazonaphthoquinone–novolac resist evolved from a printing material, but it is a workhorse of the semiconductor manufacturing owing to continuous improvement. New illumination and process modifications (e.g., phase shift and top surface imaging) are pushing the resolution limit of the diazonaphthoquinone resist into increasingly smaller dimensions. Although totally new resist systems (chemical-amplification resists; *see* Chapter 2.4) have emerged for further miniaturization of semiconductor devices, some of the conventional resist systems such as the diazonaphthoquinone resist are likely to play an important role in the industry for many more years to come.

References

1. Willson, C. G. In *Introduction to Microlithography;* Thompson, L. F.; Willson, C. G.; Bowden, M. J., Eds.; ACS Symposium Series 219; American Chemical Society: Washington, DC, 1983; Chapter 3, pp 87–159.
2. Moreau, W. M. *Semiconductor Lithography, Principles, Practices, and Materials;* Plenum: New York, 1988.
3. Bowden M. J. In *Materials for Microlithograph;* Thompson, L. F.; Willson, C. G.; Fréchet, J. M. J., Eds.; ACS Symposium Series 266; American Chemical Society: Washington, DC, 1984; Chapter 3, pp 39–117.
4. Iwayanagi, T.; Ueno, T.; Nonogaki, S.; Ito, H.; Willson, C. G. In *Electronic and Photonic Applications of Polymers;* Bowden, M. J.; Turner, S. R., Eds.; ACS Advances in Chemistry Series 218; American Chemical Society: Washington, DC, 1988; Chapter 3, pp 109–224.
5. Reichmanis, E.; Thompson, L. F. In *Polymers in Microlithography: Materials and Processes;* Reichmanis, E.; MacDonald, S. A.; Iwayanagi, T., Eds.; ACS Symposium Series 412; American Chemical Society: Washington, DC, 1989; Chapter 1, pp 1–24.
6. Reichmanis, E.; Thompson, L. F. *Chem. Rev.* **1989,** *89,* 1273.
7. Ito, H. In *Radiation Curing in Polymer Science and Technology;* Fouassier, J. P.; Rabek, J. F. Eds.; Elsevier: London, 1993; Vol. 4, Chapter 11, pp 237–359.
8. Dammel, R. In *Polymers for Microelectronics: Resists and Dielectrics;* Thompson, L. F.; Willson, C. G.; Tagawa, S. Eds.; ACS Symposium Series 537; American Chemical Society: Washington, DC, 1994; pp 252–281.
9. Haller, I.; Hatzakis, M.; Srinivasan, R. *IBM J. Res. Develop.* **1968,** 251.
10. Lin, B. J. *J. Vac. Sci. Technol.* **1975,** *12,* 1317.
11. Gupta, A.; Liang, R.; Tsay, F.; Moacanin, J. *Macromolecules* **1980,** *13,* 1696.
12. Hiraoka, H. *IBM J. Res. Develop.* **1977,** *21,* 121.
13. Moore, J. A.; Choi, J. O. In *Radiation Effects on Polymers;* Clough, R. L.; Shalaby, S. W., Eds.; ACS Symposium Series 475; American Chemical Society: Washington, DC, 1991; pp 156–192.
14. Charlesby, A. *Atomic Radiation and Polymers;* Pergamon: New York, 1960.
15. Ouano, A. C. *Polym. Eng. Sci.* **1978,** *18,* 306.
16. Allen, R. D.; Long, T. E.; McGrath, J. E. *Polym. Bull.* **1986,** *15,* 127.
17. Webster, O. W.; Hertler, W. R.; Sogah, D. Y.; Farnham, W. B.; Rajanbabu, T. Y. *J. Am. Chem. Soc.* **1983,** *105,* 5706.
18. Manring, L. E. *Macromolecules* **1988,** *21,* 488; **1989,** *22,* 2673, 4652.
19. Pittman, C.; Ueda, M.; Chen, C.; Kwiatkowski, J.; Cook, C.; Helbert, J. *J. Electrochem. Soc.* **1981,** *128,* 1758.
20. Ito, H.; Miller, D. C.; Willson, C. G. *Macromolecules* **1982,** *15,* 915.
21. Mimura, H.; Ohkubo, T.; Takeuchi, T.; Sekikawa, K. *Jpn. J. Appl. Phys.* **1978,** *17,* 541.
22. Hirai, T.; Hatano, Y.; Nonogaki, S. *J. Electrochem. Soc.* **1971,** *118,* 669.
23. Yamashita, Y.; Ogura, K.; Konishi, M.; Kawazu, R.; Ohno, S.; Mizokami, Y. *J. Vac. Sci. Technol.* **1979,** *16,* 2026.
24. Wilkins, C. W., Jr.; Reichmanis, E.; Chandross E. A. *J. Electrochem. Soc.* **1980,** *127,* 2510, 2514.
25. Hartless, R. L.; Chandross, E. A. *J. Vac. Sci. Technol.* **1981,** *19,* 1333.
26. Kawamura, Y.; Toyoda, T.; Namba, S. *J. Appl. Phys.* **1982,** *53,* 6489.
27. Kaimoto, Y.; Nozaki, K.; Takechi, S.; Abe, N. *Proc. SPIE* **1992,** *1672,* 66.
28. Moreau, W.; Merritt, D.; Moyer, W.; Hatzakis, M.; Johnson, D.; Pederson, L. A. *J. Vac. Sci. Technol.* **1979,** *16(6),* 1989.
29. Pohl, K. U.; Rodriguez, F.; Namaste, Y. M. N.; Obendorf, S. K. In *Materials for Microlithography;* Thompson, L. F.; Willson, C. G.; Fréchet, J. M. J., Eds.; ACS Symposium Series 266; American Chemical Society: Washington, DC, 1984; pp 323–338.
30. Hiraoka, H. *Macromolecules* **1977,** *10,* 719.
31. de Grandpre, M. P.; Vidusek, D. A.; Legenza, M. W. *Proc. SPIE* **1985,** *539,* 103, 250.
32. Roberts, E. D. *Appl. Polym. Symp.* **1974,** *23,* 87.
33. Lai, J. H.; Helbert, J. H. *Macromolecules* **1978,** *11,* 617.
34. Harada, K.; Kogure, O.; Murase, K. *IEEE Trans. Electron Devices* **1982,** *ED-29,* 781.
35. Tsuda, M.; Oikawa, S.; Nakamura, Y.; Nagai, H.; Yokota, A.; Tsumori, T.; Nakane, Y.; Mifune, T. *Photogr. Sci. Eng.* **1979,** *23,* 290.
36. Renaldo, A. F.; Ito, H. *Synth. Commun.* **1987,** *17(15),* 1823.
37. MacDonald, S. A.; Ito, H.; Willson, C. G.; Moore, J. W.; Gharapetian, H. M.; Guillet, J. E. In *Materials for Microlithography;* Thompson, L. F.; Willson, C. G.; Fréchet, M. J. M., Eds.; ACS Symposium Series 266; American Chemical Society: Washington, D. C., 1984; pp 179–186.
38. Ito, H.; MacDonald, S. A.; Willson, C. G.; Moore, J. W.; Gharapetian, H. M.; Guillet, J. E. *Macromolecules* **1986,** *19,* 1839.
39. Ito, H.; Renaldo, A. F. *J. Polym. Sci., Part A, Polym. Chem.* **1991,** *29,* 1001.

40. Nate, K.; Kobayashi, T. *J. Electrochem. Soc.* **1981**, *128*, 1394.

41. Brown, J. R.; O'Donnell, J. H. *Macromolecules* **1970**, *3*, 265.

42. Feldman, M.; White, D. L.; Chandross, E. A.; Bowden, M. J.; Appelbaum, J. *Proceedings of the Kodak Microelectronics Seminar;* 1975; p 40.

43. Bowden, M. J.; Thompson, L. F. *Polym. Eng. Sci.* **1974**, *14*, 525.

44. Bowden, M. J.; Chandross, E. A. *J. Electrochem. Soc.* **1975**, *122*, 1370.

45. Himics, R. J.; Ross, D. L. *Polym. Eng. Sci.* **1977**, *17*, 350.

46. Sharma, V. K.; Affrossman, S.; Pethrick, R. *Polymer* **1983**, *24*, 387.

47. Ito, H.; Hrusa, C.; Hall, H. K., Jr.; Padias, A. B. *J. Polym. Sci., Part A, Polym. Chem.* **1986**, *24*, 955.

48. Sugita, K.; Ueno, N.; Funabashi, M.; Yoshida, Y.; Doi, Y.; Nagata, S.; Sasaki, S. *Polym. J.* **1985**, *17*, 1091.

49. Higashimura, Y.; Tang, B.-Z.; Masuda, T.; Yamaoka, H.; Matsuyama, T. *Polym. J.* **1985**, *17*, 393.

50. Tsuchihara, K.; Masuda, T.; Higashimura, T. *J. Polym. Sci., Part A, Polym. Chem.* **1991**, *29*, 471.

51. Masuda, T.; Sasaki, N.; Higashimura, T. *Macromolecules* **1975**, *8*, 717.

52. Gozdz, A. S.; Baker, G. L.; Klausner, C.; Bowden, M. J. *Proc. SPIE* **1987**, *771*, 18.

53. Miller, M. D.; Michl, J. *Chem. Rev.* **1989**, *89*, 1359.

54. Nate, K.; Sugiyama, H.; Inoue, T.; Ishikawa, M. *Extended Abstracts of Papers,* 1984 Electrochemical Society Fall Meeting; Electrochemical Society: Pennington, NJ, 1984; Abstract 530.

55. West, R. In *The Chemistry of Organic Silicon Compounds;* John Wiley & Sons: New York, 1989; Part II, Chapter 19.

56. (a) Cypryk, M.; Gupta, Y.; Matyjaszewski, K. *J. Am. Chem. Soc.* **1991**, *113*, 1046. (b) Sakamoto, K.; Obata, K.; Hirata, M.; Nakajima, M.; Sakurai, H. *J. Am. Chem. Soc.* **1989**, *111*, 7641.

57. Ito, H.; Willson, C. G. *Polym. Eng. Sci.* **1983**, *23*, 1012.

58. Hatada, K.; Kitayama, T.; Danjo, S.; Yuki, H.; Aritome, H.; Namba, S.; Nate, K.; Yokono, H. *Polym. Bull.* **1982**, *8*, 469.

59. Marinero, E. E.; Miller, R. D. *Appl. Phys. Lett.* **1987**, *62*, 1394.

60. Zeigler, J. M.; Harrah, L. A.; Johnson, A. W. *Proc. SPIE* **1985**, *539*, 16.

61. Srinivasan, R.; Mayne-Banton, V. *Appl. Phys. Lett.* **1982**, *41*, 576.

62. Hatada, K.; Kitayama, T.; Danjo, S.; Tsubokura, Y.; Yuki, H.; Morikawa, K.; Aritome, H.; Namba, S. *Polym. Bull.* **1983**, *10*, 45.

63. Thompson, L. F.; Feit, E.; Heidenreich, R. D. *Polym. Eng. Sci.* **1974**, *14*, 529.

64. Tamamura, T. In *Polymers in Electronics;* Davidson, T., Ed.; ACS Symposium Series 242; American Chemical Society: Washington, DC, 1984; pp 103–118.

65. Kasai, P. H. *J. Am. Chem. Soc.* **1991**, *113*, 1539.

66. Ohnishi, Y.; Itoh, M.; Mizuko, K.; Gokan, H.; Fujiwara, S. *J. Vac. Sci. Technol.* **1981**, *19*, 1141.

67. Imamura, S. *J. Electrochem. Soc.* **1979**, *126*, 1628.

68. Liutkus, J.; Hatzakis, M.; Shaw, J.; Paraszczak, J. *Polym. Eng. Sci.* **1983**, *23*, 1047.

69. Feit, E. D.; Thompson, L. F.; Wilkins, C. W., Jr.; Eurtz, M. E.; Doerries, E. M.; Stillwagon, L. E. *J. Vac. Sci. Technol.* **1979**, *16*, 1997.

70. Sukegawa, K.; Sugawara, S. *Jpn. J. Appl. Phys.* **1981**, *20*, L583.

71. Harita, Y.; Kamoshida, Y.; Tsutsumi, K.; Koshiba, M.; Yoshimoto, H.; Harada, K. *SPSE 22nd Symposium on Unconventional Imaging Science and Technology;* 1982; p 34.

72. Thompson, L. F.; Stillwagon, L. E.; Doerries, E. M. *J. Vac. Sci. Technol.* **1978**, *15*, 938.

73. Schwalm, R.; Bottcher, A.; Koch, H. *Proc. SPIE* **1988**, *920*, 21.

74. Fréchet, J. M. J.; Tessier, T. G.; Willson, C. G.; Ito, H. *Macromolecules* **1985**, *18*, 317.

75. Reiser, A. *Photoreactive Polymers, The Science and Technology of Resists;* John Wiley & Sons: New York, 1989.

76. Iwayanagi, T.; Kohashi, T.; Nonogaki, S. *J. Electrochem. Soc.* **1980**, *127*, 2759.

77. Iwayanagi, T.; Kohashi, T.; Nonogaki, S.; Matsuzawa, T.; Douta, K.; Yanazawa, H. *IEEE Trans. Electron Devices* **1981**, *ED-28*, 1306.

78. Nonogaki, S.; Hashimoto, M.; Iwayanagi, T.; Shiraishi, H. *Proc. SPIE* **1985**, *539*, 189.

79. Toriumi, M.; Ueno, T.; Iwayanagi, T.; Hashimoto, M.; Moriuchi, N.; Shirai, S. *Proc. SPIE* **1988**, *920*, 27.

80. Iwayanagi, T.; Hashimoto, M.; Nonogaki, S.; Koibuchi, S.; Makino, D. *Polym. Eng. Sci.* **1983**, *23*, 935.

81. Niki, H.; Kobayashi, Y.; Onishi, Y. *J. Photopolym. Sci. Technol.* **1991**, *4*, 517.

82. Kobayashi, Y.; Onishi, Y.; Niki, H. *J. Photopolym. Sci. Technol.* **1990**, *3*, 215.

83. Ito, T.; Sakata, M.; Yamashita, Y. *J. Photopolym. Sci. Technol.* **1991**, *4*, 403.

84. Nakamura, K.; Sekimoto, N.; Yamada, T.; Tomascewski, G. *J. Photopolym. Sci. Technol.* **1991**, *4*, 415.

85. Hofer, D.; Kaufman, F. B.; Kramer, S. R.; Aviram, A. *Appl. Phys. Lett.* **1980**, *37*, 314.

86. Pacansky, J.; Lyerla, J. R. *IBM J. Res. Develop.* **1979**, *23*, 42.

87. Hanabata, M.; Furuta, A.; Uemura, Y. *Proc. SPIE* **1986**, *631*, 76.

88. Furuta, A.; Hanabata, M.; Uemura, Y. *J. Vac. Sci. Technol.* **1986**, *B4*, 430.

89. Templeton, M. K.; Szmanda, R.; Zampini, A. *Proc. SPIE* **1987**, *771*, 136.

90. Hanabata, M.; Furuta, A.; Uemura, Y. *Proc. SPIE* **1987**, *771*, 85.

91. Hanabata, M.; Uetani, Y.; Furuta, A. *Proc. SPIE* **1988**, *920*, 349.

92. Furuta, A.; Hanabata, M. *J. Photopolym. Sci. Technol.* **1989**, *2*, 383.

93. Urano, Y.; Miyazaki, M.; Katsuya, K.; Kikuchi, H. *J. Photopolym. Sci. Technol.* **1993**, *6*, 53.

94. Willson, C. G.; Miller, R.; McKean, D.; Clecak, N.; Thompskins, T.; Hofer, D.; Michl, J.; Downing, J. *Polym. Eng. Sci.* **1983**, *23*, 1004.

95. Ishii, W.; Kokubo, T.; Uenishi, K.; Tan, S. *J. Photopolym. Sci. Technol.* **1991**, *4*, 527.

96. Grant, B. D.; Clecak, N. J.; Twieg, R. J.; Willson, C. G. *IEEE Trans. Electron Devices* **1981**, *ED-28*, 1300.

97. Willson, C. G.; Miller, R. D.; McKean, D. R.; Pederson, L. A. *Proc. SPIE* **1987**, *771*, 2.

98. Schwartzkopf, G. *Proc. SPIE* **1988**, *920*, 51.

99. Tani, Y.; Endo, M.; Sasago, M.; Ogawa, K. *Proc. SPIE* **1989**, *1086*, 22.

100. Sugiyama, H.; Ebata, K.; Mizushima, A.; Nate, K. *Technical Papers of SPE Regional Technical Conference on Photopolymers*; Society of Plastics Engineers: Brookfield, CT, 1988; p 51.

101. Reichmanis, E.; Wilkins, C. W., Jr.; Chandross, E. A. *J. Vac. Sci. Technol.* **1980**, *19*, 1338.

102. Ito, H.; Flores, E. *J. Electrochem. Soc.* **1988**, *135*, 2322.

103. Endo, M.; Tani, Y.; Sasago, M.; Nomura, N.; Das, S. *Proc. ACS Div. Polym. Mater. Sci. Eng.* **1989**, *61*, 199.

104. Bowden, M. J.; Thompson, L. F.; Farenholtz, S. R.; Doerries, E. M. *J. Electrochem. Soc.* **1981**, *128*, 1304.

105. Ito, H.; Pederson, L. A.; MacDonald, S. A.; Cheng, Y.-Y.; Lyerla, J. R.; Willson, C. G. *J. Electrochem. Soc.* **1988**, *135*, 1504.

106. Shiraishi, H.; Isobe, A.; Murai, F.; Nonogaki, S. In *Polymers in Electronics*; Davidson, T., Ed.; ACS Symposium Series 242; American Chemical Society: Washington, DC, 1984; p 167.

107. Lin, B. J. In *Introduction to Microlithography*; Thompson, L. F.; Willson, C. G.; Bowden, M. J., Eds.; ACS Symposium Series 219; American Chemical Society: Washington, DC, 1983; pp 287–350.

108. Shaw, J. M.; Hatzakis, M.; Paraszczak, J.; Liutkus, J.; Babich, E. *Polym. Eng. Sci.* **1983**, *23*, 1054.

109. Tanaka, A.; Morita, M.; Imamura, S.; Tamamura, T.; Kogure, O. In *Materials for Microlithography*; Thompson, L. F.; Willson, C. G.; Fréchet, J. M. J., Eds.; ACS Symposium Series 266; American Chemical Society: Washington, DC, 1984; p 293.

110. Tanaka, A.; Morita, M.; Kogure, O. *1st International Polymer Conference of the Society of Polymer Science, Japan*; Society of Polymer Science, Japan: Kyoto, Japan, 1984; p 110.

111. Hayase, S.; Horiguchi, R.; Onishi, Y.; Ushirogouchi, T. In *Polymers in Microlithography*; Reichmanis, E.; MacDonald, S. A.; Iwayanagi, T., Eds.; ACS Symposium Series 412; American Chemical Society: Washington, DC, 1989; p 133.

112. Tanaka, A.; Ban, H.; Imamura, S. In *Polymers in Microlithography*; Reichmanis, E.; MacDonald, S. A.; Iwayanagi, T., Eds.; ACS Symposium Series 412; American Chemical Society: Washington, DC, 1989; p 173.

113. Watanabe, K.; Yano, E.; Namiki, T.; Fukuda, M.; Yoneda, Y. *J. Photopolym. Sci. Technol.* **1991**, *4*, 481.

114. Yamazaki, S.; Ishida, S.; Matsumoto, H.; Aizaki, N.; Muramoto, N.; Mine, K. *Proc. SPIE* **1991**, *1466*, 538.

115. Miller, R. D.; Hofer, D.; McKean, D. R.; Willson, C. G.; West, R.; Trefonas, P. T., III, In *Polymers in Microlithography*; Thompson, L. F.; Willson, C. G.; Fréchet, J. M. J., Eds.; ACS Symposium Series 266; American Chemical Society: Washington, DC, 1984; p 293.

116. Miller, R. D.; Rabolt, J. F.; Sooriyakumaran, R.; Fleming, W.; Fickes, G. N.; Farmer, B. L.; Kuzmany, H. In *Inorganic and Organometallic Polymers*; Zeldin, M.; Wynne, K. J.; Allcock, H. R., Eds.; ACS Symposium Series 360; American Chemical Society: Washington, DC, 1988; p 43.

117. Miller, R. D.; Willson, C. G.; Wallraff, G. M.; Clecak, N.; Sooriyakumaran, R.; Michl, J.; Karatsu, T.; McKinley, A. J.; Klingensmith, K. A.; Downing, J. *Polym. Eng. Sci.* **1989**, *29*, 882.

118. Wallraff, G. M.; Miller, R. D.; Clecak, N. J.; Baier, M. *Proc. SPIE* **1991**, *1466*, 211.

119. Bianconi, P. A.; Schilling, F. C.; Weidman, T. W. *Macromolecules* **1989**, *22*, 1697.

120. Kunz, R. R.; Bianconi, P. A.; Horn, M. W.; Paladugu, R. P.; Shaver, D. C.; Smith, D. A.; Freed, C. A. *Proc. SPIE* **1991**, *1466*, 218.

121. MacDonald, S. A.; Ito, H.; Willson, C. G. *Microelectronic Eng.* **1983**, *1*, 269.

122. Suzuki, M.; Saigo, K.; Gokan, H.; Ohnishi, Y. *J. Electrochem. Soc.* **1983**, *130*, 1962.

123. Reichmanis, E.; Smolinsky, G. *Proc. SPIE* **1984**, *469*, 38.

124. Novembre, A. E.; Reichmanis, E.; Davis, M. *Proc. SPIE* **1986**, *631*, 14.

125. Gozdz, A. S.; Carnazza, C.; Bowden, M. J. *Proc. SPIE* **1986**, *631*, 2.

126. Gozdz, A. S.; Shelburne, J. A., III *Proc. SPIE* **1991**, *1466*, 520.

127. Gozdz, A. S.; Ono, H.; Ito, S.; Shelburne, J. A., III; Matsuda, M. *Proc. SPIE* **1991**, *1466*, 200.

128. DeSimone, J. M.; York, G. A.; McGrath, J. E.; Gozdz, A. S.; Bowden, M. J. *Macromolecules* **1991**, *24*, 5330.

129. Tarascon, R. G.; Shugard, A.; Reichmanis, E. *Proc. SPIE* **1986**, *631*, 40.

130. Wilkins, C. W., Jr.; Reichmanis, E.; Wolf, T. M.; Smith, B. C. *J. Vac. Sci. Technol.* **1984**, *3*, 306.

131. Saotome, Y.; Gokan, H.; Saigo, K.; Suzuki, M.; Ohnishi, Y. *J. Electrochem. Soc.* **1985**, *132*, 909.

132. Smith, J. N.; Highs, H. G.; Keller, J. V.; Goodner, W. R.; Wood, T. E. *Semicond. Int.* **1979**, *2(10)*, 41.

133. Taylor, G. N.; Wolf, T. M. *J. Electrochem. Soc.* **1980**, *127*, 2665.

134. Taylor, G. N. *Solid State Technol.* **1989**, *23(5)*, 73.

135. Taylor, G. N.; Wolf, T. M.; Goldrick, M. R. *J. Electrochem. Soc.* **1981**, *128*, 361.

136. Meyer, W. H.; Curtis, B. J.; Brunner, H. R. *Microelectronic Eng.* **1983**, *1*, 29.

137. Taylor, G. N.; Stillwagon, L. E.; Venkatesan, T. *J. Electrochem. Soc.* **1984**, *131*, 1664.

138. Stillwagon, L. E.; Silverman, P. J.; Taylor, G. N. *Technical Papers of SPE Regional Technical Conference on Photopolymers*; Society of Plastics Engineers, Ridgefield, CT, 1985; p 87.

139. Nalamasu, O.; Baiocchi, F. A.; Taylor, G. N. In *Polymers in Microlithography*; Reichmanis, E.; MacDonald, S. A.; Iwayanagi, T., Eds; ACS Symposium Series 412; American Chemical Society: Washington, DC, 1989; p 189.

140. Shirai, M.; Tsunooka, M. In *Polymers for Microelectronics*; Thompson, L. F.; Willson, C. G.; Tagawa, S., Eds.; ACS Symposium Series 537; American Chemical Society: Washington, DC, 1994; p 180.

141. Coopmans, F.; Roland, B. *Proc. SPIE* **1986**, *631*, 34.

142. Lombaerts, R.; Roland, B.; Goethals, A. M.; Van den hove, L. *Proc. SPIE* **1990**, *1262*, 312.

143. Garza, C. M.; Catlett, D. L., Jr.; Jackson, R. A. *Proc. SPIE* **1991**, *1466*, 616.

144. Goethals, A. M.; Baik, K. H.; Van den hove, L.; Tedesco, S. *Proc. SPIE* **1991**, *1466*, 604.

145. Pierrat, C.; Tedesco, S.; Vinet, F.; Lerme, M.; Dal'Zotto, B. *J. Vac. Sci. Technol.* **1989**, *B7(6)*, 1782.

146. Hartney, M. A.; Rothschild, M.; Kunz, R. R.; Ehrlich, D. J.; Shaver, D. C. *J. Vac. Sci. Technol.* **1990**, *B8*, 1476.

147. Yang, B.-J. L.; Yang, M.; Chiong, K. N. *J. Vac. Sci. Technol.* **1989**, *B7(6)*, 1729.

148. McColgin, W.; Daly, R. C.; Jech, J., Jr.; Brust, T. B. *Proc. SPIE* **1988**, *920*, 260.

149. Shaw, J. M.; Hatzakis, M.; Babich, E. D.; Paraszczak, J. R.; Whitman, D. F.; Stewart, K. J. *J. Vac. Sci. Technol.* **1989**, *B7(6)*, 1709.

150. Sebald, M.; Sezi, R.; Leuschner, R.; Ahne, H.; Birkle, S. *Microelectronic Eng.* **1990**, *11*, 531.

151. Sebald, M.; Leuschner, R.; Sezi, R.; Ahne, H.; Birkle, S. *Proc. SPIE* **1990**, *1262*, 528.

152. Sebald, M.; Berthold, J.; Beyer, M.; Leuschner, R.; Nölscher, Ch.; Scheler, U.; Sezi, R.; Ahne, H.; Birkle, S. *Proc. SPIE* **1991**, *1466*, 227.

153. Brust, T. B.; Turner, S. R. *Proc. SPIE* **1987**, *771*, 102.

154. Spence, C. A.; Shacham-Diamond, Y. *Proc. SPIE* **1989**, *1086*, 207.

155. Spence, C. A.; MacDonald, S. A.; Schlosser, H. *Proc. SPIE* **1990**, *1262*, 344.

156. Dao, T. T.; Spence, C. A.; Hess, D. W. *Proc. SPIE* **1991**, *1466*, 257.

Functional Polymers for Microlithography: Chemically Amplified Imaging Systems

Hiroshi Ito

Chemical-amplification resist systems for microlithography are reviewed in this chapter, and a focus is placed on their imaging chemistries. Chemically amplified systems discussed include deprotection of polymer-bound functional groups, depolymerization, rearrangement, condensation, dehydration, and polymerization. These resist systems based on radiation-induced acid catalysis offer high sensitivity, high contrast, high resolution, and versatility. Currently, these systems are studied very heavily to support the emerging short-wavelength lithographic technologies. Several new imaging techniques such as bilayer, silylation, and all-dry imaging processes are also described.

Among the many criteria involved in the design of resist materials, resist sensitivity plays a key role. This characteristic is especially true in the case of high-resolution short-wavelength lithographic technologies such as deep-UV (<300 nm), electron-beam (e-beam), and X-ray exposure techniques. The quantum yield characterizes the efficiency of photochemical transformations and is expressed as the number of molecules transformed per photon absorbed. The quantum yield of typical diazonaphthoquinones is ca. 0.2–0.3; that is, three to five photons are required to transform a single molecule of the photoactive compound. This requirement places a fundamental limit on the photosensitivity of such systems. Increasing the quantum yield to its upper limit of unity would only result in a 3–5-fold increase in sensitivity. However, orders of magnitude of sensitivity enhancement are required to make deep-UV, e-beam, and X-ray lithography as efficient and economical as current factory exposure systems (G and I line steppers operating at 436 and 365 nm, respectively) in terms of unit area exposed per unit time.

To overcome this intrinsic sensitivity limitation imposed by quantum yield for resist systems that consume at least one photon for every productive chemical transformation, the concept of "chemical amplification" was proposed in 1982 (*1, 2*). In chemically amplified resist systems, a single photochemical (or radiation event) induces a cascade of subsequent chemical transformations and provides a gain mechanism toward the desired photochemical process.

The chemical-amplification concept is based on the generation (by irradiation) of active species that catalyze many subsequent chemical transformations (A → B) in the resist film without being consumed in one reaction (Scheme I). In general, these chemical transformations are accomplished by heating the exposed resists (postexposure bake or postbake) (*see* Figure 1 in Chapter 2.3). In principle, the active species could be either ionic or radical, but the use of photochemical acid generators (PAG) (Scheme I) [originally proposed for chemical amplification (*1, 2*)] has become the basis of an entire family of advanced resist systems (*3–8*). Various photochemical acid generators have

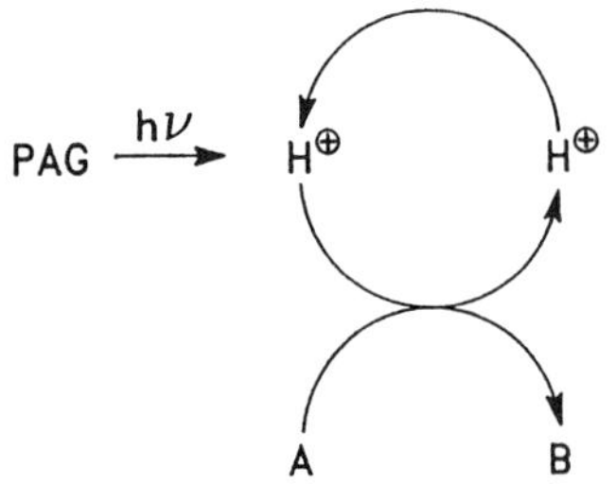

Scheme I. Schematic illustration of chemical-amplification resist (radiation-induced acid generation followed by acid-catalyzed conversion of structure A to structure B).

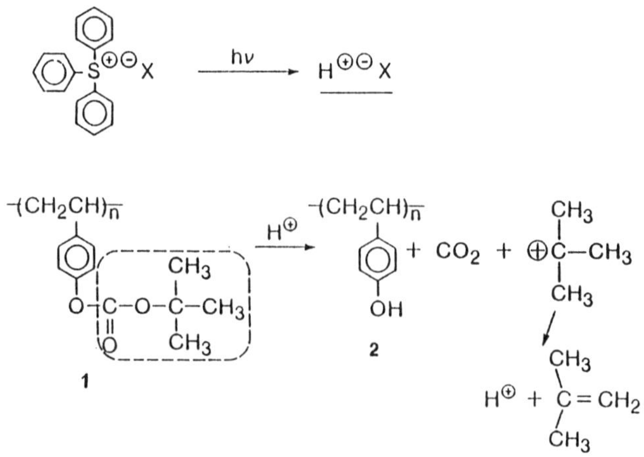

Scheme II. tBOC resist imaging mechanism (acid-catalyzed deprotection).

been studied for chemical-amplification resists (*8*, *9*). The choice of appropriate acid generators for resist formulations depends on a number of factors, including the nature of radiation, solubility, thermal stability, toxicity, and the strength and the size of the acid generated. A comprehensive review on chemical-amplification resist systems for deep-UV lithography was published recently (*8*).

In terms of lithographic processes, chemically amplified resist systems can be divided into three categories in a fashion similar to the conventional resist systems discussed in Chapter 2.3:

- positive and negative systems
- UV, e-beam, or X-ray resist systems
- two-component or three-component systems

The one-component design is not common. It is more convenient to categorize the chemical-amplification resists on the basis of imaging chemistries. In terms of chemistry, chemically amplified lithographic imaging may involve deprotection, depolymerization, rearrangement, polymerization, condensation, dehydration, etc. The topics to be discussed in this chapter include chemically amplified resist systems, their imaging chemistries, and their applications to the bilayer lithography and to top surface imaging (TSI).

Deprotection

A_{AL}-1 Acidolysis of Carbonates and Esters

Carbonates

Acid-catalyzed deprotection of polymer-bound phenyl carbonates generates base-soluble phenol and has attracted a great deal of attention. One of the first chemical-amplification resists was based on acid-catalyzed deprotection of poly(4-*t*-butoxycarbonyloxy-styrene) (**1**), in which the phenolic group of poly(4-hydroxystyrene) (**2**) is protected by acid-labile *t*-butoxycarbonyl (tBOC) group (*1*, *4*). This lipophilic carbonate polymer is transformed upon postexposure

bake at ca. 100 °C to **2** by reaction with photochemically generated acid, and carbon dioxide, isobutene, and a proton are liberated (Scheme II). The reaction is typical A_{AL}-1 acidolysis and does not require a stoichiometric amount of water. The photochemically generated acid is not consumed in one deprotection reaction, but it is regenerated to cleave many more tBOC groups to provide the desired chemical-amplification effect. The IR spectra presented in Figure 1 indicate that the film of **1** containing a small amount of PAG is relatively unchanged when exposed to UV radiation and that quantitative conversion of **1** to **2** occurs upon postbake, as evident from disappearance of the carbonate carbonyl absorption at 1755 cm^{-1} and appearance of a broad phenolic OH absorption at ca. 3600 cm^{-1}. Thus, the postbake process plays a critical role in most chemical-amplification resist systems.

The deprotection shown in Scheme II provides a large change in the polarity (and hence the solubility) of the polymer and allows dual-tone imaging depending on the polarity of the developer solvent. The lipophilic carbonate polymer (**1**) remaining in the unexposed areas is soluble in nonpolar organic solvents, which do not dissolve the phenolic polymer (**2**) produced in the exposed regions. Thus, the use of a nonpolar organic solvent such as anisole generates negative images of the mask (Figure 2). Conversely, polar solvents such as alcohol or aqueous base readily dissolve **2** but not **1** and provide positive-tone images (Figure 2). This negative imaging with an organic developer is devoid of swelling, in contrast to cross-linking negative resists, which suffer from swelling during development in organic solvents.

The positive-tone imaging of the tBOC resist with an aqueous alkaline developer has motivated a great deal of research efforts to design deep-UV resist materials in attempts to replace the novolac–diazoquinone resists, which do not function adequately in the deep-

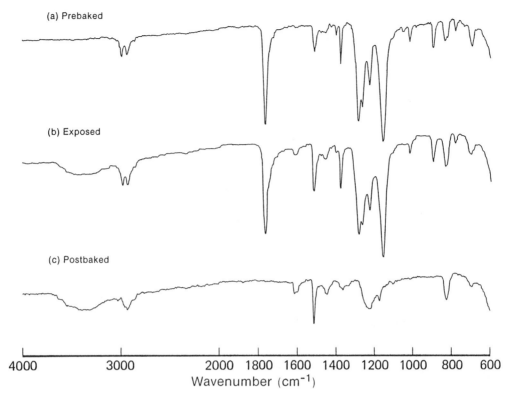

(a) Prebaked

(b) Exposed

(c) Postbaked

Wavenumber (cm⁻¹)

Figure 1. IR spectra of tBOC resist (1 + PAG) measured after prebake (a), UV exposure (b), and postbake (c). (Reproduced from reference 3. Copyright 1984 American Chemical Society.)

UV lithography (*see* Chapter 2.3). Both **1** and **2** are highly transparent at the exposing wavelength of 248 nm (KrF excimer laser) and have optical densities (OD) of 0.1 and 0.2/μm, respectively, as opposed to the highly absorbing novolac resins. Polymer **2** has a much higher glass transition temperature (T_g) than novolac resins (170–180 °C vs. 90–100 °C), and the tBOC resist is as stable in fluorocarbon plasma as the novolac–diazoquinone resists.

The tBOC resist consisting of **1** and 4.75 wt% of triphenylsulfonium hexafluoroantimonate PAG was the very first deep-UV resist used in manufacturing. Millions of 1-Mbit *dynamic random access memory* (DRAM) devices were produced at IBM on Perkin Elmer mirror-projection scanners with a numerical aperture (NA) of 0.17 in the deep-UV mode and with an excellent throughput of 100 wafers per hour (*10*).

The tBOC resist is also sensitive to e-beam (*11*) and X-ray (*12*) radiations and has shown 300-Å resolution by the use of a special e-beam system with a very small beam diameter (*13*). This phenomenon indicates that the resolution loss associated with the amplification mechanism is surprisingly small. The catalytic chain length of the tBOC resist system is estimated to be ca. 1000 under the normal lithographic conditions (*14*).

Polymer **1** can be prepared readily by radical or cationic (in liquid sulfur dioxide) polymerization of 4-*t*-butoxycarbonyloxystyrene (*15*). The polymer can

also be obtained by treating **2** with di-*t*-butyl dicarbonate in the presence of base (*16*). Monodispersed **1** was synthesized by living anionic polymerization of 4-*t*-butyl(dimethyl)silyloxystyrene, followed by desilylation with HCl and reaction with di-*t*-butyl dicarbonate (*17*). The tBOC polymer (**1**) obtained by radical polymerization has a T_g of ca. 135 °C. The polymer undergoes spontaneous thermolysis at ca. 190 °C to produce the corresponding phenolic polymer (**2**), carbon dioxide, and isobutene (Scheme II) (*15*).

The tBOC group was used to protect aqueous-base soluble polymers with high T_g such as poly-[styrene-*co*-N-(4-hydroxyphenyl) maleimide] (*18*, *19*), poly(substituted-styrene-*co*-maleimide) (*20*, *21*) poly(4-hydroxystyrene sulfone) (*22*), and poly(4-hydroxy-α-methylstyrene) (*23*, *24*) for designing thermally stable resist systems. These resist systems do not undergo image distortion due to thermal flow during high temperature lithographic processes. The tBOC group was also used for the protection of OH groups in novolac resins for resist formulation (*25*, *26*), and significant reduction in the UV absorption of the novolac resin in the deep-UV region was reported (*26*).

Poly(4-*t*-butoxycarbonyloxy-α-methylstyrene) (*23*) and poly(4-*t*-butoxycarbonyloxystyrene sulfone) (*22*) were thought to undergo main-chain degradation as well upon UV irradiation in the presence or absence of acid generators, which were assumed to assist the pos-

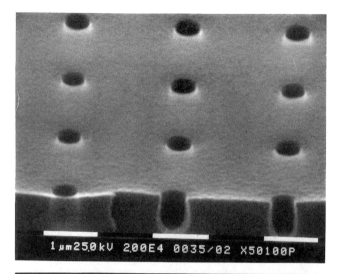

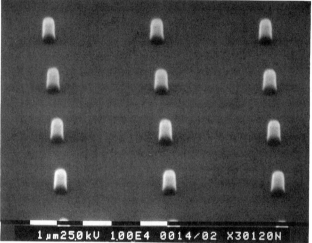

Figure 2. Positive (top) and negative (bottom) images printed in tBOC resist by synchrotron X-ray exposure. (Reproduced with permission from reference 12. Copyright 1988 American Vacuum Society.)

itive imaging (22, 24). In fact, cationically obtained poly(4-hydroxy-α-methylstyrene) and its tBOC-protected derivative undergo efficient acid-catalyzed depolymerization (27–29) (described later). Although its sensitivity to deep-UV is poor, in poly(4-t-butoxycarbonyloxystyrene sulfone) (3), X-ray-induced C–S bond scission presumably leads to generation of either sulfinic or sulfonic acid end groups, which subsequently catalyze the deprotection reaction (Scheme III) (30).

In more recent applications (31, 32), the tBOC group was employed to protect poly(hydroxyphenyl methacrylate)s, poly(N-hydroxyphenylmethacrylamide)s, and related copolymers. In contrast to 4-hydroxystyrene, the hydroxyphenyl methacrylate monomers reportedly undergo radical polymerization smoothly without protection of the phenolic hydroxy group. However, these unprotected homopolymers are very much opaque in the deep-UV region, even

though the OD of poly(4-hydroxyphenyl methacrylate) is somewhat smaller (0.48/µm) than those of novolac resins. The protection with the tBOC group reduces the UV absorption of poly(hydroxyphenyl methacrylate)s to OD = 0.18–0.29/µm, but it fails to improve the deep-UV absorption characteristics of the corresponding polymethacrylamides. Copolymers of tBOC-protected hydroxyphenyl methacrylate with methacrylic acid or maleimide were evaluated as KrF excimer laser (248 nm) positive resists (32).

Another interesting aqueous-base soluble styrenic polymer, poly[4-(2-hydroxyhexafluoroisopropyl)styrene] (4, Scheme IV) was protected with the tBOC group for chemically amplified lithographic imaging (33). The tBOC-protected polymer can be prepared by radical polymerization of a protected monomer or by reacting the fluoro-alcohol polymer (4) with di-t-butyl dicarbonate. The base-soluble fluoro-alcohol polymer (4) can also be prepared by radical polymerization of the corresponding monomer or by treating polystyrene with hexafluoroacetone in the presence of a Lewis acid catalyst (33).

The t-butyl group in tBOC has been replaced by isopropyl, substituted α-methylbenzyl, or 1-(2-tetrahydrofurfuryl)ethyl groups for greater thermal deprotection temperatures, but some of the resulting polymers may have lower sensitivity toward acidolysis (34).

Esters and Ethers

Polymers with ester functionalities that undergo A_{AL}-1 acidolysis in the absence of a stoichiometric amount of water are also highly useful in the design of dual-tone resist systems. Typical examples of such polymers are poly(4-vinylbenzoate)s (5) (11, 35) and polymethacrylates (7) (36, 37), which are converted by heating or by acidolysis to poly(vinylbenzoic acid) (6) and poly(methacrylic acid) (8), respectively (Scheme V). The ester groups in these polymers must be selected so that heterolysis of the C–O bond generates relatively stable carbocations that undergo spontaneous β-proton elimination to form olefins. For a given acid, the deprotection temperature is a good measure of the resist sensitivity of the chemically amplified systems (35, 37). The conversion of carboxylic ester to carboxylic acid provides a greater polarity or solubility change than that of phenyl carbonate to phenol and allows positive or negative imaging at lower degrees of deprotection (11, 36).

Polymers 5 and 6 are very opaque below 300 nm (OD = 1.1/µm and 3.4/µm at 248 nm, respectively), a characteristic that precludes their use as single-layer deep-UV resist materials (11, 35). However, poly(t-butyl 4-vinylbenzoate) containing a sulfonium salt PAG is a sensitive e-beam resist (11) and is attractive as a mid- or near-UV resist material (35) with a high thermal stability (T_g of 6 is 250 °C). Poly(4-vinylbenzoate)s

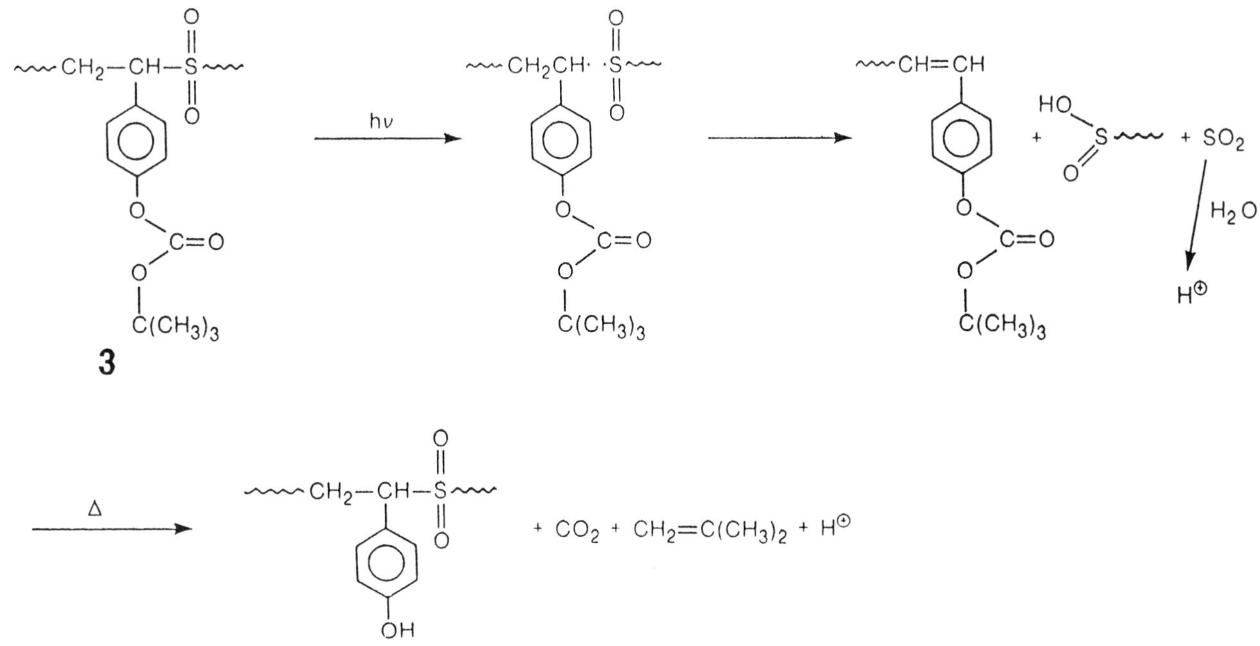

Scheme III. One-component X-ray positive resist based on tBOC deprotection.

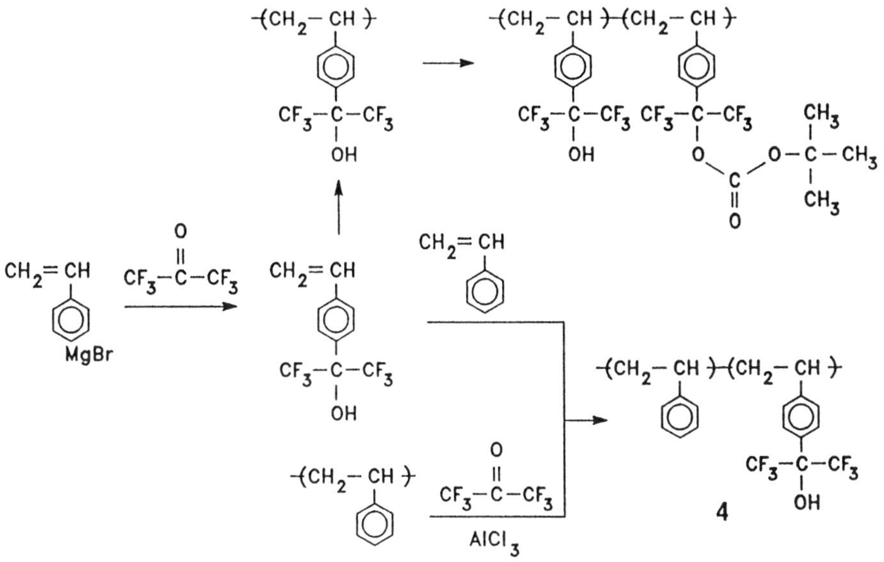

Scheme IV. Preparation of base-soluble polystyrene with a pendant hexafluoroisopropyl alcohol.

(*38*) and polymethacrylates (*37*) are typically prepared by radical polymerization; however, methacrylates also undergo anionic polymerization (sometimes living), and *t*-butyl 4-vinylbenzoate is compatible with living anionic polymerization (*39*).

The phenolic polymer (**2**) also has been protected by a *t*-butoxycarbonylmethyl group (**9**), which is converted during reaction with photochemically generated acid to a carboxylic acid (**10**) instead of phenol (Scheme VI) (*4,40*). The *t*-butyl protected polymer (**11**) has a T_g of 100–110 °C and undergoes thermal deprotection above 300 °C (*41*).

Hydrolysis

The commercialization of KrF excimer laser (248 nm) deep-UV steppers and the elegance of the tBOC deprotection chemistry have motivated considerable research efforts in the design of aqueous-base developable positive resists based on protected phenolic polymer (**2**). Hydrolysis of trimethylsilyl ether (*42, 43*) and alcoholysis of tetrahydropyranyl ether (*44, 45*) were used in the design of chemical-amplification resists based on phenolic polymers of type **2** (**12** and **13**, respectively; Scheme VII) or novolac.

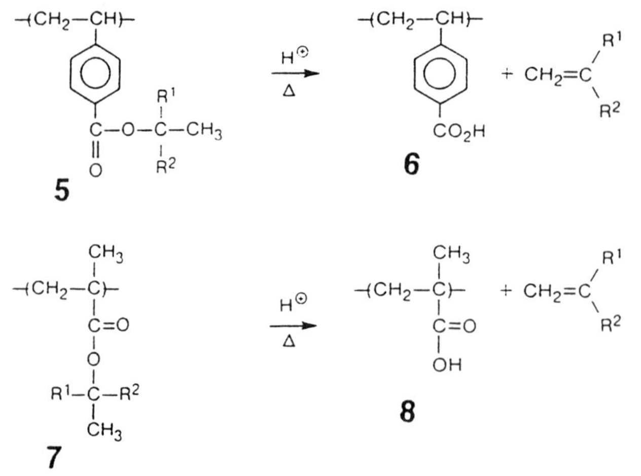

Scheme V. A_{AL}-1 acidolysis of poly(4-vinylbenzoate)s and polymethacrylates.

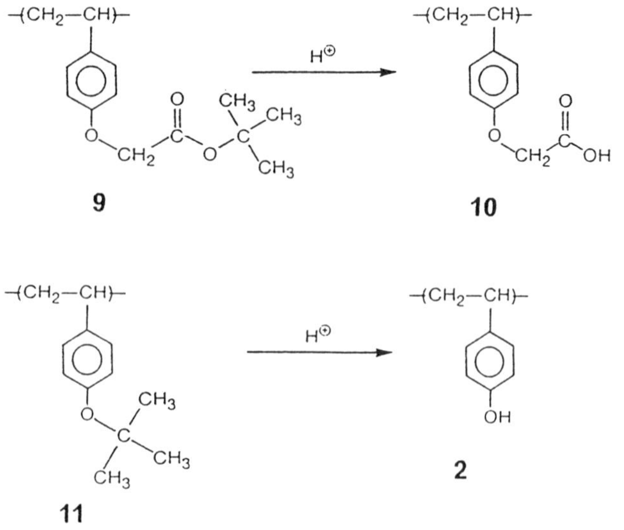

Scheme VI. Acidolysis of protected poly(4-hydroxy-styrene)s.

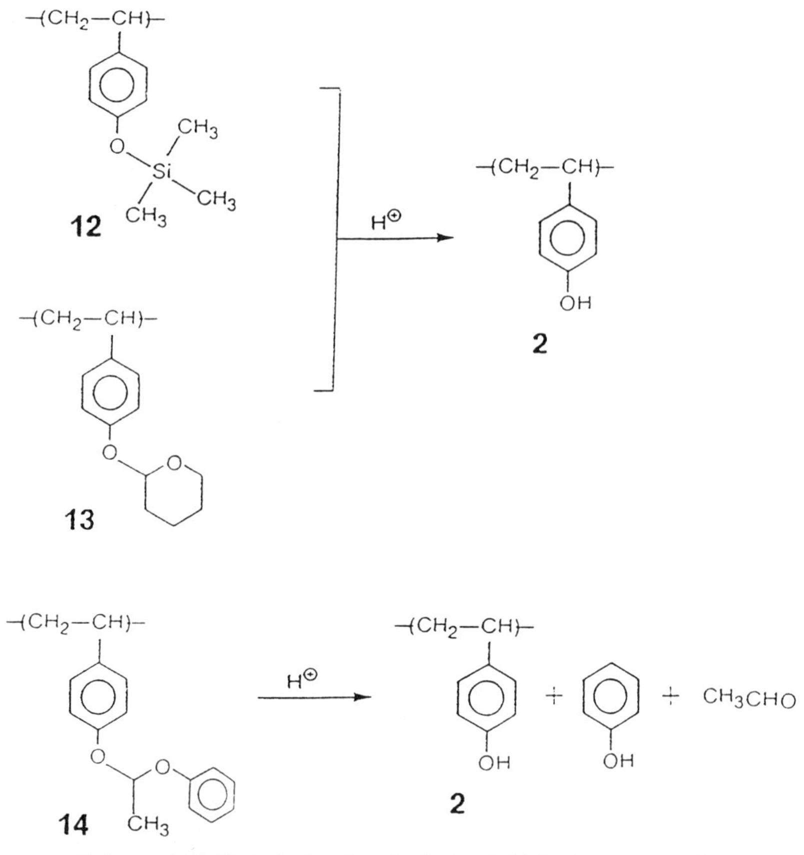

Scheme VII. Hydrolysis of protected poly(4-hydroxystyrene)s.

Acetal-protected **2** was also prepared by radical polymerization of the monomer or by chemical modification of **2**. Poly[4-(1-phenoxyethoxy)styrene] (**14**) is converted to **2** by reaction with a photochemically generated acid at ca. 100 °C (Scheme VII) (46). Development with aqueous tetramethylammonium hydroxide provides positive images.

Aqueous-Base Developable Positive Resists

Copolymers

In general, copolymers provide more flexibility for the resist design, because a single component does not have to carry all the necessary functions. One typical example is the copolymer of α,α-dimethylbenzyl

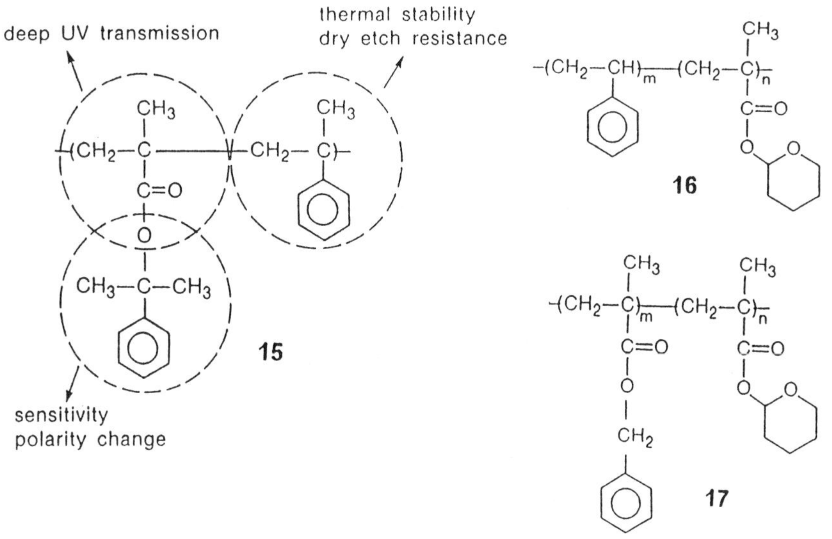

Chart I. Methacrylate copolymers for acid-catalyzed deprotection.

methacrylate with α-methylstyrene (**15**; Chart I) (*36*). The radical copolymerization of the electron-deficient methacrylate and the electron-rich α-methylstyrene tends to give an alternating copolymer, especially at high concentrations of α-methylstyrene, which does not undergo homopolymerization. The α-methyl-styrene units (ca. 50%) offer dry-etch durability and high thermal stability, and the negative images of the copolymer resist are devoid of thermal flow up to 220 °C. The high thermal stability is the result of minimizing the intramolecular dehydration through isolation of the methacrylate unit in the copolymer by alternating copolymerization by using the high T_g of poly-(methacrylic acid) (228 °C) and of poly(α-methyl-styrene) (168 °C). This copolymer concept has been extended to a copolymer of tetrahydropyranyl meth-acrylate with styrene (**16**) (*47*). It is noteworthy that these copolymer resists containing a significant amount of a lipophilic and inert styrenic unit can still be imaged in a positive mode by using aqueous base as the developer. A block copolymer of benzyl methacrylate with tetrahydropyranyl methacrylate (**17**, Chart I) also was synthesized by group-transfer polymerization for chemically amplified lithographic imaging (*48*).

4-Acetoxystyrene was incorporated into **3** by ter-polymerization (**18**, Scheme VIII) (*49*). The ester group is inert to radiation and to acidolysis and reduces the shrinkage of the exposed regions that occurs upon postbake. However, the ester group can be hydrolyzed only in the exposed regions during the base development step.

Copolymers of *t*-butyl methacrylate with adaman-tyl methacrylate were examined for chemically ampli-fied lithographic imaging at the 193 nm ArF excimer laser wavelength (*50*). The methacrylate backbone

provides a high transmission required for single-layer imaging at 193 nm, and the alicyclic group offers high dry-etch resistance needed for device fabrication. Aro-matic groups cannot be used in the ArF single-layer lithography simply because of their excessive absorp-tions. The bulky alicylcic groups increase T_g signifi-cantly (*51*).

The copolymers most commonly employed in the design of aqueous-base developable positive resists are based on **2**. When carefully prepared, **2** is very attractive as a deep-UV matrix resin due to its high transmittance at 248 nm (*11, 35*). However, the poly-mer tends to dissolve very fast in aqueous base, and its dissolution rate is difficult to control by adding

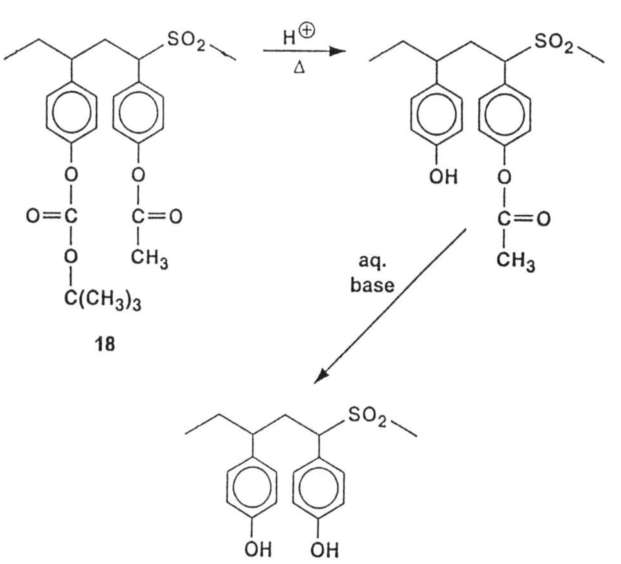

Scheme VIII. Terpolymer that undergoes photochemically induced acid-catalyzed deprotection and subsequent base-catalyzed hydrolysis during development.

inhibitors. Partial masking of the OH functionality with lipophilic acid-labile groups reduces the base dissolution rate dramatically, and the acid-catalyzed deprotection renders the exposed regions highly soluble in the base developer. This approach reduces the thermal shrinkage in the exposed areas. The tBOC group is the most popular protecting group in the copolymer approach (*52, 53*), and partially tBOC-protected polymers can be prepared by partial thermal deprotection of **1**, or more commonly by partial protection of **2** (*16*). Other protecting groups employed in the partial protection of the hydroxy group in **2** include trimethylsilyl (*54*), tetrahydropyranyl (*45*), and *t*-butylacetyl (*40*) groups. The approach based on partial protection of **2** is heavily pursued by many research groups in the design of positive deep-UV resists for replacement of the diazoquinone–novolac resist. However, because **2** is not generally miscible with its protected forms, the partial protection could results in heterogeneous products. The partially protected polymers tend to have much lower thermal-deprotection temperatures than the fully protected polymers, as demonstrated by thermogravimetric analysis curves in Figure 3 generated by using **2** protected with the tBOC group (*55*).

As mentioned earlier, the emerging ArF (193 nm) lithography cannot employ the phenolic group for base development because of the excessive absorption of the group. Terpolymers of *t*-butyl methacrylate, methyl methacrylate, and methacrylic acid provided excellent positive imaging at 193 nm by aqueous-base development (*56*). However, the terpolymer chemical-amplification resist lacks the dry-etch resistance needed for practical application.

Figure 4 shows a scanning electron micrograph of 0.25 μm line–space images printed in a positive chemical-amplification resist based on the deprotection chemistry using a KrF excimer laser stepper and aqueous tetramethylammonium hydroxide solution as the developer. These results demonstrate that features as small as the exposing wavelength can be delineated with a vertical wall profile.

Blends

Another approach to the design of positive resists that can be developed with aqueous base is to add to a phenolic matrix resin a protected dissolution inhibitor that undergoes acid-catalyzed conversion to a base-soluble form. Low molecular weight carboxylic acids or phenols were protected with acid-labile groups to be used as dissolution inhibitors for phenolic resins, particularly novolac resins (*57–59*). Small acetal compounds also inhibit dissolution of novolac resins through hydrogen-bonding (*60–62*). Strong Brönsted acid generated by irradiation cleaves acetals by catalytic hydrolysis to produce alcohols and aldehydes, which are not dissolution inhibitors, and provides positive images after development with aqueous base (Scheme IX). In contrast to the majority of the chemical-amplification resists, this three-component X-ray system does not require a postexposure bake step, because the activation energy for hydrolysis of the acetal inhibitor is sufficiently low for the reaction to proceed to near completion at ambient temperatures within 30 min.

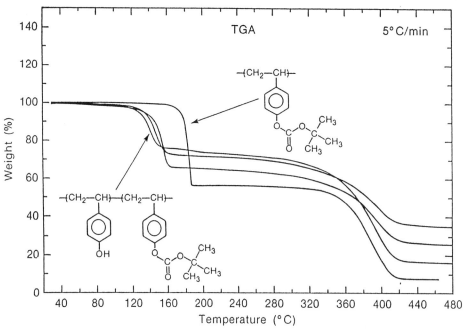

Figure 3. Thermogravimetric analysis curves of **1** and **2** partially protected with tBOC. (Reproduced with permission from reference 55. Copyright 1986 John Wiley & Sons.)

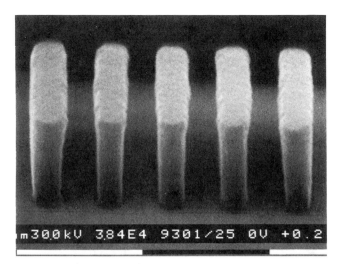

0.25 μm

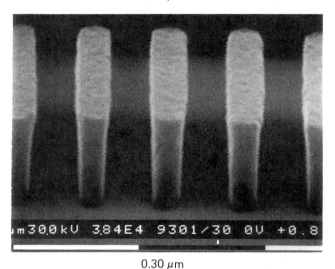

0.30 μm

Figure 4. Positive 0.25- and 0.30-μm line–space patterns delineated in a deprotection resist by KrF excimer laser stepper exposure (NA = 0.50) and aqueous-base development.

Certain acid generators such as sulfonium and iodonium salts (*63, 64*), disulfones (*65*), sulfonyl-substituted diazomethanes (*66*), and some sulfonate esters (*67*) are excellent dissolution inhibitors for novolac resins (*see* Chapter 2.3). The dissolution-inhibition effect of the sulfonium salt and the deprotection chemistry were combined (*68, 69*) in the design of a dissolution inhibitor for an aqueous-base developable positive-resist system. The sulfonium salt that carries base-solubilizing phenolic functionalities protected with the tBOC group functions as a dissolution inhibitor for **2**. Upon UV irradiation and postbake, photochemically generated acid converts the precursor sulfonium salt to a base-soluble form, and rapid dissolution of the exposed areas in aqueous base results. In a similar fashion, a *t*-butyl ester structure was incorporated into the nitrobenzyl sulfonate acid generators (*70*).

Some polymeric dissolution inhibitors have been reported. Although it is rather difficult in general to find two polymers that mix homogeneously, when **2** is partially protected with the tetrahydropyranyl or *t*-butoxycarbonylmethyl group, it is useful as a dissolution inhibitor for a novolac resin and **2**, respectively (*40, 45*). The terpolymer of *t*-butyl methacrylate, methyl methacrylate, and methacrylic acid used as a 193-nm resist is also miscible with a novolac resin (*71*). The methyl methacrylate component plays a critical role in the miscibility (*71*), but poly(*t*-butyl methacrylate) does not mix with a novolac resin (*55*). This terpolymer is only marginally miscible with **2**, but it is more compatible with C- or O-methylated forms of **2** (*71*). The ortho substitution (*72–74*) or O-alkylation (*75*) renders **2** less soluble in aqueous base for use as the matrix resin. Incorporation of a lipophilic comonomer is also useful in adjusting the dissolution rate (*75*), as is the change in the substitution position (para to ortho or meta) (*32, 75, 76*) (Chart II).

Scheme IX. Hydrolysis of dissolution-inhibiting acetal in novolac.

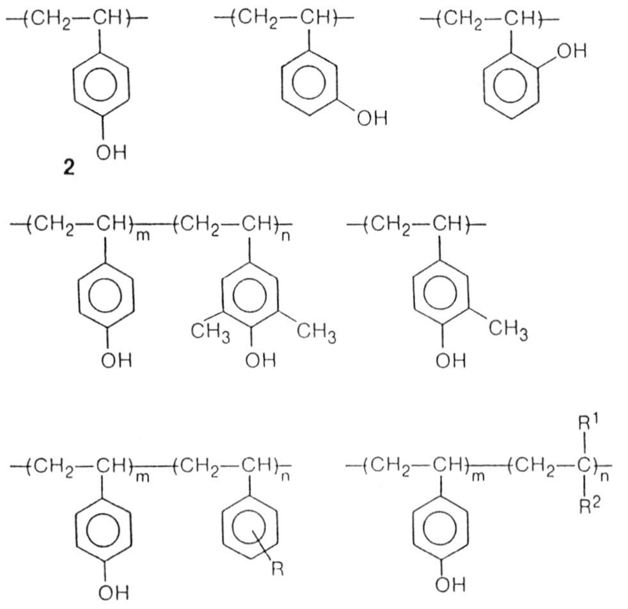

Chart II. Aqueous-base-soluble polyhydroxystyrene matrix resins.

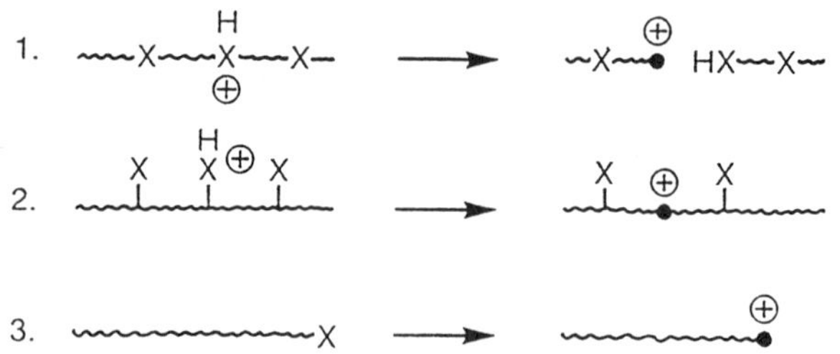

Scheme X. Initiation of acid-catalyzed depolymerization.

Three-component chemical-amplification resists consisting of a partially protected **2** a protected, small dissolution inhibitor, and a PAG have also been reported (*40, 52, 53*). The matrix polymer and the small lipophilic compound are both converted to base-soluble forms by acidolysis.

Depolymerization

Depolymerization is the ultimate form of polymer main-chain degradation, and thus it is useful in the design of sensitive positive-resist systems. Polymers that have low ceiling temperatures (T_c), including certain poly(olefin sulfone)s and polyaldehydes, are good candidates. As mentioned in Chapter 2.3, poly(olefin sulfone)s, which are prepared by alternating radical copolymerization of olefins with sulfur dioxide in an equilibrium reaction, have low T_c. Polyaldehydes are another class of polymers that are characterized by low T_c and must be end-capped because of the equilibrium nature of the anionic or cationic polymerization of aldehydes (*77*). In extreme cases for these polymers, radiation-induced chain scission, followed by thermodynamically driven depolymerization to volatile monomeric species, could produce "vapor-developed" (*78*) or "self-developed" (*2–4*) positive images without the use of any solvent.

In addition to the thermodynamically driven depolymerization of polymers with low T_c, catalytic scission of polymer backbone linkages (such as described in the earlier section) can be considered depolymerization. Furthermore, thermodynamically driven depolymerization also may be triggered by acid-catalyzed scission, which provides two stages of chemical amplification.

Thermodynamically Driven Catalytic Depolymerization

Initiation of thermodynamically driven depolymerization may be accomplished by generation of a carbocation in the polymer backbone via three alternative routes (Scheme X) (*79*):

1. protonation of the backbone heteroatom followed by scission
2. protonation and cleavage of a pendant group
3. protonation and cleavage of a polymer end group

Polyphthalaldehydes

Polyphthalaldehyde (**19**) can be prepared by anionic or cationic cyclopolymerization of phthalaldehyde, and it has a T_c of ca. –40 °C (*80*). It is an amorphous polymer and can be cast as a homogeneous film from common solvents, in contrast to highly crystalline, intractable, aliphatic polyaldehydes. Polyaldehydes consist of acetal bonds that are very labile toward acids. The C–O bond in polyphthalaldehyde is catalytically cleaved with photochemically generated acid, and this process triggers unzipping of the entire polymer chain. Positive images are produced spontaneously by exposure alone, and hence the term "self-development" is used for this process (Scheme XI) (*2–4*).

The potential contamination of exposure tools associated with self-developing resists was alleviated by suppressing depolymerization during exposure at ambient temperature. This suppression was accomplished by reducing the acid-generator concentration and the exposure dose. Complete development to the substrate was accomplished by postbaking the exposed resist to accelerate the acid-catalyzed depolymerization and to facilitate evaporation of the monomer (termed "thermal development") (*81–84*). Poly(4-chlorophthalaldehyde), which can be prepared by cationic polymerization, is particularly useful as a thermally developable resist (*81, 82*).

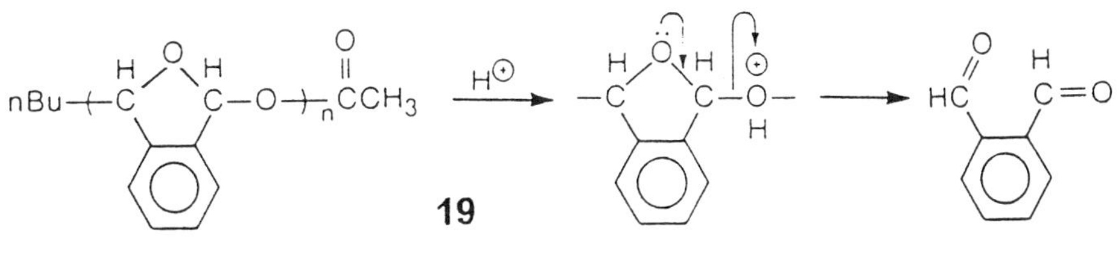

Scheme XI. Acid-catalyzed depolymerization of polyphthalaldehyde.

Poly(α-Acetoxystyrene)

In contrast to the side-chain deprotection described earlier, acidolysis of this acetophenone enol ester polymer (**20**) (Chart III) involves the polymer backbone (*85, 86*). Photochemically generated acid attacks the carbonyl oxygen of the ester group, which results in scission of the C–O bond upon postbake to form a highly stable, tertiary benzylic carbocation in the backbone and acetic acid. Polyphenylacetylene is produced when β-proton elimination occurs. At the same time, however, the backbone carbocation leads to main-chain scission to form an α-acetoxystyrene terminal carbocation, and unzipping is initiated from the scission point (*85, 86*). The degree of depolymerization amounts to 80% in solution at room temperature (*85*). Poly(α-acetoxystyrene), which can be prepared readily by radical polymerization in contrast to other α-substituted styrenes (*85, 87*), is highly resistant to common solvents due to its rigid backbone, whereas polyphenylacetylene is highly soluble. Thus, because of the solubility change due to the structural alteration and the molecular weight reduction through depolymerization, this resist system containing triphenylsulfonium hexafluoroantimonate functions as a positive resist by development with xylenes (*86*). Poly(α-acetoxystyrene) is transparent in the deep-UV region, thermally stable (no glass transition below 200 °C), and as durable in CF_4 plasma as polystyrene or a novolac resin. However, its film tends to exhibit solvent-induced cracking during development.

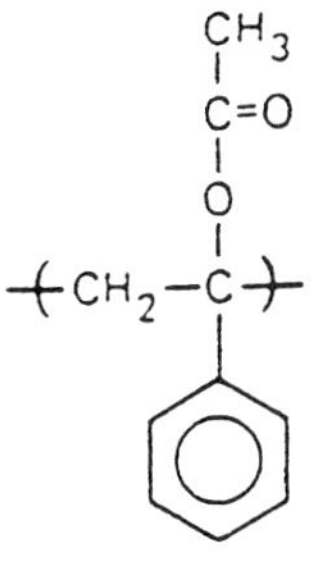

Chart III. Poly(α-acetoxystyrene) (**20**) and poly(4-hydroxy-α-methylstyrene) (**21**) capable of acid-catalyzed depolymerization.

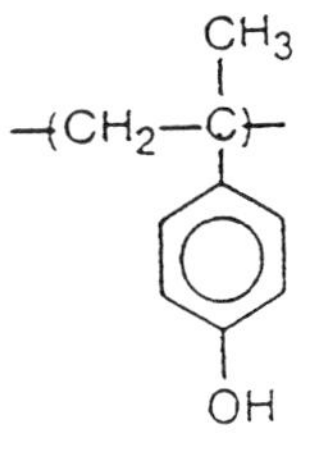

Poly(4-Hydroxy-α-Methylstyrene)

In this polymer (**21**) (Chart III), photochemically generated acid attacks the polymer end group to form a terminal carbocation, which results in depolymerization from the polymer end (*28, 29, 79*). 4-*t*-Butoxycarbonyloxy-α-methylstyrene polymerizes cationically with the acid-labile tBOC group remaining intact when liquid sulfur dioxide is used as the polymerization solvent (*23*). Heating the tBOC polymer to ca. 200 °C affords poly(4-hydroxy-α-methylstyrene) without main-chain scission, whereas the deprotection by acidolysis in solution results in extensive depolymerization even at room temperatures (*23*). A film of low molecular weight polymer **21** containing 5.4 wt% of triphenylsulfonium hexafluoroantimonate exhibited 80% shrinkage upon postexposure bake at 130 °C after exposure to 1 mJ/cm^2 (*2, 28*).

In contrast, the phenolic polymer prepared by living anionic polymerization of 4-dimethyl(*t*-butyl)silyloxy-α-methylstyrene followed by desilylation with HCl is very much inert to acidolysis. This observation suggests that depolymerization propagates from the terminal end of the polymer chain (*28, 29, 79*). The anionic polymers are expected to have acid-stable end groups (e.g., alkyl moieties derived from alkyllithium and hydrogen atoms introduced by protonation of the growing anion at termination). The cationic polymers are expected to contain acid-labile end groups, because accidental termination with trace water and termination with methanol introduce OH or OCH_3 groups. Such tertiary benzylic alcohols and ethers are highly susceptible to acidolysis and form terminal carbocations, leading to depolymerization upon heating. α-Methylstyrenic end groups introduced by deprotonation during polymerization readily react with photochemically generated acid to form polymer-end carbocations. The electron-donating *p*-OH group stabilizes the carbocation and renders the depolymerization very facile.

Repeated Main-Chain Acidolysis

Repeated cleavage of a polymer main chain to small fragments can provide another important mechanism of depolymerization, which is not thermodynamically driven but resembles catalytic cleavage of polymer pendant groups described earlier. Polymers containing tertiary, secondary allylic, or secondary benzylic carbonates, esters, or ethers (**22**) (*88, 89*) form stable carbocations by acid-catalyzed scission of the C–O bond upon mild heating. The resulting carbocations then undergo β-proton elimination to form olefins. The reaction is repeated on the polymer chain and leads to complete degradation. When the low molecular weight products evaporate out of the resist film during postbake, the degradation process is potentially suitable for thermal development. Polycarbonates synthesized from 2-cyclohexen-1,4-diol and a dihydroxy compound undergo aromatization through acidolysis to liberate benzene, a dihydroxyl compound, and carbon dioxide (*90*). These resists do not completely develop by heating alone because of the low volatility of the dihydroxyl compound. They require wet development.

Polyformals (**23**) containing secondary allylic and benzylic units decompose through acidolysis to volatile products such as an aromatic compound, formaldehyde, and water (Scheme XII) (*90*). Thus, resists consisting of polyformals and triphenylsulfonium hexafluoroantimonate develop thermally by postbake, but the positive images obtained tend to be rounded due to the low T_g of the polymers and probably due to plasticization by the aromatic degradation product before evaporation. Polyethers based on alkoxypyrimidine units (**24**) undergo an acid-catalyzed tautomeric

change of the alkoxypyrimidine to pyrimidone and release dienes (Scheme XII) (*91*).

Polymeric Dissolution Inhibitors

The lipophilic polymers that undergo acid-catalyzed decomposition to small molecules are potentially useful as polymeric dissolution inhibitors of phenolic resins for aqueous-base development of positive-resist systems. Unsubstituted polyphthalaldehyde is miscible with cresol–formaldehyde novolac resins but not with phenolic polymers of type **2** (*64, 92*). The lipophilic polyphthalaldehyde uniformly distributed in the novolac film dramatically suppresses the dissolution of the phenolic resin film in aqueous base (*64, 92*). The polymeric dissolution inhibitor is completely reverted to the monomer after postbake by reaction with photochemically generated acid, and the monomer evaporates out of the exposed regions. Consequently, the exposed regions are freed from the polymeric dissolution inhibitor and dissolve rapidly in aqueous base to provide positive images (Figure 5). Figure 6 presents dissolution curves of the three-component resist consisting of a novolac resin, 8.8 wt% of **19**, and 1.8 wt% of triphenylsulfonium hexafluoroantimonate in aqueous base. Whereas the unexposed film is insoluble in the developer for at least 3 min, the film exposed to 5 mJ/cm^2 rapidly dissolves to the substrate in ca. 60 s. In general, polymers that undergo very effi-

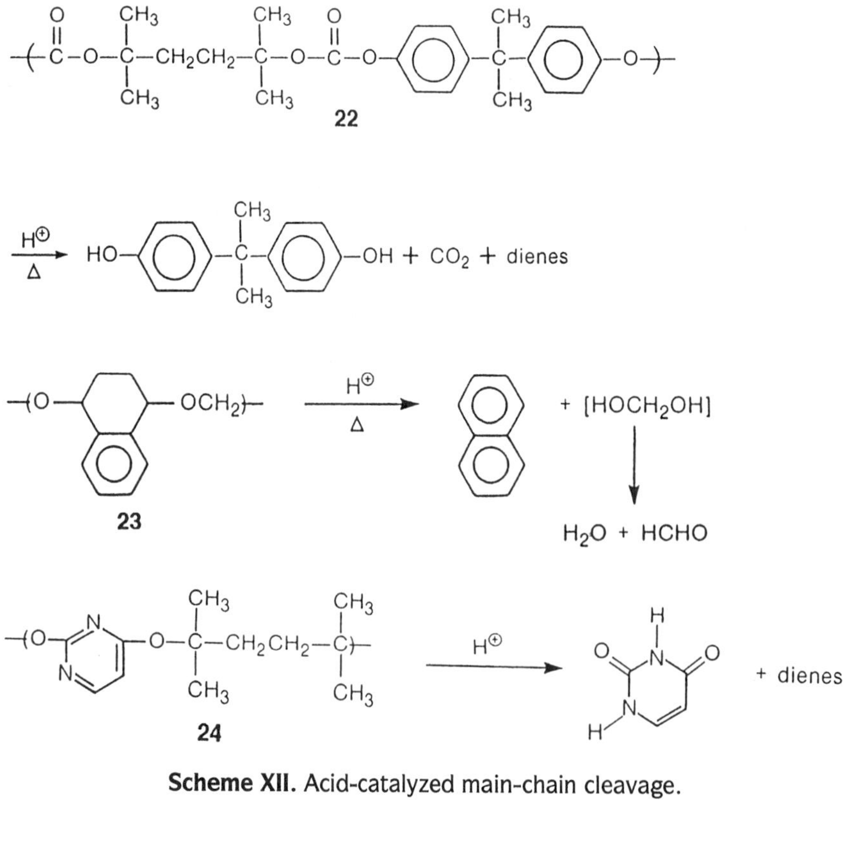

Scheme XII. Acid-catalyzed main-chain cleavage.

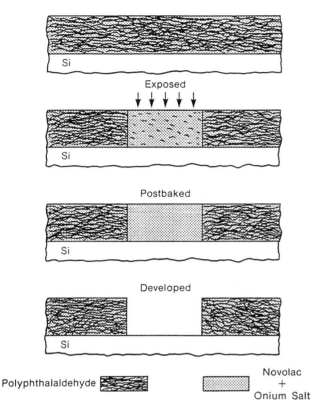

Figure 5. Acid-catalyzed depolymerization of polymeric dissolution inhibitor in novolac for positive aqueous-base development.

cient depolymerization lack the dry-etch durability needed for use in device fabrication, but the blend approach can combine the high sensitivity of the polyphthalaldehyde resist with the high dry-etch durability of the phenolic matrix resin.

Lipophilic silyl ether polymers are also useful as polymeric dissolution inhibitors for the design of chemically amplified three-component positive resists (*65, 93*). They are acid-hydrolyzed in the presence of water to alcohol and silanol, and they lose their inhibition effect.

Rearrangement

Polarity Reversal

Replacement of one of the methyl groups in poly(*t*-butyl 4-vinylbenzoate) (**5**) with a cyclopropyl residue reduces its thermal deprotection temperature by ca. 80 °C (*38*). However, in the case of poly(2-cyclopropyl-2-propyl 4-vinylbenzoate), the reaction (at 160 °C) is accompanied by ca. 10% rearrangement of the tertiary dimethyl cyclopropyl carbinol ester to a primary 4-methyl-3-pentenyl ester (Scheme XIII) (*38*). In the presence of acid, the rearrangement is more pronounced and amounts to as much as 66% (*38*). When the resist film containing triphenylsulfonium hexafluoroantimonate is postbaked at 100–130 °C, the exposed regions are not soluble in nonpolar organic solvents such as anisole due to the presence of 4-vinylbenzoic acid units (33%). This characteristic results in negative-tone imaging with anisole as the developer on the basis of polarity change.

When the resist film is postbaked at 160 °C, the unexposed regions undergo thermal deesterification to about 90%, and thus the unexposed polymer film becomes soluble in aqueous base. However, the exposed areas contain only 33% of the acidic units, and they are insoluble in the base developer. The

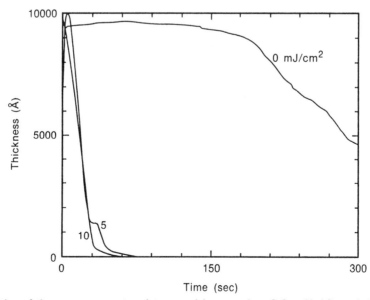

Figure 6. Dissolution kinetics of three-component resist comprising novolac, 8.8 wt% **19**, and 1.8 wt% triphenylsulfonium hexafluoroantimonate (exposed to 0, 5, and 10 mJ/cm² of 254-nm radiation) in aqueous base. (Reproduced with permission from reference 92. Copyright 1988 Electrochemical Society.)

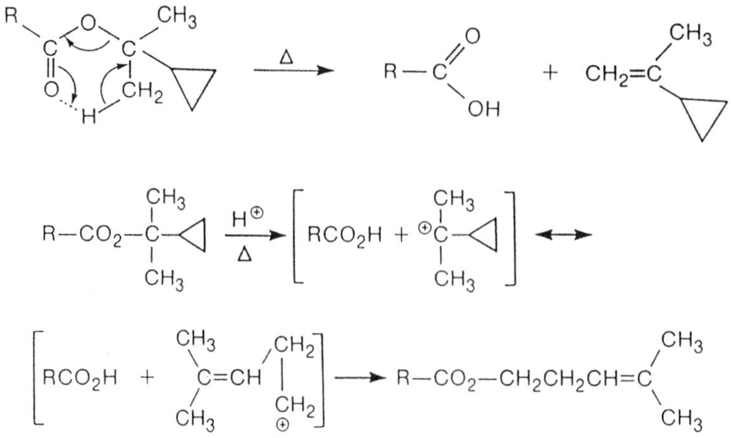

Scheme XIII. Thermal deesterification and acid-catalyzed rearrangement of cyclopropyl carbinol ester.

high-temperature postbake provides negative imaging with aqueous base on the basis of a polarity reversal (Figure 7).

Pinacol Rearrangement

The acid-catalyzed deprotection of polymer pendant groups results in a change from a nonpolar to a polar state and allows either positive or negative imaging depending on the polarity of the developer solvent. The alternative change from a polar to a nonpolar state (94) can be accomplished by the pinacol–pinacolone rearrangement, where *vic*-diols (pinacols) are converted to ketones or aldehydes with acid as a catalyst (76, 94–96). A polymeric pinacol, poly[3-methyl-2-(4-vinylphenyl)-2,3-butanediol] (25), prepared by radical polymerization of the styrenic diol monomer, can be quantitatively converted to a nonconjugated

ketone (26) when treated with an acid (Scheme XIV), as IR spectra in Figure 8 clearly demonstrate (76, 94, 96). Whereas the unexposed resist film is thermally stable to 225 °C (Figure 8c), exposure to 5.5 mJ/cm^2 of 254-nm radiation and subsequent postbake at 120 °C (Figure 8b) cleanly converts the *vic*-diol to a ketone. This phenomenon is evident from disappearance of the OH absorption and appearance of an intense carbonyl absorption. Thus, the resist consisting of the pinacol polymer and triphenylsulfonium triflate functions as a negative resist when developed with alcohol, which dissolves the polar diol polymer in the unexposed regions but cannot dissolve the less polar ketone polymer produced in the exposed areas.

In a similar fashion, the pinacol rearrangement mechanism for polarity change can be used in negative imaging with aqueous base (76, 94–97). One approach to base development involves copolymerization of the previously mentioned styrenic diol monomer with 4-acetoxystyrene, followed by base hydrolysis of the acetoxy groups to produce the aqueous-base soluble copolymer (97). Acid generated photochemically in the exposed areas converts diol groups to ketones, which form H-bonds with the phenolic residues and inhibit the dissolution of the resin in aqueous base. This process generates negative-tone images (97).

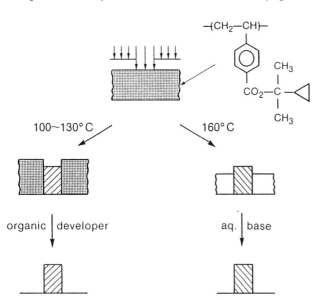

Figure 7. Negative imaging processes based on polarity change and polarity reversal of cyclopropyl carbinol ester.

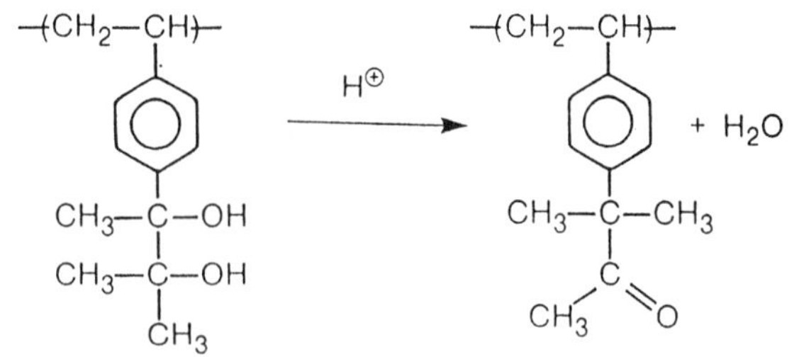

Scheme XIV. Pinacol rearrangement of polymeric *vic*-diol for polarity change.

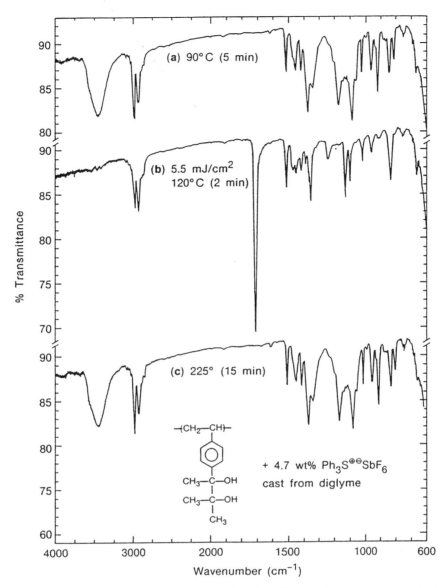

Figure 8. IR spectra of **25** containing 4.7 wt% triphenylsulfonium hexafluoroantimonate after prebake at 90 °C (a), after exposure to 5.5 mJ/cm² and postbake at 120 °C (b), and after heating at 225 °C for 15 min without UV exposure. (Reproduced with permission from reference 76. Copyright 1991 International Society for Optical Engineering.)

The polarity change in a phenolic matrix resin through the pinacol rearrangement of low molecular weight *vic*-diols to ketones or aldehydes can also provide chemical-amplification negative-resist systems that can be developed with aqueous base (*76, 94–96*). The pinacol rearrangement of *vic*-diols such as benzopinacole catalyzed by photochemically generated acid in polymer matrices (e.g., phenolic resins) occurs very efficiently, and phenyl trityl ketone produced from benzopinacole is a very strong dissolution inhibitor of phenolic resins. *meso*-Hydrobenzoin is converted to an aldehyde, which further undergoes acid-catalyzed reactions in the phenolic matrix resin. The three-component resists consisting of a novolac resin, benzopinacole or *meso*-hydrobenzoin, and a triphenylsulfonium salt were used successfully for

printing sub-half-micrometer features by aqueous-base development using KrF excimer laser steppers (*76, 94–96*).

Claisen Rearrangement

Whereas poly(4-*t*-butoxystyrene) (**11**) undergoes acid-catalyzed deprotection to form **2** (Scheme VI) and isobutene, its model compound 4-*t*-butoxytoluene is significantly realkylated in solution onto the ortho position (*41*). When the *t*-butyl group in **11** is replaced with a cyclohexenyl group, acidolysis of the ether group to afford **2** and 1,3-cyclohexadiene is only partial. Some Claisen rearrangement also takes place and results in alkylation of the ortho position (*98*). The net result of the acidolysis in this case is still a change from a nonpolar to a polar state. Poly(4-phenoxymethylstyrene) (**27**) is similarly isomerized with acid as a catalyst to produce the phenolic polymer (**28**) (Scheme XV).

Intramolecular Dehydration

Pinacol rearrangement described earlier of *vic*-diols to ketones or aldehydes involves acid-catalyzed dehydration to form a carbocation, resulting in a change from a polar to a nonpolar state. Tertiary alcohols can undergo intramolecular dehydration with acid catalysts to form olefins, thus providing another mechanism for polarity change.

Poly[4-(2-hydroxy-2-propyl)styrene] (**29**) reacts with acids to form a stable tertiary benzylic carbocation (after elimination of water), and then the carbocation undergoes β-proton elimination to produce the corresponding olefin structure (**30**) (Scheme XVI) (*99*). The net result is a conversion of a polar alcohol to a nonpolar olefin (polarity change), and this process enables negative imaging with a polar solvent such as alcohol. However, the α-methylstyrenic moieties generated by intramolecular dehydration further undergo acid-catalyzed linear and cyclic dimerizations to some degrees, concomitant cross-linking results, and positive imaging with a nonpolar solvent is precluded.

Generation of a 1,1-diphenylethylene structure instead of α-methylstyrene can eliminate the dimerization pathway because of steric hindrance (*97*). Poly[4-(1-hydroxy-1-phenylethyl)styrene] (**31**) undergoes extremely fast intramolecular dehydration when treated with photochemically generated acid, and **31** is converted to the corresponding 1,1-diphenylethylene structure (**32**) (Scheme XVI) (*97*). A model reaction indicates that the hindered olefin does not undergo any further reactions in solution in the presence of acid even at elevated temperatures (*97*). Thus, the resist based on the polymeric methyl diphenyl carbinol and 1.5 wt% of triphenylsulfonium triflate provides negative-tone images when developed with alcohol (polar) and positive-tone images when developed with xylene (nonpolar). This process is the first and only example of a dual-tone imaging material

based on this type of polarity change, and it is the most sensitive resist currently known.

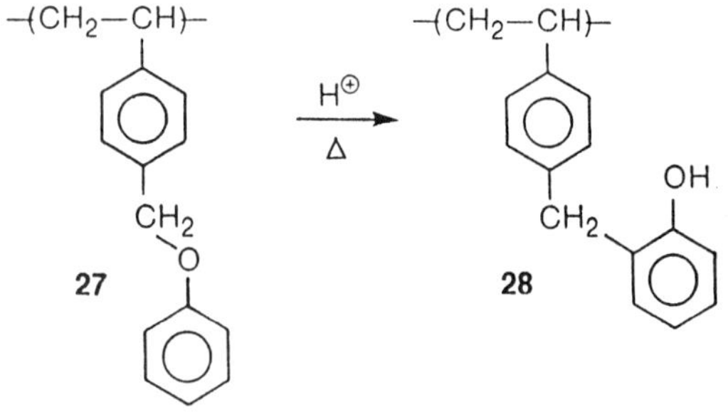

Condensation

Acid-catalyzed condensation of phenolic resins has attracted considerable attention in the design of chemical-amplification negative resists that can be developed with aqueous base, and it has provided a basis for the first commercialization of chemically amplified resists (*100–109*). These resist systems typically consist of three components:

- a base-soluble binder resin with reactive sites for cross-linking (phenolic resins)
- a radiation-sensitive acid generator
- an acid-sensitive latent cross-linking agent (electrophile)

The very first system designed on the basis of this concept employed a novolac–diazonaphthoquinone positive photoresist and an *N*-methoxymethylated melamine cross-linker (**33**) (*100*). The indenecarboxylic acid produced by photolysis of diazonaphthoquinone reacts with the melamine compound to form *N*-carbonium ions that undergo electrophilic substitution onto the electron-rich benzene rings of the novolac resin. The result is cross-linking and proton regeneration (Scheme XVII).

More recently, the novolac resin was replaced with the phenolic polymer **2** for deep-UV application along with the use of chloromethyltriazine as an HCl generator (*102–104, 107, 108*). The rate-determining step for cross-linking in this negative-resist system is

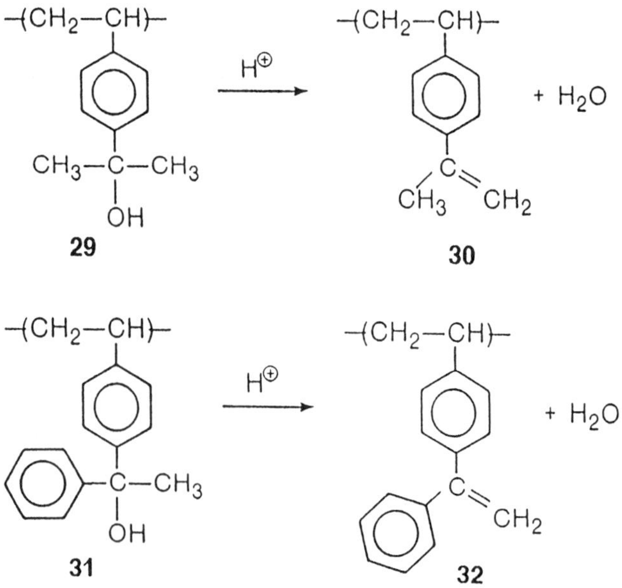

Scheme XV. Claisen rearrangement of poly(4-phenoxymethylstyrene) for polarity change.

Scheme XVI. Acid-catalyzed intramolecular dehydration of polymeric tertiary alcohol for polarity change.

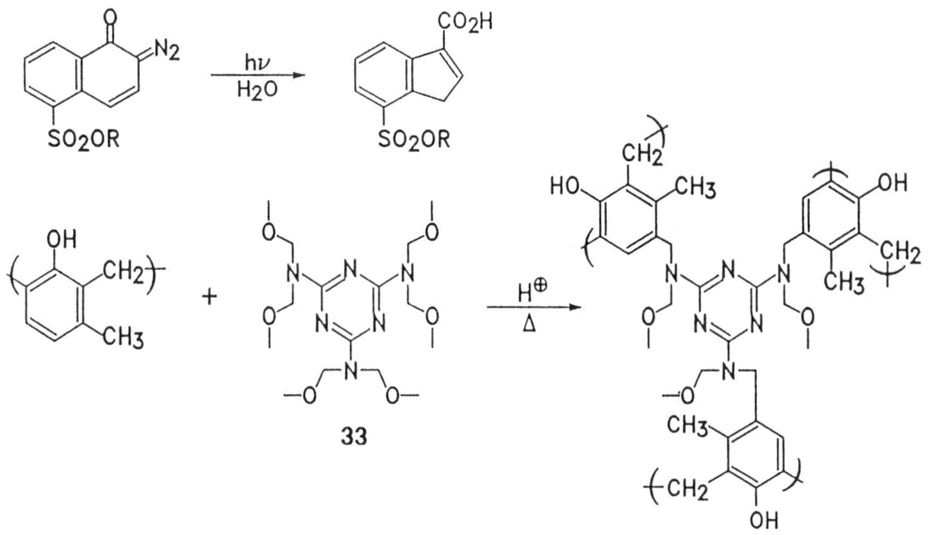

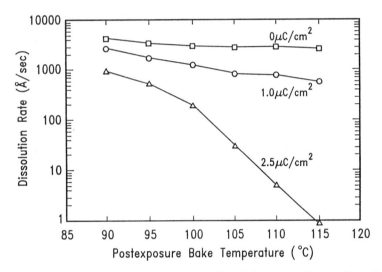

Scheme XVII. Acid-catalyzed cross-linking of novolac by condensation.

Figure 9. Effects of e-beam dose and postbake temperature on dissolution rate of a condensation resist in aqueous base. (Reproduced with permission from reference 101. Copyright 1988 American Vacuum Society.)

the formation of a carbocation from the protonated ether moiety. The cross-linking efficiency (sensitivity and contrast) and resolution of these resists are strongly influenced by the postexposure bake conditions, as is the case with most other chemical-amplification resists. Figure 9 is a plot of dissolution rates of a condensation resist in aqueous base vs. postexposure bake temperature as a function of e-beam dose. Whereas the films exposed to <1 μC/cm^2 dissolve rapidly even when baked at 115 °C, exposure to 2.5 μC/cm^2 results in dramatic reduction of the dissolution rate especially when postexposure bake is carried out at 115 °C. In this case postbake at 115 °C provides the largest dissolution rate differentiation between the unexposed and exposed regions.

Sub-half-micrometer resolution was demonstrated by KrF excimer laser, e-beam, and X-ray exposures. O-

alkylation (*108, 109*) as well as the originally proposed C-alkylation of the phenol groups are responsible for cross-linking. Figure 10 presents a scanning electron micrograph of 0.25-μm line–space patterns generated by aqueous development in a three-component condensation resist exposed on a step-and-scan deep-UV exposure system (NA = 0.50).

Other latent electrophiles also have been employed. Benzyl acetate derivatives react with photochemically generated acid to produce acetic acid and stable benzylic carbocations. These carbocations undergo electrophilic substitution reactions onto the electron-rich aromatic ring, and cross-linking results when the latent electrophile is multifunctional (*110–114*). The latent electrophile can be an additive or a monomer that can be copolymerized into the phenolic polymer. In the three-component approach, 1,4-di-

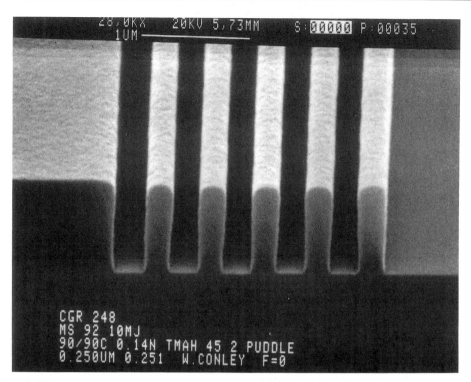

Figure 10. Negative 0.25-μm line–space patterns delineated in a three-component condensation resist by 248-nm step-and-scan exposure (NA = 0.50) and aqueous-base development (Courtesy of W. Conley, IBM Microelectronics Division).

(acetoxymethyl)benzene is added to the phenolic polymer **2**, or a novolac resin, together with triphenylsulfonium hexafluoroantimonate. In the two-component design, 4-vinylbenzyl acetate is copolymerized with the tBOC-protected monomer, followed by selective removal of the tBOC group in refluxing glacial acetic acid. The result is a copolymer that contains both the latent electrophile and the cross-linking site on the polymer chain. It is interesting to note that <5% of the acetoxy group is removed at an imaging dose of ca. 2 mJ/cm^2 (*114*).

Attachment of a bulky substituent onto the phenolic group via C- or O-alkylation can reduce the dissolution rate of the phenolic resin in aqueous base and provide negative images even when no cross-linking is involved (*75*). Monofunctional latent electrophiles such as *N*-hydroxymethyl and *N*-acetoxymethylimides provide aqueous-base developable negative-resist systems when mixed with the phenolic polymer **2** and triphenylsulfonium hexafluoroantimonate (*75*).

Aldehydes are another class of latent electrophiles for methyloting phenols with acid catalysts (*75*). The resulting phenolic resin is expected to be less soluble in aqueous base than its precursor resin, and hence the process is potentially useful for negative imaging. In addition, the methylolatedphenol can undergo further condensation with phenol. When phenolic resins are used, the second reaction results in cross-linking and hence lower dissolution rate. The chemistry is essentially the same as that of the novolac synthesis.

All the condensation resist systems discussed so far use the phenolic resins as a reaction site as well as a binder and a base solubilizer. A new aqueous-base developable negative copolymer resist was designed recently on the basis of self-condensation of polymeric furan derivatives (**34**) (Scheme XVIII) (*115*). Poly[4-hydroxystyrene-*co*-4-(3-furyl-3-hydroxypropyl)styrene] was prepared by copolymerization of the acety-protected furan monomer with the tBOC-protected monomer, followed by base hydrolysis. The product was mixed with 10 wt% of triphenylsulfonium hexafluoroantimonate (*115*). The furan methanol residue readily forms a stable carbocation by exposure to acid, and it is characterized by a high reactivity toward electrophiles (electron release from oxygen facilitating attack on the ring carbons). Thus, the furfuryl pendant groups act as both the latent electrophile and the nucleophile. Model reactions indicated that the furfuryl carbocation showed preference for reaction with the furan nucleus, rather than the less reactive phenolic moiety.

Poly[4-(1-hydroxyethyl)styrene], which carries secondary alcohol, cross-links via acid-catalyzed interchain dehydration to form ether linkages and generates negative images after development with alcohol (*99*). The negative-resist system was rendered developable with aqueous base by copolymerizing the styrenic secondary alcohol monomer with 4-acetoxystyrene followed by base hydrolysis (*97*). Model reactions indicated that acid-catalyzed intermolecular

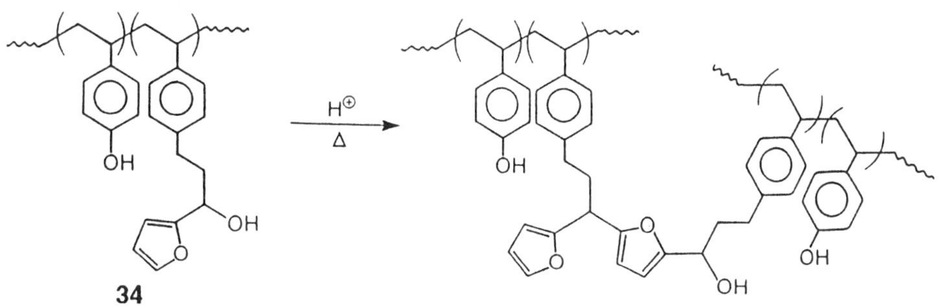

Scheme XVIII. Acid-catalyzed self-condensation of polymeric furan methanol.

dehydration results in self-condensation to di(α-methylbenzyl) ether, O-alkylation to α-methylbenzyl phenyl ether, and C-alkylation to o-(α-methylbenzyl)-phenol, all of which contribute to cross-linking (*97*).

In addition to cross-linking via condensation, self-condensation of silanol compounds in a phenolic matrix resin also has been employed as a basis for an aqueous-base-developable negative resist. This system is similar to pinacol rearrangement of small *vic*-diol to a dissolution-inhibiting compound, and it is based on a polarity change rather than cross-linking (*116–118*). This resist formulation consists of a novolac resin, an onium salt acid generator, and silanol compounds that function as dissolution promoters for the phenolic resin in aqueous base. Base-soluble silanol compounds, such as diphenylsilanediol, undergo acid-catalyzed condensation to polysiloxanes during postbake after UV exposure (Scheme XIX). This process yields negative images after development with aqueous base. However, siloxane oligomers do not inhibit the dissolution of the novolac resin at all when blended, presumably because the hydrophobic siloxane compounds are located apart from the hydrophilic phenolic OH groups. In contrast, the hydrophilic silanol compounds locate themselves in the vicinity of the phenolic OH groups. These compounds are then converted to siloxane oligomers and maintain their initial orientation relative to the phenolic OH groups. Thus, the siloxane oligomers generated from silanols in the phenolic resin can inhibit the dissolution of the resin in aqueous base due to the hydrophobic barriers surrounding the phenolic OH groups. Silsesquioxanes were also used in a similar negative-resist formulation (*119*).

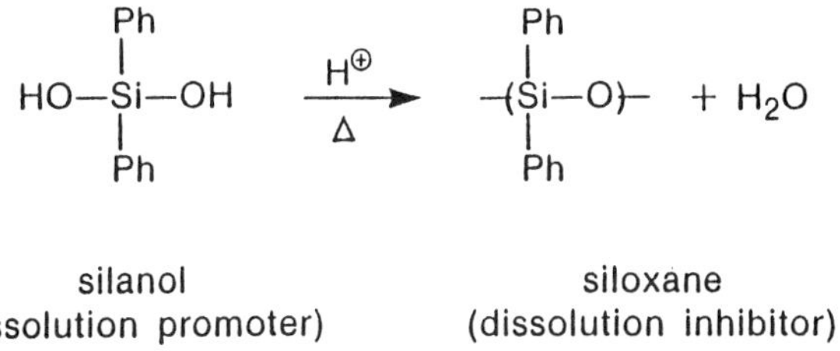

Cationic Polymerization

The onium salt cationic photoinitiators were originally developed for photochemical curing of epoxy resins, and one of the first chemical-amplification resists was based on cross-linking of epoxy resins via cationic ring-opening polymerization of pendant epoxide groups (*1–4*). Copolymers of styrene with allyl glycidyl ether (**35**) were developed for deep-UV applications (*120, 121*), and these copolymers exhibit good

Scheme XIX. Silanol condensation for polarity change.

thermal stability and low absorptivity in the deep-UV region. Polymeric episulfides (**36**) were also used as cationically polymerizable cross-linkers in resist formulations (Chart IV) (*122, 123*).

As mentioned earlier, the cross-linking mechanism does not generally provide high resolution because of swelling during development with organic solvents. The epoxy chemistry has been used in cross-linking of phenolic resins for swelling-free alkaline development by blending **2** with 15 wt% of an epoxy-novolac resin (**37**) and 10 wt% of triphenylsulfonium hexafluoroantimonate (*124*). In another formulation, a bifunctional small epoxide, bis(cyclohexene oxide), was mixed with a novolac resin containing triphenylsulfonium hexafluoroantimonate (*124*).

A copolymerization method was also used to combine the epoxy chemistry with the base development. Dicyclopentyloxy methacrylate was copolymerized with the tBOC-protected monomer, and the copolymer solution was heated to reflux (145 °C) in a casting solvent (propylene glycol methyl ether acetate) to remove the tBOC group to afford the desired copolymer (**38**) (Chart IV) (*125*). The two-component negative resist provided a high resolution by UV and e-beam exposures by using tetramethylammonium hydroxide aqueous solution as the developer (*125*).

Multilayer and Dry-Developable Resist Systems

The chemical-amplification concept was also applied to multilayer and all-dry lithographic techniques by using oxygen plasma. Two approaches are available in this case.

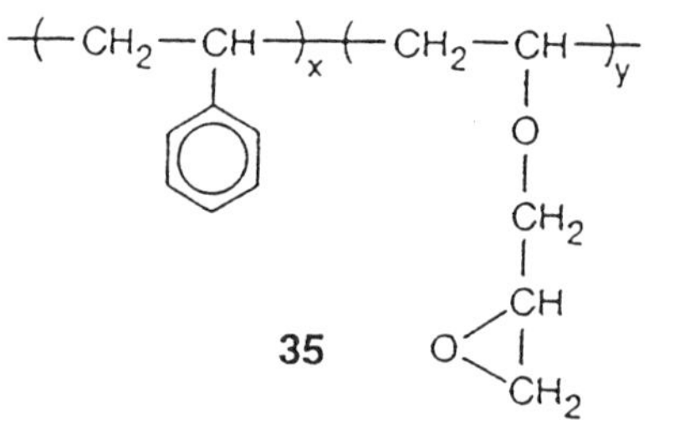

35

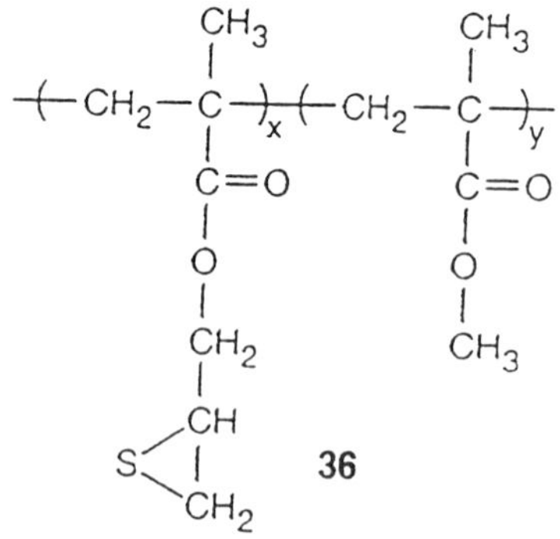

36

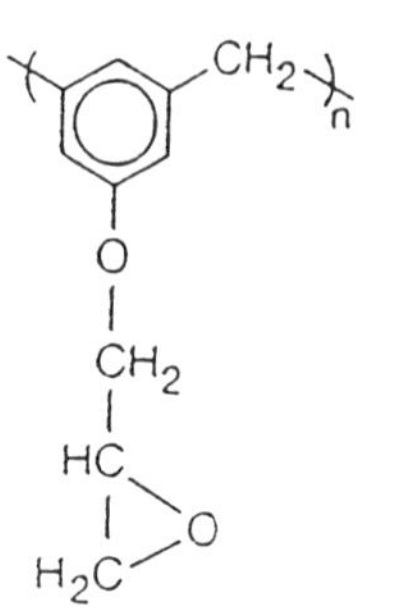

37

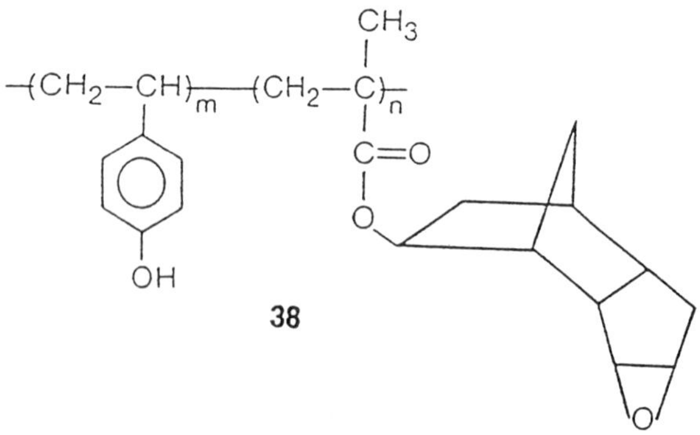

38

Chart IV. Epoxy resins for cross-linking by cationic ring-opening polymerization.

- Lithographic imaging of an organometallic resist on a thick planarizing organic polymer layer, followed by pattern transfer with oxygen reactive ion etching (RIE) (*see* Figure 14 in Chapter 2.3)
- Selective incorporation of an organometallic species into an organic resist, followed by oxygen RIE development

Silicon-Containing Resist Systems

In the bilayer scheme using silicon-containing resists, the silicon species is removed from either the exposed or the unexposed regions by solvent development, which results in positive or negative imaging, respectively. It is also possible to cleave off an Si-containing group from a polymer chain by using acid catalysis. This scission reaction can provide all-dry processes if the Si-containing small fragments can be removed from the exposed regions of the resist film without a solvent.

Semi-Dry Bilayer Lithography (Wet Development–Dry Pattern Transfer)

Acid-catalyzed silanol condensation of poly(phenylmethylsilsesquioxane) (**39**; Chart V) was used for a chemically amplified negative bilayer resist (*126*). A low molecular weight ladder polymer was selected because the condensation occurs between OH end groups on the polymer (*126*). Several acid generators were evaluated for this system. The exposed regions become cross-linked upon postexposure bake via acid-catalyzed silanol condensation, and the polymers are rendered insoluble in an organic developer (4-methyl-2-butanone/2-propanol = 1/1).

An e-beam resist consisting of poly(di-*t*-butoxysiloxane) (**40**) and an acid generator combines acid-catalyzed deprotection and silanol condensation to produce an oxygen-RIE-resistant glass (SiO_2) (*127*). The unexposed polymer is soluble in an organic solvent such as methyl isobutyl ketone, whereas the three-dimensional inorganic network formed in the exposed regions is insoluble and hence provides a negative image. The poly(di-*t*-butoxysiloxane) was synthesized by treating diacetoxy(di-*t*-butoxy)silane with triethylamine and water followed by end-capping with trimethylsilyl chloride in the presence of triethylamine (*127*).

Aqueous-base soluble silicone polymers were used (*128*) in conjunction with several acid generators to afford alkaline development systems, which undergo acid-catalyzed silanol condensation to form insoluble networks. One such polymer was synthesized by a sol–gel reaction of a mixture of phenyltrimethoxysilane and 2-(3,4-epoxycyclohexyl)ethyltrimethoxysilane. This polymer contained a high concentration of silanol groups and was soluble in aqueous base.

Polyhydroxybenzylsilsesquioxane (**41**) can be prepared by the reaction of BBr_3 on polymethoxybenzylsilsesquioxane and is soluble in aqueous base. Polymer **41** was partially protected with the tBOC group to design an aqueous-base-developable positive resist (*129*). Whereas the dissolution rate of the unexposed film is very small, acid-catalyzed deprotection renders the exposed regions highly soluble in aqueous tetramethylammonium hydroxide solution. The polymer is highly transparent at 248 nm and has a T_g of 90 °C.

All-Dry Bilayer Lithography

Acid-catalyzed desilylation can be used in all-dry bilayer positive lithography if the Si-containing acid-cleaved fragments can be removed from the exposed regions without wet development (for example, by heating). One such system was studied by using poly(4-trimethylsilyloxystyrene) and triphenylsulfonium hexafluoroarsenate (*130*). A similar approach was pursued by using copolymers (**42**) of trimethylsilyl methacrylate with styrene, methyl methacrylate, and benzyl methacrylate together with 1,2,3,4-tetrahydro-

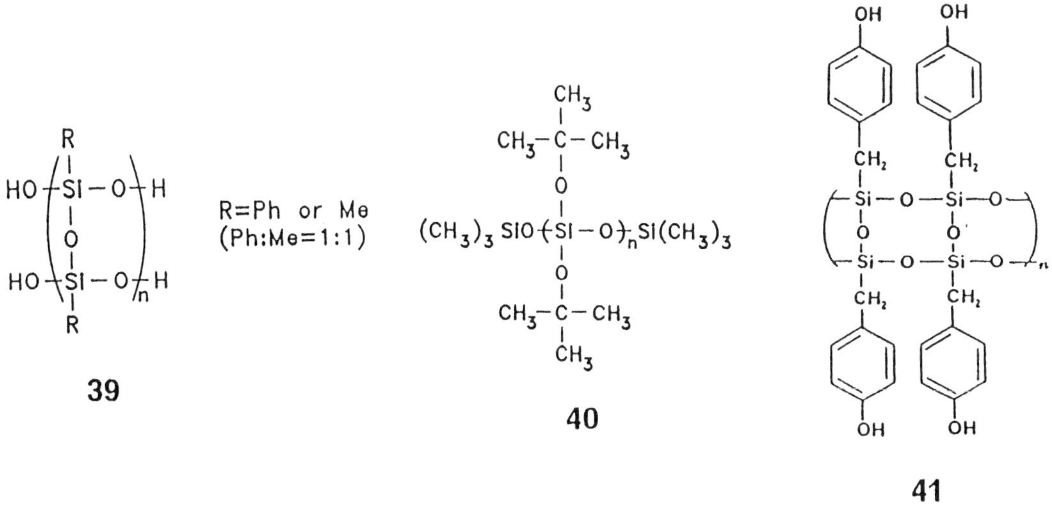

R=Ph or Me
(Ph:Me=1:1)

39

40

41

Chart V. Siloxane polymers for acid-catalyzed condensation.

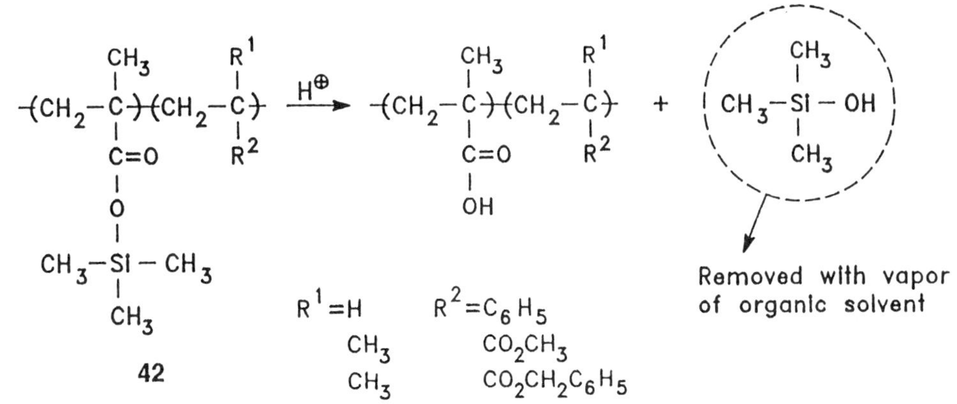

$R^1 = H$ $R^2 = C_6H_5$
    CH_3 CO_2CH_3
    CH_3 $CO_2CH_2C_6H_5$

42

Removed with vapor
of organic solvent

Scheme XX. Acid-catalyzed desilylation for O_2 RIE development.

naphthylideneamino *p*-toluenesulfonate as the acid generator (Scheme XX) (*131*). The degree of the silyl ester hydrolysis reached a limit of 85% at 390 mJ/cm² of 254 nm radiation, and only 45% of trimethylsilanol generated at this high dose escaped from the film. Complete removal of the silanol was achieved by treating the resist film, after UV exposure, with an organic vapor such as acetone, *n*-hexane, or alcohol at room temperatures. The acetone vapor treatment enhanced the acid-catalyzed hydrolysis of the silyl ester and removal of trimethylsilanol. This study indicates that removal of the Si-containing fragments produced by acidolysis of a silyl polymer is not a trivial task. If a small amount of Si-containing species remains in the exposed areas, residues result after oxygen RIE.

The very first chemical-amplification resist designed for the bilayer scheme was based on acid-catalyzed depolymerization of Si-containing polyphthalaldehyde (*82, 83*). Poly(4-trimethylsilylphthalaldehyde) (**43**), prepared by anionic polymerization and end-capped with the acetyl group, is stable thermally to ca. 200 °C, but it undergoes efficient depolymerization at ca. 100 °C by

reaction with a photochemically generated acid (*82, 83*). Thus, the resist consisting of **43** and 1.2 wt% triphenylsulfonium triflate develops to the substrate by postexposure bake (thermal development), and the positive relief images thus obtained function as an oxygen RIE mask to etch a 2-μm-thick hard-baked novolac resist film (Scheme XXI) (*82, 83*). Deep-UV sensitivity curves for the thermal development of a resist consisting of **43** and triphenylsulfonium triflate as PAG are presented in Figure 11, which indicates that the resist film is cleanly removed simply by heating at ca. 100 °C after a low dose of UV exposure.

Acid-catalyzed and thermodynamically driven depolymerization completely reverts the Si-containing polymer to the starting monomer, which evaporates during the postexposure bake step. Complete removal of the Si species from the exposed regions results. The sulfonium salt is decomposed in oxygen plasma to volatile products due to its organic nature. Sub-half-micrometer resolution has been demonstrated by e-beam and KrF excimer laser exposures. This result indicates that the thermal development is compatible

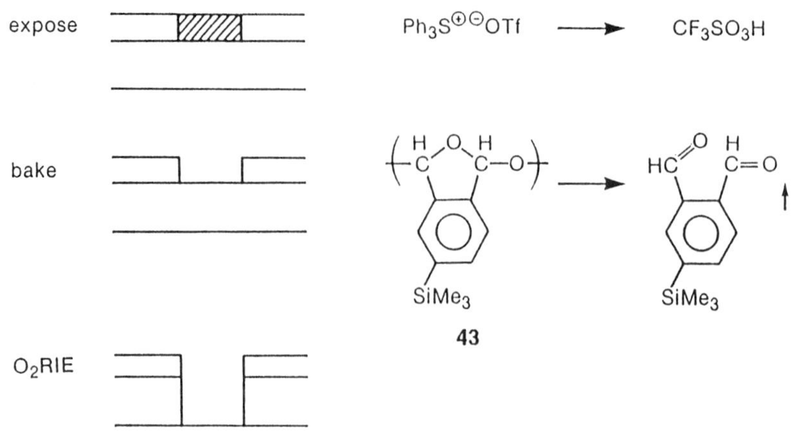

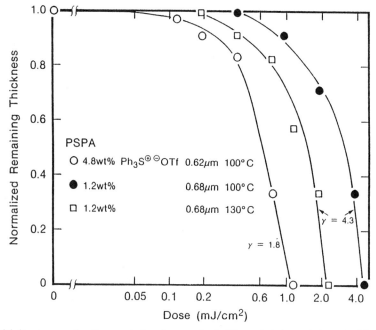

Scheme XXI. Acid-catalyzed depolymerization of polysilylphthalaldehyde for all-dry bilayer lithography (thermal development/O_2 RIE pattern transfer).

Figure 11. Deep-UV sensitivity curves for thermal development of **43** containing triphenylsulfonium triflate. (Reproduced with permission from reference 82. Copyright 1988 American Vacuum Society.)

with high resolution lithographic technologies and that the acid-catalyzed depolymerization mechanism and good dry-etch resistance are not mutually exclusive (*79, 82, 132*). Figure 12 presents sub-half-micrometer positive images printed in the resist based on **43** by thermal development followed by oxygen RIE. Poly[4,5-bis(trimethylsilyl)phthalaldehyde] was used (*133*) in the similar all-dry process to etch a ca. 2-μm-thick polyimide film.

The bottom planarizing layer for use in the bilayer scheme must satisfy a number of criteria, including the following:

- suitable for spin-coating
- resistant to common casting solvents with no interfacial mixing with the top resist

- highly opaque at the exposing wavelength to eliminate the topography effect
- not detrimental to the top resist
- etched rapidly in oxygen plasma
- high thermal stability
- resistant to fluorocarbon plasmas
- easy stripping after substrate fabrication

The bottom layer materials most commonly employed in the oxygen RIE bilayer systems are hard-baked (cross-linked) novolac resists and cured polyimide, and they generally satisfy all but the last of these requirements. To avoid interfacial mixing with the top imaging resist, the bottom layer is usually rendered insoluble by cross-linking or curing. This process enhances its thermal stability but strongly reduces

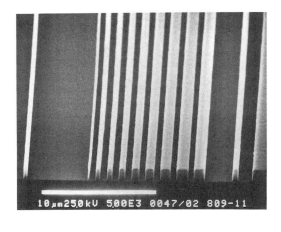
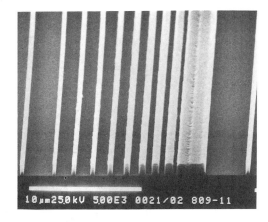
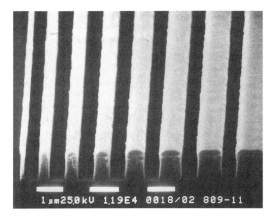
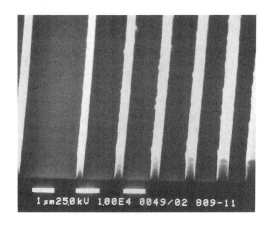

Figure 12. Positive sub-half-micrometer bilayer images delineated in polysilylphthalaldehyde resist by KrF excimer laser exposure, thermal development, and O$_2$ RIE pattern transfer.

its stripping ability. Poly(4-vinylbenzoic acid) (**6**) is reportedly highly suitable as a bottom layer for the poly(4-trimethylsilylphthalaldehyde) resist (**43**) in the deep-UV all-dry bilayer imaging, and it has good stripping ability (*132*). Polymer **6** is soluble in 2-ethoxyethanol, and the solution can be spin-coated. Polymer **6** is resistant to common casting solvents (e.g., propylene glycol methyl ether acetate and cyclohexanone) even without cross-linking. It is also very opaque in the 250-nm region and has an OD of 3.41/μm at 248 nm. The polymer has a high T_g of 250 °C and an excellent thermal stability, and it remains soluble even after prolonged heating at 230 °C. It is also resistant to fluorocarbon plasmas due to its aromatic nature, but it etches rapidly in oxygen plasma. Its high solubility in aqueous base even after high temperature treatment allows stripping with commonly available tetramethylammonium hydroxide aqueous solution.

Silylation of Organic Resists

The dual-tone resist systems based on the acid-catalyzed deprotection chemistry are uniquely suitable for selective silylation (*73, 134, 135*), because their polarity change accompanies the alteration of their reactivity (Scheme XXII). In these systems, reactive phenolic hydroxyl or carboxylic acid groups are unmasked by acid-catalyzed deprotection during postbake. The difference in chemical reactivities between the protected and unprotected areas can be converted into a differential etch rate by allowing the exposed portions of the polymer film to react with an organosilicon reagent such as hexamethyldisilazane or dimethylaminotrimethylsilane. The silylation selectivity is very high because the silylating reagents are incorporated into the exposed regions through covalent bonds and do not react with the carbonate or ester in the unexposed regions. Following gas-phase silylation, the resist is subjected to oxygen RIE for the generation of negative images (Scheme XXII). Figure 13 shows IR spectra of the tBOC resist after prebake (a), after UV exposure (10 mJ/cm² at 254 nm) and postbake (b), and after treatment with gaseous dimethylaminotrimethylsilane for 5 min at 100 °C and 26.7 kPa (c). The IR spectra clearly indicate that **2** (Figure 13b) produced from **1** (Figure 13a) is completely silylated (Figure 13c) by this treatment. The process does not

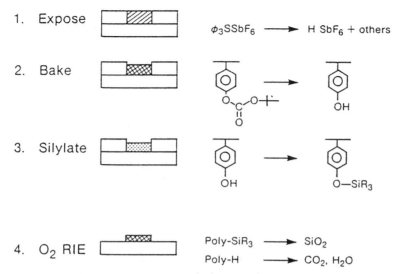

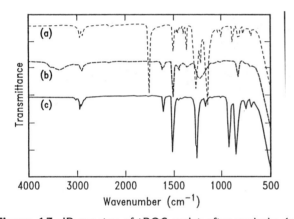

Scheme XXII. Negative-tone O_2 RIE development of tBOC resist by acid-catalyzed deprotection followed by selective silylation.

Figure 13. IR spectra of tBOC resist after prebake (a), after exposure to 10 mJ/cm^2 and postbake (b), and after treatment with gaseous dimethylaminotrimethylsilane for 5 min at 100 °C and 26.7 kPa (c). (Reproduced from reference 135. Copyright 1991 American Chemical Society.)

involve solvent development, and it can be extended to all-dry bilayer lithography.

The tone of this process can be reversed by blocking the phenolic OH groups in the exposed regions so that the polymer becomes unreactive toward silylation. The tBOC group is then removed in the initially unexposed areas by flood deep-UV exposure and postbake (Scheme XXIII) (73). This polarity reversal results in incorporation of Si in the unexposed regions and yields positive images after oxygen RIE. The masking reagent useful for the first step in this process is gaseous isocyanate, which reacts with the polymer **2** in the presence of triethylamine to form a urethane.

The tBOC resist is highly transparent in the deep-UV region; hence, the deprotection and subsequent silylation take place throughout the thickness of the resist film. Addition of an appropriate dye to the formulation renders the resist film opaque. When such opaque films are exposed, the deprotection is confined to the top surface of the resist film (top surface imaging, TSI) (Figure 14) (73, 136). Because no light penetrates to the resist substrate interface, TSI is insensitive to the variations in refractive index (reflectivity) of underlying topography and is essentially insensitive to the geometric variations in this topography.

Poly(*t*-butyl 4-vinylbenzoate) (**5**) is also highly suitable for the deep-UV TSI owing to its high absorption in the deep-UV region (132). The ester polymer is converted to its acid derivative by postbake in a fashion similar to the tBOC resist.

The diffusion-enhanced silylation resist (DESIRE) process described in Chapter 2.3 is based on limited diffusion of a silylating reagent into cross-linked regions of phenolic resists, and the process was extended to the chemical-amplification resist systems that function on acid-catalyzed cross-linking of phenolic resins (137). Such cross-linking resist systems are positive working in the silylation dry process (Figure 15) and are referred to as submicron positive dry-etch resist (SUPER).

To enhance the sensitivity of the Si-CARL bilayer process (*see* Scheme XIX in Chapter 2.3), the chemical-amplification concept was applied to a terpolymer of *N*-tBOC-maleimide, maleic anhydride, and styrene to afford an acid-catalyzed deprotection system for aqueous-base development (138). In this case, after aqueous-base development, the anhydride group in the positive relief image reacts with a Si-containing diamine, and incorporation of Si and cross-linking of the polymer result.

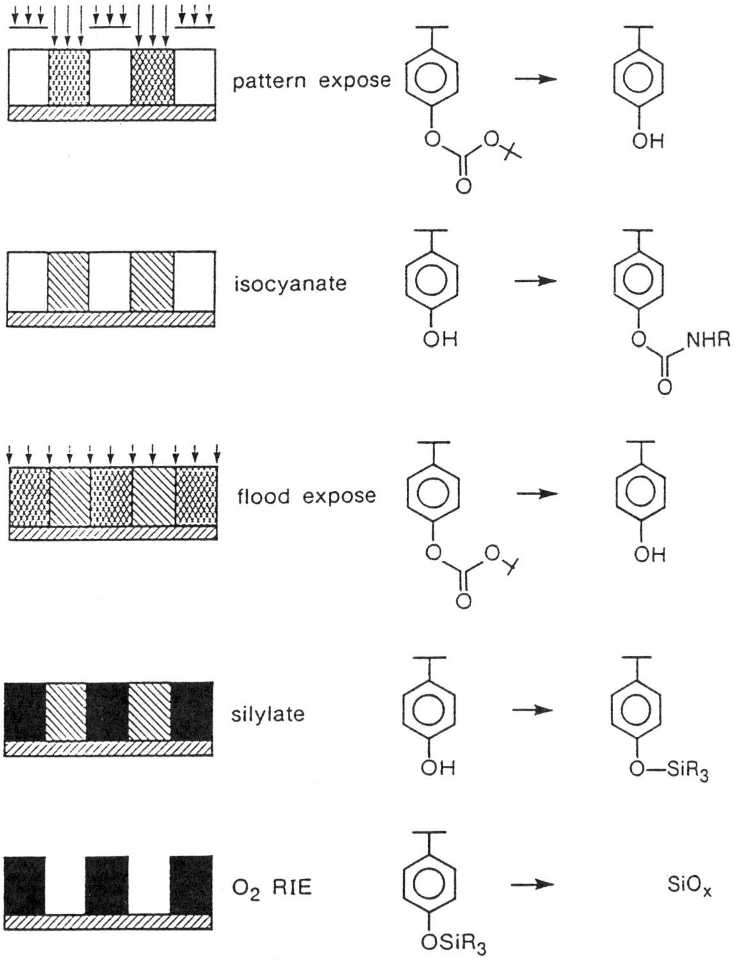

Scheme XXIII. Positive-tone O_2 RIE development of tBOC resist by selective silyaltion through polarity reversal.

Conclusions and Future Perspective

The chemical-amplification concept in microlithography offers high sensitivity, high contrast, high resolution, and versatility. The resolution capability of chemically amplified resist systems is especially remarkable, and aqueous-base developable chemical-amplification resist systems are likely to play a key role in the manufacture of 64- and 256-Mbit DRAMs by using KrF excimer laser steppers. However, these new materials may not always be plug-compatible with the processes optimized for conventional diazonaphthoquinone–novolac resists. Although resolution of <100 nm has been demonstrated, the diffusion of the photochemically generated acid during postexposure bake is not fully understood, and this phenomenon is the subject of many research activities (*14, 139–143*).

Another important issue associated with chemical-amplification resists is instability of latent images (*144–147*). Although a couple of chemical-amplification resists were used successfully for the manufacture of 1-Mbit (*10*) and 16-Mbit DRAMs by using deep-UV lithography, full implementation has been severely hampered due to the so-called "delay effect" related to the catalytic nature of the imaging chemistries. For example, a trace amount of airborne basic substances absorbed on the surface of the resist film interferes with the desired acid-catalyzed reaction and leads to the formation of "T-top" or "skin" in the positive systems. Line width shifts also result in the negative systems during standing after coating (*144, 145*). However, considerable progress has been made in solving this problem, including the use of activated-carbon filtration of the enclosing atmosphere (*144*), application of protective overcoat (*148–150*), and incorporation of additives in resist formulation (*151, 152*). More fundamental chemical approaches were proposed recently that involve reduced free volume of the resist film by annealing for maximum protection against airborne contamination (*153–155*). Availability of such environmentally stable chemical-amplification resists are expected to drive the emerging resist technology into new dimensions. We will see!

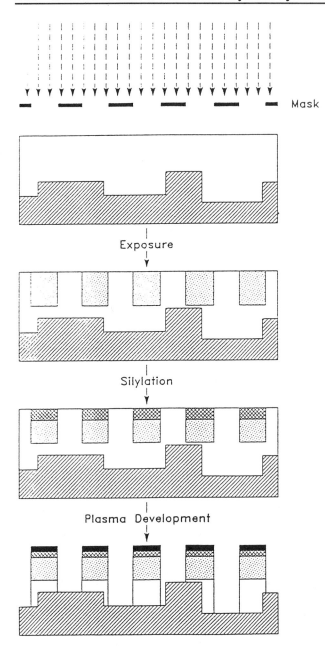

Figure 14. Negative-tone top surface imaging by shallow exposure and silylation.

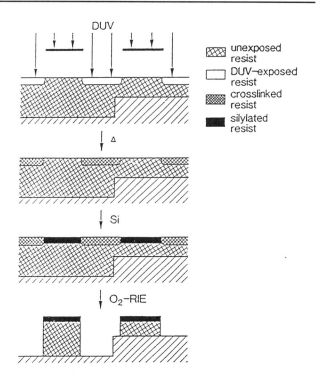

Figure 15. Positive-tone top surface imaging based on acid-catalyzed surface cross-linking of opaque phenolic resist and subsequent silylation/O$_2$ RIE.

References

1. Ito, H.; Willson, C. G.; Fréchet, J. M. J. *Digest of Technical Papers of 1982 Symposium on VLSI Technology;* 1982; p 86.
2. Ito, H.; Willson, C. G. *Technical Papers of SPE Regional Technical Conference on Photopolymers;* Society of Plastics Engineers: Brookfield, CT, 1982; p 331.
3. Ito, H.; Willson, C. G. *Polym. Eng. Sci.* **1983**, *23*, 1012.
4. Ito, H.; Willson, C. G. In *Polymers in Electronics;* Davidson, T., Ed.; ACS Symposium Series 242; American Chemical Society: Washington, DC, 1984; p 11.
5. Iwayanagi, T.; Ueno, T.; Nonogaki, S.; Ito, H.; Willson, C. G. In *Electronic and Photonic Applications of Polymers;* Bowden, M. J.; Turner, S. R., Eds.; ACS Advances in Chemistry Series 218; American Chemical Society: Washington, DC, 1988; p 107.
6. Reichmanis, E.; Houlihan, F. M.; Nalamasu, O.; Neenan, T. X. *Chem. Mater.* **1991**, *3*, 394.
7. Ito, H. In *Radiation Effects on Polymers;* Clough, R. L; Shalaby, S. W., Eds.; ACS Symposium Series 475; American Chemical Society: Washington, DC, 1991; p 326.
8. Ito, H. In *Radiation Curing in Polymer Science and Technology;* Fouassier, J. P.; Rabek, J. E., Eds.; Elsevier: London, 1993; Vol. IV, Chap. 11.
9. Pawlowski, G.; Dammel, R.; Przybilla, K.-J.; Röschert, H.; Spiess, W. *J. Photopolym. Sci. Technol.* **1991**, *4*, 389.
10. Maltabes, J. G.; Holmes, S. J.; Morrow, J.; Bar, R. L.; Hakey, M.; Reynolds, G.; Brunsvold, W. R.; Willson, C. G.; Clecak, N. J.; MacDonald, S. A.; Ito, H. *Proc. SPIE* **1990**, *1262*, 2.
11. Ito, H.; Pederson, L. A.; Chiong, K. N.; Sonchik, S.; Tsai, C. *Proc. SPIE* **1989**, *1086*, 11.
12. Seligson, D.; Ito, H.; Willson, C. G. *J. Vac. Sci. Technol.* **1988**, *B6(6)*, 2268.
13. Umbach, C. P.; Broers, A. N.; Willson, C. G.; Koch, R.; Laibowitz, R. B. *J. Vac. Sci. Technol.* **1988**, *B6*, 319.
14. McKean, D. R.; Schaedeli, U.; MacDonald, S. A. *J. Polym. Sci., Part A. Polym. Chem.* **1989**, *27*, 3927.
15. Fréchet, J. M. J.; Eichler, E.; Ito, H.; Willson, C. G. *Polymer* **1983**, *24*, 995.
16. Houlihan, F.; Bouchard, F.; Fréchet, J. M. J.; Willson, C. G. *Can. J. Chem.* **1985**, *63*, 153.
17. Ito, H.; Knebelkamp, A.; Lundmark, S. B. *Proc. ACS Div. Polym. Mater. Sci. Eng.* **1993**, *68*, 12.
18. Turner, S. R.; Arcus, R. A.; Houle, C. G.; Schleigh, W. R. *Polym. Eng. Sci.* **1986**, *26*, 1096.

19. Turner, S. R.; Ahn, K. D.; Willson, C. G. In *Polymers for High Technology*; Bowden, M. J.; Turner, S. R., Eds.; ACS Symposium Series 346; American Chemical Society: Washington, DC, 1987; p 200.
20. Osuch, C. E.; Brahim, K.; Hopf, F. R.; McFarland, A.; Mooring, C. J.; Wu, C. J. *Proc. SPIE* **1986**, *631*, 68.
21. Ahn, K. D.; Koo, D.-I.; Kim, S.-J. *J. Photopolym. Sci. Technol.* **1991**, *4*, 433.
22. Tarascon, R. G.; Reichmanis, E.; Houlihan, F. M.; Shugard, A.; Thompson, L. F. *Polym. Eng. Sci.* **1989**, *29*, 850.
23. Ito, H.; Willson, C. G.; Fréchet, J. M. J.; Farrall, M. J.; Eichler, E. *Macromolecules* **1983**, *16*, 510.
24. Houlihan, F. M.; Shugard, A.; Gooden, R.; Reichmanis, E. *Proc. SPIE* **1988**, *920*, 67.
25. Brunsvold, W.; Conley, W.; Montgomery, W.; Moreau, W. In *Polymers for Microelectronics*; Thompson, L. F.; Willson, C. G.; Tagawa, S., Eds.; ACS Symposium Series 537, American Chemical Society: Washington, DC, 1994; p. 333.
26. Gozdz, A. S.; Shelburne, J. A.; Lin, P. S. D. *Proc. ACS Div. Polym. Mater. Sci. Eng.* **1992**, *66*, 192.
27. Houlihan, F. M.; Reichmanis, E.; Tarascon, R. G.; Taylor, G. N.; Hellman, M. Y.; Thompson, L. F. *Macromolecules* **1989**, *22*, 2999.
28. Ito, H.; England, W. P. *Polym. Prepr.* **1990**, *31(1)*, 427.
29. Ito, H.; England, W. P.; Ueda, M. *Makromol. Chem., Macromol. Symp.* **1992**, *53*, 139.
30. Novembre, A. E.; Tai, W. W.; Kometani, J. M.; Hanson, J. E.; Nalamasu, O.; Taylor, G. N.; Reichmanis, E.; Thompson, L. F. *Proc. SPIE* **1991**, *1466*, 89.
31. Przybilla, K.-J.; Dammel, R.; Röschert, H.; Spiess, W.; Pawlowski, G. *J. Photopolym. Sci. Technol.* **1991**, *4*, 421.
32. Przybilla, K.; Röschert, H.; Spiess, W.; Eckes, Ch.; Chatterjee, S.; Khanna, D.; Pawlowski, G.; Dammel, R. *Proc. SPIE* **1991**, *1466*, 174.
33. Przybilla, K. J.; Röschert, H.; Pawlowski, G. *Proc. SPIE* **1992**, *1672*, 500.
34. Brunsvold, W.; Conley, W.; Crockatt, D.; Iwamoto, N. *Proc. SPIE* **1989**, *1086*, 357.
35. Ito, H.; Willson, C. G.; Fréchet, J. M. J. *Proc. SPIE* **1987**, *771*, 24.
36. Ito, H.; Ueda, M.; Ebina, M. In *Polymers in Microlithography*; Reichmanis, E.; MacDonald, S. A.; Iwayanagi, T., Eds.; ACS Symposium Series 412; American Chemical Society: Washington, DC, 1989; p 57.
37. Ito, H.; Ueda, M. *Macromolecules* **1988**, *21*, 1475.
38. Ito, H.; Ueda, M.; England, W. P. *Macromolecules* **1990**, *23*, 2589.
39. Ishizoe, T.; Hirao, A.; Nakahama, S. *Macromolecules* **1989**, *22*, 2895.
40. Onishi, Y.; Niki, H.; Kobayashi, Y.; Hayase, R. H.; Oyasato, N.; Sasaki, O. *J. Photopolym. Sci. Technol.* **1991**, *4*, 337.
41. Conlon, D. A.; Crivello, J. V.; Lee, J. L.; O'Brien, M. J. *Macromolecules* **1989**, *22*, 509.
42. Yamaoka, T.; Nishiki, N.; Koseki, K.; Koshiba, M. *Polym. Eng. Sci.* **1989**, *29*, 856.
43. Murata, M.; Takahashi, T.; Koshiba, M.; Kawamura, S.; Yamaoka, T. *Proc. SPIE* **1990**, *1262*, 8.
44. Hesp, S. A. M.; Hayashi, N.; Ueno, T. *J. Appl. Polym. Sci.* **1991**, *42*, 877.
45. Hayashi, N.; Schlegel, L.; Ueno, T.; Shiraishi, H.; Iwayanagi, T. *Proc. SPIE* **1991**, *1466*, 377.
46. Jiang, Y.; Bassett, D. R. In *Polymers for Microelectronics*; Thompson, L. F.; Willson, C. G.; Tagawa, S., Eds.; ACS Symposium Series 537; American Chemical Society: Washington, DC, 1994; p 40.
47. Kikuchi, H.; Kurata, N.; Hayashi, K. *J. Photopolym. Sci. Technol.* **1991**, *4*, 357.
48. Taylor, G. N.; Stillwagon, L. E.; Houlihan, F. M.; Wolf, T. M.; Sogah, D. Y.; Hertler, W. R. *Chem. Mater.* **1991**, *3*, 1031.
49. Kometani, J. M.; Galvin, M. E.; Heffner, S. A.; Houlihan, F. M.; Nalamasu, O.; Chin, E.; Reichmanis, E. *Macromolecules* **1993**, *26*, 2165.
50. Kaimoto, Y.; Nozaki, K.; Takechi, S.; Abe, N. *Proc. SPIE* **1992**, *1672*, 66.
51. Matsumoto, A.; Tanaka, S.; Otsu, T. *Macromolecules* **1991**, *24*, 4017.
52. Kumada, T.; Kubota, S.; Koezuka, H.; Hanawa, T.; Kishimura, S.; Nagata, H. *J. Photopolym. Sci. Technol.* **1991**, *4*, 469.
53. Kawai, Y.; Tanaka, A.; Matsuda, T. *Jpn. J. Appl. Phys.* **1992**, *31*, 4316.
54. Murata, M.; Kobayashi, E.; Yamachika, M.; Kobayashi, Y.; Yumoto, Y.; Miura, T. *J. Photopolym. Sci. Technol.* **1992**, *5*, 79.
55. Ito, H. *J. Polym. Sci., Polym. Chem. Ed.* **1986**, *24*, 2971.
56. Kunz, R. R.; Allen, R. D.; Hinsberg, W. D.; Wallraff, G. M. *Proc. SPIE* **1993**, *1925*, 167.
57. McKean, D. R.; MacDonald, S. A.; Clecak, N. J.; Willson, C. G. *Proc. SPIE* **1988**, *920*, 60.
58. O'Brien, M. J.; Crivello, J. V. *Proc. SPIE* **1988**, *920*, 42.
59. O'Brien, M. J. *Polym. Eng. Sci.* **1989**, *29*, 846.
60. Lingnau, J.; Dammel, R.; Theis, J. *Polym. Eng. Sci.* **1989**, *29*, 874.
61. Lingnau, J.; Dammel, R.; Lindley, C. R.; Pawlowski, G.; Scheunemann, U.; Theis, J. In *Polymers for Microelectronics—Science and Technology*; Tabata, Y.; Mita, I.; Nonogaki, S.; Horie, K.; Tagawa, S., Eds.; VCH Publishers: Deerfield Beach, FL, 1990; p 445.
62. Eckes, C.; Pawlowski, G.; Przybilla, K.; Meier, W.; Madore, M.; Dammel, R. *Proc. SPIE* **1991**, *1466*, 394.
63. Ito, H.; Flores, E. *J. Electrochem. Soc.* **1988**, *135*, 2322.
64. Ito, H. *Proc. SPIE* **1988**, *920*, 33.
65. Aoai, T.; Aotani, Y.; Umehara, A.; Kokubo, T. *J. Photopolym. Sci. Technol.* **1990**, *3*, 389.
66. Poot, A.; Delzenne, G.; Pollet, R.; Laridon, U. *J. Photogr. Sci.* **1971**, *19*, 88.
67. Naitoh, K.; Kanai, K.; Yamaoka, T.; Umehara, A. *J. Photopolym. Sci. Technol.* **1991**, *4*, 411.
68. Schwalm, R.; Binder, H.; Dunbay, B.; Krause, A. In *Polymers for Microelectronics—Science and Technology*; Tabata, Y.; Mita, I.; Nonogaki, S.; Horie, K.; Tagawa, S., Eds.; VCH Publishers: Deerfield Beach, FL, 1990; p 425.
69. Schwalm, R. *Proc. ACS Div. Polym. Mater. Sci. Eng.* **1989**, *61*, 278.
70. Houlihan, F. M.; Chin, E.; Nalamasu, O.; Kometani, J. M. In *Polymers for Microelectronics*; Thompson, L. F.; Willson, C. G.; Tagawa, S., Eds.; ACS Symposium Series 537; American Chemical Society: Washington, DC, 1994; p 23.

71. Allen, R. D.; Quan, P. Ly.; Wallraff, G. M.; Larson, C. E.; Hinsberg, W. D.; Conley, W. E.; Muller, K. P. *Proc. SPIE* **1993**, *1925*, 246.

72. Pawlowski, G.; Sauer, T.; Dammel, R.; Gordon, D. J.; Hinsberg, W.; McKean, D.; Lindley, C. R.; Merrem, H.-J.; Röschert, H.; Vicari, R.; Willson, C. G. *Proc. SPIE* **1990**, *1262*, 391.

73. Willson, C. G.; MacDonald, S. A.; Ito, H.; Fréchet, J. M. J. In *Polymers for Microelectronics—Science and Technology*; Tabata, Y.; Mita, I.; Nonogaki, S.; Horie, K.; Tagawa, S., Eds.; VHC Publishers: Deerfied Beach, FL; 1990; p 3.

74. McKean, D. R.; Hinsberg, W. D.; Sauer, T. P.; Willson, C. G.; Vicari, R.; Gordon, D. J. *J. Vac. Sci. Technol.* **1990**, *B8(6)*,1466.

75. Ito, H.; Schildknegt, K.; Mash, E. A. *Proc. SPIE* **1991**, *1466*, 408.

76. Sooriyakumaran R.; Ito, H.; Mash, E. A. *Proc. SPIE* **1991**, *1466*, 419.

77. Vogl, O. *J. Polym. Sci. Polym. Chem. Ed.* **1960**, *46*, 261.

78. Bowden, M. J.; Thompson, L. F. *Polym. Eng. Sci.* **1974**, *14*, 525.

79. Ito, H.; England, W. P.; Ueda, M. *J. Photopolym. Sci. Technol.* **1990**, *3*, 219.

80. Aso, C.; Tagami, S.; Kunitake, T. *J. Polym. Sci., Part A-1* **1969**, *7*, 497.

81. Ito, H.; Schwalm, R. *J. Electrochem. Soc.* **1989**, *136*, 241.

82. Ito, H.; Ueda, M.; Schwalm. R. *J. Vac. Sci. Technol.* **1988**, *B6(6)*, 2259.

83. Ito, H.; Ueda, M.; Renaldo, A. *J. Electrochem. Soc.* **1989**, *136*, 245.

84. Ito, H. *J. Photopolym. Sci. Technol.* **1992**, *5*, 123.

85. Ito, H.; Ueda, M. *Macromolecules* **1990**, *23*, 2885.

86. Ito, H.; Ueda, M.; Ito, T. *J. Photopolym. Sci. Technol.* **1990**, *3*, 335.

87. Ueda, M.; Ito, T.; Ito, H. *Macromolecules* **1990**, *23*, 2895.

88. Houlihan, F. M.; Bouchard, F.; Fréchet, J. M. J.; Willson, C. G. *Macromolecules* **1986**, *19*, 13.

89. Fréchet, J. M. J.; Eichler, E.; Stanciulescu, M.; Iizawa, T.; Bouchard, F.; Houlihan, F. M.; Willson, C. G. In *Polymers for High Technology*; Bowden, M. J.; Turner, S. R., Eds.; ACS Symposium Series 346; American Chemical Society: Washington, DC, 1987; p 138.

90. Fréchet, J. M. J.; Willson, C. G.; Iizawa, T.; Nishikubo, T.; Igarashi, K.; Fahey, J. In *Polymers in Microlithography*; Reichmanis, E.; MacDonald, S. A.; Iwayanagi, T., Eds.; ACS Symposium Series 412; American Chemical Society: Washington, DC; 1989; p 100.

91. Inaki, Y.; Matsumura, N.; Takemoto, K. In *Polymers for Microelectronics*; Thompson, L. F.; Willson, C. G.; Tagawa, S., Eds.; ACS Symposium Series 537; American Chemical Society: Washington, DC, 1994; p 142.

92. Ito, H.; Flores, E.; Renaldo, A. F. *J. Electrochem. Soc.* **1988**, *135*, 2328.

93. Aoai, T.; Umehara, A.; Kamiya, A.; Matsuda, N.; Aotani, Y. *Polym. Eng. Sci.* **1989**, *29*, 887.

94. Ito, H. In *Irradiation of Polymeric Materials*; Reichmanis, E.; Frank, C. W.; O'Donnell, J. H., Eds.; ACS Symposium Series 527; American Chemical Society: Washington, DC, 1993; p. 197.

95. Uchino, S.; Iwayanagi, T.; Ueno, T.; Hayashi, N. *Proc. SPIE* **1991**, *1466*, 429.

96. Ito, H.; Sooriyakumaran, R.; Mash, E. A. *J. Photopolym. Sci. Technol.* **1991**, *4*, 319.

97. Ito, H.; Maekawa, Y. In *Polymeric Materials for Microelectronic Applications*; Ito, H.; Tagawa, S.; Horie, K., Eds.; ACS Symposium Series 579; American Chemical Society: Washington, DC, 1994; p 70.

98. Stöver, H.; Matuszczak, S.; Chin, R.; Shimizu, K.; Willson, C. G.; Fréchet, J. M. J. *Proc. ACS Div. Polym. Mater. Sci. Eng.* **1989**, *61*, 412.

99. Ito, H.; Maekawa, Y.; Sooriyakumaran, R.; Mash, E. A. In *Polymers for Microelectronics*; Thompson, L. F.; Willson, C. G.; Tagawa, S., Eds.; ACS Symposium Series 537; American Chemical Society: Washington, DC, 1994; p 64.

100. Feeley, W. E.; Imhof, J. C.; Stein, C. M.; Fisher, T. A.; Legenza, M. W. *Polym. Eng. Sci.* **1986**, *26*, 1101.

101. Liu, H.-Y.; de Grandpre, M. P.; Feeley, W. E. *J. Vac. Sci. Technol.* **1988**, *B6*, 379.

102. Thackeray, J. W.; Orsula, G. W.; Pavelcheck, E. K.; Canistro, D. *Proc. SPIE* **1989**, *1086*, 34.

103. Thackeray, J. W.; Orsula, G. W.; Bohland, J. F.; McCullough, A. W. *J. Vac. Sci. Technol.* **1989**, *B7(6)*, 1620.

104. Berry, A. K.; Graziano, K. A.; Bogan, L. E., Jr.; Thackeray, J. W. In *Polymers in Microlithography*; Reichmanis, E.; MacDonald, S. A.; Iwayanagi, T., Eds.; ACS Symposium Series 412; American Chemical Society: Washington, DC, 1989; p 87.

105. Lingnau, J.; Dammel, R.; Theis, J. *Solid State Technol.* **1989**, *32(9)*, 105.

106. Lingnau, J.; Dammel, R.; Theis, J. *Solid State Technol.* **1989**, *32(10)*, 107.

107. Bohland, J. F.; Calabrese, G. S.; Cronin, M. F.; Canistro, D.; Fedynyshyn, T. H.; Ferrari, J.; Lamola, A. A.; Orsula, G. W.; Pavelchek, E. K.; Sinta, R.; Thackeray, J. W.; Berry, A. K.; Bogan, L. E., Jr.; de Grandpre, M. P.; Feeley, W. E.; Graziano, K. A.; Olsen, R.; Thompson, S.; Winkle, M. R. *J. Photopolym. Sci. Technol.* **1990**, *3*, 355.

108. Thackeray, J. W.; Orsula, G. W.; Rajaratnam, M. M.; Sinta, R.; Herr, D.; Pavelchek, E. *Proc. SPIE* **1991**, *1466*, 39.

109. Allen, M. T.; Calabrese, G. S.; Lamola, A. A.; Orsula, G. W.; Rajaratnam, M. M.; Sinta, R.; Thackeray, J. W. *J. Photopolym. Sci. Technol.* **1991**, *4*, 379.

110. Reck, B.; Allen, R. D.; Twieg, R. J.; Willson, C. G.; Matuszczak, S.; Stover, H. D. H.; Li, N. H.; Fréchet, J. M. J. *Polym. Eng. Sci.* **1989**, *29*, 960.

111. Fréchet, J. M. J.; Matuszczak, S.; Stover, H. D. H.; Willson, C. G.; Reck, B. In *Polymers in Microlithography*; Reichmanis, E.; MacDonald, S. A.; Iwayanagi, T., Eds.; ACS Symposium Series 412; American chemical Society: Washington, DC, 1989; p 74.

112. Fréchet, J. M. J.; Kryczka, B.; Matuszczak, S.; Reck, B.; Stanciulescu, M.; Willson, C. G. *J. Photopolym. Sci. Technol.* **1990**, *3*, 235.

113. Stöver, H. D. H.; Matuszczak, S.; Willson, C. G.; Fréchet, J. M. J. *Macromolecules* **1991**, *24*, 1741.

114. Fréchet, J. M. J.; Matuszczak, S.; Reck, B.; Stöver, H. D. H.; Willson, C. G. *Macromolecules* **1991**, *24*, 1746.

115. Fahey, J. T.; Fréchet, J. M. J. *Proc. SPIE* **1991**, *1466*, 67.

116. Ueno, T.; Shiraishi, H.; Hayashi, N.; Tadano, K.; Fukuma, E.; Iwayanagi, T. *Proc. SPIE* **1990**, *1262*, 26.

117. Shiraishi, H.; Fukuma, E.; Hayashi, N.; Ueno, T.; Tadano, K.; Iwayanagi, T. *J. Photopolym. Sci. Technol.* **1990,** *3,* 385.

118. Shiraishi, H.; Fukuma, E.; Hayashi, N.; Tadano, K.; Ueno, T. *Chem. Mater.* **1990,** *3,* 621.

119. McKean, D. R.; Clecak, N. J.; Pederson, L. A. *Proc. SPIE* **1990,** *1262,* 110.

120. Stewart, K. J.; Hatzakis, M.; Shaw, J. M.; Seeger, D. E.; Neumann, E. *J. Vac. Sci. Technol.* **1989,** *B7,* 1734.

121. Stewart, K. J.; Hatzakis, M.; Shaw, J. M. *Polym. Eng. Sci.* **1989,** *29,* 907.

122. Dubois, J. C.; Eranian, A.; Datmanti, E. *Proc. Electrochem. Soc.* **1978,** *78(5),* 303.

123. Crivello, J. V. In *Polymers in Electronics*; Davidson, T., Ed.; ACS Symposium Series 242; American Chemical Society: Washington, DC, 1984; p 3.

124. Conley, W. E.; Moreau, W.; Perreault, S.; Spinillo, G.; Wood, R. *Proc. SPIE* **1990,** *1262,* 49.

125. Allen, R. D.; Conley, W.; Gelorme, J. D. *Proc. SPIE* **1992,** *1672,* 513.

126. Sakata, M.; Kosuge, M.; Ito, T.; Yamashita, Y. *J. Photopolym. Sci. Technol.* **1991,** *4,* 75.

127. Sakata, M.; Ito, T.; Kosuge, M.; Yamashita, Y. *J. Photopolym. Sci. Technol.* **1992,** *5,* 181.

128. Kawai, Y.; Tanaka, A.; Ban, H.; Nakamura, J.; Matsuda, T. *J. Photopolym. Sci. Technol.* **1992,** *5,* 431.

129. Brunsvold, W.; Stewart, K.; Jagannathan, P.; Sooriyakumaran, R.; Parrill, J.; Muller, K. P.; Sachdev, H. *Proc. SPIE* **1993,** *1925,* 377.

130. Bonfils, F.; Giral, L.; Montginoul, C.; Sagnes, R.; Schue, F. *Makromol. Chem.* **1992,** *193,* 1289.

131. Shirai, M.; Miwa, T.; Tsunooka, M. *J. Photopolym. Sci. Technol.* **1993,** *6,* 27.

132. Ito, H. *J. Photopolym. Sci. Technol.* **1992,** *5,* 123.

133. Steinman, A. *Proc. SPIE* **1988,** *920,* 13.

134. MacDonald, S. A.; Ito, H.; Hiraoka, H.; Willson, C. G. *Technical Papers of SPE Regional Technical Conference on Photopolymers*; Society of Plastics Engineers: Brookfield, CT, 1985; p 177.

135. MacDonald, S. A.; Schlosser, H.; Ito, H.; Clecak, N. J.; Willson, C. G. *Chem. Mater.* **1991,** *3,* 435.

136. MacDonald, S. A. *Proc. ACS Div. Polym. Mater. Sci. Eng.* **1992,** *66,* 97.

137. Schellekens, J. P. W.; Visser, R. J. *Proc. SPIE* **1989,** *1086,* 220.

138. Sebald, M.; Berthold, J.; Beyer, M.; Leuschner, R.; Nölscher, Ch.; Scheler, U.; Sezi, R.; Ahne, H.; Birkle, S. *Proc. SPIE* **1991,** *1466,* 227.

139. Schlegel, L.; Ueno, T.; Hayashi, N.; Iwayanagi, T. *J. Vac. Sci. Technol.* **1991,** *B9(2),* 278.

140. Seligson, D.; Das, S.; Gaw, H. *J. Vac. Sci. Technol.* **1988,** *B6,* 2303.

141. Ban, H.; Nakamura, J.; Deguchi, K.; Tanaka, A. *J. Vac. Sci. Technol.* **1991,** *B9,* 3387.

142. Nakamura, J.; Ban, H.; Tanaka, A. *Jpn. J. Appl. Phys.* **1992,** *31,* 4294.

143. Asakawa, K. *J. Photopolym. Sci. Technol.* **1993,** *6,* 505.

144. MacDonald, S. A.; Clecak, N. J.; Wendt, H. R.; Willson, C. G.; Snyder, C. D.; Knors, C. J.; Deyoe, N. B.; Maltabes, J. G.; Morrow, J. R.; McGuire, A. E.; Holmes, S. J. *Proc. SPIE* **1991,** *1466,* 2.

145. MacDonald, S. A.; Hinsberg, W. D.; Wendt, H. R.; Clecak, N. J.; Willson, C. G.; Snyder, C. D. *Chem. Mater.* **1993,** *5,* 348.

146. Nalamasu, O.; Reichmanis, E.; Cheng, M.; Pol, V.; Kometani, J. M.; Houlihan, F. M.; Neenan, T. X.; Bohrer, M. P.; Mixon, D. A.; Thompson, L. F.; Takemoto, C. *Proc. SPIE* **1991,** *1466,* 13.

147. Schwartzkopf, G.; Niazy, N. N.; Das, S.; Surendran, G.; Covington, J. B. *Proc. SPIE* **1991,** *1466,* 26.

148. Nalamasu, O.; Cheng, M.; Timko, A. G.; Pol, V.; Reichmanis, E.; Thompson, L. F. *J. Photopolym. Sci. Technol.* **1991,** *4,* 299.

149. Oikawa, A.; Santoh, N.; Miyata, S.; Hatakenaka, Y.; Tanaka, H.; Nakagawa, K. *Proc. SPIE* **1993,** *1925,* 92.

150. Kumada, T.; Tanaka, Y.; Ueyama, A.; Kubota, S.; Koezuka, H.; Hanawa, T.; Morimoto, H. *Proc. SPIE* **1993,** *1925,* 31.

151. Röschert, H.; Przybilla, K.-J.; Spiess, W.; Wengenroth, H.; Pawlowski, G. *Proc. SPIE* **1992,** *1672,* 33.

152. Funhoff, D. J. H.; Binder, H.; Schwalm, R. *Proc. SPIE* **1992,** *1672,* 46.

153. Hinsberg, W. D.; MacDonald, S. A.; Snyder, C. D.; Ito, H.; Allen, R. D. In *Polymers for Microelectronics*; Thompson, L. F.; Willson, C. G.; Tagawa, S., Eds.; ACS Symposium Series 537; American Chemical Society: Washington, DC, 1994; p 101.

154. Hinsberg, W.; MacDonald, S. A.; Clecak, N.; Snyder, C.; Ito, H. *Proc. SPIE* **1993,** *1925,* 43.

155. Ito, H.; England, W. P.; Sooriyakumaran, R.; Clecak, N. J.; Breyta, G.; Hinsberg, W. D.; Lee, H.; Yoon, D. Y. *J. Photopolym. Sci. Technol.* **1993,** *6,* 547.

Polymer Modification by Ion Implantation: Ion Bombardment and Characterization

Y. Q. Wang, L. B. Bridwell, and R. E. Giedd

In this chapter, we review the basic principles of ion implantation In polymers and the structural characterization of ion-implanted polymers. Because of the multi-element composition and structural nonrigidity of polymers (compared with metals), the ion-beam bombardment of polymers is complicated. Ion bombardment breaks covalent bonds, releases volatile (gaseous) species, and forms carbon-rich materials In ion-implanted polymer regions. The implanted polymers are characterized by various analytical methods such as Raman spectroscopy, IR, electron spin resonance, X-ray photoelectron spectroscopy, electron energy loss spectroscopy, scanning electron microscopy, transmission electron microscopy, Rutherford back scattering, and elastic recoil detection analysis. Structural alteration In the implanted layer leads to changes in the surface electrical properties and electronic conduction behavior of the polymer. These properties and applications of ion-implanted polymers will be discussed in Chapter 2.6.

Applications for polymeric solids have become much more numerous in recent years because of the many unique advantages of the solids. These advantages include light weight, formability, versatile electronic properties, and low manufacturing cost. Current and potential uses span a wide range of applications, including exterior structural components, load-bearing elements for high-speed moving parts, electronic circuit and packaging materials, diffraction grating substrates, materials for antistatic shielding, and temperature and pressure sensor elements. The rapid growth in the field of material science and engineering drives the need for advanced polymer structures for use in composite materials.

Despite the numerous applications for advanced polymers, further use is limited by inherent softness, low thermal stability, and electrical insulating properties. The highly insulating nature of commodity polymers has caused many problems such as triboelectrifi-

cation, attraction of dust, electrostatic discharging, and electromagnetic interference. Low thermal stability, softness, poor resistance to wear, abrasion, erosion, and fatigue reduce the potential use of polymers in chemically reactive environments or at elevated temperatures. To improve the electrical and mechanical properties of polymeric materials, ion-implantation techniques have been used widely since 1980, and several comprehensive research papers and reviews of the state of the art have been published (1–14).

Ion-implanted polymers have found applications in antistatic packaging, neural networks (resistor materials), temperature sensors, pressure sensors, and hard surface coatings. The ion-implantation technique has many unique advantages compared with the conventional chemical modification of polymers. First, ion implantation is a shallow surface interaction process. The thickness of the ion-modified region can be tuned conveniently by ion energy and ion species, so a desired property on the polymer surface can be produced without altering the desired bulk physical properties of the polymer. Depending on the ion energy and mass as well as the polymer host, thickness of the implanted region can vary from ca. 100 Å to micrometers. Second, ion implantation is a much better controlled process for doping compared with chemical or thermal diffusion processes. Ion implantation can achieve greater impurity accuracy, better control of impurity uniformity, and better control of impurity location in the sample. Third, ion implantation allows doping of materials that cannot be doped chemically. Whether these compositions are stable in the long-term remains to be established.

Figure 1 is a schematic illustration of ion implantation setup at Southwest Missouri State University. Ions produced in the ion source are extracted and accelerated by a high voltage (50 kV) through an analyzing magnet. These ions are then separated according to their charge-to-mass ratios and focused onto a target wheel, where the sample is mounted. To achieve a uniform implantation, either the target wheel is mechanically moved (for this implanter) or the ion beam is electrically scanned (for most other implanters).

From a structural point of view, many polymeric materials are amorphous and have no three-dimen-

sional long-range order. The medium-range structure of polymers has been qualitatively described by the random coil model (15) and further quantitatively interpreted by the free volume theory (16) and the fractal theory (17). Figure 2 shows a two-dimensional representation of the polymers by the random coil model (15), where the bold line represents a polymer chain and the dots on the line represent polymer repeating units. Amorphous polymer structures may be visualized roughly as a bowl of spaghetti or a pile of entangled snakes.

When an ion beam bombards the surface of a polymer, great physical damage occurs on the surface. This damage includes the scissoring of polymer chains (destruction of covalent bonds), cross-linking of extended structures, chemical decomposition (release of volatile species), and substantial alteration of the molecular arrangement. As a result, this damage drastically alters the physical properties of the polymer surface.

In this chapter, we review the research and development of the polymer ion-implantation technology with an emphasis on ion-beam induced microstructure and characterization of the microstructure. The discussion about electrical conductivity enhancements, electronic conduction processes, and applica-

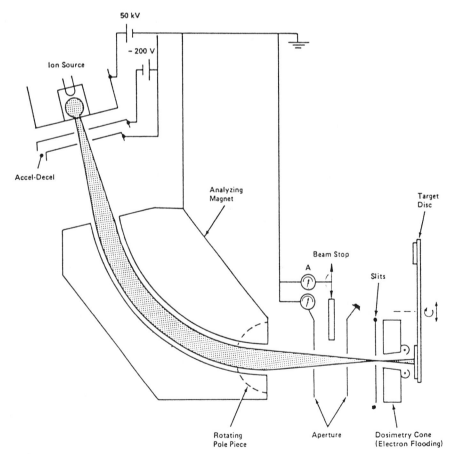

Figure 1. Schematic illustration of ion-implantation setup at Southwest Missouri State University.

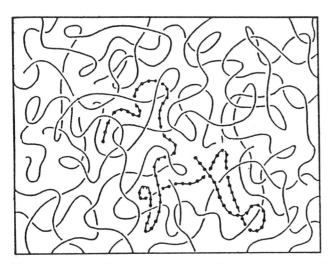

Figure 2. A two-dimensional schematic representation of polymers by the random coil model. The bold line represents a single polymer chain and the dots on the line represent polymer repeating units. (Reproduced with permission from reference 15. Copyright 1983.)

tions of ion-implanted polymers will be presented in Chapter 2.6.

Ion Bombardment of Polymers

Ion Energy Deposition Processes

When an ion penetrates a solid, it undergoes a succession of collisions with target atoms and their electrons and loses energy at each encounter. The description of energy transfer from the projectile ion to the solid is simplified by considering the electronic and nuclear collision processes separately. The principle of energy transfer from energetic ions to target atoms and their electrons is well understood in the framework of the Bethe–Born formalism (18). An important parameter characterizing the ion-to-target energy transfer is the stopping power (dE/dX), defined as the deposited energy (dE) per unit length (dX) along the ion track.

Another useful parameter is the ion mean-penetration range (R_p): that is, the most probable penetration depth. Because energy is lost by the projectile to target atoms and their electrons in a series of discrete collisions, the energy loss per collision, and hence the total ion path length, will have a statistical spread of values. This set of results leads to a near-Gaussian distribution of stopping distances, which can be evaluated by Lindhard–Scharff–Schiott theory (19) and by the transport of ions in matters (TRIM) Monte Carlo code (20). The thickness of the ion-modified layer (X) can be estimated roughly by eq 1, where σ_p is the standard deviation of the near-Gaussian distribution for the ions.

$$X = R_p + 1.18 \, \sigma_p \tag{1}$$

The TRIM code has been used successfully to simulate the ion energy loss process in simple solids such as monatomic metals, diatomic metals, and semiconductors. The use of TRIM for organic polymers is less appropriate because of the complexity of polymer structures and stoichiometries. The variation of density values for polymer films can also cause a great deal of uncertainty for both the mean penetration range and energy-loss parameters. In addition, the chemical bond effects on stopping power should be considered for low energy (keV) ion implantations (21). Nevertheless, the TRIM code is still widely used to estimate the ion energy-deposition parameters in polymers because the uncertainty of polymer film parameters such as the film density can easily overshadow the uncertainty brought in by the TRIM.

Figure 3 shows the TRIM simulated results for polyethersulfone (PES) films implanted with 50-keV He and Xe ions. The energy loss mechanism for a light helium ion is quite different from that for a much heavier xenon ion (Figure 3). In the helium implant the ratio of the energy loss through electronic collisions to that of the nuclear collisions is about 100. This ratio is only 2 for xenon implants. Each of these two processes results in a different ion energy loss and mean penetration depth and hence the formation of substantially different materials. This ratio also varies for implants with the same ions but with different energies. TRIM simulations also indicate that a high energy (MeV/amu) ion will deposit most of its energy through the electronic collisions (ionization and excitation), whereas for a low energy (keV/amu) ion, both nuclear and electronic collisions are important.

TRIM calculations show that typical amounts of deposited energy for heavy ions with acceleration energies in the kilo-electronvolt range are between 50 and 500 eV/A. This energy is deposited in a short time (10^{-13} s) and is localized in a small region around the ion track. The radial size of the ion track is important because it is critical in characterizing the structures formed near the ion track region. This radial distance is a strong function of the ion-deposited energy density and collision mechanism. For high energy ions (MeV/amu), the electronic collision mechanism dominates, and the size of the tracks is determined by the energy of ionized electrons (δ-rays). In this case, radius of the track is estimated to be about 1–10 Å for the primary knock-on ion collision events and about 50–100 Å for the secondary events caused by the aforementioned δ-rays (22). For low energy ions (keV/amu) the dominant energy-deposition process involves elastic binary collisions, and the energy localization volume coincides with the collision cascade volume. The electronic collision mechanism continues to operate for low energy ions and contributes to an increase in radius of the ion tracks. Therefore, the typ-

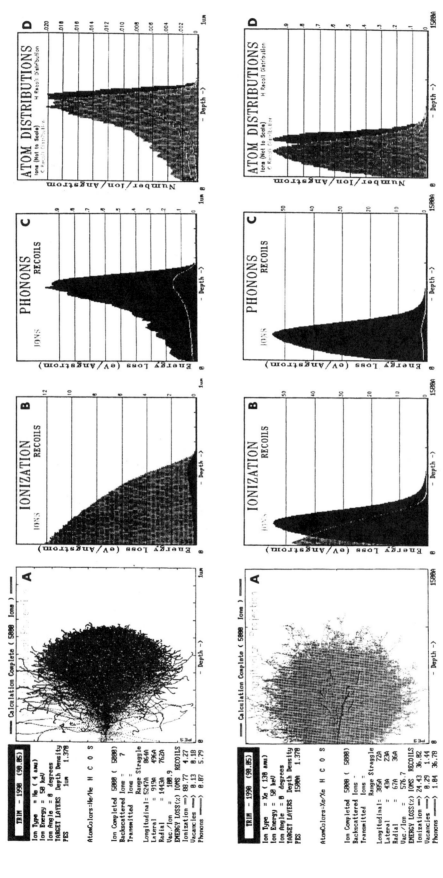

Figure 3. Simulation of 50-keV He- and Xe-implantation in polyethersulfone (PES) by the TRIM Monte Carlo code: A, ion track and collision cascade distribution; B, energy loss through ionizations; C, energy loss through phonons; D, energy loss through nuclear displacements.

ical maximum radius is 10–30 Å for the upper (keV) regime (22).

From the target point of view, the ion-track region is actually a plasma cluster of electrons, ions, excited species, and recoiled macroradicals. The target material in the cluster region experiences radiation damage as a result of the ion energy transfer. The radiation damage resulting from nuclear collisions usually refers to the displacement of target atoms, including replacement collisions and vacancies. These vacancies refer to normal interstitial vacancies as well as atoms that leave the target. In addition to these defects, electronic collisions in insulators such as organic polymers may also create defects by the Pooley processes (where δ-ray energy is converted into kinetic energy of the target atoms) or by the liberation of valence electrons and destruction of chemical bonds (both of which form free radicals) (23).

Ion implantation is a well-controlled means to introduce impurities into metals and semiconductors, and the tremendous amounts of accompanying damage, dislocations, and other defects can be removed by postimplantation annealing or by in situ annealing during implantation. However, most organic polymers, which have glass transition temperatures (T_g) <400 °C, cannot be annealed usefully following ion implantation, because chemical decomposition occurs and the displaced atoms (and molecules) cannot be re-formed to the initial polymer structure. The large number of defects around the ion tracks, which cannot be annealed, tend to distort the chemical effects of the implant ion. As a result, organic polymers, compared with metals and semiconductors, are relatively unaffected by the chemical doping effect of the impurity. Therefore, the ion-beam impact on polymers is usually referred to as the ion-created damage effect rather than the ion doping effect. In the case where the impurity level is very high (>10 atom%, which is high enough to form an alloy structure), the chemical effect of the impurity will be important.

Molecular Emission and Polymer Carbonization

Polymer materials change as a result of ion-beam radiation (24) in the following ways:

- irreversible cleavage of covalent bonds within a molecule that results in fragmentation and loss of various volatile species
- formation of chemical bonds between different molecules or different parts of the same macromolecule
- formation and disappearance of unsaturated bonds in the molecular structure

The evolution of the volatile species during ion bombardment is so prevalent that a significant change in the base pressure of the target chamber can be observed. Figure 4 shows how the target chamber pressure changes with ion-beam dose in the As implantation of PES at 50 keV (14). The sudden increase in pressure at the beginning of the implantation results from volatile gas evolution during the initial bombardment. The slow decay indicates a continuous degradation and an irreversible alteration of the PES surface. By using a residual gas analyzer (RGA), the type of gas evolved can be identified. In one experiment, Venkatesan (25) found that more than six kinds of molecules had evolved from the polystyrene (PS) films during Ar implantation at 200 keV (H_2, C_2H_2, C_6H_6, CH_4, C_8H_8, and C_3H_5 were evolved in descending abundance). Most of the ejected species in polymer implantation are small neutral molecules such as hydrogen and acetylene. The ionized components of these molecules are two orders of magnitude lower in abundance. Weak emission of monomer species indicates that polymer decomposition during the implantation process does not occur through simple pyrolysis.

The chemical yield of the bombardment (G) is defined as the number of radiolytic events or active species (ions, radicals, molecules, or bonds) changed or produced per 100 eV of absorbed energy. This parameter G has been used to more quantitatively describe the polymer degradation process during the ion implantation. The expression of G is given by eq 2, where Q_0 is the maximum emission yield, I is the current, and E is the energy deposited by the ion (5).

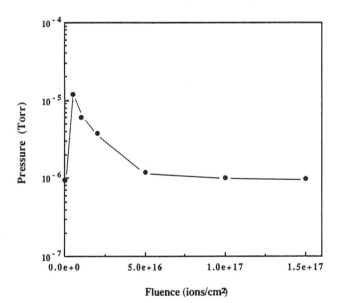

Figure 4. Pressure–dose relationship in the target chamber for As-implantation of polyethersulfone (PES) at 50 keV. (Reproduced with permission from reference 14. Copyright 1993.)

$$G = (Q_0/I)/(100/E) \qquad (2)$$

Measurements of G for 100–200-keV He- and Ar-implanted PS and polyethylene (PE) indicate that high yields for light species such as H_2 and C_2H_2 are observed (5). For a material like PS, which is considered to be a radiation-resistant material, the abundances of these light species lead to the conclusion that such light species are produced during any polymer implantation process. A similar conclusion has also been reached from ion irradiation of poly(methyl methacrylate) films (25).

Values of G for H_2 and C_2H_2 as a function of ion energy loss (dE/dX) for ion-implanted PS in the kiloelectronvolt energy range are shown in Figure 5 (5). This approximately sigmoidal (s-shaped) graph was also observed by Lewis and Lee (26) for the irradiation of PE and Kapton films with 200-keV H_2, He, and B and 1-MeV Ar. The position of the center of the sigmoid (the incremental step) occurs at approximately 18–20 eV/A for various polymer–ion-energy combinations. This universal behavior can be understood by assuming nuclear recoil plays a major role in the generation of the volatile species (26). The position of this incremental step seems to be correlated with an increment in the surface hardness of ion-irradiated PE and Kapton (Figure 6) (26). A similar sigmoid is also obtained from plotting the surface electrical conductivity as a function of average stopping power for PES films implanted by 50-keV He, B, C, N, and As (Figure 7) (27).

Ion irradiation also causes a tremendous change in the optical properties of polymers. An indication of this change is seen in the surface appearance. All experiments have indicated that the translucent gloss brown or opaque black color after implantation is indepen-

dent of the color of the unimplanted polymer. This color change is permanent and implies that an irreversible alteration has occurred in polymer structure. We have mentioned that during the polymer-implantation process, a plasma-like cluster of atoms is produced around the ion track. The evolution of volatile gases

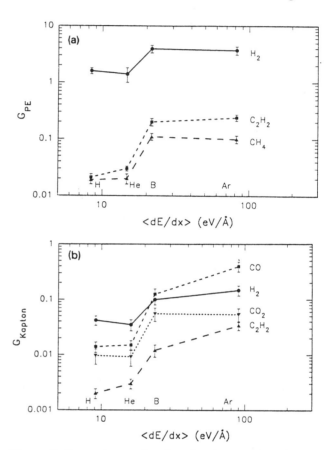

Figure 6. Dependance of surface hardness on ion energy loss for PE (a) and Kapton (b) implanted with 200-keV H_2, 200-keV He, 200-keV B, and 1-MeV Ar. (Reproduced with permission from reference 26. Copyright 1992.)

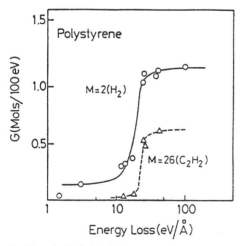

Figure 5. Chemical yields for H_2 and C_2H_2 as a function of ion energy loss for PS implanted in the kilo-electronvolt energy range. (Reproduced with permission from reference 5. Copyright 1987).

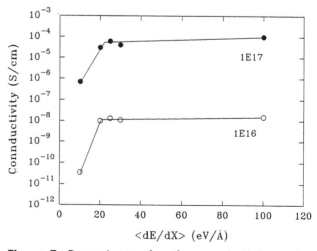

Figure 7. Dependance of surface conductivity on ion energy loss for 50-keV ion-implanted PES.

such as H_2, C_2H_2, N_2, or O_2 from the plasma cluster during ion irradiation leaves the cluster with an increased percentage of carbon atoms. At sufficient doses, these clusters will overlap and completely replace the polymer surface with a carbon-enriched surface. The existence of such carbon-enriched surfaces was observed by neutron scattering (28) and by the combination of transmission electron microscopy (TEM) and electron energy loss spectroscopy (EELS) (29).

The darkening of the surface increases linearly with dose at low and intermediate doses. Measurement of depth versus coloration by Fink et al. (30) indicated that the majority of the color change occurred near the end of the ion's mean penetration range. However, this observation was not supported by Davenas et al. (31). In the Davenas study, optical absorption measurements showed that degradation of polyimide occurred predominantly by electronic processes that occurred close to the surface. At higher doses ($>10^{15}$ ions/cm^2), opaque films are formed consistent with an increase and saturation effect in the surface conductivity. The simultaneous increase and saturation curve in optical density and electrical conductivity was observed simultaneously by Elman et al. (32) for 50-keV N-implanted polydiacetylene (PDA) thin films at the dose range of 1×10^{15} to 5×10^{16} ions/cm^2.

Ion-Induced Microstructures

We mentioned that the ion energy loss creates the plasma cluster around the ion track. Because of the effluent release of hydrogen and other volatile species from the cluster during the implantation, the cluster finally becomes hydrogen depleted and carbon enriched (termed *carbon cluster*). The model of ion-induced carbon clusters was first used by Wasserman (4) to describe the polymer insulator–conductor transformation by a percolating fractal network. This model was further developed by Davenas and Boiteaux (7) to include the formation of isolated clusters associated with polymer degradation and groups of clusters associated with the percolation network. The internal structure of the cluster as well as the number density of the clusters will determine the macroscopic properties of the irradiated polymer.

A quantitative description of the internal structure of the clusters has been very difficult because of the extreme complexity of the ion–polymer interaction involved. However, varying internal structures are assumed to account for the diverse properties of the same polymer implanted by the same dose under different energy and ion conditions. Venkatesan et al. (2) suggested that implantation at low energy (where nuclear stopping dominates) results in an amorphous carbon material with a relatively low conductivity, whereas implantation at higher energies (where electronic stopping dominates) results in a highly conducting carbon-rich material. Recent studies on poly-(ether ether ketone) (PEEK) films implanted with different ions, energies, and doses by Sofield et al. (33) provide more quantitative results. Using the combination of depth profiles on scattering spectra from a Raman microprobe and conductivity measurements, Sofield et al. concluded that a threshold energy-deposition rate exists. At energy-deposition rates above the threshold, a highly conducting material is formed; below the threshold, a less conducting amorphous material is formed.

A conducting grain model was recently suggested by Wang et al. (34) to explain this energy-deposition rate threshold. This model is based on the assumption that the intricate ion–polymer interaction process is a rapid dissipation of heat resulting from kinetic energy in the ion track region and its immediate vicinity (the cluster). This model can be used to determine the approximate internal structure of the clusters. When an energetic ion transfers its kinetic energy to a polymer matrix, δ-rays (hot electrons) as well as atomic vibrations and displacements (hot atoms) result. For high energy ions (MeV/amu), the dominant interaction results in the formation of hot electrons. The hot electron energy-deposition density (ε) is given by eq 3, where S_e is the electronic stopping power and R is radius of the ion track.

$$\varepsilon = S_e/\pi R^2 \tag{3}$$

Many of the hot electron tracks will overlap and thus deposit kinetic energy in a cylindrical region around the ion trajectory. Initially, this cylindrical region has a high temperature relative to the surrounding medium. By using the adiabatic approximation, the initial temperature T_0 of the cylindrical track is given by eq 4, where ρ is the polymer density, and c_p is specific heat of the polymer (34).

$$T_0 = \varepsilon/\rho c_p \tag{4}$$

Thus, the maximum temperature in the cylindrical region, which is of course, T_0, is determined by ε rather than the stopping power (S_e).

The magnitude of T_0 and the relaxation of temperature from T_0 to the final background temperature are critical in determining the structure of the material in the cluster. In principle, a thermodynamic relaxation of T_0 in the cylindrical cluster should form a graphite-like or even graphitic structure provided that T_0 is sufficiently high (>3000 K). At lower values of T_0 it is possible and even likely that many nonthermodynamic relaxation processes dominate and result, in a diamond-like or amorphous carbon cluster structure (8). If we assume a critical temperature (T_c) above which the thermodynamic relaxation dominates and below which the nonthermodynamic relaxation dominates, then a critical energy-deposition density (ε_c) corresponding to T_c can be found through eq 4.

The temperature gradient in the cylindrical cluster region is not uniform. In the radial direction, the closer to the center of the cylinder the greater the temperature will be. Along the axis of the cylinder (the ion trajectory) the temperature is a function of the energy loss and thus a function of the ion energy. The cooling rate in the high temperature region (usually thousands of degrees Celsius) along the ion path is slow enough so that the thermodynamic relaxation dominates. The reason for this may be that the high temperature track region is not in contact with the cold background region (near room temperature) but must conduct the heat away though the intermediate temperature region, where the thermal conductivity is poor. At large radial distances (near the wall of the cluster), the dissipation rate could be very high because of the quenching effect of the background material outside of the cluster. As a result, graphite-like or even graphitic material would be formed in the center of the cylinder, and amorphous carbon or diamond-like material would be formed around the wall of the cylinder. Figure 8 is a schematic illustration showing the formation of carbon clusters as a result of ion implantation.

For low energy (keV/amu) ion implants and for high energy implants toward the ion's mean penetration range (R_p), the nuclear energy-deposition process dominates. The hot atoms (phonons and displacements) induced by elastic nuclear collisions in the target create high temperature clusters. However, there are a number of reasons why the temperature of these clusters does not reach the critical temperature (T_c) of producing graphitic structures. First, the time constant for nuclear interactions (10^{-13} s) is much shorter than for electronic interactions (10^{-11} s), and as a result the overlapping of paths for phonons and displaced atoms will not occur often. Second, the nuclear cascade collisions encompass a larger volume and result in a significantly reduced energy-deposition density and a spherical shape. And finally, the lifetime of the temperature gradients caused by nuclear collisions may just be so small that a thermodynamic relaxation process cannot occur at all. A temperature $< T_c$ for the clusters results in the creation of the amorphous carbon or diamond-like material.

Support for the model comes from a comparison of the strength of the covalent bonds with the minimum energy-deposition density. By assuming a T_c of 3300 K for a polymeric structure (35), the minimum energy-deposition density (ε_c) is estimated (34) to be approximately 55 eV/nm^3, a result corresponding to 0.6 eV/atom. This number is close to the average strength (1 eV/atom) of a covalent bond in polymers.

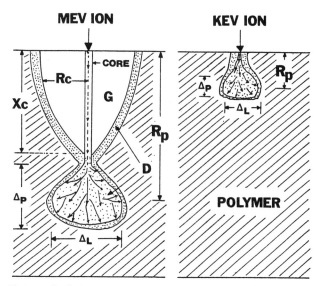

Figure 8. Schematic illustration of ion-induced carbon clusters in polymers for mega-electronvolt and kilo-electronvolt implantation: R_p, ΔR_p and ΔR_L are the ion's mean penetration range, longitudinal spreading, and lateral spreading, respectively; G is the better conducting graphite-like carbon; D is the poor conducting diamond-like carbon; R_c and X_c are the radius and length of the conducting carbon cluster, respectively.

In conclusion, an implant with kilo-electronvolt energies, induces a more spherical and less conducting amorphous (diamond-like) carbon cluster around the ion track. An implant at mega-electronvolt energies, shapes the cluster into a bullet with a spherical protrusion at the tip of the bullet, toward the ion's mean penetration range. The graphite material is in the center of the bullet. The amorphous carbon (diamond-like) material extends from the spherical section to the shells of the bullet (Figure 8).

Structural Characterization

Many polymers have been studied by ion implantation. A variety of ion-beam conditions (ion mass, energy, dose, and dose rate) have been used. Appendix 1 lists some of the polymers, their abbreviation, and the ion-beam conditions discussed in this chapter and Chapter 2.6.

Polymer Stoichiometry

The evolution of volatile species during ion irradiation alters the stoichiometry of the polymer implant layer. The main constituents of polymer materials are light elements such as hydrogen, carbon, nitrogen, and oxygen. To measure surface composition, several high-energy ion-beam analysis techniques, such as Rutherford back scattering (RBS), nuclear reaction analysis (NRA), and elastic recoil detection analysis (ERDA), have been used. The change in polymer stoichiometry can be correlated to the emission of gas molecules. Such a correlation is demonstrated in Figure 9 for 1.5-MeV He-implanted PE, where hydrogen content in the surface and hydrogen gas evolved from the surface are shown as a function of dose (*36*). The exponential emission of hydrogen as a function of dose is in agreement with the change of hydrogen in the implant layer. This relationship is also true, to a lesser extent, for other elemental constituents of the implant layer. A correlation study was also done by NRA for Teflon and Kapton films implanted with He, N, and Si ions in the energy range of 200 keV to 2 MeV (*37*). This study resulted in similar conclusions.

Studies of high-dose As-implanted polyamide–imide (PAI) films (*14*) indicate that both hydrogen content and the thickness of the hydrogen depleted region vary with ion dose. PAI films with high glass transition temperatures ($T_g \sim 400$ °C) implanted with 180-keV As in the dose range of 10^{15} to 10^{17} ions/cm^2 were analyzed for hydrogen content by ERDA with a 30-MeV Cl^{6+} beam. A double-layer hydrogen distribution is characteristic for these ion-irradiated films, compared with a uniform distribution in unimplanted, pristine polymer films. A typical spectrum for one of the PAI samples (TMPA–FDA) is shown in Figure 10 (*13*). The density change at the surface layer, a result of implant-induced carbonization and densification, is normalized by converting the ERDA-measured

areal density (atoms/cm^2) into a depth scale. The results are listed in Table I (*14*). An increase in the thickness of the implanted layer with increasing beam dose is a result of an increased number of ion collisions and hence damage at the region beyond the ion's mean penetration range. The reason that the damage at the deep tail of the ion distribution increases the thickness of the implanted layer is that the polymers are so "tender" that a little amount of irradiation can cause a significant evolution of hydrogen.

Hydrogen content of PEEK films irradiated with different ion beam conditions was measured (*13*) and is listed in Table I. A considerable amount of hydrogen still remains in the implanted polymer region after implantation with high doses or high energies. A reason for the relatively high hydrogen content after implantation may be that a balance is established between free hydrogen diffusion and hydrogen trap-

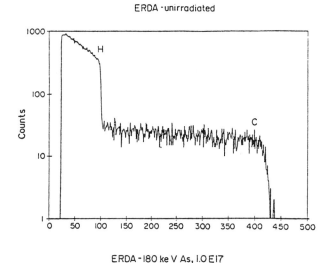

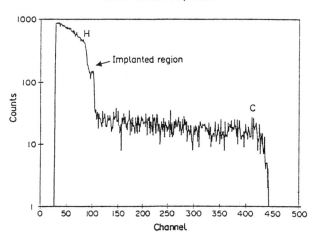

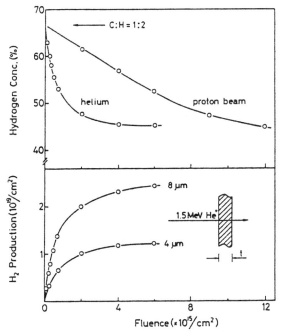

Figure 9. Hydrogen concentration as a function of dose for 1.5-MeV H- and He-implanted PE (top). Hydrogen production as a function of dose for two thicknesses of PE films (bottom) (Reproduced with permission from reference 36. Copyright 1985.)

Figure 10. Typical energy spectra collected for TMPA–FDA polymer films by ERDA: (top) unimplanted and (bottom) As-implanted at 180 keV with 10^{17} ions/cm^2. (Reproduced with permission from reference 13. Copyright 1993.)

Table I. Hydrogen Content (H/C) and Thickness of Implanted High T_g Polyamide–Imides (PAIs) and Poly(ether ether ketone) (PEEK) Films with Various Beam Conditions, Obtained by Elastic Recoil Detection Analysis (ERDA)

Sample	Ion	Energy (keV)	Dose (cm^{-2})	Layer 1 Thickness (Å)	Layer 1 H/C (at%)	Layer 2 H/C (at%)
IPMA–FDA			0		48.0	48.0
IPMA–FDA	As	180	10^{15}	1610	25.0	44.9
IPMA–FDA	As	180	10^{16}	1700	22.0	51.5
IPMA–FDA	As	180	10^{17}	2000	13.6	47.1
TMPA–BAPP	As	180	10^{17}	2000	12.4	42.9
TMPA–FDA	As	180	10^{17}	1700	15.0	47.1
BTGL–BAPP	As	180	10^{17}	2000	12.4	51.5
PDMSI–FDA	As	180	10^{17}	1700	14.9	58.7
PEEK	As	50	2.5×10^{17}		20.0	63.2
PEEK	As	180	10^{17}		15.0	63.2
PEEK	F	20,000	2×10^{16}		9.0	63.2
PEEK	Ni	41,700	10^{15}		9.0	63.2

Note: Layer 1 is the implanted region. Layer 2 is the unimplanted region.

Source: Reproduced with permission from references 13 and 14. Copyright 1993.

ping at defects or vacant sites. The remaining hydrogen plays an important role in determining the final structures and properties of implanted polymers, because it affects the cross-linkings of ion-induced radicals and thus affects the formation of three-dimensional carbon networks in the implanted region.

Measurements of other constituents such as C, N, O, F, S, and Cl are often done by RBS as well as NRA. With the exception of C, all of these constituents usually decrease with increasing ion dose. The relative carbon content increases as a result of the density change. RBS is also used to monitor the incident ion distribution in the implant layer. RBS studies of the ion distribution in 50-keV As- and Xe-implanted PAI films with high T_g values indicate a dose saturation effect (Figure 11) (13, 14). The As content of the implant layer saturated at 7.6×10^{16} ions/cm^2, even when the implant dose was as high as 2.0×10^{17} ions/cm^2. This dose saturation effect is mainly a result of the surface sputtering of As atoms at such high doses, because evidence for As diffusion out of the implanted region is not sufficient to explain this effect. Dose saturation effect for the Xe implant occurred at much smaller doses of 5.5×10^{15} ions/cm^2. This result is evidence for Xe diffusion out of the implanted region, because the Xe should only have a slightly greater sputtering rate than As for these films, according to TRIM calculations (20). RBS results indeed indicate that the Xe amount below the implant layer is as high as 20 atom% of that in the implanted layer for the implant of 2.0×10^{17} ions/cm^2. However, for the As implant at the same dose, this value is much smaller (< 0.2 atom%).

The dose saturation effect may be connected to the saturation in electrical conductivity, which will be discussed in Chapter 2.6. At relatively high doses, such as 1.0×10^{17} ions/cm^2, a significant amount of implanted species (about 35 atom%) exists in the polymer about the ion's mean penetration range. For these high dose samples, the effect of As impurities on the properties of the implant layer can be significant compared with the ion-induced damage. Recent studies by extended X-ray absorption fine structure (XAFS) analysis for 50-keV As-implanted polysulfone indicated the existence of short range (~1 nm) ordering in the vicinity of the implanted As atoms (38). This short-range local structure should be important in determining the fundamental properties of these high dose materials.

Surface Morphology

The surface morphology of ion-implanted polymers has been studied with scanning electron microscopy (SEM) and transmission electron microscopy (TEM). Early TEM studies of 2-MeV Ar-implanted 3,4,9,10-perylene-tetracarboxylicdianhydride (PTCDA) films indicated that implantation transformed the irregular surface of the original film into a smooth surface after a dose of 1.0×10^{16} ions/cm^2 (Figure 12) (39). SEM measurement on a dry-fractured cross section of a similarly implanted polymer also indicated densification and smoothness in the implant region. Figure 13 shows an SEM result for a 50-keV B-implanted polyacrylonitrile (PAN) film (40). The implant thickness measured from SEM is in agreement with that estimated by TRIM Code (20). A densification of the implant layer is consistent with the observation of surface darkening, carbonization, and electrical conductivity enhancements. Surface charging, a result of poor

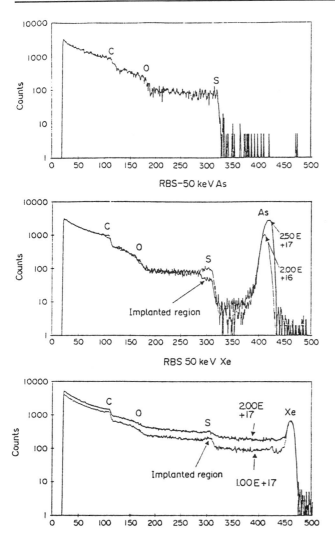

Figure 11. Typical energy spectra collected for PES films by RBS: (top) unimplanted, (middle) As-implanted at 50 keV with 2×10^{16} and 2.5×10^{17} ions/cm^2, and (bottom) Xe-implanted at 50 keV with 1×10^{17} and 2×10^{17} ions/cm^2. (Reproduced with permission from reference 13. Copyright 1993.)

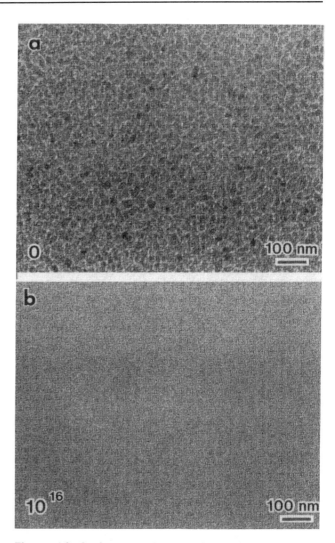

Figure 12. Surface morphology of ultra-thin ($\sim$10 nm) PTCDA films observed by TEM: (a) unimplanted surface and (b) Ar-implanted surface at 2 MeV with 1×10^{16} ions/cm^2. (Reproduced with permission from reference 39. Copyright 1984.)

electrical conduction in the implanted polymers during the SEM measurement, substantially limits the accuracy and resolution. Atomic force microscopy (AFM) has much better resolution and is not sensitive to the electrical properties of the samples. The recent use of AFM on plasma polymerized films suggests the use of AFM in this field (*41*). In the near future, AFM may provide a unique way to measure the size of ion-induced tracks in polymer films.

Bond Rearrangement

Ion irradiation breaks covalent bonds, scissors polymer chains, produces free radicals, and cross-links molecular chains and molecules. All of these modifications create a bonding rearrangement in the polymer

implant region. Earlier studies by IR transmittance in 2-MeV Ar implanted PTCDA films indicated that sharp spectral features present in the spectra of the pristine polymer gradually decrease in intensity with increasing dose (*42*). At doses $>1.0 \times 10^{14}$ ions/cm^2 the loss of molecular structural features was apparent. Studies of ion-implanted Kapton by attenuated total reflectance (ATR) IR analysis showed that ion irradiation not only damaged the sharp vibrational peaks seen in pristine materials but also formed new bonds (*10*). For example, B–N co-implanted Kapton demonstrated an absorption band near 2334 cm^{-1}, a result indicating the formation of a new B–H bond. An absorption was observed in the C≡C or C≡N band stretch region, near 2199 cm^{-1}, in 2-MeV Fe-implanted Kapton, and this result indicated the occurrence of

Figure 13. Cross-sectional morphology of dry-fractured, N-implanted PAN at 50 keV with 1×10^{17} ions/cm²: a, granularity morphology of the Au–Pd conductive coating on polymer surface; b, implanted region; c, unimplanted region; and d, glass substrate. (Reproduced with permission from reference 40. Copyright 1991.)

bond breakage and bond reformation in the aromatic chains of Kapton (*10*).

Comparison of ATR IR spectra for 50-keV He-, B-, C-, N-, and As-implanted PES films with 1.0×10^{17} ions/cm² are shown in Figure 14 (*27*). Compared with the As-implanted film, the He-implanted film clearly maintained a higher degree of spectral similarity with the pristine film (especially at a low dose of 10^{16} ions/cm²), whereas the spectral features of B-, C-, and N-implanted films are in between. This range is a result of the fundamentally different ion energy-deposition mechanisms for different ions. According to TRIM (*20*) calculations, He deposits 90% of its energy through electronic excitations–ionizations over a distance of 5000 Å, whereas As deposits 85% of its energy through nuclear events over a distance of 500 Å. The average ion energy loss in PES target along the ion track is about 10 times higher for As. Therefore, we would expect an As ion to be more destructive than an He ion at the same energy. A loss in intensity of the sulfone absorptions at 1160 cm⁻¹ was observed, in agreement with the observations of Marletta (*8*). An additional spectral absorption in the 800–1000 cm⁻¹ range in all the He- to As-implanted samples may be a result of cross-linking reactions or self-cyclizations during the implantation process. All these reactions require the loss of hydrogen, which was observed by ERDA and RGA.

Most pristine polymers do not exhibit the paramagnetic behavior that is a characteristic of free radicals and dangling bonds. Formation of these unpaired spins (paramagnetic centers) in ion-implanted polymers was observed and studied by electron spin resonance (ESR). ESR provides information on spin mobility and degree of carrier localization. It has now been accepted that the formation of free radicals results from the decay of excited molecules. Studies of

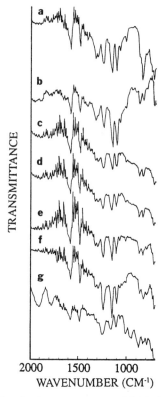

Figure 14. IR reflectance spectra of PES films implanted with 50-keV ions: a, unimplanted; b, 1×10^{16} He/cm²; c, 1×10^{17} He/cm²; d, 1×10^{17} B/cm²; e, 1×10^{17} C/cm²; f, 1×10^{17} N/cm²; and g, 1×10^{17} As/cm². (Reproduced with permission from reference 27. Copyright 1991.)

PTCDA, Ni-phthalocyanine (NiPc) (*42*), poly(phenylene oxide) (PPO), PAN (*4, 43*), and polyimide (*7*) indicated a universal behavior that the concentration of unpaired spins increases with increasing irradiation dose (ϕ) up to a certain characteristic value (ϕ_c). Above ϕ_c, the concentration slightly decreases for implants in the kilo-electronvolt energy range and rapidly decreases in the mega-electronvolt energy range. The value of ϕ_c depends on the polymer host and the implant conditions. A decrease of ESR linewidth with increasing dose is also observed for these samples. This linewidth dependence is explained by the broadening of the unpaired spin wavefunctions. The inverse correlation between hydrogen evolution and spin concentration suggests that a large number of free radicals are stabilized in the implant region due to a significant decrease in the rate of recombination. This decrease in recombination rate follows a reduction in the concentration of mobile hydrogens (*7*).

Studies of 50-keV He-implanted PES by Bridwell et al. (*27*) showed a plateau behavior in both spin susceptibility and ESR linewidth after ϕ_c (Figure 15). The radical concentrations were obtained from the integrated spin susceptibility, which was evaluated from a double integration of the ESR spectrum. The narrow-

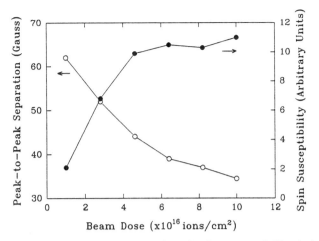

Figure 15. Dose dependencies of spin susceptibility (●) and spectral linewidth (○) of ESR spectroscopy for He-implanted PES at 50 keV. (Reproduced with permission from reference 27. Copyright 1991.)

ing of the ESR linewidth with increasing dose suggests a motional narrowing and an increased delocalization of the carriers along the conduction paths in the implant region. Both the spin concentration and the linewidth results suggest that radical–radical recombinations are important above ϕ_c of ~5 × 10^{16} ions/cm^2. Notably, the change in spectrum from a partial Gaussian–partial Lorentzian line shape at low doses (10^{16} ions/cm^2) to a pure Lorentzian shape at high doses (10^{17} ions/cm^2) suggests the existence of different degrees of spin delocalization. The broad ESR line observed at low doses is considered to be the contribution from isolated clusters of radicals below ϕ_c, whereas the narrow width of this line observed at high doses is thought to result from long-range spin mobility, which becomes possible after the formation of an infinite percolation cluster.

Carbon Configuration

Interest in the carbon bonding configuration is most important because the majority of the material in the implant region is carbon. At low doses, the effect of ion implantation principally causes the degradation along the polymer backbone. At higher doses, when most of implant region material is carbon, modification or rearrangement of the carbon structure occurs. EELS, and X-ray and UV photoelectron spectroscopy (XPS and UPS) are used to characterize the electronic band structure of ion-irradiated polymers. Optical spectroscopy provides information on electronic empty states extending above the Fermi level.

A comprehensive investigation of the carbon bonding structures by using reflective EELS (REELS), XPS, and UPS in ion-implanted (PMDA-ODA) polyimide films was performed by Davenas and Boiteaux

(7). In general, the XPS photoelectron spectrum showed three regions corresponding to three carbon bonding configurations: 284.4 eV from ring carbons (C–C/C=C), 285.6 eV from carbon in the PMDA ring (C–N or C–O), and 288.4 eV from carbon in the imide group (–C=O). The strong decrease of the carbonyl carbon component and the opening of the PMDA ring for a 100-keV Ne-implanted sample with 5.0 × 10^{16} ions/cm^2 is consistent with the rapid scission of C=O and C–O–C bonds observed from IR measurements. The UPS and REELS measurements indicate the existence of a graphite-like phase in these samples (7).

To correlate the peak intensity of these three carbon-bonding configurations with irradiation dose, 120-keV B-implanted polyimide films were studied by XPS in more detail, and the results are shown in Figure 16 (44). Figure 16 indicates the peak intensity at 284.4 eV increases with increasing dose, and this result implies the formation of the hexagonal carbon ring, C–C=C–C, as a result of irradiation. These new ring-carbon bonds result from the cleavage of bonds in C–O or C–N and –C=O, as indicated from the decrease in the height of the peaks at 285.6 and 288.4 eV. Figure 16 also indicates that the formation of the ring-carbon bonds saturates after a certain dose threshold (ϕ_c), ~10^{16} ions/cm^2, a result implying that the percolative overlapping of the ion-induced clusters in the implanted surface is complete. Further irradiation beyond this threshold probably degrades–cross-links or rearranges the hexagonal carbon rings. This further implantation results in a modified carbon configuration that modifies the microstructure of the implant region.

Raman spectroscopy is a commonly used technique to obtain information about the carbon hybrid-

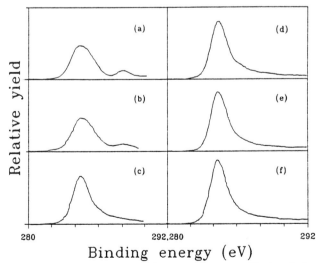

Figure 16. X-ray photoelectron spectra (C$_{1s}$) of B-implanted polyimide films at 120 keV with different doses (ions/cm^2): (a), unimplanted; (b), 1 × 10^{14}; (c), 1 × 10^{15}; (d), 1 × 10^{16}; (e), 3 × 10^{16}; and (f), 5 × 10^{16}. (Reproduced with permission from reference 44. Copyright 1992.)

ization states (sp^3, sp^2, or sp^1). The dependence on dose of the Raman spectrum for a 180-keV As-implanted TMPA–FDA polymer film is shown in Figure 17 (45). In agreement with the XPS results, the low dose irradiation destroys the sharp features of the pristine polymer. After medium and high dose irradiation, a very broad band is formed around 1550 cm^{-1} (G line). This broad, skewed band around 1550 cm^{-1} is characteristic of a diamond-like carbon structure (33, 45). In addition, a small component around 1340 cm^{-1} (D line) is just within experimental resolution at the highest dose of 10^{17} ions/cm^2. The D line is a contribution from medium-range ordered regions such as graphite-like clusters. A similar broad, skewed band is observed, at the same dose, for the other four types of PAI films with high T_g values (14). This observation seems to suggest that, independent of original composition and structure, a sufficient ion beam dose and energy converts different polymers into apparently similar carbon-enriched structures. However, widely different electrical conductivities (as much as 10 times) observed for these samples suggest significant variations in their electronic structures (45). This variation in electronic structure is related to the relative amount of the D-type carbon and the G-type carbon. By deconvoluting the Raman spectra for 120-keV B-implanted and 170-keV N-implanted polyimide films, Xu et al. (44) found that a small increase in the intensity ratio of D/G carbons corresponds to a large enhancement in electrical conductivity. A similar conclusion also can be drawn from the Raman spectra of ion-implanted PS films (11).

A distinct change in the tridimensional carbon networks of different ion-implanted PEEK surfaces was observed by Sofield et al. (33) by using a Raman microprobe. The Raman spectra of these PEEK surfaces are shown in Figure 18. Both broad G line and D line signals are clearly visible in the high energy irradiated samples (22.5 MeV, I; 41.5 MeV, Ni). This broad, twin-peak structure is a result of the existence of large graphitic crystallites in the implant region (33). In the case of the 20-MeV F-implanted film, the spectrum is similar to those shown in Figures 17c and 17d, and this result indicates a diamond-like carbon structure. These results are consistent with conductivity and hardness measurements (9, 33). For example, 20-MeV F-implanted PEEK film shows a greater mechanical hardness but a poorer electrical conductivity than a 41.5-MeV Ni-implanted PEEK film at a similar dose. In the As-implanted sample, the ion energy of 50 keV does not result in a thick enough implant layer, and the fluorescence from underlaying, undamaged PEEK dominates the signal. A depth profile of the Raman spectra along the ion track was measured for 41.5-MeV Ni-implanted PEEK (33). Results indicate that at greater depths in the polymer implant layer, the spectra are more characteristic of a diamond-like carbon (amorphous carbon) structure than a graphitic carbon structure that is seen near the surface of this sample. This observation is in agreement with the conducting grain model discussed earlier (34).

We should point out that the terminology between diamond-like carbon (DLC) and amorphous

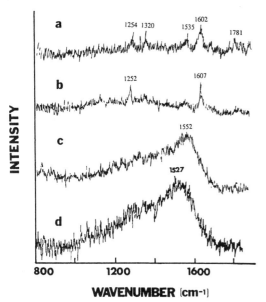

Figure 17. Raman spectra of As-implanted TMPA–FDA polymer films at 180 keV with different doses (ions/cm^2): a, unimplanted; b, 1 × 10^{15}; c, 1 × 10^{16}; and d, 1 × 10^{17}. (Reproduced with permission from reference 45. Copyright 1991.)

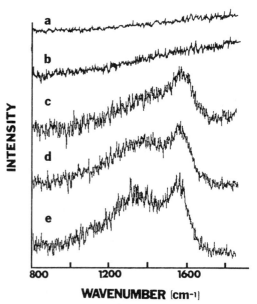

Figure 18. Raman spectra of PEEK films: a, unimplanted; b, 50-keV As, 1 × 10^{17} ions/cm^2; c, 20-MeV F, 6 × 10^{14} ions/cm^2; d, 22.5-MeV I, 4.7 × 10^{13} ions/cm^2; and e, 41.5-MeV Ni, 2 × 10^{14} ions/cm^2. (Reproduced with permission from reference 33. Copyright 1992.)

carbon is a bit ambiguous. Amorphous structures of carbon are usually distinguished into two types with relevance to hardness. Carbon phases produced by evaporation or pyrolysis are usually soft (glassy carbon), whereas carbon phases produced by radio frequency discharge or sputtering are much harder (DLC type). The hardness of the carbon phases produced by vapor deposition or plasma decomposition of hydrocarbons (hydrogenated amorphous carbon or a-C:H) depends strongly on their H content as well as their sp^3 bonding content in the films. In a recent review Robertson (46) showed that a-C:H films containing low H content (<10 atom%) and high sp^3 bonds (>90%) can be as hard as diamond (100 GPa), whereas a-C:H films containing high H content (>40 atom%) and low sp^3 bonds (<80%) are much softer (<5 Gpa). Compared with these amorphous carbon films, structures of ion-implanted polymers are more complicated. Although it is commonly accepted that ion-implanted polymers have hydrogenated amorphous carbon structures, the matches between ion-implanted polymers and a-C:H films produced by other methods are poor. To produce hard a-C:H films by other methods, both a low H content and a high sp^3 character are desirable in the film. For ion-implanted polymers, a low H content is produced by highly ionizing ions, but these ions also produce a low sp^3 content (13). There-

fore, transforming polymers into hard materials of a-C:H type by ion implantation seems difficult.

Content of sp^3 bonds in amorphous carbon films usually is determined by Raman scattering spectroscopy. However, recent studies by Lee et al. (47) showed no indication that the hardness of DLC and ion-implanted polymers is due to diamond-like sp^3 bonds. Instead, they believe that the hardness in these materials is related to the three-dimensional interconnectivity of chemical bonds (cross-linking). Cross-linking induced by ion-beam exposure is no doubt an important process for certain ion-polymer combinations. However, much work is needed in this field to understand the cross-linking process and other intricate ion–polymer interactions at the molecular level.

Acknowledgments

We would like to express our sincere gratitude to the many people who offered ideas and opinions in the course of writing this review. Specifically, C. J. Sofield was generous in his discussions of the experimental data. To M. G. Moss and S. S. Mohite, we offer thanks for supplying the polymer films and for explaining some features of polymer structures. Y. Q. Wang thanks Brewer Science, Inc., for sponsoring his research when he was at Southwest Missouri State University.

Appendix 1

Polymers and Ion Beam conditions Discussed in Chapters 2.5 and 2.6

Polymer Abbrev.	Polymer Name	Ion Species	Energy (keV)	Reference Ch. 2.5	Reference Ch. 2.6
Kapton	polyimide–Kapton	He, N, Si	200–2000	10, 37, 26	
		B, N	120–170	44	58
PAN	polyacrylonitrile	As, Br, Kr	200	1	13
		B, N	50	40, 21	
PANI	polyaniline	B, N	50		29
PAI	polyamide–imide	As, Xe	50	14, 45	23
PE	polyethylene	He, Ar	100–2000	5, 26, 36	
PEEK	poly(ether ether ketone)	He, N, F, Ni	50–42,000	33	6, 21
		As, I, Xe			23
PES	polyether sulfone	He, B, C N, As	50	27	21
PET	poly(ethylene terephthalate)	Ar, As	50–180		39, 41
PDA	polydiacetylene	N	50	32	
PMMA	poly(methyl methacrylate)	He, Ar	100–300	1, 25	
	polyacetylene	Na	50		55, 56
	polysulfone	N, As	50	38	21
PPO	poly(phenylene oxide)	N, Br	200	1	26, 42
PPP	poly(p-phenylene)	H_2	300		18
PPS	polyphenylene sulfide	As, Kr, Br	100–200		1, 14, 42
PS	polystyrene	He, Ar	100–200	5, 25	
PSA	poly(styrene–acrylonitrile)	B, N, As	50		21, 28
PVC	poly(vinyl chloride)	Ar	2000		49

Note: Beam doses are within 10^{13}–10^{17} ions/cm^2.

References

1. Dresselhaus, M. S.; Wasserman, B.; Wnek, G. E. In *Ion Implantation and Ion Beam Processing of Materials*; Hubler, G. K.; Holland, O. W.; Clayton, C. R.; White, C. W., Eds.; Matererials Research Society Symposium Proceedings 27; Pittsburgh, PA, 1983; p 413.
2. Venkatesan, T.; Levi, R.; Banwell, T. C.; Tombrello, T.; Nicolet, M.; Hamm, R.; Meixner, A. E. In *Ion Beam Processes in Advanced Electronic Materials and Device Technology*; Appleton, B. R.; Eisen, F. H.; Sigmon, T. W., Eds.; Materials Research Society Symposium Proceedings 45; Pittsburgh, PA, 1985; p 189.
3. Brown, W. L. *Radiat. Eff. Solid States* **1986**, *98*, 115.
4. Wasserman, B. *Phys. Rev.* **1986**, *B34*, 1926.
5. Venkatesan, T.; Calcagno, L.; Elman, B. S.; Foti, G. In *Ion Beam Modifications of Insulators*; Mazzoldi, P.; Arnold, G. W., Eds.; Elsevier: Amsterdam, Netherlands, 1987; Chapter 8, p 301.
6. Hersh, S. P.; Brock, S.; Grady, P. L. In *Polymers for Advanced Technology*; Lewin, M., Ed.; VCH Publishers: New York, 1988; p 52.
7. Davenas, J.; Boiteaux, G. *Adv. Mater.* **1990**, *2(11)*, 521.
8. Marletta, G. *Nucl. Instrum. Methods* **1990**, *B46*, 295.
9. Bedell, C. J.; Sofield, C. J.; Bridwell, L. B.; Brown, I. M. *J. Appl. Phys.* **1990**, *67*, 1736.
10. Lee, E. H.; Lewis, M. B.; Blau, P. J.; Mansur, L. K. *J. Mater. Res.* **1991**, *6(3)*, 610.
11. Calcagno, L.; Foti, G. *Mucl. Instrum. Methods* **1991**, *B59/60*, 1153.
12. Bridwell, L. B. In *Trends in Ion Implantation*; Rossum, V. M., Ed.; Solid State Phenomena Series 27; Trans Tech. Publications: Zurich, Switzerland, 1992; p 163.
13. Sofield, C. J.; Sugden, S.; Ing, J.; Bridwell, L. B.; Wang, Y. Q. *Vacuum* **1993**, *44(3-4)*, 285.
14. Wang, Y. Q.; Mohite, S. S.; Bridwell, L. B.; Giedd, R. E.; Sofield, C. J. *J. Mater. Res.* **1993**, *8*, 388.
15. Zallen, R. *The Physics of Amorphous Solids*; Wiley and Sons: New York, 1983; p 110.
16. Cohen, M. H.; Grest, G. S. *Phys. Rev.* **1979**, *B20*, 1077.
17. Alexander, S.; Laermans, C.; Orbach, R.; Rosenberg, H. M. *Phys. Rev.* **1983**, *B28*, 4615.
18. Bethe, H. A. *Ann. Phys.* **1930**, *5*, 325.
19. Lindhard, J.; Scharff, M.; Schiott, H. E. *Mater. Fys. Medd. Dan. Vid. Selsk.* **1963**, *33*, 14.
20. Ziegler, J. F.; Biersack, J. P.; Littmark, U. *The Stopping and Range of Ions in Solids*; Pergamon: Oxford, England, 1985.
21. Cruz, S. A.; Soullard, J. *Nucl. Instrum. Methods* **1992**, *B71*, 387.
22. Davies, J. A. In *Surface Modification and Alloying by Laser, Ions and Electron Beams*; Poate, J. M.; Foti, G.; Jacobson, D. C., Eds.; Plenum: New York, 1983; Chapter 7; p 189.
23. Biersack, J. B. In *Ion Beam Modification of Insulators*; Mazzoldi, P.; Arnold, G. W.; Elsevier: Amsterdam, Netherlands, 1987; Chapter 1, p 1.
24. Hagiwara, M.; Kagiya, T. In *Degradation and Stablization of Polymers*; Jellinek, H. G., Ed.; Elsevier: Amsterdam, The Netherlands, 1983; Vol. 1, Chapter 8, p 358.
25. Venkatesan, T. *Nucl. Instrum. Methods* **1985**, *B7/8*, 461.
26. Lewis, M. B.; Lee, E. H. *Nucl. Instrum. Methods* **1992**, *B69*, 341.
27. Bridwell, L. B.; Giedd, R. E.; Wang, Y. Q.; Mohite, S. S.; Jahnke, T.; Brown, I. M. *Nucl. Instrum. Methods* **1991**, *B59/60*, 1240.
28. Fink, D.; Ibel, K.; Goppelt, P.; Biersack, J. P.; Wang, L.; Behar, M. *Nucl. Instrum. Methods* **1990**, *B46*, 342.
29. Rao, G. R.; Wang, Z. L.; Lee, E. H. *J. Mater. Res.* **1993**, *8(4)*, 927.
30. Fink, D.; Muller, M.; Chadderton, L. T.; Cannington, P. H.; Elliman, R. G.; McDonald, D. C. *Nucl. Instrum. Methods* **1988**, *B32*, 125.
31. Davenas, J.; Xu, X. L.; Boiteaux, G.; Sage, D. *Nucl. Instrum. Methods* **1989**, *B39*, 754.
32. Elman, B. S.; Blackburn, G.; Samuelson, L.; Kenneson, D. G. *Appl. Phys. Lett.* **1986**, *49(10)*, 599.
33. Sofield, C. J.; Sugden, S.; Bedell, C. J.; Graves, P. R.; Bridwell, L. B. *Nucl. Instrum. Methods* **1992**, *B67*, 432.
34. Wang, Y. Q.; Giedd, R. E.; Bridwell, L. B. *Nucl. Instrum. Methods* **1993**, *B79*, 659.
35. Inagaki, M.; Sakamoto, K.; Hishiyama, Y. *J. Mater. Res.* **1991**, *6*, 1108.
36. Calcagno, L.; Foti, G. *Appl. Phys. Lett.* **1985**, *47*, 363.
37. Lewis, M. B.; Lee, E. H. *Nucl. Instrum. Methods* **1991**, *B61*, 457.
38. Mayanovic, R. A.; Feng, Y. P.; Groh, K. W.; Wang, Y. Q.; Giedd, R. E.; Moss, M. G. *Materials Research Society Symposium Proceedings*; Materials Research Society: Pittsburgh, PA, 1994; Vol. 316, p 75.
39. Lovinger, A. J.; Forrest, S. R.; Kaplan, M. L.; Schmidt, P. H.; Venkatesan, T. *J. Appl. Phys.* **1984**, *55*, 476.
40. Giedd, R. E.; Moss, M. G.; Craig, M. M.; Roberson, D. E. *Nucl. Instrum. Methods* **1991**, *B59/60*, 1253.
41. Li, G. F. private communication, University of Wisconsin, Madison, WI, 1993.
42. Venkatesan, T.; Forrest, S. R.; Kaplan, M. L.; Schmidt, P. H.; Murray, C. A.; Brown, W. L.; Wilkens, B. J.; Roberts, R. F.; Ruoo, L.; Schonhorn, H. *J. Appl. Phys.* **1984**, *56*, 2778.
43. Wasserman, B.; Dresselhaus, M. S.; Braunstein, G.; Wnek, G. E.; Roth, G. *J. Electron. Mater.* **1985**, *14*, 157.
44. Xu, D.; Xu, X. L.; Zou, S. C. *J. Mater. Res.* **1992**, *7*, 229.
45. Wang, Y. Q.; Giedd, R. E.; Mohite, S. S.; Jahnke, T.; Bridwell, L. B. *Mater. Lett.* **1991**, *12*, 21.
46. Robertson, J. *Surf. Coat. Technol.* **1992**, *50*, 185.
47. Lee, E. H.; Hembree, D. M.; Rao, G. R.; Mansur, L. K. *Phys. Rev.* **1993**, *B48*, 15540.

Polymer Modification by Ion Implantation: Electrical Conductivity and Applications

Y. Q. Wang, L. B. Bridwell, and R. E. Giedd

This chapter presents a review of electrical conductivity enhancements, electronic conduction mechanisms, and potential applications of ion-implanted polymers. The enhanced electrical conductivity depends on ion-beam parameters (ion species, energy, dose, and beam current) as well as on constituents and structures of the initial polymer. Large electronic energy deposition by high energy (MeV) ions transforms insulating polymers into highly conductive materials, whereas small electronic energy deposition from low energy (keV) ions forms less conductive materials. Electronic conduction is analyzed in terms of several different mechanisms and models. Potential applications of implanted polymers in the areas of electrical, mechanical, optical, chemical, and special materials syntheses are also discussed.

In Chapter 2.5, we reviewed the ion-beam modification (ion implantation) of polymers in terms of fundamental ion–polymer interactions and basic structural characterization. The significant structural modifications that occur in polymers during implantation create dramatic enhancements in the electrical conductivity of the polymer surface. This enhancement in the electrical conductivity depends on the original polymer composition, the implantation conditions, and the order–disorder of the implanted surface region (1–9).

Significant enhancements in the mechanical properties of the surface such as hardness, wear, friction, and fatigue also occur when polymers are ion implanted. A surface hardness comparable to that of stainless steel has been obtained by the implantation of high energy (MeV range) F and I ions (6) and by triple-beam implantation of B, N, and C (10). The mechanical properties of polymers enhanced by ion beams are discussed in detail in a number of other articles (6, 9–12).

In this chapter, we discuss the electrical conductivity enhancements in polymers as a result of ion-beam irradiation. The mechanisms of electronic conduction in ion-implanted polymers are analyzed, and their potential commercial applications are also discussed.

Electrical Conductivity Enhancements

Electrical conductivity of insulating polymers is increased drastically by ion implantation. A surface conductivity increase by as much as 20 orders of magnitude has been obtained depending on the original polymer structure and ion-beam conditions. For comparison, Table I lists resistivities of conductors, semiconductors, conducting polymers, and insulating polymers before and after ion implantation. Many efforts have been made to understand the mechanisms involved in the conductivity enhancement. We discuss a number of factors that affect the enhanced conductivity in implanted polymers.

3469–8/97/0387$15.00/0 © 1997 American Chemical Society

Table I. Comparison of Electrical Resistivities in Various Materials

Type	Material	Resistivity ($\Omega \cdot cm$)	Ref.
Conductors	Silver	1.59×10^{-6}	
	Copper	1.72×10^{-6}	
	Aluminum	2.82×10^{-6}	
Semiconductors	Carbon (graphite)	3.5×10^{-3}	
	Ge (pure)	46	
	Si (pure)	6.4×10^{4}	
Conducting polymers	Polyaniline (PANI)	10^{-1}	64
	Polyacetylene	10^{-5}	64
Insulating polymers	Poly(phenylene sulfide) (PPS)	10^{16}	
	Poly(vinyl chloride) (PVC)	10^{16}	
	Polyethersulfone (PES)	10^{17}	
	Polyimide–Kapton	10^{18}	
	Polyamide–imide (PAI)	10^{17}	
	Poly(ether ether ketone) (PEEK)	10^{16}	
Implanted polymers	PPS (100 keV As, 10^{16} cm^{-2})	10^{5}	1
	PVC (2 MeV Ar, 10^{16} cm^{-2})	0.3	49
	PES (50 keV As, 10^{17} cm^{-2})	10^{4}	16
	Kapton (120 keV B, 10^{16} cm^{-2})	3	58
	PAI (50 keV As, 10^{17} cm^{-2})	30	8
	PEEK (50 keV He, 10^{17} cm^{-2})	10^{3}	6, 24
	PEEK (50 keV As, 10^{17} cm^{-2})	70	6, 24
	PEEK (50 keV I, 10^{15} cm^{-2})	6×10^{4}	6, 24
	PEEK (20 MeV I, 10^{15} cm^{-2})	10^{-2}	6, 24

Ion-Beam Effect

A large number of insulating polymers have been implanted under varying conditions for electrical conductivity enhancements. The most common studies vary the ion-beam parameters such as ion species, ion energy, beam dose, and beam current (dose rate). Early studies of the surface resistivity as a function of beam dose were done for As-, Kr-, and Br-implanted polyphenylene-sulfide (PPS) and polyacrylonitrile (PAN) at energies of several hundred kilo-electron-volts (*1, 13*) and for Ar-implanted novolac resin sensitizer (HPR-204) at 2 MeV (*14*). The conductivity of the HPR-204 film increased by 14 orders of magnitude after implantation (*14, 15*). Polyethersulfone (PES) films were implanted by various ions at 50 keV (*16*). Typical results for the surface resistivity as a function of dose for these PES films are shown in Figure 1 (*16*). Similar results were also observed (*17*) for polyamide–imide films with high glass transition temperatures implanted by 50-keV As and Xe (*17*). The connecting lines in Figure 1 show the general trends for this relationship in most implanted polymers. This general trend indicates that at low and medium doses the surface resistivity decreases with increasing dose. At the high doses, the resistivity levels off to a plateau.

This saturation is believed to be a result of sputtering and nuclear damage (*6*). This sputtering and dam-

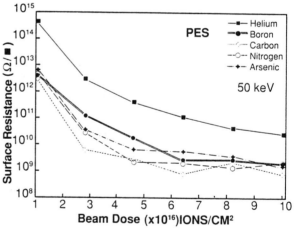

Figure 1. Surface resistivity of PES as a function of beam dose for 50-keV He, B, C, N, and As implantation. (Reproduced with permission from reference 16. Copyright 1991.)

age are expected to lead to an equilibrium conductivity when conductivity enhancement by the incoming ions balances the conductivity reduction due to sputtering or damage by nuclear elastic collisions. Variation in the saturated conductivity for implantation of different ions is caused by the differences in nuclear stopping power and sputtering rate experienced by ions. Because the nuclear stopping and sputtering rate are quite small for light ions, we expect no such saturation effect for light ions. Indeed, no saturation effect in conductivity was observed for hydrogen-implanted polymers even at much higher doses. Figure 2 shows a typical result for 300-keV H_2 ions implanted into the air-stable poly(p-phenylene) (PPP) (18).

Interestingly, a threshold dose exists in Figure 2 after which the resistivity starts to decrease rapidly. This threshold is the dose where the ion-induced conducting carbon clusters are dense enough to contact each other and form a continuous conduction network (percolative process). Different ions, energies, and polymers would exhibit different threshold doses, because the size of the carbon clusters depends on the ion–polymer interaction. This threshold behavior was studied in detail by Ruddy et al. (19). Figure 3 shows a general trend of the threshold dose (percolative dose) as a function of the ion energy loss (19). According to Figure 3, a threshold should be observed in Figure 1 at lower doses near 10^{14} ions/cm^2.

A number of experiments were conducted to determine conductivity dependence on ion dose rate (beam current). For Ni-phthalocyanine (NiPC) films implanted by 2-MeV Ar, Kaplan et al. (20) found that conductivity increased by nearly two orders of magnitude when going from low to high beam currents (20 nA to 450 nA) and leveled off at about 450 nA. This beam-current saturation effect on conductivity is generally believed to be a result of target temperature variation during the implantation.

If the background temperature of a polymer target is determined by the accumulation of temperature transients in individual ion-track regions, the implantation with a higher beam current should result in a higher target temperature. This temperature has the effect of self-annealing the carrier-trap defects in the

polymer, which may increase the conductivity. The beam-current saturation plateau corresponds to a constant target temperature, where the increment in target temperature as a result of ion implantation is balanced by the target radiation heat loss, as reported by Venkatesan et al. (14).

Dose rate effects on the resistivity of a set of 50-keV N-implanted polymers (21) also indicated a similar explanation with one exception. At very low dose rates, the resistivities of all five polymers in the study increased with increasing dose rate, (Figure 4). The five polymers were PES, poly(ether ether ketone) (PEEK), polysulfone, PAN, and poly(styrene-acrylonitrile) (PSA). This discrepancy in the low dose rate range is most likely a result of a low target temperature that cannot eliminate the carrier-trap defects by self-annealing. Figure 4 also indicates that the smaller the surface resistivity, the less sensitive the surface is to dose rate effects. The temperature-annealing effect in implanted polymers is not as obvious as in inorganic semiconductors. More studies are necessary to understand the fundamental mechanisms operating in the annealing process.

Changes in conductivity as a result of ion energy variation (2, 6, 22) indicate that surface conductivities of most polymers implanted with mega-electronvolt energies are nine to 10 orders of magnitude higher than those implanted with kilo-electronvolt ions. Even

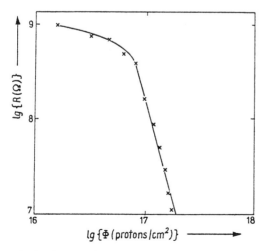

Figure 2. Surface resistivity of PPP as a function of beam dose for 300-keV H_2 implantation. (Reproduced with permission from reference 18. Copyright 1993.)

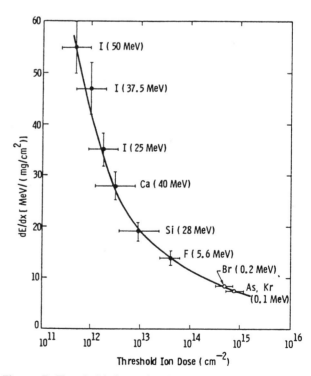

Figure 3. Threshold dose of polymers as a function of ion energy loss where the polymer conductivity starts to increase rapidly. (Reproduced with permission from reference 19. Copyright 1988.)

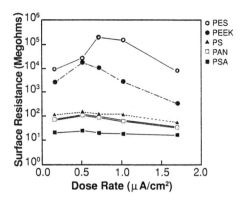

Figure 4. Surface resistivity of various polymers as a function of dose rate for 50-keV N implantation and a dose of 1×10^{16} ions/cm^2 (Reproduced with permission from reference 21. Copyright 1991.)

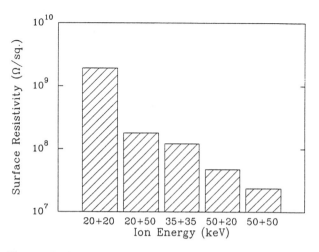

Figure 5. Surface resistivity of PSA as a function of ion energy for two-step N implantation at a dose of 1×10^{16} ions/cm^2 for each step.

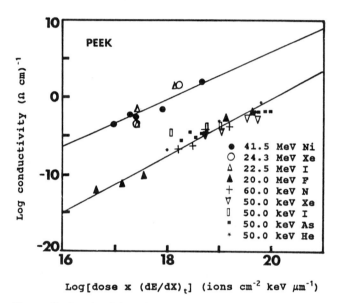

Figure 6. Conductivity of PEEK as a function of absorbed dose for various ion-beam irradiations. (Reproduced with permission from reference 24. Copyright 1992.)

in the kilo-electronvolt energy regime, the conductivity increases with increasing ion energy. Figure 5 shows surface resistivities of PSA films treated in a two-step implantation process in the 20–50 keV range with nitrogen ions. A dose of 1×10^{16} ions/cm^2 and a dose rate of 0.5 μA/cm^2 were used for each step. Figure 5 shows an apparent difference for 20 + 50-keV and 50 + 20-keV implants, where the front energy was implanted first.

This difference may be related to the variation in effective thickness of the modified surface layer. For 20 + 50-keV implant, 20 keV is implanted first and produces a carbon-rich material at the polymer surface. This carbon-rich layer is denser than the original polymer, so it reduces the mean penetration range of the following 50-keV ions. However, no such concern exists for 50 + 20-keV implant because the 50-keV implant is done first. As a result, the effective thickness of the modified surface layer is thicker for the 50 + 20-keV implant than for the 20 + 50-keV implant; therefore, the surface resistivity is expected to be lower for the 50 + 20-keV implant than for the 20 + 50-keV implant.

In terms of the ion-species (atomic mass) influence on conductivity, the limited results (2, 6, 22) seem to indicate that the heavier ions tend to produce a more conductive surface layer than light ions. This characteristic might be because heavy ions usually have a greater electronic stopping power than light ions. However, this ion-species effect may not always be true for low energies (<50 keV), where ion energy loss through nuclear collisions becomes important. Unlike the electronic ionizations, the nuclear collisions are less likely to produce graphitic clusters in polymers and instead they may destroy the conducting clusters produced by the electronic ionizations (22, 23).

The real ion-beam effect on polymer conductivity is the one in which each individual effect mentioned above functions on the polymer simultaneously. To observe this comprehensive beam effect, PEEK films

were implanted with He, N, F, Ni, As, I, and Xe in the energy range of 50 keV to 41.5 MeV and a dose range of 10^{13}–10^{17} ions/cm^2 (6, 24). The total absorbed dose (the product of dose and stopping power) was used to present the comprehensive effect of the ion beam. Despite the uncertainty about the detailed structure of the implant layer, the bulk conductivity was calculated from surface resistivity measurements, and the ion's mean penetration range was estimated by the transport of ions in matter (TRIM) code (25). Figure 6 shows the conductivity as a function of total absorbed dose for all the ion-implanted PEEK.

In Figure 6, ion-modified PEEK shows two different conductivity modes, depending on the total absorbed dose, and the conductivity difference between these two modes is as high as nine to 10 orders of magnitude. The conductivities of 20-MeV F-implanted samples belong to the lower conductivity mode. This unusual result is indicative of a threshold energy deposition rate (30 MeV cm^2/mg), above which a highly conductive graphitic material is formed and below which a less conductive material is formed. This threshold energy-deposition rate can be converted to the critical energy-deposition density (23).

All these observations suggest that ion-beam-induced conductivity depends not only on the electronic stopping powers (linear energy transfer, eV/A), but also on the energy-deposition density (eV/A^3) in polymers. As discussed in Chapter 2.5 (ion-induced microstructure), the absolute value of the electronic energy-deposition density and the dissipation rate of this energy density in the material are critical in the formation of structures around the ion track region and thus affect the physical properties of implanted polymer surface (23).

Polymer Structure Effect

Identical ion implants do not show similar conductivities on different polymer compositions or structures. Implanted polymers are, therefore, not independent of their original structure or composition. Massachusetts Institute of Technology scientists (26, 27) found that the conductivity of 200-keV Kr-implanted PPO was over two orders of magnitude lower than that of PAN implanted under identical conditions. The dose-rate effect on resistivity in Figure 4 shows that two sister polymers, PEEK and PES, exhibit an order of magnitude difference in conductivity; this difference exemplifies the significant effect of pristine structure on surface conductivity. This effect is also found for six types of polyamideimide films with high glass transition temperatures implanted with 50-keV Xe ions (17).

Despite the experimental evidence of the polymer structure effect on conductivity, there are still no clear explanations for this effect from the molecular point of view. Expectedly, the enhanced conductivities depend on original polymer structures and constituents, because the ion-induced radiolytic bond breakage, gas emission, and then cross-linking are dependent on the bonding configuration (bond type, strength, angle, and side-groups) of the original polymer. Therefore, the structures of ion-induced carbon clusters are expected to be different for different polymers. However, why one kind of polymer is more conductive than another is still not clear.

Aging Effect

The stability of the enhanced conductivity of implanted polymers is very important for implemen-

tation of the technology. For example, the use of ion-implanted polymers as sensing materials (temperature or pressure sensors) requires that the ion-implanted polymers have a long-term aging stability. It is generally believed that the enhanced conductivity in polymers by ion implantation is more stable than that by traditional chemical doping. This advantage is naturally expected and also understandable. The ion-modified polymer region is a carbon-rich amorphous material (not a polymer anymore), which has greater mechanical hardness and better chemical resistance than most ordinary polymers. Therefore, the charge carriers responsible for conductivity are expected to be more stable. However, chemically doped polymers are still polymers that often retain the weaknesses of ordinary polymers such as soft and easy chemical attack.

Surface resistivity of ion-implanted polymers increases with aging. Aging rate, defined as percent in resistivity increase over 100 hours, however, decreases with aging. Figure 7 shows the aging effect on the surface resistivity of N-implanted PSA films at 50 keV and a dose of 1×10^{17} ions/cm^2, where the time zero corresponds to the instant when the implanted sample is taken outside of the target chamber. Figure 7 indicates that a rapid increase in resistivity occurs during the initial 15 days, and then the resistivity tends to be more stable. In another example, the surface resistivity of As-implanted PEEK sheets at 50 keV and dose of 1×10^{17} ions/cm^2 increased only 2–3 times over a 1-year period (22).

Aging effect on conductivity is sensitive to implanted dose and polymer thickness. Polymers implanted at high dose (10^{17} ions/cm^2) are more stable than polymers implanted at low dose (10^{15} ions/cm^2). Conductivities of thin films tend to be more stable than those of thick ones implanted with the same

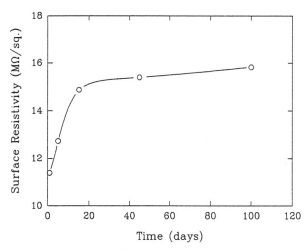

Figure 7. Surface resistivity of PSA films as a function of time in normal atmospheric conditions for 50-keV N implantation at a dose of 1×10^{17} ions/cm^2.

beam conditions. Although the exact mechanism involved in the aging effect is not understood, the de-excitation and stabilization of ion-induced carbon clusters through ambient annealing may be an important cause, because a forced annealing at elevated temperature after implantation enhances the stabilization of the conductivity. Results of the temperature dependence of conductivity, which will be discussed in a subsequent section, further indicate that the aging effect seems to be more related to carriers in a one-dimensional (1D) hopping conduction path, rather than in a three-dimensional (3D) hopping conduction network (28).

Aging effect on conductivity of conducting polymers can be more serious. For example, polyaniline (PANI) becomes conductive after being protonated. The doping level of the protonation in PANI controls its conductivity. However, the protonated sites in PANI are usually unstable, especially when exposed to humid environments, because these sites are likely occupied by diffusing water molecules. As a result, the conductivity will decrease with increasing exposure time. Figure 8 shows that the surface resistivity of protonated PANI increases more than 120% after exposure to a humid (H_2O) environment (half atmosphere) for only 15 min (29). However, conductivity of the PANI film is much more stable if the film is pre-implanted with 50-keV N ions at a dose of 1×10^{16} ions/cm^2 (Figure 8) (29). The stabilization in conductivity for implanted PANI is believed to be due to the formation of the ion-modified carbon-rich layer on the PANI surface, which serves as a hermetic sealing and prevents the external water molecules from diffusing into the PANI region.

Temperature Effect

The conductivity of ion-implanted polymers exhibits a strong temperature coefficient, and this characteristic suggests a temperature sensor application. An understanding of the conduction mechanisms in ion-implanted polymers must include an explanation of the temperature-dependent dc conductivity. The temperature-dependent dc conductivity is expressed by eq 1, where T is the absolute temperature, σ_0 is the conductivity at $T = \infty$, and T_0 is the characteristic of the conduction process.

$$\sigma(T) = \sigma_0 \exp[-(T_0/T)^m] \qquad (1)$$

Both σ_0 and T_0 are strong functions of beam dose (1) and dose rate (21). The power m depends on the conduction mechanism and has been experimentally determined to be 1/2 for a wide range of energies, doses, and polymer types (1, 3, 20, 30–32). Typical values of $\ln(R/R_0)$ as a function of temperature for one of the polyamide–imide polymer films (TMPA-BAPS) polymer films implanted with 50-keV As at a dose range of 5×10^{15} to 1×10^{17} ions/cm^2 are shown in Figure 9, where R_0 is the surface resistivity at room temperature (8). This figure suggests that the temperature dependence of the conductivity for implanted polymers is always proportional to $\exp[-(T_0/T)^{1/2}]$, but this conclusion is valid only for low dose implanted samples (8).

A number of algorithms with nonlinear least-squares fits were used to find m for high dose samples

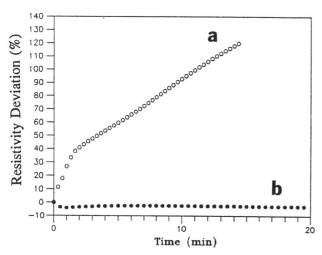

Figure 8. Surface resistivity variation of PANI as a function of time in a humid environment for (a) unimplanted and (b) 50-keV N implantation at a dose of 1×10^{16} ions/cm^2. (Reproduced with permission from reference 29. Copyright 1993.)

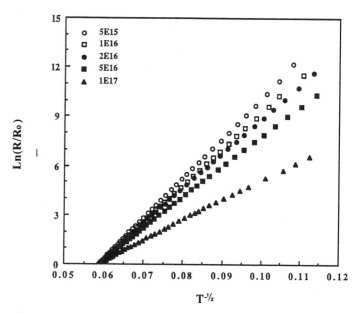

Figure 9. Surface resistivity expression $\ln(R/R_0)$ of TMPA-BAPS film as a function of $T^{-1/2}$ for 50-keV As implantation at varying doses, where R_0 is the room temperature surface resistivity. (Reproduced with permission from reference 8. Copyright 1993.)

(~10^{17} ion/cm^2). The value of *m* was determined when these different algorithms converged to the same number. For the high doses, *m* is between 1/3 and 1/4 (*8, 33*). Similar values (i.e., $m \neq 1/2$) also were reported by other investigators (*27, 34, 35*). A typical example for $m \neq 1/2$ behavior for poly[2, 4-hexadiyne-1, 6-di(*N*-carbazolyl)]polydiacetylene (DCH-PDA) films implanted with As at 150 keV and a dose of 5×10^{15} ions/cm^2 is shown in Figure 10 (*34*). It is evident from Figure 10 that $m = 1/2$ approximates the data at low temperatures, and at the high temperatures *m* approaches 1. This result was interpreted in terms of a 1D variable range hopping (VRH) conduction, which dominates at low temperatures, and a nearest neighbor hopping conduction, which dominates at high temperatures. A detailed discussion of the temperature-dependent conductivity mechanisms will be presented in a subsequent section.

AC Conductivity

Information on the conduction mechanisms can also be obtained by ac conductivity measurements. As a result of large resistances of ion-implanted polymers (usually >10 MΩ at room temperature), measuring complex impedance is quite difficult. Resistance can be made smaller by providing closer contacts. However this adaptation increases the parasitic capacitance associated with contacts and makes it very difficult to properly compensate the measurements.

The first ac conductivity measurements of ion-implanted polymers was performed by Wassermann et al. (*13*), where PAN films were implanted with Br at 200 keV and a dose of 2×10^{16} ions/cm^2. The results are in the functional form of an ac hopping conduction mechanism for amorphous materials. This functional form is expressed by eq 2, where σ is the real part of the frequency dependent conductivity, σ_{dc} is the dc conductivity, *s* is the frequency dependent exponent, ω is frequency, and *A* is usually considered to be constant.

$$\sigma(\omega, T) = \sigma_{dc}(T) + A\omega^s \qquad (2)$$

The Br-implanted PAN measurements gave an *s* value between 0 and 1.5 for frequencies between 300 Hz and 20 kHz (*13*). Studies of implanted polyimide by Aleshin et al. (*36*) gave $0.14 < s < 1.0$ for $\omega < 1$ MHz and $1.0 < s < 1.7$ for 1 MHz $< \omega <$ 10 MHz. Results for 50-keV, As-implanted PEEK at a dose of 1×10^{17} ions/cm^2 showed that *s* increased from 0 to 1.53 with increasing ω from 20 Hz to 50 kHz. Figure 11 shows the frequency dependence of the real part of the ac conductivity for this PEEK sample (*8*).

The form of eq 2 can be obtained from the Austin–Mott phonon-assisted hopping model (*37*) or from the Elliott correlated barrier hopping model (*38*). However, each model results in different expressions for *s*. In the Austin–Mott model, *s* is $1 - (n + 1)/[\ln(v_{ph}/\omega)]$, where *n* is the dimensionality of the system and v_{ph} is the phonon frequency, which is usually much larger than ω. For the Elliott model, *s* is $1 - 6kT/W_M$, where W_M is approximately equal to the band gap in the material and *k* is Boltzmann constant. Both models predict that *s* should not be greater than 1.0 for frequencies between 1 kHz and 1 MHz. Experimental results suggested that neither hopping models can satisfactorily explain the ac conductivity behavior of ion-implanted polymers in this frequency range.

Dielectric behavior of ion-implanted polymers also was studied. Studies by Tu et al. (*39*) of 180-keV

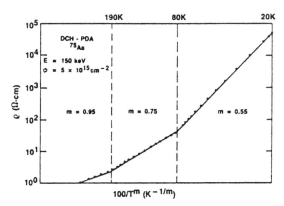

Figure 10. Resistivity of organic DCH-PDA film as a function of 100/T^m for 150-keV As implantation at a dose of 5 × 10^{15} ions/cm^2, where three temperature regions are clearly distinguished by *m* and have different scales along the horizontal axis. (Reproduced with permission from reference 34. Copyright 1985.)

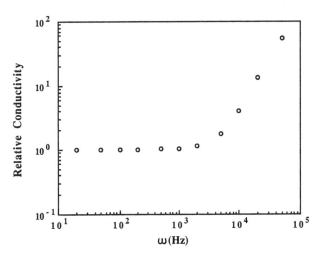

Figure 11. Relative conductivity, ln(σ/σ$_0$), of PEEK as a function of ln ω for 50-keV As implantation at a dose of 1 × 10^{17} ions/cm^2, where σ$_0$ is the conductivity at 20 Hz. (Reproduced with permission from reference 8. Copyright 1993.)

As-implanted poly(ethylene terephthalate) (PET) indicated that the ion irradiation produces a new dielectric relaxation peak in dielectric loss tangent (tanδ), which is absent for unimplanted PET with the frequency range used (10 Hz to 10 MHz) (Figure 12) (39). The capacitance (C), which is proportional to the real part of the permittivity, decreases rapidly as frequency increases around the relaxation frequency (ω_m), and this behavior represents typical lossy dielectric materials. Both ω_m and σ_{dc} of implanted PET films increased with increasing As dose and were proportional to $\exp[-(T_0/T)^{1/2}]$. The similarity in temperature dependence for ω_m and σ_{dc} leads to a conclusion that the mechanism for electric conductivity is similar to that for dielectric losses, and both temperature dependencies are believed by these authors to be due to tunneling migration of charge carriers (grain model) (40). However, this grain model cannot predict the frequency dependence of ac conductivity of ion-implanted polymers (Figure 11). Therefore, both theoretical and experimental research is needed to understand the ac conductivity mechanism involved in ion-implanted polymers.

Carrier Identification

Electric current flowing through materials is transported by charge carriers such as electrons, holes, and ions, whose behavior determines the conduction mechanisms in the materials. Conductivity measurements alone cannot identify the properties of the charge carriers. So, for a complete understanding of the conduction mechanism, other quantities such as carrier mobility, density, and type must be determined.

The Hall effect, which is an extremely valuable tool for measuring the sign and number density of charge carriers in semiconductors, is seldom used for ion-implanted polymers, mainly due to the difficulty in experimentation. Ion-implanted polymers seem to have a relatively large number of carriers resulting in very small Hall voltages, because $V_{hall} \sim 1/n_{car}$, where n_{car} is the number density of carriers. The relatively low conductivities of ion-implanted polymers are a result of poor carrier mobilities rather than a small number of carriers. Studies of the Hall effects by Venkatesan et al. (14) indicated that HPR-204 films implanted with 2-MeV Ar at a dose of 1×10^{17} ions/cm^2 had a carrier density of 10^{23} cm^{-3}. This carrier density is much larger than the densities of 10^{18}–10^{19} cm^{-3} found in highly ordered pure graphite films. However, later studies by Giedd et al. (41) for a 50-keV Ar-implanted PET film at a dose of 1×10^{17} ions/cm^2 indicated a lower carrier concentration of 10^{17} cm^{-3}. This small carrier density and the poor carrier mobility are responsible for the low electrical conductivity (0.02

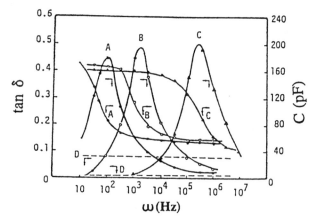

Figure 12. Dielectric loss (tanδ) and capacitance (C) of PET as a function of frequency (ω) at room temperature for 180-keV As implantation with varying beam dose: A, 5 × 10^{15}; B, 1 × 10^{16}; C, 5 × 10^{16} ions/cm^2; and D, unimplanted. (Reproduced with permission from reference 39. Copyright 1992.)

S/cm) in 50-keV Ar-implanted PET. A much better electrical conductivity (330 S/cm) was found in 2-MeV Ar-implanted HPR-204 films (14).

The number, polarity, and mobility of carriers can be determined by measuring thermoelectric power (TEP) and conductivity of the implanted materials. Early studies of TEP by Wnek et al. (42) revealed that the sign of the majority carrier in ion-implanted polymers depends on molecular structure of the polymer. PAN and poly(4-phenyl-2, 6-quiniline) exhibited n-type behavior, whereas PPO and PPS exhibited p-type behavior. Wnek et al. also showed that carrier mobility for irradiated polymers is very small [μ < 10^{-3} cm^2/(V s)] (1, 42). The magnitude of TEP also was very small (~3 μV/K) (13) and was difficult to measure when distributed across a large impedance.

Even though a larger TEP (~58 μV/K) was observed (18) in 300-keV H$_2$-implanted poly(p-phenylene) at a dose of 3 × 10^{17} ions/cm^2, TEP of ion-implanted polymers is usually similar to that of pure metals and very small compared with semiconductors. This result also suggests that the number density of carriers in implanted polymers is of the same order of magnitude as in metals. It is the very low mobility of these carriers that limits their conductivity. The low mobility is consistent with that of a magnetoresistance effect that has not been observed for ion-implanted polymers in the temperature range of 300 to 20 K (8, 41). The magnetoresistance was observed for implanted polyimide films only in the very low temperature range of 0.5–2.1 K (36).

Mechanisms of Electrical Conduction

Charge transport in a given material may be electronic, ionic, or both. In implanted polymers, electronic transport dominates. However, because most of the implanted polymer is a disordered carbon, some diffusion of the implanted ions may be expected, but no experimental evidence of this presently exists. Therefore, we will only consider electronic charge transport in implanted polymers.

Structural order and disorder are extremely important in determining the conduction mechanism in a given material, because the energetic and spatial distributions of electrons are affected by disorder in the atomic lattice. The conduction processes in the limiting cases, namely long range order in crystalline lattices and total disorder in amorphous materials, were described successfully by the band conduction model (43) and by the 3D variable range hopping (3D-VRH) model (44, 45).

Implanted polymers are evidently not at either extreme. The nature of a polymer requires that a 1D correlation should exist between the position of one repeat unit and its nearest neighbors. However, entropic and kinetic energy restraints of highly entangled chains prevent the formation of evenly spaced microscopic lattices in three dimensions. For implanted polymers, the use of band theory is questionable because they are disordered, even though the high energy implantation process may yield small graphitic regions where band theory may apply.

On the other hand, the implant layer has its own 1D structure as a result of the ion track, so application of a singular 3D-VRH mechanism is also not appropriate. We may resort to the temperature dependence of the conductivity, which distinguishes band conduction from VRH conduction. Temperature dependence of conductivity for most amorphous materials is empirically expressed as eq 1. In eq 1, the value of the exponent (m) is distinguished by the conduction mechanism involved, and the characteristic temperature (T_0) also has different physical interpretations for different conduction processes, as we will discuss.

Arrhenius-Type Conduction

For band conduction in extended periodic states, $m = 1$, and eq 1 can be written as eq 3, where the characteristic temperature (T_0) becomes the electronic activation energy if multiplied by the Boltzmann constant.

$$\sigma(T) = \sigma_0 \exp[-(T_0/T)] \qquad (3)$$

If the electronic energy states are not extended but are localized throughout the whole band, so that any mobility edge is in a higher energy band, a nearest-neighbor hopping will occur and produce a temperature dependence of the Arrhenius type given in eq 3.

Abel et al. (26) observed this type of behavior in Br- or Kr-implanted PPS. Elman et al. (34) and Sakamoto et

al. (46) found the same behavior in the 100–300 K temperature range for implanted polydiacetylene (PDA) films. Feng et al. (29) found that the Arrhenius-type conductivity presented in pristine PANI films no longer existed after the films were implanted with 50-keV N at a dose of 1×10^{16} ions/cm^2. Note that both PDA and PANI are also conductive in their pristine form.

VRH Model

For a disordered material, Mott and co-workers (44, 45) predicted that charge transport is a result of VRH between localized states. An important difference between VRH in localized states and band conduction in extended states is that in VRH, wave function overlap is incomplete. For VRH, the probability of a carrier hop is determined by the wave function overlap, and the hop is normally to a site where the activation energy (W) is lowest.

The hopping probability is expressed by v_{ph} $\exp[-2\alpha R - W/(kT)]$, where v_{ph} is the phonon frequency, R is the hopping distance, W is the energy difference between the sites, T is the absolute temperature, and α is the wave function overlap (α is also called the coefficient of exponential decay for the localized states). An optimization of the above probability will result in a temperature dependence as a function of the hopping dimensionality, which refers to the number of coincident spatial energy levels necessary for sufficient wave function overlap and hopping. This temperature dependence is given by eq 4, where n is the hopping dimensionality, A_n is a constant, and T_n is the characteristic temperature (23).

$$\sigma_n(T) = A_n T^{-2/(n+1)} \exp[-(T_n/T)^{1/(n+1)}] \qquad (4)$$

Both A_n and T_n are functions of n, α, and the density of localized states at the Fermi level, $N(E_F)$. Specifically, $T_1 = 4\alpha/kN(E_F)$, $T_2 = 48\alpha^2/4\pi kN(E_F)$, and $T_3 = 96\alpha^3/4\pi kN(E_F)$ (8).

The majority of the published results (1, 3, 32, 47) about the conduction mechanism for implanted polymers indicate a VRH where $n = 1$, and specifically, an observed conductivity of the form $\exp[-(T_1/T)^{1/2}]$. As a result of the 1D ion track, the n term in eq 4 is expected to be 1. Studies by Wang et al. (33) for 50-keV As-implanted TMPA-BAPS films indicate n in eq 4 is a function of irradiation dose. At low doses (~10^{15} ions/cm^2) n approaches 1, and this result is consistent with the 1D VRH mechanism observed by others. At high doses (~10^{17} ions/cm^2) n approaches 3, a result that implies that 3D VRH dominates. For intermediate doses of 2×10^{16} ions/cm^2, 1D VRH or 3D VRH gives equally poor fits (Figure 13) (22).

The linear correlation coefficients shown in Figure 13 seem to be very good ($R = 1.000$) for all dimen-

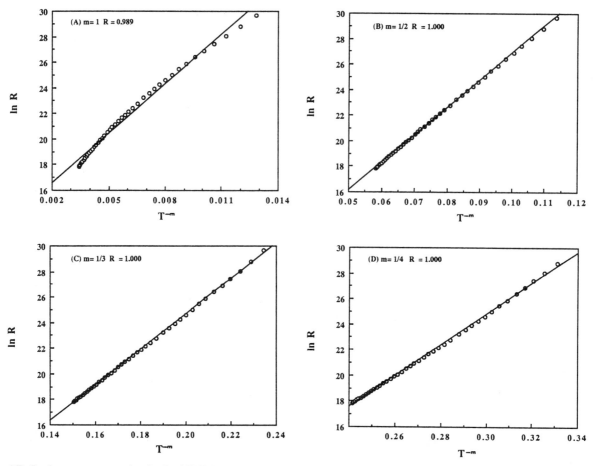

Figure 13. Resistance expression ln R of TMPA-BAPS films as a function of T^{-m} for 50-keV As implantation at a dose of 2 × 10^{16} ions/cm^2, where solid lines represent linear fits to eq 1 with $m = 1$, 1/2, 1/3, and 1/4.

sional VRHs. However, the corresponding χ^2 values are very large and are much beyond experimental error (33). In other words, neither 1D nor 3D VRH can explain the conduction process in the intermedium-dose implanted films. Consideration of the microstructure leads to a reasonable explanation of this behavior. At low doses, the carriers probably follow the disordered conductive ion tracks, which have a 1D geometry. This process should lead to 1D VRH. At higher doses the ion tracks begin to overlap and accumulate, and 3D conductivity in a disordered network results (i.e., according to VRH model).

Coulomb Gap Model

In addition to the 1D VRH, two other models predict a temperature dependence of $\exp[-(T_1/T)^{1/2}]$ to conductivity. The two models are the Coulomb gap model (48) and the charging-energy-limited tunneling (CELT) model (40). CELT, which is sometimes referred to as the grain model, will be discussed in the following section.

The Coulomb gap model assumes that interactions between localized electrons form excitons, which create a "soft gap" (called the Coulomb gap) in states near the Fermi level. The Coulomb gap model is essential for studying low temperature dc conductivity. This model is used in combination with the 3D VRH model, which dominates at higher temperatures. A crossover temperature (T_c) between the two models is defined as the low temperature limit of the VRH model. At higher temperatures, $T >> T_c$, the thermal energy that drives VRH is much larger than the Coulomb gap; therefore, the influence of the Coulomb gap can be neglected. At low temperatures, $T << T_c$, the states within the Coulomb gap energy dominate over the thermal energy, and an expression of $\exp[-(T_1/T)^{1/2}]$ for conductivity is predicted. Here $T_1 = e^2/\kappa a$, in which e is the electronic charge, κ is the dielectric constant, and a is the localization length. The crossover temperature is $T_c = e^4 a N_0(E_F)/\kappa^2$, where $N_0(E_F)$ is the unperturbed density of states near the Fermi level (48).

In amorphous carbon films, T_c is estimated to be around 15 K (48). An estimate of T_c for ion-implanted polymers is not possible, because the Fermi level, or the density of states near the Fermi level, is not known. However, we can make a reasonable assumption that T_c of implanted polymers is equal to that of amorphous carbon, because implanted polymer regions are randomly distributed carbon-rich clusters (*see* discussions in Chapter 2.5). As a result, the Coulomb gap model would not be necessary for conductivities of implanted polymers measured at temperatures greater than 15 K.

CELT Model

The conduction processes in implanted polymers are often explained by the CELT or grain model (*17, 20, 49*). This model assumes that the material consists of conducting grains embedded into an insulating matrix. The model was first proposed by Sheng and co-workers (*40*) for describing conductivity at high and low electric field limits in metal–dielectric composites.

In the high-field regime, the voltage drop between neighboring grains is much larger than kT/e. The majority of charge carriers are then created by field-induced tunneling between neutral grains, and the conductivity is strongly influenced by the external electric field. The electric field dependence of conductivity under these conditions is given by eq 5, where F is the electric field strength and B is a constant (*50*).

$$\sigma(F) = \sigma_0 \exp[B(F/T)^{1/2}] \quad (5)$$

In the low-field regime, the charge carriers are thermally activated, so charge transport is a result of tunneling between charged-conducting grains. By optimizing the carrier mobility, the temperature dependent conductivity is given by eq 6, where $T_1 = 8\chi s\varepsilon/k$, s is distance between the grains, and ε is the electrostatic energy necessary to move a charge from an uncharged grain to an adjacent uncharged grain.

$$\sigma(T) = \sigma_0 \exp[-(T_1/T)^{1/2}] \quad (6)$$

The parameter is $\chi = (2m_e\phi)^{1/2}/h$, where ϕ is the energy barrier through which the charge carrier must tunnel, h is Planck constant, and m_e is the effective mass of the charge carrier.

In the application of this model to ion-implanted polymers, the microstructure is the ion-induced carbon-rich clusters (conducting grains) embedded in the insulating pristine polymer matrix. The ratio of the size (d) of the clusters to the distance (s) between the clusters, in the case of $s/d << 1/2$, can be estimated by eq 7, where κ is the dielectric constant of the pristine polymer (*17*).

$$s/d = [\kappa h k T_1/16e^2\sqrt{2m_e\phi}]^{1/2} \quad (7)$$

Assuming $d = 100$ Å, for 50-keV As-implanted TMPA-BAPS films, s decreases from 6.2 to 3.1 Å as the dose changes from 5×10^{15} ions/cm^2 to 1×10^{17} ions/cm^2 (*17*). A rapid decrease in s with increasing dose was also observed from 2-MeV Ar-implanted 3, 4, 9, 10-perylenetetracarboxylic dianhydride films (*20*). Recent studies by Adkins (*51*) indicted that the assumption in the grain model that $s\varepsilon$ is constant is invalid for several granular metals. This assumption also may be incorrect for implanted polymers. Thus, the applicability of the grain model to ion-implanted polymers needs to be further investigated.

Composite Conduction Model

The microstructure of the ion-track region in implanted polymers, assuming that this ion track creates carriers for the conduction process, should enable conduction along paths of varying dimensionality. Ion tracks have been speculated to have a fractal geometry (*3*). For low dose implanted polymers, the ion tracks form a nearly 1D structure, where 1D VRH can occur. At higher doses, ion-track structures overlap and create a 3D disordered region where 3D VRH occurs. At intermediate doses, 1D and 3D VRH can be assumed to occur in the conduction process.

The temperature-dependent conductivity of 50-keV As-implanted TMPA-BAPS films was fitted by the empirical eq 1. The best exponent m for most samples was $1/4 < m < 1/2$ (*22*). This value corresponds to $1 < n < 3$ in eq 4, or more simply stated, as a dimensionality of between 1 and 3. This result seems to suggest that electrical conduction takes place in a fractional-dimensional space rather than the regular Euclidean space. However, it is difficult to imagine that charge carriers are excitations of the entire geometrical framework. They are probably localized carriers trapped in various geometries in their immediate vicinity. The fractal concept has been used in polymers to explain the pristine polymer structure and certain thermal excitations (*52*). Fractal excitations (or fractons) may contribute to the conduction process by scattering off the trapped or localized carriers, which are then constrained to travel in one dimension along the ion track or in three dimensions where the tracks overlap.

A singular conduction mechanism commonly has been assumed to exist in implanted polymers. This assumption may be overly simplistic because implanted polymers exhibit diverse structural characteristics. For example, conduction along the ion track region (i.e., the center of the carbon-rich cluster) is much different from conduction at the edges of the cluster (e.g., toward the ion's mean penetration range). Even conduction inside the cluster may not be uniform because of the existence of a temperature gradient within the cluster during implantation (*23*). These structural diversities suggest that more than one conduction mechanisms is involved in implanted polymers.

To explain the temperature dependence of conductivity in implanted polymers, a composite conduction model was proposed by Wang et al. (8, 23, 33). The conductivity in implanted polymers was speculated to be the sum of parallel contributions from 1D and 3D VRH conductivities. Here, 1D VRH occurs along the unbroken polymer chains in highly disordered regions, and 3D VRH occurs inside the overlapping clusters. This parallel composite conduction model (PCCM) assumes that two independent conduction carriers exist in the implanted polymers, so the total conductivity is a sum of two individual conductivities from the two different structural regions. The temperature-dependent conductivity of this model is then expressed by eq 8 (33):

$$\sigma(T) = \sigma_{1D}(T) + \sigma_{3D}(T)$$
$$= \sigma_1(T) \exp[-(T_1/T)^{1/2}] + \sigma_3(T) \exp[-(T_3/T)^{1/4}] \quad (8)$$

Equation 8 produced a much better fit compared with the singular conduction mechanism for 50-keV As-implanted TMPA-BAPS films in the dose range of 5 × 10^{15} to 2 × 10^{17} ions/cm^2 (33). Figure 14 shows the relative strength of the 1D conduction term $[\sigma_{1D}(T)/\sigma(T)]$ as a function of temperature for these As-implanted TMPA-BAPS films. The results suggest that at lower temperatures and low doses the term $\exp[-(T_1/T)^{1/2}]$ dominates, and at higher temperatures and high doses the term $\exp[-(T_3/T)^{1/4}]$ dominates. However, the better fits obtained by this model may be mislead-

ing because the function in eq 8 involves more free parameters.

The composite conduction may also be constructed from a different approach. Complex microstructures of implanted polymers may involve 1D and 3D traps for conduction carriers (e.g., electrons). This possibility implies that the same carrier would scatter off from different trap sites and would result in eq 9 for resistivity, where Matthiessen's rule (53) applies and ρ_{1D} and ρ_{3D} are resistivities due to conduction through the 1D and 3D scattering centers, respectively. Equation 9 is referred to as the series composite conduction model (SCCM).

$$\rho(T) = \rho_{1D}(T) + \rho_{3D}(T)$$
$$= \rho_1(T) \exp(T_1/T)^{1/2} + \rho_3(T) \exp(T_3/T)^{1/4} \quad (9)$$

To discriminate between SCCM and PCCM, 50-keV N-implanted PSA films at a dose of 5 × 10^{16} ions/cm^2 were systematically investigated (28, 54). These PSA films had varying thicknesses, some of which were thinner than the ion's mean penetration range.

Figure 15 shows the results of room-temperature surface resistivity as a function of film thickness. The constant resistivity observed for films thicker than 3000 Å is expected, because the thickness of the implanted layer is smaller than 3000 Å based on the TRIM calculations (25).

The temperature-dependent conductivity of these samples also was measured and fitted in terms of

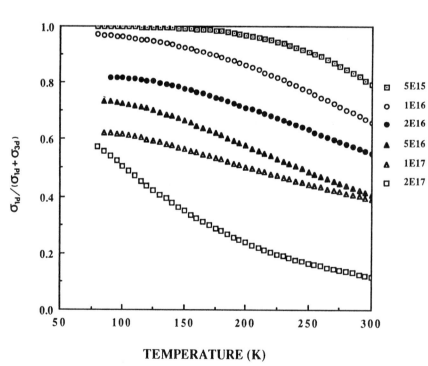

TEMPERATURE (K)

Figure 14. Contribution of the 1D conductivity $(\sigma_{1d}/\sigma_{1d} + \sigma_{3d})$ in TMPA-BAPS films as a function of temperature for 50-keV As implantation with varying beam dose. (Reproduced with permission from reference 33. Copyright 1993.)

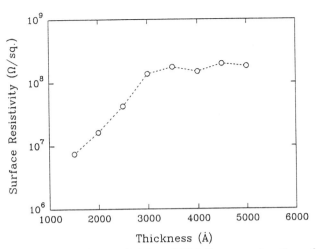

Figure 15. Surface resistivity of PSA films as a function of film thickness for 50-keV N implantation at a dose of 5×10^{16} ions/cm².

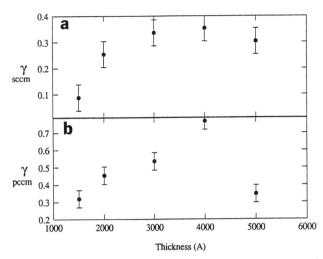

Figure 16. Overall contribution (γ) of the 1D conductivity in PSA films as a function of film thickness for 50-keV N implantation at a dose of 5×10^{16} ions/cm²: a, using SCCM; and b, using PCCM. Here γ is defined as the normalized integration of $\sigma_{1D}(T)/\sigma(T)$ for PCCM or $\rho_{1D}(T)/\rho(T)$ for SCCM over the measured temperature range. (Reproduced with permission from reference 28. Copyright 1994.)

PCCM and SCCM. Predictably, eqs 8 and 9 fit the data equally well. As mentioned previously, this phenomenon is probably a result of the number of free parameters used in the fit process. Figure 16 shows the thickness dependence of total relative contribution of the 1D term (γ) obtained from eqs 8 and 9. Here γ is defined as the normalized integration over the measured temperature range of the relative 1D terms, and the relative 1D terms are $\sigma_{1D}(T)/\sigma(T)$ for PCCM and $\rho_{1D}(T)/\rho(T)$ for SCCM. Figure 16 clearly indicates that the 1D contribution decreases with increasing film thickness for both models, a result suggesting that more 1D conduction occurs in the deeper regions close to the ion's mean penetration range.

Similar studies also were done for a set of plasma-etched PSA films that were implanted with 50-keV N at a dose of 1×10^{17} ions/cm² before etching. Results from these etched layers indicate that the ion-implanted layer is not uniformly conductive, and the more conductive region is closer to the surface, corresponding to the region where the ion energy loss is principally through the electronic ionization process (*28, 54*). All these results indicate that PCCM and SCCM support the conclusion that 1D conduction occurs toward the ion's mean penetration range and 3D conduction occurs near the surface. A saturation curve similar to Figure 15 is observed in Figure 16(a) but not in Figure 16(b), a result implying that SCCM may be more accurate than PCCM.

The 1D conductivity term in both PCCM and SCCM may result from the Coulomb gap model (*48*), provided that the crossover temperature (T_c) between the 3D VRH conduction and the Coulomb gap conduction is much higher in implanted polymers than in pure amorphous carbon ($T_c \approx 15$ K).

Any conduction mechanisms derived from the temperature dependence of conductivity alone cannot be conclusive before the characteristics of the charge carriers (generation, polarity, density, and mobility) in implanted polymers are fully understood.

Applications of Implanted Polymers

Polymers are widely used because of their low cost and easy processing compared with metals and other structural materials. The fact that polymers remain mechanically flexible after processing is important for some applications. Also, the inherent dielectric characteristics of polymers lead to their use as insulators. However, these mechanical and dielectric properties limit the use of polymers for other applications. Ion implantation provides a unique way to modify the physical and especially the surface properties of polymers for new applications.

Most of the research on ion implantation of polymers has been aimed primarily at understanding the fundamental mechanisms involved in the modification process, and less emphasis has been placed on using the modified properties and potential applications of the implanted polymers. The reason for this focus may be because ion implantation introduces irreversible chemical changes on the polymer. This effect is very different from the familiar effects of ion beams on metals and semiconductors, where the applications are based on the chemical effects of implanted impurities.

However, ion-implanted polymers have found applications both in mature technologies (*4*) such as lithography and in emerging technologies such as sensor devices. We briefly discuss some potential applications of implanted polymers under the categories of electrical, mechanical, optical, chemical, and special materials syntheses. Table II provides a partial list of such applications.

Applications Related to Electrical Properties

In all probability, the first use of implanted polymers will be for antistatic applications to reduce triboelectrification, attraction of dust, electrostatic discharging, and radio frequency (rf) interference in polymers used as packing and protective coatings (1, 4, 5). Evidence of the formation of p–n junctions in polyacetylene by implanting Na ions into a p-doped host was reported (55, 56). Field-effect transistors, based on poly(p-phenylene vinylene) doped by 30 keV iodine were recently studied (57). The practical fabrication of polymer-based p–n junctions and field effect transistors may be realized, provided that the carrier properties and conduction mechanisms in implanted polymers are clearly understood.

Ion-implanted polymers are excellent materials for the fabrication of low power consumption, high-value, small-geometry resistors. By using lithographic techniques, millimeter-size resistance arrays containing 16 mini-size megaohm resistors were produced from implanted polymers by Brewer Science, Inc. (Rolla, Missouri), and these resistors were used as memory elements in neural networks.

The most recent applications involve using implanted polymers as sensing materials. Compared with any material used for thermistors, the temperature coefficient of resistance in implanted polymers is very large. As a result, implanted polymers are good candidates for temperature sensors (47, 54, 58). By using the mature lithographic technology, miniature temperature sensing arrays were manufactured as a joint effort of Brewer Science, Inc., and Southwest Missouri State University. The implant layer can be coated on the surface of a silicon-integrated circuit and can serve as a dust and moisture protective layer, an antistatic coating, and a safety thermistor that turns off the power supply under short-circuit conditions.

Mechanical flexibility and enhanced conductivity under tension allows implanted polymers to be used as unique pressure-sensing materials (strain gauges) (54, 59). Resistance changes in the implanted layer after bending the entire polymer film are a function of

Table II. Potential Applications of Ion-Implanted Polymers Based on Their Specific Properties

Property	Potential Applications	Ref.
Electrical	Anti-static packing materials	1, 4
	Polymer-based p–n junctions	55, 56
	Field-effect transistors	
	High value small resistors	
	Temperature sensors	47, 54, 58
	Strain gauges	54, 59, 60
	Vacuum gauges	
	Infrared bolometers	
Mechanical	Protective coatings for magnetic disks	
	Exterior structural components for vehicles	
	Wear-resistant gears and bearings	
	Artificial hip and knee joints	
	Scratch-resistant plastic sunglasses	
Optical	Ion-beam microlithography	4
	Color filters	
	Optical gratings	
	Polymers as low-loss waveguides	2
	Polymer scintilators	
Chemical resistance	Reduced surface wettability	
	Reduced vapor permeability	
	Reduced environment deterioration	
	Corrosion resistant coatings	
New materials	Synthesis of hydrogenated amorphous carbon films, diamondlike carbon (DLC) films, diamond films, and doped carbon electronic devices	

the strain produced by external stress. This piezoresistivity was found in several ion-implanted polymers (*59*). The temperature dependence of the piezoresistivity also was studied (*54*). Prototype strain gauges are fabricated using ion-implanted homogeneously conductive polymer films (*60*).

Polymer films have a greater mechanical flexibility than ceramics and semiconductors, and as a result, they can undergo large flexural deformations. These characteristics imply that a wider range of applications may be possible for implanted polymers used as pressure sensors or strain gauges. A prototype of a scale using implanted polymers was fabricated at Southwest Missouri State University. This device uses a cantilever beam, coated with an implanted polymer film, where different weights can be calibrated against the resistance changes in the implanted polymer film. Potential applications of piezoresistivity in implanted polymers may include their use in the manufacture of pressure sensors for automobiles and aircraft.

Other potential applications include using the electrical properties of implanted polymers to develop vacuum gauges. The resistance of an implanted polymer film in a vacuum chamber will increase with decreasing pressure as the heat dissipation rate of the film through the surrounding gaseous molecules is reduced. Commonly used thermocouple gauges use a similar principle, but these gauges are insensitive to pressures between 101 kPa and a few hundred pascals. Newly developed Pirani-type vacuum gauges from implanted polymers (prototype) in Brewer Science, Inc., have good sensitivity in this pressure range.

The large temperature coefficient of resistance, combined with the inherent low thermal conductivity of the material and the dark color of the implant layer, make these materials good IR bolometers. Absorption of IR radiation causes a temperature rise and a decrease in resistance in the implanted layer. However, the advantages of implanted polymers over conventional IR sensing materials (semiconductors) and the solution to the technical problems in fabrication are unclear.

Applications Related to Mechanical Properties

Ion implantation significantly enhances the surface mechanical properties of polymers, such as hardness and wear resistance. Hardness increases from 0.1 to 0.5 GPa for pristine polymers to as high as 22 GPa for the implanted material (*61*). These superhard-surfaced polymers usually are achieved only by high-energy (MeV) ion irradiations. For comparison, the average values of hardness for diamond, steel, glassy carbon, and graphite are 100, 7, 2, and < 0.1 GPa, respectively.

Potential applications of such hard and wear-resistant polymer surfaces include the following: hard protective coatings for magnetic disk drives; exterior structural components for automobiles, marine vehicles, and aircraft; wear-resistant gears and bearings for machines; artificial hip and knee joints; and scratch-resistant plastic sunglasses and aircraft windows. Joint research and applications development between industrial partners and technology providers are necessary to bring products using ion-hardened polymers into the marketplace in the future.

Applications Related to Optical Properties

Color darkening of the polymer surface as a result of ion implantation is attributed to carbonization on the surface, and this darkening can be tuned by varying ion-beam parameters. This process may provide a unique way of making polymer color filters. Planar waveguides were fabricated using ion implantation of poly(methyl methacrylate) films (*62*). Fabrication of optical gratings on a polymer surface via ion implantation is another potential application. Our recent studies indicated that certain polymers become flourescent under the irradiation of MeV protons, and this result may suggest a unique way to fabricate particle scintillating detectors by using polymer films. As optical gain media, erbium-doped polymers may become very important materials to fabricate low-loss passive and active optical waveguides and to advance other photonic applications in the near future.

Applications Related to Chemically Resistant Properties

The inherent solubility of many polymers limits their use in harsh chemically reactive environments. By reducing the free volume and increasing the cross-link density of the implanted region, ion implantation can enhance the surface chemical resistance and reduce the organic vapor permeability from the polymer surface. For example, ion implantation reduces water diffusion in otherwise permeable PANI and stabilizes the polymer conductivity in a humid environment (*29*). This chemically resistant property of ion-implanted polymers may lead polymeric coatings to some important applications in corrosion science and industry where ceramic coatings are often used.

Special Materials Synthesis

Ion irradiation of polymers creates hydrogen-depleted but carbon-enriched surface structures. The desired surface properties can be tuned or tailored by choosing different ion-beam parameters. Because the main constituent of ion-implanted polymers is carbon, ion implantation of polymers may become an exotic means to produce hydrogenated amorphous carbon (a-C:H), diamondlike carbon (DLC), and even diamond films. This possibility has been enhanced by the recent research by Bianconi and co-workers (63), who found that tetrahedral plastic transforms into very hard materials (DLC type) after UV irradiation. Also, if appropriate ion species are used, the carbon films produced by ion implantation may be p- or n-type doped simultaneously. The films produced by this way may be more cost competitive in electronic industries when carbon-based semiconductor devices are needed.

Concluding Remarks

Ion implantation of polymers distorts and even decomposes the original polymer structure. It leads to evolution of volatile molecules (e.g., H_2, C_2H_2, and CH_4) and produces a carbon-rich microstructure depending on the original composition. The formation of the microstructure (carbon clusters and the network formed by the clusters) is a consequence of the radiolytic interactions between energetic ions and polymer molecules. The internal structure of the clusters can be understood from the thermal dissipation of deposited energy in the vicinity of the ion track.

The carbon-rich surface of the implanted polymer exhibits very different properties from the pristine polymer. Optical surface morphology changes from a nearly transparent state to a shiny brown or even black color. Mechanical properties improve, such as hardness and resistance to wear. Chemical resistance increases, hence the gas permeability on the surface decreases., Electrical conductivity also increases significantly. All these changes have expanded the potential applications of polymers, as discussed in the previous section.

The dose dependence of the enhanced surface conductivity suggests that the conduction is due to phonon-activated VRH between the localized electronic sites. The ion tracks are isolated from each other, and they are one-dimensional at low doses and three-dimensional when the ion tracks overlap at high doses. The temperature dependence of conductivity suggests that an either composite conduction of 1D VRH and 3D VRH or a transition between 3D VRH and Coulomb gap conduction may exist.

A definitive conclusion about the conduction mechanism is not possible at present. Carrier identification and measurements of carrier density and mobility are essential to a better understanding of the conduction process. The Hall effect, thermoelectric power, ac conductivity, magnetoresistance, and time-of-flight measurements for mobilities are needed to clarify these issues.

Penetration of super high energy (hundreds MeV) and heavy ion (e.g., Xe) penetration through thin polymer foils, which leave only electronic ionization damage in the polymer, are currently being studied. Conductivity measurements along the ion track (longitudinal) and across the ion tracks (lateral) are expected to provide further information about the conduction mechanism in implanted polymers.

Some recent comparative studies of the carbon-rich clusters in implanted polymers and the carbon structures in a-C:H films produced by other methods (e.g., chemical vapor deposition, sputtering, high-energy laser pyrolysis, and rf glow discharge) are also important. These studies are relevant because they provide a unique way to better understand the microstructure of implanted polymers and provide new techniques to synthesize DLC or even diamond materials.

Acknowledgments

We would like to express our sincere gratitude to the many people who offered ideas and opinions in the course of writing this review. Specifically, C. J. Sofield was generous in his discussions of the experimental data. To M. G. Moss and S. S. Mohite, we offer thanks for supplying the polymer films and for explaining some features of polymer structures. Y. Q. Wang thanks Brewer Science, Inc., for sponsoring his research when he was at Southwest Missouri State University.

References

1. Dresselhaus, M. S.; Wasserman, B.; Wnek, G. E. In *Ion Implantation and Ion Beam Processing of Materials*; Hubler, G. K.; Holland, O. W.; Clayton, C. R.; White, C. W., Eds.; Materials Research Society Symposium Proceedings 27; Materials Research Society: Pittsburgh, PA, 1983; p 413.
2. Venkatesan, T.; Levi, R.; Banwell, T. C.; Tombrello, T.; Nicolet, M.; Hamm, R.; Meixner, A. E. In *Ion Beam Processes in Advanced Electronic Materials and Device Technology*; Appleton, B. R.; Eisen, F. H.; Sigmon, T. W., Eds.; Materials Research Society Symposium Proceedings 45; Materials Research Society: Pittsburgh, PA, 1985; p 189.
3. Wasserman, B. *Phys. Rev.* **1986**, *B34*, 1926.
4. Venkatesan, T.; Calcagno, L.; Elman, B. S.; Foti, G. In *Ion Beam Modifications of Insulators*; Mazzoldi, P.; Arnold, G. W., Eds.; Elsevier: Amsterdam, Netherlands, 1987; Chapter 8, p 301.
5. Hersh, S. P.; Brock, S.; Grady, P. L. In *Polymers for Advanced Technology*; Lewin, M., Ed.; VCH: New York, 1988; p 52.
6. Bedell, C. J.; Sofield, C. J.; Bridwell, L. B.; Brown, I. M. *J. Appl. Phys.* **1990**, *67*, 1736.
7. Bridwell, L. B. In *Trends in Ion Implantation*; Rossum, V. M., Ed.; Solid State Phenomena, Series 27; Trans Tech Publications Ltd.: Zürich, Switzerland, 1992; p 163.
8. Wang, Y. Q.; Mohite, S. S.; Bridwell, L. B.; Giedd, R. E.; Sofield, C. J. *J. Mater. Res.* **1993**, *8*, 388.

9. Lee, E. H.; Rao, G. R.; Lewis, M. B.; Mansur, L. K. *J. Mater. Res.* **1994**, *9(4)*, 1000.

10. Lee, E. H.; Lewis, M. B.; Blau, P. J.; Mansur, L. K. *J. Mater. Res.* **1991**, *6(3)*, 610.

11. Rao, G. R.; Wang, Z. L.; Lee, E. H. *J. Mater. Res.* **1993**, *8(4)*, 927.

12. Ochsner, R.; Kluge, A.; Gyulai, J.; Bogen, S.; Ryssel, H. *Surf. Coat. Technol.* **1992**, *51*, 124.

13. Wasserman, B.; Braunstein, G.; Dresselhaus, M. S.; Wnek, G. E. In *Ion Implantation and Ion Beam Processing of Materials*; Hubler, G. K.; Holland, O. W.; Clayton, C. R.; White, C. W., Eds.; Materials Research Society Symposium Proceedings 27; Materials Research Society: Pittsburgh, PA, 1983; p 427.

14. Venkatesan, T.; Dynes, R. C.; Wilkens, B.; White, A. E.; Gibson, J. M.; Hamm, R. In *Ion Implantation and Ion Beam Processing of Materials*; Hubler, G. K.; Holland, O. W.; Clayton, C. R.; White, C. W., Eds.; Materials Research Society Symposium Proceedings 27; Materials Research Society: Pittsburgh, PA, 1983; p 449.

15. Forrest, S. R.; Kaplan, M. L.; Schmidt, P. H.; Venkatesan, T.; Lovinger, A. *J. Appl. Phys. Lett.* **1982**, *41*, 708.

16. Bridwell, L. B.; Giedd, R. E.; Wang, Y. Q.; Mohite, S. S.; Jahnke, T.; Brown, I. M. *Nucl. Instrum. Methods* **1991**, *B59/60*, 1240.

17. Wang, Y. Q.; Giedd, R. E.; Mohite, S. S.; Jahnke, T.; Bridwell, L. B. *Mater. Lett.* **1991**, *12*, 21.

18. Kuczkowski, A. *Phys. Status Solidi A* **1993**, *136*, K113.

19. Ruddy, F. H.; Barko, J.; Schoch, K. F. *J. Mater. Res.* **1988**, *3*, 1253.

20. Kaplan, M. L.; Forrest, S. R.; Schmidt, P. H.; Venkatesan, T. *J. Appl. Phys.* **1984**, *55*, 732.

21. Wang, Y. Q.; Bridwell, L. B.; Giedd, R. E.; Murphy, M. J. *Nucl. Instrum. Methods* **1991**, *B56/57*, 660.

22. Wang, Y. Q. Ph.D Thesis, Lanzhou University and Southwest Missouri State University, 1991.

23. Wang, Y. Q.; Giedd, R. E.; Bridwell, L. B. *Nucl. Instrum. Methods* **1993**, *B79*, 659.

24. Sofield, C. J.; Sugden, S.; Bedell, C. J.; Graves, P. R.; Bridwell, L. B. *Nucl. Instrum. Methods* **1992**, *B67*, 432.

25. Ziegler, J. F.; Biersack, J. P.; Littmark, U. *The Stopping and Range of Ions in Solids*; Pergamon: Oxford, England, 1985.

26. Abel, J. S.; Mazurek, H.; Day, D. R.; Maby, E. W.; Senturie, S. D.; Dresselhaus, G.; Dresselhaus, M. S. In *Ion Implantation of Materials*; Picroux, S. T.; Choyke, W. J., Eds.; Materials Research Society Symposium Proceedings 7; Materials Research Society: Pittsburgh, PA, 1982; p 173.

27. Mazurek, H.; Day, D. R.; Maby, E. W.; Abel, J. S.; Senturie, S. D.; Dresselhaus, M. S.; Dresselhaus, G. *J. Polym. Phys.* **1983**, *21*, 537.

28. Giedd, R. E.; Robey, D.; Wang, Y. Q.; Moss, M. G.; Kaufmann, J. *Materials Research Society Symposium Proceedings*; Materials Research Society: Pittsburgh, PA, 1994; Vol. 316, p 75.

29. Feng, Y. P.; Robey, D.; Wang, Y. Q.; Giedd, R. E.; Moss, M. G. *Mater. Lett.* **1993**, *17*, 167.

30. Venkatesan, T.; Forrest, S. R.; Kaplan, M. L.; Murry, C. A.; Schmidt, P. H.; Wilkens, B. J. *J. Appl. Phys.* **1983**, *54*, 3150.

31. Bridwell, L. B.; Wang, Y. Q. *SPIE* **1991**, *1519*, 878.

32. Bridwell, L. B.; Giedd, R. E.; Wang, Y. Q.; Mohite, S. S.; Jahnke, T.; Brown, I. M.; Bedell, C. J.; Sofield, C. J. *Nucl. Instrum. Methods* **1991**, *B56/57*, 656.

33. Wang, Y. Q.; Bridwell, L. B.; Giedd, R. E. *J. Appl. Phys. (Comm.)* **1993**, *73(1)*, 474.

34. Elman, B. S.; Thakur, M. K.; Sandman, D. J.; Newkirk, M. A. *J. Appl. Phys.* **1985**, *57*, 4996.

35. Bartko, J.; Hall, B. O.; Schoch, K. F. *J. Appl. Phys.* **1986**, *59*, 111.

36. Aleshin, A. N.; Gribanov, A. V.; Dobrodumov, A. V.; Suvorov, A. V.; Shlimak, I. S. *Sov. Phys. Solid State* **1989**, *31*, 6.

37. Austin, I. G.; Mott, N. F. *Adv. Phys.* **1969**, *18*, 41.

38. Elliott, S. R. *Philos. Mag.* **1977**, *36*, 1291.

39. Tu, D. M.; Wu, H. C.; Xi, B. F.; Kao, K. C. *IEEE Trans. Electr. Insul.* **1992**, *27(2)*, 385.

40. Abeles, B.; Sheng, P.; Coutts, M. D.; Arie, Y. *Adv. Phys.* **1975**, *24*, 407.

41. Giedd, R. E.; Shipman, J.; Murphy, M. In *Ion Beam Processing of Advanced Electronic Materials*; Cheung, N. W.; Marwick, A. D.; Roberto, J. B., Eds.; Materials Research Society Symposium Proceedings 147; Materials Research Society: Pittsburgh, PA, 1989; p 377.

42. Wnek, G. E.; Wasserman, B.; Loh, I.-H. In *Ion Implantation and Ion Beam Processing of Materials*; Hubler, G. K.; Holland, O. W.; Clayton, C. R.; White, C. W., Eds.; Materials Research Society Symposium Proceedings 27; Materials Research Society: Pittsburgh, PA, 1983; p 435.

43. Ziman, J. *Principles of the Theory of Solid*; Cambridge University: New York, 1964.

44. Mott, N. F. *Philos. Mag.* **1969**, *19*, 835.

45. Mott, N. F.; Davis, E. A. *Electronic Processes in Non-Crystalline Materials*; Calarendon: Oxford, England; 1979.

46. Sakamoto, M.; Wasserman, B.; Dresselhaus, M. S.; Wnek, G. E. *J. Appl. Phys.* **1986**, *60*, 2788.

47. Giedd, R. E.; Moss, M. G.; Craig, M. M.; Roberson, D. E. *Nucl. Instrum. Methods* **1991**, *B59/60*, 1253.

48. Efros, A. L.; Shklovskii, B. I. *J. Phys.* **1975**, *C8*, 149.

49. Venkatesan, T.; Forrest, S. R.; Kaplan, M. L.; Schmidt, P. H.; Murray, C. A.; Brown, W. L.; Wilkens, B. J.; Roberts, R. F.; Ruoo, L.; Schonhorn, H. *J. Appl. Phys.* **1984**, *56*, 2778.

50. Wang, Z. H.; Scherr, E. M.; MacDiarmid, A. G.; Epstein, A. *J. Phys. Rev.* **1992**, *B45*, 4190.

51. Adkins, C. J. *J. Phys.* **1987**, *C20*, 235.

52. Alexander, S.; Laermans, C.; Orbach, R.; Rosenberg, H. M. *Phys. Rev.* **1983**, *B28*, 4615.

53. Wert, C. A.; Thomson, R. M. *Physics of Solids*, 2nd ed.; McGraw-Hill: New York, 1970.

54. Wang, Y. Q.; Giedd, R. E.; Moss, M. G.; Kaufmann, J. Presented at 13th International Conference on the Application of Accelerators in Research and Industry, Nov. 7-10, 1994, Denton, Texas. To be published in *Nucl. Instrum. Methods* **1995**, *April*, B.

55. Wada, T.; Takeno, A.; Iwaki, M.; Sasabe, H.; Kobayashi, Y. *J. Chem. Soc. Chem. Commun.* **1985**, *17*, 1194.

56. Yoshida, K.; Iwaki, M. *Nucl. Instrum. Methods* **1987**, *B19*, 878.

57. Pichler, K.; Jarrett, C. P.; Fried, R. H.; Moliton, A. *J. Appl. Phys.* **1995**, *77*, 3523.

58. Xu, D.; Xu, X. L.; Zou, S. C. *Rev. Sci. Instrum.* **1992**, *63*, 202.

59. Wang, Y. Q.; Robey, D.; Giedd, R. E.; Moss, M. G. *Materials Research Society Symposium Proceedings*; Materials Research Society: Pittsburgh, PA, 1994; Vol. 316, p 349.

60. Giedd, R. E.; Wang, Y. Q.; Moss, M. G.; Kaufman, J.; Brewer, T. L. U.S. Patent 5, 505, 093, April 4, 1996.

61. Dagani, R. *Chem. Eng. News* **1995**, *Jan. 9, 24.*

62. Brunner, S.; Rück, D. M.; Frank, W. F. X.; Linke, F.; Schösser, A.; Behringer, U. *Nucl. Instrum. Methods* **1994**, *B89*, 373.

63. Visscher, G. T.; Nesting, D. C.; Badding, J. V.; Bianconi, P. A. *Science (Washington, D.C.)* **1993**, *260*, 1496.

64. Zuo, F.; Angelopoulos, M.; MacDiarmid, A. G.; Epstein, A. J. *Phys. Rev. B* **1989**, *39*, 3570.

Part Three

Optoelectronic Properties
and Applications

Photophysics
of Functional Polymers

David Phillips

This chapter provides an introduction to the topic of polymer photophysics: namely, the photophysical techniques used for studying order, subgroup motion, electronic energy transfer, and exciton migration in functional polymers. Among physical fates of excited electronic states of chromophores in macromolecules, luminescence (particularly fluorescence) provides the most convenient probe of polymer behavior. The photophysics of macromolecular chromophores is outlined stressing the photophysical processes that compete with photochemical fates. The distinction between isolated chromophores and excitons is outlined, and the consequences of high concentrations of isolated chromophores in the formation of excimer trap sites are discussed. A case history of the photophysics of poly(diacetylenes) is presented, and solvato- and thermochromism and surface effects on polymer conformation and exciton migration are discussed. Models for exciton migration in other polymers in general are also briefly described.

The photophysical properties of polymers, including functional polymers, are of diverse theoretical and practical importance. The photophysical fates of excited states are in competition with the chemical processes that may be the desired object of a functional polymer: for example, in cross-linking and other radiation effects discussed in Part Two of this book. Photophysical fates determine the nature of electronic and nonlinear optical (NLO) properties, which are the subject of the following chapters in this part of the book. Similarly, photophysical phenomena include radiative and nonradiative processes (luminescence and electronic energy transfer and migration), which are the basis of light-harvesting polymers and sensing polymers covered in Part Four. It is important, therefore, to the discussion in various parts of this volume, and in particular to the following chapters in this part, to outline some basic definitions and terms of the photophysical fates of molecules, the time scales of their decay, and some kinetic and spectroscopic considerations.

Definitions and Terms

Absorption is the term usually used to describe the radiative transition from a vibrational level of the ground electronic state to a set of vibrational levels of an excited electronic state, and it is almost exclusively confined to those transitions that occur via an electric dipole mechanism. For convenience, the zero-point level of the ground state is usually considered to be the only level populated under normal conditions, whereas the population of low-frequency higher vibrational levels may be significant and give rise to *hot-band* absorption. Transitions arising from electronically excited states may also be observed in absorption by transient spectroscopy.

Fluorescence is the spontaneous *radiative* transition between states of the same multiplicity, or spin state. The ground electronic state of most organic molecules is a spin-paired singlet state, and most (but not all) fluorescences observed occur from the first excited singlet state to the ground state. Excited electronic states

3469–8/97/0407$15.00/0 © 1997 American Chemical Society

may exhibit stimulated emission (the exact counterpart to the absorption process) if a population inversion can be achieved to the excited state with respect to the lower energy state, as in dye lasers.

Phosphorescence is the spontaneous radiative transition between two states involving a change of electron spin. This transition is most commonly observed as the emission resulting from the lowest electronically excited triplet state (*see* subsequent discussion) to the singlet ground state. Because the transition is spin-forbidden by selection rules, the transition is of low probability and the emission may have a long duration.

Internal conversion is the nonradiative decay of an electronically excited molecule to a lower state of the same multiplicity.

Intersystem crossing is the nonradiative decay of an electronically excited molecule to a lower state of different multiplicity. Most commonly, the first excited singlet decays to a lower lying triplet excited state, or the lowest triplet decays to the ground-state singlet level.

Singlet state: Each nondegenerate molecular orbital may accommodate two electrons provided the spin angular momentum (and hence magnetic moment) of one electron is opposite to that of the second (Pauli exclusion principle). The net spin on the system, S, is then 0, and the multiplicity $(2S + 1)$ is unity, giving a singlet state. Excitation of one electron to a higher orbital by an electron dipole transition gives rise to an excited electronic singlet state, and these states are usually labelled S_0, (ground), S_1, S_2, etc.

Triplet state: For two electrons in different molecular orbitals, the electrons may have parallel spins because they do not occupy the same space. In this case, the multiplicity is then $2(1/2 + 1/2) + 1 = 3$ giving a triplet state, usually labelled T_1, T_2, etc. Molecular oxygen is unusual in that the ground state is a triplet state $(^3\Sigma_g^-)$. States with one unpaired electron are doublet states, states with three unpaired electrons are quartets, and so-on.

Excimers: The word excimer comes from a coalescence of the words *excited* d*imer* and represents a sandwich dimer, which is stable only in the excited state and is repulsive in the ground state. It is characterized by a broad structureless fluorescence red-shifted with respect to emission of the monomeric species.

Exciplex: This term comes from the words *excited* com*plex* and represents an excited state complex formed from two dissimilar molecules, one of which is photoexcited. The stabilizing force in the excited state is charge-transfer in nature, and exciplexes are usually highly polar and stabilized by polar solvents. The ground-state pair is, strictly speaking, repulsive in an exciplex. Many weakly bound ground-state donor–acceptor complexes upon excitation produce a species that resembles (or is identical to) the exciplex formed by a monomeric excited-state bimolecular quenching process.

Exciton represents an excitation in an assembly of chromophores that interact strongly enough so that the excitation cannot be considered to be localized on a single chromophore. Instead, the excitation is distributed over a range of chromophores (*see* subsequent discussion).

Chromism: Molecules that show different absorption (and emission) characteristics under different physical conditions exhibit chromism (i.e., color change). A molecule may thus show solvatochromism or thermochromism.

The fates of photoexcited polyatomic molecules are conveniently outlined in the form of a Jablonskii diagram (Figure 1). For an isolated chromophore in a polymer, this description is sufficient. The competing fates following excitation to the singlet manifold will be discussed on the basis of this diagram.

Unimolecular Fates

For excitation to the first excited electronic state (S_1), the unimolecular electronic relaxation processes competing with fluorescence are intersystem crossing (ISC) to the triplet manifold, internal conversion (IC) to the ground state, and photochemical reaction. In condensed media, vibrational relaxation occurs on a picosecond time scale, and thus only photochemical processes with rate constants in excess of 10^{12} s^{-1} can enter the competition. Therefore, subsequent to excitation, vibrational relaxation is usually complete before electronic relaxation. IC is usually faster at higher excess energies, but it is not usually of great importance for lower lying vibrational levels of the first excited singlet state. Therefore, the principal process competing with fluorescence is ISC to the triplet manifold of levels. Writing a simple kinetic scheme (eqs 1–5) for the various processes permits definition of the quantum yield of fluorescence (ϕ_F) and decay time (τ_F) in terms of first-order rate constants, where h is Planck constant and ν is frequency for a molecule M.

$$M + h\nu \rightarrow {}^1M^* \qquad I_A \qquad \text{(absorption)} \qquad (1)$$

$$^1M^* \rightarrow M + h\nu \qquad k_R \qquad \text{(fluorescence)} \qquad (2)$$

$$^1M^* \rightarrow {}^3M^* \qquad k_{ISC} \qquad \text{(intersystem crossing)} \qquad (3)$$

$$^1M^* \rightarrow M \qquad k_{IC} \qquad \text{(internal conversion)} \qquad (4)$$

$$^1M^* \rightarrow \text{products} \qquad k_D \qquad \text{(dissociation)} \qquad (5)$$

From a steady-state analysis of the scheme, the quantum yield of fluorescence (ϕ_F) is given by eq 6, and the fluorescence decay time (τ_F), is given by eq 7.

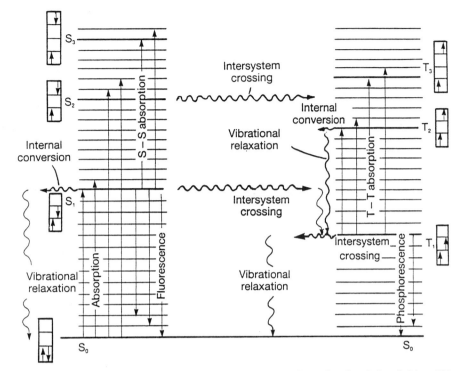

Figure 1. Fates of excited states of complex polyatomic molecules (after Jablonskii).

$$\phi_F = \frac{k_R}{(k_R + k_{ISC} + k_{IC} + k_D)} \quad (6)$$

$$\tau_F = (k_R + k_{ISC} + k_{IC} + k_D)^{-1} \quad (7)$$

The intensity of fluorescence from any molecule, even before consideration of bimolecular interactions, depends on the magnitude of the rate constant (k_R) relative to the sum ($\Sigma k = k_{ISC} + k_{IC} + k_D$). Equations 6 and 7 demonstrate that rationalization of the photophysics and photochemistry of singlet-state molecular species (i.e., absolute rate constants for the competing decay processes) cannot be obtained with the sole knowledge of quantum yields, which are merely ratios of rate constants. However, a knowledge of the excited-state lifetime (by, for example, the measurement of the fluorescence decay time) together with quantum yields provide, in principle, the absolute rate information required according to eqs 8 and 9.

$$k_R = \phi_F / \tau_F \quad (8)$$

$$k_{ISC} = \frac{\phi_{ISC}}{\tau_F} \quad (9)$$

Measurement of the fluorescence decay time is thus of immense importance. Alternatively, singlet-state decay characteristics can be monitored by singlet–singlet absorption measurements. Additional information, particularly about molecular motion, can be obtained by anisotropy measurements. By referring again to Figure 1, the population of the triplet state of a molecule can be recorded as a function of time, either by triplet–triplet absorption measurements or by monitoring phosphorescence. However, monitoring phosphorescence is often of too low a yield to be useful.

The concentration of the products of photochemical reactions is also amenable to sampling by measurement of transient absorption, which, for nonemitting species, is the conventional method. Transient grating methods and time-resolved Raman spectroscopy can yield much kinetic and spectroscopic information in this regard. To date, however, most information has been gleaned from fluorescence measurements, which are rapid, nondestructive, and relatively straightforward (*1, 2*).

Biomolecular Fates

Equations 1–9 are based on the assumption that the excited state decays only by intramolecular processes, but there are a number of bimolecular interactions involving a quencher molecule Q that lead to quenching of the excited state (eq 10). Quenching processes include enhancement of spin-forbidden nonradiative decay, often by molecular oxygen (O_2, $^3\Sigma_g^-$) and other paramagnetic species, or those containing atoms of high nuclear charge (eq 11).

$$^1M^* + Q \rightarrow products \quad (10)$$

$$^1M^* + O_2\,(^3\Sigma_g^-) \rightarrow\, ^3M^* + O_2\,(^3\Sigma_g^-) \qquad (11)$$

Charge transfer (eqs 12 and 13) and, in the limit, electron transfer (eqs 14 and 15) processes are common to all small molecules and polymers. Electronic energy transfer (eq 16) is, on the other hand, a widely observed photoexcited state in polymeric media, particularly because for some polymers (*see* subsequent discussion) there will be a high chromophore concentration.

$$^1M^* + Q \rightarrow\, ^1(M\delta^+...Q\delta^-)^{1*} \text{ (charge transfer)} \quad (12)$$

$$^1M^* + Q \rightarrow\, ^1(M\delta^-...Q\delta^+)^{1*} \text{ (charge transfer)} \quad (13)$$

$$^1M^* + Q \rightarrow M^+\bullet...Q\bullet^- \text{ (electron transfer)} \quad (14)$$

$$^1M^* + Q \rightarrow\, ^1M\bullet^-...Q^+\bullet \text{ (electron transfer)} \quad (15)$$

$$^1M^* + Q \rightarrow M +\, ^1Q^* \text{ (electron energy transfer)} \quad (16)$$

If Q is a different chromophore from M, the transfer will occur if the electronic energy of $^1Q^*$ is less than that of M. However, the situation is often met in polymers where there is a high concentration of identical chromophores, and transfer between these chromophores becomes a feasible fate of the initially excited state.

Exciton Approach

For the interacting chromophores mentioned previously, the isolated molecule approach may be insufficient, and an exciton description must be used. For example, in considering energy transfer between identical chromophores M_1 and M_2, the initial and final wavefunctions (ψ_i and ψ_f, respectively) can be written as eqs 17 and 18, where r_1 and r_2 represent electronic coordinates in M_1 and M_2, respectively. The matrix element (β) mixing initial and final states is given by eq 19, where H is the perturbation to the Hamiltonian, and $d\tau$ means $dxdydz$, the differential over the Cartesian coordinates, x, y, z. For a strong coupling, where ΔE is the electronic energy gap between zero-order states, ΔE is within the range of $\alpha \leq \Delta E \leq \beta$ (α is average coupling energy between lattice modes and molecular vibrational modes in a solid). The time probability, $P(t)$, of finding the excitation at a given time, t, is given by eq 20.

$$\psi_i = \psi_{M_1^*}(r_1)\,\psi_{M_2}(r_2) \qquad (17)$$

$$\psi_f = \psi_{M_1}(r_1)\,\psi_{M_2^*}(r_2) \qquad (18)$$

$$\beta = \psi_i H(r_1, r_2)\psi_f d\tau \qquad (19)$$

$$P(t) = (\cos^2\beta t/h)\psi_i^2 + (\sin^2\beta t/h)\psi_f^2 \qquad (20)$$

Recurrence thus occurs on a timescale of $h/4\beta$. However, $\beta \cong 1$ eV, and thus t is of the order of 10^{-15} s. Therefore, when M_1 is excited, excitation transfer to an identical neighboring chromophore takes place before the nuclei can adopt equilibrium geometry, and thus $\psi(M_1^*M_2)$ and $\psi(M_1M_2^*)$ are indistinguishable. Hence, a description must be used that treats the resonant state (or states, for an assembly of chromophores) as a single excited state. For many systems, such as poly-(diacetylenes) (*see* subsequent discussion), the exciton description is necessary. For many other polymeric systems, however, the interaction between chromophores is weak, and relaxation of the positions of nuclei can occur before transfer of energy. In this case, ψ_i and ψ_f can be taken to be independent eigenstates, and energy transfer between M_1 and M_2 follows the familiar treatments of Forster (eq 21) and Dexter (eq 22) for dipole-induced dipole and exchange interactions, respectively. Polymers with pendant chromophores or isolated chromophores can be described in terms of isolated molecules.

$$k_{DA} = \frac{9000\kappa^2 \ln 10}{128\pi^5 n^4 \tau_D R^6}\phi_D \int F_D(\bar{\upsilon})\varepsilon_A(\bar{\upsilon})d\bar{\upsilon}/\bar{\upsilon}^4 \qquad (21)$$

$$k_{DA} = JKe^{-2R/L} \qquad (22)$$

In the Forster expression (eq 21), ϕ_D is the quantum yield of emission of the donor molecule in the absence of acceptor, n is the refractive index of the medium, τ_D is the excited-state lifetime of the donor in the absence of acceptor, and R is the donor–acceptor separation distance. The integral represents spectral overlap of donor and acceptor, where $F_D(\bar{\upsilon})$ is the emission spectrum of the donor (normalized in such a way that $\int F_D(\bar{\upsilon})d\bar{\upsilon}=1$), and $\varepsilon_A(\bar{\upsilon})$ is the molar decadic extinction coefficient of the acceptor as a function of $\bar{\upsilon}$. The orientation factor κ^2 varies between 0 and 4 depending on mutual orientation of donor and acceptor electronic transition dipoles, and it is taken to be 2/3 for random orientations. For electron exchange, the rate constant k_{DA} is given by eq 22, where K is a constant, R is the donor–acceptor separation, L is the effective average Bohr radius, and J is a spectral overlap term in which both $F_D(\bar{\upsilon})$ and $\varepsilon_A(\bar{\upsilon})$ of eq 21 are normalized to unity.

Exciplex and Excimer Formation

One further consequence of a high local chromophore concentration in some polymers, particularly vinyl polymers with pendant chromophores, is the formation of exciplexes (between two different chromophores: one excited and one ground-state), and excimers (between two like chromophores: one excited and one ground-state). The energetics of excimer formation and decay are depicted in Figure 2. In general, excimer formation can occur whenever aromatic chromophores adopt a face-to-face coplanar arrangement with a separation of 0.3–0.35 nm.

In simple nonmacromolecular systems in solution, the kinetics of excimer formation and decay are straightforward. The time response to pulsed excitation of the monomeric fluorescence is given by eq 23, and that of its excimer is given by eq 24.

$$[^1M^*] = \frac{[^1M^*]_0}{(\lambda_2 - \lambda_1)} \tag{23}$$

$$\{(\lambda_2 - X)\exp(-\lambda_1 t) + (X - \lambda_1)\exp(-\lambda_2 t)\}$$

$$[^1D^*] = \frac{k_{DM}[^1M][^1M^*]_0}{(\lambda_2 - \lambda_1)}\{\exp(-\lambda_1 t) + \exp(-\lambda_2 t)\} \tag{24}$$

Here, X, λ_1, and λ_2 are functions of the rate constants (k_M, k_D, k_{MD}, and k_{DM}) in Scheme I; $X = k_M + k_{DM}[M]$,

$$\lambda_1 = \tfrac{1}{2}[X + k_D + k_M$$
$$-\{(k_D + k_M - X)^2 + 4k_{MD}k_{DM}[M]\}^{1/2}]$$

$$\lambda_2 = \tfrac{1}{2}[X + k_D + k_M$$
$$+\{(k_D + k_M - X)^2 + 4k_{MD}k_{DM}[M]\}^{1/2}]$$

From eqs 23 and 24, the monomer and excimer decays are the sum of two exponential decay terms, and the same decay constants, λ_1 and λ_2, appear in both the monomer and excimer decays. In addition, the two preexponential factors in the excimer decay are of equal magnitude but opposite sign.

In few cases are these simple kinetics in polymeric media obeyed, and debate continues as to the principal cause. In vinyl aromatic polymers, microcompositional and spatial heterogeneity, rotational and segmental motion, and electronic energy transfer and migration may all play a role in the observed nonexponential decay characteristics.

Classification of Polymer Systems

UV and visible light-absorbing chromophores may be present in synthetic polymers due to adventitious impurities such as oxidation products or initiator or termination residues (Type A), or they may be part of the repeat unit (and hence be in high concentration) (Type B). Many simple synthetic polymers, such as polyethylene and polypropylene in a pure state, will

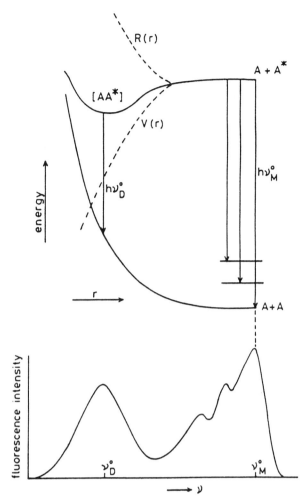

Figure 2. Excimer formation in aromatic molecule A. Excimer emission from AA* is red-shifted by virtue of stabilization of excited state and repulsion in ground state and is structureless because of dissociation in ground state.

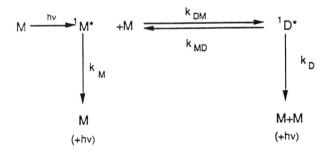

Scheme I. Birks kinetic scheme for excimer formation.

exhibit only σ–σ* absorptions in the high-energy UV region, in common with most organic molecules. Such excitations in general lead to photochemical reactions rather than luminescence, and the corresponding excited states will thus be very short-lived. Here, we focus arbitrarily on species that absorb in the spectral region from 250 nm to longer wavelengths, where the excited species may additionally take part in photophysical processes, including luminescence.

Many functional polymers are designed to perform a specific photophysical function, often mimicking a natural system. Examples of functional polymers in which photochemical excitation plays a key role include photo-cross-linking (Part Two) and NLO materials (subsequent chapters in this part) and photoconducting polymers, light-harvesting systems, and luminescent chemical sensing polymers (Part Four).

In the case of the NLO materials and in light-harvesting polymers, luminescence measurements can be particularly revealing in the study of the effect of polymer structure on the properties of the materials. Such measurements would include steady-state absorption and fluorescence spectra, fluorescence quantum yields, and time-resolved fluorescence measurements. This type of measurement would not be expected to be revealing in terms of cross-linking functional polymers.

In the case of NLO materials, the optical absorber is in general in the backbone of the polymer in the repeat unit, and thus the chromophore concentration is very high. The unit is invariably a conjugated organic system, giving rise to properties that are the consequence of extended conjugation. By contrast, in many types of light-harvesting polymers, the chromophore of interest is a substituent pendant to the polymer backbone and is thus capable of motion independent of that of the polymer chain. The range of conformations that such chromophores may adopt greatly facilitates excimer formation, and thus the photophysics of such polymers frequently is dominated by the presence of such excimer trap sites. In general, despite the very high concentration of chromophores, the preformed excimer trap site concentration is low, and population of such sites involves some local diffusion of chromophores relative to one another, either via segmental motion, or more usually, rotational motion of the chromophore. When this diffusion is restricted, as in solid polymers, the population of trap sites is dominated by electronic energy transfer and migration, and any understanding of the photophysics of such polymers must address this subject quantitatively.

The field of polymer photophysics is much too extensive to include a complete review of the photophysical processes alluded to previously. We thus confine the discussion to a detailed illustration of the criteria and methods involved in the in-depth investigation of *particular* types of *functional* polymers, namely the poly(diacetylene) (NLO and conducting polymers), and poly(vinyl) aromatic polymers (potential light-harvesting polymers) for which extensive photophysical studies have been carried out. It is hoped that the discussion may then be useful as a background model for other functional polymers in general. The discussion is

intended to impress upon the reader the wealth of detailed information about polymer conformation morphology, structure, and chain dynamics that can be obtained from photophysical measurements, both steady-state and time-resolved. No attempt is made here to discuss photochemistry because this topic is covered in Chapter 2.1.

Photophysics of Poly(diacetylenes)

Poly(diacetylenes) (PDAs; Chart I) are a class of functional polymers, the photophysics of which have been and continue to be widely studied. PDAs exhibit NLO properties and other unusual one-dimensional mechanical and electronic behavior. The polymers can be obtained as single crystals by solid-state polymerization of disubstituted diacetylene monomers (3). Recently, considerable interest and debate has been stimulated by the discovery of soluble PDAs and the striking chromism, or color changes, they display (4–6). This chromism can be produced in several ways, by adding a poor or nonsolvent (4), by altering the concentration (4), or by temperature changes (5) of the solution. Thermochromism also occurs in the solid state (7). Water-soluble PDAs (8) and other polymers such as poly(thienylenes) (9) also display chromism as a result of changes in pH and complex formation with transition metal halides.

Chromism of Poly(diacetylenes)

In these polymers chromism has been attributed to a conformational change of the skeletal backbone in

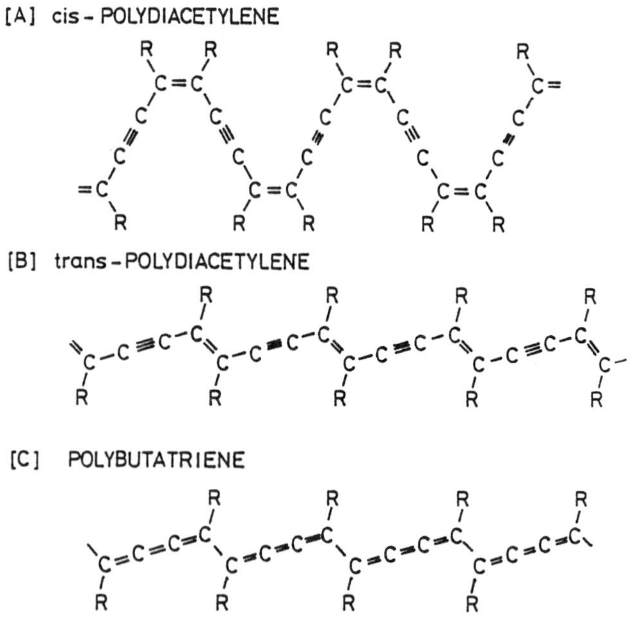

Chart I. Possible structures of polymers of diacetylenes: A, *cis*-PDA; B, *trans*-PDA; and C, poly-butatriene.

terms of the extent of delocalization of the π-bond electrons (*10–12*). Polymers can be produced that are yellow or red and sometimes blue. Yellow solutions and melts of PDAs represent statistical distributions of short conjugated segments with mean lengths of about 4–7 repeat units (*13*). Red solids and solutions have extended conjugation. For PDAs with urethane-containing side-groups, disruption of hydrogen bonds between the side groups allows the formation of short conjugated chain segments. The disruption of conjugation that leads to the short segments has been attributed to *orbital flip* defects that correspond to a 90° rotation about the single bond in the acetylenic structure (*14, 15*). Reformation of hydrogen bonds has been identified as the essential driving force behind the transformation from a random coil to an extended rigid-rod conformation in yellow and blue or red solutions of these soluble diacetylenes, respectively (*16*).

Other investigations show that PDAs without hydrogen bonds between the side groups also exhibit a chromism (*7, 17, 18*) similar to those observed for PDAs with hydrogen bonding side groups. However, the range of stabilities of the yellow solution in chloroform–hexane systems differs from one PDA to another, and chromism is thus much more likely to be a general phenomenon that does not necessarily require the formation of hydrogen bonds between side groups as the driving force.

There are differences of opinion concerning the microscopic interpretation of the color changes in PDAs. Patel and co-workers (*8, 10–12*) and Tieke (*19*) explain these in term of a nonplanar to planar transformation, which alters the backbone conjugation length involving either localized bond alternation effects or a change in backbone bonding structure and a single-chain phenomenon (Scheme II). Heeger and co-workers (*5, 20*) claimed that chromism is a single-chain effect involving cis–trans isomerization of the C=C bonds corresponding to a conformational change of random coil to rigid-rod backbone (Scheme II). They

pointed out that orbital flip defects would not affect the overall length of PDA chains, nor could the length be altered due to rotation about single bonds in polybutatriene (Figure 3), as initially proposed by the same group (*21*). Therefore, both models are incompatible with the changes in hydrodynamic radius associated with the chromic transition in PDA solutions (*5*). Wegner et al. (*6, 22*) argued that the chromic effect is a process intimately linked to the aggregation of the wormlike (Kratky–Porod) chains to form extended-chain microcrystals (*23*), rather than a single-chain phenomenon involving defects (Kuhn chain (*16, 24*)) (Figure 3).

Data on PDAs having methylene groups adjacent to the polymer backbone but with differing morphologies suggest that there are specific configurations of the backbone in different environmental conditions (*25*). For yellow PDA solutions, the absorption maxima tend to occur in the wavelength range of 460 ± 10 nm (Y phase). For red–blue solution-cast films, red–blue Langmuir–Blodgett (LB) films, and red–blue solutions and gels, the peak absorption tends to be in the wavelength range of 535 ± 10 nm for the red (R) phase and 610 ± 15 nm for the blue (B) phase, dependent on the number of methylenes adjacent to the backbone. Red LB films and red solutions have very similar fluorescence spectra with roughly equal fluorescence quantum yields (*18, 26, 27*). The R phase is usually attained for PDAs with more than three methylenes adjacent to the backbone. These three phases (Y, R, and B) suggest different (but quite specific) degrees of perturbation to the PDA backbone. Recent studies have shown that the R phase can exist as a metastable intermediate during the transition from the Y to B. The intermediate R phase occurred either as a short-lived transient (*13*) or as a long-lived intermediate (*8, 28*). Thus, dynamic behavior of the pendant groups, as well as the packing of their various conformations, has an impact on the PDA backbone and on its optical absorption and fluorescence.

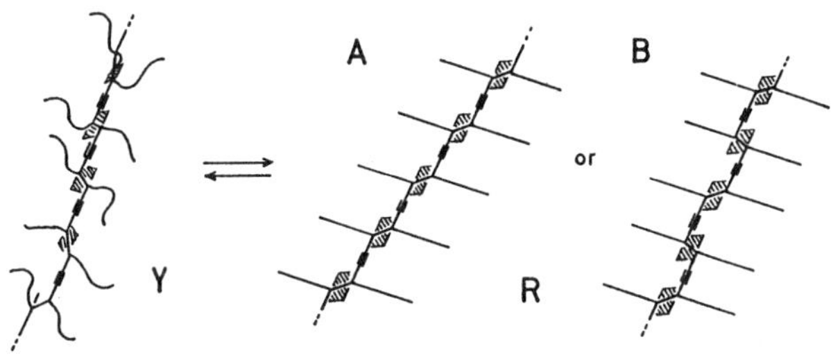

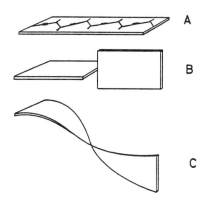

Scheme II. Proposed structure of yellow (Y) and red (R) phases for PDAs in solution (*42*): A, *trans*-transoid; B, *cis*-transoid. Hatched areas indicate planes in which the C=C bonds are extended.

Figure 3. Possible conformations of PDA chains: A, crystal (planar); B, shorter sequence lengths (dislocation); and C, wormlike chains (continuous deformations).

Despite extensive studies, the details of the structure of PDA chains in the Y, R, and B phases remain a subject of debate. Fluorescence emission was observed from monomers (29), and the efficiency of the nonradiative decay channel for perfect PDA chains is shown by the quenching of fluorescence when the monomer is converted into polymer (29). PDAs in which the extended chains are deformed in some way (e.g., in partially polymerized crystals with high defect concentrations; mechanically deformed polymer; γ-ray polymerized crystals, solutions, and glasses) *do* fluoresce (8, 18, 25, 26, 30–38). The excitonic state in PDA chains and their fluorescence were debated in several articles (18, 34–36, 39–41). The quantum yields of fluorescence (ϕ_F) for PDA solutions were ca. 0.2–0.3% (18, 27). Similar values were reported for the fluorescence from disordered films of diynoic acid polymers (26). These results indicate that *disorder* introduces a radiative decay channel that is in strong competition with the nonradiative decay mechanism operative in extended PDA chains.

Soluble Poly(diacetylenes)

Many soluble PDAs were investigated, and a class of particular interest are those with the *n*-butoxy carbonyl methyl (*n*BCMU) side groups, R (17, 27, 42) (Chart II). The *n* denotes the length of the methylene chain (spacer) between the backbone and the urethane group.

$$R = -(CH_2)_n-O-\overset{\overset{\displaystyle O}{\|}}{C}-\underset{\underset{\displaystyle H}{|}}{N}-CH_2-\overset{\overset{\displaystyle O}{\|}}{C}-O-(CH_2)_3-CH_3$$

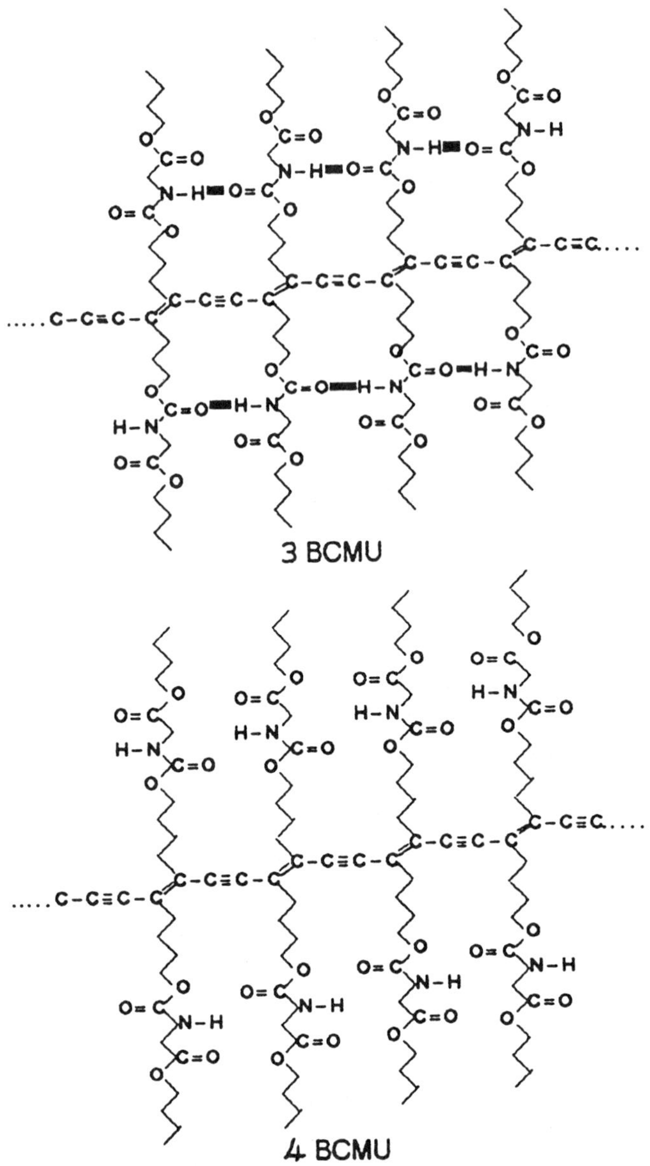

The extent of hydrogen bonding between the side is dependent on the length of the methylene chain. Thus, for $n = 3$ (3BCMU), the planar backbone is stabilized by hydrogen bonding; whereas for 4BCMU, a twisting of the backbone would be required to achieve extensive hydrogen bonding (27). As an example of the thermochromism of the PDAs, the fluorescence spectra of 4BCMU in solution at low temperature are shown in Figure 4 (42).

A detailed study of these thermo- and solvato-chromic PDAs in low-temperature glasses revealed three emitting species, A, B, and C, that have the following properties. Species A is present in both Y-phase and R-phase glasses. Its concentration is approximately independent of solution concentration. This species was identified as random PDA chains, most probably Kratky–Porod wormlike chains, frozen into the glass without significant change in molecular conformation. Results also suggested that the density of isolated chains that retain a random configuration in the glass

Chart II. Hydrogen bonding in BCMUs.

does not change significantly over the range of solution concentrations studied (10^{-3} to 10^{-6} mol^{-1}).

Species B is produced by quenching of yellow PDA solutions and is attributed to some ordering of interacting chains at concentrations $>2 \times 10^{-5}$ mol L^{-1}, because aggregation of polymer chains can occur under these conditions (5, 6, 32, 42). The degree of order is less than that achieved in slowly cooled solutions.

Species C is formed by annealing of the Y phase to produce a Y-phase glass, which appears to be identical to the ordered polymer species found in slowly cooled solutions (38). In the glasses, species A is not necessarily in intimate contact with species B or C. Long range energy transfer from species A to the other species occurs on a very short time scale, which could account for the apparent discrepancy between steady-state and time-resolved measurements.

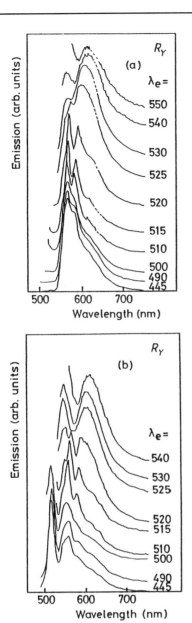

Figure 4. Fluorescence emission spectra of (4BCMU) in 2-methyl tetrahydrofuran at 77 K (first freezing of yellow solution). Fluorescence excited at different wavelengths (bandwidth, 5 nm). The spectra are arbitrarily offset for clarity: concentrations, (a) 0.064 mg/mL and (b) 0.0061 mg/mL.

Even though fluorescence spectra of glasses are consistent with the presence of a random wormlike chain in yellow solutions and more extended chains in red solutions, they do not provide direct information on the details of chain conformation. These general conclusions indicate that the inherent stiffness of the conjugated polymer chain is the main driving force behind the ordering that occurs in solution and glasses. Steric interactions and dynamics of the side groups are important perturbations even when there is no hydrogen bonding between them. Within the

constrains of these general criteria, there are two possible scenarios for the distribution of disordered regions within individual chains.

Polymer chain ends may behave differently from the interior due to their reduced steric constraints. Such an effect has been observed in *n*-alkanes and lipid bilayers (*43, 44*), where the gauche defects are unevenly distributed along the chain and the concentration is largest at the chain ends. Therefore, some of the PDA chain species observed in glasses may be characteristic of disordered chain-end regions, whereas others occur in the more ordered central regions. Exactly the opposite behavior can result from the growth of linear segments on a wormlike chain. At the chain ends, region of greater strain will grow out, but at the center of the chain these will accumulate. Whether one of these mechanisms is dominant is still an open question.

A more recent proposal (*45*) suggests that the transition of Y form to R form in these polymers may be due to chain folding and the formation of an intramolecular fringed micelle structure. This suggestion is an elegant one, which combines the features attributed to single-chain phenomenon with those of intramolecular aggregates. These intramolecular aggregates would have very similar spectral signatures to intramolecular aggregates seen at higher concentrations.

Effects of Surface on PDA Conformations

PDAs are functional polymers with an importance in technology for the production of planar optical waveguides by spin or dip coating. A luminescence spectroscopic study was performed to determine the effect of a fused-silica surface on the conformation of poly-(4BCMU) (*46, 47*). This polymer is unusual in that both Y and R forms fluoresce, facilitating the investigation by time-resolved evanescent wave-induced fluorescence (TREWIF) techniques. In this technique, which is a total internal reflection method, fluorescent solutes in solutions in contact with a surface can be probed by laser light excitation. Figure 5 shows that the EWIF spectra of 4BCMU surfaces adsorbed are identical to those of the R and Y forms measured in bulk solution, and so only two species need to be invoked to account for observed changes on presentation of a solution of either the Y form or R form to a fresh silica surface. The results are shown for the Y case in Figure 6 and for the R case in Figure 7.

The kinetics of the observed spectral changes can be analyzed. Thus, by using the 2.5 and 195 min spectra as representing early R form and late Y form, respectively, intermediate spectra were analyzed as a linear combination of these two extremes. From this preliminary kinetic study the early R form spectrum decayed with a rate constant of 0.03 min⁻¹.

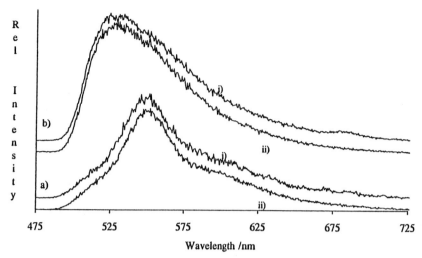

Figure 5. Comparison of a, R-form and b, Y-form fluorescence spectra measured (1) in the bulk and (2) on the silicon surface by evanescent wave-induced fluorescence (EWIF).

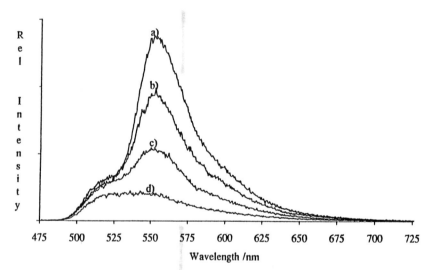

Figure 6. EWIF spectra recorded after a, 2.5 min; b, 20 min; c, 50 min; and d, 195 min following the introduction of a Y-form solution of poly(4BCMU) solution in 2-MeTHF to a fused silica surface.

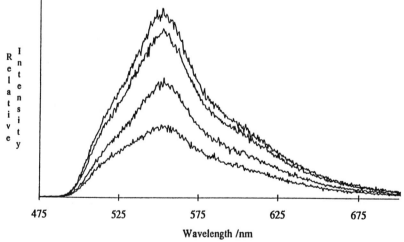

Figure 7. EWIF spectra recorded after a, 0.4 min; b, 2.7 min; c, 7.5 min; and d, 12.5 min following the introduction of an R-form solution of poly(4BCMU) solution in chloroform:hexane (7:18) to a fused silica surface.

The time resolution of this experiment is limited to ca. 2.5 min, which is the time taken to record a spectrum. At 518 nm, the emission observed is predominantly from the Y form, and a small contribution is from the R form. At 558 nm, the emission is predominantly from the R form, but the contribution from the Y form cannot be ignored. The results of monitoring these two wavelength emissions from 0–10 min are given in Figure 8 for a Y solution presented to a fresh silica surface. The inset shows the long-term behavior of the intensity at 558 nm.

EWIF spectra recorded at intervals following the introduction of an R-form solution to a cleaned and dried prism, after 0.4, 2.7, 7.5 and 12.5 min, are shown in Figure 7. Unlike the results for the Y-form solution, the spectral profiles are all similar and reach a maxi-mum intensity after ca. 15 min. Figure 9 shows a plot of intensity at 558 nm over the same time interval.

The most significant results of this study are that the emission properties for the polymer from within the evanescent region are greatly enhanced, and the time-dependent shift in the Y-R equilibrium position induced by the proximity of the glass surface. The enhanced emission within the evanescent region is best explained by a reduction in the rate of nonradiative decay of the chromophore on adsorption, a phenomenon that also was observed in the dye malachite green.

This hypothesis is interesting, because it is known from studies of crystals of poly(4BCMU) and related polymers that the fluorescence quantum yield *decreases* in this highly ordered state to an insignificant level. If the restricted motion of the adsorbed polymer,

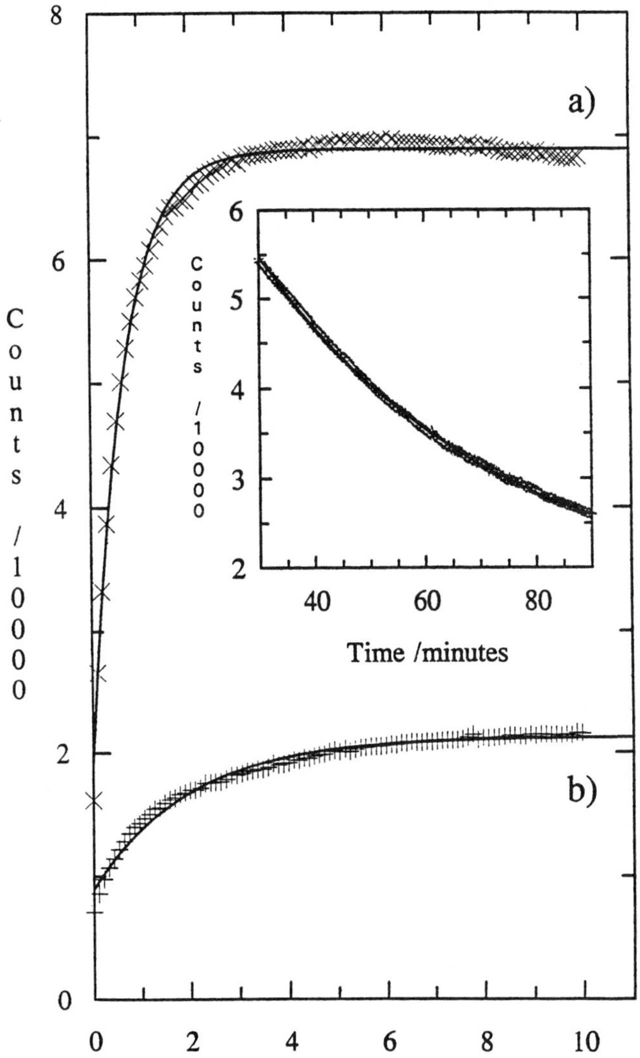

Figure 8. Time dependence of EWIF intensity at 558 nm (a) and 518 nm (b) in the time range 0–10 min following the addition of a Y-form solution to the glass surface. The inset represents the EWIF intensity at 558 nm at long times. Solid line indicates fitted exponential function.

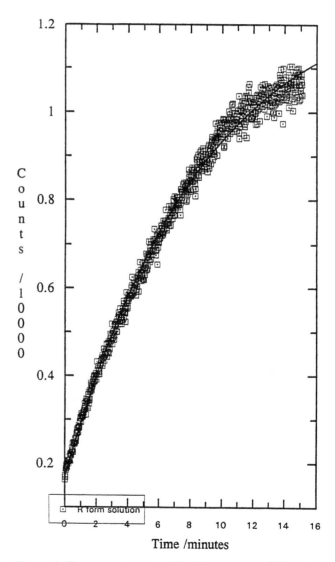

Figure 9. Time dependence of EWIF intensity at 558 nm in the time range of 0–10 min for the R-form solution. Solid line indicates fitted exponential function.

either in the Y form or in the R form, leads to an increase in the fluorescence quantum yield, then single crystals of poly(4BCMU) would be expected to be highly fluorescent, but the converse evidently is true. In the crystal, however, there are two additional factors that must be considered. First, the lower energy absorption spectrum in the crystal suggests that the average conjugation length is at least an order of magnitude larger than in either the Y or R forms of the polymer in solution or on the surface. Second, the efficiency of energy migration in a three-dimensional crystal will be far greater. In the case of the crystal, the low fluorescence quantum yield may be due to efficient energy transfer of the excitation energy to a nonfluorescent trap. In the case of the polymer adsorbed onto the surface of the glass, such efficient traps may not exist. Therefore, the reduction in the internal conversion process resulting from the restricted molecular motion may lead to the observed increase in the fluorescence quantum yield. Even if energy transfer does occur, it would simply lead to the population of fluorescent traps.

The mechanisms responsible for the very efficient deactivation of the first excited state of poly(diacetylenes), or indeed conjugated molecules in general, remain controversial. Even though time-resolved and steady-state fluorescence spectroscopy can detect the existence of nonradiative deactivation pathways, determining the mechanism is far more difficult. Consequently, the fate of the excitation energy can only be surmised.

A number of mechanisms have been proposed, which include internal conversion to the ground state followed by vibrational relaxation, intersystem crossing to the triplet state, and internal conversion to another excited state (which itself is strongly coupled to the ground state). Almost certainly a combination of all three mechanisms is extant, and the adsorption process affects only the rate of internal conversion.

The kinetics of the spectral changes were explained on the basis of the following model. In the Y-form solution at early times, the polymer chains close to the surface experience a nonfavorable environment of the nonsolvent conditioned glass surface. The dynamic equilibrium between the Y and R forms is shifted in favor of the R form, and the new equilibrium mixture then begins to adhere to the glass surface. Owing to the enhancement of the fluorescence quantum yield on the surface, the EWIF spectrum is now dominated by the adsorbed R-form favored equilibrium mixture. This process continues for up to about 1 min, after which time the good solvent (2-methyltetrahydrofuran) begins to resolvate the adsorbed polymer (and to solvate the glass surface). This process allows the equilibrium between the Y and R forms to return to normal, and the new Y-form-

dominated equilibrium mixture continues to adhere dynamically to the glass surface. Therefore, the EWIF emission is now seen to be dominated by the adsorbed Y form. At long times the Y-form dominated equilibrium mixture is now in competition with the solvent for adsorption sites on the glass surface.

Additional supporting evidence for this hypothesis is provided by the fact that if the Y-form solution is removed and the surface washed with fresh solvent, the EWIF spectrum slowly disappears. When the Y-form solution is reintroduced, no R form is observed, because the solution now encounters a glass surface that has been solvated. The initial shift of equilibrium from the Y to the R form is thus a transient response of the interfacial Y-form solution, resulting from the solution encountering a nonsolvated glass surface. The air-dried glass surface will still be well hydrated by atmospheric water, producing a poor hydrophilic environment for both the polymer and the solvent, which may explain the reason for the initial transient response.

In the case of an R-form solution, little disruption of the Y–R equilibrium mixture occurs because it already favors the R form. Therefore, all that is observed is the rate of adsorption of the R form onto the glass surface. The intensity of the EWIF spectrum grows to a constant value, at which point the rate of desorption and adsorption of the R form onto the surface is in competition with the solvent until equilibrium is reached. Because the solvent is a poor solvent, the removal rate of the R form from the surface will be slow, and thus the amount of material on the surface will be high relative to the Y-form solution case. This result could explain why the $I_{EWIF}:I_{bulk}$ ratio for the R-form solution is greater than that for the Y-form solution. The rate at which the equilibrium between the R form and Y form is disturbed supports the intramolecular fringed micelle structure for the R form.

Time-Dependent Measurements

Much of the experimental work alluded to was carried out using steady-state irradiations. The added dimension of time resolution is a potent investigative tool. On the picosecond timescale, time-correlated single-photon counting still provides the most convenient technique for such measurements (48). Figure 10 shows a typical experimental setup (1). Figure 11 presents the time scales of various intramolecular fates (above the horizontal solid line) and processes sensitive to surrounding species (below solid line), which dictate the observed lifetime or decay characteristics of any fluorophore.

The observed kinetics of decay in polymeric media are rarely simple, because chain-length distribution, tacticity, conformation, energy transfer and

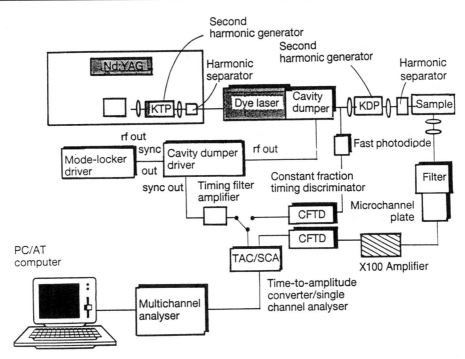

Figure 10. Schematic of time-correlated single-photon system for measurement of fluorescence decay times based on continuous wave mode-locked yttrium–aluminum–garnet–Nd laser.

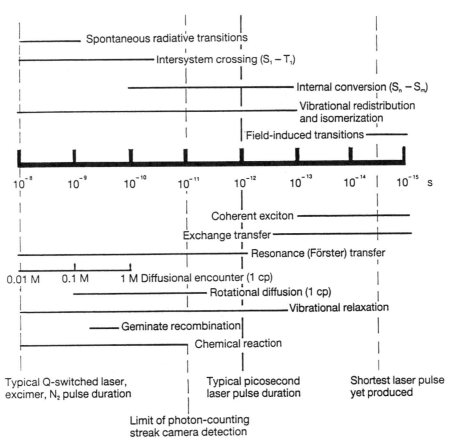

Figure 11. Representative rates of photophysical processes in complex polyatomic molecules: intramolecular processes (above horizontal line) and processes sensitive to surrounding species (below horizontal line).

migration, and rotational and segmental motion (in fluid media) can all be involved. Energy transfer and migration have been much studied by time-resolved methods. Many models have been developed involving migration to trap sites, which may be excimers or low energy conformational sites in nonexcimer forming polymers, such as the PDAs. Light-harvesting polymers rely wholly on such facile energy migration, and thus the basic physics of the process are worth discussing, even though the concept results in very complex mathematical analyses.

Random Walk Diffusion Models

Consider the diffusion of excitation through an array of repeating units. In the case of weak coupling, the excitation energy moves along the polymer chain from one chromophore to another, probably by the long-range dipole-induced dipole (Forster) mechanism. However, other energy transfer mechanisms such as the shorter range electron exchange (Dexter) mechanism may also play a part, especially in solid polymers where the chromophores are very close together. A number of different models were used to consider the time evolution of an excited state that may migrate to a trap site, and often several mathematical approaches have been employed to approximate the observable parameters for each model. A review of these complex mathematical models is beyond the scope of this chapter, but a brief summary of simple models is in order.

Random Walk Migration, Evenly Spaced Chromophores

A random walk model has been investigated for the case of evenly spaced chromophores by a number of groups, and several approximate solutions have been offered. Huber (49) solved the rate equations for the donor (monomer, M*) decay using the t-matrix approximation, resulting in eq 25 for the one-dimensional case. In the asymptotic limit (i.e., for large time), the decay can be approximated by eq 26. However, when the number of trap sites is sufficiently small, the decay reduces to a dependence of $\exp-(at + bt^{1/2})$.

$$[M^*] = A\exp(4\pi^2 q^2 Wt)\mathrm{erfc}(2\pi qW^{1/2}t^{1/2}) \quad (25)$$

$$[M^*] = \frac{A}{(4\pi q^2 Wt)^{1/2}} \quad (26)$$

where A is the preexponential term, corresponding to [M*] at time zero; q is the fraction of traps; and W is the rate of energy transfer between nearest neighbors.

In addition to the t-matrix approximation, however, a number of other methods have been used to solve the deep trapping problem. In fact, an exact

solution was offered for the one-dimensional case (50). Movaghar et al. compared the coherent potential approximation (CPA) (51) and the first-passage time approach (FPT) (52) to the exact solution; whereas Klafter and Silbey (58) found that the t-matrix approximation used by Huber is less accurate than both the CPA and FPT under all conditions. Movaghar et al. found that the FPT approach is superior to the CPA at all trap concentrations except for very high trap concentrations approaching 1 (i.e., all chromophores are traps). At long times, the FPT approach gives a solution that asymptotes to $\exp-(at^{1/2})$ similar to the result of the low trap concentration t-matrix approach, whereas the exact result asymptotes to $\exp-(at^{1/3})$.

Random Walk, Random Distribution

This more complex, but more realistic model for approximate energy migration kinetics does not require the condition of an even distribution of chromophores in the system. Such a model can involve, say, a random distribution of chromophores in three dimensions interspersed by a random distribution of traps (54, 55). In addition, a model was derived for polymers (56), which relates fluorescence decay parameters to the radius of gyration of the polymer chains.

Multiple Trap Energies

A further complication involves disorders relating to the energies as well as the position of the monomer excited states. In a polymer, the chromophores are in a range of environments, each of which may have different energies. This problem was treated theoretically by Grunewald et al. (57) and by Reichart et al. (58) in a Monte-Carlo simulation, both giving an approximate relation of the form of eq 27, and by Stein et al. (59).

$$k(t)_{DM} = b + ct^{\alpha-1} \quad (27)$$

where b and c are arbitrary constants.

The influence of dimensionality (including fractal) on the exciton dynamics and decay kinetics in a quasi-ordered PDA system, poly(diacetylene-1-hydroxy-hexadiynediol) (-1OH, Chart III) also was studied (39). PDA-1OH is an oriented fibrous polymer produced by solid-state polymerization. An exact analytical solution was obtained for the one-dimensional diffusion with randomly distributed deep traps, and this solution was assessed critically against the t-matrix approach, the coherent-potential approximation, and the first-passage time approach. For PDA-1OH, analyses showed that the MPW theory was the simplest model consistent with the experimental decay kinetics, the first ever demonstration of one-dimensionality of luminescence in these semiconducting polymers.

PDA-1OH

$$[=R'C-C{\equiv}C-CR=]_n$$

$R'=CH_2OH$ $R=CH_3$

Poly(1-vinyl naphthalene)

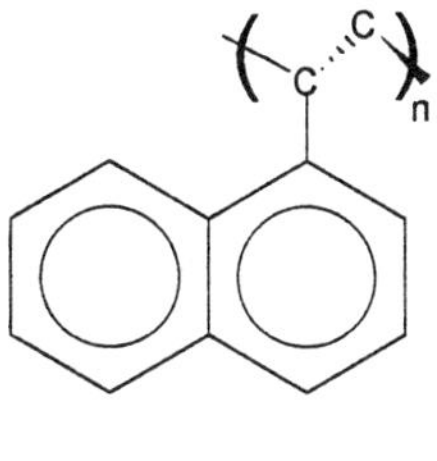

Poly(2-vinyl naphthalene)

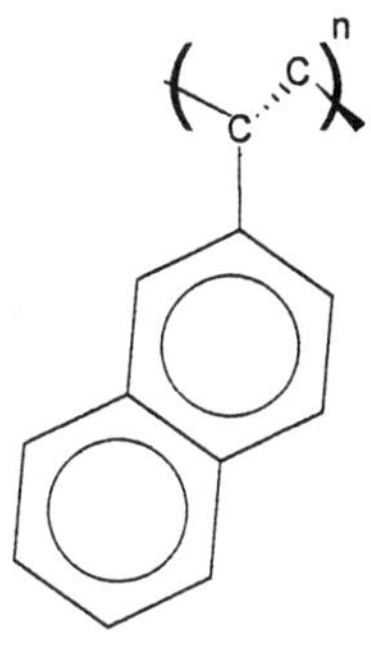

Chart III. Structures of PDA-(1OH), poly(1-vinyl naphthalene), and poly(2-vinyl naphthalene).

Excimer-Forming Polymers

Application of the previously mentioned excitation diffusional models to excimer forming polymers, where the excimer acts as the trap and can function as light-harvesting functional polymers, is bedeviled by several problems, some of which are outlined.

Reversible Excimer Formation

In the Birks kinetic scheme, back transfer is modelled simply by the rate constant k_{MD}. Weixelbaumer et al. (60) incorporated this process via an approximate method. Sienicki and Winnik (61) derived an exact method and posed the question, what happens if monomers formed by back dissociation behaved *differently* from those excited directly? The question was answered by Berberan-Santos and Martinho (62), who showed that $k_{DM}(t)$ does not necessarily decrease monotonically but can sometimes increase with time to give very complex kinetics.

Diffusion of Energy and Chromophore

Baumann and Fayer (63) considered a two-body problem in which diffusion and energy transfer occurred simultaneously. Frederickson and Frank (64) developed a simpler one-dimensional array model (FF model), which provides eq 28 for the rate of monomer

fluorescence. In this equation, $i_M(t)$ is the intensity of monomer fluorescence from the monomer (related to monomer concentration by the quantum yield of fluorescence, q_{FM}, and the rate of fluorescence decay in isolation (k_M)), q is the preformed trap fraction, and W is the rate of energy transfer between nearest neighbors on the polymer:

$$i_M(t) = q_{FM}k_M(1-q)^2 \exp\left[(4\pi^2 q^2 W - k_M - k_{rot})t\right]$$
$$\times \operatorname{erfc}(2\pi q W^{1/2} t^{1/2}) \tag{28}$$

Tao and Frank (65) found that poly(2-vinyl naphthalene) fluorescence decays fit the FF model at relatively low temperatures. However, they noted that at higher temperatures, the model breaks down, probably because of the breakdown of one of the following assumptions, which form the basis of the model.

1. The polymer is considered as a one-dimensional string of equally spaced chromophores.
2. The primary excimer forming step is energy migration, not internal rotation, and requires a number of preformed trap sites in the ground state. This characteristic means that there must be a number of sites where chromophores are in high energy configurations very close to the excimer configuration, or else there must be a low energy conformation very close to the excimer conformation.
3. The number of these preformed trap sites is low, because for high concentrations of trap sites, the t-matrix approximation becomes poor.
4. The excimer formation step is irreversible.

The FF model has been extended to high trap concentrations by using the FPT approximation. In this case, the monomer fluorescence intensity is given by eq 29, and the excimer fluorescence is determined by eq 30, where the parameters are as defined for eq 28.

$$i_M(t) = q_{FM}k_m(1-q)^2 \exp(-(t/\tau) - q$$
$$\int_0^t 2W[I_0(2Wt) + I_1(2Wt)\exp(-2Wt))dt \tag{29}$$

$$i_E(t) = q_{FE}k_E \exp(-k_E t)$$
$$-q_{FE}k_E(1-q)^2 \exp\left(\frac{-t}{\tau} - q\int_0^t 2W\exp(-2Wt)[I_0(2Wt)+I_1(2Wt)]dt - \right)du$$
$$-q_{FE}k_E(k_M - k_E)(1-q)^2 \times$$
$$\int_0^t \exp\left(-u(k_E - \tau^{-1}) - \frac{t}{\tau} - q\int_0^{t-u}\{2W\exp(-2Wt)[I_0(2Wt)+I_1(2Wt)]\}dt - \right)du \tag{30}$$

where I_0 and I_1 are modified Bessel functions.

Some of these models were tested using data from careful time-resolved fluorescence measurements on poly(1-vinyl naphthalene) and copoly(1-vinyl naphthalene-methyl methacrylate) (Chart III) in the following way. The FF function appears to have five variables: the amplitude (t), the isolated decay rate (k_M), the rate of rotation (k_{rot}), the rate of intramolecular energy transfer (W), and a number of trap sites (q). However, some of the variables cannot be treated independently, and the FF function may actually be rewritten using only three variables. This rewriting is done by substituting $\tau = 1/(k_M + k_{rot})$ and $Q = qW^{1/2}$ into eq 28 to give eq 31 and fitting the data by varying only the amplitude τ and W.

$$i_M(t) = A \exp[(4\pi^2 Q^2 - \tau)t \, \text{erfc}(2\pi Q t^{1/2})] \quad (31)$$

where τ is experimental decay time.

The efficacy of the FF model was investigated over a range of naphthalene mole fractions in the copolymer. At 290 K, fluorescence data fit the model for the copolymer containing 25% 1-vinyl naphthalene, but not for the 50% copolymer or the homopolymer. Obviously the model fits only for low naphthalene concentrations and low temperatures. The breakdown of the FF model at high temperatures and high naphthalene concentrations could be explained by the breakdown of any one of the assumptions outlined previously.

Tao and Frank (65) also found that the FF model does not adequately fit 2-vinyl naphthalene homopolymer fluorescence decay profiles at high temperatures. The FPT model should be appropriate for high trap concentrations, but Figure 12 shows the FPT model produces very similar results to the FF model and did not fit any data that did not fit the FF model (66).

In the actual interpretation of fluorescence data, models as complex as the FF model are seldom employed. Commercially available programs for fitting time-resolved fluorescence data generally cover exponential decay, the exponential of a $t\alpha$ function (or sums of these functions), but rarely anything more

complex. Consequently, knowing when data that obeys a more complex theory can also be processed adequately by simpler approximations for which fitting routines are available would be useful, so that information about a complex model can be inferred from the fit of the experimental data to a simple function. In other words, it would be useful to know when the FF model can be successfully approximated by a simpler function. If q^2W stays within certain limits, then k_{DM} in the FF model can be approximated adequately by a constant term plus a term depending on $t^{1/2}$. On integration of the rate equations, the fluorescence decay will then follow eq 32, which is commonly available in fluorescence decay fitting software.

$$i_M(t) = q_{FM} k_M (1-q)^2 \exp\left(-(k_M + k_{rot})t - \frac{4q\sqrt{Wt}}{\sqrt{\pi}}\right) \quad (32)$$

Table I shows reduced χ^2-test, $k_M + k_{rot}$, and q^2W values obtained from fits of some experimental data to the FF model and to eq 32. The last two columns of the table consist of values of $4(1 - 2/\pi)q^2W$ and $k_M + k_{rot}$. The q^2 values are equally good for both functions, but the $k_M + k_{rot}$ and q^2W parameters show some deviations that may not be explained by experimental error. The FF model consistently finds a slightly less exponential decay, indicating small inaccuracies in the approximation.

Tao and Frank (65) presented data consistent with the FF model without any reference to fitting the $\exp-(at + bt^{1/2})$ approximation. This fitting has been tested by simulation (66). Tao and Frank's data were simulated with the same amplitude as shown in their paper, from their published parameters, and Gaussian noise was added. When simulated curves were analyzed with the FF fitting program, they gave χ^2 values of 1.00 ± 0.05 (66). They were also subsequently fitted

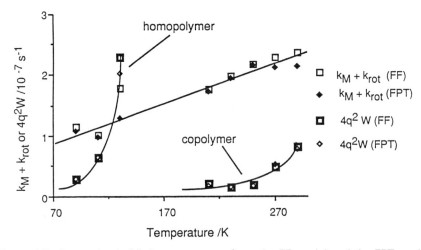

Figure 12. Comparison of fitting parameters from the FF model and the FPT model.

Table I. Quality of Fit and Some Fitting Parameters of FF Model for Copoly(1-vinyl naphthalene-methyl methacrylate) (2:75)

T (K)	χ^2(FF)	χ^2($t^{1/2}$)	q^2W (FF) ($10^7 s^{-1}$)	q^2W ($t^{1/2}$) ($10^7 s^{-1}$)	$k_M + k_{rot}$ ($t^{1/2}$) ($10^7 s^{-1}$)	$4(1-2/\pi) \times q^2W$ ($10^7 s^{-1}$)	$k_M + k_{rot}$ (FF) ($10^7 s^{-1}$)
290	1.15	1.11	0.20	0.19	2.15	0.30	2.36
270	1.11	1.05	0.12	0.12	2.16	0.18	2.30
250	1.09	1.09	0.049	0.047	2.14	0.07	2.18
230	0.99	1.06	0.041	0.038	1.95	0.06	1.99
210	1.26	1.30	0.053	0.046	1.71	0.08	1.76

to eq 32. The χ^2 values from the FF fit were then subtracted from the χ^2 values from the $t^{1/2}$ fit to give a measure of the difference in the quality of the fit. These results are presented in Table II, along with parameters extracted from the paper (65).

At low temperatures, eq 32 is well satisfied, but this characteristic is no longer true at high temperature. At 293 K, the exp–($at + bt^{1/2}$) function may possibly fit the data better than the FF model. In fact nothing in these analyses contradicts the premise that Tao and Frank's data can be fitted to eq 32 equally as well as to the FF model. *This fact means that the polymer could actually be undergoing any set of processes that approximately fit an exp–*($at + bt^{1/2}$) *function.* Therefore, very great care has to be taken in the interpretation of such time-resolved experiments.

Fluorescence Anisotropy Measurements

The fluorescence decay times of excited states are such that the fluorescence depolarization technique may be used only to examine relatively high frequency relaxation processes of polymers. Consequently, fluorescence depolarization has been primarily limited to the study of relaxation processes of polymers in solution. The anisotropy of a system, $r(t)$, is derived from measurements of the fluorescence decays with polarization parallel and perpendicular to the polarization of excitation (eq 33).

$$r(t) = [I_{\parallel}(t) - I_{\perp}(t)]/[I_{\parallel}(t) - 2I_{\perp}(t)] = D(t)/S(t)$$
$$(D = \text{difference}; \; S = \text{sum})$$
(33)

Time-resolved fluorescence anisotropy measurements (48) can provide detailed information on the reorientation dynamics of molecules in solution. Until recently, however, this information has been limited to single rotational correlation times, which are only strictly appropriate for the diffusion of spherically symmetric molecules. Improvements in instrumentation and data analysis techniques during the last decade have led to increasingly accurate measurements of fluorescence lifetimes and parallel improvements in determinations of fluorescence anisotropies.

Table II. Fitting Parameters of FF Model for Actual and Constructed Data of Tao and Frank [Poly(1-vinyl naphthalene)]

T (K)	χ^2(FF)	$\dfrac{\chi^2(t^{1/2})}{\chi^2(FF)}$	$4(1-2/\pi) \times q^2W$ ($10^7 s^{-1}$)	$k_M + k_{rot}$ ($10^7 s^{-1}$)
293	1.29	0.17	2.5	3.1
273	1.18	0.11	1.9	3.1
253	1.10	0.05	1.0	2.8
233	1.10	0.03	0.62	2.4
213	1.08	0.02	0.44	1.9
193	1.05	0.01	0.24	1.6
173	1.06	<0.01	0.09	1.5
153	1.05	<0.01	0.08	1.5
133	1.02	<0.01	0.02	1.4
113	1.03	<0.01	0.02	1.4

The advances in time-resolved techniques have fostered a reexamination of theories of the rotational motions of molecules in liquids. Models considered include the anisotropic motion of unsymmetrical fluorophores, the internal motions of probes relative to the overall movement with respect to their surroundings, the restricted motion of molecules within membranes (e.g., wobbling within a cone), and the segmental motion of synthetic macromolecules. Analyses of these models point to experimental situations in which the anisotropy can show both multiexponential and nonexponential decay. In principle current experimental techniques are capable of distinguishing between the different models. However, extracting a single average rotational correlation time demands the same precision of data and analysis as fluorescence decay experiments that exhibit dual exponential decays. Multiple or nonexponential anisotropy experiments are thus near the limits of present capabilities and generally demand favorable combinations of fluorescence and rotational diffusion times.

Conclusions and Prospects

Luminescence techniques provide a rapid, nondestructive, and in favorable cases, a very revealing probe of macromolecular behavior. In the case of polymer-bound chromophores that are essentially non- or weakly interacting, the photophysics are explained by the properties of the isolated molecular chromophores and their interactions with the surroundings, including neighboring chromophores. In such cases, the luminescence, particularly fluorescence, can be used as a probe of a variety of important functions on the polymer. Typical examples include the study of diffusion of oxygen by phosphorescence quenching, the order of a system by static fluorescence anisotropy measurements, motion of the whole molecule (or a subgroup) by time-resolved fluorescence and phosphorescence measurement, electronic energy transfer and migration by time-resolved fluorescence techniques, and in particular complex formation between chromophores studied by time-resolved fluorescence techniques.

Of what value are such studies? Fundamental questions may be addressed even though it cannot be denied that macromolecular systems present a difficult challenge in terms of understanding fundamental molecular processes. The heterogeneous nature of polymers in terms of microcomposition, tacticity, and conformational order creates a multitude of environments for the basic chromophores under study. This complexity should not, however, deter investigation, because the rewards for the investigator can be of immense practical value. Electronic energy transfer is a fundamental process in nature: in the context of functional polymers, it enables easy investigation of such important questions as the interpenetration of different polymer chains in blends, the coiling of macromolecular chains in polyelectrolytes, the mimicking of the energy transport system in green plants (i.e., channel energy to a reaction site), and as an integral part of signaling mechanisms in sensing polymers. If the photophysical questions asked are not too detailed, the answers can be of enormous use in interpretation of the behavior and its function. At the detailed molecular level, however, the complexities still defy unequivocal interpretation. Nevertheless, the application of sophisticated theoretical modelling and experimental techniques to the various processes occurring in complex polymers will one day yield tractable interpretation, as well as a rich vein of information of technological importance.

Acknowledgments

This review has depended on the recent work of Michelle Carey, to whom grateful thanks are expressed.

References

1. Rumbles, G.; Phillips, D. In *Applications of Lasers in Polymer Science and Technology*, Rabek, CRC: London, 1990; pp 1–29.

2. Beavan, S. W.; Hargreaves, J. S.; Phillips, D. J., Eds: *Adv. Photochem.* **1979**, *11*, 207–303.
3. *Polydiacetylenes*: Bloor, D.; Chance, R. R., Nijhof: The Hague, Netherlands, 1985.
4. Chance, R. R.; Patel, G. N.; Witt, J. D. *J. Chem. Phys.* **1979**, *71*, 206.
5. Lim, K. C.; Sinclair, M.; Casalnuovo, S.; Fincher, C. R.; Wudl, F.; Heeger, A. J. *Mol. Cryst. Liq. Cryst.* **1984** *105*, 329.
6. Wenz, G.; Muller, M. A.; Schmidt, M.; Wegner, G. *Makromolecules* **1984**, *17*, 837; Allegra, G.; Bruckner, S.; Schmidt, M.; Wegner, G. *Macromolecules* **1986** *19*, 399.
7. Plachetta, C.; Rau, N. O.; Schulz, R. C. *Mol. Cryst. Liq. Cryst.* **1983** *96*, 141.
8. Bhattacharjee, H. R.; Preziosi, A. F.; Patel, G. N. *J. Chem. Phys.* **1980**, *73*, 1478.
9. Rughooputh, S. D. D. V.; Hotta, S.; Heeger, A. J.; Wudl, F. *J. Polym. Sci. Polym. Phys. Ed.* **1989**.
10. Patel, G. N.; Chance, R. R.; Witt, J. D. *Polym. Sci. Polym. Phys. Ed.* **1978**, *16*, 607.
11. Patel, G. N.; Chance, R. R.; Witt, J. D. *J. Chem. Phys.* **1979**, *70*, 4387.
12. Patel, G. N.; Witt, J. D. *J. Polym. Sci. Polym. Phys. Ed.* **1980**, *18*, 1383.
13. Chance, R. R.; Washabaugh, M. W.; Hupe, D. J. In *Polydiacetylenes*; Bloor, D.; Chance, R. R., Nijhof: The Hague, Switzerland, 1985; p 239.
14. Baughman, R. H. Eds., Chance, R. R. *Ann N. Y, Acad. Sci.* **1978**, *313*, 705.
15. Baughman, R. H.; Chance, R. R. *J. Appl. Phys.* **1976**, *47*, 4295.
16. Kuhn, H. *Fortschr Chem. Org. Naturst.* **1959**, *17*, 404.
17. Rughooputh, S. D. D. V.; Phillips, D.; Bloor, D.; Ando, D. *J. Polym. Commun* **1984**, *25*, 242.
18. Rughooputh, S. D. D. V.; Phillips, D.; Bloor, D.; Ando, D. *J. Chem. Phys. Lett.* **1984**, *106*, 247.
19. Tieke, B. *Makromol Chem.* **1984**, *185*, 1455.
20. Sinclair, M.; Lim, K. C.; Heeger, A. J. *Phys. Rev. Lett.* **1983**, *51* 1768.
21. Lim, K. C.; Fincher, C. R.; Heeger, A. J. *Phys. Rev. Lett.* **1983**, *50*, 1934.
22. Muller, M. A.; Schmidt, M.; Wegner, G. *Makromol. Chem, Rapid Commun.* **1984**, *5*, 83.
23. Kratky, O.; Porod, G. *Rec. Trav. Chim.* **1949**, *68*, 1106.
24. Kuhn, H. *Fortschr. Chem. Org. Naturst.* **1958**, *16*, 169.
25. Patel, G. N.; Miller, G. G. *J. Macromol. Sci. Phys.* **1981**, *B20*, 111.
26. Olmstead, J. III; Strand, M. *J. Phys. Chem.* **1983**, *87*, 4790.
27. Rughooputh, S. D. D. V.; Phillips, D.; Bloor, D.; Ando, D. *J. Chem. Phys. Lett.* **1985**, *114*, 365.
28. Bloor, D.; Ando, D. J.; Obhi, J. S.; Mann, S.; Worboys, M. R. *Makromol. Chem., Rapid Commun.* **1986**, *7*, 665.
29. Lim, K. C.; Heeger, A. J. *J. Chem. Phys.* **1985**, *82*, 522.
30. Berlinsky, A. J.; Wudl, F.; Lim, K. C.; Fincher, C. R.; Heeger, A. J. *J. Polym. Sci. Polym. Phys. Ed.* **1984**, *22*, 847.
31. Bloor, D.; Preston, F. H.; Batchelder, D. N. *Phys. Status Solidi* **1977**, *40*, 279.
32. Eichele, H.; Schwoerer, M. *Phys. Status Solidi* **1977** *180*, 2275.
33. Tieke, B.; Bloor, D. *Makromol. Chem.* **1979**, *180*, 2275.
34. Bubeck, C.; Tieke, B.; Wegner, G. *Ber. Bunsen Ges. Phys. Chem.* **1982**, *86*, 495.

35. Sixl, H.; Warta, R. *Chem. Phys. Lett.* **1985**, *116*, 307.
36. Wong, K. S.; Hayes, W.; Hattori, T.; Taylor, R. A.; Ryan, F.; Kaneto, K.; Yoshino, Y.; Bloor, D. *J. Phys.* **1985**, *c18*, L843.
37. Koshihara, S.; Kobayashi, T.; Uchiki, H; Kotaka, T.; Ohnuma, H. *Chem. Phys. Lett.* **1985**, *114*, 446.
38. Kanetake, T.; Tokur, Y.; Koda, T.; Kotoka, Ohnuma, H. *J. Phys. Soc. Jpn.* **1985**, 4014.
39. Rughooputh S. D. D. V.; Bloor, D.; Phillips; D.; Movaghar, B. Phys. Rev. B. *Solid State* **1987**, *35*, 8103.
40. Hattori, T.; Hayes, W.; Bloor, D. *J. Phys.* **1984**, C17, L881.
41. Robins, L.; Orenstein, J.; Superfine, R. *Phys. Rev. Lett.* **1986**, *56*, 1850.
42. Rughooputh, S. D. D. V.; Bloor, D.; Phillips, D.; Jankowiak, R.; Schutz, L.; Bassler, H. *Chem. Phys.* **1988**, *125*, 355–373.
43. Maroncelli, M.; Q1, S. P.; Strauss, H. L.; Snyder, R. G. *J. Am. Chem. Soc.* **1982**, *104*, 6237.
44. Maroncelli, M.; Strauss, H. L.; Snyder, R. G. *J. Phys. Chem.* **1985**, *82*, 2811.
45. Taylor, M. A.; Odell, J. A.; Bachelder, D. N.; Campbell, A. J. **1990**, *31*, 1116.
46. Rumbles, G.; Brown, A. J.; Phillips, D.; Bloor, D. *J. Chem. Soc. Faraday Trans.* **1992**, *88*, 3313–3318.
47. Lim, K. C.; Kapitulnik, A.; Zacher, R.; Casalnuovo, S.; Wudl, F.; Heeger, A. J. In *Polydiacetylenes*; NATO ASI Series E 102; Bloor, D.; Chance, R. R., Eds.; Martinho Nijhoff: City, Holland, 1985; p 257.
48. O'Connor, D. V.; Phillips, D. *Time-Correlated Single-Photon Counting*, Academic: London, 1984.
49. Huber, D. L. *Phys. Rev. B. Solid State* **1979**, *20*, 2307.
50. Movaghar, B.; Sauer, G. W.; Wurtz, D. *J. Status Phys.* **1982**, *27*, 473.
51. Movaghar, B. *J. Phys. C. Solid State* **1980**, *13*, 4915.
52. Montroll, E. W. *J. Math. Phys.* **1969**, *10*, 753.
53. Klafter, J.; Silbey, R. *J. Chem. Phys.* **1981**, *74*, 3510.
54. Gochanour, C. R.; Andersen, H. C.; Fayer, M. D. *J. Chem. Phys.* **1979**, *70*, 4245.
55. Loring, R. F.; Andersen, H. C.; Fayer, M. D. *J. Chem. Phys.* **1982**, *76*, 2015.
56. Fredrickson, G. H.; Andersen, H. C. *Macromolecules* **1984**, *17*, 54.
57. Grunewald, M.; Pohlmann, B.; Movaghar, B.; Wurtz, D. *Philo. Mag. B* **1984**, *49*, 341.
58. Reichart, R.; Ries, B.; Bassler, H. *Philos. Mag. B* **1984**, *49*, L25–L30.
59. Stein, A. D.; Petersen, K. A.; Fayer, M. D. *J. Chem. Phys.* **1990**, *92*, 5622.
60. Weixelbaumer, W.; Burbaumer, J.; Kaufmann, H. F. *J. Chem. Phys.* **1985**, *83*, 1980.
61. Sienicki, K.; Winnik, M. A. *J. Chem. Phys.* **1987**, *87*, 2766.
62. Berberan-Santos, M. N.; Martinho, J. M. G. *J. Chem. Phys.* **1991**, *95*, 1817.
63. Baumann, J.; Fayer, M. D. *J. Chem. Phys.* **1986**, *85*, 4087.
64. Frederickson, G. H.; Frank, C. W. *Macromolecules* **1983**, *16*, 572.
65. Tao, W. C.; Frank, C. W. *J. Phys. Chem.* **1989**, *93*, 776.
66. Carey, M.; Phillips, D. *Chem. Phys.* **1994**, *185*, 75–89.

Liquid-Crystalline and Chiral Side-Chain Liquid-Crystalline Polymers

J. M. G. Cowie and T. T. Hinchcliffe

The liquid crystalline (LC) phase is a genuine thermodynamically stable state of matter displayed by molecules with a high aspect ratio or a disc-like shape. Introduction of a chiral center induces a super molecular helical structure in the nematic LC phase. This structure is capable of exhibiting selective reflection of electromagnetic radiation according to the relation $\lambda = np$, where λ is wavelength, n is the refractive index, and p is the helix pitch. This chapter presents a general overview of LC polymers and discusses examples of comb-branch polymer structures. The synthesis and properties of chiral side-chain LC polymers are described and the methods used to alter p, and hence λ, are outlined. When these chiral LC polymers form a chiral smectic C phase the possibility arises of the polymer showing ferroelectric behavior that has potential for fast switching devices. Piezo- and pyroelectric properties may also be displayed. Formation of polymer networks opens up the possible use of compression or tension to exploit these properties.

The liquid-crystalline (LC) state was first observed in 1888 by the Austrian botanist Reinitzer (1), who noted that a purified sample of cholesteryl benzoate melted to produce an iridescent opaque fluid, which on further heating cleared to give an isotropic liquid. These transitions occurred at well-defined, reproducible temperatures. Whereas this behavior was contrary to the normal observation that a crystal melted to form an isotropic liquid directly, Reinitzer's work was confirmed in the following year by Lehmann (2), who coined the term *liquid crystal* to define this intermediate phase.

Liquid crystallinity is now accepted as a thermodynamically stable phase exhibiting the properties of both a liquid and a crystalline solid. Liquid crystallinity has now been identified in a wide range of compounds that are capable of self-assembly to produce one- or two-dimensional ordering of the molecules. These molecules usually have a rigid, flat, lathe-like structure with a high aspect ratio, or alternatively they have flat disc-like shapes. The majority of those identified are *small* molecules, but these struc-

tures can be incorporated into polymeric forms that also exhibit LC phases. Liquid crystallinity can be induced either thermally (thermotropic) or on addition of a solvent (lyotropic), but we shall concentrate on the thermotropic group.

Mesomorphism

In 1922 Friedel (3) suggested that the LC phase be called the mesomorphic phase. More recent work has suggested that the LC phase is only one of three possible intermediate or mesomorphic states, the other two being plastic crystals and condis crystals. The distinction amongst these three phases relates to the type of disorder in the molecules on passing from the crystalline state into this intermediate region. Plastic crystals are said to show orientational disorder; that is, the molecules can be parallel or perpendicular to each other. Condis crystals show conformational disorder; that is, new degrees of freedom become available because bonds can now rotate more freely. Liquid crystals show only positional disordering; that is, the long axes of the molecules remain parallel to their neighbors but form structures with only one- or two-dimensional order. These distinctions tend to be overlooked, and the terms *liquid crystal* and *mesomorphic* are now used synonymously. Hence, molecules or groups that form LC phases are regularly referred to as *mesogens*.

Liquid Crystal Phases

Within the LC state there are degrees of disorder that can move progressively from the crystalline phase through to the isotropic phase as the temperature is raised. This characteristic gives rise to a range of identifiably different LC phases that can be related to the degrees of order within each phase. The two main classes are the smectic phases, which have elements of two-dimensional order, and the nematic phase, which possesses only one-dimensional order. There are several smectic phases, and these have a reasonably high degree of long range order leading to much higher viscosities than in the less regular nematic state. The anisotropic rod-like molecules that display these phases arrange themselves with the long axes parallel to each other. In the smectic phases, the centers of gravity tend to be arranged in regular layers, but this distribution is random in the nematic phase. The general orientation of these molecular axes in the LC phase is referred to as the director, and the degree of parallel alignment is estimated from the order parameter, S, which will be an average value given by eq 1.

$$S = \frac{1}{2}\langle 3\cos^2\theta - 1\rangle \tag{1}$$

Here, the angle between the long axes of any given molecule in the phase and the director (n) is θ. This order parameter is most likely to be relevant when describing the nematic phase, and it should also be remembered that the general ordering associated with a given director may refer to only one small region or domain. The whole phase will be made up of a collection of these domains whose directors may not be oriented with respect to their neighbors, but alignment of these can be achieved by applying an external force—mechanical, electrical, or magnetic—which will tend to drive the molecules to orient in one particular general direction.

Additional phases are obtained when the mesogens are optically active compounds; that is, they contain a chiral center. These phases are called *cholesteric* or *chiral* liquid crystals and they exhibit special properties. Self-assembly to form LC phases is not confined to rigid-rod type molecules. A number of stable mesophases have been identified that are formed by flat disc-shaped molecules or bowl-shaped molecules, in which the ordering is largely achieved by stacking the molecules in columns.

Smectic Phases

A large number of smectic phases have been identified, but only the most important ones will be described here. The smectic phases have the highest degree of order, but the variations arise from small changes in structural regularity within the two-dimensional layered structure that characterizes the smectic liquid crystal (Figure 1).

The smectic B (S_B) phase is the most commonly observed member of the most ordered smectic group, sometimes called *hexatic phases*, where the molecules aligned in the layers are arranged locally in a hexagonal structure. The two-dimensional order is quite pronounced in the layers, but molecules can move from layer to layer and the layers themselves can move independently of one another.

In the smectic A (S_A) phase, the arrangement of the molecules in the layers is less structured. The centers of gravity are not rigidly ordered and rotation of the molecules is easier, thereby leading to a looser layer structure. Variations on both types are obtained when the molecular axes in the layer are no longer orthogonal to the plane of the layer but are tilted with respect to this plane. The smectic C (S_C) phase is then the tilted version of the S_A phase, and smectic F (S_F) is the tilted analogue of the S_B phase. Other smectic phases have been reported, but many are closer to *soft* crystals than the liquid crystal and are less frequently observed.

Some thermotropic systems can exhibit more than one LC phase, and the natural progression on heating is for the mesogen to pass from the most ordered state to the least ordered. Thus in a hypothetical situation, S_B would appear before S_C, which would be seen before the S_A phase, on raising the temperature.

Nematic Phase

In the nematic (N) phase, the layer structures no longer exist. The centers of gravity of the molecules are as disorganized as in the isotropic liquid, but the phase retains a long range, orientational, one-dimensional order and uniaxial characteristics because of the

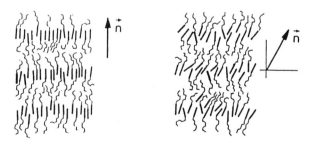

Figure 1. Schematic diagram of the general features of orthogonal and tilted smectic LC phases.

high aspect ratio of the molecules. This arrangement is represented schematically in Figure 2.

Chiral Phases

Molecules with a chiral center, such as the cholesteryl derivatives studied originally by Reinitzer, display properties that arise from the specific type of supermolecular structure that is generated in the bulk when they are optically pure. The cholesteryl derivatives, and many other optically active mesogens, form a chiral nematic (or cholesteric) phase, (N^*), that can be formally described as a series of parallel layers in which the mesogens are in the nematic phase. However, in each successive layer the director is displaced by about 10–20 minutes of arc from the neighboring layers (Figure 3). This regular displacement of the directors makes them trace out a helicoidal structure formed by the precession of the directors about an axis orthogonal to the layers.

Well-aligned samples then display selective reflection of electromagnetic radiation and act like a regular crystal lattice. The wavelength of the reflected radiation (λ_R) is related to the pitch (p) of this helical superstructure by eq 2:

$$\lambda_R = 2\bar{n}d \sin \theta = \bar{n}p \sin \theta \qquad (2)$$

where d is the distance between parallel planes, $\bar{n}$ is the refractive index, and θ is the angle between the incident beam and the N^* phase (sin θ = 1 for an angle of 90°). When the dimensions are of the same order of magnitude as visible light, the samples appear colored. This result can be manifest either as a single color or an iridescent mother-of-pearl effect. The pitch of the helix is sensitive to changes in temperature, and for the majority (but not all) of the chiral nematic phases an increase in temperature causes the helix pitch to tighten and the wavelength of selective reflection to decrease. The N^* phases also exhibit circular dichroism.

Whereas it was believed originally that optically active molecules would assume only the nematic phase, chiral modifications of the smectic C (S_C^*) and smectic A (S_A^*) phases have been identified. These molecules that form an S_C^* phase are particularly interesting because they can display ferroelectric properties. The molecular arrangement of the S_C^* phase is shown schematically in Figure 4.

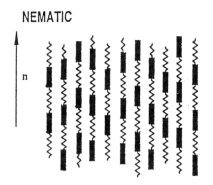

Figure 2. Schematic diagram of the nematic LC phase. The director is shown by the arrow.

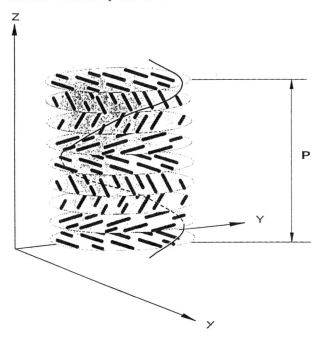

Figure 3. Schematic diagram of the chiral nematic (N^*) phase.

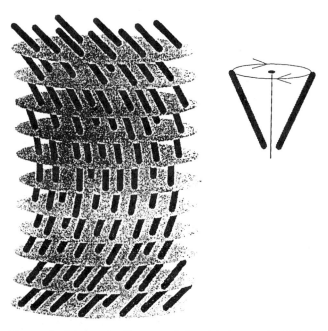

Figure 4. Schematic diagram of the chiral smectic C(S_C^*) phase.

The S_C phase is already tilted, but the chiral centers induce a precession of this tilt progressively through succeeding layers in the phase tracing out a helical distribution of the tilt direction. The S_C^* phase can show unusual dipolar effects, and using symmetry arguments Meyer et al. (4) showed that a spontaneous net polarization must exist in the S_C^* phase parallel to the layers. Because the phase only has a two-fold symmetry axis, the permanent dipole has a component parallel to this axis. Application of an electric field can reverse this permanent polarization; in this case, the molecules reverse their tilt by rotating through 180°. This process can be used in fast switching electrooptical devices.

Structural Classification of LC Polymers

The LC state was first observed in small molecules, but these molecules can be incorporated into macromolecular structures that can still exhibit LC phases and also have certain advantages over their small molecule counterparts. There is a tendency for a reduction to occur when mesogens are introduced into the polymeric structures; a small molecule mesogen with a smectic phase may display a nematic phase when in the polymeric form. However, this characteristic can be affected by different structural factors.

There are two main categories of thermotropic polymer liquid crystals; main-chain liquid-crystal polymers (MCLCPs), where the mesogenic units are linked to each other directly or through spacer units to form the main polymer backbone; and side-chain liquid-crystal polymers (SCLCPs), where the mesogens are attached pendant to the main polymer chain via an intermediate linking unit.

The incorporation of the mesogens can be achieved in so many different ways that attempts are now being made to introduce a classification protocol. An edited version of the structure classification by Brostow (5) is shown in Figure 5.

Synthesis of LC Polymers

The synthesis of LC polymers begins by recognizing that small molecule mesogens or a modification of these are ideal starting points to develop LCP structures. If we concentrate on flat, lathe-like molecules, rather than discotic structures, then some simple guiding principles can be formulated (Table I). The spacer units can vary in length and flexibility, and these differences will alter the properties of the material and the temperature range over which the LC phase is stable. The regularity of the chain formed will also be influential in controlling the melting, and hence the processing temperature, and this result is illustrated in Figure 6 for poly(aryl ester)s. Many of the important main-chain materials form a nematic phase, the low

viscosity of which aids processing. Step-growth polymerization reactions are the most commonly used methods of MCLCP synthesis, and these reactions are subject to all of the limitations inherent in this method of preparation. Thus, the formation of samples with molar masses that are large enough may be problematic, but transesterification reactions or interfacial polymerization techniques can be quite successful. Similar principles can be used in the design of SCLCPs and are illustrated in Table II.

In these structures the spacer units between the main polymer chain and the mesogenic unit assume a greater importance, because it is usually (but not always) necessary to decouple the polymer backbone motions from those of the pendant mesogens to allow

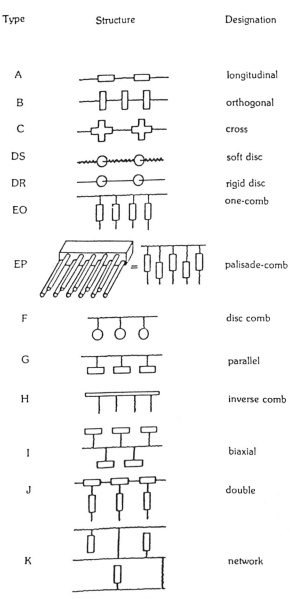

Figure 5. Structural classification of LC polymers after Brostow (5).

Table I. Group Arrangements Typical of Thermotropic MCLCPs

Cyclic Unit	Linking Group	Functional Group	Spacer
[benzene ring] x = 1 to 3	$-\overset{O}{\overset{\|}{C}}-O-$	$-\overset{O}{\overset{\|}{C}}-O-$	$(-CH_2-)_n$
[naphthalene] 1,4 1,5 2.6	$-\underset{R}{\overset{}{C}}=N-N=\underset{R}{\overset{}{C}}-$	$-\overset{O}{\overset{\|}{C}}-O-$	$(-CH_2-CHO)_n$ R
[cyclohexane]	$-CH=\underset{R}{\overset{}{C}}-$	$-O-$	$-S-R-S-$
[cyclooctane]	$-CH=N-$ $-N=N-$ $-N=N-$ $\overset{}{\underset{O}{\|}}$	$(-CH_2-)_n$	$\underset{R}{\overset{R}{\|}}{-Si-O-}$

CYCLIC UNIT	LINKING GROUP	CYCLIC UNIT	FUNCTIONAL GROUP	SPACER

MESOGENIC GROUP

formation of the LC phase. Because this chapter is concerned primarily with SCLCPs, examples of the syntheses of specific structures will be detailed later. However, two general approaches to the problem are illustrated schematically in Figure 7. Either the mesogen can be synthesized with a polymerizable group attached that allows subsequent polymer formation predominantly by addition polymerization reactions, or a polymer analogous reaction can be used where a functional terminal unit on the mesogenic structures can be attached to a preformed polymer chain. Greater control and flexibility can be achieved by using the addition approach, but the polymer analogous reaction may be easier and more direct, offering a wider range of polymer backbones as receiving sites. The relative merits of these two alternatives are more fully discussed in another chapter of this book.

Characterization of Mesophases

Precise identification of the exact nature of an LC phase is not a trivial exercise. Some are difficult to define unequivocally and a combination of methods is advisable.

Optical Textures

Perhaps the most common method used for characterizing the LC phases is the hot-stage polarizing microscope because most of the phases develop unique textures that are birefringent. The mesogenic material should be oriented as much as possible in a thin-film form. Glass slides can be rubbed in one direction with cotton, and this process will induce a homogeneous orientation of the mesogens with their long axes parallel to the surface. The sample can be enclosed by using a coverslip and then heated. Some of the patterns generated are summarized in Table III. These observations can now be coupled with thermal analysis, and the measurements run concurrently thereby aiding identification.

Thermal Analysis

Differential scanning calorimetry (DSC) is the easiest way to follow the progression of a thermotropic mesogen through the various phases. It provides an accurate measurement of the temperature range in which a phase is stable and the energy required to effect a transition from one phase to the next. These transitions are first order and so appear as endotherms in a heating cycle or exotherms in a cooling cycle. It is also advisable to investigate the sample in the cooling mode, because some LCPs are monotropic and only pass into the LC phase on cooling from the melt. A schematic diagram of typical DSC curves for a sample exhibiting multiple LC phases is shown in Figure 8.

X-ray Diffraction

Powder diffraction techniques can be used to identify phases that are not clearly defined by optical or thermal methods, but the best results are obtained when the samples can be aligned. The least ordered phases, N, S_A, and S_C, tend to show only diffuse halos, whereas the more ordered structures can exhibit Bragg reflections.

Miscibility Studies

If doubts about the nature of the LC phase still remain then the polymer can be mixed with a low molar mass mesogen that exhibits the same LC phase as that suspected for the unknown. If the two phases are identical then there will be no observable temperature difference between the two. When they do not match then two distinct LC phases will be detected that are associated with the individual components.

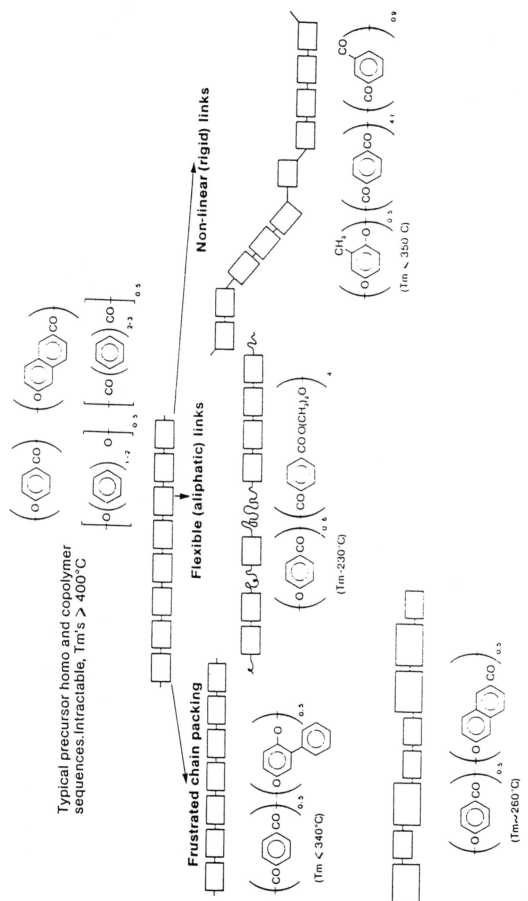

Figure 6. Methods of changing the chain regularity in thermotropic MCLCP using flexible linking units, kinks, and asymmetric monomer units. (Reproduced with permission from reference 14. Copyright 1992 Elsevier.)

Table II. Schematic Representation of Organization of an SCLCP

SCLCP	Flexible Tail	Cyclic Unit	Bridging Group	Cyclic Unit	Functional Unit	Spacer	Functional Unit	Flexible Backbone

Flexible Tail	Cyclic Unit	Bridging Group	Functional Unit	Spacer	Flexible Backbone
R	benzene ring, 1,3 or 1,4	$-\overset{O}{\underset{\parallel}{C}}-O-$	$-O-$	$+CH_2\!\!\big)_n$	$-CH-CH_2-$, OR
OR	benzene ring, X ; X = Me, Ph, Cl	$-CR=CR-$	$-\overset{O}{\underset{\parallel}{C}}-O-$	$-S-R-S-$	$-CR-CH_2-$, $\overset{C}{\underset{O}{}}{}^{OR'}$
CN	naphthalene, 1,4 or 1,5 or 2,6	$-CR=NO-$	$-O-\overset{O}{\underset{\parallel}{C}}-$	$-SiR_2-O-$	$-SiR-O-$
Chiral Unit	cyclohexane ring, n = 1,2,3	$-\overset{O}{\underset{\vert}{N}}=N-$		$+CH_2\!-CHR\big)_n$	$-P=N-$
	Cholesteryl	$-C\equiv C-$		$-NR'-R-NR'-$	
		$-CR=N-N=CR-$			

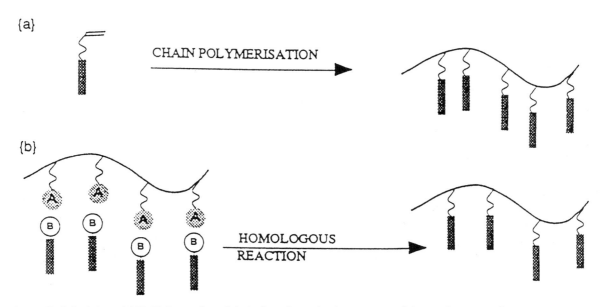

Figure 7. Principles of SCLCP formation: (a) chain polymerization process, (b) reaction on preformed backbone.

Table III. Characteristic Birefringent Patterns for Various LC Phases

LC Phase	Optical Texture
S_A	Focal-conic fan texture
S_B	Mosaic texture
S_C	Schlieren pattern (four brushes only)
S_C	Broken focal-conic fan
N	Schlieren pattern (two and four brushes)
N	Threads or marbled texture (droplets)
N^*	Planar texture with Moiré fringes; patterned cracks
N^*	Reflection colors
S_C^*	Chevron or striated texture

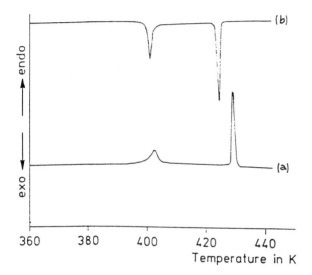

Figure 8. Schematic of a DSC measurement: (a) heating and (b) cooling cycles, showing the transition to the LC phase and then the LC to isotropic liquid transition.

General Overview

Initially work was concentrated on MCLCPs as the search for new, high performance materials intensified in the 1980s. Interest in the thermotropic LCPs centered on the production of engineering plastics, but after limited commercial success there has been a downturn in this area. Important advances have been made with materials exhibiting lyotropic behavior, leading to the development of the lyotropic aromatic polyamides such as Kevlar and associated fibers.

The literature on LCPs is extensive and benefits from regular reviewing. There are several good texts that cover the subject from different angles and can be referred to for information on nonchiral LCPs. Identification of the LC phase is of primary importance and the first method used is often optical microscopy. It is essential to have good reference books available to assist with this process, and two of note are by Demus and Richter (6), who covers all LC phases, and Gray and Goodby (7), who concentrate mainly on the smectic phases.

Whereas both main- and side-chain LCPs are normally reviewed in one volume, texts dealing mainly with the side-chain structures described here have appeared. The most widely used of these texts is edited by McArdle (8) and provides an excellent coverage of this area. The work of Russian scientists who have been particularly active in this area is not always readily accessible to Western scientists. Comprehensive reviews by Platé and Shibaev (9) and more recently by Platé (10), who compiled a volume devoted exclusively to Russian work on both MCLCPs and SCLCPs, were published.

LCPs are the subject of regular conferences, and the proceedings of some are published. Weiss and Ober (11) have edited papers from one such ACS meeting in which the physics, chemistry, and applications are dealt with in a collection of original publications.

The physical properties, phase behavior, and physics of LCPs are treated in a volume edited by Ciferri (12). The physical aspects are also emphasized by Donald and Windle (13) in their book. One of the best amongst these recently published volumes is that edited by Collyer (14), which describes characterization and physical behavior of both thermotropic and lyotropic systems and concludes with a short chapter on applications. This book is highly recommended and contains informative chapters by acknowledged experts in their field.

With the downturn in interest in MCLCPs, more emphasis has been placed on their side-chain counterparts whose potential in electronic devices and application in the information technology field has stimulated research activity on these materials. The remainder of this chapter is devoted to a review of those polymers containing a chiral center, which leads to materials with intriguing properties.

Chiral Nematic (*N**) SCLCPs

The introduction of chirality into SCLCPs led to an important class of materials that has tremendous potential for use in emerging optical technologies (15–17). The unique properties of these polymers are usually derived from either of two mesophase types: the chiral nematic (cholesteric) phase N^* or the chiral smectic C phase (S_C^*).

The liquid crystallinity in thermotropic side-chain polymers usually emanates from the ordering of the pendant groups that are linked to the polymer backbone by a flexible spacer unit (18). The chiral centers are normally located close to the mesogenic side groups.

Thermotropic SCLCPs that exhibit N^* phases are of particular interest because of their unique optical properties, which include the ability to reflect circularly polarized light in a selective manner (19–21). As mentioned previously, this effect originates from the supermolecular helical structure of the phase that causes incident light to be reflected in a Bragg-like manner from the layered twisted nematic planes.

Other properties include high sensitivity of selective light reflection to temperature changes (thermochromic materials) and an extremely high optical activity. Chiral nematic low molar mass liquid-crystal polymers (N* LMMLCPs) can be tailored to enable their use as selective wavelength notch and band-pass filters (22), highly sensitive thermo-indicators (23), IR and super high frequency radiation visualizers (24), and nonperturbing quality control of electronic devices (25).

When N* LMMLCPs are covalently attached to macromolecules, many of these properties are combined with the additional and useful properties associated with polymers (e.g., ease of processing and fiber- and film-forming ability). Of special significance is the ability to *freeze* the LC structure into the glassy matrix of a polymer, thereby retaining the optical characteristics. This ability makes N* SCLCPs ideal candidates for optical information storage materials (26). The first attempts to produce N* homopolymers with chiral centers situated in the side-chains were unsuccessful (27). The starting materials were usually cholesteryl-containing monomers (acrylates + methacrylates), which are usually liquid crystalline in themselves. Upon polymerization and incorporation into a polymer backbone, the cholesteryl systems retained their mesogenic character and smectic or highly crystalline polymers were obtained. The structures for some of these polymers are shown in Figure 9.

The most valuable route to the N* mesophase in polymers consists of introducing optically active mesogens (including cholesterol) into nematic LCPs via copolymerization. Throughout the 1980s, Soviet (27–30) and German research groups (31–34) successfully developed this approach, producing a large number of copolymers with chiral and nematogenic units. Some examples of these polymers are presented in Table IV.

The N* structure can only be frozen into the glassy matrix of a polymer providing there are no smectic or crystalline phases at lower temperatures, because these phases would tend to cause an untwisting of the helix at rates comparable to the quenching rate necessary to lock the N* phase in the glassy state below the glass transition temperature. Only a few of the copolymers derived from cholesterol produce a *solely* N* phase, and this characteristic, together with their inherent instability (photochemical and chemical) has limited their continued use and application.

A constant search for readily available non-steroidal chiral mesogens led Finkelmann and Rehage (33) to an elegant route that provided a short-lived N*

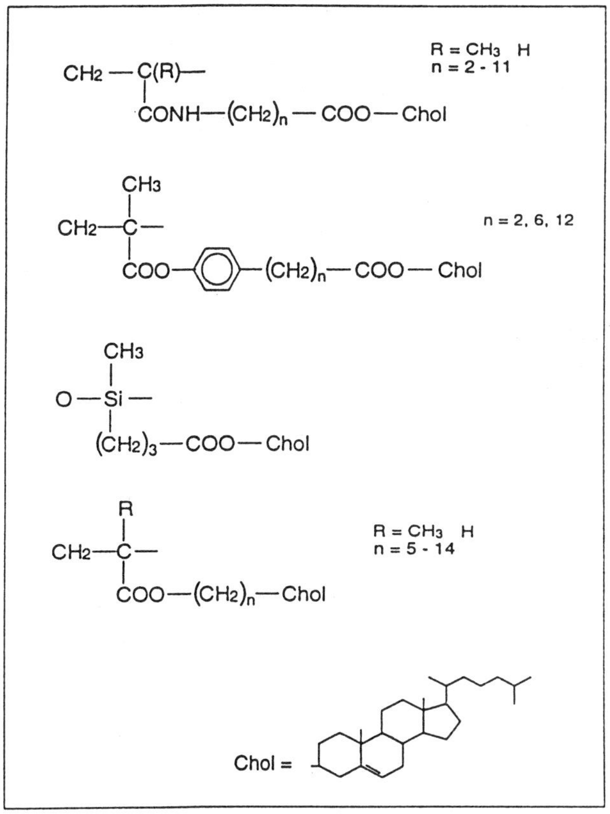

Figure 9. SCLCPs containing the cholesterol moiety in the side chain.

homopolymer (Chart I). This chiral nematic polymer has a polysiloxane backbone with chirality located in the terminal tail unit of the mesogen. The optically active center is derived from (S)-(–)-2-methyl-1-butanol, which has since been used as the principle starting material when insertion of chirality into the side chain is required.

In 1985 a new class of SCLCPs with laterally attached mesogens was reported (35). The increase in order found upon polymerization of terminally attached monomers is drastically reduced in a lateral-type system. Leube and Finkelmann (36) found that lateral attachment of an esterified cholesterol moiety led to N* polymers with elimination of undesirable smectic phases. Recently a series of novel N* homopolymers with lateral side chains were prepared by Gray and co-workers (37) using (S)-(–)-2-methyl-1-butanol as the chiral starting material; the mesogenic groups are again attached laterally to a polysiloxane backbone (Chart II).

Novel N* polymers with chirality located within the flexible spacer unit were prepared by Cowie and co-workers (38–40), who used (S)-(–)-ethyl lactate as the chiral starting material. The majority of these polymers form exclusively N* LC phases. Typical structures are shown in Chart III.

Table IV. Copolymers Exhibiting Chiral Nematic Phases and the Wavelengths of Selective Reflection from These Phases

No.	General Formula of Copolymer	Concentration of Chiral Units (mol %)	T_g (°C)	T_{cl} (°C)	λ_R Max. of Selective Reflection of Light (nm)	Ref.
1.1	*(chemical structure)*	91	70	247	1260	
1.2		84	73	229	712	19
1.3		80	77	216	562	
1.4		75	80	203	467	
2.1	*(chemical structure)*	35	48	103	495	20
3.1	*(chemical structure)*	51		209		23
4.1	*(chemical structure, $n = 3$)*	3			1700	
4.2		15			380	24
4.3	*($n = 6$)*	5			1500	
4.4		15			590	

For the chiral structures illustrated in Charts II and III suitable modifications are now possible such as reshaping of the mesogen, alteration of the polar end groups (this helps to promote N^* phase formation), and repositioning of the chiral center, all of which lead to changes in the temperature range in which the N^* phase is obtained.

The attachment of N^* LMMLCPs to preformed polymeric backbones (e.g., polysiloxanes) offers an attractive alternative that can lead to well-defined polymers with reproducible properties providing there is complete substitution of functional groups along the polymer chains (41). Adams and Gronski (42) used this synthetic pathway to produce chiral, monodisperse, AB block copolymers in which an amorphous styrene A block is connected to a 1,2-polybutadiene-type B block, to which is attached a chiral mesogenic side group. The synthetic pathway is shown in Scheme I. These types of polymers have shown very interesting morphologies as conventional block copolymers with separate LC microphases.

Another group of chiral nematic macromolecules also contains side groups; however, their helical form is derived from chirality within the backbone. Many of these materials are lyotropic in nature (e.g., poly(γ-benzyl-L-glutamate)), but there are branched polypeptides such as poly(γ-alkyl glutamates) (43, 44) or hydroxypropyl cellulose derivatives (45, 46) that

g 2°C N* 13°C I.

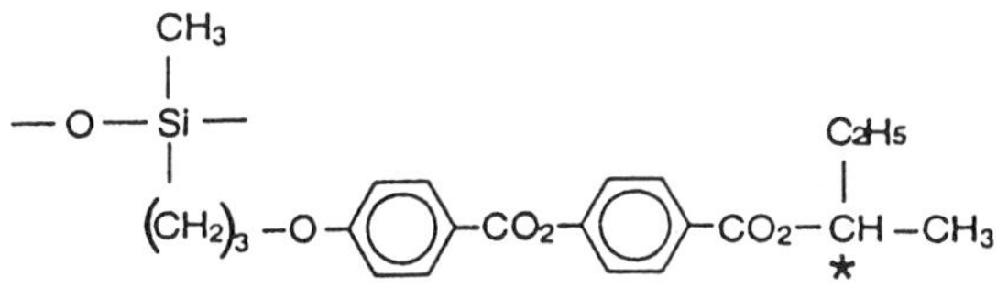

Chart I. Chiral nematic SCLCP with a polysiloxane backbone.

exhibit thermotropic N^* mesophases. Workers in Japan (47) were the first to report the thermotropic behavior of copoly(α-amino acid) derivatives based on the poly(γ-alkyl-L-glutamates) (Chart IV).

These types of branched polypeptides give rise to anisotropic melts characterized by selective reflection of visible light at 50–80 mol% of the shorter side-chain substituent. The aliphatic side chain appears to play the role of solvent molecules, and in this respect the thermotropic systems can be regarded as very highly concentrated solutions.

Optical Properties of Chiral Nematic Copolymers

The development of chiral nematic polymers has been impeded by a scarcity of information relating molecular structures to relevant optical properties. The optical properties of induced N^* phases mainly have been studied in low molar mass liquid crystals that consist of mixtures of chiral and nematic forming molecules. The notion of the twisting force of the chiral additive and the capability of a nematic matrix to be twisted is often expressed in terms of helical twisting power (HTP). The HTP is defined as the slope of the inverse helical pitch (p) of the chiral nematic phase plotted as a function of the mole fraction (X_{ch}) of the added chiral guest molecules that will induce the chiral nematic pitch in the nematic host at a constant reduced temperature, $T^* = T_m/T_i$; (T_m = measuring temperature, T_i = isotropic clearing temperature) (eq 3):

$$\text{HTP} = \bar{n}(dp^{-1}/dX_{ch})_{T^*} \quad X_{ch} << 1 \qquad (3)$$

By using eq 2, the slope of the function is then $\lambda_R (X_{ch})$ and is directly proportional to HTP by eq 4:

$$\text{HTP} = \bar{n}(d\lambda_R^{-1}/dX_{ch})_{T^*} \qquad (4)$$

The formation of the N^* mesophase can be rationalized on the basis of theories proposed by Goosens (48), van der Meer and Vertogen (49), and Finkelmann and Stegemeyer (50). These theories argue that the

g 26°C N* 123°C I

g 6°C N* 25°C I

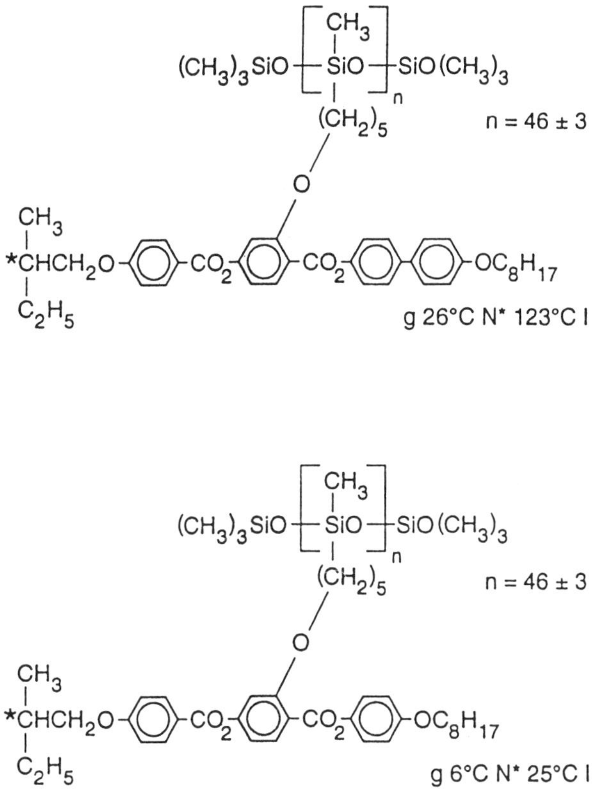

Chart II. Polysiloxanes with laterally attached chiral mesogens.

chiral dopant molecules induce the most significant helical twist in the mesophase when it is highly asymmetric and has a planar shape. Under such conditions dipole–quadropole interactions between chiral and achiral molecules composing the system are increased. Subsequently, this process leads to an increasingly hindered rotation of chiral molecules about the long molecular axis resulting in a helical twisting effect.

Finkelmann and Rehage (51) have shown that the hindered rotation of the nematic molecules also contributes significantly to the HTP of induced N^* phases. This contribution was discovered by studying the HTP in a series of N^* copolysiloxanes in which the flexible spacer of the nematogenic component was gradually increased from 3 to 6 units. The induced N^* copolymer structures and their relative HTP values are shown in Chart V and Table V.

Finkelmann found that the value of the HTP decreased as the spacer length of the nematogenic component was increased. Obviously the hindered rotation of the methoxy phenyl benzoate mesogens are alleviated as they move farther away from the polysiloxane backbone. This result is accompanied by a decrease in dipole–quadropole interactions between chiral and achiral molecules, which results in the observed decrease of the HTP.

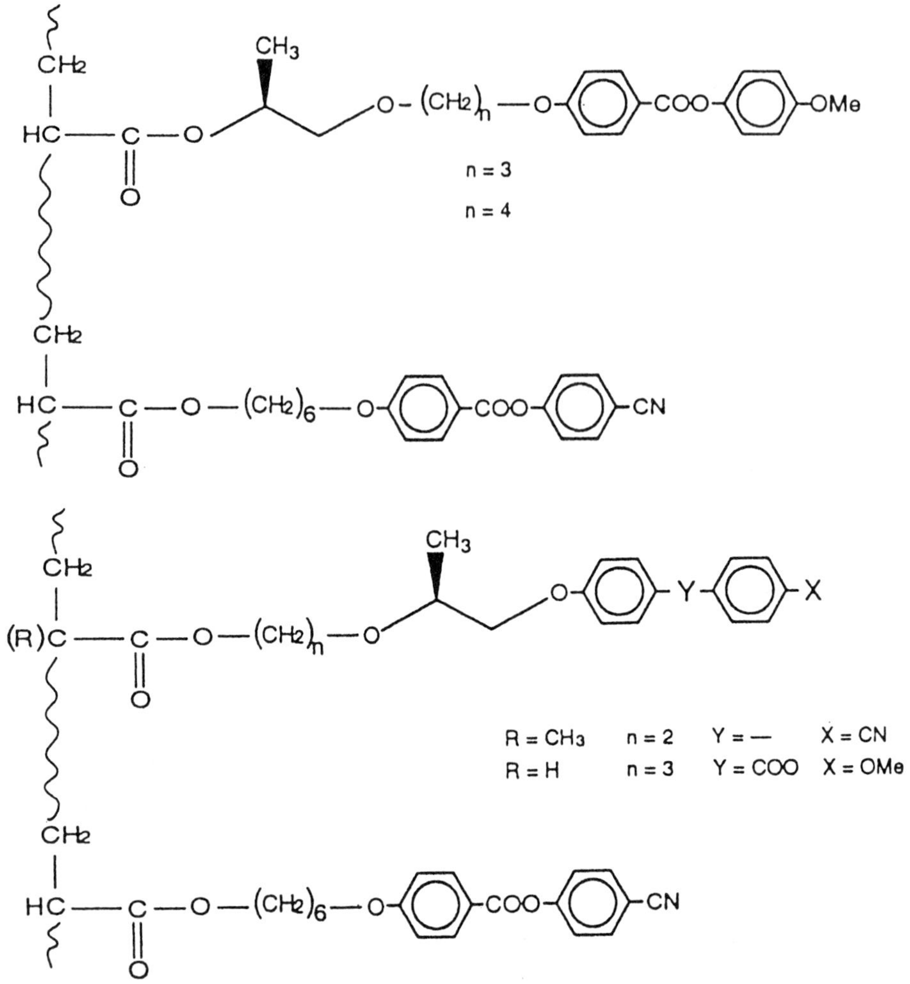

Chart III. Side-chain LC copolymers with the chiral unit located in the spacer.

Recently, we (40) assessed the HTP for chiral nematic copolymers in which the chiral center was repositioned in a flexible spacer of varying length. The chiral copolyacrylates and the relative HTP values are shown in Figure 10. Location of the chiral center away from the mesogenic moieties (mBA-6) gave the weakest helical twisting effect. These results are in agreement with the theories of Goosens, Vertogen, and Finkelmann and can be explained also on the basis of hindered rotation of chiral and achiral species.

The helical sense and twisting power of new thermotropic N^* copolymers containing (S)-(–)-1-phenylethanol[1] and (R)-(–)-methylmandelate[2] has been studied by Chen and Krishnamurthy (52). Instead of optical rotation or absolute configuration, chiral nematic molecular interactions of a steric nature distorted the configurational characteristics that are responsible for the twist sense. Furthermore, an increased chiral nematic structural similarity appeared to contribute to an enhanced HTP. The representative

copolymers and their relative HTP values are illustrated in Figure 11.

Chiral Smectic C (S_C^*) SCLCPs

Chiral modification of the smectic C phase (S_C^*) has led to the synthesis of an important class of materials exhibiting ferroelectric behavior. The molecular arrangement for the S_C^* phase is shown in Figure 4. In the S_C^* phase the molecules are parallel to one another and are arranged in layers, with the long axis tilted with respect to the normal to layer planes. However, the structure is helicoidal formed by a precession of the director about the axis of the layers. The symmetry plane and the inversion center of the ordinary achiral smectic C phase are absent, which stems directly from the chirality of the compounds.

Meyer and co-workers (4), after reporting that on symmetry arguments a ferroelectric liquid crystallinity should exist, proved this phenomenon by synthesiz-

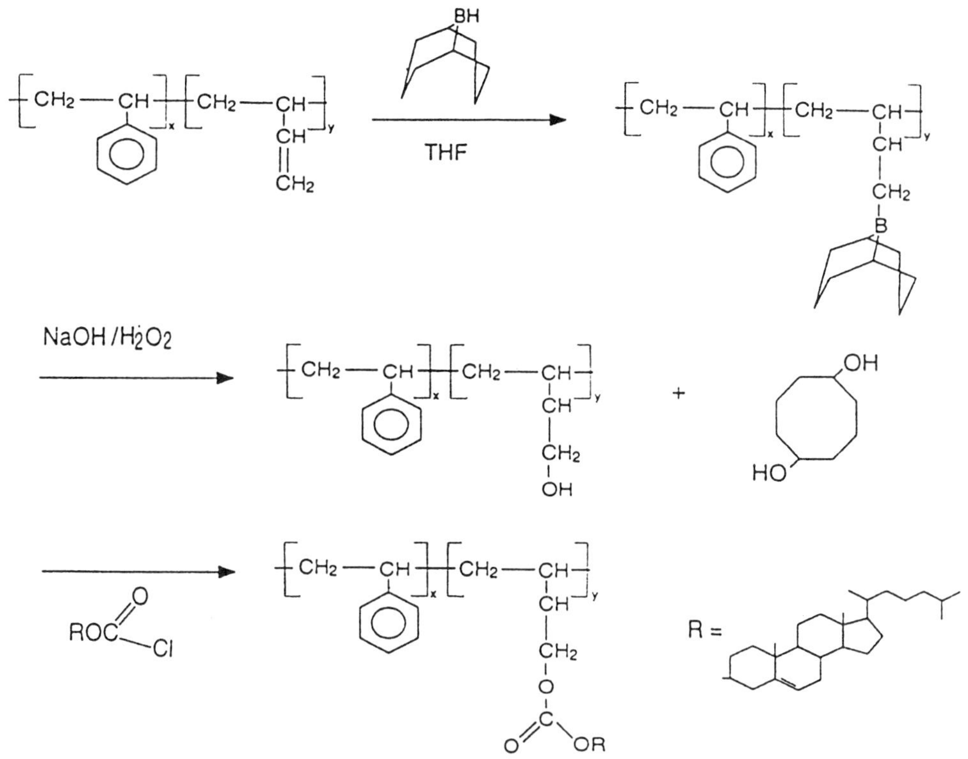

Scheme I. Synthetic route used by Adams and Gronski (*42*).

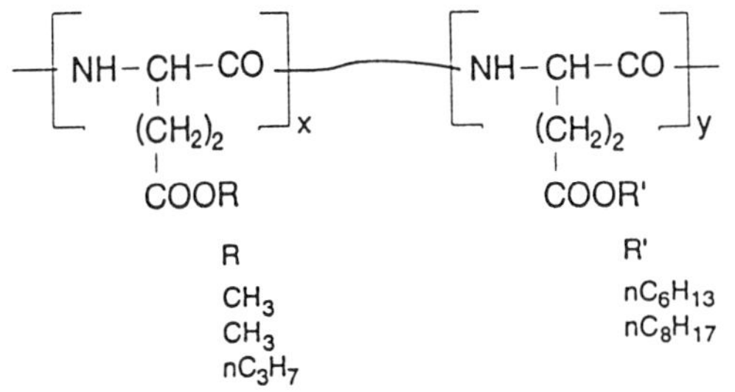

Chart IV. Structure of poly(α-amino acid) copolymers.

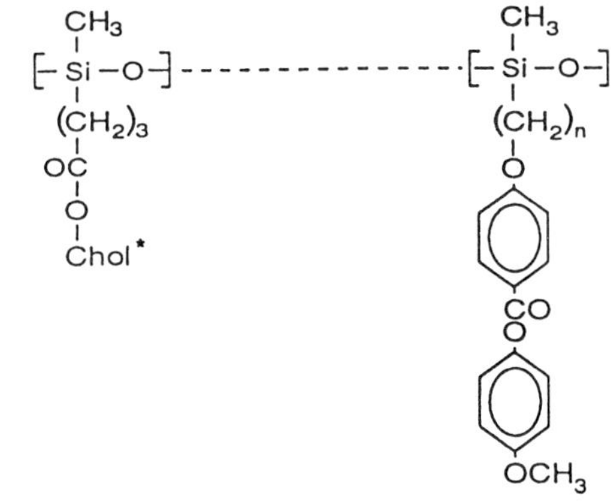

R	R'
CH_3	nC_6H_{13}
CH_3	nC_8H_{17}
nC_3H_7	

Chart V. Structure of the polysiloxane copolymers whose helical twisting power is shown in Table V.

Table V. Helical Twisting Power (htp) of Induced Chiral Nematic Phases

Induced Chiral Nematic Copolymer	$10^2-(d\gamma_R/dXch)nm^{-1}$
$n=3$	1.71
$n=4$	1.43
$n=6$	1.28

ing a compound that exhibited such a phase. To exhibit $S_C{}^*$ phases the majority of LC compounds have to comply with the following conditions:

1. They should have a lamellar structure.
2. The molecular long axis must be twisted within the layers.
3. The molecules must possess a transverse dipole moment.
4. The molecules must be chiral.

In almost all LC compounds a transverse dipole moment exists; however, for $S_C{}^*$ compounds (due to the symmetry of the phase) the component parallel to the axis cannot be averaged to zero, and this result leads to a spontaneous polarization in the molecule. When an electric field, E, is applied normal to the helicoidal axis, the matrix becomes distorted; above a critical field, E_c, it is completely unwound and the sample is poled with the molecules tilted along a preferred direction normal to E.

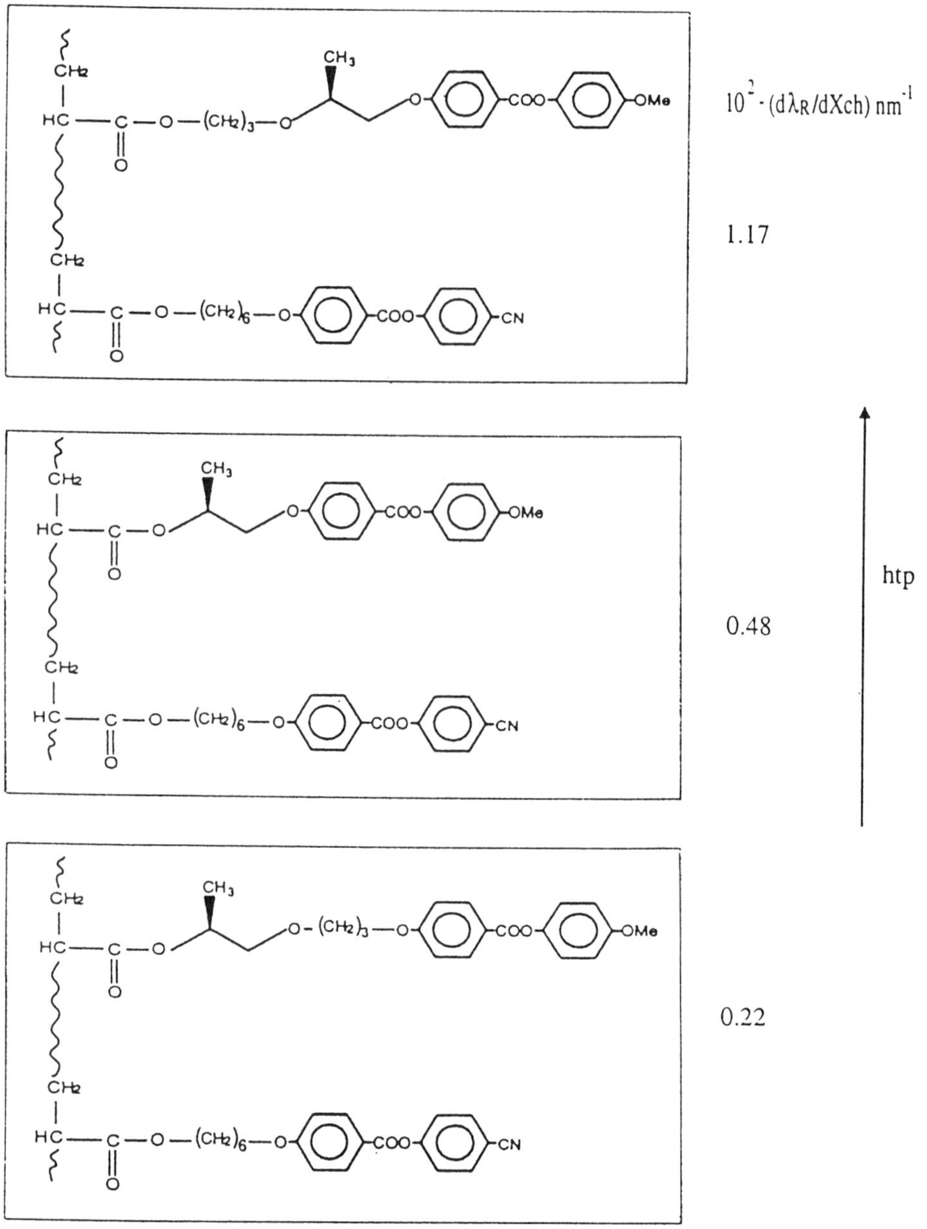

Figure 10. Relative helical twisting power (HTP) in relation to the positioning of the chiral center in the spacer unit.

The $S_C{}^*$ phase has shown promising potential for use in electrooptical displays because of the ferroelectric properties (4) and the very fast switching speeds between the two stable states of the molecules (53). At present there is no widespread use of devices employing $S_C{}^*$ LCPs; however, some effective applications have emerged that employ LMM $S_C{}^*$ LCPs or their mixtures (54).

Ferroelectric LC polymers (FLCPs) are not likely to have switching times as fast as their LMM counterparts because of the much higher viscosity of the polymer as opposed to low molar mass material. Many questions still remain unanswered with the development of fast electrooptical switching FLCPs

now occupying a major part of LC research. The hierarchy of response times for different materials is as follows: nematic LCPs, several seconds; nematic LMMLCPs, milliseconds; ferroelectric LMMLCPs, microseconds. A question mark remains over the FLCP group.

The switching time is normally defined as the time of the change in transparency from 10% to 90% in an oriented sample in a surface-stabilized LC cell. The switching time of a ferroelectric device is inversely proportional to the spontaneous polarization as expressed by eq 5, where τ is the reorientation time, η is the viscosity, P_s is the spontaneous polarization, and E' is the applied electric field.

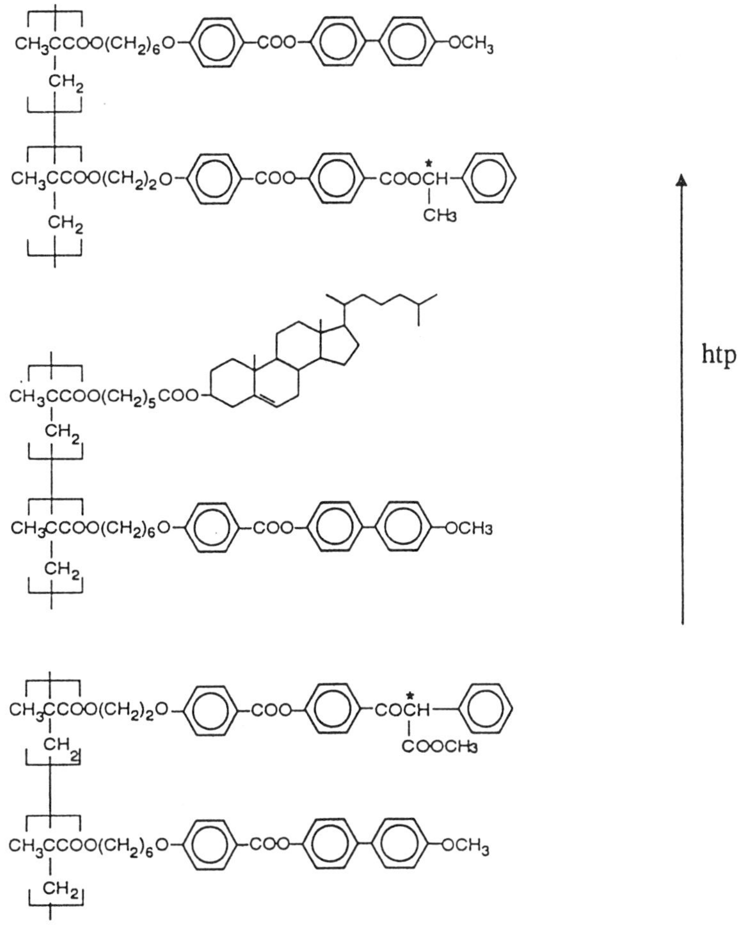

Figure 11. Change in helical twisting power (HTP) as a function of copolymer structure.

$$\tau = \eta/(P_s E') \qquad (5)$$

The first LMM $S_C{}^*$ LCP found was (S)-4-n-decyl-oxybenzylidene amino-2'-methyl butyl cinnamate (DOBAMBC) (4), Chart VI, but it was not until 1984 that an FLCP was synthesized by Shibaev and co-workers (55). The polymer was a methacrylate with the chiral center (derived from (S)-(−)-2-methyl-1-butanol) located in the terminal group, Chart VII, which passed through two mesophases between the glass and the isotropic liquid, one of which was $S_C{}^*$.

Japanese workers (56) have since produced a set of FLCP structures similar to those of Shibaev, and these compounds are shown in Chart VIII, where the spacer length, n, is >10, and the core moiety is either a biphenyl unit or ester-linked phenyl groups. These FLC polyacrylates have been used successfully to make large area FLCP switching displays with dimensions of 15×40 cm.

Scherowsky (57) made substantial contributions toward the production of FLCPs and synthesized a series of novel chiral acrylate side-chain polymers that show great promise as fast switching optical displays. The main advantages of these materials is their better

DOBAMBC

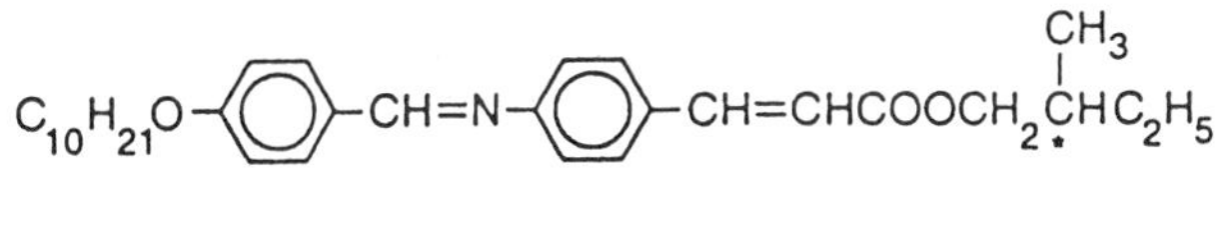

Chart VI. Structure of (S)-4-n-decyloxybenzylidene amino-2'-methyl butyl cinnamate.

shock resistance and applicability as flexible display devices when laminated with plastic substrates. Some of the polymers are shown in Chart IX, and the location of chirality within the flexible spacer is also shown to be effective toward the production of $S_C{}^*$ phases as well as N^* ones.

The most common FLCPs consist of relatively long spacers containing 10 or more methylene groups. Very little work has been carried out to study the influence of molecular weight and the length of the flexible spacer upon the properties of these polymers. Some work has been carried out by Kühnpost and Springer (58), who have studied the influence of spacer length ($n = 2, 6, 8,$ and 11 methylene groups) on the polymorphism and ferroelectric properties of chiral SCLC polyacrylates. Various molecular weights were obtained

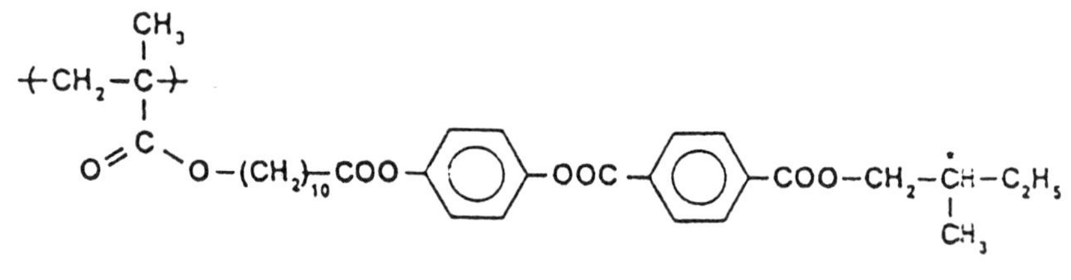

G 20 - 30 S$_c$$\cdot$ 73 - 75 S$_A$ 83 - 85 L

Chart VII. Structure of polymethacrylate chiral side-chain material that exhibits ferroelectric behavior (46).

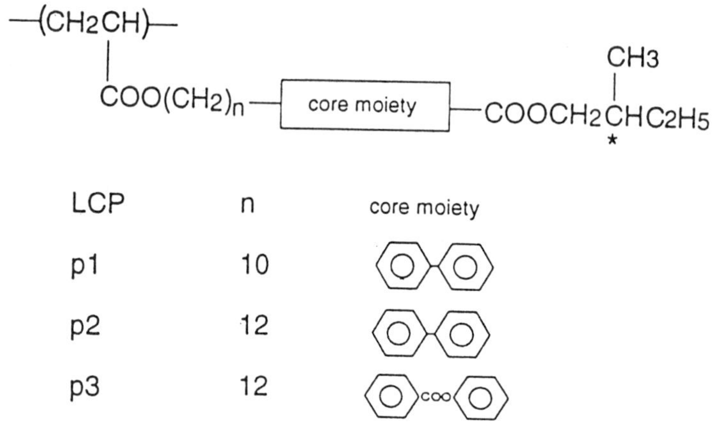

LCP	n	core moiety
p1	10	
p2	12	
p3	12	

Chart VIII. SCLCPs possessing an S_C^* phase and exhibiting ferroelectric properties (56).

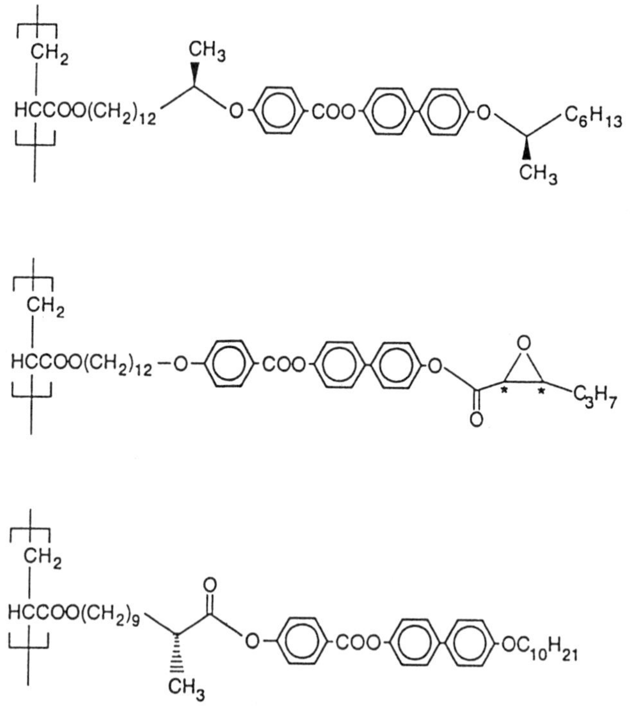

Chart IX. Some polymers exhibiting ferroelectric behavior prepared by Scherowsky (57).

from the polymers $n = 8$ and $n = 11$, whose structures are shown in Chart X.

These initial studies are of importance to new workers in the field and illustrate the fact that polymorphism of complex S_C^* phases can strongly depend on the molecular weight and spacer lengths of polymers. What effect the various viscosities of the samples have on the behavior of these materials is not yet clear.

Kumar and co-workers (59) recently adopted a novel approach to the preparation of FLCPs through the thermodynamically favorable route of using self-assembly of a suitable hydrogen bond donor polymer with a hydrogen bond acceptor molecule. The concept of introducing ferroelectricity through hydrogen bonding was realized with the assemblies shown in Chart XI. The FLCPs in question consist of polysiloxanes with pendant *p*-alkoxybenzoic acid groups complexed with an optically active (S)-(–)-*trans*-4-(2-methoxypropyloxy)-4'-stilbazole.

Zentel and Kapiza (60) prepared chiral nematic combined MC/SC polymers by a polymer-analogous esterification reaction with the help of dicyclohexylcarbodiimide (DCCI). The synthetic route to these materials is shown in Scheme II. By using a polymer-analogous reaction the molecular preconditions for high spontaneous polarization (P_s) can be created by employing a wide selection of chiral acids, of the type shown, and attaching these acids as terminal units for mesogenic structures (Chart XII).

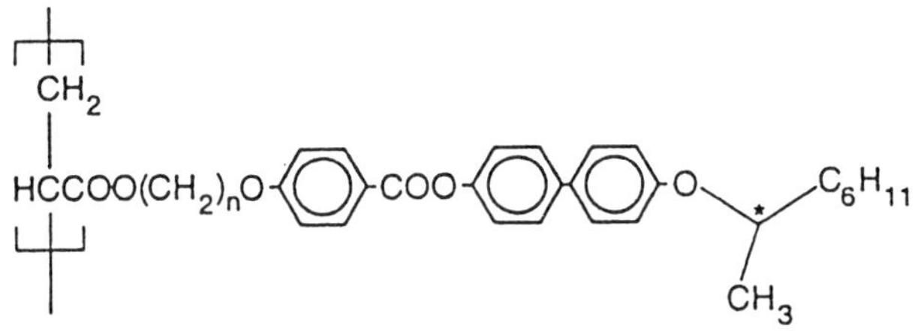

	n	Mw (g /mol)	Mw/Mn
1a	2	117 000	2.19
1b	6	83 000	2.12
1c	8	27 000	1.42
1d	8	68 000	1.12
1e	8	406 000	1.57
1f	11	15 000	1.50
1g	11	38 000	2.44
1h	11	240 000	3.07

Chart X. Structure of materials prepared by Kuhnpost and Springer (*58*).

The interest in these materials is related to the possibility of *locking in* different orientations below the glass transition temperature of the polymeric structure to produce optically recording media (*8*) or to introduce additional dye molecules, preferably with nonlinear optical (NLO) properties (*61*). To use side-chain polymers for frequency doubling (second harmonic generation) in NLO applications, a copolymer was prepared in which 6% of the mesogenic groups were esterified with a group that had an appreciable hyperpolarizability component perpendicular to its long axis. The molecule used was 2-nitro-5-dimethylaminobenzoic acid (Chart XIII), and the introduction of this NLO chromophore had virtually no effect on the LC phase transitions of the copolymer when compared with the homopolymer.

The orientational effects of the $S_C{}^*$ phase on the guest NLO molecules have led to materials that are effective at second harmonic generation combined with high spontaneous polarization.

Recently a tristable switching effect was discovered in an LMM ferroelectric LC (*62*). The third state of the tristable switching corresponds to a novel antiferroelectric mesophase ($S_C{}^*A$). Antiferroelectric LCPs exhibited sharp threshold characteristics (*63*). Consequently, a prototype of a matrix display device with high contrast was developed (*64*). To study these effects in polymers, Scherowsky and co-workers (*65*) attached mesogenic chiral heterocycles to polyacrylate backbones (Chart XIV). A third switching state was discovered in the $S_C{}^*$ phase, but whether this third state consists of an antiferroelectric phase or a helical arrangement with a short pitch in the $S_C{}^*$ phase has not yet been firmly established.

As discussed previously, copolymerization of chiral and nematogenic monomers is a useful strategy for the synthesis of N^* SCLCPs. As already seen in this section, the majority of $S_C{}^*$ polymers are homopolymers with a chiral moiety situated in every group. Kozlovsky and co-workers (*66*) recently realized a new approach to the creation of ferroelectric LC copolymers based on the preparation of copolymers of the chiral homopolymer (having no tilted smectic phases) and the tilted smectic F homopolymer (P6M) (Chart XV).

Introduction of 37% or less of the chiral side chains into the macromolecule of the smectic F polymer did not destroy this tilted structure. Copolymers with spacer lengths of C_8–C_{13} showed a pyroelectric effect that was accompanied by spontaneous polarization of the phase. By using this approach, many of the concepts developed in the design of ferroelectric LMMLC mixtures can be applied to the preparation of FLCPs with high values of spontaneous polarization.

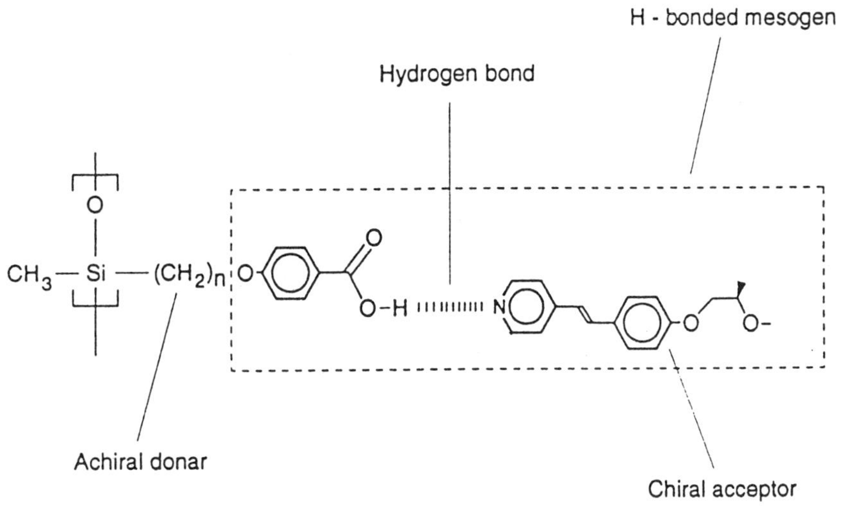

Chart XI. Formation of chiral SCLCPs through hydrogen bonding.

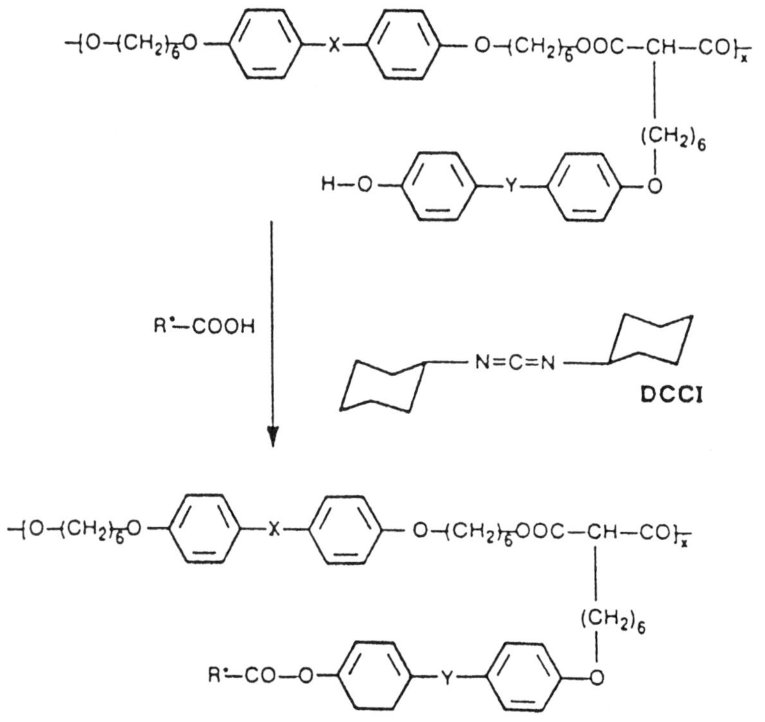

Scheme II. Combined MC and SCLCPs exhibiting a chiral nematic phase (60).

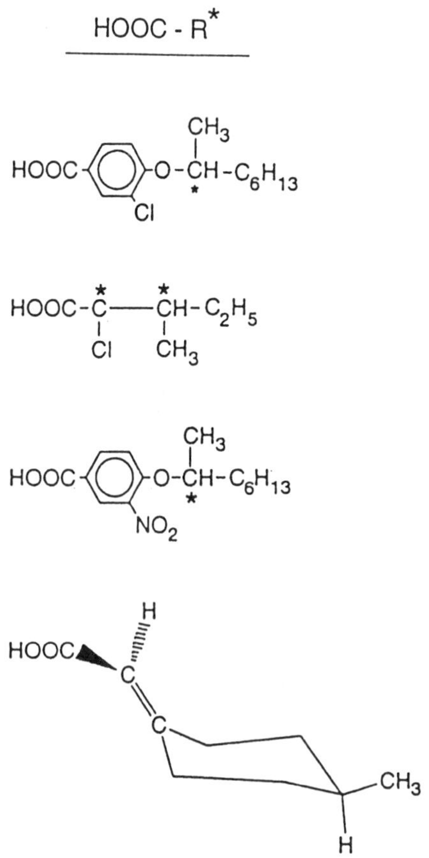

HOOC - R*

Chart XII. Selection of chiral acids capable of being used as terminal units for the synthesis of chiral mesogens.

Chiral LC Networks

In 1969 de Gennes suggested that an elastomeric network could trap LC order within a polymeric structure (67). The formation of networks by the cross-linking of linear polymer chains is well known, and if these methods are applied to linear polymers that contain mesogenic groups the generation of LC networks may ensue. Cross-linking prevents only the macro-Brownian motion (mobility of polymer chains), whereas the micro-Brownian motion (mobility of chain segments) is not impaired except close to the network junctions. Thus, depending on the length and flexibility of the cross-linking agents, the formation of LC networks is possible that will exhibit contrasting mechanical properties. Early research in this field demonstrated that the preparation of both flexible and rigid networks is feasible, some structures of which are shown schematically in Figure 12. A brief overview of this topic is given subsequently, followed by full details of monodomain LC networks by in situ polymerization in Chapter 3.3.

Finkelmann and co-workers (68) were the first to report on lightly cross-linked elastomeric networks prepared from linear polysiloxanes (Scheme III). These materials were produced by cross-linking the chains with a flexible difunctional siloxane carrying no mesogenic groups.

The glass transition and the mesophase-to-isotropic clearing temperatures of the elastomers were

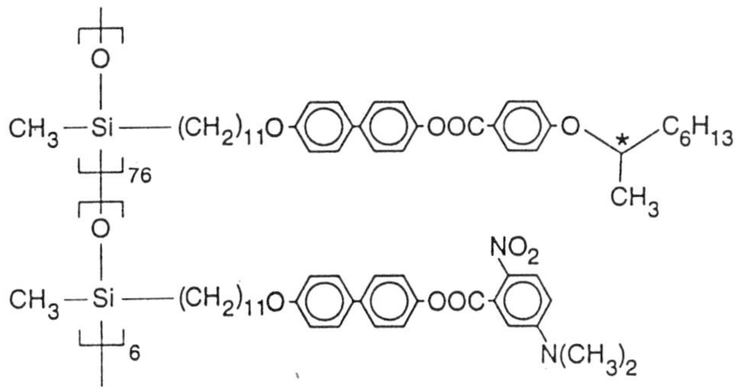

Chart XIII. Copolymers used for NLO application (*61*).

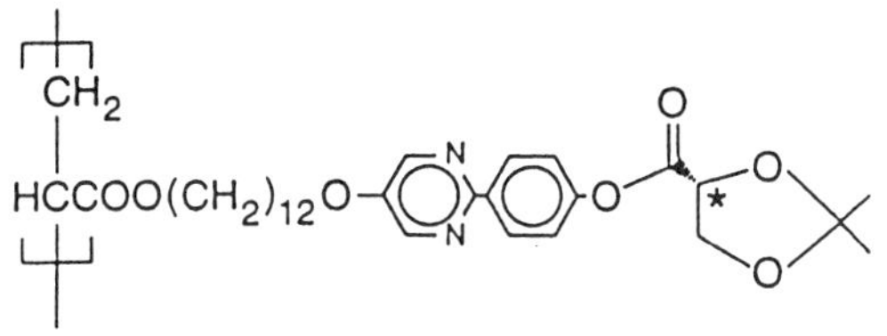

Chart XIV. Structure of polymer showing antiferroelectric behavior.

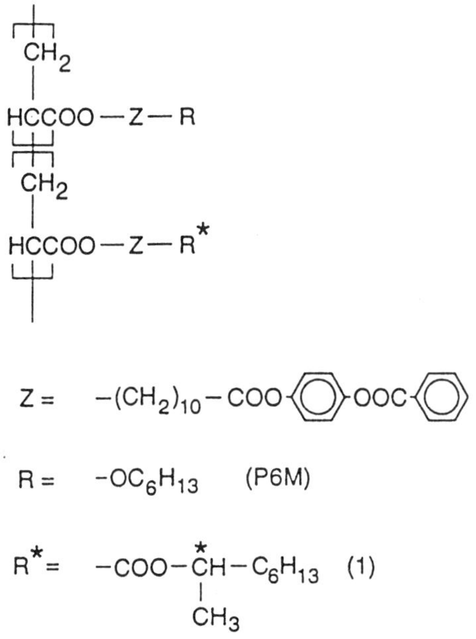

$$Z = \quad -(CH_2)_{10}-COO\text{—}\langle O \rangle\text{—}OOC\langle O \rangle$$

$$R = \quad -OC_6H_{13} \quad (P6M)$$

$$R^* = \quad -COO-\overset{*}{C}H-C_6H_{13} \quad (1)$$
$$\qquad\qquad\qquad | \\ \qquad\qquad\quad CH_3$$

Chart XV. Ferroelectric LC copolymer structures prepared by Kozlovsky et al. (*66*).

lower than the corresponding linear polymers, whereas the range over which the LC phases were observed remained constant. The unstretched rubbery elastomers in the LC state behave as monomeric LC melts and are turbid in appearance; upon stretching, the mesogenic groups become macroscopically ordered as indicated by a uniform transparency of the samples.

These effects in LC elastomers have opened up a new technological interest in switched optical devices. LC elastomers with polyacrylate and polymethacrylate

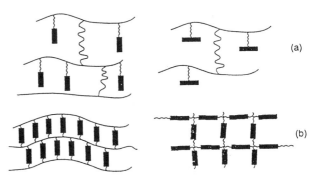

Figure 12. Schematic diagram of different types of network structures: (a) flexible, (b) rigid.

main chains were synthesized (*69*) by using synthetic routes similar to those employed for the polysiloxane samples. If an acrylic mesogenic monomer is copolymerized with an acrylic hydroxy functional monomer, the resulting copolymer is suitable for cross-linking with a diisocyanate component (Scheme IV).

Elastomeric networks comprising chiral LC components have recently opened up completely new applications (e.g., piezosensors and piezotransducers) that could not be covered by LMMLC systems. The special optical or electrical properties of the N^* and S_C^* phases are closely related to their helical superstructure. For these materials the presence of an electrical or magnetic field can give rise to untwisting within the mesophase. Zentel et al. (*70*) demonstrated that for an N^* elastomer these effects can be induced by using only mechanical fields. Thus, for an N^* LCP that is selectively reflecting visible light such an unwinding would result in a nematic structure without selective reflection. Materials with these properties may be used as mechanical/optical devices. For an S_C^* elastomer (without macroscopic polarization) mechanical unwinding would result in a smectic C structure (with macroscopic polarization); thus, one has the ability to transform mechanical signals into electrical signals (the basis of a piezoelectric device).

The materials used to demonstrate the helical unwinding effects of chiral LC elastomers consist of lightly cross-linked, combined MC/SC polymers that exhibit broad N^* and S_C^* phases (Scheme V). In the N^* phase, strains of 150% generated oriented X-ray patterns typical of nematic phases, thus proving that the helical superstructure of an N^* phase can be untwisted upon strain. After removing the strain an unorientated X-ray pattern was obtained, presumably due to restoration of the N^* phase. Similar results were obtained for chiral smectic phases at strains of approximately 300%.

In 1989 Brand predicted (*71*) that it should be possible to observe the piezoelectric response for a compression parallel to the helical structure in the N^* phase. These elastomers would seem to be the ideal

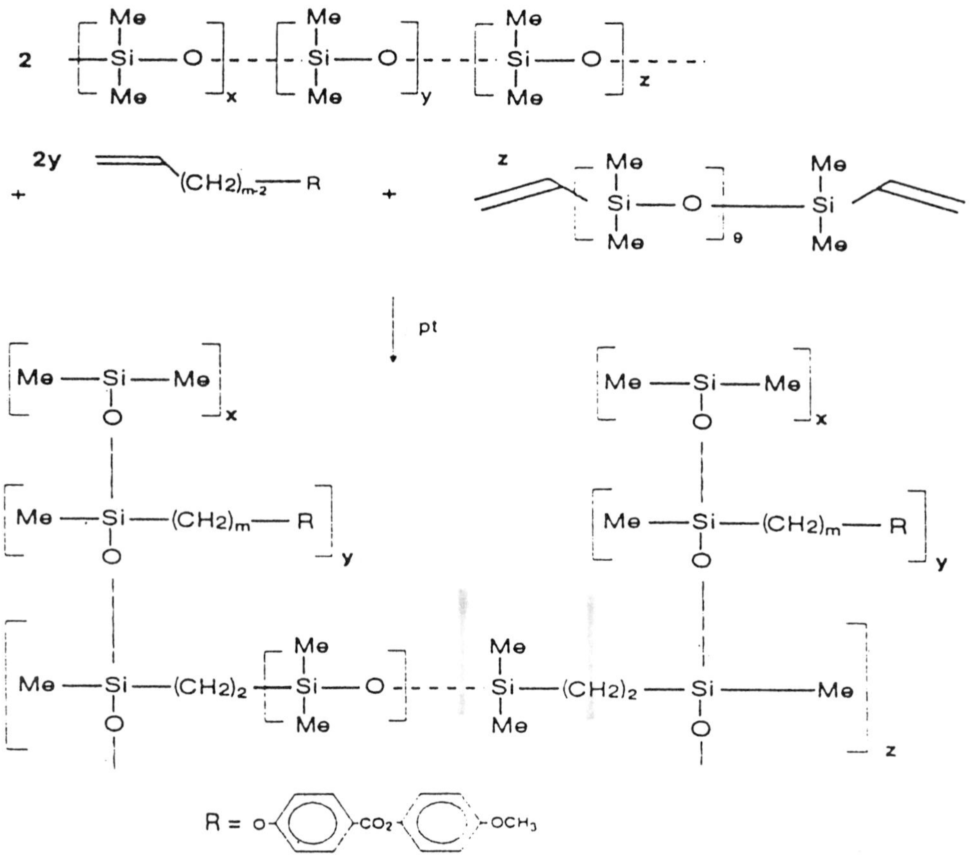

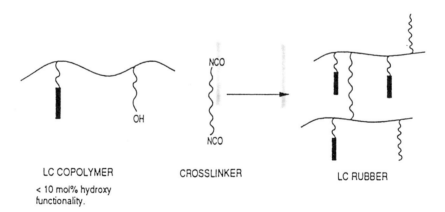

Scheme III. Method of preparing flexible polysiloxane LC networks.

LC COPOLYMER
< 10 mol% hydroxy
functionality.

CROSSLINKER

LC RUBBER

Scheme IV. Schematic of formation of a cross-linked network using a diisocyanate cross-linking agent.

materials to display this effect, because they are capable of supporting uniaxial compression, whereas LMMLCPs are not able to sustain this deformation. These effects were also predicted for S_C^* elastomers. In 1990 experimental proof of piezoelectricity was found in both S_C^* and N^* phases of elastomers derived from very similar materials to those shown in Scheme V. Finkelmann (72) at the same time prepared an N^* elastomer by swelling a nematic elastomer with a LMMLC chiral dopant; the resulting materials produced piezoelectric voltages of magnitudes comparable with those of quartz, a material that is also piezoelectric and nonferroelectric.

Workers at Philips research laboratories made piezoelectric networks (73) by photopolymerization of LC diacrylate molecules doped with chiral mesogenic molecules by using the molecules shown in Chart XVI. The mixtures are suitable for photopolymerization in both the N^* and S_C^* phases. The networks obtained by polymerizing poled mixtures in the S_C^* phase

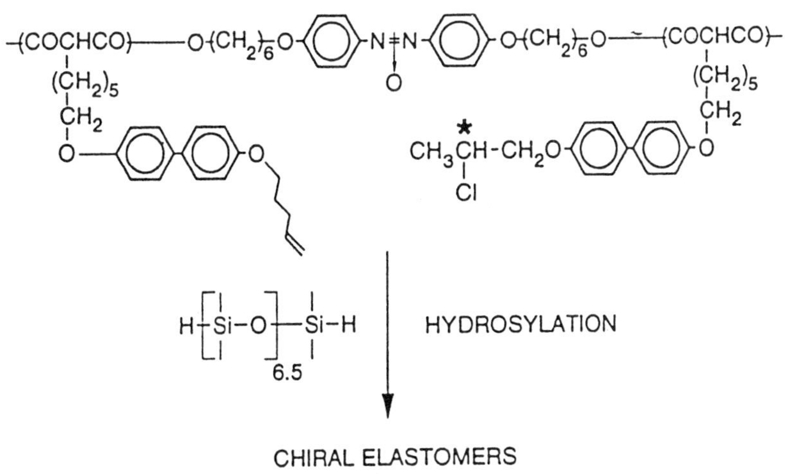

Scheme V. Structure of chiral elastomers combining main-chain and side-chain mesogenic groups (*70*).

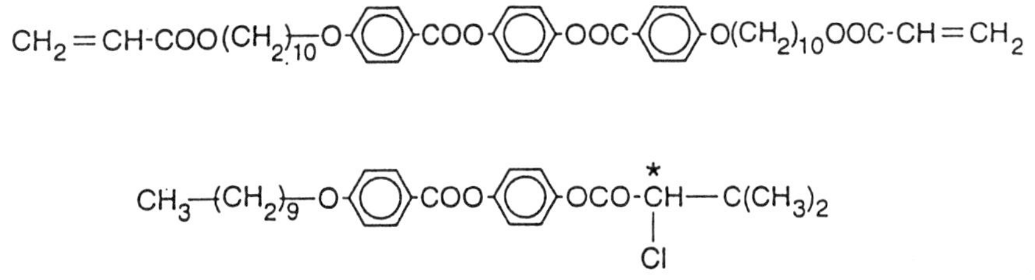

Chart XVI. Molecules used to synthesize piezoelectric active networks (*65*) by photo-cross-linking.

showed piezoelectricity, as discussed more fully in Chapter 3.3.

Cross-linked networks displaying chiral thermotropic LC phases were prepared from hydroxypropyl cellulose (*74*). The cellulose-based LC elastomers exhibit the phenomenon of mechanically induced molecular switching, although it is less marked than in typical cross-linked chiral SCLC systems.

We (*40*) prepared chiral acrylic elastomeric networks that display only N^* mesophases (Scheme VI). The elastomers exhibit selective reflection of light in the visible region; different reflection colors were produced upon application of stress to the networks. This type of material may show promise as colored strain gauges.

Tailor-made LC networks exhibiting an S_C^* mesophase also were produced by Percec and Zheng (*75*) using the technique of living cationic copolymerization. The copolymers shown in Scheme VII contain reactive methacryloyl side groups that can be thermally cross-linked to produce the desired S_C^* phases.

Conclusions and Prospects

Much of the research to date that has lent itself to commercial exploitation has been directed toward MCLCPs, where strength and stiffness are the desirable properties in materials applications. These high performance engineering plastics characteristics are of much less importance when studying SCLCPs, where attention now tends to be focused on the electro- and thermooptical properties.

The effort expended on the development of synthetic routes to these SCLCP materials has been considerable and so far has exceeded the work on assessment of their properties. This focus should now change, and work centered on the application of SCLCPs will become progressively more important now that a number of sound synthetic strategies have been established that lead to polymers with interesting properties. This does not mean that all the synthetic problems have been solved. Many of the polymerization processes used do not really produce polymers with significantly long chain lengths and are often more likely to be oligomeric. The molecular weight distribution of the samples is also poorly controlled, and variations in these characteristic properties can affect the viscosity, the electrooptical response, and the temperature range over which the LC behavior is observed. Thus, it is essential that to ensure reproducibility of the sample properties and behavior greater attention should be given to improving the control over these factors and establishing optimum preparative conditions. Only then will the full potential of SCLCPs as useful materials be realized.

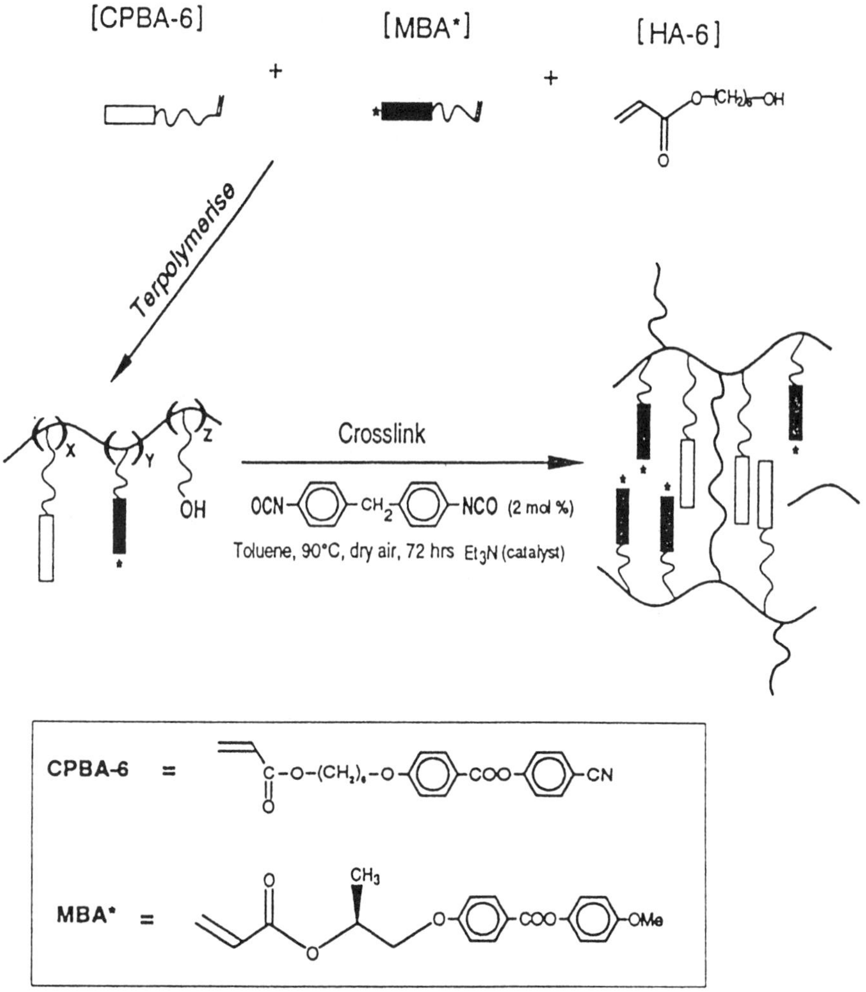

Scheme VI. Synthesis of networks exhibiting only N^* mesophases (40) and color changes under stress.

Whereas small molecule LCPs have been used widely in electrooptical devices because of their fast switching characteristics, SCLCPs suffer from having much larger molar masses with correspondingly higher viscosities, such that the switching and response times are much slower. Consequently, other areas of application may be more appropriate and work on network structures has suggested more promising possibilities. Chiral LCPs have been incorporated into flexible networks and have potential as optical strain gauges. These materials can also show piezoelectric behavior because the polymers have no inversion center.

The property of selective reflection of electromagnetic radiation by chiral SCLCPs could be exploited in the visible region to produce stable colored films, but a more significant possibility is to make use of SCLCPs that are more likely to reflect the shorter wavelengths. By designing the polymer such that the pitch of the helical supermolecular structure corresponds to a reflection of radiation in the UV region, it may be possible to produce materials suitable for use as films or coatings that have some protection from degradation by UV radiation. One problem is that for a particular chiral SCLCP, only 50% of the radiation will be reflected: the right-hand polarized component of the radiation is reflected by a right-handed helix but the left-hand component is transmitted. This result could be improved by having alternate layers of chiral SCLCPs with the opposite optical rotation, which should then reflect all components of the selected wavelength.

Recent theory has predicted that rigid LC networks formed by connecting mesogenic units into a highly cross-linked network structure will produce a material of very high strength. Incorporation of such structures into a composite could lead to new high strength materials. The ability of LCPs to improve hardness and impact strength can be regarded as a key discovery of importance to the coatings industry.

Also, increasing evidence supports the proposition that some thermotropic SCLCPs may have a microheterogeneous morphology. Because this morphology could have an important bearing on their physical

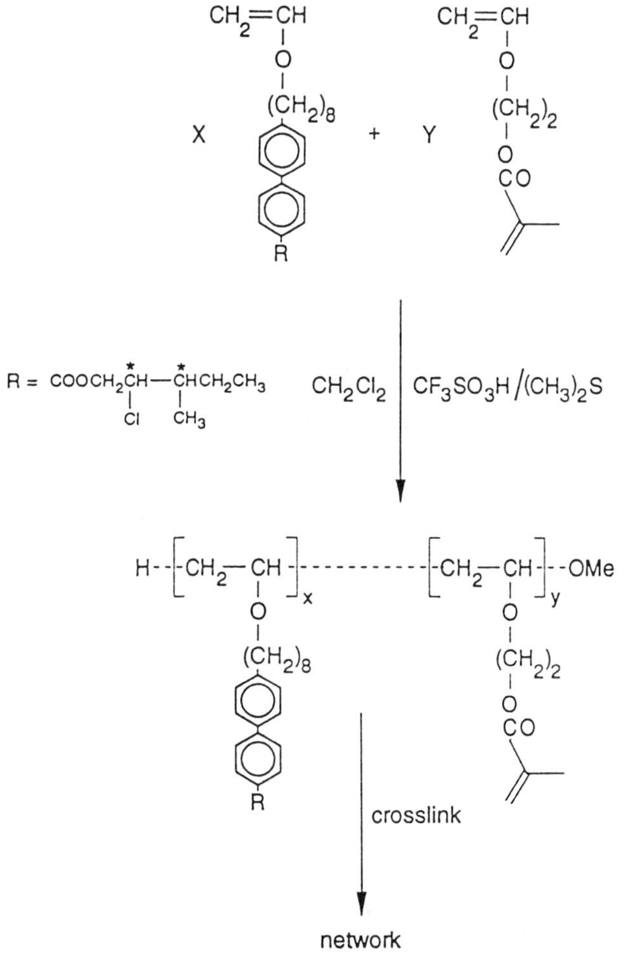

$CH_2{=}CH$ $CH_2{=}CH$
O O
$(CH_2)_8$ $(CH_2)_2$
X   +   Y O
     CO
R

$R = COOCH_2\overset{*}{C}H{-}\overset{*}{C}HCH_2CH_3$ CH_2Cl_2 $CF_3SO_3H/(CH_3)_2S$
         Cl CH_3

$H{-}[CH_2{-}CH]_x{-}\cdots\cdots{-}[CH_2{-}CH]_y{-}OMe$
     O O
     $(CH_2)_8$ $(CH_2)_2$
                       O
                       CO
     R

crosslink

network

Scheme VII. Synthesis of $S_C{}^*$ networks (*75*) using cationic polymerization methods.

properties it is an area that merits further investigation. This area, coupled with the development of new block copolymers containing LCP units, which also undergo microphase separation, could open up a new area of research on materials with unique properties. Thus, new applications in both the areas of devices and materials remain to be explored that will make use of the self-assembling properties of LCPs.

References

1. Reinitzer, F. *Monatsh. Chem.* **1888**, *9*, 421.
2. Lehmann, O. *Z. Phys. Chem.* **1889**, *4*, 462.
3. Friedel, M. G. *Ann. Phys.* **1922**, *18*, 273.
4. Meyer, R. B.; Liebert, L.; Strzelecki, L.; Keller, P. *J. Phys. (Paris) Lett.* **1975**, *36*, 69.
5. Brostow, W. In *Liquid Crystal Polymers*; Collyer, A. A., Ed.; Elsevier: Amsterdam, Netherlands, 1992.
6. Demus, D.; Richter, L. *Textures of Liquid Crystals*; Verlag Chemie: New York, 1978.
7. Gray, G. W.; Goodby, J. W. *Smectic Liquid Crystals*; Leonard Hill: London, 1984.
8. *Side-Chain Liquid Crystal Polymers*; McArdle, C. B., Ed.; Blackie: Glasgow Scotland, 1989.
9. Platé, N. A.; Shibaev, V. P. *Comb-Shaped Polymers and Liquid Crystals*; Plenum: New York, 1987.
10. *Liquid Crystal Polymers*; Platé, N. A., Ed.; Plenum: New York, 1993.
11. *Liquid Crystalline Polymers*; Weiss, R. A.; Ober, C. K., Eds.; ACS Symposium Series 435; American Chemical Society: Washington, DC, 1990.
12. *Liquid Crystallinity in Polymers: Principles and Fundamental Properties*; Ciferri, A., Ed.; VCH Publishers: New York, 1991.
13. Donald, A. M.; Windle, A. H. *Liquid Crystalline Polymers*; Cambridge Solid State Science Series; Cambridge University Press: Cambridge, England, 1992.
14. *Liquid Crystal Polymers: From Structure to Applications*; Collyer, A. A., Ed.; Elsevier: Barking, UK, 1992.
15. Ortler, R.; Brauchle, C.; Miller, A.; Reipl, G. *Makromol. Chem., Rapid Commun.* **1989**, *10*, 189.
16. Eich, M.; Wendorff, J. H. *Makromol. Chem., Rapid Commun.* **1987**, *8*, 467.
17. Nakamura, T.; Uerio, T.; Taris, C. *Mol. Cryst. Liq. Cryst.* **1989**, *169*, 167.
18. Finkelmann, H.; Happ, N.; Portugal, M.; Ringsdorf, H. *Makromol. Chem.* **1978**, *79*, 2541.
19. Finkelmann, H.; Rehage, G. *Adv. Polym. Sci.* **1984**, *60/61*, 99.
20. Chiellini, E.; Galli, G. In *Recent Advances in Liquid Crystalline Polymers*; Chapoy, L. L. Ed.; Elsevier: Amsterdam, Netherland, 1985.
21. Shibaev, V. P.; Platé, N. A. *Adv. Polym. Sci.* **1984**, *60/61*, 173.
22. Jacobs, S. D.; Cergua, K. A.; Marshall, K. L.; Schmid, A.; Guardalben, M. J.; Skerrett, K. J. *J. Opt: Soc. Am. B* **1988**, *5*, 1962.
23. Parker, R. U.S. Patent 3,822,594, 1974.
24. Brown, G. H. U.S. Bureau of Radiological Health, Seminar Program, selected papers. No. 9, 1970, CA76, 19069v; 21 pp.
25. La Marr Sabourin; NASA Accession No. N66–34137, Rept. No. NASA–TM–TX–X–57823, 1966, 12 pp., CA66, 106583v.
26. Nakamura, T.; Ueno, T.; Tari, C. *Mol. Cryst. Liq. Cryst.* **1989**, *169*, 167.
27. Freinzon, Ya; Shibaev, V.; Platé, N. A. *Advances in Liquid Crystal Research and Application*; Bata, L., Ed.; Pergamon: Oxford, England; Akademiai Kiado: Budapest, Hungary, 1950; p 899.
28. Mousa, A.; Freidzon, Ya; Shibaev, V.; Platé, N. A. *Polym. Bull.* **1982**, *6*, 485.
29. Wedler, W.; Talroze, R.; Korobeynikova, I.; Freidzon, Ya; Shibaev, V.; Platé, N. A. *Krystallographiya* **1987**, *32*, 1222.
30. Mousa, A. M.; Freidzon, Ya; Shibaev, V.; Platé, N. A. *Polym. Bull.* **1982**, *6*, 485.
31. Finkelmann, H.; Koldehoff, J.; Ringsdorf, H. *Angew. Chem.* **1978**, *B90*, 92.
32. Finkelmann, H.; Ringsdorf, H.; Siol, W.; Wendorff, S. *Makromol. Chem.* **1978**, *179*, 829.
33. Finkelmann, H.; Rehage, G. *Makromol. Chem., Rapid Commun.* **1980**, *1*, 733; *ACS Polym. Prepr.* **1980**, *24(2)*, 277.

34. Finkelmann, H.; Koldenhoff, J.; Ringsdorf, H. *Angew. Chem. Int. Ed. Engl.* **1978**, *17*, 935.
35. Hessel, F.; Finklemann, H. *Polym. Bull.* **1985**, *14*, 375.
36. Leube, H.; Finkelmann, H. *Polym. Bull.* **1988**, *20*, 53.
37. Gray, G. W.; Toyne, K. J.; Lewthwaite, J. *Mater. Chem.* **1992**, *2*, 119.
38. Cowie, J. M. G.; Hunter, H. W. *Makromol. Chem.* **1990**, *191*, 1393.
39. Cowie, J. M. G.; Hunter, H. W. *Makromol. Chem.* **1991**, *192*, 143.
40. Cowie, J. M. G.; Hinchcliffe, T. T. *Polymer*, in press.
41. Gray, G. W.; Hawthorne, W. D.; Hill, J. S.; Lacey, D.; Lee, M. S. K.; Nestor, G.; White, M. S. *Polymer* **1989**, *30*, 964.
42. Adams, J.; Gronski, W. In *Liquid Crystalline Polymers*; Wiess, R. A.; Ober, C. K., Eds.; ACS Symposium Series 435: American Chemical Society: Washington, DC, 1990.
43. Watanabe, J.; Nagase, T. *Macromolecules* **1988**, *21*, 171.
44. Watanabe, J.; Nagase, T.; Itoh, H.; Istill, T.; Saroh, T. *Mol. Cryst. Liq. Cryst.* **1988**, *164*, 135.
45. Yamagishi, T.; Fukuda, T.; Miyamoto, T.; Watanabe, J. *Mol. Cryst. Liq. Cryst.* **1989**, *172*, 17.
46. Yamagishi, T.; Fukuda, T.; Miyamoto, T.; Ichizuka, T.; Watanabe, J. *Liq. Cryst.* **1990**, *7*, 155.
47. Uematsu, J.; Uematsu, Y. *Adv. Polym. Sci.* **1984**, *59*, 37.
48. Goosens, W. J. A. *J. Phys. Colloq.* **1979**, *40*, 158.
49. Van der Meer, B. W.; Vertogen, G. *Phys. Lett.* **1976**, *A59*, 279.
50. Finkelmann, H.; Stegemeyer, H. *Ber. Bunsen Ges, Phys. Chem.* **1978**, *82*, 1302.
51. Finkelmann, H.; Rehage, G. *Makromol. Chem., Rapid Commun.* **1980**, *1*, 31.
52. Krishnamurthy, S.; Chen, S. H. *Macromolecules* **1991**, *24*, 3481.
53. Clark, N. A.; Lagerwall, S. T. *Appl. Phys. Lett.* **1980**, *36*, 899.
54. Lagerwall, S. T.; Otterholm, B.; Skarp, K. *Mol. Cryst. Liq. Cryst.* **1987**, *152*, 503.
55. Shibaev, V.; Kozlovsky, M.; Beresnev, L.; Binov, L.; Platé, N. A. *Polym. Bull.* **1984**, *12*, 299.
56. Yuasa, K.; Uchida, S.; Sekiya, T.; Hashimoto, K.; Kawasaki, K. *Polym. Adv. Technol.* **1992**, *3*, 205.
57. Sherowsky, G. *Polym. Adv. Technol.* **1992**, *3*, 219.
58. Kühnpost, K.; Springer, J. *Makromol. Chem.* **1992**, *193*, 3097.
59. Kumar, U.; Fréchet, J. M. J.; Kato, T.; Ujiie, S.; Timura, K. *Angew. Chem. Int. Ed. Engl.* **1992**, *31*, 1531.
60. Kapiza, H.; Zentel, R. *Makromol. Chem.* **1991**, *192*, 1859.
61. Prasad, P. N.; Ulrich, D. R. *Nonlinear Optical and Electroactive Polymers*; Plenum: New York, 1987.
62. Chandani, A. D. L.; Hagiwara, T.; Suzuki, Y.; Ouchi, Y.; Takezoe, H.; Fukuda, A. *Jpn. J. Appl. Phys.* **1988**, *27*, L 729.
63. Hiji, N.; Chandani, A. D. L.; Nishiyana, S.; Ouchi, Y.; Takazoe, H.; Fukuda, A. *Ferroelectrics* **1989**, *85*, 99.
64. Yamakawa, M.; Yamada, Y.; Yamamoto, N.; Mori, K.; Hagishi, H.; Suzuki, Y.; Negi, Y. S.; Hagiwara, T.; Kawamura, J.; Orihara, H.; Ishiboshi, Y. *Japan Display '89, Conference Proceedings*; Tokyo, Japan, 1989; Abstract No. 26.
65. Scherowsky, G.; Kühnpost, K.; Springer, J. *Makromol. Chem. Rapid Commun.* **1991**, *12*, 381.
66. Kozlovsky, M. V.; Fodor-Csorba, K.; Bata, L.; Shibaev, V. *Eur. Polym. J.* **1992**, *28*, 901.
67. de Gennes, P. S. *Phys. Lett.* **1969**, *A28*, 725.
68. Finkelmann, H.; Hock, H. J.; Rehage, G. *Makromol. Chem., Rapid Commun.* **1981**, *2*, 317.
69. Zentel, R.; Benaria, M. *Makromol. Chem.* **1987**, *188*, 665.
70. Zentel, R.; Reckert, G.; Bualek, S.; Kapitza, H. *Macromol. Chem.* **1989**, *190*, 2869.
71. Brand, H. R. *Makromol. Chem., Rapid Commun.* **1990**, *11*, 599.
72. Meier, W.; Finkelmann, H. *Makromol. Chem. Rapid Commun.* **1990**, *11*, 599.
73. Hikmet, R. A. M. *Macromolecules* **1992**, *25*, 5759.
74. Mitchell, G. R.; Gus, W.; Davis, F. J. *Polymer* **1992**, *33*, 68.
75. Percec, V.; Zheng, Q. *Polym. Bull.* **1992**, *29*, 501.

Monodomain Liquid-Crystalline Networks by In Situ Photopolymerization

D. J. Broer

Photoinitiated polymerization of pre-oriented mesogenic polyfunctional monomers yields films of densely cross-linked polymers with a monodomain liquid-crystalline molecular order. The morphology of the monomer mesogenic phase is fixed by the rapid photocross-linking, which enables the formation of either nematic or one of the various smectic structures, depending on the phase behavior of the monomers. The desired macroscopic molecular order in the monomeric state can be accomplished by using techniques such as external fields or surface-induced orientation. Surface structures are also produced by replication polymerization, and combinations of these techniques enable the modulation of the orientational director in the plane of the films and also into the third dimension perpendicular to the film surface. Photocross-linking enables lithographic techniques to fix the director selectively by local polymerization, before permanently fixing the total structure. The introduction of chiral centers in the liquid-crystalline monomers produces helicoidally ordered networks, whose pitch of the molecular helix can be adjusted accurately by composition or by polymerization temperature. This fine tuning of 3-D molecular orientation within thin polymeric films makes this process very unique in the world of oriented polymers.

Oriented polymers are of interest because of their anisotropic properties. Well-known features along the axis of orientation are the following:

- high tensile modulus and strength
- low or negative thermal expansion
- high refractive index
- high absorption if anisotropic dye molecules are present

In addition, the well-ordered structures may give rise to nonlinear optical, ferroelectric, and piezoelectric properties. Currently, oriented polymers are obtained by mechanical deformation during or after their processing. However, in applications where the polymer forms an integrated part of a device, such as a thin layer on or between two substrates, the conventional techniques to obtain oriented structures are difficult to use. This difficulty is especially apparent when devices are involved that need more than one localized director orientation within one layer.

The scope of this chapter concerns the control over the molecular orientation into three dimensions by using the photoinitiated bulk polymerization of liquid-crystalline (LC) monomers (1–3), also sometimes denoted as reactive liquid crystals. A schematic two-dimensional representation of the process is shown in Figure 1. The mesomorphic behavior of the monomers designed for this purpose provides the unique monomer features of long range alignment. These monomers are designed by using well-known techniques such as pretreated substrates or external fields. The

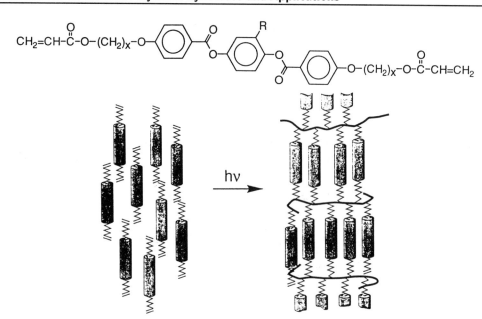

Figure 1. Schematic presentation of the photoinitiated polymerization and cross-linking of LC diacrylates. The substituent R in the chemical formula can be hydrogen *(1, 3)*, methyl *(2, 3)*, chlorine, or other relatively small side-groups *(4)*.

reactive groups enable fixation of the oriented structure in a thin polymer film that will become permanently fixed if monomers with a higher functionality are used to form an LC polymer network.

The history of LC networks dates back to the 1960s, when various authors suggested (5, 6) cross-linking liquid crystals in their mesophase to obtain highly ordered polymers. Thermosetting LC diacrylates yielded 3-D cross-linked polymer networks with strong optical anisotropy and molecular order that was frozen up to the decomposition temperature (3, 7–9). However, the use of high curing temperatures often conflicts with the temperature ranges of the LC phases of many reactive LC monomers. During heating above the melting temperature and processing of the LC phase, local polymerization and cross-linking may start before the desired phase and orientation are achieved; hence, defects become permanently frozen in the resulting network. Therefore, photoinitiation is highly preferable for bulk polymerization and network formation. The monomers can be processed in their LC state until the desired order has been obtained. From this moment on, the molecular arrangement can be fixed very rapidly by exciting a dissolved photoinitiator with light of appropriate wavelength.

The first reports on the bulk photopolymerization of reactive LC monomers were on monoacrylates forming linear LC side-chain polymers (10–13). Fixation of molecular order in most cases does not occur because the formed polymer still exhibits various mesophases in temperature ranges different from those of the initial monomer. Therefore, phase transitions from one LC phase into another, or from the LC phase to the crystal phase, take place during the poly-

merization; consequently, the initial molecular order is lost. Also, phase separation will take place if the LC polymer formed is not soluble in the initial monomer. Many optical defects may be observed in the expected and initially established monolithic order. For this reason, bulk photopolymerization of polyfunctional LC monomers (photosetting LC monomers) became so important. For instance, the free-radical photopolymerization of monolithically ordered nematic LC diacrylates resulted in a stable polymer with the same texture and almost the same degree of molecular orientation (1–3). The use of LC diepoxides (14, 15) and LC divinylethers (16–18) polymerized by photocationic mechanisms essentially led to the same results.

This chapter will review the various possibilities of making structured, oriented polymer films by in situ photopolymerization of pre-ordered LC monomers. I will concentrate on the formation of uniaxially oriented films and their basic process parameters, followed by the formation of superstructures with high degrees of director control.

Uniaxially Oriented Networks

Materials and Principles

A very convenient class of monomers to be photopolymerized in their LC state are the bis[(4-ω-acryloyloxy) alkyloxy] benzoates shown in Figure 1 (1–3). The specific advantages of this class of materials are as follows:

1. a broad mesogenic temperature range (both nematic and smectic LC phases can be achieved)

2. stability at the processing temperatures when small amounts of inhibitor are added

3. very high rate of photopolymerization that forms polymers on a time scale of seconds depending on light intensity and photoinitiator concentration (3)

The LC phases and transition temperatures of examples of these monomers are summarized in Table I.

All molecules have nematic phases at elevated temperatures. In analogy with conventional LC phases, the smectic phase is stabilized by an increased length of the alkylene spacer between the central aromatic unit and the reactive moieties and is destabilized by lateral side-groups. The processing window is, in general, at elevated temperatures; for viscosity reasons the nematic or isotropic phase is most suited for thin film formation, and the nematic or smectic state is used for molecular orientation and successive polymerization. The addition of small amounts of photoinitiator (typically up to a few percent by weight) and inhibitor (50–100 ppm) has no important effect on the transition temperatures. Optimization on the processing properties for specific applications can be done by blending of the various monomers (2). Other formulation possibilities are available, such as blending with lower concentrations of LC or non-LC monoacrylates (20) or acrylates with a higher functionality to tailor the (thermo)mechanical, optical, or electrical properties of the final polymer. Finally, the monomer or polymer properties can be tailored by the addition of nonreactive LC materials to produce plasticized networks or, at higher concentrations of nonreactive LC materials, to obtain LC gels (21, 22).

Macroscopic alignment of the monomers into an LC monodomain can be induced by an external mag-netic field or by pretreated substrates. The temperature-dependent order parameter of the aligned nematic monomers is comparable to that of conventional LC monomers and can be fitted to one master curve when plotted against the reduced temperature [which is the actual temperature divided by the isotropic transition temperature (T_c) expressed in kelvin] as shown in Figure 2. Typical values of the order parameter are between 0.4 and 0.7, and a steep drop to zero occurs near the first order transition at T_c.

When the desired molecular organization has been achieved, the LC monomers are polymerized. A convenient photoinitiator to initiate the free-radical polymerization is α,α-dimethoxydeoxybenzoin (Irgacure 651, CIBA-Geigy). Other free-radical initiators can be applied, but precaution should be taken with respect to evaporation or thermal degradation under the processing conditions. The quantum yield for the chain propagation polymerization and cross-linking reaction is very high, which means that the UV-induced polymerization proceeds on a short time scale despite the use of relatively low-intensity UV lamps such as fluorescent lamps (Philips TL08, TL09, or PL10). Figure 3 shows a typical polymerization curve measured by photo differential scanning calorimetry.

The eventual influence of the molecular organization within the LC phases on the kinetics of polymerization is difficult to analyze because of the very large temperature effects that are common for bulk polymerization. At temperatures below 90 °C the polymerization behavior is dominated by viscosity effects, especially at higher conversions, and these effects give

Table I. Mesomorphic Behavior of Bis[(4-ω-acryloyloxy)-alkyloxy]benzoate LC Diacrylate Monomers

R	x	Monomeric Phase Transitions	Ref.
H	4	Cr 107 N 165 I	3
H	5	Cr 92 N 170 I	3
H	6	Cr 108 (SmC 88) N 155 I	1–3
H	8	Cr 82 SmC 108 N 148 I	3
H	10	Cr 87 SmC 112 N 137 I	3
H	11	Cr 81[a] SmC 114 N 134 I	3
CH₃	4	Cr 80 N 120 I	3
CH₃	5	Cr 93 N 124 I	3
CH₃	6	Cr 86 N 116 I	2, 3
Cl	6	Cr 100 N 109 I	19
OCH₃	6	Cr 66 N 80 I	19
CO·CH₃	6	Cr 60 N 87 I	19

Note: Cr is crystalline, N is nematic, SmC is smectic C, and I is isotropic state. The numbers between the phases are transition temperatures (°C). The phases between brackets are observed in the supercooled state.

[a]Smectic-to-smectic transitions are also observed at 87 and 95 °C.

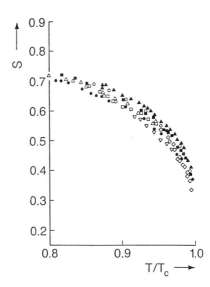

Figure 2. Temperature-dependent order parameters as determined from refractive indices of the various monomers studied. Symbols: R = H with x = 4 (□), x = 5 (△), x = 6 (○), x = 8 (▽), x = 10 (◊); R = CH₃ with x = 4 (■), x = 5 (▲), x = 6 (●).

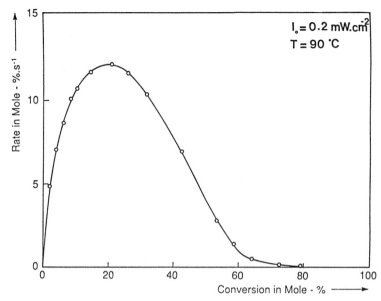

Figure 3. Photoinitiated polymerization of LC diacrylate (Table I: R = H, x = 6) at 90 °C.

the steep increase of the polymerization rate with temperature. At high temperatures, that is, >120 °C, the polymerization is dominated by thermodynamic parameters that ultimately lead to a practically zero rate around 220 °C. In between 90 and 220 °C, and fortunately corresponding to the LC ranges of most monomers, one finds the most optimum conditions for bulk polymerization. At these conditions, the Trommsdorff effect gives high rates almost directly from the onset of the reaction, and conversions of the available acrylate double bonds are ≤80% within seconds and >90% at prolonged irradiation (minutes) or during a short postbake at the same temperatures.

During the process of photoinitiated polymerization, the initial texture of the monomer (monodomain nematic or smectic, depending on the monomer and temperature) is preserved. X-ray measurements confirmed (23) this macroscopic observation by showing the formation of nematic and smectic networks after polymerization from nematic or smectic monomers, respectively. When the polymerization was carried out in the nematic state close to the transition to the smectic state, some local smectic molecular ordering (denoted as cybotactic) was deduced from the small angle X-ray scattering (SAXS) diffraction pattern. The change of the order parameter in the nematic state during polymerization was recorded by dichroic measurements of the various monomer samples (3, 24).

Figure 4 shows that at relatively high reduced temperatures close to T_c the order parameter increases during polymerization, whereas at lower temperatures the order parameter does not change. When polymerization occurred above T/T_c no long range ordering was found. In the case of a substituent in the central ring, even a small decrease in the order param-

eter might then be observed at a high initial degree of orientation. In general the films that are formed show an appearance equal to that of conventional aligned liquid crystals: transparent in the oriented state and high birefringence when observed between crossed polarizers. However, the films differed in the thermal stability of the orientation due to the dense cross-linking by the polyacrylate chains formed. In fact, the samples can be heated up to their pyrolysis temperature without a significant loss of orientation.

On the basis of this principle many variations can be generated. For instance, instead of acrylate functional groups, other reactive moieties can be used with essentially the same behavior but with differences in

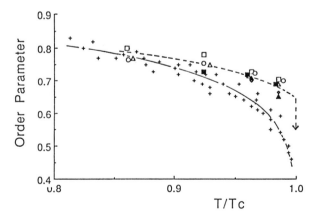

Figure 4. Change of the order parameter during polymerization: (+) from the master curve of the monomers; the other symbols are from the polymerized samples corresponding to R = H with x = 4 (□), x = 5 (△), x = 6 (○), x = 8 (▽), x = 10 (◊); R = CH₃ with x = 4(■), x = 5 (▲), x = 6 (●).

mesogenic properties and polymerization rate. As indicated previously, the use of vinylethers (*16–18*) and epoxides (*14, 15*) was described. In addition to oriented nematic or smectic reactive mesogens, other types of LC order can be frozen-in by using cholesteric (*25–27*), chiral smectic (*28*), or discotic monomers.

Process Features

Because of their low viscosity (0.05 to 1 Pa•s at the temperatures of interest) the LC monomers can be processed in the absence of solvents. A convenient application process is suction between two heated plates, distanced at 1–100 µm by spacers to make use of the capillary forces. Alternatively, other film-forming techniques can be used such as squeezing between two substrates, spreading with a doctor blade, wire-wound rod coating, and screen printing.

For the formation of a monodomain oriented LC structure with a low defect concentration, use can be made of the techniques developed for the conventional LC monomers. The homogeneous, planar, uniaxial, oriented monodomains with the director parallel or almost parallel to the substrate can be obtained at (or preferably between) substrates that are coated with a thin film of tissue-rubbed polyimide. Perpendicular or so-called homeotropic orientation is established at or between substrates that are modified with a monolayer of surfactant that directs its apolar tails perpendicular to the substrate surface (*29*). Director angles between 0 and 90° can be obtained by rotation of the flat substrate in a magnetic field or by specially designed polyimides yielding a planar orientation with a high pretilt. Because of the small negative dielectric anisotropy ($\Delta\varepsilon$) of the monomers presented in Table I, electrical field alignment can only be applied after modification of the monomer formulation with an amount of LC material (reactive or nonreactive) that has a high positive $\Delta\varepsilon$, such as a cyanobiphenyl-based material (*20*). For such a material, substrates can also be used that are provided with a UV-transparent conductive coating such as indium tin oxide to orient the monomer mixture by an electrical field. This freedom in the selection of one of the available orientation techniques provides the flexibility in the choice of the orientation direction. A molecular orientation can be controlled not only within the plane of a film as with conventional techniques but also into every other desired direction.

The use of stable UV-sensitive initiators permits processing and orientation without premature polymerization, which may diminish the desired molecular organization. As soon as the desired orientation is achieved, a short UV exposure freezes the structure into a stable oriented network. Fast polymerization proved to be essential for the whole process because orientation disturbing effects such as phase transitions

or separations could be suppressed on kinetic grounds. Moreover, the use of UV light or other radiation provides the possibility to polymerize only a localized area in the film by using masking or interference techniques. This process means that the orientation can still be altered in the nonpolymerized area and can be fixed successively by a second (or third) irradiation step.

Materials Properties

The mechanical properties of the LC networks obtained by the photopolymerization process were described previously (*8*). An increase in both strength and modulus along the orientation direction was reported, but in contrast to the LC main-chain polymers, these increases do not seem to be high enough to render these compounds suitable for demanding mechanical applications. More interesting is the typical thermomechanical behavior with an almost zero thermal expansion along the orientation direction over a broad temperature range (*30*). This property is demonstrated in Figure 5 and can be used for adhesives and encapsulates especially when thermal stresses between a sensitive inorganic substrate and the organic adhesive or encapsulate should be avoided. For the encapsulate applications a low thermal expansion is needed into the two directions in the plane of the film, and this expansion could be obtained by using the cholesteric materials (*26*), as discussed subsequently.

The anisotropic optical properties of the LC networks in combination with their transparency is of interest for the manufacturing of various optical elements (*3, 15*). The large difference in refractive index parallel and perpendicular to the orientation director, denoted as n_e and n_o, respectively, is demonstrated in Figure 6. This difference is particularly suited for vari-

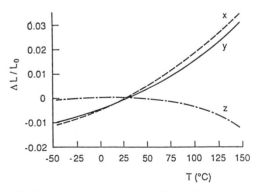

Figure 5. Dimensional change of a film of LC network (R = H, x = 6) measured in three directions relative to the values at room temperature: z is along the orientation in the plane, y is perpendicular to the orientation in the plane, and x is perpendicular to the orientation but perpendicular to the plane of the film. (Reproduced with permission from reference 30. Copyright 1991.)

ous types of polarizers, retarder plates, anisotropic waveguides, and diffraction and selective wavelength filters. Depending on the molecular structure and the polymerization temperature of the monomers, the birefringence of the LC networks is roughly between 0.1 and 0.2 and is subjected only to small reversible changes when heated up to of 200 °C.

The oriented structures obtained from photo-cross-linking of LC monomers are very stable and are insensitive to temperature variation or to the presence of external electric or magnetic fields. In case a switching behavior of the networks is desired for such processes as the manufacturing of nonmechanical shutters based on light scattering, blends of reactive and nonreactive LC materials were proposed for fast partial rotation of the nonbounded molecules by the application of an electric field.

Director Control in Three Dimensions

Combined Orientation Techniques

In the preceding paragraph, uniaxially oriented LC networks were described which are obtained from an orientated monomer structure by using one single orientation technique. What I demonstrate in this section is that by using a combination of orientation techniques one can build interesting structures with an additional control over the orientation direction. There are two different possibilities:

- using two interfaces with a different sense of orientation, such as two differently treated substrates
- using two different techniques, pretreated substrates in combination with an external field, where one technique locally overrules the other

A basic example of the first possibility is the use of two polyimide-coated substrates that are rubbed unidirectionally with the mutual rubbing directions under an angle in the plane of the film. When the LC monomer is applied as a thin film between these substrates the molecules near the two interfaces are aligned following the direction of rubbing in the differently rubbed substrates. In between the two interfaces and over the cross-section of the layer, there will be a gradual twist from one orientation direction to the other. To prevent different handedness of the molecular rotation, which might result in domain (and therefore defect) formation, a small amount of chiral liquid crystal can be added as a dopant to direct the rotational twist. Figure 7 shows a scanning electron microscope picture of a cryofractured surface of such a nematic structure, which is twisted over 90°. The fracture texture is determined by the molecular orientation, which enables visualizing the orientation direction and changes therein. By this technique, the director of orientation is observed to be rotated over 90°. If the rotation takes place over a film thickness that is not too small, the resulting film rotates the plane of polarization of transmitted polarized light, which in fact means that the film works as an achromic half-wave retarder. At smaller thicknesses the wave function becomes more complicated, but the molecular rotation still can be proven by using the

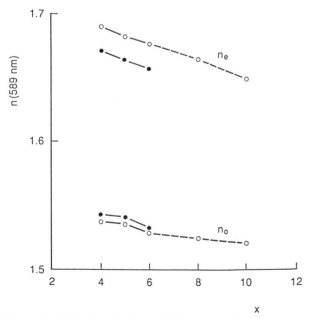

Figure 6. Refractive indices of LC network films as a function of spacer length x (compare Table I). The LC networks are based on monomers with R=H (○) and R=CH₃ (●). (Reproduced with permission from reference 3. Copyright 1991.)

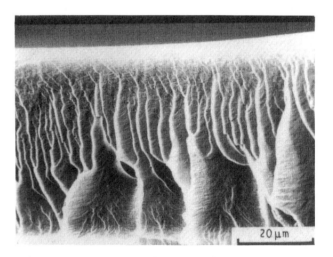

Figure 7. Scanning electron micrograph of a cryofractured cross-section of an LC network aligned at the two interfaces with a planar interfacial alignment under a mutual angle of 90° within the plane of the film. (Reproduced with permission from reference 29. Copyright 1992.)

Gooch and Tarry technique (*31*) to record UV-vis spectra between two polarizers (*25, 26*).

In this first example the director twist is established in the plane of the film. Another interesting structure is obtained when a planar orienting substrate is combined with a homeotropic orienting substrate. In this case the molecules also rotate over 90°, but this time in the plane of the cross-section. This result is demonstrated in Figure 8, which shows the perpendicular orientation at the surfactant-treated bottom substrate and the homogeneous planar orientation at the top obtained from rubbed polyimide. Figure 8 shows that the actual rotation is less gradual than is the case in Figure 7 and takes place over a thickness of less than 10 μm in the 60-μm thick film. Obviously, structures demonstrated in Figures 7 and 8 are very hard to produce by any other polymer-orienting technique.

Instead of combining the rubbed-polyimide substrate with a homeotropic orienting second substrate, a comparable effect can be achieved by using a surfactant-like LC additive (*32*). In this case the planar orientation is obtained at the rubbed-polyimide interface, and a homeotropic orientation is obtained at the interface with air. The surfactant LC additive can be simply a cyanobiphenyl with an alkyl tail directed into the air interface. Because the transition from 0 to 90° cannot take place over an infinite small distance determined by the elastic constants of the LC mixture, intermediate angles of rotation can be found at smaller thicknesses. At these thicknesses, the planar orientation at the polyimide interface is fixed and the director at the air interface varies with thickness. The thickness can

be adjusted easily by using spin-coating techniques from solutions with an adjusted concentration.

The second possibility is the use of two different orientation-inducing methods, where one method locally overrules the other. One simple example, which in fact is also used in display applications based on nonreactive LCs, is based on glass substrates that are (partially) covered with a transparent indium tin oxide electrode and subsequently coated with rubbed polyimide. The rubbed polyimide provides the homogeneous planar alignment. However, on the location of an applied electrical field, the molecules are aligned homeotropically (Figure 9). This structure is established in the monomer phase and is fixed by one single irradiation step. An example of such a film is shown in Plate 1. At angles below 45° between crossed polarizers, the anisotropic homogeneously aligned area is transparent. On the location of the characters, the monomer is aligned along the field lines of an applied electrical field. When observed head-on, the homeotropic structure becomes optically isotropic and black between the polarizers. The film can be heated to high temperatures without losing its stored information.

Cholesterically Ordered Networks

Another type of 3-D control over the molecular orientation can be achieved by the use of chiral additives (*25, 26*) or by modification of the reactive LC materials with chiral centers (*27*). The enantiomeric molecules induce a molecular rotation perpendicular to the long molecular axes and produce structures known as cholesteric or chiral nematic phases. When such materials are applied between rubbed substrates, LC networks are formed with a superstructure describing a molecular helix perpendicular to the film surface. The structure (i.e., the pitch of the molecular helix, *p*, or the cumulative angle of molecular rotation, ψ) can be very

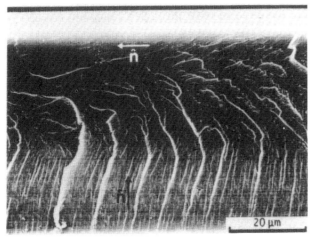

Figure 8. Scanning electron micrograph of a cryofractured cross-section of an LC network aligned at the two interfaces, one planar and the other perpendicular, resulting in a tumbling of the molecules over 90° within the cross-section of the film. (Reproduced with permission from reference 29. Copyright 1992.)

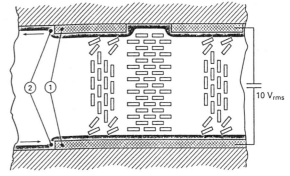

1. ITO electrodes
2. Rubbed polyimide (arrows indicate rubbing direction)

Figure 9. Schematic presentation of the director modulation in LC networks oriented by a combination of rubbed polyimide and, locally, an electrical field.

controlled by the combination of film thickness, amount of chiral molecules, and the angle in the rubbing directions of the two substrates. Adding relatively small amounts (few percent) of dopants to the molecules listed in Table I produces networks with relatively large pitch. Figure 10 shows a scanning electron microscope image of a cross-section of a helicoidally ordered network where the helix has a pitch of 20 μm. An apparent layer structure is formed, and each layer has a thickness of $p/2$ corresponding to a 180° rotation of the molecules. During polymerization the pitch decreases somewhat and the decrease corresponds to the one-dimensional polymerization shrinkage. During heating or cooling, p changes corresponding to the one-dimensional thermal expansion perpendicular to the film surface (27). During polymerization, heating, and cooling, ψ remains constant. The analogy with an expanding or contracting spring representing the molecular helix was proposed (26).

By using larger amounts of chiral materials, preferably by modification of the spacers of the LC molecules presented in Table I, one can obtain a pitch of the molecular helix on the order of 0.2 μm. Therefore, by blending with the nematic monomers, every pitch between 0.2 μm and infinity can be obtained. A particularly interesting range of pitches are those that have a value corresponding to the wavelength of the light in the network medium. In this case selective reflection of circular polarized light will occur with a wavelength of $\lambda = \langle n \rangle \times p$, where $\langle n \rangle$ is the average refractive index in the plane of the film [i.e., $(n_e + n_o)/2$]. The sense of polarization of the reflected light corresponds with the sense of molecular rotation. An example of such a film is shown in Figure 11 for a cholesteric network with $p = 300$ nm.

Light Modulation

A third way to produce ordered LC networks with a 3-D superstructure is by taking advantage of the photopolymerization enabling fixation of the structure only at a localized area by using lithographic or interference techniques. The most simple example is illustrated in Figure 12. The monomer is oriented homogeneously in a magnetic field and irradiated through a mask. In the next step the sample is rotated with respect to the field lines, and the mask is removed. Only the molecules in the nonpolymerized area are able to follow the field lines. The subsequent flood irradiation also fixes these molecules by yielding a network pattern with a modulated director. Many variations are possible on this theme:

1. The nonpolymerized area can be removed after the first irradiation step to yield an oriented surface relief that, if desired, can be covered again by either a new layer of LC monomer or any other isotropic polymer.
2. The nonpolymerized area can be heated to the isotropic phase and polymerized into nonoriented structures yielding a modulated orientation in an isotropic environment.
3. Rotation in the magnetic field before the second irradiation step is possible into all directions to give no limitation to the angle between the directors of the two different area types.
4. Instead of a magnetic field, treated substrates may be used for the first irradiation step, after which the second irradiation step might be carried out either under the action of magnetic forces or in the isotropic state.

The number of possibilities to control the director into all direction are unlimited. The only limitation is set by the resolution of the process determining the minimal size of the oriented structures formed. Although the limitations have not yet been determined, a definition on a micrometer scale seems feasible. The only intrinsic limitation that is typical for this kind of material seems to be gradient of the director at the boundary between polymerized and nonpolymerized material. Apart from that limitation, the common lithographic limitations such as internal reflections of the UV light, masker definition, and diffraction at mask edges play a role.

An interesting alternative to the masking technique is the use of holographic techniques. Interference determines the spatial distribution of UV light during polymerization yielding local polymerization

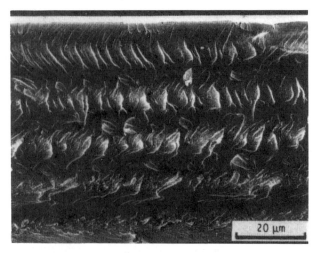

Figure 10. Scanning electron micrograph image of a cryofractured cross-texture of a helicoidally ordered network with a pitch of 20 μm. (Reproduced with permission from reference 29. Copyright 1992.)

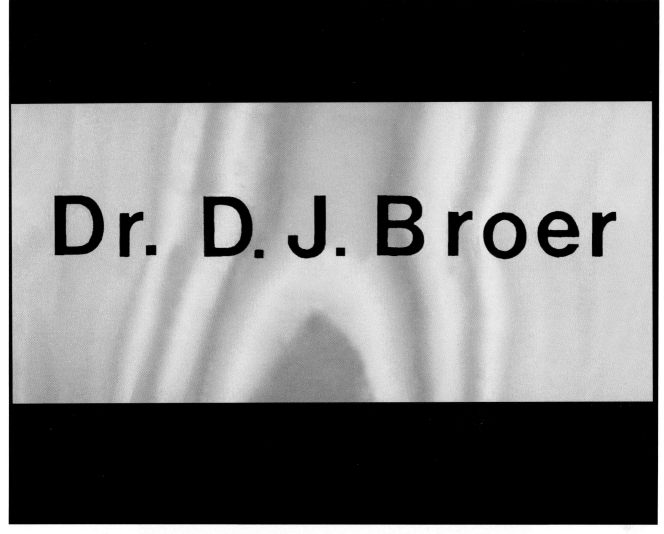

Plate 1. An example an LC network produced by the technique shown in Figure 9 visualized between two crossed polarizers. The polymerized monomer mixture contained the diacrylate from Figure 1 with R = H and x = 6 and modified with a cyanobiphenyl mixture (E7-Merck) to obtain the positive Δε.

at the area of constructive interference either as layers in the plane of the film or as planes over the cross-section (*33*). The resulting strong phase diffraction after switching of the nonpolymerized area by means of an electrical field points to a pattern resolution in the micrometer region.

Another interesting possibility is to use modulated light to polymerize cholesteric monomers. Because the pitch of the molecular helix can be adjusted by means of the polymerization temperature, the area can be fixed with different pitches next to each other. This ability opens the possibility to make color patterns with different reflection wavelengths when pitches are selected in the visible region. Figure 13 is an example of a cholesteric monomer blend that was UV irradiated through a mask with 100 μm lines at 30 °C and subsequently flood irradiated at 60 °C to yield lines with a pitch of 346 and 385 nm, respectively. The

respective reflection wavelengths are, therefore, in blue and green.

Replication Polymerization

As is the case with photoinitiated polymerization of isotropic acrylates (*34, 35*), molding techniques can be used to structure the surface of the anisotropic LC networks. In a practical example, both the mold and the substrate are provided with an orientation-inducing polyimide layer. The monomer is pressed in the nematic state between mold and substrate and subsequently photopolymerized. In the next step the mold can be removed leaving an exact, but negative, replication of its surface in the surface of the LC network. Plate 2 shows a surface tunneling microscopy picture of such a structure, where the orientation direction of the molecules is perpendicular to the saw-tooth. A

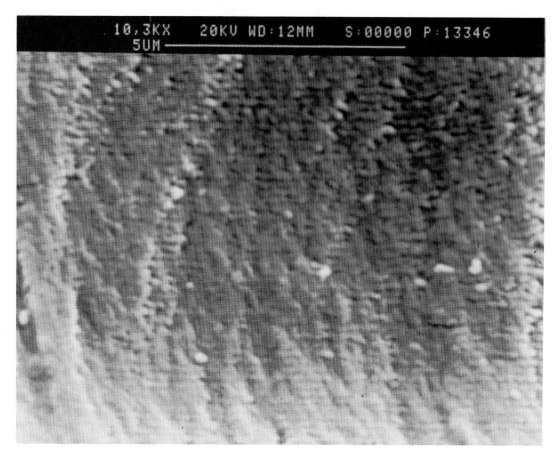

Figure 11. Scanning electron micrograph image of a cryofractured cross-section of a helicoidally ordered network with a pitch of 300 nm.

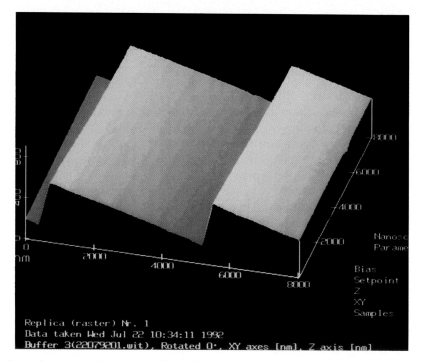

Plate 2. Surface tunneling microscope pictures of oriented LC network polymerized against brass mold treated with rubbed polyimide.

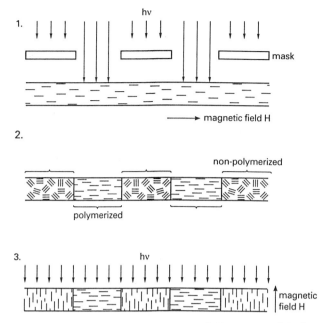

Figure 12. Schematic presentation of director modulation by means of localized irradiation: 1, orientation in a magnetic field and irradiation through a mask; 2, sample rotation with respect to the field lines and mask removal; 3, subsequent flood irradiation.

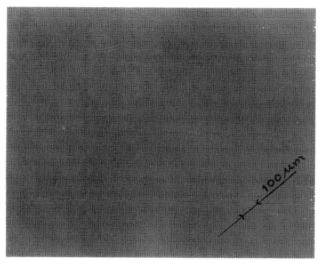

Figure 13. Lines of different reflection color obtained by local polymerization at different temperatures. The alternating lines reflect 540 and 620 nm, respectively.

typical application of structures like this is refraction-based beam splitting of light into its polarization components in a Fresnel-type configuration. If desired the structured, oriented LC network might be covered with a new layer of isotropic or anisotropic monomer, which can be polymerized against a structured or nonstructured mold. Alternatively, a structured isotropic network can be formed by replication, and the network subsequently is covered by a thin orientation-inducing polyimide layer and the anisotropic LC network. Both techniques yield complicated laminates, however, with an exact control over the positioning of the molecules into three dimensions.

Conclusions

In situ photopolymerization of polyfunctional LC monomers is a versatile technique that is able to produce oriented structures with a spatial control over the molecular orientation. The molecular organization within the networks corresponds to those found in the current low-molar mass liquid crystals; however, the mesogenic parts are completely immobilized by the polymer chain grown at both sides of the molecules and therefore have become thermally stable up to the degradation temperature. Macroscopic molecular alignment can be established by using known techniques such as rubbed substrates, surfactant-treated substrates, magnetic fields, and, after a small adaptation of the formulation composition, electrical fields.

These techniques can be combined to make complicated structures that are hard to make by other techniques. The use of light to induce polymerization gives an additional spatial control by using masking or interference techniques. Sub-micrometer structures with modulated directions of molecular orientation seem to be feasible. Replication of specially treated molds introduces additional possibilities to build up laminated structures with different types of molecular organization. Chemical modification of the monomers brings in the helicoidal type of order where the pitch of the molecular helix can be varied locally by means of polymerization temperature and masking irradiation. This combination of possibilities to control the molecular organization in three dimensions makes the technique unique in the world of oriented polymers. The films produced by this process combine a high optical transparency with anisotropic optical, (thermo)-mechanical, and electrical properties; therefore, these films are useful in a variety of applications, especially in the fields of optics and electro-optics.

Acknowledgment

I wish to express my thanks to I. Heynderickx, R.A.M. Hikmet, J. Lub, G.N. Mol, and C.M.R. de Witz for valuable discussions, advice, and many experimental contributions.

References

1. Broer, D. J.; Boven, J.; Mol, G. N. *Makromol. Chem.* **1989**, *190*, 2255.
2. Broer, D. J.; Hikmet, R. A. M.; Challa, G. *Makromol. Chem.* **1989**, *190*, 3201.

3. Broer, D. J.; Mol, G. N.; Challa, G. *Makromol. Chem.* **1991**, *192*, 59.
4. Wendorff, J. H. *Liquid Crystalline Order in Polymers;* Blumstein, A., Ed.; Academic: Orlando, Fl, 1967.
5. DeGennes, P. G. *Phys. Lett.* **1969**, *28A*, 725.
6. Strzelecki, L.; Liebert, L. *Bull. Soc. Chim. Fr.* **1973**, *2*, 597, 605.
7. Bouligand, Y.; Cladis, P.; Liebert, L.; Strzelecki, L. *Mol. Cryst. Liq. Cryst.* **1974**, *25*, 233.
8. Clough, S. B.; Blumstein, A.; Hsu, E. C. *Macromolecules* **1976**, *9*, 123.
9. Arslanov, V. V.; Nikolajeva, V. I. *Vysokomol. Soedin. Ser. B* **1984**, *26*, 208.
10. Shannon, P. J. *Macromolecules* **1984**, *17*, 1873.
11. Broer, D. J.; Finkelmann, H.; Kondo, K. *Macromol. Chem.* **1988**, *189*, 185.
12. Hoyle, C. E.; Chawla, C. P.; Griffin, A. C. *Polymer* **1989**, *30*, 1909.
13. Hoyle, C. E.; Chawla, C. P.; Griffin, A. C. *Mol. Cryst. Liq. Cryst.* **1988**, *157*, 639.
14. Broer, D. J.; Lub, J.; Mol, G. N. *Macromolecules* **1993**, *26*, 1244.
15. Jahromi, S.; Lub, J.; Mol, G. N. *Polymer* **1994**, *35*, 621.
16. Hikmet, R. A. M.; Lub, J.; Higgins, J. A. *Polymer* **1993**, *34*, 1736.
17. Andersson, H.; Gedde, U. W.; Hult, A. *Polymer* **1992**, *33*, 4014.
18. Johnson, H.; Anderson, H.; Sundell, P. E.; Gudde, U. W.; Hult, A. *Polym. Bull.* **1991**, *25*, 641.
19. Geibel, K.; Hammerschmidt, A.; Strohmer, F. *Adv. Mater.* **1993**, *5*, 107.
20. Hikmet, R. A. M.; Broer, D. J. *Polymer* **1991**, *32*, 1627.
21. Hikmet, R. A. M. *Liq. Cryst.* **1991**, *9*, 405.
22. Hikmet, R. A. M. *Mol. Cryst. Liq. Cryst.* **1991**, *198*, 357.
23. Hikmet, R. A. M.; Broer, D. J. *Integration of Fundamental Polymer Science and Technology;* Kleintjes, L.A.; Lemstra, P.J., Eds.; Elsevier: Amsterdam, Nethelands, 1989.
24. Heynderickx, I.; Broer, D. J.; van de Boom, H.; Teesselink, W. J. D. *J. Polym. Sci. Polym. Phys. Ed.* **1992**, *30*, 215.
25. Broer, D. J.; Heynderickx, I. *Macromolecules* **1990**, *23*, 2474.
26. Heynderickx, I.; Broer, D. J. *Mol.Cryst. Liq. Cryst.* **1991**, *203*, 113.
27. Lub, J. Broer, D. J.;Hikmet, R. A. M.; Nierop, K. G. J. *Liq. Cryst.* **1995**, 18, 319.
28. Hikmet, R. A. M. *Macomolecules* **1992**, *25*, 5759.
29. Heynderickx, I.; Broer, D. J.; Tervoort-Engelen, Y. *J. Mater. Sci* **1992**, *27*, 4107.
30. Broer, D. J.; Mol, G. N. *Polym. Eng. Sci.* **1991**, *31*, 625.
31. Gooch, C. H.; Tarry, H. A. *J. Phys. D.* **1975**, *8*, 1557.
32. Hikmet, R. A. M.; de Witz, C. .M. R. *J. Appl. Phys.* **1991**, *70*, 1265.
33. Zhang, J.; Sponsler, M. *J. Amer. Chem.Soc.* **1992**, *114*, 1506.
34. Zwiers, R. J. M.; Dortant, G. C. *Appl. Optics* **1985**, *24*, 4483.
35. Kloosterboer, J. G.; Lippits, G. J. M. *J. Radiat. Curing* **1984**, *11*, 10.

Conducting Polymers

Richard A. Pethrick

Over the last decade a considerable volume of research has been published on electrical conductivity in polymeric materials, and more recently on the application of these materials to a range of technological problems. This chapter reviews the synthesis, structure, and electrical characteristics of different types of conducting polymers. Conventionally filled polymer systems are still a very important group of materials, but in recent years they have been surpassed by intrinsically conducting polymers, which are moving from being an academic curiosity to having real practical potential. Understanding the theories of conduction in intrinsically conducting polymers is crucial if attempts to improve their properties are to be fruitful, and hence a brief discussion of this problem is also included.

Polymers have traditionally been used as insulators. However, in recent years polymers with high electric conductivity were generated and became technologically important. Conducting polymers are used in a range of electronic, medical, and defense applications. Polyethylene, polyvinylchloride, polystyrene, and polytetrafluoroethylene are used as electrical insulators; however, incorporation of conducting fillers in the outer layer of a cable can effectively reduce problems associated with static buildup in insulators. Electrostatic charge generation occurs when materials with different surface potentials or conductivities are brought into contact. Failure to discharge static can lead to personal discomfort, failure of electronic equipment, and even explosions. Charge buildup can be reduced by increasing the surface conductivity. Too slow a rate of discharge leads to high potentials. Too fast a rate of discharge leads to ignition, generation of fire, or less hazardous but equally troublesome, radio frequency emissions.

Surface conductivity can be increased by compounding a conductive carbon-black or metallic filler into the insulating polymer matrix at a level such that particle–particle contacts are infrequent. Carbon black is a very useful filler; however, it suffers two serious disadvantages. First, the curve controlling the composition–conductivity relationship is often so sharp that the intermediate conductivities required for safe charge dissipation are not possible in all cases. Second, carbon black is a very effective pigment, and therefore the resulting filled product is always black. The increase in the conductivity with carbon-black content is associated with the filler achieving a percolation threshold at which the carbon-black particles aggregate to form conducting chains through the matrix. Metallic loaded polymers also suffer from problems of corrosion, generating an insulating oxide, and consequent aging of the electrical performance over a period of time.

Although the concept of intrinsically or inherently conductive polymers (ICPs) was proposed over 40 years ago, only in the past 15 years has their existence been demonstrated practically. Theoretically, a polymer having a conjugated backbone has the capability of exhibiting high electronic mobility and electronic conductivity. NMR studies performed 40 years ago indicated the existence of a magnetic field associated with these mobile electrons in conjugated structures. The key breakthrough in ICPs came as a result of a synergistic interaction between the groups of Shirakawa, interested in synthesis of polyacetylene, and of MacDiarmid, concerned with doping (1–7). ICPs are very different from metallic conductors in that they are highly anisotropic and possess a quasi 1-D structure.

However, the large majority of conducting polymeric materials currently in use are based on the more traditional approach of incorporation of conducting fillers (carbon, silver, or aluminum) into a polymer matrix (8). These materials are technologically more important than the ICPs, although they have received less attention.

To place conducting polymers in context, copper has a conductivity of about 5×10^5 S/cm (equivalent to Ω^{-1} cm^{-1}), and polystyrene has a value of 10^{-18} S/cm. Silicon has a value of 10^{-4}, which on doping can be increased to a value of 1–100 S/cm. Nylon, with its slightly higher protonic mobility, has a value of 10^{-14} S/cm; and mercury has a value of 10^4 S/cm. Undoped conducting polymers have values comparable with those of other insulating polymers ($\approx 10^{-12}$ S/cm), which on doping are increased to $\approx 10^2$ S/cm.

General Classification of Conducting Polymers

Conducting polymers can be classified according to the scheme shown in Figure 1. As a chapter in a book on functional polymers, this review focuses on ICPs. However, a brief description of filled conducting polymers is also included for the sake of completeness.

Filled Polymers

Carbon black, silver, aluminum, and other conducting particulate materials dispersed in the form of fibers, flakes, beads, or needles in thermoplastics and ther-

mosets raise the conductivity to a level at which facile charge migration occurs. Filled polymers can be used for electrostatic screening and a variety of charge-dissipation applications. Highly filled materials, with loading of particles in excess of 20% w/w, produce high levels of conductivity and in certain cases a *positive temperature coefficient* of conductivity (9, 10).

Carbon-black particles can form chained structures in which interparticle contacts allow currents to be carried. A construction generated by filling the gap between two metallic conductors with a conducting carbon-black matrix can be used as a heating tape. When the matrix is heated, it expands and the number of particle–particle contacts will be reduced and a corresponding increase in the resistivity and reduction will occur in the current carried; hence, heating will cease. A current-limited condition is produced and determined by the temperature at which the loss of contacts produced by expansion of the matrix is balanced by the increased conductivity generated by the cooling of the matrix, which increases the contacts. This type of self regulatory or *smart structure* is used commercially in heating pipes and vessels where spot heating could produce hazardous conditions.

The loading level at which conductivity is imparted to a polymer matrix depends on the shape, size, and distribution of particles. The critical conductivity has little to do with the conductivity of the filler, but rather with its size, geometry, and density. For two fillers of equal particle size and shape, the one with the lower density will have a lower critical composition or volume. The higher the aspect ratio, the lower the critical composition: needle-shaped materials are better than flakes, which are in turn more effective

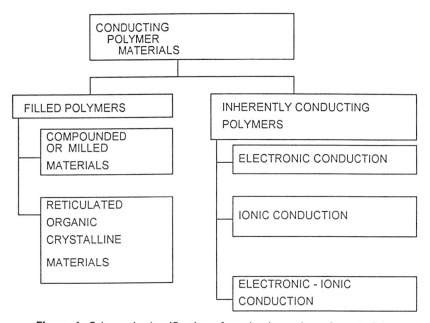

Figure 1. Schematic classification of conducting polymeric materials.

than spheres. In practice, carbon black, which is constituted from spherical submicrometer particles, will tend to aggregate into chains and increase conductivity. Particle–particle contact is not essential if the filler particles are less than 100 nm in diameter, because quantum tunnelling effects may well become operative. A number of authoritative publications on carbon-black-loaded material exist (*11–15*).

Reticulated Organic Crystalline Materials

Molecules such as tetrathiafulvalene (TTF) and tetracyanoquinodimethane (TCNQ) can be crystallized into needle-shaped structures that have a very high intrinsic conductivity (*16*). Crystals are typically of submillimeter dimensions and are of little practical use. Kryszewski (*17*) and Berlin (*18*) grew dendritic structures in polymer matrices that exhibited conductivity levels of 1–10 S/cm at 1–2% w/w loading levels. The TTF–TCNQ molecules form stacks in which charge transfer between adjacent molecules is assisted by their donor–acceptor nature (Chart I).

A wide range of organic molecules are capable of donor–acceptor behavior (*11*). The TTF–TCNQ complex dispersed in polycarbonate at 1% loading is a light-green transparent solid, making it attractive for applications where conductivity, flexibility, and transparency are all important. These materials had not attracted the attention they deserved until relatively recently. By using similar concepts, nonlinear optically active polymer films were generated, and this approach is likely to find greater utility in the future for the generation of materials with conductivities in the range of 10 to 10 S/cm (*see* earlier chapter in this part).

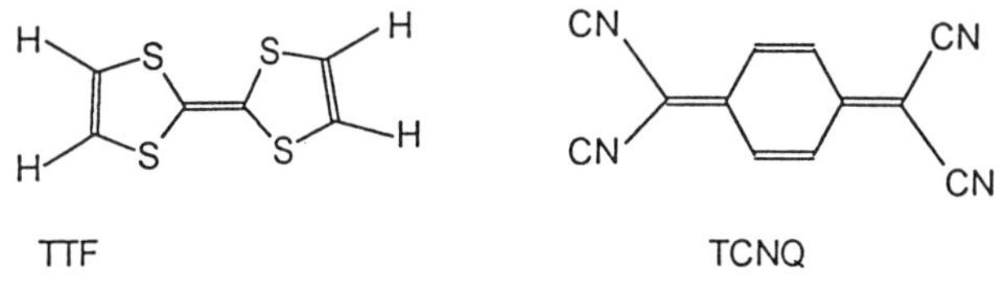

Inherently Conducting Polymers

ICPs can be divided into three types:

- electronically conducting materials (electronic conductivity is dominant)
- ionically conducting materials (ionic conductivity is dominant)
- mixed conductivity materials (ionic and electronic processes both play a significant role)

Because of their application in areas such as batteries and smart window technology, ionic conducting materials have been labelled *polymer electrolytes*. Polyethylene oxide is the most widely studied of the polymer electrolytes. Doped with suitable ionic species, it is capable of exhibiting ionic conductivities of the order of 10^{-4} S/cm at room temperature and 10^{-2} S/cm at 100 °C. Polymer electrolytes have allowed the development of solid-state devices and form an important

Chart I. Molecular structures of tetrathiafulvene (TTF) and tetracyanoquinodimethane (TCNQ).

component in batteries used in heart implants, smart windows, and a range of applications where ionic rather than electronic conductivity is important. Ionically doped polyethylene oxide, polyphosphates, and polytungstates all have applications in this area. Conductivity is achieved by migration of ionic dopants in a manner analogous to a molten electrolyte. A considerable volume of research on this topic was compiled by Vincent (*19*).

Electronically conducting materials can be divided into two groups:

1. Polyacetylene, polypyrrole, polythiophene, polyphthalocyanines, polyphenylvinylene, and polyphenylsulfide are all dominantly electronically conducting and rely on their conjugated structure to facilitate electronic mobility.
2. Polyaniline and one or two related systems behave in a somewhat different manner by exhibiting conductivity that is best described in terms of both ionic and electronic mechanisms.

In both these systems the conductivity is a consequence of a highly conjugated backbone structure that is capable of being doped to yield a high bulk conductivity on doping.

Theory of ICPs

Polyacetylene (PA) is the simplest conducting organic material, and central to understanding conductivity in organic polymers is the *soliton* model. This mathematical description of the energy levels in conducting polymers can be applied with suitable modification to any system in which the backbone forms an extended conjugated structure (*20–24*). In the case of polyaniline it is necessary to consider the role of proton migration; however, even in this case the underlying concepts of the soliton model are still correct.

PA is the simplest conjugated polymer conductor, and each unit has three of its four valence electrons in carbon sp^2 hybridized orbitals: Two generate σ-bonds and form the 1-D lattice, and the third is used to form a bond with hydrogen (*25–31*). The remaining π-electron forms the conjugated structure that is a prerequisite of organic conductors. PA can exist in cis or trans isomeric structures, and these structures have nonde-

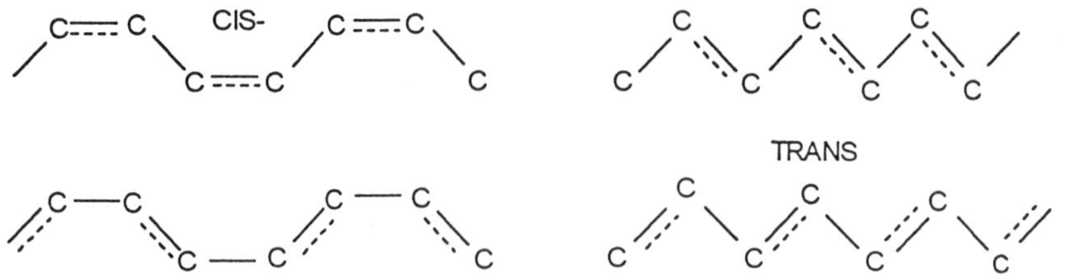

Chart II. Structures and bonding in polyacetylene.

generate energy levels and different conductivities (Chart II).

Pristine films of PA will contain a mixture of the less conducting cis and the more conducting trans forms (30). The trans form is the thermodynamically more stable structure, and isomerization is achieved by thermal annealing at high temperature. Semiconducting and metallic properties are achieved by doping the films by using either gaseous diffusion or electrochemical techniques. p-$[(D^+)_y(CH)]_x$ or n-$[(A^-)_y(CH)]_x$ type material can be obtained depending on whether A and D dopants are acceptor or donor molecules. Semiconducting properties are achieved at approximately 1% dopant levels; higher levels lead to metallic conductivity (32). Doping may increase the conductivity by as many as 10 orders of magnitude, and understanding this process is fundamental to the development of more effective conducting polymers.

Soliton and Related Models

Conductivity in conjugated polymers relies on the generation of accessible energy levels within the band gap, formed by delocalized bonding and antibonding electron sites and associated with conformationally excited states (i.e., *solitons*). PA is a broad-band quasi 1-D semiconductor with an overall band width (W) associated with electronic motion along the chain. The value of W can be estimated from tight-binding theory by eq 1, where z is the number of nearest neighbors and t_0 is the intercarbon transfer energy for an electron.

$$W = 2zt_0 \qquad (1)$$

The t_0 parameter was obtained from spectroscopic studies and has a value of 2–2.5 eV, indicating that W is approximately 9–10 eV. However, because of the effects of bond alternation, *trans*-$(CH)_x$ is a semiconductor with a band gap of about 1.5 eV. The ability of stable *trans*-$(CH)_x$ to sustain free stable solitons as natural nonlinear excitation leads to the importance of PA as a prototype conducting polymer (33–35).

In an idealized PA structure, all the C–C bonds would be identical; however, as a consequence of the Peierls instability, pairs of CH groups move toward each other to form alternating shorter (pseudo-dou-ble) and longer (pseudo-single) bonds. The resulting lowering of the energy and creation of an energy gap at the Fermi level allows conductivity to be achieved. This process can be described theoretically in terms of changes in the resonance integrals associated with the π-σ* excitation and generation of an energy gap. The gap (φ) separates the highest occupied and lowest unoccupied molecular orbital states and corresponds to a difference in the transfer (resonance) integrals for the single and double bonds. Change in the bond structure from cis to trans leads to changes in the electron distribution and energy gap separation.

For the trans structure, the two forms are isoenergetic, and both forms are thermodynamically stable and degenerate. The two cis forms are nonisoenergetic. The *cis*-transoid has a lower energy than the *trans*-cisoid form; therefore, only the lower energy form is thermodynamically stable, and the ground state is nondegenerate. X-ray studies (26, 36) of the trans form have confirmed the effects of the Peierls instability. The basic crystal structure conforms to a $P2_1/n$ space group, where $a = 0.425$ nm, $b = 0.737$ nm, $c = 0.246$ nm, and $\beta = 91.5$ (where β is the angle of the unit cell, which contains two chains). The instability leads to a distortion of the a–c plane of approximately 0.003 nm, which is sufficient to break the symmetry and generate an energy gap. Theoretically, the energy gap is given by eq 2:

$$\phi = 8\beta u_0 \qquad (2)$$

where β is the electron–phonon coupling constant describing the modulation of the π-electron transfer integral, and u_0 is the energy distortion resulting from changes caused by the Peierls instability.

Analysis of optical absorption spectra indicates that φ is between 1.6 and 1.8 eV. Substitution of this value in eq 2 yields a value of 7.1 eV for the electron–phonon coupling constant (β). Good agreement between measured (37) and predicted (34) energy gaps implies that effective electron–electron interaction is relatively weak and does not dominate the conduction process. The degenerate ground state in *trans*-PA leads to the possibility of stable phase domains and the presence of domain walls within the material. As the temperature is raised, a *cis*-$(CH)_x$ chain begins to isomerize, but because each carbon is identical, the isomerization will commence randomly along the chain. The ground state is degenerate, different phases will be produced at random, and where they meet a nonlinear topological excitation, a soliton, will occur (Figure 2, top).

These solitons give rise to localized states in the center of the band gap. The nature of these sites and their interaction with other soliton sites are strongly dependent on the occupancy of these localized states. The soliton is negative and has a spin of 1/2 when the state is singly occupied. If the state becomes unoccupied or doubly occupied, the soliton has no spin. Theoretical studies (*37–40*) indicated that two neutral solitons simply combine and leave no deformation, two identically charged solitons will tend to repel each other, but a charged and neutral soliton will be stable when in close proximity. This neutral–charged soliton pair is called an antisoliton pair, or a *polaron*. The presence of a polaron generates two states in the band gap (a bonding and nonbonding state) that are spaced equally on either side of the Fermi level (Figure 2, bottom).

Analysis of the spatial distribution of these structures (*40*) suggests that the soliton is spread over about 15 CH units and involves a slow variation in both bond length and charge. Whereas the point location on the chain is arbitrary, the motion of the soliton observed by external nuclear double resonance (*41*) involves two discrete states. These two states (Figure 2, top) have different charge locations and result in the soliton being distributed over different lengths of chain segment. The moving soliton may be treated (*24*) rather like a particle that has a kinetic mass of m_g given by eq 3, where M_{CH} is the mass of the C–H fragment, U_o^2 is a potential function, l is the half width of the soliton wave function in units of the primary C–C chain (found theoretically to be about 7 units), and a is the primary bond length.

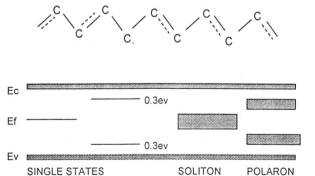

$$m_g = M_{CH}U_o^2/al \qquad (3)$$

When an anion or cation was brought close to the PA chain, it was originally expected to generate a charged soliton of the opposite sign on the chain. This simplistic approach did not account for the generation of the alternate phase. It is now widely believed that at low doping levels, when the centers are widely separated, polarons are generated on the chain. As the concentration is increased the polarons begin to interact, allowing annihilation of two neutral solitons and the liberation of two charged solitons. These two charged solitons move apart and give *free* solitons on the chain (Figure 3). This hypothesis is supported by electron spin resonance (ESR) data, which indicate that at low doping levels the spin of the charge carriers can be detected, but as the doping concentration rises the number of spin carriers fall although the conductivity rises. Therefore, the dopant plays not only a role in the generation of solitons but also in interchain charge migration. In this context even PA may be suggested to have an ionic contribution to its conductivity.

The doping behavior of the system depends on the relative magnitude of the energy required to add an electron or hole, or gap (ϕ), and the energy (E_s) required for creation of the distortion. If $\phi < E_s$, charge-transfer doping will occur by creating free bond excitations; if $\phi > E_s$, then distortion formation would be favored. The continuum model yields eq 4.

$$E_s = 2\phi/\pi \qquad (4)$$

The fact that $E_s < \phi$ ensures that topological distortions are the primary electronic excitations of the *trans*-(CH)$_x$ lattice. The topological nature of these defects makes them very different from the states involved in conduction in conventional 3-D semiconductors. The soliton mid-gap state is generated with no dependence on the nature of the dopant, although there is evidence from external nuclear double resonance measurements that the half-width of the soliton wave function is dependent on the nature of the

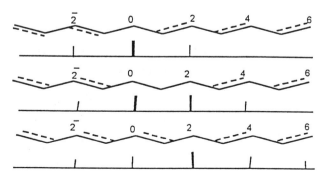

Figure 2. Bonding distribution in a soliton in polyacetylene (top). Energy levels of the soliton and polaron states in polyacetylene (bottom), where E_c, E_f, and E_v are, respectively, the conduction, effective doping, and the valence band levels.

Figure 3. Change of electron distribution with motion of a soliton down the polymer chain.

dopant and may be as long as 46 units at 4.2 K for an ReF$_6^-$ anion (42).

The process by which charge is transmitted in PA was initially believed to be dominated by solitons, but studies have shown temperature (T) dependencies of the form $T^{-1/4}$. These dependencies are consistent with the Mott random range hopping model (43). These results are not consistent with the soliton-dominated hypothesis. Kivelson (44) suggested that at low doping levels phonon-assisted hopping between soliton band states is the predominant mechanism, whereas at higher doping levels soliton transport along the chain is more important. Pulsed photoconductivity measurements on trans-(CH)$_x$ suggest a carrier lifetime of <3 ns (45) and conclude that if solitons do exist, they are merely short lifetime fluctuations of the chain order.

Other optical and photoconductivity data, however, are interpreted as supporting the proposed soliton model. Calculations based on photoexcitation of an electron–hole pair suggest that it evolves into a soliton–antisoliton pair in a time of the order of the reciprocal of the optical photon frequency. Sethna and Kivelson (46) suggested that photon energies less than the Peierls gap can give rise to the direct production of a soliton–antisoliton pair via a standard dipole electronic transition accompanied by a highly nonlinear lattice motion associated with fluctuation of the chain. This process gives rise to photoconduction at energies smaller than the band gap; indeed, the threshold for trans-(CH)$_x$ normally lies at about 1.0 eV. Free carrier generation efficiency rises exponentially above this threshold and then changes to a slow increase with the onset of direct gap transitions. In the case of the cis-(CH)$_x$ the lack of ground state degeneracy means that the soliton pairs are inherently highly unstable and recombine rapidly.

A pair of kinks would require a total energy of E_{tot}, given by eq 5, where E_s is the energy for creation of a single soliton, n is the number of CH units separating the two kinks, and E_o is the energy difference between cis-transoid and trans-cisoid configurations, which is expected to be a small fraction of the full gap [$2\Delta = 2.0$ eV in cis-(CH)$_x$].

$$E_{tot} = 2E_s + n\Delta E_o \qquad (5)$$

As a consequence, two solitons would be confined or bound into a polaron-like entity: the farther apart the greater the energy. As a result, soliton photogeneration does not make a significant contribution to the photoconduction; indeed, there is little generation of free carriers even for photon energies 1 eV greater than the band gap, because the photoexcited carriers are confined by carrier site interactions (47–49).

These photoconduction properties are consistent with the observed luminescence in the respective systems. In cis-(CH)$_x$, a relatively broad luminescence structure peaking at 1.9 eV near the interbond absorption edge, together with a number of multiple Raman lines, is observed, a result consistent with the confinement of photogenerated carriers by interaction with the site. Further, the luminescence turns on sharply for excitation energies greater than 2.05 eV and implies a Stokes shift of about 0.15 eV. In trans-(CH)$_x$, there is the complete absence of any hard edge luminescence, even at the lowest temperature. This absence is consistent with the generation and separation of the soliton–antisoliton pairs.

Doping and High Conductivity Polymers

Semiconducting and metallic properties are only achieved when PA is doped with donor or acceptor materials. A wide range of materials have been used as doping agents. Transition metal halides such as TiCl$_4$, ZrCl$_4$, HfCl$_4$, NbCl$_5$, TaCl$_5$, TaBr$_5$, TaI$_5$, MoCl$_5$, and WCl$_3$ and also nontransition metal halides such as TeCl$_4$, TeBr$_4$, TeI$_4$, SnCl$_4$, and SnI$_4$ all act as dopants (50). In the initial studies, iodine, bromine, and AsF$_5$ were used as dopants, and the maximum conductivity in the trans-PA was higher than in cis-PA. The efficiency of doping is observed to be a function of solvent. Doping in a good solvent indicates that the conductivity is independent of the metal halide but is a function of the ratio of dopant to CH units. At a mole ratio of 0.1:1, the conductivity is 10^3 S/cm. Mössbauer spectroscopy and extended X-ray absorption fine structure (EXAFS) studies of FeCl$_3$ indicated (26) that on doping:

$$2FeCl_3 + e^- \rightarrow FeCl_4^- + FeCl_2$$

Several structural factors may influence the electronic properties; the ideal PA crystal has the chains ordered periodically. This periodicity is perturbed or destroyed on a molecular scale by stacking faults, dislocation vacancies in dopant occupation, chain defects associated with chain ends, irregularities in backbone conformation, and substitution (Figure 4). Doping in PA is assumed to occur randomly; however, clustering is favored because it minimizes the strain energy associated with insertion into the matrix. Iodine is isovolumetric with the CH unit, and insertion of a column of iodine requires less energy than a random distribution of dopant, as indicated by X-ray studies (50–53). The temperature and field dependence of the electrical conductivity and dielectric constant is reproduced if dopant is assumed to be in the form of sheets that are 1.27-nm thick and are intercalated layers (54–59). Maximum conduction is obtained when the structure consists of alternating layers of PA chains; lower conductivities arise when the conducting layers are interspersed with undoped PA chains.

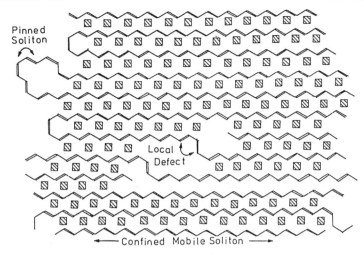

Figure 4. Defect structure of polyacetylene: pinned soliton, local defect, and confined mobile soliton.

PA chains not constrained into crystalline regions are susceptible to chemical reaction. Extensive spectroscopic studies including time-resolved ESR (spin echo and saturation recovery) and NMR (magic-angle spinning sample, cross-polarization, and magic-angle spin echo) experiments identified three different types of defect structures (*59*): mobile spins (neutral solitons), fixed distributed spins (pinned solitons or polarons), and chain defects (*see* Figure 4). The oxidation occurs predominantly in the amorphous regions and can be associated with the accessible pinned solitons (*60, 61*). Local defects will lead to a lowering of the crystalline perfections, and within the highly ordered regions, percolation of oxygen is assumed to be very low.

Environmental Susceptibility

Oxygen is capable of acting as a dopant at low concentrations and leads to an increase in conductivity. This initial doping is also observed to be reversible and indicates that no chemical reaction has occurred (*62*) (Figure 5). However, further increase in the oxygen content has the reverse effect of lowering the conductivity. Oxygen reacts with the localized states in the amorphous regions to form highly reactive peroxides and leads to the generation of furan or related structures. The intrinsic reactivity with oxygen has caused many device applications to fail in the final product stage. However, Naarman (personal communication) developed a method of synthesis for PA that produces a material on doping with iodine. This material has a conductivity about half that of copper, is stable in air for several weeks, and has an initial conductivity of 147×10^3 S/cm. These conductivities are much higher than those that were obtained previously, typically 200 S/cm. The BASF material is more crystalline than that produced by the alternative methods and contains less than 1% of sp^3-hybridized carbon defects. The

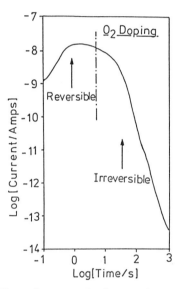

Figure 5. Effect of oxygen doping on the conductivity of polyacetylene: reversible and irreversible doping as a function of time.

improved stability of the polymer composite is associated with its low amorphous content. These highly crystalline materials are difficult to process and are somewhat of academic interest, but they indicate that the PA story is far from being complete.

Mechanism of Charge Migration

A detailed analysis led to the conclusion that the inter-soliton-hopping model is the best representation of the processes that occur (*63–66*). The charged defect migration process at low doping levels is controlled by the concentration of the dopant. The population of the free charges is mainly governed by the interaction energy (E_{ip}) of the ionized-dopant–injected-charge ion pair. The conductivity (σ_{cm} based on charge migration) can be described by eq 6, where D_{ch} is the diffusion coefficient of the charged defects, k is the Boltzmann constant, T is temperature (K), e is the electron charge, n is number of charged defects introduced per chain by each dopant ($n < 1$), x is number of possible states corresponding to E_{ip}, and N_a is the density of carbon atoms.

$$\sigma_{cm} = \frac{e^2}{kT} N_a D_{ch} \left(\frac{1-n}{n} \right) x \exp\left(\frac{-E_{ip}}{kT} \right) \quad (6)$$

In the absence of disorder, the diffusion coefficient of the charged defects (D_{ch}) may be approximated by the diffusion constant of the neutral radicals (D_{neut}), which may in turn be estimated from the expression of a 1-D Brownian motion of domain walls in contact with thermal phonons. Calculations lead to $D_{neut} \cong 2 \times 10^{-2}$ cm²/s. If a_c is the lattice constant, this diffusion coefficient corresponds to a diffusion rate D_n/a_c^2 equal to 10^{14} s⁻¹ (67). This value is in good agreement with the NMR relaxation result of 4×10^{14} s⁻¹. The thermoelectric power (T_{cm}) is given by eq 7.

$$T_{cm} = \frac{e}{k}\left[\frac{E_{ip}}{kT} + \ln\frac{(1-n)x}{n} \right] \quad (7)$$

In the process involving charge transfer between neighboring chains (intersoliton hopping), the energy necessary for electron hopping depends on whether a dopant molecule is near the neutral radical defect. If so, the hopping energy is lowered considerably. The corresponding expression for the conductivity is given by eq 8, where Y_n and Y_{ch} are, respectively, the molar concentrations of neutral and charged defects, N is the number of carbon atoms per conjugated sequence, and $W_{ih}(T)/N$ is a frequency proportional to the fraction of time that the defects are situated within kT of each other.

$$\sigma_{in} = 0.45 \frac{c^2 W_{ih}(T)}{kTN} \frac{\Psi_d}{d_o^2} \frac{Y_n Y_{ch}}{(Y_n + Y_{ch})} \exp\left(\frac{2.78 d_o}{\Psi_d} \right) \quad (8)$$

where σ_{in} is intrinsic conductivity, and c is velocity of light.

The exponential factor reflects the electronic overlap between two defects separated by the mean distance between the dopant molecules (d_o) (eq 9).

$$[\text{dopant}] = (4/3\pi d_o^3)^{-1} \quad (9)$$

The Ψ_d variable is the 3-D averaged electronic decay length according to eq 10, where $\Psi_{d=}$ and $\Psi_d\perp$ are, respectively, the parallel and perpendicular resonance integrals.

$$d_o = (\Psi_{d=} \Psi_{d\perp}^2)^{1/3} \quad (10)$$

At every hop, a spin and charge are transported in opposite directions. Minimization of the ion-pair binding energy (E_{ip}) is achieved by increasing the local dielectric constant near the doping molecule or the distance between the dopant and the charge carrier. A maximum conductivity increase of 10^4 is observed at low sodium doping levels; an estimation of E_{ip} is possible from eq 11, leading to a value of 0.23 eV, which is a lower limit for the ion-pair binding energy (66, 68).

$$\exp(-E_{ip}/kT) = \sigma_{unsolvated}/\sigma_{solvated} = 10^{-4} \quad (11)$$

where $\sigma_{unsolvated}$ and $\sigma_{solvated}$ are conductivities of unsolvated and solvated ions, respectively.

Electronic Characteristics

Conjugated polymers in their pure undoped state are electrical insulators. The conductivity of *cis*-PA is 10^{-9} S/cm and increases to 10^{-5} S/cm on isomerization to the trans form. The conductivity (σ) is proportional to the product of the free carrier (n) and the carrier mobility (μ) according to eq 12, where e is the unit electronic charge (1.6×10^{-19} C).

$$\sigma = en\mu \quad (12)$$

For intrinsic conductivity, the carrier concentration decreases exponentially with increasing band gap and is very low at normal temperatures (68). Doping of conjugated polymers generates high conductivity primarily by increasing the carrier concentration. The term doping is misleading, because the concentrations used are exceptionally high compared with those in inorganic semiconductors, where levels of parts per million are used. In organic systems the dopant can constitute 50% of the weight, and a more appropriate description for these materials might be charge-transfer complexes rather than doped polymers.

Synthesis of Polyacetylene

Initially PA was generated by Zeigler–Natta, Freidel–Crafts, or nucleophilic displacement methods and produced thin smooth films (19, 20). Exposure of smooth surfaces wetted with a solution of the Zeigler–Natta polymerization catalyst, $Al(C_6H_5)_3$:$Ti(OC_4H_9)_4$ (4:1), in toluene (0.1–0.2 M in Ti) at –78 °C to 80 kPa (600 mm Hg) of acetylene for 20 min produces a coating with a metallic sheen, which is easily removed providing a free standing PA film that is not very mechanically stable. Isomerization to the trans form occurs on heating to 170 °C for about 20 min. According to calorimetric measurements the trans isomer is about 3.56 kJ/mol more stable than the cis isomer.

An entirely different approach to the synthesis of PA uses a retro Diels–Alder reaction shown in Scheme I (*69, 70*). The PA is predominantly cis and can be isomerized to the trans form. Derivatives of PA have proliferated; methyl-, phenyl-, and trimethylsilyl-substituted analogues of PA were prepared. The cyclopolymerization of 1,6-heptadieneyne yields a bridged PA derivative comprising alternating endo- and exocyclic double bonds (Scheme I) (*71*).

Although their chemical and morphological properties differ from each other greatly, substituted PAs consistently display conductivities on doping several orders of magnitude less than those of the parent unsubstituted polymer. The substituted PAs, however, do have the advantage of greater flexibility. Copolymers of acetylene in which acetylene units alternate with a second electron-rich unit have been produced and exhibit conductivity (Scheme I). A wide variety of synthetic techniques were developed that have contributed to the evolution of a new generation of conducting polymers (*72*).

Molecular Weights and Morphology of Polyacetylene

Because PA is insoluble in most common solvents, determination of its molecular weight is difficult. Chemical control of the conjugation length was attempted (*73–81*). Transmission electron microscopy indicates that PA exhibits a fibrillar gel structure. The fibers are between 10 and 30 nm in diameter and are independent of the concentration of catalysts used in the synthesis (*81, 82*). Long fibers are obtained when the catalysts are freshly prepared, and the concentration of PA is slightly lower than that normally used to produce the gel-type materials. These extended structures have a *poker-chip* lamellae-type structure. With an aged catalyst, lamellae overgrowths are observed and the chain axis is parallel to the fibril axis. Resonance Raman studies of materials prepared by using a variety of different conditions indicate that the conjugation lengths vary according to the synthetic route used (*26*).

The correlation of molecular weight with conductivity is important, and material with a molecular weight of 10,500, as determined by ^{14}C and ^{3}H radiolabels, was investigated. Conjugation lengths, defined as the number of carbon atoms or monomer units composing an uninterrupted segment of conjugated olefin or aromatic units, were studied (*82*) on a ^{13}C-enriched sample of *cis*-PA. The conjugation lengths indicated that the π-orbital of the delocalized electron is extended over 6.0 nm or 50 carbons (25 monomer repeat units) (*82*). Studies of morphology of both neutral and doped PA indicated a dependence on the conditions of synthesis. Fibrillar and rod morphologies are typically obtained from the Shirakawa process, as indicated by scanning and transmission microscopy. PA obtained by using the Luttinger catalyst, $Co(NO_3)_2$–$NaBH_4$, was characterized as fibers of 5–25-nm diameter incorporated into a fibrillar net and as loose conglomerates of platelets tens of nanometers in diameter and 10.0-nm thick. When grown under shear conditions, PA assumes the form of thin films of highly aligned fibers that consist of lamellae with a periodicity of 11.0 nm. The chain axis appears to be oriented along the fiber direction in the film plane.

On doping with AsF_5, neutron diffraction indicates a new structure is produced with a lattice structure. On a macroscopic level, PA films are composed of bundles of fibers with about 50% void volume; for continuous carrier transport, these interparticle voids

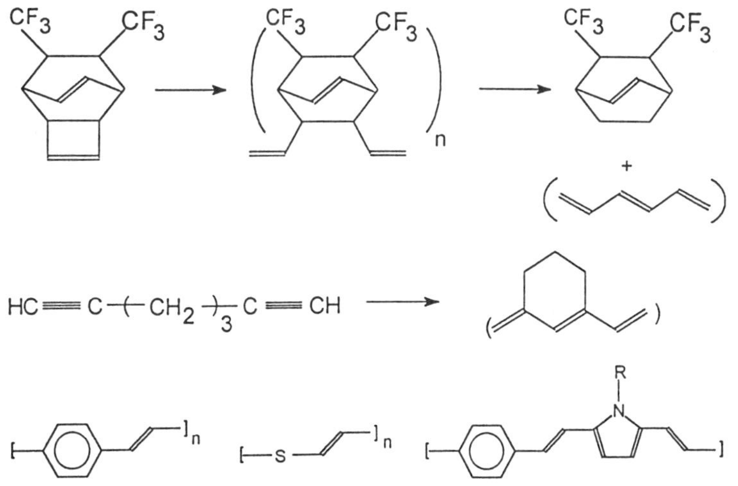

Scheme I. Retro Diels–Alder synthesis of polyacetylene (top), cyclization reaction to form polyacetylene precursor (middle), and the structures of various copolymers of polyacetylene that have been prepared (bottom).

have to be traversed (72). On a microscopic level, the PA chain interrupted by a defect cross-link arising from oxidation or other chemical impurity approximately every 100 carbon atoms necessitates hopping of carriers. Experimental determinations of mobility in PA confirm that interchain and interparticle transport are the limiting factors for conduction. The macroscopic mobility derived from the measurement of PA doped with I_2 is 1.7 cm^2/(V·s) ($n = 1.8 \times 10^{21}$ cm^{-3} and $\sigma = 500$ S/cm) (see eq 12). The microscopic mobility obtained from optical measurements is about 100 cm^2/V·s, a value that was confirmed by magnetoresistance measurements, which allow interparticle effects to be removed from the measurement. The potential conductivity of a perfectly oriented, PA-doped single crystal devoid of defects and interparticle effects should be about 10^5 S/cm, which is comparable with that observed in many metals (83).

Effects of Chain Alignment and Interchain Order

PA is the prototype of a high performance conducting polymer to which all other systems are ultimately compared. PA is capable of being aligned when subjected to elongation. PA films are typically brittle and lack mechanical strength; however, an improved method of tensile drawing of PA films prepared by bulk polymerization using a thermally pretreated variation of the Shirakawa catalyst was developed by Cao and co-workers (84, 85). After equilibration in organic liquids, which act as plasticizers in the deformation process, the films can be stretched to a maximum draw ratio of $\lambda \approx 15$ (80). Although the conductivity parallel to the draw direction increases dramatically, the conductivity perpendicular remains essentially constant. The anisotropy ratio $\sigma_{par}/\sigma_{perp} \approx 250$. The mechanical properties also are improved. Even though the relatively high mobility associated with intrachain transport is increased, one must avoid problems associated with interchain charge transfer to prevent localization inherent to 1-D conductors (85). The electrical conductivity of perfectly oriented doped PA would be approximately 2×10^5 S/cm; however, the practical value is likely to be limited by such things as structural defects and -sp^3 carbons rather than by intrinsic phonon scattering.

In real PA the macromolecules are often coiled and tangled with high volume fraction of amorphous material. Even in polymers with relatively high crystallinity, the structural coherence lengths are rarely greater than 10 nm. This disordered secondary structure leads to local deviations in bond lengths, bond angles, and interchain interactions and limits the effective conjugation length and localization of the π-wave functions (86). There have also been attempts to increase the

order by gel processing; however, the entanglement present in the original gel is never removed by mechanical processing, and the conductivity levels achieved still remain lower than those predicted theoretically. Fibers have been drawn to eight times their initial length, and values of the Young's modulus as high as 35 GPa and tensile strength of 0.7 GPa were reported (81). Doping of the fibers leads to a moderate reduction of the modulus and essentially no loss of tensile strength. Even in polymers with relatively high crystallinity, the structural coherence lengths are rarely greater than 10 nm (80). This process of enhancement of mechanical and electrical properties by drawing also was applied to other conducting polymers with similar significant improvement in their properties with increased morphological perfection.

Other Conducting Polymers

In the 1980s it became apparent that the essential feature for conduction in organic polymers was the existence of an extended conjugated structure capable of being doped. In principle, any of the structures shown in Chart III could exhibit conductivity.

Kossmehl (87) indicated in a review the range of chemistry that can be encompassed by this simple concept, and the number of possible structures runs into the thousands. Selection between different systems relies on the identification of a design philosophy that needs to identify the critical parameters for a particular system and balance these with other factors. For instance, many of these systems are important because of their potential nonlinear optical properties. In this case, optical clarity is an important factor in the total design. These factors will be exemplified in a subsequent section.

Polyphenylene

Poly(p-phenylene) is among the very few polyenic systems where the degree of delocalization along the polymer chain may be varied by small conformational changes (88). In model compounds, the angle between

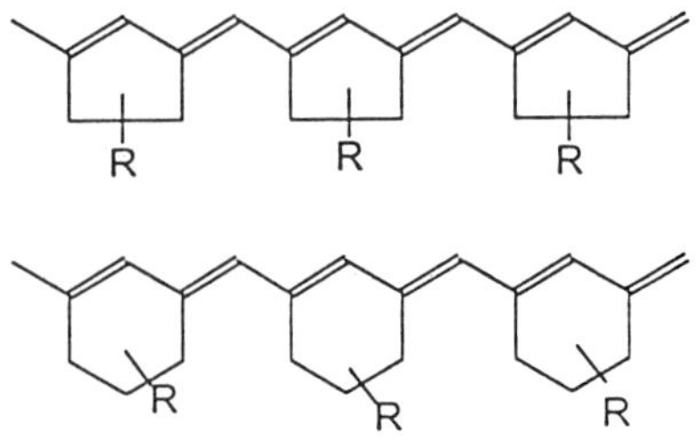

Chart III. Schematic structures for various conducting polymer types where R is any structure that can form the pendant ring and may also include heteroatoms.

the phenyl rings is about 22°. The width of the highest occupied band is strongly affected by this rotational angle and varies from 3.5 eV for 22° to a value of 3.9 eV for coplanar phenyls and to 0.2 eV for perpendicular phenyls. Polyphenylene may be prepared by treating benzene with $AlCl_3$–$CuCl_2$ and an oxidant; and the polymer, unlike *trans*-PA, can exist in two resonance forms (benzenoid and quinoid) (Chart IV) (*89, 90*). These structures are energetically nonequivalent or degenerate, and therefore no mid-gap states are formed.

Stable deformations, however, can be formed when a pair of neutral solitons are energetically constrained to recombine. A neutral and a charged soliton (a polaron) is stable and has a binding energy of 0.03 eV and is spread over 4–5 rings. This deformation is relatively weak and gives rise to two states: a bonding state about 0.2 eV above the valence band and an antibonding state equidistant above. The band gap in poly(*p*-phenylene) is 3.4 eV, twice that of PA (*91, 92*). A bipolaron is also stable but causes a stronger deformation than the polaron and is more compact. The probability of finding the polaron charge within the rings is 0.79, whereas that for bipolaron is 0.91. The bonding energy is also substantially higher at about 0.4 eV. The presence of numbers of polarons or bipolarons on the chains will cause these single levels to broaden out into bands. At high densities the bands will merge with the valence and conduction bands, respectively, but unlike PA, even at the highest densities, they never close the band gap.

Modelling of the interaction of Li and Na atoms with poly(*p*-phenylene) indicates that charge transfer toward the chain is predicted to be about 0.64 eV for lithium and 0.93 eV for sodium, and this charge transfer gives rise to band gaps of about 0.45 and 0.7 eV, respectively (*93*). If this approach is extended to a longer chain with higher doping levels, then the bipolaron bands just merge with the valence and conduction bands. Increasing the doping concentration gives rise to bipolarons that were detected by ESR. Poly(*p*-phenylene) is oxidatively more stable than PA but requires the use of very reactive dopants to achieve semiconducting or metallic properties.

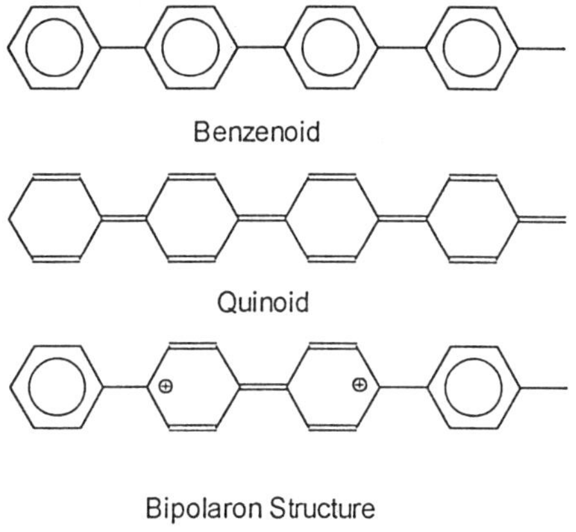

Polypyrrole

Interest in polypyrrole arose primarily because it could be produced electrochemically as a highly conducting polymer film (*94–96*). The original electrochemical synthesis of polypyrrole by Dall'Olio and Dascola (*97*) involves the monomer dissolved with a suitable salt, and polymer is formed at the anode and incorporates anion in a ratio of anion:monomer between 1:3 and 1:4. A number of solvents were tested, including benzonitrile and propylene carbonate, but the most widely used are water and acetonitrile.

A wide range of salts were examined, particularly because the anion has a substantial effect on both the structure and the physical properties of the resulting polymeric films (*98–105*). In general, salts giving facile cathodic reactions such as $AgClO_4$ and $LiBF_4$ are most widely used, but materials such as tetrabutylammonium *p*-toluenesulfate are used, despite the complexity of the cathodic reactions. The use of *n*-alkyl sulfonates of varying chain length has dramatic effects on the resultant conductivity (*99*). The anode and cathode must remain inert under growth and in some cases reduction, and this requirement is satisfied by materials like Pt, Au, and SnO_2 conducting films on glass. Stainless steel, graphite, and germanium for use in attenuated total reflection Fourier transform IR spectroscopy also were used and allowed examination of in situ growth of conducting films (*101–103*). Because the polymers are sensitive to impurities arising from both the original material and the atmosphere, careful control of growth conditions is essential. Electrochemical syntheses of a wide range of related polymers are discussed in relation to their nonlinear optical behavior in this part.

Electrochemical Synthesis of Polypyrrole

This method of synthesis produces films that are doped simultaneously. A wide range of electrolytes have been used, including quaternary ammonium salts, R_4NX, where R = alkyl or aryl, and X = Cl^-, Br^-, I^-, ClO_4^-, BF_4^-, PF_6^-. Also, metal salts, MX, were used, where M = Li^+, Na^+, and As^+, and X = BF_4^-, ClO_4^-, PF_6^-, $CF_3SO_3^-$, AsF_6^-, and $CH_3C_6H_4SO_3^-$ (*106*). The advantages of the electrochemical polymerization of conducting polymers are as follows.

1. Reactions are carried out at room temperature.
2. Thickness of the films can be controlled by adjusting either the potential or current.

Benzenoid

Quinoid

Bipolaron Structure

Chart IV. Various resonance structures for polyphenylene.

3. Polymer films are formed directly at the electrode surface.
4. It is possible to produce homogeneous films.
5. The salts provide conductivity and allow incorporation of an anion into the matrix.
6. It is possible to obtain copolymers and graft copolymers.

Polypyrrole electrochemically doped has been the subject of extensive study over the last decade (107–111). Much of the early research was concerned with attempts to resolve the apparent variability of the values of conductivity that were possible. More recently, better understanding of various factors influencing the synthesis has led to concerns over the extent to which the simple chemistry of the polymer adequately describes the real situation. Even when the polymer structure is well defined the problem of location and nature of the doping species is important. As with other conducting polymers, morphological variations have a profound effect on the conductivity developed in the material.

Variation of Electrolyte Concentration

Films grown with a constant potential and with the salt concentration varying between 0.005 and 0.8 M indicate that the conductivity increases markedly until a limiting value is achieved (Figure 6). In most cases the optimum ratio of salt to pyrrole approaches a value of 0.33 and is slightly dependent on the anion (111).

Variation of Growth Potential

The largest effect on the conductivity of the final film is achieved by variation of the potential used in the growth of the film (111–115). The correct potential can be selected by measurement of the oxidation–reduction potential for the particular electrode system (Fig-

ure 7). The conductivity of the film has a maximum when the preparation potential equals the redox value.

Extensive electrochemical measurements on the pyrrole system indicate that the polymerization process follows Scheme II and involves the formation of a radical cation that combines with a second radical liberating two protons, a process that is electrochemically reversible. The liberation of protons gives rise to a steady fall in the pH of the medium, and the pH may be used to indicate the extent to which the reaction has proceeded. The polymerization process continues because the oxidation potential of the oligomers is lower than that of the monomer, a fact that is easily demonstrated by polymerizing from a solution of dimers or trimers of the original monomer.

The limiting factors on chain length are steric hindrance, which prevents the monomer reaching the end of the chain, and the generation of the α,β-link-

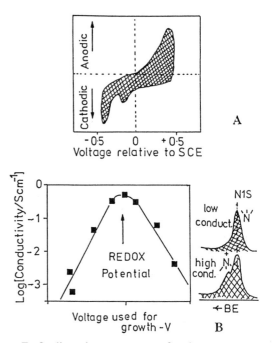

Figure 7. Cyclic voltammogram of polypyrrole grown on gold (A), and variation of conductivity with the growth voltage (B). The maximum conductivity coincides with the redox potential.

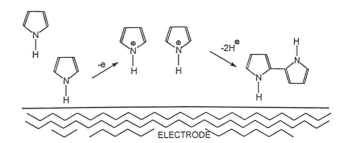

Scheme II. Schematic representation of electrochemical polymerization of pyrrole.

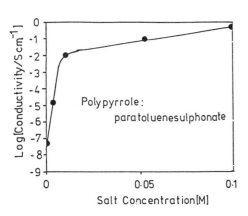

Figure 6. Effect of doping level on the conductivity of polypyrrole films.

ages and saturated α,α-linkages (i.e., when the proton fails to leave the site of the link), which impede the transport of charge along the chain that is itself a fundamental prerequisite for continued polymerization. The charge involved in this process is measured experimentally at between 2.2 and 2.4 and indicates that, in addition to the polymerization that involves 2 charges per monomer, a further positive charge is developed for every three to four monomer units (*115*).

X-ray photoelectron spectroscopic (XPS) and secondary ion mass spectroscopic analysis of films with varying degrees of conductivity indicate that the N 1s peak is split into a doublet (*116, 117*). The intensity of the two peaks changes with conductivity. The lower conductivity sample has a diminished higher bonding energy contribution in comparison with that observed in the more highly conducting sample. The higher binding energy N 1s peak corresponds to that of a quaternized nitrogen and indicates the presence of a greater concentration of the oxidized monomer in the more highly conducting material. The highly conducting state corresponds to one in which approximately one pyrrole ring in three is quaternized (*see* Figure 7).

Growth Rate and Surface Structure

Variation of the anion has major effects on the surface morphology; the surface structure varies from being mirror-smooth to a highly fibrous form (*111*). Electron microscopic examination failed to show any significant difference in these materials. The growth rate of the film is very sensitive to the potential used in the deposition of the polymer (Figure 8). At the edge of electrodes, where an overpotential may exist, significant thickening of the film is often observed. The rate of deposition increases markedly once the oxidation–

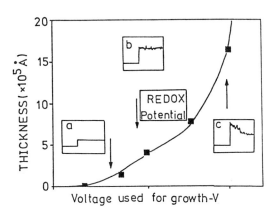

Figure 8. Variation of the rate of growth with growth voltage. Insets: surface profile traces showing the film thickness across an electrode edge (abscissa = distance; ordinate = thickness): a, smooth surface; b, matt surface; c, edge profile resulting from rapid growth as a consequence of the overpotential.

reduction potential has been exceeded, and the material thus deposited will be less conductive. The rapidly deposited and less-conductive films tend to be rough and fibrous. Glossy films are deposited at voltages around or just below the redox potential and under conditions where the current is limited.

Electrochemical Properties

Oxidation of the polymer, even though the process is not electrochemically reversible, is coulombically reversible and may be taken through many cycles with no apparent deterioration in the properties. The oxidation process is characterized by a single peak on the cyclic voltammogram, typically about 150-mV wide. This width is somewhat broader than would be predicted for an ideal process and suggests electroactive site interaction or that all sites are not electrochemically equivalent. As the film thickness increases, the breadth of the peak increases. This result indicates that all sites are not electrochemically equivalent. The ion current (i_p)/ramp rate relationship is consistent with the reaction of surface localized species. The oxidation potential is slightly dependent on the ionic species, but the potential is much more dependent on the material being polymerized; thus, poly(*N*-methylpyrrole) (*118,119*) has an oxidation peak that is 0.65 V more aniodic, relative to that for pyrrole, whereas poly(3-methylpyrrole) (*118, 119*) is slightly less aniodic than pyrrole. Salmon and Carbajal (*120*) examined a series of *N*-ortho-substituted phenylpyrrole polymers and observed a steady change in oxidation potential. Inoue, Yamase, and co-workers found (*120–122*) that at room temperature the polymerization gives a brittle fibrillar structure; whereas at 40 °C a smooth, dense, and flexible polymer was obtained.

Reduction

The reductive part of the cycle is substantially more complex than the oxidative part and may consist of a single broad asymmetric peak or two separate peaks, depending on the monomer and anion (*123, 124*). In systems where two distinct peaks exist, only partial reduction is achieved if the cycle is stopped between the peaks; therefore, reduction may involve at least two processes. One possible process involves deprotonation of the oxidized pyrrole rings. The variation of i_p with sweep rate is linear and indicates reaction of surface localized species. Analysis of the total charge passed and the weight of material are in close agreement with the anions being removed. A polymer with a lower level of π-electron delocalization and a higher resistivity is left, and the second peak may be associated with this process. The color changes from the steel blue/black of the oxidized form to transparent yellow in the reduced form leads to the suggestion that these materials could be used in electrochromic devices, although the changes tend to be slow.

Reduced films are sensitive to chemical oxidation. Even short exposure of neutral polypyrrole films to oxygen results in a reduction in the height and an increase in the width of the anodic peak. Over longer periods of time an irreversible color change to dark blue–black occurs (*109*). A study of the electrochemical properties of polypyrrole by using electron energy loss spectroscopy led to Scheme III (*124*). Nucleophilic attack of OH⁻ on oxidized pyrrole units leads to en-amine-like structure elements in which the π-system of the polymer chain is partially interrupted. The OH⁻ groups are not removed and may react to form hetero groups in the subsequent oxidation cycle destroying the conjugation.

Acid and Alkali Treatment

Treatment with alkali can produce a significant reduction in the conductivity, which can be regained by subsequent treatment with acid (Figure 9). XPS measurements indicate that the conductivity is closely related to the number of oxidized quaternized nitrogens (*111, 117*). Examination of a series of films with different thicknesses indicates that the process is diffusion controlled and is a linear function of the film thickness. Initial deprotonation at the surface will have little effect on the bulk conductivity. Only when the deprotonation reaction permeates throughout the bulk of the sample does the total conductivity of the films decrease markedly. Similarly, the rapid rise in the conductivity on treatment with acid is consistent with surface reprotonation increasing the conductivity. Even though the reprotonation occurs initially only at the surface, the process is sufficiently near to the saturation value of the film for the conductivity to have returned to its original value.

The decrease in the amplitude of change in conductivity on deprotonation for repeated cyclization can be explained in several ways. Exchange of the initial anion, *p*-toluenesulfonate, with chloride anion may lead to a reduction in the conductivity. Oxidation of the polymer by dissolved oxygen will be more effective once the films are reduced and may cause an irreversible loss in the conductivity. Diffusion processes can also irreversibly change the morphology.

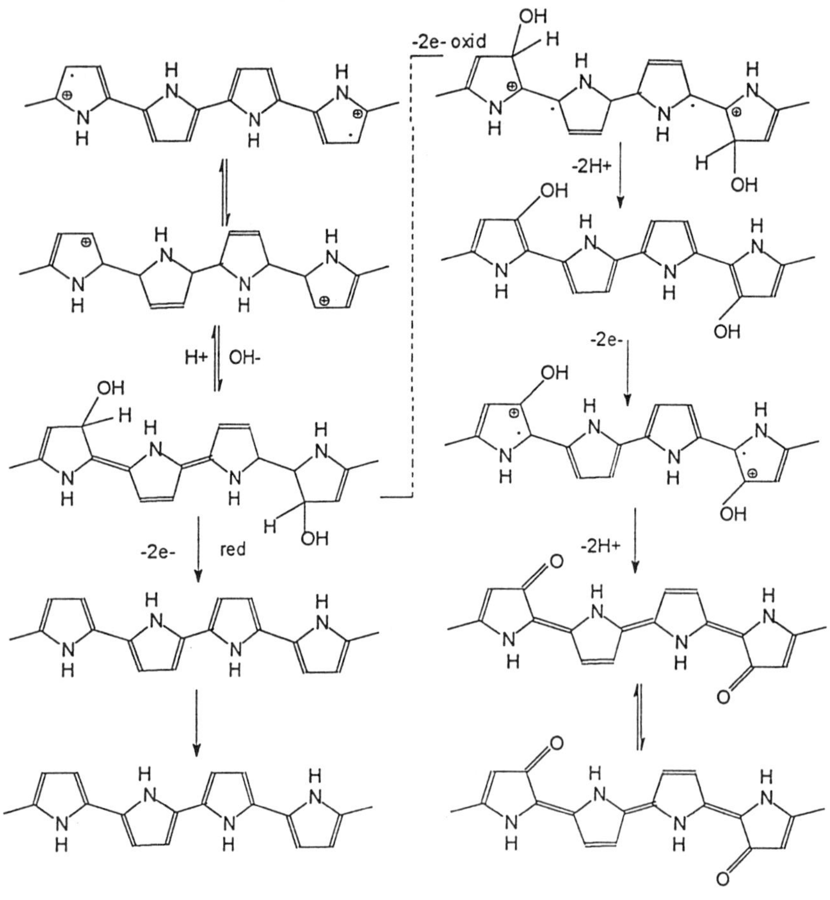

Scheme III. Chemistry of the reduced state of polypyrrole.

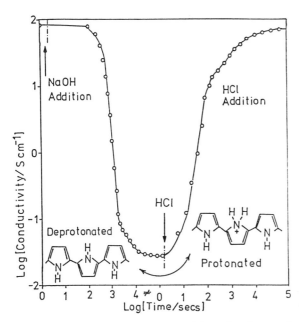

Figure 9. Variation of the conductivity of polypyrrole on deprotonation and reprotonation.

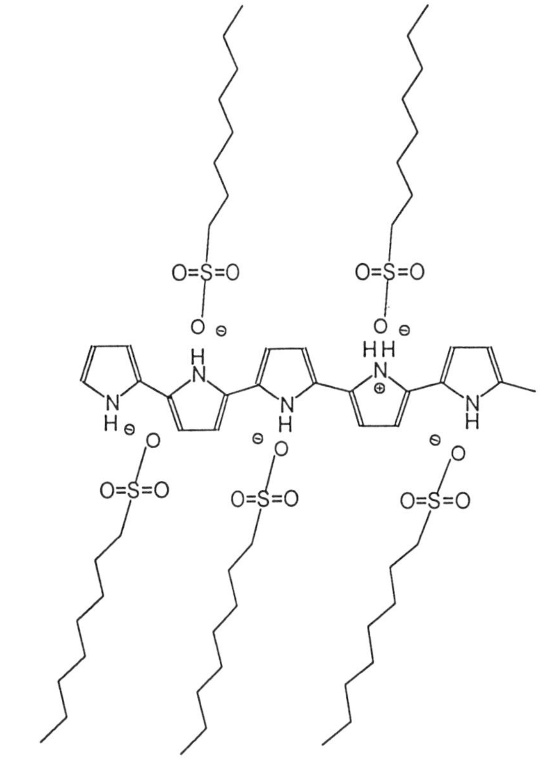

Morphology and Microstructure

Polypyrrole shows little evidence of crystalline structure when examined by X-ray and electron diffraction techniques and is characterized by diffuse amorphous scattering (*118, 119*). Examination of poly(3,4-dimethylpyrrole) by using electron diffraction indicates small crystalline regions in a larger amorphous structure (*123–125*). Analysis of the diffraction pattern shows that the chains are oriented parallel to the plane of the film and conform to a monoclinic cell structure with a chain spacing of 0.365 nm. A value of 0.341 nm was reported for polypyrrole (*94*). A study of polypyrrole generated in the presence of *n*-alkylsulfonates indicates organization of polypyrrole chains (*124*). Increasing the length of alkyl chains by one unit leads to a change in the X-ray scattering spacing of 0.19 nm. The contour length of a single alkyl chain, however, increases by 0.125 nm if one CH_2 unit is added. These results indicate that the anions are present as a tiled double layer (Chart V) and that in α,ω-disulfonates the polymer chains have the form of a single layered structure.

Microanalysis of polypyrrole films reveals an imbalance of the hydrogen, nitrogen, and carbon attributed to the incorporation of water (*126–129*). XPS results indicate the presence of both α- and β-linked carbons, as well as disordered structures that may result from chain or incomplete hydrogen elimination. These results indicate a lower anion:ring ratio than obtained from macroscopic analysis, and the XPS values are typically between 1:6 and 1:8 (*129–131*). The presence of excess hydrogen has made molecular

Chart V. Structure of a polypyrrole alkylsulfonate film.

weight determination very difficult because the polymer is insoluble. Nazzi and Street (*132*) elegantly solved the problem by producing an α-tritiated monomer and analyzing the tritium content of the resultant polymer. This study indicated a molecular weight of about 80,000 (i.e., about 1500 rings of pyrrole).

^{1}H and ^{13}C NMR studies (*133–140*) of poly(3-methylpyrrole) indicated the occurrence of trans and cis forms of the polymer. Also, a small but significant number of β-linkages were detected. Doping causes a Knight shift of approximately 30 ppm (*137*). However, the alternating ring structure clearly predominates, as determined by ESR performed on polypyrrole and related molecules. Both neutral and oxidized films were investigated (*138*). Exposure of the films to oxygen leads to a broadening of the ESR signal that is only partially reversible, accompanied by a reduction in the maximum obtainable conductivity by an order of magnitude. This observation indicates that the interactions occurring are a combination of physical and chemical interactions. The data are interpreted in terms of highly mobile spins at a concentration of 1 spin per 3000 rings for lightly doped materials, and as the doping level increases the spin concentration first rises then falls.

The transport properties of polypyrrole were investigated using the temperature dependence of direct current and 35-GHz thermopower measurements (*141*). The Mott variable-range hopping model extended to a system of mobile localized states appears appropriate. In this model, a parameter, l, is introduced that defines the mean distance that a mobile localized state carrying an electron can be displaced before the electron hops away. In the case of polypyrrole, the states are referred to as those localized on polarons and bipolarons, and these ideas are similar to those proposed to explain the ESR data (*142*). Clearly, the conduction process in polypyrrole is somewhat more complex than that in PA.

Polythiophene and Its Derivatives

Unsubstituted polythiophene is prepared directly either by electrochemical or chemical oxidation (*143–145*). The polymerization process is similar to that for polypyrrole (Scheme IV). The neutral polymer is an insulator but can be oxidized and doped to yield mobile charge carriers that lead to conductivities exceeding 100 S/cm. The formation of these charge carriers (radical cations and dications) modifies the geometry and leads to absorption bands in the visible and near-IR electrochromism.

Because of relatively long and flexible substituents in the 3-position, a rigid-rod structure that is highly insoluble can be converted into a material that is soluble in chloroform, tetrahydrofuran, or toluene. Poly(3-alkylthiophene)s have a relatively high oxidation potential of 0.95 V, indicative of relatively poor stability of the polymer in its oxidized form. The solubility of the material however, has led to considerable interest and speculation of technological application in

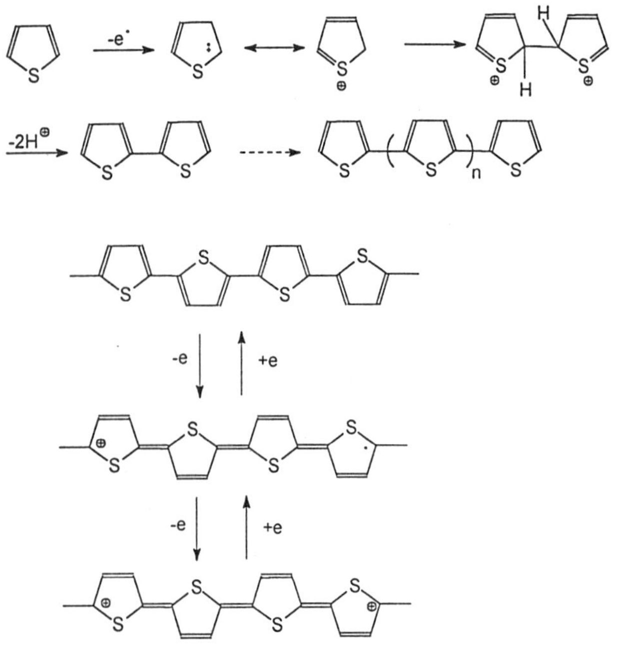

Scheme IV. Synthesis and oxidation–reduction of polythiophene.

rechargeable batteries, electrochromic and thermochromic devices, light-emitting diodes, and biochemical and chemical sensors (*146–149*).

The polymers exhibit interesting thermochromic, solvatochromic, and piezochromic properties. The temperature dependence of the UV-vis absorption spectrum of poly(3-octyloxy-4-methylthiophene) in the solid state exhibits an absorption maximum at 545 nm at 25 °C. However, on heating to 150 °C, the polymer undergoes a thermally induced transition indicated by the absorption shifting to 395 nm (*143*). This remarkable 150-nm blue shift is believed to be the result of a loss of planarity of the conjugated backbone and a decrease in effective conjugation length. The observation of an isobestic point comes from steric interactions between substituents on the backbone. Disorder in the side chain also must be playing an important part in the overall process. The flip of the first repeat unit induces a twisting of the next repeat unit, and a domino effect follows.

The distinctive optical properties of these materials attracted the attention of a number of researchers (*146–149*). The morphology depends on the growth conditions, and the degree of order increases with film thickness. Transmission electron microscope studies reveal a fibrillar structure with fibrillar diameter of 20 nm in the undoped and 80 nm in the doped conducting state. The solvent used as the supporting electrolyte influences the degree of aggregation, and the fibrillar structure can increase to several micrometers in acetonitrile (*150*).

Copolymers of Thiophene and Pyrrole

Copolymers of N-substituted pyrroles and thiophene are conductive. A Schottky-type diode with low-work-function indium was constructed whose I–V characteristics obey a simple barrier model (*151*). Copolymers using monomers such as styrene also were prepared (*152*). Styrene was first copolymerized with chloromethylstyrene by using a radical initiator in the electrochemical cell and then electro-oxidatively coupling with pyrrole to form a conducting composite (*143*). The resultant films exhibit a widely varying degree of conductivity (from 50 S/cm to 5×10^{-2} S/cm) and enhanced mechanical properties. The nature of the cosolvent used in the electrochemical preparation of the polymer strongly influences both the conductivity and mechanical properties of the films (*113*).

In an attempt to obtain more processible conducting polymers, a range of polythiophene-related structures (Chart VI) were produced. These structures have good electrical properties, molecular weights of the order of 35,000, and glass transition temperatures of 58 °C (*153*, *154*). A major problem with many of these simple conducting polymer films is processibility. Ideally, it

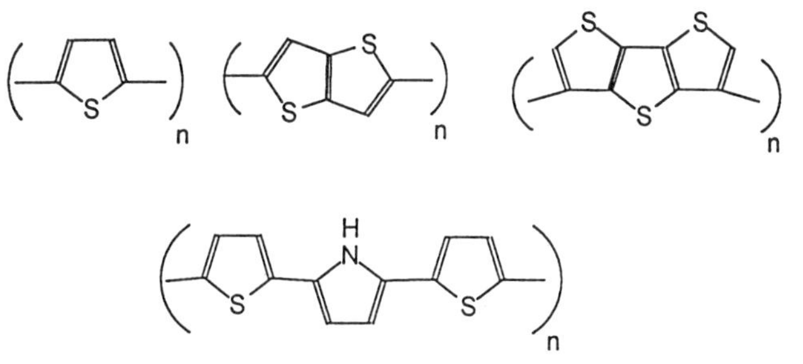

Chart VI. Structures of poly(2,5-dithienylpyrrole).

would be desirable to be able to dissolve the polymer in a solvent and then cast the solution to yield a conducting polymer film. The electrochemical polymerization of 2,5-dithienylpyrrole yields a polymer (Chart VI) of modest conductivity, but more importantly the polymer is soluble in polar solvents (153).

Solubility is a consequence of a low molecular weight (10,000) and the mixture of monomer–monomer interactions present. Polymerization shifts the absorption maximum from 477 nm for the trimer to 516 nm, a result that is indicative of an increased degree of conjugation in the polymer. The polymer exhibits the same type of electrochromic behavior found in many thiophene materials. and thin films can be switched from blue to green color.

A number of copolymers were synthesized electrochemically. Kumar et. al (155) reported that phenylene oxide and pyrrole units form a material that has better mechanical properties than polypyrrole but a slightly lower conductivity. A copolymer of copoly[(*N*-*p*-nitrophenylpyrrole)pyrrole] also was reported and was derivatized to produce highly conducting polymer films (156). Certain of these copolymers are amorphous, and have glass transition temperatures that, in their rubbery phase, exhibit greatly enhanced mobilities (157–160).

Polyaniline

Considerable interest has been shown in polyaniline because it can be grown by using aqueous and nonaqueous routes (154, 161–163). The product has the typical emeraldine black color and can be considered as being derived from alternating reduced and oxidized forms (Scheme V). The averaged oxidation state can be varied continuously from zero to give a completely reduced polymer. Studies of polyaniline as a composite electrode with $LiClO_4$ in propylene carbonate as electrolyte indicate polyaniline has very good recyclability and essentially 100% coulombic efficiency. Derivatives of aniline are polymerized readily and present the possibility of changing the oxida-

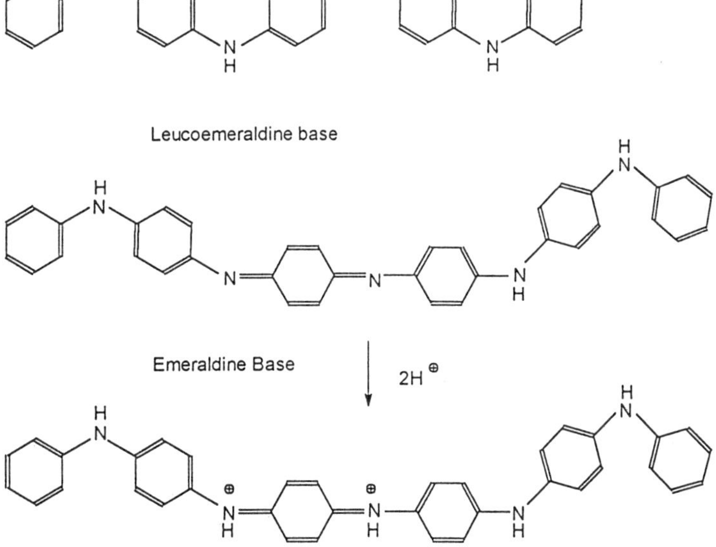

REDUCED   OXIDIZED

Leucoemeraldine base

Emeraldine Base $\downarrow$ $2H^{\oplus}$

Scheme V. Structure of polyaniline and the process of acid doping.

tion–reduction characteristics of the resultant polymer to match particular electrochemical applications.

The base form of polyaniline (Scheme V) corresponds to the emeraldine oxidation state and is converted from an insulator ($\sigma \approx 10^{-10}$ S/cm) to a conductor ($\sigma \approx 5$ S/cm) by treatment with 1 M aqueous HCl. The corresponding salt, emeraldine hydrochloride, is formed (154). In this type of doping the number of electrons associated with the polymer does not change as a result of doping, in contrast to the processes discussed previously. The conductivity mechanism in polyanaline is, therefore, rather different from that discussed for polyacetylene, polypyrrole, and polythiophene.

The metallic emeraldine hydrochloride is believed to be a delocalized poly(semiquinone radical cation) having a polaron conduction band with most of the positive charge residing on the nitrogen atom. It exhibits a finite density of states at the Fermi energy, as identified by ESR (164–167). The observed transition is proposed to be an isolated bipolaron-to-polaron lattice transition and corresponds to the disproportion between the protonated imine plus amine to form two semiquinones (168, 169). By assuming a 1-D tight binding model for a polaron band, a polaron band width of $W_p = 0.37$ eV is obtained. Using eq 13 from the continuum theory, where Δ_o is the half gap 2 eV, R is the polaron–polaron separation, and ξ_o is the polaron decay length, the band width (W_p) is given by:

$$W_p = 4\sqrt{2}\Delta_o \exp(R/2\sqrt{\xi_o}) \qquad (13)$$

If the polaron length is twice the nitrogen–nitrogen distance (N–N), then $\xi_o = 0.4$ N–N and charge is expected to be spread out considerably along the polyaniline chain (170–174).

Polyquinolines

An interesting group of conducting polymers that were studied to a limited extent are the polyquinolines (175) (Chart VII). High number-average molecular weight ($M_n = 5 \times 10^5$) polyquinolines can be prepared by an acid-catalyzed reaction of aromatic α-amino ketones with ketomethylene compounds. The route is easily modified to yield a wide range of structures.

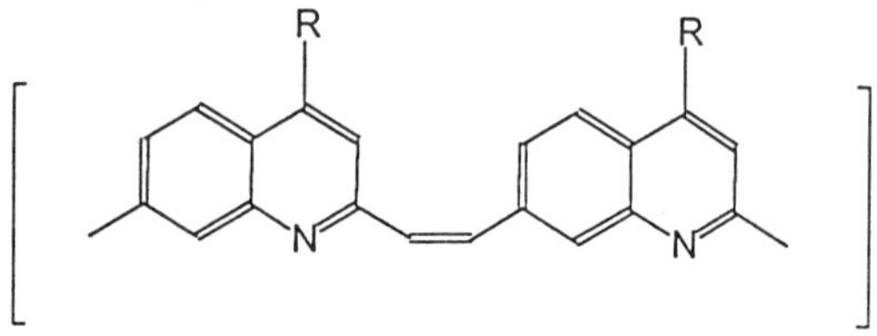

Chart VII. Structure of polyquinoline.

Rigid-rod polyquinolines are highly crystalline (melting temperature ≈ 500 °C and are soluble in the polymerization dope (m-cresol/di-m-cresyl)phosphate in which they form anisotropic solutions. Fibers spun from the polymerization dope containing 5–18% dopants have good tensile strength (8–10 g/denier) and high Young's modulus.

The X-ray structure of a crystalline fiber of poly[2,2'-p,p'-biphenylene-6,6'-bis(4-phenylquinoline)] showed that the rod-like chains are orientated in the direction of the fiber axis and stacked in parallel, nearly coplanar sheets above one another. Thus, this polyquinoline resembles the structure of graphite. Several features of these materials can be identified.

- incorporation of a vinyl group into the backbone lowers the conductivity
- conductivity is very sensitive to the thermal history of the fiber
- annealing the fibers decreases the conductivity associated with the generation of crystalline structure

These materials have interesting processing characteristics, and changes in the structure by ring substitution lead to systematic changes in the oxidation–reduction potential. Also, these materials exhibit a high degree of stability to oxygen.

Polycarbazoles and Polyphenothiozones

Similar types of polymer were studied based on polycarbazoles and polyphenothiozones (176–178). Two specific examples are poly(N-methyl-3,6-carbazolyl) and poly(N-methyl-3,7-phenothiozinyl) (Chart VIII). When doped with iodine, conductivities of these polymers are 1–10 S/cm. The attraction of these polymers is that they are relatively stable to air and can be processed by standard solution techniques.

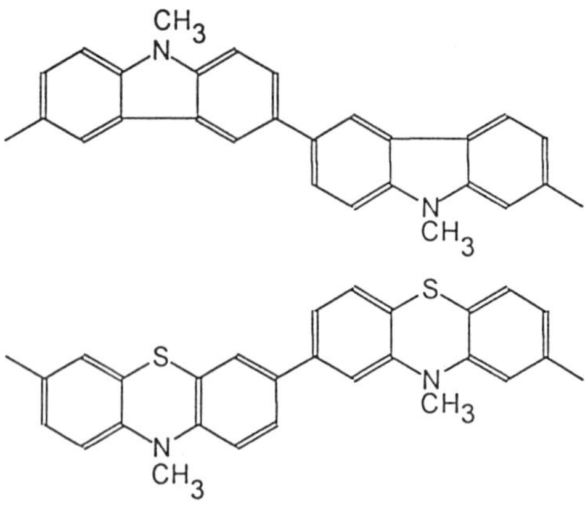

Chart VIII. Structures of poly(N-methyl-3,6-carbazoyl) and poly(N-methyl-3,7-phenothiazinyl) conducting polymers.

Polyazulene

In the case of azulene (*179*) (Scheme VI), the macromolecule produced electrochemically is linked via a five-membered ring, and this structure implies that the resulting radical cation, in principle, has two options.

1. It can act as an electophile and attack neutral azulene.
2. It can act as a radical and dimerize.

Electrochemically, the reaction occurs almost exclusively through the 1,3-position of the cyclopentadienyl ring, and the resulting conducting films are remarkably stable when subjected to a range of different environmental conditions. A common feature of conducting polymers is their insolubility in acids, bases, and organic solvents. The thick films of polyazulene are stable up to 80 °C, but thin films display a sensitivity to oxidation.

When kept for 3 h in atmospheric conditions, the electrical conductivity decreases from 0.4 to 0.2 S/cm, whereas heating for about 20 h results in a 10-fold decrease. The loss in conductivity is probably a result of irreversible changes in the degree of conjugation as a result of degradation of the polymer structure.

Optical and Electronic Conduction Properties

All conducting polymers exhibit complex absorption spectra (*180–187*). At low doping levels polypyrrole exhibits four absorption peaks at 0.7, 1.18, 2.3, and 3.2 eV, which can be associated with transitions from below the valence band edge to bonding and antibonding levels of polarons (for the 0.7- and 1.18-eV transitions i.e., situated in the band gap), bonding to antibonding levels of polarons (for the 2.3-eV transition: i.e., intraband), and valence band to conduction band (for the 3.2-eV transition: i.e., interband).

Interestingly, the 1.18-eV transition decreases in intensity, eventually disappearing as a consequence of increased doping. This transition is interpreted as being a result of the formation of bipolaronic levels (which are either fully occupied or empty) due to combination of free polarons brought about by the increasing dopant concentration. Bipolaronic defects can be observed even at low doping levels in certain organic structures [e.g., poly(dimethylpyrrole)], which inherently have an increased degree of ordering compared with the amorphous structure of polypyrrole. This increased order results in a lower probability of polaronic recombination into bipolarons.

Polythiophene exhibits transitions at 0.65, 1.3, 1.5, 1.8, and 2.6 eV, which are critically dependent on the

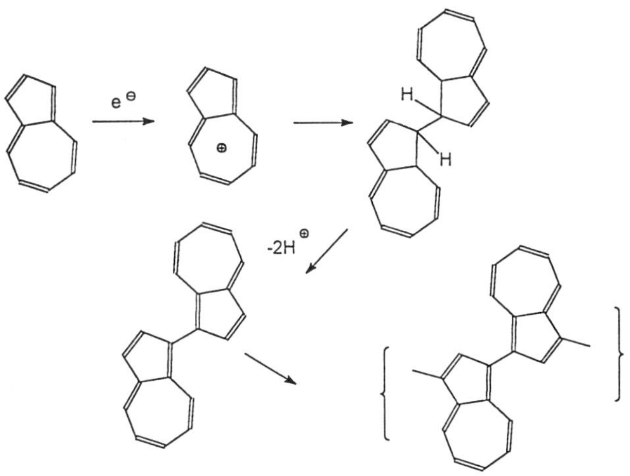

Scheme VI. Electrochemical synthesis of polyazulene.

dopant concentration. These transitions have been assigned to valence band bonding and antibonding levels of shallow and deep polarons, and the conduction band, respectively. The absorption peak at 1.3 eV disappears at high doping levels (~33–66%). A deep polaron exists in polythiophene as a result of interaction between the dopant ion and the nearest thiophene ring, whereas a shallow polaron has been thought to result from the interaction of the dopant ion and the nearest ring. The difference in behavior of the thiophene and pyrrole rings in part can be explained by the presence of *d*-orbitals on the sulfur atom enhancing certain of the through-space interactions. Similar characteristics were reported for other substituted polythiophenes such as poly(3-methylthiophene) and poly(dimethylthiophene) and also poly(heteropyrrole)s.

Potential Applications of Conducting Polymers

A variety of applications were investigated, and the following selection is not intended to be a comprehensive review of the literature, but rather to illustrate the types of potential devices that might be developed. Initially there was considerable interest in the use of these materials for electrical conductivity and in battery applications. In all cases, oxidative sensitivity and the ability to carry significant currents are critical, and in this context conducting polymers appear to have limited application. Conducting polymers are most likely to be formulated as composite materials based on carbon black, ICPs, and a binder (*188–190*).

Potentially one of the most interesting areas of development is in the construction of modified conductive electrode coatings (*191, 192*). In the majority of such applications the conducting polymer provides a source for electrons to balance the redox process that is used as a sensor (*193*). The type of format used is often conceptually simple. Because the electrical conductivity of the polymer depends on the oxidation state of the polymer, the oxidation state can also be used as the active element of a sensor. This use can be achieved provided that the sandwich construction allows for supply or removal of the associated counter ions; for example, poly[osmium(2,2-bipyridine)$_2$(4-vinylpyridine)$_2$] perchlorate on a platinum electrode covered by a porous gold film allows counter-ion permeation (194). The electrochemical potentials of the metallic electrodes can be controlled independently with respect to a reference electrode immersed in the contacting electrolyte.

Applying the same potential relative to the reference electrode to both electrodes of the sandwich defines the oxidation state of the electroactive polymer. The current established between the two connecting electrodes when their potentials are not the same ($\Delta E = 5$ mV) gives the conductivity of the polymer at the oxidation state. If polypyrrole or other conducting polymers that have the capacity of reversible reduction processes are used, then this structure can be used as a sensor. Polypyrrole reduced at –0.5 V relative to sodium chloride–saturated calomel reference electrode (SSCE) is a poor conductor, but when it is oxidized at potentials more positive than about –0.1 V compared with SSCE it is a relatively good conductor. Films of oxidized polypyrrole under a given potential difference are the same whether they are dry or wetted with solvent. Also, the conductivity is ohmic up to 100 mV.

The polypyrrole composite electrode has a capacity to store charge that is directly related to the electrode potential. The peaks in the storage are centered at the standard potentials of the polymer: +0.72, –1.33, and –1.53 V relative to SSCE. Reduction of the Os(III) to Os(II) is primarily a metal-centered reduction, whereas the subsequent reductions to Os(I) and Os(0) involve the bipyridine ligands. As in the case of polypyrrole, the conductivity is a function of the electrode potential and has a maximum value when the film is in mixed valence state. It is only then that electrons on reduced sites can hop to oxidized sites, thereby allowing electron conduction. This self-exchange process between neighboring sites can be written as follows:

$$Os(III) + OS(II) \xrightarrow{k_c} Os(II) + Os(III)$$

where k_c is the electron self-exchange rate constant. The ClO_4^- counterion migrates within the polymer to eliminate the applied field in the interior of the polymer, and the driving force for electron hopping is the concentration gradient of the reduced sites. This self-exchange process is believed (195) to be general to all electrochemically active polymer films that conduct electrons. At high temperatures, the electron flow between occupied and unoccupied sites is by electrolytic development of a concentration gradient. At low temperatures, the ions in the polymer are essentially immobile, and the electron flow is driven by a voltage gradient rather than a concentration gradient. Lowering the temperature has a marked effect on the conductivity, as would be expected for a thermally activated hopping process.

If a sufficiently large potential is applied to the film, all the polymer contacting one electrode becomes oxidized to the Os(III) state and the rest becomes reduced to Os(II). The current established is the largest that can be supported by the Os(II)/Os(III) mixed valence state of the sandwich. This limiting current (I_{lim}) is given by eq 14, where F is the Faraday constant, A is the sandwich electrode area, $D_{e(III/II)}$ is the Os(III)/Os(II) electron diffusion constant, C_T is the total concentration of Os sites (1.4 M), and d is the film thickness. The value of D_e corresponds to the mean squared displacement of the electron per unit time and is given by the ratio of the redox conductivity to the electrochemical capacity of the material.

$$I_{lim} = nFAD_{e(III/II)}C_T/d \qquad (14)$$

In general there will be three different values of D_e corresponding to $D_{e(III/II)}$, $D_{e(II/I)}$, and $D_{e(I/0)}$. The absolute values of these coefficients will depend on the mechanism involved in the conduction process. The electrons that reduce Os(III) complex to Os(II) complex are about 20 times less mobile than those that reduce Os(I) complex to Os(0) complex. In other words, the self-exchange process rate constants differ for Os(III)/Os(II), Os(II)/Os(I), and Os(I)/Os(0) complexes. This difference reflects the difference in electron mobility in a delocalized band material such as polypyrrole and in materials like osmium powder, where the electron faces a significant kinetic barrier between layers of adjacent monomer sites.

One of the principal attractions of using conducting polymers is the possibility of using microfabrication methods. Reduction in size of the electrodes can give a significant increase in the sensitivity to the reagent it is designed to detect. Consider a one-terminal 10-nm disc microelectrode that has an area of only 8×10^{-7} cm^2. A 100-nm thick coating will contain about 10^9 monomer sites, which is a relatively small number of chemically reactive sites. Such devices have been fabricated and tried as in vivo electrochemical probes in neurochemistry.

A number of reviews have appeared listing systems with potential as ion-selective probes, and it is probable that the first commercial applications of conducting polymers will be in the area of electroanalytical probes (190, 193). In these structures, the current carried by the conducting polymer is relatively low, and hence the stress on the material is low, compared with the stress incurred when batteries are used.

The electro- and photochromic properties of thiophene-containing conducting polymers were recognized by many workers. These polymers and the related phenylenevinylene polymers were used to generate light-emitting diodes. Electroluminescence through charge injection under a high applied field has been known in organic semiconductors for some time. A considerable level of interest has occurred in the properties of anthracene, initially as bulk crystal (*196–198*) and later as thin films (*199*). Poly(phenylenevinylene) was used as the emissive layer in electroluminescent devices (*200*).

Structures for electroluminescence are fabricated with polymer film formed on a bottom electrode deposited on a suitable substrate–glass and a top electrode formed on to the fully converted poly(phenylenevinylene) film. Electrode materials are chosen with a low work function for use as the negative electron-injecting contact and with a high work function as the positive hole injecting contact. At least one of these layers must be semitransparent for light emission normal to the plane of the device, and both indium oxide and thin aluminum, typically at 7–15 nm, have been used in this role. The schematic structure of these devices is shown in Figure 10. Application of voltages as low as 5 V across the 100-nm thick film with calcium yields a device that is 1% efficient (photons per electron injected).

The spectrum of the luminescence produced in these devices is essentially the same as that of the photoluminescence. Poly(phenylenevinylene) has a low concentration of extrinsic charge carriers and exhibits a resistivity above 10^{12} $\Omega \cdot$cm, and charge injection only proceeds as the field is raised above 10^5 V/cm. These devices operate by double charge injection of electrons and holes from, respectively, the negative and positive electrodes. The high field is required to enable tunnelling across barriers due to work function mismatch or interfacial insulating layers.

The efficiency in these electroluminescent devices is determined by a number of factors:

1. The quantum yield for luminescence from a singlet exciton is usually well below 100%, because there are nonradiative decay mechanisms present. Charges present as polarons or bipolarons are probably the principal sources of quenching sites. There is evidence that photoluminescence is quenched through chemical doping through photogeneration of charged excitations and through generation of a charge-accumulation layer. These processes can reduce the number of quenching sites or limit the mobility of the excitons. The quantum yield for devices fabricated with aluminum for both electrode layers and with a copolymer with a varying ratio of the starting materials shows a very strong dependence on copolymer composition (Figure 11).

2. Double charge injection processes followed by a combination of positive and negative charges to form excitons will give singlets and triplets in the ratio of 1:3. The triplet excitons are relatively long-lived, a result indicating that they are considerably lower in energy than singlets. Delayed red-shifted luminescence due to direct triplet to ground-state transitions have not been reported; indeed, the spin orbit coupling is small. However, there are possible decay routes that allow radiative emission through processes such as triplet–triplet collisions that result in the formation of singlet excitons.

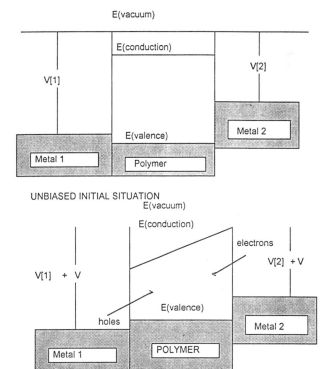

Figure 11. Schematic presentation of the energy states in a poly(phenylenevinylene) luminescence device.

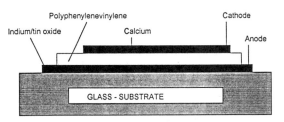

Figure 10. Schematic presentation of an electroluminescent device based on poly(phenylenevinylene).

3. The requirement for injection of both electrons and holes is best met if the positive and negative electrodes have high and low work functions, respectively. In general it is easier to match the ionization potential (top valence band) for the polymer with a metal with a high work function than it is to match the electron affinity (bottom of conduction band) with a metal with a low work function. If there is imbalance in the injection rates for the two carrier types, the more readily injected carrier passes through the polymer film, possibly combining with the other carrier type at the far electrode. However, a loss of electroluminescence efficiency certainly results.

Conducting polymers are also being studied for other applications, and there is every reason to believe that they will be used in composite battery structures (201–203), microelectroanalytical probes, and possibly some photo- or electrochromic (204–208) and electromagnetic screening applications. As electromagnetic screening applications, conducting polymers are effective screens for the electric component but are ineffective for the magnetic component of high frequency electrical radiation. A variety of modified electrode systems were listed in reviews (204–209) and include porphyrin-containing films, porphorins coupled with cobalt-amine complexes capable of converting oxygen to peroxide, and the controlled release of dopant from polypyrrole films (210–214).

Conclusions and Prospects

A wide range of chemical structures were shown to exhibit electronic conduction. Certain of these have produced soluble systems, but the problem of stability of the films to atmospheric modification still remains an area of concern. The very nature of the conduction process leads to the possibility of irreversible chemistry being promoted; however, polyaniline is the least susceptible of the systems to this problem and hence is the most likely to provide viable devices in the future. The intrinsic instability of conducting polymers due to irreversible oxidation has limited their potential application to high voltage or current applications, but their mechanical and widely varying electrochemical properties have indicated a significant potential for these materials as part of electrochemical probes. Conducting polymers are also attractive as part of environmentally friendly battery systems and are currently becoming commercially available.

Recent publications would suggest that it may be possible to solve the problem of stability and in turn enhance the conductivities. The conducting polymer story is by no means completed, and new materials are likely to emerge in the near future. Conducting polymers have moved from being theoretical concepts to a practical reality, and the initial potential of these systems now appears a little more optimistic.

References

1. Etemad, S.; Heegar, A. J.; MacDiarmid, A. G. *Ann. Rev. Phys. Chem.* **1982,** *33,* 443.
2. Su, W. P.; Schrieffer, J. R.; Heeger, A. R. *Phys. Rev. Lett.* **1979,** *42,* 1968.
3. Rice, M. J. *Phys. Lett. A.* **1979,** *71,* 152.
4. Takayama, H.; Lin-Lin, Y. R.; Maki, K. *Phys. Rev. B.* **1980,** *21,* 2388.
5. Brazovskii, S. *JETP Lett.* **1980,** *78,* 677.
6. *The Physics and Chemistry of Low Dimensional Solids;* Alcacer, L., Ed.; Reidel: Dordrecht, Netherlands, 1980; pp 353, 393.
7. Etemad, S.; Heeger, A. J. *Mol. Cryst. Liq. Cryst.* **1981,** *77,* 43.
8. Kathirgamanathan, P. In *Royal Society of Chemistry;* Fawcett, A. H., Ed.; Specialist Publication No. 87; Royal Society of Chemistry: London, 1990; pp 175–205.
9. Doijack, F. A. *IEEE Trans. Compon. Hybrids Manuf. Technol.* **1981,** CHMT-4(4), 372.
10. Sherman, R. D.; Middleman, L. M.; Jacobs, S. M. *Polym. Eng. Sci.* **1983,** *23,* 36.
11. *Inherently Conducting Polymers;* Aldissi M. Noyes Data Corporation: Park Ridge, NJ, 1989.
12. Margolis, J. M. *Conducting Polymers and Plastic;* Chapman and Hall: London, 1989.
13. Degussa Company *Technical Bulletin on Pigments;* No. 69.
14. Cabot Corporation *Technical Report, S. 39;* Cabot Corporation.
15. Eley, D. D.; Willis, M. R. *Symposium on Electrical Conductivity in Organic Solids;* Kallman, H.; Silver, M., Eds.; Interscience: New York, 1961; p 257.
16. Eley, D. D. *J. Polym. Sci.* **1967,** C17, 73.
17. Kryszewski, M. *Semiconducting Polymers;* Polish Scientific Publishers: Warsaw, Poland, 1980.
18. Berlin, A. A. *IUPAC Symposium on Macromolecular Chemistry;* International Union of Pure and Applied Chemistry, Wiley: New York, 1965; Preprint 281.
19. Vincent, C. A. *Electrochemical Science and Technology of Polymers;* Linford, R. G., Ed.; Elsevier Applied Science: Amsterdam, Netherlands, 1990; Chapter 2, p 149.
20. Ito, T.; Shirakawa, H.; Ikeda, S. *J. Polym. Sci. Polym. Chem. Ed.* **1974,** *12,* 11.
21. Ito, T.; Shirakawa, H.; Ikeda, S. *J. Polym. Sci. Polym. Chem. Ed.* **1974,** *13,* 1943.
22. Karasz, F. E.; Chien, J. C. W. *Nature (London)* **1979,** *282,* 286.
23. Akaishi, Y.; Miyasaka, K. *J. Polym. Sci. Polym. Phys. Ed.* **1980,** *18,* 745.
24. Finder, C.; Chen, C. E. *Phys. Rev. Lett.* **1982,** *48,* 100.
25. Park, Y. W.; Heeger, A. J. *J. Chem. Phys.* **1980,** *75,* 946.
26. Nigrey, P. J.; McInnes, D. *J. Electrochem. Soc.* **1981,** *128,* 1651.
27. Nigrey, P. J.; MacDiarmid, A. G. *Mol. Cryst. Liq. Cryst.* **1981,** *77,* 81.
28. Ozaki, M.; Peebles, D. *J. Appl. Phys.* **1980,** *51,* 4252.

29. Toni, T.; Grant, P. M.; Gill, W. D. *Synth. Met.* **1980,** *1,* 301.
30. Chen, S. N.; Heeger, A. J. *Appl. Phys. Lett.* **1980,** *36,* 96.
31. Su, W. P.; Schrieffer, J. R. *Phys. Rev. B* **1981,** *22,* 2099.
32. Brazovskii, S. A.; Kirov, N. N. *JETP Lett.* **1981,** *33,* 6.
33. Simon, J.; Andre, J-J. *Molecular Semiconductors;* Springer Verlag: Berlin, Germany, 1985; p 166.
34. Baeriswyl, I. *Electronic Properties of Polymers and Related Compounds;* Kuzmany, H.; Mehring, M.; Roth, S., Eds.; Springer Verlag: Heidelberg, Germany, 1985; p 85.
35. Gill, W. D.; Clarke, T. C.; Street, G. B. *Appl. Phys. Commun.* **1983,** *2,* 211.
36. Wegner, G. *Electronic Properties of Polymers and Related Compounds;* Kuzmany, H.; Mehring, M.; Roth, S., Eds.; Spinger Verlag: Heidelberg, Germany, 1985; p 18.
37. Kohler, B. E. *Electronic Properties of Polymers and Related Compounds;* Kuzmany, H.; Mehring, M.; Roth, S., Eds.; Springer Verlag: Heidelberg, Germany, 1985; p 100.
38. Peyrard, M.; Remoissenet, M. *Phys. Rev. B* **1982,** *26,* 2886.
39. Bredas, J. L.; Chance, R. R. *Phys. Rev. B* **1982,** *26,* 5843.
40. Boudreaux, D. S.; Chance, R. R. *Phys. Rev. B* **1982,** *26,* 6927.
41. Heeger, A. J.; Schreiffer, J. R. *Solid State Commun.* **1983,** *48,* 207.
42. Baker, G. J.; Rayner, J. B. *J. Chem. Phys.* **1984,** *80,* 5250.
43. Mott, N. F.; Davies, E. A. *Electronic Processes in Non-Crystalline Materials;* Clarendon: Oxford, England, 1977.
44. Kivelson, S. *Mol. Cryst. Liq. Cryst.* **1981,** *47,* 65.
45. Yacoby, Y.; Roth, S. *Solid State Commun.* **1983,** *47,* 869.
46. Sethna, J. P.; Kivelson, S. *Phys. Rev. B* **1982,** *26,* 3513.
47. Etemad, S.; Ozalei, M. *Mol. Crystal. Liq. Crystal.* **1981,** *77,* 121.
48. MacDiarmid, A. G.; Maxfield, M. *Electrochemical Science and Technology of Polymers;* Linford, R. G., Ed.; Elsevier Applied Science: London, 1987; Vol. 1, p 67.
49. Nigery, P. J.; MacDiarmid, A. G. *J. Chem. Soc. Commun.* **1979,** 594.
50. Shirakawa, H.; Kabayaski, T. *J. Phys.* **1983,** *C3,* 3.
51. Baughman, R. H.; Bredas, J. L.; Chance, R. R. *Chem. Rev.* **1982,** 82.
52. Monkenbusch, M.; Morva, B. S. *Macromol. Chem. Rapid Commun.* **1982,** *3,* 69.
53. Hasslin, H. W.; Rickel, C. *Synth. Met.* **1982,** *5,* 37.
54. Hoffman, D. M.; Tanner, D. B. *Mol. Cryst. Liq. Cryst.* **1982,** *83,* 143.
55. Epstein, A. J.; Gibson, H. W. *Phys. Rev. Lett.* **1981,** *47,* 1549.
56. Boils, D.; Schue, F. *J. Phys.* **1983,** *C3,* 189.
57. Reynolds, J. R.; Chein, J. C. W. *J. Phys.* **1983,** *C3,* 171.
58. Billard, D.; Kulsziwioz, I. *J. Phys.* **1983,** *C3,* 29.
59. Schacklette, L. W.; Chance, R. R. *Synth. Met.* **1979,** *1,* 307.
60. Heeger, A. J. *Nature (London)* **1987,** *327,* 403.
61. Zurer, P. *Chem. Eng. News* **1987,** 453.
62. Pochan, J. M.; Pochan, D. F.; Rommelmann, F.; Gibson, H. W. *Macromolecules* **1981,** *14,* 110.
63. Kivelson, S. *Phys. Rev. Lett.* **1981,** *46,* 1344; *Phys. Rev. B* **1982,** *25,* 3798.
64. Park, Y. W.; Heeger, A. J.; Druy, M. A.; MacDiarmid, A. G. *J. Chem. Phys.* **1980,** *73,* 946.
65. Nechtschein, M.; Devreaux, F.; Greene, R. L.; Clarke, T. C.; Street, G. B. *Phys. Rev. Lett.* **1980,** *44,* 356.
66. Andre, J-J.; Bernard, M.; Mathis, F. *J. Phys.* **1983,** *C3,* 44, 189.
67. Street, G. B.; Clarke, T. *Solid State Chemistry;* Holt, S. L.; Milstein, J. B.; Robbins, M., Eds.; Advances in Chemistry Series 186; American Chemical Society: Washington, DC, 1980; p 177.
68. Simon, J.; Andre J-J. *Molecular Semiconductors;* Springer Verlag: Berlin, Germany, 1985; p 182.
69. Edwards, J. H.; Feast, W. J. *Polymer* **1984,** *25,* 395.
70. Edwards, J. H.; Feast, W. J. *Polymer* **1980,** *21,* 595.
71. Bott, D. C.; Chai, C. K. *J. Phys.* **1983,** *C3,* 143.
72. Gibson, H. W.; Pochan, J. M. *Electrical and Electronic Properties of Polymers: A State of the Art Compendium;* Kroschwitz, J. I., Ed.; John Wiley: New York, 1988; p 35.
73. Drury, M. A.; Tsang, C. H. *J. Polym. Sci. Phys. Ed.* **1980,** *18,* 429.
74. Ripathy, S. E.; Rubner, M. *J. Phys.* **1983,** *C3,* 37.
75. Bott, D. C.; Chai, C. K. *J. Phys.* **1983,** *C3,* 45.
76. Fincher, C. R.; Moses, D. *Synth. Met.* **1983,** *6,* 243.
77. Wegner, G. *Angew. Chem. Int. Ed.* **1981,** *20,* 361.
78. Hocker, J.; Schreiber, G. *J. Phys.* **1983,** *C3,* 147.
79. Galvin, M. E.; Wnok, G. E. *Polym. Commun.* **1982,** *23,* 795.
80. Galvin, M. E.; Wnok, G. E. *J. Phys.* **1983,** *C3,* 151.
81. Baker, G. L.; Bates, F. S. *J. Phys.* **1983,** *C3,* 11.
82. Enkelman, V.; Halim, J.; Fischer, H.; Wegner, G.; Albouy, P. A. *Synth. Met.* **1992,** *51,* 1.
83. Korner, F.; Thummer, G.; Kotzler, J. *Synth. Met.* **1992,** *51,* 31.
84. Cao, Y.; Smith, P.; Heeger, A. J. *Polymer* **1991,** *32,* 1210.
85. Heeger, A. J. *Conjugated Polymers and Related Materials;* Salaneck, W. R.; Lundstrom, I.; Ranby, B., Eds.; Oxford Science Publications: Oxford, England, 1993; p 38.
86. Tokito, S.; Smith, P.; Heeger, A. J. *Polymer* **1991,** *32,* 464.
87. Kossmehl, G. *Ber Bunsenges Phys. Chem.* **1979,** *83,* 417.
88. Baghman, R. H.; Bredas, J. L.; Chance, R. R.; Fleenbaumer, R. L.; Shacklette, L. W. *Chem. Rev.* **1982,** *82,* 209.
89. Chaing, C. K.; Druy, M. A.; Gau, S. C.; Heeger, A. J.; Louis, E. J.; MacDiarmid, A. G.; Park, Y. W.; Shirakawa, H. *J. Am. Chem. Soc.* **1978,** *100,* 1013.
90. Speight, J. G. *J. Macromol. Sci.* **1971,** *C5,* 295.
91. Hotta, S.; Hosaka, T. *Synth. Met.* **1983,** *6,* 317.
92. Bredas, J. L.; Chance, R. R. *Mol. Cryst. Liq. Cryst.* **1981,** *77,* 319.
93. Bredas, J. L. *J. Mol. Struct. (Theochem.)* **1984,** *107,* 169.
94. Kanazawa, K. K.; Diaz, A. F.; Geiss, R. H.; Gill, W. D.; Kwak, J. F.; Logan, J. A.; Rabolt, J. F.; Street, G. B. *J. Chem. Soc. Chem. Commun.* **1974,** 854.
95. Kanazawa, K. K.; Diaz, A. F.; Gardini, G. P.; Gill, W. D.; Grant, P. M.; Pfluger P.; Scott, J. C.; Veiser, G. *Synth. Met.* **1979** *1,* 329.
96. Street, G. B.; Clarke, T. C.; Krounei, M.; Kanazawa, K. K.; Lee, V.; Pfluger, P.; Scott, J. C.; Veiser, G. *Mol. Cryst. Liq. Cryst.* **1982,** *83,* 253.
97. Dall'Olio, A.; Dascola, Y. *Acad. Sci. Ser. C* **1968,** *267,* 433.
98. Pickup, P. G.; Osteryoung, R. A. *J. Am. Chem. Soc.* **1984,** *106,* 2294.
99. Wegner, W.; Mokenbusch, M. *Macromol. Chem.* **1984,** *5,* 157.
100. Irebeau, P. *J. Phys.* **1983,** *C3,* 44, 579.
101. Wnek, G. E.; Chein, J. C. W. *Polymer* **1979,** *20,* 1441.
102. Neugebauer, H.; Nauer, G. *J. Phys. Chem.* **1984,** *88,* 652.
103. Frank, A. J. *Mol. Cryst. Liq. Cryst.* **1982,** *83,* 341.

104. Noufi, R.; Frank, A. J. *J. Am. Chem. Soc.* **1981**, *103*, 1849.
105. Bargon, J.; Mohmand *IBM J. Res. Dev.* **1983**, *27*, 330.
106. Monstedt, H. *Electronic Properties of Polymers and Related Compounds*; Kuzmany, H.; Mehring, M.; Roth, S., Eds.; Springer Verlag: New York, 1985; p 8.
107. Baughman, E. H. *Mol. Cryst. Liq. Cryst.* **1985**, *118*, 1.
108. Mahoubian-Jones, M. G. B.; Pethrick, R. A. *Polymer Yearbook II*; Gordon and Breach: London, 1985; p 225.
109. Street, G. B.; Lindsay, S. A.; Nassal, A. I.; Wynne, K. J. *Mol. Cryst. Liq. Cryst.* **1985**, *118*, 137.
110. Wernet, W.; Mikenbush, M.; Wegner, G. *Mol. Cryst. Liq. Cryst.* **1985**, *118*, 193.
111. Arrayed, F. M.; Benham, H. L.; McLeod, G. G.; Mahoubian-Jones, M. G. B.; Pethrick, R. A. *Mater. Forum* **1986**, *9*, 209.
112. Satoh, M.; Daneto, K.; Yoshino, K J. *Appl. Phys. Jpn.* **1985**, *24*, 423.
113. Inganas, O.; Erlandsson, R.; Nylander, C.; Lundstrom, I. *J. Phys. Chem. Solids* **1984**, *45*, 427.
114. Diaz, A. F.; Hall, B. *IBM J. Res. Dev.* **1983**, *27*, 342.
115. Selaneck, W. R.; Erlansson, R.; Prejza, J.; Lunstrom, I.; Inganas, O. *Synth. Met.* **1983**, *5*, 125.
116. Eaves, J. G.; Munro, H. S.; Parker, D. *Polym. Commun.* **1987**, *28*, 38.
117. McLeod, G. G.; Jeffreys, K.; McAllister, J. M. R.; Mudell, J.; Affrossman, S.; Pethrick, R. A. *J. Phys. Chem. Solids* **1987**, *48*, 921.
118. Kanazawa, K. K.; Diaz, A. F. *Synth. Met.* **1979**, *1*, 327.
119. Tourillon, G.; Garneir, F. *J. Electroanal. Chem.* **1984**, *161*, 51.
120. Salmon, M.; Carbajal, M. E. *J. Chem. Soc. Chem. Commun.* **1983**, 1532.
121. Inoue, T.; Yamase, T. *Bull. Chem. Soc. Jpn.* **1983**, *56*, 985.
122. Tourillon, G.; Garnier, F. *J. Electroanal. Chem.* **1984**, *161*, 407.
123. Geiss, R. H.; Street, G. B.; Voksen, W.; Econcomy, J. *IBM J. Res. Dev.* **1983**, *27*, 321.
124. Wegner, G.; Wernet, W.; Glatschafer, D. T.; Vlanskj, J.; Krohnske, C. L.; Mohammad, M. *Synth. Met.* **1987**, *18*, 1.
125. Diaz, A. F.; Castillo, J. I. *J. Electroanal. Chem.* **1981**, *129*, 115.
126. Zotti, G.; Schiavon, G. *J. Electroanal. Chem* **1984**, *163*, 233.
127. Schiavon, G.; Zotti, G. *J. Electroanal. Chem.* **1984**, *161*, 323.
128. Watanabe, A.; Tanaka, M. *Bull. Chem. Soc. Jpn.* **1981**, *54*, 2278.
129. Pfluger, P.; Street, G. B. *J. Chem. Phys.* **1984**, *80*, 544.
130. Pfluger, P.; Krounbi, M. *J. Chem. Phys.* **1983**, *78*, 3213.
131. Street, G. B.; Clarke, T. C. *J. Phys.* **1983**, *C3*, 599.
132. Nazzai, A.; Street, G. B. *J. Chem. Soc. Chem. Commun.* **1984**, 83.
133. Amer, A.; Zimmer, H. *J. Polym. Sci. Polym. Lett. Ed.* **1984**, *22*, 77.
134. Hotta, S.; Hosaka, T. *J. Chem. Phys.* **1984**, *80*, 954.
135. Kume, K.; Mizuno, K. *Mol. Cryst. Liq. Cryst.* **1982**, *83*, 385.
136. Clarke, T. C.; Scott, J. C. *IBM J. Res. Dev.* **1983**, *27*, 313.
137. Duke, C. B.; Paton, A. *Mol. Cryst. Liq. Cryst.* **1982**, *83*, 177.
138. Tourillon, G.; Gourier, D. *J. Phys. Chem.* **1984**, *88*, 1049.
139. Berlin, A.; Bradamente, S.; Ferracciolz, R.; Pagani, G. A.; Annicolm, F. S. *Synth. Met.* **1987**, *18*, 117.
140. Shen, Y.; Carneiro, K.; Jacobsen, C.; Qian, R.; Qiu, J. *Synth. Met.* **1987**, *18*, 77.
141. Devreux, F.; Genovo, F.; Nechtschein, M.; Villeret, B. *Synth. Met.* **1987**, *18*, 89.
142. Vardeny, Z.; Ehrenfreund, E.; Brafman, O.; Nowak, M.; Schaffer, J. R.; Hegger, A. J.; Wudl, F. *Phys. Rev. Lett.* **1986**, *56*, 671.
143. Roux, C.; Faid, K.; Leclarc, M. *Polym. News* **1994**, *19*, 6.
144. Kanatzidis, M. G. *Chem. Eng. News* **1990**, *68*, 36.
145. Roncali, J. *Chem. Rev.* **1992**, *92*, 711.
146. Leclarc, M.; Diaz, F. M.; Wegner, G. *Makromol. Chem.* **1989**, *190*, 3105.
147. Roux, C.; Leclarc, M. *Macromolecules* **1992**, *25*, 2141.
148. Roux, C.; Bergeron, J. Y.; Leclarc, M. *Makromol. Chem.* **1993**, *194*, 869.
149. Kaneko, M.; Nakamore, H. *J. Chem. Soc. Chem. Commun.* **1985**, 346.
150. Tourillon, G.; Garnier, F. *J. Electroanal. Chem.* **1982**, *135*, 173; *J. Phys. Chem.* **1983**, *87*, 2289; *J. Polym. Sci. Polym. Phys. Ed.* **1984**, *22*, 33.
151. Keozuka, H.; Etoh, S. *J. Appl. Phys.* **1983**, *54*, 2511.
152. Elsenbaumer, R. L.; Jen, K. Y.; Miller, G. G.; Shacklette, L. W. *Synth. Met.* **1987**, *18*, 277.
153. McLeod, G. C.; Mahoubin-Jones, M. G. B.; Pethrick, R. A.; Watson, S. D. *Polymer* **1986**, *27*, 455.
154. MacDiarmid, A. G.; Chiang, J. C.; Richter, A. F.; Epstein, A. J. *Synth. Met.* **1987**, *18*, 285.
155. Kumar, N.; Malholtra, B. D.; Chandra, S. *J. Polym. Sci. Polym. Lett.* **1985**, *23*, 57.
156. Veloazquiz Rosenthal, M.; Skotheium, T. A.; Mela, A.; Florit, M. I.; Salmon, A. *J. Electroanal. Chem.* **1985**, *185*, 297.
157. Malholtra, B. D.; Ramanathan, R. *Phys. Lett. A.* **1985**, *108*, 153.
158. Malholtra, B. D.; Pethrick, R. A. *Macromolecules* **1985**, *16*, 1175.
159. Rolland, M.; Abadie, M. J.; Cadene, M. *Rev. Phys. Appl.* **1984**, *19*, 187.
160. Kaneto, K.; Yoshima, K.; Inuishi, Y. *Jpn. J. Appl. Phys.* **1982**, *21*, L567.
161. MacDiarmid, A. G. *Conjugated Polymers and Related Materials*; Salaneck, W. R.; Lundstrom, I.; Ranby, B., Eds.; Oxford Science Publications: Oxford, England, 1993; p 73.
162. Bredas, J. J. *Conjugated Polymers and Related Materials*; Salaneck, W. R.; Lundstrom, I.; Ranby, B., Eds.; Oxford Science Publications: Oxford, England, 1993; p 187.
163. Whang, Y. E.; Miyata, S. *Conjugated Polymers and Related Materials*; Salaneck, W. R.; Lundstrom, I.; Ranby, B., Eds.; Oxford Science Publications: Oxford, England, 1993; p 149.
164. Vachon, D.; Angus, R. O.; Lu, F. L.; Nowak, M.; Liu, Z. K.; Schaffer, H.; Wudl, F.; Heeger, A. J. *Synth. Met.* **1987**, *18*, 297.
165. Salaneck, W. R.; Lundstrom, I.; Hjerlberl, T.; Duke, C. B.; Conwell, E.; Paton, A.; MacDiarmid, A. G.; Somasiri, N. L. D.; Huane, W. S.; Richter, A. F. *Synth. Met.* **1987**, *18*, 291.
166. Chaing, J. C.; MacDiarmid, A. G. *Synth. Met.* **1986**, *13*, 193.
167. Kaya, M.; Kitani, M.; Sasaki, K. *Chem. Lett.* **1986**, 147.

168. Wack, G. E. *Polym. Prepr. (Am. Chem. Soc. Div. Polym. Symp.)* **1986**, *27*, 277.
169. McManus, P. M.; Yang, S. C.; Cushman, J. *Chem. Soc. Chem. Commun.* **1985**, 1556.
170. Onodera, B. *Phys. Rev. B* **1984**, *30*, 775.
171. Kirdson, S.; Hegger, A. J. *Phys. Rev. Lett.* **1985**, *55*, 308.
172. Chance, R. R.; Boudreaux, J. *Synth. Met.* **1987**, *18*, 329.
173. Epstein, A. J.; Ginder, J. M.; Zoo, F.; Bigelow, R. W.; Woo, H. S.; Tanner, D. B. *Synth. Met.* **1987**, *18*, 306.
174. Richter, A. F.; Huang, W. S.; MacDiarmid, A. G. *Synth. Met.* **1987**, *18*, 306.
175. Tunney, S. E.; Suenaga, J.; Stille, J. K. *Macromolecules* **1983**, *16*, 1398.
176. Wellinghoff, S. T.; Kedrowski, T.; Jenekje, S. A.; Ishida, H. *Mol. Cryst. Liq. Cryst.* **1984**, *105*, 175.
177. Jenekje, S. A.; Wellinhoff, S. T.; Reed, J. F. *Mol. Cryst. Liq. Cryst.* **1984**, *105*, 175.
178. Wellinghoff, S. T.; Deng, Z.; Kedrowski, T. J.; Dick, S. A.; Jenekje, S.A.; Ishida, H. *Mol. Cryst. Liq. Cryst.* **1984**, *105*, 208.
179. Kaneto, K.; Kohno, Y.; Yoshino, K. *Solid State Commun.* **1984**, *51*, 267.
180. Horowitz, G.; Tourillon, G.; Garnier, F. *J. Electrochem. Soc.* **1984**, *131*, 151.
181. Audebert, P.; Bidan, G. *J. Electroanal. Chem.* **1985**, *190*, 129.
182. Kaneto, K.; Yoshima, K. *Synth. Met.* **1987**, *19*, 133.
183. Bredas, J. L.; Themans, B.; Fripiat, J. G.; Andre, J. M.; Chance, R. R. *Phys. Rev. B* **1984**, *29*, 6781.
184. Mo, Z.; Lee, K. B.; Moon, Y. B.; Kobayashi, B.; Heeger, A. J.; Wudl, F. *Macromolecules* **1985**, *18*, 1972.
185. Yong, C.; Renezuan, Q. *Solid State Commun.* **1985**, *54*, 211.
186. Taliani, C.; Daniels, R.; Zameoni, R.; Ostaja, P.; Porzio, W. *Synth. Met.* **1987**, *18*, 177.
187. Malholtra, B. D.; Kumar, N.; Chandra, S. *Prog. Polym. Sci.* **1986**, *12*, 179.
188. Munstedt, H. *Electronic Properties of Polymers and Related Compounds*; Kuzmany, H.; Mehring, M.; Roth, S., Eds.; Springer Verlag: Heidelberg, Germany, 1985; p 8.
189. Chidsey, C. E. D.; Murray, R. W. *Science (Washington, D.C.)* **1986**, *231*, 25.
190. Dim, A. F.; Kanazawa, K. K.; Gardini, C. P. *J. Chem. Soc. Chem. Commun.* **1979**, 635.
191. Tourillon, G.; Garnier, F. *J. Phys. Chem.* **1983**, *87*, 2289.
192. Hillman, A. R. *Electrochmical Science and Technology of Polymers*; Linford, R. G., Ed.; Elsevair Applied Science: London, 1987; Vol. 1, p 241.
193. Jarnigan, J. C.; Chidsey, C. E. D.; Murray, R. W. *J. Am. Chem. Soc.* **1985**, *107*, 2824.
194. Murray, R. W. *Chem. Eng. News* **1987**, 28.
195. Pope, M.; Swenberg, C. E. *Electronic Processes in Organic Crystals*; Clarendon: Oxford, England, 1982.
196. Pope, M.; Kallmann, H.; Magnante, P. *J. Chem. Phys.* **1963**, *38*, 2024.
197. Helfich, W.; Schneider, W. G. *Phys. Rev. Lett.* **1965**, *14*, 229.
198. Vincent, P. S.; Barlow, W. A.; Hann, R. A.; Roberts, G. G. *Thin Solid Films* **1982**, *94*, 476.
199. Friend, R. A. *Conjugated Polymers and Related Materials*; Salaneck, W. R.; Lundstrom, I.; Ranby, B., Eds.; Oxford Science Publications: Oxford, England, 1993; p 285.
200. Peramunage, D.; Tomkiewicz, M.; Ginley, D. S. *J. Electrochem. Soc.* **1987**, *134*, 1384.
201. Jow, J. R.; Schacklette, J.; Maxwell, M.; Vermick, D. *J. Electrochem. Soc.* **1987**, *184*, 1731.
202. Corradini, A.; Mastragostino, M.; Panero, A. S.; Prosperi, P.; Scrosati, B. *Synth. Met.* **1987**, *18*, 625.
203. Genies, E. M.; Pornaut, J. M. *J. Electroanal. Chem.* **1985**, *191*, 111.
204. Skotheim, T. A. *Synth. Met.* **1986**, *14*, 31.
205. Quian, B.; Qiu, J.; Yan, B. *Synth. Met.* **1986**, *14*, 81.
206. Hanh, S. J.; Gajoa, W. S.; Vogerhut, P. O.; Zellor, M.V. *Synth. Met.* **1986**, *14*, 89.
207. Macer, K. A.; Spiro, T. G. *J. Am. Chem. Soc.* **1988**, *105*, 5601.
208. Amrers F. C.; Wi, C. L.; Saveart, J. M. *J. Am. Chem. Soc.* **1985**, *107*, 3442.
209. Zehger, B.; Miller, L. L. *J. Am. Chem. Soc.* **1985**, *106*, 6861.
210. Waltman, C.; Berger, J. J. *Can. J. Chem.* **1986**, *64*, 76.
211. Frommer, J. E. *Acc. Chem. Res.* **1986**, *19*, 2.
212. Naitoh, S. *Synth. Met.* **1987**, *18*, 237.
213. Pethrick, R. A. *Electrochemical Science and Technology of Polymers*; Linford, R. G., Ed.; Elsevier Applied Science: London, 1990; Vol. 2, p 149.

Transparent Polymers for Optical Applications

B. Boutevin, D. Bosc, and A. Rousseau

Transparent polymers offer numerous interesting features for the fabrication of optical fibers and waveguides. In this chapter the possibilities of using halogenated monomers to prepare transparent polymers are discussed. A method to estimate the contribution of halogen atoms to the refractive index to control the numerical aperture of such polymeric optical fibers is described. The need to reduce intrinsic absorption in the polymer, the synthesis, and the characteristics of high halogenated polymers are discussed. Heat and moisture resistant and highly transparent polymers for optical applications can be made by using poly(halogenated acrylate).

Polymers have begun to attract considerable attention from chemists and technologists as challenging systems for basic medium of optical devices. However, a number of difficulties must be overcome for polymers to find their place in the optical arena. First, polymers come after well-established inorganic materials, especially glass and silica, as optical devices. Second, most engineers in optics are only familiar with commodity polymers that are not exactly suitable for the manufacture of optical fibers.

Thus, it seems that polymers hide their great potential for the desired property and performance. But as polymer chemists, we know that the art of organic chemistry allows us to synthesize polymers with the right characteristics for the desired application. The purpose of this chapter is to show how polymer materials can be adapted to improve their optical properties and to render them suitable for the building of polymeric optical components. The required improvements depend on the type of application.

For example, if we consider the optical fiber network, it is well known that the interconnection of conventional diameter fibers is very expensive. This characteristic is especially important for short links in multimode domestic area networks. In this case, the only solution is in the use of very thick fibers, more than 200 μm, for which connections and splices can be very simple. However, connections and splices for 50-μm silica optical fibers need great accuracy. Large diameter makes the silica optical fiber very brittle, impractical to bend, and very difficult to handle, generally.

For silica fibers, the minimum radius of curvature is 100 times the diameter, at least in respect with the bend modulus. On the other hand, polymer optical fibers (POF) have the advantages of flexibility and high elasticity, which make POFs much more convenient for short multimode networks. In comparison with glass fibers, the radius of curvature for POF is as low as 20 times the diameter of the fiber. This characteristic makes possible the use of POFs with diameters larger than 500 μm. Another advantage of POF versus silica fibers is the low density of polymers, around 1.0–1.3 g/cm^3, compared to more than 2 for silica.

However, in the case of polymer optical fibers, high transparency is necessary. This claim is very difficult to fulfill. It is easy to make transparent material for lenses because the thickness or the optical path is very short, but it so much more tricky to have the light crossing a large thickness of tens of meters with little or no damping.

In this chapter we illustrate the importance of new polymers with low light absorption, with emphasis on the middle way between the full transparency of glass and that of the usual nontransparent polymers. At the same time, we also emphasize that it is necessary to keep the customary useful properties of polymers and not to make their cost inhibitory.

Polymer materials offer a number of interesting features for the fabrication of planar channel optical waveguides because of their possibilities to make films of various thicknesses and compositions, such as those already used in photoresist processes. In the optical area we review the various materials and the fabrication processes that have been studied for optical interconnects and integrated optic applications. Here, the requirements are different from those of POFs. The transparency is less critical for planar channel optical waveguides because the optical path is much shorter. The main problems here are the difficulty of processing and interconnecting with other optical waveguides, and the possibility of adding other materials to realize optically active functions, such as modulators and amplifiers.

A planar channel optical waveguide requires finding a process to print an optical path in a film. For that purpose, we have to choose a method to realize a refractive index difference, and we discuss different ways to achieve this. We explain how we can correlate the monomer structure to the refractive index of the polymer. The choice of the polymer that has to be processed for the waveguide is important. Another route is optical printing by the change of chemical structure inside the polymer film. For example, one of these methods is photobleaching. Many methods require the use of functional polymers to fulfill all of these process requirements. Recently there has been considerable interest in polymers with high nonlinear optical (NLO) properties. It was shown in the 1980s that NLO properties of organic molecules may be better than those of usual single crystals such as lithium niobate or potassium dehydrogenophosphate. These results open a wide field of applications, and functional polymers have the potential for the development of components much less costly than existing products. For example, we can expect to realize light modulators based on new grafted copolymers. Presently there is discussion about the emergence of full optical data processing and telecommunication networks, as well as full optical computers. The challenge rests with the synthesis of polymers with the right dye molecules anchored to the macromolecular chains.

These three topics—POFs, polymer for planar channel waveguide, and polymers with high and stable NLO properties—are the main features of a future all-organic optical systems for many applications. These systems require the control of a few basic prop-erties such as the refractive index of polymers and the light attenuation. These two points will be elaborated in this chapter. The more specific topics of NLO polymers and their third-order nonlinearity are the subjects of Chapters 2.5 and 2.6.

Refractive Index in Optical Waveguides

There are two kinds of construction design for optical waveguides: cylindrical and planar (Figure 1). The cylindrical design is used for long distance transmission applications, and the planar design is used for very short distances (less than 0.1 m). The materials requirements are expressed in terms of attenuation of light and refractive index difference between the core and the optical cladding of the guide. The attenuation of light will be discussed in the next section. The refractive index difference (RID) enables the medium to guide the light. The refractive index of the core, where the light travels, must be higher than the refractive index of the surrounding material (optical cladding). The RID also controls the number of propagation modes. It is linked to the numerical aperture (NA) criterion. From the value NA, one can calculate the acceptance angle, Θ_A, the light-gathering ability of

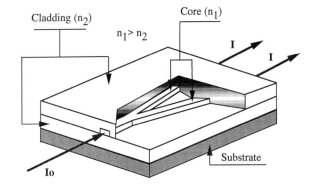

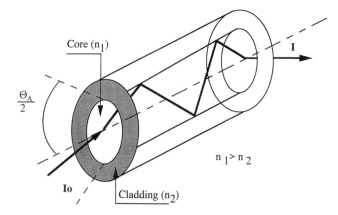

Figure 1. Schematic diagrams of planar and cylindrical waveguides: n_1 and n_2 are refractive indexes; I_o and I are, respectively, the input and output power.

the fiber, according to eq 1, where n_1, n_2, and n_0 are the refractive indexes, respectively, of the core, optical cladding, and the medium where the light is coming from.

$$NA = \sin\left(\frac{\Theta_A}{2}\right) = \frac{n_1}{n_0}\sqrt{1 - \frac{n_2^2}{n_1^2}} \qquad (1)$$

Fortunately, we will see in this chapter that the substitution of hydrogen atoms in the polymer by halogen atoms can afford the control of the refractive index as well as the decrease of the intrinsic absorption. However, the control of the refractive index is of primary importance to build a light waveguide, and we will see the wide abilities of polymers for producing materials with the desired refractive indexes. We also explain a new method of ours that is well suited for predicting the refractive index of materials. We begin with the chemical structure.

This section is by no means a report about waveguide technologies, but it is important to know about the different methods available for making POFs and planar polymer waveguides and the requirements for the guiding process. For both cases, many methods are based on coating the core of the guide by optical cladding polymers (1, 2). Recently, a practical way for making two-dimensional waveguides was described based on modifying the refractive index of the polymer by photobleaching (3). In this way, it is quite possible to find a more efficient route to build a waveguide than with the classical mineral materials. Other examples are the process implemented by B. L. Blooth (4), injection molding (5), and photolithography and dry-etching processes (6). However, for each process one needs to know and to tune the refractive index of the polymer.

Structure–Refractive Index Relationship in Monomers and Organic Liquids

The refractive index (n) of a medium is defined as the ratio of the velocity of the light in vacuum (c) to the velocity in the medium (v) (eq 2). The well-known Lorentz–Lorenz equation (eq 3) correlates n of an organic compound to the density (ρ), the molecular weight (M), and the molar refractivity (R) (7).

$$n = c/v \qquad (2)$$

$$\frac{n^2 - 1}{n^2 + 2}\frac{M}{\rho} = R \qquad (3)$$

For a very large number of compounds, the assessment of the molar refraction can be achieved by adding the contributing value of each group of atoms

present in the molecule. However, for the determination of the refractive index from eq 3, one needs to know the density of the compound.

Hoshino et al. worked out a method to calculate theoretically some of the physical characteristics of organic compounds such as refractive index of aliphatic hydrocarbons (8), liquid molar volume (9), and the latent heat of vaporization at normal boiling point (10). Their method is called the "group contribution method". It enables calculation of n from the chemical formula of the organic compound, according to eq 4, where n^{25} is the refractive index at 25 °C, and Δd_i is the increment of the ith functional group. This method yields good assessments of the refractive index of aliphatic hydrocarbons, but with halogen compounds it is inaccurate.

$$n^{25} = \frac{M}{\sum_i \Delta d_i} \qquad (4)$$

More recently W. Groh et al. (11) created a method to calculate the refractive index n_D of unknown amorphous organic polymers below the glass transition temperature (T_g). This method is based on the Lorentz–Lorenz equation, which can be written in the form of eq 5, where n_D is the refractive index at 589 nm, k_j is the number of increments of the substructure j in the repeating unit, R_j and V_j are, respectively, the molar refraction increment (cm³/mol) and the molar volume (cm³/mol) of the substructure j, and J is the total number of individual substructures. However, very few data are available concerning halogenated alkanes and acrylic monomers.

$$\frac{n_D^2 - 1}{n_D^2 + 2} = \left(\sum_{j=1}^{J} k_j R_j\right)\left(\sum_{j=1}^{J} V_j\right)^{-1} \qquad (5)$$

For this reason, we have proposed a new method enabling the calculation of the refractive index of halogenated alkanes and halogenated acrylic monomers with good accuracy at several temperatures. According to this method, the refractive index is written as an empirical rule given by eq 6, where n_D^T is the calculated refractive index at λ = 589 nm, and T is temperature (n_D^T is denoted n^T in the rest of this chapter). The index, i, is related to the nature of the atoms or groups of atoms in the molecules. The contribution of each group, i, in the whole refractive index is represented by K_i^T at temperature T, and N_i is the number of atoms or groups (i) in the molecule.

$$n_D^T = \frac{\sum_i K_i^T N_i}{\sum_i N_i} \qquad (6)$$

Refractive Index of Halogenated Alkanes

In the case of halogenated alkanes (12) containing carbon, fluorine, chlorine, and bromine atoms, eq 6 can be written in the form of eq 7.

$$n^T = \frac{K_C^T N_C + K_F^T N_F + K_{Cl}^T N_{Cl} + K_{Br}^T N_{Br}}{N_C + N_F + N_{Cl} + N_{Br}} \quad (7)$$

The determination is carried out in two steps. First, the constants K_i are determined from model molecules, and then the viability of the method is checked with compounds having known refractive indexes. A number of perhalogenated molecules and their telomers were prepared (13). Thus, chlorotrifluoroethylene and tetrafluoroethylene were telomerized with carbon tetrachloride, bromotrichloromethane, and dibromodichloromethane (14–16). Chemical formulas of perhalogenated alkanes and their refractive indexes at several temperatures are reported in Table I.

From the data of Table I, the contribution K_i at different temperatures can be calculated. The accuracy of calculations is better than 5%, and the linear correlation coefficient is higher than 0.99999. The results are reported in Table II.

On the one hand the influence of a halogen on the refractive index is larger as the atom is bigger. This characteristic is in agreement with the results previously published (11). On the other hand, when the temperature dependence of K_i is linear, the correlation coefficient is in all cases better than 0.99. Thus, the

contribution of halogen and carbon at 20 °C can be calculated. Then, the K_i values are K_C^{20}, 2.622; K_F^{20}, 0.7215; K_{Cl}^{20}, 1.161; and K_{Br}^{20}, 1.429. To check the method, the refractive indexes of halogenated alkanes were calculated, and the results are reported in Table III. The accuracy is quite good and has less than 1% error. This result means that the method is well suited for predicting the refractive index of simple halogenated molecules. Therefore, expanding the method for the determination of refractive indexes of halogenated acrylic monomers is interesting.

Refractive Indexes of Halogenated Acrylic Monomers

Halogenated acrylates and methacrylates are prepared by esterification of the corresponding acid or the acid chloride with halogenated alcohols (20, 21). The esters of α-chloroacrylic acid are synthesized from the 1,2-dichloropropionic acid as shown in Scheme IA. Halogenated alcohols can be obtained by reaction of oleum with the halogenated telomers prepared by telomerization of trifluorochloroethylene with carbon tetrachloride, followed by reduction by lithium aluminium hydride (Scheme IB) (22). We also used lithium aluminium hydride to transform $C_7F_{15}CO_2H$ into $C_7F_{15}CH_2OH$ (23). Another way to prepare halogenated alcohols is telomerization, by a redox catalyst, of allylic alcohol and carbon tetrachloride (Scheme IC) (24).

Acrylic monomers used for the calculation of the group contributions are gathered in Table IV. As with halogenated alkanes, the refractive indexes of halogenated acrylic monomers can be calculated by eq 6, re-written as eq 8, where K_A, K_C, K_H, K_F, and K_{Cl} are, respectively, contributions of acrylic group (C=CCO$_2$), carbon, hydrogen, fluorine, and chlorine atoms.

After calculation (correlation coefficient higher than 0.99999), the contributions of each group and atoms are determined at 20 °C:

$$n^T = \frac{K_A^T N_A + K_C^T N_C + K_H^T N_H + K_F^T N_F + K_{Cl}^T N_{Cl}}{N_A + N_C + N_H + N_F + N_{Cl}} \quad (8)$$

where $K_A^{20} = 3.558$, $K_C^{20} = 2.579$, $K_H^{20} = 0.8489$, $K_F^{20} = 0.6699$. and $K_{Cl}^{20} = 1.147$. The viability of the method

Table I. Refractive Indexes of Halogenated Alkanes at Various Temperatures

Halogenated Alkane	n^{25}	n^{30}	n^{35}	n^{40}
CCl$_3$CF$_2$CFCl$_2$	1.4360	1.4341	1.4319	1.4290
CCl$_3$(CF$_2$CFCl)$_2$Cl	1.4325	1.4310	1.4290	1.4275
CCl$_3$(CF$_2$CFCl)$_3$Cl	1.4320	1.4307	1.4286	1.4270
CCl$_3$CF$_2$CF$_2$Cl	1.4000	1.3980	1.3950	1.3921
CCl$_3$CF$_2$CFClBr	1.4621	1.4600	1.4575	1.4550
CCl$_2$BrCF$_2$CFClBr	1.4857	1.4835	1.4810	1.4789
CCl$_3$CF$_2$CCl$_3$	1.4775	1.4752	1.4730	1.4701
CClF$_2$CFCl$_2$	1.3558	1.3535	1.3505	1.3475

Source: Reproduced with permission from reference 12. Copyright 1984 Elsevier.

Table II. Group Contributions K_i^T and Function $K_i = f(T)$

K_i^T	25 °C	30 °C	35 °C	40 °C	$K_i = f(T)$
K_C	2.639	2.650	2.664	2.685	$3.035 \times 10^{-3}\,T + 2.561$
K_F	0.7109	0.7051	0.6944	0.6840	$-1.826 \times 10^{-3}\,T + 0.758$
K_{Cl}	1.151	1.144	1.136	1.124	$-1.773 \times 10^{-3}\,T + 1.196$
K_{Br}	1.421	1.413	1.404	1.395	$-1.691 \times 10^{-3}\,T + 1.463$

Source: Reproduced with permission from reference 12. Copyright 1984 Elsevier.

Table III. Experimental and Calculated Values of Refractive Index for Halogenated Molecules at 20 °C

Halogenated Alkane	n^{20}_{exp}	n^{20}_{calc}	Error (%)
$CF_2BrCFClBr$	1.4270	1.4285	0.10
$CFCl_2CF_2CCl_3$	1.4389	1.4395	0.04
$CF_2ClCFClCCl_3$	1.4392	1.4395	0.02
$CFCl_2CFClCFCl_2$	1.4387	1.4395	0.05
$CCl_3CF_2CF_2Cl$	1.3961	1.3396	0.25
$CF_2ClCFClCFCl_2$	1.3960	1.3996	0.25
$CF_2ClCCl_2CF_2Cl$	1.3958	1.3996	0.38
$CF_3CFClCCl_3$	1.4000	1.3996	0.03
$CF_2ClCFClCF_2Cl$	1.3512	1.3596	0.62
$CFCl_2CFClCF_3$	1.3530	1.3596	0.48
$CF_2BrCFBrCClF_2$	1.4028	1.4084	0.40
$CF_2BrCClBrCClF_2$	1.4471	1.4484	0.09
CCl_4	1.4664	1.4532	0.90
CCl_3Br	1.5061	1.5068	0.05
CBr_3F	1.5256	1.5268	0.08

Note: Values compiled from references 7, 17–19.

Source: Reproduced with permission from reference 12. Copyright 1984 Elsevier.

was assessed for different acrylic monomers, and the results are reported in Table V.

The contributions of K_i for the monomers are quite different from the contributions obtained from the halogenated alkanes. This difference is because the polarizations are not the same in halogenated alkanes and in halogenated acrylic monomers. Therefore, two different K_i values are available for the refractive index calculation: one for halogenated alkanes and one for halogenated acrylates.

Practical Examples

For the design of a planar or fiber waveguide, one needs to take into account the core and the cladding refractive indexes to enable efficient light transmission. Two examples are chosen to illustrate this point.

1. The implementation of a planar waveguide showing the use of materials with different refractive index is illustrated in Scheme II (3). In this special case only one polymer is used: the poly(vinylcin-

A)

$$CH_2ClCHClCO_2H \xrightarrow[\text{2) Base}]{\text{1) SOCl}_2} CH_2=CClCOCl \xrightarrow{ROH} CH_2=CClCO_2R$$

B)

$$Cl-(CFClCF_2)_n-CCl_3 \xrightarrow{\text{oleum}} Cl-(CFCl-CF_2)_n-CO_2H \xrightarrow{AlLiH_4} Cl-(CFClCF_2)_n-CH_2OH$$
$(n = 1, 2, 3)$

C)

$$CH_2=CHCH_2OH + CCl_4 \longrightarrow CCl_3CH_2CHClCH_2OH$$

Scheme I. Syntheses of halogenated acrylates and alcohols.

Table IV. Acrylic Monomers for Calculation of K_i

Acrylic Monomer	n^{20}
$CH_2=CHCO_2CH_3$	1.3984
$CH_2=C(CH_3)CO_2CH_3$	1.4140
$CH_2=CClCO_2CH_3$	1.4420
$CH_2=C(CH_3)CO_2CH_2CH_3$	1.4128
$(CH_2=C(CH_3)CO_2CH_2CH_2)_2-$	1.4565
$CH_2=CClCO_2CH_2CHClCH_2CCl_3$	1.503
$CH_2=CHCO_2CH_2CH_2C_6F_{13}$	1.338
$CH_2=CHCO_2CH_2C_7F_{15}$	1.328

Source: Reproduced with permission from reference 23. Copyright 1988 Elsevier.

Table V. Theoretical and Experimental Refractive Indexes of Acrylic Monomers

Acrylic Monomer	$n^{20}exp$	$n^{20}calc$	Error (%)
$CH_2=CHCO_2CH_2CHClCH_2CCl_3$	1.493	1.485	0.5
$CH_2=CHCO_2CH_2-(CF_2CFCl)_3Cl$	1.418	1.403	1.06
$CH_2=CHCO_2CH_2-(CF_2CFCl)_2Cl$	1.416	1.408	0.6
$CH_2=CClCO_2CH_2CH_2C_6F_{13}$	1.351	1.349	0.15
$CH_2=CClCO_2CH_2C_7F_{15}$	1.338	1.337	0.07
$CH_2=CHCO_2(CH_2)_3CH_3$	1.417	1.415	0.14
$CH_2=CClCO_2CH_2CH_3$	1.438	1.437	0.07
$CH_2=CHCO_2CH_2CH_3$	1.405	1.409	0.3
$(CH_2=CHCO_2CH_2)_3CCH_2CH_3$	1.474	1.487	0.9

Source: Reproduced with permission from reference 23. Copyright 1988 Elsevier.

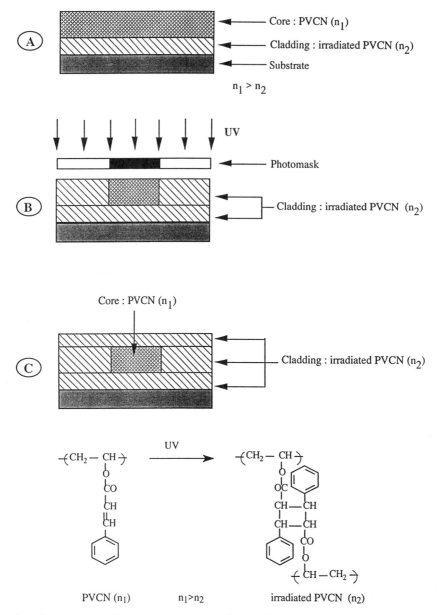

Scheme II. Example of a planar waveguide processing: A, coating a substrate with PVCN on a layer of irradiated PVCN; B, photobleaching for refractive index modification through a photomask; C, coating the core waveguide with a top layer.

namate) (PVCN). The cinnamate group (C_6H_5-CH=CHCO$_2$R) undergoes [2 + 2] photocycloaddition under UV irradiation. Therefore, the PVCN polymer is used to obtain a localized optical waveguide in one step. Under UV irradiation the [2 + 2] photocycloaddition between the double bonds leads to a bleaching process and a cross-linked polymer. This photodimerization provides a decrease in the refractive index and allows the drawing of an optical pattern on the polymer material with cross-linking.

2. The abacus of a numerical aperture (NA) controls the admittance of more or less light energy depending on the refractive indexes of the core-cladding (Figure 2) (25). Thus, to obtain a well-

defined NA this abacus provides a match between the core and cladding materials. In the Figure 2, the cladding material is a copolymer of methyl methacrylate and methacrylate bearing fluorinated groups, and its refractive index is 1.40. As can be seen from the abacus, one can choose either poly(methyl methacrylate) (PMMA) or polystyrene (PS) as the core material, depending on other materials already used in the optical device where the waveguide will be located.

These examples show how important it is, by monomers syntheses, to have a wide range of refractive indexes of available materials. For this reason, we devoted this section to the synthesis of monomers hav-

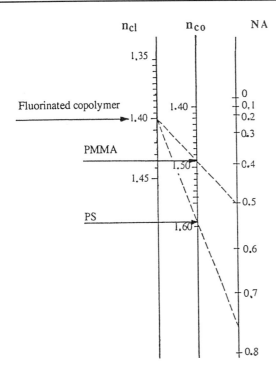

Figure 2. Abacus of numerical aperture (NA) and refractive index for the design of optical core (n_{co}) and cladding (n_{cl}) in a waveguide. The cladding is a copolymer of methyl methacrylate, $CH_2=C(CH_3)CO_2CH_3$, and methacrylate bearing fluorinated group, $CH_2=C(CH_3)CO_2C_2H_4C_6F_{13}$. The positions of poly(methyl methacrylate) (PMMA) and polystyrene (PS) are indicated. (Reproduced with permission from reference 25. Copyright 1988.)

Table VI. Physical Characteristics of Amorphous Polymers PMMA, PS, and PC

Property	PMMA	PS	PC
Refractive index	1.491	1.590	1.586
Glass transition temperature (°C)	105	100	150
Max. usuable temperature (°C)	80	70	120
Thermal expension coefficient (10^{-6}/°C)	63	80	70
Density (g · cm^{-3})	1.17	1.06	1.20

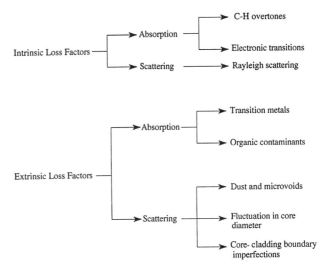

Figure 3. Loss factors in polymer optical fibers.

Highly Transparent Polymers

A large number of organic polymers are well known to exist as fully amorphous glasses at room temperature (Table VI). They show good transparency in the visible region, and some of them are candidates for fabrication of POFs (26–29). For about a decade, POFs were made and used for short length optical transmission data links and lighting (30). Indeed, POFs have many advantages, including good ductility, ease of handling, the ability to connect together (because of their large core diameter), and high numerical aperture. Usually, PMMA (31), PS (32, 33), and polycarbonate (34) are used as core materials, and polyfluorinated polymers are used as cladding materials.

ing the refractive indexes needed in the range of 1.34–1.5. The corresponding refractive indexes for the halogenated polyacrylates are about 0.2–0.5 higher than those of the monomers, depending on the polymer structure. Thus, a method was developed to build the suitable material with the required refractive index. The method is based on the contribution of chemical groups to the refractive index of the molecules with great accuracy for the acrylate monomers family.

However, light loss in POFs is much higher than for silica optical fibers. The total attenuation loss (β in decibels per kilometer and with an error of 5%) of a POF is defined in eq 9, where I_0 and I are, respectively, the input and output power, and L (km) is the length of the POF.

$$\beta = (10/L)\log(I_0/I) \qquad (9)$$

The loss factors of POF are shown in Figure 3 (35). The attenuation curve in the range of 500–750 nm for a PMMA POF is drawn in Figure 4 (36).

The damping of light in POF is caused by both intrinsic and extrinsic losses (37, 38). Extrinsic losses in POF play a more significant role than in silica optical fibers (39). At present, the best reported result closest to the theoretical limit for PMMA is 60 dB/km at 570 nm (37). For PMMA in the visible and near IR (NIR) regions, losses are dominated by overtones of C–H IR vibration. The sixth overtone of C–H stretching vibration occurs at 620 nm and has an intensity of 300–400 dB/km (Figure 4). To use the optical properties of amorphous polymers to their best (high numerical and good flexibility), it is a challenge to reduce these high absorptions.

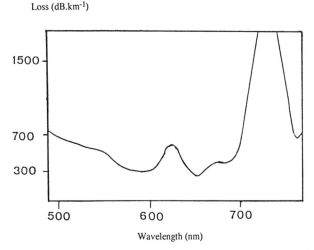

Loss (dB.km⁻¹)

Figure 4. Transmission loss spectrum for a PMMA optical fiber core. (Adapted and reproduced with permission from reference 1. Copyright 1991.)

Many attempts were made during the 1980s to produce POFs with new polymers with weak absorption in the visible and NIR regions, and fully deuterated polymers were examined for this purpose (*40*, *41*). The first poly(octadeutero methyl methacrylate) used as a POF exhibited a low attenuation of 20 dB/km in the red region (*42*), and the ultimate loss limit of POF is expected to be about 6 dB/km with poly (1,1,3-trideuterohexafluorobutyl pentadeuteromethylmethacrylate) (*43*).

However, POFs made with deuterated polymers have high water absorption, and moreover, a high cost of synthesis. Unfortunately, fully halogenated (especially fluorinated) polymers cannot be used as POF core, because of their high degree of crystallinity and low refractive index. Meanwhile, for cladding, poly-(tetrafluoroethylene-*co*-hexafluoropropylene) (FEP) can be used (*44*). Other polymers from halogenated acrylic monomers have been made with partially halogenated polymers. They seem to be a good compromise between optical and physical properties on the one hand and ease of synthesis on the other. However, before we can appreciate the nature of such balanced properties, we need to study the basic optical properties of polymers.

Absorption Phenomena in Polymers

Intrinsic loss phenomena in polymers are mainly due to electronic transitions in the visible region as Rayleigh scattering absorption of overtone of C–H IR vibration.

The electronic transition absorption (α_e) is due to UV absorption. There are two kinds of electronic absorptions: the $\pi \rightarrow \pi^*$ transition of the phenyl group in the case of polystyrene (PS), and the $n \rightarrow \pi^*$ transi-

tion of the carbonyl group in PMMA. Thus, electronic absorption is located in the UV region with its tail in the visible region and obeys the Urbach rule (*45*). The values of α_e (dB/km) for PMMA and PS are given by eqs 10 and 11, where λ is the wavelength of the absorbed light (*46*). Using these equations for $\lambda = 500$ nm, one can calculate an electronic absorption of less than 1 dB/km for PMMA and 98 dB/km for PS.

$$\alpha_e(\text{PMMA}) = 1.58 \times 10^{-12} \exp\left(\frac{1.15 \times 10^4}{\lambda}\right) \quad (10)$$

$$\alpha_e(\text{PS}) = 1.10 \times 10^{-5} \exp\left(\frac{8.0 \times 10^3}{\lambda}\right) \quad (11)$$

The Rayleigh scattering, based on the fluctuation of density in the isotropic material within the core of optical fibers, has a value around 10–20 dB/km (*47*) at 633 nm.

In wavelength regions longer than 700 nm, intrinsic losses such as electronic transition and Rayleigh scattering have little influence on the attenuation loss. Thus, the overtones of C–H bonds are the most important loss factor in the visible and NIR regions. High overtone absorptions occur at approximate multiples of the fundamental frequency, and the intensity of these overtones decreases by one order of magnitude with each successive overtone.

Energy of absorption depends on the strength of the oscillator C–H. If a diatomic molecule can be visualized as a harmonic oscillator, the fundamental vibration wavenumber ($\bar{\nu}_0(\text{cm}^{-1})$), the radiation wavenumber emitted or absorbed) is given by eq 12, where k is the restoring force and μ is the reduced mass (*48*).

$$\bar{\nu}_0 = \frac{1}{2\pi c}\sqrt{\frac{k}{\mu}} \quad (12)$$

But in many cases, diatomic vibration in a polyatomic molecule cannot be considered as a harmonic oscillator but as an anharmonic one. Therefore, the position of the fundamental wavenumber ($\bar{\nu}_1$) (observed in the IR spectra) is given by eq 13, where x is the anharmonicity constant. In a similar manner, the wavenumber ($\bar{\nu}_n$) of the nth overtone in stretching vibration is given by eq 14 (*49*):

$$\bar{\nu}_1 = \bar{\nu}_0 - 2\bar{\nu}_0 x \quad (13)$$

$$\bar{\nu}_n = \frac{\bar{\nu}_1 n[1 - x(n+1)]}{1 - 2x} \quad (14a)$$

$$x = \frac{n\bar{v}_1 - \bar{v}_n}{n(n+1)\bar{v}_1 - 2\bar{v}_n} \quad (14b)$$

Table VII lists the fundamental and overtones of O–H, C–H, C–D, C=O, C–F, and C–Cl bonds.

When $\bar{v}_1$ and $\bar{v}_2$ are known from the absorption spectra, x can be calculated. For example, for a C–H bond, $\bar{v}_1 = 2950$ cm^{-1} and $\bar{v}_2 = 5783$ cm^{-1}; thus, from eq 14a and $n = 2$: $x_{C-H} = 1.9 \times 10^{-2}$. In a similar manner, x values for other main bonds can be obtained: $x_{C-D} = 1.46 \times 10^{-2}$; $x_{O-H} = 1.95 \times 10^{-2}$; $x_{C-F} = 4 \times 10^{-3}$; $x_{C=O} = 6.5 \times 10^{-3}$; $x_{C-Cl} = 5.9 \times 10^{-3}$.

To reduce absorption in the visible and NIR regions, the reduced mass μ must be increased. For example, for C–D bonds in polymer materials, the absorption due to stretching vibration is shifted toward higher wavelengths. For C–F bonds the shift is more pronounced, and the fundamental vibration occurs around 8 μm instead of 4.4 μm for C–D bond. This shift would allow the use of halogenated organic materials at longer wavelength optical windows than is possible with deuterated polymers.

Table VII. Fundamental and Overtone Oscillations of Main Chemical Bonds in Visible, near IR, and IR Regions

Oscillation	Fundamental and Overtone for Main Bonds (nm) (wavenumber, cm^{-1})					
	O–H	C–H	C–D	C=O	C–F	C–Cl
v_1	2818	3390	4484	5417	8000	12987
	(3548)	(2950)	(2230)	(1846)	(1250)	(770)
v_2	1438	1729	2276	2727	4016	6533
	(6952)	(5783)	(4393)	(3668)	(2490)	(1531)
v_3	979	1176	1541	1830	2688	4381
v_4	750	901	1174	1382	2024	3306
v_5	613	736	954	1113	1626	2661
v_6	523	627	808	934	1361	2231
v_7	—	549	704	806	1171	1924
v_8	—	492	626	710	1029	1694
v_9	—	447	566	635	919	1515
v_{10}	—	—	519	576	830	1372
v_{11}	—	—	—	527	758	1256
v_{12}	—	—	—	—	698	1158
v_{13}	—	—	—	—	647	1076
v_{14}	—	—	—	—	603	1006
v_{15}	—	—	—	—	565	945
v_{16}	—	—	—	—	532	892
v_{17}	—	—	—	—	503	845
v_{18}	—	—	—	—	—	803
v_{19}	—	—	—	—	—	766
v_{20}	—	—	—	—	—	732
v_{21}	—	—	—	—	—	702

Note: v_1 is the fundamental and v_n, $n>1$, are the overtones. Dashes mean that values are not calculated and not given in reference 49 because they are out of visible range.

Source: Adapted and reproduced with permission from reference 49. Copyright 1988 Hüthig & Wepf Verlag.

Halogenated Monomers and Polymers

Preliminary studies reported in Figure 5 *(50)* show the absorption bands (in dB/cm) versus wavelength for halogenated monomers used in the refractive index studies. At the wavelength of 1.14 μm, absorption bands are due to the third overtone of CH=C vibration; at a wavelength of 1.17 μm, the bands are those of the third overtone of C–H. At 1.17 μm, and in terms of absorption strength, the lesser the number of C–H bonds per cubic centimeter the lower the absorption. The very important losses introduced by stretching vibration of N–H, C–H, and O–H bonds were studied by Takezawa et al. *(51)*. They demonstrated that the variation in the absorption strength of the second overtone of C–H, N–H, and O–H vibrations is directly proportional to the concentration of these bonds. For an equal concentration, the order of absorption is O–H > N–H > C–H.

According to the Beer–Lambert law, the absorption coefficient (ε^n) for a given bond at a wavelength corresponding to the nth overtone is expressed in eq 15, where I_0 and I are the input and output power, c is the concentration (mol/cm^3) of the absorbing species, and l is the path length of the sample:

$$\log\frac{I_0}{I} = \varepsilon^n \cdot c \cdot l \quad (15)$$

If l is in kilometers, eq 15 can be rewritten as eq 16, according to eq 9, where β^n (dB/km) is the total absorption of the sample, and N is the concentration (mol/cm^3) of the absorbing bonds; N is defined in eq. 16a, where ρ is the density (g/cm^3) of the compound, M is the molecular weight (g/mol), and n_a is the number of the given bond in the molecule or in the monomer unit in case of polymer.

$$\beta^n = \varepsilon^n N \quad (16a)$$

$$N = \rho n_a / M \quad (16b)$$

For example, the number, N, of C–H bonds per cubic centimeter for a PMMA sample is $N = 94 \times 10^{-3}$, and $\rho = 1.17$ g/cm^3, $n_a = 8$, and $M = 100$ g/mol.

From model compounds, absorption coefficients (ε^n) of fundamentals and overtones can be measured from eq 16. The linear relationships are given in Table VIII and are also drawn in Figure 6 *(52)*.

As an example, for an organic compound with an equal number of C–H and C–F bonds per cubic centimeter, the total absorption due to C–H and to C–F at a wavelength close to 1 μm (third overtone of C–H ($n = 3$) and the eighth of C–F ($n = 8$)) can be calculated as follows, and values are given in Table VIII.

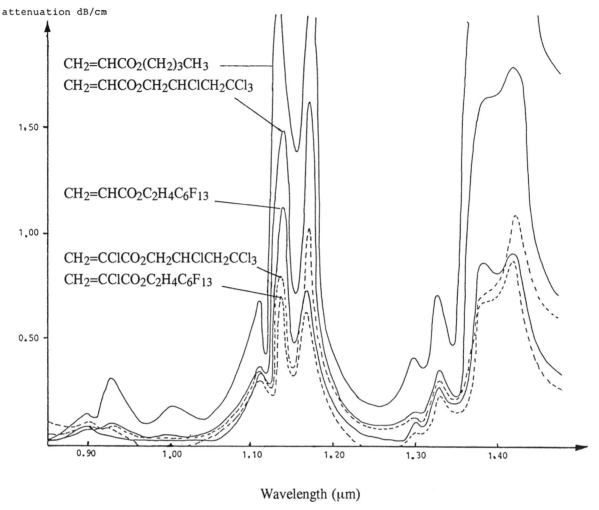

attenuation dB/cm

$CH_2=CHCO_2(CH_2)_3CH_3$ ————
$CH_2=CHCO_2CH_2CHClCH_2CCl_3$

$CH_2=CHCO_2C_2H_4C_6F_{13}$

$CH_2=CClCO_2CH_2CHClCH_2CCl_3$
$CH_2=CClCO_2C_2H_4C_6F_{13}$

Wavelength (μm)

Figure 5. Attenuation spectra of ordinary (hydrogenated) and halogenated acrylates in the NIR region. (Reproduced with permission from reference 50. Copyright 1984.)

$$\log\varepsilon_{C-F}^8 = -2.68 \text{ and } \log\varepsilon_{C-H}^3 = 6.8 \therefore \frac{\beta_{CF}^8}{\beta_{CH}^3} = 3.3\times10^{-10}$$

Thus, in the visible and NIR regions, the absorptions of C–F stretching vibrations are negligible compared with those of C–H bonds. On the basis of these studies, we prepared new monomers containing very few C–H bonds for the development of POFs.

To reduce the number of C–H bonds in the molecule, we have prepared α-chloroacrylates, α-fluoroacrylates, halogenated vinyl carbonates, and the trideuteromethyl trifluoroacrylate (a monomer without any C–H bonds). α-Fluoroacrylates were prepared according to the procedure described by Molines et al. (53, 54), as indicated in Scheme IIIA (55). Recently Nguyen and Wakselman (56) proposed a new convenient three-step method to synthesize α-fluoroacrylates from 2,3-fluorinated propanols. The overall yield on acid fluoride is 84% (Scheme IIIB). Trifluoroacrylate

also was synthesized from 1,2-difluorotetrachloroethane (Scheme IIIC) (57, 58). A monomer with 2,2,2,1-tetrachloroethyl group was prepared by reaction of methacryloyl chloride and chloral (Scheme IIID) (59). Vinyl carbonate monomers also were synthesized by reaction between vinylchloroformiate and alcohols (Scheme IIIE) (60). The monomer $CCl_2=CHOCO_2C_6F_5$ was prepared by reaction of phosgene with chloral.

The refractive index and the number of C–H bonds per cubic centimeter (N_{CH}) of the new monomers prepared are listed in Table IX (61). The values of N_{CH} of halogenated monomers vary from 0 to 53.5 × 10^{-3} compared with 75 × 10^{-3} C–H bonds per cubic centimeter for the methyl methacrylate. The variation of refractive index is from 1.3450 to 1.4975 for acrylic monomers and 1.3441 to 1.4667 for vinyl carbonate monomers. Because of their absorption, all the new monomers have an N_{CH}, and therefore a β^n, lower than the N_{CH} of methyl methacrylate. The lowest values are obtained with $CF_2=CFCO_2CD_3$, $CH_2=CXCO_2$-

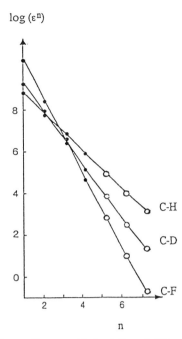

log (ε^n)

n

Figure 6. Absorption coefficient (log (ε^n)) of C–H, C–D, and C–F bonds related to the quantum number (*n*). Experimental values (●) and calculated values (○) from equations given in Table VIII. (Reproduced with permission from reference 52. Copyright 1992 John Wiley & Sons, Inc.)

Table IX. Refractive Index at 20 °C and Number C–H Bonds per cm³ (N_{C-H}) of New Halogenated Monomers

Monomer	n^{20}	N_{C-H} (10^{-3})
Halogenated Acrylic		
$CF_2=CFCO_2CD_3$	1.3667	0
$CH_2=CClCO_2C_6F_5$	1.4555	11
$CH_2=CFCO_2C_6F_5$	1.4305	12
$CH_2=C(CH_3)CO_2CH_2(CF_2CFCl)_3Cl$	1.4257	22
$CH_2=CFCO_2C_2H_4C_6F_{13}$	1.3450	23
$CH_2=CFCO_2CH_2CCl_3$	1.4636	26
$CH_2=CClCO_2CH_2CCl_3$	1.4950	26.5
$CH_2=C(CH_3)CO_2CH_2(CF_2CFCl)_2Cl$	1.4217	27.5
$CH_2=C(CH_3)CO_2C_6F_5$	1.4413	28
$CH_2=CFCO_2CH_2CF_3$	1.3450	31
$CH_2=C(CH_3)CO_2C_2H_4C_6F_{13}$	1.3480	31
$CH_2=C(CH_3)CO_2CHClCCl_3$	1.4856	34
$CH_2=CFCO_2CH_2CF_2CF_2H$	1.3600	35
$CH_2=C(CH_3)CO_2CH_2CF_2CFCl_2$	1.4209	37.5
$CH_2=C(CH_3)CO_2CH_2CCl_3$	1.4725	43
$CH_2=C(CH_3)CO_2CH_2CHClCH_2CCl_3$	1.4975	49
$CH_2=CFCO_2C_6H_5$	1.5030	50
$CH_2=(CH_3)CO_2CH_2CF_3$	1.3613	50
$CH_2=CFCO_2CH_3$	1.3915	53.5
$CH_2=C(CH_3)CO_2CH_3$	1.4140	75
Vinyl Carbonates		
$CCl_2=CHOCO_2C_6F_5$	1.4663	6
$CH_2=CHOCO_2C_6F_5$	1.4288	18
$CH_2=CHOCO_2C_2H_4C_6F_{13}$	1.3441	23
$CH_2=CHOCO_2CH_2CCl_3$	1.4667	30
$CH_2=CHOCO_2CD_3$	1.3990	31
$CH_2=CHOCO_2CH_2CF_3$	1.3552	37

Source: Adapted and reproduced with permission from reference 61. Copyright 1994 Taylor & Francis.

Table VIII. Linear Relation Between Absorption Coefficient, Log (ε^n), and Quantum Number *n*

Bond	$Log\ (\varepsilon^n) = f(n)$
C–H	$-(n-1) + 8.8$
C–D	$-1.3\,(n-1) + 9$
C–F	$-1.84\,(n-1) + 10.2$

Source: Reproduced with permission from reference 52. Copyright 1992 John Wiley & Sons.

A)

$CH_2BrCFBrCOBr + ROH \longrightarrow CH_2BrCFBrCO_2R \xrightarrow{Zn} CH_2=CFCO_2R$

$R = C_6F_5,\ C_2H_4C_6F_{13},\ CH_2CCl_3.$

B)

$HCFXCF_2CH_2OH \xrightarrow{2R'Li} CFX=CFCH_2OH \xrightarrow{H^+} CH_2=CFCOF \xrightarrow{ROH} CH_2=CFCO_2R$

$X = Cl\ or\ F$

C)

$CCl_2FCCl_2F \xrightarrow{Zn} CFCl=CFCl \xrightarrow{CFCl_3\ /\ AlCl_3} CF_2ClCFClCCl_3$

$CF_2ClCFClCCl_3 \xrightarrow{oleum} CF_2ClCFClCOCl \xrightarrow[2)\ Zn]{1)\ CD_3OD} CF_2=CFCO_2CD_3$

D)

$CH_2=C(CH_3)COCl + CCl_3CHO \longrightarrow CH_2=C(CH_3)CO_2CHClCCl_3$

E)

$CH_2=CHOCOCl + ROH \longrightarrow CH_2=CHOCO_2R$

$R=C_2H_4C_6F_{13},\ CH_2CCl_3,\ CD_3,\ CH_2CF_3,\ C_6F_5$

Scheme III. Syntheses of monomers with a few number of C–H bonds per cubic centimeter.

C_6F_5 (X = Cl, F), and $CCl_2=CHOCO_2C_6F_5$. The pentafluorophenyl esters have five fluorine atoms, and therefore their refractive indexes are higher than the refractive index of methyl methacrylate; this difference is due to the aromatic ring. From Table IX, two monomers seem to be very interesting for optical applications: $CF_2=CFCO_2CD_3$ and $CCl_2=CHOCO_2$-C_6F_5. However, these two monomers do not undergo free-radical polymerization under normal conditions. The homopolymer of $CF_2=CFCO_2CD_3$ is formed only under severe conditions (*63*), and all attempts to homopolymerize $CCl_2=CHOCO_2C_6F_5$ by a free-radical process have failed. All other halogenated monomers polymerize readily under free-radical conditions. Optical and physical properties of halogenated polymers prepared are reported in Table X.

The T_g values of polyacrylates, given in Table X, vary from 31 to 180 °C, and these values are mostly

Table X. Physical and Optical Characteristics of New Halogenated Polymers

Homopolymer	n^{20}	Tg (°C)	N_{CH} (10⁻³)	Homopolymer	n^{20}	Tg (°C)	N_{CH} (10⁻³)
Acrylic				**Acrylic—continued**			
$-(CH_2-CCl)-$ | $CO_2C_6F_5$	1.5000	120	12	$-(CH_2-C(CH_3))-$ | $CO_2CH_2CCl_3$	1.5251	134	49
$-(CH_2-CF)-$ | $CO_2C_6F_5$	1.4650	160	14	$-(CH_2-CF)-$ | $CO_2C_6H_5$	1.5600	180	56
$-(CH_2-C(CH_3))-$ | $CO_2CH_2(CF_2CFCl)_3Cl$	1.4438	31	26	$-(CH_2-C(CH_3))-$ | $CO_2CH_2CF_3$	1.4146	80	60
$-(CH_2-CCl)-$ | $CO_2CH_2CCl_3$	1.5342	140	28	$-(CH_2-C(CH_3))-$ | $CO_2CH_2CF_2CF_2H$	1.422	71	61
$-(CH_2-CF)-$ | $CO_2CH_2CCl_3$	1.4995	124	30	$-(CH_2-CF)-$ | CO_2CH_3	1.460	140	68
$-(CH_2-C(CH_3))-$ | $CO_2C_6F_5$	1.4873	125	32	$-(CH_2-C(CH_3))-$ | CO_2CH_3	1.489	105	92
$-(CH_2-C(CH_3))-$ | $CO_2CH_2(CF_2CFCl)_2Cl$	1.4551	47	34	**Vinyl Carbonate**			
$-(CH_2-C(CH_3))-$ | $CO_2C_2H_4C_6F_{13}$	1.3800	49	37	$-(CH_2-CH)-$ | $OCO_2C_2H_4C_6F_{13}$	1.3342	37	26
$-(CH_2-C(CH_3))-$ | $CO_2CHClCCl_3$	1.5179	165	37	$-(CH_2-CH)-$ | $OCO_2CH_2CCl_3$	1.4965	74	36
$-(CH_2-CF)-$ | $CO_2CH_2CF_3$	1.385	123	39	$-(CH_2-CH)-$ | OCO_2CD_3	1.4619	55	37
$-(CH_2-CF)-$ | $CO_2CH_2CF_2CF_2H$	1.398	95	42	$-(CH_2-CH)-$ | $OCO_2CH_2CF_3$	1.4045	57	47
$-(CH_2-C(CH_3))-$ | $CO_2CH_2CF_2CFCl_2$	1.4639	67	45	$-(CH_2-CH)-$ | $OCO_2C_6F_5$	—	65	—

Note: n^{20} is the refractive index at 20 °C, T_g is the glass transition temperature, N_{CH} is the number of C–H bonds per cm³. Values compiled from references 61, 62, and 64. Dashes indicate there are no experimental values.

Source: Adapted and reproduced with permission from reference 61. Copyright 1994 Taylor & Francis, Inc.

higher than the T_g of PMMA, except when the alkyl chain is very long such as $-CH_2(CF_2CFCl)_nCl$ ($n = 1,2$, or 3), $-C_2H_4C_6F_{13}$, $-CH_2CF_3$, and $-CH_2CF_2CF_2H$. The T_g values of polycarbonates are lower than the T_g values of the corresponding polymethacrylates and are higher than the T_g values of polyacrylates with the same side chain. In general, α-halogenated acrylates have higher T_g values than methacrylates. The thermal stability of α-fluoroacrylates are superior to those of α-chloroacrylates because of dehydrochlorination

that occurs in the case of α-chlorinated polymers around 200 °C.

Pentafluorophenyl esters of methacrylate, its α-chloro and α-fluoro analogs, and also trichloroethyl esters of α-chloro and α-fluoroacrylate seem to be good polymers for optical transparent materials because of their high T_g and low N_{CH}. For example, the absorption spectra of a rod of PMMA (4 × 1 cm) and of poly(pentafluorophenyl methacrylate) (PMAPF) (5 × 1 cm) are given in Figure 7. The absorption in the 800–1200-nm

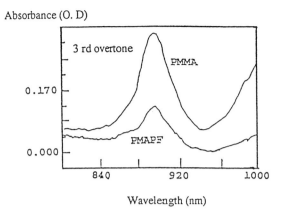

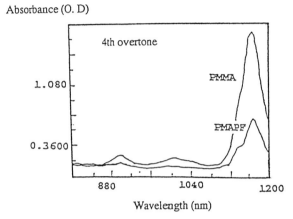

Figure 7. Optical density of C–H 4th and 3rd overtone vibration in PMMA (4 × 1 cm) and in PMAPF (5 × 1 cm). (Reproduced with permission from reference 52. Copyright 1992 John Wiley & Sons, Inc.)

range for PMAPF is about 30% of that of PMMA. This reduction of absorption is in the same order as the N_{CH} ratio of the monomers, MAPF and MMA:

$$\frac{N_{C-H}^{MAPF}}{N_{C-H}^{MMA}} \cong \frac{N_{C-H}^{PMAPF}}{N_{C-H}^{PMMA}}$$

Fluorinated acrylates already have a wide domain of applications (65). We have seen that introducing fluorine atoms into polymers, in the main chain or in the side chain, improves both heat and moisture resistance and substantially reduces the intrinsic absorption in the visible and NIR regions. Thus, poly(fluoroalkyl α-fluoroacrylate)s exhibit very interesting properties for optical components, and they should have a considerable impact on the development of optical devices.

Related Polymers for Optical Applications

In the area of planar waveguides, tremendous possibilities are created by grafting optical active functions onto the polymer backbone. In this area we are now achieving organic materials with high and steady electro-optic coefficients (66, 67). Besides the planar waveguides, NLO effects are being exploited to fully develop polymer single-mode optical fibers (68, 69). Other polymeric optical components are also being studied. Thus, active research brings to light the ability of electroluminescent polymers for building LED (70–72). In this way the electroluminescent yields of polymers are in progress, and it is possible now to achieve 1000 cd/cm² of luminance. Theoretical concepts are well established to improve the performance, especially the response time (73).

Researchers are able to construct a polymer interferometer on a silicon wafer with a well-adapted buffer layer. The device can withstand the baking of the resist up to 150 °C (74). Polymer waveguides are also seen for the interconnection of electronic circuits of large scale integration and high bit rate to take advantage of the lightwave high bandwidth (75).

Concluding Remarks and Future Prospects

In this chapter we have elaborated what can be done to control the two main characteristics needed to develop cylindrical and planar polymeric waveguides. These two characteristics are the refractive index and the absorption due to molecular vibrations in the visible and NIR regions. First, we have shown that it is possible to tune the monomer structures to get the desirable refractive index of the resulting polymers. Thus, with polyacrylates we can vary the refractive index from 1.35 to 1.55. This range affords a wide scope to build waveguides with different numerical apertures. Also, the substitution of hydrogen by halogens in acrylates and other monomers offers a real means of reducing the intrinsic absorption. The attenuation is the main drawback of POFs for transmission over long distances. The control of these two features is the key to the future development of organic components for integrated optics. The use of fluorinated polymer, as core or cladding materials, reduces the total attenuation of POF, especially the intrinsic absorption factor. Also, fluorinated polymers should be a very promising alternative to PS and PMMA in terms of attenuation and thermal stability.

These new developments in combination with the versatility of polymers can provide some real advantages over glass. So, for the planar optical waveguide, relatively simple polymer fabrication processes can overcome the drawbacks of other materials that need etching, laying, and annealing. By direct UV modification of polymer films bearing appropriate side groups, we have printed a channel optical waveguide without the loss or adding of materials.

Therefore, it is realistic to imagine that all the optical components needed for integrated optical communications (including transmitter, receiver, modulator, couplers, multiplexing systems, and data links) could be manufactured from specifically tailored functional polymers in the not too distant future.

References

1. Bosc, D.; Froyer, G.; Clarisse, C. *Écho Rech.* **1991**, *143*, 27.
2. Booth, B. L. *J. Lightwave Technol.* **1989**, vol7No(10), 1445.
3. Moutonnet, D.; Guilbert, M.; Bosc, D.; Leroux, M. *Mater. Lett.* **1992**, *15*, 220.
4. Booth, B. L. In *Polymers for Lightwave and Integrated Optics. Technology and Applications*: Hornak, L. A., Ed.; Marcel Dekker: New York, 1992; pp 231–266.
5. Neyer, A.; Knoche, T.; Müller, L. *Electron. Lett.* **1993**, *29*, 399.
6. Imamura, S.; Yoshimura, R.; Izawa, T. *Electron, Lett.* **1991**, *27*, 1342.
7. *Handbook of Chemistry and Physics,* 64th ed.; Weast, R. C., Ed.; CRC: Boca Raton, FL, 1983–1984; p E-362.
8. Hoshino, D.; Nagahama, K.; Hirata, M. *J. Jpn. Pet. Inst.* **1979**, *22*, 218.
9. Hoshino, D.; Nagahama, K.; Hirata, M. *J. Jpn. Pet. Inst.* **1979**, *22(1)*, 32.
10. Hoshino, D.; Nagahama, K.; Hirata, M. *J. Jpn. Pet. Inst.* **1981**, *24(3)* 197.
11. Groh, W.; Zimmerman, A. *Macromolecules* **1991**, *24*, 6660.
12. Bosc, D.; Pietrasanta, Y.; Rigal, G.; Rousseau, A. *J. Fluorine Chem.* **1984**, *26*, 369.
13. Boutevin, B.; Pietrasanta, Y. In *Comprehensive Polymer Science*; Allen, G.; Bevington, J. C., Eds.; Pergamon: New York, 1989; Vol. 3, Part 1, pp 185–194.
14. Boutevin, B.; Pietrasanta, Y. *Tetrahedron Lett.* **1973**, *12*, 887.
15. Boutevin, B.; Cals, J.; Pietrasanta Y. *Eur. Polym. J.* **1975**, *12*, 225.
16. Boutevin, B.; Cals, J.; Pietrasanta, Y. *Tetrahedron Lett.* **1974**, *12*, 939.
17. Kakac, B.; Hudlicky, M. *Coll. Czech. Chem. Commun* **1965**, *30*, 745.
18. Hazeldine, R. N. *J. Chem. Soc.* **1951**, *1*, 588.
19. Miller, W. T.; Fainberg, A. H. *J. Am. Chem. Soc.* **1957**, *79*, 4164.
20. Pittman, A. G.; Sharp, D. L.; Ludwig, B. A. *J. Polym. Sci.* **1968**, *6*, 1729.
21. Boutevin, B.; Pietrasanta, Y.; Sideris, A. *J. Fluorine Chem.* **1982**, *20*, 727.
22. Boutevin, B.; Pietrasanta, Y. *Eur. Polym. J.* **1975**, *12*, 231.
23. Boutevin, B.; Rigal, G.; Rousseau, A.; Bosc, D. *J. Fluorine Chem.* **1988**, *38*, 47.
24. Boutevin, B.; Hugon, J. P.; Pietrasanta, Y. *Makromol. Chem.* **1981**, *182*, 2927.
25. Bosc, D. In *Les Acrylates et Polyacrylates fluorés. Dérivés et Applications*; Boutevin, B.; Pietrasanta, Y., Eds; EREC Ed: France. 1988; pp 159–181.
26. *Polymers for Lightwave and Integrated Optics:* Hornak, L. A., Ed.; Marcel Dekker: New York, 1992.
27. Emslie, C. *J. Mater. Sci.* **1988**, *23*, 2281.
28. Herbrechtsmeier, P. *Kunstoffe* **1989**, *79*, 1040–1044.
29. Glen, R. M. *Chemtronics* **1986**, *1*, 98.
30. Kaino, T. *J. Polym. Sci., Part A: Polym. Chem. Ed.* **1987**, *25*, 37.
31. Kaino, T.; Fujiki, M.; Oikawa, S.; Nara, S. *Appl. Opt.* **1981**, *20*, 2886.
32. Kaino, T.; Fujiki, M.; Nara, S. *Appl. Phys.* **1981**, *52*, 7061.
33. Oikawa, S.; Fujiki, M.; Katayama, Y. *Electron. Lett.* **1979**, *15*, 829.
34. Tanaka, A.; Sawada, H.; Wakatsuki, N. *Fujitsu Sci. Tech. J.* **1987**, *23*, 166.
35. Kaino, T.; Fujiki, M.; Jinguji, K. *Rev. Electr. Commun. Lab.* **1984**, *32*, 478.
36. Bosc, D. Ph.D. thesis, University Montpellier II, France, 1992.
37. Kaino, T. *Jpn. J. Appl. Phys.* **1985**, *24*, 1661.
38. Crist, B.; Marhic, M. E.; Raviv, G.; Epsein, M. *J. Appl. Phys.* **1980**, *51*, 1160.
39. Bosc, D.; Guilbert, M.; Toinen, C. *Proceedings of OPTO 89;* ESI Publishers: Paris, 1989; pp 15–19.
40. Schleinitz, H. M. *26th Int. Wire Cable Symp.* **1977**, *26*, 352.
41. Kaino, T.; Jinguji, K.; Nara, S. *Appl. Phys. Lett.* **1982**, *41*, 802.
42. Kaino, T.; Jinguji, K.; Nara, S. *Appl. Phys. Lett.* **1983**, *42*, 567.
43. Kaino, T.; Katayama, Y. *Polym. Eng. Sci.* **1989**, *29*, 1209.
44. Takezawa, Y.; Tanno, S.; Taketani, N.; Ohara, S.; Asano, H. *J. Appl. Polym. Sci.* **1991**, *42*, 3195.
45. Urbach, F. *Phys. Rev.* **1953**, *92*, 1324.
46. Kaino, T. In *Polymers for Lightwave and Integrated Optics. Technology and Applications*; Hornak, L. A., Ed.; Marcel Dekker: New York, 1992; pp 1–38.
47. Tanaka, A.; Sawada, H.; Takoshima, T.; Wakatsuki, N. *Fiber Intern Opt.* **1987**, *7*, 139.
48. Apert, L.; Keiser, W. E.; Szymanski, H. A. *I.R. Theory and Practice of Infra Red Spectroscopy*, 2nd ed.; Plenum: New York, 1970.
49. Groh, W. *Makromol. Chem.* **1988**, *189*, 2861.
50. Boutevin, B.; Pietrasanta, Y.; Rigal, G.; Rousseau, A. *Ann. Chim. Fr.* **1984**, *9*, 723
51. Takezawa, Y.; Taketani, N.; Tanno, S.; Ohara, S. *J. Polym. Sci. Part B: Polym. Phys. Ed.* **1992**, *30*, 879.
52. Boutevin, B.; Rousseau, A.; Bosc, D. *J. Polym. Sci., Part A: Polym. Chem. Ed.* **1992**, *30*, 1279.
53. Molines, H.; Nguyen, T.; Wakselman, C. *Synthesis* **1985**, *8*, 754.
54. Molines, H.; Nguyen, T.; Wakselman, C. *Synthesis* **1985**, *8*, 755.
55. Boutevin, B.; Pietrasanta, Y.; Bosc, D.; Rousseau, A. Fr. Patent 87 16671, 1987.
56. Nguyen, T.; Wakselman, C. *J. Org. Chem.* **1989**, *54*, 5640.
57. Paleta, O.; Posta, A.; Liska, F. *So. Vys. Sk. Chem.—Technol. Praze, Org. Chem. Technol.* **1978**, *C25*, 105.
58. Boutevin, B.; Pietrasanta, Y.; Rousseau, A.; Bosc, D. *J. Fluorine Chem.* **1987**, *37*, 151.
59. Hrabak, F.; Picova, H. *J. Polym. Sci. Chem. Ed.* **1972**, *10*, 3125.
60. Boutevin, B.; Pietrasanta, Y.; Laroze, A.; Rousseau, A. *Polym. Bull.* **1986**, *16*, 391.
61. Boutevin, B.; Rousseau, A.; Bosc, D. *Fiber Int. Opt.* **1994**, *13(3)*, 309.

62. Boguslavskaya, L. S.; Panteleeva, I. Yu.; Morozova, T. V.; Kartashov, A. V.; Chuvatkin, N. N. *Russ. Chem. Rev.* **1990**, *59*, 906.

63. Weise, J. K. *Polym. Prep. Am. Chem. Soc. Div. Polym. Chem.* **1971**, *12*, 512

64. Wuensche, P.; Mueller, S.; Lorkowski, H. J. *J. Polym. Sci. Part A: Polym. Chem. Ed.* **1988**, *26*, 2669.

65. *Les Acrylates et Polyacrylates Fluorés. Dérivés et Applications*; Boutevin. B.; Pietrasanta, Y., Eds.; EREC Ed.: France. 1988.

66. Foll, F.; Bosc, D.; Liang, J.; Rousseau, A.; Boutevin, B. Fr. Patent 93 105 72, Sept. 6, 1993.

67. Levenson, R.; Liang, J.; Rossier, C.; Van Beylen, M.; Samyn, C.; Foll, F.; Rousseau, A.; Zyss, J. In *Organic Thin Films for Photonic Applications*; Optical Society of America, Washington, DC, 1993; Vol. 17, pp. 81-84.

68. Bosc, D.; Toinen, C. *Photonics Technol. Lett.* **1992**, *4*, 749.

69. Bosc, D.; Toinen, C. *Polym. Compos.* **1993**, *14*, 410.

70. Brown, A. R.; Greenham, N. C.; Bradley, D. D. C.; Friend, R. H.; Burn, P. L.; Kraft, A.; Holmes, A. B. *Conduction Polymers, Conduction Polymers Conference* Rapra Technology Shrewbury, U.K., 1992; paper 7.

71. Cacialli, F.; Friend, R. H.; Moratti, S. C.; Holmes, A. B. *Synth. Met.* **1994**, *67(1–3)*, 157.

72. Baigent, D. R.; Greenham, N. C.; Gruener, J.; Marks, R. N.; Friend, R. H.; Moratti, S. C.; Holmes, A. B. *Synth. Met.* **1994**, *67(1–3)*, 3.

73. Schott, M.; Le Barny, P. In *Optoelectronique Moleculaire*; OFTA-Masson, Ed.; Paris, 1993; pp 61–76.

74. Holland, W. R. In *Polymers for Lightwave and Integrated Optics. Technology and Applications*; Hornak, L. A., Ed.; Marcel Dekker: New York, 1992; p 397.

75. Sullivan, C. *Proc. SPIE* **1988**, *994*, 92.

Functionalized Polymers for Second-Order Nonlinear Optics

Douglas R. Robello

Functionalized polymers may provide the combination of properties required for the construction of practical photonic devices for communication and imaging by using second-order nonlinear optical (NLO) effects. These materials hold the potential for the efficient production of hitherto inaccessible frequencies of light and better control of these and existing frequencies. In this chapter, an introduction to the field of second-order nonlinear optics is presented along with a survey of leading research in the area. Diverse polymers with covalently attached NLO-active moieties are presented and categorized according to chemical structure, means of synthesis, and intended use. Some of the main topics discussed are poled polymers, (side-chain, main-chain, and cross-linked materials), surface-oriented systems (Langmuir–Blodgett films and self-assembled multilayers), and photorefractive polymers.

Because increasing demands for information transmission, storage, and processing have begun to outstrip the capabilities of electronic and magnetic technologies, attention has turned to systems that depend on light for these functions. To cite two obvious examples, fiber-optic cables and optical memory (e.g., compact disks) have represented such major technological improvements that they each rendered older systems obsolete within just a few years. (For information on polymeric light carrying materials with potential application to fiber optics, *see* reference 1.) The benefits of optical technology are evident. Today, fiber-optic telecommunications and compact disks are the foundations of huge industries that employ thousands of people and provide products and services for millions more. The situation is poised for future growth. Such futuristic concepts as interactive television and video telecommunications are already becoming realities. There has even been discussion in the U.S. political arena of promoting an "information superhighway" based on fiber-optic transmission lines. (For a provocative discussion of the implications for U.S. governmental policy in the area of optical technology, *see* reference 2.) It is conceivable that we are witnessing passage from the "Electron Age" to the "Photon Age".

Photons do not interact strongly with each other or with external fields. This feature is an important advantage over electrons when considering high-density circuitry and transmission lines. However, because the interactions are weak, it is difficult to do much with light waves. An as yet unfulfilled need exists to manipulate photons with the facility that is now possible for electrons. The prospect for controlling light in this way has given rise to a new technological area, dubbed *photonics* to emphasize the analogy with electronics. (It is important to differentiate photonics, which uses relatively low power light sources and where photons carry information, from applications such as laser cutting and drilling, laser-induced nuclear fusion, and laser weaponry, which obviously employ much higher power light sources, and where photons are used only for their energy content.) In a related development, researchers have been attempt-

3469–8/97/0505$15.75/0 © 1997 American Chemical Society

ing to create miniature photonic devices, leading to the new field of *integrated optics*. The most likely outcome of these technologies is the production of hybrid optoelectronic modules ("chips") that employ both electrons and photons for information acquisition, processing, transmission, and storage.

Description and Theory

Key elements of these prospective photonic devices are nonlinear optical (NLO) processes. (A more detailed discussion of the physical basis of nonlinear optics is beyond the scope of this chapter. The interested reader should consult references 3–7). Nonlinear optics concerns interactions of light with matter that lead to new optical fields that are altered in amplitude, frequency, phase, or other characteristics. In the creation of these new fields, manipulation of photons becomes possible.

The basis for nonlinear optics lies in the polarization, P, of charge carriers in a material responding to an applied electric field, E. This relationship can be approximated by the power series given by eq 1 in which P_0 is the unperturbed polarization, and the χ factors describe the interactions. Note that, because of the three-dimensional (3D) nature of the interactions, the coefficients in eq 1 are not simple quantities. Variables P and E are vectors, and the χ factors are tensors: $\chi^{(1)}$ is the linear polarizability and deals with the familiar phenomena of refraction, reflection, and absorption; $\chi^{(2)}$ and $\chi^{(3)}$ are the first and second nonlinear susceptibilities, respectively. Because $\chi^{(2)}$ and $\chi^{(3)}$ ordinarily assume relatively small values, it is only at the high field strength attainable in lasers that E becomes large enough for the nonlinear terms to be significant. Also, because of symmetry considerations, $\chi^{(2)}$ (as well as all other even-rank terms in eq 1) is zero in a medium with a center of symmetry, a feature that has significant consequences for the design and fabrication of practical second-order NLO materials. In any case, the overall goal of NLO research has centered on ways to maximize the susceptibilities, χ.

$$P = P_0 + \chi^{(1)}E + \chi^{(2)}E^2 + \chi^{(3)}E^3 + \dots \qquad (1)$$

The subject of nonlinear optics can be divided into two broad areas based on second-order ($\chi^{(2)}$) and third-order ($\chi^{(3)}$) phenomena. For practical purposes, second-order and third-order phenomena are concerned with interactions between two and among three electric fields, respectively. These fields can be high frequency (from light waves) or low frequency (provided by electronic devices). This chapter will dwell on materials intended for $\chi^{(2)}$ applications, whereas Chapter 3.7 (8) is concerned with $\chi^{(3)}$ issues.

Applications and Devices

Before turning to specific examples, mention should be made of the concept of guided wave optics. Integrated optical circuits require the optical equivalent of wires, which usually take the form of fibers carrying light between modules, and waveguides on the modules themselves. These optical wires confine light by using a core of material with a higher refractive index than the cladding (total internal reflection condition). Waveguides are thin films (ca. 1 µm) fabricated to carry light at or near the surface of a planar substrate. Such films can cover the entire substrate, in which case they are known as slab waveguides. However, for most proposed integrated optical circuitry, thin ribbons a few micrometers wide known as channel waveguides are required. The waveguides can be passive, meaning that they serve only to transmit the light, or they can be active and contain NLO components that manipulate the light. Of course, the chip can hold electronic circuitry and provide convenient interconnections between the optical and electrical domains.

The most widely known $\chi^{(2)}$ process is second-harmonic generation (SHG, also known as frequency doubling), which is a special case of frequency combination. Here, the electric fields of two photons interact to produce a new photon that possesses twice the frequency of the original photons. (When dissimilar photons interact, it is possible to create new photons with either the sum or difference of the original frequencies.) SHG has been widely employed by researchers as a tool for assaying the nonlinear susceptibility of a material.

However, the need for SHG materials has been driven by one practical consideration: Efficient and inexpensive miniature lasers that emit wavelengths shorter than about 600 nm do not exist and may be infeasible to construct directly. On the other hand, solid-state (usually semiconductor diode) lasers that emit at near-IR wavelengths are commonplace, and twice the frequency of these existing lasers could produce blue and green wavelengths needed (in combination with red) for color-imaging applications. A second, perhaps more important application of SHG lies with optical disk storage. The working size of the digital markings on optical disks is proportional to the wavelength used to read them (currently near-IR). If a frequency-doubled laser were employed, the markings could be reduced by a factor of two in both length and breadth, a reduction leading to a four-fold increase in information density on the optical disk (Figure 1).

In practice, SHG materials must exhibit extremely low optical absorption at both the fundamental and SHG wavelengths, otherwise the material would

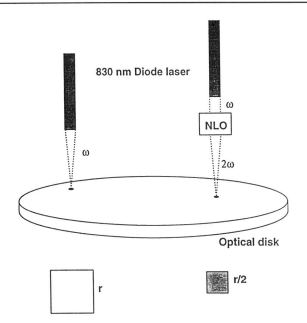

Figure 1. Second harmonic generation applied to optical disk memory.

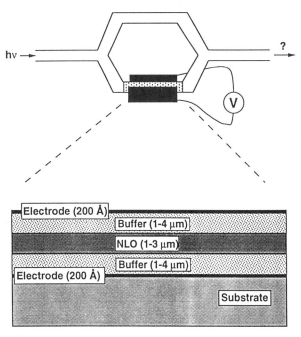

Figure 2. Mach–Zehnder interferometer for electrooptic modulation. The lower part of the figure shows the enlarged cross section of a typical waveguide package.

quickly overheat and be destroyed by the very large power present in operating devices. Finding materials with simultaneously large susceptibility and transparency at blue and green wavelengths has been a particularly difficult challenge, especially for organic materials (*9, 10*). Many measurements of $\chi^{(2)}$ by SHG in the literature were carried out on thin-film samples perpendicular to the plane of the film, conditions under which absorption is negligible. (NLO susceptibility by SHG is often reported in the literature as *d* value, which is simply $\frac{1}{2}\chi^{(2)}$). Many of these materials are actually strongly absorbing below 600 nm and are completely unsuitable for practical SHG applications in which the interaction distance would be much longer. Also, because of stronger coupling between the light and material, greatly increased susceptibility is observed at absorptive wavelengths. This phenomenon is known as resonance enhancement, and it must be considered an artifact of the experimental technique. Much data on second-order NLO susceptibility reported in the literature is unrealistically inflated because of this enhancement.

Unfortunately, NLO chromophores with the largest hyperpolarizabilities are strongly colored. The nonlinearity is diminished severely in compounds in which the absorption maximum is shifted well below the visible region of the spectrum. Strong colorless NLO chromophores represent an unsolved (and possibly unsolvable!) challenge to synthetic chemists.

A second device concept for second-order NLO materials is the electrooptic effect (also known as the Pockels effect). Here, the electric field of a light wave interacts with a relatively low frequency field generated by electrons. The result is a change in the refrac-

tive index (and therefore in the speed of light) in the NLO material, an effect that can be put to work. For electrooptic processes, $\chi^{(2)}$ is usually expressed in terms of the magnitude of the change in refractive index with applied voltage, known as the electrooptic coefficient (*r*). For comparison, the largest tensor element of *r* for the well-studied inorganic material LiNbO$_3$ is ca. 30 pm/V for off-resonance measurements. This benchmark figure has been achieved for polymeric materials in only a few rare instances. There is clearly room for much improvement. Note that *r* can be conveniently measured for laboratory samples by using a reflection technique developed independently by Schildkraut and Clays (*15, 16*) and by Teng and Man (*17*).

To give one simple example, Figure 2 depicts a device known as a Mach–Zehnder interferometer for electrooptic modulation. The guided input beam is divided into two paths, at least one of which is NLO-active. The beam is then recombined. The light will arrive at the second junction in or out of phase depending on the relative times needed to traverse the two paths. If the appropriate voltage (*V*) is applied to the path containing the NLO material, the resulting refractive change will alter the phase relationship at the second junction. This alteration will lead to, for example, complete constructive or destructive interference. Therefore, the light output will be controlled by *V*.

A second related example deals with electrooptic switching (Figure 3). This device, known as a direc-

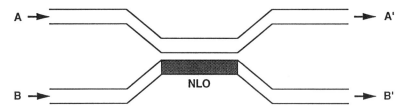

Figure 3. Electrooptical directional coupler. Electrodes around the NLO waveguide are omitted for clarity.

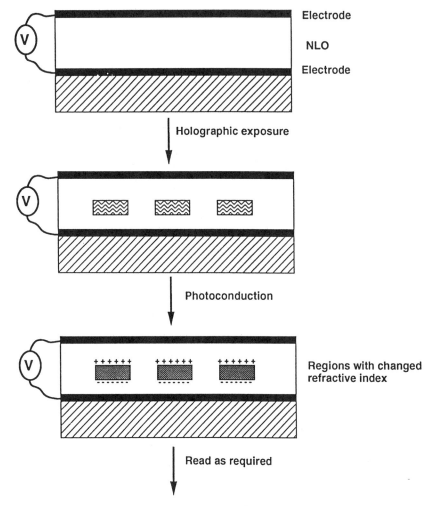

Figure 4. Photorefractive optical information storage. Note that the hologram is actually written in three dimensions.

tional coupler, provides a method for steering a light beam down one channel or another according to electronic inputs. The light guides (A and B) are brought close enough to each other so that light energy "leaks" between the two channels and oscillates back and forth along the interaction distance. When light is injected into A, it may emerge at A' or at B' (or be divided between the two output channels), depending on the relative refractive indexes of the two paths and on the interaction distance. If one (or both) of the channels is NLO-active, then its refractive index can be changed by application of a voltage, and the light can therefore be directed out at either A' or B' at will.

A third device concept in second-order NLO is photorefractivity, which is a consequence of simultaneous second-order NLO response and photoconductivity, and could be useful for 3D optical storage (Figure 4). A thick sample of NLO material is irradiated by using an interference pattern (hologram) that encodes the information to be stored. The irradiated regions conduct charge, which sets up local electric fields. The refractive index in these regions changes because of the electrooptic effect, and a 3D diffraction pattern is produced that can be read later by a probe beam.

Many similar NLO devices have been described, and even complicated large-scale integrated optoelectronic circuits are envisaged based on repetition of simpler NLO elements. A number of working devices based on molecular NLO materials have already been fabricated and tested (*18–22*), and research in this area is accelerating as the performance of materials improves.

Functional Polymers

Several inorganic crystalline materials, most notably $LiNbO_3$ and KH_2PO_4, have been discovered to possess large $\chi^{(2)}$ values. However, complicated thin-film (waveguide) devices are inconvenient to fabricate using such bulk materials. In addition, bulk inorganic materials possess relatively slow NLO responses and large dielectric constants (compared to organics), factors that limit their use in ultra-high speed optical devices. Also, many inorganic crystals are subject to damage at high optical intensities, although solutions to this problem are being researched.

Because of these limitations, researchers turned to molecular materials as an alternative. The term "molecular materials" is meant to represent any solid composed of discrete molecules, including organometallic compounds and molecules containing no carbon atoms at all. Nevertheless, the great majority of molecular materials developed for NLO is composed of traditional organic molecules containing C, H, N, S, etc.

The polarization of molecule, *p*, can be expressed in a power series given by eq 2, which is analogous to eq 1 where the terms α, β, γ, etc., are the molecular equivalents of the χ parameters in eq 1. The hyperpolarizability, β, is the molecular quantity of interest for second-order NLO molecules. As before, even-ranked terms in eq 2 such as β are zero for centrosymmetric molecules.

$$p = p_0 + \alpha E + \beta E^2 + \gamma E^3 + \dots \qquad (2)$$

Unlike bulk materials, the NLO susceptibility of molecular materials arises directly from the properties of their constituent molecules. Hence, to a good approximation, $\chi^{(2)}$ is defined by eq 3, where N/V is the concentration of NLO-active molecules, and $\mathscr{L}$ is an orientational factor describing the degree of non-centrosymmetry in the material. Values of $\mathscr{L}$ can be between 0 (for centrosymmetric systems) and 1 (perfect polar molecular alignment). Therefore, if β of a molecule is known, $\chi^{(2)}$ can be predicted based on structural factors (such as concentration, crystal habit, and poling conditions) and other simple quantities (*6, 23–25*). This calculation ignores interactions among the NLO-active molecules; nevertheless, fairly accurate results have been obtained.

$$\chi^{(2)} \propto N/V \beta \mathscr{L} \qquad (3)$$

Structure 1. 4-Dimethylamino-4´-nitrostilbene, a prototypical NLO molecule.

The substantial NLO properties of molecular materials were first recognized in the 1970s. For second-order NLO, the strongest molecules contain loosely bound electrons with large unsymmetrical spatial distributions. In general, strong electron-donating and withdrawing groups separated by a conjugated π-framework are required. Because many such compounds are strongly colored, the donor–π-acceptor moiety is often referred to as the NLO chromophore. 4-Dimethylamino-4'-nitrostilbene (DANS, **1**) is a prototypical example of an organic compound with a large β value.

The NLO properties of a great many compounds have been measured and the results correlated with structure (*12, 26–33*). Unfortunately, because of differing conventions and treatment of data, hyperpolarizability measurements from different laboratories cannot be compared directly. Standardization of the measurement technique is being pursued cooperatively among several organizations and would be a substantial boon to the NLO field. With the tools available from modern synthetic chemistry, unlimited variations can be synthesized to tailor the properties of the material as required. All of the polymers to be described in this review share one common feature: a donor–π-acceptor chromophore is present and covalently attached to a polymer backbone.

The NLO response in molecular materials is based on polarization of the π-electron cloud, one of the fastest processes in nature that can be harnessed. The timescale has been estimated to be femtoseconds (10^{-15} s). From the perspective of optoelectronic circuitry, this rate means that the electronics would be the slow step. The molecular optical material responds essentially instantaneously on the electronic timescale. With this inherently fast response combined with low dielectric constants, molecular NLO materials are suitable for ultra-fast optoelectronic circuits.

Polymers provide the facility to form waveguides. Thin films of polymers on a substrate can be fabricated easily and inexpensively under mild conditions by methods such as spin- or dip-coating. These methods are compatible with current electronic integrated-circuit fabrication technology; therefore, existing techniques and equipment can be used, and hybrid optoelectronic chips can be constructed.

In addition to their film-forming ability, polymers may provide other benefits such as mechanical toughness, adhesion to the substrate, good optical transmission characteristics, and resistance to thermal and optical damage. Also, complicated structures such as channels and junctions can be constructed by using conventional microlithography.

To impart NLO properties to a thin film, a polymer containing an NLO-active chromophore and a means for achieving a noncentrosymmetric orientation of these chromophores are required. To obtain the largest NLO susceptibility, eq 3 calls for the highest possible concentration (N/V) of chromophores with the largest possible molecular hyperpolarizability (β) ordered in a structure as highly polar as possible (L). Obviously, the molecular alignment, once attained, must be preserved for long periods of time under operating conditions. This stability must also extend to the rather drastic temperature extremes inherent to device manufacture, including soldering, metal deposition, and chip bonding. The development of "orientationally robust" polymers has been an important area of NLO research. In this respect, polymers with very high glass transition temperatures (T_g) are desirable, because movement of oriented chromophores would be inhibited even at elevated temperatures. However, high temperatures are then required to process and orient these materials, so some research has been directed toward discovery of thermally stable donor–acceptor chromophores.

Many other practical factors must be considered. For example, the polymer must adhere well to its substrate; resist damage from thermal and mechanical stresses; be nearly perfectly transparent at the wavelengths present in the device; and be free of defects such as cracks, particles, and crystal- or phase-domain boundaries that can cause light scattering. The transparency constraints virtually rule out the application of crystalline and liquid crystalline polymers, despite their desirable order. It is exceedingly difficult to produce single-domain, poled, semicrystalline polymers. In addition, the polymer should have a low dielectric constant, low conductivity (i.e., nearly ideal insulator behavior), and low cost.

The most expeditious method for producing an NLO-active material is to dissolve an appropriate small molecule in a conventional polymer. Even though numerous studies have dealt with such host–guest systems, there are three major disadvantages:

1. The amount of dopant than can be dissolved is usually quite low, and because $\chi^{(2)}$ is proportional to the concentration of NLO-active species (N/V, eq 3), the susceptibility is therefore limited.
2. The presence of a low molecular weight dopant in a polymer acts as a plasticizer and lowers the T_g of

the mixture. This effect leads to an increase in the rate of chromophore disorientation, thereby degrading the initial susceptibility over time.
3. Even well below the T_g of the mixture, the NLO chromophore can experience significant free volume and can disorient at a relatively rapid rate.

The solution to these problems is straightforward: the NLO-active chromophore must be covalently attached to the polymer backbone; hence, functional polymers are required. In this way, very high concentrations of chromophore can be achieved, in many cases exceeding half of the total volume of the material. Covalent attachment also avoids plasticization effects and requires that disorientation be accompanied by substantial molecular motions of the polymer itself (34, 35). The result is that the rate of chromophore disorientation is considerably reduced when the chromophores are an integral part of the polymer (36).

An ensemble of randomly oriented chromophores, as would be the case in an amorphous polymer, is net centrosymmetric; therefore, no second-order NLO behavior would be shown. Two principal techniques for achieving noncentrosymmetric arrangements of chromophores have been developed: electric field poling and surface orientation. In addition, considerable effort has been devoted to methods for preserving the induced orientation under the rigorous conditions anticipated during manufacture and use of the devices that would contain these materials.

For the purpose of discussion, functional polymers in the literature will be divided into two main groups according to the method used for chromophore orientation, and they will be further subdivided by structural features. These divisions are arbitrary, and it is possible for certain examples to fall within more than one class. A comprehensive review of the field would be prohibitively long; therefore, only representative examples of the various classes of second-order NLO polymers will be presented. Also, the NLO-inactive components (i.e., comonomers) are omitted in many of the structures drawn for brevity and clarity.

Poled Polymers

In this technique, a sample of the NLO-active polymer is exposed to a strong direct-current electric field while above T_g. The poling field can be provided by two closely spaced electrodes or by a corona discharge (36–38). The constituent chromophores, which possess fairly large ground-state dipole moments, are mobile and align with the applied field. Then, the alignment is "locked in" by either cooling the material to well below its T_g or by rigidifying the matrix by using a thermal or photochemical reaction. Once the locking process is complete, the poling field can be removed. The degree of orientation that can be achieved by poling depends on the strength of the field applied (E_p), the dipole moment of the chromophores (μ), and the absolute temperature (T). The orientation factor ($\mathscr{L}$, eq 3) for poling is a Langevin function of these three parameters, but for the rather low degrees of orientation that are typical of the poling process ($0.1 < \mathscr{L} < 0.3$), $\mathscr{L}$ can be approximated by eq 4.

$$\mathscr{L} \propto \mu E_p / T \qquad (4)$$

There are practical limitations to E_p and T. The poling voltage cannot be much higher than about 200 V/μm without the probability of cataclysmic dielectric breakdown in the sample. The temperature during poling must be at or above T_g of the material, which in turn must be well above ambient to prevent subsequent relaxation of the induced order. Typical poling temperatures range from 100 to 250 °C.

However, μ can be increased by clever chemical synthesis. Several researchers have produced materials with enhanced dipole moments (*39–43*), and progress to date in this area is quite encouraging.

The main advantage of poling lies in its simplicity. Ordinary laboratory equipment can be employed, and the whole process can be done quickly. Any nonconductive, NLO-active polymer can be used, provided that the molecular dipoles can be given sufficient mobility. In addition, many samples can be poled simultaneously, a real advantage for eventual manufacture. The main disadvantages of poling are the relatively low degree of order that can be obtained with practical poling fields (the orientation process is opposed by thermal fluctuations), and the locked material is in a metastable state that tends to decay over time.

Different types of polymers oriented by this technique are discussed in the following sections.

Side-Chain Systems by Polymerization of Functional Monomers

In these systems, an NLO-active chromophore is functionalized with a polymerizable group (or groups), and a polymer is formed by conventional means. The NLO-active moiety is thereby attached to the backbone, often via a flexible spacer group. This method has the advantage that the concentration of chromophore can be readily controlled at the point of synthesis, provided that the chromophore does not interfere with the polymerization reaction.

The basic architectural features of these polymers were derived from earlier extensive work on side-chain liquid-crystalline polymers (*44*). The first published examples of side-chain polymers specifically designed for second-order NLO were the liquid-crys-

talline polyacrylates (**2**) of LeBarny et al. (*45*) and polyesters (**3**) of Griffin et al. (*46*). In these two cases, the authors employed or emulated the structures of known liquid-crystalline polymers in attempts to achieve high degrees of orientation. However, it has since become clear that light scattering from such materials would be unacceptably high.

Since the publication of these two structural classes, a host of similar materials have been reported. For example, I (*47*) polymerized a series of related monomers by free-radical techniques and showed that, although homopolymers were liquid crystalline, copolymers with methyl methacrylate were amorphous. Difficulties caused by cross-linking during free-radical polymerization of stilbene-containing

2

3

Structures 2 and 3. The first reported functionalized side-chain liquid crystalline polymers designed for NLO.

monomers were encountered, but an azobenzene-containing monomer polymerized uneventfully. Problems in the free-radical polymerization of related monomers were also reported by Griffin and Bhatti (48).

At about the same time, a large variety of similar side-chain (meth)acrylate NLO polymers were disclosed by DeMartino, East, and co-workers at Hoechst-Celanese Corp. The chromophores contained biphenyl (4, 5) (49, 50), naphthyl (6) (51), stilbene (7, 8) (52–54), azobenzene (9) (55), and diphenylbutadiene (10) (56) push–pull structures. Homopolymers and copolymers with conventional acrylate monomers were synthesized from the substituted monomers. Liquid-crystal and isotropic polymers were reported.

Additionally, a series of polymers analogous to 3 were synthesized by DeMartino (57–59) (e.g., 11) by using condensation polymerization. One noteworthy

feature of polymers such as 11 is the *lack* of any spacer group. The donor nitrogen atom is bound directly in the polymer backbone, a feature that may slow reorientation processes. Similar structures based on linear epoxy polymers (e.g., 12) were synthesized by workers at IBM (60–62). In one case, (60) this architecture was referred to as "main-chain", but this classification is probably better applied to polymers in which substantial portions of the rigid, aromatic moieties are bound within the backbone. DeMartino and co-workers (63) also disclosed a rather rare example (13) in which cationic polymerization of NLO-functionalized monomers was employed (Scheme I).

Unfortunately, in much of the early work, only sparse descriptions of the NLO measurements and properties were published. This situation has begun to change. For example, van der Vorst and van Rheede

Structures 4–10. Examples of the acrylic side-chain polymers for NLO synthesized by Hoechst-Celanese Company.

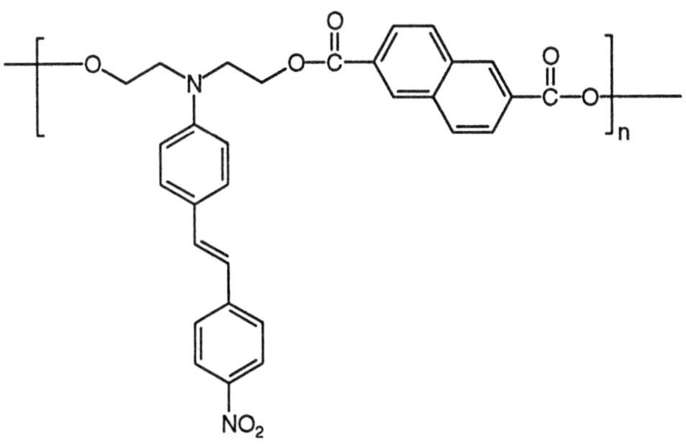

Structures 11 and 12. Side-chain NLO polymers without spacer groups.

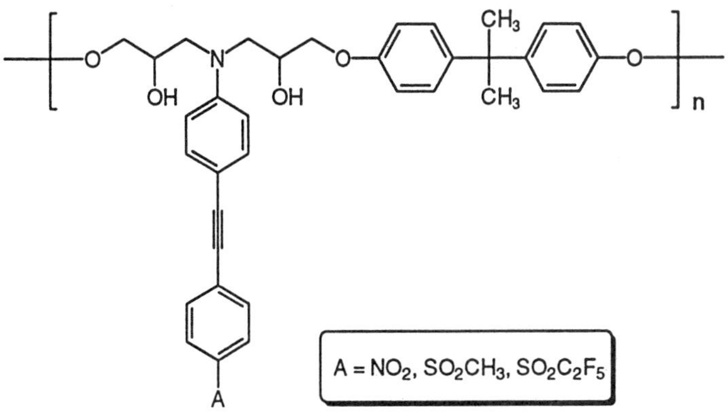

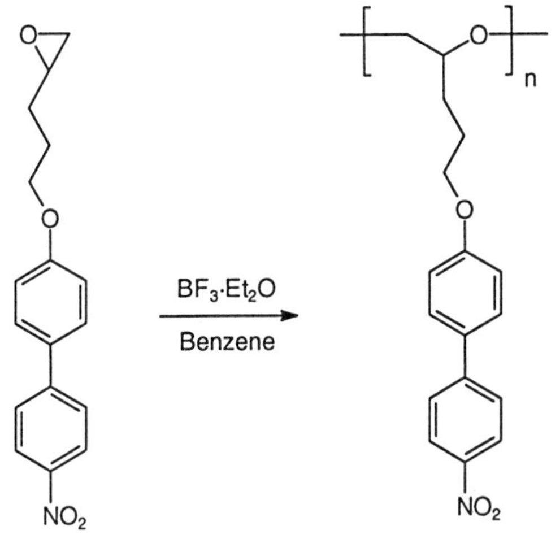

Scheme I. Side-chain NLO polymer via cationic polymerization.

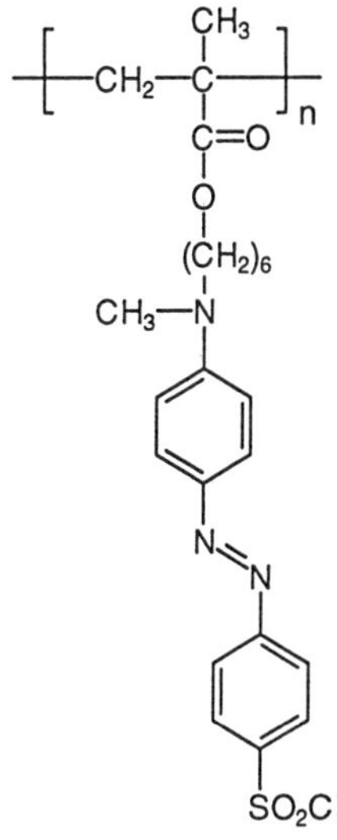

Structures 14 and 15. Side-chain NLO polymers with sulfonyl acceptor groups.

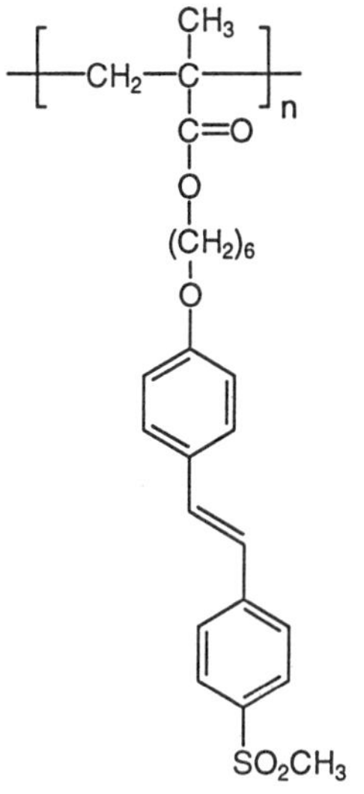

(*64*) published a detailed description of the poling and NLO properties of polymers apparently similar to **7**. An electrooptic coefficient of 34 pm/V was claimed for a poled polymer functionalized with DANS (**1**). Papers from the IBM group (*60, 61, 65*) also described the NLO properties of their materials in commendable detail. Similarly, a synthesis of a series of side-chain polymers (e.g., **14**) was reported by myself and co-workers (*25*), along with a complete description of fabrication, poling, electrooptic measurements, prototype device operation, temporal stability, and a predictive model. In the best case, a nonresonant electrooptic coefficient of 12.5 pm/V was found in this study. Hopefully, the trend toward more complete reporting of NLO materials will continue.

Ulman et al. (*12*) demonstrated that the sulfonyl group is an attractive acceptor because of improved transparency in the visible region compared with other common acceptor groups. This feature was put into practice by Nijhuis et al. (*13*), who synthesized color-

less NLO polymers (e.g., **15**) potentially useful for frequency doubling. The trade-offs between transparency and NLO properties were evident from this study. The susceptibility of **14** after poling was rather weak (d_{33} of 1–4 pm/V by SHG at 1064 nm fundamental).

All of the polymers discussed thus far have chromophores bound to the backbone via the donor atom (usually O or N), obviously for synthetic convenience. However, two examples have appeared in which the chromophore was bound to the opposite end by using the acceptor group: an ester acceptor (**16**) by Ford et al. (*66*) and a sulfone acceptor (**17**) by Dalton et al. (*67, 68*). In polymer **17**, the donor group is substituted with two hydroxyl groups that are available for cross-linking during poling, and cross-linking markedly improves the orientational stability. For example, a highly cross-linked, poled sample of **17** showed very little decay of NLO properties even at 125 °C for many days. Polymer

Structures 16 and 17. Side-chain NLO polymers bound through the acceptor groups.

17 could also be classified under cross-linked systems, and it illustrates the importance of chemical flexibility in the design of complex NLO functionalized polymers. A trend toward polymers bearing extended chromophores such as **18** (69, 70) has been noted. Even longer molecules are being studied (71).

The final example in this category is the polyimide (**19**) synthesized by Marks et al. (72), which displayed

remarkable orientational stability after poling, even at 85 °C. The T_g of this polymer was estimated to be 236 °C. Because of the recognized stability of polyimides, further examples of NLO polymers with this backbone have begun to appear (73–75).

Side-Chain Systems by Polymer Modification Reactions

Because of the possible incompatibilities involved with polymerization reactions and the complexities in synthesizing complicated functional chromophores, many researchers have formed the polymer first and then appending the chromophore as a side-chain in a polymer-modification reaction. In most cases, the polymer is electrophilic, and the chromophore is nucleophilic. For example, Marks and co-workers (76, 77) attached chromophores to preformed poly(vinylbenzyl chloride) and to brominated poly(phenylene oxide) to form polymers **20** and **21**, respectively. Polymer **21** displayed the desirable attributes of unusually large loading of chromophore and high T_g.

Poly(acryloyl chloride) was used by Choe (78), and poly(methacryloyl chloride) was used by Müller et al. (79), in similar routes, to produce polymers **22** and **23**, respectively. Even the inorganic polymer poly(phosphazene) was derivatized by Dembek et al. (80) to obtain the mixed fluoroalkyl-NLO chromophore polymer **24** (Scheme II). However, the phosphazene backbone is extremely flexible, and it was not possible to

Structure 18. Side-chain NLO polymer with extended chromophore.

Structure 19. The first reported functionalized NLO polyimide.

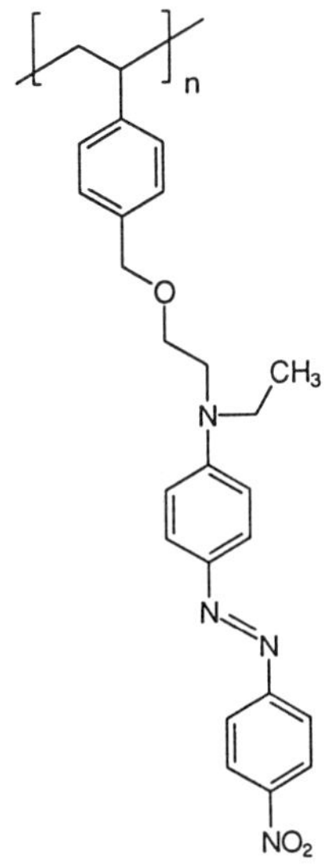

Structures 20 and 21. NLO polymers via chemical modification using benzyl halides.

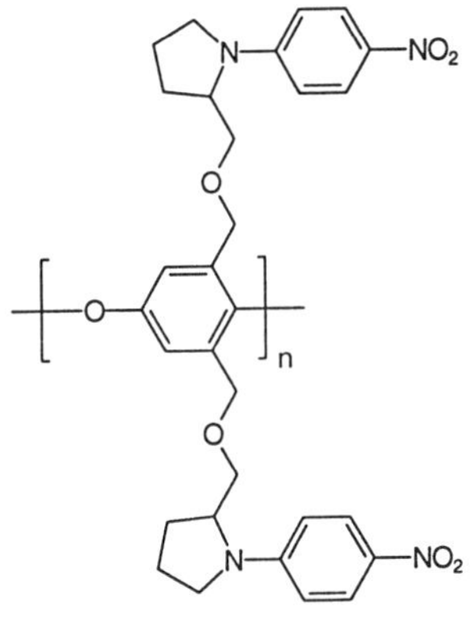

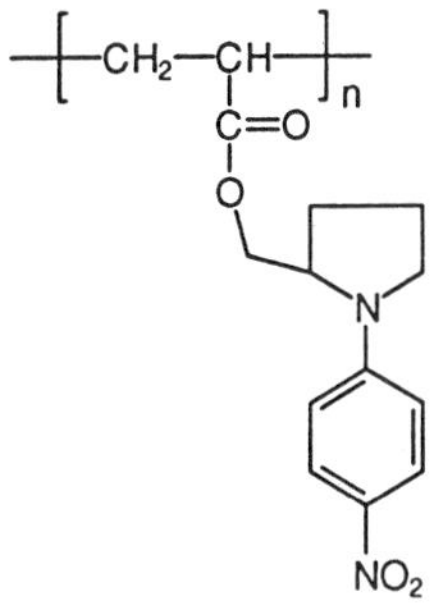

Structures 22 and 23. NLO polymers via chemical modification using acid chlorides or anhydrides.

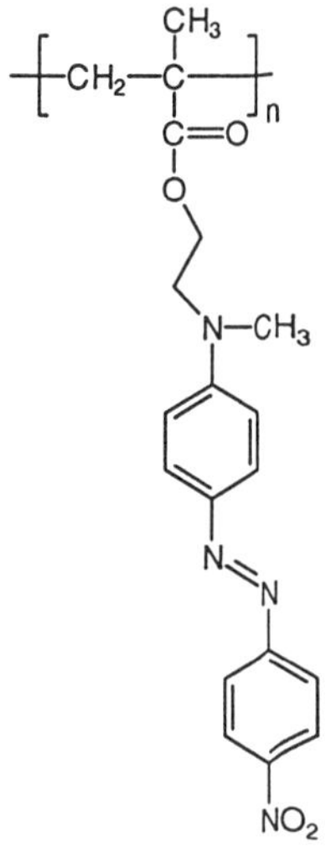

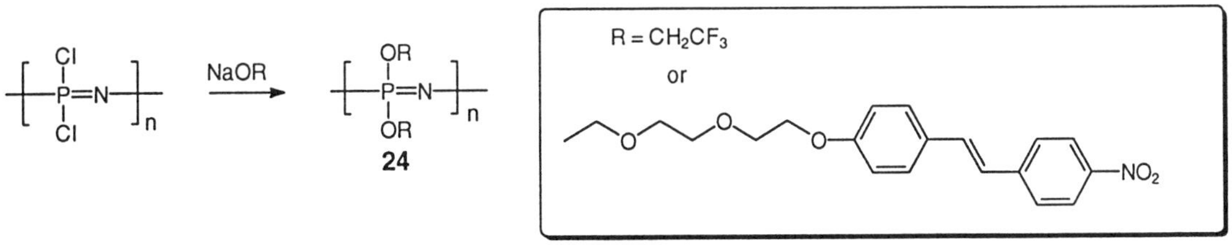

Scheme II. NLO polymers via chemical modification using acid chlorides or anhydrides.

prepare polymers with sufficiently high T_g for the preservation of poling-induced order.

Leslie (*81*) and Ford et al. (*82*) used hydrosilation of poly(siloxane) backbones for the connection of NLO chromophores to create structures **25** and **26**, respectively. Again, the high degree of flexibility inherent to the siloxane backbone leads to very low T_g.

The mechanistic roles of the polymer and chromophore were reversed in a series of materials synthesized by Marks et al. (*83–85*) in which poly(4-hydroxy styrene) was used to produce an NLO-active polymer (**27**, Scheme III). The chromophores were substituted with an electrophilic group for attachment to the nucleophilic polymer. Because only a fraction (<50%) of the repeat units were converted to ether linkages, the residual phenols were able to participate in hydrogen bonding and cross-linking reactions that stabilized poling-induced alignment (*84, 85*).

Co-workers and I (*86*) were able to achieve high degrees of substitution of NLO chromophores onto a styrenic backbone by using Heck carbonylation chemistry to produce polymers such as **28** (Scheme III). Advantages of this approach were the inertness of the polymeric precursor, poly(4-bromo styrene), and the compatibility of the reaction to variations in the chromophore structure.

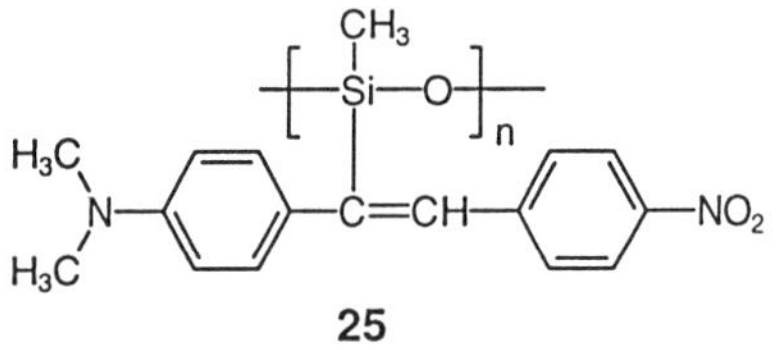

Structures 25 and 26. NLO polymers via chemical modification using hydrosilylation.

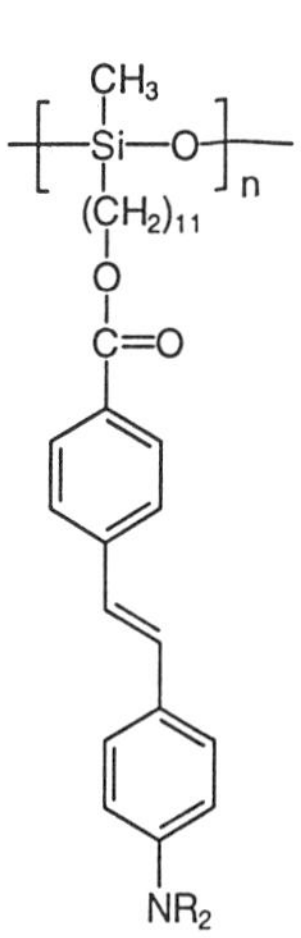

Interesting variants of the grafting technique for azobenzene-containing materials were investigated by Schilling et al. (*87*) and by Feringa et al. (*88*). The NLO-active chromophore was synthesized in stages on the polymer backbone. In the first case, a polymethacrylate containing aniline moieties was synthesized by polymerization of a partially functionalized monomer

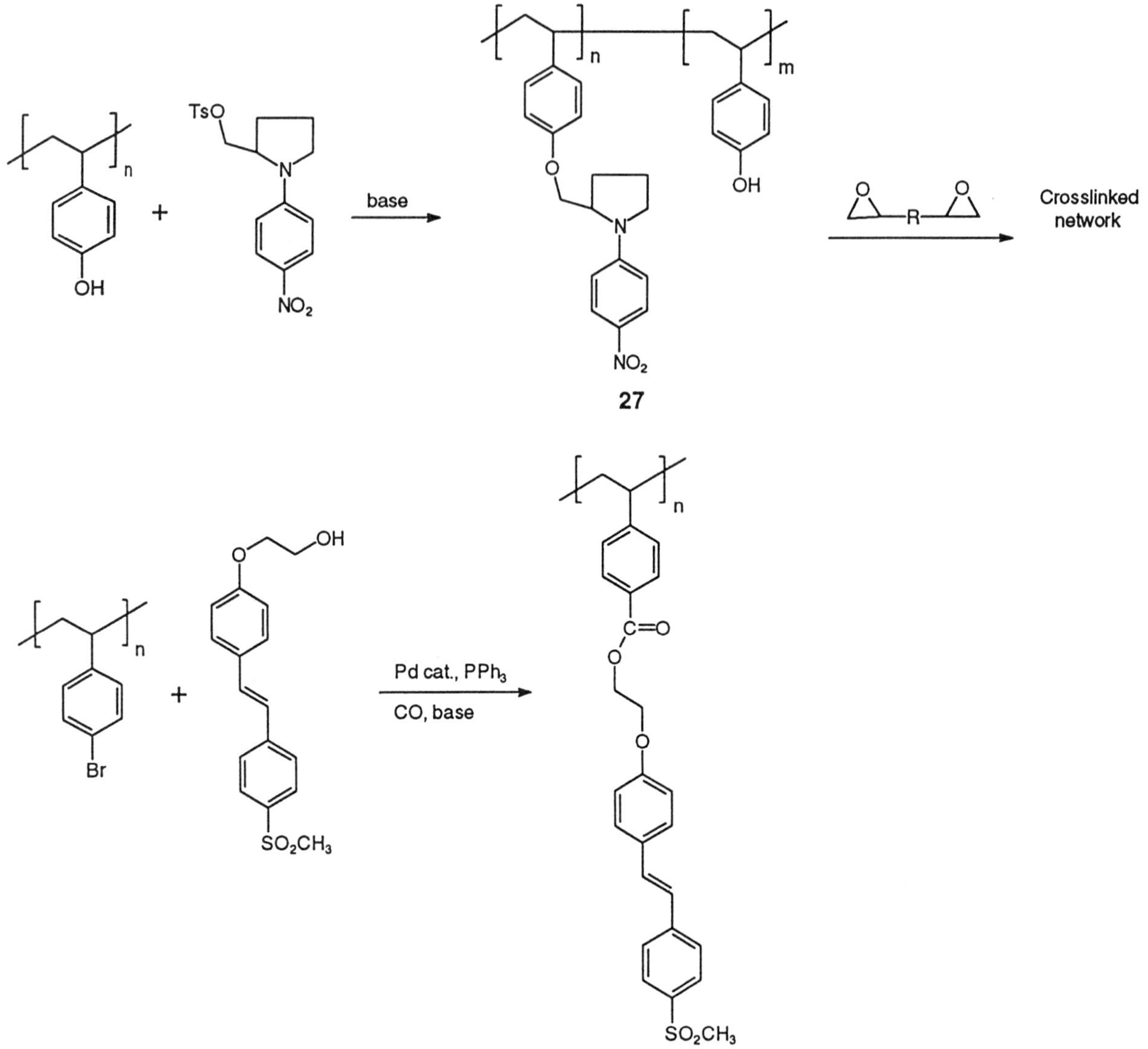

Scheme III. NLO polymers via chemical modification of polystyrene derivatives.

(89), and then a dicyanovinyl acceptor group was appended by diazo coupling to produce an NLO polymer (**29**). Excellent results were obtained. More than 90% of the available aniline groups were transformed to push–pull dyes; consequently, strong NLO properties were found (electrooptic coefficient, r, of 18 pm/V off-resonance) (90).

Feringa et al. (88) carried out a similar reaction on a styrenic polymer to produce **30**. The difference in their approach was that the aniline group itself was formed by a polymer modification reaction on poly-(vinylbenzyl chloride).

A final example serves to illustrate the ease by which NLO-active polymers can be synthesized by using grafting reactions. Sen et al. (91) reacted two simple, commercially available materials, poly(allyl-

amine) hydrochloride and 4-fluoronitrobenzene, and obtained in one step a polymer with an unusually high concentration of p-nitroaniline chromophores (**31**, Scheme IV). In a subsequent, detailed presentation of the properties of this polymer (92), these researchers reported a high nonlinearity, d_{33}, of 31 pm/V (measured by SHG at 1064 nm fundamental), although, unfortunately, this value decreased to 19 pm/V after storage at room temperature for 5 days.

Main-Chain Systems

Two forces drive the incorporation of the NLO-active chromophore within the polymer backbone. First, Williams and Willand (39) predicted that it may be possible to gain cooperativity during poling of an

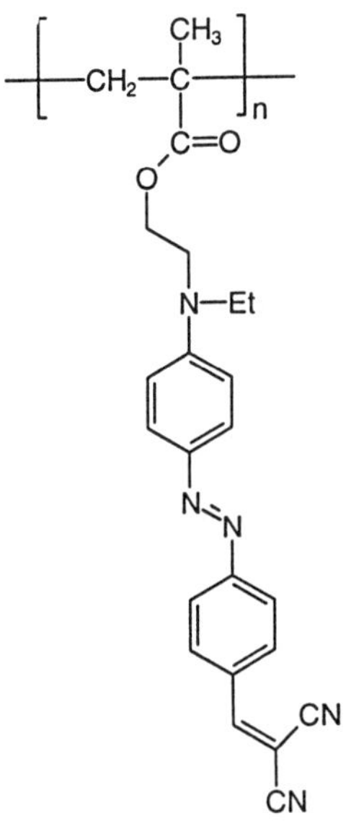

Structures 29 and 30. NLO polymers by diazo coupling onto polymeric precursors.

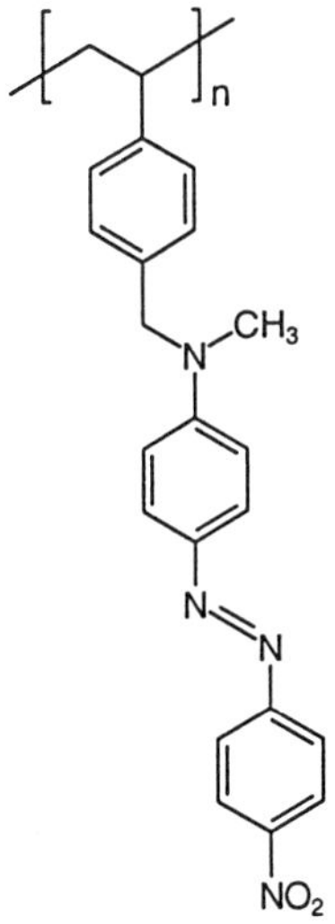

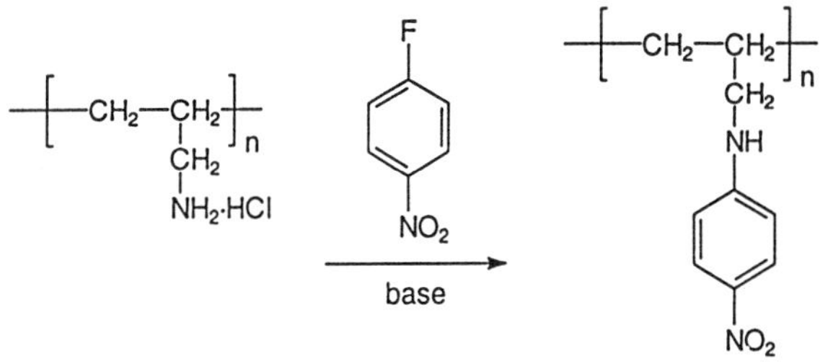

Scheme IV. NLO polymers by aromatic nucleophilic substitution.

isoregic (i.e., head-to-tail) polymer that has a significant component of the chromophore dipole along the backbone. In this way, the effective dipole moment of the NLO chromophores would be increased (93), and a more complete orientation would be achieved within a given poling field; therefore, the orientation

factor, $\mathscr{L}$ (eq 3), would be increased. Second, the close connection between chromophore and backbone in a main-chain system would require a relatively large displacement of the polymer to allow for chromophore movement. Main-chain systems, therefore, should be slower to orient, but, more important, slower to disorient compared to analogous side-chain systems.

The first reports in this area were those of Green et al. (94, 95), in which NLO polyesters (32, 33) were synthesized by the self-condensation of appropriately functionalized quinodimethane and benzene NLO chromophores. The copolymer of 32 containing a flexible comonomer (for solubility) exhibited strong enhancement of NLO properties in solution but only slight enhancement in the solid state (96).

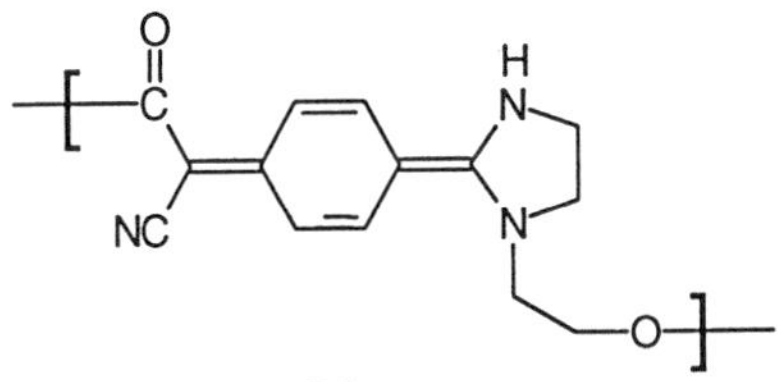

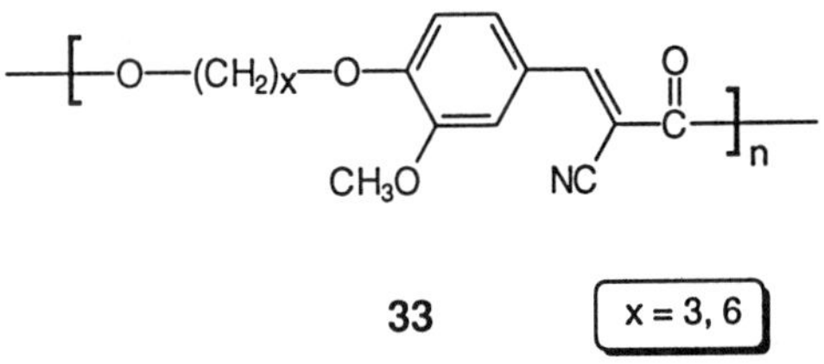

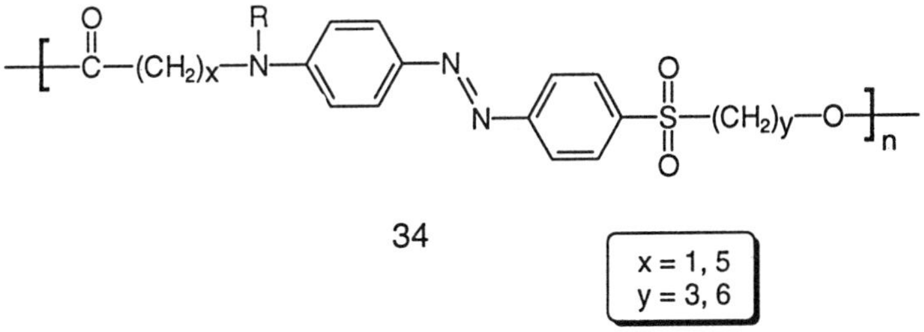

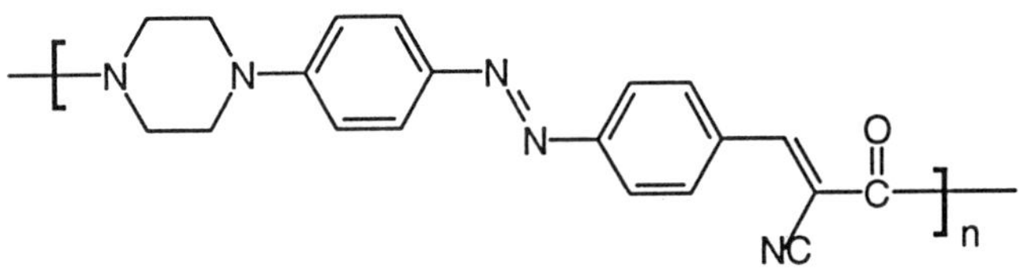

Structures 32–35. Main-chain NLO polyesters.

Köhler et al. (*34, 35*) studied isoregic polymers, **34**, that exploited the bifunctional sulfone acceptor group for main-chain architecture. Although no dipole enhancement was found, their measurements confirmed the prediction that main-chain polymers would disorient more slowly than related side-chain polymers. In fact, the loss of orientation appeared to begin slightly above the T_g in these systems.

In a related study on discreet *p*-aminosulfone oligomers, Mitchell et al. (*97*) demonstrated that even a small number of flexible bonds between chromophores can preclude cooperative orientation in an electric field. When rigid spacer groups are used, it is essential to maintain an extended, linear geometry to observe dipole additivity (*43*). Schilling and Katz (*40, 98*) synthesized oligomers, **35**, in which this concept was put in practice. Dimers to tetramers with partially reinforcing dipole moments were investigated, and a modest enhancement was noted.

Because of the difficulties in obtaining soluble, rigid, extended structures, other groups attempted to gain dipole additivity with syndioregic (head-to-head) structures designed to fold in a manner that would lead to parallel dipoles. This architecture, originated by workers at the China Lake Naval Weapons Center, was dubbed *accordion* polymers. For example, Lindsay et al. (*99*) synthesized a polyamide (**36**) with relatively rigid spacers in an attempt to induce chain folding. However, no information on dipole enhancement in accordion polymers has yet appeared.

An intriguing variation on this theme was synthesized by Wright and Sigman (*100*), in which the chromophore structure contained both ferrocene and silane groups (**37**). Whether this structural class has value other than novelty has yet to be demonstrated. The use of organometallic NLO chromophores has been explored in several other reports, but there seems to be no particular advantage in incorporating transition metals in the structures considered thus far. A theoretical treatment predicting relatively weak

NLO performance for most simple organometallic compounds has appeared (*101*).

Cross-Linked Systems

The reason for interest in cross-linked NLO materials is obvious. Rigidification of the matrix as occurs on cross-linking is expected to lower the mobility of the active chromophores, thereby reducing the rate of disorientation after poling. Naturally, because immobilization occurs, it is necessary to carry out the cross-linking reaction simultaneously with poling. Here, instead of cooling the poled system below its T_g, the process essentially raises the T_g of the matrix above the poling temperature. However, the material must still be heated during poling to achieve the high T_g needed for orientational stability. This point was apparent in the first NLO networks (**38**) reported by co-workers and myself (*102, 103*). These networks exhibited rather poor orientational stability (Scheme V). When simultaneous poling and photo-cross-linking were carried out at room temperature, only relatively low T_g values were obtained, presumably because the reaction becomes very slow as a cross-linked matrix is formed, and the cross-link density is not very high.

Eich et al. (*104*) reported the first example of a thermosetting NLO material with the epoxy system (**39**, Scheme VI). A T_g as high as 150 °C was obtained when the sample was processed at 140 °C, and the NLO susceptibility of this material exhibited short-term stability at 85 °C.

A similar concept was used by Tripathy et al. (*105*), who used sol–gel chemistry to produce a network polymer with covalently bound azo-benzene chromophores (**40**, Scheme VII). This system was reasonably stable even at 105 °C.

Rather than producing a functional monomer capable of cross-linking, subsequent researchers concentrated on preformed polymers that could be cross-linked. A system based on poly(*p*-hydroxystyrene) (**27**)

Structures 36 and 37. *Accordion* main-chain NLO polymers.

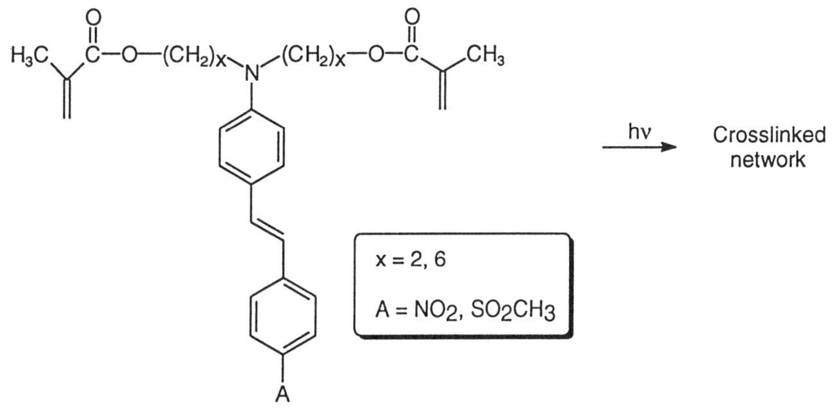

Scheme V. Polyfunctional acrylic NLO chromophores designed for photo-cross-linking.

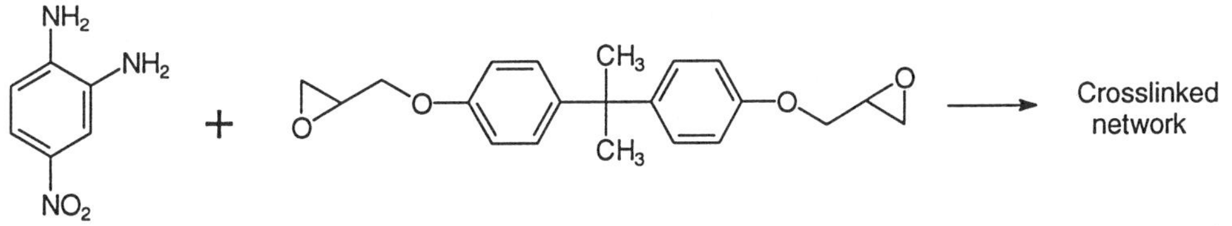

Scheme VI. Thermally cross-linked NLO polymer based on epoxy chemistry.

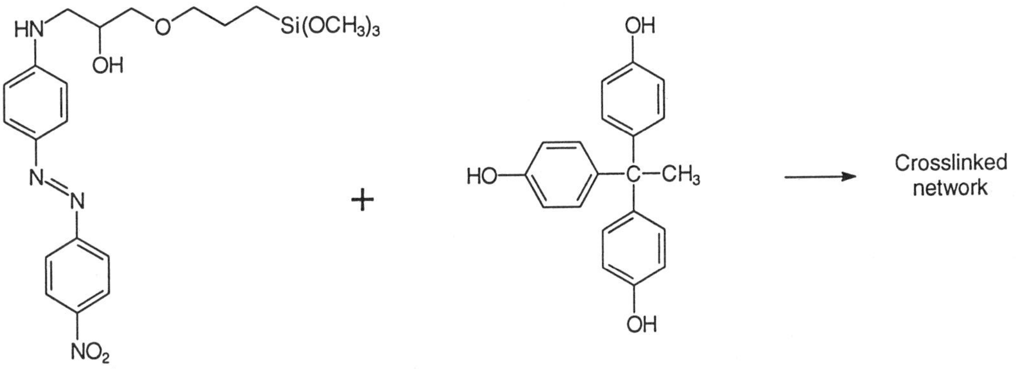

Scheme VII. Cross-linked NLO polymer based on sol–gel chemistry.

was discussed previously. Similar systems were based on photochemical cross-linking. For example, a polymer bearing both NLO chromophores and cinnamate esters on separate side chains (**41**) was synthesized by Tripathy et al. (*106*) (Scheme VIII). This polymer could be rendered insoluble by irradiation, but some photochemical decomposition of the chromophores was noted. Even a simple 4-alkoxynitrobenzene chromophore suffered significant damage during irradiation of a related polymer synthesized by Hashidate et al. (*107*). Light absorption by the chromophore is a severe problem in these systems. Even if the chromophore resists photo-induced decomposition, it still

can act as a powerful photon collector. Only a small fraction of the incident light may remain available for photocross-linking reactions (*103*). Also, wasteful energy transfer from the photo-cross-linking system to the chromophore can easily occur, depending on the relative electronic states of the components. These limitations were overcome by Beecher et al. (*108, 109*) with a judicious combination of polymer-bound chromophore (**42**), cross-linking group, and photoinitiator (**43**) (Scheme IX). The ferrocenyl photoinitiator absorbs at longer wavelengths than the stilbene chromophore, bleaches after irradiation, does not transfer energy to the chromophore, and is thermally stable.

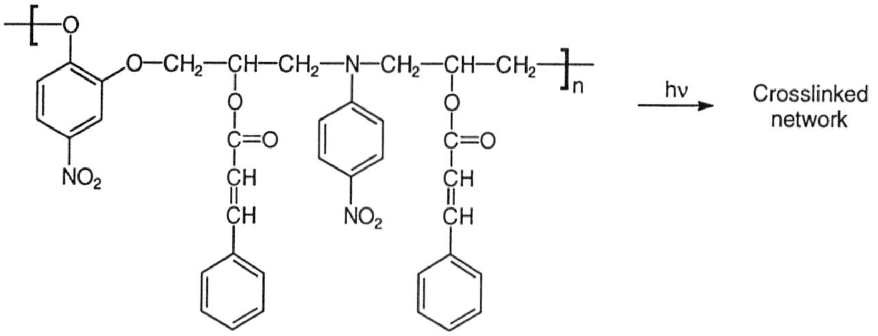

41

Scheme VIII. NLO polymer cross-linkable by cinnamate photodimerization.

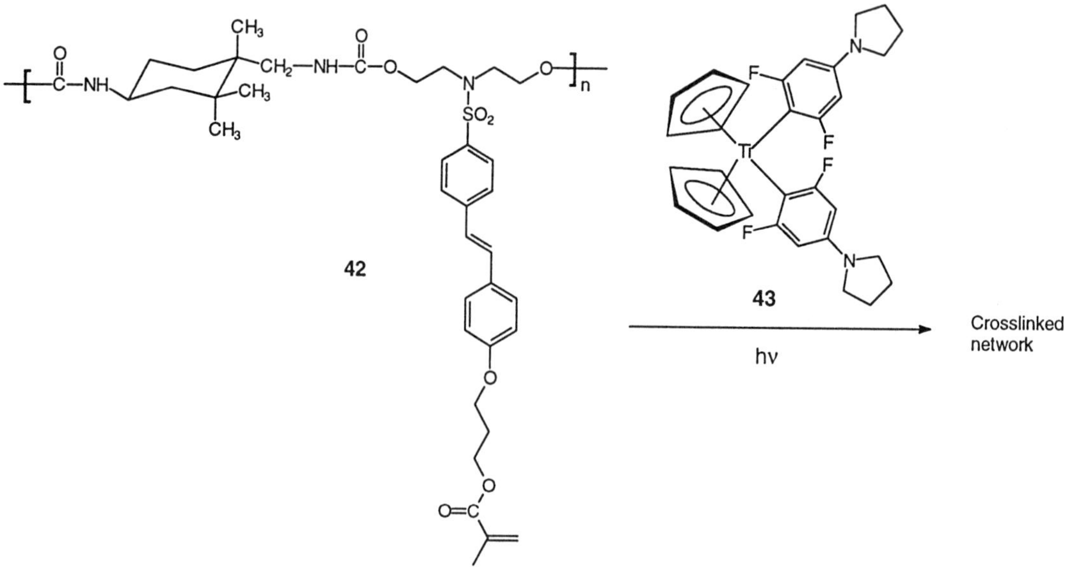

42

43

Scheme IX. Photo-cross-linkable functionalized NLO polyurethane system.

Cross-linking was effected by irradiation above the absorption band of the stilbene-chromophore. These features allowed direct photolithography of the NLO-active polymer, a process that may permit the creation of channel waveguides simultaneously with poling.

Photorefractive Polymers

In early work, the required photoconductivity was obtained by doping an NLO polymer with appropriate charge-transfer agents (*110, 111*). Later, more sophisticated polymers containing both the photoconductive and NLO-active groups bound to the same backbone were synthesized and studied. For example, Tamura et al. (*112*) partially fuctionalized poly(*N*-vinylcarbazole) by using tetracyanoethylene to produce an intrinsically photorefractive polymer (**44**).

Yu et al. (*113*) synthesized an even more advanced polymer (**45**) that contained charge-generating, charge-transport, and NLO-active chromophores. Two-beam

coupling was performed, an experiment that confirms the photorefractive properties of this polymer.

Surface-Oriented Systems

Surface Alignment

A surface is inherently noncentrosymmetric. Two related techniques take advantage of this dissymmetry in an attempt to create useful structures: Langmuir-Blodgett (L–B) deposition and self-assembly (*114*). In both techniques, macroscopic films are built up from single-molecule thick layers, with, potentially, a high degree of control over the structure within and among the layers. These techniques have obvious applications in second-order NLO, in which control of molecular orientation is important. For example, one might envisage structures with each layer containing chromophores all pointing directly away from the original surface, a highly noncentrosymmetric arrangement

Structure 44. Photorefractive poly(*N*-vinylcarbazole) derivative.

(obtained without poling). The main disadvantage lies in the tedium in building up hundreds of monomolecular layers into a film of sufficient thickness for waveguiding applications. Also, L–B deposition requires specialized equipment that may be difficult to adapt to a commercial manufacturing environment.

Langmuir–Blodgett Films

To form an L–B film (*114*), a sample of a surfactant molecule is floated on a water surface in a large, flat compartment (trough) and is compressed mechanically to form a two-dimensional "solid" film. A substrate with a suitable surface preparation is then passed through the interface, and the film adheres, as shown greatly simplified in Figure 5. The process is repeated many times by depositing one monomolecular layer on top of another until a film of sufficient thickness is reached. Hundreds of passes are typically required in practice.

NLO-active chromophores present in each layer may assume a polar orientation with respect to the surface of the substrate, in which case the film can exhibit second-order susceptibility (*115*). However, there is one caveat: each layer must deposit in a fashion that reinforces the susceptibility of the preceding

layers. Unfortunately, in the most common occurrence, the direction of the susceptibility alternates and leads to a cancellation of nonlinearity for each pair of layers, so each bilayer is net centrosymmetric. Several methods for overcoming this situation have been developed, most notably alternate deposition of two materials chosen to complement rather than cancel each other in bilayers.

Small molecules have been used for L–B films for nearly 60 years, but recently polymeric systems have begun to attract interest (*116*). For second-order NLO applications, a polymer with amphiphilic character and functionalized with NLO-active chromophores is required. Polymeric L–B films show significant improvements in optical transparency and robustness compared with their low molar mass counterparts. The structures used are varied; examples of polymers synthesized via functional monomers, grafted chromophore polymers, and main-chain polymers will be presented.

In most cases, the polymer backbone comprises the hydrophilic portion of the surfactant needed to form L–B films. One example is the stilbazolium polyether (**46**) of Saperstein et al. (*117*), who grafted chromophores to a polyether backbone. Polar multilayers of **46** were constructed by alternate dipping with calcium behenate, and the expected increase in NLO properties with increasing number of layers was observed. Unfortunately, the assembly was rather fragile; it rearranged noticeably on heating to 60 °C.

Researchers at Eastman Kodak Company synthesized L–B materials based on surfactant-like polymers that incorporated NLO chromophores **47** (*118, 119*). When interleaved with poly(butylacrylate), relatively thick assemblies with hundreds of layers could be produced with strong NLO properties. A sample of **47** produced significant quantities of green or blue frequency-doubled light as Cerenkov radiation (*119*) in the glass substrate. The waveguide quality of **47** was quite good for an NLO film, but unfortunately, the chromophore in **47** is too absorptive to pass fully-

Structure 45. Highly functionalized photorefractive polyurethane.

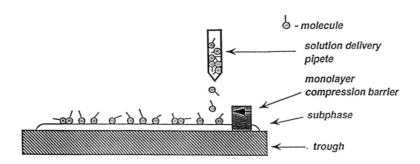

Figure 5. Schematic of Langmuir–Blodgett deposition.

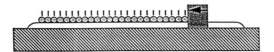

46

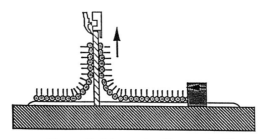

47

48

Structures 46–48. Functionalized side-chain NLO polymers designed for Langmuir–Blodgett deposition.

guided visible light, and it is difficult to harness Cerenkov radiation in a useful form.

These authors went on to demonstrate blue light SHG in a fully waveguided format by using a colorless, functionalized polymer **48** (*120, 121*). Using precise control of the thickness and molecular architecture, these researchers were able to achieve phase-matched SHG from 820 to 410 nm within a waveguide of **48**, a particularly noteworthy accomplishment. The assembly appeared to be unchanged over 4 months at ambient conditions. With foreseeable improvements in chromophore concentration and with the development of channel waveguides, practical amounts of blue light might be produced by this system.

As a final example, even main-chain NLO polymers can be applied to L–B films. The fluorinated "accordion" polymers, **49**, of Hoover et al. (*122*) are designed to fold when compressed on the water surface and lead to polar structures. Interestingly, even nominally symmetric bilayer films of **49** showed SHG activity and indicated that, possibly, asymmetry from the dipping process was involved. However, the films were not very stable even at ambient temperature.

Self-Assembled Multilayers

The field was pioneered by Sagiv and co-workers (*123*). The concept of self-assembled films is closely related to L–B films (*114*), in that a material is constructed by piling up many monomolecular layers. However, as the name implies, self-assembled films do not require a trough for construction, but instead, given the appropriate structure, they spontaneously form a layered structure, stepwise, with its associated asymmetry. The process involves covalent bonding of a molecule to a surface that carries reactive functional groups as well as covalent bonding within the plane of the layer. Therefore, the self-assembled system resembles a layered network polymer. The molecule contains, at opposite ends, a group for bonding to the surface and a latent functional group that can be chemically transformed to reconstruct the reactive

surface. The functional group serves to glue the chromophoric layers together both laterally and between layers. In addition, for certain structures, the molecules within a layer are capable of forming densely packed structures resembling a crystal lattice with the accompanying high order and physical stability.

The main advantage of the self-assembly technique lies in its simplicity. The formation and transformation of each layer is accomplished merely by dipping the substrate in the appropriate chemical reagent. Obviously, automation of this technique would be straightforward and would make large-scale manufacture feasible.

Two remarkable approaches to polar NLO materials via self-assembly have appeared. In the first, Katz et al. (*124*) emulated the layer structures known for zirconium diphosphonates in constructing self-assembled multilayers (**50**) containing an azo chromophore. The phosphonate moiety has a dual function by acting as an electron-acceptor group and as a linking group between layers. More than 30 layers were deposited by dipping the substrate in solutions of the phosphonate-substituted chromophore alternately with solutions of $ZrOCl_2$, and excellent consistency was achieved from one layer to the next. The assembly was orientationally stable at 150 °C for short periods, although it began to decompose chemically over prolonged periods at this temperature.

Marks et al. (*125, 126*) constructed analogous multilayer assemblies, **51**, that comprised three distinct sublayers: a coupling layer for attachment to the surface, a chromophore layer for NLO properties, and a capping layer for tying together the layers and for regenerating the reactive surface. In addition to knitting the layer structure, the capping layer served to greatly increase the NLO response from layers of **51**. The intensity of SHG with an increasing number of layers indicated good consistency, but only five and nine layers of the two structures were reported, respectively. The construction and performance of multilayer assemblies thick enough to permit waveguiding are yet to be demonstrated. Nevertheless, self-

49

Structure 49. Main-chain NLO polymers designed for Langmuir–Blodgett deposition.

Structures 50 and 51. Self-assembled multilayer polymers with NLO chromophores. An idealized two layer structure is shown.

assembly is an exciting and promising alternative to poled systems.

Concluding Remarks and Future Outlook

Functionalized polymers designed for second-order nonlinear optics have grown substantially in complexity and sophistication over the past few years, especially as the requirements for practical device fabrication and operation have become clearer. Recent developments have included the application of polymers stable at high temperatures to NLO and careful studies of optical losses. However, optimized systems combining very large susceptibility, orientational stability, and other desired characteristics have not yet been disclosed. Certainly, there is still much room for improvement in existing NLO materials. The proof of the pudding is in the tasting; despite approximately 10 years of intense worldwide investigation, not a single commercial NLO device based on molecular materials has been produced.

Surface-oriented materials offer the promise of improved orientation compared with poled polymers. However, the monolayer-based systems have the additional hurdle of adaptation to manufacturing practice, which may be particularly difficult for L–B

films. In addition, the thermal stability of these systems needs to be addressed.

Almost all of the proposed NLO devices depend on waveguided optics, but this technology is also in its infancy. For commercialization of integrated guided optics to be accepted, there must be an overriding reason for engineers to abandon the bulk optical systems that are currently in place. Alternatives to molecular NLO devices may supersede. For example, advances in inorganic crystal technology have continued apace, and a material such as $LiNbO_3$ may be superior to any molecular material for blue light generation. Naturally, frequency doubling itself may be obviated completely by the development of a practical semiconductor laser that emits primary blue light. This area is being vigorously pursued, although the feasibility of constructing a short-wavelength-emitting diode laser is by no means certain. Similarly, electrooptic modulation using any kind of NLO material must compete with direct modulation of the laser source itself.

Even if adequate NLO materials are produced, the proper auxiliary technologies must also be in place. For example, because of phase-matching considerations, the efficiency of SHG depends strongly on the wavelength of the source. Even a small drift of the fundamental laser can drastically reduce the intensity of the second-harmonic light. Frequency-stabilized semiconductor lasers have been made, but they are not commercial. A catch-22 exists. At the moment, SHG materials and frequency-stabilized lasers have no reason for existence without each other. Both must be developed in parallel and, most likely, by separate companies.

In the area of photorefractive materials, a pressing need exists for materials that would permit 3D optical information storage. For example, the existing compact disk has remarkably large (planar) storage capacity, yet it can only hold a few minutes of high-definition television video information. Practical photorefractive polymers would provide an ideal high-density storage medium, but only if considerable improvements are made, especially in charge-generation efficiency.

Perhaps second-order nonlinear optics is a case in which the technology has run ahead of commercial demand. Particularly, the necessarily large investments in research, development, and eventual manufacturing technology of integrated optical components are not justifiable if only a few highly specialized machines are to be produced. A large base of applications must develop concurrently, as has occurred in the small computer industry. Perhaps the appearance of high-definition television, or the demands of very fast microprocessors, or rapid expansion of fiber-optic telecommunications will catalyze the growth of integrated optical technology.

Nevertheless, there is good reason to believe that advances in NLO materials will continue to accelerate, as will the demands to manipulate photons faster and more efficiently. A large and growing application of guided wave optics is already in place: fiber optics. Whether second-order NLO molecular materials will play a key role in the expansion of photonic technology is not certain. Much depends on cost and efficiency issues and the developments in competing technologies. The best chance for the application of NLO materials to devices is not as a substitute for bulk materials such as $LiNbO_3$, but rather in components in which the ease of forming complicated structures integrated with electronic components is paramount. Certainly great potential exists for interesting basic and applied research in the future.

References

1. Boutevin, B. Chapter 3.5 in this volume.
2. Strenberg, E. *Photonic Technology and Industrial Policy, U.S. Responses to Technological Change*; State University of New York: Albany, NY, 1992.
3. Shen, Y. R. *Principles of Nonlinear Optics*; Wiley: New York, 1984.
4. Prasad, P. N.; Williams, D. J. *Introduction to Nonlinear Optical Effects in Molecules and Polymers*; Wiley: New York, 1991.
5. Williams, D. J. *Angew. Chem. Int. Ed. Eng.* **1984**, *23*, 690–703.
6. *Nonlinear Optical Properties of Organic Molecules and Crystals*; Chemla, D. S.; Zyss, J., Eds.; Academic: Orlando, FL, 1987.
7. *Materials for Nonlinear Optics: Chemical Perspectives*; Marder, S. R.; Sohn, J. E.; Stucky, G. D., Eds.; ACS Symposium Series 455; American Chemical Society: Washington, DC, 1991.
8. Chandrasekhar, P. Chapter 3.7 in this volume.
9. Ulman, A.; Willand, C. S.; Köhler, W.; Robello, D. R.; Williams, D. J.; Handley, L. *J. Am. Chem. Soc.* **1990**, *112*, 7083–7090.
10. Nijhuis, S.; Rikken, G. L. J. A.; Havinga, E. E.; tenHoeve, W.; Wynberg, H.; Meijer, E. W. *J. Chem. Soc. Chem. Commun.* **1990**, 1093–1094.
11. Schildkraut, J. *Appl. Opt.* **1988**, *27*, 2839–2841.
12. Clays, K.; Schildkraut, J. S. *J. Opt. Soc. Am. B* **1992**, *9*, 2274–2282.
13. Teng, C. C.; Man, H. T. *Appl. Phys. Lett.* **1990**, *56*, 1734–1736.
14. Lytel, R.; Lipscomb, G. F.; Stiller, M.; Thakara, J. I.; Ticknor, A. J. *Proc. SPIE* **1988**, *971*, 218–229.
15. Lipscomb, G. F.; Lytel, R.; Ticknor, A. J.; Van Eck, T. E.; Kwiatkowski, S. L.; Girton, D. G. *Proc. SPIE* **1990**, *1337*, 23–34.
16. Lipscomb, G. F.; Lytel, R. *Mol. Cryst. Liq. Cryst. Sci. Technol. Ser. B: Nonlin. Opt.* **1992**, *3*, 41–60.
17. Möhlmann, G. R. In *Nonlinear Optics. Fundamentals, Materials, and Devices*; Miyata, S., Ed.; Elsevier Science: Amsterdam, Netherlands, 1992; pp 415–430.
18. Van Tomme, E.; Daele, P. P.; Baets, R. G.; Lagasse, P. E. *IEEE J. Quant. Electron.* **1991**, *27*, 778–787.
19. Meredith, G. R.; VanDusen, J. G.; Williams, D. J. *Macromolecules* **1984**, *17*, 2228–2230.
20. Singer, K. D.; Kuzyk, M. G.; Sohn, J. E. *J. Opt. Soc. Am. B* **1987**, *4*, 968–976.
21. Robello, D. R.; Dao, P. T.; Phelan, J.; Revelli, J.; Schildkraut, J. S.; Scozzafava, M.; Ulman, A.; Willand, C. S. *Chem. Mater.* **1992**, *4*, 425–435.
22. Katz, H. E.; Singer, K. D.; Sohn, J. E.; Dirk, C. W.; King, L. A.; Gordon, H. M. *J. Am. Chem. Soc.* **1987**, *108*, 6561–6563.
23. Cheng, L.-T.; Tam, W.; Stevenson, S. H.; Meredith, G. R.; Rikken, G.; Spangler, C. W.; Marder, S. R. *J. Phys. Chem.* **1991**, *95*, 10631–10643.
24. Cheng, L.-T.; Tam, W.; Marder, S. R.; Steigman, A. E.; Rikken, G.; Spangler, C. W. *J. Phys. Chem.* **1991**, *95*, 10643–10652.
25. Stiegman, A. E.; Graham, E.; Perry, K. J.; Khundkar, L. R.; Cheng, L.-T.; Perry, J. W. *J. Am. Chem. Soc.* **1991**, *113*, 7658–7666.
26. Barzoukas, M.; Fort, A.; Klein, G.; Serbutoviez, C.; Oswald, L.; Nicoud, J. F. *Chem. Phys.* **1992**, *164*, 395–406.
27. Cheng, L.-T.; Tam, W.; Feiring, A. E. *Mol. Cryst. Liq. Cryst. Sci. Technol. Ser. B: Nonlin. Opt.* **1992**, *3*, 69–71.
28. Moylan, C. R.; Tweig, R. J.; Lee, V. Y.; Swanson, S. A.; Betterton, K. M.; Miller, R. D. *J. Am. Chem. Soc.* **1993**, *115*, 12599–12600.
29. Marder, S. R.; Gorman, C. B.; Tiemann, B. G.; Cheng, L.-T. *J. Am. Chem. Soc.* **1993**, *115*, 3006–3007.

30. Köhler, W.; Robello, D. R.; Willand, C. S.; Williams, D. J. *Macromolecules* **1991**, *24*, 4589–4599.

31. Köhler, W.; Robello, D. R.; Dao, P. T.; Willand, C. S.; Williams, D. J. *J. Chem. Phys.* **1990**, *93*, 9157–9166.

32. Singer, K. D.; Kuzyk, M. G.; Holland, W. R.; Sohn, J. E.; Lalama, S. J.; Comizzoli, R. B.; Katz, H. E.; Schilling, M. L. *Appl. Phys. Lett.* **1988**, *53*, 1800–1802.

33. Comizzoli, R. B. *J. Electrochem. Soc.* **1987**, *134*, 424-429.

34. Dao, P. T.; Williams, D. J.; McKenna, W. P.; Goppert-Berarducci, K. *J. Appl. Phys.* **1993**, *73*, 2043–50.

35. Willand, C. S.; Williams, D. J. *Ber. Bunsenges. Phys. Chem.* **1987**, *91*, 1304–1310.

36. Katz, H. E.; Schilling, M. L. *J. Am. Chem. Soc.* **1989**, *111*, 7554–7557.

37. Kelderman, E.; Derhaeg, L.; Heesink, G. J. T.; Verboom, W.; Engbersen, J. F. J.; van Hulst, N. F.; Persoons, A.; Reinhoudt, D. N. *Angew. Chem. Int. Ed. Eng.* **1992**, *31*, 1075–1077.

38. Heesink, G. J. T.; vanHulst, N. F.; Bölger, B.; Kelderman, E.; Engbersen, J. F. J.; Verboom, W.; Reinhoudt, D. N. *Appl. Phys. Lett.* **1993**, *62*, 2015–2017.

39. Katz, H. E.; Lavell, W. T. *J. Org. Chem.* **1991**, *56*, 2282–2284.

40. *Side Chain Liquid Crystal Polymers*; McArdle, C. B., Ed.; Blackie: Glasgow, 1989.

41. LeBarny, P.; Ravaux, G.; Dubois, J. C.; Parneix, J. P.; Njeumo, R.; Legrand, C.; Levelut, A. M. *Proc. SPIE* **1987**, *682*, 56–64.

42. Griffin, A. C.; Bhatti, A. M.; Hung, S. L. *Proc. SPIE* **1987**, *682*, 65–69.

43. Robello, D. R. *J. Polym. Sci., Part A: Polym. Chem.* **1990**, *28*, 1–13.

44. Griffin, A. C.; Bhatti, A. M. *Spec. Pub. R. Soc. Chem.* **1989**, *69*, 295–300.

45. DeMartino, R. N.; Khanarian, G.; Leslie, T. M.; Sansone, M. J.; Stamatoff, J. B.; Yoon, H. N. *Proc. SPIE* **1989**, *1105*, 2–13.

46. DeMartino, R. N. U.S. Patent 4 913 844, 1990.

47. East, A. J. U.S. Patent 4 913 836, 1990.

48. DeMartino, R. N.; Yoon, H.-N. U.S. Patent 4 801 670, 1989.

49. DeMartino, R. N.; Yoon, H.-N. U.S. Patent 4 808 332, 1989.

50. DeMartino, R. N.; Yoon, H.-N. U.S. Patent 4 822 865, 1989.

51. Allen, D.; Lee, C.; DeMartino, R. N. U.S. Patent 5 041 510, 1991.

52. Feuer, B. I. PCT Intl. Appln. WO 91/03683, 1991.

53. DeMartino, R. N. U.S. Patent 4 57 30, 1988.

54. DeMartino, R. N. U.S. Patent 4 795 664, 1989.

55. DeMartino, R. N. U.S. Patent 4 829 950, 1989.

56. Teraoka, I.; Jungbauer, D.; Reck, B.; Yoon, D. Y.; Twieg, R.; Wilson, C. G. *J. Appl. Phys.* **1991**, *69*, 2568–2576.

57. Jungbauer, D.; Teraoka, I.; Yoon, D. Y.; Reck, B.; Swalen, J. D.; Twieg, R.; Wilson, C. G. *J. Appl. Phys.* **1991**, *69*, 8011–8017.

58. Ebert, M.; Lux, M.; Smith, B. A.; Twieg, R.; Wilson, C. G.; Yoon, D. Y. *Polym. Prepr. (Am. Chem. Soc. Div. Polym. Chem. Symp.)* **1991**, *32(3)*, 130–131.

59. van der Vorst, C. P. J. M.; van Rheede, M. *Proc. SPIE* **1992**, *1775*, 186–197.

60. Page, R. H.; Jurich, M. C.; Reck, B.; Sen, A.; Twieg, R. J.; Swalen, J. D.; Bjorkland, G. C.; Wilson, C. G. *J. Opt. Soc. Am. B* **1990**, *7*, 1239–1250.

61. Ford, W. T.; Bautista, M.; Zhao, M.; Reeves, R. J.; Powell, R. C. *Mol. Cryst. Liq. Cryst.* **1991**, *190*, 351–356.

62. Xu, C.; Wu, B.; Dalton, L. R.; Shi, Y.; Ranon, P. M.; Steier, W. H. *Macromolecules* **1992**, *25*, 6714–6715.

63. Xu, C.; Wu, B.; Todorova, O.; Dalton, L. R.; Shi, Y.; Ranon, P. M.; Steier, W. H. *Macromolecules* **1993**, *26*, 5303–5309.

64. Amano, M.; Kaino, T. *J. Appl. Phys.* **1990**, *68*, 6024–6028.

65. Kaino, T.; Amano, M.; Shuto, Y. In *Nonlinear Optics Fundamentals, Materials and Devices*; Miyata, S., Ed.; Elsevier Science: Amsterdam, Netherlands, 1992; pp 163–178.

66. Wong, K. Y.; Jen, A. K.-Y.; Rao, V. P.; Drost, K.; Mininni, R. M. *Proc. SPIE* **1992**, *1775*, 74–84.

67. Lin, J. T.; Hubbard, M. A.; Marks, T. J.; Lin, W.; Wong, G. K. *Chem. Mater.* **1992**, *4*, 1148–1150.

68. Miller, R. D.; Burland, D. M.; Dawson, D.; Hedrick, J.; Lee, V. Y.; Moylan, C. R.; Twieg, R. J.; Volksen, W. *Polym. Prepr. (Am. Chem. Soc., Div. Polym. Chem. Symp.)* **1994**, *35(2)*, 122–123.

69. Peng, Z.; Yu, L. *Macromolecules* **1994**, *27*, 2638–2640.

70. Yu, D.; Gharavi, A.; Yu, L. *Macromolecules* **1995**, *28*, 784–786.

71. Ye, C.; Marks, T. J.; Yang, J.; Wong, G. K. *Macromolecules* **1987**, *20*, 2324–2326.

72. Dai, D.-R.; Marks, T. J.; Yang, J.; Lundquist, P. M.; Wong, G. K. *Macromolecules* **1990**, *23*, 1891–1894.

73. Choe, E. W. U.S. Patent 4 755 574, 1988.

74. Müller, H.; Nuyken, O.; Strohriegl, P. *Makromol. Chem. Rapid Commun.* **1992**, *13*, 125–133.

75. Dembek, A. A.; Kim, C.; Allcock, H. R.; Devine, R. L. S.; Steier, W. H.; Spangler, C. W. *Chem. Mater.* **1990**, *2*, 97–99.

76. Leslie, T. M. U.S. Patent 4 855 078, 1989.

77. Bautista, M. O.; Ford, W. T.; Duran, R. S.; Naumann, M. *Polym. Prepr. (Am. Chem. Soc., Div. Polym. Chem. Symp.)* **1992**, *33(1)*, 1172–1173.

78. Ye, C.; Minami, N.; Marks, T. J.; Yang, J.; Wong, G. K. *Macromolecules* **1988**, *i*, 2899–2901.

79. Park, J.; Marks, T. J.; Yang, J.; Wong, G. K. *Chem. Mater.* **1990**, *2*, 229–231.

80. Jin, Y.; Carr, S. H.; Marks, T. J.; Lin, W.; Wong, G. K. *Chem. Mater.* **1992**, *4*, 963–965.

81. Robello, D. R.; Perry, R. J.; Urankar, E. J. *Macromolecules* **1993**, *26*, 6940–6944.

82. Schilling, M. L.; Katz, H. E.; Cox, D. I. *J. Org. Chem.* **1988**, *53*, 5538–5540.

83. Feringa, B. L.; de Lange, B.; Jager, W.; Schudde, E. P. *J. Chem. Soc., Chem. Commun.* **1990**, 804–805.

84. Katz, H. E.; Bohrer, M. P.; Mixon, D. A.; Alonzo, J.; Markham, J.; Sohn, J. E.; Cox, D. I. *J. Appl. Polym. Sci.* **1990**, *40*, 1711–1715.

85. Singer, K. D.; Holland, W. R.; Kuzyk, M. G.; Wolk, G. L.; Katz, H. E.; Schilling, M. L. *Proc. SPIE* **1989**, *1147*, 233–244.

86. Sen, A.; Eich, M.; Twieg, R. J.; Yoon, D. Y. *Nonlinear Optical Properties of Organic Molecules and Crystals*; Chemla, D. S.; Zyss, J., Eds.; Academic: Orlando, FL, 1987; pp 250–257.

87. Eich, M.; Sen, A.; Looser, H.; Bjorklund, G. C.; Swalen, J. D.; Tweig, R.; Yoon, D. Y. *J. Appl. Opt.* **1989**, *66*, 2559–2567.

88. Levine, B. F.; Bethea, C. G. *J. Chem. Phys.* **1976**, *65*, 1989–1993.

89. Green, G. D.; Hall, H. K.; Mulvaney, J. E.; Noonan, J.; Williams, D. J.; *Macromolecules* **1987**, *20*, 716–722.

90. Green, G. D.; Weinschenk, J. I.; Mulvaney, J. E.; Hall, H. K.; *Macromolecules* **1987**, *20*, 722–726.

91. Willand, C. S.; Feth, S. E.; Scozzafava, M; Williams, D. J.; Green, G. D.; Weinschenk, J. I.; Hall, H. K. In *Nonlinear Optical and Electroactive Polymers*; Prasad, P. N.; Ulrich, D. R., Eds.; Plenum: New York, 1988; pp 107–120.

92. Mitchell, M. A.; Tomida, M.; Padias, A. B.; Hall, H. K.; Lackritz, H. S.; Robello, D. R.; Willand, C. S.; Williams, D. J. *Chem. Mater.* **1993**, *5*, 1044–1051.

93. Katz, H. E.; Schilling, M. L.; Fang, T.; Holland, W. R.; King, L.; Gordon, H. *Macromolecules* **1991**, *24*, 1201–1204.

94. Lindsay, G. A.; Henry, R. A.; Stenger-Smith, J. D. *Proc. SPIE* **1992**, *1775*, 425–431.

95. Wright, M. E.; Sigman, S. *Polym. Prepr. (Am. Chem. Soc., Div. Polym. Chem. Symp.)* **1992**, *33(1)*, 1121–1122.

96. Kanis, D. R.; Ratner, M. A.; Marks, T. J. *J. Am. Chem. Soc.* **1992**, *114*, 10338–10357.

97. Robello, D. R.; Ulman, A.; Willand, C. S. U. S. Patent 4,796,971, 1989.

98. Robello, D. R.; Willand, C. S.; Scozzafava, M.; Ulman, A.; Williams, D. J. In *Materials for Nonlinear Optics: Chemical Perspectives*; Marder, S. R.; Sohn, J. E.; Stucky, G. D., Eds.; ACS Symposium Series 455; American Chemical Society: Washington, DC, 1991; pp 279–293.

99. Eich, M.; Reck, B.; Yoon, D. Y.; Wilson, C. G.; Bjorklund, G. C. *J. Appl. Phys.* **1989**, *66*, 3241–3247.

100. Jeng, R. J.; Chen, Y. M.; Chen, J. I.; Kumar, J.; Tripathy, S. K. *Macromolecules* **1993**, *26*, 2530–2534.

101. Mandal, B. K.; Jeng, R. J.; Kumar, J.; Tripathy, S. K. *Makromol. Chem., Rapid Commun.* **1991**, *12*, 607–612.

102. Hashidate, S.; Nagasaki, Y.; Kato, M.; Okada, S.; Matsuda, H.; Minami, N.; Nakanisi, H. *Polym. Adv. Technol.* **1992**, *3*, 145–149.

103. Beecher, J. E.; Fréchet, J. M. J.; Willand, C. S.; Robello, D. R.; Williams, D. J. *J. Am. Chem. Soc.* **1993**, *115*, 12216–12217.

104. Beecher, J. E.; Durst, T.; Fréchet, J. M. J.; Godt, A.; Willand, C. S. *Macromolecules* **1994**, *27*, 3472–3477.

105. Schildkraut, J. *Appl. Phys. Lett.* **1991**, *58*, 340–342.

106. Moerner, W. E.; Walsh, C.; Scott, J. C.; Ducharme, S.; Burland, D. M.; Bjorklund, G. C.; Tweig, R. J. *Proc. SPIE* **1991**, *1560*, 278–289.

107. Tamura, K.; Padias, A. B.; Hall, H. K.; Peyghambarian, N. *Appl. Phys. Lett.* **1992**, *60*, 1803–1805.

108. Yu, L.; Chan, W.; Bao, Z.; Cao, S. X. *Macromolecules* **1993**, *26*, 2216–2221.

109. Ulman, A. *An Introduction to Ultrathin Organic Films: From Langmuir-Blodgett to Self-Assembly*; Academic: Orlando, FL, 1991.

110. Allen, S. *Inst. Phys. Conf. Ser.* **1989**, *103*, 163–174.

111. Tredgold, R. H. *Thin Solid Films* **1987**, *152*, 223–230.

112. Saperstein, D. D.; Rabolt, J. F.; Hoover, J. M.; Stroeve, P. In *Macromolecular Assemblies in Polymeric Systems*; Stroeve, P.; Balazs, A. C., Eds.; ACS Symposium Series 493; American Chemical Society: Washington, DC, 1992; pp 104–112.

113. Penner, T. L.; Armstrong, N. J.; Willand, C. S.; Schildkraut, J. S.; Robello, D. R.; Ulman, A. *Nonlin. Opt.* **1993**, *4*, 191–209.

114. Clays, K: Penner, T. L.; Armstrong, N. A.; Robello, D. R. *Proc. SPIE* **1992**, *1775*, 326–339.

115. Clays, K.; Armstrong, N. J.; Ezenyilimba, M. C.; Penner, T. L. *Chem. Mater.* **1993**, *5*, 1032–1036.

116. Penner, T. L.; Motschmann, H. R.; Armstrong, N. J.; Ezenyilimba, M. C.; Williams, D. J. *Nature (London)* **1994**, *367*, 49–51.

117. Hoover, J. M.; Henry, R. A.; Lindsay, G. A.; Nadler, M. P.; Nee, S. F.; Seltzer, M. D.; Stenger-Smith, J. D. In *Macromolecular Assemblies in Polymeric Systems*; Stroeve, P.; Balazs, A. C., Eds.; ACS Symposium Series 493; American Chemical Society: Washington, DC, 1992; pp 94–103.

118. Sagiv, J. *J. Am. Chem. Soc.* **1980**, *102*, 92–98.

119. Katz, H. E.; Scheller, G.; Putvinski, T. M.; Schilling, M. L.; Wilson, W. L.; Chidsay, C. E. D. *Science (Washington, D.C.)* **1991**, *254*, 1435–1437.

120. Li, D. Q.; Ratner, M. A.; Marks, T. J.; Zhang, C.; Yang, J.; Wong, G. K. *J. Am. Chem. Soc.* **1990**, *112*, 7389–7390.

121. Kakkar, A. K.; Yitzchaik, S.; Roscoe, S. B.; Kubota, F.; Allan, D. S.; Marks, T. J.; Lin, W.; Wong, G. K. *Langmuir* **1993**, *9*, 388.

Polymers for Activated Laser Switching

P. Chandrasekhar

The impelling need for improved materials for photonic and optoelectronic applications in areas such as optical computing has produced much work in polymeric materials. Although rigid-rod, ladder, and pendant-modified inert polymers have been center stage, considerable recent interest has also focussed on conducting polymers (i.e., conjugated polymers, CPs). A general overview of CPs is presented in another chapter. I focus on the use of CPs for practical laser switching and other photonic applications and on the more promising recent developments in third-order nonlinear optical effects. My work on novel poly(aromatic amines) and ring-S-heteroatom polymers is discussed, in which appreciable off-resonant third-order nonlinearities were observed: namely, third-order nonlinear bulk susceptibility ($\chi^{(3)}_{xxxx}$), where xxxx denotes all four beams of identical polarization in the measurement method used, of 1.3–2.7×10^{-10} esu at 1.06 μm and time response <40 ps. The highlight of the work is the unique semiconductor–CP (SC–CP) interfaces, in which the inorganic SC is excited by an ultrafast laser, transfers charge to the CP, and causes it to switch in a fast or ultrafast regime. Thus, the SC provides both the trigger and the energy for the CP switching. Applications for passive laser limiting for military use are discussed.

The increasing need for improved materials for photonic, optoelectronic, and related applications (especially optical computing) has produced much activity in the development of improved nonpolymeric as well as polymeric materials (*1–3*). Among the polymeric materials, several classes such as rigid-rod and ladder polymers, pendant-modified linear polymers, σ-conjugated materials based on bond-alternant polysilanes, metal-containing macrocycles (e.g., Pt and Pb phthalocyanines), and conducting polymers (i.e., conjugated polymers, CPs) have shown much promise (*4–6*). Of these materials, CPs, which possess appreciable conductivities ($>10^{-7}$ S/cm), have received special attention, primarily because of the possibility of tailoring of their properties such as processibility, doping level, conductivity, and optical transparency in combination with photonic properties.

Applications

Third-Order Nonlinear-Optical Materials

The primary focus for third-order nonlinear-optical (NLO) and related properties of CPs has been in all-optical signal processing and "optical computing", where the drawbacks of conventional electromagnetic interference would be absent. The principal components of an optical computing system are, in order: light sources, light beam steering materials, and light modulators. In this respect, semiconducting lasers are widely presumed to be the light sources of choice in these systems. Thus, a particular need has been created for beam steering and other light modulation with sub-watt switching thresholds, combined with a passive switching mode and good optical transparency. CPs especially have attracted interest in this respect, primarily because other systems do not appear to meet these specifications. Other applications of third-order NLO materials have been in optoelectronic interconnects (forming the interface between transmission of optical and electric signals) and spatial light modulators, which are specialized modulators that modify the phase, amplitude, polarization, or intensity of light passing through them and are used in image processing and pattern recognition (7–9).

The very fast switching times of nearly all CP-based third-order NLO materials, typically in the picosecond or sub-picosecond region, and their real-time imaging capabilities make them very attractive. Unfortunately, third-order nonlinearities of CPs as well as all other materials thus far studied remain two to three orders of magnitude lower than required for them to effectively compete commercially with electronic devices.

CPs as Laser-Blocking Materials

Another important application of CPs (especially from the military point of view) is "laser limiting": passive attenuation of potentially hazardous laser radiation. This blocking is applied to both friendly and unfriendly lasers used in the battlefield. This application is most effectively addressed by CPs incorporated into unique semiconductor (SC)–CP interfaces.

Specific Photonic Applications

For illustrative purposes, some typical photonic device applications are cited. Etemad and Soos (10) described the operation of an all-optical directional coupler based on the off-resonant third-order NLO effect in CPs (Figure 1). A pair of waveguides coupled through the overlap of evanescent tails of their respective guided modes composes the device. Depending on the input pulse intensity, the output pulse switches between the two guides. Switching can occur in less than 10^{-14} s.

Applications in all-optical signal processing include devices described by Stegeman et al. (7) and Caulfield et al. (11) based on functional third-order NLO polymers for parallel and serial all-optical signal processing. For parallel processing, 2D arrays of all-optical logic functions are used with arrays of parallel NLO elements, each of which acts as a small etalon. In this case, much larger etalon thicknesses of up to 1 cm are required for achieving even outputs comparable with GaAlAs devices, and thus the only advantage of CPs for parallel processing appears to be their speed. For serial devices, however, CP-based materials offer clear advantages. Because long propagation distances, necessitated by the large nonlinear phase shifts

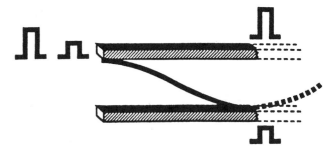

Figure 1. Optical directional coupler employing CPs (*10*). (Input, left; output, right). With light coupled into the upper guide input, power is transferred periodically between guides as wave propagates. At low intensities, with the guides' optical modes coupled, intensity oscillates between guides with period *L*, the coupling length. At high intensity, refractive index changes cause light to stay in upper guide. (Reproduced with permission from reference 10. Copyright 1991 J. Wiley & Sons, New York.)

required (π to 4π), are necessary, guided wave methods are required. Optical fibers and planar and channel waveguides are some of the waveguiding options available. Stegeman's group (*7, 8*) used NLO organics, including CPs, in planar waveguides for the first time, and these materials reduced the switching beam power requirements from the kilowatt regime to the sub-watt regime. This research group has discussed the design of a number of specific devices, including nonlinear distributed couplers, distributed feedback gratings, and Mach–Zehnder interferometers.

CPs are in principle also usable in opto-electronic interconnects of the local (point to point), regional (point to many points), and global (all points to all points) varieties (*12*). However, the impediment to this use has mainly been the lack of functional polymers with high optical quality, low threshold, and large area.

Specific Laser-Limiting Applications

Perhaps the most rigorous photonic application envisioned for CPs is the protection of sensors, including visible-region sensors, IR-region sensors, and cameras and human eyes from random incidence of hazardous laser radiation in the battlefield. This basic optical limiting application requires passive operation, sub-nanosecond switching risetimes, a virtual 100% rejection of intensities above the laser tolerance threshold (which is very low especially for human eyes), and broad-band transmission above ca. 75% (luminous) in the unactivated state. This application has been addressed, but with little success (*9, 12*), with third-order NLO materials (including CPs). This failure primarily has resulted because the magnitudes of the nonlinearities required are at least two orders of magnitude higher than those currently achieved, and no material appears to meet the high unswitched-state transmission requirements. Such large nonlinearities can in principle be emulated by large thicknesses, that is, stacking of many laser limiting units, but this modification leads to high opacities. Co-workers and I addressed this problem through the use of unique SC–CP interfaces, as discussed at length in the final section of this chapter.

Third-Order NLO CPs and Current State of the Art

In a dipole approximation, the bulk polarization or susceptibility of condensed material is given by eq 1:

$$P = P_0 + \chi^{(1)}E + \chi^{(2)}E^2 + \chi^{(3)}E^3 + \ldots \quad (1)$$

where P is the bulk polarization, E is the electromagnetic field vector, and the $\chi^{(1)}$, $\chi^{(2)}$, $\chi^{(3)}$ are, respectively, the first- second- and third-order bulk susceptibilities of the material. Even though these parameters in eq 1 are correlated to the corresponding molecular polarizability, hyperpolarizability, and second hyperpolarizability (α, β, and γ, respectively), exact relationships between the bulk and molecular parameters are still not established, and much work continues. In spite, or perhaps because of this problem, much of the current experimental work has concentrated on empirical enhancement of observed third-order nonlinearities in CPs with optimal processibility, transparency, and high damage thresholds needed for device application. However, the highest off-resonant third-order nonlinearity achieved is in the range of $\chi^{(3)} = 10^{-8}$ esu (2), which is about two orders of magnitude less than what would be required for efficient device application (9). Furthermore, the linear absorption–scattering losses at the wavelengths of interest must be less than about 0.1 dB/cm, and this level currently is difficult to achieve.

Prasad at the State University of New York, Buffalo (13), and Druy and Lusignea at Foster-Miller Labs, Boston (14), developed poly(*para*-phenylene-vinylene) (PPV) (Chart I) in a high optical quality, highly oriented, uniaxial fiber form suitable for many device applications. By using femtosecond degenerate four wave mixing (DFWM) at 602 and 580 nm in an oriented film, $\chi^{(3)}$ values of 4×10^{-10} esu and sub-picosecond response times were demonstrated. Druy and Glatkowski (15) also demonstrated third-order NLO polymers related to PPV, such as poly(dimethoxy-PPV). Pang and Prasad (16) recently processed semi-soluble poly(dodecylthiophene) (PDDT) into optical-quality thin films and measured their resonant third-order nonlinearities at 620 nm by using DFWM with 60-fs pulses. They observed effective values for $\chi^{(3)}$ of 5.5×10^{-11} esu. Jenekhe et al. (17) reported

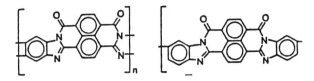

Chart I. Structures of benzimidazo- and benzophenanthroline-type ladder polymers: full-ladder polymer, BBL (left), and semi-ladder polymer, BBB (right).

preparation of poly(thiophene) (PT) derivatives coprocessed with poly(2,5-bismethene thiophene), the quinoid form of PT. These derivatives had reported bulk resonant susceptibilities, ($\chi^{(3)}$), measured via DFWM at 532 nm, of about 10^{-7} esu.

The advantageous feature in all the cited work has been the facile processing of the CPs, either from soluble precursors followed by in-situ doping (in the case of PPV) or from the inherent partial solubility in organic solvents of long-chain alkyl-substituted polymers (in the case of PDDT). This feature permits thin-film preparation via standard polymer-coating techniques such as spin- or dip-coating, which are cost effective for ultimate device application. The drawback of the work, however, has been that the polymers frequently require orientation to maximize the effective nonlinearity. For instance, orientational anisotropy for PPV, measured as the ratio $\chi^{(3)}(\| / \perp)$, is ca. 37 (13, 18). Indeed, some of the highest off-resonant susceptibilities observed for polymers, ($\chi^{(3)}$ ca. 10^{-8} esu) have been for polyacetylene (PA) polarized along the polymer axis (19). However, such PA films are air-sensitive. Apart from simple, nonvacuum methods such as spin- and dip-coating, Langmuir–Blodgett and Shear methods also were used for fabrication of very thin CP films in devices. An example that comes to mind is use of polydiacetylenes (PDA) in waveguides (20).

Among other works of note for third-order NLO polymers in practical device applications, Etemad and Soos (10) and others (21) characterized third-order NLO properties of combinations of PDA with 4-butoxycarbonylmethylurethane. The materials were spin-coated or solution-processed for waveguide applications over macroscopic distances. The Dalton group (22) reported a new variety of ladder polymer formed by derivatizing vinylamine with dichloroquinone with vinylamine groups. Off-resonant bulk susceptibilities ($\chi^{(3)}$) were claimed to be ca. 10^{-9} esu. In some of the first work of its kind, Lindle et al. (23) studied third-order NLO properties of benzimidazo-benzophenanthroline-type ladder polymers based on the well-known "BBL" and "BBB" (Chart I) skeletons and obtained $\chi^{(3)}_{xxxx}$ values at 1.064 µm of ca. 1.5×10^{-11} esu. These polymers had high tensile strengths and were readily spin-coated or solution-processed into thin films at room temperature.

Relation of CP Conductivity and Third-Order NLO Properties

The relation between conductivity and third-order NLO effects in CPs has been dealt with at some length by many workers, although primarily from theoretical points of view (24–27). The extended electron delocalization along the CP chain, deemed essential for high conductivity, is also important for third-order NLO properties. Garito and co-workers (27–29) have shown that a delocalization along about 8 monomer units of a CP chain is sufficient for realizing the maximum NLO response, and further extension of the delocalization does not enhance this response. Delocalization encompasses charged entities such as polarons, bipolarons, and solitons, all essential in determining CP conductivity. Observed, macroscopic conductivity of CPs, however, is dependent on two factors: the intrachain charge mobility and the interchain charge mobility, called "charge hopping". The interchain mobility, however, has no known direct relation with intrinsic third-order NLO effects in the CP.

Structure–Property Correlations

Most of the work in NLO polymers has been accomplished without a systematic understanding of the relationships between the known molecular structure and morphologies of the polymers and the observed bulk third-order susceptibilities. Garito's group pioneered the study of structure–property relationships in this area and showed, in some of their first detailed work (29), that large third-order susceptibilities require only that polymer chains be of intermediate length of ca. 10 nm. This finding is in quite explicit contradiction to earlier theories of increasing chain lengths leading to increasing nonlinearities (30) based on infinite-chain electron delocalization models. Dalton and co-workers (22, 30) subsequently substantiated the findings of Garito in large measure in experimental work with ladder polymers, heterocyclic aromatics, and other CPs.

More recently, Jenekhe's group at the University of Rochester (31) attempted a systematic structure–property study of rigid-rod poly(quinolines) and related polymers (Chart II) by using systematic variation in structure and third-order NLO measurements primarily via third-harmonic generation at many different wavelengths. In my considered opinion, however, this very extensive work appears to have simply generated a large amount of data with no systematic or clear trends in properties observed, due to in part a large number of seeming contradictions between different polymers in the observed effects of particular structural features. Other attempts at systematic structure–property studies have been those of Bryne and Blau (32); Heeger (33), especially as applied to devices; and other work reviewed recently by Muellen (34).

Recently, our group (35, 36) initiated a systematic study of the structure–property relationships of a group of CPs comprising mainly poly(aromatic amines) and polymers containing ring-S-heteroatoms (Table I),

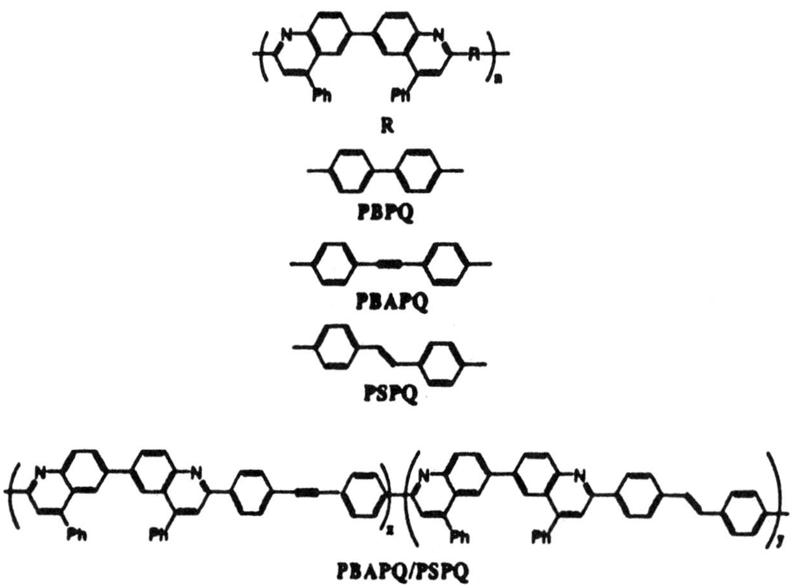

Chart II. Representative rigid-rod poly(quinolines) studied by Jenekhe and co-workers (31). Structure with R as shown for the polymers poly[2,2'-p,p'-biphenylene)-6,6'-bis(4-phenylquinoline)] (PBPQ), poly[2,2'(p,p'-biphenyl-acetylene)-6,6'-bis(4-phenylquinoline)] (PBAPQ), and finally poly[2,2'-(p,p'-stilbene)-6,6'-bis(4-phenyl quinoline)] (PSPQ). The PBAPQ–PSPQ co-polymer is also shown.

Table I. Comparative Nonlinearities, $\chi^{(3)}_{xxxx}$(pure), for Selected CPs Studied at Identical Doping (*ca.* 2.0% with BF_4^-) and Concentration (20 mM) in DMF

Polymer Symbol, Name	Monomer Structure	$\chi^{(3)}_{xxxx}$(pure) (esu $\times 10^{-10}$)	TCL[a]
Poly(aromatic amines)			
P(DPA), poly(diphenylamine)		1.3	10–40
P(4ABP), poly(4-aminobiphenyl)		1.4	10–40
P(NNPheBz), poly(*N,N'*-diphenylbenzidine)		0.9	10–40
P(1APyre), poly(1-aminopyrene)		2.7	10–40
P(3APAz), poly(3-aminopyrazole)		0.40	10–40
P(NPhe2NaA), poly(*N*-phenyl-2-naphthylamine)		<0.04	10–40
P(Ani) (poly(aniline), emeraldine base form)		0.11	up to 10^5
Poly(amino quinolines)			
P(3AQ), poly(3-aminoquinoline)		0.37	*ca.* 220
P(5AQ), poly(5-aminoquinoline)		0.19	45–56
P(6AQ), poly(6-aminoquinoline)		0.26	45–56
P(8AQ), poly(8-aminoquinoline)		0.23	45–56
P(3APy), poly(3-aminopyridine)		0.22	45–56
S-Heterocycle Polymers			
P(ITN), poly(isothianaphthene)		5.0	45–56
P(2ATAz), poly(2-aminothiazole)		<0.01	45–56
P(2ABTAz), poly(2-aminobenzothiazole)		<0.01	45–56
P(4A213BTDAz), poly(4-amino-2,1,3-benzothiadiazole)		<0.01	45–56

[a]TCL is typical chain length in monomer units.

which were originally developed in unrelated work (37–40) for less esoteric applications such as IR–radar signature control and electromagnetic impulse (EMI) shielding.

Ashwin-Ushas Work on Third-Order NLO Polymers

Polymers Studied

A series of polymers developed initially for IR–radar signature control, stealth, and EMI shielding applications primarily for the U.S. Defense Advanced Research Projects Agency and U.S. Naval Sea Systems Division (39, 40) were subsequently adapted for far-IR signalling and camouflage applications in a joint project with Martin Marietta Laboratories (41). The monomer precursors for these polymers are depicted in Table I. The dopants used in our work were the following: BF_4^- (tetrafluoroborate), ClO_4^- (perchlorate), PF_6^- (hexafluorophosphate), poly(vinyl sulfate), and p-CH_3-C_6H_5-SO_3^- (tosylate). These doped polymers were produced by an electrochemical synthesis (37–40), wherein a stepped potential algorithm was used successively to generate dimer, tetramer, and so forth to yield short polymers or oligomers with chain lengths of 12–35 monomer units. These polymers yield the optimal match of processibility, conductivity, optical, and other properties. Notably, the electrochemical synthesis is not a procedure producing a definitive chain length, but rather it gives a range of chain lengths. All polymers have chain lengths of 10–40 monomer units with the exception of the poly(aminoquinolines), which have chain lengths of ca. 45–56 monomer units for P(5-, 6-, and 8-AQ), and ca. 222 monomer units for P(3AQ). The polymers have low doping levels (1–12%) and exhibit appreciable solubilities of up to 20 w/w% (1.0 M) in organic solvents such as dimethylformamide (DMF) and 1-methyl-2-pyrrolidinone. This solubility enabled efficacious determination of NLO properties in solution with minimal detrimental effects of the low laser damage thresholds frequently found in thin films. We employed a number of new techniques to improve conductivity, solubility, processibility, and other properties. Conductivities ranged from 10^{-8} to 10^{-1} S/cm for the poly(aminoquinolines) and from 10^{-2} to 50 S/cm for the poly(aromatic amines).

An understanding of structural features of the polymers is useful for appreciation of structure–property relationships. In the case of P(DPA) and P(4ABP), structural analysis (35, 37) revealed that in addition to the normally expected head-to-tail linkages, these polymers differ from each other in additional chain branching at the N atom for P(DPA) and at the ortho position to the amino group for P(4ABP). They differ

from previously synthesized forms (42) in these structural features and in their high solubility and stability. For the poly(aminoquinolines), structural analysis revealed probable linkages at the NH group and at either ortho or para positions in the α-ring, except for P(8AQ), where data indicate possible linkage at the 5-position in the β-ring. P(3APAz) has probable linkages at the NH group and the 5-position (36). The probable structures of P(ITN) and P(1APyre) were established elsewhere (43, 44).

For potential processing of these polymers for NLO applications, especially for EMI shielding, we developed techniques for co-processing of these CPs with thermoplastics such as acrylonitrile–butadiene–styrene (ABS) or polycarbonate. We obtained composites that had little change in conductivities from those of the "virgin" CP at CP loadings as low as 20 w/w%. For instance, original conductivities of poly(3-methylthiophene (P3MT) of ca. 30 S/cm were lowered to 10 S/cm in 80–20% (w/w) ABS–P3MT composites. Similarly, original P(DPA) conductivities of 5 S/cm were lowered to ca. 2 S/cm in 80–20% ABS–P(DPA) composites. EMI shielding effectiveness for these composites was above 40 dB for most frequency regions of interest. In another application, we demonstrated high and controllable dynamic signature in the mid- and far-IR regions for a number of CP–dopant combinations (41).

NLO Measurements and Data

Filtered solutions of the CPs in DMF (normally ca. 20 mM) were used for all NLO measurements. The vis–near-IR spectra of the solutions as used for the NLO measurements had optical densities (ODs) of less than 0.1 through the 800–1400-nm region. Absorptions at 532 nm were substantial only for P(ITN) (OD ca. 0.32) and P(4ABP) (OD ca. 1.7). Figure 2 shows typical data for P(DPA) and, in the inset, representative vis–near-IR absorption spectra of P(DPA) and P(4ABP). For the NLO measurements, we employed DFWM (Figure 2) supplemented by Z-scan (45) and 2-photon absorption measurements. These data together with the linear absorption spectra served to confirm the purely off-resonant, electronic character of the nonlinearities recorded.

The DFWM procedure, which used the apparatus depicted in Figure 3, was described at length elsewhere (35, 36). Briefly, DMF solutions of the polymer and CS_2 reference samples were placed in matched 1-mm path quartz cuvettes. A yttrium–aluminum–garnet (YAG):Nd laser (L) at 1.064 μm, 2 Hz, 5 mJ, and 40-ps pulse was used. In Figure 3, $\omega_{(1,2,3,4)}$ denote, respectively, the pump beam, the back reflection from mirror (M) placed behind sample (S), the probe beam at 10° to ω_1 and temporally synchronized with it by using the moveable delay line (DL) and frequency-

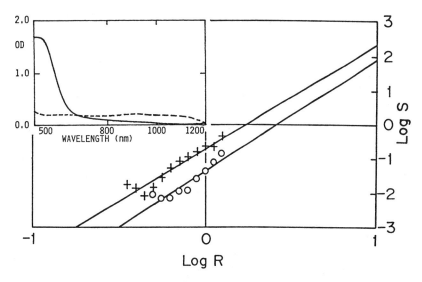

Figure 2. (Inset): Vis–near-IR absorption spectra for BF_4^--doped, 20-mM solutions in DMF of P(4ABP) (solid line) and P(DPA) (dotted line) as used for NLO measurements. Main Figure: Typical DFWM data for P(DPA). S = signal; R = reference.

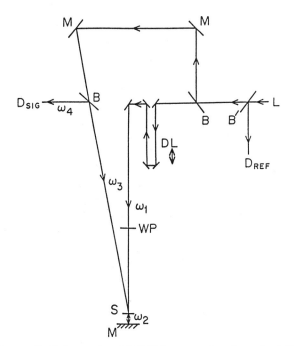

Figure 3. Schematic of DFWM apparatus (*see* text for abbreviations).

doubling crystal, and the phase conjugate (signal) reflection detected using the photodiode D_{SIG}. Output of diodes D_{SIG} and D_{REF} was fed to a digital oscilloscope and computer processed shot-to-shot. At the sample, beam power density was up to 20 GW/cm². The half-wave plate (WP) was used to adjust pump beam polarization: that is, with all four beams of same polarization (*xxxx*) and pump and probe beams of orthogonal polarization (*yyxx*). Errors due to beam fluctuation were estimated to be less than 30%.

The magnitudes of $\chi^{(3)}$ with respect to the known $\chi^{(3)}$ value of CS_2, 4.0×10^{-13} esu, were calculated according to eqs 2 and 3:

$$\chi^{(3)}{}_{soln.} = \chi^{(3)}{}_{CS_2} \times [S_{soln.}/S_{CS_2}]^{1/2} [n_{soln.}/n_{CS_2}]^2 A \quad (2)$$

where $A = \alpha l/\{\exp(-\alpha l)[1 - \exp(-\alpha l)]\}$, S = signal intensity, and $\alpha l = 2.3 \times$ OD. The refractive index, n, of CS_2 is taken as 1.625, and n of solutions of all polymers is 1.431, which is the pure DMF value. For the pure materials:

$$\chi^{(3)}{}_{pure\ polymer} = [N_{pure\ polymer}/N_{soln.}] \times \chi^{(3)}{}_{soln.} \quad (3)$$

where N is the effective density.

Structure–Property Correlations

For a meaningful comparison of nonlinearities of different CPs in relation to their structure, we used identical doping type and level (Table I).

To summarize the structural variations, P(DPA) and P(4ABP) are analogs of P(Ani), but with an additional unfused phenyl ring along the polymer chain. P(NPhe2NaA) is an analog with two additional fused phenyl rings. P(1APyre) represents a multiple fused ring poly(aromatic amine). In the poly(aminoquinolines), an *N*-heteroatom is introduced into the ring and the amino substituent position is varied; P(3APy) represents the single aromatic ring analog of the poly(aminoquinolines). P(ITN) in Table I is a fused ring analog of poly(thiophene), whereas the other polymers contain multiple *N*- or *S*-heteroatoms.

The nonlinearities of P(DPA) and P(4ABP) are significantly higher than those of the soluble emeraldine base form of P(Ani). Our results for P(Ani) are in agreement with those obtained recently by MacDiarmid (private communication). The substitution of unfused multiple aromatic rings for a single ring in the polymer backbone [i.e., in P(DPA) and P(4ABP) as compared with P(Ani)] significantly increases the nonlinearity, possibly due to increasing delocalization of π-electron density between two aromatic rings. With multiple fused rings, as in P(1APyre), however, the increase is dramatic. This effect is also observed for P(ITN). The lack of observable nonlinear response in P(NPhe2NaA) may be due to breaking of the π-electron delocalization along the chain by increased quinonoid character introduced by the second fused aromatic ring and absence of head-to-tail linkage.

The incorporation of an N-heteroatom into the aromatic polymer backbone in poly(aminoquinolines) and P(3APAz) generally appears to lower the nonlinear susceptibility. Molecular third-order hyperpolarizabilities, γ_{ijk}, calculated by us for the monomers parallel these data: Ani, 8.67; DPA, 29.72; 1APyre, 41.15; 8AQ, 12.92 (all values are $\times 10^{-36}$ esu). The data also show the absence of substitution position effects in the case of the poly(aminoquinolines) and a similar nonlinearity for the single-ring analog, P(3APy).

In the S-containing polymers, when N- and S-heteroatoms are coupled in the same aromatic ring in the polymer backbone, the nonlinearity practically disappears, as observed for P(2ATAz), P(2ABTAz), and P(4A213BTDAz). However, with identical N-heteroatoms in the same aromatic ring, the nonlinearity is somewhat reduced.

Dopant Type and Level

We consistently observed that lower doping levels gave nonlinearities that were larger than those for more doped polymers and were well above experi-

mental error. For example, the nonlinearity was nearly three times higher in the case of specific polymers such as P(DPA) when doping level was reduced from 4.5 to 1%. These results are shown in Figures 4 and 5. One proposed explanation is possible intrinsic lattice defects occurring in mid-gap states, such as bipolarons produced on greater doping and interrupting the electron delocalization along the polymer chain, which ultimately is responsible for the nonlinearity.

Dopant type did not have an appreciable influence on the nonlinearities among the dopants we examined. Typical $\chi^{(3)}_{xxxx}$ values that can be cited for P(DPA) at a constant 4.5% doping level and 40-mM concentration in DMF are as follows (all values are $\times 10^{-10}$ esu): BF_4^-, 0.72; PF_6^-, 0.74; ClO_4^-, 0.35; tosylate, 0.37. Concentration effects were discussed elsewhere (35, 36).

Chain Length Effects

Many of the trends observed, such as the effect of doping level, may be interrelated with polymer chain length, which may vary for differing dopant type or level. Because of the general experimental difficulty of controlling polymer chain lengths for short chain polymers, the effect of this factor on nonlinearities was not studied by us. However, work by the Garito group (29) indicated that the effect of chain lengths beyond ca. 12 monomer units is minimal. Our bulk susceptibilities ($\chi^{(3)}_{xxxx}$) are appreciable and may be compared to the data previously cited and to other off-resonant data given for single-crystal poly(diacetylenes), which yielded values for $\chi^{(3)}_{zzzz}$ (z is polymer chain axis) of 1.6×10^{-10} esu at 2.62 μm and 8.5×10^{-10} esu at 1.89 μm. These values are among the largest off-resonant values hitherto observed (31). Additionally, we may qualify our results with the caveat that they represent a

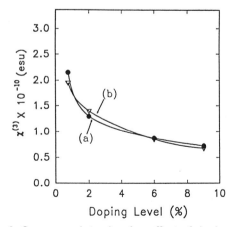

Figure 4. Summary data showing effect of doping level on third-order nonlinearity for BF_4^--doped P(DPA) (a) and P(4ABP) (b).

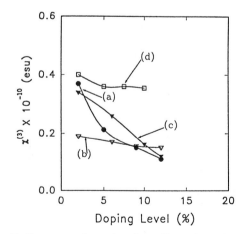

Figure 5. Summary data showing effect of doping level on third-order nonlinearity for BF_4^--doped P(3AQ) (a), P(5AQ) (b), P(6AQ) (c), and P(3APAz) (d).

preliminary systematic study, and more conclusive interpretation of the observed relationships awaits more detailed studies.

Time Response, Z-Scan, and Two-Photon Measurements

The time response of signals obtained by varying the delay line had the time dependence of the laser pulse for all polymers studied. This relationship shows that the nonlinear responses for all polymers were faster than 40 ps and discounted long-lived thermal contributions.

The DFWM technique produces a vector sum of the real and imaginary parts of the bulk third-order susceptibility. The Z-scan method on the other hand, as first propounded by Sheik-bahae et al. (45), measures the real part of the nonlinearity and determines its sign. In this method, the transmittance (T) of a single tightly focused laser beam through a small aperture in the far field is measured as the sample is translated from one axis ($-z$) to the opposite axis ($+z$). The z-translation (abscissa) is plotted versus the T values (ordinate). The negative lens effect of nonlinear materials [negative refractive index, n_2 and $\chi^{(3)}$], yields a T peak followed by a valley, whereas corresponding positive values yield a valley–peak sequence. The two-photon absorption similarly measures the imaginary part of the third-order nonlinearity.

Z-scan and two-photon absorption data are shown for the representative P(5AQ) doped with BF_4 in Figures 6A–6C. The DFWM measurement yields a value for P(5AQ) of 1×10^{-13} esu. Figures 6A and 6B depict Z-scan data for CS_2 (A, reference) and P(5AQ) (B). The sign of the measurement is the same for both and corresponds to a positive $\chi^{(3)}$. The two signals (T) are equal in size; the signal from CS_2 was obtained with an incident intensity one-third that for P(5AQ). By using eq 3,

$$\chi^{(3)}_{P(5AQ)} = \chi^{(3)}_{CS_2} \times \{[n \times T_{P(5AQ)}]/[n \times T_{CS_2}]\}$$

and appropriate reference values, we obtain $Re(\chi^{(3)})$ for P(5AQ) = 1.2×10^{-13} esu, equal to the value from the DFWM experiment; therefore, $Re(\chi^{(3)})$ dominates the total nonlinearity observed, and the DFWM experiment is indeed nonresonant at 1.064 μm.

The imaginary part of $\chi^{(3)}$ is measured by the two-photon experiment. For the P(5AQ) sample, high power irradiation (800 GW/cm^2) yielded a 15% beam attenuation (Figure 6C). The parameters for the particular experiment were as follows: beam waist, 45 μm; diffraction length, 6 mm; energy, 2 mJ in 40-ps pulse; incident power density, 800 GW/cm^2. By using values for β, the two-photon absorption coefficient, and L, the path length (0.1 cm), we obtained for the absorbed

fraction, $\beta I_0 L$, a value of 0.15. Values of β and the imaginary component (Im) of the nonlinearity could then be computed according to eqs 4 and 5:

$$\beta = 0.15/(0.1 \text{ cm} \times 800 \text{ GW/cm}^2)$$
$$= 1.8 \times 10^{-3} \text{ cm/GW} = 1.8 \times 10^{-19} \text{ cm/erg} \quad (4)$$

$$Im(\chi^{(3)}) = n_0^2 c^2 \beta/8\pi^2\omega = 3 \times 10^{-15} \text{ esu} \quad (5)$$

where n_0 is the solvent refractive index, c is the velocity of light, and ω is the frequency of the radiation. This value is 30 times smaller than both the real and total components of the nonlinearity measured. Thus,

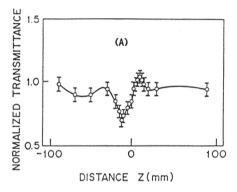

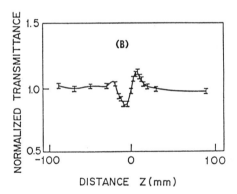

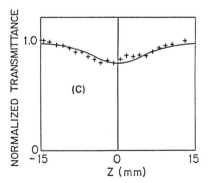

Figure 6. Representative Z-scan data (transmittance vs. translation) for CS_2 (reference, A) and for BF_4^--doped P(5AQ) (B). Representative two-photon absorption data for BF_4^--doped P(5AQ) (C).

the Z-scan and two-photon measurements cited for the representative polymer P(5AQ) clearly demonstrate that the nonlinearity measured in our studies was truly off-resonant.

Semiconductor and Conducting Polymer Interfaces

Military Laser Limiting and Other Requirements

To understand the needs and the driving force for the development of SC–CP interface technology, it is pertinent to briefly overview the contemporary military requirements in this area. Both pulse and continuous wave (CW) lasers are increasingly used in the battlefield for range finding, target designation, communication, and similar uses. Although most lasers in use have been of fixed wavelength, there is an increasing desire for use of continuously variable wavelength lasers primarily for enemy-evasive purposes. Compounding this problem from friendly lasers, the use of destructive lasers by unfriendly forces is an increasing possibility.

Lasers currently known to be in friendly use range from the ruby rangefinder laser (694 nm) with a typical 30-ns 10-J pulse and 0.2-GW peak power, a standard Nd:YAG laser (532 nm) with a 250-mJ 10-ns pulse and 100-MW peak power, to tunable dye lasers with 50-mJ 10-ns pulse and 10-MW peak power. The GaAlAs, Ti:Al$_2$O$_3$, excimer, and Cu vapor lasers are believed to be implementable (*46*). Even though battlefield sensors and sensitive cameras can be seriously damaged by laser contact, the target of most concern remains the eyes of human personnel. A typical danger can be visualized with the standard 694-nm ruby rangefinder laser. The laser has an energy of 100 mJ/pulse, a 30-ns pulsewidth, a 0.25-mrad divergence, and an exit diameter of 2 cm. The beam diameter at 1 km is 25 cm (it is 1 m for a 1.0-mrad divergence); thus, the likelihood of contact is greatly increased.

Table II lists typical medical maximum permissible exposures (MPEs) (a term denoting the highest allowed exposure before visible, irreversible damage to the retina or other ocular component occurs) for eye contact for standard Class IV battlefield lasers (*46*). The current requirements for laser eye protection are most stringent: passive action, with rest-state scotopic (night vision) transmission in the 50–75% range; dynamic (i.e., switchable), broad-band (400–700 nm) action in the fast (sub-nanosecond) or ultrafast (10–100 ps) mode; high laser damage thresholds (i.e., high resistance before damage occurs) and preferable ballistic protection (i.e., protection against shattering by projectiles, usually provided by incorporation with polycarbonate); and no angular dependence (i.e., no

Table II. Maximum Permissible Exposures (MPEs) for 1-s Laser Exposure to Pulsed and CW Lasers for 3-mm Pupil Diameter (i.e., Photopic)

Wavelength (nm)	Rate (Hz)	Pulse Width	MPE (μJ)
1064	CW	—[a]	2500
1064	10	10 ns	50.0
514.5	CW	—[a]	750
514.5	10	10 μs	0.8
530	5	15 ns	5.0
441.6	CW	—[a]	600
441.6	1	18 ms	10.0
694	1	6 ps	0.2
694	1	30 ps	1.3
694	1	30 ns	2.1
694	1	100 ms	27.0

Note: MPE is a term denoting the highest allowed exposure before visible, irreversible retinal or other ocular component damage. MPE gives total intraocular energy.

[a]Dash indicates not applicable.

dependence of protection capacity on angle of laser incidence). Other technologies offered have various drawbacks. Optical notch or holographic filters have angular dependence or fixed wavelength action. Liquid crystals and VO$_2$ devices have very slow switching times. Absorbing dyes suffer from low transmission, and NLO materials still have nonlinearities at least two orders of magnitude below those required.

SC–CP Interfaces

The well-known electrochromic behavior of CPs means that they can be switched electrochemically from fairly transparent (generally reduced) states to highly opaque or colored (generally oxidized) states by the application of a suitable potential (bias) through a conductive electrode. This process occurs in the millisecond regime. Figure 7 shows the electrochromism for poly(diphenyl benzidine) [P(DPBz)]. This electrochromism represents the electrochemical oxidation and reduction of the polymer.

However, this electrochromic switching also can be affected through the medium of an inorganic SC electrode interfaced to the CP and activated by a laser, which would transfer or accept charge from the CP. This SC electrode activation and CP switching would then occur in a fast or ultrafast regime, and the SC would provide both the trigger and the energy for the switching. This mechanism is the basis for the action of SC–CP interfaces, as represented schematically in Figure 8. The steps involved are as follows: photoexcitation of the SC and charge transfer to the CP, both occurring in the sub-nanosecond regime; oxidation or reduction of the initial CP layers to a depth of ca. 10–50 nm, which we have characterized as also occurring in the sub-nanosecond regime (*see* subsequent

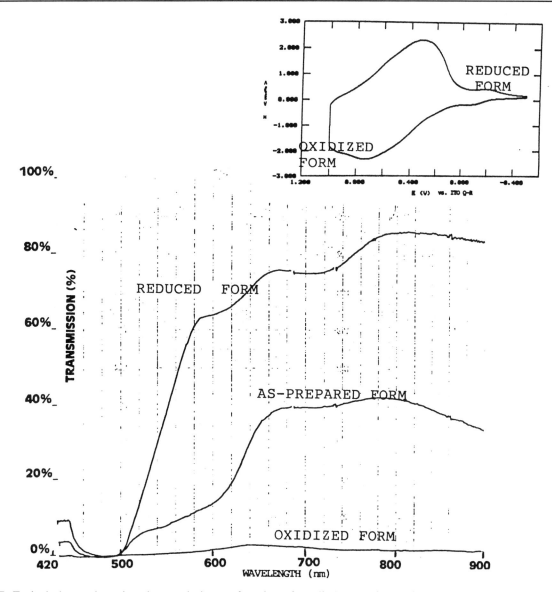

Figure 7. Typical electrochromism (transmission as function of applied potential vs. indium–tin-oxide) for the CP, a ring-substituted P(DPBz), in hermetically sealed device. Top right inset: Typical voltametric behavior corresponding to electrochromism

discussion); and the final step, charge transport within the bulk CP. This final step appears to be a comparatively slow process from the data we have obtained.

Interface Construction

The SC–CP interface can be constructed in a number of ways. Commercial thin-film SCs are unavailable at reasonable cost except for indium–tin-oxide. Single-crystal SCs cannot be used except in a cumbersome reflective mode due to their high absorptions. We initially prepared thin-film SCs on quartz substrates via thermal evaporative deposition in a dedicated vacuum deposition apparatus by adapting literature procedures. For instance, CdSe deposition involved controlled-temperature deposition from three raw materials: Cd, Se, and CdSe. We found that for many SCs, if a near-monolayer was first deposited and annealed, subsequent depositions of very thin films less than ca. 70 nm were epitaxial. SC films were characterized by X-ray diffraction, Rutherford backscattering, and other appropriate methods.

The semiconductors we tested include CdSe, CdS, AlSb, Bi_2S_3, and Se. Initial results with interfaces employing thin-film SCs were encouraging and validated the feasibility of the technology. However, even with modest incident laser energies (in the high nano-Joule–square centimeter per pulse region) adhesion of the SC films to and their ablation from the quartz or polycarbonate substrates were serious problems. Electrochemical procedures for SC deposition on the substrates, especially for the III–V SCs, yielded

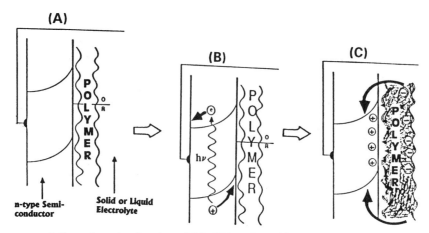

Figure 8. Schematic representation of mode of action of SC–CP interface. The *n*-type semiconductor bands are bent upward at the polymer–SC interface, as shown. O/R is the electronic level corresponding to the O/R redox potential: rest state (A); laser excitation (B); charge transport in polymer (i.e., polymer switching) (C). The processes occurring in chronological order are: laser excitation of the SC (sub-nanosecond) (B); charge transfer to polymer (sub-nanosecond) (B); oxidation–reduction of initial polymer layers (sub-nanosecond risetime; falltime from ca. 10 ns to microseconds) (C); charge transport in bulk polymer (slow process) (C).

no epitaxy and frequent problems with the stoichiometry itself. Subsequently, pre-ball-milling of the SCs into very fine particulates of nanometer dimension was tried. The SCs were then intimately mixed and co-processed with similarly prepared fine-particulate CPs and then immobilized in inert transparent polymer matrices such as poly(ethyl methacrylate) or poly(carbonate) (47). The SCs were selected on the basis of matching of their bandgaps, band locations, and other properties to the properties of the CP (redox potentials, doping levels, and expected midgap states) and consideration of the laser wavelengths to be attenuated.

Test Procedure

Two simple procedures for testing the laser shielding efficiency were devised in relation to the ultimate practical use of the technology. In the first procedure, a standard pump-and-probe methodology with a nanosecond-pulsed laser was used to establish charge transfer and to measure switching rise and fall times. Typical laser parameters were a 5-ns 532-nm pulse at 10 Hz and energies from 1 nJ/cm^2 per pulse to several milli-Joule–square centimeter per pulse. Following the pump, a Xe flash lamp probe for the 200–900-nm window was used to generate transient spectra at delay times from a nominal 0 ns to 100 μs, and the output was collected at a photodiode array connected to a personal computer where the data were processed. Each experiment consisted of an average of over 200 shots generally at 10 Hz; thus, any phenomenon measured had to be highly reversible within that time frame to be observable at all. For all SC–CP combinations tested,

the risetime was observed to be in the sub-nanosecond region (instrumentally limited), but the fall times were observed to vary from ca. 10 ns to 0.5 ms. Figures 9 and 10 show some of the first pump-and-probe data obtained for epitaxial, thin-film SC samples.

In the second procedure, the rest-state unswitched OD values (OD_{rest}) of the sample were first measured in a spectrophotometer at the laser wavelengths to be used. Subsequently, the apparatus depicted schematically in Figure 11 was used to obtain the laser-induced OD, OD_{laser}, through $\log_{10}(I_0/I)$, where I_0 is the incident laser intensity without sample, and I is the intensity through the sample. All measurements were done at 532 nm (frequency double Nd:YAG). A measure of the laser-shielding efficiency was thus obtained by eq 6:

$$\Delta OD = OD_{laser} - OD_{rest} \qquad (6)$$

Figures 12–14 show representative data obtained with the previously described method. As observable in these figures and as we found for nearly all samples, higher laser-incident energies frequently gave rise to lower shielding efficiencies. This result was ascribed to a combination of two effects: sample ablation effects (even at low energies) and saturation effects. Both effects also gave some indication of damage thresholds.

Switching threshold energies were in the 0.01 mJ/cm^2 per pulse range for most samples. Accompanying NLO studies of the CPs, as described previously, and of the interfaces appeared to indicate that the actual laser shielding was the cumulative effect of three processes:

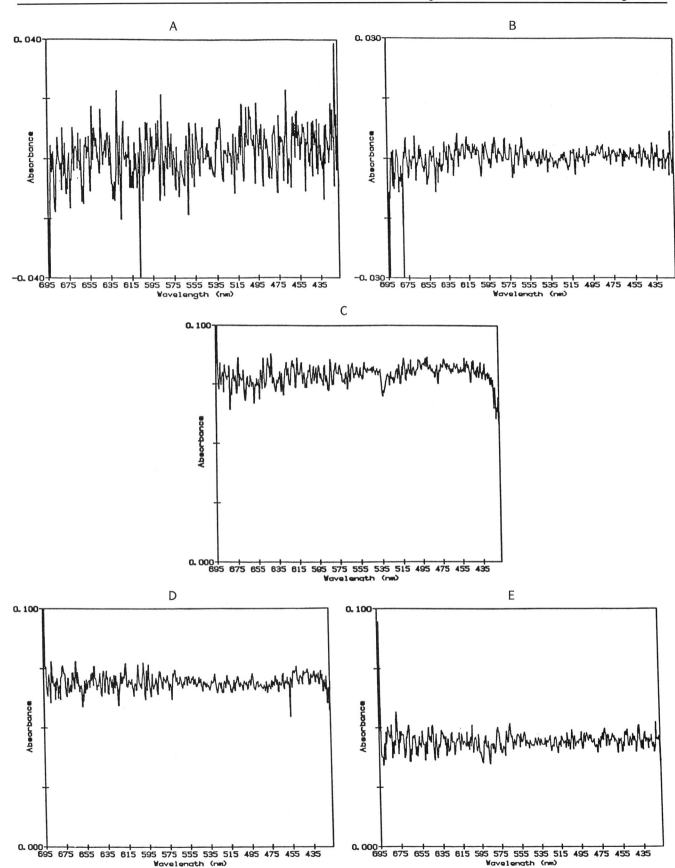

Figure 9. Initial transient absorption (TA) data for the Se/P(DPA) SC–CP system: (A) and (B), references; (C), (D), and (E), TA decay from 20 ns to 100 ns. In all SC–CP systems studied where a nominal 0-ns delay measurement was possible, the TA was already well developed and indicated sub-nanosecond risetimes in all cases.

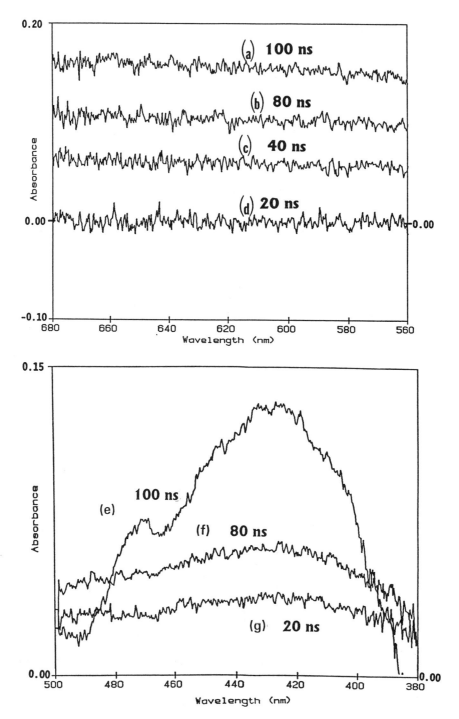

Figure 10. Initial TA data, as in Figure 9: for (a)–(d), the Se/P(3AQ) system shows TA decay from 100 ns; for (e)–(g), the P(DPBz)–CdSe system shows TA decay from 100 ns.

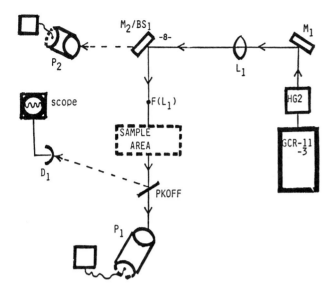

Figure 11. Schematic of "total-photon-count" method used to characterize laser shielding efficiency of SC–CP interfaces. GCR-11-3, Spectra-Physics Q-switched pulsed Nd:YAG laser; HG2, frequency doubling and tripling crystals; M_1, 45° high power laser mirror; L_1, focussing lens, 50-cm FL; M_2/BS_1, 45° high power laser mirror or beam splitter; PKOFF, 1% reflection coated optic or cover glass pickoff; D_1, ultrafast (ca. 400-ps risetime), large-area photodiode; P_1, digital power meter; P_2, second, reference, digital power meter.

- SC → CP charge transfer
- small intrinsic third-order nonlinear effects in the CP
- intrinsic third-order nonlinearity of the inorganic SC, if any

Of these, the first process is, to us, unquestionably the most important.

Another observation in our work was that when CP thicknesses greater than 100 nm were employed, no increase in laser attenuation was observed even at the lowest energies. This result again led us to the hypothesis that actual charge transfer and switching of the CPs was occurring only in the first CP chains directly at the interface, to a depth of ca. 10–15 nm: it appeared to be a surface phenomenon. This dimension appeared to tally well with the known CP chain lengths, and it may indicate that charge conduction within a polymer chain is rapid but very slow between chains. This theory is again in line with presently accepted theories of charge conduction in CPs (*48*). This dimension, however, led us to conclude that the proposed technology could not come near to meeting the specifications for an efficient military laser-shielding device, and the broad-band switched OD approached 4.0.

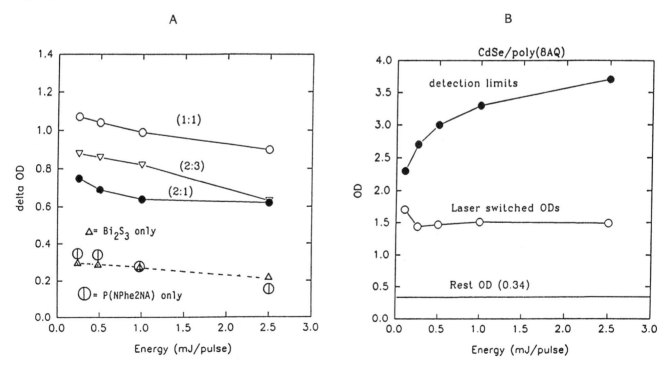

Figure 12. (a) Laser shielding efficiency, ΔOD, as a function of incident laser energy for the $Bi_2S_3/P(NPhe2NaA)$ system, also showing data for SC-only and CP-only references. Molar ratios, CP–SC, are in parentheses. All data in this and subsequent figures are for immobilized fine-particulate configuration. (b) Total laser-induced OD as function of incident energy for the CdSe/P(8AQ) system (1:1 molar proportion). The instrumental detection limits and the sample rest (unswitched) OD are also shown.

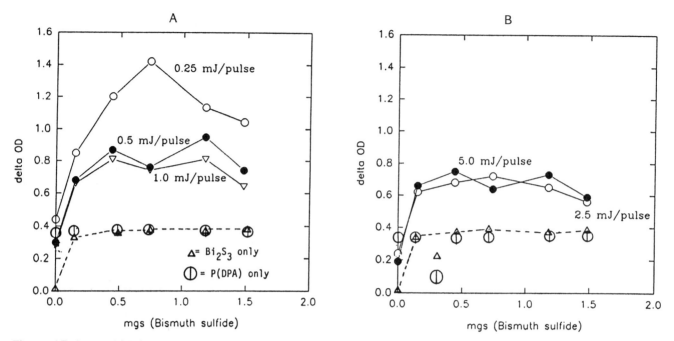

Figure 13. Laser shielding efficiency ΔOD, as function of proportion of Bi_2S_3, for the Bi_2S_3/P(DPA) system. The apparent higher efficiencies at lower incident energies are a combined artifact of sample ablation and saturation effects, as discussed in text: (A) and (B), respectively, for lower and higher incident energies.

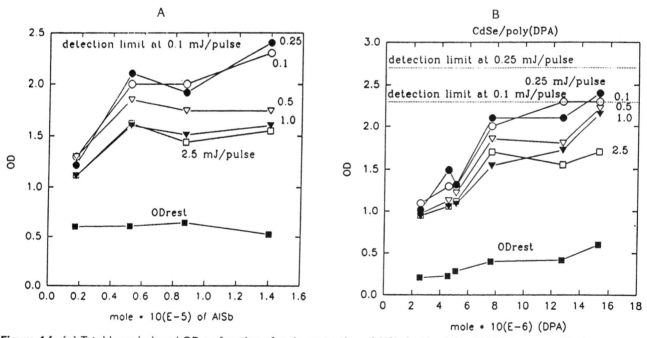

Figure 14. (a) Total laser-induced OD as function of mole proportion of AlSb for the AlSb–P(8AQ) system. The instrumental detection limits (ordinate limit) and the sample rest (unswitched) OD are also shown. (b) Total laser-induced OD as function of mole proportion of P(DPA) for the CdSe–P(DPA) system. The instrumental detection limits and the sample rest (unswitched) OD are also shown. For this system, the reduction in efficiencies with increasing incident energy is especially pronounced due to very high ablation effects associated with CdSe.

Concluding Remarks and Future Prospects

The study of third-order NLO polymers has progressed significantly in recent years, and an understanding of fundamental structure–property relationships has just begun. However, much remains to be done before practical applications of the technology can be realized. First, a more thorough understanding of structural features responsible for high experimental bulk susceptibilities needs to be established and translated into the structural design of high-nonlinearity systems. A true, complete understanding of structure–property relationships for bulk NLO properties of CPs remains elusive. Practical aspects of CP processibility and isotropic durability (i.e., high tensile strengths along all axes, not just one) also need to be addressed at the design stage. Practical problems of achieving high transparency and the associated high laser-damage threshold (i.e., high resistance to damage by lasers) also need to be addressed. Ulrich (49) suggested the tailoring of bipolaron band positions in relation to the main valence–conduction band as one possible solution for this problem. Most fundamentally, the *off-resonant* third-order nonlinearities need to be at least in the 10 esu range for optical devices to be viable.

As for the SC–CP interfaces, a number of variables affecting interface performance remain to be studied. For instance, whereas the use of epitaxial SC thin films was obviated in part by immobilized fine-particulate SCs, the precise effects of SC particle size, morphology, impurity doping, and related parameters are unknown. Similarly, the effects of CP dopant type and level have not been fully studied. The use of epitaxial or fine-particulate, visible-transparent SCs, such as doped SnO_2, is also worthy of investigation. Finally, confirmation of the surface (i.e., one-polymer-chain) nature of the SC → CP charge transfer and ways to circumvent this for higher laser-shielding efficiencies need to be addressed by, perhaps, the design of longer chain polymers or modification of polymer morphology. Multidisciplinary efforts between laser spectroscopists, physicists, materials scientists, and polymer chemists appear imperative for success.

References

1. Blau, W. J. *Nonlinear Optical Properties of Conjugated Polymers*; Springer Series of Solid-State Science; Springer: Heidelberg, Germany, 1992; Vol. 107, pp 183–189.
2. Eaton, D. F. *Science (Washington, D.C.)* **1991**, *253*, 281.
3. Glass, A. M. *Science (Washington, D.C.)* **1984**, *226*, 657.
4. *Nonlinear Optical Properties of Polymers*; Heeger, A. J.; Orenstein, J.; Ulrich, D. R., Eds.; Materials Research Society Symposium Proceedings; Materials Research Society: Pittsburgh, PA, 1988; Vol. 109, Part I, Applications and Device Requirements, pp 3–90.
5. *Conjugated Polymeric Materials: Opportunities in Electronics, Optoelectronics and Molecular Electronics*; Bredas, J. L.; Chance, R. R., Eds.; NATO ASI Series E: Applied Sciences; Kluwer: Dordrecht, Netherlands, 1990; Vol. 182.
6. *Nonlinear Optical Effects in Organic Polymers*; Messier, J.; Kajzar, F.; Prasad, P.; Ulrich, D., Eds.; NATO ASI Series E: Applied Sciences; Kluwer: Dordrecht, Netherlands, 1989; Vol. 162.
7. Stegeman, G. I.; Zanoni, R.; Rochford, K.; Seaton, C. T. *Nonlinear Optical Effects in Organic Polymers*; Messier, J.; Kajzar, F.; Prasad, P.; Ulrich, D., Eds.; NATO ASI Series E: Applied Sciences; Kluwer: Dordrecht, Netherlands, 1989; Vol. 162, pp. 257–270.
8. Stegeman, G. I.; Zanoni, R.; Seaton, C. T. *Nonlinear Optical Properties of Polymers*; Heeger, A. J.; Orenstein, J.; Ulrich, D. R., Eds.; Materials Research Society Symposium Proceedings; Materials Research Society: Pittsburgh, PA, 1988; Vol. 109, Part I, Applications and Device Requirements, pp 53–64.
9. Ulrich, D. R. In *Nonlinear Optical Effects in Organic Polymers*; Messier, J.; Kajzar, F.; Prasad, P.; Ulrich, D., Eds.; NATO ASI Series E: Applied Sciences; Kluwer: Dordrecht, Netherlands, 1989; Vol. 162, pp 299–326,.
10. Etemad, S.; Soos, Z. G. In *Spectroscopy of Advanced Materials*; Clark, R. J. H.; Hester, R. E., Eds.; Advances in Spectroscopy Series; J. Wiley and Sons: New York, 1991; Vol. 19, pp 87–132.
11. Caulfield, H. J.; Horovitz, S.; Tricoles, G. P.; von Windle, W. A. *Proc. IEEE* **1985**, *72*, 755 (Special Issue on Optical Computing).
12. Kushner, B. G.; Neff, J. A. *Nonlinear Optical Properties of Polymers*; Heeger, A. J.; Orenstein, J.; Ulrich, D. R., Eds.; Materials Research Society Symposium Proceedings; Materials Research Society: Pittsburgh, PA, 1988; Vol. 109, Part I, Applications and Device Requirements, pp 3–18.
13. Prasad, P. N. "Final Report to AFOSR Contract No. F49620–87–C–0097," 1989, and refs. therein. Available from U.S. National Technical Information Service (NTIS), Washington, DC, or U.S. Defense Technical Information Center (DTIC), Arlington, VA.
14. Druy, M.; Lusignea, R. "Final Report to AFOSR Contract No. F49620–88–C–0065", 1990, and refs. therein. Available from U.S. National Technical Information Service (NTIS), Washington, DC, or U.S. Defense Technical Information Center (DTIC), Arlington, VA.
15. Druy, M. A.; Glatkowski, P. J. *Proceedings of the 38th Sagamore Army Materials Research Conference*; Available from U.S. National Technical Information Service (NTIS): Washington, DC, or U.S. Defense Technical Information Center (DTIC): Arlington, VA, 1991.
16. Pang, Y.; Prasad, P. N. *J. Chem. Phys.* **1990**, *93*, 2201–2204, and refs. therein.
17. Jenekhe, S. A.; Chen, W.-C.; Lo, S.; Flom, S. R. *Appl. Phys. Lett.* **1990**, *57*, 126–131.
18. Gagnon, D. R.; Karasz, F. E.; Thomas, E. L.; Lenz, R. W. *Synth. Met.* **1987**, *20*, 85–95.
19. Heeger, A. J.; Moses, D.; Sinclair, M. *Synth. Met.* **1986**, *15*, 95–104.
20. Thakur, M.; Verbeek, B.; Chi, G. C.; O'Brien, K. J. *Nonlinear Optical Properties of Polymers*; Heeger, A. J.; Orenstein, J.; Ulrich, D. R., Eds.; Materials Research Society Sympo-

sium Proceedings; Materials Research Society: Pittsburgh, PA, 1988; Vol. 109, Part I, Applications and Device Requirements, pp 41–52.

21. Townsend, P. D.; Baker, G. L.; Schlotter, N. E.; Klausner, C. F.; Etemad, S. *Appl. Phys. Lett.* **1988**, *53*, 1782–1784.

22. Dalton, L. R. *Nonlinear Optical Properties of Polymers;* Heeger, A. J.; Orenstein, J.; Ulrich, D. R., Eds.; Materials Research Society Symposium Proceedings; Materials Research Society: Pittsburgh, PA, 1988; Vol. 109, Part I, Applications and Device Requirements, pp 301–312, and refs. therein.

23. Lindle, J. R.; Bartoli, F. J.; Hoffman, C. A.; Kim, O.-K.; Lee, Y. S.; Shirk, J. S.; Kafafi, Z. H. *Appl. Phys. Lett.* **1990**, *56*, 712.

24. Stegeman, G. I.; Tortuellas, W. In *Electrical, Optical and Magnetic Properties of Organic Solid State Materials;* Garito, A. F.; Jen, A. K.-Y.; Lee, C. Y.-C.; Dalton, L. R., Eds.; Materials Research Society Symposium Proceedings 328; Materials Research Society: Pittsburgh, PA, 1994; pp 397–412.

25. Twieg, R. J.; Burland, D. M.; Hedrick, J. W.; Lee, V. Y.; Miller, R. D.; Moylan, C. R.; Volksen, W.; Walsh, C. A. In *Electrical, Optical and Magnetic Properties of Organic Solid State Materials;* Garito, A. F.; Jen, A. K.-Y.; Lee, C. Y.-C.; Dalton, L. R., Eds.; Materials Research Society Symposium Proceedings 328; Materials Research Society: Pittsburgh, PA, 1994; pp 421–432.

26. Kaino, T.; Ooba, N.; Tomaru, S.; Kurihara, T.; Yamamoto, T. In *Electrical, Optical and Magnetic Properties of Organic Solid State Materials;* Garito, A. F.; Jen, A. K.-Y.; Lee, C. Y.-C.; Dalton, L. R., Eds.; Materials Research Society Symposium Proceedings 328; Materials Research Society: Pittsburgh, PA, 1994; pp 449–460.

27. Yamada, S.; Cai, T. M.; Shi, R. F.; Wu; M. H.; Chen, W. D.; Qian, Q. M.; Garito, A. F. In *Electrical, Optical and Magnetic Properties of Organic Solid State Materials;* Garito, A. F.; Jen, A. K.-Y.; Lee, C. Y.-C.; Dalton, L. R., Eds.; Materials Research Society Symposium Proceedings 328; Materials Research Society: Pittsburgh, PA, 1994; pp 523–528 and references therein.

28. *Electrical, Optical and Magnetic Properties of Organic Solid State Materials;* Chiang, L. Y.; Garito, A. J.; Sandman, D. J., Eds.; Materials Research Society Symposium Proceedings 247; Materials Research Society: Pittsburgh, PA, 1992; and references therein.

29. Grossman, C.; Heflin, J. R.; Wong, K. Y.; Zamani-Khamiri, O.; Garito, A. F. "Final Report to AFOSR Contract No. F49620–85–C–0105," 1988, and refs. therein.

30. Dalton, L. R. "Annual Technical Reports to AFOSR Contracts Nos. F49620–87–C–0100 and F49620–88–C–0071," 1988, 1989, and refs. therein.

31. Agrawal, A. K.; Jenekhe, S. A.; van Herzeele, H.; Meth, J. S. *Chem. Mater.* **1991**, *3*, 765 and refs. therein.

32. Bryne, H. J.; Blau, W. *Spec. Publ. R. Soc. Chem. (Org. Mater. NLO Opt.)* **1991**, *91*, 242–248.

33. Heeger, A. J. *Gov. Rep. Announce. Index (U. S.)* **1992**, *92(202)*, 573.

34. Muellen, K. *Pure Appl. Chem.* **1993**, *65*, 89–96.

35. Chandrasekhar, P.; Thorne, J. R. G.; Hochstrasser, R. M. *Appl. Phys. Lett.* **1991**, *59*, 1661–1663.

36. Chandrasekhar, P.; Thorne, J. R. G.; Hochstrasser, R. M. *Synth. Met.* **1989**, *53*, 175–191.

37. Chandrasekhar, P.; Gumbs, R. W. *J. Electrochem. Soc.* **1991**, *138*, 1337–1346.

38. Chandrasekhar, P.; Masulaitis, A. M.; Gumbs, R. W. *Synth. Met.* **1990**, *36*, 303–326.

39. Chandrasekhar, P. "Final Annual Reports to U.S. DARPA Contracts Nos. DAAH01–90–C–0556, DAAH01–92–C–R120, DAAH01–91–C–R151," 1990, 1992, 1992. Available from U.S. National Technical Information Service (NTIS), Washington, DC, or U.S. Defense Technical Information Center (DTIC), Arlington, VA.

40. Chandrasekhar, P. "Final Report to U.S. Naval Sea Systems Division Contract No. N00024–91–C–4045," 1992.

41. Chandrasekhar, P. "Final Reports to Contracts Nos. MML–TSC–93–01 and MML–TSC–93–05," Ashwin-Ushas Corp., Inc.: Freehold, NJ; Martin Marietta Labs.: Baltimore, MD, 1993.

42. Guay, J.; Dao, L. H. *J. Electroanal. Chem.* **1989**, *274*, 135–142 and refs. therein.

43. Yashima, H.; Kobayashi, M.; Lee, K.-B.; Chung, D.; Heeger, A. J.; Wudl, F. *J. Electrochem. Soc.* **1987**, *134*, 46–52 and refs. therein.

44. Ohsaka, T.; Hirabayashi, K.; Oyama, N. *Bull. Chem. Soc. Jpn.* **1986**, *59*, 3423–3429 and refs. therein.

45. Sheik-bahae, M.; Said, A. A.; van Stryland, E. W. *Opt. Lett.* **1989**, *14*, 955.

46. Chisum, G. T. "Laser Eye Protection for Flight Personnel;" Report No. NADC–78158–60, Naval Air Development Center (now Naval Air Warfare Center): Warminster, PA, 1986.

47. Chandrasekhar, P. "Final Report to U.S. Navy Contract No. N62269–90–C–0234," Naval Air Development Center (now Naval Air Warfare Center): Warminster, PA, 1992, and refs. therein.

48. *Handbook of Conducting Polymers;* Skotheim, T. A., Ed.; Marcel Dekker: New York, 1986; Vols. 1, 2.

49. Ulrich, D. R. *Proc. SPIE* **1989**, *1104*, 201.

Part Four

Chemical and Physicochemical Applications

Light-Harvesting Functional Polymers

Diana M. Watkins and Marye Anne Fox*

Organized polymeric assemblies are investigated for the unique functional properties induced by the absorption of light. These arrays, bearing pendant photoactive groups in their excited state, are useful for such diverse functions as solar-energy conversion, photoconductivity, optical storage, and other functions thought to be attainable only in nature and in small-scale electronic devices. The successful design and synthesis of such polymers relies on the judicious assembly of functional components in an organized array. Light-induced electron and energy transfers that occur in these arrays constitute the fundamental events that are discussed in detail in this chapter.

Recent developments in science and technology have provided access to macromolecules capable of performing physical functions that were once limited to natural systems (1–5). Among the most intriguing new functional macromolecular systems are polymeric assemblies. Such assemblies usually consist of a polymer backbone with appended photoactive functional groups that can be assembled in an organized manner to perform useful chemical and physicochemical processes such as photochemical oxidation and reduction or photoinitiated electrical switching. Nature provides the best example of such a process in photosynthesis, where the absorption of a photon initiates the transfer of an electron through a directed pathway and results in spatial charge separation. The production of carbohydrates by the reduction of carbon dioxide and the oxidation of water are ultimately accomplished.

The principles operative in photosynthesis have been extended to a wide range of artificial photopowered electronic or mechanical devices. This phenomenon of collecting light to separate charge, and hence, to power chemical redox reactions or to control electri-

cal switches has prompted photochemists, polymer scientists, and material scientists to devise macromolecular light-harvesting systems. These arrays have been used to construct memory storage devices (6), small-scale photochemical energy conversion devices (7), and molecular-based electronic components (8, 9). As in the intricately composed natural photosynthetic unit (10) or in artificial, multicomponent solar cells, small-scale molecular systems that collect light are composed of strategically placed units with defined individual functions that operate in unison to produce desired physicochemical processes. However, these angstrom-sized components are a challenge to manipulate on a molecular level, and therefore, fine tuning the desired function on a macroscopic level is difficult. Many researchers have taken on the challenge of preparing such molecular assemblies (1–5), but the small-scale systems prepared to date are still relatively simple and are composed of only a few basic units.

As such, a polymeric light-harvesting system generally contains a unit or set of units that absorb light, a corresponding unit or set of units to accept and store the energy (excitons) from the absorbed light, and a set of intervening units through which the energy (excitons) can be transferred. The stored energy can

*Corresponding author

3469–8/97/0549$15.00/0 © 1997 American Chemical Society

then be used to perform a variety of functions, some of which are discussed in detail in this chapter. More fundamental aspects of photochemistry and photophysics underlying these functions are discussed in other chapters.

The idea of harvesting light also applies directly to the utility of light-sensitive polymers in solar-energy conversion. In most polymer systems, the transformation of light energy to chemical, mechanical, or electrical energy entails the absorption of light and its subsequent conversion into a graded flow of energy or electrons along the polymer chain. Many applications ensue from this process, and the production of chemical fuels represents one important use. The conversion of light energy into electrical signals in sensors is another objective that can be approached with light-harvesting polymers. For example, photoinduced charge separation over micrometer distances that results in a positive electrical current or trapped charge is required to achieve photoelectric effects in electrooptical material or lithographic films. A number of polymers that potentially can be used in such photovoltaic or photoconducting devices and for information storage will be discussed in this chapter.

Clearly, the continuing development of new synthetic methods, along with an improved understanding of molecular photophysics, has enabled the construction of macromolecular systems capable of performing functions reminiscent of those found in nature, solid-state semiconductors, or microelectronic devices. Accordingly, functionalized macromolecules, once identified as the goals for the future, are today becoming accessible scientific tools. The objective of this chapter is to overview such polymer systems, to discuss their potential applications, and to propose ideas for new systems that still need to be developed.

Required Components in Photon-Harvesting Polymers

The preparation of functionalized, supramolecular, light-harvesting systems is accomplished by the assembly of functionalized molecular components. The excited states of single molecules have been studied extensively throughout the past two decades (11). With the extensive progress that has been made toward understanding the synthesis and photophysics (12, 13) of individual molecules, the stage is now set for the preparation of macromolecular, multicomponent systems capable of performing an additional function characteristic of the assembly itself, such as light harvesting. We can clearly learn much about the requirements necessary for the proper assembly of molecules from natural systems (10). Nature has taught us that individual chromophoric units must be held rigidly and in the correct relative

orientation for maximally efficient communication between successive units. This organization in the natural system allows for a series of energy- and electron-transfer events between absorptive units and relays that result in an assembly that functions as a template for chemical oxidation and reduction. In an attempt to mimic such a system, or more practically, the functions that compose the system in operation, the individual functions must be clearly defined.

We can begin by defining the operative components required for the design of a variety of photon-harvesting polymers (5, 14–16). In polymeric systems with multiple photoresponsive units, the potential exists for four different photophysical processes to occur:

- energy transfer within a homopolymer between identical chromophores
- direct electron transfer between two adjacent complementary redox-active groups on a polymer
- mediated electron transfer through a conjugated or nonconjugated polymer backbone between two complementary redox-active groups
- energy storage resulting from a physical or chemical change to the pendant photoactive species following light absorption

This fourth process, although not directly involved in most natural systems, is a method by which large amounts of light can be stored or harvested and subsequently released to perform work. A detailed description of these four processes is paramount to a clear comprehension of their significance to their operation in an assembly as a whole.

Energy Migration Within a Homopolymer

Energy migration along a polymer chain occurs through the sequential interaction of pairs of adjacent chromophores attached to the backbone. These interactions consist of a series of isoenergetic excitation-energy transfers from an excited chromophore (C^*) to an adjacent ground-state chromophore (C), ultimately resulting in a relaxation of the first molecule (sensitizer) to its ground state and production of the excited-state of the second. This process can be repeated several times, and the excitation energy hops to the next adjacent ground-state chromophore and so forth along the polymer (Scheme I).

These transfers can continue until the excited state is quenched by radiative or nonradiative decay, by energy transfer (i.e., interaction with a photoactive species whose excitation energy is lower than that of the originally excited chromophore), or by electron transfer as discussed subsequently. The efficiency of energy migration along a polymer chain also depends

Scheme I. Energy migration in a homopolymer bearing an array of chromophores (C).

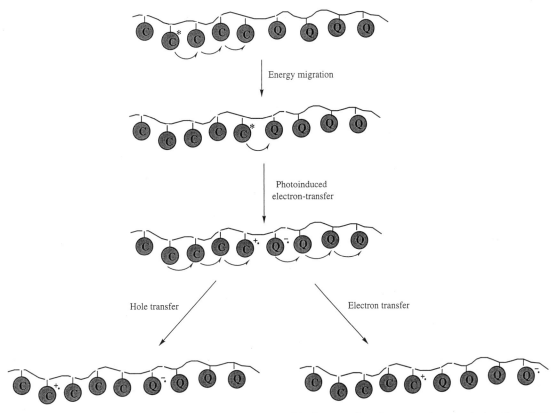

Scheme II. Macroscopic separation of charge by electron and hole transfer along a polymer chain. C and Q represent chromophores and quenchers, receptively.

greatly on the distance and orientation between adjacent photoactive species. These distances can be varied by synthetically manipulating the loading of the pendant species, the morphology of the polymer backbone, and the rigidity of the polymer chain. Hence, nonrigid polymers are poor candidates for energy-migration systems because dissipative energy wasting through excimer formation can occur readily as a result of folding of the flexible polymer chain.

Electron Migration in Chromophore and Quencher-Containing Copolymers

Excited states are inherently more reactive toward electron transfer than their ground-state precursors by virtue of population of a higher lying orbital (making oxidation more favorable) while creating a vacancy in a lower lying orbital (making reduction more favorable). When coupled with energy migration (*see* Scheme I), photoinduced electron transfer can generate an oxidized and reduced radical ion pair at a site remote from the initial absorption of light (Scheme II).

When coupled with electron or hole migration, the separation between these redox-activated species can be enhanced further.

Electron transfer along a polymer chain is governed by the same spatial and geometrical requirements as energy migration. In this step, an electron hops from site to site from the interface where the radical ion pair was produced by photoinduced electron transfer between a donor and adjacent acceptor with a lower lying vacant orbital. In a simple molecular collision, the charges produced by photoinduced electron transfer quickly recombine to regenerate the neutral ground-state species from which they derive. In a polymer, however, the electron and hole can be sequentially transferred, and the electron can migrate to an adjacent molecule isoenergetically. A greater spatial separation of charge is thus attained, provided that the system is rigid. Such polymeric arrays provide a clear advantage over low molecular weight donor–acceptor pairs insofar as charge separation over an extended framework provides a method by which polymers can perform work, such as that encountered in photosynthesis or in the operation of microscale devices.

Electron Transfer Through Conjugated or Nonconjugated Polymer Backbones

Electron transfer between donors and acceptors attached at opposite ends of a rigid polymer chain can occur by two different pathways depending on whether the matrix coupling element for the connecting bridge is substantial; for example, whether the polymer linking them is conjugated or nonconjugated. For species attached to a conjugated polymer chain, electrons between the donor and acceptor are completely delocalized through the π-network of the polymer (Figure 1a). For species attached at the opposite ends of a nonconjugated polymer, electrons can flow between donor and acceptor through the σ-framework of the polymer (Figure 1b). If the polymer is conformationally flexible, as in Figure 1c, charge separation between the donor and acceptor is quenched quite rapidly, because the polymer ends have the potential to come in contact with one another by conformational equilibration either before or after photoexcitation.

Energy Storage from Photochemical Transformations of Photoactive Species

The storage of light energy as a photochemical product with a higher heat for formation of a chromophore bound to or incorporated within the polymer backbone requires the same rigidity and conformational restrictions as those for energy migration and electron transfer. Photoactive species able to undergo transformations such as isomerization, dimerization, or cycli-

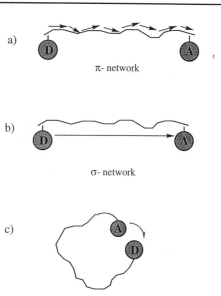

Figure 1. Electron transfer in a rigid polymer bearing a pendant donor (D) and acceptor (A) through a π-network (a) and a σ-network (b). Electron transfer in a nonrigid polymer (c) when donor and acceptor come within spatial contact of one another.

zation during light absorption were studied extensively (*16*) for their ability to store light energy in thermally stable higher-energy products (Figure 2). Attachment of these moieties to a polymer backbone creates macroscale, photoresponsive polymers with the ability to induce modifications of bulk optoelectronic properties of the material, such as optical density or refractive index. More importantly these systems act as data storage devices, reading information via light signals and storing the signal via a structural modification. These systems potentially may be able to retrieve that information thermally.

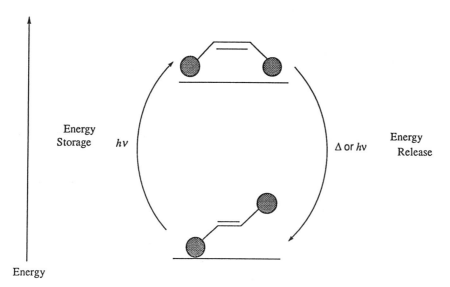

Figure 2. Energy storage via a photoinitiated change in a molecule (i.e., azobenzene).

Functional Light-Harvesting Polymers

In the past decade, many investigators assembled light-responsive molecules into complex supramolecular arrays (*4*) to elucidate and mimic some of the functional components in natural solar-energy conversion. The objective of preparing and studying these artificial assemblies was to learn how to control, by chemical design, processes such as intramolecular electron and energy transfer that occur quite efficiently in nature. Much has been learned about the complexity and intricacy of combining multiple functions into a single supramolecular structure from studying the crystal structure of the photosynthetic bacteria *Rhodopseudomonas viridis* (*10*). This multicomponent biological unit consists of several subunits that act in unison to absorb light and mediate the process that results in the transfer of electrons and energy through directed pathways, ultimately creating a charge-separated species capable of chemical work. In an attempt to analyze the complexity of this array by examining the essential functionality of the natural system, photophysicists and molecular biologists have begun to apply photophysical principles that are well-established in simple molecular systems to extended polymers (*13–16*).

Solar-Energy Conversion in Polymers

Polymer-based assemblies play an important role in the design of multicomponent, light-absorbing networks capable of transmitting electrons or energy in one direction over distances of 1 μm (10,000 Å) or more (*5*). Functional components that permit the preparation of useful solar-energy conversion systems include a rigid polymer backbone, a chromophore for absorption of light, a relay component to transmit the light energy, and a trap or quencher to store the incident energy until it is used for an electrical or chemical purpose. Moreover, the extended chains characteristic of polymers provide a means by which the spatial and distance requirements for acquiring and maintaining light-induced charge separation essential to artificial photosynthesis can be addressed. The lack of rigidity or regularity in most nonabsorptive organic polymers, however, has made the study of chromophore-labeled organic polymers challenging, although a few successful systems have been described (*17–25*).

Accordingly, a major effort has been directed toward the study of the photophysical properties of polymers bearing several precisely positioned light-absorbing species (*13–25*)., For example, Webber and co-workers (*26, 27*) investigated homopolymers bearing pendant naphthyl groups and alternating copolymers with pendant anthracene and carbazole groups to obtain information about the course of singlet energy migration along the polymer chain. Their investigations showed that energy transfer over long distances is inhibited by excimer formation in random copolymers, whereas increased energy-migration efficiency along the polymer can be achieved by preparing copolymers with regular alternating attachment of donor–acceptor pairs. Unlike random copolymers, alternating copolymers had a smaller number of energy traps, and this characteristic allowed for efficient energy migration over longer distances. Such a picture of the factors that control energy-migration efficiency enhances our understanding of the complicated photophysics in polymers and permits better design for the ultimate preparation of functionalized photon-harvesting supramolecular systems.

Closs, Miller, and co-workers (*12, 28*) studied distance and driving-force dependence of intramolecular electron transfer in a series of small-length donor–spacer–acceptor molecules, and the spacers had an average length of ~10 Å. Their investigations showed that several factors influence the rate of electron transfer: donor–acceptor orientation and distance, solvent polarity, and the nature of the covalent spacer intervening between the donor and the acceptor.

Verhoeven, Paddon-Row, and co-workers (*29, 30*) prepared a series of nonconjugated donor–spacer–acceptor assemblies (Structure 1). In these systems, intramolecular charge separation results from photoexcitation of the donor, followed by long-distance electron tunneling through the saturated hydrocarbon spacer to the acceptor. In this rigid nonconjugated system, the optical transitions that produce the charge-separated state occur on a different time-scale than in a conjugated system. For example, the time required for charge separation is partially determined by competition between electron tunneling and the natural decay of the locally excited donor or acceptor. As with the molecules studied by Closs, Miller, and co-workers, these systems show that the electron transfer can

Structure 1. A rigidly linked nonconjugated donor–spacer–acceptor assembly.

be mediated by rigid σ-bonded bridges; π-conjugated linkages are not necessarily required. The rigidity in the spacer separating the donor and acceptor is once again important, regardless of the pathway used for electron transfer. In addition, energy-wasting charge recombination in these systems is inhibited because electron tunneling to the ground state is slower than forward charge separation as a result of differences in driving force. Clearly, such systems have potential applications as solar-energy conversion devices in which charge separation might be maintained long enough to perform oxidative or reductive tasks.

Verhoeven, Warman, and co-workers (*31*) also described a similar but flexible donor₁–donor₂–acceptor assembly that forms a folded loop with a charge transfer after illumination (Structure 1). Photoinduced charge separation produces a separated ion pair (in the extended conformation of the molecule) that rapidly pulls the ends of the extended molecule together via electrostatic attraction of opposite charges; this process is called *harpooning*. In these systems, the dipole created by the initial charge separation is neutralized in the folding step. This type of photoinitiated change in physical conformation of a molecule that creates on and off switching of a dipole moment (and ultimately conductivity) meets the requirements of a sensing device (*32*).

Another approach to artificial photosynthetic systems involves the construction of oligomeric arrays of rigidly tethered light-absorbing molecules. Harriman (*33*) studied the photophysical properties of a series of covalently linked metalloporphyrin dimers and pen-

tamers. In the systems studied, asymmetrical dimers showed excitation-energy transfer, whereas symmetrical dimers showed only a small degree of electronic coupling (excitation) between the two porphyrin rings. The pentameric array shows light-harvesting features and energy transfers that occur efficiently from antenna-zinc porphyrins to a central free-base porphyrin. The nature of the connecting chain, solvent, and central metal cation are factors that directly influence the extent of excitation coupling or transfer in the arrays described.

Lindsey and co-workers (*34*) recently probed organization effects on energy transfer and migration processes using similar assemblies (Structure 2). The pentameric array, consisting of four zinc-based porphyrins attached to a free-base porphyrin core, harvests light in a method similar to that in the natural system. For example, the four peripheral porphyrins act as antennae that funnel the absorbed light via energy migration across the diphenylethyne spacers to the porphyrin free-base core, the efficiency of this energy transfer is estimated to be ~90% based on fluorescence quenching of the zinc porphyrins. As in natural photosynthetic systems, this carefully architectured array shows that rigidity and orientation play an important role in facilitating energy transfer between chromophores and acceptors. Such criteria are essential to the preparation of macromolecular systems for future applications in molecular devices.

Many functional polymeric assemblies have been devised to serve as media for a direct imaging or lasing application. The coumarin-substituted poly(meth-

Array	R	R'	R"
8	CH₃O	H	H
9	CH₃	CH₃	CH₃

Structure 2. A pentameric array of five covalently linked porphyrins with peripheral zinc (M = zinc) porphyrins and a free base (M = hydrogen) porphyrin core.

acrylic acid) prepared by Jones and Rahman (*35*) is used to partition and organize mixtures of co-bound laser dyes so that energy migration between them is maximized (Structure 3). This organization allows for a shift in fluorescence emission to longer (more useful) wavelengths without complications stemming from intermolecular aggregation. The crucial importance of orientation and distance between the appended dyes is evident. Moreover, by changing the pH of the surrounding aqueous medium, the rate of energy-migration efficiency in these water-soluble, methacrylic, acid-based polymers can be adjusted. For example, methacrylic acid-based polymers at pH 3 show 70% transfer efficiency, whereas the same polymers at pH 7 show a decreased transfer efficiency because of ionization of the polymer, and the loss of dye substituents results.

Thus, this functionalized polymer can be tuned to maximize electron- or energy-migration efficiency by adjusting pH of the medium. At a pH greater than 7 the polymer is fully ionized, and as a result it has a rod-like shape because of charge repulsion of the carboxylate side chains. Conversely, at a pH less than 4 the polymer is uncharged and becomes highly coiled. An important aspect of these devices is the distance separating the respective donors or donor–acceptor pairs along the polymer chain. Förster theory predicts that chromophores must be *within* a maximum distance (dictated by orbital overlap) for energy transfer to occur efficiently between groups on a polymer (*13*). Polymer systems bearing chromophores spaced at distance larger than this maximum fail to transfer energy through the polymer chain across extended distances.

Meyer and co-workers (*36–41*) also studied energy transfer over long distances in styrene-based polymers. Recently, they prepared polymers bearing a series of randomly distributed sensitizers [ruthenium(II) tris(bipyridine[bpy])], quenchers [osmium(II) tris(bipyridine)], and intermediate energy relays (anthracene) (Scheme III). In such an array, the excitation energy stored in the metal-to-ligand triplet charge-transfer state of $[Ru(bpy)_3]^{2+}$ is transferred through a series of anthryl groups to $[Os(bpy)_3]^{2+}$ and results in long-range energy transfer along the polymer chain.

Solar-energy conversion devices incorporating biopolymers also have been described (*42, 43*). The inherent conformational rigidity and helicity of most peptides make them prime candidates for facilitating long-range photoinduced electron transfer. Jones and Weiss-Farahat (*42*) investigated photoinduced charge migration along a polymer surface between pendant

Abbreviated structure for poly(methacrylic acid)

(ionized form)   (fully potonated)

Coumarin laser dyes cobound with poly(methacrylic acid) for allignment

Structure 3. Methacrylic acid polymers in their ionized form and coumarin laser dyes.

Multi-functional Polystyrene Copolymer

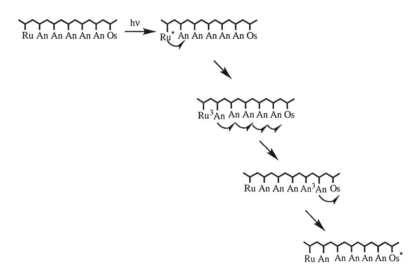

Energy Transfer and Migration with a Triplet Relay

Scheme III. Energy migration and transfer in a styrene-based polymer with pendant $Ru(bpy)_3^{2+}$ sensitizer, $Os(bpy)_3^{2+}$ quencher, and anthracene as a relay system for triplet energy migration.

electroactive donors and acceptors. Their studies show that electron and hole migration along the chain depend on peptide conformation and the interaction of neighboring functional groups. Research in our group (*43*) using chromophore-quencher functionalized polypeptides has also confirmed the importance of helical rigidity for efficient electron transfer through these types of biopolymers. Currently, investigations focus on the preparation of sequentially ordered, redox-active, functionalized polypeptides, which should potentially improve the ability of these systems to serve as solar-energy conversion devices that can transfer energy over extended distances.

The utility of polymer-based devices for solar-energy conversion can be improved with functionalized solid-state materials. In this respect, Wamser and co-workers prepared (*7, 44–46*) chemically asymmetric polyporphyrin films (Figure 3) that exhibit directional photocurrents when irradiated while sandwiched

between semitransparent electrodes. During the formation of these films, two monomers approach the polymerization reaction at the interface from different directions because of their confinement to separate, immiscible solutions.

The direction of the photoinduced electrical current derives from the difference in redox potentials between the two porphyrins on opposite sides of the film surface. During irradiation, the surface contacting tetrakis-μ-(*p*-chlorocarbonylphenyl)porphyrin develops a more negative potential than the opposite surface containing tetrakis-μ-(*p*-hydroxyphenyl)porphyrin (Figure 4).

A series of rigid and highly organized polymers bearing sequentially ordered, pendant, redox-active groups (Structure 4) show promise as materials for photoinduced, directional energy migration and electron transfer (*47*). Directional energy migration in several of the diblock copolymers (**II**) bearing pendant

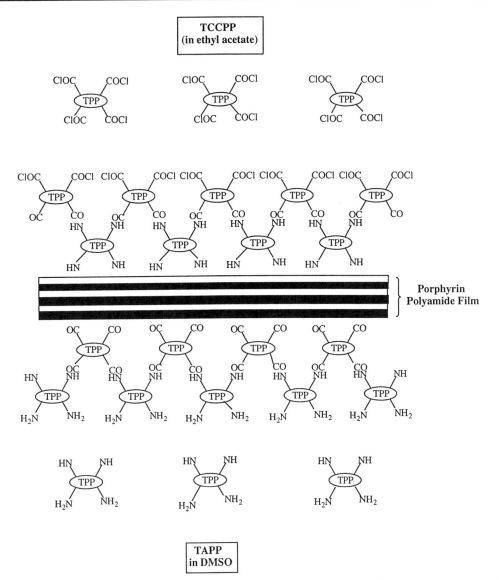

Figure 3. Tetrakis(4-aminophenyl)porphyrin (TAPP) and tetrakis[4-(chlorocarbonyl)phenyl]porphyrin (TCCPP) form an asymmetric polymer film.

dimethylaniline and naphthalene chromophores is shown by a greater than expected fluorescence quenching of naphthalene by electron transfer at the copolymer interface. The rigidity in the prepared polymers is demonstrated by the lack of excimer emission in the naphthalene homopolymer (**I**) and by the absence of exciplex emission in a naphthalene–pentamethylbenzene–dimethylaniline triblock copolymer (**III**).

Photoconduction in Polymeric Devices

Photoconduction is defined as the flow of electrons (current) in a material upon the application of a potential during irradiation with light (*48*). Materials that display this property are useful as luminescence detectors in spectrophotometric devices, as components in xerographic films, and in various optical computer applications, to name a few applications. Polymers have played a major role in the preparation of such materials because they provide extended π-networks over which electrons can flow. Doped polyacetylenes and polythiophenes traditionally have been used for these applications (*49*), although a variety of more complex and intricate polymers have been prepared and studied (*48*). For a polymer to prove useful as a light-harvesting photoconductor, broad absorption of light by the entire polymer is required. To achieve an uninterrupted flow of electrons, rigidity is once again a necessity.

With these criteria in mind, Meier and Albrecht (*9*) studied the photoconductivity of bridged polymeric phthalocyanines without any dopants or additives. The uniquely rigid and one-dimensional polymers shown in Structure 5 exhibit a range of photoconduc-

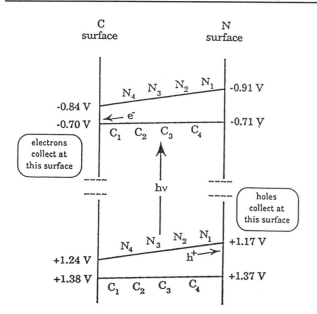

Figure 4. Representation of the redox potential gradient predicted based on the different patterns of porphyrin substituent expected in a TCCPP/TAPP interfacial polymer film. (Reproduced from reference 44. Copyright 1991 American Chemical Society.)

the functionality that can be achieved when a system is designed with the appropriate features. Similar polymers were first described by Marks (50), who established and studied electrical conductivity in several phthalocyanine-based macrocyclic assemblies.

Photoconduction in polymers can also be probed by studying non-linear optical (NLO) effects (51). Optical communication devices based on NLO effects have drawn worldwide attention (52–55). Since the invention of lasers and the discovery of second-harmonic generation, progress in the fabrication of materials for NLO for information processing has been rapid (54, 55). NLO materials have been used directly for frequency doubling of low-energy-wavelength light and indirectly for electrical switching. Photoconductive and NLO polymers are discussed in other chapters, but these materials also relate directly to photon-harvesting polymers and are hence briefly covered here for the sake of completeness.

A photoconductive material is capable of harvesting light and converting it into an electrical signal in an optical switch or optical waveguide. Polymers in particular have commanded significant attention because of their ability to provide both an NLO response and photoconductivity. Relatively few examples of such dual-use polymers exist, but the structural requirements for optical nonlinearity or photoconductivity are similar in several aspects. For example, an NLO material requires an array of photoactive species aligned so as to enforce a positive macroscopic dipole, which in turn interacts with linearly polarized light (55). Similarly, a photoconductive material requires a series of precisely aligned chromophores to interact

tivities depending on the choice of the central metal atom and of the bridging ligand, and the efficiency of charge transfer depends directly on the structure and chemical composition of the polymer. Moreover, a polymer's photoelectric properties can be tailored to a specific wavelength response by using various combinations of bridging ligands and central atoms. Rigid linear assemblies such as these provide examples of

Naphthalene substituted polynorbornene
homopolymer (**I**)

Naphthalene-dimethylaniline
substituted polynorbornene diblock (**II**)

Naphthalene-pentamethylbenzene-dimethylaniline
substituted polynorbornene triblock (**III**)

Structure 4. Rigid polynorbornene-based naphthalene homopolymer, naphthalene–dimethylaniline diblock copolymer, and naphthalene–pentamethylbenzene–dimethylaniline triblock copolymer prepared by ring-opening metathesis polymerization techniques. (Reproduced from reference 44. Copyright 1994 American Chemical Society.)

= Phthalocyanine =

L = Bridging Ligand =   N   N   or,   N   N

or, NC—⟨ ⟩—CN   or,   N N N N

M = Metal atom =   Ge, Fe, Co, Ru, Cr, or Mn

Structure 5. Two classes of polymers: M = Ge, L = oxygen or sulfur; and M = transition metal, L = bidentate π-electron containing ligands pyrazine, 4,4'-bipyridine, 1,4-diisocyanobenzene, or tetrazine cyanide.

after absorption of light. In NLO materials, this interaction results in a nonlinear output of light, whereas in a photoconductor the interaction results in the conversion of light energy into an electrical current. Conductivity in a photoconductor can only result from the interaction between adjacent chromophores, as described previously. NLO materials, on the other hand, may possess, but do not require, this interaction.

One of the few examples of an NLO material that also possesses the interchromophoric interactions necessary for photoconductivity was prepared by Tripathy and co-workers (52). The dual-function polymer 3-cinnamoyloxy-4-[4-(*N*,*N*-diethylamino)-2-cinnamoyloxyphenylazo]nitrobenzene (CNNB-R) (Structure 6) is prepared by photocross-linking an excited cinnamoyl group of one chain with a ground-state cinnamoyl group of another chain. The polymer doped with an azo dye displayed a photoconductivity of $4.7 \times 10\ \Omega$ cm when sandwiched between glass slides covered with indium–tin-oxide and a thin layer of gold (53). Third-order optical nonlinearity was also observed. This observed photoconductivity is comparable with that of a similar polymer that required the addition of photosensitizers and photocharge transport agents to induce conductivity (49).

The experimental setup shown in Figure 5 for the photoconductivity measurement displays the use of the previously mentioned material as a photoconducting device. Initial absorption by the CNNB-R at 518 nm is the origin of both photocharge generation and transport.

Information Storage in Polymeric Devices

It has been said that "progress in information storage is directly related to the ability to store and recall large amounts of data in the smallest possible space" (55). In this context, the ever-increasing need for new technologies capable of efficient data storage and information retrieval has spurred interest in polymers for optical information storage. Photon-harvesting polymers play a prime role in the design and synthesis of such devices. For example, polymers with pendant, photoactive, functional groups or functional groups that are incorporated into the backbone are capable of reversible or irreversible storage of light signals upon excitation at a defined wavelength. The optical energy is "stored" upon excitation of the pendant chromophore. The energy is released when the higher-energy-state molecule reverts, either thermally or photochemically, to its pre-excitation state. This phenomenon (the use of light to cause a physical or chemical change) is the basis of optical recording. Polymers bearing photochromic dyes (such as azobenzene or spiropyrans) that undergo cis–trans isomerization or ring closure upon excitation of light can be directly applied as recording layers in these devices.

Polymer-bound photoactive norbornadienes (NBD) that are converted to quadricyclanes (QC) upon excitation were also studied (56–58) for use as memory storage systems. For example, Nishikubo and co-workers (56, 57) prepared photoresponsive polymers bearing NBD moieties by cationic polymerization of 2-[(3-phenyl-2,5-norbornadienyl-2-carbonyl)oxy]ethyl vinyl ether (PNVE). This photochemical transformation (Scheme IV) converts an appended NBD to a higher energy but thermally stable QC. The energy can be released in a controlled manner by the reversion of the ring-strained QC to NBD either thermally, catalytically, or photochemically. Nishikubo and co-workers (57) demonstrated that polymers with pendant NBD groups can release as much as 92 kJ/mol upon rever-

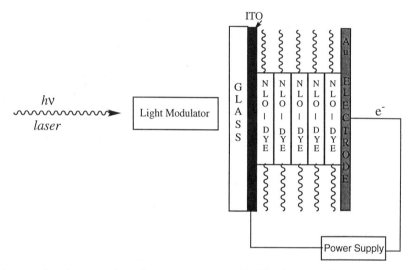

Structure 6. A polymer doped with an azo dye.

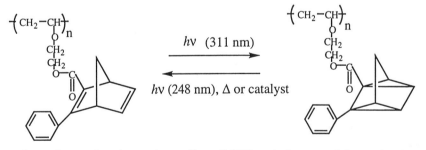

Figure 5. Experimental setup for photoconductivity measurement of bridged polymeric phthalocyanines. (Reproduced with permission from reference 50. Copyright 1991 American Association for the Advancement of Science.)

Scheme IV. Photoresponsive polymers bearing norbornadiene (NBD) moieties store information in the form of energy via photoisomerization.

sion of the QC to NBD following initial photochemical isomerization. Homopolymers of poly(vinyl ether)s bearing only NBD showed higher rates of photochemical reaction of pendant NBD in homopolymers than copolymers bearing NBD and either 2-phenoxy-ethyl vinyl ether, 2-isobutyl vinyl ether, or 2-chloroethyl vinyl ether. This result is consistent with energy migration between NBD units in homopolymers that can expedite photochemical reaction along the homopolymer chain. Copolymers hinder this "communication" by placing a spatial barrier between respective NBD units. Unfortunately, the chemical reversibility of these systems is not complete and polymer degradation begins to appear after only a few cycles.

Another potential memory storage device is a rigid-rod polymer incorporating photoactive groups within the backbone. Sahyun and co-workers (59) used this approach to prepare rigid-rod polymers that incorporate anthracene units in their backbone. These anthracene-containing polymers (Scheme V) tend to "preorganize" (as seen by interstrand excimer formation) in the solid state to allow for light-induced photodimerization between polymer strands. As discussed previously, dimerization of photoactive chromophores provides a means by which light energy can be captured and stored in a higher-energy state (a dimer in this case). The amount of energy that can be stored is increased several fold in this polymer compared with that attainable in solution, because at least 10 anthracene units per polymer strand can associate with 10 from another strand. Incorporation of the photoactive anthracene unit into the backbone provided enhanced reactivity over polymers with pendant anthracene units for the photo-cross-linking reaction because of the nonrigidity of polymers with pendant anthracenes and the reduced efficiency with which anthracene units linked at the 9- and 10-positions can photo-cross-link.

Optical information storage using azo-derivatized polymers was studied extensively over the past two decades (55). Polymers that incorporate azo-functionalities in the backbones or pendant to the chain undergo cis–trans photoisomerization during capture of light energy (60). Spiess and co-workers (61) investigated the use of liquid-crystalline polymers with azobenzene-derivatized side groups for information storage. Laser-induced transformations (n-π*; excitation at 460 nm) in the azobenzene cause changes in the local orientation of the liquid-crystalline matrix and create changes in the macroscopic refractive index. These changes persist when the irradiation has ceased.

New azobenzene-based monomers and polymers can serve as pH indicators (62), and reversible optical storage has been accomplished in azobenzene-bound amorphous poly(methyl methacrylate), poly(ethyl methacrylate), polystyrene, and polyamide (63). This work demonstrates that liquid crystallinity is not a necessary condition for effective optical storage.

Other new materials using a different photoresponse are also being investigated. Fréchet and Cameron (64) investigated novel photoresponsive polymers bearing photocyclizable 2,4,6-triisopropyl benzophenone moieties. After the absorption of 313-nm light, a Norrish type II photocyclization converts a pendant ortho-functionalized benzophenone group into the corresponding benzocyclobutenol (Scheme VI). Light energy is stored along the polymer chain until heat is used to reverse the cyclization. High loadings of the chromophore on the polymer result not only in intramolecular photocyclization but also in potentially useful intermolecular cross-linking.

A sophisticated scheme for memory storage was proposed by Hopfield and co-workers (6) in a shift register based on electron transfer between molecules on a polymer chain (Figure 6). The polymer, consisting of three redox or photoactive residues per repeat unit, would be capable of reading and storing data as light-activated electronic transitions in the orbitals of the pendant molecules. Electron shifts are initiated by excitation of the donor α (indicated by solid arrow pointed up) in the first repeat unit. The presence of an electron is represented by the presence of a small up arrow. The curved arrows indicate electron transfers that shift the electron one unit down the chain following photexcitation. Back reactions that decrease the efficiency of the device are indicated by dashed arrows. The information is shifted along the chain each time a new light signal is absorbed by the first chromophore in the chain.

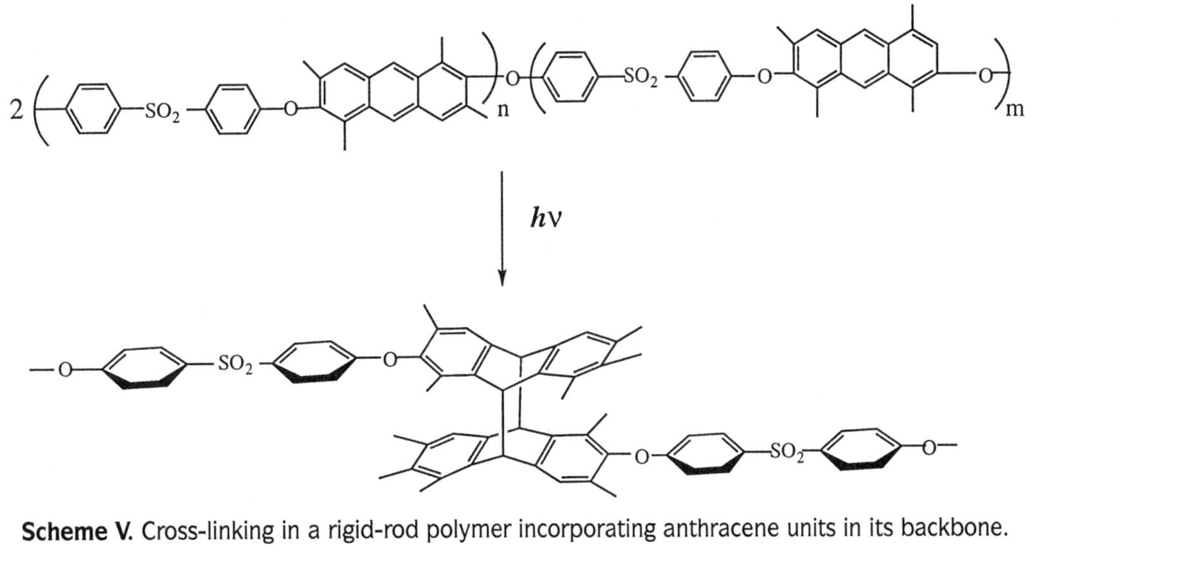

Scheme V. Cross-linking in a rigid-rod polymer incorporating anthracene units in its backbone.

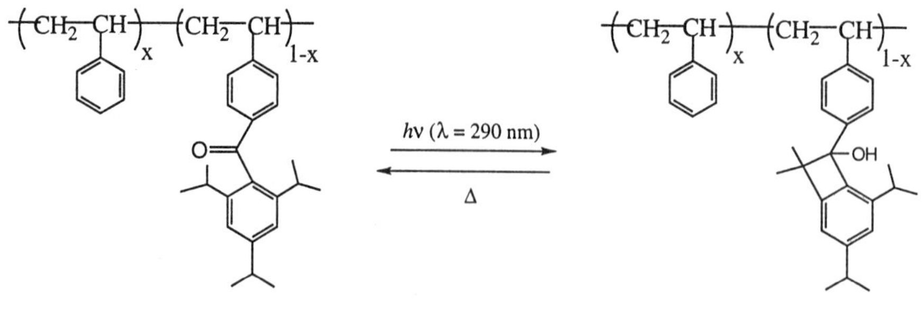

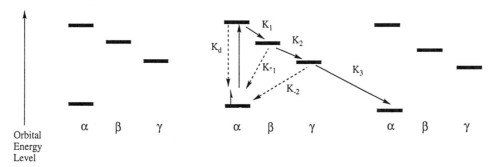

Scheme VI. Photoresponsive polymers involving the Norrish type II photocyclization of pendant 2,4,6-triisopropylbenzophenone groups (**1**) into the corresponding benzocyclobutenols (**2**).

Figure 6. Proposed design of a molecular-level memory shift register based on electron transfer. (Reproduced with permission from reference 6. Copyright 1988 American Association for the Advancement of Science.)

Other Applications

Although most light-harvesting polymers have so far been studied for their electro-optical properties, they are also capable of directly photocatalyzing chemical processes or influencing the solubility of a particular type of polymer chain. Photoresponsive polymers are particularly useful for these applications because the pendant photoactive group is often able to change its physical properties or perform a particular physical function during photoirradiation. For example, Shinkai and co-workers (65) prepared polymers bearing pendant, photoresponsive, crown ethers able to extract metal ions from aqueous solution. The idea is illustrated in Scheme VII. Two adjacent, pendant, azo-crown-ether groups on a polystyrene chain are able to bind a metal cation between them when the azo moiety is in its trans or extended form. The metal ion can then be released when the azo-crown ether is irradiated to the more compact cis form. This extraction of a metal ion from aqueous solution works only when the polymer backbone is extended or rigid, which is observed when the azo-crown ethers are in their trans form in a nonpolar solvent; nonrigid backbones prevent organized uptake and release of metal ions.

A unique biomimetic approach to photochemical energy conversion using membrane-bound, donor–acceptor pre-organization in lipid bilayers (66) was described recently by O'Brien and co-workers. Polymerization-induced domain formation of the lipid bilayers has been used to tightly organize the membrane-bound donor (a tetracationic porphyrin) and acceptor (cyanine dye), and this organization results in enhanced energy transfer between the two species (Figure 7). The polymerizable lipids BAPC and Bis-SorbPC bind to either side of the nonpolymerizable lipid dioleoylphosphatidic acid. This reaction creates two outer domains that are inaccessible to the cofactors (P^{4+} and Cy^{3+}, which then have a shorter average distance of separation) that bind only to the nonpolymerized domain in the center. After polymerization, energy transfer was enhanced 8-times more efficiently than in the unpolymerized form.

Unique methods for preparing highly controlled polymer systems using well-defined organometallic catalysts are being investigated by Schrock and co-workers (67–73). These methods have been used to make phenothiazine- or ferrocene-labeled polynorbornenes with a pyrene endcap. Photophysical studies on these polymers show that emission from the pyrene endgroup that is adjacent to blocks containing ferrocene or phenothiazine is quenched by a factor of 30 or 110, respectively. These results indicate that energy or electrons can transfer through the ferrocene or phenothiazine units to effectively quench the pyrene emission.

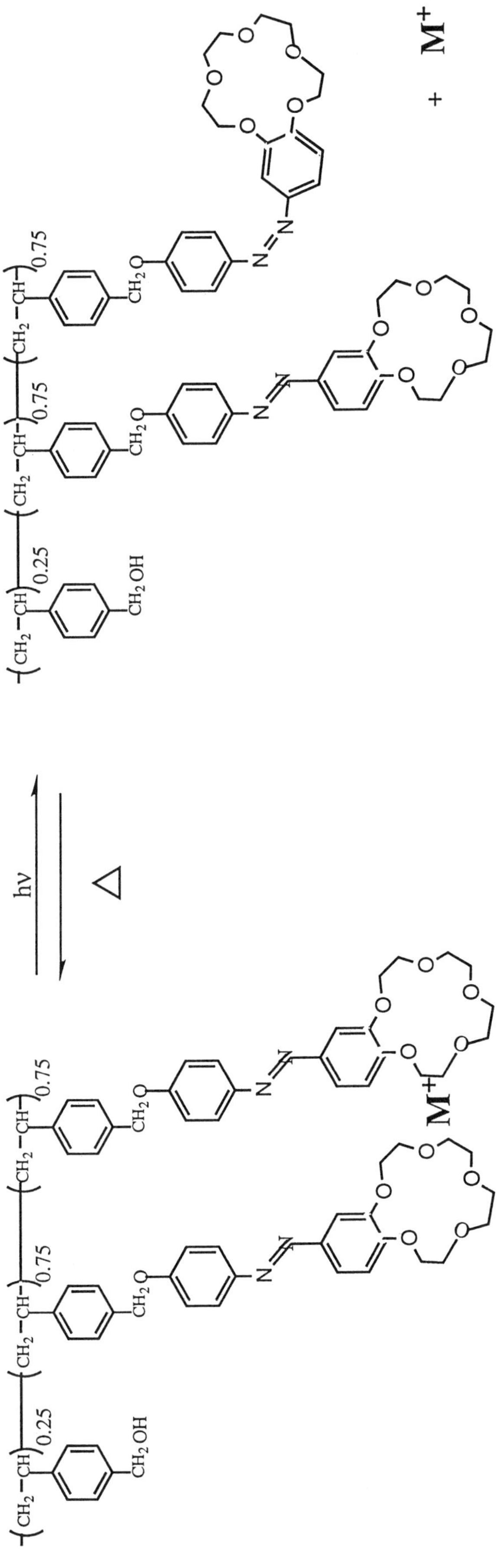

Scheme VII. Polystyrene derivative bearing photoresponsive 4'-azo(benzo-15-crown-5) used for the extracting metal ions from aqueous solution.

Electron Transfer in Polymerizable Lipid Media

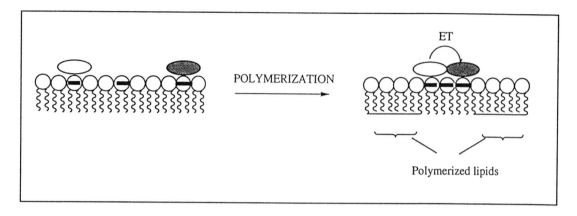

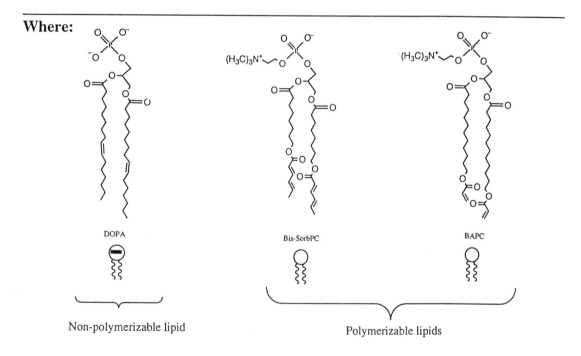

Figure 7. Biomimetic energy conversion in lipid bilayers.

Concluding Remarks

The exposure of organized polymeric assemblies to photons permits solar-energy conversion, photoconductivity, and memory storage. The successful execution of such functions can be attributed to the skillful design and synthesis of novel polymeric systems. The systems described in this chapter judiciously control energy and electron transfer to produce electro-optical effects or photoinduced chemical transformations of photoactive molecules pendant to the polymer to store light energy. Improved methods for the synthesis of rigid polymers and better understanding of photochemical events are required to design and access more highly organized systems that can perform even more sophisticated functions.

Progress in the synthesis of rigid polymers through controlled polymerization mechanisms is key to designing functional polymeric arrays. In addition, organized incorporation of multiple functional units onto a single rigid polymer provides maximum potential for preparing highly functionalized polymeric assemblies. Effective use of such systems will also require an increased understanding of photophysics in macromolecules. Advances in this area are likely to take place with a multidisciplinary approach in which the design, preparation, and photophysical characterization of highly organized and complex polymeric arrays is a highly sought goal.

Acknowledgments

We gratefully acknowledge the financial support of the Office of Basic Energy Sciences, Fundamental Interactions Branch of the U.S. Department of Energy. Diana M. Watkins also thanks the Department of Education for a graduate fellowship award.

References

1. O'Brien, D. F.; Thauming, K.; Ulrich, L. *Polym. J.* **1991**, *23*, 619–628.
2. Fox, M. A. *Accts. Chem. Res.* **1992**, *25*, 569–574.
3. Wrighton, M. *Com. Inorg. Chem.* **1985**, *4*, 269–294.
4. Balzani, V. *Tetrahedron* **1992**, *48*, 10443–10514.
5. Fox, M. A.; Jones, W. E.; Watkins, D. M. *Chem. & Engr. News* **1992**, *70*, 38–48.
6. Hopfield, J. J.; Onuchic, J. N.; Beratan, D. N. *Science (Washington, DC)* **1988**, *241*, 817–820.
7. Wamser, C. C.; Bard, R. R.; Denthilathipan, V.; Anderson, V. C.; Yates, J. A.; Lonsdale, H. K.; Rayfield, G. W.; Friesen, D. T.; Lorenz, D. A.; Stangle, G. C.; van Eikeren, P.; Baer, D. R.; Randsdell, R. A.; Golbeck, J. H.; Babcock, W. C.; Sandberg, J. J.; Clarke, S. E. *J. Am. Chem. Soc.* **1989**, *111*, 8485–8491.
8. Meier, H.; Albrecht, W. *Mol. Cryst. Liq. Cryst.* **1991**, *194*, 75–83.
9. Deisenhofer, J.; Epp, O.; Miki, K.; Huber, R.; Michel, H. *J. Mol. Biol.* **1988**, *180*, 385–98.
10. Turro, N. *Modern Molecular Photochemistry*; University Science: Mill Valley, CA, 1991; p 261.
11. Closs, G. L.; Calcaterra, L. T.; Green, N. J.; Penfield, K. W.; Miller, J. R. *J. Phys. Chem.* **1986**, *90*, 3673–83.
12. Fox, M. A.; Chanon, M. *Photoinduced Electron Transfer*; Elsevier: Amsterdam, Netherlands, 1989.
13. Guillet, J. *Polymer Photophysics and Photochemistry*; Cambridge University: New York, 1985.
14. Liu, G.; Guillet, J. E. *Macromolecules* **1990**, *23*, 1388–1392.
15. White, B.; Nowakowska, M.; Vansco, G. J.; Guillet, J. E. *Macromolecules* **1991**, *24*, 2903–2906.
16. Hallensleben, M. L.; Weichart, B. *Polym. Bull.* **1989**, *22*, 553–556.
17. Bai, F.; Chang, C.-H.; Webber, S. E. *Macromolecules* **1986**, *16*, 2484–2494.
18. Itoh, Y.; Webber, S. E. *Macromolecules* **1989**, *22*, 2766–2775.
19. Byers, J. D.; Parsons, S. W.; Friesner, R. A.; Webber, S. E. *Macromolecules* **1990**, *23*, 4835–4844.
20. Cao, T.; Webber, S.E. *Macromolecules* **1991**, *24*, 79–86.
21. Hargreaves, J. S.; Webber, S. E. *Macromolecules* **1984**, *17*, 235–240.
22. Chatterjee, P. K.; Kamioka, K.; Batteas, J. D.; Webber, S. E. *J. Phys. Chem.* **1991**, *95*, 960–965.
23. Shand, M. A.; Rodgers, M. A. J.; Webber, S. E. *Chem. Phys. Lett.* **1991**, *177*, 11–16.
24. Sowash, G. G.; Webber, S. E. *Macromolecules* **1988**, *21*, 1608–1611.
25. Webber, S. E. *Chem. Rev.* **1990**, *90*, 1469–1482.
26. Itoh, Y.; Nakada, M.; Satoh, H.; Hachimori, A.; Webber, S. E. *Macromolecules* **1993**, *26*, 1941–1946.
27. Closs, G. L.; Calcaterra, L. T.; Green, N. J.; Penfield, K. W.; Miller, J. R. *J. Phys. Chem.* **1986**, *90*, 3673–3683.
28. Oliver, A. M.; Craig, D. C.; Paddon-Row, M. N.; Kroon, J.; Verhoeven, J. W. *Chem. Phys. Lett.* **1988**, *150*, 366–373.
29. Warman, J. M.; Haas, M. P. D.; Oevering, H.; Verhoeven, J. W.; Paddon-Row, M. N.; Hush, N. S. *Chem. Phys. Lett.* **1986**, *128*, 95–99.
30. Warman, J. M.; Smit, K. J.; Haas, M.; P. D.; Jonker, S. A.; Paddon-Row, M. N.; Oliver, A. N.; Kroon, J.; Oevering, H.; Verhoeven, J. W. *J. Phys. Chem.* **1979**, *93*, 1979–1987.
31. Brouwer, A. M.; Verhoeven, J. W.; Warman, J. M. *Chem. Phys. Lett.* **1991** *186*, 481–489.
32. Harriman, A. *NATO ASI Ser., Ser. C* **1987**, *214*, 207–203.
33. Prathapan, S.; Johnson, T. E.; Lindsey, J. S. *J. Am. Chem. Soc.* **1993**, *115*, 7519–7520.
34. Jones, G.; Rahman, M. A. *Chem. Phys. Lett.* **1992**, *200*, 241–250.
35. Worl, L. A.; Strouse, G. F.; Younathan, J. Y.; Baxter, S. M; Meyer, T. J. *J. Am. Chem. Soc.* **1990**, *112*, 7571–7578.
36. Olmsted, J.; McClanahan, S. F.; Danielson, E.; Younathan, J. N.; Meyer, T. J. *J. Am. Chem. Soc.* **1987**, *109*, 3297–3301.
37. Meyer, T. J. *Coord. Chem. Rev.* **1991**, *111*, 47.
38. Younathan, J. N.; McClanahan, S. F.; Meyer, T. J. *Macromolecules* **1989**, *22*, 1048–1054.
39. Baxter, S. M.; Jones, W. E., Jr.; Danielson, E.; Worl, L. A.; Strouse, G. F.; Younathan, J.; Meyer, T. J. *Coord. Chem. Rev.* **1991**, *111*, 47–71.
40. Jones, W. E., Jr.; Chen, P.; Meyer, T. J. *J. Am. Chem. Soc.* **1992**, *114*, 387–389.

41. Jones, III, G.; Weiss-Farahat, C. *Proc. DOE Solar Photochem. Res. Conf., 14th* **1990**, 91–6.
42. Meier, M. S.; Fox, M. A.; Miller, J. R. *J. Org. Chem.* **1991**, *56*, 5380–5383.
43. Wamser, C. C.; Bard, R. R.; Denthilathipan, V.; Anderson, V. C.; Yates, J. A.; Lonsdale, H. K.; Rayfield, G. W.; Friesen, D. T.; Lorenz, D. A.; Stangle, G. C.; van Eikeren, P.; Baer, D. R.; Randsdell, R. A.; Golbeck, J. H.; Babcock, W. C.; Sandberg, J. J.; Clarke, S. E. *J. Am. Chem. Soc.* **1989**, *111*, 8485–8491.
44. Wamser, C. C. *Mol. Cryst. Liq. Cryst.* **1991**, *194*, 65–73.
45. Wamser, C. C.; Londsdale, H. K. *Energy Res. Abstr.* **1991**, *16*, 1.
46. Watkins, D. M.; Fox, M. A. *J. Am. Chem. Soc.* **1994**, *116*, 6441.
47. Lessard, R. A. *Proc. Int. Soc. Opt. Eng.* **1991**, *1559*, 1–499.
48. Schildkraut, J. S. *Appl. Phys. Lett.* **1991**, *58*, 340–342.
49. Marks, T. *Science (Washington, DC)* **1985**, *227*, 881–889.
50. Li, L.; Jeng, R. J.; Lee, J. Y.; Kumar, J.; Tripathy, S. K. *SPIE, Nonlinear Opt. Prop. Org. Mater.* **1991**, *1560*, 243–250.
51. Mandal, B. K.; Kumar, J.; Huang, J.-C.; Tripathy, S. *Makromol. Chem. Rapid. Commun.* **1991**, *12*, 63–68.
52. *Nonlinear Optical Properties of Organic Molecules and Crystals*; Chemla, D. S., Zyss, J., Eds.; Academic: Boston, MA, 1987; Vol 1.
53. *The Principles of Nonlinear Optics*; Shen, Y. R., Ed.; John Wiley & Sons: New York, 1984.
54. Mittal, K. *Polymers in Information Storage Technology*; Plenum: New York, 1988.
55. Kamagawa, H.; Yamada, M. *Macromolecules* **1988**, *21*, 918–922.
56. Nishikubo, T.; Kameyama, A.; Kishi, K.; Kawashime, T. *Macromolecules* **1992**, *25*, 4469–4475.
57. Nishikubo, T.; Kawashima, T.; Inomata, K.; Kameyama, A. *Macromolecules* **1992**, *25*, 2312–2318.
58. Bruss, J. M.; Sahyun, M. R. V.; Schmidt, E.; Sharma, D. K. *J. Polym. Sci., Part A, Polym. Chem.* **1993**, *31*, 987–993.
59. Kumar, G. S. *Azo Functional Polymers*; Technomic: Hyderabad, India, 1992.
60. Wiesner, U.; Antonietti, M.; Boeffel, C.; Spiess, J. W. *Makromol. Chem.* **1990**, *191*, 2133–2149.
61. Natansohn, A.; Rochon, P.; Gosselin, J.; Xie, S. *Macromolecules* **1992**, *25*, 2268–2273.
62. Haitjema, H. J.; v. Ekenstein, G. O. R. A.; Werkman, P. J.; Koldijk, A. J.; Boer, T. S. *Polym. Prepr.* **1991**, *32*, 132–133.
63. Cameron, J. F.; Fréchet, J. M. J. *Macromolecules* **1991**, *24*, 1088–1095.
64. Shinkai, S.; Ishihara, M.; Manabe, O. *Polym. J.* **1985**, *17*, 1141–1144.
65. Armitage, B.; Klekotka, P. A.; Oblinger, E.; O'Brien, D. F. *J. Am. Chem. Soc.* **1993**, *115*, 7920–7921.
66. Bazan, G. C.; Schrock, R. R.; Cho, H.-N.; Gibson, V. C. *Macromolecules* **1991**, *24*, 4495–4502.
67. Bazan, G. C.; Khosravi, E.; Schrock, R. R.; Feast, W. J.; Gibson, V. C.; O-Regan, M. B.; Thomas, J. K.; Davis, W. M. *J. Am. Chem. Soc.* **1990**, *112*, 8378–8387.
68. Bazan, G. C.; Oskam, J. H.; Cho, H.-N.; Park, L. Y.; Schrock, R. R. *J. Am. Chem. Soc.* **1990**, *113*, 6899–6907.
69. Albagli, D.; Bazan, G.; Wrighton, M. S.; Schrock, R. R. *J. Am. Chem. Soc.* **1992**, *114*, 4150–4158.
70. Albagli, D.; Bazan, G.; Wrighton, M. S.; Schrock, R. R. *J. Am. Chem. Soc.* **1993**, *115*, 7328–7334.
71. Schrock, R. R. *Accts. Chem. Res* **1990**, *23*, 158–165.
72. Albagli, D.; Bazan, G.; Wrighton, M. S.; Schrock, R. R. *J. Am. Chem. Soc.* **1993**, *97*, 10211–10216.

Polymers for Solar-Energy Devices

Gary Jorgensen, John Pern, Stephen Kelley, Al Czanderna, and Paul Schissel

The functional role of polymeric materials for a wide range of applications in solar-energy conversion systems is summarized. These materials include metallized polymer reflectors, photovoltaic encapsulants, electrochromic devices, desiccant materials, and others. Metallized, flexible polymeric film reflectors can be used in solar concentrator technologies to focus sunlight to generate thermal heat for process applications or for conversion to electrical power. Transparent elastic polymers such as ethylene vinyl acetate have been used as encapsulant materials for photovoltaic modules. Developmental electrochromic devices consist of a multilayer stack of thin films mounted onto a suitable substrate. Polymers are used in the ion-conducting layer and are planned for use as substrates for electrochromic windows in buildings applications. The use of cationic and anionic polymers as desiccants in cooling systems that are regenerated at temperatures below 353 K is briefly described. Recent work on metals deposited onto the organic functional end groups of self-assembled monolayers is also summarized. Problems, research needs, and future prospects for polymeric materials used in solar-energy systems are identified and discussed.

Solar insolation represents an attractive source of clean, renewable, sustainable, and low-cost energy. Practical use requires efficient collection and conversion of the intermittent, yet abundant, terrestrial solar resource in an economical manner. The level of solar energy arriving at the Earth's surface exceeds humankind's present energy usage by four to five orders of magnitude. Decentralized solar-energy systems are particularly attractive for remote applications in third-world nations. The potential benefit associated with the use of polymeric materials in solar-energy devices has been recognized internationally (1, 2).

Incorporation of polymeric materials into solar-energy devices offers a number of economic and performance advantages. Polymeric materials are typically lighter weight, lower cost, easier to process, and more conveniently assembled in the field than most inorganic and metallic materials. Even with these numerous advantages, polymeric materials are still not widely used for many solar-energy applications. There are several reasons for this limited acceptance. These reasons include a decline in the properties of polymeric materials following exposure to solar irradiation, thermal and humidity cycles, and mechanical stress.

Despite these limitations, polymer materials increasingly are used in solar-energy devices. Examples of recent developments in the field were summarized (1–3). Acrylates, especially polymethyl methacrylate (PMMA), are good candidates for encapsulants and as mirror superstrates. Polycarbonate can withstand the combined effects of temperature and moisture cycles. However, when UV irradiation also occurs, microcracks rapidly form on the surface of the polymer. Fluoropolymers, silicone rubbers, and ethylene propylene rubbers are all used as sealants and adhesives in solar collectors. This chapter outlines some of the general strengths and limitations of polymeric materials in solar-energy applications and provides a detailed discussion of the use of different polymers in specific devices.

3469–8/97/0567$15.50/0 © 1997 American Chemical Society

Properties of Polymers for Solar-Energy Devices

Bulk Properties of Polymers

For solar applications, the bulk properties of greatest concern include mechanical strength, toughness, permeability, UV stability, and optical properties. Each of these bulk properties is affected in different ways by solar irradiation, thermal and humidity cycles, and mechanical stress. Mechanical properties are important in allowing the polymer to support both long-term (e.g., creep resistance) and short-term (e.g., wind loads) design loads at high and low temperatures. Toughness and ductility are important in allowing the material to sustain thermal expansion or mechanical tension without brittle failure. Permeability affects the movement of plasticizers, UV stabilizers, and degradation products out of the polymer and the movement of permeative gases and moisture through the polymer. Finally, the optical clarity and absorbency of polymers can be influenced by changes in polymer structure and morphology.

The bulk mechanical properties of most commercially important polymers were studied extensively (4). Changes in the stiffness or modulus of a polymer in response to an applied load or thermal stress can be explained by a number of viscoelastic models (5, 6). The effects of temperature over extended periods of time, or temperature cycles on stiffness, can be predicted with short-term experiments and the application of the principle of time–temperature superposition (TTSP). For example, an increase in temperature results in a decrease in stiffness, leading to distortion or creep in a polymer under applied load. More recent TTSP models have been expanded to include the effects of humidity and humidity cycles on stiffness. Failure due to cyclic loading or fatigue may be caused by growth of a crack tip or concentration of stress in an imperfection in the polymer.

UV irradiation of polymers causes two primary types of reactions: chain scission and cross-linking (7). With sufficient exposure to either long-term moderate irradiation or short-term intensive irradiation, both of these reactions eventually result in changes in the mechanical properties of the polymers. Chain-scission reactions lead to a reduction in the molecular weight of the polymer, eventually causing a decrease in mechanical properties. The ultimate mechanical properties, such as elongation to break, are usually very sensitive to a reduction in molecular weight. Cross-linking reactions increase polymer stiffness and decrease the elongation to break. High levels of cross-linking lead to embrittlement of the polymer, and this embrittlement is particularly important for rubbery polymers used as adhesives or sealants. Cross-linking of rubbery polymers can decrease the polymer ability to recover from cyclic stress.

The permeability of polymers is influenced by both the mobility of the polymer chains and the strength of the interaction between the polymer and the penetrant. The mobility of the polymer chains can be affected by prolonged exposure to environmental stresses. The rate of gas permeation increases with increasing temperature for glassy polymers, but the rate is less sensitive to temperature for rubbery polymers. Migration of plasticizers and stabilizers through glassy polymers is also accelerated by increases in temperature. A buildup of plasticizers or stabilizers on the polymer surface is often manifested in the appearance of a hazy layer or *bloom*. For hydrophilic polymers, an increase in the relative humidity can result in increased permeability of both water vapor and other permeative gases.

Chain scission or cross-linking caused by UV irradiation can have a substantial influence on permeability. Chain scission increases the free volume of the polymer, which can lead to an increase in permeability. Cross-linking restricts the mobility of the polymer chains, which generally leads to a decrease in permeability. Oxidation of the polymer backbone with the formation of relatively polar groups can also increase the water sorption and water permeation properties of polymers. For example, the formation of hydroxyl and carbonyl groups in oxidized polyethylene can lead to an increase in water sorption (8).

Bulk optical properties of polymers typically are not influenced by changes in temperature or cyclic stress. However, if these two environmental factors can cause a change in polymer morphology (i.e., crystallization), there can be a dramatic decrease in the clarity of the polymer. The simultaneous application of stress and temperature can induce crystallization and the growth of spherulitic structures, which scatter light and decrease clarity.

UV irradiation will typically have the greatest influence on a polymer's clarity or absorption spectrum. Chain-scission reactions can lead to the formation of chromophores, which can drastically increase absorption in either the UV or visible range. For example, polydimethylsiloxane and butyl rubbers both show a decrease in optical properties when exposed to high levels of UV irradiation (9).

Surface Properties of Polymers

Environmental stresses can have a dramatic influence on the surface properties of polymers, including the adhesive interface between two materials. The formation of a durable bond between dissimilar materials that are expected to withstand exposure to solar irradiation, thermal and humidity cycles, and mechanical stress is a particularly difficult problem. Two types of bonds are of particular interest in solar applications: the bond between a polymer support and a metallic overlay, and the bond between a polymer (or an inorganic material) and a polymeric adhesive or sealant. Each of these bonds presents a unique problem for the selection of appropriate materials.

Selection of a polymeric material for use as a support for a metallic overlay is a complex problem. First, the bulk mechanical properties of the polymer must be adequate. Then, the surface properties of the polymer must be considered. The surface and bulk properties of polymers may differ for several reasons. Surfaces are in a higher energy state than the polymer bulk. This relatively higher energy provides a driving force for the polymer system to reorganize itself and locate low-energy components at the surface. The phenomenon is particularly important for rubbery block copolymers and polymer systems with high loadings of plasticizers or stabilizers. Metals are well-known to adhere much more strongly to a properly cleaned polymer surface than an untreated surface (*10*).

Processing techniques used to prepare the material (i.e., solvent casting or thermal processing) can also influence the surface composition of the polymer. Differences in the thermal expansion coefficients and humidity responses of two materials being bonded can generate significant internal stresses that the adhesive layer must resist. Interactions of water and metal oxide layers can cause the formation of a weak interface even after the adhesive bond is formed. The detrimental effects of UV irradiation can be accentuated at the interface between an organic polymer and a metallic surface. The metallic surface has the potential to act as a catalyst for degradation of the polymer especially if certain metal cations are formed and diffuse into the polymer. Degradation in the first one or two molecular layers of the polymer surface can rapidly lead to failure of the joint.

Metallized Polymer Reflectors

The economic viability of solar concentrator technologies depends on the availability of inexpensive and high-performance reflector materials capable of a lifetime of extended service in outdoor environments. In the mid 1980s, the state-of-the-art solar reflectors were back-silver-coated low-iron glass. However, glass mirrors were considered to be too heavy, too fragile (requiring special handling), and too expensive for solar concentrator applications.

Metallized polymeric reflectors are significantly lighter in weight and potentially much less expensive than conventional glass mirrors. Additionally, metallized polymers offer greater system design flexibility. In particular, polymer reflectors allow the use of a stretched-membrane concept (an innovative, low-cost concentrator design) for concentrating collectors. The structure of these reflectors is shown in Figure 1. The top polymer film is intended to protect the metal reflective layer from adverse environmental factors such as moisture, temperature, ambient pollutants, airborne particulates, hail, and harmful (typically UV) radiation. Special additives (e.g., UV absorbers and antioxidants) are typically incorporated into the bulk polymer to improve the durability of the overall construction. An adhesive layer allows lamination to a substrate material for structural integrity. Abrasion-resistant top coatings have been proposed to allow ease of cleaning without scratching the polymer surface. Adhesion-promoting layers between the polymer film and the metal layer have been tried; such interlayers can also provide barrier capability. The application of back protective layers has also resulted in increased corrosion resistance.

Candidate Polymers

A number of candidate polymers have been evaluated for use as metallized reflector substrate and superstrate materials, including acrylic, silicone, fluoropolymer, polyacrylonitrile (PAN), polycarbonate (PC), and polyester. The general structures of these common polymers are shown in references 11 and 12. Both silver and aluminum reflective layers have been consid-

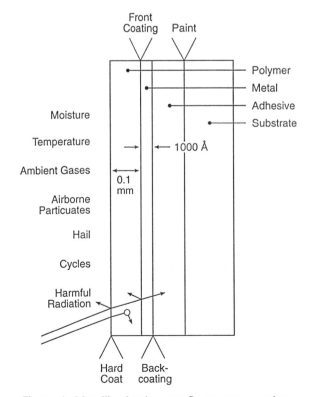

Figure 1. Metallized polymer reflector construction.

ered. On the basis of accelerated screening tests, acrylic (primarily PMMA) was determined to be the most promising polymer candidate performer. Because silver metallization of polymer films provides the highest level of reflectance, the main thrust of recent development efforts has focused on silvered PMMA reflector materials.

Optical Measurements

Optical performance of metallized polymers is characterized in terms of specular reflectance—the degree to which a mirror is capable of transferring directed radiation to a target receiver surface. Microroughness of a mirror surface or crazing of protective top coats can result in scattering (loss) of light outside a specified acceptance angle, defined as the half angle (ψ) subtended by the receiver as viewed from the reflector surface. At each wavelength (λ), the level of specular reflectance (ρ_s) is a function of both the hemispherical reflectance ($\rho_{2\pi}$) and the half-width (σ) of the (assumed Gaussian) distribution of scattered light, as defined in eq 1.

$$\rho_s(\psi, \lambda) = \rho_{2\pi}(\lambda)[1 - e^{-[\psi/\sigma(\lambda)]^2/2}] \qquad (1)$$

Measurements of hemispherical reflectance throughout the solar spectrum (250–2500 nm) are made with a dual-beam UV–vis–near-IR spectrometer with an integrating sphere attachment. The value of ρ_s is routinely measured at wavelengths of interest (selected by bandpass filtered light) as a function of collection angle by using a specially designed specular reflectometer (13).

Exposure Testing

The optical durability of a mirror is its ability to resist losing specular reflectance during real-world service (14). Such a loss can result from corrosion of the reflective layer and subsequent loss in $\rho_{2\pi}$, from increased scattering and widening of the reflected beam (σ) caused by soiling or abrasion of the mirror surface, or a variety of other mechanisms. To screen candidate reflectors on a comparative basis, accelerated weathering chambers are used in which various stress factors can be isolated and emphasized under controlled laboratory conditions. Real-time outdoor testing is also carried out at a number of exposure sites throughout the United States. As new, advanced reflector materials demonstrate increased lifetimes, correlations between outdoor and accelerated exposures are needed to predict service lifetimes and to avoid having to wait too long to determine material lifetimes.

Development Efforts at the National Renewable Energy Laboratory

On the basis of the results of screening experiments, PMMA was selected as the most promising candidate polymer film for solar-reflector applications. PMMA film itself exhibits high optical quality and is inherently stable against outdoor weathering. However, unmodified PMMA provides inadequate protection against silver corrosion at the silver–PMMA interface during real-world exposure (11). Consequently, an extensive effort was made at the National Renewable Energy Laboratory (NREL) to stabilize silver–PMMA mirrors.

Test mirrors were made by using Corning 7809 glass as a substrate onto which a thin, reflective silver film was deposited, and polymeric films were then cast from solution onto the silver. The reflective layer consisted mainly of silver either deposited onto glass substrates by the wet-chemical process or sputtered onto glass substrates by vacuum metallization as a precursor to casting polymeric films.

When polymer formulations are altered to improve durability, the solution-casting technique is convenient to use to make the polymeric films. In this case, the polymer is coated onto the silver in contrast to the use of extruded films where silver is deposited onto the polymer. Neidlinger and co-workers studied a series of additives (UV screens, quenchers, antioxidants, and antipermeants) that can enhance the effectiveness of PMMA. Commercially available additives (Table I) were tested, and UV screens were emphasized. Additional polymeric stabilizers were synthesized specifically to protect silvered polymers (List I).

Polymerizable stabilizers were copolymerized with MMA to increase not only the light stability and permanence but also the compatibility of the stabilizer with the substrate. Attention was focused on the synthesis and characterization of polymeric UV-absorbing stabilizers of the 2-hydroxybenzophenone and 2-hydroxy phenylbenzotriazole classes, which are methacrylic, vinyl, or isopropenyl derivatives. Other work concentrated on the synthesis and polymerization of 2,2,6,6-tetramethylpiperidine derivatives, which can act as non-UV-absorbing stabilizers.

For example, 3,5-[di(2H-benzotriazole-2-yl)]-2,4-dihydroxyacetophenone (DBAP) was synthesized in good yield. DBAP was transformed successfully in four steps to 2-(2,4-dihydroxy-5-vinylphenyl)-1,3-2H-dibenzotriazole (DBDH-5V) and to 2-(2,4-dihydroxy-5-isopropenylphenyl)-1,3-2H-dibenzotriazole (DBDH-5P). The synthesis essentially followed the procedure of Vogl and his associates (15). However, some reaction solvents and conditions were modified to improve yields.

DBDH-5V and DBDH-5P were copolymerized with MMA in the presence of azobisisobutyronitrile

Table I. Commercially Available Stabilizers

Product	Supplier
Tinuvin P	
2-(2'-hydroxy-5'-methylphenyl)benzotriazole	Ciba-Geigy
Uvinul 408	
2-hydroxy-4-*n*-octoxybenzophenone	BASF-Wyandotte
4-methacryloxy-2-hydroxybenzophenone	Aldrich
National Starch 78–6121	
developmental copolymer of a 2-hydroxybenzophenone	National Starch
Tinuvin 292	
bis(1,2,2,6,6 pentamethyl-4-piperidinyl) sebacate	Ciba-Geigy
Irganox 1010	
tetrakis[methylene (3,5-di-*tert*-butyl-4-hydroxyhydrocinnamate]	
urethane	Ciba-Geigy
Goodrite UV 3034	
1,1'-(1,2-ethanedioyl)bis(3,3,5,5-tetramethylpiperazinone)	Goodrich
Goodrite UV 3125	
3,5-di-*tert*-butyl-5-hydroxyhydrocinnamic acid triester with	
1,3,5-tris(2-hydroxyethyl)-*S*-triazine-2,4,6-(1*H*,3*H*,5*H*)-trione	Goodrich
Irgastab 2002	
nickel bis[*O*-ethyl(3,5 di-*tert*-butyl-4-hydroxybenzyl)]phosphonate	Ciba-Geigy
Chimassorb 944	
polymeric hindered amine light stabilizer (HALS)	Ciba-Geigy
Spinuvex A36	
polymeric HALS	Borg-Warner

List I. Polymeric Stabilizers Used at NREL

p-MMA-*co*-MHB
Copolymers of methyl methacrylate and 4-methacryloxy-2-hydroxybenzophenone

p-MMA-*co*-2H5V
Copolymers of methyl methacrylate and 2-hydroxy-5-vinylphenylbenzotriazole

p-MMA-*co*-DBDH-5V
Copolymers of methyl methacrylate and 2-(2,4-dihydroxy-5-vinylphenyl)-1,3-2*H*-dibenzotriazole

p-MMA-*co*-DBDH-5P
Copolymers of methyl methacrylate and 2-(2,4-dihydroxy-5-isopropenylphenyl)-1,3-2*H*-dibenzotriazole

p-MMA-*co*-BDHM
Copolymers of methyl methacrylate and 2-(2-hydroxy-4-methacryloxyphenyl)-2*H*-benzotriazole

p-MMA-*co*-MAP
Copolymers of methyl methacrylate and 4-methacrylamido-2,2,6,6-tetramethylpiperidine

p-MMA-*co*-MAP-E
Copolymers of methyl methacrylate and *N*-methacryloxyethyl-*N*'-(2,2,6,6-tetramethyl-4-piperidyl)-urea

(AIBN) as initiator in a mixture of toluene and dimethylacetamide as solvents. Yields above 90% were obtained for the *p*-MMA-*co*-DBDH-5P series, and yields were 72–76% for the *p*-MMA-*co*-DBDH-5V series. Details of the syntheses are available (*16–19*).

Weathering, the term used to describe the deterioration of polymer properties in an outdoor environment, involves a complex interaction of physical and chemical processes; photooxidation is recognized as being one of the most important of these processes. The characteristics of photodegradation permit application of stabilization methods to prevent degradation by involving photon absorption, transfer of energy, and antioxidant mechanisms. For the protection of solar reflectors, however, the stabilization of the polymer glazing is just one facet of the total problem. Other aspects include the stabilization of the polymer–silver interface, as well as the stabilization of a potential polymeric backing if no precautions are taken to avoid the UV transparency of the silver layer. Because UV absorbers decrease the amount of solar light reflected from a mirror, the necessity of the presence of such additives to protect and extend the lifetime of the reflector is of high practical interest.

Emphasis at first was on using stabilizers in solution-cast thin glazings (1–20 µm). Tests performed with these thin glazings were meant to guide later formulations for thicker extruded films (ca. 90 μm). Accelerated weathering tests demonstrated an enhanced degradation of unstabilized silver–PMMA mirrors, whereas stabilizers added to PMMA impeded the degradation. Researchers identified a benzotriazole derivative as a leading stabilizer for silver–PMMA mirrors. Three other stabilizer combinations were comparable in performance in accelerated tests (*16–19*).

Physical aspects of the silver surface (bulk plasmon and surface plasmon absorption) and silver surface degradation phenomena (photodarkening and corrosion) interfere with the direct evaluation of thin polymer glazings on silver mirrors (*18*). Therefore, to evaluate changes in the polymer, transparent non-metallized PMMA films were studied.

Free-standing PMMA films were prepared by casting toluene solutions (10 wt% solids) of polymer–stabilizer (98.5 wt%/1.5 wt%) combinations onto glass substrates. PMMA (from Polysciences) was used in most cases as received without further purification. Changes in specular transmittance are one measure of degradation of the polymer films. Accelerated laboratory tests showed that unstabilized films lose transmittance more rapidly than films with stabilizers (*19*).

PMMA, if not unique, is unusual among polyalkylacrylates in that it is not cross-linked during photolysis of its films in air. PMMA is also unusual in that the number of chain scissions per molecule tends to be linear with irradiation time. The effect of stabilizers on chain scission was studied by gel permeation chromatography (GPC) after irradiation. Values of molecular weight showed a rapid decrease in the case of PMMA without stabilizer. This molecular weight decrease is considerably retarded through addition of stabilizers. Exact molecular weight changes for the system stabilized with polymeric additives are difficult to elucidate from the GPC traces because of the superposition of guest and host traces. However, no noticeable degradation for the stabilized films occurs during accelerated weathering.

The permanence of stabilizers in the PMMA films is another important factor for preventing long-term degradation of silver–PMMA mirrors and PMMA films. Despite the high inherent UV stability of the selected stabilizers, their concentrations do fall steadily during UV irradiation. The change in the absorption maximum of the UV-screening additives in thinner films (13–25 μm) was chosen to monitor this factor. Polymeric stabilizers decompose, in general, in a fashion similar to their low molecular weight analogs, and this process is evident through losses in their characteristic absorption bands. However, no noticeable molecular weight changes of the polymeric stabilizers can be observed during their consumption. This result may indicate that only cleavage and/or

modification of the pendant stabilizer groups are involved in the consumption process.

To demonstrate the expected increase in physical permanence of the polymeric stabilizers in contrast to their low molecular weight analogs, selected accelerated extraction experiments were performed. In the Soxhlet-extraction mode, the samples are washed repetitively with a constant volume of warm water (ca. 75 °C) for 1 week. NREL tests did not indicate any polymeric stabilizer leaching in the samples tested. These results are in contrast to a noticeable loss of low molecular weight stabilizers incorporated in other films. Similar observations were made in a more severe extraction test in ethanol at 75 °C for 24 h. UV spectra of samples taken from the supernatant ethanol revealed low molecular weight stabilizers, whereas none of the NREL polymeric stabilizers leached out of the films under these extreme conditions.

Delamination Failure

During outdoor service, silvered-PMMA mirrors have demonstrated an unexpected and sporadic failure mode termed delamination or *tunneling*. During tunneling the polymer delaminates from the silver metal in a characteristic pattern. Tunneling usually occurs when the mirror is exposed to high humidity, and it almost always begins at the edges of the polymer. Tunnel failures are unpredictable. If water is allowed to accumulate on a large mirror, tunneling may begin within a few days. Typically, a pattern of tunnels propagates rapidly over the complete mirror surface, rendering the material unusable.

Several promising solutions that substantially reduce delamination failures have been identified and demonstrated. These include (a) use of Tedlar edge tape, (b) thermal treatment of laminated reflector–substrate constructions, and (c) application of silver to the polymer film through an alternative deposition process. Procedures a and b offer readily available engineering solutions to the delamination problem. Solutions b and c provide tunnel resistance over the entire surface of the reflector material (including edges).

Non-PMMA Candidates

During the early phase of the program, a number of candidate non-PMMA polymer reflectors were evaluated. Fourier transform IR (FTIR) spectroscopy was employed to investigate a number of polymer film–metal substrate systems (*20*). Non-PMMA polymers included bisphenol-A PC, polyvinylidene fluoride (PVF), atactic polypropylene, and PAN. A controlled environment exposure chamber allowed FTIR reflection–absorption measurements to monitor functional changes (indicative of degradation) within polymers as a function of in situ weathering. Results from these studies supported other work that suggested that metallized PMMA mirrors were the most promising candidates for solar applications.

Although dramatic improvements have been achieved in the optical performance and lifetime of silvered PMMA reflector materials, concerns exist with regard to the cost of these materials. Consequently, many new alternative advanced solar reflector materials have been identified and are being tested. One of the most promising ways to reduce the cost of solar mirrors is to metallize an inexpensive substrate material such as a poly(ethylene terephthalate) (PET) thermoplastic co-polyester film, and then overcoat the reflector with an abrasion-resistant, durable, protective top layer. Some reflectors having this generic construction have demonstrated promising results in accelerated durability tests at NREL. Such materials may ultimately be capable of achieving the cost and performance goals of the program. Candidate top layers can be either organic (such as organosilicones, polyurethanes, and acrylics) or inorganic (such as Si_3N_4, diamond-like carbon, SiO_x, Al_2O_3, and other oxides). Organic–inorganic composite coatings also have been suggested.

NREL explored a number of top coat options. Compared with other protective coatings having Si–O bonds, experimental mirrors provided by industry have performed remarkably well in accelerated weatherometer (WOM) exposure. Polyurethane also was evaluated as a protective top coat for silver reflector materials. Such painted mirrors (mirrors with organic coatings applied in a painted fashion, such as solution casting or dip coating) are adaptable to current production techniques and can be inexpensive, but their optical durability has not been demonstrated. Silver protected by polyurethane paint having UV absorber additives generally has not maintained hemispherical reflectance in accelerated WOM testing.

Collaborative Efforts with Industry

Early tests of the reactivity of polymers with metal reflective materials were carried out with NREL support by SRI International. Polymer–metal combinations were subjected to accelerated WOM exposure and ranked on the basis of interfacial stability. PMMA, PVF, and PET showed promise as protective coatings for solar mirrors (21).

In parallel with the development of silvered PMMA solar reflectors, PAN has been extensively explored as well. Both PAN and PMMA are commonly used in industry, and PAN ranked close to PMMA in early screening tests in its suitability as a baseline polymer for this application. Researchers at the University of Denver, in collaboration with NREL, studied the photodegradation and photostabilization of silver-backed PAN as well as PC, PVF, and PET (22).

During the mid 1980s, a collaborative effort was initiated between NREL and the 3M Company (a leading manufacturer of PMMA films) to develop a commercial metallized PMMA film for solar applications. Before 1984, improvements in specular reflectance were emphasized. A highly specular, biaxially oriented, extruded PMMA film formed the basis for excellent optical properties. Use of a silvered version of the 3M aluminum reflector material resulted in further improvement in reflectance because of the higher reflectance of silver relative to aluminum.

Between 1984 and 1991, improvements in optical durability were stressed. On the basis of results from NREL experiments, UV absorber type and quantity were optimized. A commercial product that survived for approximately 1 year during outdoor exposure in Phoenix, AZ (ECP-300A in Figure 2) was improved further by the elimination of additives within the bulk polymer that may have reduced the longevity of the material. This improved material (ECP-305) exhibited dramatically increased corrosion resistance during outdoor exposure in Arizona (Figure 2) and Florida. Accelerated exposure tests at NREL demonstrated greater improvements in optical durability of silvered PMMA film reflectors having protective back coatings of copper (Figure 3). This prototype construction also lead to the latest commercial silvered PMMA reflector from the 3M company (ECP-3054).

Recently, a number of collaborative, cost-shared research and development efforts were initiated with industrial partners (23). These efforts include a directly deposited reflective surface, a metal–polymer multilayer stack construction, a metallized fluoropolymer reflector, and an all-polymeric solar reflector material.

Science Applications International Corporation investigated directly deposited solar reflector materials. Polymeric levelizing layers and hard top coats, including epoxies and silicon-based materials, were tested.

The Battelle Memorial Institute Pacific Northwest Laboratory explored the encapsulation of silver reflective layers by vacuum flash evaporation of monomer material to prevent corrosion. Deposition of the silver and polymer multilayers can be done without breaking vacuum. This process may yield extremely high line speeds and low production costs.

Industrial Solar Technology, Inc., worked toward developing a silvered Teflon solar reflector material. Although the unweathered specular reflectance of metallized fluoropolymer films is generally poor, progress has been made in understanding this problem and in identifying ways to improve specularity.

Dow Chemical Company is developing initiated development of an all-polymeric solar reflector mate-

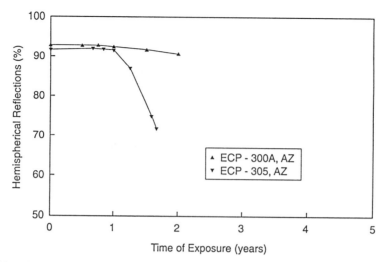

Figure 2. Solar-weighted hemispherical reflectance of silvered PMMA reflectors on painted aluminum substrates as a function of outdoor exposure in Phoenix, AZ.

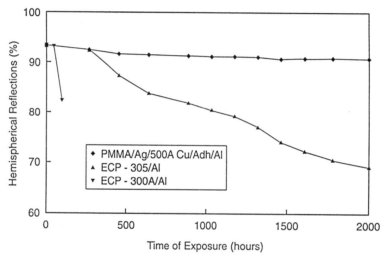

Figure 3. Solar-weighted hemispherical reflectance of silvered PMMA reflectors as a function of accelerated exposure in a solar simulator chamber (80 °C, 75% RH, 50× light intensity).

rial. This approach uses alternating coextruded layers of low-cost, commercially available transparent thermoplastics. Abrasion-resistant top coatings with UV protection are also used. Efficient, broad-band solar reflectors are envisioned that are fabricated from a high-speed, low-cost technique that was commercially demonstrated.

Photovoltaic Encapsulants

In the development of flat-plate photovoltaic (PV) modules for large scale, long-term terrestrial applications, a critical component in addition to the solar cells is the encapsulation material that provides structural support, optical coupling, electrical isolation, physical isolation–protection, and thermal conduction for the solar cell assembly (24–27). Although a few different encapsulation methods were conceived, the most

common method today employs polymer encapsulants for a variety of PV modules that range from the most technologically mature monocrystalline Si (c-Si) to polycrystalline Si (pc-Si), and to the recently emerging thin-film modules of amorphous Si (a-Si), polycrystalline $CuInSe_2$ (CIS), and CdTe. Although the exact configuration of the encapsulation scheme may vary depending on the kind and design of PV modules, the need for a suitable insulating encapsulant remains paramount.

To be economically and practically useful, the encapsulant materials have to meet the requirements of low cost, good processability, high optical transmission, high dielectric constant, low water absorptivity and permeability, high resistance to UV degradation and thermal oxidation, good adhesion, mechanical strength, and chemical inertness. The encapsulant materials are usually made of a visibly transparent polymer to provide good optical transmission in a prescribed spectral region. The transparent copolymer of ethylene-vinyl acetate (EVA), formulated with stabilizers, was developed by the Jet Propulsion Laboratory (JPL) and Springborn Laboratories in the early 1980s (*24–26*) and is used today predominantly by the PV industry. An earlier encapsulant, poly(vinyl butyral) (PVB), is no longer used.

Researchers at NREL conducted a series of systematic investigations aimed at understanding the stability issues of the EVA encapsulants. These studies include examinations of the UV-excitable chromophores inherent in EVA, effects of thermal processing, degradation mechanisms, fluorescence analysis of the structural changes of chromophores by field and simulated degradation, photodecomposition of Cyasorb UV 531 and its stabilization effectiveness by antioxidants, efficiency loss of solar cells due to EVA discoloration, destructive analysis of weathered PV modules, and efforts on the modification of current EVA formulations. The results are summarized subsequently.

EVA Copolymer Encapsulants and Formulations

EVA is a random copolymer of ethylene with 33% (by weight) vinyl acetate. One major source of the raw copolymer is Elvax 150 manufactured by DuPont. It was selected by JPL and Springborn Laboratories from a list of several polymer pottant candidates, including ethylene-methyl acrylate copolymer, poly(*n*-butyl acrylate), and aliphatic poly(ether urethane), on the basis of reasonably low cost ($0.60/lb to $0.70/lb in 1980), elas-

tomeric (thermoplastic) properties, film extrudability, transparency, weatherability, and good module processability by dry-film lamination methods (*25*).

Two widely used EVA films, designated by Springborn Laboratories as A9918 and 15295 (*24–26*), are formulated with three stabilizers and two curing agents. The three stabilizers are Cyasorb UV 531 (2-hydroxy-4-*n*-octyloxybenzophenone), Tinuvin 770 [bis(2,2,6,6-tetramethyl-4-piperidinyl) sebacate], and Naugard P [tri(monononylphenyl)phosphite]. The curing agents are Lupersol 101 (2,5-bis(*tert*-butyldioxy)-2,5-dimethylhexane) for EVA A9918 and Lupersol TBEC [*O,O-t*-butyl-*O*-(2-ethylhexyl)monoperoxycarbonate] for EVA 15295. The two EVA films are made by first compounding the Elvax 150 pellets with the ingredients, and then extruding as *uncured* sheets at a die temperature well below 120 °C (typically, ~100 °C) to avoid initiating any premature curing reaction. The typical thickness of an extruded uncured EVA film is ~0.46 mm (18 mil).

These two EVA formulations have most commonly been used by the PV industry since the 1980s to the present for the encapsulation of c-Si and pc-Si modules, although there are other experimental *advanced* formulations such as EVA 16870 and 18170 developed by Springborn Laboratories (*24*). A simplified encapsulation scheme using a double vacuum-bag laminator is shown in Figure 4. EVA is also being used in the encapsulation of thin film modules (*28*). For example, Siemens Solar uses EVA for lamination of thin-film CIS and CIS–a-Si solar cells between two glass plates (*28a*), and Advanced Photovoltaic Systems also uses EVA to laminate a-Si thin-film solar cells between two glass plates (*28c*).

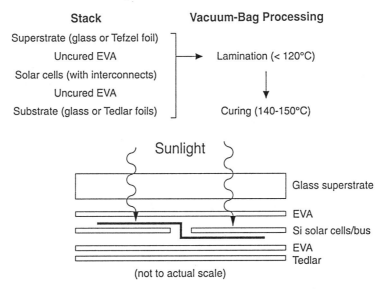

Figure 4. A typical processing scheme and conditions for encapsulation of solar cells with the uncured EVA films (top) and a simplified diagram showing the configuration commonly featured in monocrystalline and polycrystalline Si modules (bottom).

Polymer Stabilization and EVA Formulations

Degradation of polymer materials via either thermo-chemical or photothermochemical pathways, or both, is a topic that has been widely studied for many years. There are many sources that lead to polymer degradation. For example, chain-scission and cross-linking reactions are known to proceed in poly(vinyl acetate) via Norrish type I and II or radical recombination mechanisms in the presence of carbonyl groups during illumination by UV light. The α,β-unsaturated carbonyl groups, –C=C–CO– or –C=C–CO–C=C– [or (–C=C)$_2$–CO–] in enone or enal form, frequently found in commercial polyolefins, can be UV-excited leading to photodegradation of polyolefins (29). Short polyenes and metal ions are also known to facilitate or catalyze the degradation of polymers (29, 30), and peroxides on the polymeric chains produced from the peroxide-type curing agent may assist the oxidation of polymers. All these degrading parameters can act separately or synergistically and result in a reduced weatherability of polymer materials; artificial stabilization is therefore needed to ensure a prolonged service life.

In general, polymer photooxidation involves four basic mechanistic steps: chain initiation of free radicals, chain propagation, chain branching, and chain termination. To effectively stabilize a polymer against photooxidation, the first three chaining steps should be reduced or blocked by the use of adequate physical or chemical means. Formulation is a chemical approach to stabilize a polymer with the addition of various antioxidants. The stabilization effectiveness depends on the adequacy of the formulation and the stability of the individual antioxidant. For EVA, the functions of the stabilizers used in the formulations are

- UV absorption and quenching of the excited states by Cyasorb UV 531
- free-radical scavenging by Tinuvin 770
- hydroperoxide decomposition by Naugard P

The adequacy of the EVA formulation is questionable, however, and JPL noted several potential limitations that range from the rapid physical loss and limited storage life of Lupersol 101 to the low molecular weight and rapid physical depletion of Tinuvin 770 and Cyasorb UV 531 leading to a gradual loss of EVA weathering protection (25). Photothermal degradation of unstabilized and stabilized EVA was reported previously (31, 32), including some examples by JPL researchers in the 1980s (32). Despite these problems, the EVA encapsulants formulated in this way have continued to be widely used by the PV industry for more than a decade.

EVA Raw Material and Extruded Uncured Films

An important factor that may determine or even dictate the photostability of a polymer material is the presence of numerous chromophores and the excitation of these chromophores by UV light leading to the degradation of the polymer or destruction or depletion of embedded stabilizers. Therefore, verifying the existence and determining the basic structure of the UV-excitable chromophores in the EVA films are important. From the results of UV-vis absorption measurements and fluorescence analysis, Pern and Czanderna (33) showed that the raw material, Elvax 150 pellets, is contaminated with a photosensitizer that decomposes during UV irradiation, and that the extruded uncured EVA films have a mixture of short α,β-unsaturated carbonyl groups that are excitable by UV light. The presence of Cyasorb UV 531 can effectively shield these short chromophores and quench the excitation. The uncured EVA films have a thermal stability lower than the cured EVA films as evaluated by thermal gravimetric analysis by Pern and Czanderna (33a). Yellowing of the uncured EVA films during storage in the dark at ambient temperature was observed but appears to be a noncritical problem, because the yellow color disappears when the aged films are exposed to white light or heated at curing temperature (33b).

Thermal Encapsulation Process

The typical module lamination-curing process in a double vacuum-bag laminator involves the following steps (26).

1. Both chambers are under vacuum to remove air after the module stack is loaded.
2. The entire assembly of the module stack is heated to melt EVA.
3. At about 120 °C, while the bottom chamber remains under vacuum all the time, the top chamber is gradually released from vacuum to 1 atm over 5–8 min to initiate the lamination.
4. The temperature is increased to and held at 145–150 °C for the curing process.
5. After the desired curing period [which depends on which EVA formulation is used (25)], the chamber temperature is gradually decreased and the module is removed at less than 50 °C.

Effects of Lamination, Curing, and Curing Agent

To investigate the effects of lamination and curing on EVA chromophores, Pern and Glick (34) used fluorescence analysis to study samples of uncured EVA films that were subsequently laminated and cured between two quartz slides. Because the extruded uncured film sheets were produced from an extruder at <120 °C, little change was observed for the uncured films being laminated between quartz slides at ~115 °C. The curing step at ~145 °C produces new UV-excitable chromophores in the EVA, however. The concentration of the new chromophores is significantly greater for the slow-cured EVA A9918 formulation than for the fast-cured EVA 15295 formulation, primarily due to the typically longer curing time for the EVA A9918 formulation (*24–26, 34*). Pern and Glick (*34*) showed that on a 16-cm long cured A9918 film strip the concentration of chromophores excitable by ≤322 nm light varied from ~20% to ~37%, a result clearly indicating that a significant nonuniformity was produced by the curing process. A likely consequence will be nonuniform EVA photothermal degradation in encapsulated modules.

The generation of chromophores by curing increases as the curing time increases. Whereas Lupersol peroxide is the main oxidizing reagent that promotes the generation of chromophores, the oxidative strength of Lupersol TBEC is greater than Lupersol 101 and the presence of air during curing is not as critical (*34*). The slow-curing Lupersol 101 increases the concentration of the chromophores as the curing temperature increases from 130 °C to 175 °C for a fixed curing period of 30 min, but the fast-curing Lupersol TBEC produces almost no net increase in the chromophores in the same temperature range for a curing time of 15 min. Therefore, to achieve a desired gel content, the process can be completed in less time with Lupersol TBEC than with Lupersol 101. The consequence of generating large amounts of UV-excitable chromophores by curing with Lupersol 101 in A9918 is their incomplete shielding by the UV absorber, Cyasorb UV 531. Accordingly, protection of the cured EVA A9918 from UV degradation is well below what is expected from its formulation, and EVA A9918 is expected to degrade and discolor earlier and faster than EVA 15295. A detailed comparison of EVA A9918 and 15295 is available (*34*).

Loss of Solar Cell Efficiency by EVA Encapsulation

A small percentage efficiency loss (ca. 1–2%) is expected for encapsulated modules because of the transmission losses from light absorption by Cyasorb UV 531 (below 350 nm), cured EVA, and low-iron plate glass (superstrate). Efficiency loss of c-Si and pc-Si solar cells due to direct physical contact with the EVA laminate has not been reported. However, an ~15% recoverable power loss resulted from EVA encapsulation of CdS–CIS modules, and an irreversible loss also

occurred in CdS–CdTe modules (*28*). Further studies are needed to understand these phenomena. The effect of EVA encapsulation on the aluminum back contact of a-Si modules has not been reported to date.

Long-Term Reliability of EVA

Encapsulant stability is required for PV modules deployed in the field where a service lifetime of 20–30 years and a consistent power output are needed. In reality, after a number of years of weathering, some c-Si and pc-Si modules show a reduced power output degradation as the EVA discolors from yellow to dark brown, depending on location and operating temperature of the module. EVA discoloration has been reported for modules deployed over 3–12 years in a variety of regions ranging from Florida and the southwestern United States, to Australia, countries in Africa, and Israel, where hot, dry and hot, humid climates prevail (*35*). Therefore, this discoloration is no longer an isolated incident, as some first believed in the case of the power plant in Carrisa, California, where EVA browning was accelerated by concentrating mirrors (*36*). Yellowing and browning of PVB has also occurred on some aged rooftop modules. The discoloration of EVA and PVB apparently results primarily from weathering-induced, photothermal degradation of the polymers.

Mechanisms of EVA Degradation

Thermal and photothermal stability of unstabilized and stabilized EVA was studied previously by several groups (*24, 31, 32*). Pern and Czanderna (*35*) also investigated the general degradation mechanisms of the stabilized, cured EVA through a series of simulated degradation experiments. Their results indicate that the EVA discoloration results from the formation of polyconjugations by multistep deacetylation and that the acetic acid produced by thermal and photothermal decomposition exhibits an autocatalytic effect on the EVA yellowing. Pern and Czanderna (*33a*) also reported that the UV absorber, Cyasorb UV 531, is gradually depleted in yellow to brown EVA in field-deployed modules. As the color darkens, depletion of the Cyasorb increases. The gel content increases in the exposed EVA and is greater in the yellowed or browned regions than in the clear portions around the cell edges. Acetic acid and other volatile organic components are present in the yellowed and browned modules. Any losses in Tinuvin 770 and Naugard P by photodegradation or other mechanisms in the weathering-degraded EVA has yet to be determined. However, no apparent discoloration was observed for the films exposed in the presence of air due to photobleaching reactions that involves the oxidative breakdown of the polyconjugations by both UV and visible light (*34*).

Stabilizer Effectiveness

The effectiveness of stabilizers used in the EVA formulations was recently investigated by Pern (*37*). The results are shown in Figure 5. The Cyasorb UV 531 dissolved in cyclohexane solutions or, when embedded in Elvax 150 thin films, photodecomposed rapidly into undetermined aromatic compounds. Addition of the hydroperoxide decomposer, Naugard P, only slightly decreases the Cyasorb photodecomposition. A greater stabilization is obtained when Tinuvin 770, a free-radical scavenger, is added. These results are seen for the cyclohexane solutions (Figure 5a) and Elvax 150 thin films (Figure 5b) separately exposed to a concentrated Xe light. In addition, all the Elvax 150 thin films that contained only Cyasorb or other kinds of UV absorbers in the study became cross-linked after exposure to the light. No cross-linking was observed when Tinuvin 770 was present in the thin films. Therefore, cross-linking in the Elvax 150 thin films or EVA encapsulant is primarily initiated by a mechanism involving photogenerated free radicals. In the absence of a free-radical scavenger, the free radicals attack and cause the photodecomposition of Cyasorb UV 531 or other UV absorbers.

Pern (*37*) also found that there is an important concentration dependence among Cyasorb UV 531, Tinuvin 770, and Naugard P when studied for thin films of Elvax 150. The study results in an improved photothermal stability of current EVA formulations when the concentrations of the three key stabilizers are properly balanced (*35*). Similar improvements can also be obtained by using other light stabilizers and antioxidants with Cyasorb UV 531 (*37*).

EVA Discoloration and Loss of Module Efficiency

Loss of module efficiency due to EVA browning and other electrical problems has resulted in about 47% loss between 1986 and 1990 in the annual energy output (or annual capacity factor) from the initially rated, 5.2-MW PV system of the Carrisa, California, power plant (*36*). EVA browning alone accounts for at least 10% power loss in the Carrisa modules (*36d*). A further study on the module performance revealed that performance degradation is highly nonuniform from module to module and that the *average* module power output is 35.9% below that of the initial rating. Mismatching between neighboring modules due to nonuniform EVA degradation contributes an additional power loss of 11.1% (*36c*). Optical characterization of the discolored EVA films taken from the Carrisa

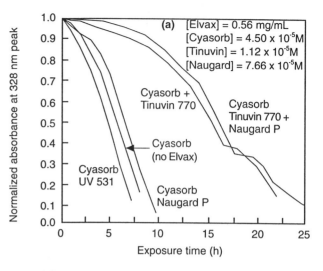

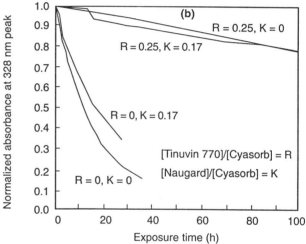

Figure 5. Photodecomposition of the UV absorber Cyasorb UV 531 and its stabilization by addition of Naugard P, Tinuvin 770, and both Naugard P and Tinuvin 770. The samples are (a) cyclohexane solutions and (b) Elvax 150 thin films. The thin film samples in (b) were prepared in cuvettes by solvent evaporation from the cyclohexane solutions in (a). The samples were exposed individually at a black panel temperature of 55–60 °C to an Oriel 1-KW Xe light beam that was condensed to ~2.5-cm diameter and filtered by an aqueous $CuSO_4$ solution filter and a 305-nm LP cut-off filter. Absorption spectra were measured with an HP-8452A UV-vis spectrophotometer.

modules was reported by Pern and co-workers (*33, 38*). The 10-kW system at Sede Boqer, Israel, had a 4–5% loss in current and a 6–7% loss in the maximum power for the mirror-enhanced, sun-tracking modules when the EVA browned from 1987 to 1992 (*39*).

There has been a lack of a clear correlation between browning of EVA and the degradation of electrical components such as at the gridlines–cells and gridlines–buslines interfaces. However, Pern (*37*) found in Carrisa modules that severe oxidation and corrosion occurred on the Pb–Sn alloy-coated Cu rib-

bon buslines that are in direct contact with browned EVA (Figure 6). The effects on the contact resistances at the gridlines–Si cell interfaces and gridlines–buslines interfaces have not been determined because of experimental difficulties. In short, EVA discoloration is only *partially* responsible for module efficiency loss,

and its association with other electrical problems has yet to be established experimentally. Figure 7 shows the decreasing spectral responses (absolute quantum efficiency) as the EVA film gradually turns brown in the solar cell exposed to a RS4 UV light source at 85 °C for 198 days (*40*).

EPMA Composition Analysis for the New and Exposed Buses

Sample Source PV Module No.	Solar Cell Location	Side of Bus-line	EVA Color	Pb-M	Sn-L	Cu-L	O-K
New bus				39.52	60.48		
Carrizo 1B	L9, R2	EVA	Light yellow	10.96	68.57	5.35	15.12
Carrizo 309259	L9, R2	EVA	Brown	41.13	5.80	13.27	39.80
Carrizo 309262	L9, R2	EVA	Dark brown	15.91	30.60	20.41	33.08
Carrizo 309262	L7, R3	EVA	Dark brown	46.56	4.62	8.08	40.74

Auger Depth Profile Analysis

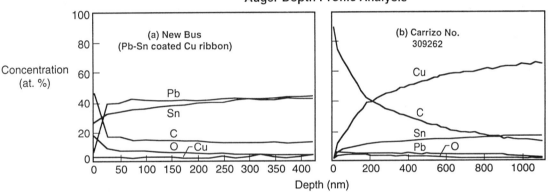

Figure 6. Results of electron probe microanalyzer (EPMA) compositional analysis and Auger depth profiling analysis for new and degraded buslines (Pb–Sn alloy-coated Cu ribbons) revealing the severe oxidation, loss, and interdiffusion of the metal components on the surfaces of the ribbons in direct contact with the yellowed and browned EVA films in three Carrisa modules.

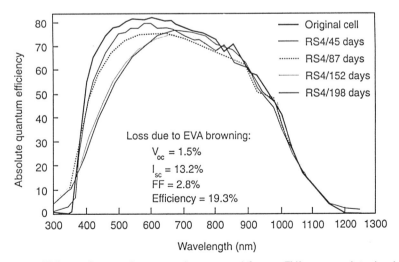

Figure 7. Absolute quantum efficiency (spectral response) measured for an EVA-encapsulated solar cell that was exposed in a chamber to a 3 × 100 W RS4 (UV) light source at 85 ± 2 °C for 198 days. The EVA film turned brown and reduced the light transmission, and reduced the final efficiency by 19.3%. No noticeable series resistance change was observed from the dark I–V measurements: V_{oc}, open-circuit voltage; I_{sc}, short-circuited current; FF, fill factor.

Electrochromic Devices

Considerable interest has developed about materials and devices that can be used for optical switching of large-scale glazings, especially for those based on electrochromism (41). In this section, we strongly emphasize the potential applications of electrochromism for buildings, but we will also mention other potentially interesting applications in vehicles, aerospace, eyeglasses, and information displays. Electrochromic (EC) devices provide the potential for controlling heat gain through windows while enhancing the use of daylight for the interior lighting of buildings. By a process of ion insertion into an EC material, the essential optical properties of the device are changed in a way so that they provide a net energy savings and can also reduce the peak power demand for air conditioning (including the size of the units).

EC devices are extremely desirable for four reasons. First, EC windows exhibit a memory and maintain a colored (or bleached) state for long periods; this technology is in contrast to liquid crystals. Second, the optical properties of EC windows can be changed at will, in contrast to devices based on thermochromism and photochromism (41). Third, EC glazings can be automated. Finally, they consume low amounts of power for switching and minimal amounts of power to maintain a given optical state (42).

Although EC windows have great potential, they also have a number of technical uncertainties and (potential) problems. A few prominent degradative effects have been observed for many different devices, including residual coloration, loss of dynamic range, and loss of stability of the device during long-term cyclic switching. The origins of these effects are only partially understood at this time. The economics of EC devices and their acceptance by those who design and construct buildings are also uncertain, which is part of the reason NREL scientists have proposed using EC devices supported on polymeric substrates for retrofit applications (43). The durability requirements of polymer-supported devices are considerably less stringent and the economics are considerably improved by reducing the costs for fabrication, installation, and replacement.

EC devices function by the reversible injection or ejection of both ions and electrons in or out of an EC material, as illustrated in Figure 8 for a cathodically colored device. To color the EC film, a voltage source is connected between the transparent conductive electrodes (e.g., indium tin oxide, ITO). With a negative voltage on one ITO electrode, electrons from the ITO electrode and protons from an ion storage (IS) layer are injected into the EC film. This process continues until the EC layer is converted to the desired color. The process can be stopped at lower color values than

the maximum needed for normal change. Under open-circuit conditions, the color remains for some time, which provides a long-term memory device. To bleach the film, the polarity is reversed so that electrons return to the ITO electrode from the EC layer and protons return to the IS layer. Current flows until the entire film is restored to its original color.

A typical EC reaction for a cathodic coloring material is as follows:

$$a\text{-WO}_3 + x\text{M}^+ + xe^- \leftrightarrow a\text{-M}_x\text{WO}_3 \qquad (2)$$

in which a-WO$_3$ is amorphous WO$_3$; M is typically H$^+$, Li$^+$, or Na$^+$; WO$_3$ is transparent; and M$_x$WO$_3$ is blue.

A typical anodic coloration reaction is as follows:

$$\text{Ni(OH)}_2 \leftrightarrow \text{NiOOH} + \text{H}^+ + e^- \qquad (3)$$

in which Ni(OH)$_2$ is transparent, and NiOOH is bronze. EC devices are designed to store ions that will be transported back and forth into the EC layer when a potential (typically, 2–3 V) is applied. Even though an EC device may consist of four or five active layers sandwiched between substrates and superstrates (42), the five-layered device shown in Figure 8 is most typical.

The multilayered device begins and ends with a transparent electronic conductor layer on a glass (or polymer) substrate. Typically, the conductor is ITO, or tin oxide doped with antimony or fluorine. The second layer is the EC layer with both ionic and electronic conduction. The typical EC layer can be amorphous or crystalline tungsten oxide, molybdenum oxide, iridium oxide, or nickel hydroxide; however, the most used material is tungsten oxide. The third layer is an ion conductor (IC), which is a liquid electrolyte in many prototype EC devices but is expected to be a solid elec-

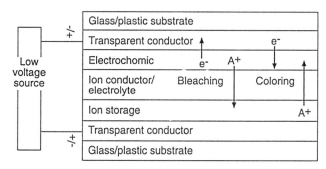

Figure 8. Schematic structure of a cathodic EC device in which the transparent conductor is typically ITO, and the electrochromic material is tungsten oxide. The ion storage (IS) and ion conducting (IC) materials are less well defined, but the IS layer serves as a reservoir for A$^+$ (typically, H$^+$ or Li$^+$) inserted into WO$_3$ when a potential is applied (coloring) or returned across the IC layer to the IS material when the potential is reversed (bleaching). Electrons (e$^-$) flow into or out of the EC layer as shown.

trolyte for buildings glazing applications. Polymers may have significant potential as the IC layer. The fourth layer is the counter electrode (CE) material, and this layer may be a complementary EC material or a material with only IS capability. The device with an IS layer is the most commonly fabricated device. In the complementary device design, the CE and EC electrodes are both EC materials, in which one colors anodically and the other cathodically. In this way, they color and bleach together. Polyorganodisulfide has been used as an IS layer, and polyethyleneoxide–Li salt complexes have been used in a recent EC device (44).

A major problem with today's EC window technology is the electrolyte or IC layer. Both inorganic and polymeric electrolytes can be considered, but the material used must not compromise the economics of manufacturing and potential durability of the units. More details about the need for an IC layer are available (42).

For complete devices, the superstrate and substrate will both be glass, a polymer, or a combination of the two. A retrofit EC device will likely use polymers for the superstrate and substrate. PET has been used as a substrate in early prototype devices. The IC layer is the layer most likely to be formed from a polymeric material.

Evaluation and Performance Criteria for EC Windows

The evaluation criteria for complete EC windows, which were summarized in depth (42), include the design function, aesthetics, economics, performance expectations, construction and preparation methods, and durability. Polymeric retrofit units are expected to improve the economics, if they pass the required preparation and durability criteria. The preparation methods must be compatible with gaining economies of scale, and the durability criterion is now thought to be 5 years or more for meeting the performance expectations. The IC layer, which serves as the medium for ion transport between the EC and CE layers, should be colorless in the visible spectrum, thermally stable, and UV stable. The transport rate and temperature dependence for the insertion ion used in the device must be known so an appropriate layer thickness can be chosen. Knowledge of the transport mechanism may help avoid choosing materials with inherent durability problems.

The performance criteria for complete EC windows were listed (42), and they include transparency in the visible spectrum and optical switching. Important criteria related to polymer degradation might include any of the optical properties from photothermal or oxidative degradation of polymeric superstrate and substrates, memory, cycle lifetime, lifetime, and operating temperatures. The physical property requirements are not likely to be as stringent as for

EVA discussed previously, but the durability factor may prove challenging.

Experience with Polymers Used for IC Layer

In the so-called aprotic systems, a-WO_3 corrosion is slowed by the low ionizability of large organic molecules and inhibited tungsten ion formation. After 10,000 cycles of coloration and bleaching of a-WO_3 electrodes cycled in $LiClO_4$–propylene carbonate (PPC), a decrease in injected charge density is noted. Similar declines in the charge per unit area, and hence optical density, have been seen in many devices after long cycling. After cycling, lithium ions were bound into the film structure and were judged responsible for the electromotive force shift, but these ions were not active in coloration. Meanwhile, as a result of lithium incorporation, the film in the bleached state becomes more ion conductive. Further details about the degradation modes in this system are available (45, 46). Investigators also detected degradation in the response time (cyclic switching) for cells cycled at 22 °C, 50 °C, 60 °C, and 90 °C.

Thermal stability may also be important for devices with polymer electrolytes, as concluded from studies that were also performed on $LiClO_4$ and PPC (47). Decomposition was noted in 1,000 h at 60 °C, 300 h at 70 °C, 200 h at 80 °C, and 15 h at 90 °C. Gas bubbles usually formed in the device during decomposition.

The organic viologen system has always been attractive for switching devices, because its absorption spectrum can be adjusted at any visible wavelength; it charges very efficiently and switches fast. The problem with viologens is that they show unexpected irreversible secondary reactions with use (48).

With the importance of lithium insertion ions to EC devices, it is not surprising that researchers working with polymeric electrolytes in battery systems have also made EC devices with a polymeric IC layer. An example of using 2-acrylamido-2-methyl-1-propane sulfonic acid (polyAMPSA) and subjecting the device to cyclic voltage and current stresses was reported by Cogan et al. (49).

More recently, perfluorosulfonate (PFS) (50a), PMMA-poly(propylene glycol $LiClO_4$) (50b), and PMMA-$LiClO_4$ (50c,d) were used for the IC layer; stability studies have been completed but not published for PFS. The operation of devices with these polymers (50), those previously studied (48, 49), and other polymers was compared and contrasted (51).

Durability of EC Devices with Polymeric Components

No published information is available, except for early prototypes, on EC devices that have survived extensive accelerated life testing or abbreviated life testing. Polymers probably will be used for the IC layer or for superstrate or substrate materials in these devices. For retrofit applications, the EC device will most likely be placed on the inside of a glass pane in a heated room, thus reducing considerably the temperature and UV stresses. An alternative design might be a layer between two glass panes, with or without an inert gas in the enclosed space.

In either case, the UV stresses will be limited to radiation that is transmitted by conventional glass windows, and the temperature extremes will most likely be limited to 20–40 °C above or below room temperature. Furthermore, relative humidity is also likely to be between 40 and 50% at a temperature of 22 °C, and air pollutants will be comparable to those found in most indoor building environments. A test matrix for EC windows has been developed for the various stages of research, development, and commercialization from prototype to full-sized windows (42). The report also addresses the key technical issues for evaluation criteria, test methods, and determination of the durability of EC windows. The issues are addressed under the categories of practical windows, durability testing, and fundamental mechanisms.

Polymers as Advanced Desiccant Materials

Desiccant cooling systems (DCS) include a desiccant regeneration process (i.e., desorption of water) that can use heat provided by flat plate collectors of solar energy or from some other source. The typical temperature supplied by flat plate collectors is not high enough for conventional inorganic desiccants such as silica gel or zeolites. As potential advanced desiccant materials, polymers have been investigated for their water adsorption properties and ability to be regenerated at about 80 °C or below.

Solid polymeric materials can serve as desiccants in DCS that process water vapor in an atmosphere to produce net cooling (52). Systems with solid desiccants most commonly used in industrial air-drying applications and DCS were described (53). For the typical DCS, a desiccant-laden wheel is used in which air may flow in the axial direction only. The solid desiccant (the state-of-the-art material is silica gel) is mounted onto the wheel and the air to be dried flows through one side of the wheel, while the desiccant on the other side of the wheel is being dried by an externally heated air stream. These two air streams must be kept physically separated to maintain the distinctly separate functions of air dehumidification and desiccant regeneration. The historical development of the present commercial systems was summarized (52, 53).

Important differences exist between the design philosophies of regenerative DCSs and commercial desiccant dehumidifiers. Most importantly, it is apparent from thermodynamic analyses of recirculation cycles that the adsorption characteristics of the desiccant can have a large influence on the cooling capacity and coefficient of performance (COP) of the cycle (52). Silica gel has been considered for years to be the most important candidate material for DCS. Extensive optimization studies of the engineering design and systems analysis studies of DCS units for estimating the achievable thermal and electrical COP have used the properties of silica gel, which have been documented extensively over several decades.

Because the water sorption properties of the desiccant are crucial to the performance of any commercial DCS, desiccant materials research was undertaken in a search for polymers that could become better desiccants than silica gel (54). The purpose of studying polymers as an advanced desiccant material is to identify new materials with optimal performances in the temperature range used in DCS. The performances of these new polymeric materials were then ranked and compared with the performance of the industrial standard, silica gel. Some of the important parameters for the water solid-desiccant-material system include the isotherm shape, sorption capacity, heat of sorption, rate of sorption at or near the desiccant bed temperature (20 to 55 °C), the rate of desorption at an elevated temperature (>80 °C), the physical and chemical stability of the desiccant, and the cyclic repeatability of the sorption amount and rates. Important conclusions and recommendations about these parameters were reached in a modeling study (55), and these results were used as a basis for studies to identify new materials with optimal desiccant properties (56–58). The life-cycle cost (i.e., initial cost, performance, and durability) has a direct impact on the cost-effective deployment of any regenerative DCS, as has been summarized in detail (53).

Water Vapor Sorption–Desorption by Desiccant Materials

The sorption of water vapor by a desiccant is given by the simplified scheme:

$$D(s) + H_2O(v) \rightarrow D{\cdot}H_2O \text{ surface} \qquad (4)$$

$$D(s) + D{\cdot}H_2O \text{ surface} \rightarrow D{\cdot}H_2O \text{ internal} + D(s) \quad (5)$$

in which D(s) is a solid desiccant adsorption site. Water vapor adsorbs onto the surface (eq 4) and permeates into the solid by several possible diffusion processes (eq 5) to provide the total sorption. Empty surface sites are filled by further adsorption (eq 4).

Desorption of water vapor from the desiccant occurs by reversing the sequence shown for adsorption and permeation.

Even though the details of the molecular processes are much more complex than illustrated by eqs 4 and 5, it is evident that the sorption capacity is gained by at least two kinetic steps involving the rate of adsorption (eq 4) and the rate of permeation into the solid (eq 5). The reverse steps of desorption and diffusion provide the *solid-side* resistance in the sorption–regeneration sequence. A combination of isotherm shape, sorption capacity, sorption heat, sorption rates at the bed temperature, evolution rate at a minimally elevated temperature, cyclic stability of the sorption properties, and stability of the desiccant itself is required for suitable desiccant materials. Practically, the sorption and evolution rate processes will be reduced by pore volume diffusion and intraparticle diffusion in a packed bed of inorganic materials, but organic polymers offer great potential for reducing the diffusion limitations by properly using optimized polymer synthetic chemistry and engineering.

Why Polymers?

By considering polymers as advanced desiccant materials in a DCS, we note that polymers have the potential for being

- capable of sorbing water from 5% to 5000% or more of their own weight (compared with ca. 35% for silica gel)
- readily fabricated into shapes required for DCS (e.g., a honeycomb structure having approximate dimensions of $25 \times 0.10 \times 0.15$ cm)
- modified so sorption isotherms of the desired shape and heats of adsorption of about 10.8 kcal/mol are both obtained
- fabricated to provide rapid diffusivities of water vapor through the materials
- regenerated by using hot air for the thermal desorption of sorbed water
- capable of maintaining long-term stability through thousands of sorption–desorption cycles
- commercially available at costs comparable to that of commercial-grade silica gel

Accordingly, polymeric materials could not only serve as both the desiccant and support structure in a desiccant wheel, but could also be replaced easily and inexpensively if their water sorption capacity degrades for any reason. Two sets of criteria were developed and applied for ranking potential polymers as advanced desiccants, as was detailed (*56*).

The six key research issues in identifying acceptable polymers for DCS are as follows:

1. Will any commercially available polymer have the properties required for use in a DCS?
2. Can laboratory-prepared polymers be synthesized with suitable sorption, desorption, and stability properties?
3. Do impurities in a gas-fired air stream result in the loss of sorption performance of candidate polymers?
4. Are there other degradation processes that cause a candidate polymer to lose sorption performance?
5. Can the polymeric material be modified to improve performance properties such as isotherm shape, heat of adsorption, regeneration temperature, cyclic stability, and diffusivity?
6. Do the performance properties depend on the sample size and configuration?

Progress to Date

The scientific literature is incomplete for elucidating the properties needed to characterize polymers as suitable for DCS applications. For those found potentially suitable in the initial search, measurements were made with a quartz crystal microbalance system for isotherms, rates of adsorption and desorption, and cyclic stability (*56*). Evaluation of the measurements reduced the candidate materials to a number of anionic polymers: alkali salts of polystyrene sulfonic acid (PSSA), including lithium salt (LS), sodium salt (SS), potassium salt (KS), ammonium salt (AS), and cellulose sulfate sodium salt (*53, 57–60*). A detailed listing of 24 potential candidate commercially available polymers, their sources, and pertinent specific comments is given in reference 58 (Table 1 in reference 58). The water vapor sorption for these polymers is also summarized (Table 2 of reference 58) for isotherm type, adsorption capacity at 80% RH at 22.1 °C, hysteresis, sorption–desorption kinetics, and preliminary comments about cyclic stability. The PSSA itself has an initial capacity of over 400% at 60% relative humidity, but the capacity decreases rapidly over time.

Six water vapor isotherms taken at 22.1 °C are shown in Figure 9 for all but PSSAAS (the VERSA TL materials are made by the National Starch Co.). These are all Brunauer, Emmett, and Teller (BET) type 2 isotherms and are close to the ideal type 1M isotherms described by Collier et al. (*55*). The PSSAAS data are not shown because of marked hysteresis between the adsorption and desorption legs at isothermal conditions, but the isotherms are also available (*60*). The PSSALS and the PSSASS remain the two best candidate materials for DCS, although other cationic and anionic ionomers offer some potential with further study. The structures of the anionic alkali salts of PSSA are shown in Chart I, as well as for those discussed in the next paragraph.

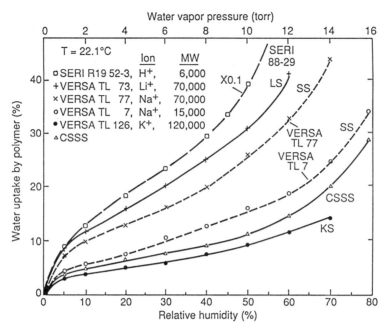

Figure 9. Water-vapor isotherms at 22.1 °C for PSSA (SERI R19-52-3), PSSALS (VERSA TL 73), PSSASS (VERSA TL 7 and 77), and PSSAKS (VERSA TL 126). An isotherm for CSSS (Scientific Polymer) is also included (Data from reference 60).

Recently, additional alkali-exchanged anionic polymers of AMPSA monomers and sodium salts of AMPSA were prepared with copolymers of acrylamide sodium hydrogel (ASH), sodium methacrylate (SM), dimethylaminoethyl methacrylate (DEM), sodium acrylate (SA), and dimethylacrylamide (DA) (61). With 10% water uptake at 60% relative humidity given as the minimum needed, the DEM formulations are rejected and the SA copolymers are just above the minimum. As expected, the potassium and cesium salts (CS) of polyAMPSA were also at or below the minimum capacity needed. PolyAMPSA is unstable with time and its degradation is parallel to that of PSSA, except the initial capacity of polyAMPSA is not nearly as great as for PSSA. Although the initial measurements of 17 different formulations show that the weight percent uptake at 60% relative humidity and the change in water uptake between 5 and 60% relative humidity are favorable, a loss in performance of 10 to 60% occurred after 1 year of storage of the raw material. The causes of the performance losses are not known, and further study of the materials has not been possible.

Four cationic polymers, poly(vinylbenzyltrimethyl ammonium chloride) (VBTAC), *n*-vinylpyrrolidonedimethylaminoethyl methacrylate copolymer quaternized (VPDAMQ), (trimethyl ammonium)propyl methacrylamide methylsulfate, and (trimethyl ammonium)ethyl methacrylate methylsulfate, were also measured (61). The capacity of the latter two was less than 7% at 60% relative humidity, but capacity of VPDAMQ was 57% and of VBTAC was 34% at 60% relative humidity. The structures of VPDAMQ and of

VBTAC are shown in Chart II. After 1 year, the capacities decreased to 15% and 20%, respectively; as with the other materials in this study (61), the cause of performance loss has not been determined. In all cases, it is not clear whether performance loss would have occurred if the polymers were used in a DCS, so further study is merited.

The materials that exhibit the best combination of sorption properties studied at 22.1 °C, including isotherm shape (most are BET type 2 and two are type 1M), sorption capacity at 60% relative humidity, sorption capacity change between 5 and 60% relative humidity, lack of hysteresis, and fast kinetics (over 90% of the total change in 3 min or less) of adsorption and desorption are in the order as follows: PSSALS and PSSASS; polyAMPSASS copolymers with SM and ASH, with the ratios to be optimized; the cationic polymers VPDAMQ and VBTAC; and polyAMPSASS copolymer with DA and polyAMPSA in water.

Interactions at Metal–Self-Assembled Organic Monolayer Interfaces

The multilayer devices described in previous sections have extensive interfaces that in most cases will be between inorganic materials and polymers. As part of our long-range studies to underpin the multilayer-based technologies, we have begun studies on model systems as precursors to future work on inorganic oxide–self-assembled monolayer (SAM) interfaces.

The key properties of SAMs, which have been summarized in a recent chapter (62), are forming a thi-

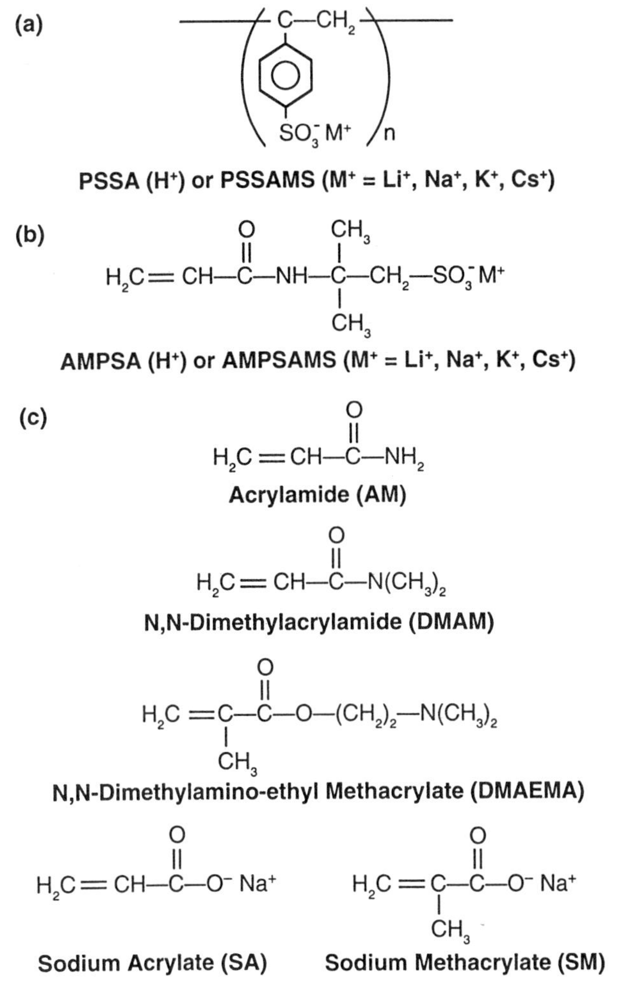

(a)

PSSA (H⁺) or PSSAMS (M⁺ = Li⁺, Na⁺, K⁺, Cs⁺)

(b)

AMPSA (H⁺) or AMPSAMS (M⁺ = Li⁺, Na⁺, K⁺, Cs⁺)

(c)

Acrylamide (AM)

N,N-Dimethylacrylamide (DMAM)

N,N-Dimethylamino-ethyl Methacrylate (DMAEMA)

Sodium Acrylate (SA) **Sodium Methacrylate (SM)**

Chart I. Polymers that deserve further study as promising candidates as advanced, water-sorbing materials for desiccant cooling applications.

olate–Au bond of 44 kcal/mol between alkane thiol molecules and gold substrates, an S superlattice on Au of $\sqrt{3} \times \sqrt{3}$ R30°, a 0.499-nm S–S nearest neighbor on Au, a molecule tilt angle of the methylene chain of 25 to 30°, and a twist angle of 52° (62). Disordering of the hexagonal close-packed chains begins at about 380 K, and C_{22} thiolate monolayers decompose or desorb at about 500 K (62).

We recently completed a critical review chapter on studies of metal (M)–SAM interfaces (63). The purpose of research on M–SAM interfaces (Figure 10) is to understand the interactions between metals on well-ordered organic substrates. Application of M–SAM and inorganic–SAM research results to understand real inorganic–organic interfaces in vacuum and under environmental conditions can potentially play a key role in the development of advanced devices with stable interfacial properties. The M–SAM approach to interface research is delineated as a new subfield in surface science in the context of other approaches to inorganic–organic interface research. In the critical review (63), current issues in M–SAM research were outlined that include chemical compound formation, the morphology (spreading, clustering, or penetration) of the metal species, the kinetics of the metal morphology, the effect of the metal on the degree of order in the SAM, and the rate of metal penetration into the SAM. Progress in studying metals deposited onto alkane thiol SAMs [HS(CH₂)ₙX], in which the terminal group X = CH_3, OH, COOH, $COOCH_3$, and CN, has been summarized for Ti, Cr, Ni, Cu, Al, K, Na, and Ag overlayers (63). The work to date on these M–SAM combinations was reviewed and ranked according to reactivity and penetration (63). The

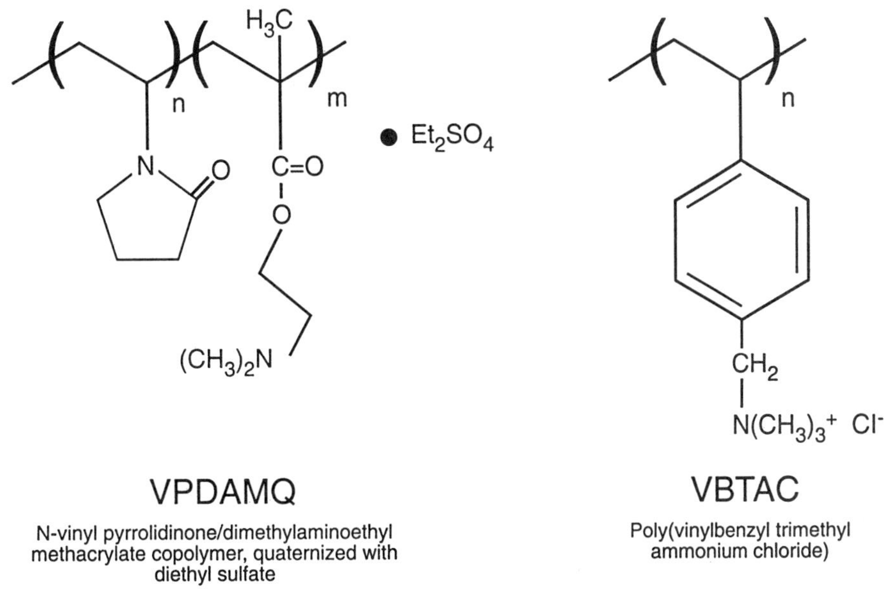

VPDAMQ

N-vinyl pyrrolidinone/dimethylaminoethyl methacrylate copolymer, quaternized with diethyl sulfate

VBTAC

Poly(vinylbenzyl trimethyl ammonium chloride)

Chart II. Structures of cationic polymers VPDAMQ and VBTAC.

Arrangement of a Deposited Metal on an Organic Organized Molecular Assembly Attached to a Substrate

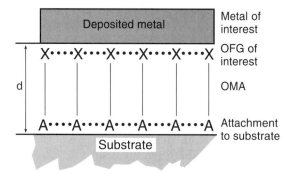

Figure 10. Arrangement of a deposited metal on an organic organized molecular assembly (OMA) attached to a substrate.

future of this subfield of surface science lies in its expansion from M–SAM interfaces in vacuum, to other inorganic–SAM interfaces in vacuum, and then under environmental conditions (*63*).

Concluding Remarks

Significant progress has been demonstrated in the development of metallized polymer solar reflectors during the past 10 years, but additional work is needed to achieve newly proposed goals based on the aspirations and recommendations of the solar thermal manufacturing industry. Polymer-based mirrors that maintain high specular reflectance for extended lifetimes (typically at least 10 years) under outdoor service conditions and whose present cost is reduced by half to roughly $11.00/m^2 are desired. To achieve these aggressive goals, an expanded, more robust program in reflector material development has been initiated that includes close collaborative efforts with industrial partners.

For at least the past 13 years, the PV industry has used EVA encapsulants almost exclusively. EVA browning and other related issues have become a serious concern with regard to the acceptance of PV modules by the utilities industry. Although improved photothermal stability can be achieved with modified formulations (*37*), EVA is a thermally and photochemically unstable material. The future prospects for warranting a 30-year service life of the flat-plate PV modules thus seem to depend on the successful development of new encapsulant materials using more stable polymers, and probably in combination with innovative encapsulation processes and some physical methods like using UV-filtering superstrates.

Glazings with EC coatings have great potential for window applications in buildings, but a number of

technical concerns still exist. Several problems that degrade performance are known, including residual coloration, loss of dynamic range, and loss of stability of the device during long-term cyclic switching. The origins of these effects are only partially understood at this time and need further study. A good deal of uncertainty is likewise associated with the cost and acceptance level by building designers of EC devices.

Additional research is needed to fully evaluate candidate polymeric desiccants. Performance loss with time has been observed for a large number of materials. It is not clear whether such performance losses would have occurred in actual service conditions, so further research and development is warranted.

Self-assembled monolayers offer a wide spectrum of potential benefits including molecularly engineered chemical interfaces and coatings designed for optimal material properties. To fully realize the advantage of such material systems, investigations must be expanded from metal–SAM interfaces studied in vacuum to other inorganic–SAM interfaces in vacuum and eventually to conditions representative of real-world environments.

Acknowledgments

In support of reflector material development, optical characterization capabilities were developed at NREL by Keith Masterson and Ingo Susemihl, with help from Richard Pettit and Rodney Mahoney of Sandia Laboratories. Hermann Neidlinger and Peter Gomez led the developments in stabilizer chemistry. John Webb developed and carried out the FTIR activities. Burton A. Benson of the 3M Company provided a great deal of help (including materials). Edwin Tracy helped develop the magnetron sputtering capabilities. Sample preparation, weathering, and optical measurements were accomplished with the aid of Rita Goggin, Cheryl Kennedy, Yvonne Shinton, Betsy Hovermale, and Mary Steffeck. We are indebted to other members of the Materials Branch at NREL for their help and encouragement.

Technical assistance in the EVA encapsulation work has been provided at NREL by Stephen H. Glick, Keith A. Emery, Ramesh G. Dhere, Alice Mason, and David Niles. The managerial support of Richard DeBlasio is also appreciated.

The interactions with S. K. Deb, D. Benson, E. Tracy, and J.-G. Zhang and C. Lampert on electrochromic devices and R. K. Collier, T. R. Penney, A. Pesaran, T. M. Thomas, H. H. Neidlinger, G. C. Herdt, and N. T. Tilman on desiccants are deeply appreciated. The U.S. Department of Energy provided continuing support for all of these efforts under Contract No. DE–AC36–83CH10093, which is gratefully acknowledged.

References

1. Rabek, J. F. *Prog. Polym. Sci.* **1988**, *13*, 83–188.
2. Neidlinger, H. H.; Schissel, P. *Polymers in Solar Technologies, International Symposium on Polymers for Advanced Technologies, International Union of Pure and Applied Chemistry;* VCH Publishers, Inc.: New York, 1988; pp 34–51.
3. Carroll, W. F.; Schissel, P. In *Polymers in Solar Energy Utilization;* Gebelein, C. G.; Williams, D. J.; Deanin, R. D., Eds.; ACS Symposium Series 220; American Chemical Society: Washington, DC, 1983; pp 3–19.
4. Brandrup, J.; Immergut, E. H. *Polymer Handbook;* John Wiley and Sons: New York, 1989.
5. Ferry, J. D. *Viscoelastic Properties of Polymers;* John Wiley and Sons: New York, 1980.
6. Christensen, R. M. *Theory of Viscoelasticity, an Introduction;* Academic: Orlando, FL, 1982.
7. Mendelsohn, M. A.; Navish, F. W.; Luck, R. M.; Yeoman, F. A. In *Polymers in Solar Energy Utilization;* Gebelein, C. G.; Williams, D. J.; Deanin, R. D., Eds.; ACS Symposium Series 220; American Chemical Society: Washington, DC, 1983; pp 39–79.
8. Sfirakis, A.; Rogers, C. E. *Polym. Eng. Sci.* **1980**, *20*, 294.
9. Luck, R. M.; Mendelsohn, M. A. In *Polymers in Solar Energy Utilization;* Gebelein, C. G.; Williams, D. J.; Deanin, R. D., Eds.; ACS Symposium Series 220; American Chemical Society: Washington, DC, 1983; pp 81–98.
10. Schonhorn, H.; Ryan, F. W. *J. Polym. Sci.* **1969**, *7*, 105.
11. Schissel, P.; Czanderna, A. W. *Solar Energy Mater.* **1980**, *3(1–2)*, 225–245.
12. Saunders, K. J. *Organic Polymer Chemistry;* Chapman and Hall: London, 1973.
13. Susemihl, I.; Schissel, P. *Solar Energy Mater.* **1987**, *16*, 403–421.
14. Jorgensen, G. J.; Schissel, P. In *Metallized Plastics 1;* Mittal, K. L; Susko, J. R., Eds.; Plenum: New York, 1989; pp 79–92.
15. Li, S.; Albertson, A. C.; Gupta, A.; Bassett, W., Jr.; Vogl, O. *Monatshefte Chem.* **1984**, *115*, 853.
16. Neidlinger, H. H.; Schissel, P. In *Proceedings of the Society of Photo-Optical Instrumentation Engineers (SPIE);* International Society for Optical Engineering: Bellingham, WA, 1987; Vol. 823, pp 181–189.
17. Schissel, P.; Neidlinger, H. H.; Czanderna, A. W. *Energy* **1987**, *12(3,4)*, 197–202.
18. Neidlinger, H. H.; Schissel, P. *Polymer Synthesis and Modification Research During FY 1985;* SERI/TR–255–2590; National Renewable Energy Laboratory: Golden, CO, July 1985.
19. Schissel, P.; Neidlinger, H. H.; *Polymer Reflectors Research During FY 1986;* SERI/PR–255–3057; National Renewable Energy Laboratory: Golden, CO September 1987.
20. Webb, J. D.; Czanderna, A. W.; Schissel, P. In *Degradation and Stabilization of Polymers;* Jellinek, H. H. G., Ed.; Elsevier: New York, 1989; Vol. 2, pp 373–430.
21. Brauman, S. K.; MacBlane, D. B.; Mayo, F. R. In *Polymers in Solar Energy Utilization;* Gebelein, C. G.; Williams, D. J.; Deanin, R. D., Eds.; ACS Symposium Series 220; American Chemical Society: Washington, DC, 1983; pp 125–141.
22. Welch, W. F.; Graham, S. M; Chughtai, A. R.; Smith, D. M.; Schissel, P. *Appl. Spectrosc.* **1987**, *41(5)*, 853–860.
23. Kennedy, C.; Jorgensen, G. In *Proceedings of the Eighth International Conference on Vacuum Web Coating;* Bakish Materials Corporation: Englewood, NJ, 1994; pp 33–44.
24. *Polymers in Solar Energy Utilization;* Gebelein, C. G.; Williams, D. J.; Deanin, R. D., Eds.; ACS Symposium Series 220; American Chemical Society: Washington, DC, 1983; Chapters 22–24.
25. Cuddihy, E.; Coulbert, C.; Gupta, A.; Liang, R. *Flat-Plate Solar Array Project Final Report, Vol. VII: Module Encapsulation;* JPL Publication 86–31, DOE/JPL–1012–125; Jet Propulsion Laboratory: Pasadena, CA, October 1986.
26. Willis, P. B. *Investigation of Materials and Processes for Solar Cell Encapsulation;* final report of JPL Contract No. 954527 S/L Project 6072.1 to JPL, DOE/JPL–954527–86/29; Springborn Laboratories, Inc.: Enfield, CT, October 1986.
27. Rabek, J.F. In *New Trends in the Photochemistry of Polymers;* Allen, N. S.; Rabek, J. F., Eds.; Elsevier: New York, 1985; Chapter 16, pp 265–288.
28. (a) Gay, R.; Chesarek, W.; Fredric, C.; Knapp, K.; Pier, D.; Tarrant, D.; Willett, D. In *Proceedings of the PV Performance and Reliability Workshop;* National Renewable Energy Laboratory: Golden, CO, 1992; Report No. SERI/CP–411–5184, p 197. (b) Ackerman, B.; Albright, S. P. In *Proceedings of the Photovoltaic Module Reliability Workshop;* National Renewable Energy Laboratory: Golden, CO, June 21, 1989; Report No. SERI/CP–213–3553, pp 149–169. (c) *Photovoltaic Insider's Report;* Curry, R., Ed.; 1993; Vol. XII, No. 2, p 2.
29. *Photochemistry of Man-Made Polymers;* McKellar, J. F.; Allen, N. S. Eds, Applied Science Publishers: London, 1979.
30. (a) Owen, E. D.; Williams, J. I. *J. Polym. Sci.* **1974**, *12*, 1933–1943. (b) Osawa, Z.; *Polym. Deg. Stab.* **1988**, *20*, 203–236.
31. (a) Andrei, C.; Hogea, I.; Dobrescu, V. *Proceedings of the IUPAC Macromolecular Symposium;* International Union of Pure and Applied Chemistry: Oxford, England, 1982; Vol. 28, p 329. (b) *Rev. Roum. Chim.* **1988**, *33*, 53–58.
32. (a) Stefeno, S. D.; Gupta, A. *Polym. Prepr. (Am. Chem. Soc. Div. Polym. Chem.)* **1980**, *21*, 178–179. (b) Liang, R. H.; Chung, S.; Clayton, A.; DiStefeno, S.; Oda, K.; Hong, S. D.; Gupta, A. *Polym. Sci. Technol.* **1983**, *20*, 267–278. (c) Gallagher, B. D. In *Proceedings of the Flat-Plate Solar Array Project Research Forum, PV Metal Systems;* JPL Publication 83–93; Jet Propulsion Laboratory: Pasadena, CA, 1983; pp 147–158.
33. (a) Pern, F. J.; Czanderna, A. W. *Solar Energy Mater. Solar Cells* **1992**, *25*, 3–23. (b) Pern, F. J.; *Polym. Deg. Stab.* **1993**, *41*, 125–139. (c) Pern, F. J. In *Proceedings of the 23rd Institute of Electrical and Electronics Engineers Photovoltaic SC;* The Institute of Electrical and Electronics Institute, Inc.: Piscataway, NJ, 1993; pp 1113–1118.
34. Pern, F. J.; Glick, S. H. In *AIP Conference Proceedings for the 12th NREL Photovoltaic Program Review Meeting;* Noufi, R.; Ullal, H., Eds.; American Institute of Physics: New York, 1994; pp 573–585.
35. Pern, F. J.; Czanderna, A. W. In *AIP Conference Proceedings for the 11th NREL Photovoltaic AR&D Project Review*

Meeting; Noufi, R., Ed.; Book No. 268; American Institute of Physics: New York, 1992; pp 445–452.

36. (a) Gay, C. F.; Berman, E. *CHEMTECH* **1990**, *March*, 182–186. (b) Schaefer, J. In *Proceedings of the Photovoltaic Module Reliability Workshop*; SERI/CP-4079; National Renewable Energy Laboratory: Golden, CO, 1990; pp 75-78. (c) Rosenthal, A. L.; Lane, C. G. ibid., pp 217–229. (d) Kusianovich, J. ibid., pp 241–245. (e) Wenger, H.; Schaefer, J.; Rosenthal, A.; Hammond, B.; Schlueter, L. In *Proceedings of the 22nd Institute of Electrical and Electronics Engineers Photovoltaic SC*; Institute of Electrical and Electronics Engineers, Inc.: Piscataway, NJ, 1991; pp 586–592.

37. Pern, F. J. In *Proceedings of the Photovoltaic Performance and Reliability Workshop*; Report No. SERI/CP-410-6033, NREL; National Renewable Energy Laboratory: Golden, CO, 1993; pp 358–374.

38. Pern, F. J.; Czanderna, A. W.; Emery, K. A.; Dhere, R. G. In *Proceedings of the 22nd Institute of Electrical and Electronics Engineers Photovoltaic SC*; Institute of Electrical and Electronics Engineers, Inc.: Piscataway, NJ, 1991; p 557.

39. Faiman, D.; Slonim, M. *Efficiency Measurements on the 189 Solarex SX-146 Modules of the PAZ Photovoltaic System at Sede Boqer*; R&D Division, Ministry of Energy and Infrastructure: Ben Gurion University: Sde Boger Campus: Ben Gurion, Israel, October 1992.

40. Pern, F. J. In *Proceedings of the Photovoltaic Performance and Reliability Workshop*; SERI/CP–411–5184; National Renewable Energy Laboratory: Golden, CO, 1992; pp 326–344.

41. *Large Area Chromogenics: Materials and Devices for Transmittance Control*; Lampert, C. M.; Granqvist, C., Eds.; The International Society for Optical Engineering: Bellingham, WA, 1990; Vol. IS4.

42. Czanderna, A. W.; Lampert, C. M.; *Evaluation Criteria and Test Methods for Electrochromic Windows*; SERI/TP–255–3537; National Renewable Energy Laboratory: Golden, CO, 1990.

43. Benson, D. K.; Tracy, C. E. *Electrochromic Sun Control Coverings for Windows*; SERI/TP–212–3786; National Renewable Energy Laboratory: Golden, CO, 1990.

44. Doeff, M.; Lampert, C. M.; Visco, S. J.; Ma, Y. P. *SPIE* **1992**, *1728*, 223

45. Nagai, J.; Kamimori, T.; Mizuhashi, M. In *Proceedings of the SPIE*; The International Society for Optical Engineering: Bellingham, WA, 1984; p 59. *Solar Energy Mater.* **1986**, *502(13)*, 279; **1986**, *14*, 175.

46. Kamimori, T.; Nagai, J.; Mizuhashi, M. *Solar Energy Mater.* **1987**, *16*, 27.

47. Masumi, T.; Normura, K.; Nishioka, K.; Deguchi, H.; Ono, H. *SID Dig.* **1982**, *13*, 100.

48. Kase, T.; Kawai, M.; Ura, M. In *Proceedings of the Society of Automotive Engineers (SAE) Meeting*; Society of Automotive Engineers: Warrendale, PA, 1986.

49. Cogan, S. F.; Plante, T. D.; McFadden, R. S.; Rauh, R. D. *Proc. SPIE* **1987**, *823*, 1060.

50. (a) Shabrang, M.; Sealed Electrochromic Device, U.S. Patent 5, 136, 419, August 4, 1992. (b) Anderson, A. M.; Granqvist, C. G.; Stevens, J. R. *Appl. Opt.* **1989**, *28*, 3295. (c) Bohnke, O.; Rousselot, C.; Gillet, P.; Truche, C. *J. Electrochem. Soc.* **1992**, *139*, 1862. (d) Zhang, J.; Benson, D. K.; Deb, S. J. National Renewable Energy Laboratory, 1993, private communication.

51. Granqvist, C. G.; *Solid State Ionics* **1992**, *53–56*, 479.

52. Collier, R. K.; Barlow, R. S.; Arnold, F. H. *J. Solar Energy Engr.* **1982**, *104*, 28.

53. Pesaran, A. A.; Penney, T. R.; Czanderna, A. W. *Desiccant Cooling State-of-the-Art Assessment*; NREL/TP–254–4147; National Renewable Energy Laboratory: Golden, CO, 1992.

54. Parent, Y. National Renewable Energy Laboratory, Golden, CO, has extensive literature about the properties of silica gel.

55. Collier, R. K.; Cale, T. S.; Lavan, Z. *Advanced Desiccant Materials Assessment: Final Report*; GRI–86/0181, Gas Research Institute: Chicago, IL, 1986.

56. Czanderna, A. W.; Thomas, T. M.; *Advanced Desiccant Materials Research*; SERI/PR–255–2887; National Renewable Energy Laboratory: Golden, CO, 1986.

57. Czanderna, A. W.; Thomas, T. M. *J. Vac. Sci. Technol.* **1986**, *A5*, 2412.

58. Czanderna, A. W. *Am. Soc. Heat. Refrig. Air Cond. Eng. Trans.* **1989**, *95*, (part 2), 1098.

59. Czanderna, A. W.; Neidlinger, H. H. *Am. Soc. Heat. Refrig. Air Cond. Eng. Trans* **1991**, *97*, (part 2), 615.

60. Czanderna, A. W. *Polymers as Advanced Materials for Desiccant Applications: Progress Report for 1989*; SERI/TP–213–3608; National Renewable Energy Laboratory: Golden, CO, 1990.

61. Czanderna, A. W.; Herdt, G. C.; Tillman, N. T. *Polymers as Advanced Materials for Desiccant Applications Part 3: Alkali Salts of PSSA and PolyAMPSA and Copolymers of PolyAMPSASS*; ASHRAE Transactions, Vol.1; American Society of Heating, Refrigerating and Air Conditioning Engineers: Atlanta, GA, 1995, Part I, pp. 697–712.

62. Dubois, L. H.; Nuzzo, R. G.; *Ann. Rev. Phys. Chem.* **1992**, *43*, 437.

63. Jung, D. R.; Czanderna, A. W. In *Critical Review of Solid State Material Science*; Green, J.; Holloway, P., Eds.; CRC: Boca Raton, FL, 1994; Vol. 19, pp 1–54.

Photoluminescent Polymers for Chemical Sensors

Toru Ishiji and Masao Kaneko*

A general review of photoluminescent chemical sensors based on various luminescent probes and different polymer matrixes or carriers is presented. The review is based mainly on our work, but related developments in other laboratories around the world are also highlighted. The photoluminescent probe–polymer combinations covered include fluoresceinamine–cellulose, morin–cellulose, ruthenium complex–Nafion, ruthenium complex–silk, and pyrenebutyric acid–silicone. The effect of the structure and morphology of the polymer on the sensor is also briefly discussed.

Optical sensors are based on fiber optic technology and are attracting much attention for the measurement of physical and chemical properties (1–27). The main advantage of optical sensors over electrochemical sensors is their reliability against electromagnetic disturbance. They can also be employed for microdevice fabrication (3, 4). Many kinds of optical properties such as absorption intensity, luminescence, absorption or emission maxima, emission decay, and polarization can be measured by optical sensors. Most of these optical parameters are strongly affected by external variables such as temperature, solvent viscosity, pH, and concentration of other molecules. Among optical properties, photoluminescence is particularly suited for analysis by optical sensing because of its simplicity, high sensitivity, and ease of fabricating photoluminescent sensors (2, 3). A photoluminescent sensor device can be constructed with a photoluminescent probe and an optical fiber (Figure 1).

Among photoluminescent sensors, those used for oxygen, pH, and metal ions have been more widely studied. Photoluminescent oxygen sensors have been reported to a greater extent because there is much demand for measuring oxygen in industry, clinic and research laboratories, and the environment. Almost all optical sensors for oxygen are based on the change of photoluminescence: that is, quenching of photoluminescence by O_2.

This type of photoluminescent probe can be advantageously incorporated into a polymer film for practical use, and various polymers have been reported for this purpose. In designing a polymer matrix for a luminescent probe, it is important to understand that the interaction of the probe with analytes is strongly affected by its microenvironment (i.e., the polymer).

The luminescence resulting from a probe incorporated in a polymer film reflects the microenvironment around the probe and gives information about the location of the probe in the polymer matrix. Therefore, the nature of a polymer microenvironment can be studied by a luminescent probe. Such a study on the microenvironment around a luminescent probe is also important for constructing photoluminescent sensors.

When compared with conventional electrical or electrochemical sensors, photoluminescent chemical sensors offer the following advantages (3, 4):

*Corresponding author.

3469–8/97/0589$15.00/0 © 1997 American Chemical Society

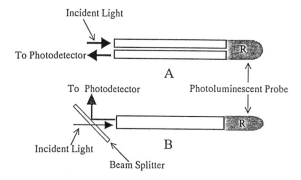

Figure 1. Schematic diagrams of a photoluminescent oxygen sensor: (A) sensor based on bifurcated fiber optic in which separate fibers carry probe and detector radiation; (B) single optic with a beam splitter separating probe and detector radiation.

- stable against electromagnetic disturbance
- usable in flammable gases or liquid without anti-explosion design
- applicable to human body
- rapid response
- possibility of designing microdevices by using of optical fiber
- inexpensive designs can be used as disposable sensors
- wider temperature range than electrochemical sensors

Photoluminescent sensors also have some disadvantages such as interference by ambient light. However, these advantages make photoluminescent sensors suitable for a number of new applications. One such application is in situ detection. For example, important substances such as gases (O_2 and CO_2), pH, and ions in the blood were detected recently by photoluminescent sensors (3) as will be discussed in a subsequent section.

In this chapter, we review the principles of photoluminescent sensors and their application to sensing oxygen, pH, and metal ions by using polymer matrixes. We also discuss our own work on oxygen photoluminescent probes such as tris(2,2'-bipyridine)ruthenium(II) [Ru(bpy)$_3^{2+}$] complex or pyrene derivatives incorporated into different polymer matrixes.

Several other groups also investigated photoluminescent sensors. Since Lübbers and Opitz proposed (10) to use a photoluminescent polymer containing a pyrene derivative for the measurement of oxygen, many researchers applied pyrene derivatives as O_2, pH, and CO_2 probes. Wolfbeis et al. (1, 13, 23) were the first to report on photoluminescent pH and oxygen sensors and their use in biology and medicine. These and subsequent works by Seitz (2, 20–22, 27), Leiner (4), Peterson et al. (12, 19), and others are also highlighted throughout the review. For related topics covered in this book, see also Chapter 3.5.

Principles of Photoluminescent Chemical Sensors

Electronic Transitions

Figure 2 illustrates transitions of an electron between the ground and excited states of a probe. An electron is excited from a ground state (S_0) to an excited state (S_1 to S_n) by absorbing light ($h\nu$, where h is the Planck constant and ν is wavenumber). The electron excited to a higher S_n state returns to S_1 state by internal conversion followed by various other processes from S_1. The electron can return to the original S_0 by a nonradiative process. In this nonradiative process, excess energy is released as heat. Luminescence is observed when excess energy in an excited state is released in the form of light. The electron in S_1 can transit to a triplet excited state (T_1) by intersystem crossing. Similar transition can also take place from T_1. The emission is called fluorescence ($h\nu'$) or phosphorescence ($h\nu''$), depending on whether its excited state is singlet (S_1) or triplet (T_1), respectively.

When an electron or energy is transferred to a different molecule or species present near the excited state, the emission is quenched (5). By using this emission quenching, analysis of these species is possible. Quenching can also take place when an electron is transferred from an electron-donating molecule to the excited probe molecule.

Quenching and Chemical Analysis

Most optical photoluminescent sensors are based on emission quenching of a suitable photoluminescent probe. When there is a quenching species around the excited probe, the observed photoluminescence intensity decreases, depending on the quencher concentration. Accordingly, quencher concentration can be determined by measuring the intensity of the luminescence.

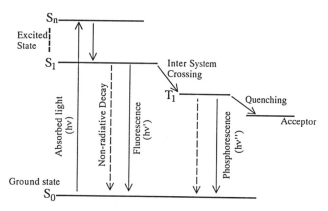

Figure 2. Principle of electronic transitions including quenching by a quencher: S_0, ground state; $S_1 \sim S_n$, excited singlet state; T_1, excited triplet state.

The relationship between the emission intensity and the quencher concentration is expressed by the Stern–Volmer equation (6) (eq 1), where I_0 and I are the luminescence intensity in the absence and presence, respectively, of the quencher, Q, and k_{SV} is the Stern–Volmer quenching constant.

$$I_0/I = 1 + k_{SV}[Q] \qquad (1)$$

When quenching follows a dynamic mechanism that takes place by diffusion and collision of an excited probe with a quencher molecule, the relative emission lifetime of the excited probe is also represented by a similar Stern–Volmer equation (eq 2), where τ_0 and τ are the emission lifetimes in the absence and the presence of the quencher, respectively.

$$\tau_0/\tau = 1 + k_{SV}[Q] \qquad (2)$$

After plotting a calibration line for I_0/I against $[Q]$, the concentration of the quencher can be determined by measuring the emission intensity.

Examples of Photoluminescent Chemical Sensors

Oxygen Probe

Oxygen is an important compound in general, and it is essential for supplying life energy by respiration in particular. It is, therefore, hardly surprising that oxygen sensors have attracted much attention. Various types of oxygen sensors and their characteristics are listed in Table I. Electrochemical oxygen sensors, called Clark electrodes, are widely used as an alarm for O_2 deficiency in working environments such as tunnel constructions and chemical plants. Although conventional oxygen sensors are satisfactory for most applications, they are not suitable for miniature size and high speed devices. Photoluminescent sensors can overcome these problems, and hence these sensors can be used for many applications requiring such devices.

Although emission quenching by oxygen has been known from the 1930s (8), its application to oxygen sensors was first reported in 1968 by Bergman and co-

workers (9) by using fluoranthene in a porous Vycor glass matrix, Lübbers and Opitz reported (10) oxygen measurement in 1983 by using pyrene and pyrenebutyric acid (PBA) adsorbed into a silicone film. The photoluminescence intensity of PBA was dependent on the concentration of oxygen (Figure 3) (11). More recently, photoluminescent oxygen sensors were studied by Peterson et al. (12), Wolfbeis (13), and others (14, 15).

Almost all photoluminescent probes reported for oxygen analysis are aromatic compounds that are excited only by UV light. The use of these materials is limited because of the hazardous effect of UV light on the measurement system. Metal complexes often absorb and emit visible light and are therefore excellent candidates for photoluminescent oxygen probes (16–18). Such visible light probes are especially useful for in situ measurement of dissolved oxygen in biological systems, and the possibility of using inexpensive plastic optical fibers as waveguides exists (see Chapter 3.5). We investigated photoluminescent and quenching properties of tris(2,2'-bipyridine)ruthenium(II) complex, as described in the section on polypyridine membranes.

pH Probe

Continuous measurement of pH is required in many practical fields such as the chemical industry and environmental pollution control. Sensors also were developed for in vitro and in vivo monitoring of pH, oxygen, and CO_2. At present, a potentiometric glass electrode is most widely used as a pH sensor. Although a pH microsensor is highly desirable in the clinic, it is difficult to fabricate a microdevice based on a glass electrode. This problem can be overcome by an optical pH sensor using an optical fiber. Optical pH sensors are based on the fact that optical properties of the sensing probe such as absorbance, reflectance, and luminescence vary with pH. The first optical pH sensor for in vivo study was developed by Peterson et al. (19) using phenol red as an indicator. Here, the color (absorbance) of the indicator changes with pH. Subsequently, a pH sensor based on the photoluminescence of fluoresceinamine was published (20). However, more recently, luminescent pH sensors are preferred

Table I. Characteristics of Various Types of Oxygen Sensors

Name	Composition	Mechanism	Applications
Galvanic cell (Clark electrode)	Electrolyte solution	O_2 reduction (Noble metal-Pb electrode)	Oxygen deficiency alarm
Polarography	Electrolyte solution	O_2 reduction (Noble metal-Ag–AgCl electrode)	Medical use, research
Concentration cell	Solid electrolyte (ZrO_2)	O_2 equilibrium (Potentiometry)	Automobile
Limiting current	Solid electrolyte (ZrO_2)	O_2 reduction (Amperometry)	Automobile
Magnetism	Dumbbell ball	O_2 paramagnetism	Industrial
Photoluminescence	Photoluminescent probe	O_2 quenching	General

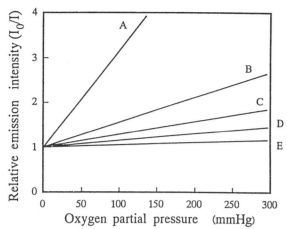

Figure 3. Effect of oxygen partial pressure on relative emission intensity for (A) pyrenebutyric acid (PBA) in silicone film, (B) pyrene in dioctylphthalate, (C) PBA in dioctylphthalate, (D) PBA in water, and (E) perylene in dioctylphthalate.

because of their advantages in comparison with absorption pH sensors. The sensing device is usually constructed with a polymer matrix incorporating a photoluminescent probe. Several types of luminescent pH sensors are described subsequently. Because optical pH sensors can also be applied to CO_2 detection, recent studies on CO_2 sensors are also briefly reviewed.

A variety of luminescent pH indicators are known, but only a few meet the requirements of pH sensors because of their excitation wavelength, luminescent intensity, and stability. Typical luminescent indicators include fluoresceinamine, coumarins, and the trisodium salt of 8-hydroxyl-1,3,6-pyrene trisulfonic acid (HOPSA). A photoluminescent pH sensor relies on an indicator dye whose acid-base equilibrium (i.e., its acid (HA^z) and base (A^{z-1}) forms (eq 3)) have different luminescent properties.

$$HA^z \leftrightarrow H^+ + A^{z-1} \tag{3}$$

Saari and Seitz reported (20) a pH sensor based on the fluorescence of fluoresceinamine. The fluorescent dye was immobilized on cellulose, and the sensing system was fabricated as shown in Figure 4. The emission intensity of the acid form of the dye (HA^z) is higher than that of its dissociated form, and the fluorescent intensity of immobilized dye probe is dependent on pH. The intensity increases between pH 3–8 and then decreases above pH 9. The probe response to pH change was demonstrated by adding KOH into a solution of 0.1 M acetic acid and fluorosceiamine (Figure 5). The dependence of emission on pH indicated that the fluorosceinamine immobilized on a polymer matrix can be used as a pH sensor.

Zhujun and Seitz proposed (21, 22) a new type of sensor by using immobilized HOPSA. The photolumi-

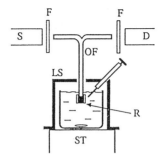

Figure 4. Schematic diagram of a pH sensor based on photoluminescence: S, light source; F, interference filter; OF, optical fiber; LS, light tight shield; R, probe layer; ST, stirrer; D, detector.

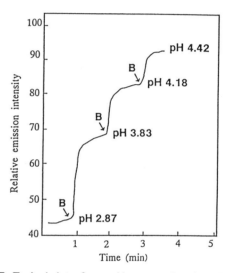

Figure 5. Typical data for a pH sensor showing changes in emission intensity of fluoresceinamine–cellulose in response to added base: B, base addition point.

nescent probe was adsorbed on an anion-exchange membrane. They employed an excitation source that selectively excited either HOPSA or its dissociated base ($OPSA^-$). The advantage of this sensor is its independence on the light source intensity.

One of the most attractive applications of optical pH sensors is considered to be the measurement of CO_2. The principle of a CO_2 sensor is based on the pH change of a solution in contact with CO_2 through a permeable membrane. Many kinds of optical sensors for CO_2 have been reported; we describe a multisensing device (23) for simultaneous detection of both O_2 and CO_2. The mechanism of oxygen detection is based on luminescence quenching from the excited $Ru(bpy)_3^{2+}$ by oxygen as described previously. The CO_2 sensing probe consists of a pH-dependent dye (HOPSA) and a gas permeable membrane. Both O_2- and CO_2-sensitive probes were incorporated into a cellulose layer, which was in turn mounted on the top of an optical fiber.

Absorption and emission spectra of the two probes, Ru complex and HOPSA, are shown in Figure 6. The sensor gives two different emission maxima from the two probes that can be quenched by O_2 and CO_2. The change in luminescence signal versus CO_2 concentration is shown in Figure 7. The emission intensity decreased with the increase of CO_2 concentration. The sensor also showed good response to oxygen concentration. Other polymer matrixes incorporating a luminescent probe were investigated for the purpose of sensing CO_2 (24, 25).

Metal-Ion Probe

Detection of metal ions is similar to that of the hydrogen ion (i.e., pH). A probe indicator incorporated into a polymer matrix forms a complex with the metal ion and hence changes its optical properties. Several types of optical sensors for metal ions were reported (26). Absorptions of transition metal ions are known to be related to their complexing ligands. Molar absorptivities of transition metal ions are not very high, but the luminescence of their complex is attracting attention for detection of metal ions.

A well-known photoluminescent probe for metal ions is 3,5,7,2',4'-pentahydroxyflavone (morin) for the detection of Al(III) (27, 28). When morin was immobilized on cellulose powder at the top of an optical fiber, and this fiber was placed in an Al^{3+} solution, photoluminescence was observed. Complex formation between the polymer-bound probe and the metal ion is depicted in Scheme I. The relationship between emission intensity and concentration of Al^{3+} was studied by Saari and Seitz (Figure 8). Morin is a weakly fluorescent compound by itself, but its complex with Al^{3+} is highly fluorescent, so that morin is a suitable mate-

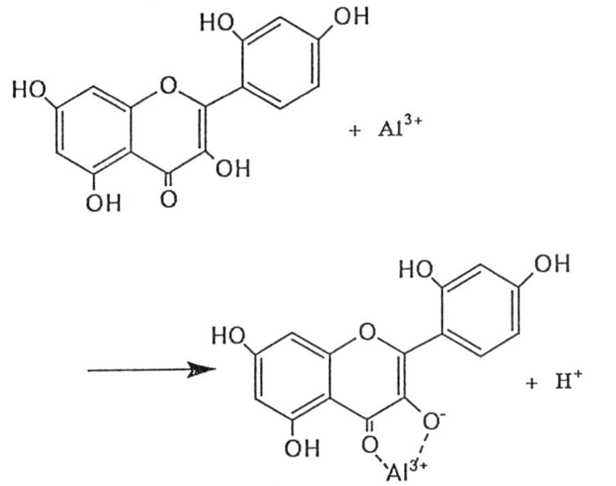

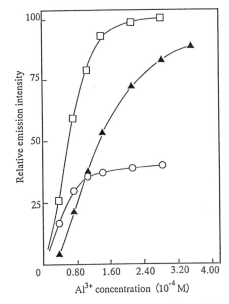

Scheme I. Complexation of Al^{3+} combined with 3,5,7,2',4'-pentahydroxyflavone (morin).

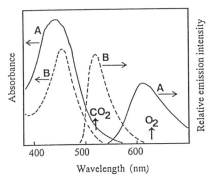

Figure 6. Absorption and emission spectra of $Ru(bpy)_3^{2+}$ (A) and HOPSA (B) on cellulose paper with internal buffer at pH 8. The arrows indicate the wavelengths used for measurement.

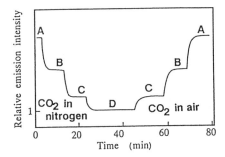

Figure 7. Typical response of a double (O_2-CO_2) sensor to CO_2 (probe; HOPSA) in nitrogen (left) and air (right) by measuring fluorescence at 520 nm. Concentration of CO_2: (A) 2.5%, (B) 5.0%, (C) 15%, and (D) 100%.

Figure 8. Relative emission intensity of immobilized morin as a function of Al^{3+} concentration at pH 3.8 (▲), pH 4.8 (□), and pH 5.9 (○): excitation wavelength, 420 nm; emission wavelength, 488 nm. *See* Scheme I for interaction of morin with Al^{3+}.

rial as an Al^{3+} sensor. Emission intensity of morin also is dependent on pH.

Several research groups reported selective photoluminescent ion sensors (*28, 29*). An *ionophore* is considered to be an interesting ligand for selective metal-ion complexation. Ionophores have been very important in the development of ion-selective membrane electrodes in analytical chemistry. An ionophore incorporated into a polymer membrane can be used for a photoluminescent ion sensor as well as an ion-selective electrode. One example is the detection of Ca^{2+} and Mg^{2+} by a polytetrafluoroethylene membrane incorporating crown ethers (Chart I) (*28*). This sensor could detect about 0.5 mM concentrations of Ca^{2+} and Mg^{2+} selectively.

Polypyridine Membranes as Oxygen Sensors

Many earlier probes reported for oxygen detection were aromatic compounds that are excitable only by UV light. However, some metal complexes can be used as photoluminescent probes, and [Ru(bpy)$_3^{2+}$] (Chart II) is regarded as an excellent candidate for a visible

light photoluminescent oxygen probe (*16*). An inexpensive plastic optical fiber can be used as a waveguide in this probe. Because this complex is a cation, it can be easily bound to polyanionic materials such as Nafion film (Chart II) by ionic interaction. We studied photoluminescence behavior of this combination for a photoluminescent oxygen sensor (*30–33*). Figure 9 shows the absorption and emission spectra of the Ru complex incorporated into a Nafion film.

Another approach for a polymer incorporating the complex is attaching the complex to the polymer by a covalent bond. Conventional polymers such as cellulose (*34, 35*) and silk fibroin (*36–38*) were useful for incorporating photoluminescent probes. However, in the design of polymeric luminescent probes, the quenching efficiency of the luminescence is affected by the microenvironment. In the case of dissolved oxygen in a liquid phase, for example, interaction of oxygen molecule with the polymer-bound photoluminescent probe strongly depends on the polymer structure. In this section we discuss our studies on the microenvironment of Ru(bpy)$_3^{2+}$ incorporated into styrene copolymer, polysiloxane, and Nafion and its quenching by oxygen.

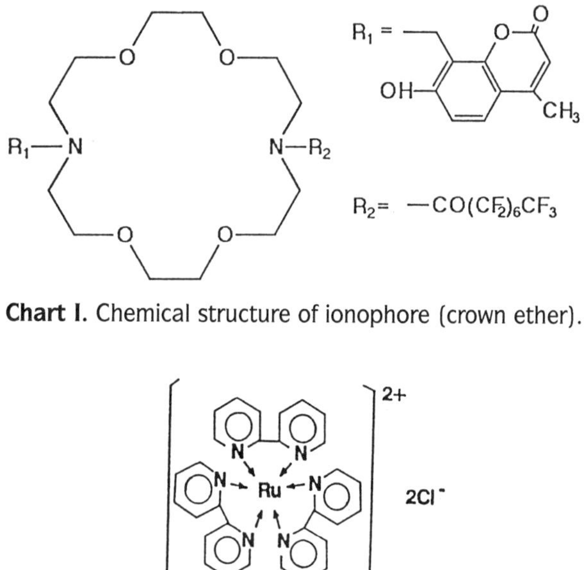

Polymeric Bipyridines Membrane

Various copolymers carrying Ru(bpy)$_3^{2+}$ (Chart III) were prepared by treating copolymers of 4-methyl-4'-vinyl-2,2'-bipyridine with *cis*-Ru(bpy)$_2$Cl$_2$ (Chart III), followed by the study of their photoluminescent properties (*37–39*). The emission from the hydrophobic polymer based on styrene (Chart III) was not quenched by oxygen in water but was quenched in methanol (*38*). This result shows that the compatibility of the polymer with the medium is crucial in the photoluminescence quenching. The hydrophobic styrene copolymer is not solvated by water, and hence the incorporated probe is not accessible to the dissolved O$_2$ molecules; however, it is solvated in methanol. The polymer-pendant probe is useful from a practical point of view because leaking of the probe molecule is prohibited because of the covalent bonding.

Chart I. Chemical structure of ionophore (crown ether).

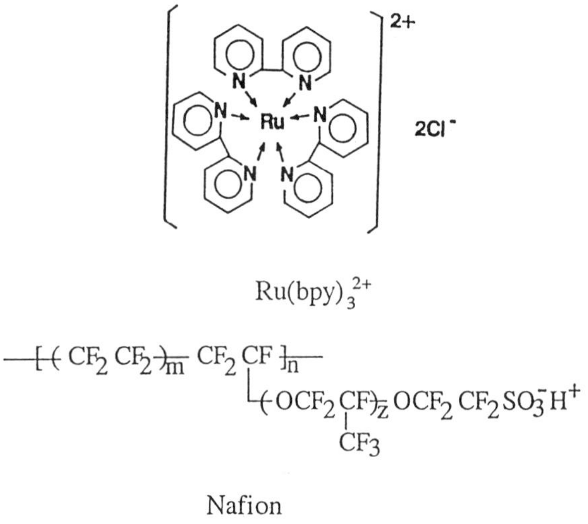

R$_1$ =

R$_2$ = —CO(CF$_2$)$_6$CF$_3$

Ru(bpy)$_3^{2+}$

2Cl$^-$

—[(CF$_2$CF$_2$)$_{\overline{m}}$ CF$_2$CF]$_{\overline{n}}$—
    |
    [OCF$_2$CF]$_{\overline{z}}$ OCF$_2$CF$_2$SO$_3^-$H$^+$
        |
        CF$_3$

Nafion

Chart II. Chemical structures of Ru(bpy)$_3^{2+}$ and perfluoroalkylsulfonate (Nafion) used as oxygen sensor.

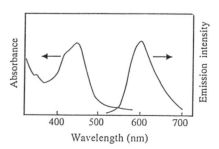

Figure 9. Absorption and emission spectra of Ru(bpy)$_3^{2+}$ incorporated into Nafion film.

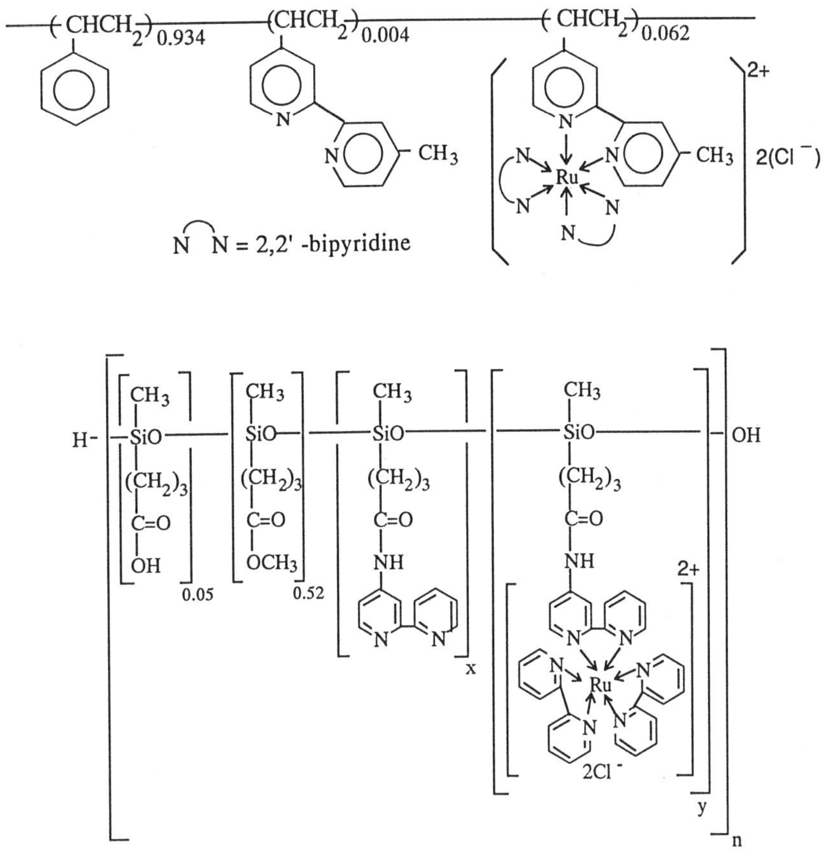

Chart III. Structures of copoly(styrene-Ru(bpy)$_3$$^{2+}$) and polysiloxane carrying Ru(bpy)$_3$$^{2+}$.

Polysiloxane-pendant Ru(bpy)$_3$$^{2+}$ (Chart III) was prepared from the corresponding polysiloxane bpy (*40*). The photoexcited polymer probe was quenched by O$_2$ in a methanol solution. The quenching rate constant of 10^9 dm^3 mol^{-1} s^{-1} for the polymer in methanol was in the same order of magnitude as that of the low molecular weight Ru(bpy)$_3$$^{2+}$ in methanol. The emission quenching of the polymer film by O$_2$ in the gas phase was also studied and gave a linear Stern–Volmer plot.

Ru(Bpy)$_3$$^{2+}$ Ionically Bound to Nafion Membrane

A Nafion film incorporating Ru(bpy)$_3$$^{2+}$ can be prepared easily by dipping the polymer in an aqueous solution of Ru(bpy)$_3$$^{2+}$ (*30, 31*). The luminescence of the resulting film decreased with increasing oxygen concentration in a gas phase. The relative emission intensities (I_0/I) showed good linear relationships against [O$_2$], and the slope (k_{SV}) was 0.14 atm^{-1}. The film system can be used to measure the concentration of dissolved oxygen in liquids. Typical emission spectra of Ru(bpy)$_3$$^{2+}$ in a Nafion film in methanol as a function of [O$_2$] are shown in Figure 10. The relative emission intensity (I_0/I) at the emission maximum (ca.

605 nm) versus [O$_2$] in methanol shows a linear relationship (Figure 11).

Linear relationships of the Stern–Volmer plots for the previously mentioned system were obtained in water, methanol, and benzene. These results indicate that the Nafion–Ru(bpy)$_3$$^{2+}$ film can be used as a photoluminescent oxygen sensor in these media (*41*). The quenching efficiency by oxygen was high in alcohol, which has both polar and nonpolar groups but does not contain bulky alkyl groups.

When the Nafion–Ru(bpy)$_3$$^{2+}$ sensor was left in a liquid medium for several hours, no desorption of the complex from the film was observed, a result indicating strong ionic interactions between the anionic polymer and the cationic probe. Similarly, hydrophobic interactions between the hydrophobic bpy ligands of the complex and the fluorocarbon chain of the Nafion is expected to prevent leaking of the complex in a hydrophilic medium. A polymer–metal complex that has both hydrophobic and hydrophilic characters is present at the boundary region between hydrophilic and hydrophobic groups in the film. The complex bound in such a boundary region by both hydrophilic and hydrophobic interactions can migrate to, and be stabilized by, either hydrophobic or hydrophilic regions of the film, depending on the polarity of the medium.

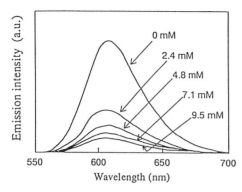

Figure 10. Quenching of Nafion-bound Ru(bpy)$_3^{2+}$ by oxygen in methanol. Each value in the figure is oxygen concentration in methanol.

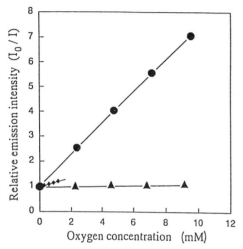

Figure 11. Stern–Volmer plots of relative emission intensity of Ru(bpy)$_3^{2+}$–Nafion film against oxygen concentration in water (♦), methanol (●), or benzene (▲). Solubility of 1 atm O$_2$ in water is 1.16 mM.

In the present study, because the Nafion film contains both hydrophobic and hydrophilic residues, emission quenching by oxygen was observed both in water and benzene. However, quenching efficiency in benzene was very low ($k_{SV} = 13.2$ M^{-1}) as compared with that in water ($k_{SV} = 165$ M^{-1}). This result indicates that in water oxygen can approach the complex through hydrophilic regions of the Nafion film, whereas in benzene oxygen hardly approaches the complex.

Ru(Bpy)$_3^{2+}$ Adsorbed on Silk Membrane

Silk fibroin can be cast into a transparent film from its aqueous solution containing high concentrations of lithium bromide. Silk fibroin film is biocompatible, and we studied this material as matrix for possible biomedical use (34–36).

A silk fibroin film was treated with methanol to make it insoluble in water and other liquids by forming

a beta-structure on its surface, followed by dipping in an aqueous solution of Ru(bpy)$_3^{2+}$. The photoexcited silk fibroin–Ru(bpy)$_3^{2+}$ film was not quenched by O$_2$ in water but was quenched in methanol (34). An electron spin resonance study of a spin-labelled silk fibroin film containing the Ru complex indicated that the molecular mobility of tyrosine residues is strongly suppressed (35). These results may indicate that the probe is bound to hydrophobic clusters of tyrosine phenolic residues. The relatively long excited-state lifetime (about 1000 ns) of the probe in the silk fibroin film, compared with about 600 ns in aqueous solution, supports the view that molecular mobility of the probe is reduced by being bound to the film.

Although the emission is not quenched by O$_2$ in water when the probe complex is adsorbed into a precast silk fibroin film, it is quenched by O$_2$ in water when the film is cast from the mixed solution of the silk fibroin and the Ru complex (35). Such casting procedures may disrupt the ordered structure of silk and produce a more porous membrane accessible by water molecules.

Copolymer-pendant Ru(bpy)$_3^{2+}$ was also grafted onto silk fibroin by attaching copoly(4-methyl-4'-vinyl-2,2'-bpy-methylmethacrylate) on nonwoven silk fabric, followed by treatment with *cis*-Ru(bpy)$_2$Cl$_2$ (36). Photoluminescence of this sample and its quenching by O$_2$ were studied in the gas phase as well as in methanol and water. The relative emission intensity and the emission lifetime of the sample showed two kinds of Ru complex sites, a major and longer lifetime component (77%, 1070 ns under argon gas) and a minor and shorter one (23%, 288 ns). The shorter lifetime species were quenched by O$_2$ more efficiently than the longer ones. The results may indicate a heterogeneous distribution of the complex with the exposed and nonexposed regions of the polymer. The quenching was studied in terms of dynamic and static mechanisms that depend on the medium (Table II).

Silicone–Pyrene Membrane as Oxygen Sensor

Many kinds of polymer films have been reported as solid matrixes for photoluminescent sensors. Silicone is widely used as an excellent polymer matrix, because its affinity with oxygen is high. Also, in mass production of an oxygen sensor, casting of the luminescent probe–silicone mixture is economical for coating an optical fiber. However, the use of solvents for silicones is very much limited, and it has been difficult to prepare a silicone film containing the probe molecule by a simple casting of the mixture (polymer + probe). Therefore, almost all of the reports on the use of silicone as a matrix are based on adsorption of the probe molecules on preformed silicone materials (42, 43).

Table II. Quenching of Excited Silk-Bound and Free Ru Complex ($Ru(bpy)_3^{2+}$)

Sample	Medium	O_2 Conc. (mM)	Mechanism	k_q ($M^{-1}s^{-1}$)
Polymer-bound	Argon	0.0–40.8	For τ_1; dynamic and static	Not determined
			For τ_2; dynamic	2.28×10^7
	Methanol	0.0–9.5	Dynamic and static	9.96×10^7
	Water	0.0–1.2	Nearly dynamic	5.40×10^8
Free (solid, solution)	Argon	0.0–40.8	No quenching	0
	Methanol	0.0–9.5	Dynamic	1.91×10^9
	Water	0.0–1.2	Dynamic	3.23×10^9

Note: K_q is second order quenching rate constant. τ_1 is longer and major lifetime. τ_2 is shorter and minor lifetime.

Recently a soluble silicone was developed, and we adopted this polymer for a simple mixture-casting method (*44*) using its solution to prepare a film. We prepared a photoluminescent film by the soluble silicone powder R-910 (Dow Corning Toray) and pyrenebutyric acid (PBA) (Figure 12) by mixture casting using ethanol, *n*-butanol, 1,2-dichloroethane, or toluene. R-910 is a poly(dimethyl siloxane) containing 3 wt% Si–OH groups.

The relative emission intensity ratios (I_0/I) against oxygen partial pressure for the resulting dry films (dry film) are given in Figure 12 and show good linear relationships. The quenching efficiency of the silicone–PBA films is much higher than neat PBA powder. Only slight differences in quenching efficiency were observed among the films prepared by different casting solvents. The linear relation of the Stern–Volmer plots is a desirable characteristic for the development of practical oxygen sensors. The emission quenching of PBA incorporated in the soluble silicone film was primarily dependent on the chemical structure of the silicone. Luminescent oxygen sensor can thus be fabricated by the present simple mixture casting method.

The luminescence from a probe incorporated in a polymer film reflects the environment around the probe and gives information about its microenvironment in the polymer matrix. Therefore, the nature of the polymer microenvironment can be studied by luminescent probing. When the film is obtained by solvent casting, different casting solvents produce different microenvironments around the probe. The study of such microenvironments is important for understanding the characteristics of the probe, such as quenching efficiency, sensitivity, and stability, which are affected by the microenvironment in the support matrix.

Because the fine luminescence spectra of pyrene and its derivatives are strongly dependent on the microenvironment (*33, 45–47*), pyrene is often used to study microenvironment effects as well as the nature of the matrix. Because the soluble silicone used to incorporate the PBA probe has both polar and nonpolar groups, the texture of the cast film is strongly affected by the polarity of the casting solvent.

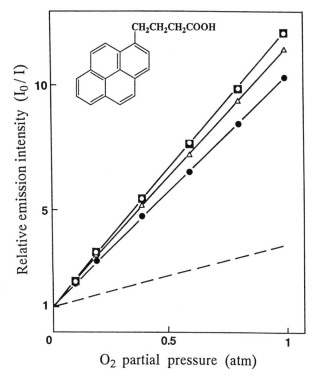

Figure 12. Chemical structure of pyrenebutylic acid (PBA) and Stern–Volmer plots for the quenching of neat PBA (– – –) and silicone–PBA films against oxygen partial pressure. Films were cast in toluene (○), 1,2-dichloroethylene (■), *n*-butanol (△), or ethanol (●).

The luminescent characteristics of the silicone–PBA films were studied by absorption and emission spectra. The absorption spectrum of the films (Figure 13) resembled that of the PBA solution. When comparing the emission spectra of films cast from toluene and ethanol, the maximum absorption wavelength (λ_{max}) and maximum emission wavelength (E_{max}) were independent of the evaporation conditions used to remove the solvent. The effect of the casting solvent was not reflected in λ_{max} or E_{max}, but the fine structure of the emission spectra was dependent on the casting solvent.

Three peaks were observed in the emission spectrum of PBA in the cast silicone film: I (376 nm), II (383

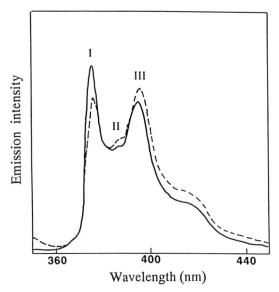

Figure 13. Emission spectra of PBA incorporated in silicone film (film dried by slow evaporation). Casting solvent: ethanol (—) toluene (– – –).

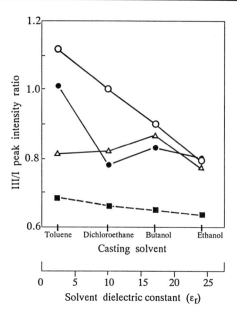

Figure 14. III/I peak intensity ratio of silicone–PBA films cast from different solvents (*see* Figure 13). Film-drying conditions: (○) predry, (●) dry at 60 °C, and (△) well-dried for 3 months; (■) 10 μM PBA in each solvent.

nm), and III (393 nm). The III/I intensity ratios are plotted in Figure 14. Significant dependence of the III/I intensity ratio on both dielectric constant and drying conditions was observed. Before drying, the III/I intensity ratio for the film clearly decreased with increasing polarity of the casting solvent. After drying at 60 °C in vacuo, the III/I intensity ratio changed substantially, except for ethanol. These results indicate that residual casting solvent in the film strongly affects the emission spectrum. Because the silicone used in the present work is hydrophobic, its affinity with the nonpolar solvent is high, and hence nonpolar solvents are more difficult to remove from the film.

After these films were kept in dry air for 3 months (well-dried film), the III/I intensity ratio became nearly equal for all of the casting solvents. Thus, the relative III/I intensity ratio of the emission spectrum is inferred to be sensitive to the presence of trace solvent, and hence information on the drying process can be obtained by using the peak ratio of the emission spectrum of PBA.

Concluding Remarks

Photoluminescent chemical sensors based on photoluminescent polymers are attracting considerable interest because they offer a number of practical advantages over conventional sensors based on electrochemical or magnetic devices. For example, polymer-based photoluminescent probes are more generally stable and applicable to gases, liquids, and solid as well as in the human body. They can also be designed in the form of micrometer-size disposable devices of particular interest for biomedical and environmental use. All these

attributes, and possible effects of polymer microenvironment (macromolecular structure and morphology) render polymer-based photoluminescent chemical sensors a highly attractive field of multidisciplinary research with many potential applications in research and industry.

List of Abbreviations

TDI	toluene diisocyanate
MDI	diphenylmethane diisocyanate
CHD	1,4-cyclohexane diisocyanate
FDI	fluorinated aliphatic diisocyanate (Asahi Glass)
EDA	ethylene diamine
XDA	xylylene diamine
1,4-BD	1,4-butanediol
DED	diethylene diol
PEA	polyethylene adipate
PPD	polypropylene diol
POPD	polyoxypropylene diol
PEO	polyethylene oxide
PTMO	polytetramethylene oxide
PCL	polycaprolactone

References

1. Wolfbeis, O. S. *Trends Anal. Chem.* **1985,** *4,* 184–188.
2. Seitz, W. R. *Anal. Chem.* **1984,** *56,* 16A–34A.
3. *Fiber Optic Chemical Sensors and Biosensors;* Wolfbeis, O. S., Ed.; CRC; London, 1992; Vol. I.
4. Leiner, M. J. P. *Anal. Chim. Acta* **1991,** *255,* 209–222.
5. Timpson, C. J.; Carters, C. C.; Olmsted, J. O. *J. Phys. Chem.* **1989,** *93,* 4116–4120.

6. Stern, V. O.; Volmer, M. *Phys. Zeitschr.* **1919**, *20*, 283–288.

7. Clark, L. C. Jr. U.S. Patent 2,913,386, 1959.

8. Bowen, E. J.; Norton, A. *Trans. Faraday Soc.* **1939**, *35*, 44–48.

9. Bergman, I. *Nature (London)* **1968**, *218*, 396.

10. Lübbers, D. W.; Opitz, N. *Sens. Actuators* **1983**, *4*, 641–654.

11. Opitz, N. *Chem. Ind.* **1984**, *XXXVI*, 742–744.

12. Peterson, J. I.; Fitzgerald, R. V.; Buckhold, D. K. *Anal. Chem.* **1984**, *56*, 62–67.

13. Wolfbeis, O. S. *Fiber Optic Chemical Sensors and Biosensors*; Wolfbeis, O. S. Ed.; CRC: London, 1992; Vol. II, pp 19–53.

14. Li, P. Y. F.; Narayanaswamy, R. *Analyst* **1989**, *114*, 663–666.

15. Li, P. Y. F.; Narayanaswamy, R. *Analyst* **1989**, *114*, 1191–1195.

16. Juris, A.; Balzain, V.; Barigelleti, F.; Campagna, S.; Belser, P.; Zelewsky, A. V. *Coord. Chem. Rev.* **1988**, *84*, 85–277.

17. Krausz, E. *Comments Inorg. Chem.* **1988**, *7*, 139–158.

18. Demas, J. N.; Degraff, B. A. *Anal. Chem.* **1991**, *63(17)*, 829A–837A.

19. Peterson, J. I.; Goldstein, S. R.; Fitzgerald, R. V. *Anal. Chem.* **1981**, *52*, 864–869.

20. Saari, L. A.; Seitz, W. R. *Anal. Chem.* **1982**, *54*, 821–823.

21. Zhujun, Z.; Seitz, W. R. *Anal. Chim. Acta* **1984**, *160*, 47–55.

22. Zhujun, Z.; Seitz, W. R. *Anal. Chim. Acta* **1984**, *160*, 305–309.

23. Wolfbeis, O. S.; Weis, L. J.; Leiner, M. J. P.; Ziegler, W. E. *Anal. Chem.* **1988**, *60*, 2028–2030.

24. Orellana, G.; Moreno-Bondi, M. C.; Segovia, E.; Marazuela, M. D. *Anal. Chem.* **1992**, *64*, 2210–2215.

25. Mills, A.; Chang, Q. *Analyst* **1993**, *118*, 839–843.

26. Seitz, W. R. *Fiber Optic Chemical Sensors and Biosensors*; Wolfbeis, O. S., Ed.; CRC: London, 1992; Vol. II, pp 1–17.

27. Saari, L. A.; Seitz, W. R. *Anal. Chem.* **1983**, *55*, 667–670.

28. Blair, T. L.; Cynkowski, T.; Bachas, L. G. *Anal. Chem.* **1993**, *65*, 945–947.

29. Bakker, E.; Simon, W. *Anal. Chem.* **1992**, *64*, 1805–1812.

30. Kaneko, M.; Hayakawa, S. *J. Macromol. Sci. Chem.* **1988**, *A25*, 1255–1261.

31. Ishiji, T.; Kaneko, M. *Kobunshi Ronbunshu* **1990**, *47*, 869–873.

32. Kaneko, M.; Yamada, A. *Photochem. Photobiol.* **1981**, *33*, 793–798.

33. Rajeshwar, K.; Kaneko, M. *Langmuir* **1989**, *5*, 255–260.

34. Yoshimizu, H.; Asakura, T.; Kaneko, M. *Makromol. Chem.* **1991**, *192*, 1649–1654.

35. Kaneko, M.; Iwahata, S.; Asakura, T. *J. Photochem. Photobiol., A* **1991**, *61*, 373–380.

36. Kaneko, M.; Takekawa, T.; Asakura, T. *Makromol. Chem., Macromol. Symp.* **1992**, *59*, 183–197.

37. Kaneko, M.; Yamada, A.; Tsuchida, E.; Kurimura, Y. *J. Polym. Sci, Polym. Lett. Ed.* **1982**, *20*, 593–597.

38. Kaneko, M.; Nakamura, H. *Makromol. Chem.* **1987**, *188*, 2011–2017.

39. Hou, X-H.; Kaneko, M.; Yamada, A. *J. Polym. Sci., Polym. Chem. Ed.* **1986**, *24*, 2749–2756.

40. Nagai, K.; Nemoto, N.; Ueno, Y.; Ikeda, K.; Takamiya, N.; Kaneko, M. *Makromol. Chem., Macromol. Symp.* **1992**, *59*, 257–266.

41. Ishiji, T.; Kudo, K.; Kaneko, M. *Sens. Acuators B* **1994**, *22*, 205–210.

42. Bacon, J. R.; Demas, J. N. *Anal. Chem.* **1987**, *59*, 2780–2785.

43. Opitz, N.; Graf, H. J.; Lübers, D. W. *Sens. Actuators* **1988**, *13*, 159–163.

44. Ishiji, T.; Kaneko, M. *Analyst* **1995**, *120*, 1633–1638.

45. Blatt, E.; Launikonis, A.; Mau, A. W. H.; Sasse, W. H. F. *Aust. J. Chem.* **1987**, *40*, 1–12.

46. Kalyanasundaram, K.; Thomas, J. K. *J. Am. Chem. Soc.* **1977**, *99*, 2039–2044.

47. Lianos, P.; Georghiou, S. *Photochem. Photobiol.* **1979**, *30*, 355–362.

Functional Polymers
for Chemical Sensors

R. Zhou, K. E. Geckeler*, and W. Göpel

This chapter provides a review of functional polymers suitable for the development of chemical sensors. Functional polymers may be used for different types of chemical sensors, including acoustic wave sensors (bulk acoustic wave, surface acoustic wave, and flexural plate wave sensors), electronic conductance sensors (semiconductive and capacitance sensors), and calorimetric sensors. These polymers are discussed in terms of their physical and chemical properties, as well as their sensitivity and selectivity toward the detection of certain analytes. Partition coefficient constants, thermodynamics, and kinetics of analyte sorption in polymer coatings for acoustic wave devices are discussed. Future prospects of functional polymers for the development of more sensitive and selective chemical sensors are also discussed.

Demand is increasing for monitoring various hazardous compounds in environmental, medical, and technological fields. Chemical sensors provide a strong tool for such detection by transducing chemical signals into appropriate electrical signals. Chemical sensors can be fabricated as small, robust, and inexpensive devices for process automation. Therefore, chemical sensors have been studied extensively during the past two decades to detect hazardous compounds in both gas and liquid phases (1–7).

Sensors should not only be designed to detect sub-parts-per-million concentrations, but they should also be small, portable, and inexpensive. Sensor devices that can be adapted easily to many different detection problems would also be advantageous. Few devices have the necessary sensitivity and selectivity and retain the small size needed in the field. Microelectronic chemical sensors, such as acoustic wave devices and semiconductive, optical, thermometric, electrical, and electrochemical devices meet this size requirement, and they can also be very sensitive. Generally, such chemical sensors have been fabricated by modification of the transducer surface with a nonvolatile coating that can physically or chemically interact with the desired analyte to improve sensitivity. Functional polymers are especially suitable for this purpose because of their film-forming properties, good chemical and thermal stability toward many environmental conditions, and versatility of functional groups and backbone structures. They also show relatively high sensitivity and selectivity and are inexpensive. For these reasons, functional polymers have been used to develop different types of chemical sensors, and many studies now focus on the preparation of new sensors with a variety of functional polymers for different transducers (1–8).

In this chapter, chemical sensors based on functional polymers are discussed according to the type of sensor transducer: namely acoustic wave, electronic conductance, and calorimetric transducers. Sorption processes and physical and chemical properties of the polymer and analyte are also discussed. The use of functional polymers for photochemical sensors is covered in another chapter.

*Corresponding author

Classification of Chemical Sensors

Chemical sensors are devices that characterize a chemical state by transforming chemical information (e.g., a single concentration or total composition) into an analytically useful signal (Figure 1). A chemical state (or environment) is brought about by the various concentrations, pressures, or activities of analytes such as atoms, molecules, ions, or biological species to be detected in the gas, liquid, or solid phase. The majority of scientists in the field agree that sensors should operate reversibly in a thermodynamic sense, but many biosensors do not fulfill this requirement. Various chemical sensors may be classified according to the operating transducer principles illustrated in Table I. Chemical sensors studied in our laboratory include bulk acoustic wave (Δm), conductive ($\Delta \sigma$), capacitance (ΔC), and calorimetric sensors (ΔT), whose oscillators are shown in Figure 2.

Functional Polymers Used for Chemical Sensors

Polymer sensors are classified here according to the transduction method used for measuring the effect and not the primary effect they sense. Polymer sensors may also be classified according to polymer properties, such as elasticity, crystallinity, chemical properties, polarity, polarizability, dipolarity, functionality, and the coating morphology (porous or unporous). Classification according to the mode of analyte detection (e.g., sensor for pH or for toxic gas) or the mode of application (e.g., for use in vivo or for process monitoring) also was attempted. Either of these classifications may be adapted as long as they are based on clearly defined and logical principles. Certain proper-

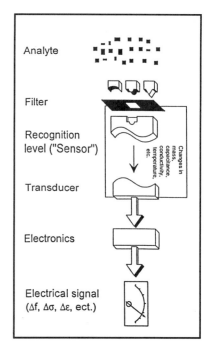

Figure 1. Schematic presentation of a chemical sensor.

ties of the polymers used as chemical sensors are discussed briefly because they may be directly relevant to sensor preparation or the sensing process. However, the range of polymers used is so extensive that a fuller discussion of the polymers cannot be included.

Sensor Preparation

Functional polymer sensors are usually prepared by depositing a thin film of the polymer on a substrate surface. The method of coating the polymer on the sensor surface is very important, because sensor stabil-

Table I. General Classification and Description of Chemical Sensors

Sensor Type	Sensor Principle
Electronic	Measurements of conduction and capacitance, where no electrochemical process takes place, but signals arise from the change of electrical properties including conductivity and electric permittivity caused by the interaction of the analyte with polymer
Electrochemical	Transformation of the effect of electrochemical interaction between the analyte and the electrode into electronic signals
Mass sensitive	Transformation of mass change on the electrode surface into electronic signals
Magnetic	Change in paramagnetic properties
Optical	Transformation of changes in optical phenomena (absorbance, reflectance, luminescence, fluorescence, refractive index, optothermal, light scattering) produced by interaction of the analyte with chemically sensitive sites
Thermometric	Transformation of thermal effects of a chemical reaction or sorption of the analyte into electronic signals

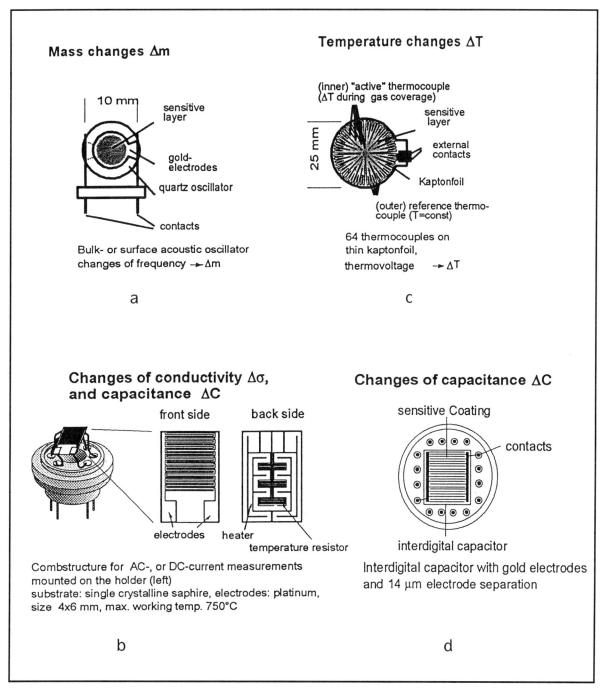

Figure 2. Schematic representation of four types of transducers used in our laboratory: a), 10-MHz quartz oscillator with gold electrodes to measure Δm effects; b), interdigital comb structure with platinum electrodes and 100-µm electrode separation on a single crystalline Sapphire to measure $\Delta\sigma$ and ΔC; c), planar thermopile with 64 Cu–Cu–Ni thermocouples integrated in a thin kapton foil to measure ΔT effects [thermovoltage (V_{therm}) generated by 32 inner active thermopiles, while temperature of the 32 outer reference thermocouples is kept constant]; d), interdigital capacitor with gold electrodes and 14-µm electrode separation on a glass substrate to measure ΔC effects.

Table II. Different Methods of Coating Sensors by Functional Polymers

Coating Technique	Ref.
Dip coating	26
Spin coating	42, 43, 85
Spraying	16, 42, 43, 59
Langmuir–Blodgett	82
Chemical immobilization	65
Plasma polymerization	54, 81
Electropolymerization	13
Vapor deposition	80

ity strongly depends on the adhesion of the polymer to the surface. Several coating methods described for the preparation of sensors are given in Table II.

Simple coating methods include dipping, spraying, and spin coating. More elaborate techniques involve Langmuir–Blodgett, physical vapor deposition, or chemical vapor deposition. In general, most polymers coated on sensor device substrates form weak physical bonds. For more stable polymer surfaces it is preferable to use chemical immobilization (i.e., chemical attachment to the substrate). Vapor deposition (sublimation) techniques are possible only for thermally stable materials.

Acoustic Wave Chemical Sensors

Sensor Principles

Acoustic wave sensors were first introduced as detectors in gas chromatography on the basis of the bulk acoustic wave (BAW) by King in 1964 (9). Since then a variety of other acoustic wave devices have been used to create acoustic wave chemical sensors. These devices include surface acoustic wave (SAW) (10), flexural plate wave (FPW) (6), shear horizontal acoustic plate-mode (SH-APM) (6), and electrochemical quartz microbalance (EQMB) (6) devices. Some of these devices are illustrated in Table III.

In each case, metal transducer on the piezoelectric substrate converts the electrical energy into mechanical energy in the form of an acoustic wave. For sensor applications, the device is usually placed in an oscillator circuit, where it functions as the resonant element. Changes in the frequency are observed as the mass on the transducer surface of the sensor changes by, for example, sorption of analyte molecules onto the surface. Interaction of analyte with sensitive coating on different BAW devices is shown in Figure 3.

The function of BAW sensor is based on frequency changes (Δf) in the fundamental oscillation frequency (f_0) as a result of analyte sorption. To a first approximation, Δf relates to changes in the oscillating mass (Δm) according to eq 1 (11), where ρ_q and μ_q are the density

and shear modulus of quartz, respectively, and A is the piezoelectric active area.

$$\Delta f = -\frac{2f_0^2 \Delta m}{A\sqrt{\rho_q \mu_q}} \qquad (1)$$

If the surface film is soft and nonconducting, the response of SAW devices to Δm can be expressed by eq 2 (10), where k_1 and k_2 are constants for the piezoelectric substrate, h is coating thickness, and ρ is coating density.

$$\Delta f = (k_1 + k_2)f_0^2 h\rho \qquad (2)$$

Equation 2 predicts a linear decrease in Δf with increasing mass per unit area, because $h\rho$ has the same unit as $\Delta m/A$. However, f_0 can be affected by changes in elastic properties or resistance of the surface coating. The influence of mass and elastic properties of a thin, isotropic, nonconducting coating on the frequency of an SAW device is expressed by eq 3 (10), where v_0 is the Rayleigh wave velocity in the piezoelectric substrate, and λ and μ are the first Lamé constant and shear modulus of the coating material, respectively.

$$\Delta f = (k_1 + k_2)f_0^2 h\rho - k_2 f_0^2 \frac{4\mu(\lambda+\mu)}{v_0^2(\lambda+2\mu)} \qquad (3)$$

The first term in eq 3 represents the mass effect, and the second term represents the effects of elastic properties of the surface coating on the frequency of the surface acoustic wave. Wohltjen pointed out (10) that for certain polymers, the second term becomes negligible when the ratio $4\mu/v_0^2 < 100$.

The frequency change in FPW or SH-APM can also be obtained from mass change on the active surface. The relationships for FPW and SH-APM are given by eqs 4 and 5, respectively.

$$\Delta f = \frac{f_0 \Delta m}{2\rho h A} \qquad (4)$$

$$\Delta f = \frac{f_0 \Delta m}{\rho h A} \qquad (5)$$

The mass–frequency relation for the fundamental mode of SH-APM is identical to that of the flexural mode devices, but the SH-APM mode operates at high harmonics and has a better sensitivity.

The mass per unit area of the polymer-coated FPW in liquid phase can be determined by a liquid-loading method according to eq 6 (12), where f_0 is the

Table III. Schematic Presentation of Typical Acoustic Wave Devices

Mode and Format	Type of Acoustic Wave	Phase Velocity
Bulk acoustic wave Quartz / Electrodes (Au, Ag, or Al)	Longitudinal volume waves	5000...12,000 m/s
	Transverse volume waves	2500...6000 m/s
Surface acoustic wave Absorber / Transmitter IDT's / Receiver IDT's / Piezoelectric substrate	Surface acoustic wave: Raleigh wave	2500...6000 m/s
	Horizontally polarized shear wave	
Flexural plate wave Sensitive coating / Si-substrate / ZnO/Al/Si₃N₄ Plate Thickness = 3 μm / IDT	Lamb wave: symmetric	5000...12,000 m/s
	Lamb wave: asymmetric	100...5000 m/s
Shear horizontal acoustic plate-mode Input IDT / Output IDT / Quartz plate Thickness = 300 mm / Cross-section displacement	Horizontally polarized plate wave-mode in cross-section	2500...6000 m/s

fundamental oscillator frequency in air, f_{Load} is oscillator frequency with mass loading ($f_0 + \Delta f$), M is mass per unit area of the bare coating, ρ_L is density of the liquid, and δ is depth of the evanescent disturbance.

$$\frac{f_{Load}}{f_0} = \left(\frac{M}{M + \rho_L \delta}\right)^{\frac{1}{2}} \quad (6)$$

Mass changes of fluxural thin rod acoustic wave (FTRAW) devices can be calculated by determining the change of the phase angle ($\Delta\phi$) of the thin rod according eq 7 (*13a*), in which l is the distance between the two glass horn tips, a is the radius of the thin rod, and λ is the wavelength of the flexural wave.

$$\Delta\phi = \frac{180l}{\rho a \lambda} \Delta m \quad (7)$$

FTRAW devices have, potentially, the same advantage of high mass sensitivity as plate-mode sensors. Because their rod may be produced directly from the coating materials normally used for other sensors, very high mass sensitivity could be obtained due to the potentially large ratio of $\Delta m/m$. Li et al. developed (*13b*) such a device for chemical sensors and showed that it has a higher sensitivity than other devices and that it may give improved results for water analysis.

In most studies, the quartz frequency is the only measurable quantity, and frequency changes are directly transformed to mass changes by the well-

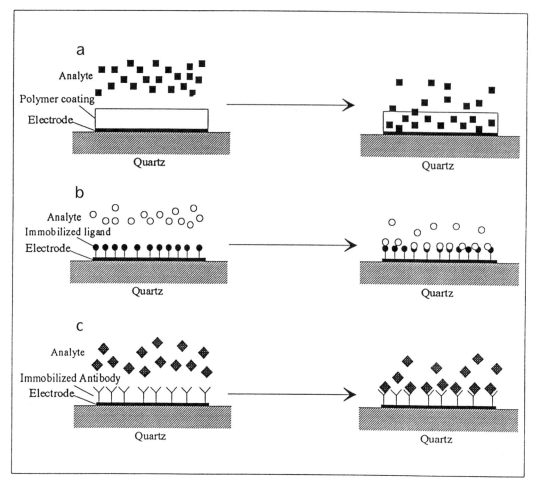

Figure 3. Schematic representation of interaction of an analyte with a) polymer coating, b) immobilized ligands, and c) immobilized antibodies on a BAW device.

known Sauerbrey equation (*11*). However, this equation is not justified when the viscoelastic properties of the contacting medium (liquid or film) change during the experiment. This situation can be resolved on the basis of the quartz equivalent circuit. The EQMB provides correlation between electrochemically induced mass change at the electrode surface and the charge consumed in the process. For liquid loadings, frequency change can be expressed by eq 8 (*14*), where f_s is the series resonance frequency, N is the overtone or harmonic number, and η_L and ρ_L are viscosity and density of the liquid, respectively.

$$\Delta f_s = -\frac{f_s^{\frac{3}{2}}}{N}\left(\frac{\rho_L \eta_L}{\pi \mu_q \rho_q}\right)^{\frac{1}{2}} \qquad (8)$$

Under ideal conditions $\rho_L \eta_L$ is constant in a typical BAW experiment and eq 8 (Sauerbrey equation) is applicable. However, commonly used polymers are viscoelastic, and the contributions of the coating viscosity (η_f), density (ρ_f), and elasticity (μ_f) to the reso-

nant frequency must be considered. One simple way is to treat the polymer as a liquid and use eq 9 (*15*), although the elasticity changes are ignored.

$$\Delta f_s = -\frac{f_s^{\frac{3}{2}}}{N}\left(\frac{\rho_L \eta_L}{\pi \mu_q \rho_q}\right)^{\frac{1}{2}}\left(\frac{\Delta m}{A} + \frac{\rho_f \eta_f}{4\pi f_s}\right)^{\frac{1}{2}} \qquad (9)$$

Applications of Acoustic Wave Sensors

Applications of acoustic wave sensors were reviewed by several authors (*1–7*), but acoustic wave sensors coated with polymers have not been separately reviewed in detail. Early application examples include the detection of hydrocarbons by means of BAWs coated with chromatographic substrates (*9, 16*). Detection of gaseous pollutants such as ammonia, formaldehyde, hydrogen sulfide, ozone, sulfur dioxide, and hydrocarbons have been the most extensively investigated, but there are a number of reports for the detection of liquids and solids.

Acoustic Wave Sensors for Gases

Several inorganic gases such as CO_2 and SO_2 can be detected by polymer-coated acoustic wave sensors, especially polymer-coated BAW sensors ([1-7, 13, 16-30]). These studies were based on the sorption of gas in polymer coatings. An overview of polymer-coated acoustic gas sensors used for inorganic analytes is presented in Table IV.

Interaction among the sensor surface and inorganic analytes is mostly a chemical reaction or chemisorption. For example, CO_2 sensors are based on acid–base reactions ([17, 18]), and humidity sensors are based on H-bonding among water molecules and the polymer ([18, 20]). As a typical example, the time-dependent response of a BAW sensor coated with 3-aminopropylethoxysilane–propylmethoxysilane copolymer for different concentrations of CO_2 and the interaction mechanism are shown in Figure 4 ([17]).

The signals in Figure 4 are completely reversible, and response times are less than 60 s. However, the sensor baseline for some polymers is not always stable. To eliminate the baseline drift, a differential method for response measurement was proposed ([26]). The differential signals of a polymer-coated BAW sensor for NO_2 are shown in Figure 5, in which the differential baseline is stable and sensor responses are reversible.

Acoustic Wave Sensors for Organic Vapors

Polymers can absorb (or adsorb) organic solvents in which the polymer is soluble or swellable. Therefore, polymers have been widely used as sensitive chemical sensors for the detection of organic solvents (Tables V and VI).

Although many authors have reported sensor response for organic vapors (Table V), it is very difficult to find a selective polymer coating for a given ana-

Table IV. Overview of Polymer-Coated Acoustic Wave Gas Sensors Used for Inorganic Analytes

Analyte	Polymer	Method	Ref.
CO_2	3-Aminopropyltrimethoxysilane–propyltrimethoxysilane copolymer	BAW	17, 91
	3-Aminopropyltrimethoxysilane–octadodecyltriethoxysilane copolymer	BAW	17, 91
	Poly(ethyleneimine)	BAW, SAW	18, 91
HCl	Nonylphenylpolyethoxylate	BAW	19
H_2O	Cellulose acetate butyrate	SAW	19
	$CoCl_2$–polyethylene glycol	BAW	21
	Poly(ethyleneimine)	SAW	18
	Polyimide	SAW	20
	2-Hydroethylmethacrylate–2-acrylamido-2-methylpropanesulfonic acid copolymer	BAW	22
	Cellulose acetate	BAW	22
	Modified epoxy resin	BAW	22
	Electropolymerized polypyrrole	FTRAW	13
NH_3	Poly(vinylpyrrolidone)	BAW	23
	Polypyrrole	BAW	24
	Pyridoxine hydrohydride and nonyl phenylpolyethoxylate	BAW	25
NO_2	Polystyrene-bound metal phthalocyanines	BAW	26
	Poly(4-vinylpyridine-*co*-styrene)-bound metal phthalocyanines	BAW	26
O_3	Poly(1,4-butadiene)	BAW	27
SO_2	Carbowax 400 and 20M	BAW	28
	Poly(styrene-*co*-dimethylaminopropylmaleimide)	BAW	29
	N-Dimethyl-3-aminopropylmethoxysilane–propylmethoxysilane copolymer	BAW	30

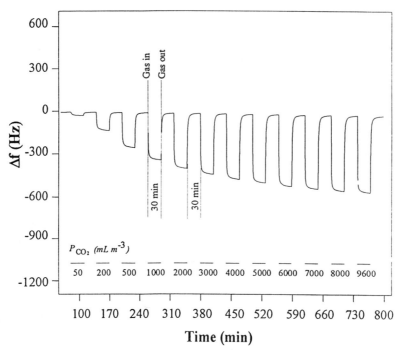

Figure 4a. Typical Δf responses of a BAW sensor coated with 3-aminopropyltriethoxysilane–propyltrimethoxysilane copolymer as a function of time for stepwise exposure to different CO_2 concentrations in synthetic air at 70 °C (r.h. = ~0%) (17).

Figure 4b. Presumable acid–base interaction between CO_2 and aminopolymer.

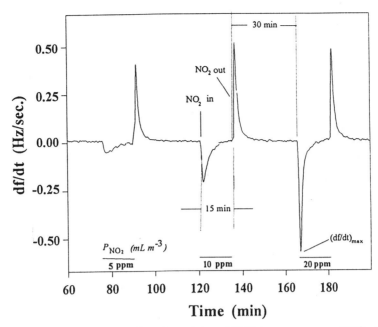

Figure 5. Typical results of frequency changes as a function of time (*df/dt*) in a sequence of NO_2 exposures and pump-down experiments. In all experiments, exposure times and pump-down times were the same. (Polymer: poly(4-vinylpyridine-*co*-styrene) carrying dicarboxylphthalocyanatocopper(II)) (*26*).

lyte. However, by employing a specific chemical reaction between analyte and the coating, a high selectivity can be achieved. For example, a poly(ethylene maleate)-coated SAW sensor for cyclopentadiene vapor (*35*) is based on a Diels–Alder reaction (Scheme I), and the sensor response is irreversible. In this work, it was possible to differentiate the mass change caused by the reaction of poly(ethylene maleate) with cyclopentadiene from the physical absorption of non-reacting solvent molecules. In addition, sensor arrays have been used to improve sensor selectivity and sensitivity (*38, 47, 51, 53, 56*).

Acoustic Wave Sensors for Liquids

Acoustic wave sensors have been used for detection of analytes and bioactive compounds in the liquid phase. Most of these sensors are BAWs (Table VII).

Okahata and Ebato reported (*64a*) water analysis by means of multi-bilayer-coated BAW devices. They claimed that a special homemade oscillator was used to drive the quartz at resonance frequency in aqueous solution and that one side of the BAW sensor was covered to avoid an electrical short circuit in the electrolyte solution. Some immunosensors are based on immunological reactions using polymer-bound antibodies for antigen detection, and they have high selectivity because of the specific bioreactions involved (*65–67*).

Several groups reported (*6, 14, 60*) FPW, SH-APM, and EQMB sensors for liquid chemical sensors.

Recently, in our laboratory, polymer-coated EQMB sensors were investigated (*64b, 91*) for the detection of hydrocarbons in water, and results were satisfactory. However, SAW sensors are not adaptable in liquid phase (*7*).

Polymer-coated acoustic wave sensors can be used to study polymers properties such as glass transition temperature (T_g), film shear storage, and moduli. Similarly, interface phenomena can be monitored by EQMB (*71*), and viscosity can be monitored by SH-APM devices (*72*).

Partition Coefficient in Acoustic Wave Sensors

Equilibrium sorption of analyte vapor into the sensor coating represents a partition of the analyte between the gas phase and the polymer at sensor surface. The distribution can be quantified by a partition coefficient (*K*), which is the ratio of the vapor concentration in the polymer phase to that in the vapor phase (C_p/C_g). The value of *K* can be determined easily by acoustic wave devices, because it can be directly related to the frequency change when vapor is sorbed into the polymer coating. The calculation is given by eq 10, where Δf_s is the frequency change caused by vapor sorption, Δf_p is the frequency change cause by the coating on the sensors, ρ is the density of the coating, *t* is the operation temperature in degrees Celsius, *Mg* is the molecular mass (weight) of the vapor molecule (g is gas), and *Y* is the vapor concentration (mL/m³).

Table V. Polymer-Coated BAW Gas Sensors for Organic Analytes

Analyte	Polymer Coating	Ref.
Acetone	Carbowax 20M	31
Anesthetic gas	Silicon rubbers	32
	Apiezon N vacuum grease	33
	Apiezon M vacuum grease	33
	Dri-Film SC-87	33
	Silicon oil 200	33
	Silastic medical grade elastomer #382	33
	Silastic adhesive sealant #738 RTV	33
	Silastic medical grade adhesive type A	33
	Silicon oil (DC-190)	34
Amines	Polydimethylsiloxane	7
Chloroform	Carbowax 20M	31
Dimethylhydrazine	Butadiene copolymer	7
Ethanol	Polydimethylsiloxane	7
Ethylbenzene	Pluronic L-64	31
Hexane	Pluronic L-64	3
Nitrobenzene	Carbowax 100 etc.	36, 37
	PEG 400–1540	37
	Poly(ethyleneimine)	36
Organophosphorus compounds	Fluoropolyol	47
	Poly(vinylpyrrolidone)	38
	Poly(epichlorohydrin)	38
	Poly(butadiene–acrylonitrile)	38
	XAD-4-Cu^{2+}-amines	39
	XAD-4-Cu^{2+}-amines–poly(hexadecyl methacrylate)	39
	Tetramethylethylenediamine Cu^{2+}chloride–poly(vinylpyrrolidone)	40
	Tetramethylethylenediamine Cu^{2+}chloride–poly(vinylbenzyl chloride)	40
Propylene glycol dinitrate	Poly(cyanopropylphenylsiloxane)	41
	Nitrile silicone gum (X6-6)	41
Styrene	Polystyrene	4
	PEG-400	4
Tetrachloroethylene	Poly(dimethylsiloxane)	42, 43
Toluene	Silicone oil DC-190	4
	Carbowax 550	45
	Pluronic F68	45
Toluene diisocyanate	PEG 400	4, 46
	Carbowax 550	45, 47
	Silicone fluid DC high vacuum	46
	Silicone grease	46
	Silastic LS-420	46
Trinitrotoluene	Carbowax 1000	37

Table VI. Different Modes of Polymer-Coated Acoustic Wave Sensors Used for Various Organic Vapors

Polymer Coating	Method	Ref.
Carbowax 550, 1000, 20M	BAW, SAW	48
Polyethylene oxide	SAW	49
Poly(2,6-dimethyl-*p*-phenylene oxide)	BAW	50
Poly(ethylene maleate)	BAW, SAW	38, 51, 53
Poly(vinyl stearate)	BAW	53
Poly(butadiene methacrylate)	BAW	53
Poly(methyl methacrylate)	BAW	53, 91
Plasma-polymerized tetraphenylborate	BAW	54
Polyphosphazene	BAW, SAW	38, 52
Poly(ethylene succinate)	BAW	50
Polycarbonate	BAW	50
Poly(ethylene phthalate)	SAW	51
Ethylene–vinyl acetate copolymer	SAW	49
Styrene–butylmethacrylate copolymer	SAW	49, 91
Polycarbonate resin	SAW	49
Poly(acrylic acid)	SAW	49
Octadecyl vinyl ether–maleic anhydride copolymer	SAW	38
Poly(vinyl acetate)	FPW	57
Poly(*t*-butyl acrylate)	FRW	55
Poly(1,4-butadiene adipate)	BAW	50
Poly(butadiene–acrylonitrile)	BAW	53
Hydroxy-terminated poly(butydiene)	BAW	53
Poly(butadiene), hydroxylated	SAW	51
Poly(1,2-butadiene)	SAW	49
Styrene–butadiene copolymer	SAW	49
Acrylonitrile–butadiene copolymer	SAW	38
Polystyrene	BAW, EQMB	53, 54, 61
Styrene–acrylonitrile copolymer	SAW	49
Poly(vinyl isobutyl ether)	BAW	53
Polyethylene	BAW	50
Polyisobutylene	SAW, FPW	51, 55
Polyisoprene	SAW	38
Poly(vinylphenol)	BAW	53
Phenoxy resin	BAW	50
Poly(caprolactonediol)	SAW	49
Poly(caprolactonetriol)	BAW	53
Poly(vinyl alcohol)	FPW	55
Collodion	BAW	53
Ethylcellulose	BAW ,SAW	51, 53, 91
Hydroxybutyl methylcellulose	SAW	49
Cellulose acetate	BAW	51b, 91
Cellulose propionate	BAW	51b, 91
Poly(vinyl chloride)	BAW	53, 54
Plasma-polymerized phthalocyanine	BAW	54
Polystyrene-bound phthalocyanines	BAW	44
Poly(4-vinylpyridine-*co*-styrene)-bound phthalocyanines	BAW	44
Plasma-polymerized naphthene	BAW	54
Poly(ethyleneimine)	BAW, SAW	51, 56
Polycaprolactone	BAW	50
Polysulfone	BAW, SAW	49, 50
Fluoropolyol	SAW	38, 51, 56, 58
Polyepichlorhydrin	BAW, SAW	38, 51, 54
Poly(vinylpyrrolidone)	SAW, EQMB	38, 51, 61

continued on next page

Table VI. Different Modes of Polymer-Coated Acoustic Wave Sensors Used for Various Organic Vapors (continued)

Polymer Coating	Method	Ref.
Poly(2-vinylpyridine)	EQMB	61
Polyamide resin	SAW	49
Polyamidoxime	SAW	38
Poly(vinylformal)	SAW	49
Hypalon 20	BAW	52
Thiokol LP-31	BAW	52
Poly(ether urethane)	BAW	91
Poly(dimethylsiloxane) (PDMS)	BAW, FPW	42, 43, 59, 60
Aminoalkyl-PDMS	BAW	42b, 91
Carboxyalkyl-PDMS	BAW	42b, 91
Hydroxyalkyl-PDMS	BAW	42b, 91
Glycidylalkyl-PDMS	BAW	42b, 91
Poly(cyanopropylmethylsiloxane)	BAW	42b, 52, 59
Poly(methyltrifluoropropylsiloxane)	BAW	52
Poly(diphenylphenylmethylsiloxane)	BAW	42b, 91
Poly(isopropanoic acid methylsiloxane)	BAW	59
Poly(phenylmethylsiloxane)	BAW	59

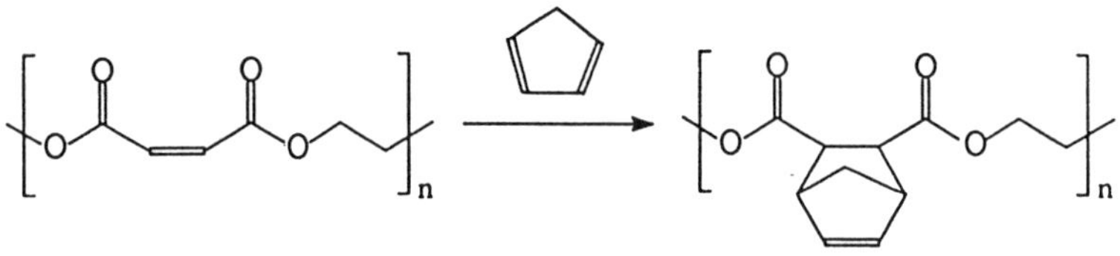

Scheme I. Diels–Alder reaction between cyclopentadiene and poly(ethylene maleate) (35).

Table VII. Polymer-Coated Bulk Acoustic Wave (BAW) Sensors Reported for Detection in the Liquid Phase

Analyte	Polymer Coating	Ref.
pH	Amphoteric polymer film	62
Cu^{2+}	Poly(2-vinylpyridine)	63
Alcohols	Lipid membrane	64
Bitter substances, odorants, anaesthetics, surfactants, antibiotics (duramycin)	$2C_{18}N^+2Cl/PSS^{-a}$	65
Human immunoglobulin	Antibody immobilized on polymers	65, 66
Polynucleotide	Other polynucleotide	7
Human serum albumin	Anti-human serum albumin	68
Protein	Phospholipid	69
Organic solvents in water	Polysiloxanes, poly(epichlorohydrin)	64b, 91
Organic acid in organic solvents	Poly(ethyleneimine)	70

$^a 2C_{18}N^+2Cl$ is dioctadecyldimethylammonium chloride; PSS^- is poly(styrene-sulfonate) ion.

$$K = \frac{C_p}{C_g} = 2.24 \times 10^{10} \left(1 + \frac{t}{273}\right) \frac{\Delta f_g \rho}{\Delta f_p M_g Y} \qquad (10)$$

The partition coefficient (K) is strongly temperature-dependent, because sorption normally decreases with increasing temperature. The more strongly the vapor is sorbed, the more K decreases with increasing temperature. Calculated partition coefficients for several organic vapors using BAW sensors coated with different polysiloxanes are given in Table VIII (59).

In addition to gas sorption, frequency change is dependent on polymer properties (e.g., swelling), which can be dominant in certain cases (73). Accordingly, calculated partition coefficients may deviate from observed values.

Table VIII. Partition Coefficient (eq. 10) of Different Solvents in Polysiloxanes at 30 °C

Solvent	PAPMS	PDMS	PCMS	PIPAMS	PPMS
Alkanes					
Octane	1683	1537	819	$\cong 0$	1227
Methylcyclohexane	527	534	227	$\cong 0$	338
Chlorinated Hydrocarbons					
Chloroform	1033	157	740	188	273
Tetrachloromethane	1608	554	660	$\cong 0$	790
Tetrachloroethylene	724	1220	1248	645	1675
Toluene		878	1469	455	1135
Ether					
Diisopropyl ether	238	188	120	$\cong 0$	148
Ketones					
Acetone	185	64	342	176	98
Alcohols					
Ethanol	276	$\cong 0$	579	724	136
1-Propanol	768	291	1656	2050	383
2-Propanol	251	73	563	699	101
Amines					
1-Butylamine	(1319)	598	(2776)	(1319)	(612)

Notes: PDMS is poly(dimethylsiloxane); PCMS is poly(cyanopropylmethylsiloxane); PIPAMS is poly(isopropanoic acid methylsiloxane); PPMS is poly(phenylmethylsiloxane); PAPMS is poly(aminopropylmethylsiloxane).

Source: Data are taken from reference 59.

Thermodynamics of Sorption

The standard Gibbs free energy (ΔG_s^0) for a gaseous solute can be calculated from the partition coefficient (K) by eq 11, where the standard states are unit concentration in the gas phase and unit concentration in solution, R is the proportionality constant, and T is temperature (K).

$$\Delta G_s^0 = -RT \ln K \tag{11}$$

From the temperature dependence of ΔG_s^0, the differential enthalpy of the sorption is given by eq 12. According to the Helmholtz equation ($\Delta G_s^0 = \Delta H_s^0 - T\Delta S_s^0$), the corresponding differential entropy of the sorption is calculated by eq 13, where ΔH_s^0 and ΔS_s^0 are the standard enthalpy and entropy, respectively. Other thermodynamic quantities for mixing 1 mol of analyte with an infinite quantity of the polymer coating can be derived by eqs 14 and 15, where ΔH_m^0 and ΔS_m^0 are excess enthalpy and entropy of mixing, respectively, and ΔH_e^0 and ΔS_e^0 are evaporation enthalpy and entropy of analyte, respectively. Values of ΔH_e^0 are available (74), and ΔS_e^0 can be calculated by eq 16, where T_b is the boiling temperature of the analyte.

$$\Delta H_s^0 = R \frac{\partial \ln K}{\partial (1/T)} \tag{12}$$

$$\Delta S_s^0 = \frac{\left(\Delta H_s^0 - \Delta G_s^0 \right)}{T} \tag{13}$$

$$\Delta H_m^0 = \Delta H_s^0 + \Delta H_e^0 \tag{14}$$

$$\Delta S_m^0 = \Delta S_s^0 + \Delta S_e^0 \tag{15}$$

$$\Delta H_e^0 = T_b \Delta S_e^0 \tag{16}$$

According to these principles, Schierbaum and coworkers calculated (42, 43) different enthalpies of sorption of tetrachloroethylene in poly(dimethylsiloxane) coating as follows: $\Delta H_e^0 = 34.1$ kJ/mol, $\Delta H_s^0 = -32.7$ kJ/mol for the total interaction process, and $\Delta H_m^0 = 1.4$ kJ/mol. Further details of the sorption thermodynamics were discussed by Zhou and co-workers (42b, 51b, 91). Brace and Sanfelippo studied (20) the thermodynamics of hygroscopic polymer–water system by an SAW device, and they illustrated the general equation of sorption thermodynamics on the basis of Huang's results (75). However, the excess enthalpy of mixing was not discussed.

Kinetics of Sorption

Interaction of the polymer coating with analyte molecules is time-dependent, and the changes can be estimated from sensor signals. Brace and Sanfelippo (20) described the sorption kinetics of water vapor in a hygroscopic polymer (glassy polymer) by using SAW devices and eq 17, where Δf_t is the frequency change, and m_t is the coating mass at time t during the sorption, and n' is diffusion order.

$$\frac{\Delta f_t}{f_0} \cong m_t = kt^{n'} \tag{17}$$

This study showed that, as expected, a multistage sorption process occurs for the glassy polymer examined. Sorption kinetics of polymer coating exposed to a liquid also were discussed by Charlesworth et al. (52) by using BAW devices and Frick's diffusion law. The results showed that sorption curves are not in agreement with Frick's law for some glassy cross-linked polymers. This result arises probably because many polymers below their T_g are unable to relax during the time scale involved (76a).

Electronic Conductance Chemical Sensors

Chemically sensitive solid-state devices are based on changes in electronic or mixed electronic and ionic conductance (conductance sensors) or capacitances (capacitance sensors). These changes are recorded as electric responses for the detection of the analytes in the gas phase. Different terminology commonly in use for these sensors includes homogeneous and heterogeneous semiconductors, Schottky diode, and dielectric sensors (7).

Semiconductive Sensors

Most of the conductance sensors described in the literature are based on inorganic semiconducting oxides with different metals or oxide additives and dopants. However, some organic semiconductors such as phthalocyanines or a variety of polymers also show surprisingly high sensitivities to certain gases. A number of semiconductive sensors coated with conducting polymers are listed in Table IX.

Operating temperature for these sensors usually must be very high to obtain a reversible sensor response. To decrease the operating temperature, BAW sensors with a polymer-bound metal phthalocyanine coating were studied (26). The results showed that the response of these sensors is reversible, and that they can be used for the detection of NO_2 at room temperature in the air.

Capacitance Sensors

Interdigital capacitors (IDCs) are widely used basic devices for capacitive chemical sensors because of their simple technology (see Figure 2d). An IDC consists of two planar lumped electrodes manufactured by conventional thin-film technology on an insulating substrate. The electrodes are coated with a thin gas-sensitive film that changes its permittivity by adsorption of gases. The adsorption must be reversible to reach a stable equilibrium in the analyte concentration between the gas phase and the polymer coat.

The total capacitance of a polymer-coated IDC can be separated into contributions from the gas phase, the substrate, and the polymer (eq 18) (30, 83, 85, 91). The three capcitances (C_i) have three different permittivities ε_i ($i = 1, 2, 3$) that are characteristic of the gas phase, the substrate, and the polymer, respectively. The gas phase permittivity (ε_1, corresponding to C_1), is practically constant even for large variations of analyte concentration (for analytes with high dipole moments). The substrate capacitance, C_2, remains constant for inert substrates like glass. However, sorption of the analyte leads to a change in the permittivity of the polymer (ε_3) and hence to a sensor response according to eq 19.

$$C_{tot} = C_1 + C_2 + C_3 \tag{18}$$

$$\Delta C = \Delta C_3 \tag{19}$$

Typical sensor signals for chloroform and toluene from a capacitive sensor coated with poly(ether urethane) are shown in Figure 6.

Table IX. Semiconductive Gas Sensors Coated with Conducting Polymers

Polymer Coating	Gas	Ref.
Polyacetylene, polypyrrole, polyphenylene, polythiophene, poly(ethylene sulfide)	NO_2, H_2S, and NH_3	77
Electropolymerized polypyrrole	NO_2	77
Polystyrene	NO_2	79
μ-Cyano(tetrakiscarboxylphthalocyaninatocobalt(III))	NO_2	26
μ-Fluoro(phthalocyaninatoaluminium(III))	NO_2	80
Plasma-polymerized manganese acetylacetone	CO	81
Metal tetrakis(cumylphenoxy)- phthalocyanines/fluoropolyol	Organophosphorous compounds	82

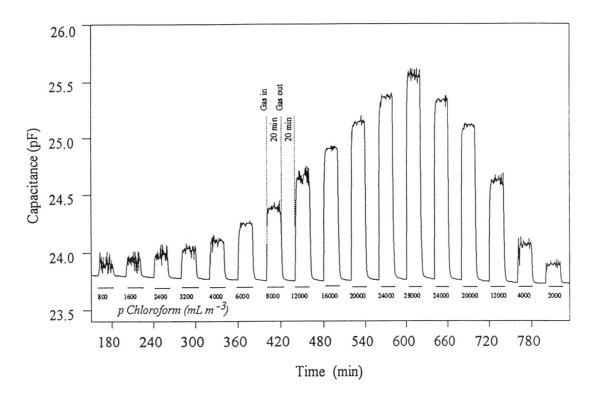

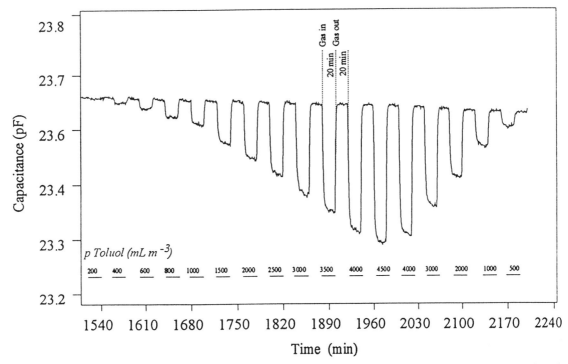

Figure 6. Typical capacitance responses (ΔC) for capacitance sensors coated with poly(ether urethane) as a function of time for stepwise exposure to different concentration of chloroform (top) and toluene (bottom) in air (*35*).

The results show that the polar solvent leads to an increased capacitance, and nonpolar solvent leads to a decreased capacitance. These results can be interpreted by assuming that the polar solvent increases permittivity of the sensor coating, but the nonpolar one decreases the permittivity as a result of polymer swelling by solute molecules (85, 91). A number of other polymers studied as capacitance sensors are summarized in Table X. Another application of capacitance sensors is the monitoring of curing of the polymer coating (89). This application is made possible because of the change of dielectric properties of the polymer as a result of cross-linking (curing).

Calorimetric Chemical Sensors

Recently, polymer-coated calorimetric chemical sensors (see Figure 2b), in which temperature changes are recorded by a thermovoltage, were proposed in our laboratory (42, 43, 91). In this type of sensor, analyte concentration is determined by measuring temperature changes produced by the interaction of the polymer coating with the analyte under chopped flow condition. Only changes in equilibrium concentrations of adsorbed or absorbed analyte lead to (time-dependent nonequilibrium) heat generation. As an example, a time-dependent signal for toluene from a poly(ether urethane)-coated calorimetric sensor is shown in Figure 7.

For the same sensor, temperature changes are positive during exposure to different concentrations of tetrachloroethylene in air and negative for subsequent exposure to pure air. Kinetics of temperature changes during the sorption process also were studied in detail (42). Very polar polymers (e.g., cellulose-acetate and celluloce-propianate) cannot be used for this type of chemical sensors, because no or only small sensor responses were observed (91).

Effects of Polymer Properties in Chemical Sensors

Physical Properties

It is essential for chemical sensors that physical properties of the coating material should not change or lead to hysteresis during sensor measurements. These requirements can be met by nonvolatile liquids, amorphous oligomers, or polymers. Amorphous polymers should be elastomeric; that is, their operating temperature should be above their T_g. Under these conditions, the high motion of the polymer chains leads to rapid diffusion of the analyte, and the material remains essentially unchanged during the sorption-desorption process (90). Under similar operating conditions, sorption (and hence sensitivity) in elastomers is greater than in glassy polymers. The swelling of elastomers is also important for the sensor response, because some sensors (e.g., BAW and SAW) are sensitive to the changes of polymer moduli (71, 73).

The slower diffusion of analyte in glassy polymers leads to slower sensor response and recovery times (90). Retention of analyte in glassy polymers may lead to changes in their physical properties (as a result of plasticization, flow, or crazing) and hence cause partial sensor irreversibility. Glassy polymers show poor reproducibility from sensor to sensor, or from vapor to vapor, and some also show poor recovery characteristics (20, 38). However, some glassy polymers produce well-behaved BAW and SAW vapor sensors (90, 91). Partially crystalline polymers also show irreversible sensor response, proba-

Table X. Polymer-Coated Capacitance Sensors Reported for Gas Detection

Polymer	Gas	Ref.
Various polymers[a]	Organic vapors	42, 43, 84
Ethylcellulose	Organic vapors	51b, 91
Poly(epichlorohydrin)	Organic vapors	91
Poly(ether urethane)	Organic vapors	91
Aminopropyltrimethoxysilane–propyltrimethoxysilane copolymer	CO_2	87
Polyimide	H_2O	7
N,N-Dimethyl-3-aminotrimethoxysilane–propyltrimethoxysilane copolymer	SO_2	30, 83, 86
Poly(phenylacetylene)	CO, CO_2, CH_4, H_2O	84
3-Aminopropyltrimethoxysilane–n-butyltrimethoxysilane copolymer-bound antibody layer	Immunosensor	88

[a]Poly(dimethylsiloxane), poly(cyanopropylmethylsiloxane), poly(methyltrifluoropropylsiloxane), poly(cyanopropylmethyl–dimethylsiloxane), poly(vinylpyrrolidone), polystyrene, poly(vinyl acetate), polyethylene glycol, poly(vinyl chloride), poly(vinylpyrrole), and poly(acrylonitrile).

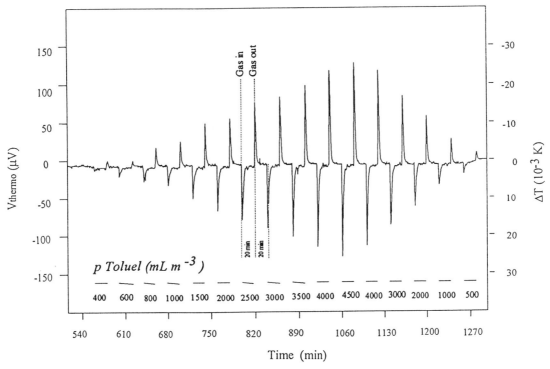

Figure 7. Typical V_{thermo} and ΔT responses of thermocouples of calorimetric thermopile sensors coated with poly(etherurethane) as a function of time for stepwise exposures to different concentration of toluene: V_{thermo} is the signal of three thermopiles connected in series (91).

bly due to changes in crystallinity after the sorption process.

The thickness and film morphology of the polymer coat also affect sensor performance. Thicker coatings may have relatively slower response and slower recovery. In terms of morphology, porous polystyrene coating gives relatively good sensor signals, whereas nonporous polystyrene coating shows only irreversible response with BAW devices (91).

Chemical Properties

Chemical structure of the coating material determines analyte solubility in the coat, and hence sensor sensitivity and selectivity. Nonpolar polymers strongly interact with nonpolar analytes and polar polymers interact with polar analytes, as is evident by the partition coefficient data in Table VIII and the corresponding normalized vectors in Figures 8 and 9 (59).

The relatively nonpolar polymer poly(dimethylsiloxane) has a higher partition coefficient for the aliphatic solvent (octane), whereas the more polarizable polymer poly(cyanopropylmethylsiloxane) interacts with the polarizable solvent (toluene). Only the more highly polar solvent (ethanol) strongly interacts with poly(isopropanoic acid-methylsiloxane) and poly-(aminopropylmethylsiloxane). However, octane, methylcyclohexane, and tetrachloromethane do not interact

with poly(isopropanoic acid-methylsiloxane). The interaction of alcohols with poly(isopropanoic acid-methylsiloxane) is mainly the result of hydrogen bonding (Chart I). In contrast, polar solvents, like chloroform. that are unable to form hydrogen bonding selectively interact with poly(aminopropylmethylsiloxane).

However, polymer–analyte interaction is not dependent merely on a single property. For a better understanding of these interactions, Abraham, Zhou, and co-workers compared (42b, 58, 91, 92) data of sensor coatings with those of chromatographic packings. Results obtained from sensor arrays also indicate that performance of polymer-coated sensors depends on the chemical structure of the polymer and its solubility power for the analyte (38).

Conclusions and Future Prospects

Functional polymers have been applied successfully for the development of chemical sensors based on a variety of transducers including acoustic wave, electronic conductance, and calorimetric transducers. Today, mass-sensitive acoustic wave chemical sensors are particularly promising. Such acoustic wave sensors can be used for detection in gaseous and liquid states. The majority of acoustic wave sensors studied so far for liquid phase are bulk acoustic wave biochemical sensors. Flexural plate wave, and especially horizontal

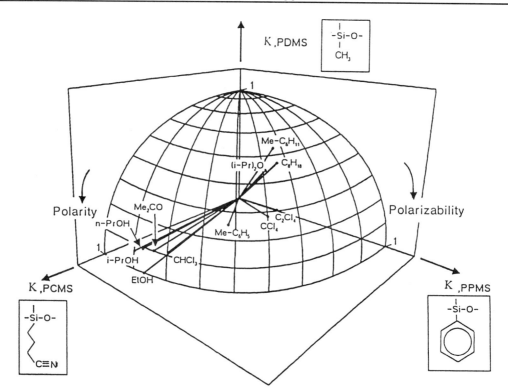

Figure 8. Three-dimensional plot of normalized partition coefficient vectors (K_{PCMS}, K_{PPMS}, K_{PDMS}) of different solvents in poly(cyanopropylmethylsiloxane) (PCMS), poly(phenylmethylsiloxane) (PPMS), and poly(dimethylsiloxane) (PDMS) on the basis of their poplarities and polarizability. The endpoint of each vector is positioned at a spherical surface represented by azimuthal and longitudinal lines and may be characterized by their polar coordinates (91).

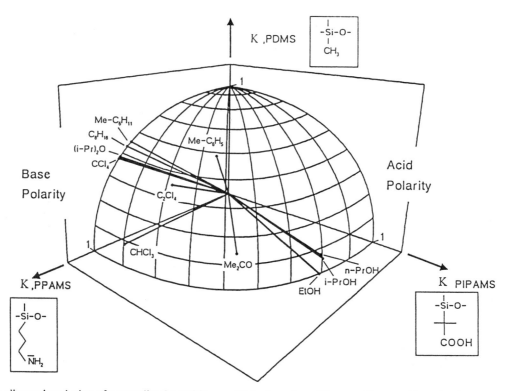

Figure 9. Three-dimensional plot of normalized partition coefficient vectors (K_{PAPMS}, K_{PIPAMS}, K_{PDMS}) of different solvents in poly(aminopropylmethylsiloxane) (PAPMS), poly(isopropanoic acid methylsiloxane) (PIPAMS), poly(dimethylsiloxane) (PDMS) on the basis as their acid–base polarity. The endpoint of each vector is positioned at a spherical surface represented by azimuthal and longitudinal lines and may be characterized by their polar coordinates (59).

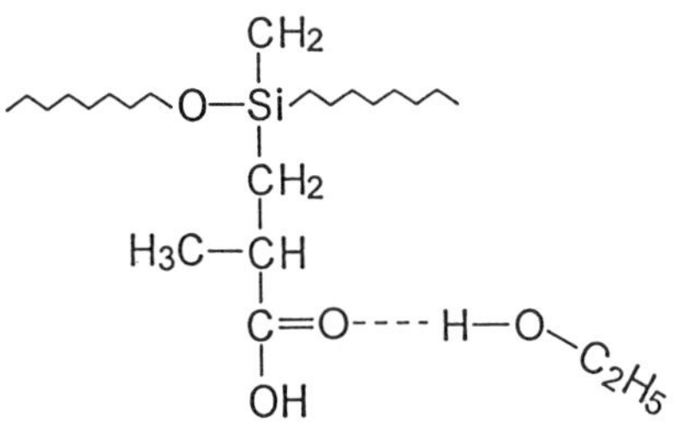

Chart I. Schematic presentation of hydrogen bonding between poly(isopropanoic acid methylsiloxane) and ethanol.

shear acoustic plate-mode devices, have also been developed. Electrochemical quartz microbalance can be used as a chemical sensor to measure analytes in the gas or liquid phase, but their most useful application is the determination of polymer properties such as T_g, moduli, and interfacial phenomenon. Semiconductive sensors (i.e., conducting polymers) are mainly used for detection of inorganic gases, but they also were used for organic vapors. Capacitance and calorimetric sensors based on the change of permittivity and thermal conduction, respectively, are useful for detecting large concentrations of analytes, but calorimetric sensors have no selectivity.

Because most functional polymers studied have poor selectivity for different analytes, the development of selective polymer-based chemical sensors remains the main challenge in the future of sensor development. Polymer selectivity may be improved by designing more efficient polymers, by derivatizing analytes, by preconcentrating analytes, by chromatographic separation, or by using selectively permeable membranes or tubings before detection. The development of sensor arrays may also improve selectivity and sensitivity.

References

1. Hlavay, J.; Guilbault, G. G. *Anal. Chem.* **1977**, *49*, 1890–1898.
2. Alder, J. F.; McCallum, J. J. *Analyst* **1983**, *108*, 1169–1189.
3. Mierzwinski, A.; Witkiewicz, Z. *Environ. Pollut.* **1989**, *57*, 181–198.
4. McCallum, J. J. *Analyst* **1989**, *114*, 1173–1189.
5. Fox, C. G.; Alder, J. F. *Analyst* **1989**, *114*, 997–1004.
6. Ward, M. D.; Buttry, D. A. *Science (Washington, D.C.)* **1990**, *249*, 1000–1007.
7. *Sensors: A comprehensive Survey*; Göpel, W.; Hesse, J.; Zemel, J. N., Eds.; VCH Verlag: Weinheim, Germany, 1991; Vol. 2.
8. Geckeler, K. E.; Müller, B. *Naturwissenschaften* **1993**, *80*, 18–24.
9. King, W. H., Jr. *Anal. Chem.* **1964**, *36*, 1735–1740.
10. Wohltjen, H. *Sens. Actuators* **1984**, *5*, 307–325.
11. Sauerbrey, G. *Z. Phys.* **1959**, *155*, 206–222.
12. White, R. M.; Wenzel, S.W. *Appl. Phys. Lett.* **1988**, *52*, 1653–1655.
13. (a) Jen, C. K.; Olivera, J. E. B.; Yu, J. C. H.; Dai, J. D.; Bussiere, J. *Appl. Phys. Lett.* **1990**, 56, 2183–2185. (b) Li, P. C. H.; Stone, D. C.; Thompson, M. *Anal. Chem.* **1993**, *65*, 2177–2180.
14. Buttry, D. A.; Ward, M. D. *Chem. Rev.* **1992**, 1355–1379.
15. Kanazawa, K. K.; Gorden, J. G., II *Anal. Chem.* **1985**, *57*, 1770–17771.
16. Ho, M. H. In *Application of Piezoelectric Quartz Crystal Microbalances*; Lu, C.; Czeanderna, W., Eds.; Elsevier: New York, 1984; pp 351–388.
17. Zhou, R.; Vaihinger, S.; Geckeler, K. E.; Göpel, W. *Sens. Actuators* **1994**, *B19*, 415–420.
18. Nieuwenhuizen, M. S.; Nederlof, A. J. *Sens. Actuators B* **1990**, *B2*, 97–101.
19. Lai, C. S. I.; Moody, G. J.; Thomas, J. D. R. *Analyst* **1986**, *111*, 511.
20. Brace, J. G.; Sanfelippo, T. S. *Sens. Actuators* **1988**, *14*, 47–68.
21. McCallum, J. J.; Fielden, P. R.; Volkan, M.; Alder, J. F. *Anal. Chim. Acta* **1984**, *162*, 75–84.
22. Randin, J.-P.; Züllig, F. *Sens. Actuators* **1987**, *11*, 319.
23. Fraser, S. M.; Edmonds, T. E.; West, T. S. *Analyst* **1986**, *111*, 1183–1188.
24. Slater, J. M.; Watt, E. J. *Analyst* **1991**, *116*, 1125–1130.
25. Moody, G. J.; Thomas, J. D. R.; Yarmo, M. A. *Anal. Chim. Acta* **1983**, *155*, 225–229.
26. (a) Zhou, R.; Haug, M.; Geckeler, K. E.; Göpel, W. *Sens. Actuators B* **1993**, *B15–16*, 312–316. (b) Zhou, R.; Geckeler, K.E.; Göpel, W. *Makromol. Chem.* **1994**, *195*, 2409–2421.
27. Fog, H.; Rietz, B. *Anal. Chem.* **1985**, *57*, 2634–2638.
28. Guilbault, G. G.; Lopez-Roman, A. *Environ. Lett.* **1971**, *2*, 35–45.
29. Frechte, M. W.; Fasching, J. L. *Environ. Sci. Technol.* **1973**, *7*, 1135–1137.
30. Endres, H.-E.; Mickle, L. D.; Kösslinger, C.; Drost, S.; Hutter, F. *Sens. Actuators B* **1992**, *6*, 285–288.
31. Edmonds, T. E.; West, T. S. *Anal. Chim. Acta* **1980**, *117*, 147–157.
32. Kindlund, A.; Lundström, I. *Sens. Actuators* **1982/83**, *3*, 63–77.
33. Cooper, J. B.; Edmonds, J. H.; Joseph, D. M.; Newbower, R. S. *IEEE Trans. Biomed. Eng.* **1981**, *BME-28*, 459–466.
34. Kindlud, A.; Sundgen, H.; Lundström, I. *Sens. Actuators* **1984**, *6*, 1–17.
35. Snow, A.; Wohltjen, H. *Anal. Chem.* **1984**, *56*, 1411–1416.
36. Sanchez-Pedreno, J.A. C.; Drew, P. K. P.; Adler, J. F. *Anal. Chem. Acta* **1986**, *182*, 285–291.
37. Tomita, Y.; Ho, M. H.; Guilbault, G. G. *Anal. Chem.* **1979**, *51*, 1475–1478.
38. Ballantine, D. S., Jr.; Rose, S. L., Grate, J. W.; Wohltjen, H. *Anal. Chem.* **1986**, *58*, 3058–3066.
39. Guiltbault, G. G.; Affolter, J.; Tomita, Y.; Kolesar, E. S., Jr. *Anal. Chem.* **1981**, *53*, 2057–2060.
40. Guiltbault, G. G.; Kristoff, J.; Oven, D. *Anal. Chem.* **1985**, *57*, 1754–1156.
41. Turnham, B. D.; Yee, L. K.; Luoma, G. A. *Anal. Chem.* **1985**, *57*, 2120–2124.

42. (a) Schierbaum, K. D.; Gerlach, A.; Haug, M.; Göpel, W. *Sens. Actuators A* **1992**, *31*, 130–137. (b) Zhou, R.; Weimar, U.; Schierbaum, K. D.; Geckeler, K. E.; Göpel, W. *Sens. Actuators*, in press.

43. Haug, M.; Schierbaum, K. D.; Gauglitz, G.; Göpel, W. *Sens. Actuators B* **1993**, *11*, 383–391.

44. Zhou, R.; Geckeler, K. E.; Schierbaum, K. D.; Göpel, W. *J. Macromol. Sci., Pure Appl. Chem.* **1994**, *A31,Suppl. 6–7*, 1291–1298.

45. Ho, M. H.; Guilbault, G. G.; Rietz, B. *Anal. Chem.* **1983**, *55*, 1830–1832.

46. Mirrison, R. C.; Guiltbault, G. G. *Anal. Chem.* **1985**, *57*, 2342–2344.

47. Ho, M. H.; Guilbault, G. G.; Rietz, B. *Anal. Chem.* **1980**, *52*, 1489–1492.

48. Woltjen, H.; Dessy, R. *Anal. Chem.* **1979**, *51*, 1465–1470.

49. Amati, D.; Arn, D.; Blom, N.; Ehrat, M.; Saunois, J.; Widmer, H. M. *Sens. Actuators B* **1992**, *7*, 587–591.

50. Yokoyama, K.; Ebisawa, F. *Anal. Chem.* **1993**, *65*, 673–677.

51. (a) Pehrsson, S. L.; Grate, J. W.; Ballantine Jr., D. S.; Jurs, P. C. *Anal. Chem.* **1988**, *60*, 2801–2811. (b) Zhou, R.; Hierlemann, A.; Schierbaum, K. D.; Göpel, W. *Sens. Actuators*, in press.

52. Charlesworth, J. M.; Riddel, S. Z.; Mathews, R. J. *J. Appl. Polym. Sci.* **1993**, *47*, 653–665.

53. Carey, W. P.; Beebe, K. R.; Kowaslki, B. P.; Illman, D. L.; Hirschfield, T. *Anal. Chem.* **1986**, *58*, 149–153.

54. Krosawa, S.; Kamo, N.; Matsui, D.; Kobatake, Y. *Anal. Chem.* **1990**, *62*, 353–359.

55. Grate, J. W.; Wenzel, S. W.; White, R. M. *Anal. Chem.* **1992**, *64*, 413–423.

56. Grate, J. W.; R-Pehrsson, S. L.; Venezky, D. L.; Klusty, M.; Wohltjen, H. *Anal. Chem.* **1993**, *65*, 1866–1881.

57. Grate, J. W.; Wenzel, S. W.; White, R. W. *Anal. Chem.* **1991**, *63*, 1552–1561.

58. Grate, J. W.; Snow, A.; Ballantine Jr., D. S.; Wohtjen, H.; Abraham, M. H.; McGill, R. A.; Sasson, P. *Anal. Chem.* **1988**, *60*, 869–875.

59. Schierbaum, K. D.; Hierlermann, A.; Göpel, W. *Sens. Actuators B* **1994**, *19*, 448–452.

60. Wenzel, S. W.; White, R. M. *Sens. Actuators A* **1990**, *21–23*, 700–703.

61. Hauptmann, P.; Lucklum, R.; Hartmann, J.; Auge, J. *Sens. Actuators A* **1993**, *37–38*, 309–316

62. Wang, J.; Ward, M. D.; Ebersole, R. C.; Foss, R. P. *Anal. Chem.* **1993**, *65*, 2553–2562.

63. Nomura, T.; Sakai, M. *Anal. Chem. Acta* **1986**, *183*, 301–305.

64. (a) Okahata, Y.; Ebato, H. *Trends Anal. Chem.* **1992**, *11*, 344–354. (b) Zhou, R.; Patskovsky, S.; Noetzel, G.; Schierbaum, K. D.; Göpel, W. *1. Dresdner Sensor Symposium, Forschungsgesellschafte für Meß- und Sensortechnik e.v. Dresen*, 14–16. December 1993, Dresen, Germany; Schierbaum, K. D.; Hierlemann, A.; Zhou, R.; Göpel, W. *7th International Symposium on Synthetic Membrane in Science and Technology*, August 29–September 1, 1994, Tübingen, Germany.

65. Roederer, J. E.; Bastiaans, G. J. *Anal. Chem.* **1983**, *55*, 2333–2336.

66. Thomopson, M.; Arthur, C. L.; Dhaliwal, G. K. *Anal. Chem.* **1986**, *58*, 1206–1209.

67. Muramatsu, H.; Dicks, J. M.; Tamiya, M.; Karube, I. *Anal. Chem.* **1987**, *59*, 2760–2763.

68. Muratsugu, M.; Ohta, F.; Miya, Y.; Hosokawa, T.; Kurosawa, S.; Kamo, N.; Ikeda, H. *Anal. Chem.* **1993**, *65*, 2933–2937.

69. Ebara, Y.; Okahata, Y. *Langmuir* **1993**, *9*, 574–576.

70. Charlesworth, J. M. *Anal. Chem.* **1990**, *62*, 76–81.

71. Martin, S. J.; Frye, G. C. *Ultroasonics Symp.* **1991**, *15(8)*, 393–398.

72. Ricco, A. J.; Martin, S. J. *Appl. Phys. Lett.* **1987**, *50*, 1474–1476.

73. Grate, J. W.; Klusty, M.; Mcgill, R. A.; Abraham, M. H.; Whiting, G.; Andonian- Haftvan, J. *Anal. Chem.* **1992**, *64*, 610–624.

74. Synowietz, C. *D'Ans-Lax, Taschenbuch für Chemiker und Physiker*; Springer-Verlag: Berlin, Germany, 1983; Band II.

75. Huang, P. H. *Sens. Actuators* **1985**, *8*, 23–28.

76. *Diffusion in Polymers*; Crank, J.; Park, G. S., Eds.; Academic: Orlando, FL, 1968; Chapter 1.

77. Miasik, J. J.; Hooper, A.; Tofield, B. C. *J. Chem. Soc., Faraday Trans. I* **1986**, *82*, 1117–1126.

78. Hanawa, T., Kuwabata, S.; Yoneyama, H. *J. Chem. Soc., Faraday Trans. I* **1988**, *84*, 1587–1592.

79. Christensen, W. H.; Sinha, D. N.; Agnew, A. F. *Sens. Actuators B* **1993**, *10*, 149–153.

80. Berthet, G.; Blanc, J. P.; Germain, J. P.; Laribi, A.; Maleysson, C.; Robert, H. *Synth. Met.* **1987**, *18*, 715–720.

81. Inagaki, N.; Tasaka, S.; Kobayashi, M. *Polym. Bull.* **1991**, *25*, 273–278.

82. Grate, J. W.; Klusty, M.; Barger, W. R.; Snow, A. W. *Anal. Chem.* **1990**, *62*, 1927–1934.

83. Endres, H.-E.; Drost, S. *Sens. Actuators B* **1991**, *4*, 95–98.

84. Hermanns, S. C. M. *Sens. Actuators* **1984**, *5*, 181–186.

85. Haug, M.; Schierbaum, K. D.; Endres, H.-E.; Drost, S.; Göpel, W. *Sens. Actuators A* **1992**, *32*, 326–332.

86. Lin, J.; Obermeier, E. *Sens. Acuators B* **1993**, *15-16*, 319–322.

87. Lin, J.; Heurich, M.; Schlichting, V.; Obermeier, E. *Technical Digest of the 4th International Meeting on Chemical Sensors*; September 13–17, 1992, Tokyo, Japan.

88. Saby, C.; Jattrezic-Renault, N.; Martelet, C.; Coin, B.; Carles, M.-H.; Delair, T.; Mandrand, B. *Sens. Actuators B*, **1993**, *15–16*, 458–462.

89. Bruckman, H. W. L.; de Goeje, M. P. *Prog. Org. Coat.* **1992**, *20*, 501–16.

90. Elias, H-G. *Makromoleküle*; Hüthig und Wepf Verlag Basel: Heidelberg, Germany, 1992; Band 2.

91. Zhou, Thesis, University of Tuebingen, Germany, 1994

92. Grate, J. W.; Abraham, M. H. *Sens. Actuators B* **1991**, *2*, 85–111.

Polymeric Stabilizers and Antioxidants

Wayne W. Y. Lau and Pan Jiang Qing

Over 2×10^8 tons of polymers are produced every year, and a large portion needs various types of stabilizers against degradation for longer service life. There are many effective stabilizers to choose from, and addition of stabilizers to polymers remains the most convenient and effective way of enhancing polymer life and performance. In this chapter the principles of four generations of stabilizers (screeners, UV absorbers, quenchers, and free-radical scavengers) are reviewed, and special attention is focused on hindered amine light stabilizers. Recent trends are discussed in the development of polymeric stabilizers, antioxidants, and multifunctional stabilizers for higher stabilizer effectiveness and stability. In each case, the background and synthetic polymer chemistry are discussed, and the performance of the resulting stabilizers is outlined. A brief comparison between conventional and polymeric stabilizers is also provided.

Polymeric materials exposed to sunlight undergo degradation that shortens their service life, mainly as a consequence of a process called photooxidation. There are several ways to combat photooxidation in polymers. The addition of light stabilizers to the polymer is the most convenient and effective, because such addition does not alter processing conditions to any significant extent and many effective stabilizers exist. In the rubber industry a stabilizer is often referred to as an antioxidant, whereas in the plastics industry a stabilizer is an antioxidant as well as a light stabilizer. Antioxidants include primary antioxidants (mainly sterically hindered phenols and aromatic amines) and secondary antioxidants, which are hydroperoxide decomposers containing organic phosphite and sulfur compounds. Light stabilizers, on the other hand, can be classified into screeners, UV absorbers, excited energy quenchers, and free-radical scavengers.

Polymeric materials have very wide applications in industry, agriculture, construction, medicine, defense, and other high performance technologies. More than 2 $\times 10^8$ tons of polymers are produced every year. A large proportion of these polymer products need light stabilizers against photodegradation, and a wide range of stabilizers are used for this purpose. However, as demand on stabilizer performance and stabilization mechanism increases, conventional stabilizers may not meet all the requirements of certain applications. For example, for products needing high temperature processing, conventional stabilizers may be volatile, unstable, or toxic. Under these circumstances polymeric stabilizers appear particularly attractive. In addition, small concentrations of one, two, or more different stabilizers may be covalently bound to polymers for increased effectivity. Polymeric stabilizers also overcome the problem of stabilizer exudation. In this review a general discussion of polymer stabilization is presented with special emphasis on recent advances in research and development, synthesis, and potential applications of polymeric light stabilizers. A brief discussion on polymeric antioxidants will also be included to provide a more complete coverage of polymeric stabilizers.

Photooxidation, Photodegradation, and Photostabilization

Polymeric materials such as transparent plastics for greenhouses, mulching films in agricultural applications, fishing nets, cables, pipes, surface coatings, and plastic construction materials undergo degradation to various extents under sunlight and artificial light. The main cause is photooxidation (1–5). The ozone layer in the atmosphere absorbs much of the solar irradiation, but light that reaches the earth of wavelength longer than 290 nm, particularly UV light in the range of 290–400 nm, can be quite damaging to polymeric materials (6). UV light carries sufficient energy to cause chain scissions in macromolecules and to produce free radicals, which in turn enhance the propagation of chain reactions (2, 4, 5).

According to principles of photochemistry, two prerequisites are needed to promote a photoreaction in a polymeric material. One prerequisite is that either the material itself, or something in the material, must be able to absorb light; the other requirement is that the number of chain scissions occurring per quantum absorbed must be sufficiently large. If the quantum yield per unit time is low, it may take a long time (for example a thousand years) for a photoreaction to become observable.

Certain polymeric materials, such as polymethylethylketones and polymers used in the Ecolyte process, contain UV-absorbing functionalities (Chart I). These polymers can undergo photolysis or photo-cross-linking in sunlight to produce free radicals and to initiate photodegradation reactions. For example, poly(1,2-butadiene) can undergo photocyclization, which renders the material brittle. These photodegradable materials have been widely used as packaging materials, mulching films, and products for which environmental pollution is of special consideration (2, 7, 8).

Materials like polypropylene (PP) and polyethylene (PE), which have a paraffinic structure, do not absorb light of wavelength longer than 186 nm (9–13). In theory these materials should not undergo photodegradation, yet they do degrade quite fast in sunlight. For example, under the summer outdoor weather condition of Beijing, PP can last for about 14 days; even in winter, it can only last for about 3 months (10). The rea-

son for the deterioration of these materials is that they contain impurities that are light-absorbing. Polyolefins may contain chromophoric impurities such as catalyst residues like TiO_2, Al_2O_3, Fe_2O_3, and Cl^- and by-products like polymeric carbonyl and hydroperoxides generated by high temperature processing (9, 11–13). Through their process history many polymeric materials can acquire minute quantities of light-absorbing chemicals necessary for degradation through photooxidation. Table I lists wavelengths to which some common polymers are susceptible because of light-absorbing impurities they inherit from various fabrication and thermal treatments (14).

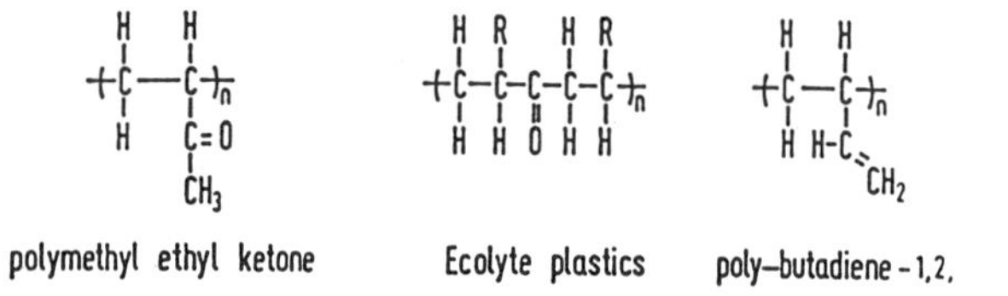

Free-Radical Autooxidation

After nearly 20 years of research on photooxidation in polymeric materials, a better understanding of the mechanisms involved has been reached. As in thermal oxidation, photooxidation in polymers is also a free-radical autooxidation reaction (Scheme I) (1–6, 9–26).

Oxymacroradicals, such as the polypropylene oxide free radical, can break down to smaller free radicals, carbonyls, and unsaturated compounds, which can cause further photooxidation and degradation in the polymer backbone (Scheme II). Photooxidation in a polymer can be retarded, or even inhibited, when the chain reaction is stopped.

Norrish-Type Photolysis

Norrish-type photolysis (1, 2, 4, 26) occurs in chemical compounds containing carbonyl groups (Scheme III). Free radicals are generated in the type I reaction, which can initiate chain photooxidation reactions. The type II reaction causes chain scission, in which ketones and fragments carrying a double bond are produced. Ketones are chromophoric and thus can initiate photooxidation. Plastics made by the Ecolyte process contain carbonyl groups that cause the material to degrade through Norrish-type photolysis. These

Table I. Wavelengths to Which Common Polymers Are Sensitive

Polymer	λ nm
Polyester	325
Polypropylene	310
Polyethylene	300
Polystyrene	318
Poly(vinyl chloride) (PVC)	310
PVC–poly(vinyl acetate) copolymer	364
Polycarbonate	295
Poly(methyl methacrylate)	290, 315
Polyformaldehyde	300, 320

Source: Adapted from reference 14.

Chart I. Examples of materials containing UV-absorbing functionalities.

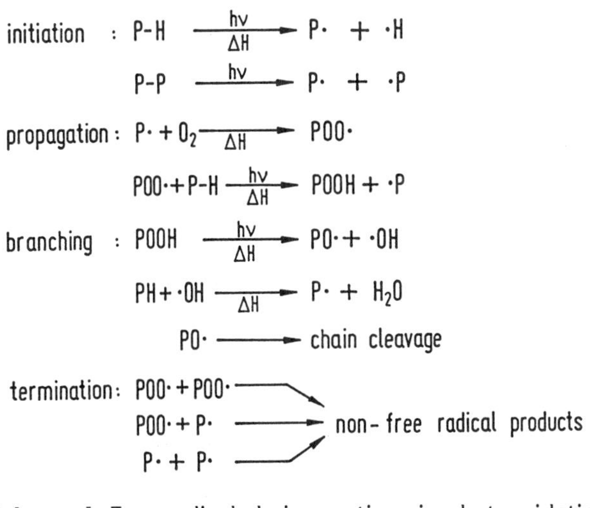

initiation : P-H $\xrightarrow[\Delta H]{h\nu}$ P· + ·H

P-P $\xrightarrow{h\nu}$ P· + ·P

propagation : P· + O$_2$ $\xrightarrow{\Delta H}$ POO·

POO·+P-H $\xrightarrow[\Delta H]{h\nu}$ POOH + ·P

branching : POOH $\xrightarrow[\Delta H]{h\nu}$ PO·+ ·OH

PH+ ·OH $\xrightarrow{\Delta H}$ P· + H$_2$O

PO· $\longrightarrow$ chain cleavage

termination: POO·+ POO·

POO·+ P· $\longrightarrow$ non-free radical products

P· + P·

Scheme I. Free-radical chain reactions in photooxidation of polymers.

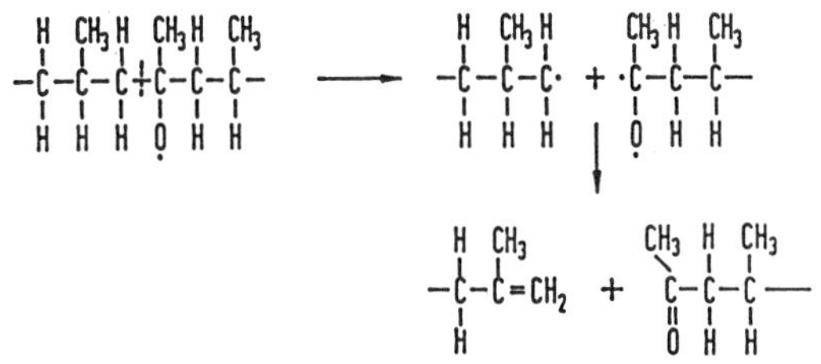

Scheme II. Breakdown of macro oxy-radicals producing smaller free radicals that can cause further photooxidation and degradation in a polymer backbone.

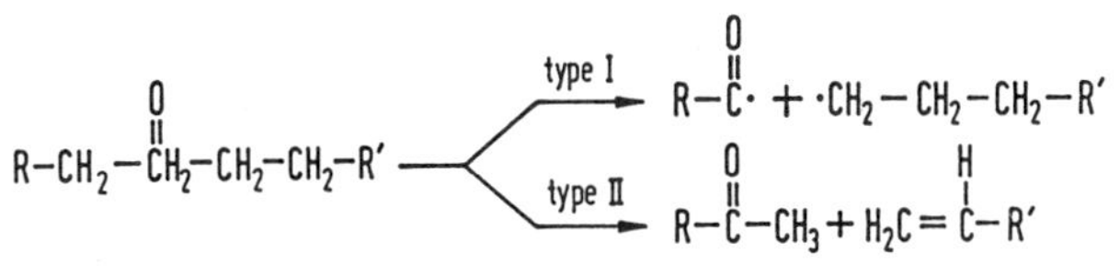

Scheme III. Norrish type photolysis of carbonyl compounds.

materials are regarded as environmentally friendly for mulching films and in packaging applications.

Oxidation by Singlet Oxygen

Singlet oxygen (2, 27), $^1O_2$, is active oxygen at a higher energy state. It can react with many organic compounds that carry double bonds. Polymers such as polybutadiene, polyisoprene, and natural rubber are prone to oxidation by $^1O_2$. In general, this type of oxidation cannot be inhibited by phenolic antioxidants, and only $^1O_2$-quenchers can stop it. $^1O_2$ can be generated by microwave discharge and by certain special chemical reactions. For example, some dye molecules in solution exist in an excited state and react with atmospheric oxygen to yield $^1O_2$ through energy transfer. Trozzolo and Winslow (28) were the first to propose the $^1O_2$ oxidation mechanism shown in Scheme IV.

Saturated polymers are not sensitive to $^1O_2$; however, compounds containing many double bonds (like polydienes) are very sensitive to $^1O_2$, which attacks a hydrogen atom at the double bond to form a peroxide by a mechanism known as ene-photooxidation. The generated peroxide may trigger the formation of free radicals, which in term lead to degradation by photooxidation.

The three mechanisms outlined previously have profound implications for photooxidative polymeric materials. Free-radical chain autoxidation plays an important role in all degradation reactions in polymers, whereas Norrish photolysis takes place mainly in polymers carrying carbonyl groups. Oxidation by $^1O_2$ occurs mainly to polydienes.

Photostabilization of Polymers

There are many ways to prevent photooxidation in polymeric materials. For example, the polymer may be

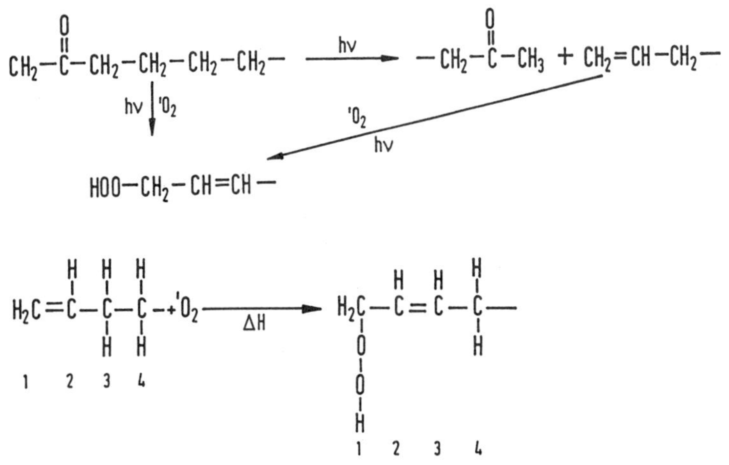

Scheme IV. Oxidation by singlet oxygen (top); Ene-photooxidation (bottom).

coated so that it is not exposed to light, the polymer can be modified by grafting and compositing, and high purity monomers can be used. Reduction in catalyst residue in the polymer and less branching in the polymer structure also help to reduce the probability of photooxidation. However, addition of stabilizer remains the most effective and most popular way of combating photodegradation (1–5, 11–26). The practical reason for this method is the fact that nowadays there are many effective stabilizers available to meet various needs, and stabilizer addition to the polymer need not require or cause any significant change in existing polymer processing technologies. Research and development on stabilizers and their applications have enabled polymeric materials to enjoy a service life between 20 to 50 years. For example, a 2% carbon black can render 20 years of outdoor life to PE, 50 years to PE–PP copolymer roofing, and over 5 years to the highly photooxidation-susceptible PP in outdoor applications (29, 30).

Research on photooxidation and photostabilization of polymeric materials has been very extensive during the past 20 years, and a great deal has been learned about the mechanisms involved. The first journal on polymer degradation and stabilization, *Polymer Degradation and Stability*, appeared in 1981 and was followed by the publication of two book series: *Development in Polymer Degradation*, Vols. 1–7, N. Grassie, Ed., Applied Science Publishers, 1977–1987, and *Development in Polymer Stabilization*, Vols. 1–8, G. Scott, Ed., Elsevier Applied Science Publishers, 1979–1987. In addition, several international conferences were held. These activities indicate that research on polymer degradation and stabilization has reached maturity, and, as a result, four generations of highly effective photostabilizers have evolved, including screeners, UV absorbers, quenchers, and free-radical scavengers.

Screeners

Screeners are the first generation of photostabilizer. They act by shielding the substrate from light. This mechanism is, in principle, the simplest, because a screener interferes with the first step in a photooxidation process. Reflective or opaque pigments like TiO_2, Cr_2O_3, Pb_3O_4, ZnO, Fe_2O_2, and Fe_3O_4 are used. Organic pigments like azo compounds, anthraquinone, and thioindigo compounds also find wide application in the plastics industry (2, 3). Carbon black is a very effective UV screener; it can also act by other mechanisms such as UV absorption and radical scavenging (4).

UV Absorbers

Most screeners are colored compounds, a characteristic that excludes their application in noncolored products.

Inorganic screeners also pose effectiveness and compatibility problems with organic substrates. These shortcomings encouraged the development of the second generation of stabilizers, namely UV absorbers (1–3). These stabilizers have excellent UV-absorption properties. Among popular UV absorbers are *o*-hydroxybenzophenone (UV-531) and *o*-benzotriazole (UV-327) (Scheme V). These compounds absorb UV energy and then dissipate this energy in a harmless way, perhaps through intramolecular proton transfer between the keto and the enol tautomers. Benzotriazoles are believed to exhibit a similar behavior. Salicylates and benzoates under the effect of light can undergo Fries rearrangement (1, 2) to *o*-hydrobenzophenone structure.

Quenchers

Quenchers belong to the third generation of stabilizers. They include many nickel chelates related to those shown in Scheme VI. When a polymer molecule (P_0) absorbs light energy it may be raised to an excited state (P^*). A quencher is able to absorb the excited-state energy (E) from the excited macromolecule and return it back unharmed to its ground state (P_0). Oth-

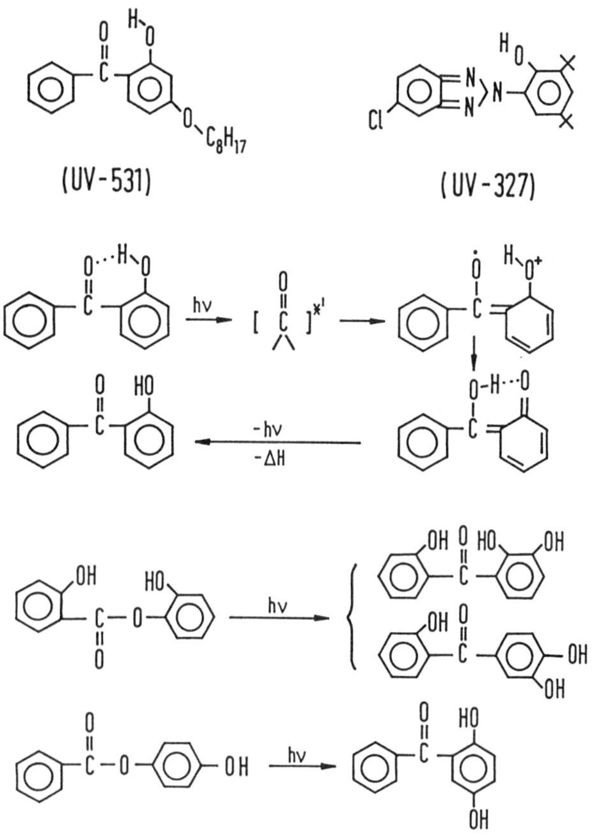

Scheme V. Examples of UV-absorber (top); "keto" and "enol" tautomers (middle); Fries rearrangement (bottom).

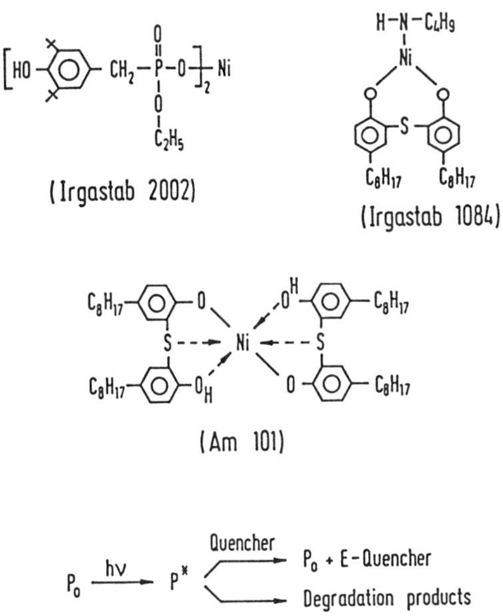

(Irgastab 2002)

(Irgastab 1084)

(Am 101)

Scheme VI. Examples of quenchers and their mechanism of action.

R = Alkyl, P = Polymer

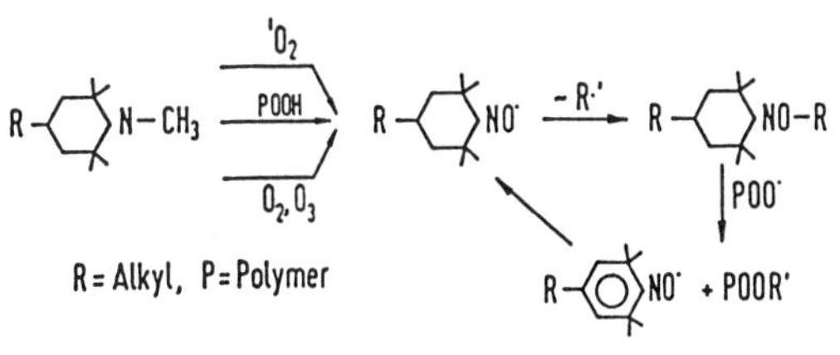

Scheme VII. Free-radical scavenging by a HALS.

erwise, the excited polymer may undergo photodegradation to degradation products (Scheme VI).

Unlike UV absorbers, the effectiveness of quenchers is not affected by thickness of the substrate material. Hence, quenchers are particularly suitable for thin film products. Recent work indicated that photostabilizing mechanisms of nickel chelates are rather complex. In addition to quenching the excited state, they may also act as hydroperoxide decomposers and free-radical scavengers (1, 2, 23, 26). However, quenching is probably their main function.

Free-Radical Scavenger

Sterically hindered phenols and hindered amine light stabilizers (HALS) are popular fourth generation photostabilizers. They act by scavenging free radicals (1–3) formed by autooxidation in a polymer backbone. In the course of photooxidation, a steric phenol can form molecular structures that can exert photosensitizing effects (37–40); therefore, the photostabilizing effectiveness of steric phenols is low. At high concentration, steric phenols can even sensitize a photooxidation reaction. They are not photostabilizers but they are indeed very effective thermal antioxidants. Hindered amines are truly effective light stabilizers. They protect the substrate by a combination of several mechanisms, including free-radical scavenging, hydroperoxide decomposition, and excited-state quenching (1–3). However, radical scavenging appears to be their main function (Scheme VII).

HALS are very effective photostabilizers even at as low a concentration as 0.02%, and their effectiveness does not appear to change with thickness of the

substrate. This capability may be because HALS have a special affinity for hydroperoxide. The concentration of HALS in the immediate surrounding of a hydroperoxide can be as high as 25 times higher than in other regions of the same matrix (41). Furthermore, its stable nitroxy radical can be regenerated to take on more harmful radicals. Autosynergism among the generated hydroamine and nitroxy radicals also is believed to occur and further reinforces the effectiveness of HALS. Synergism also exists among different types of stabilizers, for example between HALS and UV absorbers and between primary antioxidants and secondary antioxidants (1–3).

Commercial Photostabilizers

The ability of a compound to inhibit a photooxidation reaction does not necessarily make a chemical compound a useful photostabilizer. A useful photostabilizer must meet the following requirements: high enough thermal and photostability, high resistance to extraction, compatibility with the polymer substrate, high effectiveness, colorlessness, nontoxic nature, and low cost. On the basis of these criteria, the following trends in the development of polymer stabilizers have been pursued in recent years.

1. Multifunctional stabilizers (Chart II) are an obvious preference. For example GW-540 is a free-radical scavenger as well as a hydroperoxide decomposer. Tinuvin-144 scavenges free radicals and also acts like an antioxidant; whereas nickel chelates can decompose hydroperoxide, scavenge free radicals, and quench excited states (42).

2. Autosynergistic stabilizers carry functional groups that reinforce the effect of each other. For example, the effectiveness of GW-540 is believed to be due to a synergism between the hindered amine and the phosphite group. The thiol group in 2,6-di(*tert*-butyl)4-propanylthiolphenol (DBPT, Chart II) reinforces the effect of its phenolic group, a compounded effect that makes it a hydroperoxide decomposer as well as a free-radical scavenger.

3. High thermal stability, compatibility, and resistance to extraction are also important. A stabilizer of higher molecular weight and compatibility with the substrate exhibits higher thermal stability and

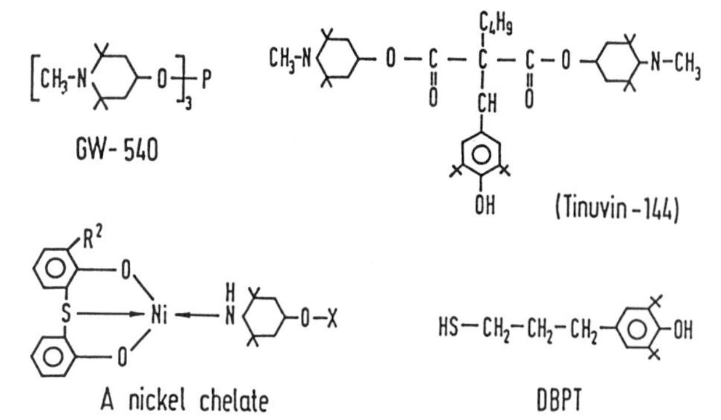

GW-540

(Tinuvin-144)

A nickel chelate   DBPT

Chart II. Examples of multifunctional (GW-540, Tinuvin-144, and nickel chelate) and autosynergistic (DBPT) stabilizers.

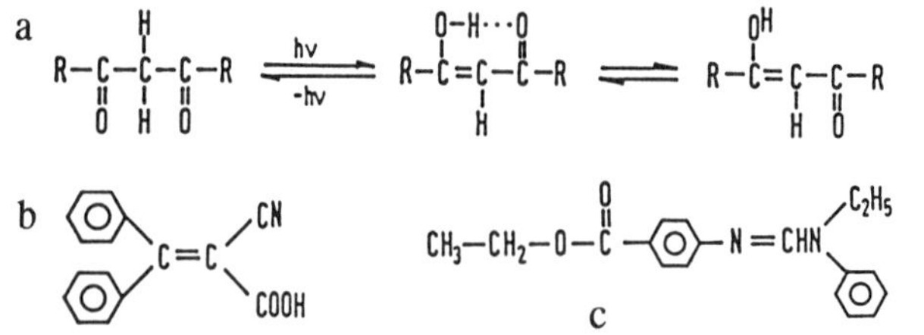

Scheme VIII. Tautomorism in a β-diketone (a); 2-cyano-3,3-diphenyl acrylate, which can undergo cis–trans-isomerization (b); phenylformamidine (c).

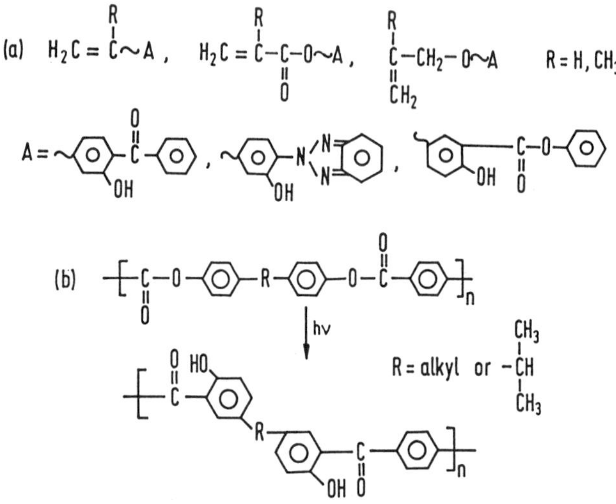

Scheme IX. Example of vinyl monomers carrying UV-absorbing functional groups (a); photo-Fries rearrangement of an aromatic polyester producing a polymeric UV-absorber (b).

durability. A stabilizer can be made more durable by increasing its molecular weight or lengthening its alkyl chain (43–45). Long alkyl side chains help promote better compatibility with the substrate.

4.   Polymeric stabilizers with high molecular weight have several desirable properties. Copolymerization of suitable monomers with different stabilization functionalities can lead to combinations of desirable properties in the copolymers.

Polymeric Stabilizers

The molecular weight of polymeric stabilizers brings higher thermal stability, more resistant to extraction, and lower toxicity. Therefore, the effectiveness of polymeric stabilizers is more lasting. Polymeric photo-stabilizers have been used successfully in applications where polymers have to go through heat processing involving large surfaces such as in fiber and film making (33). Among the four types of photostabilizers, screeners are colored, and nickel-chelate-type quenchers are toxic. Therefore, research and development efforts are focused mainly on polymeric UV absorbers and HALS.

Polymeric UV Absorbers

Classic UV absorbers (1, 2, 45) include o-hydroxy-benzophenone, o-hydroxybenzotriazole, phenyl salicylate, and phenylbenzoates. They function mainly through fast proton transfer in tautomorism by which excited state energy in a polymer is dissipated. β-Diketone (Scheme VIIIa) can undergo rapid tautomorism to produce a photostabilizer effect (46). Excited state energy also can be dissipated by cis–trans isomerization of some double bond compounds such as 2-cyano-3,3-diphenyl acrylate (Scheme VIIIb) (1, 2). Certain phenylformamidines (Scheme VIIIc) protect their substrate by absorbing UV radiation and quenching excited state energy (47). In principle, these structures

can be built into monomers and then polymerized to produce polymeric light stabilizers (48, 49).

Polymerization of vinyl monomers carrying photo-stabilizer functionality (Scheme IXa) can be carried out by free-radical polymerization (7, 48, 49). Aromatic polyesters can also be subjected to UV radiation to produce o-hydroxybenzophenone segments via photo-Fries rearrangement (Scheme IXb).

Monomers carrying UV-absorbing functional groups can also be bound to a polymer molecule by grafting through a free-radical mechanism, or a macromolecular chain radical can be end-capped with UV-absorbing functionality (50, 51). Grafting can also be carried out during thermal processing of a polymer (50, 51). Poly-o-hydroxybenzophenone can be prepared through polycondensation (with dicarboxylic-diol) of the monomer, a derivative of o-hydroxybenzophenone containing dicarboxylic acid, or diol groups (45, 52–54).

Polymeric UV absorbers offer high thermal stability and high resistance to extraction. Luston et al. (55) demonstrated that with a molecular weight higher than 1600, a UV absorber would not be extracted with trichloroethylene. Substrate–stabilizer compatibility can be improved by incorporating monomers with long alkyl chains through copolymerization (*44, 45*). Usually low molecular weight UV absorber would exudate to the substrate surface when used at concentration near 1% (*45*). Polymeric stabilizer can be applied at concentration far higher than 1% without exudation. In fact, a polymeric stabilizer can be added as required if the mechanical properties of the substrate are not affected. This major advantage for a UV absorber occurs because the more light that is absorbed by the higher concentration of UV absorber, the less the substrate would be affected. Higher thermal stability and higher resistance to extraction enable the use of polymeric UV absorbers in applications that require high temperature processing, long exposure to liquids, and exposure to an outdoor environment (*48*).

Polymeric UV absorbers have, at the same time, some inherent shortcomings:

1. It is expensive to make a monomer with UV absorbing functionality (*1*).
2. Their effectiveness is low when used at low concentrations.
3. When used in high dosage they may cause the substrate molecules to crystallize, with adverse effect on the mechanical properties of the product.

Moreover, their effectiveness declines with increase in molecular weight (*56–60*), often to below that of low molecular weight UV absorbers (*52, 61*). When UV-absorbing functional groups are grafted to the surface layer of a substrate, the amount of grafting is usually small and the thin grafted layer is not able to completely absorb harmful UV light that strikes the surface. Photooxidation and degradation can still occur in the subsurface layers. In addition, effectiveness of UV absorbers is generally far below that of HALS. As a result few companies would invest in research on polymeric UV absorbers, and very few polymeric UV absorbers are in commercial use (*1, 2*). One example of an available product is Permasorb MA (*2*).

Polymeric HALS

Background

HALS (*1–3*) refer to derivatives based on 2,2,6,6-tetramethylpiperidine (Scheme X). Developed around 1970, these hindered amine derivatives have proved to be the most effective photostabilizers. Neiman (*62*) reported his finding of stable nitroxy free radical in 1962, and subsequently the application of this type of stabilizers to commercial products was mainly developed in Japan. Japanese researchers worked extensively on 2,2,6,6-tetramethylpiperidines; they screened 1200 related compounds (*63*) and finally selected two (*see* subsequent discussion) to be used as effective photostabilizers. The HALS were commercially produced

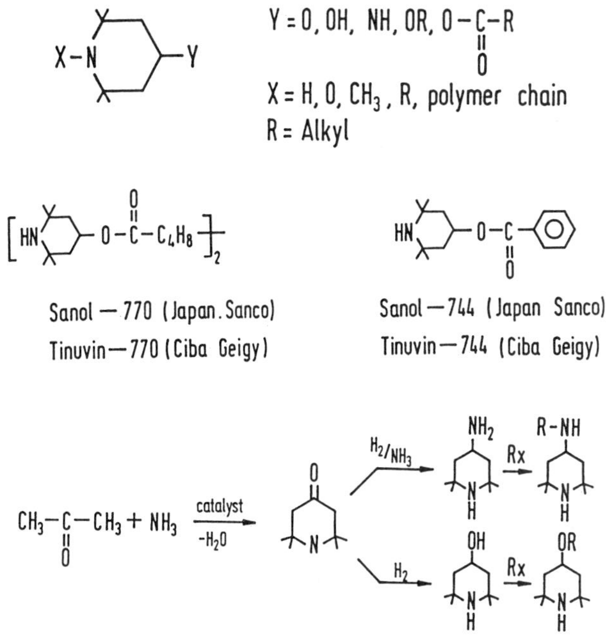

Scheme X. 2,2,6,6-Tetramethyl piperidine (top) and two commercial HALS derived from it (middle); synthesis routes for making these HALS (bottom).

in 1969–1973 in Japan, and later Ciba-Geigy joined this venture to distribute the HALS products to the international market.

From 1970 to 1985, there were 600 patents taken out on HALS (*33*); 80 companies and research institutes participated in research and development of HALS and averaged 50 patents each year. In addition, a large number of research publications have appeared on the mechanisms and effectiveness of HALS (*33, 64, 65*). 2,2,6,6-Tetramethylpiperidine is the main intermediate for the synthesis of HALS, and it can be produced through the reaction of acetone with ammonia (Scheme X).

The initially developed HALS were low molecular weight derivatives that had high volatility and low resistance to extraction. Higher molecular weight HALS are a trend in recent HALS development (*57*). Table II lists a number of recently developed commercial HALS, including high molecular weight HALS and polymeric HALS.

Synthesis of Polymeric HALS

Vinyl monomers carrying a hindered amine undergo copolymerization or homopolymerization to yield polymeric HALS (*1–3, 57, 61*) like PDS in Chart III (*1, 3*). Other polymeric HALS like Tinuvin-622 and Chimassorb-944 (Chart III) are obtained by polycondensation (*1, 3, 57, 62*). Alternatively many polymeric piperidinyl HALS can be prepared through functionalization of preformed polymers (Scheme XI) (*66*).

In certain polymerization processes such as the free-radical polymerization of styrene or methyl

Table II. Examples of Commonly Available HALS, Including Polymeric HALS

Trade Name	Chemical Structure	MW	T_m °C	Marketed
GW 540 Ruian, China	$\left(CH_3-N\langle\bigcirc\rangle-O\right)_3P$	541	120–122	March 1980
PDS Beijing	(styrene–piperidinyl methacrylate copolymer structure)	3000	122	1982
Tinuvin 770 Ciba-Geigy	$\left(HN\langle\bigcirc\rangle-O-\underset{O}{\overset{}{C}}-C_4H_8\right)_2$	480	81–86	1974
Tinuvin 144	$CH_3N\langle\bigcirc\rangle-O-C(=O)-C(C_4H_9)(CH_2\text{-}Ar\text{-}OH)-C(=O)-O-\langle\bigcirc\rangle NCH_3$	685	146–150	1980
Sanol 2626	$HO-\langle\bigcirc\rangle-C_2H_4-C(=O)-O-C_2H_4-N\langle\bigcirc\rangle-O-C(=O)-C_2H_4-\langle\bigcirc\rangle-OH$	722	135–140	March 1981
Mark LA-57	$CH_2-CH-CH-CH_2$ tetra(C=O–O–R); $R=\langle\bigcirc\rangle NH$	790	137–140	Sept. 1983
Mark LA-52	$CH_2-CH-CH-CH_2$ tetra(C=O–O–R); $R=\langle\bigcirc\rangle NCH_3$	846	139–141	March 1987
Mark LA-67	$CH_2-CH-CH-CH_2$ tetra(C=O–O–R); $R=\langle\bigcirc\rangle NH$ and, $-C_{13}H_{27}$	900	(liquid)	June 1986
Tinuvin 622 Ciba-Geigy	$H\left[O-\langle\bigcirc\rangle N-C_2H_4-O-\underset{O}{\overset{}{C}}-C_2H_4-\underset{O}{\overset{}{C}}\right]_n O-CH_3$	3000	130–146	March 1981
Chimassorb 944	$\left[N-(CH_2)_6-N-\langle\text{triazine}\rangle\right]_n$ with piperidinyl NH groups and C_8H_{17}	>2500	100–300	July 1984

Continued on next page

Table II. Examples of Commonly Available HALS, Including Polymeric HALS—*continued*

Trade Name	Chemical Structure	MW	Tm °C	Marketed
Cyasorb UV 3346 A.C.C.	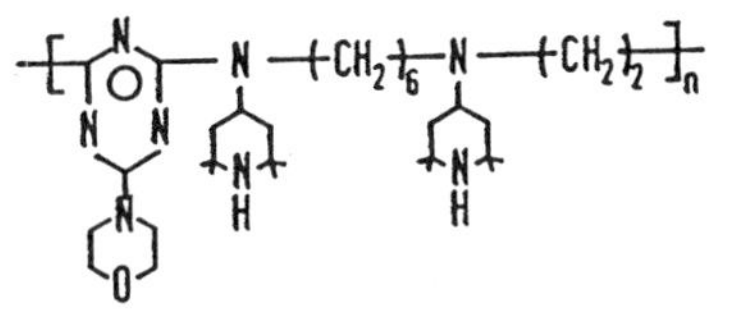	>2000	110–130	Feb. 1987
Chimassorb 119	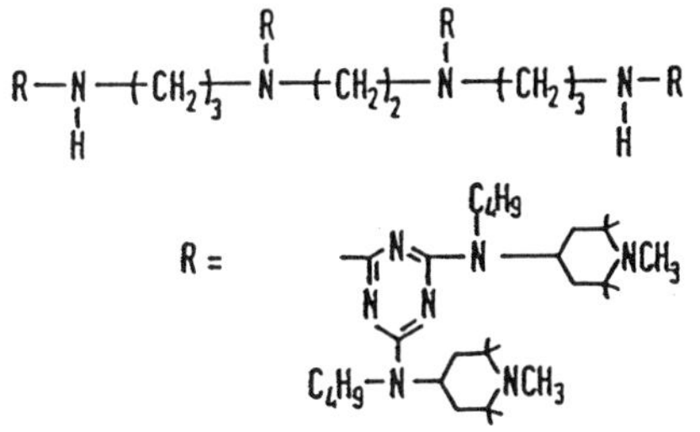	2282	—	—
HALS-4	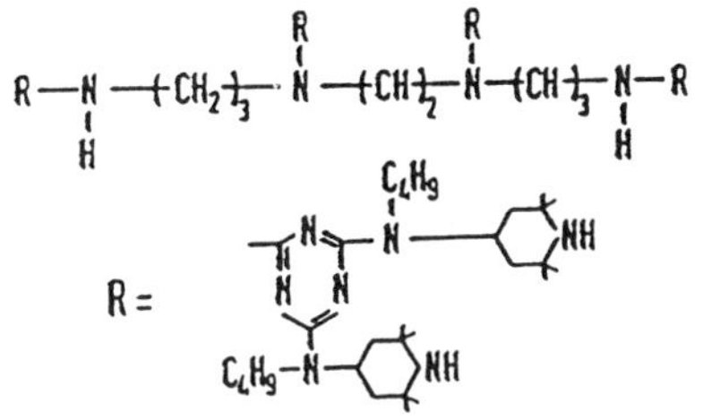	2170	—	—
Spinuvex A36 Borg-Werner	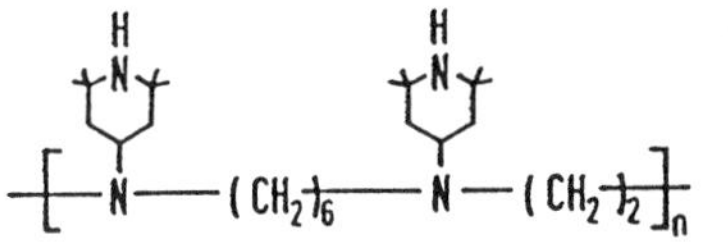		—	—

Note: Dashes mean data are not available.

Sources: Data are tkaen from references 3, 57, and 61.

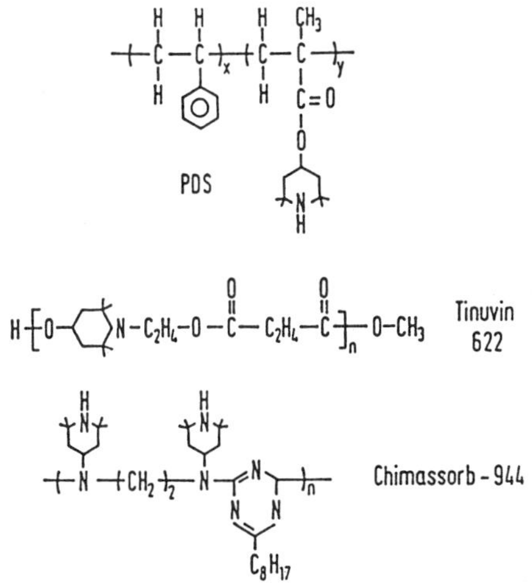

Chart III. Examples of polymeric HALS obtained by polymerization (PDS) or polycondensation (Tinuvin-662 and Chimassorb-994).

methacrylate, vinyl monomers carrying hindered amine groups can be added at a late stage of polymerization so that they do not adversely affect the polymerization (*1*, *50*). Alternatively, HALS monomers can be added during thermal processing of a polymer, during which time segmental free radicals produced from chain scissions will react with the added monomers to produce polymer-bound HALS (*1*, *50*, *67*).

In principle, grafting of HALS functionality onto a polymer surface should be an effective way of applying a HALS, because a small quantity of HALS can be distributed over a substrate surface at the frontline attack by photooxidation (*1*, *2*, *11*). Figure 1 shows the higher performance characteristics of HALS grafted onto PP film surface compared with those of HALS blended with PP (*68*).

However, this type of monomer is expensive, and the grafting operation can lead to an environmental pollution problem. The grafted surface may also lose its effectiveness when it rubs against other surfaces. Therefore, this method of HALS application is not yet commercialized.

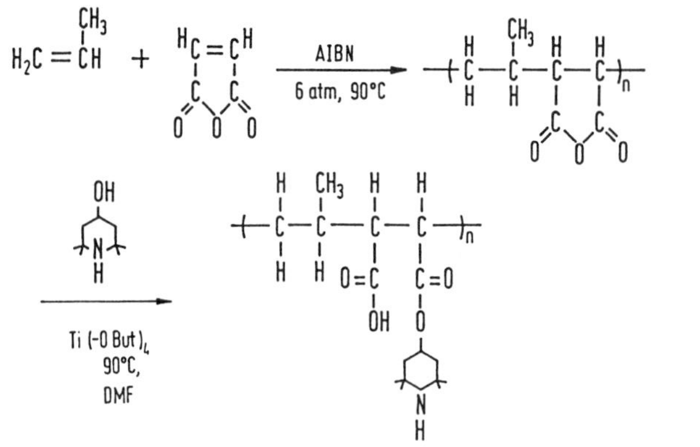

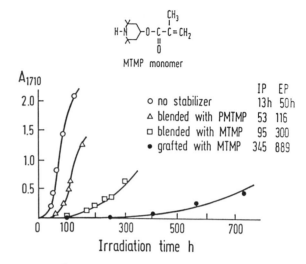

Scheme XI. Example of synthesis of polymeric HALS by polymer functionalization.

In addition to polymerization, oligomeric photostabilizers can be prepared through simple reactions of multifunctional organic compounds to yield photostabilizers of near optimum molecular weight. Some of these routes are polycarboxylation [e.g., Mark LA-57, Mark LA-52, and Mark LA-67 in Table II (57, 61)], esterification of pentaerythritol (e.g., A1010, Table III), condensation of multifunctional amines [e.g., Chimassorb 119 in Table II (57, 61)], condensation with triisocyanates, hydrosilylation (90), and triazination (57, 61). For example, triphenyl methane triisocyanate (M_r, 367.38) reacts with 2,2,6,6-tetramethyl piperidinol-4 to yield a HALS of M_r of 838. Star oligomeric HALS have been prepared through hydrosilylation (87, 90). Monomers such as 2,2,6,6-tetramethyl-4-piperidinyl-*m*-isopropenyl-α,α-dimethylbenzyl carbamate (Scheme XII) possess high photostabilizing effectiveness (88, 89). Such monomers undergo hydrosilylation to yield photostabilizers in the molecular weight range of 1000 to 2000 (90).

Triazine ring can be derivatized to carry functional groups of rather high reactivity. Compounds containing hindered amine and bifunctional groups were made to react with such triazines to make polymeric HALS such as Cyasorb UV3346, Chimassorb-119, and Chimassorb-944 (57, 61; Table II). However, controlling M_r of HALS made via reaction with triazine is difficult.

Figure 1. Performance of HALS grafted onto PP films compared with HALS blended into PP film: IP, induction period (h); EP, embrittlement period (h).

Also, cross-linking may occur, which is undesirable, because cross-linked HALS are difficult to disperse in a substrate. Compounds containing hindered amine and mono-functional groups, on the other hand, can be made to combine with triazines to yield HALS of M_r in the range of 1000 to 2000 (Chart IV).

Low molecular weight HALS is prone to evaporate or sublime. This characteristic is why many low molecular weight piperidine compounds cannot be effective photostabilizers (63). Therefore, increasing its molecular weight is a very important way to enhance the effectiveness of a HALS.

Figure 2 shows a general profile of service life of some HALS in relation to their molecular weight. Table IV also shows effect of molecular weight of polymeric HALS on the mechanical strength (tensile) of PP film in a thermal oxidation environment.

However, higher molecular weight in HALS does not necessarily always lead to higher effectiveness because bulkier and heavier molecules have lower

Table III. Effect of Molecular Weight of Steric Phenols on Their Antioxidant Effectiveness

Antioxidant Trademark	Structure	Phenolic Oxygen (%)	MW	Induction Period at 140 °C (h)
BHT	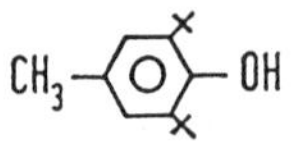	7.3	220	6
A1010	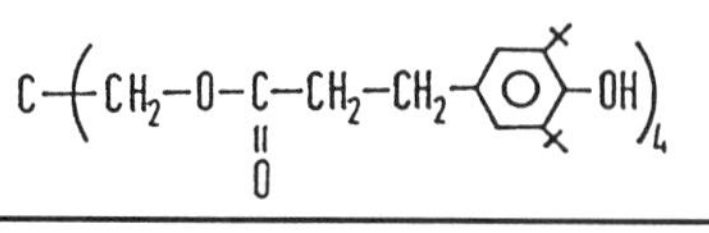	5.5	1176	198

Source: Adapted from reference 10.

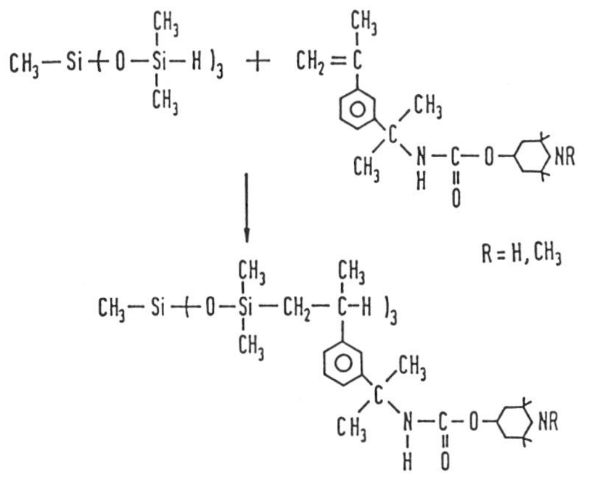

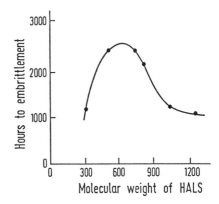

Scheme XII. Synthesis of oligomeric HALS via silylation of monomeric HALS carrying a vinyl group.

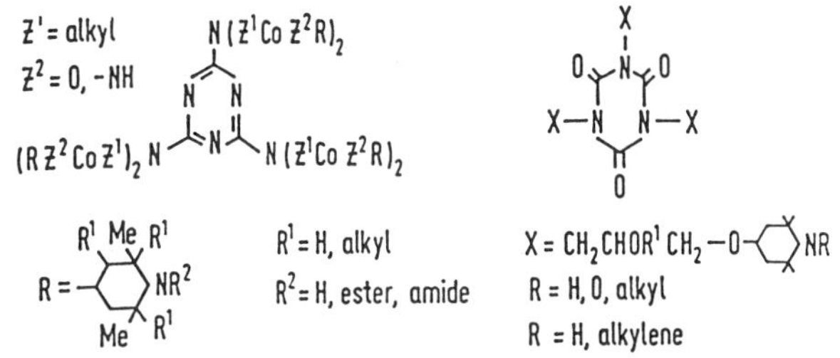

Z¹ = alkyl
Z² = 0, -NH

R^1 = H, alkyl
R^2 = H, ester, amide

X = CH₂CHOR¹CH₂—O—NR
R = H, 0, alkyl
R = H, alkylene

Chart IV. Oligomeric HALS based on triazine.

Figure 2. Effect of molecular weight of HALS (Mark AO-50 with carboxyl groups) on its performance.

Table IV. Effect of MW on Stabilizing Activity of Polymeric HALS in PP films

	Stabilizing Effect on PP	
MW of HALS	50% Tensile Strength Retention (days)	Thermooxidation Stability (days)
1800	267.5	125
2720	280.0	127
6800	172.5	98
13,200	139.6	88
23,400	79.7	51
No HALS	28.8	39

Note: 0.2% poly(1,2,2,6,6-pentamethylpiperidine 4-acrylate).

Source: Adapted from reference 72.

mobility in the substrate and they become more difficult to disperse evenly in the matrix (*57, 58, 60, 72, 82–85*). An optimum molecular weight of a HALS exists at which its effectiveness is at its highest. The optimal molecular weight for polymeric HALS is different from that for nonpolymeric HALS. An optimum molecular weight for polymeric HALS is around 2700 (*72*); for ordinary, HALS this value is around 600 (*57*). This optimum value also varies with the structure of a HALS. Furthermore, the effectiveness of a HALS depends on its applied dosage, the type of polymer substrate, its molecular weight, and test (or usage) conditions (*58–60, 72, 74*).

Among currently available HALS Tinuvin-770 with a molecular weight of 480 shows very high effectiveness and sufficiently high thermal stability. It is a white powder that facilitates application. However, it has low resistance to extraction by water, and this characteristic excludes its usage for fibers. On the other hand polymeric HALS like PDS and Tinuvin-622 (Table II) possess high enough thermal stability and high resistance to extraction. Therefore, they are suitable for fibers and films even though they are otherwise less effective compared with Tinuvin-770.

Polymerization remains an important route for the preparation of polymeric photostabilizers. Because of difficulties in solving environmental pollution problems arising from grafting processes, grafting technology for making polymeric photostabilizers cannot be

realized commercially yet. Grafting on a substrate surface is not long lasting because of mechanical abrasion of the surfaces (*50*). One potentially attractive route for attaching photostabilizing functionality to a substrate matrix is a method of chemical binding developed by Scott (*50*), in which suitable monomers are added to a polymerization system at a late stage of polymerization or are added during a melt processing operation. On the other hand, polymeric photostabilizers made through formation of macromolecules has attracted much research interest (*50*).

Polymeric Antioxidants

Background

The chief function of an antioxidant is to protect a polymer substrate against thermal oxidation (*75*). Such protection is required and is necessary for polymers during thermal processing, fabrication, storage, and even for room temperature applications. For instance, without the addition of antioxidants certain polymers cannot be melt spun to produce fibers because of rapid degradation by thermal oxidation.

Antioxidants are classified into primary and secondary antioxidants. Primary antioxidants consist mainly of steric phenols and aromatic amines, whereas secondary antioxidants (also called hydroperoxide decomposers) include organic phosphites and organic thio-compounds. Primary antioxidants act mainly through the steric hydroxyl function (–OH) by scavenging free radicals in the substrate matrix and hence breaking up chain reactions. The stable radicals so generated can couple to form molecules to terminate a chain reaction. Aromatic amines act through a similar mechanism.

Secondary antioxidants act through a non-free-radical mechanism, in which polymeric hydroperoxides are decomposed so that autooxidation in macro free radicals cannot take place. The various mechanisms and effects of antioxidants have been reviewed (1, 2, 4, 5, 32–40, 75). When primary and secondary antioxidants are suitably coupled in usage they often form synergistic systems, which significantly enhance their overall effectiveness against oxidation.

Because antioxidants are often used in thermal processing of polymers and in polymer finished products for high temperature applications, thermal stability of an antioxidant becomes a more stringent requirement than for a photostabilizer. Again, increasing the molecular weight of an antioxidant as an effective means of increasing its thermal stability (Table III) is a recent trend of antioxidant development. Also, using nontoxic, noncolored polymer-bound steric phenols in place of aromatic amines (which are toxic and give off colored pollutants) is another trend in recent antioxidant development; hence, steric phenols have become an active area of research.

The active phenol concentration in BHT is higher than in A1010 (Table III), but its effectiveness is far below that of A1010. This difference is attributable to the higher M_r of A1010, which makes it thermally more stable (10). However, the effectiveness of an antioxidant does not increase linearly with its M_r (Figure 3) for similar reasons as with HALS. For antioxidants, the optimum M_r appears to be 1000–2000 (57).

Synthesis of Polymeric Antioxidants

Polymeric antioxidants can be prepared (1, 2) through homopolymerization and copolymerization of monomers carrying antioxidant functionality. These polymeric antioxidants possess very high thermal stability and resistance to extraction (77). Examples of monomers used are shown in Chart V. These monomers also can be grafted onto chains of a polymer substrate by a free-radical mechanism. The introduction of antioxidant capability by grafting was reviewed extensively (50). Antioxidant functionality can also be attached to a polymer through attachment of steric phenols to polymers carrying functional groups (50, 76). Such attachment can also be made through end-capping of growing chains with monofunctional molecules such as those shown in Chart V. Certain phenolic resins (novolac resins, Chart V) carry the necessary phenolic functionality, and so they themselves are effective antioxidants. Because of their high thermal stability, novolacs possess quite high antioxidant capability (78–81).

Merits and Demerits of Polymeric Photostabilizers

As pointed out previously, polymeric photostabilizers carry many merits, such as high thermal stability, high resistance to extraction, high performance, and relative nontoxicity. They are also available with a broad range of properties to suit various applications. This range is made possible, for example, through copoly-

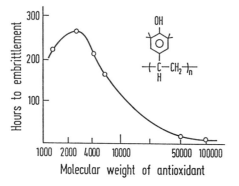

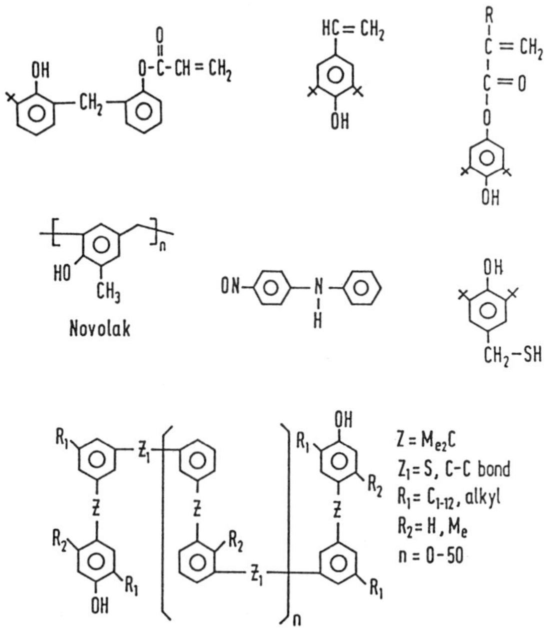

Figure 3. Effect of molecular weight of polymeric antioxidant on its performance (adapted from reference 57).

Chart V. Examples of monomers carrying antioxidant functionality, monofunctional molecules suitable for end-capping growing chains, and novalac resins.

merization to give the stabilizers dual functionality of UV absorber–HALS and HALS–antioxidant. Their compatibility with the substrate also can be enhanced through copolymerization with monomers that give long alkyl branching. However, polymeric photostabilizers also have a number of demerits:

1. Their photostabilizing effectiveness declines with increasing molecular weight (57–61). Monomeric HALS always show higher effectiveness than their polymeric counterparts (69, 74), a consequence attributable to the low mobility and low dispersibility of higher molecular weights in the substrate matrix (57–59).
2. Synthesis of a monomer involves additional labor and cost.
3. Photostabilizers made by copolymerization always bear a dilution effect by the incorporation of a co-monomer.

Regarding molecular weights of polymeric stabilizers, one faces a problem of where to strike a good balance of their properties. For nonpolymeric HALS the optimum molecular weight lies between 500 and 600 (57), but for polymeric HALS this value is around 2700 (72). However, this optimum molecular weight changes with the chemical structure of the stabilizer, test conditions, and purpose of its application. A good polymeric photostabilizer, therefore, should be one that strikes a good balance on its effectiveness, thermal stability, compatibility with its substrate, and high resistance to extraction. For all these characteristics, molecular weight stands out to be a very important factor, because an increase in molecular weight enhances thermal stability and resistance to extraction, but at the same time molecular weight lowers effectiveness when it becomes too high (57–61). High resistance to solvent extraction can be achieved (55) in a polymeric UV absorber when its molecular weight reaches 1600. For example antioxidant A1010, with an M_r of 1178, shows high thermal stability and very high resistance to extraction (86). In fact, commercial photostabilizers currently available are oligomers of molecular weight less than 3000.

References

1. Rabek, J. F. *Photostabilization of Polymers*; Elsevier: New York, 1990.
2. Ranby, B.; Rabek, J. F. *Photodegradation, Photooxidation, and Photostabilization of Polymers*; John Wiley & Sons: London, 1975.
3. Pan, J. Q. *Polym. Bull.* 1992, 3, 138–146.
4. Hawkins, W. L. *Polymer Stabilization*; Interscience: New York, 1972.
5. Hawkins, W. L. *Polymer Degradation and Stabilization*; Springer Verlag: Berlin, Germany, 1984.
6. Guillet, J. E. *Pure Appl. Chem.* 1972, 30(1), 135–144.
7. *Degradability of Polymers and Plastics*; Plastics Institute: London, 1993.
8. Gilead, D. *Polym. Degrad. Stab.* 1990, 29(1), 65–71.
9. Cicchetti, O. *Adv. Polym. Sci.* 1970, 7, 70–112.
10. Pan, J. Q. *Polym. Degrad. Stab.* 1991, 33(1), 67–75.
11. Carlsson, D. J.; Wiles, D. M. *J. Macromol. Sci. Rev. Chem.* 1976, C14(1), 65–105; 1976, C14(2), 155–192.
12. Allen, N. S. *Degradation and Stabilization of Polyolefins*; Applied Science: London, 1986.
13. Carlsson, D. J.; Carton, A.; Wiles, D. M. In *Developments in Polymer Stabilization—1*; Scott, G., Ed.; Applied Science: London, 1979; 219–260.
14. Hirt, R. C.; Searle, N. Z. *Soc. Plast. Eng. Trans.* 1991, 1, 1–15.
15. Padron, A. J. C. *J. Photochem. Photobiol. Chem.* 1989, 49(1), 1–39.
16. Padron, A. J. C. *J. Macromol. Sci. Rev. Chem.* 1990, C30(1), 107–154.
17. Grassie, N. *Polymer Degradation and Stabilization*; Cambridge University: Cambridge, England, 1985.
18. Gugumus, C. *Polym. Degrad. Stab.* 1989, 24(4), 289–301.
19. Allen, N. S. *Polym. Degrad. Stab.* 1985, 13(1), 31–76.
20. Scott, G. *Polym. Degrad. Stab.* 1985, 10(2), 97–125.
21. Scott, G. *Mechanism of Polymer Degradation and Stabilization*; Elsevier: London, 1990.
22. Hahim, H. *Handbook of Polymer Degradation*; Marcel Dekker: New York, 1992.
23. Allen, N. S. *Fundamentals of Polymer Degradation and Stability*; Elsevier: New York, 1992.
24. Carlsson, D. J.; Carton, A.; Wiles, D. M. *Macromolecules* 1976, 9(5), 695–701.
25. Carlsson, D. J.; Wiles, D. M. *Macromolecules* 1969, 2(6), 597–606.
26. Allen, N. S. *Chem. Soc. Rev.* 1986, 15(3), 373–404.
27. Rabek, J. F. *Singlet Oxygen 4*; CRC: Boca Raton, FL, 1985.
28. Trozzolo, A. M.; Winslow, F. H. *Macromolecules* 1968, 1(1), 98–100.
29. Dyogir, I. *Soc. Plast. Eng. J.* 1965, 21(3), 248–252.
30. Pan, J. Q. *Age Commun.* 1977, 4, 9–18.
31. Yang, X. Z.; Chien, J. C. W. *Polym. Degrad. Stab.* 1988, 20, 1–35.
32. Allen, N. S.; Rabek, J. F. *New Trends in Photochemistry of Polymers*; Elsevier: New York, 1985.
33. Müller, H. K. In *Polymer Stabilization and Degradation*; Klemchuk, P. P., Ed.; ACS Symposium Series 280; American Chemical Society: Washington, DC, 1985; pp 55–68.
34. Geuskens, G. *Degradation and Stabilization of Polymers*; Elsevier: London, 1975.
35. *Stabilization and Degradation of Polymers*; Allara, D. L.; Hawkins, W. L., Eds.; Advances in Chemistry 169; American Chemical Society: Washington, DC, 1978.
36. Kamal, M. R. *Weatherability of Plastic Materials*; J.A.P.S. Symposium No. 4; Interscience: New York, 1969.
37. Pospisil, J. In *Developments in Polymer Photochemistry—2*; Allen, N. S., Ed.; Applied Science: London, 1981; pp 53–134.
38. Pospisil, J. *Polym. Degrad. Stab.* 1988, 20(4), 181–202.
39. Pospisil, J. *Polym. Degrad. Stab.* 1991, 34(1), 85–109.
40. Pospisil, J. *Polym. Degrad. Stab.* 1990, 27(3), 227–255.

41. Chan, K. H.; Carlsson, D. J.; Wiles, D. M. *J. Polym. Sci., Polym. Lett. Ed.* **1980**, *18(9)*, 607–612.

42. Fukuoka, N. *Jpn. Kokai Tokkyo Koho* 87–68831, 1987; *Chem Abstr.* **1987**, *108(6)*, 39004.

43. Hawkins, W. L.; Worthington, M. A.; Matreyer, W. *Ind. Eng. Chem. Prod. Res. Dev.* **1962**, *1*, 241–251.

44. Tocker, S. Br. Patent 893507, 1962; *Chem Abstr.* **1962**, *57*, 13955e.

45. Bailey, D.; Vogl, O. *J. Macromol. Sci. Rev. Chem.* **1976**, *C14-2*, 267–293.

46. Wu, S. K. *Polym. Degrad. Stab.* **1986**, *16(2)*, 169–186.

47. Pan, J. Q. *Polym. Degrad. Stab.* **1992**, *37 (3)*, 195–199.

48. Vogl, O. In *Polymer Stabilization and Degradation*; Klemchuk, P. P., Ed.; ACS Symposium Series 280; American Chemical Society: Washington, DC, 1985; pp 197–210.

49. Vogl, O. In *New Trends in the Photochemistry of Polymers*; Allen, N. S.; Rabek, J. F., Eds.; Applied Science: New York, 1985; pp 247–264.

50. Munteanu, D. In *Developments in Polymer Stabilization—8*; Scott, G., Ed.; Applied Science: London, 1987; pp 179–208.

51. Mukherjee, A. K.; Gupta, B. D. *J. Macromol. Sci. Chem.* **1983**, *A19(7)*, 1069–1099.

52. Balaban, L.; Rysavy, D.; Uhlir, M. Czech. Patent 122,619, 1967; *Chem. Abstr.* **1968**, *68*, 79132e.

53. Strobel, A. F.; Catino, S. C. Fr. Patent 1,334,971, 1963; *Chem. Abstr.* **1964**, *60*, 731ac.

54. Coleman, R. A. U.S. Patent 3,391,110, 1968; *Chem. Abstr.* **1968**, *69*, 36767u.

55. Luston, J.; Schubertova, N.; Manasek, D. *J. Polym. Sci. Polym. Symp.* **1973**, *40*, 33–42.

56. Osawa, Z.; Suzuki, M.; Ogiwara, Y. *J. Macromol. Sci. Macromo. Chem.* **1971**, *A5(2)*, 275–285.

57. Motonobu, M. *Polym. Degrad. Stab.* **1989**, *25(2)*, 121–141.

58. Chmela, S.; Hrdlovic, P. *Polym. Degrad. Stab.* **1985**, *11(3)*, 233–241.

59. Chmela, S.; Hrdlovic, P. *Polym. Degrad. Stab.* **1985**, *11(4)*, 339–348.

60. Vassova, G. *Thermochim. Acta* **1985**, *93*, 175–178.

61. Gugumus, F. *Angew. Macromol. Chem.* **1991**, *190*, 111–136.

62. Neiman, M. B. *Nature (London)* **1962**, *196*, 472–474.

63. Kurayama, K. *J. Synth. Org. Chem. (Jpn.)* **1973**, *31(3)*, 198–201.

64. *Polymer Stabilization and Degradation*; Klemchuk, P. P., Ed.; ACS Symposium Series 280; American Chemical Society: Washington, DC, 1985.

65. Gugumus, F. Presented at the Symposium on Polymer Degradation Stability, Manchester, UK, 1985.

66. Pan, J. Q. *Polym. Acta* **1987**, *3*, 234–237.

67. Thompson, R. E. Eur. Patent 293253, 1989; *Chem Abstr.* **1989**, *111(2)*, 8365.

68. He, M.; Hu, X. *Polym. Degrad. Stab.* **1987**, *18(4)*, 321–328.

69. Pan, J. Q.; Chen, W. H. *Polym. Degrad. Stab.* **1993**, *39(1)*, 85–91.

70. Rody, J.; Rasberger, M. U.S. Patent 4,233,412, 1978; *Chem. Abstr.* 1978 *88*, 74945.

71. Gillies, M. T. In *Chemical Technology Review No. 199*; Noyes Date Corporation: Park Ridge, NJ, 1981.

72. Gugumus, F. *Res. Discl.* **1981**, *209*, 357–358.

73. Chmela, S.; Hrdlovic, P. *Polym. Degrad. Stab.* **1990**, *27(2)*, 159–167.

74. Pan, J. Q.; Lau, W. W. Y. *Polym. Degrad. Stab.* **1993**, *44(1)*, 85–91.

75. Terman, L. M.; Kochneva, L. S. *Russ. Chem. Rev.* **1972**, *41(10)*, 1876–1902.

76. Lin, S. A. *Additives for Rubber and Plastics*; Chemical Industry: Beijing, China, **1983**.

77. Evans, B. W.; Scott, G. *Eur. Polym. J.* **1974**, *10(6)*, 453–458.

78. Thiagarajan, R.; Sridher, S.; Ratra, H. C. *Polym. Degrad. Stab.* **1990**, *28(2)*, 153–171.

79. Samud, E. Ger. Offen DE 3,430,735, 1985; *Chem Abstr.* 1985 *103*, 7268.

80. Spivack, J. P. Eur. Patent 137329, 1986; *Chem Abstr.* **1986**, *104*, 20248.

81. Shigeo, S. *Jpn. Kokai Tokkyo Koho* 62,227988 (87–227988) 1987; *Chem Abstr.* **1988**, *108*, 113563.

82. Pavol, C.; Martoe, P. *Angew. Macrol. Chem.* **1985**, *137*, 249–260.

83. Bauer, D. R.; Gerlock, J. L. *Polym. Degrad. Stab.* **1990**, *28(1)*, 39–51.

84. Hrdlovic, P.; Chmela, S. *J. Polym. Mater.* **1990**, *13(1-4)*, 245–254.

85. Gugumus, F. *Kunststoff* **1987**, *77(10)*, 1065–1069.

86. Pan, J. Q. *Synth. Fiber Ind.* **1978**, *1*, 17–25.

87. Zhou, G.; Smid, J. *J. Polym. Sci., A. Chem.* **1991**, *29(8)*, 1097–1105.

88. Pan, J. Q.; Lau, W. W. Y. *Polym. Degrad. Stab.* **1993**, *41(3)*, 275–281.

89. Lau, W. W. Y.; Pan, J. Q. *Proceeding of the 4th Asian Chemical Congress*; Chinese Chemical Society: Beiging, China, 1991; pp 644–645.

90. Pan, J. Q.; Lau, W. W. Y.; Lee, C. S. *J. Polym. Sci., A. Chem.* **1993**, *32(5)*, 997–1000.

Functional Polymers for Selective Flocculation of Minerals

Vincenzo Bertini and Marco Pocci

This chapter presents a review of functional polymers useful for recovering valuables from ultrafine mineral dispersions by selective flocculation. Principles of selective flocculation and the main features of polymer flocculants are briefly discussed. Significant examples of selective polymer flocculants (natural and synthetic) and their properties and applications are reported. Common features of selective flocculation processes based on complementary interactions between the polymer and a dispersant are discussed, and suggestions on the rational design of selective flocculating chemical systems are presented.

In the field of provisioning raw materials, selective flocculation of minerals is an important tool for the recovery of ultrafine valuables. Many ore deposits afford composite (often micrograined) rocks, where the valuables in the form of small crystals are embedded in siliceous or other hard matrixes (Table I). The treatment of such ores requires a stage of fine grinding for liberating the valuables, so inevitably a significant portion of the comminution products undergoes overgrinding and ultrafines (below 20 μm) result. These ultrafines are generally untreatable by the usual separation techniques and are discarded. In situations where the valuable is distinctly softer than the gangue minerals, this process may lead to the loss of up to nearly half the total valuable present in the deposit.

An effective procedure for the recovery of such ultrafines is selective flocculation, a technique that acts on aqueous dispersions of the ultrafine material. The procedure causes the aggregation of the particles of one component into *flocks*, which are more readily separable from the gangue.

Flocculation in its general definition is an established technology widely used for clarifying water, treating effluents, dewatering (minimizing the water content in solid sediments), and facilitating solid–liquid separations. The process is based on the action of substances like inorganic salts and natural or synthetic functional polymers (generally termed flocculants or coagulants) on small suspended particles or colloids.

A peculiarity of polymer flocculants is their selective action in stabilizing or flocculating multicomponent systems (selective flocculation), depending on their functional groups and their application mode. The less encountered but conceptually equivalent process of selective dispersion also may be promoted by suitably designed functional polymers. Numerous functional-polymer flocculants are commercially available for total flocculation, but the real problem for the development of selective flocculation is the scarcity of functional polymers specifically tailored to the different species present in ores. This scarcity has created an extensive research for functional polymers with selective flocculating activity and has stimulated the study of theoretical principles and practical variables that influence the process of flocculation.

Selective Flocculation

To understand the overall physicochemical problem of selective flocculation of minerals by functional poly-

Table I. Composition of Various Minerals

Mineral	Formula	Mineral	Formula
Apatite	$Ca_{10}(PO_4)_6(OH)_2$	Ilmenite	$FeTiO_3$
Barite	$BaSO_4$	Kaolinite	$Al_4Si_4O_{10}(OH)_8$
Bauxite	$Al_2O(OH)_4$	Magnesite	$MgCO_3$
Bornite	Cu_5FeS_4	Magnetite	Fe_3O_4
Calcite	$CaCO_3$	Malachite	$CuCO_3 \cdot Cu(OH)_2$
Cassiterite	SnO_2	Molybdenite	MoS_2
Chalcocite	Cu_2S	Pyrite	FeS_2
Chalcopyrite	$CuFeS_2$	Pyrolusite	MnO_2
Chrysocolla	$CuSiO_3 \cdot 2H_2O$	Quartz	SiO_2
Covellite	CuS	Rhodochrosite	$MnCO_3$
Cuprite	Cu_2O	Rutile	TiO_2
Dolomite	$CaMg(CO_3)_2$	Scheelite	$CaWO_4$
Feldspar	$KAlSi_3O_8$	Sepiolite	$Mg_2Si_3O_8 \cdot 2H_2O$
	$NaAlSi_3O_8$		
	$CaAl_2Si_2O_8$		
Fluorapatite	$Ca_{10}(PO_4)_6F_2$	Sphalerite	ZnS
Fluorite	CaF_2	Sylvinite	KCl
Galena	PbS	Tourmaline	$NaMg_3Al_6(OH)_4$
			$(BO_3)_3Si_6O_{18}$
Hematite	Fe_2O_3	Wolframite	$FeWO_4 + MnWO_4$

mers, a brief survey of the general principles governing the flocculation process is presented. Such principles generally relate to liquid suspended particles having conventional colloidal size of 0.001–1 μm, but in many cases the principles also satisfactorily apply to larger particles up to the size of ultrafine minerals (about 20 μm). However, our discussion is limited to aqueous dispersions relevant to the flocculation of minerals.

Stability of Aqueous Dispersions

Aqueous dispersions of fine solids may be thermodynamically stable or unstable, depending on particle surface properties. Aqueous dispersions of water-insoluble or poorly soluble materials are, for example, thermodynamically unstable. However, they settle very slowly (or not at all), and hence they are defined as metastable. Settling of these systems corresponds to the attaining of a lower energy state, but it can take place only by overcoming an energy barrier whose height determines the effective stability of the dispersion.

The metastable state is treated by assuming that the dispersed particles are submitted to two opposite forces, namely attractive and repulsive interactions. The main components of these forces are van der Waals attraction and electrical repulsion, which form the basis of the DLVO theory of colloid stability (from the names of Deryagin, Landau, Verwey, and Overbeek) (1). In addition, other minor forces, either attractive or repulsive and connected with the effects of various "interposition agents" like hydrated ions, hydrophobic surfaces, and adsorbed polymers (2), may also play important roles in flocculation by polymers.

In conformity with van der Waals forces, colloidal particles develop a reciprocal attraction through a dipole–dipole-induced mechanism: for example, nonpolar molecules in the liquid state that generate interactions by tuning the motion of their electrons through a reciprocal induction. In the case of colloids, such attractions are conditioned by the solvent and the chemical nature of the particles, and they show direct dependence on particle size and inverse dependence on particle separation (i.e., they are strong only at very short distances).

Electrical repulsion comes from the fact that solid surfaces become charged in electrolyte solution. This phenomenon, mainly due to ionization of surface groups (e.g., SiOH on silica), means that all the dispersed particles acquire a net positive or negative charge, and therefore that they repel each other and perturb the surrounding media. For a given colloid, the electrical repulsion shows a direct dependence on the particle size and an inverse dependence on the particle-separation distance. The charge on each particle is exactly balanced by oppositely charged ions (counterions) in solution, which, due to their attraction toward the particle surface, reciprocal repulsion, solvation, and thermal energy, cannot be expected to be either concentrated around the particles or uniformly distributed in solution.

A widely accepted model, to be attributed to Stern, assumes the formation of an electrical double layer, where the main portion of the counterions is confined in a layer close to the particle surface (Stern layer). The estimated thickness corresponds to the diameter of a hydrated counterion. The remainder of the counterions, endowed with much higher freedom, are distributed in a much broader layer (Gouy–Chapman or diffuse ionic layer) (*1*) (Figure 1).

Within these assumptions, the potential at the Stern layer is of interest in determining the stability of the colloidal system. In practice, the electrostatic repulsion and van der Waals attraction dominate at distances, respectively, greater and smaller than the thickness of the Stern layer. This difference creates a potential energy barrier: The particles have difficulty breaching the boundary of the Stern layer, but if they collide with enough kinetic energy to overcome the barrier, they interact to form an aggregate (Figure 2). The potential at the Stern layer decreases, and the aggregation of the particles becomes favorable when the ionic strength is increased.

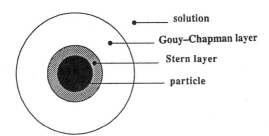

Figure 1. Schematic representation of Stern and Gouy–Chapman layers.

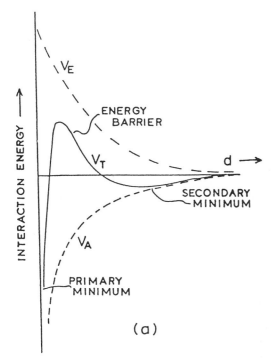

Figure 2. Potential energy diagram for the interaction of colloidal particles. The van der Waals attraction (V_A), electrical repulsion (V_E), and the total energy (V_T) are shown as a function of the particle separation (*d*). (Reproduced with permission from reference 2. Copyright 1989 CRC.)

In addition to van der Waals attraction and electrical repulsion, in a colloid system containing various ions and counterions on the particle surface or close to the surface in the Stern layer, a significant role may be played by water molecules that hydrate such ions. Because contact between the particles needs the expelling of water, work against the hydration forces must be done in addition to the work needed for DLVO repulsion. Sometimes, large deviations may be observed from the prediction of DLVO theory (*2*).

A further extra DLVO interaction may exist as a result of hydrophobic (non-wettable) areas on the particle surface. Such attractive forces also can be generated artificially by addition of adsorbing reagents to the system (*2*). The non-wettable surfaces cannot establish hydrogen bonds with water molecules, and they enhance the possibility of particle aggregation. Such an extra DLVO attractive force (hydrophobic effect) extends far beyond the Stern layer and may be quite substantial. An interesting case of extra DLVO interactions, either repulsive or attractive, arises from the presence of certain substances (especially polymers present in the system such as humic acids or those added intentionally) that adsorb on the particle surface and influence inter-particle contacts (*2*).

Examples of repulsive interactions for colloid stabilization are those of hydrated polymers adsorbed at the particle surface (*3*). The hydrated polymer segments protruding from the particle surface into the solution oppose the approach of different particles with the energy corresponding to their interpenetration (osmotic effect) and distortion by compression (volume restriction effect) (Figure 3). In either case, the

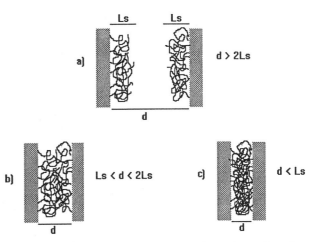

Figure 3. Schematic representation of the three domains of close approach for sterically stabilized flat plates (*d* = distance between the plates; *Ls* = average conformational thickness of the steric layer): (a) the noninterpenetrational domain; (b) the interpenetrational domain; (c) interpenetrational plus compressional domain. (Reproduced with permission from reference 3. Copyright 1983 Academic.)

process is generally referred to as "steric stabilization". The opposite effect, that is, examples of attractive interactions caused by polymer adsorption for flocculation, will be discussed in a subsequent section.

Destabilization of Aqueous Dispersions

The DLVO theory of colloidal stability predicts that the electrically charged dispersed particles (metastable system) may form an aggregate (stable system) by overcoming an electrical potential-energy barrier. Such a process may be possible if the dispersed particles collide with sufficient kinetic energy (1). The slow settling of colloidal dispersions occurs because the thermal energy of the system generates random displacements (Brownian motions) of the particles and hence collisions and aggregation. From a kinetic point of view, the frequency and efficiency of collisions (summarizing viscous retarding effects) determine the rate of this aggregation process, which is known as *perikinetic flocculation*. This process may be accelerated if the barrier is lowered by the addition of the so-called destabilizers or flocculants.

Mechanical stirring of a colloidal dispersion gives rise to another aggregation process, known as *orthokinetic flocculation*, and is normally employed with destabilized colloids to increase the flocculation rate. In this process, the fluid motion imparts different rates to particles of different sizes (or density) and results in enhanced particle collision and aggregation. The flocculation rate generally increases with particle concentration and size. On the basis of the same principles, particle collision and aggregation may be enhanced by applying convection and diffusion motions.

The electrical potential at the Stern layer is sensitive to the ionic strength of the medium. Accordingly, a simple method to reduce the electrical repulsion between the particles and enhance their aggregation is the addition of a salt to the system. The effectiveness of salts as destabilizers increases with the charge of the counterion. For negative particles, for example, Ca^{2+} is much more effective than Na^+. In general, however, various salts produce two different effects: double layer compression and specific ion adsorption. The double layer compression (4) is produced by strong electrolyte salts without specific chemical interaction with the particle surface (indifferent electrolytes). This phenomenon facilitates flocculation without overdosing or restabilization. The optimum flocculant concentration is independent of the particle concentration. Specific ion adsorption (5) is afforded by certain hydrolyzable salts, which dramatically enhance flocculation at relatively low concentrations but restabilize the dispersion at higher concentrations by reversing the electrical charge of the particles. In this case, particle concentration affects the optimum flocculant concentration attaining a nearly linear relationship in the presence of strong specific adsorptions. For the same flocculation effect, the optimum flocculant concentration by specific ion adsorption is noticeably smaller than that by double-layer compression.

The use of hydrolyzable salts as destabilizers may also yield flocculation by another mechanism known as sweep flocculation or enmeshment (6). When hydrolysis of the salt at the operating pH leads to precipitation in the form of a voluminous gel (e.g., $Al(OH)_3$ at neutral pH), the dispersed particles may be enmeshed by the precipitate without any specific ion adsorption. Sometimes enmeshment and specific ion adsorption may coexist. For the enmeshment mechanism, in contrast with specific ion adsorption, the optimum flocculant concentration is relatively independent of particle concentration, so the sweep flocculation process is particularly appropriate for very dilute particle dispersions (e.g., in water clarification). Sweep flocculation also may be promoted by polymer flocculants under precipitating conditions. The effects of counterionic, nonionic, and isoionic polymers on colloidal dispersions will be discussed in a subsequent section.

Effects of Polymeric Flocculants

Despite extensive empirical information, our knowledge of the influence of polymers on the stability of colloidal dispersions does not enable a comprehensive rationalization of the phenomenon. Polymers may act either as stabilizers or flocculants. Highly hydrophilic polymers in colloidal dispersions can cause repulsion between the dispersed particles (steric stabilization, as discussed previously). Flocculating polymers are generally water-soluble polymers and are classified as ionic and nonionic. Their interactions with colloidal dispersions are interpreted on the basis of two main mechanisms: electrostatic patch and polymer bridging.

When a colloidal dispersion interacts with a water-soluble counterionic polymer, a flocculation can occur possibly according to the electrostatic patch mechanism (7). Thus, the polymer molecules are adsorbed onto the particle surface through multiple electrostatic interactions but without quantitative neutralization of the ions on the particles and the polymer. In this way, each adsorbed polymer molecule on the particle surface forms a "patch" with an excess of counterionic charge surrounded by a region where the original particle charge is maintained. Moreover, the adsorbed polymer molecules may, depending on molecular weight, form flattened or extended loops on the particle surface and lead to particle aggregation. The flattened polymer patches bring about areas of positive and negative charges on the surface of the particles, and hence their aggregation occurs by direct electrostatic attraction (Figure 4).

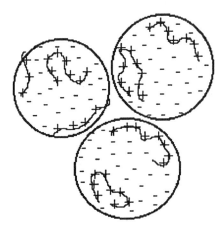

Figure 4. Schematic representation (not to scale) for the flocculation of colloidal particles by electrostatic patch mechanism for polycations adsorbed on negatively charged particles.

The electrostatic-patch mechanism is in agreement with the following observations:

1. Particles and polymer flocculant have charges of opposite sign.
2. Optimum polymer dosage corresponds approximately to the stoichiometric neutralization of the particle charge by the polymer.
3. Excess polymer can reverse the sign of the particle charge and restabilize the dispersion.
4. The flocculation efficiency of the polymer is affected by its charge density and its molecular weight and by particle concentration.

When the adsorbed counterionic polymer forms extended loops, especially in the presence of high particle concentrations, flocculation may occur through electrostatic interactions between particle-bond polymer chain segments in solution and unoccupied surface areas of other particles. Hence, polymeric bridges are formed between the particles.

In the case of water-soluble nonionic polymers or low charge density isoionic polymers, flocculation may occur through a polymer-bridging mechanism (*8*). In general, polymer molecules adsorb on the particle surface not only by electrostatic bonds but also through other interactions like hydrogen, van der Waals, hydrophobic bonding, and even coordination and covalent bonds. Thus, polymer adsorption is made possible even when the polymer contains ions of the same sign as the particle.

Adsorption of nonionic or isoionic polymers can destabilize colloidal dispersions through a dynamic process involving polymer–particle interactions by random attachment of polymer segments at the particle surface, and contemporaneous formation of polymer loops and tails in the solution occurs (Figure 5).

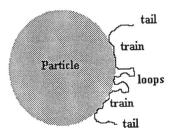

Figure 5. Schematic representation (not to scale) of polymer segments adsorbed on the particle surface.

Figure 6. Schematic representation (not to scale) for the flocculation of colloidal particles by polymer-bridging mechanism.

When the particles approach the electrostatic repulsion sphere, the loops and tails from different particles come into contact and produce aggregates (Figure 6).

This polymer-bridging mechanism is in agreement with the fact that flocculation improves as polymer concentration increases up to a certain value, and an inverse effect follows. This agreement indicates that as the particle coverage by the polymer increases, the number of free sites for bridge formation decreases. Flocculation is enhanced by increasing polymer molecular weight and particle concentration, both of which lead to increased loop and tail (and hence bridge) formation. The flocs formed through polymer bridging are stronger than those obtained by other mechanisms. Notably, in the case of nonelectrostatic adsorption of polymer on particle surfaces, either stabilization or flocculation is possible depending on the hydrophilicity of the polymer (*9*).

Selective Action of Polymer Flocculants

The achievement of selective flocculation of minerals (aggregation and separation of one or more desired components from a complex mixture of ultrafine minerals) is only possible by the use of polymer flocculants on the basis of polymer–particle interactions (Figure 7).

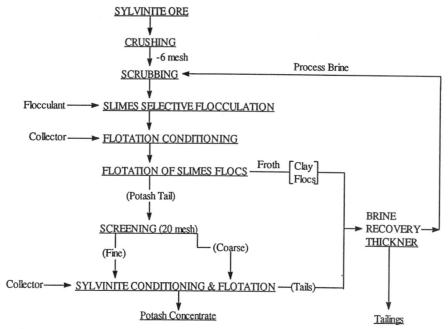

Figure 7. Flow chart of selective flocculation flotation of crude sylvinite slimes. (Reproduced with permission from reference 10. Copyright 1987 Elsevier Science.)

Therefore, the ideal selective polymer flocculant should show optimal adsorption to the target particles and zero adsorption to all other components of the mixture. These requirements mean that selective polymer flocculants must be designed on the basis of specific chemical interactions such as coordination or covalent bonds, rather than nonspecific interactions like ionic or hydrogen bonds. On this basis, selective flocculations may then be achieved by the polymer-bridging mechanism using water-soluble nonionic or low charge density isoionic polymers that interact specifically with the active sites at the particle surface. Such polymers are characterized by the presence of hydrophilic and hydrophobic functions adequately balanced (*11*).

Several studies on functional groups potentially suitable for specific adsorptions on different types of minerals are available (*12, 13*). However, such indications are deduced from the chemistry of free ions; thus, the information is useful only as a general guide owing to different behavior of free ions and ions of a crystal lattice located at the particle surface. For hydrophilic minerals, cyclic structures forming coordination bonds in axial direction with surface sites may be preferred. These cyclic structures shield the particle surface with its large cross-sectional area, and they influence the free energy of the adsorption process by displacing water molecules from the particle surface (*14, 15*).

Granular starch, which plays a key role in the industrial separation of iron oxides (*see* subsequent discussion), deserves some attention. It cannot form polymeric bridges between the oxide particles in the

usual sense, but because the granules are larger than the Stern layer thickness, they may bind several particles and bring about the desired aggregation process.

The physical state of the dispersion also has a strong influence on flocculation selectivity. Important features are low concentration of solids, absence of particles forming aggregates with valuables or gangue, and stability of the dispersion to avoid settling of undesired minerals during the selective flocculation process. The absence of these conditions may lead to substantial loss of selectivity. For example, many polymers produce excellent flocculation of certain isolated minerals and are thus selective flocculants, but they fail to promote the selective flocculation of the same minerals when they are in natural or artificial mixtures. This loss of selectivity has been attributed to factors such as mechanical entrapment within flocks, entrainment, heterocoagulation, ion activation, and heteroflocculation (*16*).

Entrapment of unflocculated particles and contemporaneous settling of unflocculated particles and flocks of other particles (entrainment) depend on the degree of stabilization of the particles. These processes may not lead to serious loss of selectivity.

Heterocoagulation occurs when dispersed particles of different minerals attract each other by electrostatic interactions. Heterocoagulation leads to the loss of selectivity, but its occurrence is limited and occasionally prevented by changing the dispersion conditions.

Low concentration of ions (added, present in the process water, or dissolved from the dispersed minerals) may dramatically influence the flocculation behavior of the dispersion (ion activation) by changing, for example, polymer conformation or particle surface activity. This effect may be favorable or unfavorable; under favorable conditions, a small concentration of an added ion may act as a useful modifier.

Heteroflocculation is also a very common problem in selective flocculation. It is observed when a selective polymer flocculant tailored for the target particles is also weakly adsorbed by the nontarget mineral particles (*17*). In this case the polymer promotes the heteroflocculation of the target and nontarget minerals, because the polymer tails and loops extending from the gangue particles may adsorb strongly enough to form bridges with the available active sites on the target particles (*18*).

A further complication arises from the fact that selective flocculation involves aqueous dispersions of ultrafines produced from the grinding of mineral mixtures in the form of natural ore deposits. This grinding process often modifies the particle surface to such an extent that its behavior may be quite different from that of particles of the corresponding pure mineral.

In principle, a selective flocculation process may be based on the selective flocculation and removal of the valuables leaving the gangue minerals suspended, or vice versa, depending on convenience and economy of the process. In practice, a selective polymer flocculant may by itself not bring about the selective flocculation of a multicomponent dispersion. This limitation explains why many selective flocculation processes use a selective polymer flocculant in combination with a suitable "dispersant". In fact, when a polymeric flocculant adsorbs strongly on the target particles and weakly on particles of other minerals, it is necessary to prevent the weak adsorption to avoid heteroflocculation. This prevention may be achieved by the use of a suitable dispersant capable of competing with the flocculant for adsorption to nontarget particles.

Selective flocculation of a multicomponent dispersion by the combined action of a polymer flocculant and a dispersant is actually a function of the relative selectivity of the two reagents. Thus, with a poorly selective flocculant a more selective dispersant must be employed, and vice versa. Commercially available dispersants are hydrophilic organic or inorganic compounds (small molecules, oligomers, or polymers). Common examples of polymeric dispersants are alkali polystyrenesulfonates, polyacrylates, polyphosphates, and silicates. However, in the search for selective flocculation processes, selective polymer flocculants have received much more attention than selective dispersants.

A wide range of functional polymers, both natural or synthetic and intact or modified, have been examined for selective flocculation of ultrafine mineral dispersions. The most commonly used natural polymers

for flocculation of mineral dispersion are polysaccharides, polypeptides, and tannins. Polysaccharides include various starches, guar and locust bean gums, seaweed extracts (alginates), cellulose, and lignin. However, only starches, alginates, cellulose, and their derivatives have been applied profitably to selective flocculation.

Synthetic polymers most commonly used for flocculation may be divided into two groups: commodity and specialty. The first group is derived from a few monomers characterized by good polymerizability, low price, hydrophilicity, and the presence of functional groups capable of ionic or hydrogen bonding. Typical examples of such polymers include acrylics, polyamines, polyethers, and polystyrenesulfonates. The second group of polymers, as their name implies, are those tailor-made for specific applications requiring specific chemical functions, electrical charge, and molecular weights. These polymers are obtained via transformation from those of the first group or from the corresponding functional monomers. Representative examples of these polymers studied for selective flocculation of minerals are discussed subsequently, but no attempt was made to provide a comprehensive coverage of the field.

Naturally Occurring Polymers

As mentioned, among functional polymers of natural origin, only polysaccharides (starches, alginates, cellulose, and their derivatives) have been applied to selective flocculation of ultrafine mineral dispersions. In all these polymers, the high number of OH groups capable of hydrogen bond formation and interaction with the surface of water-suspended mineral particles is the dominating factor. However, polysaccharides show different organization levels regarding the type and sequence of monomeric units, tridimensional structure (conformation), and association of two or more chains in their supramolecular forms. Therefore, a given polysaccharide or its derivative may develop particular interaction patterns with the solvent and the particles and hence alter the specificity and modality of flocculation.

Starches of different origin and their derivatives are used as flocculants. Alginates are obtained from seaweeds in a variety of water-soluble forms with different structure and composition, and they are usually employed without further chemical modifications. Cellulose, which is insoluble in its native form, is invariably used in the form of various derivatives.

Starches

Natural starches generally occur as granules (2–150 μm) composed of two types of polysaccharides: linear amylose and branched amylopectin. The ratio of the two polysaccharides and their, molecular weights and physical properties vary with the vegetal source (corn, potato, tapioca, wheat, sorghum, rice, etc.). The amylose content may range from less than 2% to about 85%, but it is usually about 25%. Both amylose and amylopectin are homopolymers composed of D-glucose units: that is, cyclic pyranose α-anomeric forms joined together by α-glucosidic (1 → 4) linkages to form linear chains (amylose, **1**) or α-D-(1 → 6) branch points (amylopectin, **2**) (Chart I) (*19*).

Depending on the natural source, amylopectin molecules contain a highly variable number of dihydrogen orthophosphate groups as D-glucose-6-ester (*20*). The ratio of phosphate to α-D-glucopyranosyl units varies between zero and about 0.17. This difference may have a substantial influence on the flocculating properties of the starch. Amylose molecules also contain some α-D-(1 → 6) branching, but many of the properties of amylose fit well into a model of linear molecules capable of adopting different conformations in solution as random coil in neutral water. Amylose also forms helical structures (with about six α-D-glucopyranosyl units per helical turn) in the presence of complexing agents like iodine or polar organic solvents, which are enclosed in the helix. Insoluble, intertwined double helices may also be formed when the polysaccharide undergoes "retrogradation": that is, autoseparation from solution. However, starch granules show peculiar properties not deducible from those of the two separate components of starch, amylose and amylopectin.

Starch granules swell reversibly in cold water and irreversibly in hot water. Swelling involves hydration of the molecules with breaking of some hydrogen bonds between glucose residues. Some amylose molecules are released into solution. Despite this process, the bulk of the α-D-glucopyranosyl units remain shielded inside the swollen granule, and only the surface units can develop interactions with, for instance, the surface of ultrafine mineral particles. In particular, the hydroxy and acetal groups may play a determinant role in these interactions, depending on the preferred conformation of the α-D-glucopyranosyl units. Calculations for the monomer α-D-glucopyranose indicate an equilibrium in solution between the two more stable chair forms **3** and **4** (Chart I) corresponding to a percent content of 96–99.9 of conformer **3** characterized by four out of five substituents in the equatorial position (*21, 22*).

In general, mono- and polysaccharides form weak complexes (*23*). Interesting complexes are formed with oxyanions like borate (*24*) and vanadate (*25*), or with cations (trivalent > divalent > monovalent) having ionic radius >0.8 Å (*24*). Complex formation may involve one or more hydroxy groups or acetal oxygen, and usually stable structures are attained when three or more sites are involved. Substantial complex stability is attained when three consecutive hydroxy or acetal oxygens are in an axial–equatorial–axial

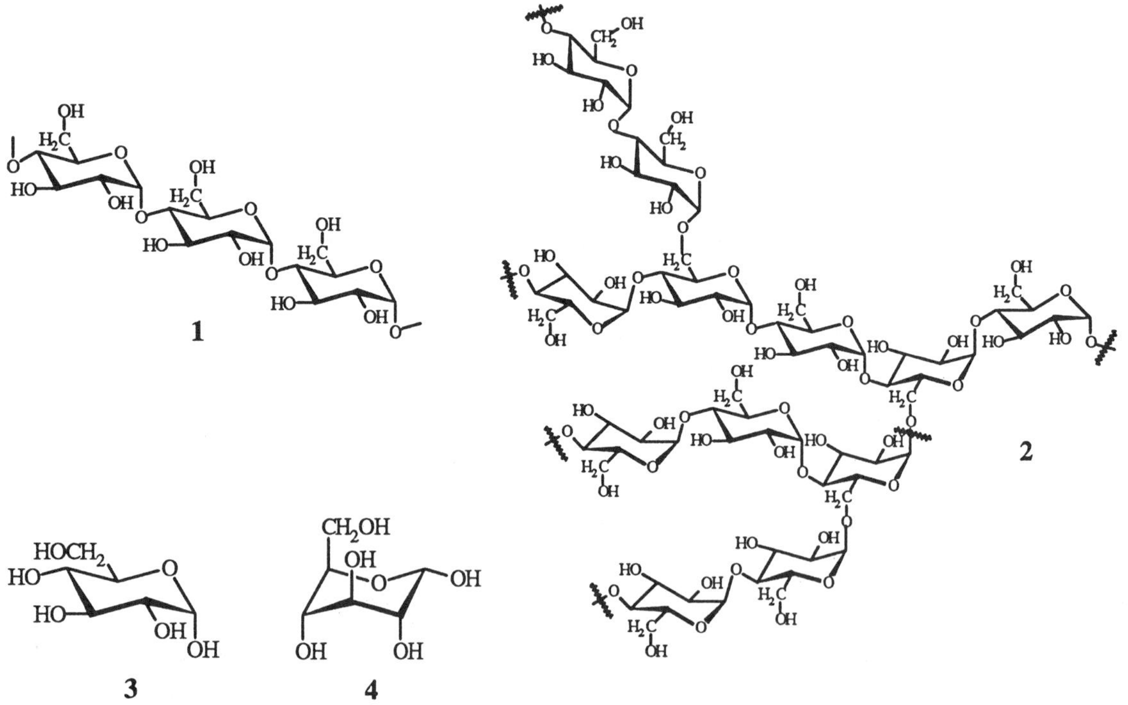

Chart I. Structures of starch polymers amylose (**1**) and amylopectin (**2**) and of the more stable chair forms of α-D-glucopyranose (**3** and **4**).

sequence (23). Other configurations involving three or fewer oxygen atoms in the monosaccharide units may also be active but less effective. In conclusion, the extent of complex formation with mono- and polysaccharides varies greatly with their configuration. For instance, allose, talose, and ribose readily form complexes, whereas glucose and glucose units in polysaccharides are much less active.

The interaction of the mineral particle surfaces with starch in various forms (e.g., amylose content, granule size, gelatinized, or partially hydrolyzed pastes or sols) is likely to involve favorably oriented hydroxy or acetal oxygens. Such interactions may yield hydrogen or coordinative bonds with surface metal ions and ester bonds with oxyanions. These interactions afford a "multilink" association with the particle in a relatively selective manner, depending on density and nature of the sites on the particle surface.

Nonionic starches in natural form (e.g., tapioca flour or causticized) or in various modified forms (26–34) are used for the selective flocculation and separation of iron oxides (mostly hematite) from gangue (mostly silica) obtained by wet grinding (Table I). These examples include important industrial processes, in which starch selectively flocculates iron oxides from the dispersed pulp in the presence of $CaCl_2$, which accelerates the formation and settling of the flocks. Unmodified corn starch selectively flocculates phosphate particles from slimes (35, 36). The presence of calcium, magnesium, or sodium ions improves the settling rate, probably by reducing repulsive forces. Causticized starch causes the selective flocculation of coal from dispersions of mixtures of coal and bentonite (37). Starch also promotes the selective flocculation of manganese minerals from ore pulps (38). Similarly causticized starch and amylopectin enable the selective flocculation of tin oxide (cassiterite) from calcite, quartz, or kaolin suspension (39, 40).

Starch Derivatives

Various starch derivatives (including nonionic, anionic, and cationic) are widely used for flocculation of ultrafine mineral dispersions as flocculants or dispersants, but they are seldom mentioned as selective flocculants. Chemical transformations are based on the hydroxy and acetal groups, and include oxidation, hydrolysis,

and the formation of esters and ethers. Generally, the granular structure of starch is preserved.

Oxidation with sodium hypochlorite affords carbonyl and carboxyl groups, whereas hydrolysis with acids gives rise to the scission of D-glucosyl linkages and formation of new hemiacetal-reducing groups and lowering of the molecular weight. Common starch esters are alkanoates (**5**), phosphates (**6**), and xanthates (**7**). Typical starch ethers include hydroxyalkyl ethers (**8**), tertiary aminoalkyl ethers (**9**), and quaternary ammonium alkyl ethers (**10**) (Chart II). These derivatives are produced in basically the same way as those described for cellulose and chitin in other chapters in this book.

Various starch xanthates (soluble or cross-linked) enable the selective flocculation of different dispersed mineral sulfides in mixture with quartz or calcite, such as chalcopyrite (41), chrysocolla (42), covellite, bornite, pyrite, or sphalerite (43). The interaction of cross-linked starch xanthate with covellite, bornite, pyrite, and sphalerite appears to involve covalent bond formation between the xanthate group and the metal components on the surface of the mineral particles.

Alginates

Alginates are produced by alkaline extraction from various seaweeds and are available in the form of soluble salts of different cations (ammonium, sodium, potassium, and couples of ammonium and sodium with calcium). They are also available as propylene glycol esters of partially esterified alginic acid.

Alginic acid is a polysaccharide composed of β-D-(1 → 4)-mannuronic acid (**11**) and α-L-(1 → 4)-guluronic acid (**12**) units in the form of crystalline blocks of each of the homopolymers and amorphous blocks of the copolymer (44, 45) (Chart III). The α-L-guluronic acid blocks in the alginic acid may give a cross-linked network with divalent cations, especially calcium ions (46).

Ammonium alginate enables selective flocculation in a complex ultrafine mineral dispersion for the separation of two different components at two different pHs (47). For example, the aqueous dispersion of a finely ground fraction of natural ores containing mainly barite, calcite, fluorapatite, fluorite, and clays treated with ammonium alginate in the presence of a small amount of sodium silicate as dispersant, selec-

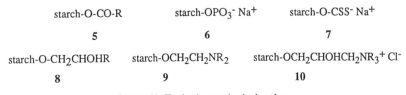

Chart II. Typical starch derivatives.

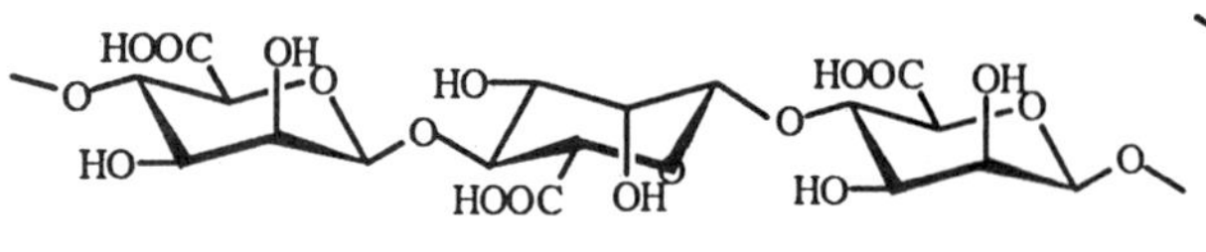

11 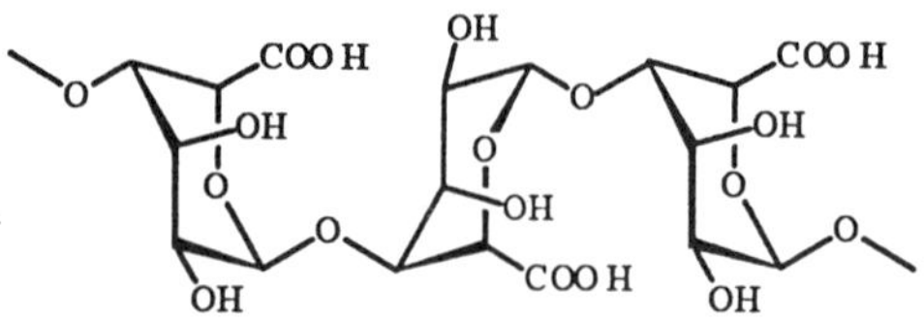   **12**

Chart III. Homopolymers of β-D-(14)-mannuronic acid (**11**) and α-L-(14)-guluronic acid (**12**) components of alginic acid.

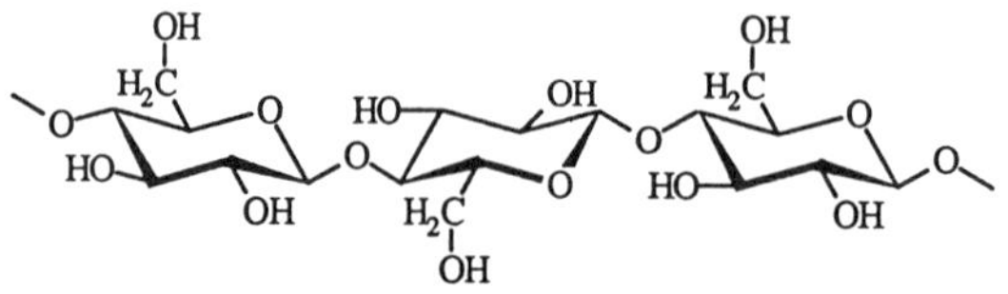

13

cellulose-OH + NaOH +ClCH$_2$COO$^-$Na$^+$ → cellulose-OCH$_2$COO$^-$Na$^+$ + NaCl + H$_2$O

14

cellulose-OH + NaOH + CS$_2$ → cellulose-O-CSS$^-$Na+ + H$_2$O

15

Scheme I. The chain structure of cellulose (**13**) and typical cellulose derivatives (**14** and **15**) used in selective flocculation of minerals.

tively flocculates barite at pH 8.2. In a second stage, fluorite is flocculated at pH 6.2.

Cellulose

Cellulose is a homopolymer composed of (1 → 4) linked β-D-glucopyranosyl units, and all five substituents are in equatorial position in the chair conformation **13** (Scheme I). Native cellulose usually exists in the form of highly organized microfibrils with inter- and intramolecular hydrogen bonds. Unmodified cellulose is insoluble in most solvents (including water), and only some of its derivatives have found applications in the treatment of ultrafine mineral dispersions.

The main cellulose derivatives commercially available are esters and ethers. However, only the anionic derivatives carboxymethylcellulose (**14**) and cellulose xanthate (**15**) (Scheme I) are used for selective flocculation.

The equatorial position of all the substituents in the β-D-glucopyranosyl units enhances the tendency of the polysaccharide to give ordered supramolecular aggregates, thus rendering cellulose less active than starch for forming stable complexes with cations or oxyanions (*see* previous discussion). Among possible interactions between cellulose derivatives and ultrafine mineral particles, hydrogen bonding may be dominant, but more specific coordinative bonds between anion carboxylate or xanthate with surface active sites are also possible. For example, addition of

carboxymethylcellulose to a suspension of pyrolusite and quartz (Table I) selectively flocculates the manganese mineral and enables the separation of quartz by flotation (*48*). Cellulose xanthate also effects a modest selective flocculation of ultrafine dispersion of copper mineral chrysocolla (*42*).

Synthetic Polymers

Synthetic functional polymers are usually employed as general flocculants, and under suitable conditions they are also used as selective flocculants. Typical examples of these polymers studied for flocculation of minerals are polyacrylamide, partially hydrolyzed polyacrylamide, poly(acrylic acid), and poly(ethylene oxide).

Polyacrylamide

Polyacrylamide (PAM, **16**, Scheme II) is the polymer most widely used for general and selective flocculation owing to its physicochemical properties, ready availability, and low cost. It is prepared by radical polymerization with molecular weights in the range 1.5 million to 30 million Da. Poly(β-alanine) (**17**) is an isomer of PAM and is easily obtainable by anionic polymerization of acrylamide; however, the absence of flocculation studies with **17** impedes an interesting comparison between these two isomeric polymers.

In the flocculation literature (including patents), the term polyacrylamide often includes partially

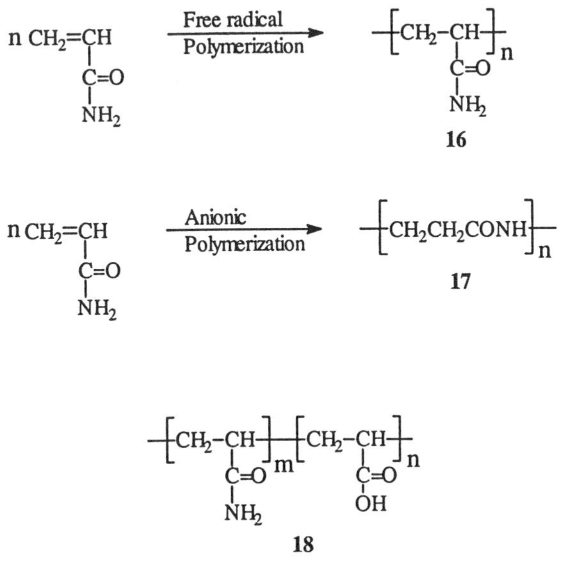

n CH$_2$=CH $\xrightarrow[\text{Polymerization}]{\text{Free radical}}$ $\left[\text{CH}_2-\text{CH}\right]_n$
　　|　　　　　　　　　　　　　　　|
　　C=O　　　　　　　　　　　　C=O
　　|　　　　　　　　　　　　　　　|
　　NH$_2$　　　　　　　　　　　　NH$_2$

16

n CH$_2$=CH $\xrightarrow[\text{Polymerization}]{\text{Anionic}}$ $\left[\text{CH}_2\text{CH}_2\text{CONH}\right]_n$
　　|
　　C=O
　　|
　　NH$_2$

17

$$\left[\text{CH}_2-\text{CH}\right]_m\left[\text{CH}_2-\text{CH}\right]_n$$
　　　　|　　　　　　|
　　　　C=O　　　　C=O
　　　　|　　　　　　|
　　　　NH$_2$　　　　OH

18

Scheme II. Synthesis of polyacrylamide (**16**) and poly(β-alanine) (**17**). Copoly(acrylamide-acrylic acid) (**18**) obtainable by direct copolymerization or hydrolysis of polyacrylamide.

hydrolyzed polyacrylamides (**18**) containing various percentages of carboxylic or carboxylate groups, which are obtained either by hydrolysis of PAM or direct copolymerizations of acrylamide with acrylic acid or its salts. A peculiar feature of high molecular weight PAMs ($\sim 1.5 \times 10^6$ Da) is the instability of their aqueous solutions leading to a decrease of viscosity without any variation in molecular weight. This phenomenon

is interpreted as a conformational transition from a stiffer intramolecularly hydrogen-bonded and partially ordered coil to a more flexible disordered one. Such conformational transition causes a reduction in the hydrodynamic volume as confirmed by gel permeation chromatographic measurements. Also, a parallel decrease in the flocculation activity is evidenced by experiments on kaolin suspensions (*49*). Various examples of flocculation of minerals by polyacrylamide and its hydrolyzed derivatives are summarized in Table II. Interestingly, the recovery of sylvinite by PAM (*50–52*) is an important industrial process.

The effects of grinding on surface properties of high grade natural hematite were studied by the use of electron spectroscopy for chemical analysis (ESCA) (Figure 8), electrophoresis, electron microscopy, potentiometric titration, and flocculation tests with various PAMs. These studies show that coarser crystallites (5 μm) have surface properties significantly different from the very fine particles (0.1–1 μm), and this result is in agreement with their different behavior toward flocculation with PAMs. Also, a release of cations from hematite impurities into solution leads to deactivation of anionic polyacrylamides (*87*).

Studies on the interactions between PAM or partially hydrolyzed PAM and particles of synthetic hematite and silica at various pHs have established that PAM adsorbs to a much lesser extent on silica than on ferric oxide (*88*). Adsorption probably occurs through hydrogen bonding between proton-donor groups at the particle surface (OH or OH$_2^+$) and the

Table II. Examples of Selective Flocculations with Polyacrylamide and Its Hydrolyzed Derivatives

Flocculant	Minerals Flocculated	Minerals Not Flocculated	Ref.
B	Clay gangue	Sylvinite	50–52
B+E	Bauxite	Siliceous ferruginous gangue	53–55
A+O, B+E, B+G, B+H	Rhodochrosite, pyrolusite	Siliceous gangue	56–58
A	Kaolinite	Ilmenite gangue	59, 60
B+E, B+G, B+E+M	Ilmenite gangue	Clay	61–64
B+D	Ilmenite	Feldspar	65
C+E+K, C+G, C+M, C+I	Hematite	Siliceous gangue	66–71
C+E	Siliceous gangue	Hematite	72, 73
A+E	Gangue	Sepiolite	74
C	Magnesite	Silica	75
A+E+L, C+N	Chrysocolla	Quartz	76, 77
C+E	Apatite	Calcite, clay	78
C+G	Apatite	Quartz	79
C+J	Apatite	Kaolinite	80
B	Wolframite	Fluorite, quartz	81
C+F	Calcite	—	82
C	Calcite	—	83–84
C	Dolomite	—	85
C	Kaolinite	—	86

Note: A is polyacrylamide; B is partially hydrolyzed polyacrylamide; C is anionic acrylamide-acrylic acid copolymers; D is oxidized paraffin soap collector; E is sodium hexametaphosphate; F is sodium triphosphate; G is sodium silicate; H is water glass; I is silicic acid; J is sodium polyacrylate; K is sodium fluoride; L is sodium chloride; M is calcium chloride; N is magnesium sulfate; O is aluminum sulfate.

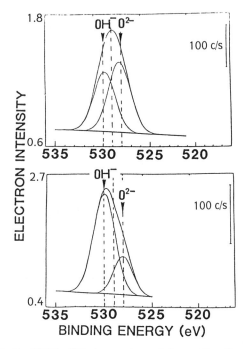

Figure 8. ESCA (01s) spectra of hematite fines: (a) coarser black crystallites (<5 μm); (b) colloidal red particles. (Reproduced with permission from reference 87. Copyright 1987 Elsevier Science.)

Table III. Properties of Polyacrylamide Samples Adsorbed on TiO_2 at Equilibrium Concentration of 300 ppm and pH 6

M_n (Da)	M_w/M_n	Thickness of Adsorbed Layer (nm)	Density of Adsorbed Layer (mg/m²)
84,500	1.30	28.6	1.28
120,400	1.25	37.4	1.38
227,000	1.21	50.6	1.58
343,000	1.15	69.9	1.90

Note: M_w is weight-average molecular weight.

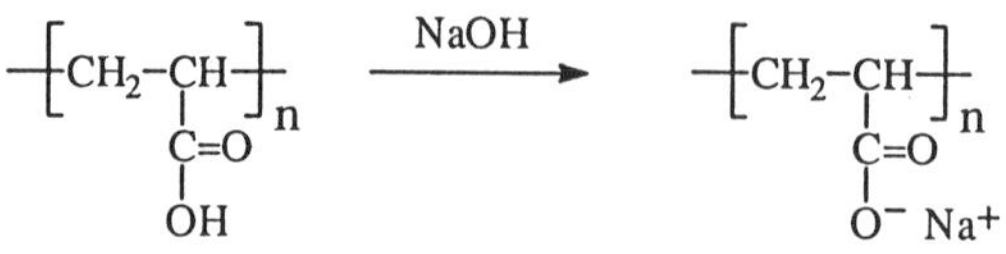

$$\left[CH_2-CH\right]_n \xrightarrow{NaOH} \left[CH_2-CH\right]_n$$

Scheme III. Poly(acrylic acid) (**19**) and its sodium salt (**20**).

amide oxygen in PAM. In partially hydrolyzed polyacrylamides (**18**), in addition to hydrogen bonding, electrostatic attractions between positively charged sites of the oxide and negatively charged carboxylate groups on the polymer also probably effect adsorption.

Adsorption of monodispersed polyacrylamides and polyacrylic acids of various molecular weights onto synthetic titanium dioxide particles also was studied (*89, 90*). The highest levels of adsorption were observed with the highest molecular weights at acidic pH values (Table III), and these levels were interpreted on the basis of interactions between amide groups or undissociate carboxylic groups with hydroxy residues on titania (TiOH, $TiOH_2^+$) (*90*).

Poly(Acrylic Acid)

Poly(acrylic acid) (PAA, **19**) (Scheme III) is obtained by free-radical polymerization of acrylic acid. PAA and highly hydrolyzed PAM are polyelectrolytes whose charge density is easily controlled by pH. When PAA is titrated with an aqueous base such as NaOH, it transforms into the polyelectrolyte **20**, and the flexible polymeric chain "expands" as a result of electrostatic repulsion among carboxylate groups (*91*). The addition of salts due to the presence of positive ions has the effect of shielding the carboxylate anions, and the polymer chains return from rigid rods to flexible coils (*91*).

At low pH, PAA adsorbs on silica only after addition of ferric chloride, probably owing to strong adsorption of the hydrolyzed ferric ions at the silica surface (*88*). In the case of hematite, adsorption takes place without any modifier, probably through electrostatic attraction and hydrogen bonds between OH and OH_2^+ surface sites and carboxylic oxygen on the polymer. At pH > 8.8, hematite and silica particles are negatively charged, so the repulsion from the negatively charged polymer precludes any hydrogen bonding. A similar pH dependence was also observed for the system PAA–rutile (*90*).

PAA enabled upgrading of the $Ca_3(PO_4)_2$ content in a synthetic mixture with quartz (*92*). Fourier transform IR analysis on the $Ca_3(PO_4)_2$/PAA system suggested that adsorption occurred mostly through interaction between the carboxylate groups on the polymer and Ca^{2+} ions at the phosphate particles. Studies on possible correlation between the conformation of poly(acrylic acid) adsorbed onto alumina and flocculating properties of the polymer were performed by using PAA labelled with pyrene residues. Such a technique is based on the observation that the extent of the excimer formation has a direct bearing with the coiled or expanded polymer conformation, because it depends on the interaction possibility of different pyrene pendant groups in the polymer, respectively, in the excited and ground state. Furthermore, the dependence of coiled or expanded PAA conformation on pH values is very effective for polymers adsorbed at the particle surface (Figure 9). With PAA at low concentrations, an increase in pH resulted in stretched polymer chains, more efficient interparticle bridging, and hence improved flocculation (*93–95*). At higher polymer concentrations, the crowding of polymer chains on the particles probably prevents conformational transitions to the stretched form, and this characteristic has a negative effect on flocculation.

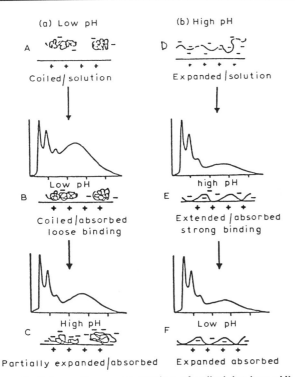

Figure 9. Schematic representation of coiled (at low pH) or expanded (at high pH) pyrene-labelled poly(acrylic acid) in solution and adsorbed on alumina, connected with fluorescence emission spectra at various pH: (A) coiled polymer in solution at low pH, (B) coiled polymer adsorbed on alumina particles at low pH, (C) only partially expanded polymer adsorbed on alumina when pH is brought to high values, (D) expanded polymer in solution at high pH, (E) expanded polymer adsorbed on alumina, and (F) expanded polymer adsorbed on alumina when pH is brought to low values.

Poly(Ethylene Oxide)

The term poly(ethylene oxide) (PEO) indicates polymers of high molecular weight ($>100,000$ Da) obtained by anionic polymerization of ethylene oxide (oxirane). The term poly(ethylene glycol) (PEG) is used to indicate low molecular weight samples (2000–20,000 Da) of the same polymer. Interest in PEO for the selective flocculation of minerals is stimulated by its favorable properties. PEO is miscible with water in all proportions, and it is available in various molecular weight grades. PEO also provides high concentration of ethereal oxygens useful as electron donors for interaction with coordination sites on mineral particles, and it contains methylene groups useful as hydrophobic functions.

PEO efficiently flocculates dolomite in synthetic mixture with alumina. Flocculation is also satisfactory in a mixture with quartz, but it is not satisfactory in a mixture with apatite. These results are interpreted on the basis of the observed adsorption of the polymer on

these minerals in the order of dolomite $>>$ apatite $>$ quartz $>$ alumina. Under the experimental conditions applied (*16*), phenomena such as heterocoagulation, ion activation, physical entrapment, and entrainment were ruled out. However, "heteroflocculation" was thought to be a possible reason for the loss of selectivity in the apatite–dolomite system (*16*). PEO can be used to selectively flocculate copper minerals, including chrysocolla and malachite from quartz, calcite, and dolomite by first hydrophobizing their particle surfaces through the selective adsorption of a "flotation collector" such as amyl xanthate (*97*).

PEO in conjunction with silicic acid (as dispersant) has been used for upgrading iron oxide in slimes (*69*). In the presence of poly(vinyl alcohol), a competitor for the active sites at the particle surface, PEO enables the selective flocculation of dolomite in mixture with apatite (*98*). PEO is also useful in coal recovery (*99*).

Studies on the flocculation of oxides such as silica gel, quartz, hematite, alumina, and kaolinite with PEO have shown that the adsorption of PEO is governed by hydrogen bonding and the degree of hydration of the mineral. Thus, PEO was a good flocculant at acidic pH for sodium kaolinite and silica gel, whereas natural quartz was flocculated only at pH < 3. At higher pH values, the presence of oxyanions at the quartz surface precludes hydrogen bonding with PEO. Synthetic alumina and hematite did not adsorb PEO, probably for the inability of the polymer to replace the high concentration of water molecules bound to these minerals (*100, 101*).

Tailor-Made Polymers

Important examples of tailor-made functional polymers used for selective flocculation of ultrafine mineral dispersions include derivatives of polyacrylamide, polyacrylonitrile, poly(acrylic ester), poly(acrylic acid), and a number of other polymers.

Sulphomethylated Polyacrylamide

Sulphomethylation of polyacrylamide (PAMS, **21**) is usually carried out by treatment with formaldehyde and sodium bisulfite (Scheme IV) (*102, 103*). PAMS is a good flocculant for extremely fine tin-ore cassiterite, but it is inactive for quartz and calcite (*104*). It also enables the selective flocculation of wolframite while being inactive for fluorite and quartz (*81*). PAMS also adsorbs strongly on ilmenite and feldspar in aqueous suspensions and leads to flocculation of the first mineral but stabilization of the second. UV-vis spectra suggest that the polymer–particle interactions occur through coordination between SO_3^- residues on the polymer and Ti^{4+} or Fe^{2+} on the mineral (*105, 106*).

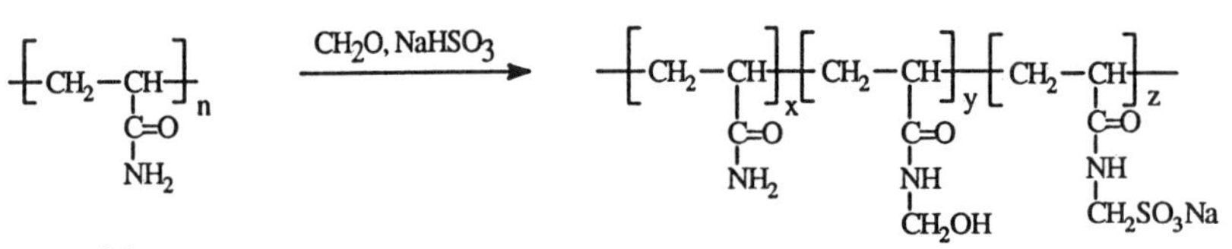

Scheme IV. Preparation of sulphomethylated polyacrylamide (PAMS, **21**).

Glyoxal Bis(2-Hydroxyanil)Polyacrylamide

Reaction of polyacrylamide (**16**) with glyoxal bis(2-hydroxyanil) (**22**) and formaldehyde or with 2-aminophenol, glyoxal, and formaldehyde forms water-soluble glyoxal bis(2-hydroxyanil)polyacrylamide (PAMG, **23**) (Scheme V) (*107, 108*). PAMG in the presence of sodium hexamethaphosphate or low molecular weight carboxylic polymers (as dispersant) enables selective flocculation of copper minerals from natural ore (*108*) and from various synthetic mixtures containing calcite, feldspar, quartz, and dolomite (*107*). Resins containing glyoxal bis(2-hydroxyanil) residues, obtained by condensation of glyoxal and 2,4,6-triaminophenol, are also known to show high affinity for ions such as Cu^{2+}, Ni^{2+}, and UO^{2+} (*109*).

Polymers Containing Hydroxamic Acid Residues

Polymers bearing hydroxamic acid residues (CON-HOH) have been considered as possible selective flocculants for the known ability of the hydroxamic functions to form complexes particularly with Fe^{3+} (*110*). The synthesis of vinyl monomers containing hydroxamic acid residues is difficult (*111, 112*). Therefore, polymers containing hydroxamic acid residues are usually prepared from polyacrylamide, poly(acrylic esters), and polyacrylonitrile. Polyacrylamide reacts with hydroxylamine at room temperature at pH 10 to afford a polymer (**24**) composed of 70% hydroxamic acid units, 25% unreacted amide groups, and less than 5% carboxylic groups (*113*) (Chart IV).

Hydroxamic acid polymers prepared from PAM selectively flocculate iron oxides from kaolin (*114*) and cassiterite from quartz (*115*). They are also active toward the flocculation of ilmenite and feldspar (*116*).

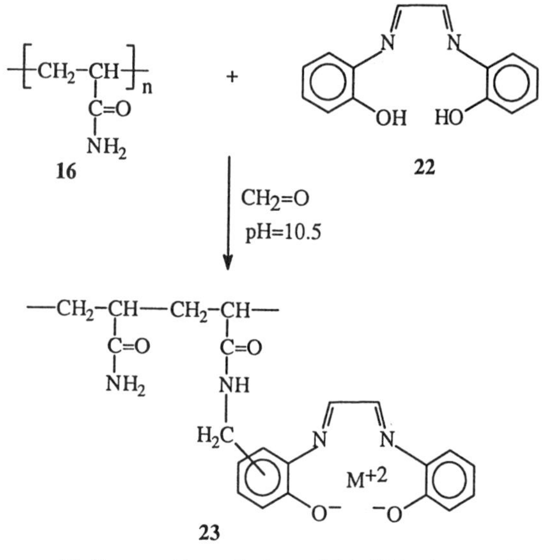

Scheme V. Preparation of glyoxal bis(2-hydroxyanil)polyacrylamide (PAMG, **23**).

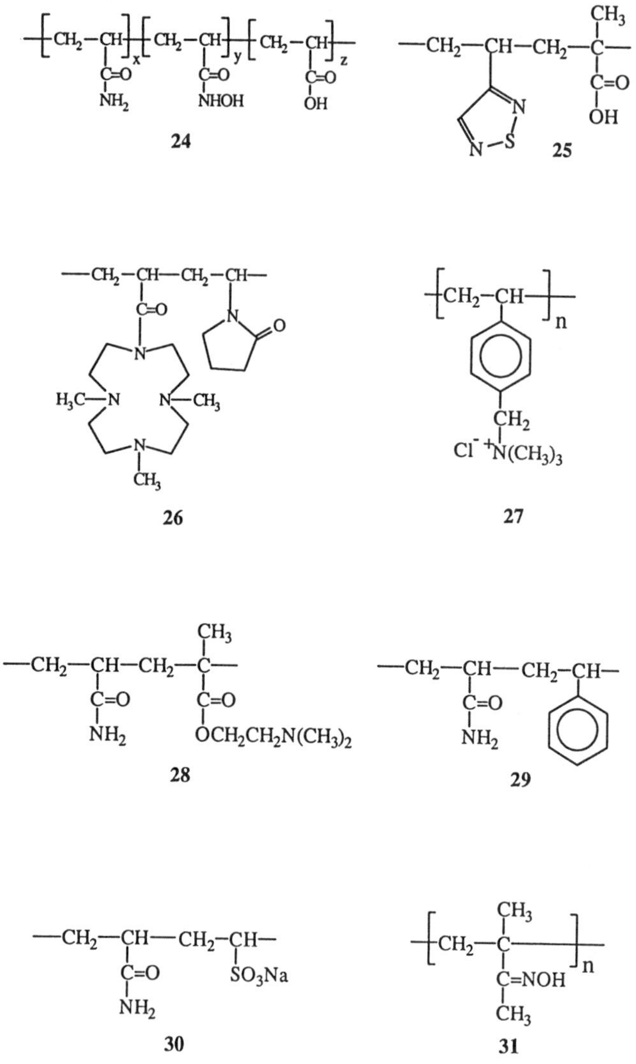

Chart IV. Chemical structures of some tailor-made selective polymeric flocculants.

Poly(Acrylic Acid) with Xanthate Groups

This type of polymer is reported as a selective dispersant for ultrafine pyrite in the recovery of coal, whereas coal itself is flocculated with a general flocculant (*117*).

Vinyl Copolymers Containing 1,2,5-Thiadiazole Residues

These polymers are obtained by polymerization of 3-vinyl-1,2,5-thiadiazole or its copolymerization with comonomers of different polarity such as acrylic acid, methacrylic acid, 4-vinylpyridine, styrene, and N-vinyl-pyrrolidone (*118*). An interesting feature of 1,2,5-thiadiazole is its complexing ability toward a number of cations such as Zn^{2+}, Cd^{2+}, Hg^{2+}, and Ag^+.

Copolymers of 3-vinyl-1,2,5-thiadiazole with methacrylic acid (25) (Chart IV) are good flocculants for finely ground chalcocite dispersions at acidic pH values. However, these copolymers are dispersant at basic pH and selective for the gangue minerals calcite and quartz (*119*). Poly(methacrylic acid) itself shows only a modest flocculation power for chalcocite and gangue minerals around neutral or acidic pH values. The IR analysis of the copolymer 25 adsorbed on chalcocite particles showed batochromic shifts of the thiadiazole ring bands at 528, 606, 796, and 885 cm^{-1} that are attributable to coordinative interactions between ring nitrogens and copper ions at the particle surface (*119*).

Other Vinyl Polymers

The known complexing ability of azamacrocycles toward many cations (*120, 121*) and their application in various fields (*122*) has stimulated the study of new vinyl monomers and water-soluble copolymers containing such chemical structures for use as selective flocculants (*123*). Tests with hydrosols of gold, silver, copper ferrocyanide, arsenic sulfide, silica, and sulfur at different pH values have shown that the copolymers 26 (Chart IV) promote the indiscriminate flocculation of all these hydrosols at pH 4–6. The results are interpreted as nonspecific electrostatic interactions between the hydrosols particles and positively charged polymer molecules. At pH 11.2, only gold (and to a lesser extent silver) particles are flocculated by a selective adsorption of the polymer on the particles. No flocculation is observed in the presence of the azacrown monomer, and this result confirms the importance of the polymeric structure in the flocculation process (*123*).

Poly(4-vinylbenzyltrimethylammonium chloride) (27) in the presence of sodium triphosphate and seaweed extracts (as nonionic surfactants) allows the selective flocculation of phosphate–carbonate ore fines for the enrichment of P_2O_5 (*124*).

Copoly(acrylamide-styrene) (28) in the presence of sodium hexamethaphosphate affords the selective flocculation of ultrafine bituminous coal (*125*). Copolymers of acrylamide with dimethylaminoethyl methacrylate (29) and sodium vinyl sulfonate (30) are effective flocculants for barite dispersions at pH 4.5–11 (*126*).

Polyoximes (31) are obtained by reacting either poly(methyl vinyl ketone) with hydroxylamine hydrochloride or commercial anionic polyacrylamide with formaldehyde oxime (*127*). Polyoximes, when used together with sodium polyacrylate as a dispersant, selectively flocculate cassiterite with respect to quartz and tourmaline. The modified polyacrylamide enables the upgrading of cassiterite tailings from a tin content of 0.2 to 1.99%, although low recovery yields result.

Copolymers Containing Catechol Residues

Catechol systems are characterized by good complexing ability toward titanium ions especially at acidic pH (*128*). Adsorption of catechols at the surface of titanium mineral particles through coordination bonds, especially for derivatives having condensed bicyclic penta-atomic 1,3-benzodioxole or hexa-atomic 1,4-benzodioxane structures, may occupy a large, flattened cross-sectional area on the particle surface displacing a significant number of water molecules. A wide choice of monomers containing catechol systems (32–36) (Chart V) (*14, 129–131*) were polymerized and copolymerized with acrylic (AA), methacrylic, 4-vinylbenzoic acid, and 4-vinylpyridine (4VP) to yield the corresponding polymers. These polymers showed the expected selective flocculation properties for rutile and ilmenite. They also showed the "crossed" selective flocculations of the two titanium oxides.

Copolymers of 33a with AA (number-average molecular weight: M_n = 35,000–476,000 Da) produced stable, well-separable flocks with rutile and ilmenite at acidic pH, but these copolymers did not flocculate gangue minerals such as quartz (*11*). Copolymers of 34b with AA (M_n = 150,000–350,000 Da) enabled selective flocculation of rutile with respect to ilmenite (*132*). Copolymers of 32b with AA (M_n = 80,000 Da) showed high selective flocculating power toward ilmenite with respect to rutile, but copolymers of 32a (M_n = 420,000 Da) and 35b (M_n = 154,000 Da) with AA were less effective for the same process (*133*). Good selective flocculants for rutile were also found among the copolymers of 32c, 32d, and 34c with AA (*14*). All of these polymers have the distinctive feature of being flocculants at acidic pH but dispersants at basic pH, and this feature adds great advantages for the efficiency of the flocculation process. In all the examined cases, poly(acrylic acid) had little flocculation ability. Copolymers of 32a with 4VP proved a good flocculant for rutile at neutral pH (*14*). A summary of the floccu-

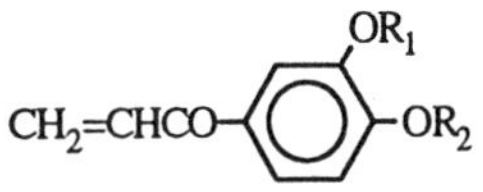

32a $R_1 = R_2 = H$
32b $R_1 = R_2 = COCH_3$
32c $R_1 = H, R_2 = CH_3$
32d $R_1 = COCH_3, R_2 = CH_3$
32e $R_1, R_2 = -CH_2-$
32f $R_1, R_2 = -CH_2CH_2-$

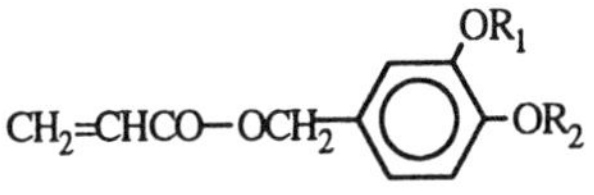

33a $R_1, R_2 = -CH_2-$
33b $R_1, R_2 = -CH_2CH_2-$

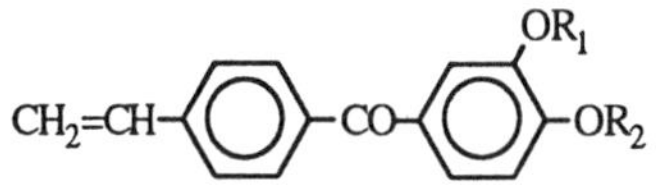

34a $R_1 = R_2 = H$
34b $R_1 = R_2 = CH_2OCH_3$
34c $R_1, R_2 = -CH_2-$

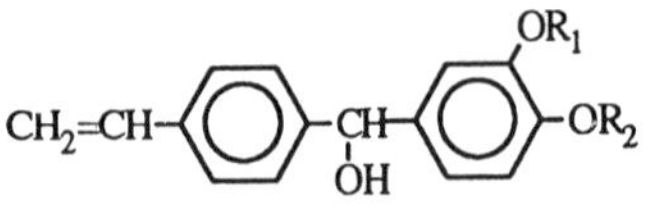

35a $R_1 = R_2 = CH_2OCH_3$
35b $R_1, R_2 = -CH_2-$

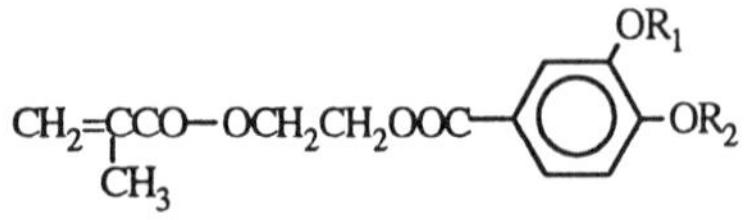

36a $R_1, R_2 = -CH_2-$
36b $R_1, R_2 = -CH_2CH_2-$

Chart V. Various monomers containing catechol residues.

Table IV. Flocculating or Dispersing Power of Catechol Copolymers and PAA at 50 ppm pH = 2.5

Copolymer Composition			*Flocculating (F) or Dispersing (D) Power (%)*		
Catechol Monomer (%)	*Acrylic Acid (%)*	M_n *(Da)*	*Rutile*	*Ilmenite*	*Quartz*
32b (5)	95	80,000	D (75)	F (97)	Inert
33a (18)	82	48,000	F (96)	F (72)	D (8)
34b (5)	95	343,000	F (97)	D (57)	F (3)
—	100	360,000	D (3)	F (14)	Inert

lating properties of the various catechol polymers are given in Table IV.

Conclusions

The key to the selective flocculation of a multicomponent ultrafine mineral dispersion is the availability of a selective flocculating material (i.e., a water-soluble nonionic or low charge density isoionic polymer), which may be used alone or in combination with a suitable competitor or dispersant. A selective polymer flocculant must adsorb strongly on the surface of the target particles and not at all or very weakly on the other particles to avoid co-flocculation or heteroflocculation. However, the creation of such specific polymer flocculants is not an easy task. So in practice, the question is how to achieve selective flocculation with partially selective polymer flocculants. When approached from this point of view, various polymer flocculants can be tailored to carry interactive functions complementary to those present on the surface of the target particles. In this way, either one or a combination of hydrogen bonding, coordination, and even covalent bonds may be envisaged to devise relatively efficient flocculating processes based on specifically designed functional probes. Suitable dispersants, in combination with modifiers if necessary, may complement inadequate polymer selectivity by counteracting the polymer adsorption onto undesired particles.

References

1. Bleier, A. V. In *Kirk–Othmer Encyclopedia of Chemical Technology;* Krochwitz, J. I.; Howe-Grant, M., Eds.; John Wiley & Sons: New York, 1993; Vol. 6, pp 820–822.
2. Gregory, J. In *Critical Reviews in Environmental Control;* CRC: Roca Baton, FL, 1989; Vol. 19, No. 3, pp 185–230.
3. Napper, D. H. In *Polymeric Stabilization of Colloidal Dispersions;* Academic: London, 1983; pp 18–30.
4. Sennet, P.; Olivier J. P. *Ind. Eng. Chem.* **1965**, *57*, 32.
5. Matijevic, E. *J. Colloid Int. Sci.* **1973**, *43*, 217.
6. Packham, R. F. *J. Colloid Int. Sci.* **1965**, *20*, 81.
7. Gregory, J. *J. Colloid Int. Sci.* **1973**, *42*, 448.
8. Kitchener, J. A. *Br. Polym. J.* **1972**, *4*, 217.
9. Miller, C. A.; Neogi, P. *Colloidal Dispersions, Interfacial Phenomena, Equilibrium and Dynamic Effects;* Surfactant Science Series 17; Marcel Dekker: New York, 1985; pp 125–126.
10. Yu, S.; Attia, Y. A. In *Flocculation in Biotechnology and Separation Systems: Part 8. Flocculation in Mineral Processing;* Attia, Y. A., Ed.; Process Technology Proceedings 4; Elsevier: Amsterdam, Netherlands, 1987; pp 601–637.

11. Bertini, V.; Pocci, M.; Marabini, A.; Barbaro, M.; De Munno, A.; Picci, N. *Colloids Surf.* **1991**, *60*, 413.

12. Attia, Y. A. In *Flocculation in Biotechnology and Separation Systems: Part 3. Design and Synthesis of Selective Flocculants*; Attia, Y. A., Ed.; Process Technology Proceedings 4; Elsevier: Amsterdam, Netherlands, 1987; pp 227–246.

13. Warren, L. J. In *Metallurgy*; Fielding L. E.; Gordon A. R., Eds.; 13th Congress of the Council of Mining and Metallurgical Institutions; The Australian Institute of Mining and Metallurgy: Victoria, Australia, 1986; Vol. 4, pp 249–257.

14. Bertini, V.; Marabini, A.; De Munno, A.; Pocci, M.; Picci, N.; Barbaro, M. In *Flocculation in Biotechnology and Separation Systems: Part 3. Design and Synthesis of Selective Flocculants*; Attia Y. A., Ed.; Process Technology Proceedings 4; Elsevier: Amsterdam, Netherlands, 1987; pp 247–261.

15. Yehia, A.; Ateya, B. G.; Youssef, A. A. In *Advances in Fine Particle Processing: Part 3. Surface and Colloidal Phenomena in Fine Particle Processes*; Hanna, J.; Attia, Y. A., Eds.; Elsevier: New York, 1990; pp 171–180.

16. Moudgil, B. M.; Shah, B. D. In *Flocculation in Biotechnology and Separation Systems: Part 8. Flocculation in Mineral Processing*; Attia, Y. A., Ed.; Process Technology Proceedings 4; Elsevier: Amsterdam, Netherlands, 1987; pp 729–739.

17. Moudgil, B. M.; Vasudevan, T. V. *J. Colloid Int. Sci.* **1989**, *127*, 239.

18. Moudgil, B. M.; Behl, S. *J. Colloid Int. Sci.* **1991**, *146*, 1.

19. Zimm, B. H. *J. Am. Chem. Soc.* **1952**, *74*, 1111.

20. Tabata, S.; Nagata, K.; Hizukuri, S. *Staerke* **1975**, *27*, 333.

21. Kildeby, K.; Melberg, S.; Rasmussen, K. *Acta Chem. Scand.* **1977**, *A31*, 1.

22. Angyal, S. J. In *The Carbohydrates—Chemistry and Biochemistry*; Pigman, W.; Horton, D., Eds.; Academic: Orlando, FL, 1972; Vol. IA, pp 202–206.

23. Angyal, S. J. *Chem. Soc. Rev.* **1980**, *9*, 415.

24. Angyal, S. J.; McHugh, D. J. *J. Chem. Soc.* **1957**, 1423.

25. Tracey, A. S.; Gresser, M. J. *Inorg. Chem.* **1988**, *27*, 2695.

26. Frommer, D. W.; Colombo, A. F. U.S. Patent 3 292 780, 1966; *Chem. Abstr.* **1967**, *66*, P58118y.

27. Frommer, D. W. *Mezhdunar. Kongr. Obogashch. Polezn. Iskop. [Tr.] 8th* **1968**, *1*, 464; *Chem. Abstr.* **1970**, *72*, 57919x.

28. Frommer, D. W. *Eng. Min. J.* **1969**, *170*, 92.

29. Colombo, A. F.; Jacobs, H. D. *Rep. Invest. U.S. Bur. Mines* **1976**, *RI*, 8180.

30. Jacobs, H. D.; Colombo, A. F. *Rep. Invest. U.S. Bur. Mines* **1980**, *RI*, 8482.

31. Rao, K. H.; Narasimhan, K. S. *Int. J. Miner. Process.* **1985**, *14*, 67.

32. Singh, B. P. *Hymalayan Chem. Pharm. Bull.* **1990**, *7*, 4; *Chem. Abstr.* **1991**, *114*, 27574z.

33. Singh, B. P.; Singh, R. *J. Mines Met. Fuels* **1992**, *40*, 127.

34. Lien, H. O.; Morrow, J. G. *CIM Bull.* **1978**, *71*, 109; *Chem. Abstr.* **1979**, *90*, 26714u.

35. Haseman, J. F. U. S. Patent 2 660 303, 1953; *Chem. Abstr.* **1954**, *48*, P1930h.

36. Ibrahim, S. S.; Abdel-Khalek, N. A. *Miner. Eng.* **1992**, *5*, 907.

37. Klunder, H.; Koopmans, K. *Delft Prog. Rep.* **1991**, *6*, 190.

38. Semioshko, V. M.; Safronov, A. F. *Obogashch. Rud.* **1975**, *20*, 20; *Chem. Abstr.* **1976**, *84*, 124726v.

39. Liu, P. *Kuangye Gongcheng* **1991**, *11*, 41; *Chem. Abstr.* **1992**, *116*, 63923n.

40. Schulz, G. *Freiberg. Forschungsh. A* **1973**, *513*, 95.

41. Sresty, G. C.; Somasundaran, P. *Int. J. Miner. Process.* **1980**, *6*, 303.

42. Baudet, G.; Morio, M.; Rinaudo, M.; Nematollahi, H. *Ind. Miner. (St. Etienne, Fr.)* **1978**, *1*, 19.

43. Termes, S. C.; Wilfong, R. L.; Richardson, P. E. *Rep. Invest. U.S. Bur. Mines* **1983**, *RI*, 8819.

44. Atkins, E. D. T.; Nieduszynski, I. A.; Mackie, W.; Parker, K. D.; Smolko, E. E.; *Biopolymers* **1973**, *12*, 1865.

45. Atkins, E. D. T.; Nieduszynski, I. A.; Mackie, W.; Parker, K. D.; Smolko, E. E.; *Biopolymers* **1973**, *12*, 1879.

46. Bryce, T. A.; McKinnon, A. A.; Morris, E. R.; Rees, D. A.; Thom, D. *Faraday Discuss.* **1974**, *57*, 221.

47. Carta, M.; Alfano, G. B.; Del Fa, C.; Ghiani, M.; Massacci, P.; Satta, F. *Proceedings of the 11th International Mineral Processing Congress*; Cagliari, Italy, 1975; p 1187.

48. Yousef, A. A.; Arafa, M. A.; Boulos, T. R. *Int. Min. Metall. Sect. C*, **1971**, *80*, 223.

49. Kulicke, W. M.; Kniewske, R.; Klein, J. *Prog. Polym. Sci.* **1982**, *8*, 373.

50. Brogoitti, W. B.; Howald, F. P. Ger. Offen. 2 309 583, 1973; *Chem. Abstr.* **1974**, *80*, P49901k.

51. Thompson, P.; Huiatt, J. L.; Seidel, D. C. U.S. Patent Appl. 942 833, 1979; *Chem. Abstr.* **1979**, *91*, 143995t.

52. Thompson, P.; Huiatt, J. L.; Seidel, D. C. U.S. Patent Appl. 4 192 737, 1980; *Chem. Abstr.* **1980**, *92*, 218669w.

53. Volova, M. L.; Lyubimova, E. I.; Akopova, K. S.; Yakubovich, I. A.; Kotov, A. M. *Isvetn. Met.* **1974**, *78*; *Chem. Abstr.* **1975**, *82*, 75813u.

54. Lifirenko, V. E.; Volova, M. L. *Isvetn. Met.* **1977**, *78*; *Chem. Abstr.* **1977**, *86*, 158750e.

55. Volova, M. L.; Lyubinova, E. I.; Shishkova, L. M.; Bershitskii, A. A. *Isvetn. Met.* **1977**, *72*; *Chem. Abstr.* **1977**, *86*, 174709d.

56. Gogitidze, T. A.; Okromchedlidze, I. V.; Shekriladze, N. Sh.; Papavadze, N. S. *Soobshch. Akad. Nauk Gruz. SSR*; **1972**, *65*, 401; *Chem. Abstr.* **1972**, *77*, 104182s.

57. Sergo, E. E.; Korchagin, L. V. *Obogashch. Polezn. Iskop.* **1973**, *12*, 35; *Chem. Abstr.* **1973**, *79*, 148308h.

58. Sergo, E. E.; Korchagin, L. V.; Temchenko, O. I. *Obogashch. Polezn. Iskop.* **1973**, *13*, 58; *Chem. Abstr.* **1974**, *81*, 40075g.

59. Chapman, J. H.; Anderson, D. Brit. Patent Appl. 2 025 913, 1980; *Chem. Abstr.* **1980**, *93*, 9841c.

60. Ermakov, A. J.; Medvedev, M. I. *Khim. Teckhnol. Vody* **1985**, *7*, 61; *Chem. Abstr.* **1985**, *102*, 224962c.

61. Maynard, R. N. Ger. Offen. 2 329 455, 1974; *Chem. Abstr.* **1974**, *80*, 124072w.

62. Sheridan, J. J., III U.S. Patent 3 837 482, 1974; *Chem. Abstr.* **1974**, *81*, 175148d.

63. Mercade, V. V. U.S. Patent 3 826 365, 1974; *Chem. Abstr.* **1974**, *81*, 157755v.

64. Mercade, V. V. U.S. Patent 3 826 027, 1975; *Chem. Abstr.* **1975**, *82*, 144166z.

65. Chen, J.; Chen, W.; Huang, L. *Youse Jinshu* **1983**, *35*, 23; *Chem. Abstr.* **1984**, *100*, 71775a.

66. Read, A. D.; Whitehead, A. *Erzmetall* **1972**, *25*, 64.

67. Bogdanova, I. P.; Bodnarashik, L. G. *Fiz. Tekh. Probl. Razrab. Polezn. Iskop.* **1975**, 117; *Chem. Abstr.* **1976**, *85*, 146207z.

68. Balasubramani, K. J.; Raju, G. B.; Khangaonkar, P. R. *Bull. Electrochem.* **1990**, *6*, 836.

69. Shrader, E. A.; Puchkova, O. E.; Ratmirova, L. I. *Soversh. Metodov Pererab. Miner. Syr'ya* **1984**, 66; *Chem. Abstr.* **1985**, *102*, 65312x.

70. Bagster, D. F.; Mc Ilvenny, J. D. *Int. J. Miner. Process.* **1985**, *14*, 1.

71. Jin, R.; Hu, W.; Hou, X. *Colloids Surf.* **1987**, *26*, 317.

72. Read, A. D. *Inst. Min. Metall. Trans. Sect. C* **1971**, *80*, 24.

73. Read, A. D. *Br. Polym. J.* **1972**, *4*, 253.

74. Cai, R. *Kuangchan Zonghe Liyong* **1990**, 1; *Chem. Abstr.* **1992**, *116*, 238344d.

75. Yarar, B.; Ozgur, T. *Filtr. Sep.* **1976**, *13*, 443.

76. Rubio, J.; Goldfarb, J. *Inst. Min. Metall. Trans. Sect. C* **1975**, *84*, C123-C127.

77. Ye, Y.; Fuerstenau, M. C. In *Advances in Fine Particles Processing: Proceedings of the International Symposium: Part 5. Processing of Fine Particles by Flocculation and Dispersion*; Hanna, J.; Attia, Y. A., Eds.; Elsevier: New York, 1990; pp 285–297.

78. Pradip *Proceedings of the International Conference on Progress in Metallurgical Research: Fundamental Application Aspects*; Mehrotra S. P.; Ramachandran, T. R., Eds.; Tata Mc Graw Hill: New Delhi, India, 1985; pp 45–50.

79. Rubio, J.; Marabini, A. M. *Int. J. Miner. Process.* **1987**, *20*, 59.

80. Pradip; Kulkarni, R. A.; Gundiah, S.; Moudgil, B. M. *Int. J. Miner. Process.* **1991**, *32*, 259.

81. Jin, H.; Li, B. *Kuangye Gongcheng* **1983**, *3*, 21; *Chem. Abstr.* **1984**, *100*, 142744c.

82. Friend, J. P.; Kitchener, J. A. *Chem. Eng. Sci.* **1973**, *28*, 1071.

83. Yarar, B.; Kitchener, J. A. *Inst. Min. Metall. Trans. Sect. C* **1970**, *79*, C23–C33.

84. Kuz'Kin, S. F.; Nebera, V. P. *Sb. Nauchn. Tr. Inst. Tsvetn. Met. im. M. I. Kalinina* **1960**, *33*, 202; *Chem. Abstr.* **1962**, *57*, 10827i.

85. Janusz, W.; Palak, T.; Szczypa, J. *Rudy Met. Niezelaz.* **1986**, *31*, 269; *Chem. Abstr.* **1987**, *107*, 157277w.

86. Clark, A. Q.; Herrington, T. M.; Petzold, J. C. *Colloids Surf.* **1990**, *44*, 247.

87. Pugh, R. J.; Lundstroem, H. *Flocculation in Biotechnology and Separation Systems: Part 8. Flocculation in Mineral Processing*; Attia, Y. A., Ed.; Process Technology Proceedings 4; Elsevier: Amsterdam, Netherlands, 1987; pp 673–694.

88. Drzymala, J.; Fuerstenau, D. W. In *Flocculation in Biotechnology and Separation Systems: Part 1. Colloid and Surface Science of Flocculation*; Attia, Y. A., Ed.; Process Technology Proceedings 4; Elsevier: Amsterdam, Netherlands, 1987; pp 45–60.

89. Chibowski, S. *Powder Technol.* **1990**, *63*, 75.

90. Chibowski, S. *J. Colloid Int. Sci.* **1990**, *140*, 444.

91. Kay, R. J.; Treloar, F. E. *Makromol. Chem.* **1974**, *175*, 3207.

92. Pradip; Moudgil, B. M. *Int. J. Miner. Process.* **1991**, *32*, 271.

93. Tjipangandjara, K. F.; Somasundaran, P. In *Advances in Fine Particle Processing: Part 5. Processing of Fine Particles by Flocculation and Dispersion*; Hanna, J.; Attia, Y. A., Eds.; Elsevier: New York, 1990; pp 259–268.

94. Tjipanngandjara, K. F.; Huang, Y.; Somasundaran, P.; Turro, N. J. *Colloids Surf.* **1990**, *44*, 229.

95. Tjipangdjara, K. F.; Somasundaran, P. *Colloids Surf.* **1991**, *55*, 245.

96. Somasundaran, P.; Kunjappu, J. T. *Colloids Surf.* **1989**, *37*, 245.

97. Rubio, J.; Kitchener, J. A. *Trans. Inst. Min. Metall. Sect. C* **1977**, *86*, 97.

98. Behl, S.; Moudgil, B. M. *Miner. Metall. Process.* **1992**, *9*, 4.

99. Baichenko, A. A.; Baichenko, A. A. *Izv. Vyssh. Uchebn. Gorn. Zh.* **1980**, 100; *Chem. Abstr.* **1981**, *94*, 194711s.

100. Koksal, E.; Ramachandran, E.; Somasundaran, P.; Maltesh, C. *Powder Technol.* **1990**, *62*, 253.

101. Rubio, J.; Kitchener, J. A. *J. Colloid Int. Sci.* **1976**, *57*, 132.

102. Schiller, A. M.; Suen, T. J. *Ind. Eng. Chem.* **1956**, *48*, 2132.

103. Bakalik, D. P.; Kowalski, D. J. *ACS Div. Polym. Mater. Sci Eng. Proc.* **1987**, *57*, 845.

104. Zhong, H.; Zhu, D. *Kuangye Gongcheng* **1988**, *8*, 21; *Chem. Abstr.* **1989** *110*, 99250f.

105. Chen, J.; Chen, W.; Xu, Z. *Zhongnan Kuangye Xueyuan Xuebao* **1986**, 86; *Chem. Abstr.* **1987**, *107*, 138159t.

106. Chen, J.; Chen, W.; Xu, Z. *Zhongnan Kuangye Xueyuan Xuebao* **1987**, *18*, 223; *Chem. Abstr.* **1987**, *107*, 180649c.

107. Attia, Y. A. *Int. J. Miner. Process.* **1977**, *4*, 191.

108. Attia, Y. A. I.; Kitchener, J. A. *Int. J. Miner. Process.* **1977**, *4*, 209.

109. Bayer, E. *Angew. Chem. Int. Ed. Engl.* **1964**, *3*, 325.

110. Yale, H. L. *Chem. Rev.* **1943**, *33* 209.

111. Becke, F.; Mutz, G. *Chem. Ber.* **1965**, *98*, 1322.

112. Narita, M.; Teramoto, T.; Okawara, M. *Bull. Chem. Soc. Jpn.* **1972**, *45*, 3149.

113. Domb, A. J.; Carvalho, E. G.; Langer, R. *J. Polym. Sci. Chem. Ed.* **1988**, *26*, 2623.

114. Shankar Pradip, S.; Ravi, A.; Deo, M. G.; Kulkarni, R. A.; Gindiah, S. *Bull. Mater. Sci.* **1988**, *10*, 423.

115. Clauss, C. R. A.; Appleton, E. A.; Vink, J. J. *Int. J. Miner. Process.* **1976**, *3*, 27.

116. Huang, L.; Chen, W.; Chen, J. *Zhongnan Xueyuan Xuebao* **1982**, 55; *Chem. Abstr.* **1983**, *99*, 74452q.

117. Attia, Y. A. *Prepr. Pap. Am. Chem. Soc. Div. Fuel Chem.* **1985**, *30*, 19.

118. Bertini, V.; De Munno, A.; Pocci, M.; Picci, N.; Lucchesini, F. *J. Chem. Res.* **1986** (Synopsis) *358*, (Miniprint) 3060.

119. Bertini, V.; Pocci, M.; Marabini, A.; Barbaro, M.; De Munno, A. *Part. Sci. Technol.* **1986**, *4*, 203.

120. Pedersen, C. J.; Frensdorff, H. K. *Angew. Chem.* **1972**, *84*, 16.

121. Christensen, J. J.; Etaough, D. J.; Izatt, R. M. *Chem. Rev.* **1974**, *74*, 351.

122. Smid, J.; Sinta, R. *Top. Curr. Chem.* **1984**, *121*, 105.

123. Psheshetsky, V. S.; Lishinski, V. L.; Kokorin, A. I.; Tsarkova, L. A.; Rakhnianskaya, A. A.; Pertsov, N. V. *Makromol. Chem. Macromol. Symp.* **1992**, *59*, 163.

124. Baichenko, A. A.; Baichenko, A. A. *Flotatsionnoe Obogashch. Rud Ochiska Stochnykh Vod.* **1980**, 75; *Chem. Abstr.* **1982**, *96*, 203008p.

125. Daiichi Kogyo Seiyakui Co., Ltd.; Jpn. Kokai Tokkyo Koho Jpn. Patent 58 142 984, 1983; *Chem. Abstr.* **1984**, *100*, 71125p.

126. Paprotny, J.; Sobel, P. *Zesz. Nauk. Politech. Slask. Gorn.* **1987**, *160*, 57; *Chem. Abstr.* **1989**, *111*, 81725a.

127. Attia, Y. A.; Sinclair, R. G.; Markle, R. A.; Cousin, M.; Krishnan, S. V.; Keys, R. O. In *Flocculation in Biotechnology and Separation Systems: Part 3. Design and Synthesis of Selective Flocculants*; Attia, Y. A., Ed.; Process Technology Proceedings 4; Elsevier: Amsterdam, Netherlands, 1987; pp 263–275.

128. Sommer, L. *Z. Anorg. Allgemeine Chem.* **1963**, *321*, 191.

129. Bertini, V.; Pocci, M.; Picci, N.; Lucchesini, F.; Falbo, A.; De Munno, A. *J. Polym. Sci. Chem. Ed.* **1987**, *25*, 2625.

130. Bertini, V.; Pocci, M.; De Munno, A.; Picci, N.; Lucchesini, F. *Eur. Polym. J.* **1988**, *24*, 467.

131. Pocci, M.; Picci, N.; Bertini, V.; De Munno, A.; Lucchesini, F. *Eur. Polym. J.* **1990**, *26*, 15.

132. Pocci, M.; Bertini, V.; Picci, N.; Fabiani, F.; Marabini, A.; Barbaro, M.; De Munno, A. *Atti Soc. Toscana. Sci. Nat. Mem. Ser. A* **1991**, *98*, 271.

133. Bertini, V.; Pocci, M.; Marabini, A.; Barbaro, M.; Picci, N.; De Munno, A. *Part. Sci. Technol.* **1991**, *9*, 191.

Part Five

Biomedical Applications

Biocompatible Polymer Surfaces

Ih-Houng Loh, Min-Shyan Sheu, and Alan B. Fischer

The increasing use of polymeric materials in medical devices has placed new demands on the performance of these materials. In particular, these polymers need to be biocompatible. Several methodologies have been developed for modifying the surfaces of polymeric materials to enhance or inhibit their interactions with biological environments. High energy techniques, such as glow discharge, have been used to impart specific chemical functionality to the biomaterial surfaces or to deposit new polymer films with desired properties. Alternatively, chemical agents have been grafted onto the surfaces or activated surfaces have been used to initiate polymerization of new polymer overlayers directly on the substrate. The development of new materials and the use of surface-modification procedures leading to polymers with improved biocompatibility and use as biomedical devices are discussed in this chapter.

The increasing use of polymeric materials in medical devices has placed new demands on the performance of these materials: in particular, the need to be biocompatible. The exact meaning of biocompatibility depends on how and where the device will be used and the tissues with which the device will come in contact. Fortunately, polymeric materials are amenable to both chemical and physical modifications to enhance their biocompatibility. Appropriately chosen modifications can result in inert biocompatible surfaces or surfaces that act to promote beneficial interactions with surrounding tissues.

Many types of polymers are used for medical applications, ranging from soft hydrogels for contact lenses to high strength materials for orthopedic applications. Chemical formulation and processing variables help to determine both chemical and morphological surface characteristics. Tools such as scanning electron microscopy, atomic force microscopy, and X-ray photoelectron spectroscopy help relate surface properties to interactions with biological tissues and fluids.

Proteins and cells are known to interact with the surfaces of polymers through hydrophobic, ionic, and chemical means. In many cases, surfaces are rapidly covered by a biofilm of adsorbed proteins that mediates interactions with tissues or blood cells. The surface properties of the polymeric material can strongly influence the constituents of the biofilm and thereby influence the action of surrounding cells. A great deal of research has been directed at understanding how to specifically and selectively influence protein and cell binding reactions and how to induce cell-growth spreading at polymer surfaces.

Materials used for in vivo applications involving extensive blood contact have additional biocompatibility constraints placed on them. Blood-contacting surfaces must not activate the blood clotting cascade or lead to long-term inhibition of clotting. Control of the interaction of platelets and incorporation of specific nonthrombogenic and antithrombogenic coatings are some of the approaches used to improve performances of such biomaterials.

Several methodologies have been developed for modifying the surfaces of polymeric materials to enhance their interactions with biological environments. High energy techniques such as radio-frequency (rf)-plasma have been used to impart specific chemical functionality to surfaces or to deposit new polymer films with desired properties. Alternatively, chemical agents have been grafted onto the surfaces, or activated surfaces have been used to initiate polymerization of new polymer overlays directly on the existing surface.

Polymeric Biomaterials

Polymeric materials have a wide variety of applications for biomedical devices or prostheses (1–4) because of their desirable physical properties and versatile manufacturing capabilities. In biomedical applications (either for long-term implant or temporary use), polymeric materials typically come in contact with biological systems such as cells, proteins, tissues, organs, and organ systems. Table I lists a variety of polymers that are now being specifically used for medical applications.

Polymer Materials Used for Medical Devices

In general, the polymeric biomedical materials have five principal applications. They are cardiovasculars, orthopedic implants, wound dressing, ophthalmic prostheses, and controlled-release devices.

Cardiovasculars

Polymeric biomaterials used in cardiovascular applications come in contact with blood, so it is important

that these materials are made of thrombo-resistant materials. However, the perfect material, which resists degradation and fatigue and is nontoxic, noncarcinogenic, and totally thrombo-resistant, has yet to be developed. Table II lists some of the most common polymeric medical devices and materials that are currently used in the cardiovascular fields.

Orthopedic Implants

The orthopedic implant market has grown at an annual rate of 10% over the past few years, reaching approximately $580 million in 1992 (4). While the basic technology has remained the same for several decades, new materials and configurations are now driving the market. Although metals, ceramics, and specialty alloys dominate the orthopedic materials markets, new polymeric materials and advanced composites are slowly gaining in this growing market. The most common polymeric orthopedic implant materials are summarized in Table III.

Wound Dressing

Scientific advances in wound management are occurring rapidly, and newer products are displacing conventional dressings made from gauze. New film, foam, and gel products that control pharmaceutical delivery, regulate wettability, and promote wound healing rates have been developed. These materials

Table I. Polymeric Biomedical Materials and Applications

Polymeric Material	*Applications*
Non-Degradable Polymers	
Acrylics	Housing materials, maxillofacial prostheses, dentures, bone cements
Fluorocarbons	Vascular grafts, shunt, catheter coatings
Hydrogels	Contact lenses, ocular prostheses, catheter coatings
Polyacetal	Heart valve structures
Polyamide	Sutures
Polycarbonate	Housing materials for extracorporeal devices
Polyesters	Vascular grafts, soft tissue replacements, angioplasty balloons
Poly(ether ketones)	Heart valve and artificial heart structures
Polyimides	Heart valve and artificial heart structures
Polyolefins	Sutures, catheters, tubing, artificial heart bladder
Polystyrene	Cell culture substrates
Polysulfones	Heart valve and artificial heart structures
Polyurethanes	Catheters, artificial heart bladder, wound dressing
Poly(vinyl chloride)	Tubing, blood bags
Silicone rubbers	Tubing, catheters, soft tissue reconstruction, heart valve poppets
Polyethylene	Acetabular cup, total hip, knee, joints
Biodegradable Polymers	
Poly(amino acids)	Drug controlled release devices, cell adhesion peptides
Polyanhydride	Drug controlled release devices
Polycaprolactones	Drug controlled release devices, bone plates
Poly(lactic–glycolic acid)	Drug controlled release devices, sutures, bone plates
Polyhydroxybutyrates	Drug controlled release devices
Polyorthoesters	Drug controlled release devices, bone plates

Table II. Polymeric Medical Devices in Cardiovascular Fields

Medical Devices	Materials
Arterial catheters	PU, PE, PTFE
Blood bags	PVC
Blood tubing	SR, PVC, polyolefin rubber
Central venous catheters	SR, PVC, PU
Cardiovascular sutures	Nylon, PP
Extracorporeal devices	Housing: PC, acrylic
	Membranes: PP, SR, PSF
Hybrid artificial organs (pancreas, liver)	Housing: ABS, PC
	Membranes: PP, SR, PSF
Intra-aortic balloon pump	PU
Sutures	PGA, PLA
Total artifical hearts and left ventricular assist devices	PU
Vascular grafts	PET, PTFE

Note: PU, polyurethane; PE, polyethylene; PTFE, poly(tetrafluoroethylene) (Teflon); PVC, poly(vinyl chloride); SR, silicone rubber (poly(dimethylsiloxane)); PP, polypropylene; PC, polycarbonate; PSF, poly(sulfone); CA, cellulose acetate; ABS, acetylene–butadiene–styrene copolymer; PET, poly(ethylene terephthalate); PGA, poly(glycolic acid); PLA, poly(lactic acid).

Table III. The Most Common Polymeric Orthopedic Implant Materials

Medical Devices	Materials
Knee joint	Acrylic, ultrahigh molecular weight PE
Bone fixation device (bone plates, staples, screws)	Advanced composites: nondegradable, poly(ether ether ketone), and PSF; degradable, PLA
Ligaments	Expanded PTFE, woven polyester, braided PP, polyester elastomers

are generally different types of polyurethane (PU), hydrogels, silicon rubber (SR), and poly(tetrafluoroethylene) (PTFE) materials.

Ophthalmic Prostheses

Contact lenses and intraocular lenses are two of the most popular examples of biomaterials on ophthalmologic devices. Historically, poly(methyl methacrylate) (PMMA) was the polymer of choice to replace glass as the lens material because of its high refractive index, hardness, and biocompatibility. However, PMMA has its own disadvantages such as hydrophobicity and low oxygen transmissibility. Silicone lens materials offer excellent oxygen transfer but lack surface hydrophilicity. Hydrophilic polymers such as poly(hydroxyethyl methacrylate) and poly(N-vinylpyrollidone) have excellent wettability but possess low rigidity. The most promising polymers for ophthalmic applications should offer these advantages and eliminate these disadvantages. Functional polymers with efficient surface wettability and excellent physical properties will be the ones chosen for future ophthalmic uses.

Controlled-Release Devices

Controlled-release technologies are probably one of the most exciting biomedical developments in the past few decades. Controlled-release technology has thera-

peutic and economic benefits for a wide range of diseases, drugs, and delivery modes. It depends critically on advanced materials having special properties, such as biocompatibility, selective permeability, site-specific affinity, and in some cases, biodegradability. These materials include various natural and synthetic polymers. Some of the common biodegradable polymer materials are listed in Table I.

Polymers in Biological Environments

The surface properties of a material being introduced into the body are of critical importance because they help determine the initial response of the body's own defense. Whereas the bulk properties of a device such as compressibility or tensile strength of a material may be critical for the intended use, the surface properties will set the course for the interaction with the tissues of the host environment.

The earliest metallic implants were inadvertently modified by layers of native oxides that served to present a suitable surface to the surrounding tissues and to passivate the metal to further degradation (corrosion). Deliberate oxidation is now often used for preparation of metallic implant surfaces (5). Polymeric materials, unlike metals, do not have an appreciable oxide surface but can be modified by chemical and physical means to change both the chemical and physical structure of the surface. The ability to modify and tailor the surface properties of polymeric materials gives the medical device designer a much freer hand in selecting materials for a particular application. Materials can be chosen on the basis of their desired bulk properties and then surface modified to impart desired surface characteristics and render the materials biocompatible.

The ways in which the surface of a polymeric material interacts with the biological environment are described subsequently. The knowledge of these interactions has been used to develop many different surface-modification techniques. Examples of the techniques and the methods of characterization illustrate the power of this approach.

Biocompatible Surfaces

Operational Definition of Biocompatibility

The field of medicine has seen dramatic advances in the ability to detect and diagnose disease states and the ability to repair and restore damaged function. New types of invasive probes including catheter-delivery imaging dye injections, ultrasound transducers, and endoscopes allow the clinician to examine problems where they occur. In situ repair or replacement of diseased or damaged tissues is now possible not only by surgery using synthetic grafts but also by catheter-guided delivery, as in the case of percutaneous angioplasty. These advances have been aided by the use of polymer-based biomaterials that are compatible with these invasive applications.

The term *biocompatibility* has been used widely in discussions of biomaterials (6). Definitions for the term vary, and the specific criteria used to judge biocompatibility depend on the type of application. Thus, a more general definition of biocompatibility is that the material must not cause any harm to the host environment or trigger deleterious responses. Depending on the specific use, different material properties may be important for biocompatibility. In all cases, the physical and chemical structure of the surface form the immediate interface with the host and are hence critical for initiating acceptance of the biomaterial.

Interaction of Biomaterial Surfaces with Proteins and Cells

Implanted materials can provoke immune and inflammatory responses with detrimental effects on the organism (7, 8). Responses at the molecular and cellular levels lead to inflammation, blood coagulation, and complement activation, and if uncontrolled, these responses can lead to serious complications. The surface properties of the implanted material influence the initial response. When a material is placed in a physiological solution, a large variety of proteins reach the surface on a time scale of milliseconds to seconds (9, 10). Biomolecules continuously approach the surface, absorb and denature on the surface, or displace proteins in the adsorbed layer (11, 12). The extent, type, and orientation of proteins that adsorb on the surface may depend on the location within the body and on the material surface. Thus, a hydrophobic surface such as silicone will more strongly adsorb hydrophobic proteins such as fibrinogen and immunoglobulin compared with a hydrophilic surface. The proteins that adsorb onto the surface serve as the interface for subsequent processes that occur at the cellular level. Adsorbed protein molecules can serve to either promote or inhibit the binding and activity of cells. Cells eventually come into contact with the protein-covered surface, and their responses will depend on the composition, organization, and conformation of the protein layers: some cells may be induced to adhere to the surface, whereas others may be repelled from it.

The surface properties of materials in contact with cells can have a significant impact on cellular metabolic function. Surface composition, charge, polarity, mobility, and other properties have all been shown to be important, although there are significant differences among cell types and growth conditions (13–18). Proteins and other biomolecules adsorbed under in vitro and in vivo conditions are very important for cell attachment, growth, and activation (19–21). The character of the surface influences the deposition and binding characteristics of the intermediating proteins and thereby influences cellular responses to the surface. The process of adhesion of cells to substrates involves the following main steps (20):

- adsorption of biomolecules onto the substrate
- contact of cells with the protein layer
- attachment of the cells to the surface
- spreading of the cells on the surface

In most cases of specific adhesion, cells adhere to the surface by relatively specific interactions between integrin-type receptors and *adhesive proteins*, which are already adsorbed onto the substrate (20). Several extracellular matrix (ECM) and plasma proteins have been shown to mediate cell substrate attachment. Fibronectin, a 440-kd glycoprotein that is found in both plasma and the ECM, has been the most thoroughly investigated adhesive protein, and much information has been gained on the mechanisms of fibronectin-mediated cell adhesion. Fibronectin has a significant role in endothelial cell and osteoblast attachment (22). However, many other biomolecules play a role as well.

In addition to its role in mediating cell attachment to artificial substrates, fibronectin is responsible for other functions such as bone development (23–25). Fibronectin mediates the migration, attachment, and therefore the interaction of mesenchymal cells with collagenous extracellular matrices. Pluripotent mesenchymal stem cells (26) (which can give rise to bone, cartilage, tendon, and ligament) appear to be induced to differentiate into chondroblasts and osteoblasts and, thus, to form cartilage and bone by this interaction. Thus, surface properties of polymer biomaterials not only influence protein and cellular attachment but can have a long-range impact on cellular response and function.

Blood Compatible Surfaces

Materials used in in vivo applications involving extensive blood contact must meet several criteria to be biocompatible. Blood-containing surfaces must not activate the blood clotting cascade or lead to long-term clotting inhibition. Control of the interaction of platelets and incorporation of specific nonthrombogenic and antithrombogenic coatings are some of the approaches used to improve performance.

Polymer–Blood Interactions

When a foreign biomaterial contacts blood, the first measurable response in the initial seconds to minutes is the adsorption of biomolecules, mainly proteins. This response is followed in the next minutes to hours by cellular interactions from platelets and white blood cells (or leukocytes). These interactions involve the intrinsic blood coagulation cascade, and they end with the formation of thrombi on the implant surface. In addition, platelets may adhere and aggregate to the foreign materials when deposited fibrin or adsorbed blood proteins are present. The adhesion of platelets to a thin layer of deposited fibrin may cause platelet aggregation and the release of phospholipids, which accelerate the blood coagulation cascade and fibrin formation. On the other hand, physical and chemical properties of the biomaterial can be altered during the implantation. Important responses following the interaction of polymeric biomaterials with blood are summarized in Table IV.

Important biological responses in the blood in contact with the polymer include activation of the complement and coagulation cascades leading to neutropenia and fibrin formation, platelet adhesion and aggregation, thrombus deposition, embolization, and endothelial cell proliferation. On the other hand, leaching of impurities or additives, degradation, calcification, cracking, and fatigue are often seen in the cardiovascular implants. Failure of an implant may be due to the following (27):

- its material nature (chemical and physical incompatibility)

Table IV. Responses in Blood and Polymeric Biomaterials upon Interactions

Parameter	Responses
Blood	Protein adsorption
	Coagulation system activation and fibrin deposition
	Kallikrein system activation
	Complement system activation
	Fibrinolysis
	Platelet adhesion, release, and aggregation
	Thrombosis
	Embolization
	Endothelial proliferation
	Thrombocytopenia
	Neutropenia
	Thrombophlebitis
	Infection
	Anastomatic hyperplasia and stenosis (in blood vessels)
Polymeric materials	Swelling
	Leaching
	Delamination
	Degradation
	Absorption
	Mineralization (calcification)
	Cracking
	Fatigue
	Oxidation
	Cross-linking

- design of the implant
- fabrication method
- sterilization technique
- surgical implantation and clinical procedures
- the patient

The fibrin thrombus can either be removed by the fibrinolytic system or go into the circulation (embolization). After removal of a thrombus from the implant because of embolization, the biomaterial surface continues to stimulate the thrombosis process and to form a new thrombus. Embolization of thrombi, including white thrombi, may block the distal circulation and cause strokes if such blockage occurs in the brain. Strokes resulting from embolization are considered the major clinical roadblock to the use of the artificial heart.

Polymers with Enhanced Blood Compatibility

Over the past few decades, a number of methods have been developed to obtain blood-compatible surfaces. Generally recognized hemocompatible surfaces are

- hydrophilic and nonfouling polymers, [e.g., polyethylene oxide (PEO)]
- pre-adsorbed proteins (e.g., albumin)
- immobilized drug or enzymes that inhibit the blood coagulation process (e.g., heparin and urokinase)
- cell lining (e.g., endothelium)

Nonfouling PEO Surface. Because of its unusual solubility characteristics in water and most organic solvents (28), PEO has been used in a number of applications in biomaterials and biotechnology (29). Surfaces modified with PEO exhibit resistance to fouling by protein adsorption and platelet adhesion and suppression of immunogenic reactions (30). Some of the explanations proposed for the low affinity of PEO surfaces for proteins and cells have been proposed and are summarized:

- non-ionic hydrophilic characteristics (high water content) of the PEO molecule
- large excluded volume of the PEO molecule (31)
- high mobility of PEO chains in water (32)
- osmotic repulsion of PEO surfaces (33)
- lack of a protein binding site on the PEO molecule (34)
- ability of PEO to complex metal ions (35)

Such observations have stimulated great interest in surface modification of biomaterials (especially hydrophobic polymers) with PEO. Polymer surfaces have been modified with PEO in many ways, such as physical adsorption (36–38), surface entrapment (39),

chemical immobilization (40, 41), chemisorption (42, 43), surface grafting (44–46), plasma (or glow discharge) polymerization (47), and glow discharge immobilization (48, 49). In addition, further attachment of biomolecules such as heparin (50) onto PEO surfaces was considered to enhance their blood compatibility.

Surface Passivation of Albumin. Albumin, the predominant plasma protein, is a globular, amphipathic, negatively charged protein of modest size (M_r ~66,000) (51). Albumin probably plays a number of physiological roles, such as the binding and transport of normal catabolites (e.g., bilirubin). Albumin may also be important to the vascular endothelium. Normal vascular surfaces are covered with a continuous film of plasma proteins. Even though it is not clear whether endothelium possesses a specific receptor for albumin, there is evidence for a high affinity interaction of albumin with the endothelial surface (52). These observations suggest that the selective adsorption of albumin to the surface of implantable devices might yield materials with desirable blood compatibility.

Biomaterials to which albumin has been passively adsorbed are markedly less procoagulant and discourage platelet and phagocyte adherence (53–56). Alkylate surfaces that display at least five or six CH_2 groups (intended to act as a ligand for the fatty acid binding domain(s) of albumin) preferentially bound (57–60) albumin and improved the thrombo-resistance of a biomaterial in an acute canine ex vivo experiment (61). In addition, an albumin-affinity coating containing a cibacron blue dye was developed (62).

Heparinized Surfaces. Heparin, a water-soluble heteropolysaccharide that contains 2–3 sulfate groups in each disaccharide unit, has demonstrated anticoagulation ability. The action of heparin in anticoagulation is due to its ability to form strong complexes with a variety of blood-clotting factors such as thrombin, fibrinogen, prothrombin, and Factors IX–XII (63). Thus, heparin neutralizes the action of the factors. However, the greatest contribution to the anticoagulant activity of heparin is through its enhancement of the activity of anti-thrombin III, which is normally found in plasma to inhibit the activation of clotting factors and other proteases (64).

Heparinized biomaterial surfaces have been obtained by ionic or covalent immobilization of heparin and by synthesis of block copolymers (65). A biomaterial surface with a cationic group can attract heparin binding through ionic interactions. Such cationic surfaces can be introduced by physical deposition with a cationic detergent (66), cationic salt (67, 68), or anion-exchange resins (69) or by radiation grafting with ionogenic monomers (70, 71). In general, surfaces coated with ionically bound heparin can prevent thrombosis after the implantation. However, this heparin coating is often unstable and can be eluted eventually from the biomaterial surface, which limits long-term applications of the medical implants. Immobilization of heparin, on the other hand, becomes more favored because of its long-term stability (72–77).

Endothelialization. The anticoagulatory function of the blood vessel surfaces (endothelium) was used to develop a hemocompatible surface (78). A biomaterial surface is rapidly covered with endothelial cells before and/or after the implantation, and the process is called endothelialization (endothelial seeding). The hemocompatibility of the composite would then depend largely on the function of the adherent cells. Successful endothelialization of biomedical implants may improve biocompatibility and aid in reducing antigenicity and thrombogenicity of vascular grafts. Therefore, many surface-modification methods have been developed to achieve enhanced endothelialization on a biomaterial surface, and the methods are generally categorized as nonspecific and specific endothelialization. The nonspecific method can be morphological or chemical modifications to a polymer surface (e.g., microfiber or flocked surfaces) (79), whereas the specific method is surface plasma treated with oxygen–acetone mixtures (80).

Endothelialization involves cell adhesion molecules (CAMs), such as fibronectin, laminin, and collagen. Surfaces precoated with CAMs can enhance the receptor-mediated cell adhesion of anchorage-dependent cells such as endothelial cells. Recently, the attachment domain (binding site) present in these CAMs was determined as a tripeptide, Arg-Gly-Asp (RGD). Massia and Hubbell (81–83) covalently immobilized RGD peptides on a glass substrate. The peptide-modified surfaces showed the support for focal contact and stress fiber formation within fibroblasts, indicating cell adhesion and spreading.

Evaluation of Biocompatibility

Cytotoxicity

Obviously, medical devices must satisfy safety and efficacy requirements before they are applied to humans. Implanted biomaterials must not be toxic to cells or tissue directly or through degradation products. Therefore, both short-term and long-term evaluations generally are required during the development of biomaterial implants. Table V lists the major methods used to evaluate cytotoxicity of medical implants. Acute cytotoxicity tests in cell cultures are always considered as an initial screening test of biocompatibility for both tissue and cardiovascular implants. The results from cell culture evaluations are usually obtained within 1 week. Mouse

fibroblasts and chick embryo cells are normally used in theses cell culture assays. To pass the initial screen of biocompatibility, materials or extracts of materials in contact with cells under controlled conditions should not cause cell death or inhibit cell growth.

When the biomaterials are compatible with cells, in vivo implantation is then performed to evaluate their compatibility with tissue. Such implantation evaluations include the following:

- materials (shape, size, films, fabric, or extracts)
- location (skin, subcutaneous tissue, muscle, or intraperitoneum)
- application (contact, injection, or implantation)
- animal selection (mice, rats, guinea pigs, and rabbits)
- periods (few days to several months)

Biomaterials can be placed in a cage and then implanted beneath the skin. Tissue fluid is then periodically removed by a syringe and examined to observe the cells present. The material should be implanted for at least 4 days to evaluate the acute inflammatory responses of the tissue. Longer implantation may be conducted to examine chronic inflammation mainly due to biodegradation. In general, inflammatory responses of tissue to the implant are examined by direct microscopic observation of fixed and stained explanted material (84) and/or analyses of enzymes specific to macrophages and polymorphonuclear leukocytes (85–87).

Hemocompatibility

As described previously, biomaterials may interact with blood components and provoke the blood coagulation cascade, which is different form material–tissue interactions. The consequence of these blood–material interactions (e.g., thrombosis and embolization) usually causes the failure of implantation. Therefore, in addition to acute cytotoxicity tests using cell culture, it is particularly important to examine hemocompatibility of the cardiovascular implants. The major evaluations of hemocompatibility of implants are summarized in Table VI.

Table V. Evaluation of Cytotoxicity

Test	Method
Acute cytotoxicity tests (cell culture)	Agar overlay assay Fluid medium assay Culture with material extracts
In vivo toxicity tests	Skin contact Subcutaneous implantation Cage implantation Muscle implantation Intraperitoneal implantation

Table VI. Evaluation of Hemocompatibility

Method	Information
In vitro tests	Protein adsorption (fibrinogen)
	Complement activation
	Platelet adhesion, aggregation, and release
	Thrombus deposition
	Whole blood clotting time
Ex vivo shunts	Platelet adhesion, aggregation, and release
	Platelet survival, turnover, and consumption
	Thrombus deposition
	Emboli formation
In vivo implantation	Platelet adhesion, aggregation, and release
	Platelet survival, turnover, and consumption
	Thrombus deposition
	Emboli formation

Table VII. Surface Modifications of Polymers

Type	Method	Ref.
Physical	Physical adsorption	92, 93
	Langmuir–Blodgett film	92
Chemical	Acid etching	95–98
	Alkaline etching	94
	Chemical infusion	99
	Ozone	100–102
	Flame treatment	94, 103
Radiation	Plasma (glow discharge)	104, 105
	Photo-activation (UV)	106–108
	Corona	
	Electron beam	109, 110
	Ion beam	111
	Laser	112

In vitro tests of biomaterials are usually performed as an initial screening test in the early stage of hemocompatibility evaluation. Biocompatibility of biomaterials is believed to be strongly influenced, if not dictated, by a layer of host proteins spontaneously adsorbed to the surfaces after their implantation. For example, adsorbed albumin may decrease platelet adhesion and activation, whereas fibrinogen may enhance the thrombus formation. Adsorbed proteins can be determined by enzyme-linked immunoassay (88), X-ray photoelectron spectroscopy, attenuated total reflection Fourier transform IR spectroscopy (89), ellipsometry (90), fluorescent-labeling, and radio-labeling (91). Adhered platelets, on the other hand, are usually measured or observed by using radio-labeling or microscopy.

Several hematological tests also were applied to in vitro hemocompatibility evaluation of biomaterials. These tests include blood-clotting time, clot weight, thrombin time, prothrombin time, and partial thromboplastin time. Platelet activity (aggregabililty) can be measured by release factors such as serotonin (dense granule), adenosine 5'-diphosphate (dense granule), Platelet Factor IV (a-granule), β-thromboglobulin (α-granule), and thromboxane.

Although in vitro tests may provide rapid screening of hemocompatible materials, the results are only used to predict their in vivo performance of hemocompatibility. An ex vivo shunt study is usually performed to eliminate unnecessary sacrifice in in vivo tests. In ex vivo tests, the blood of an animal is bypassed via a cannulate vessel though a test chamber or section of shunt containing testing materials, and the blood is then returned to the animal. Baboons, dogs, sheep, or rabbits have been considered for ex vivo shunt studies. Emboli formation may be observed using Kusserow ring or laser scattering.

Finally, suitability of the biomaterials after implantation can be determined only by in vivo evaluation.

An in vivo animal model can be used to evaluate both short-term and long-term hemocompatibility. After sacrifice of the animal, the implanted material is then removed for further histological examination.

Surface Modification of Polymers for Biocompatibility

Various technologies were developed in the past decade for the functionalization of polymer surfaces (Table VII) (92–94). The classification of these methods is arbitrary because of the overlap of some modification techniques.

Polymer surfaces can be modified by physical deposition or adsorption of amphiphilic molecules or macromolecules, or surfaces can be chemically modified by liquid- or gas-phase reactions. For example, strong oxidizing acids were used to introduce carbonyl (95, 96) and carboxylic acid (97, 98) groups at polyolefinic surfaces. Treating polyolefins with ozone yields surfaces with hydroxyl, carbonyl, and carboxylic acid groups (100). Flame and thermal treatments induce radicals and chain scission on polymer surfaces (103). Furthermore, ion beam treatments are used to induce chemical and morphological changes of the surfaces, and these treatments can also be used for the incorporation of various elements into surfaces (111). Laser treatments combining thermal and UV effects were used to induce chemical and morphological changes at the polymer surface (112).

These surface-modification methods can also induce further surface functionalization via grafting methods. For this purpose, radicals or peroxides in the surface region of polymers are activated by different types of activating regents, and γ-radiation, electron beam (109), UV, plasma discharge, chemical (113), ozone (101, 102), and mechanochemical activation (114) methods were reported. The active sites created are then used as surface-bound initiators to start graft reactions with unsaturated compounds. The use of compounds containing functional groups often yields functionalized surfaces. For example, acrylic acid has been grafted onto poly(ethylene terephthalate) by γ-radiation (102) or by plasmas discharge (104) and onto polyethylene by electron beam (110), UV (106), γ-radiation (115), and plasma discharge (105) treatments.

It is impractical to describe all of the surface-modification techniques in detail here, but two of the most recently developed surface-functionalization technologies—photoactivation and plasma surface modification—are discussed at some length. For a full discussion on plasma surface modification, *see* Chapter 1.7.

Photoactivation

Photoactivation is normally performed on the surface by photoradiation to directly dissociate backbone C–C bonds, and for our purposes to create polymer free radicals (116). In this process, a hydrogen-abstracting photoinitiator is normally used to generate the polymer radicals. In general, photoactivation can be accomplished in three ways:

1. pre-irradiation, where the backbone polymer is irradiated in vacuum to produce radicals before exposure to the reacting monomer
2. per-oxidation, where the polymer is irradiated in the presence of oxygen to produce peroxy- and hydroperoxy radicals, which can initiate grafting
3. the simultaneous process

A promising photoactivation surface-modification process (PhotoLink) is chosen to illustrate this photoactivation process.

The PhotoLink Process

The PhotoLink process uses photochemistry to covalently bond reagents to the surfaces of biomaterials (117). The process consists of first covalently coupling the molecule of interest to a photoreactive moiety, usually a benzophenone derivative. Next, this photoreagent is applied to the device surface in an aqueous solution. After exposure to UV radiation (1–2 min), the benzophenone group (1) is promoted to the excited state (2), followed by a rapid intersystem crossing to the triplet excited state (3). This species then abstracts a hydrogen from the substrate surface, resulting in a pair of radicals (4) that rapidly collapse to form a covalent bond to the surface (5) (Scheme I). In Scheme I, the R group can be a synthetic polymer, a biomolecule, or an organic tether for immobilizing biomolecules. This process is capable of covalent coupling to most polymers possessing C–H bonds on their surface. Also, because most of the photoreagents have multiple photogroups per molecule, exposure to light produces coupling to the device surface and to adjacent photoreagents. This coupling allows the generation of coatings that range from widely dispersed molecules to coating that are several molecules thick.

The major advantages for PhotoLink in biomaterials surface modification are that it does not require vacuum, elevated temperatures, harsh solvents, or long reaction times. A wide variety of hydrophilic polymers and biomolecules were immobilized to polymeric biomaterials surfaces by using this PhotoLink process, examples of which include poly(ethylene glycols), polyacrylamides, poly(vinylpyrrolidone), heparin, hirudin, antibodies, collagens, other ECM proteins, and ECM peptides.

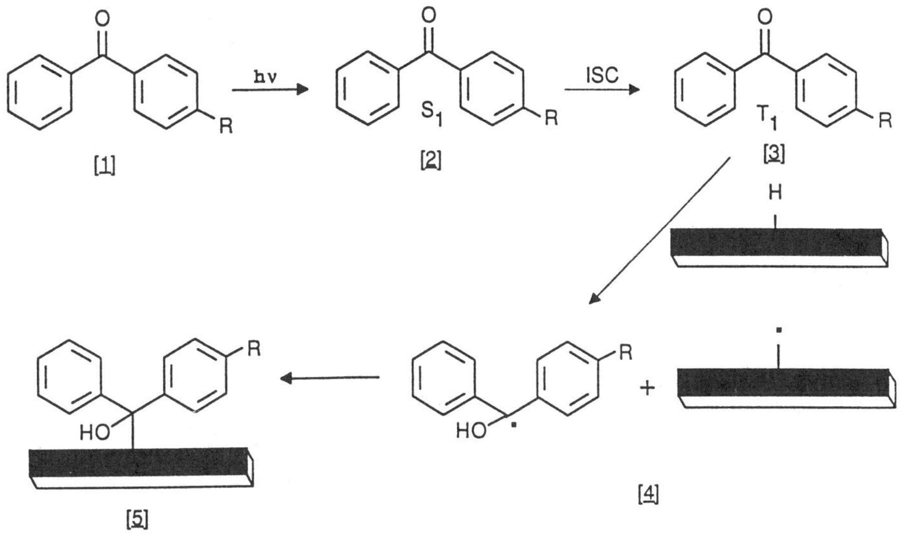

Scheme I. General scheme and chemical processes involved in the PhotoLink process. (Reproduced with permission from BSI Corporation.)

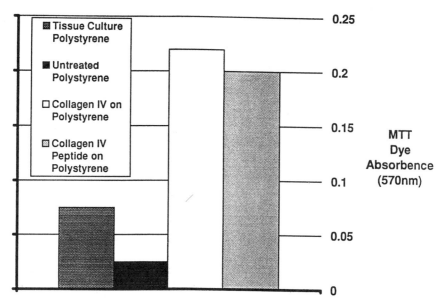

Figure 1. Growth of human umbilical vein endothelial cells on surface-modified PS 24-well plates. The plates were seeded with 1500 cells per well, cultured for 8 days, and relative cell numbers were quantitated with a tetrazolium metabolic dye method. Each bar is the average of 4 determinations, and the standard error of the mean is shown. (Reproduced with permission from BSI Corporation).

Biocompatible Polymer Surfaces

Figure 1 presents in vitro data from a study designed to improve cell growth on polystyrene (PS). In this study, PS tissue culture plates were coated with type IV collagen (COL IV) and a peptide derived from COL IV. Low-passage human-umbilical-vein endothelial cells were added and cultured for 8 days. Relative cell numbers were quantified by using a tetrazolium metabolic dye. The data show that cell numbers were about eightfold greater on surfaces coated with COL IV or COL IV peptide than on untreated PS. Also, cell numbers were about 3.5-fold greater on the COL IV and COL IV peptide coated surfaces than on standard tissue culture PS.

This effect also was demonstrated in vivo in several applications. For example, a combination of fibronectin plus COL IV produced a threefold increase in lumenal endothelial cell coverage on 4-mm expanded PTFE (ePTFE) and PU vascular grafts that were implanted in dog femoral arteries (118). Each protein immobilized individually was less effective than when immobilized in combination.

In another example, a coating of type I collagen was applied to a PMMA intracorneal lens and implanted in rabbit corneas for 15 months (119). Compared with uncoated controls, the coated lenses promoted bonding of stromal tissue, reduced inflammation adjacent to the device, and greatly reduced necrosis of corneal tissue over the device. These lenses have been implanted into two patients who could not be treated by other methods and who had each been

blind for over 5 years. Their respective vision, 5 and 6 months postoperative, was 20/50 and 20/100.

Also, a coating of COL IV was applied to polydimethylsiloxane (SR) breast implants and implanted subcutaneous in pigs for 4 months (120). Compared with uncoated controls, the coated model breast implants showed greater bonding of tissue to the device surface and a 48% thinner fibrous capsule.

These in vitro and in vivo results demonstrate that photochemistry is a gentle process that can immobilize fragile biomolecules, such as collagen and ECM peptides onto a variety of biomaterials including SR, ePTFE, and PS with retention of critical activities. The resultant coated surfaces greatly improve the culture of low passage cells in vitro and tissue integration and blood compatibility with implant devices in vivo. The potential commercial values of the photoactivation techniques currently being evaluated include hydrophilic–lubricious coatings such as photoactive derivatives of heparin (indwelling catheters, pacemaker leads, contact lenses, etc.), blood compatibility coatings (e.g., indwelling catheters, stents, and shunts), and photoactive derivatives of ECM proteins and carbohydrates (e.g., vascular grafts, breast implants, and artificial corneas).

Plasma Discharge Surface Modification

Plasma discharge techniques have been used extensively to functionalize polymer surfaces (121–126). Surface modification is achieved by exposing the surface to a partially ionized gas (plasma). Depending on

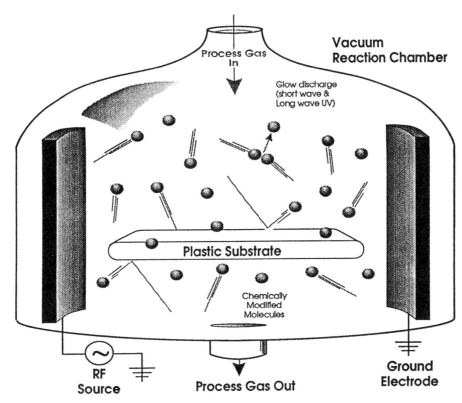

Figure 2. Schematic diagram of the glow discharge plasma.

the operating pressure, two types of processes can be distinguished: corona discharge (atmospheric pressure) and glow discharge (reduced pressure). Corona treatment is commonly chosen to oxidize a polymer surface yielding several oxygen-containing groups such as hydroxyl, carbonyl, and carboxylic groups. Although corona treatments is a simple and economical method to introduce some selected functional groups, this technique is not particularly suitable to specifically introduce one type of functional group. Because of the requirements of high electrical voltage, corona treatment has limited use for delicate biomaterials. Therefore, the discussion in this chapter will only focus on the glow discharge techniques.

A glow discharge plasma is created by evacuating a vessel, usually quartz (for inertness), then refilling with a low pressure gas (Figure 2). System pressures typically range from 13.3 to 66.7 kPa. The gas is excited by electrical energy, usually in the rf range, such as 13.56 MHz. As the gas ionizes, it emits a characteristic glow (which is why a plasma is sometimes referred to as a glow discharge).

The energetic species in a gas plasma—including ions, electrons, radicals, metastables, and photons (short-wave UV)—collide with the surfaces polymer in the plasma chamber and excite the molecules in the surface layers. The energies of the plasma constituents are equal to or exceed the bond energies of many common chemical bonds. Hence, a low pressure plasma has the ability to break chemical bonds. Although the precise mechanisms of bond scission are only just beginning to be understood, scientists generally assume that as the bonds on the surface break, new bonds are formed: some with the gaseous species (if the gas is reactive) and some within the polymer itself. Because the plasma treatment process is conducted in a vacuum, it tends to be completely pervasive and is capable of treating surfaces of complex shape.

Gas plasma treatments are typically carried out in a batch process, but continuous on-line treatment of fibers, tubing, membranes, fabrics, and films is also common. In many cases a custom reactor design is necessary to maintain a high level of throughput, maximizing the economic feasibility of the process. Once a process is developed and the reactor is set up, the treatment levels are extremely stable and predictable. Reproducibility is excellent due to the inherent stability and control of the reaction parameters: gas flow, pressure, electrical power input, and residence time.

One major concern to everyone is the environmental impact of any chemical process. This area is where gas plasma treatment excels. The byproducts of the plasma process are mostly gaseous and consist mainly of the original gas species introduced into the reaction chamber. This form is possible because when the gas is no longer under the direct effect of the electrical field it recombines into its nonexcited state. In the case of a nonreactive gas such as argon there are very few environmental concerns except that the process vent is constructed in a safe and responsible manner. Reactive gases are obviously changed by the process into something different than was initially introduced. However, the process is far from stoichiometric; a comparatively small amount of the gas used actually reacts with the surface. Therefore, the major constituent of the effluent gas is still the primary species. Many of these species can be condensed in a cold trap installed in the vacuum line and recycled. In almost all cases the amount of environmental impact is so small and the effluent so innocuous that regulation is usually unnecessary.

Plasma Treatment

Plasma treatment can be performed by any reactive gas (all noninert gases are defined as reactive gases). Oxygen-containing gases, such as CO_2, O_2, acetone, water, or mixtures of the gases, are commonly used to create carbonyl or hydroxyl groups, whereas ammonia is often used to create amine groups. Also, fluorinated gases such as Freon can be used to implant fluorine atoms in some polymers to render them hydrophobic. In a similar way, almost any reactive gas can be used to tailor the surface properties of a polymer.

When inert gases are used to treat polymer surfaces, a different type of mechanism is in place (127). Because the molecules of the plasma gas are reluctant to form bonds with those of the surface, most of the new bonds are formed with donor molecules from within the polymer itself. This formation of bonds often results in polymer chain scission (degradation) and re-splicing along with the formation of cross-links between chains (cross-linking). Predomination of either process, degradation or cross-linking, depends on the chemical properties of the treated polymer. The cross-linking mechanism is usually of greatest interest for inert gas plasma treatment. By using this type of treatment, a highly cross-linked surface can be created on most polymers. Generally speaking, this treatment can impart such qualities as increased hardness, improved solvent resistance, decreased permeability, reduced surface tack, and improved biological properties.

Plasma Polymerization

Another method of surface modification by plasma is coating by plasma polymerization (122). When the gas used to create the plasma is a polymerizable molecule and the reaction parameters are properly controlled, a polymer may form on the surfaces in the reaction chamber. Because of the action of the plasma, these polymers are typically highly cross-linked and strongly adhered to the surface of the substrate.

The high cross-linking density in these plasma polymers makes them quite brittle, which precludes the possibility of their being used as flexible coatings, particularly in thickness above 15 μm. However, a layer of plasma polymer (a few hundred angstroms thick) can form an extremely tenacious primer for other coatings. In addition, the surface of the plasma polymer can be designed as chemically active, enabling the subsequent coating to be very strongly adhered to the plasma polymer.

Recent Developments

Several articles extensively reviewed plasma discharge technologies for polymer surface modification (122–126). In this section, only the most recent progress in the biomedical applications will be discussed.

Plasma Discharge Immobilization for Nonfouling Surfaces.

Recently, Sheu et al. (48, 49, 128) developed a novel plasma discharge process that can covalently immobilize PEO surfactants to polymer biomaterials for nonfouling biomaterials applications. This development addresses some critical problems facing biomedical devices, separations media, and other surface-sensitive applications. This process produces a hydrophilic, low adhesion PEO surface that resists the fouling by proteins, cells, or microorganisms. In addition, PEO-coated surfaces show a significant reduction in the amount of bacterial adhesion (132) and thrombus formation (45). As a result, this process can be instrumental in enhancing a variety of medical and biotechnological devices including medical implants, cardiovascular or urinary catheters, blood contact devices, ophthalmic lenses, biosensors, and separation membranes.

The BioPhilic process consists of two steps: coating the substrate with non-ionic PEO surfactants via organic solvents or aqueous solutions and cross-linking the hydrophobic tails of the surfactants to the substrate surface by means of an rf-plasma of an inert gas (Figure 3). The obtained PEO surface is a strongly adherent, thin coating and shows reduced water contact angles, protein adsorption, and cell adhesion. This plasma discharge immobilization process has several advantages:

- a fast and simple two-step process
- possibility of simultaneous sterilization
- covalent immobilization
- no change in the bulk properties
- low dependency on substrate surface functionality
- high surface coverage

This novel process also was applied to immobilize ionic surfactants [e.g., sodium dodecyl sulfate (133)] for enhancing surface wettability or to functionalize polymer surfaces [e.g., with primary amine groups when alkyl amines were used (134)].

The polyethylene surface coated with a plasma-immobilized-PEO surfactant demonstrated significant reduction of fibrinogen and platelet binding by 10-fold and fourfold, respectively (Figure 4). This process is quite versatile and can be applied to a wide variety of polymeric biomaterials (Figure 5). Such surfaces modified by the BioPhilic process are stable over time and retain greater hydrophilicity compared with oxygen plasma treated surfaces (Figure 6). These characteristics are particularly important for very mobile chain polymers on which the effects of thin surface treatments are often short-lived.

Plasma Deposited Polymers That Enhance Cell Growth.

As discussed in the previous section, most cells require strong interactions with the substrate surfaces to effectively stimulate proliferation and strong adhesion. Unfortunately, most of the polymeric implant surfaces lack the desired functional groups to support the cell attachment. Therefore, several biological coatings such as collagen, polylysine, laminin, or fibronectin typically are used to mimic in vivo conditions and enhance anchorage of the cells.

An alternative method, low temperature gas plasma treatment, has been used to induce uniform changes in the chemical functionality of tissue cultureware surface. This modification makes the surface more conducive to cell attachment and proliferation. Air plasma (*135*) or corona discharge are commonly

used to modify PS for anchorage-dependent cell growth. Ramsey et al. (*136*) reported increased growth of monkey kidney cells on O_2 plasma-treated PS as well as with other wet chemical treatments. They attributed this growth to an increase in carboxyl groups on the surface. Such an effect was demonstrated by several independent researchers for a variety of plasma-treated cultureware cell lines (*137*).

Recently, plasma-deposited coatings have been developed to a variety of substrates (including polymer, ceramic, and metal substrates) for the promotion

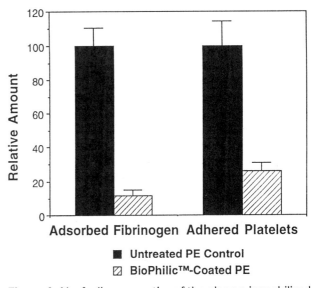

Figure 4. Nonfouling properties of the plasma-immobilized-PEO surfactant (BioPhilic coated) PE surfaces. Samples were incubated in 0.2 mg/mL ^{125}I-fibrinogen or 1×10^8 ^{111}In-platelets/mL solutions for 2 h at 37 °C. (Reproduced with permission from reference 48. Copyright 1992.)

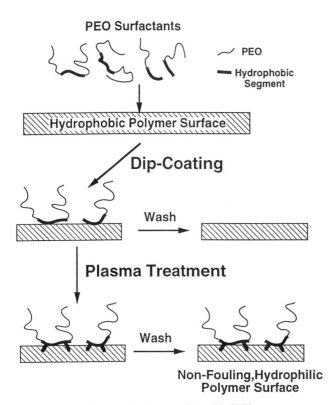

Figure 3. General scheme of the BioPhilic process.

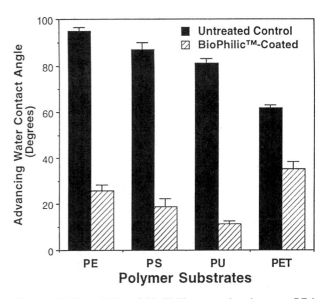

Figure 5. Wettability of BioPhilic coated polymers: PE is polyethylene; PET is poly(ethylene terephthalate).

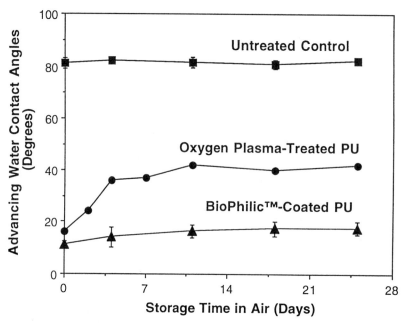

Figure 6. Stability of BioPhilic coating versus oxygen plasma treatment on PU. (BioPhilic coated samples have been washed in isopropyl alcohol before the stability study.)

of cell attachment and proliferation in vivo, also called CellStage by AST. This enhancement is due to increased adsorption of the adhesive protein fibronectin to the plasma-deposited layer. A recent study demonstrated (80) that in a medium containing 10% calf serum, cloning efficiency of 3T3 Swiss mouse fibroblasts adhered to surfaces that were treated with acetone plasma (60%) was increased approximately sevenfold compared with uncoated, nonadhesive substrates (8%). By examining fibronectin adsorption from the same medium, approximately seven times as much fibronectin adsorbed to the acetone-plasma-coated (2.1 ng/cm^2) versus untreated (0.3 ng/cm^2) materials. Plasma-deposited polymer coatings also were shown to accelerate cell spreading (129). Overall, the increased fibronectin-adsorbing capacity of plasma-deposited substrates correlates well with increases in cell attachment, spreading, and growth.

Osteoblast Responses to Plasma-Deposited Coatings.

Because of the role of fibronectin in osteoblast attachments and the importance of fibronectin–collagen interactions in bone induction, a fibronectin-adsorbing biomaterial (e.g., plasma-coated substrate) implanted into a collagenous matrix (e.g., bone) may be able to promote bone growth and bone–implant bonding. Therefore, implant fixation would be enhanced.

Increasing numbers of orthopedic implants are being used to restore function to diseased and damaged joints. Because of the "graying" population, the need for joint replacement is expected to continue increasing. Loosening of implants, however, remains a significant problem. Processes capable of enhancing bone formation have applications in treating orthopedic implants to promote bone–implant bonding. Surface-modification processes are most desirable because the treatment, depending on the specific process being employed, will not alter the bulk properties of the substrate. This point is of particular importance, because such technologies can be applied to any material without altering its bulk mechanical properties. Hence, plasma-deposited coatings that enhance osteoblast responses and, consequently, bone formation have commercial applications in modifying orthopedic implants of complex shape for the purpose of stabilizing the bone–implant interface.

Plasma Polymerization with Controllable Compositions.

Most scientists have known that plasma-deposited organic films are unpredictable and have complex, irregular composition. These characteristics are due to the complicated reaction mechanisms of the plasma and the many possible reactions (both plasma-phase and surface reactions). An organic compound (precursor) can be dissociated, rearranged, and deposited on the substrate of interest by a much more different chemistry than from the precursor. Recently, Lopez and a co-worker developed (130, 131) a unique plasma deposition technique by lowering the substrate temperature related to the plasma environments. This novel approach produced films with predictable chemical features (e.g., elemental composition and functional group constitution). Specifically, plasma-deposited 2-hydroxylethyl methacrylate and tetraethylene glycol dimethyl ether thin films were

Table VIII. Biomedical Applications of Plasma Surface Modification

Applications	Materials	Purposes of Treatment	Biological Interfaces	Ref.
Adhesion promotion	PE, PET, PP, PTFE, PU, PVC, SR	Chemical bonding, interfacial adhesion, surface grafting	Blood, tissue	138
Bioabsorbable polymers, controlled release systems	PGA, PLA	Hydrophobicity enhancement, surface cross-linking	Blood, body fluids, central nervous system	139–141
Biocompatibility enhancement	PET, PP, PS, PE	Endothelial cell attachment, surface grafting of anticoagulant agent	Blood, plasma	141, 143
Catheter, vascular grafts	PE, PET, PTFE	Biocompatility enhancement, reduction of blood clotting	Blood, tissue fluids	144–147
Cell culture substrates	PS	Surface charging, surface functionalization	Cell attachment, cell growth	80, 129, 135, 136
Diagnostic devices, biosensors	PP, PS	Protein–antibody immobilization	Blood	128, 148, 149
Ophthalmic lenses	PMMA, PHEMA	Reduction of protein deposits, wettability enhancement	Tear fluids, tissue fluids	150–153
Bioseparation membranes	Cellulose acetate (CA), PSF	Surface grafting, wettability enhancement	Blood, plasma	154

prepared and their suitability for nonfouling biomedical applications was demonstrated.

Biomedical Applications

Plasma surface-modification techniques are being used or investigated to control tissue adhesion, reduce drag, provide thrombo- or infection resistance, and act as immobilizing agents for biomolecules. Major applications and specific products based on plasma technologies are listed in Table VIII.

Probably the most common use of plasma surface modification is in the area of adhesion promotion (*138*). Using reactive gases, plasma scientists can create functional groups on the surface of a polymer to enable an adhesive to form strong molecular bonds with the surface. For example, ammonia plasma has been used to create amine groups that can bond heparin to the surface of vascular grafts for decreased thrombogenicity. Also, this type of treatment is often used to enhance the bond strength of some epoxies to substrate materials. Oxygen is often used in a similar role, forming active oxidized sites on the surface of the substrate to enhance adhesive bonding.

Another common application of plasma surface modification is the tailoring of surface energies. Both hydrophilic and hydrophobic surfaces can be created on polymers by reaction with gas plasma. For example, plasma surface modification of PS cell culture-ware surfaces is widely recognized as a significant improvement on cell culture technology. The use of oxygen to create hydroxyl functionality will increase the wettability of the surface. This process has been used to enhance the performance of a catheter by the creation of a wettable surface on the polymer tubing. In a similar way, surfaces can be specifically engineered to modify protein binding and improve blood compatibility.

In addition to surface functionalization, surface cross-linking is often used to enhance the performance of polymers. The activity of the plasma creates a higher cross-linking density within the material to a depth of a few thousand angstroms. The resulting increase in hardness and chemical resistance can be used to great advantage in many applications. For example, SR components treated in an inert gas plasma can be modified to form a hard skin on the surface. This modification results in a substantial decrease in surface tack and coefficient of friction.

Plasma polymers have been used in conjunction with parylene coating to provide tough barrier coatings that strongly adhere to the substrate. Many other types of coatings can be enhanced in a similar manner. These plasma surface-modification techniques have been used in many applications. For example, the treatment of intraocular prostheses such as contact lenses and intraocular lenses demonstrated improved wettability and decreased bacterial and protein adhesion, which would enhance wearer comfort.

With current systems for sustained drug release, zero-order release kinetics and tolerance by the human body are difficult to maintain. Plasma surface modification of known controlled-release delivery systems can overcome both problems by creating a tightly cross-linked barrier membrane (*139–141*), which is biocompatible and regulates the dispersion rate within a narrow range.

Biomaterials used in contact with blood or protein require special surface treatments to enhance biocompatibility. Amine functional groups attached by ammonia plasma treatment act as hooks for anticoagulants such as heparin, thereby decreasing thrombogenicity (*142, 143*). Plasma-deposited PTFE coatings on small-diameter Dacron vascular grafts demonstrated a dramatic improvement in patency and decreased embolization (*144, 145*).

Synthetic polymeric implant materials can be surface activated using rf-plasma techniques to enable covalent immobilization of cell-binding peptides derived from the ECM proteins fibronectin and laminin. The resulting grafted peptides can promote complete coverage of a surface with a monolayer of intact, healthy endothelial cells (the natural blood compatibility surface). As such, the successful endothelialized bioimplants will have improved biocompatibility and reduced antigenicity and thrombogenicity.

The use of rf-plasma permits the formation of reactive functional groups on just about any polymer of interest for use in medical devices. These reactive groups enable peptides to be coupled to the surface by using well-established immobilization reactions. Specifically, poly(ethylene terephthalate) and PTFE surfaces have been activated using rf-plasma modification to enable covalent coupling of adhesive binding peptides. Surfaces modified in this way enhanced the adhesion and proliferation of bovine pulmonary endothelial cells in vitro.

Conclusions and Future Outlook

Synthetic biomaterials have become widely accepted for in vivo and in vitro medical applications. However, interaction of blood components or biological media with the biomaterial surface often limits their performance. Various coatings and surface derivatizations are being used or investigated to control tissue adhesion, reduce tissue drag, provide thrombo-resistance or infection resistance, and act as immobilizing agents for chemotherapy and the removal of particular proteins or cells.

Among various surface-modification methods, plasma and plasma polymerizable coating offer favorable flexibility, effectiveness, safety, and environmental consciousness. For example, the plasma is effective at near ambient temperature without damage on most heat-sensitive biomaterials. Plasma modifies only the near surface of the treated substrates and does not change the bulk material properties. Plasma can modify almost any kind of substrate geometry without line-of-sight problem. The most important feature of plasma process is the ability of surface functionalization, which chemical process cannot offer. Some of the disadvantages of plasma surface treatment include vacuum setups, high cost, poorly defined chemistry, and irregularity of the modified surface. In particular, it is extremely difficult to generate plasma in very small pores or within long, small diameter tubing. For further details of plasma surface modification, *see* Chapter 1.7.

Presently, there are relatively few commercial uses for plasma depositions and reactions in the biomedical field. One of the major obstacles confronting the industry is the lack of clearly defined U.S. Food and Drug Administration guidelines as to how surface modifications will be regulated. In addition, although plasma processes and applications have been proposed and studied extensively in academic settings, process reproducibility has often been overlooked. There is a pressing need for the biomaterials industry to work with research institutions for process optimization and scale-up economics. On the whole, however, advantages of plasma surface treatment far outweigh its disadvantages for many types of modifications that simply cannot be generated by other methods.

The biomedical industry is expanding rapidly with new medical products, devices, materials, and surgical procedures that are improving current health care practice. Many of these products and procedures are designed around polymeric devices that must meet certain clinical requirements. Plasma surface-modification technologies can improve biocompatibility and biofunctionality, and they are currently being considered for enhancing biocompatibility of devices for cardiovascular, orthopedic, dental, and ophthalmic use.

Acknowledgments

We appreciate Eric J. Simso at BSI Corporation for providing PhotoLink information. In addition, we thank Gena M. Patch for performing some of the studies presented in this chapter.

References

1. *Biomaterials: An Introduction*; Park, J. B.; Lakes, R. S., Eds.; Plenum: New York, 1992.
2. *Biomaterials Science and Engineering*; Park, J. B., Ed.; Plenum: New York, 1984.
3. Helmus, M. N. *MRS Bull.* **1991**, *16*, 33–38.
4. Helmus, M. N. *DR Reports*; Decision Resources, Inc.: Burlington, MA, 1990.
5. *Annual Book of ASTM Standards: Medical Devices, 13.01*; American Society for Testing and Materials: Philadelphia, PA, 1986.
6. Bush, R. B. *J. Appl. Biomater.* **1993**, *4*, 195–209.
7. Anderson, J. M. *Trans. ASAIO* **1988**, *34*, 101–107.
8. Park, J. B.; Lakes, R. S. In *Biomaterials: An Introduction*, 2nd ed.; Plenum: New York, 1992; pp 223–244.
9. Kasemo, B.; Lausmaa, J. *CRC Crit. Rev. Biocomp.* **1986**, *2*, 223–244.
10. Andrade, J. D. In *Surface and Interfacial Aspects of Biomedical Polymers*; Andrade, J. D., Ed.; Plenum: New York, 1985; Vol.2, pp 1–80.

11. Vroman, L. In *Biocompatible Polymers, Metals, and Composites*; Szyncher, M., Ed.; Technomic: Lancaster, PA, 1983; p 81.

12. Horbett, T. A.; Brash, J. L. In *Proteins at Interfaces: Physicochemical and Biochemical Studies*; Brash, J. L.; Horbett, T. A., Eds.; ACS Symposium Series 343; American Chemical Society: Washington, DC, 1987; pp 1–31.

13. Scharenradd, J. M.; Busscher, H. J.; Wildevuur, C. R. H.; Arends, J. *J. Biomed. Mater. Res.* **1986**, *20*, 773.

14. Lyndon, M. J.; Minett, T. W.; Tighe, B. J. *Biomaterials* **1985**, *6*, 396.

15. Salthouse, T. N.; Matlaga, B. F. In *Biomaterials in Reconstructive Surgery: Some Cellular Effects Related to Implant Shape and Surface*; Rubin, L. R., Ed.; Mosby: St. Louis, MO, 1983; pp 40–45.

16. Petit, D. K.; Horbett, T. A.; Hoffman, A. S.; Chan, K. Y. *Invest. Opthalmol. Visual Sci.* **1990**, *31*, 2269.

17. Baier, R. E.; Meyer, A. E.; Natiella, J. R.; Natiella, R. R.; Carter, J. M. *J. Biomed. Mater. Res.* **1984**, *18*, 337.

18. Dewez, J. L.; Doren, A.; Schneider, Y. J.; Legras, R.; Rouxhet, P. G. *Surf. Interface Anal.* **1991**, *17*, 449.

19. Furcht, L. T. *Mod. Cell Biol.* **1983**, *1*, 53.

20. Grinnell, F. *Int. Rev. Cytol.* **1978**, *53*, 65–129.

21. Klebe, R. J.; Bently, K. L.; Hanson, D. P. In *Proteins at Interfaces: Physicochemical and Biochemical Studies*; Brash, J. L.; Horbett, T. A., Eds.; ACS Symposium Series 343; American Chemical Society: Washington, DC, 1987; pp 615–628.

22. Puleo, D. A.; Bizios, R. *Bone* **1991**, *12*, 271–276.

23. Weiss, R. E.; Reddi, A. H. *Proc. Natl. Acad. Sci. U.S.A.* **1990**, *77*, 2074–2078.

24. Weiss, R. E.; Reddi, A. H. *Exp. Cell Res.* **1981**, *133*, 247–254.

25. Sampath, T. K. In *Extracellular Matrix: Structure and Function*; Reddi, A. H., Ed.; Alan R. Liss: New York, 1985; pp 398–408.

26. Caplan, A. I. *J. Orthop. Res.* **1991**, *9*, 641–650.

27. Hoffman, A. S. *CHEMTECH* **1986**, *16*, 426.

28. Merrill, E. W.; Salzman, E. W. *ASAIO J.* **1983**, *6*, 60.

29. *Poly(Ethyene Glycol) Chemsitry: Biotechnical and Biomedical Applications*; Harris, J. M., Ed.; Plenum: New York, 1992.

30. Lee, W. Y.; Sehon, A. H. *Nature (London)* **1977**, *276*, 618–619.

31. Hermans, J. *J. Chem. Phys.* **1982**, *77*, 2193–2203.

32. Friedrich, C.; Lauprete, F.; Noel, C.; Monnerie, L. *Macromolecules* **1980**, *13*, 1625–1629.

33. Klein, J. *Makromol. Chem., Macromol. Symp.* **1986**, *1*, 125–137.

34. Gombotz, W. R.; Wang, G.; Horbett, T. A.; Hoffman, A. S. In *Protein Adsorption to and Elution from Polyester Surfaces*; Harris, J. M., Ed.; Plenum: New York, 1992; pp 247–261.

35. Balasubramanian, D.; Chandani, B. *J. Chem. Ed.* **1983**, *60*, 77–78.

36. Cohen Stuart, M. A.; Waajen, F. W.; Cosgrove, T.; Vincent, B.; Crowley, T. L. *Macromolecules* **1984**, *17*, 1825–1830.

37. O'Mullane, J. E.; Davison, C. J.; Petrak, K.; Tomilson, E. *Biomaterials* **1988**, *9*, 203–204.

38. Lee, J. H.; Kopecek, J.; Andrade, J. D. *J. Biomed. Mater. Res.* **1992**, *23*, 351.

39. Desai, N. P.; Hubbell, J. A. *Biomaterials* **1991**, *12*, 144–153.

40. Gombtz, W. R.; Wang, G. H.; Hoffman, A. S. *J. Appl. Polym. Sci.* **1989**, *37*, 91–107.

41. Kim, S. W.; Jacobs, H.; Lin, J. Y.; Nojori, C.; Okano, T. *Ann. N.Y. Acad. Sci.* **1988**, *516*, 116–130.

42. Prime, K. L.; Whitesides, G. M. *Science (Washington, D.C.)* **1991**, *252*, 1164.

43. Pale-Grosdemange, C.; Simon, E. S.; Prime, K. L.; Whitesides, G. M. *J. Am. Chem. Soc.* **1991**, *113*, 12.

44. Mori, Y.; Nagoaka, S.; Takiuchi, H.; Kikuchi, T.; Noguchi, N.; Tanzawa, H.; Noishiki, Y. *Trans. ASAIO* **1982**, *28*, 459.

45. Nagaoka, S.; Nakao, A. *Biomaterials* **1990**, *11*, 119–121.

46. Sun, Y. H.; Gombotz, W. R.; Hoffman, A. S. *J. Bioact. Compat. Polym.* **1986**, *1*, 316.

47. Lopez, G. P.; Ratner, B. D.; Tidwell, C. D.; Haycox, C. L.; Rapoza, R. J.; Horbett, T. A. *J. Biomed. Mater. Res.* **1992**, *26*, 415.

48. Sheu, M.-S.; Hoffman, A. S.; Terlingen, J. G. A.; Feijen, J. *J. Adhes. Sci. Tech.* **1992**, *6*, 995–1009.

49. Sheu, M.-S.; Hoffman, A. S.; Terlingen, J. G. A.; Feijen, J. *Clin. Mater.* **1993**, *13*, 41–45.

50. Llanos, G. R.; Sefton, M. J. *Biomaterials* **1992**, *13*, 421–424.

51. Peters, T. *Adv. Protein Chem.* **1985**, *1*, 39–54.

52. Shasby, D. M.; Shasby, S. S. *Cir. Res.* **1986**, *57*, 903–908.

53. Zucker, M. B.; Vroman, L. *Proc. Cos. Exp. Biol. Med.* **1969**, *131*, 318–320.

54. Lee, R. G.; Adamson C.; Kim, S. W. *Thromb. Res.* **1974**, *4*, 458–490.

55. Horbett, T. In *Biomaterials: Interfacial Phenomena and Applications*; Cooper, S. L.; Peppas, N. A., Eds.; Advances in Chemistry 199; American Chemical Society: Washington, DC, 1982; pp 233–244.

56. Absolom, D. R.; Zingg; Neumann, A. W. *J. Biomed. Mater. Res.* **1987**, *21*, 161–171.

57. Munro, M. S.; Quattrone, A. J.; Ellsworth, S. R.; Kulkarni, P.; Eberhart, R. C. *Trans. ASAIO* **1981**, *27*, 499–503.

58. Frautschi, J.; Munro, M. S.; Lloyd, D. R.; Eberhart, R. C. *Trans. ASAIO* **1983**, *24*, 242–244.

59. Eberhart, R.; Munro, M.; Frautischi, J.; Lubin, M.; Clubb, F.; Miller, C.; Sevastianov, V. *Ann. N.Y. Acad. Sci.* **1987**, *516*, 78–95.

60. Pitt. W. G.; Cooper, S. L. *J. Biomed. Mater. Res.* **1988**, *22*, 359–382.

61. Grasel, T. G.; Pierce, J. A.; Cooper, S. L. *J. Biomed. Mater. Res.* **1987**; *21*, 815–842.

62. Keogh, J. R.; Velander, F. F.; Eaton, J. W. *J. Biomed. Mater. Res.* **1992**, *26*, 441–456.

63. Gentry, P. W.; Alexander, B. *Biochem. Biophys. Res. Commun.* **1973**, *50*, 500–509.

64. Sas, G.; Jako, J.; Doman, J.; Laszlo, C.; Padar, J. *Thromb. Diat.* **1971**, *25*, 555.

65. Plate, N. A.; Valuev, L. I. *Adv. Polym. Sci.* **1986**, *79*, 95–137.

66. Gott, V. L.; Whiffen, J. D.; Dutton, R. S. *Science (Washington, D.C.)* **1963**, *142*, 1297.

67. Grode, G. A.; Falb, R. D.; Crowley, J. P. *J. Biomed. Mater. Res.* **1972**, *6*, 77.

68. Leininger, R. I.; Crowley, J. P.; Falb, R. D.; Grode, G. A. *Trans. ASAIO* **1972**, *18*, 312–315.

69. Kammermeyer, K. *CHEMTECH* **1971**, *1*, 719.

70. Falb, R. D.; Leininger, R. I.; Grode, G. A.; Crowley, J. P. *Adv. Exp. Med. Biol.* **1975**, *52*, 365.

71. Falb, R. D.; Grode, G. A.; Takahachi, M. T.; Leininger, R. I. *Interac. Liq. Solid Substr.* **1968**, *87*, 181.

72. Schmer, G. *Trans. ASAIO* **1972**, *18*, 321–324.

73. Goosen, M. F. A.; Sefton, M. V. *J. Biomed. Mater. Res.* **1983**, *17*, 359–373.

74. Heyman, P. W.; Cho, C. S.; McRea, J. C.; Olsen, D. B.; Kim, S. W. *J. Biomed. Mater. Res.* **1985**, *19*, 419–436.

75. Park, K. D.; Okano, T.; Nojiri, C.; Kim, S. W. *J. Biomed. Mater. Res.* **1988**, *22*, 977–992.

76. Solomon, D. D.; McGary, C. W.; Pascarella, V. J. U.S. Patent 4,521,564, 1985

77. Hoffman, A. S.; Schmer, G.; Harris, C.; Kraft, W. G. *Trans. ASAIO* **1972**, *18*, 10–17.

78. Burkel, W. E.; Graham, L. M.; Stamley, J. C. *Ann. N.Y. Acad. Sci.* **1988**, *516*, 131–144.

79. Schultz, L.; Bull, B. S.; Braunwald, N. S. *Surgery* **1971**, *69*, 557–562.

80. Ertel, D. I.; Ratner, B. D.; Horbett, T. A. *J. Biomed. Mater. Res.* **1990**, *24*, 1637.

81. Massia, S. P.; Hubbell, J. A. *Anal. Biochem.* **1990**, *187*, 292.

82. Massia, S. P.; Hubbell, J. A. *Trans. Soc. Biomat.* **1990**, *13*, 240.

83. Massia, S. P.; Hubbell, J. A. *J. Biomed. Mater. Res.* **1991**, *25*, 223–242.

84. Yam, L. T.; Li, C. Y.; Crosby, W. H. *Am. J. Clin. Pathol.* **1971**, *55*, 283–290.

85. Himmelhoch, R. S.; Evans, W. H.; Mage, M. G.; Peterson, E. A. *Biochemistry* **1969**, *8*, 914–921.

86. Torres, J. L.; Rush, R. S.; Main, A. R. *Arch. Biochem. Biophys.* **1988**, *267*, 271–279.

87. Wright, D. G; Bralove, D. A.; Gallin, J. I. *Am. J. Pathol.* **1977**, *87*, 273–283.

88. Elwing, H.; Nilsson, B.; Svensson, K. E.; Askendahl, A.; Nilsson, U. R.; Lundstrom, I. *J. Colloid Interface Sci.* **1988**, *125*, 139–145.

89. Lenk, T. J.; Ratner, B. D.; Gendreau, R. M.; Chittur, K. K. *J. Biomed. Mater. Res.* **1989**, *23*, 549–569.

90. Vroman, L.; Adams, A. L. *Surf. Sci.* **1969**, *16*, 438.

91. Brash, J. L.; Davidson, V. J. *Thromb. Res.* **1976**, *9*, 249–259.

92. Ward, W. J.; McCarthy, T. J. In *Encyclopedia of Polymer Science and Engineering*; Kroschwitz, J. I., Ed.; Wiley: New York, 1990; Vol. 16, pp 674–689.

93. *Surface Modification Engineering*; Kossowsky, R., Ed.; CRC: Boca Raton, FL, 1989; Vols. 1–2.

94. *Handbook of Surface Preparation*; Snogren, R. C., Ed.; Palmerton: New York, 1974.

95. Kato, K. *J. Appl. Polym. Sci.* **1976**, *20*, 2451–2460.

96. Blais, P.; Carlsson, D. J.; Csullof, G. W.; Wiles, D. M. *J. Colloid Interface Sci.* **1974**, *47*, 636–649.

97. Holmes-Farley, S. R.; Whitesides, G. M. *Langmuir* **1987**, *3*, 62–76.

98. Holmes-Farley, S. R.; Bain, C. D.; Whitesides, G. M. *Langmuir* **1988**, *4*, 921–937.

99. Duchane, D. V. *Surface Modification Engineering*; Kossowsky, R., Ed.; CRC: Boca Raton, FL, 1989; Vol. 1, pp 285–312.

100. Briggs, D.; Brewis, D. M.; Konieckzo, M. B. *Eur. Polym. J.* **1978**, *14*, 1–4.

101. Yamauchi, J.; Yamaoka, A.; Ikemoto, K.; Matsui, T. *J. Appl. Polym. Sci.* **1991**, *43*, 1197–1203.

102. Dasgupta, S. *J. Appl. Polym. Sci.* **1990**, *41*, 233–248.

103. Briggs, D.; Brewis, D. M.; Konieckzo, M. B. *J. Mater. Sci.* **1979**, *14*, 1344–1348.

104. Meichsner, J.; Poll, H. U.; Steinrucken, A. *Thin Solid Film* **1984**, *112*, 369–380.

105. Iwata, H.; Matsuda, T. *J. Membr. Sci.* **1988**, *38*, 185–199.

106. Yao, Z. P.; Ranby, B. *J. Appl. Polym. Sci.* **1990**, *40*, 1647–1661.

107. *Radiation Chemsitry of Polymeric Systems*; Chapiro, A., Ed.; Wiley Interscience: New York, 1962.

108. Garnett, J. L. *J. Rad. Phys. Chem.* **1979**, *14*, 79–99.

109. Heger, A. *Textiltechnik* **1988**, *38*, 359–363.

110. Ishigaki, I.; Sugo, T.; Senoo, K.; Okada, T.; Okamoto, J.; Machi, S. *J. Appl. Polym. Sci.* **1982**, *27*, 1033–1041.

111. Loh, I. H.; Lu, P. H.; Hirvonen, J. K. *Polym. Mater. Sci. Eng.* **1988**, *59*, 1028–1032.

112. Occhiello, E.; Gardassi, F.; Malatesta, V. *Angew., Makromol. Chem.* **1989**, *169*, 143–151.

113. Baszkin, A.; Ter-Minassian-Saraga, L. *J. Colloid Interface Sci.* **1973**, *43*, 473–477.

114. Lerchenthal, C. H.; Brenman, M.; Yits'haq, N. *J. Polym. Sci.: Polym. Chem. Ed.* **1975**, *13*, 737–740.

115. Hsiue, G. H.; Wang, C. C. *J. Appl. Polym. Sci.* **1990**, *40*, 235–247.

116. Chapiro, A. *Radiation Chemistry of Polymeric Systems*; Wiley Interscience: New York, 1962.

117. Amos, R. A.; Guire, P. E. *Trans. Soc. Biomater.* **1991**, *14*, 175.

118. Clapper, D. L.; Hagen, K. M.; Hupfer, N. M.; Anderson, J. M.; Guire, P. E. *Trans. Soc. Biomater.* **1993**, *16*, 42.

119. Kirkhman, S. M.; Dangle, M. E. *Ophth. Surg.* **1991**, *22*, 455–461.

120. Clapper, D. L.; Anderson, A. B.; Duquette, P. H.; Guire, P. E. BSI Corporation: Minneapolis, MN, unpublished results.

121. Loh, I. H. *Spectrum: Materials and Manufacturing*; Decision Resources, Inc.: Burlington, MA, 1991; pp 15–19.

122. *Plasma Polymerization*; Yasuda, H., Ed.; Academic: Orlando, FL, 1985.

123. Gombotz, W. R.; Hoffman, A. S. *CRC Crit. Rev. Biocompat.* **1987**, *4*, 1–42.

124. *Plasma Deposition, Treatment, and Etching of Polymers*; D'Agostino, R., Ed.; Academic: Orlando, FL, 1990.

125. *Glow Discharge Processes—Sputtering and Plasma Etching*; Chapman, B., Ed.; Wiley: New York, 1980.

126. *Techniques and Applications of Plasma Chemistry*; Hollan, J. R.; Bell, A. T., Eds.; Wiley: New York, 1974.

127. Schonhorn, H.; Hansen, R. H. *J. Appl. Polym. Sci.* **1976**, *11*, 1461–1474.

128. Sheu, M. S.; Hoffman, A. S.; Terlingen, J. G. A.; Feijen, J. *Polym. Prepr. (Am. Chem. Soc. Div. Polym. Symp.)* **1991**, *32*, 239.

129. Chinn, J. A.; Horbett, T. A.; Ratner, B. D.; Schway, M. B.; Haque, Y.; Hauschka, S. D. *J. Colloid Interface Sci.* **1989**, *127*, 67–87.

130. Lopez, G. P.; Ratner, B. D. *Langmuir* **1991**, *7*, 766–773.

131. Lopez, G. P. Ph. D. Dissertation, University of Washington, Seattle, WA, 1991.

132. Desai, N. P.; Hossainy, S. F. A.; Hubbell, J. A. *Biomaterials* **1992**, *13*, 417–420.

133. Terlingen, J. G. A.; Feijen, J.; Hoffman, A. S. *J. Colloid Interface Sci.* **1993**, *155*, 55–56.

134. Terlingen, J. G. A.; Brenneisen, L. M.; Super, H. T. J.; Pijpers, A. P.; Hoffman, A. S.; Feijen, J. *J. Biomater. Sci.* **1993**, *4*, 165.

135. Amstein, C. F.; Hartman, P. *J. Clin. Microbiol.* **1975**, *2*, 46–54.

136. Ramsey, W. S.; Hertl, W.; Nowlan, E. D.; Binkowski, N. J. *In Vitro* **1984**, *20*, 802–808.

137. Technical Data; Tekmat Corp.: Ashland, MA, May 1988.

138. Kaplan, S. L.; Rose, P. W. *Int. J. Adhes. Adhes.* **1991**, *11*, 109–113.

139. Lin, H. L.; Chu, C. C.; Loh, I. H. *J. Appl. Biomater.* **1992**, *3*, 131.

140. Zhang, L.; Chu, C. C.; Loh, I. H. *J. Biomed. Mater. Res.* **1993**, *27*, 1425–1441.

141. Colter, K. D.; Shen, M.; Bell, A. T. *Biomat. Med. Dev. Art. Org.* **1977**, *5*, 13–24.

142. Yuan, S.; Szakalas-Gratzl, G.; Ziats, N. P.; Joacobsen, D. W.; Kottke-Marchant, K.; Marchant, R. E. *J. Biomed. Mater. Res.* **1993**, *27*, 811–819.

143. Hollahan, J. R.; Stafford, B. B.; Falb, R. D.; Payne, S. T. *J. Appl. Polym. Sci.* **1969**, *13*, 807.

144. Yasuda, H. K.; Matsuzawa, Y. *J. Appl. Polym. Sci., Appl. Polym. Symp.* **1984**, *38*, 65–74.

145. Fowler, B. C.; Bohnert, J. L.; Horbett, T. A.; Hoffman, A. S. *Trans. Third World Biomater. Congr.* **1988**, *XI*, 99.

146. Ratner, B. D.; Yoon, S. C.; Mateo, N. B. In *Polymer Surfaces and Interfaces*; Feats,W. J.; Munro, H. S., Eds.; John Wiley & Sons: New York, 1987.

147. Triolo, P. M.; Andrade, J. D. *J. Biomed. Mater. Res.* **1983**, *17*, 129–147, 149–165.

148. Hsu, T. T.; Wang, M. T.; Hsiung, K. P.; Hsiue, G. H.; Sheu, M. S. U.S. Patent 5,028,657, 1991; 5,171,779, 1992.

149. Fujimotot, K.; Inoue, H.; Ikada, Y. *J. Biomed. Mater. Res.* **1993**, *27*, 1559–1567.

150. Ho, C. P.; Yasuda, H. *J. Biomed. Mater. Res.* **1988**, *22*, 919–937.

151. Sipehia, R.; Garfinkle, A. *Biomater. Art. Cells, Art. Org.* **1990**, *18*, 643–655.

152. Koziol, J. E.; Peyman, G. A.; Yasuda, H. *Arch. Ophthal.* **1983**, *101*, 1779–1781.

153. Mateo, N. B.; Ratner, B. D. *Invest. Ophthalmol. Visual Sci.* **1989**, *30*, 853–860.

154. Chawla, A. S. *Artif. Organs* **1979**, *3*, 92–96.

$$2 \ O=C=N-R-N=C=O + HO-(R')_n-OH \rightarrow O=C=N-R-NH-CO-O-(R')_n-O-OC-HN-R-N=C=O$$

Diisocyanate   Polyol   Prepolymer (macrodiisocyanate)

$$O=C=N-R-NH-CO-O-(R')_n-O-OC-HN-R-N=C=O \quad + \quad HO-R''-OH \quad \rightarrow$$

Prepolymer   Diol

$$-[-O-R''-O-OC-HN-R-NH-CO-O-(R')_n-O-OC-HN-R-NH-CO-O-R''-O-]_x-$$

Polyurethane

$$O=C=N-R-NH-CO-O-(R')_n-O-OC-HN-R-N=C=O \quad + \quad H_2N-R''-NH_2 \quad \rightarrow$$

Prepolymer   Diamine

$$\rightarrow \quad -[HN-R''-NH-CO-NH-R-NH-CO-O-(R')_n-O-OC-HN-R-NH-CO-HN-R''-NH-]_x-$$

Polyurethaneurea

Scheme I. General scheme for synthesis of polyurethanes.

along the chain. Alternating and block copolyurethanes can be obtained in the two-step process, in which diisocyanate and polyol are reacted first, and then a chain-extension reaction is carried out by adding a diol or a diamine.

A number of polyurethanes for demanding applications are produced by a two-step solution polymerization. The polymerization solvent should act as a nonreactive medium in which the diisocyanate, the polyol, the diol or diamine, and the resulting polymer are soluble. Dimethylacetamide, dimethylformamide, dimethyl sulfoxide, N-methylpyrrolidone, tetrahydrofuran, and mixed solvents are used for synthesis of polyurethanes (1–17).

Polyurethane Monomers

Monomers used for the syntheses of polyurethane elastomers are polyols with a molecular weight usually in the range of 500 to 5000, aromatic or aliphatic diisocyanates, and a range of diols or diamines. The monomers used for preparation of biomedical polyurethanes should be of high purity because traces of metals, salts, and nucleophilic organic reagents could lead to an acceleration of side reactions (1–33). The catalysts should be nontoxic and preferably removed from the raw polymer together with the low molecular components. These compounds, when left in the material, may affect biocompatibility and in vivo stability of the final devices.

Polyols

Low molecular weight polyesters, polyethers, polyetheresters, hydrocarbon polymers, and poly(dimethylsiloxane)s with hydroxy end groups are used for the

synthesis of polyurethane elastomers (Table I). Polyester diols are prepared by polyesterification of dicarboxylic acids such as adipic, sebacic, oxalic, maleic, dimerized linoleic, or phthalic anhydrides with diols of ethylene, propylene, 1,2-butylene, and 1,6-hexane. Polyesters derived from ε-caprolactone or hydroxy-terminated aliphatic and cycloaliphatic polycarbonates also have been used for synthesis of polyurethanes. Polyurethanes based on poly(caprolactone) (11) and especially on 1,6-hexandiol-polycarbonate (18, 57) exhibit improved hydrolytic stability as compared with polyesterurethanes from other polyols (57). This hydrolytic stability can be improved further by using a diisocyanate containing carbodiimide groups. Polyether diols are usually synthesized by the addition of alkylene oxides (ethylene oxide, propylene oxide, and tetramethylene oxide) to diols, polyols, diamines, or polyamines. They also can be produced by ring-opening polymerization of tetrahydrofurane. Polyetherurethanes are claimed to have better hydrolytic stability than polyesterurethanes although there are deviations to this trend. The use of hydroxy-terminated homo- and copolymers of butadiene and polyisobutylene and copolymers of butadiene with styrene leads to elastomers with high resistance to light, thermal, and hydrolytic degradation (22, 53).

Diisocyanates

The most frequently used aromatic diisocyanates (Table II) are 4,4'-diphenylmethane diisocyanate (MDI) and toluene diisocyanate (TDI). MDI in its pure form contains two equally reactive isocyanate groups. TDI is available as pure 2,4-isomer or as an 80/20 mixture of the 2,4- and 2,6-isomers. At room temperature, the –NCO group in the 4 position is about eight times as

Biomedical Polyurethanes

Sylwester Gogolewski

Biomedical polyurethanes have found applications in total artificial hearts, heart valves, intraaortic balloons, mammary implants, wound dressings, pacing leads insulation, and angioplasty balloons, to mention but a few. Limited long-term molecular stability in vivo, relatively high thrombogenic activity of surfaces, and the susceptibility to calcification could explain why this class of biomaterials has not found broader biomedical application. Various techniques have been and are being used to enhance both the molecular stability and blood compatibility of biomedical polyurethanes with varying degrees of success. However, it does not seem very probable that there will ever be a biomedical polyurethane developed that is stable in vivo for an indefinite time and whose blood-contacting surface has ultimate properties, such as enhances formation of the functional neo-intima in vascular prostheses and does not induce clot formation or calcification in heart valves.

This chapter provides a survey of biomedical polyurethanes (both commercial and laboratory synthesized). Trends are outlined in syntheses of commercial materials with improved in vivo functionality. Their biocompatibilty and hemocompatibility, medical applications, and stability in vitro and in vivo are also outlined.

Polyurethanes are polymers with the characteristic –NH–CO–O– linkage in the chain, although in urethane copolymers, the urethane group is one of several groups containing nitrogen in the polymer. Thermoplastic polyurethanes are mainly linear elastomeric block copolymers of the $(AB)_n$ type. One block is derived from polydiols with molecular weights in the range of 500 to 5000 and consists of relatively flexible-chain *soft segments*. The second stiff block, referred to as *hard segment*, is formed by the reaction between diisocyanates and low molecular weight diols or diamines used as chain extenders (1–36). It has been suggested, however, that these materials are not simple $(AB)_n$ copolymers and may contain significant amounts of dimeric soft and hard repeat segments, depending on the relative reactivities of the polydiol and the diisocyanate. The presence of these dimeric segments will affect the overall structure and properties of these materials (15).

Polyurethanes used initially in various biom devices were commercial polymers design fibers, coatings, tubes, and other nonmedical a[tions. For the past two decades, however, much has been made to synthesize polyurethane elast with a wide range of physical and chemical pro to be used in contact with body fluids and t (37–126).

Syntheses of Polyurethane Elastome

Thermoplastic linear polyurethane elastomer polyurethanes) are produced by the reaction of molecular weight hydroxyl-terminated aliphatic ester, polyether, polyetherester, polyalkene, or (dimethylsiloxane)s with diisocyanates and low cular weight diols or diamines used as chain exte (Scheme I) (1–45). When polyfunctional reagent used, highly cross-linked polymer networks formed. Polyurethanes can be produced in a one or a two-step process, either by bulk or solution merization. In the one-step process all the reac and catalyst are mixed together, and an approp temperature is maintained in the reaction vessel the reaction is complete. The resulting copolymer cally has a random distribution of monomer t

Table I. Typical Polyols Used for Preparation of Polyurethane Elastomers

Chemical Structure	Name
$HO-(CH_2CH_2-O-)_nH$	Polyethylene oxide (PEO)[a]
$HO-(CH_2CH_2CH_2CH_2-O-)_nH$	Poly(tetramethylene oxide) (PTMO)
$(HO)-(CH_2-\overset{\overset{\displaystyle CH_3}{\mid}}{CH}-O-)_nH$	Poly(propylene oxide) (PPO)
$HO-[C-(CH_3)_2-CH_2]_n-OH$	Poly(isobutylene) (PIB)
$HO-[(CH_2)_4-O-OC(CH_2)_2CO-O-]_nH$	Poly(ethylene adipate) (PEA)
$HO-[(CH_2)_5-CO-O-]_nH$	Poly(caprolactone)
$HO-(CH_2)_4-[Si(CH_3)_2-O-]_xSi(CH_3)_2-(CH_2)_4-OH$	Hydroxybutyl terminated poly(dimethylsiloxane) (PDMS)–OH
$HO-CH_2-CH=CH-CH_2-CH-CH_2C\ H=CH-CH_2-OH$ $\overset{\mid}{\underset{H_2C=HC}{}}$	Polybutadiene (PBD)

[a]Low molecular weight PEO is also known as polyethylene glycol (PEG).

reactive as the –NCO group in the 2 position, which may enhance side reactions. Polymers based on MDI and TDI tend to yellow under prolonged exposure to sunlight. The methylene group in MDI is also susceptible to oxidation via a proton abstraction mechanism, involving autooxidation of the aromatic urethane groups to a form of quinoneimide (*1–4, 12–14, 16*). Paraphenylene diisocyanate (PPDI) and 1,5-naphthalene diisocyanate (NDI), although used occasionally for commercial polyurethanes, are not suitable for biomedical materials due to their high toxicity.

Polyurethanes that are claimed to exhibit better resistance to light, hydrolysis, and thermal degradation are based on aliphatic diisocyanates (*12–14*). The most common commercial aliphatic diisocyanates are 1,6-hexamethylene diisocyanate, 4,4'-dicyclohexylmethane diisocyante (H_{12}MDI), isophorone diisocyanate, *trans*-1,4-cyclohexane diisocyanate (CHDI), *m*-tetramethyl xylylene diisocyanate (TMXDI), *m*-isopropenyldimethylbenzyl, dimeryl diisocyanate derived from dimerized linoleic acid, xylylene diisocyanate (XDI), and 1,1,6,6-tetrahydroperfluoro-hexamethylene diisocyanate. Hexamethylene diisocyanate is available in its basic form and as isocyanate-terminated biuret (triisocyanate) or an isocyanurate ring-containing isocyanate (*1, 18–20*). It has been used in preparation of commercial polyurethane coatings and in experimental biomedical polyurethane materials. Dicyclohexylmethane diisocyante is available commercially as a mixture of isomers (trans–trans, trans–cis, and cis–cis). Isomeric hard segments formed from this monomer lack regularity. This characteristic may limit their close packing within adjacent chains and, in consequence, lead to a polymer network that can be easily disrupted by heat and water.

Polyurethanes based on isophorone diisocyanate, polyether, or polyester diols and chain extended with 1,4-butanediol or diamines show good hydrolytic sta-

bility (*18, 57*). The chemical reactivity of CHDI allows the preparation of polyurethanes with or without catalysts (*21, 22, 46*). This preparation may lead to materials with improved thermal and hydrolytic stability, because catalyst residues enhance the rate of depolymerization. Polyurethanes based on CHDI are semicrystalline materials with excellent thermal and mechanical properties that are retained at elevated temperatures, excellent solvent resistance, and hydrolytic stability (*21, 22*). Similarly to CHDI, no catalyst is needed for preparation of polyurethanes from XDI with hydroxy-terminated polyols. Polyurethanes based on TMXDI and *m*-isopropenyldimethylbenzyl diisocyanate (TMI) (unsaturated aliphatic) show good toughness, weatherability, resistance to yellowing, abrasion, and action of chemicals (*34*). Because TMXDI and TMI were not designed for medical applications, nothing is known about their properties in vivo. Polyurethanes from tetrahydroperfluoro-hexamethylene diisocyanate are claimed to exhibit low thrombogenic activity (*38*). In general, aliphatic diisocyanates are less reactive than aromatic diisocyanates, as illustrated by their relative reactivities (*22*): PPDI (1.85) > MDI (1.00) > NDI (0.37) > CHDI (0.28) > IPDI (0.15) > H_{12}MDI (0.13). Table II shows diisocyanates used for syntheses of biomedical polyurethanes.

Chain Extenders

Various compounds with active hydrogen atoms such as aliphatic and aromatic diols, diamines, dicarboxylic acids, or water are used as chain extenders or crosslinkers in the synthesis of polyurethanes (Table III). In general, polyurethanes produced with aromatic chain extenders are stiffer than those prepared with aliphatic ones. Diamine chain extenders lead to polyurethanes with usually better physical properties than those chain-extended with diols. Polyurethaneureas

Table II. Typical Diisocyanates Used in Syntheses of Polyurethane Elastomers

Chemical Structure	Name
	4,4'-Diphenylmethane diisocyanate (MDI)
	2,6- and 2,4-Toluene diisocyanate (TDI)
	4,4'-Dicyclohexylmethane diisocyanate ($H_{12}MDI$)
	trans-1,4-Cyclohexane diisocyanate (CHDI)
	Isophorone diisocyanate (IPDI)
	Tetramethyl xylene diisocyanate (TMXDI)
	Dimeryl diisocyanate (DDI)
$OCN-(CH_2)_6-NCO$	Hexamethylene diisocyanate (HMDI)
$OCN-CH_2(CF_2)_4CH_2-NCO$	1,1,6,6-Tetrahydroperfluoro-hexamethylene diisocyanate (THFDI)

are, however, more difficult to process from solution and from the melt. Chain extenders commonly used in the syntheses of polyurethanes include ethylene-, diethylene-, dipropylene-, 1,4-butane-, and 1,6-hexanediols; aliphatic ethylene-, hexamethylene-, and propylene diamines; and cycloaliphatic isophorone- and 1,4-cyclohexyldiamines and water (1–22). Table III shows chain extenders used in syntheses of polyurethanes.

Catalysts

Dibutyltin dilaurate, tetrabutyltin, dibutyltin dicaprylate, stannous octoate, ferric chloride, zinc chloride, manganese oxide, and lithium acetate are commonly used as catalysts in syntheses of polyurethanes, espe-

cially with aliphatic diisocyanates, which are usually less reactive than aromatic ones. Aliphatic diisocyanates in the presence of metallic salt catalysts react faster than aromatic diisocyanates. Reactivities of both the aliphatic and aromatic diisocyanates are comparable in the presence of organometallic catalysts (16).

Side Reactions in Polyurethanes

Aromatic and aliphatic diisocyanates can enter into a number of side reactions with themselves or other molecules that contain active hydrogen (Scheme II). Dimerization of aromatic diisocyanates is one such process. This equilibrium process, which proceeds easily at low temperatures especially in the presence of acidic or basic catalysts, introduces a thermally unsta-

Table III. Typical Chain Extenders Used in Syntheses of Polyurethane Elastomers

Chemical Structure	Name
HO–$(CH_2)_4$–OH	1,4-Butanediol
HO–$(CH_2)_6$–OH	Hexanediol
HO–$(CH_2)_2$–OH	Ethylene diol
HO–$(CH_2)_2$–O–$(CH_2)_2$–OH	Diethylene
H_2N–$(CH_2)_2$–NH_2	Ethylene diamine
H_2N–CH_2–$\overset{\overset{\displaystyle CH_3}{\vert}}{CH}$–$NH_2$	Propylene diamine

ble bond in the polymer chain. Dimerization may, in addition, hinder the formation of high molecular weight linear urethane copolymers because it disturbs the reactant balance and introduces branching or cross-linking sites in the polymer chain (1–16). Aliphatic diisocyanates do not appear to form dimers, but both aliphatic and aromatic diisocyanates form trimers. The presence of acidic and basic catalysts enhances trimerization. Cyclic trimers are very stable to thermal and hydrolytic attack. Similarly to dimerization, trimerization of diisocyanates also leads to cross-linking and disturbances in the balance of reactants.

Diisocyanates can also form carbodiimides. The carbodiimide linkage is very reactive toward chemical groups containing active hydrogen, originating from impurities such as water, monofunctional alcohols,

amines, amides, or carboxylic acids. In consequence, this reaction may lead to highly branched, cross-linked, or low molecular weight materials (1–16). Diisocyanates can also react with urethane or urea groups in the polymer chain forming allophanate and biuret linkages, which lead to cross-linking or branching of the polymer. Traces of metals and salts in the chemicals used for synthesis accelerate all these side reactions. The by-products formed in polyurethanes will affect their physical properties and performance in vivo. Thus, the use of high purity reagents is a must for synthesis of high molecular weight linear biomedical polyurethanes (1–16, 37–51).

Structure–Property Relations in Polyurethanes

Hydrogen Bonding in Polyurethane Elastomers

Hydrogen bonding in polyurethanes results from the presence of the –NH group, which is the proton donor, and the carbonyl or ether oxygen groups, which are hydrogen bond acceptors (Scheme III). The hydrogen bond acceptor may be either the hard segment (the carbonyl of the urethane group) or the soft segment (an ester carbonyl or ether oxygen). Relative amounts of the two types of hydrogen bonds are determined by the extent of microphase segregation. Increased phase segregation favors interurethane

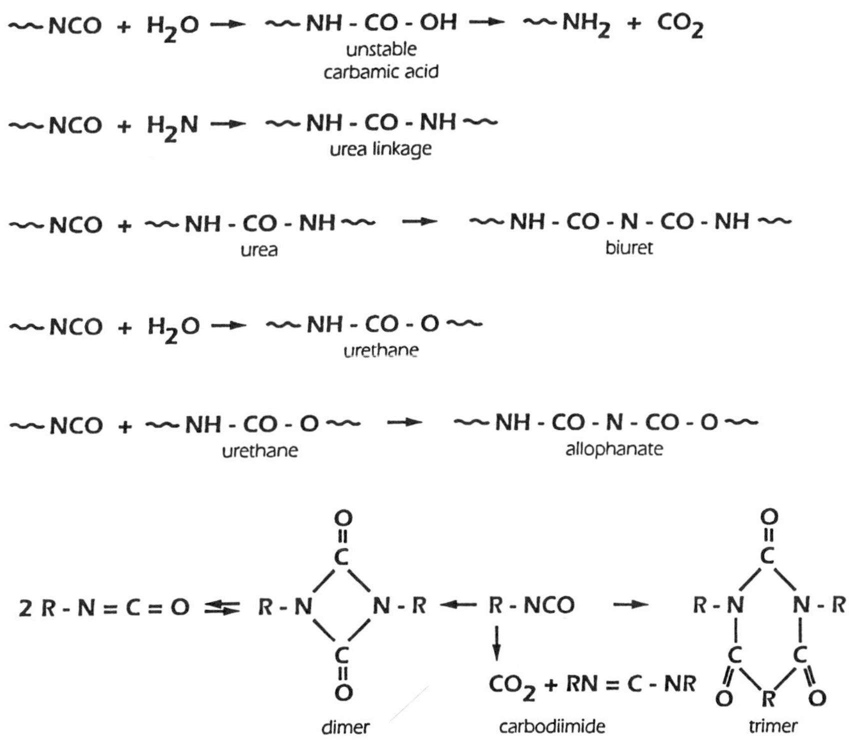

Scheme II. Typical side reactions in the polyurethane synthesis.

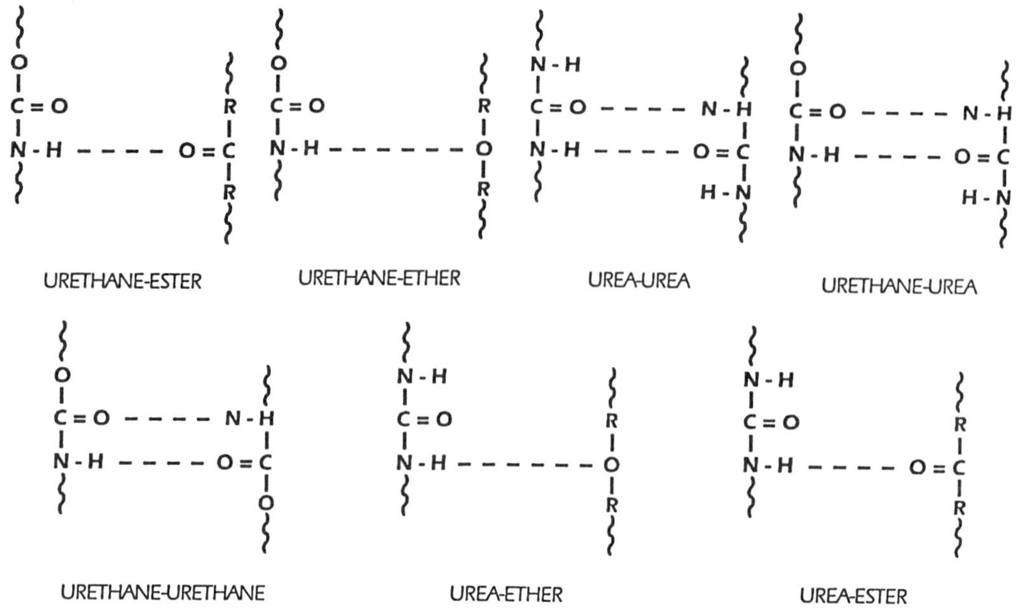

Scheme III. Possible hydrogen bonding in polyurethanes.

hydrogen bonding. At room temperature about 80–90% of the –NH groups in the hard segment are hydrogen bonded. The extent of interurethane hydrogen bonding was higher in the materials having better phase segregation (longer soft and hard segments). Hydrogen bonding contributes to domain cohesiveness, but it is not directly responsible for the unique mechanical properties of polyurethanes (23–33).

Phase Separation in Polyurethanes

The soft segments in polyurethane elastomers are polyesters, polyethers, poly(dimethylsiloxane)s, or polybutadienes having a glass transition temperature (T_g) below the temperature of use, and they are therefore rubbery. The hard segments can be either aliphatic or aromatic diisocyanates, extended with diols or diamines. Their T_g softening or melting temperatures are above normal use temperatures and, in consequence, the hard segments are rigid, glassy, or crystalline. Incompatibility between the segments leads to their segregation into hard- and soft-segment domains. The hard-segment domains with strong dipole–dipole interaction and hydrogen bonding act as physical cross-links for the rubbery soft-segment matrix (23–33). The two-phase domain structure of polyurethanes is a basic determinant of their properties. The factors that affect phase segregation in polyurethanes include segment polarity, length and crystallizability, tendency for interaction between the hard and soft segments (e.g., via hydrogen bonding), overall material composition, and molecular weight. Longer blocks undergo higher degrees of phase segregation and

more perfect hard-segment domains. Higher hard-segment content results in higher amounts of hard segments mixed into the soft-segment phase. Polar soft segments that form strong hydrogen bonding with the hard segments exhibit a higher extent of phase mixing. Thus, polyester soft segments are usually more compatible with urethane hard segments than polyether soft segments. This compatibility is because the strength and degree of hard-segment–soft-segment hydrogen bonding is higher for the ester carbonyl group than for the ether oxygen. This tendency is even more pronounced for polyurethanes with polybutadiene or polyisobutylene soft segments, which do not form hydrogen bonding. These polymers form highly phase-segregated systems. Polyurethanes based on aromatic reactants generally show better phase segregation than those synthesized from aliphatic ones.

Phase segregation in polyurethanes is strongly affected by temperature. Raising the temperature of a polymer system induces phase mixing. Subsequent cooling causes phase segregation, leading to the original morphology. Stretching of polyurethanes results in the orientation of hard segments in the elongation direction. This orientation enhances the formation of a maximum number of hydrogen bonds between the –NH and –CO groups in adjacent segments, and at high elongations, it may lead to the stress-induced crystallization of the soft segments that in turn enhances the tensile properties of the material (21–33). Because both the soft- and hard-segment phases are interpenetrating, large deformation of the one phase results in similar deformation of the other.

Physical Characteristics

Physical properties of polyurethane elastomers are directly related to the extent of the hard- and soft-segment domain segregation. Hence, such factors as symmetry of monomers, chemical composition, molecular weight and polydispersity of hard and soft segments, type of interaction between these segments, and crystallinity will affect the final properties of the polyurethanes. Soft-segment chemical composition, molecular weight, and degree of phase mixing determine the low T_g of the system, elongation at break, initial modulus, and tensile strength. The volume fraction of hard segments and their symmetry control moduli, strength, and high softening temperatures. Increasing the length of the soft segment at a fixed length of the hard segment leads to polyurethanes with lower tensile strength and modulus and higher elongation at break. Increasing the hard-segment content at constant soft-segment length results in increased T_g, crystallinity, and crystalline hard-segment melting temperature; tendency of the materials to form morphology with a hard-segment continuous phase and isolated soft segment domains; enhanced mixing of the continuous and dispersed phases; increased interfacial area, tensile strength, and modulus; and decreased ultimate elongation. These effects seem to be similar for both aliphatic and aromatic hard segments.

A decrease of the hard-segment polydispersity increases modulus, tensile strength, and elongation at break, whereas a decrease of the soft-segment polydispersity causes a slight increase in modulus, a moderate increase in tensile strength and elongation, and a decrease in T_g. Polyurethanes subjected to multiple stretching show hysteresis caused by a disruption of the hard segments. This characteristic results in reduction of their ability to reinforce the rubbery matrix. Polyurethanes under ultimate tensile stress undergo heterogeneous fracture via propagation of cracks through areas of defects acting as sites of stress concentration (26–33).

Criteria for Selection of Biomedical Polyurethanes

In the early 1960s, commercial polyurethanes were used in implantable devices with mixed results. It was soon realized that the negative results might be due to the quality of the commercial materials, which contained unspecified impurities, additives, catalyst residues, and structural irregularities. Since that time, extensive research and development work was undertaken to develop biomedical polyurethanes with good physical and biological characteristics. Since 1975, pacing leads with polyurethane insulation have been used in humans. In 1982 the Jarvik 7 total artificial heart with blood sacs from polyurethanes (Biomer, Cardiothane 51) was first implanted in a human (58).

Biomedical polymers, and especially those for intracorporeal devices, should not induce adverse inflammatory or foreign body reaction; be carcinogenic, mutagenic, teratogenic, or toxic; or have any other adverse effect in the body. When in contact with blood, they should not induce thrombosis or abnormal intima formation; interfere with the normal clotting mechanism; activate the complement system; alter the configuration or stability of any cellular or soluble materials in the blood that would lead to cell changes or allergic, hypersensitive, or toxic reactions.

Biomedical polyurethanes should be reproducibly produced as pure materials; processed into the desired form and sterilized without changes in properties or form; have the required chemical, physical, and mechanical properties for performing their functions; and not adversely change their physical, chemical, or mechanical properties in vivo (50, 56). No polyurethane products are available that fulfill all these requirements.

Physical properties of biomedical materials should be adequate for their particular applications. Orthopedic surgery requires materials with optimal yield strength, modulus, fatigue resistance, wear resistance, and friction. Modulus, mechanical strength (tear and burst), and fatigue resistance are essential for polymers to be used in reconstructive surgery of soft tissue and cardiovascular surgery. Thermal expansion, thermal conductivity, wear and abrasion resistance, mechanical strength, and modulus affect the performance of dental materials. The required properties of polymers are usually achieved by using additives (e.g., antioxidants, stabilizers, plasticizers, and opacifiers), molecular orientation, crystallization, cross-linking, pendant groups, variation of molecular weight, blending, or copolymerization. All these factors will affect, to a greater or lesser extent, polymer biocompatibility.

Tissue Compatibility

Materials in contact with soft or hard body tissues usually cause inflammatory responses (acute or chronic) or fibrous capsule formation, which may be seen over various time periods following implantation. In the worse case, there may be systemic effects to tissues remote from the site of implantation (60–65). The ideal foreign body reaction of soft tissue to a polymeric implant is one in which the acute inflammatory response is minimal, and fibrosis and eventual encapsulation of the implant by fibrous tissue occurs rapidly. Body reactions to an implant material can be affected by implant geometry, configuration, size, and chemical characteristics, as well as positional stability of the implant at the implantation site.

The formation of a thin, permeable capsule around the implant or a surface lining with cells is a normal and desirable body response to an implant. In some cases, however, there can be a chronic inflammatory reaction, extensive fibrosis, or excessive capsule thickening. These processes may involve adjacent organs or tissues and lead to adverse reactions. In the extreme, solid thick implants with large, smooth, uninterrupted surfaces and sharp edges may lead to solid-state carcinogenesis (66, 67). Additives such as plasticizers, catalysts, opacifiers, curing agents, residues of ethylene oxide used for sterilization, unreacted monomers, or products of polymer degradation when leached out from the implant into the host tissue environment may elicit serious toxic or inflammatory response.

Various in vitro and in vivo biocompatibility and blood compatibility tests provide useful tools for qualifying candidate polymers for extracorporeal and intracorporeal devices. Such tests include characterization of physical and chemical properties, resistance to various techniques of sterilization, aging in vivo, aging in buffers, or aging in pseudoextracellular fluid, tissue culture, cytotoxicity of extracts, Ames mutagenicity, U.S. Pharmacopeia (USP) Class VI procedure, hemolysis, and intramuscular or intercerebral animal implantations. (The Ames test utilizes histidine auxotrophs of *Salmonela typhimurium* to assess mutagenicity of leachable components. The histidine auxotrophs will not grow in medium lacking histidine (68).) Most chemically stable polymers in a configuration acceptable for implants will pass bicompatibility tests (60–65, 68).

Blood Compatibility

Thrombogenic events in the cardiovascular system proceed via extrinsic and intrinsic mechanisms. In the extrinsic pathway, the thrombogenic events are initiated by chemical releases at the damaged vessel site. The intrinsic coagulation process occurs within the blood itself (e.g., on the surface of aggregating platelets) and can be initiated by the contact of circulating blood components with the foreign surface. Exposure of blood to foreign surfaces leads to a rapid adsorption of plasma proteins (mainly fibrinogen) to the material surface, thus its modification, and subsequent interactions with various blood elements will be determined by this modified surface. Adsorption of plasma proteins is affected by the type of surface, nature of blood proteins, hemorheological parameters, and other compounds present in plasma (e.g., ions). This stage of blood–material interaction is sometimes called *passivation*, although not all proteinated surfaces become passive toward blood components. In fact, they can induce platelet activation, adhesion, and aggregation leading to thrombosis.

The adsorbed proteins can be enzymatically degraded, replaced by other proteins, or undergo conformational and configurational changes and denaturation. These processes play a dominant role in the activation of plasma coagulation factors and affect the blood elements, especially the platelets. Platelets that have adhered to a proteinated surface become activated, release adenosine 5'-diphosphate and other platelet constituents, and irreversibly aggregate. The coagulation process initiated on the surface of aggregating platelets leads to transformation of soluble fibrinogen to an insoluble fibrin network. Individual platelets lose their integrity, fuse with each other, and together with entrapped blood cells, form a thrombus. Thus, any material to be used in blood-contacting devices has to be carefully tested for its blood compatibility, defined as the inability of an artificial surface to activate the intrinsic blood coagulation system, to attract or alter platelets or leukococytes. It should be kept in mind that a material that does not show adherent clots after contact with blood may induce clot formation elsewhere in the body and thus is not blood compatible, although it may be called *thromboresistant*.

Tests for the evaluation of blood–polymer interaction usually involve the following:

- in vitro measurements of whole blood clotting time in tubes of test material
- whole blood clotting time on sheets or films of test material
- centrifugal force tests for cell adhesion
- bead column tests
- closed loop tests
- static and dynamic flow chambers
- spinning disc
- other rheological test chambers

Parameters measured in these tests are

- formation of a blood clot
- adhesion of erythrocytes, leukocytes, or platelets
- clot size
- adherence of cells in native blood
- detection of radiolabelled proteins
- thrombus–unclotted blood ratio
- platelet changes

In vivo tests include extracorporeal evaluation of initial blood–material interaction in canulae or shunts, stagnation point flow measurements, annular axial tests, extracorporeal evaluation of chronic blood–material interactions using an arterio-venous shunt, intracorporeal evaluation of initial blood–material interactions with intravascular catheters, vena cava ring tests, and renal embolus tests. The parameters measured are thrombus rate, morphology of adherent cellular elements, occlusion time, extent of the activation mechanism, influence of surface irregularities, radiolabelled platelet count, deposition of fibrinogen, and renal emboli (*69–84*).

Surface Properties

The introduction of a foreign material into a living organism creates an interface between the material and tissue. This result means that such material characteristics as surface tension, surface free energy, morphology (roughness, texture, and porosity), electrical properties, hydrophilicity, presence of ionic groups, and surface contamination will all affect the tissue–material interaction (*69–129*). Materials with critical surface tension within a zone of biocompatibility in the range of 20 to 30 dynes/cm should show minimum bioadhesion and thrombus formation (*85*). A number of polyurethanes including commercial biomedical polyurethanes have surface tension values in the range of 30 to 70 dynes/cm, and critical surface tension in the range of 27 to 29 dynes/cm (*53*): that is, within the zone of biocompatibility.

Smooth polymer surfaces exposed to blood in the short-term tests seem to be more blood compatible than rough ones. Such surfaces, however, especially when produced on the lumen side of vascular prostheses, are particularly prone to embolization due to low fibrin–polymer bioadhesion forces. Rough, textured surfaces enhance cell adhesion and, in contact with blood they may be initially more thrombogenic than smooth ones, but they seem to enhance more rapid and firm attachment of the fibrin lining and thus endothelialization (*86, 87*). Medical polyurethanes in textured forms with rough surfaces perform favorably in contact with soft tissue and blood (*88–107*).

The presence of pores of an adequate size in the implants enhances tissue ingrowth. This effect has been used for fixation and stabilization of electrodes, artificial heart valves, artificial larynx, bladders, skins, tendons, and ligaments and in orthopedic surgery (hard tissue ingrowth). Tissue ingrowth–ongrowth has been used to create a blood compatible surface in vascular prostheses (formation of the neo-intima), blood vessel replacements (formation of neo-artery), nerve regeneration conduits, prevention of epidermal downgrowth by mature connective tissue ingrowth, artificial skins, percutaneous implants, bone regeneration membranes, and paradental membranes (*88–106*).

Electrical properties of solid surfaces such as surface charge or streaming (zeta) potential may affect thrombogenic events at the blood–polymer interface. Negatively charged surfaces have less tendency for thrombosis than positively charged ones (*72–74, 107,*

108). This result can be easily explained because under normal conditions the walls of the blood vessels and the platelets also have a negative surface charge. Negatively charged polyurethanes prepared by the electret effect or by admixing to the polymer a conductive carbon black (Bioelectric Polyurethane) showed good performance in contact with blood (*111, 112*).

Hydrophilic (wettable) surfaces are believed to be more blood compatible than hydrophobic ones, although it has also been suggested that a good hydrophilic–hydrophobic balance is required for optimal blood compatibility (*53, 112–116*). Surface wettability influences adhesion and proliferation of different types of mammalian cells. Cell adhesion occurs preferentially to water-wettable substrates.

Anionically charged surfaces show good blood compatibility, because repulsive interactions are acting between the surface and the platelets, which also possess an anionic charge. On the other hand, the Hageman factor XII (an intrinsic blood clotting factor) was observed to be activated by a number of negatively charged surfaces including collagen. In general, however, the anionic polyurethanes were more thromboresistant than neutral or cationic polyurethanes (*53*).

Impurities on the polymer surface affect the polymer–tissue interaction. Impurities introduced in a polymer during synthesis (catalyst residue, solvents, monomers, and oligomers) tend to migrate to the surface. Impurities such as airborne particulates, dust, water, solvents, and mold-release agents can be deposited on the surface in the processing–storage stage. Such impurities may make the surface highly thrombogenic or irritant, although the material itself is not. Leachable impurities and residual solvents in the polyurethane may cause necrosis of the neointima, prolonged inflammation, and cell death (*72–74*). Ultrasonic cleaning or extraction with solvents may improve the surface quality. Polyurethanes for biomedical applications should possibly be free from stabilizers, plasticizers, antioxidants, particulate fillers, and fibrous or textile fillers. Extraction of polyurethanes enhances their blood- and biocompatibility (*110*). Physical defects on the surface such as cracks, pinholes, fissures, and gas nuclei may trap microemboli and disturb the laminar flow of blood, which frequently results in thrombus formation not to mention the negative effects on the mechanical properties of implants.

Polymer surfaces are usually characterized by measuring surface tension and surface free energy, contact angle, X-ray photoelectron spectroscopy, secondary ion mass spectrometry, ion scattering spectroscopy, attenuated total reflection IR spectroscopy, and the zeta (streaming) potential (*79–85*).

Polymer Surface Modification and Blood Compatibility

Numerous approaches have been tried to enhance the blood compatibility of polymeric surfaces. These approaches involve developing negative charges on the surface by admixing a conductive carbon black (*111*), using the electret effect (*112*), attaching anionic groups (*53*), or producing neutral polymers with negative zeta potential (*113*). Binding of heparin by ionic or covalent linking to a polymer or coating with heparin-poly-(vinyl alcohol) hydrogels seems to improve their thromboresistance because of the inhibition of the surface-induced activation of the intrinsic coagulation (*117–122*). Albumin coating of surfaces may improve their blood compatibility because of reduced platelet adhesion, thrombogenesis, faster endothelial healing, and a reduced tendency for embolization (*123–125*).

Polyurethanes based on hydrophilic polyethylene oxide adsorb albumin faster than those prepared with hydrophobic poly(propylene oxide). Albumin adsorption can be enhanced by derivatization of the surface to carry alkyl groups (*53, 124–126*). Certain polyetherurethane-ureas and a block copoly(dialkyl siloxane)–urethane show a high albumin and a low fibrinogen adsorption (*126*). The reason for this specific protein adsorption behavior of polyurethanes is not clear, although it may be due to phase segregation in these polymers. Minimal phase segregation in a polymer reduces the rate of factor XII activation (lower platelet and fibrinogen deposition) (*107*), although an opposite behavior was reported in another study with similar polymers (*109*). These trends were not observed in polyurethanes synthesized with poly(dimethylsiloxane) or polyethylene oxide (*109*).

Binding prostaglandins on polymer surfaces or entrapping them in a polymer matrix inhibits platelet aggregation and significantly reduces platelet adhesion (*53*). Covalent grafting of polymer surfaces with hydrogels such as acrylamide or poly(hydroxyethyl methacrylate) enhances blood compatibility, which may be controlled by the extent of the protein adsorption–desorption process on the surface (*53, 72–74*). Other surface modifications to improve the surface blood compatibility include deposition of glow-discharge polymers such as hexamethyldisiloxane or tetrafluoroethylene and the radiofrequency glow discharge treatment (*127*) (*see also* Chapters 1.13 and 5.4). Blood compatibility and biocompatibility of polymers is significantly enhanced by biolization of a polymer surface by treatment with a solution of albumin or gelatin followed by cross-linking with gluteraldehyde or formaldehyde (*128*).

Coating of polymers with heparin-like materials such as sulfonated water-soluble polymers, heparinoid systems based on unsaturated poly(*cis*-1,4-iso-prene) and *N*-chlorosulfonyl isocyanate, or with resins prepared by binding sulfonate or amide groups onto cross-linked polystyrene and polysaccharides, leads to more nonthrombogenic surfaces (*129*). Deposition of low-temperature-isotropic (LTI) carbon on polymeric surfaces produces materials with good blood compatibility. This improved compatibility may be associated with the ability of these materials to maintain a native protein layer at the carbon–blood interface, which prevents the sequence of reactions leading to thrombus formation (*72–74*). Sulfonation or modifications of polyurethanes with polyethylene oxide improve their biocompatibility (*130, 131*).

Commercial Biomedical Polyurethanes

During the past two decades a number of medical-grade polyurethanes have been developed and used in various extracorporeal and intracorporeal devices (Table IV). After successful animal trials with a commercial DuPont polyetherurethane (Lycra) (Scheme IV), a segmented aromatic polyetherurethane-urea based on MDI, poly(tetramethylene oxide) (PTMO), and ethylenediamine was developed by Ethicon, under license from DuPont. This polymer with the trade name Biomer (Scheme V) is distributed as 30 wt% solution in dimethylacetamide, which can be processed by dip-coating, solution-casting, spraying, or by precipitation from dilute solutions. The polymer cannot be processed by melt extrusion or injection molding without degradation, because it contains a relatively high concentration of urea groups. Removal of trace solvent residues from the product is critical for good performance in vivo. Production of an extrusion-grade Biomer based on MDI, PTMO, and water has been abandoned. Biomer has been used in various blood-contacting devices such as bladders and coatings for left ventricular assist pumps, catheters, and heart valves (*53*). The polymer showed acceptable mechanical properties, blood biocompatibility, and long-term in vivo stability (*132–134*).

Cardiothane 51 (Scheme VI) distributed by Kontron as 16 wt% solution in tetrahydrofuran-1,4-dioxane mixture is a block copolymer of 90% polyurethane and 10% poly(dimethylsiloxane) (*135*). The polymer contains reactive acetoxy groups in the chain, and on exposure of the solution to air it undergoes cross-linking by moisture. Cardiothane exhibits an acceptable blood compatibility and has been used in intra-aortic balloons, catheters, artificial hearts, and blood tubings. The polymer can be processed by solution casting, dip coating, spraying, or glueing. Other polyurethanes from the same group are Cardiomat 610 and Cardiomat 40. Cardiomat 610 is similar to Biomer in properties. Cardiomat 40 is a silicon-containing, silica-free, polyetherurethane that vulcanizes at room tempera-

Table IV. Mechanical Properties of Medical Polyurethanes

Trade Name	Tensile Strength (Mpa)	Young's Modulus (Mpa)	Hardness Shore Durometer (%)	Ultimate Elongation (%)	Creep
Biomer	28–41	2.8–12.4	75A	600–800	—
Cardiothane 51	43	6.5–37.9	72A	525	—
Cardiomat 610	28.1	8.2	80A	500	—
Pellethane 2363-75A	32.1	2.2–6.2	73A	650	—
Pellethane 2363-80AE	31.0	4.8–8.9	83A	650	1.2
Pellethane 2363-80A	37.9	3.8–11.0	83A	550	—
Pellethane 2363-90AE	37.2	6.3–15.9	87A	560	—
Pellethane 2363-90A	44.8	7.7–23.0	93A	460	—
Pellethane 2363-55DE	41.2	10.6–29.9	52D	450	—
Pellethane 2363-55D	45.8	13.1–34.0	55D	390	—
Pellethane 2363-65D	43.2	16.3–31.4	62D	450	—
Tecoflex EG-80A	33.6	2.8–10.7	80A	740	—
Tecoflex EG-85A	44.3	—	85A	600	—
Tecoflex EG-93A	47.7	—	93A	430	—
Tecoflex EG-100A	50.1	—	48D	400	—
Tecoflex EG-60D	50.3	15.0–42.2	60D	370	—
Tecoflex EG-65D	49.3	—	65D	310	—
Tecoflex EG-72D	55.2	15.4–24.7	72D	250	—
Thoratec BPS-215	38	10.3	75A	700	—
Thoratec BPS-105	34.5	4.1	70A	870	—
Eurothane 2003	—	6.8	—	—	1.2
Eurothane 2005	—	11.5	—	—	2.6
Corethane 80A	45–52	5.3–8.6	—	400–490	—
Corethane 55D	48–59	12.8–15.2	—	365–440	—
Corethane 75D	48–63	36.5–39.3	—	255–320	—
Rimplast PYUA 102	21	28	70A	700	—
Rimplast PYUA 103	10	13.8	60A	1000	—
Tyndale Plains-Hunter	2.7–38	1.6	—	200–880	—
Biothane	18	8.6–12.0	44D	136	—
Erythrothane	43	4.8	83A	520	—

Note: Dashes indicate that data are not available.

ture. The polymer manifested good in vivo performance (*136*).

Pellethane 2363 (Scheme VII) is a polyetherurethane based on MDI, PTMO, and 1,4-butanediol that was developed by Upjohn, at present produced and supplied by Dow Chemical. Pellethanes contain residues of processing aids and catalysts, which should be removed from the polymer by extraction with organic solvents (*53*). Pellethane is supplied in a granular form in both injection-molding (2363-80A) and extrusion (2363-80AE) grades. The polymer is biocompatible and reasonably blood compatible (*51–55*). Pellethanes have been used as insulation in pacing leads, pacemaker connectors, blood bags, gas therapy tubings, cardiac assist devices, endotracheal tubings, and as a substrate for interpenetrating network hydrogels (*54*). Because Pellethanes undergo surface cracking on implantation (*137–141*), the producer intends to withdraw these materials as unsuitable for long-term use in the body (*142*).

Mitrathane was produced by Mitral Medical International. The polymer is polyetherurethaneurea with a chemical structure similar to that of Biomer, supplied as 20 wt% solution in dimethylacetamide. It is claimed to be particulate free with good potential for blood-contacting devices such as vascular prostheses or pericardial patches (*96, 97*). Recent studies have shown, however, that vascular prostheses produced from Mitrathane and implanted in dogs were occluded by thrombosis after six months and degraded to varying degrees on the external surface (*143, 144*).

Toyobo TM5 segmented polyurethaneurea developed by Toyobo Co. is based on MDI, PTMO, and propylene diamine. The polymer is supplied as a 15 wt% solution in dimethylformamide. The polymer has higher strength and ductility than Biomer, but lower static and dynamic flexibility. Toyobo polyurethaneurea exhibits acceptable blood compatibility and is intended for blood-pump applications (*145, 146*).

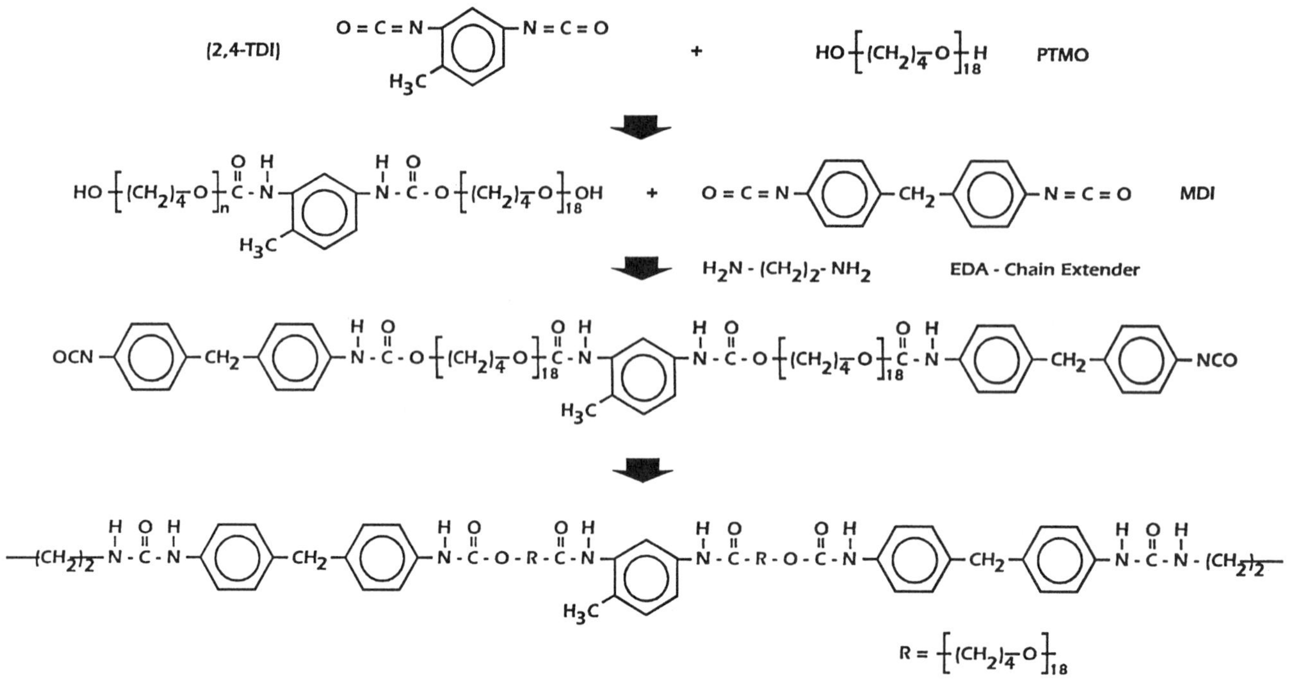

Scheme IV. Reaction scheme for the synthesis of Lycra.

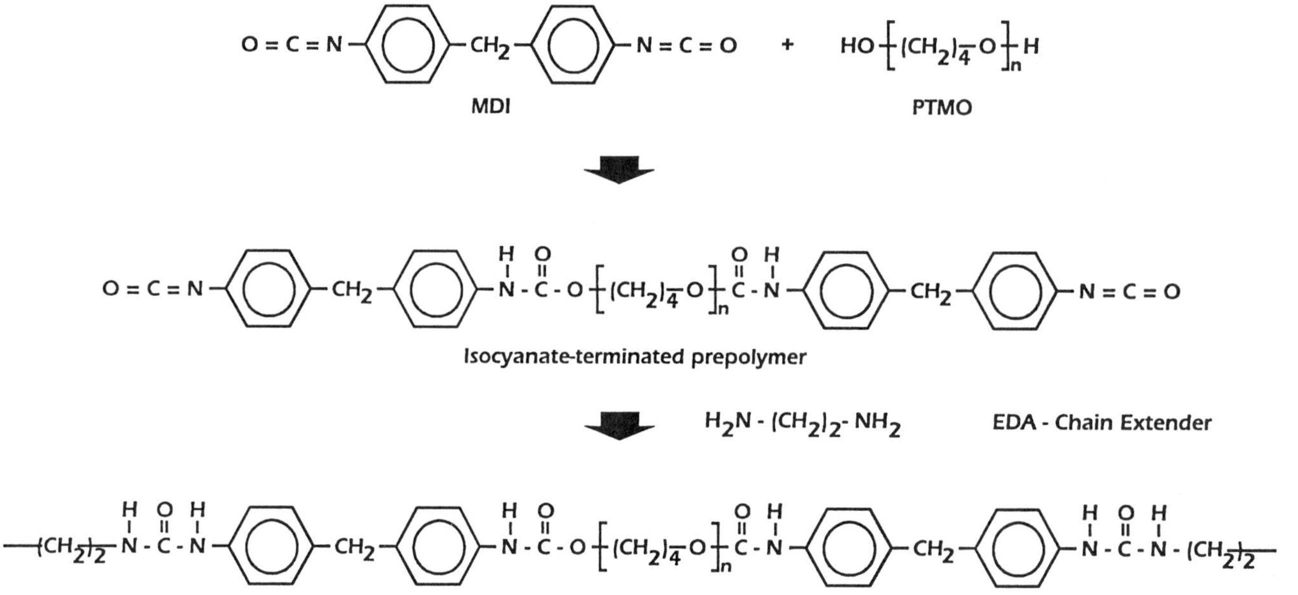

Scheme V. Reaction scheme for the synthesis of Biomer.

Tecoflex (Scheme VIII) is a cycloaliphatic polyetherurethane produced by Thermedics from $H_{12}MDI$, PTMO, and 1,4-butanediol. This polymer is supplied in various grades as melt-processable and solution-processable materials. Tecoflex is claimed to be hemocompatible, thromboresistant, biocompatible, noncarcinogenic, and nonmutagenic (39). However, the polymer is reported to be less water and heat stable than its aromatic analog Pellethane.

Texin 985-AM, 85 Shore A and Texin 865DM, 65 Shore D are polyether-based, MDI-type thermoplastic urethane elastomers. Texin 4210-M, 70 Shore D, is a polyester-based, MDI-type polyurethane–polycarbonate blend. All three grades have passed the USP XXI biological test, meeting class VI standards for biocompatibility in tissue contact. Texin polymers produced by Mobay Company are melt processable and are claimed to have excellent chemical, physical, and mechanical properties for implantation in the human body (147).

The Thoratec BPS series of solution-processable polyurethanes are based on MDI, PTMO, and ethyl-

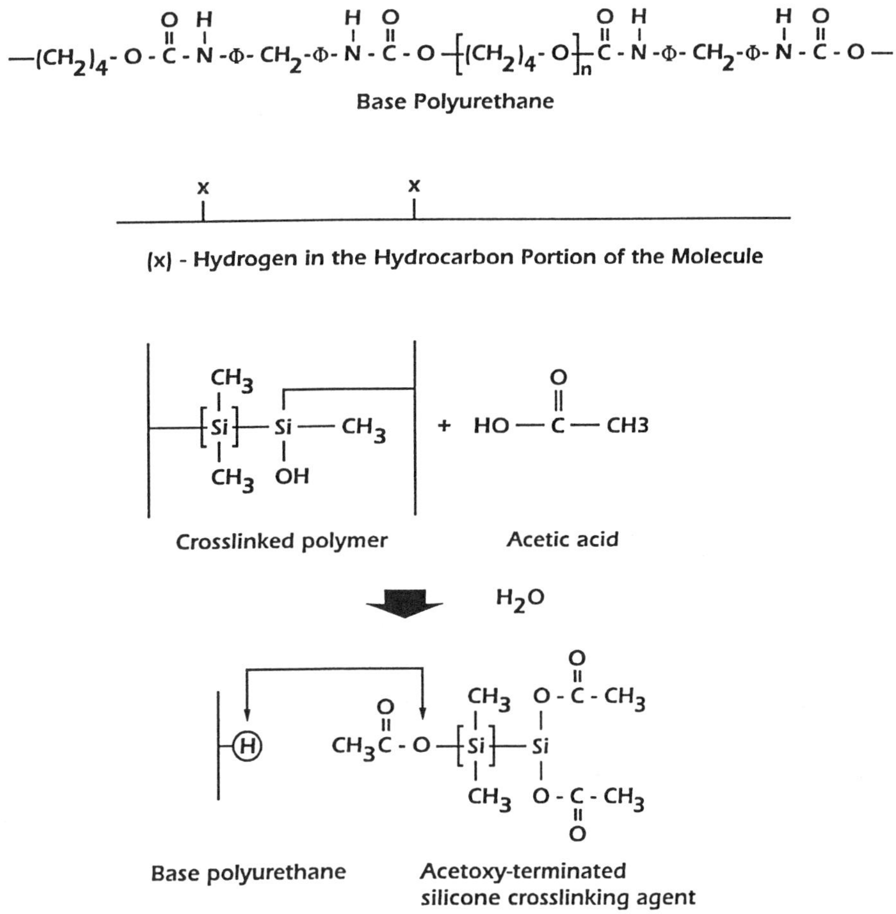

$$—(CH_2)_4 - O - \overset{O}{\underset{||}{C}} - \overset{H}{\underset{|}{N}} - \Phi - CH_2 - \Phi - \overset{H}{\underset{|}{N}} - \overset{O}{\underset{||}{C}} - O - \left[(CH_2)_4 - O \right]_n \overset{O}{\underset{||}{C}} - \overset{H}{\underset{|}{N}} - \Phi - CH_2 - \Phi - \overset{H}{\underset{|}{N}} - \overset{O}{\underset{||}{C}} - O —$$

Base Polyurethane

(x) - Hydrogen in the Hydrocarbon Portion of the Molecule

Crosslinked polymer   **Acetic acid**

H_2O

Base polyurethane   **Acetoxy-terminated silicone crosslinking agent**

Scheme VI. Reaction scheme for the synthesis of Cardiothane.

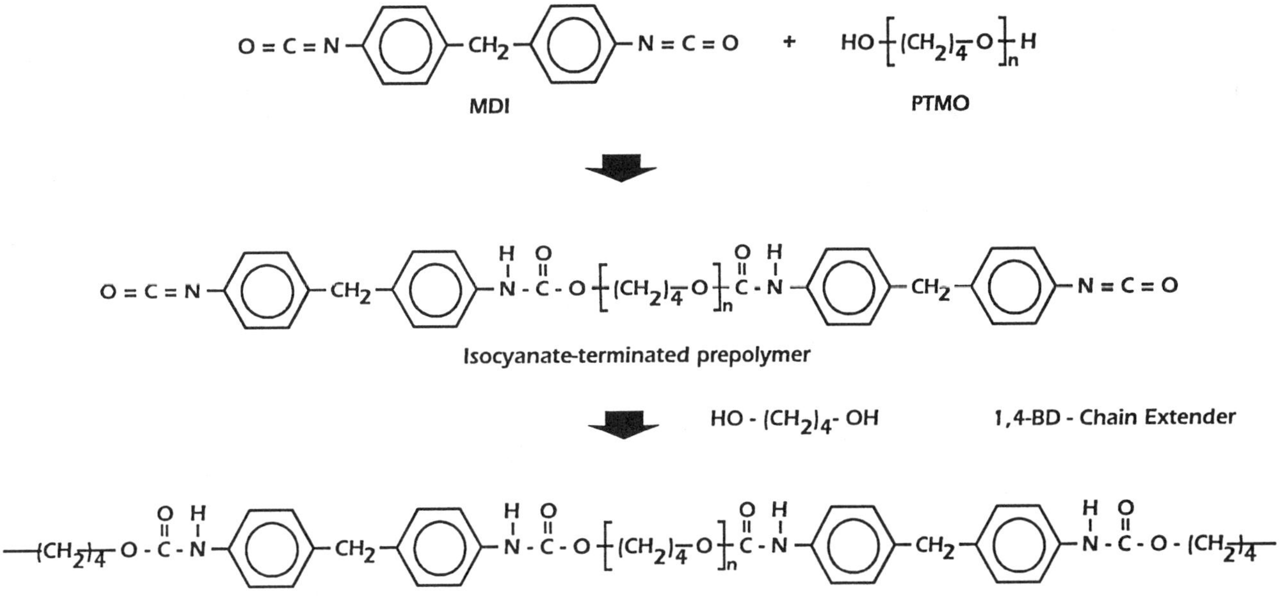

MDI   **PTMO**

Isocyanate-terminated prepolymer

$HO - (CH_2)_4 - OH$ **1,4-BD - Chain Extender**

Scheme VII. Reaction scheme for the synthesis of Pellethane.

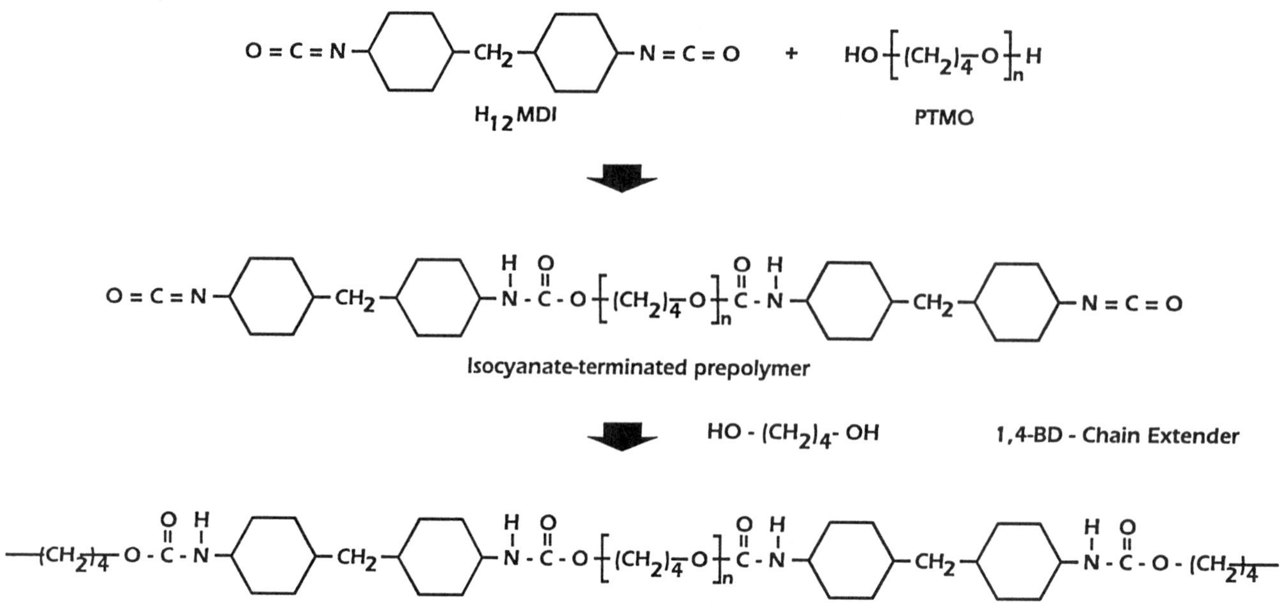

Scheme VIII. Reaction scheme for the synthesis of *Tecoflex*.

enediamine they have a chemical structure similar to that of Biomer. The medical-grade polymers BPS-215 M and BPS-105 developed by Thoratec are claimed to be unique in their combination of high tensile strength, elongation, and flex life (45). The materials are suggested to be used in artificial hearts, ventricular assist devices, prosthetic heart valves, and vascular grafts. The BPS-215 M polymer has gained a U.S. Food and Drug Administration investigational device exemption for the substitution of Biomer in the clinical use of the Pierce–Donachy ventricular assist device (148).

Toray Industries solution-processable polyetherurethane-urea is the former DuPont Lycra T-127 polymer and has been used for sac-type artificial hearts. The polymer showed good survival time in goats (149).

Eurothane 2003 and 2005 polyurethanes were developed by Beam Tech, England. The polymers are chemically similar to Biomer, but the length of the hard blocks in the chain is reduced. The Eurothane 2003 is being used in the development of a three leaflet valve for short-term use in a left ventricular assist device (148).

Corathane, polycarbonate urethane elastomer developed by Corvita, is based on MDI, poly(1,6-hexyl-1,2-ethyl carbonate)diol and 1,4-butanediol (Scheme IX). Corethane was designed as an improved replacement for Pellethane and was reported to be biostable (47).

The Biothane polyurethane systems of Cas Chem, Inc., are two-part liquid-coating polymers prepared from polyols derived from castor oil and non-TDI-based liquid prepolymers. The polymers are used in hemodialysis and blood oxygenator devices. Biothane is claimed to be hydrolytically stable and resistant to steam, heat, and ethylene oxide treatment (ETO) sterilization; it can be applied for potting, encapsulation, adhesives, sealants, and coatings (40).

TPH-Hydrophilic Polyurethanes produced by Tyndale Plains-Hunter are described as water absorptive, hydrogen-bonding, hydrogel systems. The chemical composition of the polymers is not disclosed. These polymers are supplied in durometers 78-99A, dry (40-90A, wet). Suggested applications include coatings (catheters, feeding tubes, and guide wires), cannulae, absorbents, wound dressings and drapes, controlled release, tendons, tissue expanders, contraceptives, etc. (41).

Erythrothane, a specially formulated thermoplastic polyurethane fabricated by Biosearch, has passed biocompatibility tests and is suitable for thermal processing, solvent welding, etc. It can also be delivered as tubes for blood, pharmaceuticals, and biological fluids (42). Hydromer is a hydrophilic polymer produced by interpolymerization of isocyanate prepolymers with high molecular weight polyvinylpyrrolidone. Hydromer exhibits low platelet adhesion and thrombus formation and can be used for coating of catheters, burn wound dressings, and sutures (54, 100).

Rimplast, of Petrarch Systems, are interpenetrating networks of aliphatic–aromatic and ester–ether polyurethanes modified with silicones. Rimplast PYUA 102 and PYUA 103 Tecoflex-based polyurethane networks are recommended by the producer for biomedical applications (43, 150).

Hypol hydrophilic polyurethanes are derived from foamable Hypol FHP isocyanate-terminated polyethers based on TDI (FHP 2002), MDI (FHP 4000, FHP 5000), and on aliphatic IPDI (FHP 6100) diiso-

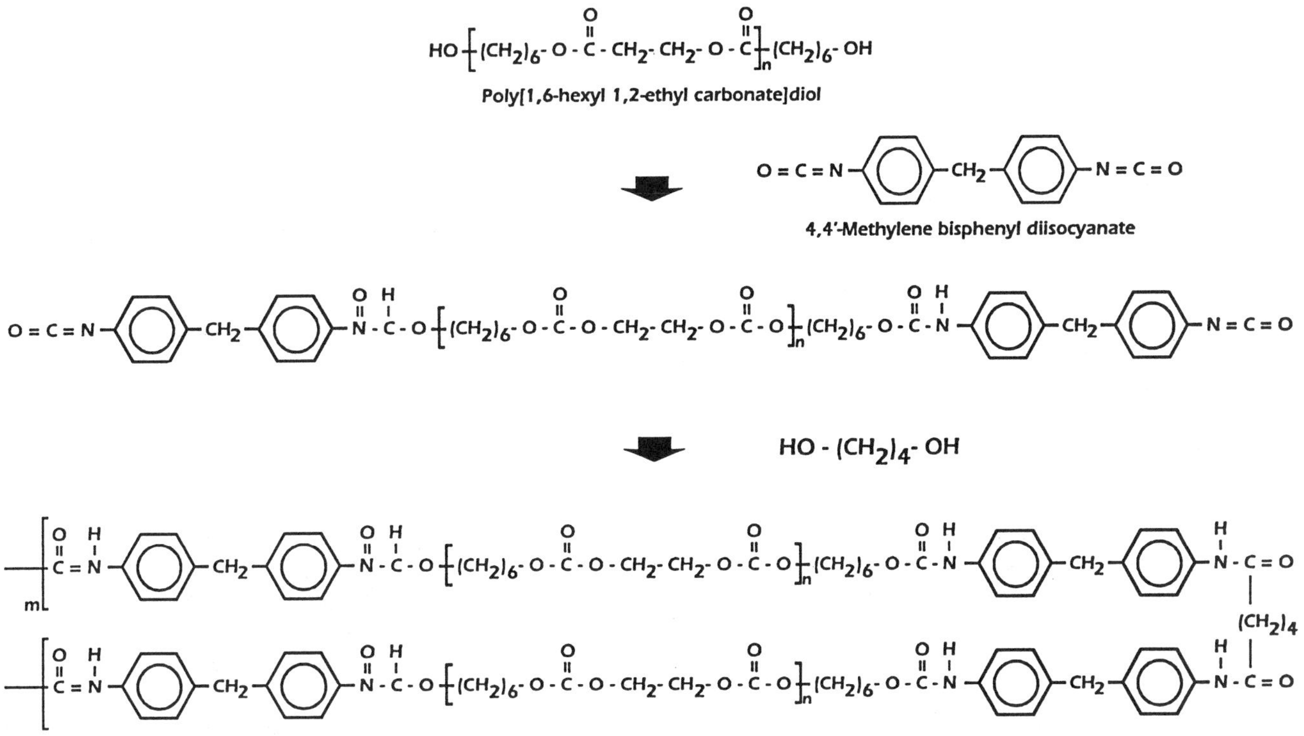

$$HO-[(CH_2)_6-O-\overset{\overset{O}{\|}}{C}-CH_2-CH_2-O-\overset{\overset{O}{\|}}{C}]_n-(CH_2)_6-OH$$

Poly[1,6-hexyl 1,2-ethyl carbonate]diol

$$O=C=N-\langle\rangle-CH_2-\langle\rangle-N=C=O$$

4,4'-Methylene bisphenyl diisocyanate

$$O=C=N-\langle\rangle-CH_2-\langle\rangle-\overset{\overset{O}{\|}}{\underset{H}{N}}-\overset{\overset{O}{\|}}{C}-O-[(CH_2)_6-O-\overset{\overset{O}{\|}}{C}-O-CH_2-CH_2-O-\overset{\overset{O}{\|}}{C}-O]_n-(CH_2)_6-O-\overset{\overset{O}{\|}}{C}-\overset{H}{N}-\langle\rangle-CH_2-\langle\rangle-N=C=O$$

$$HO-(CH_2)_4-OH$$

Scheme IX. Reaction scheme for the synthesis of Corethane.

cyanates. Produced by W. R. Grace & Co., these polymers can be used in wound dressings, drug delivery, and molded prosthetic devices (*44*).

Chronoflex, produced by PolyMedica, the composition of which is not disclosed, is based on aliphatic diisocyanate. The polyurethane is claimed to be biostable.

Surethane of Cardiac Control Systems, is prepared from Lycra (Spandex) of DuPont by removal of surface lubricants and dissolution of the polymer in dimethylacetamide. The material contains 3–5% poly(dimethylsiloxane) (blended and partially covalently bonded). Surethane is considered to be a candidate for pacing leads insulation.

Stable Experimental Polyurethanes

Polyurethaneurea based on MDI, polyoxypropylenediol, and ethylene diamine from the University of Utah was used for the development of vascular prostheses and hemispherical artificial heart (Table V) (*89, 90, 151*).

Polyetherurethane-ureas based on MDI, PTMO, and ethylene diamine from Stanford Research Institute were used for the construction of intra-aortic balloons, heart massage cups, and left ventricular assist device bladders. Their in vivo performance was comparable to that of Biomer. The development work on these polyurethanes was discontinued (*152*).

Hydrophilic polyurethaneureas based on MDI, polyoxyethylene diamine, and diethylenediol from

Ontario Research Foundation can absorb up to 62 wt% of water, show good biocompatibility, and are intended for percutaneous and subcutaneous implants. Elastic moduli of these polymers approach that of human cartilage (*54*).

Hydrophilic polyetherurethane-ureas prepared at Massachusetts Institute of Technology from *trans*-1,4-cyclohexane diisocynate, polyethylene oxide, and ethylene diamine swell up to 10-fold in water and show good blood compatibility (*54*).

Hydrophilic and hydrophobic polyether or polyesterurethanes from Enka AG (*46*) are based on *trans*-1,4-cyclohexane diisocyanate. These polymers have passed USP XX biocompatibility tests. Potential applications include vascular prostheses (*94, 95*), heart valves, blood pumps for extracorporeal assist devices, catheters, and blood tubings. The polymers are melt and solution processable.

Fluorine-containing, solution-processable aliphatic polyurethanes prepared by Asahi Glass from a fluorinated aliphatic diisocyanate, PTMO or PEO, and ethylene diamine or *m*-xylylene diamine are claimed to exhibit low thrombogenic activity (*38*).

Degradable Experimental Polyurethanes

Block copolyurethanes with peptide links in the main chain were synthesized from hexamethylene diisocyanate or MDI, PTMO 1000, and various amino acids or dipeptides as rigid-chain extenders. The increase in

Table V. Stable Experimental Polyurethanes

Origin	Composition	Suggested Use	Remarks
University of Utah	MDI–PPD–EDA	Vascular prostheses, artificial heart	Not available
Stanford Research Institute	MDI–PTMO–EDA	Intra-aortic balloons, heart massage cups, left ventricular assist device bladders	Not available
Ontario Research Foundation	MDI–PODA–DED	Percutaneous–subcutaneous implants, cartilage	Not available
Massachusetts Institute of Technology	CHD–PEO–EDA	Hydrogels, wound dressings, vascular prostheses	Not available
Enka AG	CHD-7-7	Vascular prostheses, heart valves, blood pumps, catheters, blood tubings	Available in small batches
Asahi Glass	FDI–PTMO–EDA FDI–PEO–XDA	Vascular prostheses, surface coatings, catheters	Not available

peptide length was shown to improve the microphase segregation of soft and hard segments in both aliphatic and aromatic polyurethanes. Polyurethanes with peptide links possess, in principle, all the properties typical for regular segmented polyurethanes, but at the same time they show a tendency of specific degradation by enzymes (*153*).

Polyurethanes synthesized from hexamethylene diisocyanate, PTMO 1000, and various sugar derivatives showed a high susceptibility to specific enzymatic degradation as compared with regular polyurethanes (*154*). Aromatic and aliphatic polyetherurethanes containing L-serine dipeptide in the main chain, prepared from MDI or hexamethylene diisocyanate, PTMO or PEO, and a linear or cyclic dipeptide of L-serine chain extender showed low thrombogenic activity and were permeable to oxygen and uremic toxins (*155, 156*). Physical mixtures of polyurethanes with polylactides were used for the preparation of degradable vascular prostheses and artificial skin. Both types of implants performed satisfactorily in rats and guinea pigs (*157*). Biodegradable lysine diisocyanate-based poly(glycolide-*co*-ε-caprolactone) urethane network was suggested as a dermal analogue in bilayer artificial skin (*158*). Cross-linked biodegradable composites for medical applications were also fabricated from poly(D,L-lactide-urethane) matrix based on L-lysine diisocyanate (ethyl-2,6-diisocyanatohexanoate) and poly(D,L-lactide) triol (*159*).

Biomedical Applications of Polyurethanes

Various polyurethanes were used alone or as components of various medical devices such as total artificial hearts, heart valves, ventricular assist devices, vascular prostheses, pericardial patches, roller pump tubings in artificial heart or blood pumps, vascular stents, angioplasty balloons, endovascular embolization, pacing leads insulation, mammary implants, oesophageal and tracheal prostheses, ureteral prostheses, fallopian tubings, endotracheal tubings, gastric balloons and feeding tubings, catheters and cannulae, sutures, ligaments, wound dressings and drapes, blood bags, peripheral nerve repair devices, implants for craniofacial and maxillofacial reconstruction, liners in dentistry and paradental membranes, shock absorbing elements for root implants, adhesives, orthopedic casting tapes, drug delivery devices, enveloping membranes for soft organ fixation, dialysis membranes, filters in blood oxygenators, meniscus reconstruction membranes, catheters, artificial kidneys, implants for craniofacial and maxillofacial reconstruction, and a number of other devices.

Criteria of biomedical polyurethanes to be used for various devices will depend on their intended applications. Materials for total artificial heart (TAH) and heart valves should have good bio- and blood compatibility, long-term molecular stability, high mechanical strength, and resistance to calcification and infection. Polyurethanes used in TAH were Biomer, Cardiothane 51, Toyobo, and Thoratec BPS-205M. Artificial heart valves produced from Biomer, Cardiothane 51, and bioelectric polyurethanes were used with limited success, mainly due to dystrophic calcification of the devices on the blood-contacting surfaces. There were also problems with long-term in vivo molecular stability of devices produced from Biomer and other polyurethanes. Vascular prostheses based on various aromatic polyether-urethane-ureas, aliphatic polyetherurethanes, and polyester-urethanes were used successfully in experimental animals (*88–101*). Except for numerous trials, the polyurethane vascular prostheses are not as yet used in the clinics.

Processing of Polyurethanes

Linear thermoplastic polyurethanes are processable from the melt or from solution on conventional thermoplastic processing equipment. Typical processing includes extrusion molding, injection molding, rotational molding, blow molding, compression molding, reaction injection molding, calendring, casting, coating (blade coating, brushing, spraying, and dipping), laminating, filament winding, vacuum forming, and thermoforming. Polyurethaneureas for biomedical applications are frequently processed from solution by such methods as coating, casting, rotational molding, and eventually filament winding. In solution casting, the quality of mold surface, solvent characteristics (e.g., boiling point), solution and environmental temperatures (if different from solution temperature), humidity and purity of air, and concentration are important for the quality of the resulting product.

A contaminated mold surface may result in highly thrombogenic implants with poor mechanical properties. The rate of solvent evaporation affects phase segregation in polyurethanes and thus their mechanical and surface properties. The completeness of solvent removal has to be carefully controlled, because a number of solvents suitable for dissolution of polyurethanes are either toxic, carcinogenic, mutagenic, or teratogenic. Polyurethanes for melt processing should be carefully dried to reduce the risk of hydrolytic degradation on heating. Processing temperatures should be kept at the lowest possible level, whereas the residence of a polymer in a processing machine should be short to diminish thermal degradation. All the processing aids and release agents have to be efficiently removed from the fabricated devices. Usually, solvent extraction and plasma etching will improve the surface properties. Removal of low molecular weight impurities and oligomers enhances biocompatibility, blood compatibility, and molecular in vivo stability.

Porous polyurethanes can be produced by using blowing (foaming) agents using replamineform from molds (e.g., porous coral), admixing additives (low or high molecular weight compounds) that are subsequently washed out from the product, precipitation of a polymer solution on a suitable support with a nonsolvent, controlled evaporation of solvent–nonsolvent systems from a polymer solution, and nonwoven spinning. (Replamineforms: implantable devices, e.g., vascular prostheses, are produced using a template machined from the microporous calcite spines of sea urchins. Polymer solution is injected in the template, solvent is removed, and subsequently the calcite is dissolved in hydrochloric acid, leaving behind a microporous structure (a replica of the calcite structure).) In general, the use of more concentrated polymer solutions leads to smaller pores that are irregular in shape, whereas precipitation of dilute polymer solutions gives materials with large, interconnected pores. Porosity of materials produced in the presence of inorganic salts depends on the size and size distribution of the additive crystals. Pore size in the material produced in the presence of organic salts that are soluble in solvents used for dissolution of polymers depends on the concentration of the additive in the polymer solution. Generally, pores sizes increase with an increase in the additive concentration.

Sterilization of Polyurethanes

Dry Heat

Sterilization at 125 °C for 6 h has no significant effect on the mechanical properties of polyurethanes but will result in a slight discoloration. The soft-grade polyurethanes may cloud because of the migration of processing aids to the surface (52). In porous materials, dry heat sterilization may cause deformation of the pore structure and diminish the pore sizes.

Steam Autoclaving and Boiling in Water Methods

Steam autoclaving (125 °C, 1.5 h) and boiling water sterilization are not recommended for polyurethanes because they lead to polymer hydrolysis. In the case of MDI-based polyurethanes, hydrolysis gives rise to the formation of 4,4'-diphenylmethanediamine, which has been recognized as a carcinogen in the part-per-billion range.

Ethylene Oxide Treatment

ETO sterilization is performed at 21–66 °C and a relative humidity of 30–60%. Effective gas concentrations are in the range of 400–1600 mg/L. Ethylene oxide is used in its pure form or as 12/88% or 20/80% mixtures with freon or carbon dioxide, respectively. The sterilization time depends on sterilizer size, load, material mass, and material density (porosity). The drawbacks of ETO sterilization are related to the possibility of by-products formation in the sterilized material and the necessity for material degassing to remove the gas residue. By-products in ETO sterilized polymers are ethylenediol and ethylenechlorohydrin. Too high a temperature and humidity applied during sterilization of porous articles may affect the pore size and structure. Polyurethanes can be sterilized with ETO at a low temperature and humidity. Sterilized materials have to be carefully evacuated, possibly in combination with flushing by an inert gas.

High Energy Radiation

Radiation sterilization (λ-^{60}Co and β high energy electrons) requires doses in the range of 1.5–2.5 Mrad. A standard dose used in Europe for the sterilization of medical devices is 2.5 Mrad. Polyurethanes can be sterilized without significant changes in physical properties. A slight increase in tensile strength and no changes in ultimate elongation are often observed. Multiple sterilization may cause cross-linking or degradation of the polymer (*160*).

Molecular Stability of Polyurethanes

The molecular stability of polyurethanes may be affected by prolonged action of water (and other liquids), heat, photochemical and high energy radiation, and the biologically active environment of the living body. Water, hydrogen-bonding liquids, and acids cause hydrolysis of ester, ether, and urethane linkages in polyurethanes. Hydrolysis of ester bonds is an autocatalytic process due to the formation of carboxylic groups, whereas hydrolysis of ether bonds seems to be catalyzed by metal ions and peroxides. Temperatures exceeding 100 °C, especially in the presence of air and moisture, may cause nonreversible, destructive changes in polyurethanes. The products of thermal degradation are basically the monomers, although in the case of polyurethanes based on 4,4'-diphenylmethane diisocyanate, 4,4'-diphenylmethanediamine may also be formed. This compound has been recognized as a carcinogen. Visible light and UV radiation, when acting for a prolonged time, cause degradation of polyurethanes, especially those free from light stabilizers and antioxidants. Beta and gamma radiations at the standard dose of 2.5 Mrad used for sterilization of medical devices do not affect the mechanical properties of polyurethanes.

The biologically active environment of the living body may degrade polyurethanes, mainly through hydrolytic chain scission within ester and urethane linkages of a polymer and oxidative attack within polyether segments. In vivo degradation often manifests itself in implant fragmentation or surface cracking. Phagocytosis and -lysis of polymer fragments by macrophages and giant cells is the second essential stage of the in vivo degradation of polyurethanes. In vivo degradation of polyurethanes appears to be accelerated by the action of cell enzymes (Table VI), metal ions, lipid pickup, calcification, surface defects in the implants, organotin (used in the preparation of polyurethanes as catalyst), and stress accumulated in the implants. Polyetherurethane-ureas and polyetherurethane–poly(dialkylsiloxane) copolymers are more stable in the biological environment than polyetherurethanes and regular commercial polyesterurethanes (*137–141, 161–173*). A survey of degradation processes in polyurethanes was given elsewhere (*173*).

Toxicity from Polyurethanes

Some early publications on carcinogenicity testing of nonmedical grade, industrial polymers contaminated

Table VI. Enzymes Affecting Molecular Structure of Polyurethanes

	Polyurethane Type[a]						
Enzyme	1	2	3	4	5	6	7
Bromelain	+	–	X	X	X	X	X
Cathepsin C	+	–	X	X	X	X	X
Cholesterol esteraze	X	X	–	X	X	X	+
Chymotrypsin	+	–	+	+	+	X	X
Collagenase	–	–	X	X	X	X	X
Cytochrom C oxidase	–	–	X	X	X	X	X
Esterase	+	–	X	X	X	X	X
Ficin	+	–	X	X	X	X	X
Horseradish peroxidase	X	X	X	X	X	X	+
Leucine aminopeptidase	X	X	X	+	+	+	X
Lysosomal enzymes from a liver	+	–	X	X	X	X	X
Papain	+	–	+	+	+	+	X
Trypsin	+	–	X	X	X	X	X
Xantine oxidase	X	X	–	–	X	X	X
Urease	X	X	+	X	X	X	X

Note: +, affected; –, unaffected; X indicates that data is not available.

[a]1, TDI+MDI+EDA+PTMO 1000 (*Lycra* type); 2, TDI+MDI+EDA+PTMO 650; 200 (*Lycra* type); 3, MDI+EDA+PTMO 1300 (*Biomer*); 4, MDI+1,4-BD+PTMO (*Pellethane*); 5, MDI+1,4-BD+PEA (*Esthane*); 6, MDI+EDA+PPD 625; 2000; 7, TDI+PCL+EDA.

Source: Adapted from references 160–172.

with residual monomers, catalysts, curing agents, plasticizers, etc., raised the issue of their potential mutagenicity and carcinogenicity (*174–179*). It was suggested that mutagenicity and carcinogenicity from polyurethanes might eventually be of chemical origin: that is, caused by the leachable impurities introduced into the polymer during the preparation process (catalysts, plasticizers, pigments, unreacted monomers, etc.) or by-products formed in the polyurethane upon its degradation (e.g., 4,4'-diphenylmethanediamine).

Malignant tumors were induced in rats after implantation of 17 chemically different polyurethanes. About 90% of all the cases were fibrosarcomas, and 3–4% were carcinomas. The carcinogenic response was much less pronounced for aliphatic polyurethanes than for aromatic ones. It was speculated that at least a few cases in this series might be due to chemical carcinogenesis.

However, polyethylene samples of the same size implanted as a reference produced tumor as well. In another set of experiments, no mutagenicity or evidence of tumor was found for commercial polyurethane based on MDI implanted in rats up to 2 years (*140*). No tumors were observed in rats and pigs implanted up to 1 year with degradable vascular prostheses based on aliphatic polyesterurethanes (*93, 98–100*). Even though the data on carcinogenic potential of some polyurethanes have been gathered using rodents as the experimental model (animals especially susceptible to tumors), so far there is no evidence of a carcinogenic response to polyurethanes in humans (to implantable pacing leads with an aromatic polyurethane insulation or to degradable polyurethane KL-3 tissue adhesive) (*164–168*).

Concluding Remarks

Polyurethanes, with their unique morphology resulting from the hard-segment–soft-segment microphase segregation, are candidates materials for bio- and blood-compatible devices for extracorporeal and intracorporeal applications. Polyurethanes with acceptable physical properties and bio- and blood compatibility can preferably be produced by two-step solution polymerization processes from high purity reactants in high purity polymerization media. The purity of raw biomedical polyurethanes and the final devices determine to a great extent their in vivo performance (biocompatibility, blood compatibility, and molecular stability). Chemical and biological modification of surfaces contacting blood may greatly improve blood compatibility and biocompatibility of polyurethane devices. None of the commercial or experimental polyurethane elastomers available at present fulfills all the requirements for the ultimate biomedical material: that is, the material with optimal blood compatibility, biocompatibility, mechanical properties, and long-term molecular stability in vivo. The main shortcoming of polyurethanes is their long-term molecular instability in vivo. In addition, the blood compatibility of nonmodified surfaces of polyurethanes is insufficient.

Various experimental, scientifically interesting approaches used to enhance the blood compatibility of polyurethanes did not solve this problem satisfactorily. This shortcoming is especially true if the implantable devices have to contact blood for prolonged time. The polyurethanes available at present are far from being perfect; however, they can be materials of choice for various biomedical applications providing that implantable devices produced from these polymers are used only for a limited period of time, regularly checked, and replaced with new ones when needed. As with available monomers it is hard to synthesize biomedical polyurethanes with ultimate properties, the efforts should be undertaken to design new classes of biomedical polymers. Elastomeric copolyester ethers may potentially supplement polyurethanes. These copolyester ethers available at present are, however, general-use polymers that were not designed for medical applications and especially not for implantable devices. Hence, extensive studies are needed to evaluate their long-term stability in vivo and bio- and blood compatibility.

References

1. Bayer, O. *Angew. Chem.* **1947**, *A59*, 257–288.
2. Dombrow, B. *Polyurethanes*; Reinhold: New York, 1957.
3. Saunders, J. H.; Frisch, K. C. *Polyurethanes: Chemistry and Technology*; Interscience: New York, 1962; Vols. 1, 2.
4. Saunders, J. H. *Polymer Chemistry of Synthetic Elastomers*; Interscience: New York, 1969.
5. Phillips, L. N.; Parker, D. B. V. *Polyurethanes*; Iliffe Books: London, 1964.
6. Wright, P.; Clumming, A. P. C. *Solid Polyurethane Elastomers*; MacLaren: London, 1969.
7. Reegan, S. L.; Frisch, K. C. *Adv. Urethane Sci. Technol.* **1971**, *1*, 1–15.
8. Entelis, S. G.; Nesterov, O. V.; Tiger, R. P. *J. Cell. Plast.* **1967**, *3*, 360–365.
9. Frisch, K. C.; Dieter, J. A. *Polym. Plast. Technol. Eng.* **1975**, *4*, 1–17.
10. Hepburn, C. *Polyurethane Elastomers*; Applied Sciences: London, 1982.
11. Frisch, K. C.; Kordomenos, P. *Advance in Urethanes: Elastomers, Coatings, Adhesives, Sealants*; Seminar Materials; Technomic: Westport, CT, 1985; pp 1–80.
12. Schollenberger, C. S.; Stewart, F. D. *J. Elastoplast.* **1971**, *3*, 28–56.
13. Schollenberger, C. S.; Dinbergs, K. *J. Elastoplast.* **1973**, *5*, 222–251.
14. Schollenberger, C. S.; Dinbergs, K. *J. Elastoplast.* **1979**, *11*, 58–91.
15. Wang, T. L. D.; Lyman, D. J. *J. Polym. Sci., Part A–1.* **1993**, *31*, 1983–1995.
16. Lyman, D. J. *Rev. Macromol. Chem.* **1966**, *1*, 191–237.
17. Lyman, D. J. *J. Polym. Sci.* **1960**, *45*, 49–59.
18. Goyert, W. *Swiss Plast.* **1982**, *4*, 7–18; Bayer AG; "Vulkolan 2020 Polycarbonate Diol;" Information Brochure: Bayer AG: Bayerwerk, D-5090 Leverkusen, Germany, 1983.
19. Schauerte, K. In *Kunstoff-Hanbuch, Polyurethane*; Carl Hauser Verlag: München, Germany, 1983; pp 42–120.
20. Bayer AG. *Desmodur/Desmophen Bayer Coating Materials Bulletin*; Bayer AG: Leverkusen, Germany, 1979; No. 5, p 79.
21. Dieter, J. W.; Byrne, C. A. *Polym. Eng. Sci.* **1988**, *27*, 673–683.
22. Cohen, D.; Siegman, A. *Polym. Eng. Sci.* **1987**, *27*, 286–293.
23. Seymour, R. W.; Cooper, S. L. *Macromolecules* **1970**, *3*, 579–583.
24. Estes, G. M.; Seymour, R. W.; Cooper, S. L. *Macromolecules* **1971**, *4*, 452–457.
25. Srichatrapimuk, V. W.; Cooper, S. L. *J. Macromol. Sci. Phys.* **1978**, *B15*, 267–311.
26. Kimura, I.; Ishihara, H.; Ono, H.; Yoshihara, N.; Nomura, S.; Kawai, H. *Macromolecules* **1974**, *7*, 355–363.
27. Estes, G. M.; Seymour, R. W.; Huh, D. S.; Cooper, S. L. *Polym. Eng. Sci.* **1969**, 383–387.
28. Gibson, P. E.; Valance, M. A.; Cooper, S. L. *Development in Block Copolymers*; Goodman, I., Ed.; Applied Science Publishers Ltd.: Barking, England, 1984; Vol. 1. pp 217–259.
29. Harrell L. L., Jr. *Macromolecules* **1969**, *2*, 607–612.

30. Bonart, R.; Morbitzer, L.; Hentze, G. *Macromol. Sci. Phys.* **1969**, *B3*, 337–356.

31. Seefried, C. G., Jr.; Koleske, J. V.; Critchfield, F. E. *J. Appl. Polym. Sci.* **1975**, *19*, 2493–2502.

32. Seefried, C. G., Jr.; Koleske, J. V.; Critchfield, F. E. *J. Appl. Polym. Sci.* **1975**, *19*, 2503–2513.

33. Seefried, C. G., Jr.; Koleske, J. V.; Critchfield, F. E. *J. Appl. Polym. Sci.* **1975**, *19*, 3185–3191.

34. American Cyanamid Company *Plast. Eng.* **1988**, *44*, 55.

35. General Mills Corp. *Tech. Bull.* **1965**, *CDS 10*, 65.

36. Henkel Corp. *Tech. Bull.* **1965**, *CDS 3*, 65.

37. Stenzenberger, H.-D.; Hummel, D. O. *Angew. Makromol. Chem.* **1979**, *82*, 129–148.

38. Takaura, T.; Kato, M.; Yamabe, M. *Antithrombogenic Fluorine Containing Segmented Polyurethanes*; Reports Research Laboratory, Asahi Glass Co., Ltd.: Yokohama, Japan, 1984; Vol. 34, pp 35–42.

39. Thermedics, Inc. Tecoflex, *Catalog Suppl.*; Thermedics, Inc.: Waltham, MA, 1985.

40. Cas Chem, Inc. *Biothane Polyurethane Systems*; Publication TB102, 1985; Vol. 2/85, p 177.

41. Tyndale Plains Hunter *Handbook of Hydrophyilic Polymers*; Princeton, NJ, 1986: pp 1–47.

42. Biosearch, Inc. *Erythrothane Biomedical Tubing*; Technical Literature; Biosearch, Inc.: Somerville, NJ, 1973.

43. Petrarch Systems, Inc. *Medical Applications of Rimplast Silicone Urethanes*; Promotional Bulletin; Petrarch Systems, Inc.: Bristol, PA, 1983.

44. W. R. Grace & Co. *Hypol Foamable Hydrophylic Polymers, Laboratory Procedures and Foam Formulation*; Production Bulletin; W. R. Grace & Company: Lexington, MA, 1986.

45. Mercor, Inc. *Catalog of Biomaterials, Speciality Polymers and Reactive Intermediates*; Production Bulletin; Mercor, Inc.: Berkeley, CA, 1987.

46. Enka, AG. *Polyurethanes for Medical Applications*; Production Bulletin; Enka, AG: Wuppertal, Germany, 1986.

47. Corvita Company *Corethane, Polycarbonate Urethane Elastomers*; Production Bulletin; Corvita Company: Miami, FL, 1993, CC–304241.

48. Beam Tech, Ltd. *Eurothane, Low Elastic Modulus Polyurethanes*; Technical Note; Beam Tech, Ltd.: Tarvin, England, 1992.

49. Lee, H.; Neville, K. *Handbook of Medical Plastics*; Pasadena Technology: Pasadena, CA., 1971.

50. Lyman, D. J. *Rev. Macromol. Chem.* **1966**, *1*, 355–391.

51. Wilkes, G. L. *Polymers in Medicine and Surgery*; Kronenthal, R. L.; Oser, Z.; Martin, E., Eds.; Plenum: New York, 1975; pp 45–75.

52. Ulrich, H.; Bonk, H. W. *Polyurethanes in Biomedical Engineering*; Planck, H.; Egbers, G.; Syre, I., Eds.; Elsevier: Amsterdam, Netherlands, 1984; pp 165–179.

53. Lelah, M. D.; Cooper, S. L. *Polyurethanes in Medicine*; CRC: Boca Raton, FL., 1986.

54. Coury, A. J.; Cobian, K. E.; Cahalan, P. T.; Jevene, A. H. *Adv. Urethane Sci. Technol.* **1984**, *9*, 130–168.

55. Coury, A. J.; Slaikeu, P. C.; Cahalan, P. T.; Stokes, K. B. *Prog. Rubber Plast. Technol.* **1987**, *3*, 24–36.

56. Gogolewski, S. *Colloid Polym. Sci.* **1989**, *267*, 757–785.

57. Gogolewski, S. "In Vivo Molecular Stability of Vascular Prostheses Produced from Various Thermoplastic Polyurethanes," *Medinvent Internal Report* **1984**, 1–13;

Walpoth, B. H.; Rheiner, P.; Cox, J. N.; Faidutti, B.; Mégevand, R.; Gogolewski, S. *Helv. Chir. Acta* **1988**, *55*, 157.

58. Olsen, D. *Transactions of the 13th Annual Meeting of the Society for Biomaterials*; Society for Biomaterials: New York, 1987; p 168.

59. Pierce, W. S. *Polymers in Medicine and Surgery*; Kronenthal, R. L.; Oser, Z.; Martin, E., Eds.; Plenum: New York, 1975; pp 263–286.

60. Bruck, S. D.; Rabin, S.; Ferguson, R. J. *Biomaer Med. Devices Artif. Organs* **1973**, *1*, 191–222.

61. Ocumpaugh, D. E.; Lee, H. L. *J. Macromol. Sci. Chem.* **1970**, *A4*, 595–613.

62. Bruck, S. D. *Biomaer Med. Devices Artif. Organs* **1973**, *1*, 79–88.

63. Park, J. B. *Biomaterials Science and Engineering*; Plenum: New York, 1984.

64. Williams, D. F. *Med. Prog. Technol.* **1976**, *4*, 31–42.

65. *Techniques of Biocompatibility Testing*; Williams, D. F., Ed.; CRC: Boca Raton, FL, 1986; Vols. 1, 2.

66. Oppenheimer, B. S.; Oppenheimer, E. T.; Stout, A. P. *Surg. Forum* **1953**, *4*, 672–676.

67. Oppenheimer, B. S.; Oppenheimer, E. T.; Stout, A. P. *Proc. Soc. Exp. Biol. Med.* **1952**, *79*, 366–369.

68. *Guidelines for Physicochemical Charaterization of Biomaterials*: NIH Publ. No 80–2186; National Institutes of Health, U.S. Department of Health and Human Services: Washington DC, 1980.

69. Lyman, D. J.; Muir, W. M.; Lee, I. J. *Trans. Am. Soc. Artif. Intern. Organs* **1965**, *11*, 301–306.

70. Lyman, D. J.; Muir, W. M.; Lee, I. J. *Trans. Am. Soc. Artif. Intern. Organs* **1968**, *14*, 250–255.

71. Baier, R. E.; Dutton, R. C. I. *J. Biomed. Mater. Res.* **1969**, *3*, 191–206.

72. Bruck, S. D. *J. Biomed. Mater. Res. Symp.* **1977**, *8*, 1–21.

73. Bruck, S. D. *Biomat. Med. Devices Artif. Organs* **1978**, *6*, 341–349.

74. Bruck, S. D. *J. Polym. Sci., Polym. Symp.* **1979**, *66*, 283–312.

75. Forbes, C. D.; Prentice, C. R. M. *Br. Med. Bull.* **1978**, *34*, 201–207.

76. Merill, E. W. *Ann. N. Y. Acad. Sci.* **1987**, *516*, 196–203.

77. Colman, R. W.; Scott, Ch. F.; Schmaier, A. H.; Wachtfogel, Y. T.; Pixley, R. A.; Edmunds, L. H. Jr. *Ann. N. Y. Acad. Sci.* **1987**, *516*, 253–267.

78. Chenoweth, D. E. *Ann. N. Y. Acad. Sci.* **1987**, *516*, 307–311.

79. Owen, J.; Kaplan, K. L. *Ann. N. Y. Acad. Sci.* **1987**, *516*, 621–631.

80. Cooper, S. L.; Fabrizius, D. J.; Grasel, T. G. *Ann. N. Y. Acad. Sci.* **1987**, *516*, 572–585.

81. Leininger, R. I.; Hutson, T. B., Jakobsen, R. J. *Ann. N. Y. Acad. Sci.* **1987**, *516*, 173–183.

82. Dewanjee, M. K. *Ann. N. Y. Acad. Sci.* **1987**, *516*, 541–571.

83. *Guidelines for Blood–Material Interactions*; NIH Publ. No. 80–2185; National Institutes of Health, U.S. Department of Health and Human Services: Washington, DC, 1980.

84. Kohwa, S.; Litwak, R. S.; Rosenfeld, R. E. *Ann. N. Y. Acad. Sci.* **1977**, *283*, 457–472.

85. Baier, R. E. *Applied Chemistry at Protein Interfaces*; Baier, R. E., Ed.; Advances in Chemistry 145; American Chemical Society: Washington, DC, 1975; pp 1–33

86. White, R. A. *Transactions of the 10th Annual Meeting of the Society for Biomaterials*; Society for Biomaterials: Washington, DC, 1984; p 163.

87. White, R. A. *Trans. Am. Soc. Artif. Intern. Organs* **1988**, *34*, 95–100.

88. Mirkovitch, V.; Akutsu, T.; Kolff, W. *Trans. Am. Soc. Artif. Intern. Organs* **1962**, *8*, 79–84.

89. Kardos, J. L.; Mehta, B. S.; Apostolou, S. F.; Thies, C.; Clark, R. E. *Biomaer Med. Devices Artif. Organs* **1974**, *2*, 387–396.

90. Lyman, D. J.; Kwan-Gett, C.; Zwart, H. H. J.; Bland, A.; Eastwood, N.; Kawai, J.; Kolff, W. J. *Trans. Am. Soc. Artif. Intern. Organs* **1971**, *17*, 456–463.

91. Lyman, D. J.; Albo, D., Jr.; Jackson, R.; Knutson, K. *Vascular Prosthes Trans. Am. Soc. Artif. Intern. Organs* **1977**, *23*, 253–261.

92. Annis, D.; Bornat, A.; Edwards, R. O.; Higham, A.; Loveday, B.; Wilson, J. *Vascular Prosthes Trans. Am. Soc. Artif. Intern. Organs* **1978**, *24*, 209–214.

93. Pollock, E.; Andrews, E. J.; Lentz, D.; Sheikh, K. *Trans. Am. Soc. Artif. Intern. Organs* **1981**, *27*, 405–409.

94. Gogolewski, S.; Pennings, A. J. *Makromol. Chem., Rapid Commun.* **1983**, *4*, 213–219.

95. Hess, F.; Jerusalem, C.; Braun, B. *J. Cradiovasc. Surg.* **1983**, *24*, 509–515.

96. Hess, F.; Jerusalem, C.; Grande, P.; Braun, B. *J. Biomed. Mater. Res.* **1984**, *18*, 745–755.

97. Gilding, D. K.; Reed, A. M.; Askill, I. N.; Briana, S. *Trans. Am. Soc. Artif. Intern. Organs* **1984**, *30*, 571–576.

98. Ives, C. L.; Zamora, J. L.; Eskin, S. G.; Weilbaecher, D. G.; Gao, Z. R.; Noon, G. P.; DeBakey, M. E. *Trans. Am. Soc. Artif. Intern. Organs* **1984**, *30*, 587–589.

99. Gogolewski, S.; Galletti, G. *Colloid Polym. Sci.* **1986**, *264*, 854–858.

100. Gogolewski, S.; Galletti, G.; Usia, G. *Colloid Polym. Sci.* **1987**, *265*, 774–778.

101. Galletti, G.; Gogolewski, S.; Ussia, G.; Farruggia, F. *Ann. Vasc. Surg.* **1989**, *3*, 236–243.

102. MacGregor, D. C.; Wilson, G. J.; Klement, P.; Weber, B. A.; Binnington, A. G.; Pinchuk, L. *Transactions of the 17th Annual Meeting of the Society of Biomaterials*; Society for Biomaterials: Scottsdale, AZ, 1991; p 145.

103. Gogolewski, S.; Walpoth, B.; Rheiner, P. *Colloid Polym. Sci.* **1987**, *265*, 971–977.

104. Gogolewski, S.; Pennings, A. J. A. *Makromol. Chem., Rapid Commun.* **1983**, *4*, 675.

105. Warrer, K.; Karring, T.; Nyman, S.; Gogolewski, S. *J. Clin. Periodontol.* **1992**, *19*, 633. Farsφ Nielsen, F.; Karring, T.; Gogolewski, S. *Acta Orthop. Scand.* **1992**, *63*, 66.

106. Wang, F.; Nieuwenhuis, P.; Gogolewski, S.; Pennings, A. J.; Wildevuur, Ch. R. H. *Polyurethanes in Biomedical Engineering*; Planck, H.; Egbers, G.; Syre, I., Eds.; Elsevier: Amsterdam, Netherlands, 1984; pp 317–332.

107. Hepp, W.; Syre, I.; Planck, H.; Schrioy, N.; Wasmuth, C. *Polyurethanes in Biomedical Engineering*; Planck, H.; Egbers, G.; Syre, I., Eds.; Elsevier: Amsterdam, Netherlands, 1984; pp 333–361.

108. Picha, G. J.; Gibbons, D. F.; Auerbach, R. A. *J. Bioeng.* **1978**, *2*, 301–311.

109. Lelah, M.; Lambrecht, L. K.; Young, B. R.; Cooper, S. L. *J. Biomed. Mater. Res.* **1983**, *17*, 1–22.

110. Grasel, T. G.; Stuart, S. L. *Biomaterials* **1986**, *7*, 315–328.

111. Grasel, T. G.; Lee, D. C.; Okkema, A. Z.; Slowinski, T. J.; Cooper, S. L. *Biomaterials* **1988**, *9*, 383–392.

112. Sharp, W. V.; Taylor, B. C.; Wright, J.; Finelli, A. F. *J. Biomed. Mater. Res. Symp.* **1971**, *1*, 75–81.

113. Murphy, P. V.; La Croix, A.; Bernhard, W. *J. Biomed. Mater. Res. Symp.* **1971**, *1*, 59–74.

114. Leininger, R. I.; Epstein, M. M.; Falb, R. D.; Grode, G. A. *Trans. Am. Soc. Artif. Int. Organs* **1966**, *12*, 151–154.

115. Engbers, G. H. M.; Feijen, J. *Intern. J. Artif. Organs* **1991**, *14*, 199–215.

116. Takahara, A.; Okkema, A. A.; Cooper, S. L.; Coury, A. J. *Biomaterials* **1991**, *12*, 324–334.

117. Takahara, A.; Jo, N. J.; Kajiyama, T. *J. Biomater. Sci. Polym. Ed.* **1989**, *1*, 17–29.

118. Nojiri, Ch.; Jacobs, H.; Kim, S. W. *IEEE Eng. Med. Biol. Mag.* **1989**, *6*, 17–21.

119. Barbucci, R; Magnani, A.; Albanese, A.; Tempesti, F. *Int. J. Artif. Organs* **1991**, *14*, 499–507.

120. Azzuli, G.; Barbuci, R.; Benvenuti, M.; Ferruti, P.; Nocentini, M. *Biomaterials* **1987**, 61–66.

121. Mazid, M. A.; Scot, E.; Li, N. H. *Clin. Mater.* **1991**, *8*, 71–80.

122. Leininger, R. I.; Crowley, J. P.; Alb, R. D.; Grode, G. A. *Trans. Am. Soc. Artif. Intern. Organs* **1972**, *18*, 312–315.

123. Evangelista, R. A.; Sefton, M. V. *Biomaterials* **1986**, *7*, 206–211.

124. Sipehia, R.; Chawla, A. S. *Biomater. Med. Devices Artif. Organs* **1982**, *10*, 229–246.

125. Eberhart. R. C. *IEEE Eng. Med. Biol. Mag.* **1989**, 26–28.

126. Eberhart, R. C.; Munro, M. S.; Williams, G. B.; Kulkarni, P. V.; Shannon, W. A., Jr.; Brink, B. E.; Fry, W. J. *Artif. Organs* **1987**, *11*, 375–382.

127. Lyman, D. J.; Metcalf, L. C.; Albo, D., Jr.; Richards, K. F.; Lamb, J. *Trans. Am. Soc. Artif. Intern. Organs* **1974**, *20*, 474–479.

128. Yasuda, H.; Gazicki, M. *Biomaterials* **1982**, *3*, 68–77.

129. Kambic, H. E.; Murabayashi, S.; Nose, Y. *Biocompatible Polymers, Metals, and Composites*; Szycher, M., Ed.; Technomic: Lancaster, PA, 1983; pp 179–198.

130. Jozefowicz, M.; Josefonovicz, J. *Pure Appl. Chem.* **1984**, *56*, 1335–1344.

131. Hergenrother, R. W.; Wabers, H.; Cooper, S. L. *Transactions of the 17th Annual Meeting of the Society for Biomaterials*; Society for Biomaterials: Scottsdale, AZ, 1991; p 298.

132. Okkema, A. Z.; Grasel, T. G.; Zdrahala, R. J.; Solomon, D. D.; Cooper, S. L. *J. Biomater. Sci., Polym. Ed.* **1989**, *1*, 43–62.

133. Boretos, J. W.; Detmer, D. E.; Donachy, J. H. *J. Biomed. Mater. Res.* **1971**, *5*, 373–387.

134. Boretos, J. W. *J. Biomed. Mater. Res.* **1972**, *6*, 473–476.

135. Kontron Cardiovascular Inc.; Cardiothane 51 (formerly Avcothane 51), Cardiomat 610 (formerly Avcomat 610), and Cardiomat 40 (Avcomat 40) Polymers; *Technical Data Sheet*; 1983.

136. Boretos, J. W.; Pierce, W. S.; Baier, R. E.; Leroy, A. E.; Donachy, H. J. *J. Biomed. Mater. Res.* **1975**, *9*, 327–340.

137. Ashar, B.; Turcotte, L. R. *Trans. Am. Soc. Artif. Intern. Organs* **1981**, *27*, 372–379.

138. Stokes, M. *Polyurethanes in Biomedical Engineering*; Planck, H.; Egbers, G.; Syre, I., Eds.; Elsevier: Amsterdam, Netherlands, 1984; pp 243–255.

139. Szycher, M.; McArthur, W. A. *Corrosion and Degradation of Implant Materials;* Fraker, A. C.; Griffin, C. D., Eds.; ASTM STP 859; American Society for Testing and Materials: Philadelphia, PA, **1985;** pp 308–321.

140. Parins, D. J.; McCoy, K. D.; Horvath, N.; Olson, R. W. *Corrosion and Degradation of Implant Materials:* Fraker, A. C.; Griffin, C. D., Eds,; ASTM STP 859; American Society for Testing and Materials: Philadelphia, PA, **1985;** pp 322–339.

141. Stokes, K.; Cobian, K. *Biomaterials* **1982,** *3,* 225–231.

142. Angeline, B. D.; Hiltner, A.; Anderson, J. M. *Degradable Materials;* Barenberg, N., Ed.; CRC: Boca Raton, FL, **1991;** 295–312.

143. Duncan, E. *Biomater. Forum* **1990,** *12,* 4.

144. Martz, H.; Paynter, R.; Forest, J. C.; Guidoin, R. *Biomaterials* **1987,** *8,* 3–11.

145. Therrien, M.; Guidoin, R.; Adnot, A.; Paynter, R. *Biomaterials* **1989,** *10,* 517–520.

146. Hayashi, K.; Matsuda, T.; Takano, H.; Umezu, M.; Taenaka, Y.; Nakamura, T. *Biomaterials* **1985,** 82–88.

147. Matsuda, T.; Takano, H.; Hayashi, K.; Taenaka, Y.; Takaichi, K.; Umezu, M.; Nakamura, T.; Iwata, H.; Nakatani, T.; Tanaka, T.; Takatani, S.; Akutsu, T. *Trans. Am. Soc. Artif. Intern. Organs* **1984,** 353–358.

148. Miles Inc. *Plast. Eng.* **1994,** *50,* 38. (Polymers Marketing Communications Dept., Miles Inc., Mobay R., Pittsburgh, PA 15205-9731.)

149. Farrar, D. J.; Litwak, P.; Lawson, J. H.; Ward, R. S.; White, K. A.; Robinson, A. J.; Rodvien, R.; Hill, J. D. *Thorac. Cardiovasc. Surg.* **1988,** *95,* 191–200.

150. Imachi, K.; Fujimasa, I.; Miyake, H.; Takido, N.; Nakajima, M.; Kauno, A.; Ono, T.; Mori, Y.; Nagaoka, S.; Kewase, S.; Kikuchi, T.; Atsuma, K.; *Trans. Am. Soc. Artif. Intern. Organs* **1981,** *27,* 118–123.

151. Arkles, B. C. *Med. Device Diagn. Ind.* **1983,** *6,* 66–67.

152. Lyman, D. J.; Fazzio, F. J. *U.S. Patent* 4 173 689, 1979.

153. Heller, J.; Brauman, S. K. "Development of Materials for Circulatory Assist Devices;" Contract No. 1–HV–3–2959, NTIS, SRI Report, 1979.

154. Lipatova, T. E.; Phhakadze, G. A.; Vsilchenko, D. V.; Vorona, V. V.; Shilov, V. V. *Biomaterials* **1983,** *4,* 201–204.

155. Lipatova, T. E.; Pkhakadze, G. A.; Snegirev, A. I.; Vorona, V. V.; Shilov, V. V. *J. Biomed. Mater. Res.* **1984,** *18,* 129–136.

156. Sosa, K.; Mori, A.; Sisiodo, M.; Imanishi, Y. *Biomaterials* **1985,** *6,* 312–324.

157. Mori, A.; Imanishi, Y.; Ito, T.; Sakaoku, K. *Biomaterials* **1985,** *6,* 325–337.

158. Gogolewski, S.; Pennings, A. J. *Makromol. Chem., Rapid Commun.* **1982,** *3,* 839–845.

159. Bruin, P.; Veenstra, G. J.; Nijenhuis, Pennings, A. J. *Makromol. Chem., Rapid Commun.* **1988,** *9,* 589–594.

160. Storey, R. F.; Wiggins, J. S.; Mauritz K. A. *Polym. Compos.* **1993,** *14,* 17–25.

161. Landfield, H. *Biocompatible Polymers, Metals, and Composites;* Szycher, M., Ed.; Technomic; Lancaster, PA, **1983;** pp 975–999.

162. Ratner, B. D.; Gladhill, K. W.; Horbett, T. A. *J. Biomed. Mater. Res.* **1988,** *22,* 509–527.

163. Phua, S. K.; Castillo, E.; Anderson, J. M.; Hiltner, A. *J. Biomed. Mater. Res.* **1987,** *21,* 231–246.

164. Smith, R.; Williams, D. F.; Oliver, C. *J. Biomed. Mater. Res.* **1987,** *21,* 1149–1166.

165. Lipatova, T. E.; Vesolovski, R. A. *Vysokomol. Soedin* **1969,** *A11,* 1459–1464.

166. Lipatova, T. E.; Loos, S. M.; Momuzhai, M. M. *Vysokomol. Soedin* **1970,** *A12,* 2051–2056.

167. Kovalev, M. M.; Tereschenko, T. L.; Suslow, E. I.; Pchakadze, G. A. *Polym. Med.* **1974,** *4,* 277–283.

168. Lipatova, T. E.; Novikova, O. E.; Pchakadze, G. A.; Novikova, T. *J. Polym. Med.* **1974,** *4,* 313–321.

169. Lipatova, T. E. *J. Appl. Polym. Sci., Polym. Symp.* **1979,** *66,* 239–257.

170. Takahara, A.; Hergenrother, R. W.; Coury, A. J.; Cooper, S. L. *J. Biomed. Mater. Res.* **1992,** *26,* 801–818.

171. Tyler, B. J.; Ratner, B. D. *J. Biomed. Mater. Res.* **1993,** *27,* 327–334.

172. Meijs, G. F.; McCarthy, S.; Rizzardo, E.; Chen, Y. Ch.; Chatelier, R. C. *J. Biomed. Mater. Res.* **1993,** *27,* 345–356.

173. Santerre, J. P.; Labow, R. S.; Duguay, D. G.; Erfle, D.; Adams, G. A. *J. Biomed. Mater. Res.* **1994,** *28,* 1187–1199.

174. Gogolewski, S. *In Vitro and In Vivo Molecular Stability of Medical Polyurethanes: Review, Trends in Polymer Sciences;* Menon., Ed.; Council Science Integration: Trivandrum, India, 1991; Vol. 1, pp 47–56.

175. Autian, J.; Singh, A. R.; Turner, J. E.; Hung, G. W. C.; Nunez, L. J.; Laurence, W. H. *Cancer Res.* **1975,** *35,* 1591–1596.

176. Darby, T. D.; Johnson, H. J.; Northup, S. J. *Toxicol. Appl. Pharmacol.* **1978,** *46,* 449–453.

177. Hueper, W. C. *Am. J. Clin. Pathol.* **1960,** *34,* 328–333.

178. Hueper, W. C. *J. Natl. Cancer Inst.* **1964,** *33,* 1005–1027.

179. Rigdon, R. H. *South. Med. J.* **1974,** *67,* 1459–1463.

180. Gogolewski, S. *Life Support Syst.* **1987,** *5,* 41–46.

Homopolymers and Copolymers of 2-Hydroxyethyl Methacrylate for Biomedical Applications

Jean-Pierre Monthéard, Jaroslav Kahovec, and Daniel Chappard

This chapter presents a general review of the chemistry of 2-hydroxyethyl methacrylate (HEMA) (a commercially available monomer), its polymerization and copolymerization, chemical transformations of the resulting homopolymer (PHEMA) and cross-linked PHEMA gels, and various biomedical applications of these polymers. The main property of PHEMA is its swelling in aqueous media giving hydrogels with numerous applications in the biomedical field. The most important uses of HEMA polymers and copolymers such as histological embedding, dental cements, immobilization of cells and enzymes, controlled release of drugs, prostheses, and optical lenses are summarized briefly.

The chemistry of biomedical polymers has received a great deal of attention in recent years due to the numerous applications of the polymer. Among the macromolecules used in the biomedical field poly(2-hydroxyethyl methacrylate) (PHEMA) is one of the most important for the following reasons.

1. PHEMA is a nontoxic, biocompatible polymer, and hence it is of potential interest for use in biomedical technologies. The pioneering work of Wichterle (1) showed that PHEMA gives hydrogels. This discovery opened the way to many new developments in the technology of soft optical lenses.
2. The monomer, 2-hydroxyethyl methacrylate (HEMA) can be easily polymerized or copolymerized (like the majority of methacrylic derivatives), and its primary alcohol function allows the formation of new HEMA derivatives leaving the double bond unchanged. These new monomers are generally able to be polymerized and copolymerized.
3. Similar substitutions are also possible with PHEMA to provide a wide range of PHEMA derivatives for various biomedical applications.

Therefore, PHEMA and its copolymers have many uses as materials for immobilization of cells, drugs, and enzymes; contact lenses and numerous ophthalmic devices; dentistry (root canal fillers and soft denture liners); prostheses (ureter prostheses, ligament prostheses, etc.); and for hemocompatible materials.

All these properties and applications explain why the literature contains thousands of articles and patents dealing with HEMA, PHEMA, and related copolymers. This chapter gives an overview of the syntheses and purification of HEMA, its polymerization and copolymerization, chemical modification of the monomer and copolymer, and applications in the biomedical field. Several books and reviews were published in the general area of biopolymers and should be referred to for further information (2–5).

Monomer Syntheses

Because HEMA is now a commercial product with a relatively low cost, its method of synthesis will be cov-

$$CH_2=C(CH_3)COOCH_3 + HOCH_2CH_2OH \longrightarrow CH_2=C(CH_3)COOCH_2CH_2OH + CH_3OH$$

HEMA

$$CH_2=C(CH_3)COOCH_2CH_2OH + CH_2=C(CH_3)COOCH_3 \longrightarrow$$

$$CH_2=C(CH_3)COOCH_2CH_2OCOC(CH_3)=CH_2$$

EDMA

$$CH_2=C(CH_3)COOCH_3 + CH_2 - CH_2 \longrightarrow CH_2=C(CH_3)COOCH_2CH_2OH$$

O

Scheme I. Preparations of HEMA by different methods.

ered only briefly. HEMA can be prepared in a single step by transesterification of methyl methacrylate (MMA) with ethylene glycol (6) (Scheme I). During this reaction a second transesterification occurs providing a divinyl monomer, ethylene glycol dimethacrylate (EDMA), and a low percentage of methacrylic acid resulting from the hydrolysis of HEMA. Recently, a Japanese patent described (7) a preparation of HEMA in the presence of 1-naphthol-2-carboxylic acid, which prevents the formation of EDMA.

The reaction of methacrylic acid with ethylene oxide is also a common method used to prepare HEMA (Scheme I) (8). In this method, too, EDMA and methacrylic acid are found in HEMA as low percentage impurities. Other methods also were described (9), but they are less common today. Recent patents described methods of preparation of HEMA by using the two previous reactions, and the yields and the purity were slightly improved. Because many applications need pure HEMA, we present a survey of the purification methods.

Purification of HEMA is based on the solubility of the monomer in water or in diethyl ether and its insolubility in petroleum ether or hexane (EDMA is soluble in hexane). Preliminary assessment of the purity of commercial HEMA was made by thin layer chromatography (TLC) by using n-hexane-diethylether (1/1, v/v) as the solvent, silica gel as the absorbent, and n-hexane as the counter-current solvent. EDMA was removed by liquid–liquid continuous extraction of an aqueous solution of HEMA (water–HEMA, 1/4, v/v) containing 0.3% w/v t-butylcatechol as the polymerization inhibitor. However, a small amount of thermal polymerization was noted during prolonged extractions. The thermal polymerization was overcome by circulating cold water around the extraction vessel. Extraction for about 14–15 h generally sufficed to remove EDMA as evidenced by monitoring with TLC. HEMA was then recovered from aqueous solution by salting out with NaCl, after which the separated HEMA was diluted again with 10 volumes of water

and the salting out was repeated. This process ensured the maximum possible extraction of EDMA.

Methacrylic acid was removed by dissolving the HEMA in aqueous sodium hydrogenocarbonate followed after 90 min by a salting-out procedure. The HEMA was extracted with diethylether, the solution was dried, and the monomer was distilled under vacuum. A further purification was made by column chromatography by using silica gel (60–120 mesh) as the adsorbent and diethyl ether as the eluant (10).

Methacrylic acid also can be removed by soaking technical grade HEMA with 15% by weight of anhydrous sodium carbonate for 3 h at 24 °C, then vacuum filtering through No. 50 Whatman filter paper. EDMA is then removed by repeated extractions with hexane (11) after dissolving HEMA in distilled water. The aqueous solution is salted with NaCl then extracted as in the previous procedure. Purification of HEMA with an ion-exchange resin (Amberlyst A21) seems like a simple method for elimination of methacrylic acid (12), but the yields are relatively low. N,N'-Dicyclohexylcarbodiimide (DCCI) also is used to condense methacrylic acid with HEMA to form EDMA, but variations in the quality of the reagent often outweigh the value of the method (13). HEMA is a colorless liquid; its boiling points (°C/mmHg) are as follows: 85/5, 64/0.5, 69/0.1. The ^{1}H NMR spectrum is given in reference 14.

Polymerization of HEMA

HEMA can be polymerized by means of free-radical initiators like many methacrylic derivatives. Anionic initiators also were used to prepare isotactic PHEMA via the benzoyl esters of HEMA (15). The radical polymerization is strongly exothermic ($\Delta H_p = 50$ kJ/mol.) and should be carried out in a medium that removes the heat of reaction. The early studies dealt with the polymerization of HEMA containing minor proportions of methacrylic acid and EDMA. For example, by using benzoyl peroxide as the initiator, HEMA was polymerized in bulk (*16*). The resulting product swelled in water after a lengthy exposure, and the amount of water in the polymer was 37.5%. Polymerization of HEMA also was carried out with a redox system (ammonium persulfate and sodium metabisulfite) producing a hydrogel (*16*). The formation of hydrophilic gels of PHEMA was studied (*17*) by polymerization of HEMA in organic solvents with very low concentrations of EDMA (0.31%) in the presence of diisopropylperoxycarbonate as the initiator. Azobisisobutyronitrile (AIBN) also was used as an initiator for the polymerization of HEMA (*6*) containing a small amount of EDMA, and the kinetics of the polymerization were studied by dilatometry.

Emulsion polymerization of HEMA has been studied relatively little due to its solubility in water. However, because one of the application of PHEMA is devoted to the release of drugs, suspension polymerization giving microspheres of PHEMA has been studied more often. A recent example is given (*18*): for producing microspheres of PHEMA–EDMA, 45 mL of a 20% aqueous solution of NaCl was mixed with 2 g of $MgCl_2 \cdot 6H_2O$. The temperature was increased to 90 °C. This solution was supplemented with 5 mL of 4 N NaOH solution until a fine gelatinous precipitate of $Mg(OH)_2$ was formed. The suspension was supplemented with 10 mL of a mixture composed of EDMA–HEMA at different molar ratios (EDMA–HEMA, 0.033, 0.065, 0.13, and 0.26) and 0.1 g of AIBN. The suspension was maintained at 90 °C for 2 h. The product was cooled, and HCl was added until a neutral pH was achieved and an $Mg(OH)_2$ precipitate was dissolved. The microspheres obtained were collected by filtration, washed with water, desiccated to a constant weight, and stored in a desiccator. The microspheres were subsequently sieved, and the fraction between 315 and 420 μm was loaded by soaking in water in the presence of various drugs to study the controlled release.

A number of other procedures of polymerization of HEMA are also known. Microwaves (without any radical initiator) polymerize HEMA and give an insoluble PHEMA (*19*). In this case the starting HEMA was distilled and probably contained EDMA. ^{60}Co γ-irradiation was an effective technique for the synthesis of hydrophilic functional microspheres with 0.3–3 μm diameter (*20*). Plasma-initiated solution polymerization of HEMA in water, ethanol, or dimethylsulfoxide also was reported (*21*). The monomer was exposed to radiofrequency cold plasma for 1 min followed by postpolymerization at 25 °C.

Solutions of HEMA, water, and biological substances such as amino acids, enzymes, or microbial cells were polymerized with ^{60}Co γ-irradiation at various temperatures (*22*). Electropolymerization of HEMA on carbon fibers was studied to provide submicrome-

ter coatings of PHEMA (*23*). HEMA also was photopolymerized in the presence of sensitizers in aqueous solution (*24*). HEMA was grafted onto poly(vinyl alcohol) (*25*) by using Ce^{4+} ions as initiators and onto hydroxypropylcellulose by photoinitiation (*26*). Radiation-induced grafting onto polyethylene membranes (*27*) or silicone rubber filters (*28*) also was studied.

Profiles of reaction rate as a function of time were obtained for free-radical copolymerization of HEMA and EDMA (or other dimethacrylates) by using differential scanning calorimetry (Figure 1) (*29*). Copolymerization of HEMA with small amounts of dimethacrylate cross-linking agents was studied in addition to homopolymerization of dimethacrylates. In the copolymerization case, the effects of the cross-linking agent, initiator, solvent, and pendant glycol chain length were investigated. Typically, a sharp increase in the reaction rate was observed and was attributed to autoacceleration. The time period before the onset of autoacceleration increased and the maximum reaction rate decreased as the reaction solvent or the pendant chain was increased or the cross-linking agent concentration was decreased. These results were explained by an increase in the mobility of the growing polymer leading to a decrease in the magnitude of the gel effect (*29*).

Characterization of PHEMA

Numerous studies have been devoted to the physical properties of PHEMA. However, because the biomedical applications generally need a cross-linked

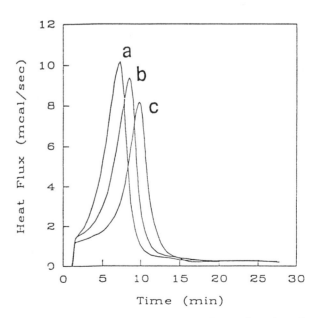

Figure 1. Heat flux as a function of reaction time for HEMA illustrating the effect of EDMA concentration with 0.5 wt% azobisisobutyronitrile: a, 2 mol% EDMA; b, 1 mol% EDMA; c, 0 mol% EDMA.

PHEMA, only few studies deal with the properties of soluble PHEMA, which can be prepared from pure HEMA or from HEMA containing a very low percentage of EDMA. Uncross-linked polymer is insoluble in water but soluble in dimethylsulfoxide, dimethylformamide, alcohols, or diethylene glycol (1). Molecular weights of PHEMA can be measured (30) by viscometry (Mark–Houwink parameters are known) (31), by osmometry, and by light scattering. The cross-linked polymer swells in water, and the degree of swelling (ratio of bulk swollen volume to dry volume) shows that PHEMA swells better in polar organic solvents such as alcohols or ethylene glycol (5). A ^{13}C NMR study of the cross-linked gels of PHEMA in water showed that carbon resonances are the same as those of the uncross-linked polymer in solution (32).

To remove residual monomers and oligomers from PHEMA gels obtained by the cross-linking polymerization, the gels are washed with distilled water (33). Thermal properties of PHEMA hydrogels also were studied. The results indicate that the internal structure of the hydrogels can be described as that of an elastic solution in which water molecules are distributed continuously in all possible orientations for interaction with the polymer (34). A correlation was established between the percentage of EDMA and methacrylic acid in the gel and its water uptake. PHEMA at cross-linker content >0.15 mol% does not swell above an equilibrium value of 39–42% (35). The permeability of PHEMA hydrogels (containing 38% of water in weight) when used as a membrane for oxygen was compared to other polymers such as a polyelectrolyte complex or poly(dimethylsiloxane) (36). The diffusion of water through low cross-linked PHEMA hydrogels was studied in connection with their applications in the biomedical field (37). Mechanical and viscoelastic properties of PHEMA were summarized in two previous reviews (38, 39), and the influences of the percentage of EDMA on the tensile strength and strain at break are given in Table I.

Table I. Effect of Cross-Linker Concentration During Copolymerization of HEMA with EDMA upon Tensile Strength (σ_b) and Strain (ε_b) at Break

EDMA Concentration $10^4 (mole \times cm^{-3})$	Swelling Ratio	Strain to Break			
		$\sigma_b (kg \times cm^{-2})$		$\varepsilon_b (\%)$	
		5 °C	25 °C	10 mm/s	1 mm/s
0.0855	0.525	7.4	3.4	480	480
0.225	0.528	6.6	4.1	390	300
0.435	0.533	7.8	2.8	305	130
0.845	0.544	10.3	4.1	225	110
1.570	0.559	12.0	3.2	160	50
3.000	0.583	35.5	7.4	120	60
5.650	0.696	49.0	18.6	55	50

Source: Reproduced with permission from reference 38. Copyright 1973 Marcel Dekker.

Copolymerization of HEMA

HEMA has been copolymerized mainly with acrylic and methacrylic derivatives by using various initiators and solvents. The majority of these works give the reactivity ratios of HEMA (M_1) and the monomers (M_2). The Alfrey–Price parameters Q and e of HEMA are, respectively, 1.75 and –0.39 (40). The values of the reactivity ratios r_1 and r_2 are given in Table II.

The procedures of the copolymerization are very often similar to those of the polymerization (bulk, solution, or suspension), and the resulting copolymers are soluble when pure HEMA is used. Emulsion polymerization of HEMA with styrene as a comonomer was studied (49, 50). An example of a recipe of emulsion terpolymerization of HEMA, ethyl acrylate (EA), and MMA to produce microcapsules is given in the subsequent paragraph (51).

A monomer mixture (433 g in total) was used in each polymerization. The mole ratio of EA, MMA, and HEMA was 12/6/9. The monomer mixture (150 g) was poured in distilled water (1300 g) containing sodium dodecyl sulfate (4 g) and emulsified in a mixer. The

Table II. Examples of Copolymerization of HEMA (M_1) with Other Monomer (M_2) and Values of Their Reactivity Ratios (r_1 and r_2)

M_2	Conditions (solvent, initiator)	r_1	r_2	Ref.
Vinyl acetate	H_2O, Mn [(acetylacetonate)]$_3$	—	—	41
Methacrolein	DMF, AIBN	0.36	0.77	42
Methacrylic acid	^{60}Co γ-radiation bulk	0.32	1.02	43
Acrylonitrile	DMF, AIBN	0.63	0.02	44
Methyl acrylate	Bulk, BP	7.14	0.01	45
Furfuryl acrylate	DMF, AIBN	1.42	0.92	46
Methyl methacrylate	Bulk, BP	1.05	0.30	45
4-Methacryloyloxyacetanilide	DMF, AIBN	0.90	2.15	47
Styrene	Toluene, AIBN	1.65	0.50	48

Notes: — is value not given. AIBN is azobisisobutyronitrile, BP is benzoylperoxide.

emulsion was kept at 80 °C in the presence of nitrogen, and 1 mL of aqueous ammonium peroxodisulfate (2 g/30 mL) was introduced to initiate polymerization. The remainder of the monomer mixture was slowly dropped into the reactor over 3 h. The reaction was further continued for 2 h. During the 5-h reaction, 1 mL of ammonium peroxodisulfate was introduced every 30 min.

The reaction product was passed through an 80-mesh sieve to remove coarse solid masses and used for coating. A known weight of dispersion was dried at 110 °C to constant weight. The yield was almost 100% as total polymer and 88–97% as dispersed polymer. The solid content of the dispersion was 21–23%.

Chemical Modifications of HEMA

Most chemical modifications of HEMA have been carried out on the hydroxy function so as to leave the double bond unchanged. Generally the resulting monomers can be polymerized or copolymerized giving functionalized macromolecules. Modification of HEMA can rise to various extents of polymerization if reaction conditions are not carefully controlled (9). Table III gives some examples of the most recent reactions of chemical modification of HEMA.

Earlier literature on the following reactions of HEMA was reviewed previously (9):

- synthesis of ether by reaction of HEMA with hexamethylolmelamine to prepare nonscratchable polymer coatings
- preparation of esters by reaction of HEMA with acid chlorides or anhydride (e.g., phthalic anhydride) to incorporate in polymeric adhesives
- preparation of carbamates by reaction of HEMA with various diisocyanates (e.g., tolylene diisocyanate) used in adhesives

Numerous applications other than biomedical applications were found, and the most recent examples are as follows:

- HEMA with phosphorous substituent to impart fire retardancy to polymeric systems (54)
- Photo-cross-linkable monomer for nonlinear optics synthesized by reaction of HEMA with cinnamoyl chloride (55)
- HEMA ester with 4-(*N,N*-dimethylamino)benzoic acid or 4-azidobenzoic acid providing a new polymeric dye for nonlinear optics (57, 58)
- HEMA containing 1,4-benzodioxane group used as flocculant for titanium minerals (59)

Chemical Modifications of HEMA Polymers

Most chemical modifications described here were performed on cross-linked polymers and copolymers of HEMA. The reason is practical: for various applications of these polymers (such as ion exchangers, reactive polymers, and polymer carriers) the polymers must be insoluble in the working medium. This requirement is why cross-linked polymers of HEMA (i.e., copolymers of HEMA with a cross-linker), both gel type and porous, are used. However, it can be safely assumed that the same reactions occur with homopolymers of HEMA in solution. Moreover, the modifications of PHEMA performed in homogeneous solutions may show higher conversions but lower yields (cross-linked polymers are easier to recover).

Polymers and copolymers of HEMA with EDMA (both lightly and highly cross-linked) are commercially available today, and due to their hydrophilicity, they have numerous applications in biomedical fields as sorbents and ion-exchangers. Because these copolymers contain primary alcohol function, many new functional polymers may be synthesized by chemical

Table III. Examples of Chemical Modifications of HEMA According to the Following Scheme: $CH_2=C(CH_3)COOCH_2CH_2OH + reagent \rightarrow CH_2=C(CH_3)COOCH_2CH_2OR$

Reagent	R	Yield (%)	Ref.
Hexamethyldisilazane	$-Si(CH_3)_3$	72	52
Diketene	$-COCH_2COCH_3$	—	53
Diethyl chlorophosphate	$-PO(OC_2H_5)$	48	54
Cinnamoyl chloride	$-COCH=CH-C_6H_5$	73	55
2,3,6-Trichlorophenyl isocyanate	$-CONHCOCH_2C_6H_2Cl_3$	92	56
N,N-Dimethylaminobenzoic acid	$-COC_6H_4N(CH_3)_2$	68	57
3,4-Ethylenedioxylbenzoyl chloride	$-COC_6H_4$ (benzodioxane ring with two O and $(CH_2)_2$)	83	58

Note: — is value not given.

transformations of the hydroxy polymers. Both practical performance and conversions of the modification reactions on HEMA polymers differ in dependence on what type of polymers is used: soluble (noncrosslinked), gel-type, or porous. Modifications of soluble polymers are carried out in an appropriate solvent such as acetone, chloroform, or pyridine. The same modification applies to the gel-type polymers where the solvents act as good swelling agents. In both cases, the conversions are generally high and converge to those obtained in the corresponding low molecular weight organic reactions.

The choice of reaction medium is not critical with porous polymers. Here, however, the accessibility of hydroxy groups determines the attainable degree of conversion in chemical transformations and plays a significant role. Generally, a substantial portion of the hydroxy groups present are not accessible to a chemical reagent, because they are buried in the heavily cross-linked mass of microparticles. The recovery of modified polymers is easy with the cross-linked polymers by removing soluble substances by washing with an appropriate solvent. Modified soluble polymers are recovered by precipitation into a nonsolvent. Some examples of chemical modifications are now given for cross-linked HEMA–EDMA resins containing 25% and 40% of EDMA. (HEMA polymers are generally represented by $\circledP$, and only functional groups of interest are shown, that is, $\circledP$–OH or $\circledP$–CH_2CH_2OH.)

Esters of PHEMA with inorganic and sulfonic acids are readily available by reaction of the hydroxy copolymers with the corresponding acid chlorides. Thus, with chlorosulfonic acid, phosphorus oxychloride, and aromatic sulfonylchlorides, the corresponding polymeric alkylsulfuric acid (60), monoalkyl phosphate (after hydrolysis) (61), and arenesulfonate (62), respectively, are obtained (Scheme II). Halogen derivatives of PHEMA can be obtained by reaction with hydrohalogenic acids (e.g., HCl or HBr) or with halogenating agents (e.g., $SOCl_2$ or $SOBr_2$) (63).

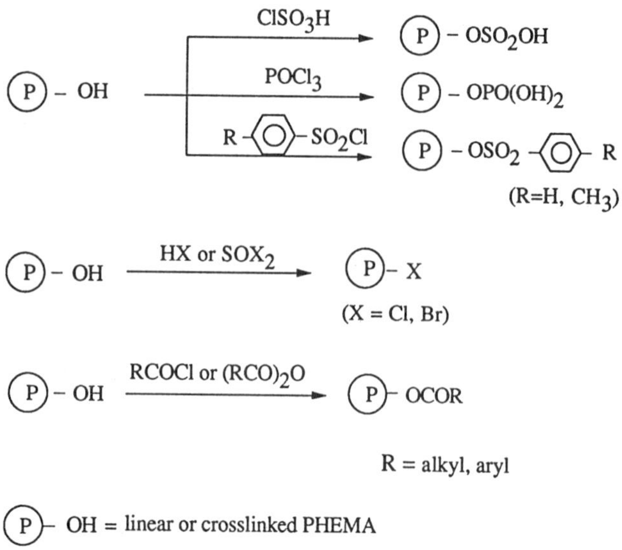

Scheme II. Modification of PHEMA with inorganic, sulfonic acid, hydracids, and halogenating agents.

Carboxylic esters are obtained by reaction with the corresponding acid chlorides or anhydrides, especially in pyridine as the medium (Scheme II). Thus, acetate, butyrate, caprylate, laurate, stearate, 4-nitro and 3,5-dinitrobenzoate (64), and 3-acetylamino-2,4,6-triiodobenzoate (65) were obtained from the corresponding acid chlorides. The succinate (Scheme II; R = CH_2CH_2COOH) similarly was prepared by reaction with succinic anhydride (66). The methoxycarbonyl derivative of mixed carbonic acid ester (R = OCH_3) was obtained by reaction with methyl chloroformate (67). PHEMA-based chelating agents can be prepared by reaction of the polymer with ethylenediaminetetraacetic acid anhydride (68). An activated carbonate of HEMA–EDMA and N-hydroxy-5-norbornene-2,3-dicarboximide also is described that may be used for the immobilization of amino acids and enzymes (69).

Sulfur derivatives of HEMA–EDMA were prepared by the reactions shown in Scheme III (60, 70,

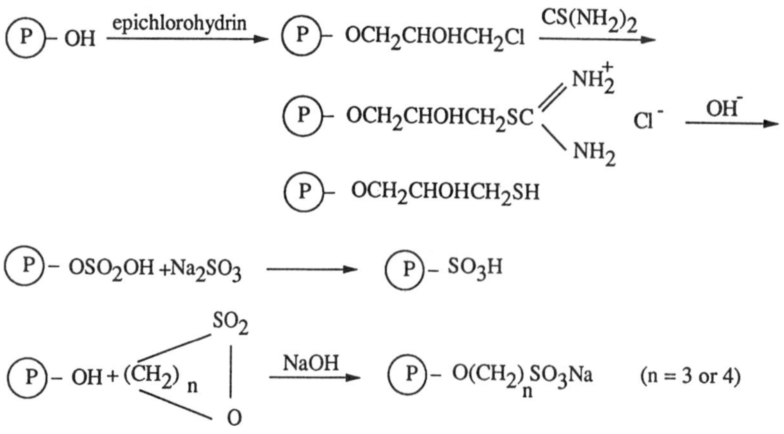

Scheme III. Formation of sulfur derivatives.

71). The polymeric alkylsulfuric acid, Ⓟ–OSO₂OH, is an alkylating reagent, and hence it can also be used for the preparation of sulfonic acid *(72).* (NOTE: Propane and butanesulfone are carcinogenic.)

Carboxylic acids and esters of HEMA–EDMA can be obtained by oxidation of the CH₂OH groups with potassium permanganate in an acid medium *(66)* (Scheme IV). Alternative methods for the preparation of carboxylic acid derivatives of PHEMA are carboxymethylation of hydroxy groups with chloroacetic acid or acylation with dicarboxylic acid anhydrides (e.g., maleic anhydride or phthalic anhydride) as shown in Scheme IV *(66).* From these carboxylic acid derivatives, esters may be prepared by acid-catalyzed esterification *(72),* by a reaction with diazomethane *(73),* or by condensation with an alcohol in the presence of DCCI *(74).* By the DCCI method, active esters of *N*-hydroxysuccinimide and of substituted phenols were obtained *(74)* (Scheme IV).

Aldehyde derivatives of the polymers may be obtained by mild oxidation of primary alcohol groups by selective reagents, such as dimethylsulfoxide or pyridine-chromium trioxide. However, the conversions to aldehyde groups are very low *(75)* (Scheme V).

Aliphatic amino derivatives of PHEMA are most often prepared by reaction of an alkylating derivative of HEMA–EDMA such as halide, tosylate, or sulfate with ammonia or amine *(68)* (Scheme V). The 2,3-epoxypropyl derivative of HEMA–EDMA is particularly convenient for this preparation. By using the last derivative, HEMA–EDMA polymers carrying ethylamine, ethylenediamine, tetramethylenediamine and hexamethylenediamine, ethanolamine, 6-aminocaproic acid, imidazole, L-lysine, and peptide groups *(76–78)* as well as branched polyethylenimine *(68)*

were prepared. Immobilized amines can also be prepared by reaction of the amine with HEMA–EDMA after activation of hydroxy groups with cyanogen bromide *(79)* or *N,N'*-carbonyldiimidazole *(80).* In all of these modifications, reaction of diamines with the polymer leads to extensive cross-linking, and the amine content of the polymer is generally lower than what may be expected theoretically depending on various details of the preparation. Similarly, in alkylation reactions, a mixture of primary, secondary, and tertiary amines may be produced depending on the amine used. If the primary amine (Ⓟ–NH₂) without secondary and tertiary amino group contaminants is required, reduction of the corresponding azide derivative or the Delepine reaction (i.e., alkylation of hexamethylenetetramine and subsequent hydrolysis of the resulting hexamethylenetetraminium salt) is advantageous *(68)* (Scheme V). The 2-(diethylamino)ethyl derivative is prepared by reaction of PHEMA with 2-(diethylamino)ethyl chloride *(81).*

Aromatic amino groups can be introduced into the polymer in three ways. The first method is analogous to the functionalization of cellulose fibers by using reactive dyes *(82)* (Scheme VI). The conversion of hydroxy groups is rather low; the greater part of the reagent is hydrolyzed in the alkaline medium. According to the second method, a polymeric carboxylic acid is condensed with an aromatic diamine (e.g., benzidine) in the presence of *N,N'*-dicyclohexylcarbodiimide (Scheme VI). Diazotization of these amines and azo coupling with various reagents (e.g., phenols, aromatic amines, or other π-electron systems) leads to immobilized analytical reagents bearing *o*-hydroxyazoformazane and other moieties *(82, 83)* (Scheme VI). Another method consists of the reaction of PHEMA with 4-nitrobenzenesulfonyl isocyanate and reduction of the resulting sulfonylcarbamate to the amine (Scheme VI). Treatment of

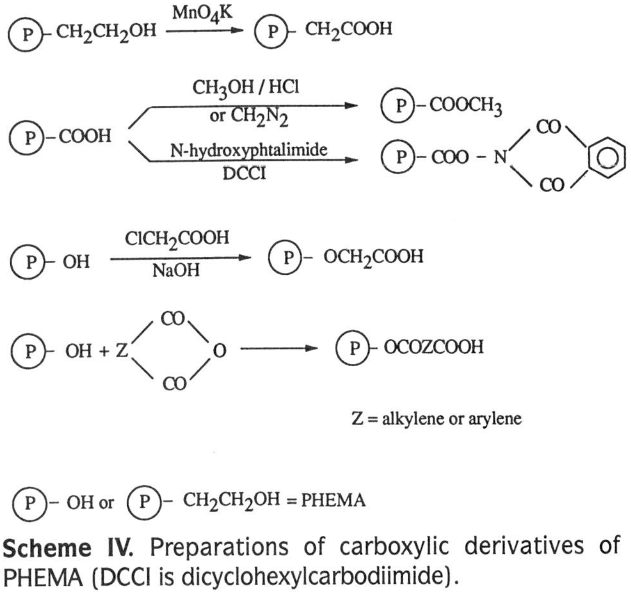

Scheme IV. Preparations of carboxylic derivatives of PHEMA (DCCI is dicyclohexylcarbodiimide).

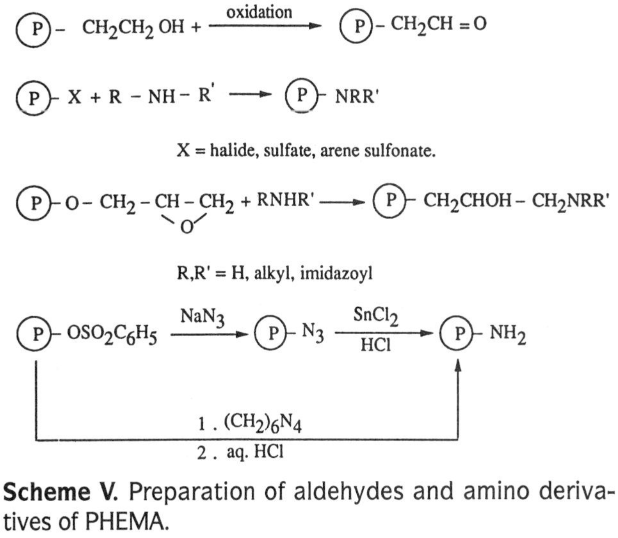

Scheme V. Preparation of aldehydes and amino derivatives of PHEMA.

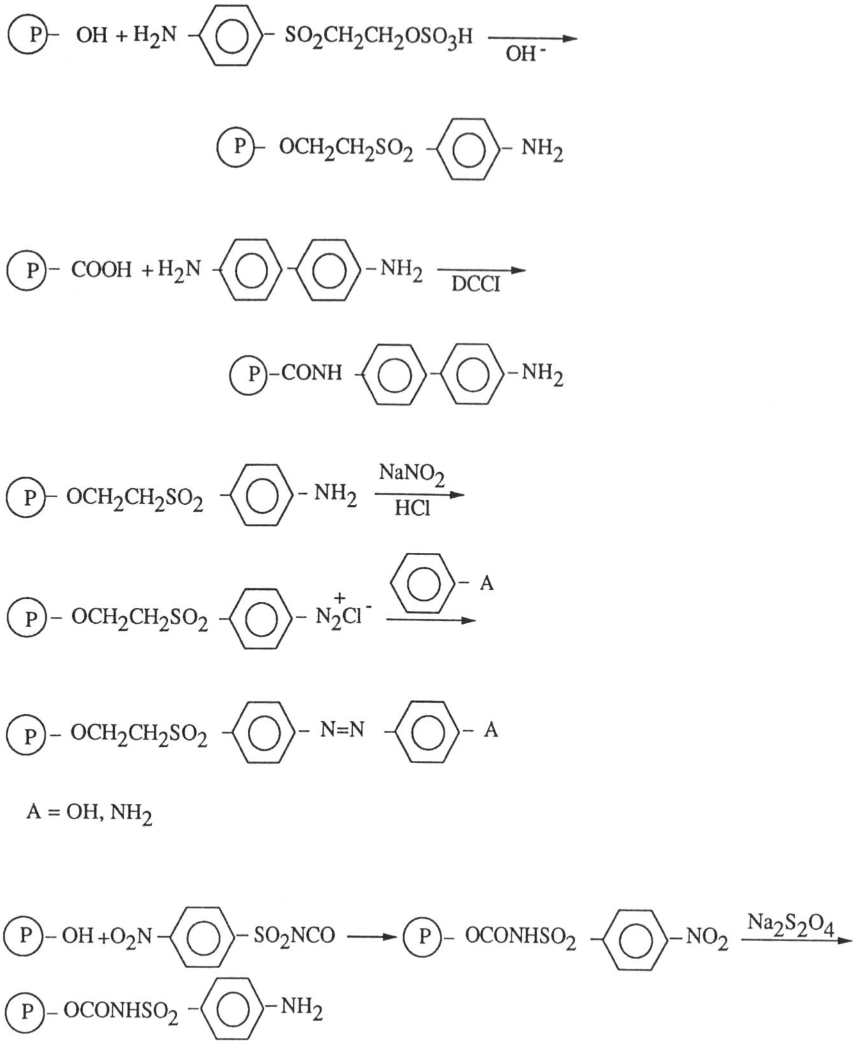

(P)– OH + H₂N –⟨O⟩– SO₂CH₂CH₂OSO₃H →(OH⁻)

(P)– OCH₂CH₂SO₂ –⟨O⟩– NH₂

(P)– COOH + H₂N –⟨O⟩–⟨O⟩– NH₂ →(DCCI)

(P)–CONH –⟨O⟩–⟨O⟩–NH₂

(P)– OCH₂CH₂SO₂ –⟨O⟩– NH₂ →(NaNO₂ / HCl)

(P)– OCH₂CH₂SO₂ –⟨O⟩– N₂⁺Cl⁻   ⟨O⟩– A →

(P)– OCH₂CH₂SO₂ –⟨O⟩– N=N –⟨O⟩– A

A = OH, NH₂

(P)– OH + O₂N –⟨O⟩– SO₂NCO ⟶ (P)– OCONHSO₂ –⟨O⟩–NO₂ →(Na₂S₂O₄)

(P)– OCONHSO₂ –⟨O⟩–NH₂

(DCCI = Dicyclohexylcarbodiimide)

Scheme VI. Preparation of aromatic amino derivatives of PHEMA.

the amine with glutaraldehyde produced supports capable of binding trypsin (*84*).

Other nitrogen derivatives such as hydrazide derivative of HEMA–EDMA may be obtained by ester hydrazinolysis (*72, 73*) (Scheme VII). Polymer hydrazide may also be obtained by direct hydrazinolysis of HEMA–EDMA, in which case hydrazide units are directly attached to the polymer backbone (*67*). Hydrazinolysis of the polymeric methoxycarbonyl derivative leads to half-carbohydrazide (–OCONHNH₂) group (*67*) (Scheme VII).

An important derivative of HEMA–EDMA is 2,3-epoxypropyl or glycidyl ether, which is easily obtainable by a reaction of the basic polymer with epichlorohydrin (*76*) (Scheme VII). The epoxy polymer is also produced by direct copolymerization of glycidyl methacrylate and EDMA (*85*). The epoxide ring is readily opened by various nucleophiles (AH) (Scheme VII), and hence this derivative is useful as a reactive polymer in the preparation of other derivatives of HEMA–EDMA (*60, 76–78, 86*). In contrast, the acid-catalyzed (BF₃/Et₂O) reaction of HEMA–EDMA with epichlorohydrin leads to the 3-chloro-2-hydroxypropyl derivative, (P)–OCH₂CH(OH)CH₂Cl (*70*) (Scheme III). HEMA–EDMA ethers can be prepared by alkylation of the basic polymer with alkyl or aryl halides (*64–81*) (Scheme VII).

Glycosidic derivatives of HEMA–EDMA are readily obtainable by an acid-catalyzed condensation of the polymer with monosaccharides (Scheme VII). The

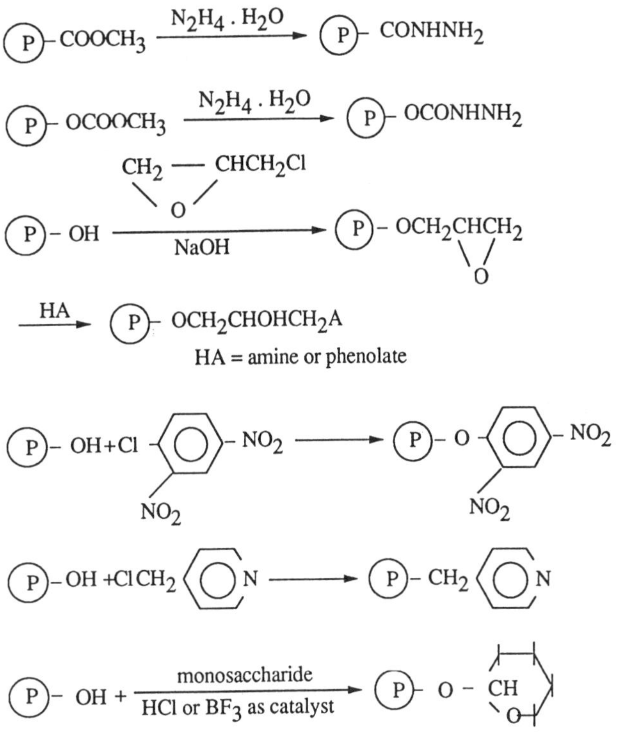

(P)–COOCH$_3$ $\xrightarrow{\text{N}_2\text{H}_4 \cdot \text{H}_2\text{O}}$ (P)–CONHNH$_2$

(P)–OCOOCH$_3$ $\xrightarrow{\text{N}_2\text{H}_4 \cdot \text{H}_2\text{O}}$ (P)–OCONHNH$_2$

(P)–OH $\xrightarrow[\text{NaOH}]{\text{CH}_2\text{—CHCH}_2\text{Cl} \diagdown \text{O} \diagup}$ (P)–OCH$_2$CHCH$_2$

$\xrightarrow{\text{HA}}$ (P)–OCH$_2$CHOHCH$_2$A

HA = amine or phenolate

(P)–OH + Cl—⟨O⟩—NO$_2$ (with NO$_2$) ⟶ (P)–O—⟨O⟩—NO$_2$ (with NO$_2$)

(P)–OH + ClCH$_2$—⟨O⟩N ⟶ (P)–CH$_2$—⟨O⟩N

(P)–OH + $\dfrac{\text{monosaccharide}}{\text{HCl or BF}_3 \text{ as catalyst}}$ ⟶ (P)–O–CH

Scheme VII. Preparation of various nitrogen and other derivatives of PHEMA.

resulting polymers may be used as carriers in affinity chromatography (*87*). Diphenylphosphine moiety was anchored to HEMA–EDMA by the reaction with diphenylchlorophosphine (*88*). The oligo (*N*-acetylimi-noethylene) grafts, (N(COCH$_3$)CH$_2$CH$_2$)$_n$, may be introduced into HEMA–EDMA by the reaction of an alkylating derivative of HEMA–EDMA (bromide, tosy-late, or glycidyl ether) with 2-methyloxazoline. By hydrolysis, a chelating polymer with oligo(iminoethyl-ene) units was obtained (*68*). Acetylsalicylic acid (aspirin) was immobilized on PHEMA, and its anti-thrombogenic activity was evaluated (*89*).

Biomedical Applications

A complete survey of biomedical literature on HEMA appears impossible, so we present some of the most recent biomedical uses of HEMA and PHEMA. We included some of the earlier literature in a recent review (*90*) and in others referenced in the introduc-tory paragraphs. Various biomedical applications of HEMA and PHEMA include the following: irritant and toxic effects, use as a histological embedding medium, use in dentistry, possibility to immobilize molecules (e.g., enzymes or drugs) and cells, biocom-patibility in cell culture systems or as implants or artifi-cial organs, blood compatibility, and realization of ocu-lar lenses.

Irritant and Toxic Effects

The low toxicity of HEMA is generally accepted, but few reports are available on its potent irritant effects. Intradermal injection of crude HEMA monomer at low concentrations in saline solution (~1%) induced a very mild irritation in the rat, whereas higher concen-trations (up to 20%) resulted in pronounced reactions. Similar findings were observed with sodium benzoate (an end-product of degradation of dibenzoyl peroxide often used as initiator in solution polymerization) and emphasized the irritant role of residues (*91*). PHEMA hydrogels implanted in muscle tissue of rats released residual irritants continuously but at a very low rate, and thus no cellular reaction was induced. HEMA used at 0.01–1% altered the fine structure of cultured cells as determined by quantitative video microscopy. On the other hand, numerous clinical trials (discussed subsequently under specific organ description) have found little irritant reactions.

Histological Embedding

The use of HEMA in the histological practice (i.e., the study of living tissues and cells at the microscopic level) was proposed in 1960 (*92*). The hydrophilic properties of the monomer made it useful as a com-bined dehydrating agent for the tissues and as an embedding medium for electron microscopy. How-ever, blocks of pure PHEMA appeared difficult to sec-tion and had poor resistance under the electron beam. Also, the quality of commercially available HEMA was reported to vary considerably until 1965. Copolymers with butyl methacrylate or styrene also were less satis-factory than the epoxy resins. During the last decades, HEMA found a new interest in light microscopy (*93*).

An extensive review on this topic was presented by Bennett et al. (*12*). Briefly, HEMA embedding is favored for light microscopy because of the following reasons:

1. The embedding duration is shorter than classic methods, even when embedding very large specimens.
2. Preservation of tissular and cellular structures is by far superior to other classic methods. This dif-ference is due to the adherence of tissue sections onto the microscopic glass slides and because the resin is not removed before staining.
3. Sectioning is easier and semi-thin sections (i.e., 2–3 μm) can be obtained with conventional micro-tomes with steel or Ralph's glass knives. Further-more, once cut, the sections spread on water and do not shrink.
4. Numerous staining methods can be performed on PHEMA sections. Classic stains (excepted for those having a hydro-alcoholic vehicle, which makes

the sections swell) work well, sometimes after minor modifications (*94*) (Figure 2).

Enzymological studies readily can be done and a large amount of enzymes are preserved. The enzymes of calcified tissue cells were demonstrated on undecalcified sections, whereas decalcification hampers their identification (*95*).

At the present time, several HEMA-based commercial kits are available (e.g., Historesin or JB4). However, the slow hydrolysis of the resin makes it difficult to obtain regular results from these kits. The methacrylic acid formed de novo appears to combine with basic stains, and even small amounts (1.5% or less) of methacrylic acid impair correct staining by strongly obscuring the background (*12*). Several purification methods specially devoted to histotechnology have been designed. The HEMA monomer alone was repeatedly found to be a poor medium for calcified tissues, because the relatively large size of the molecule makes it difficult to infiltrate such tissues. Combined with MMA (*96*) or various types of alkyl methacrylates or acrylates, HEMA provided suitable embedding mediums. HEMA is usually polymerized by a redox reaction (dibenzoyl peroxide and *N,N*-dimethylaniline) reported in numerous papers. Various concentrations of both polymerization initiator and accelerator have been proposed and adapted to different tissues, and the method has been used to embed specimens in the cold to preserve enzyme activities (*95*). AIBN, bar-

biturate cyclic compounds, and butazolidine have been proposed as initiating systems. PHEMA produced better sections when small amounts of crosslinkers were used (*97*). Recently, we showed that HEMA embedding is a nonhomogeneous process, and polymerization rates and polymer quality vary according to the volume of monomer to be polymerized in bulk (*98*) (Figure 3).

Dentistry

Synthetic apatitic calcium–phosphate cements were prepared with a PHEMA hydrogel containing tetracalcium phosphate and dicalcium phosphate. PHEMA was a highly biocompatible and resorbable material for primary-teeth endodontic filling. However, because of its hydrophilicity, PHEMA appeared more useful in dentistry as a bonding reagent between dentine and other types of restorative resin. Various mixtures of HEMA and glutaraldehyde were investigated. HEMA was a suitable vehicle for dentine self-etching primers (such as acid monomers) (*99*). Other clinical trials were done with an antiseptic (Chlorhexidine) entrapped in a HEMA–MMA copolymer membrane to develop a controlled-release delivery system (*100*).

Immobilization of Enzymes, Cells, and Drugs

Immobilization implies the entrapment within a polymeric network of a 'foreign' compound (e.g., an

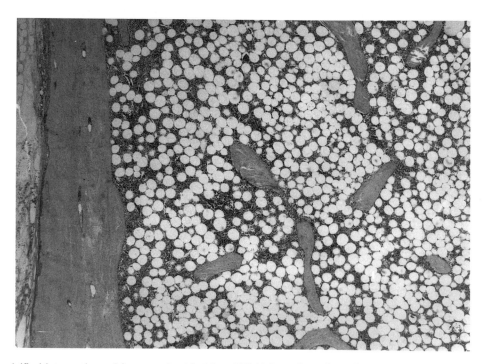

Figure 2. Undecalcified human bone biopsy embedded in a HEMA-based medium. Section, 7-μm thickness. Note the well-preserved bone architecture and bone marrow distribution Goldner's trichromic stain. Magnification, ×25.

enzyme, a cell, or a drug), whether it is simply confined or grafted onto the polymeric chains.

Immobilization of various enzymes on solid supports has found a number of biotechnology applications because enzyme molecules become reusable (*101*). To preserve enzyme activity, radiation-induced polymerization is often reported: cellulase was well preserved in HEMA polymerized by γ–radiations at low temperature. Trypsin bound covalently on a composite material made of an alginate copolymerized with HEMA and glycidyl methacrylate. The loss of enzyme activity was only 7% after five successive uses. Glucose oxidase was readily immobilized in PHEMA membranes, and a glucose-sensitive electrode was prepared by immobilizing the enzyme in PHEMA or polyurethane (*102*). More recently, a needle-type electrode with immobilized glucose oxidase was proposed for glucose monitoring in diabetic patients (*103*). The location of the enzyme lipase within a PHEMA hydrogel has been studied. The distribution of glucoamylase labeled with fluorescein isothiocyanate was investigated with fluorescence microscopy. The enzyme was located on the interface between polymer membrane and pore structures and partly in the polymer itself (*104*). HEMA–EDMA copolymer (Separon HEMA) was used to study the covalent immobilization of various enzymes. The type and concentrations of added salts modified yields (*105*).

Several types of microbial cells or yeasts known to possess biotechnologically interesting enzymes have been entrapped into PHEMA hydrogels. Cell systems studied include *Streptomyces phaechromogenes* (containing glucose isomerase), *Mortierella vinacea* (containing a galactosidase), *Saccharomyces cerevisae* (which synthesizes ethanol at high rates), and others (Figure 4). Immobilization in PHEMA appears more satisfactory than on tuff for continuous alcoholic fermentation (*106*). Spores of fungi (*Humicola lutea*) were entrapped in a PHEMA hydrogel and produced more acid proteinase than free cells (*107*).

Pancreatic islets enclosed in a PHEMA hydrogel synthesized and released insulin in vitro (*108*). The biocompatibility of such pancreatic islets was excellent when implanted into animals. Diffusion chambers made of a PHEMA hydrogel were successfully used in vivo after immobilization of rabbit embryos, the chamber was implanted in the peritoneal cavity of male mice, and early developmental stages were followed (*109*). A hydrogel of pure PHEMA has no effect on spermatozoid motility, but the copolymer HEMA–methacrylic acid has been proposed as a male contraceptive technique when injected into the vas deferens.

Composites of alginate and HEMA have been used to prepare microspheres for the microencapsulation of cells. A detailed method for the encapsulation of fibroblast from Chinese hamster ovary was reported (*110*). Disks of poly(ethylene terephthalate) coated with HEMA–2-diethylaminoethyl methacrylate copolymer selectively separated both activated and helper T-lymphocytes from peripheral blood of rheumatoid patients

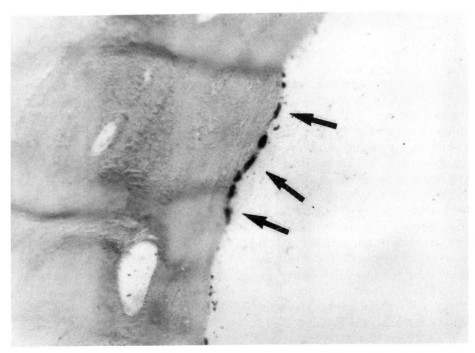

Figure 3. Histochemical identification of osteoclasts (the bone-resorbing cells) by staining their specific enzyme, the tartrate resistant acid phosphatase. Only osteoclasts are stained (arrows); bone marrow is unstained. Bone matrix is lightly counterstained: α-Naphylphosphate, fast violet B blue salt method, ×100 magnification.

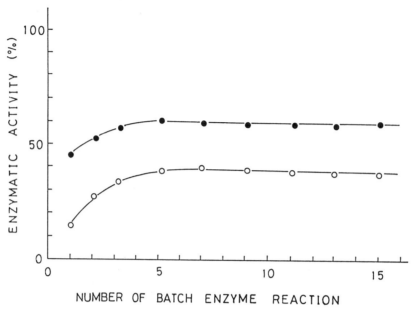

Figure 4. Variation of enzymatic activity of *Mortierella Vinacea* cells with repeated batch reaction. HEMA monomer concentration: ● 15% HEMA, ○ 50%. HEMA. Cell concentration, 0.5%. (Reproduced with permission from Kumakura et al. *J. Appl. Polym. Sci.* **1983**, *28*, 295.)

(111). The ability of drugs to diffuse out of polymers may be used in various types of biotechnologies such as membrane separation and drug-delivery devices. Numerous drugs have been entrapped (or immobilized) in radiation-polymerized HEMA (ergotamine, salicylic acid, and hormones) to produce drug-delivery devices. Chloramphenicol immobilized in PHEMA hydrogels that were cross-linked with EDMA was released during swelling of the gel in water, and the diffusion obeyed Fick's second law *(112)*. The kinetics of thiamine (vitamin B₁) and theophylline diffusion from previously loaded PHEMA beads were studied *(113, 114)*. PHEMA membranes were favored as transdermal delivery systems for long-term constant drug delivery (e.g., nitroglycerine and hormones) *(115)*. Vidarabine (an antiviral agent) was entrapped into PHEMA membranes and used as transdermal patches *(116)*. The concentrations of the blood drugs could be predicted and the permeability coefficient of the membranes could be adjusted by controlling hydration. Synthetic organ substitutes having the capacity to slowly release hormones have been designed. Diffusion of insulin and luteinizing hormone–releasing hormone analogues *(117, 118)* through PHEMA membranes was studied. Because PHEMA hydrogels are slowly degraded in vivo, entrapment in a blend of PHEMA–albumin resulted in a slowly degraded matrix with continuous release of the drug. Testicular prosthesis releasing testosterone were reported *(119)*.

Anticancer drugs (methotrexate and 3'5'-dibromoaminopterin) entrapped in matrices of PHEMA provide a hard material that can be implanted into the tumor to deliver higher amounts of the drug. 5-Fluorouracil was embedded in HEMA–EDMA copolymer in beads (3 mm in diameter) that could be implanted subcutaneously *(120)*. The effect of cross-linking on the swelling of PHEMA gels (and the drug-diffusion coefficient through these gels) were studied *(121)*.

Finally, various substances were immobilized in PHEMA to prepare diagnostic tools. An antiserum raised against methotrexate was entrapped in PHEMA during polymerization. The lyophilized powder was used for radio-immunoassay of this anticancer drug. The entrapment of immunoglobulins was used for immunochemical studies to provide immunoaffinity sorbents for isolation of proteins *(122)*.

Biocompatibility of HEMA

Biocompatibility of PHEMA was studied at cell and tissue levels. Cell cultures on PHEMA-coated slides or on PHEMA hydrogels were used to investigate the intimate mechanisms of cellular compatibility. Implanting pieces of gel in an animal by a surgical procedure allows the study of the adverse responses of the whole organisms to the resin. Because implantations in the eye or in direct contact with blood induces specific problems, these two aspects of the biocompatibility will be treated separately.

Cell Culture

The hydrophilicity of PHEMA is thought to be favorable for cell culture. Cellular adherence to PHEMA has

been recognized since 1975. Lydon et al. (*123*) found that spreading of cells was higher on modified PHEMA than on polystyrene because of the hydrophilic properties of PHEMA. Other authors presented the opposite results: leukocyte locomotion was suppressed on PHEMA-coated glass plates (*124*), malignant melanoma cells grown onto PHEMA formed aggregates of round cells and generated polykaryons, and adrenal tumor cells showed decreased steroidogenesis secondary to altered cell shape. The time required for rat peritoneal macrophages to adhere to HEMA–ethyl methacrylate copolymers was higher than adherence to hydroxystyrene–styrene copolymers due to the high hydrophilicity of HEMA–ethyl methacrylate copolymers. Decreasing rates of adhesion of staphylococci onto MMA–HEMA copolymers parallel the increasing HEMA content (i.e., the hydrophilicity). Adhesion of 3 strains of *Escherichia coli* on PHEMA and PHEMA containing methacrylic acid was compared with other methacrylates: homopolymers having a positive radical ζ potential retained more bacteria (*125*).

Simple PHEMA gels do not permit cell spreading. When ionizable groups are entrapped, cells spreading is no longer inhibited. When collagen is added, cell proliferation occurs. Peritoneal macrophages adherence decreased as a function of the hydrophilicity of the polymer. Cellular adherence on PHEMA is favored by absorption of proteins of the extracellular milieu. Albumin, fibronectin, and immunoglobulins G favor cellular adherence, whereas fibrinogen, elastin, and blood plasma copolymerized with PHEMA hamper this phenomenon (*126*). Alternative modifications of PHEMA allowing cell proliferation have been incorporation of methacrylic acid or 2-ethylaminoethyl methacrylate or treating the polymer with concentrated sulfuric acid (creating surface carboxylic groups) (*127*) (Figure 5).

Implants and Artificial Organs

PHEMA is a suitable biomaterial for implantation because of its nontoxicity and high resistance to degradation. Numerous composite biomaterials based on PHEMA and collagen blends have been proposed. By using different additives, the mechanical properties of PHEMA hydrogels can be adjusted to various biomedical applications. HEMA–methacrylic acid copolymers were more biocompatible than PHEMA alone, which induced an inflammation with a giant cells granuloma (*128*). When collagen was entrapped in PHEMA gels, the composite was highly biocompatible after subcutaneous implantation in rats. Composites with a low collagen content were better preserved in long-term implantation studies, whereas those containing higher amounts of collagen exhibited calcification at the early stages followed by full biodegradation (*129*).

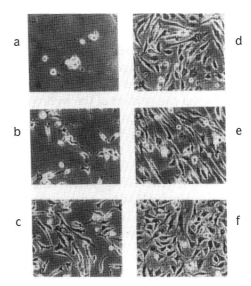

Figure 5. Appearance of cells 24 h after attachment to pHEMA etched with sulfuric acid. Etching times were a, 0 s; b, 2.0 s; c, 5.0 s; d, 10.0 s; e, 15.0 s; and f, control polystyrene. The fibroblastoid morphology of cells (e) appears to be the result of forms of surface channeling by prolonged acid treatment of PHEMA and was not due to a subpopulation of cells. Reproduced with permission from reference 128. Copyright 1987 John Wiley & Sons.)

Composites of collagen and cross-linked PHEMA were degraded after a 3-month implantation period in the rat. The collagen matrix appeared fully resorbed, but small, residual, round-shaped particles (1–15 mm diameter) composed of PHEMA were still present (*130*).

Calcification of a synthetic biomaterial implies poor biocompatibility; although the chemical composition appears important, the macroscopic structure and surface characters of a PHEMA implant play a key role. Extensive calcium accumulation in cell mitochondria in close contact with the gels was proposed as the primary mechanism of calcification. In addition, hydrogels of HEMA and methacrylic acid copolymers picked up large amounts of Ca^{2+} when exposed to aqueous solutions of calcium. This effect was taken into account when porous sponges of PHEMA were compared with demineralized bone for inducing ectopic bone formation (*131*). When implanted into bone, these collagen–PHEMA composites were resorbed even when collagen concentration was low. The presence of collagen could be a prerequisite for osteoblastic stimulation (*132*).

Hydrogels of PHEMA have excellent biocompatibility but present poor mechanical properties. The mechanical and hydration properties of PHEMA and other poly(hydroxyalkyl) methacrylate membranes were studied (*133*). Radiation grafting of HEMA was done on polyurethane films (with good mechanical properties) to increase hydrophilicity and tolerance (*134*). Hemodialysis membranes of PHEMA cross-linked with ethylene dimethacrylate were prepared (*135*). The interaction of urea (the end product of protein catabolism) with PHEMA hydrogels revealed that small amounts of methacrylic acid may dramatically increase the swelling properties of the gel. Membranes made of PHEMA and silicone were proposed as skin substitutes for modelization of permeability of drugs (*136*). Composite membranes made of a poly(ethylene terephthalate) mesh coated by a PHEMA hydrogel were used in vivo as pericardial substitutes without significant epicardic reaction (*137*).

Prosthetic Vascular Implants and Blood Compatibility

In addition to biocompatibility, a very interesting property of PHEMA-based hydrogels is their high hemocompatibility. The thrombus formation is delayed when the blood is in contact with PHEMA. Because blood is a complex milieu, we consider all the relationships of PHEMA with blood cells, endothelial cells (i.e., the inner cells of the blood vessels), and blood components. Because of the hydrophilicity of PHEMA, films of styrene–butadiene–styrene when grafted to PHEMA had a better blood compatibility. Copolymers of HEMA–styrene or HEMA–dimethylsiloxane suppress platelet adhesion and aggregation (and thus reduce the thrombus formation) by creating hydrophilic–hydrophobic microdomains (*138*) (Figures 6 and 7).

Similar findings were obtained with HEMA–polyethylene oxide and HEMA–poly(propylene oxide) copolymers. A HEMA–polyamine copolymer induced neither blood platelet adherence nor activation. Also, this copolymer was used to separate T lymphocyte from B lymphocyte sub-populations by using its

hydrophilic–hydrophobic microdomain composition. An artificial lung containing hollow fibers of PHEMA in contact with blood and surrounded with air was constructed (*139*).

Vascular tubes of polyethylene blended with 14% PHEMA have a very low thrombogenicity due to the hydrophilicity of PHEMA. Radiation grafting of HEMA and *N*-vinylpyrrolidone onto silicone rubber was used to improve the hydrophilicity of artery-to-vein shunts and thus to reduce thrombus formation (*140*). HEMA was used as a compound for polymerization within detachable balloons used in vascular neurosurgery and especially for aneurysmal occlusion. The rationale is that PHEMA hydrogel eliminates balloon deflation. However, the hydrogel was somewhat incompatible with the polyisoprene rubber and led to balloon failure (*141*). Intracarotid perfusion of mongrel dogs with HEMA was followed by severe cerebral infarctions that could be attributed to intravascular polymerization (*142*).

Another important aspect of blood compatibility is that a biomaterial should not activate the complement system (a complex system of plasma proteins activated in a cascade fashion and leading to inflammation). Intra-ocular lenses made of PHEMA were ineffective in vitro in activating the serum complement system (C3a, C4a, C5a) (*143*). On the other hand, copolymers of HEMA–EMA were reported to activate the complement system when the polymer contained 60% or more HEMA. When HEMA was used as a surface modifier for polystyrene, it activated the complement via the classic pathway (*144*).

The hemocompatibility of PHEMA has led to the development of a medical method used to remove

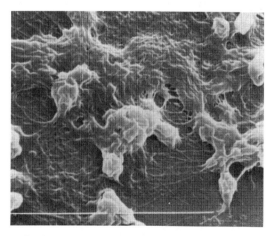

Figure 6. Scanning electron microscopic views of blood platelets adhered to the surface of polystyrene (× 4700). (Reproduced with permission from reference 140. Copyright 1990 John Wiley & Sons.)

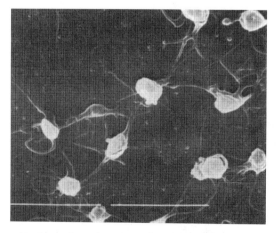

Figure 7. Scanning electron microscopic views of blood platelets adhered to the surface HEMA–Styrene–ABA-type block copolymer containing 0.608 mol fraction of HEMA (×4700). Note well-preserved morphology of platelets. (Reproduced with permission from reference 140. Copyright 1990 John Wiley & Sons.)

endo- or exotoxins from blood. Hemoperfusion takes the advantage of activated charcoal to bind such toxics (barbiturates and tricyclic antidepressants). Activated carbon particles have been encapsulated with PHEMA for the construction of hemoperfusion columns.

Heparinized blood is purified by adsorption of toxic molecules on the entrapped charcoal particles, and the cleaned blood is then returned to the patient (*145*). Composites of PHEMA, polyethylene glycol, and activated carbon were useful for other blood-perfusion applications (Figure 8).

Another important application of PHEMA is the occlusion of blood vessels in various organs, principally in tumors (which are always hypervascularized). Spherical particles of PHEMA of regular shape were produced by suspension polymerization. When injected in a vessel close to the tumor, the small beads act as emboli and obliterate the smaller vessels. Thus, tumor vascularization is stopped and endovascular embolization is followed by tumoral-cell necrosis and size reduction of the tumor. The swelling in water of PHEMA beads made them suitable for close obliteration of vessels (*146*). Beads can be loaded or coupled with an X-ray contrasting agent (iodine), which helps radiographic tracing. Adsorption of hemostatic drugs such as ethamsylate, aminocaproate, thrombin, and 4-nitrophenyl-chloroformate of PHEMA beads used for vascular embolization favors hemostasis (*147*) (Figures 9 and 10).

Optical Lenses

The main application of PHEMA hydrogels is the preparation of contact and intra-ocular lenses used after cataract extraction (*148*). Black-pigmented

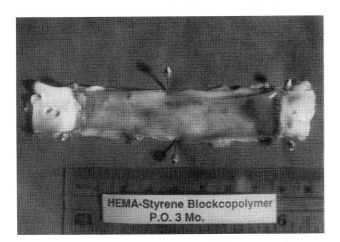

Figure 8. Luminal view of HEMA–Styrene block copolymer coated graft 3 months after implantation. The surface is bare without detectable thrombi. (Reproduced with permission from reference 140. Copyright 1990 John Wiley & Sons.)

PHEMA was used to prepare light-occluding lens after ophthalmic surgery. Gentamicin-soaked contact lenses retained bactericidal concentrations of the antibiotic for up to 3 days of eye contact. Deep corneal stromal opacities were seen in PHEMA contact lenses and were related to chronic corneal anoxia (*149*). Deposits are sometimes observed within contact lenses. They occur after 12 months of daily-lens wear and may lead to a decrease in vision acuteness.

The protein deposits on contact lenses change according to the copolymer. With HEMA–methacrylic acid copolymers, lenses adsorb large amount of lysosyme, whereas HEMA–MMA copolymers preferentially adsorb albumin. These protein deposits can be detected by using the dye Coomassie Blue R in a sensitive staining method detecting ≥ 2 $\mu g/cm^2$ (*150*). Contact lenses of copolymer of HEMA with methacrylic acid or various silanes adsorbed less lysosyme than unsilanized lenses. Coating the lens with phosphorylcholine and using a glow-discharge technique that modified surface hydrophilicity reduced deposits (*151*). Deposits of calcium in contact lens of PHEMA were reported.

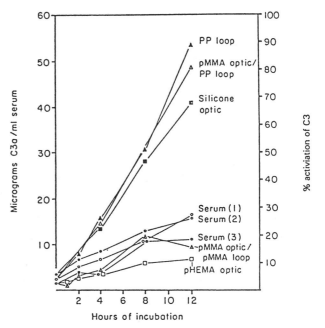

Figure 9. Quantitation of C3a in serum incubated either alone or with intra-ocular lenses. Levels of C3a were quantitated by radio immunoassay of serum incubated with intra-ocular lenses and/or loops. Each point is the average of 2 samples; the standard error of each point is <10% of the mean. The concentration of C3a at 4-, 8-, and 12-h time points was significantly higher in serum incubated with polypropylene loops ($\triangle$), PMMA optic with polypropylene loops ($\blacktriangle$), and silicone optic ($\blacksquare$) ($p < 0.05$) than in serum incubated alone. (Reproduced with permission from reference 144. Copyright 1987 Butterworth.)

(a)

(b)

Figure 10. Electromicrograph of SAC particles (a) without poly-HEMA coating and (b) with poly-HEMA coating (×5000). (Reproduced with permission from reference 146. Copyright 1990 John Wiley & Sons.)

Intra-ocular strips of PHEMA hydrogels containing small amounts (1.2–1.4%) of methacrylic acid were tolerated favorably in vivo due to the high water and carboxylic group content. Intra-ocular PHEMA lenses were better tolerated than conventional amino-polyamide-based implants, but the presence of microvilli on corneal cells suggested the release of impurities from the lenses (152). PHEMA-based intra-ocular lenses were well preserved after yttrium–aluminum–garnet–Nd laser surgery. Various drugs (e.g., chloramphenicol, pilocarpine, and dexamethasone) had long wash-out periods when entrapped in intra-ocular lenses. The bioclinical results of PHEMA intra-ocular lenses were most: favorable: 92% of implanted patients experienced a recovery of visual acuity (153).

Concluding Remarks and Future Prospects

Among the synthetic hydrogels [polyacrylamide, poly(N-vinyl pyrrolidone), poly(glyceryl methacrylate) and PHEMA], PHEMA is a very accessible, highly versatile polymer and is useful for numerous biomedical applications. Some of the main uses of HEMA polymers and copolymers with acrylic, methacrylic monomers, or N-vinyl pyrrolidone include vascular grafts, cell culture substrates, contact lenses, breast and other soft tissue substitutes, carriers of antibodies, various prostheses, antibiotic and antitumor drug-delivery devices, denture lines, and dentures. However, most of these current and potential applications have emerged during the past 20 years, and the long-term viability of some of them remains to be proved.

At the same time, and despite extensive studies, applied PHEMA research, fundamentals of polymer chemistry, and polymer properties are not yet fully explored. Thus, from both academic and technological view points, PHEMA research is most likely to be pursued very actively in the future. But how rapidly biomedical technology of PHEMA is likely to grow is more difficult to predict.

References

1. (a) Wichterle, O.; Lim, D. *Nature (London)* **1960**, *185*, 118. (b) Wichterle, O. In *Encyclopedia of Polymer Technology*; Mark H. F.; Gaylord N. G.; Bikales, N., Eds.; Interocienie Publisher: New York, 1971; Vol. 15, p 273.
2. Ratner, B. D. In *Comprehensive Polymer Science*; Allen, G.; Bevington, J. C.; Aggarwal, S. L., Eds.; Pergamon: Oxford, Great-Britain, 1989; Vol 7, p 201
3. Wisniewski, S. J.; Gregoris, D. E.; Kim, S. W.; Andrade, J. D. *Hydrogels for Medical and Related Applications*; Andrade, J. D., Ed.; ACS Symposium Series 311; American Chemical Society: Washington, DC, 1976; pp 80–87.
4. Horbell, T. A. *Hydrogels in Medicine and Pharmacy*; Peppas, N. A., Ed.; CRC: Boca Raton, FL, 1986; Vol. 1, p 127.
5. Singh, J.; Agrawal, K. K *J. Macromol. Sci., Rev. Macromol. Chem.* **1992**, *C32*, 521.
6. Kopecek, J.; Jokl, J.; Lim, D. *J. Polym. Sci. Part C* **1968**, *16*, 3877.
7. Sumimoto Kagaku Kogyo, Jpn. Patent 04066554 A2, 1992; Okamura, H.; Mizuno, T.; Shiozaki, T. *Chem. Abstr.* **1992**, *117*, 292052.
8. Casella Farbwerke Mainkur. A.G. Dutch Patent 6603510, 1966; *Chem. Abstr.* **1967**, *66*, 55059
9. *Hydroxy Monomers*; Yocum, R. H.; Nyquist, E. B., Eds.; Marcel Dekker: New-York, 1973; Vol. 1.

10. Al-Issa M. A.; Davis, T. P.; Huglin, M. B.; Yahia, I. B.; Yip, D. C. F. *Eur. Polym. J.* **1984**, *20*, 947.

11. Pinchuk, L.; Eckstein, E. C.; Van de Mark, M. R. *J. Appl. Polym. Sci.* **1976**, *29*, 1749.

12. Bennet, H. S.; Wyricks, A. D.; Lee, S. W.; Mc Neil, H. J. *Stain Technol.* **1976**, *51*, 71.

13. Frater, R. B. *Stain Technol.* **1979**, *54*, 241.

14. *The Aldrich Library of NMR Spectra*; Pouchert, C.; Campbell, J. R., Eds.; Aldrich Society, 1976; Vol. 3, p 50.

15. Gregoris, D. E.; Russel, G. A.; Andrade, J. D.; De Visser, A. C. *Polymer* **1978**, *19*, 1279.

16. Refojo, M. F.; Yasuda, Y. *J. Appl. Polym. Sci.* **1965**, *9*, 24.

17. Wichterle, O.; Chromecek, R. *J. Polym. Sci., Part C: Polym. Chem.* **1969**, *16*, 1677.

18. Orienti, I.; Gianasi, E.; Zecchi, V. *Farmaco* **1993**, *48*,1577.

19. Teffal, M.; Gourdenne, A. *Eur. Polym. J.* **1983**, *19*, 543.

20. Renbaum, A.; Yen, S. P. S.; Cheong, E.; Wallage, S.; Molday, R. S.; Gordon, I. L.; Dreyer, W. J. *Macromolecules* **1976**, *9*, 328.

21. Osada, Y.; Irimaya, Y.; Takase, M.; Imo, Y.; Ohte, M. *J. Appl. Polym. Sci., Appl. Polym. Symp.* **1984**, *38*, 45.

22. Kumakura, M.; Kaetsu, I. *J. Polym. Sci., Polym. Lett.* **1983**, *21*, 609.

23. Zinger, B.; Shkolnik, S.; Höcker, H. *Polymer* **1989**, *30*, 628.

24. Allen, N. S.; Catalina, F.; Green, P. N.; Green, W. A. *Eur. Polym. J.* **1986**, *22*, 871.

25. Gonen, S.; Kohn, D. H. *J. Polym. Sci. Polym. Chem. Ed.* **1981**, *19*, 2215

26. Da Silva, M. A; Gil, M. H; Lapa, E. *J. Appl. Polym. Sci.* **1987**, *34*, 871.

27. Yul, E. F.; Tianyi, S.; Wengfeng, W.; Fang, S. *J. Appl. Polym. Sci.* **1989**, *38*, 821.

28. Fang, Y.; Shi, T. Y.; Liu, W. S.; Shi, F. *J. Macromol. Sci. Chem.* **1990**, *A27(1)*, 117.

29. Scranton, A. B.; Bowman, C. N.; Klier, J.; Peppas, N. A. *Polymer* **1992**, *33*, 1683.

30. Fort, R. J.; Polyzoidis, T. M. *Eur. Polym. J.* **1976**, *12*, 685.

31. Bohdanecky, M.; Tuzar, Z.; Stol, M.; Chromecek, R. *Collect. Czech. Chem. Commun.* **1968**, *33*, 4104.

32. Yokota, K.; Abe, A.; Hosaka, S.; Sakai, I.; Saito, H. *Macromolecules* **1978**, *11*, 95.

33. Brynda, E.; Stol, M.; Chytry, U.; Cifkova, I. *J. Biomed. Mater. Res.* **1985**, *19*,1169.

34. Roorda, W. E.; Bourwstra, J. A.; de Vries, M. A.; Junginger, H. E. *Biomaterials* **1988**, *9*, 494.

35. Pinchuk, L.; Eckstein, E. C.; Van der Mark, M. R. *J. Biomed. Mater. Res.* **1984**, *18*, 671.

36. Yasuda, H. *J. Polym. Sci., Part A: Polym. Chem.* **1967**, *15*, 2952.

37. Spacek, P.; Kubin, M. *J. Polym. Sci., Part C: Polym. Symp.* **1967**, *16*, 705.

38. Janacek, J. *J. Macromol. Sci., Rev. Macromol. Chem.* **1973**, *C9*, 1.

39. Acierno, D.; Nicolais, L.; Vojta, V.; Janacek, J. *J. Polym. Sci., Polym. Phys. Ed.* **1975**, *13*, 703.

40. Greenley, R. Z. *J. Macromol. Sci. Chem.* **1980**, *14A*, 427.

41. Lavrov, N. A.; Nikolaev, A. F.; Sergeeva, E. P. *Zh. Prikl. Khim.* **1991**, *64*, 2004; *Chem. Abstr.* **1991**, *118*, 169672.

42. Schauer, J.; Houska, M.; Kalal, J. *Makromol Chem.* **1980**, *181*, 367.

43. Ramakrishna, M. S.; Desphande, D. D.; Babu, G. N. *J. Polym. Sci., Part A: Polym. Chem.* **1988**, *26*, 445.

44. Bajaj, P.; Jain, P. C.; Gandopadhyay, D. *J. Polym. Sci. Polym. Chem. Ed.* **1979**, *17*, 595.

45. Varma, I. K.; Patnaik, S. *Eur. Polym. J.* **1976**, *12*, 259.

46. Zaldivar, D.; Peniche, C.; Bulay, A.; San Roman, J. *J. Polym. Sci., Part.A: Polym. Chem. Ed.* **1993**, *31*,625.

47. San Roman, J.; Levenfeld, B.; Madruga, E. L.; Vairon, J. P. *J. Polym. Sci., Part A: Polym. Chem. Ed.* **1991**, *29*,1023.

48. Lebduska, J.; Snuparek, J.; Kaspar, K.; Cermak, V. *J. Polym. Sci., Part A: Polym. Chem.* **1986**, *24*, 777.

49. Chen. S. A.; Chang, H. S. *J. Polym. Sci., Part A: Polym. Chem.* **1990**, *28*, 2547.

50. Kamei, S.; Okubo, M.; Matsumoto, T. *J. Polym. Sci., Part A: Polym. Chem.* **1986**, *24*, 3109.

51. Fukumori, Y.; Yamaoka, Y.; Ichikawa, H.; Takeuchi, Y.; Fukuda, T.; Osako, Y. *Chem. Pharm. Bull* **1988**, *36*, 3070.

52. Hirao, A.; Kato, H.; Yamaguchi, K.; Nakahama, S. *Macromolecules* **1986**, *19*, 1294.

53. Moszner, N.; Salz, U.; Rheinberger, V. *Polym. Bull.* **1994**, *32*, 411

54. Regunadhan Nair, C. P.; Clouet, G.; Brossas, J. *J. Polym. Sci., Part A, Polym. Chem.* **1988**, *26*, 1791.

55. Bosc, B.; Boutevin, B.; Granier-Azema, D.; Rousseau, A. *Polym. Bull.* **1992**, *29*, 289.

56. Feld, W. A.; Bombick, D. W.; Friar, L. L. *J. Polym. Sci. Polym. Chem.* **1981**, *19*, 263.

57. Hayeshi, A.; Goto, Y.; Nakayama, M.; Sato, H.; Watanabe, T.; Miyata, S. *Macromolecules* **1992**, *25*, 5094.

58. Kato, M.; Hirayama, T.; Matsuda, H.; Minami, N.; Okada, S.; Nakanishi, H. *Makromol. Chem., Rapid. Commun.* **1994**, *15*, 741

59. Bertini, V.; Pocci, M.; De Munno, A.; Picci, N.; Lucchesini, F. *Eur. Polym. J.* **1988**, *24*, 467.

60. Mikes, O.; Strop, P.; Hostomska, Z.; Smrz, M.; Slovakova, S.; Coupek, J. *J. Chromatogr.* **1984**, *301*, 93.

61. Mikes, O.; Strop, P.; Hostomska, Z.; Smrz, M.; Coupek, J.; Frydrychova, A.; Bares, M. *J. Chromatogr.* **1983**, *261*, 363.

62. Kuhn, M.; Kirstein, D. East Ger. Patent 157 340, 1982; *Chem. Abstr.* **1982**, *98*, 144433.

63. Smrz, M.; Hradil, J. Czech. Patent 183 508, 1980; *Chem. Abstr.* **1981**, *94*, 140469

64. Pecka, K.; Smidl, P.; Vavra, M.; Havel, Z.; Hala, S. *Sb. Vys. Sk. Chem. Technol. Praze Technol. Paliv.* **1982**, *46D*, 49; *Chem. Abstr.* **1982**, *98*, 60434.

65. Horak, D.; Metalova, M.; Svec, F.; Drobnik, J.; Kalal, J.; Borovicka, M.; Adamyan, A. A.; Voronkova, O. S.; Gumargalieva, K. Z. *Biomaterials* **1987**, *8*, 142.

66. Mikes, O.; Strop, P.; Smrz, M.; Coupek, J. *J. Chromatogr.* **1980**, *192*, 159.

67. Klyashitskii, B. A.; Mezhova, I. V.; Pozdnev, V. F.; Slavik, E. P.; Shvets, V. I. *Bioorg. Khim.* **1986**, *12*, 141.

68. Kahovec, J.; Jelinkova, M.; Coupek, J. *Polym. Bull.* **1987**, *18*, 495.

69. Buettner, W.; Becker, M.; Rupprich, C.; Boeden, H. F.; Henklein, P.; Loth, F.; Dautzenberg, H. *Biotechnol. Bioeng.* **1989**, *33*, 26.

70. Slovak, Z.; Smrz, M. S.; Docekal, B.; Slovakova, S. *Anal. Chim. Acta* **1979**, *111*, 243.

71. Hradil, J.; Smrz, M. Czech. Patent 175 255, 1978; *Chem. Abstr.* **1979**, *90*, 138570.

72. Smrz, M.; Slovakova, S.; Vermousek, I.; Viska, J.; Kiss, F. Czech. Patent 176 607, 1979; *Chem. Abstr.* **1979**, *91*, 58092.

73. Smrz, M.; Slovakova, S.; Vermousek, I.; Viska, J.; Kiss, K. Czech. Patent 169 316, 1977; *Chem. Abstr.* **1978**, *88*, 153558.

74. Smrz, M.; Vermousek, I.; Slovakova, S. Czech. Patent 172 044, 1978; *Chem. Abstr.* **1978**, *89*, 198500.

75. Smrz, M.; Slovakova, S.; Vermousek, I.; Kiss, F.; Viska, J. Czech. Patent 169 326, 1977; *Chem. Abstr.* **1978**, *88*, 153565.

76. Turkova, J.; Blaha, K.; Horacek, J.; Vajcner, J.; Frydrychova, A.; Coupek, J. *J. Chromatogr.* **1981**, *215*, 165.

77. Hadrabova,V.; Coupek, J.; Majer, J. *Chem. Zvesti* **1983**, *37*, 225

78. Kühn, M.; Mohr, P.; Coupek, J. East Ger. Patent 136 269, 1979; *Chem. Abstr.* **1979**, *91*,158548

79. Strop, P.; Coupek, J.; Mikes, O. Ceskoslovenska Akademie Ved. Ger. Patent 2505 892, 1975; *Chem. Abstr.* **1976**, 122 812.

80. Development Finance Corp. of New Zealand. U.S. Patent 4 330 440, 1982; Ayers, J. S.; Bethell, G. S.; Hancock, W. S.; Hearn, M. T. W. *Chem. Abstr.* **1982**, *97*, 88257.

81. Mikes, O.; Strop, P.; Zbrozek, J.; Coupek, J. *J. Chromatogr.* **1979**, *180*, 17

82. Smrz, M.; Slovakova, S.; Slovak, Z.; Pribyl, M. Czech. Patent 169 321, 1977; *Chem. Abstr.* **1978**, *88*, 192034

83. Smrz, M.; Slovakova, S.; Docekal, B.; Slovak, Z. Czech. Patent 217 841, 1984; *Chem. Abstr.* **1985**, *102*, 132956

84. Daly, W. H.; Shih, F. *Biological Activities of Polymers*; Carraher, C. E., Jr.; Gebelein, C. G., Eds.; ACS Symposium Series 186; American Chemical Society: Washington, DC, 1982; pp 133–148.

85. Svec, F. *Acta Polym.* **1980**, *31*, 68

86. Janak, K.; Janak, J. *Collect. Czech. Chem. Commun.* **1986**, *51*, 650.

87. Filka, K.; Coupek, J.; Kocourek, J. *Biochim. Biophys. Acta* **1978**, *539*, 518

88. Berglund, M.; Andersson, C. *J. Mol. Catal.* **1986**, *36*, 375.

89. Sato, H.; Kojima, J.; Nakajima, A.; Morita, T.; Noishiki, Y.; Gu, Z.; Li, F.; Feng, X. *J. Biomater. Sci. Polym. Ed.* **1991**, *2*, 1

90. Chappard, D.; Monthéard, J. P.; Chatzopoulos, M.; Alexandre, C. *Innov. Technol. Biol. Med.* **1992**, *13*, 322.

91. Stol, M.; Cifkova, I.; Brynda, E. *Biomaterials* **1988**, *9*, 273.

92. Rosenberg, M.; Bartl, P.; Lesko, J. *J. Ultrastr. Res.* **1960**, *4*, 298.

93. Rudell, C. L. *Stain Technol.* **1967**, *2*, 253.

94. Litvin, J. A. *Prog. Histochem. Cytochem.* **1985**, *16(2)*, 1.

95. Chappard, D.; Alexandre, C.; Riffat, G. *Basic Appl. Histochem.* **1983**, *27*, 75.

96. Chappard, D.; Alexandre, C.; Camps, M.; Monthéard, J. P.; Riffat, G. *Stain Technol.* **1993**, *58*, 299.

97. Gerrits, P. O.; Van Leeurven, M. B. M. *Stain Technol.* **1987**, *62*, 181.

98. Chappard, D.; Benascar, Z.; Alexandre, C.; Monthéard, J. P. *J. Histotechnol.* **1989**, *12*, 89.

99. Uno, S.; Asmussen, R. *Acta Odontol. Scand.* **1991**, *49*, 297.

100. Mirth, D. B.; Bartkiewickz, A.; Shern, R. J.; Little, W. A. *J. Dent. Res.* **1990**, *68*, 1285.

101. Keatsu, I.; Kumakura, M.; Yoshida, M. *Biotechnol. Bioeng.* **1979**, *21*, 847.

102. National Research Development Corp. (U.K.) Patent CooperationTreaty, Int. Appl. World, 9013021 A1, 1990; Pickup, J. C.; Claremont; D. J. *Chem. Abstr.* **1991**, *115*, 67971.

103. Shaw, G. W.; Claremont, D. J.; Pickup, J. C. *Biosens. Bioelectron.* **1991**, *6*, 401.

104. Yoshida, M.; Kaetsu, I. *J. Appl. Polym. Sci.* **1981**, *26*, 687.

105. Smalla, K.; Turkova, J.; Coupek, J; Hermann, P. *Biotechnol. Appl. Biochem.* **1988**, *10*, 2.

106. Carenza, M.; De Alteriis, E.; Lora, S.; Parascandolla, P.; Scardi, V. *Ann. Microbiol. Enzimol.* **1989**, *39*, 257.

107. Aleksieva, P.; Petricheva, E.; Konstantinov, Ch.; Robeva, M.; Mutafov, S. *Acta Biotechnol.* **1991**, *11*, 255.

108. Roneil, S. H.; D'Andreas, M. J.; Hashiguchi, H.; Klomp, G. F.; Dobelle, W. H. *J. Biomed. Mater. Res.* **1983**, *17*, 855.

109. Pollard, J. W.; Pineda, M. H. *In Vitro Fert. Embryo Tranfer.* **1975**, *5*, 207.

110. Dawson, R. M.; Broughton, R. L.; Stevenson, W. T.; Sefton, M. V. *Biomaterials* **1987**, *8*, 360.

111. Asahi Medical Company Ltd. Jpn. Patent, A2 0403 6656T., 1992; Nishimura, T.; Ijichi, S. *Chem. Abstr.* **1992**, *116*, 251650.

112. Yean, L.; Bunel, B.; Vairon, J. P. *Makromol. Chem.* **1990**, *191*, 1119.

113. Lee, P. I. *Polym. Commun.* **1983**, *24*, 45.

114. Ivan, B.; Kennedy, J. P.; Mackey, P. W. *Polym. Prepr. (Am. Chem. Soc. Div. Polym. Chem.)* **1990**, *31*, 217.

115. Seki, T.; Sugibayashi, K.; Juni, K.; Morimoto, Y. *Drug Des. Delivery* **1989**, *4*, 69.

116. Miyajima, M.; Okano, T.; Kim, S. W.; Higushi, W. I. *J. Controlled Release* **1987**, *5*, 179.

117. Domb, A.; Davidson, G. W. R.; Sanders, L. M. *J. Controlled Release* **1990**, *14*, 133.

118. Sefton, M. V.; Nishimura, E. *J. Pharm. Sci.* **1980**, *69*, 208.

119. Yoshida, M.; Asano, M.; Kaetsu, I.; Imai, K.; Mashimo, T.; Yuasa, H.; Yamanaka, H.; Kawaharada, U.; Suzuki, K. *Biomaterials* **1987**, *8*, 124.

120. Kaetsu, I.; Yoshida, M.; Yamada, A. *J. Biomed. Mater. Res.* **1980**, *14*, 185, 197.

121. Canal, T.; Peppas, N. A. *J. Biomed. Mater. Res.* **1989**, *23*, 1183.

122. Prisyazhnoy, V. S.; Fusek, M.; Alakhov, Y. B. *J. Chromatogr.* **1988**, *424*, 243.

123. Lydon, M. J.; Minett, T. W.; Tighe, B. J. *Biomaterials* **1985**, *6*, 396.

124. De Boisfleury-Chevance, A.; Rapp, B.; Gruler, H. *Blood Cells* **1989**, *15*, 315.

125. Harkes, G.; Feijen, J.; Dankert, J. *Biomaterials* **1991**, *12*, 853.

126. Lentz, A. J.; Horbertt, T. A.; Hsu, L.; Ratner, B. D. *J. Biomed. Mater. Res.* **1985**, *19*, 1101.

127. McAuslan, B. R.; Johnson, G. J. *J. Biomed. Mater. Res.* **1987**, *21*, 921.

128. Smetana, K.; Sulc, J.; Krcova, Z. *Exp. Mol. Pathol.* **1987**, *47*, 271.

129. Cifkova, I.; Brynda, E.; Holusa, R.; Adam, M. *Biomaterials* **1987**, *8*, 30.

130. Stol, M.; Cifkova, I.; Tyrakova, V.; Adam, M. *Biomaterials* **1991**, *12*, 454.

131. Ericksson, C. *J. Biomed. Mater. Res.* **1985**, *19*, 833.
132. Smetana, K., Jr.; Stol, M.; Korbelar, P.; Novak, M.; Adam, M. *Biomaterials* **1992**, *13*, 639.
133. Kumakura, M. *J. Biomed. Mater. Res.***1986**, *20*, 521.
134. Jansen, B.; Elling Horst, E. *J. Biomed. Mater. Res.* **1985**, *19*, 1085.
135. Lai, J. Y.; Shih, C. Y.; Tsai, S. M. *J. Appl. Polym. Sci.* **1991**, *43*,1431.
136. Hatanaka, T.; Inuma, M.; Sugibayaski, K.; Morimoto, Y. *Int. J. Pharm.* **1992**, *79*, 21.
137. Blue, M. A.; Guilbeau, E. J.; Brandon, T. A.; Walker, A. S.; Bjotvedt, G.; Fish, R. L.; Leighton, R. *ASAIO Trans.* **1991**, *37*, M152–M153; *Chem. Abstr.* **1992**, *117*, 157573.
138. Okano, T.; Aoyagi, T.; Kataoka, K.; Abe, K.; Sakurai, Y.; Shimadada, M.; Shinohara, I. *J. Biomed. Mater. Res.* **1986**, *20*, 919.
139. Nojiri, C.; Okano, T.; Jacobs, H. A.; Park, K. D.; Moham-mad, S. F.; Olsen, D. B.; Kim, S. W. *J. Biomed. Mater. Res.* **1990**, *24*,1157.
140. Seifert, L. M.; Green, R. T. *J. Biomed. Mater. Res.* **1985**, *19*, 1043.
141. Nonent, M.; Laurent, A.; Merland, J. J.; Huguet, J.; Vert, M. *J. Mater. Sci.: Mater. Med.* **1991**, *2*, 89.
142. Purdy, P.; Charles, C. L.; Batjer, H.; Brewer, K.; Hodges, K. D.; Samson., D. *J. Neurosurg.* **1990**, *73*, 756.
143. Gobel, R. J.; Janatova, J.; Googe, J. M.; Apple, D. J. *Biomaterials* **1987**, *8*, 285.
144. Goelander, C. G.; Lassen, B.; Nilsson-Ekdahl, K.; Nils-son, U. R. *J. Biomater. Sci. Polym. Ed.* **1992**, *4*, 25.
145. Lee, C. J.; Hsu, S. T. *J. Biomed. Mater. Res.* **1990**, *24*, 243.
146. Horak, D.; Svec, F.; Kalal, J.; Gumargalieva, K.; Adamyan, A.; Skuba, N.; Titova, M.; Trostenyuk, N. *Biomaterials* **1986**, *7*, 188.
147. Horak, D.; Svec, F.; Adamyan, A.; Titova, M.; Skuba, N.; Voronkova, O.; Trostenyuk, N. *Biomaterials* **1992**, *13*, 521.
148. Menace, R.; Skorpik, C.; Juchem, M.; Scheidel, W.; Schranz, R. *J. Cataract Refract. Surg.***1989**, *15*, 264.
149. Remeijer, L.; Van-rij, G.; Beekhuis, W. H.; Polak, B. C.; Van-nes, J. *Ophthalmology* **1990**, *97*, 281.
150. Goldenberg, J. M. S.; Beckman, A. C. *Biomaterials* **1991**, *12*, 267.
151. Sunny, C. M.; Sharma, C. P. *Biomater. Artif. Cells Immobi-lization Biotechnol.* **1991**, *19*, 599.
152. Yalon, M.; Blumenthal, M.; Goldberg, E. P. *J. Am. Intraocul. Implant Soc.* **1984**, *10*, 315.
153. Barrett, G. D.; Constable, I. J.; Steward, A. D. *J. Cataract Refract. Surg.* **1986**, *12*, 623.

Molecularly Designed Dental Polymers

Joseph M. Antonucci and Jeffrey W. Stansbury

Among a wide range of polymers examined for dental applications, acrylics have played the major role in preventive and restorative dentistry because of their unique combination of properties such as facile free-radical polymerization, excellent aesthetics, and good mechanical strength. However, major shortcomings of conventional acrylic dental polymers include excessive polymerization shrinkage, less than optimal conversion, susceptibility to air inhibition, inadequate wear resistance, poor physicochemical stability in the oral environment, and lack of durable adhesion to tooth structure. These critical deficiencies limit the service life and applicability of current acrylic-based dental materials. Newly designed monomers, oligomers, and polymers are under development to enhance the properties and range of applications of polymeric dental materials. The new materials include hydrophobic, low-shrinking, multifunctional oligomers; cyclopolymerizable and ring-opening monomers and oligomers; and more effective adhesion-promoting systems.

Dentistry is concerned with the prevention, diagnosis, and treatment of diseases and traumas that adversely affect oral tissues and structures. A primary goal of modern dentistry is the replacement of defective or lost teeth with materials that will restore function and aesthetics. Because the regeneration of tooth tissue is presently not feasible, dental practitioners have turned to essentially bioinert materials (metals, ceramics, polymers, and their various combinations) for restorative purposes.

Polymers are widely used for dental applications such as artificial teeth, denture bases, composite restoratives, temporary crowns, veneers, cements, sealants, adhesives, root-canal fillers, impression materials, and maxillofacial prostheses. Examples of natural polymers used in dentistry are agar (1) and alginic acid (2), both used as hydrocolloidal impression materials, and *trans*-1,4-polyisoprene or gutta-percha (3), a material used to obturate root canals.

Historically, dentistry has always been quick to exploit new developments in synthetic polymers.

Goodyear's discovery of vulcanized rubber in 1839 soon led to the development of the denture-base material vulcanite, which was used until the advent of dental acrylics in 1937. Since then, a wide spectrum of other synthetic polymers (e.g., acrylics, polyvinyls, polyesters, polyamides, polycarbonates, polyethers, epoxides, polysulfides, and polysiloxanes) has been examined for a variety of dental uses.

Acrylic polymers have maintained a predominant position in dentistry (1) because of their unique combination of favorable properties pertaining to polymerization and processing, aesthetics, and mechanical and physicochemical behavior. However, conventional acrylic monomers and polymers have several critical deficiencies that limit their clinical performance. Important shortcomings of acrylic systems include excessive polymerization shrinkage, incomplete polymerization, inhibition by oxygen during cure, less than optimal stability in the oral environment, and inadequate wear resistance and adhesion to tooth structure (especially to dentin). These and other deficiencies of

Structures 1–3. Examples of natural polymers used in dentistry: agar (**1**), alginic acid (**2**), and gutta percha (**3**).

current dental polymeric materials are being addressed by searching a wide range of other monomers, oligomers, and polymers. This chapter delineates significant new directions in the design of monomers, oligomers, and polymers for dental and related biomedical applications as we enter the twenty-first century.

Early acrylic denture-based and restorative materials were made by the ambient polymerization of a composite dough derived from mixing methyl methacrylate (MMA) and fine beads of poly(methyl methacrylate) (PMMA) by using an initiator system consisting of benzoyl peroxide and a tertiary aryl amine such as N,N-dimethyl-p-toluidine. This type of organic composite has a number of desirable properties: ease of manipulation and fabrication, excellent optical properties, low solubility, smooth surface texture, and a tough ductile nature. Unfavorable properties of MMA–PMMA composites that limit their use as a permanent filling material include a coefficient of thermal expansion significantly greater than tooth structure, a relatively low modulus, high polymerization shrinkage, and questionable biocompatibility due to the volatility and leachability of MMA (1). The MMA–PMMA system is still widely used as a denture base material and as an orthopaedic bone cement.

The modern era of dental acrylic materials can be traced to the pioneering research of R. L. Bowen (2–5), who originally conceived the use of epoxies such as the diglycidyl ether of bisphenol A (**4**, Scheme I) as the polymer matrix for mineral-filled restorative composites. However, this early use of epoxides failed, mainly because of the poor ambient curing characteristics of this class of monomers with the initiator systems then available (2). These studies, however, paved the way for the synthesis of a thermosetting acrylic derivative of **4**, 2,2-bis[p-(2'-hydroxy-3'-methacryloxypropoxy)phenyl]-propane (commonly referred to as BIS-GMA, **6**, Scheme I). The introduction of this monomer in turn led to the

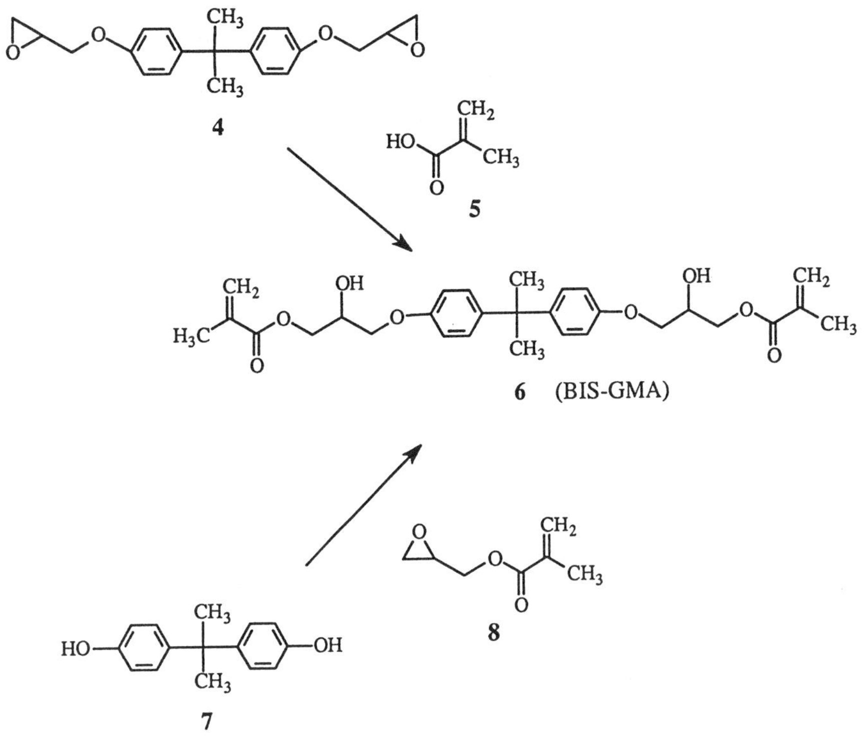

Scheme I. Synthesis of BIS-GMA (**6**) from the diglycidyl ether of bisphenol A (**4**) and methacrylic acid (**5**), or alternatively, from bisphenol A (**7**) and glycidyl methacrylate (**8**).

development of the modern polymer-matrix, restorative-dental composites, resin cements, sealants, adhesives, and prostheses that have revolutionized the practice of restorative and preventive dentistry (*3–5*).

Although BIS-GMA represents a major advance over MMA, it still has several deficiencies as a dental monomer. Its shortcomings include high polymerization shrinkage, relatively low cure efficiency at ambient temperatures, high viscosity, and plasticization of its polymers by oral fluids. These shortcomings lead to accelerated degradation. Intermolecular hydrogen bonding renders BIS-GMA an extremely viscous monomer, and the use of considerable amounts of less-bulky diluent monomers are required for obtaining dental resins with tractable viscosities. The use of diluent monomers further augments the degree of polymerization shrinkage and the attendant problems of internal strains, gap formation, and microleakage. **9–14** are typical diluent monomers that can be used with BIS-GMA (and other viscous monomers) to formulate resin systems for dental use.

The hydroxyl groups of BIS-GMA and the ethylene oxide segments of diluent monomers such as triethylene glycol dimethacrylate **9** (commonly referred to as TEGDMA) contribute to the relatively high water-sorption values of the resulting polymeric materials. Replacement of **9** with more hydrophobic diluent monomers such as hexamethylene 1,6-dimethacrylate **10**, decamethylene 1,10-dimethacrylate **11**, dodeca-

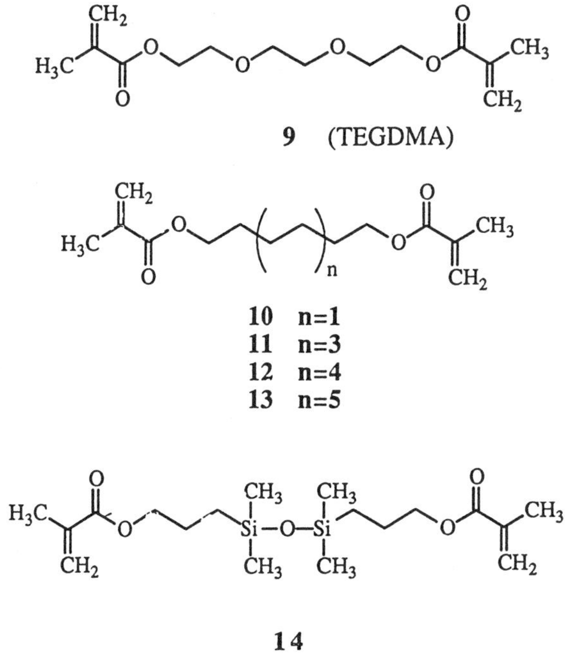

Structures 9–14. Diluent monomers used in the formulation of dental resin systems.

methylene 1,12-dimethacrylate **13**, tetradecamethylene 1,14-dimethacrylate **13**, or bis(methacryloxypropyl-tetramethyl) disiloxane **14** reduces water uptake. Finally, the more rigid structure of BIS-GMA typically results in monomer systems with relatively low degrees of cure or conversion (*6*). In an attempt to overcome the shortcomings of BIS-GMA, several alternative types of acrylic and nonacrylic monomers have been synthesized and evaluated for dental applications and are described subsequently.

New Acrylic Monomers and Oligomers

Hydrophobic Hydrocarbon-Based Acrylics

To enhance the hydrophobicity, flexibility, and cure efficiency of dental polymers, nonhydroxylated BIS-GMA-type monomers such as 2,2-bis(*p*-methacryloxyphenyl)propane **15**, 2,2-bis[*p*-(2'-methacryloxyethoxy)phenyl]propane **16**, and 2,2-bis[*p*-(3'-methacryloxypropoxy)phenyl]propane **17** were synthesized (*7–9*). Dimethacrylates made from ethoxylated derivatives of bisphenol A are relatively hydrophobic (if degree of ethoxylation is relatively low) and have relatively flexible structures with low viscosities. Dental monomer systems based on these flexible derivatives of bisphenol A yield polymers with relatively high degrees of cure and low polymerization shrinkage (*6*). With higher degrees of ethoxylation in **16**, hydrophilicity and flexibility increase in a manner analogous to that observed for aliphatic dimethacrylates based on polyethylene glycols. In contrast, dimethacrylates based on long alkyl-chain diols increase in hydrophobicity with an increase in the bulk of the hydrocarbon spacer

group (**10** versus **13**). Excellent polymeric composites having low water sorption were developed based on the tricyclic aliphatic hydrocarbon acrylates and methacrylates, **18**, derived from tricyclo[5.2.1.0^{2,6}]-decanediol (*9*).

Urethane- and Urea-Acrylics

Various urethane dimethacrylates and oligomeric urethane methacrylates have been employed as resin binders for dental composites and related materials (*10–16*). The facile synthesis of urethane- and urea-acrylics has the option of using a variety of hydroxylated or aminated acrylates (such as 2-hydroxyethyl methacrylate **19** [HEMA] or *N-tert*-butylaminoethyl methacrylate **20**, respectively) and isocyanates and permits the introduction of a wide range of properties into these monomers and oligomers and their resulting polymers. Both monomeric and oligomeric urethane derivatives (**21** and **22**) of BIS-GMA were synthesized and are finding increased use in a variety of dental applications.

Properties of oligomeric urethane derivatives of BIS-GMA can be significantly affected by the choice of the diisocyanate component. Thus, the oligomeric derivative **22** based on 1,6-hexamethylene diisocyanate **23** is a stiff solid, whereas the oligomer derived from the high molecular weight (600 daltons), isomeric mixture of diisocyanates (of which **24** is a typical

Structures 15–18. Hydrophobic hydrocarbon dimethacrylates used in dental resin systems.

Structures 19–24. Examples of functionalized methacrylates (**19** and **20**) that can be combined with diisocyanates to form simple urethane and urea dimethacrylates. The reaction of BIS-GMA with isocyanates or diisocyanates (such as **23** and **24**) provides monomeric (**21**) or oligomeric (**22**) urethane derivatives, respectively.

example) derived from dimer acid is a viscous liquid. When prepared in dichloromethane, oligomer **22** (obtained from chain-extending BIS-GMA with diisocyanate **24**) can have average molecular weights as high as 60,000 daltons (*17*). Dental polymers based on this oligomer or prepolymer yield tough, strong materials with high degrees of cure under ambient photopolymerization (Table I) (*17*). Composites based on this oligomer have excellent mechanical strength and are expected to have low water sorption and low polymerization shrinkage. Urea-acrylics have also been prepared by similar routes. Thus, reaction of **20** with diisocyanate **23** yields a solid product (mp 50 °C), and reaction with **24** yields a viscous liquid (*16, 17*).

Fluorocarbon- and Siloxane-Acrylics

The polymeric binder of dental composites can be regarded as the first line of defense against the con-

stant challenges of a hostile oral environment. Excessive absorption of oral fluids by the polymer matrix leads to plasticization, dimensional instability, accelerated wear, and ultimately failure of composite fillings. Similar adverse effects occur with cements, sealants, and adhesives that are based on conventional acrylic resins. Dental polymers with low surface energy and solubility parameters highly different from those of the oral fluids provide a means of enhancing the durability of polymeric dental materials (*18, 19*).

A wide range of fluorinated acrylics and methacrylates (**25–29**) have been synthesized from the polyfluorinated telomer alcohols and from alcohols obtained from hexafluoroacetone (*16, 18–20*). Polyfluorinated acrylic oligomers (**31–33**) have been prepared from the poly(fluoroprepolymer polyol) **30**. Composites based on these low-surface-energy oligomers exhibited acceptable mechanical strength, low polymerization shrinkage, and extremely low water sorption (Table II).

Table I. Degree of Cure and Diametral Tensile Strength of Oligomeric Urethane-Based Composites

Resin Composition	Initiator (wt%)	Cure (%)	Tensile Strength (MPa)
BIS-GMA/**9**	0.20/0.80	74.4	52.4
22/9	0.34/1.40	82.2	50.6
22/10	0.32/0.82	85.0	43.7
22/11	0.39/1.33	75.7	47.2
22/12	0.33/1.39	82.9	43.3
22/13	0.47/1.49	89.2	35.7
22/14	0.28/1.02	90.8	38.6

Note: Resin composition was 1:1 by weight. Photoinitiator system was camphorquinone/ethyl 4-*N*,*N*-dimethylaminobenzoate (amine). Urethane oligomer was based on BIS-GMA + diisocyanate **24**. Composites contained silanized particulate glass fillers at 83.3%.

Table II. Properties of Glass-Filled Resin Composites Based on Polyfluorinated Acrylic Oligomers

Resin Composition	Weight (%)	Tensile Strength (MPa)	Water Uptake (mg/cm²)	Polymerization Shrinkage (%)	Cure (%)
32	70	41	0.17	—	—
11[a]	30				
33	68	47	0.21	2.8	69
10[b]	32				
33	70	45	0.20	2.3	71
14[b]	30				
BIS-GMA	70	45	0.73	3.9	65
9[b]	30				

Note: Composites contained silanized particulate glass fillers at 75 wt%.

[a]Composite chemically cured with benzoyl peroxide–*N*,*N*-dihydroxyethyl-*p*-toluidine initiator.

[b]Composite photochemically cured with camphorquinone–ethyl 4-*N*,*N*-dimethylbenzoate.

Structures 25–33. Structures of polyfluorinated acrylic monomers (**25–29**) and multifunctional fluorinated acrylic oligomers (**31–33**) derived from a polyfluoropolyol **30**.

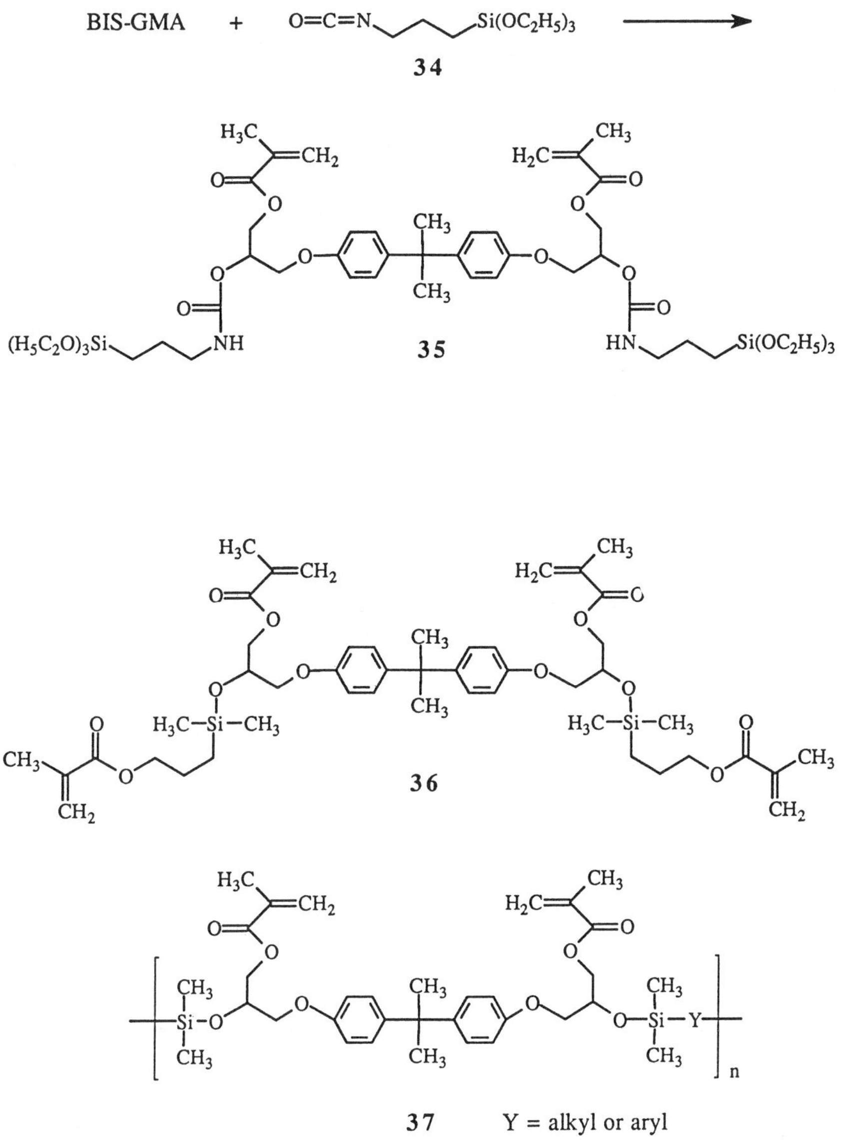

BIS-GMA   +   O=C=N⁀⁀⁀Si(OC₂H₅)₃

34

35

36

37   Y = alkyl or aryl

Scheme II. Synthesis of a multifunctional silane coupling agent (**35**) derived from BIS-GMA and examples of silyl ether derivatives (**36** and **37**) of BIS-GMA.

Siloxane monomers such as bis(methacryloxypropyl)tetramethyldisiloxane (**14**) also have been shown to be effective in reducing the water absorption of polymeric composites and enhancing the degree of cure (*21*). Multifunctional silane coupling agents such as **35** have been synthesized by using 3-isocyanato-propyltriethoxysilane (**34**, Scheme II) with hydroxylated monomers in an effort to improve the interfacial phase in dental composites (*22*). Several unique silyl ether derivatives (**36** and **37**) of BIS-GMA also have been synthesized. These compounds appear to be surprisingly stable despite the presence of the potentially hydrolyzable silyl ether groups in their structures (*22*).

Composite samples (Table III) exhibited no evidence of hydrolytic degradation during prolonged storage in water. The observed hydrolytic stability of composites based on silyl ether polymers is probably

Table III. Diametral Tensile Strength of Siloxane- and Silyl-Ether-Containing Composites

Resin Composition	Weight (%)	Glass/Resin Ratio	Tensile Strength (MPa)
BIS-GMA	70	4	45
14	30	5	49
36	100	6	42
37	71	4	35
10	29		
37	58	5	42
10	42		
BIS-GMA	48	5	43
36	48		
MPDES	4		

Note: Composites were photochemically cured with camphorquinone–ethyl 4-*N,N*-dimethylbenzoate. MPDES is 3-methacryloxypropyldimethylethoxysilane.

due to the sterically hindered silyl ether groups and the highly cross-linked structure of the matrix. Self-healing (curing) mechanisms such as those shown in Scheme III also may act to maintain the structural integrity of this type of composite. Preliminary studies (Table III) indicate that these composites have acceptable mechanical strength.

Cyclophosphazene Acrylics

Hexafunctional and octafunctional methacrylates derived from the base-catalyzed condensation of hexachloro- and octachlorocyclophosphazenes with HEMA **19** have been synthesized (**39** and **40**, Scheme IV) for composite and other dental applications (*18, 23*).

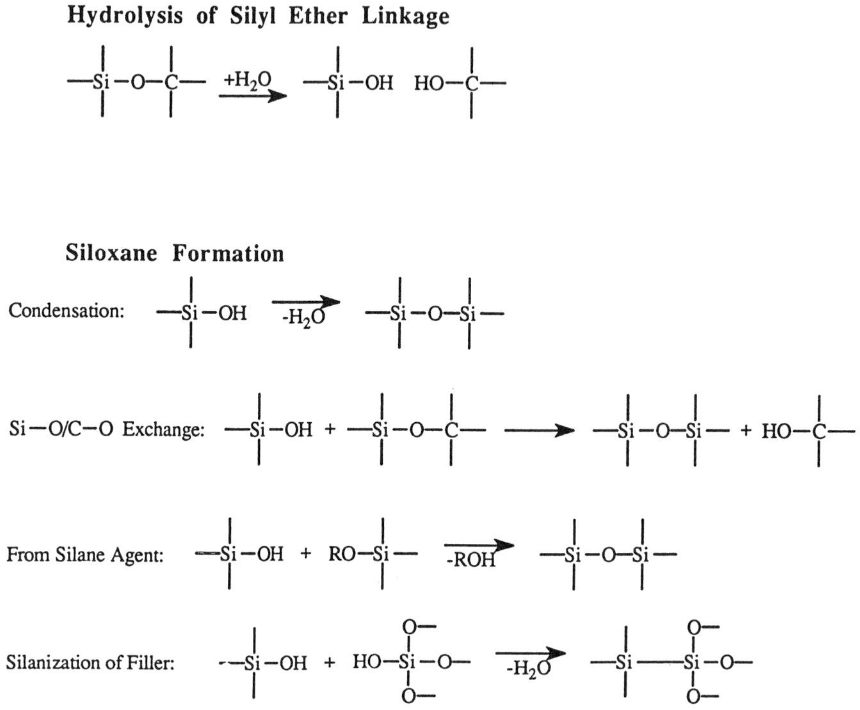

Hydrolysis of Silyl Ether Linkage

Siloxane Formation

Scheme III. Possible self-healing mechanism for glass-filled composites based on silyl ether monomers.

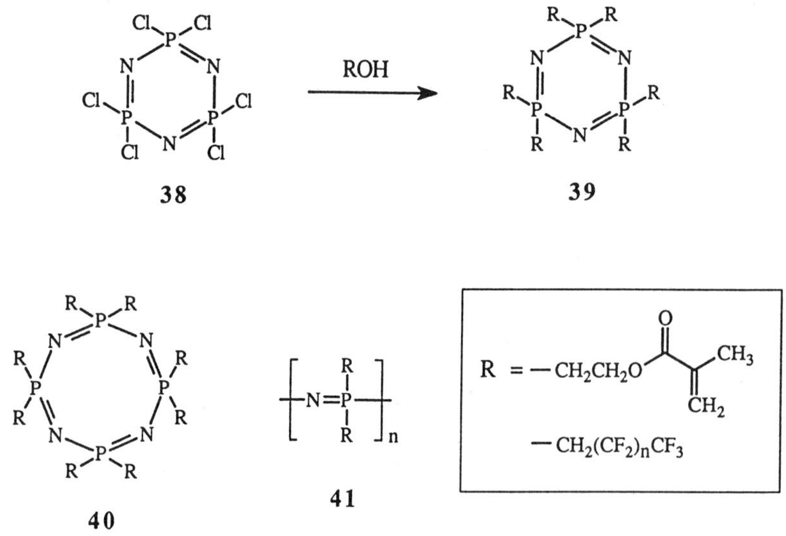

38 → ROH → **39**

40   **41**

$R = -CH_2CH_2O$ ... CH_3, CH_2

$-CH_2(CF_2)_nCF_3$

Scheme IV. General structures of acrylic cyclophosphazenes derived from the trimer (**39**) and tetramer (**40**) forms of the corresponding phosphonitrilic chlorides and 2-hydroxyethyl methacrylate (**19**); other ring substituents such as those derived from fluoroalcohols can replace some of the HEMA substituents. General representation of a linear ring-opened polyphosphazene (**41**).

These cyclophosphazene rings, with their semi-inorganic structures, confer their polymers and copolymers several interesting properties such as low coefficients of thermal expansion, exceptional surface hardness and compressive strength, and low polymerization shrinkage. However, they have relatively low tensile strength and high water-sorption values. The relatively brittle nature of these cyclophosphazene polymers is in contrast with the elastomeric ring-opened phosphazenes **41**. The replacement of some of the oxyethyl methacrylate substituents of the cyclophosphazene monomers by the more hydrophobic oxyfluorocarbon substituents reduces the water sorption of their polymers, but this replacement also reduces their cross-link density and moduli (*23*).

Methylene Lactone Monomers

α-Methylene lactones have been examined as alternatives to conventional acrylic monomers. The simplest example of this class of monomer is α-methylene butyrolactone (MBL), which can be described as the cyclic analog of MMA. As a monomer, MBL is a colorless, nonvolatile liquid of low viscosity and a boiling point of 210 °C (more than 100 °C higher than that of MMA) (*24*). In free-radical polymerizations, MBL is significantly more reactive than MMA and produces polymers with high glass-transition temperatures and improved solvent resistance (*25, 26*).

The use of MBL as a reactive diluent comonomer in photocured BIS-GMA resins demonstrated that incorporation of small amounts of MBL in the formulation provides polymers with improved mechanical properties and reduced levels of residual unsaturation (Table IV) (*27*). The polar nature of MBL makes this monomer an excellent solvent for polymers and for the introduction of a variety of surface-active bonding agents used in dentistry. The synthesis of substituted α-methylene lactones and difunctional bis(methylene lactone)s was reported (*28–30*). Cross-linked polymers derived from multifunctional lactone monomers are expected to exhibit high conversion and strong solvent resistance, and they provide interesting candidates for composite and adhesive dental materials.

Cyclopolymerizable Monomers

Cyclopolymerization (ring-closing polymerization) refers to a polymerization process in which intermolecular addition between monomer units alternates with an intramolecular cyclization step (*31*). Cyclization also occurs to a minor degree in the polymerization of conventional dimethacrylates (*32*). Since the recent discovery that monomers that are efficiently cyclopolymerizable can be produced simply from the amine-catalyzed coupling between acrylates and formaldehyde

Table IV. Properties of Light-Cured Unfilled Resins

Parameter	BIS-GMA/9	BIS-GMA/9/ MBL
Resin composition		
Weight (%)	70:30	63:27:10
Mole (%)	54:46	37:32:31
Tensile strength (MPa ± SD)	42.2 ± 3.6	46.6 ± 3.7
Transverse strength (MPa ± SD)	75.3 ± 4.3	82.4 ± 2.4
Conversion of C=C (%)		
Methacrylate	57	71
Methylene lactone	—	96
Overall conversion	57	75

Note: SD is standard deviation.

(*33–36*), developments with this new class of free-radical polymerizable monomers have advanced rapidly. 1,4-Diazabicyclo[2.2.2]octane (DABCO) is used as the amine catalyst in the reaction between an acrylate and formaldehyde (generally as paraformaldehyde) to yield an α-hydroxymethylacrylate **42** (Scheme V). This functionalized intermediate undergoes further DABCO-catalyzed condensation to provide the cyclopolymerizable oxybismethacrylate **43**. These ether-linked 1,6-diene monomers show a strong propensity toward intramolecular cyclization during polymerization that results in high degrees of conversion and reduced polymerization shrinkage (Table V) (*37–39*). Bulky ester groups on the oxybismethacrylate monomers further enhance polymerization via internal cyclization when compared with polymerization via 1,2-addition (*see* Scheme V) (*40, 41*).

Completely cyclized polymers (**44**, *x* = 100%) are obtained from suitably dilute-solution polymerizations of oxybismethacrylate monomers, and depending on the monomer structure, cyclization efficiencies of 65% to >99% can be achieved in bulk polymerizations (*40, 42*). Therefore, the resulting cyclopolymers **44** often contain few cross-links and may not be suitable for most dental material applications. However, diacrylates can also be chain-extended via the formaldehyde condensation reaction to yield multifunctional oligomers (**45**, Scheme VI) that have several discrete cyclopolymerization points per chain (*43, 44*). This process gives access to highly cross-linked polymers and polymerization shrinkage values approximately 30% lower than those of conventional resin monomers (Table VI). The cyclopolymerizable oligomers are generally quite viscous and require the addition of diluent comonomers to produce workable resins for dental-material applications. Recently, completely cyclopolymerizable dental-resin formulations were prepared (*45, 46*) in a single-step process that combines mono- and diacrylate starting materials in the reaction with formaldehyde. A wide variety of acrylates and diacrylates can be used to obtain resins

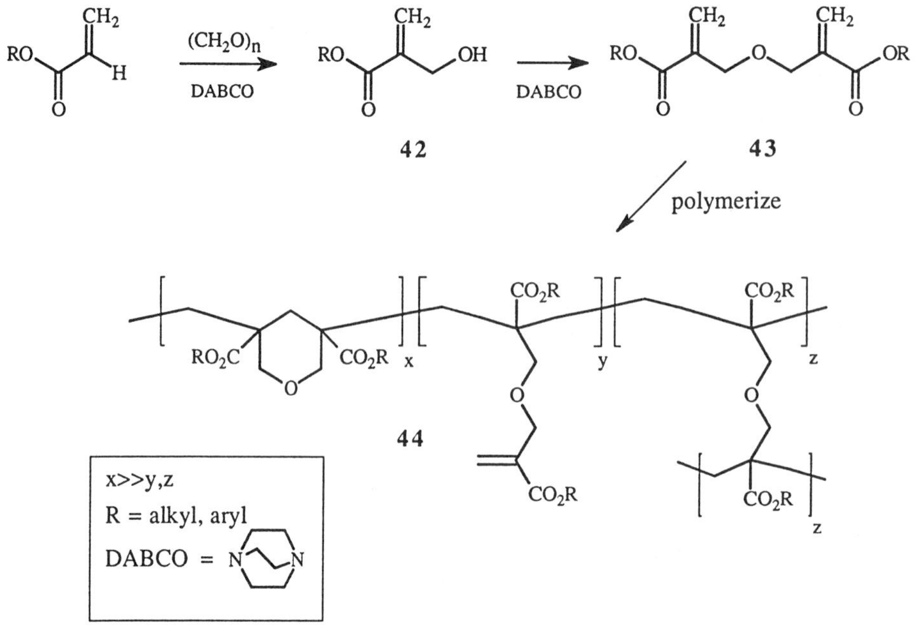

Scheme V. Synthesis of cyclopolymerizable oxybismethacrylate monomers (**43**) by the amine-catalyzed reaction of acrylates with paraformaldehyde.

Table V. Oxybismethacrylate and Dimethacrylate Homopolymers

Monomer	Polymerization Temp. (°C)	Cure (%)	Volumetric Shrinkage (%)
43	60[a]	100	12.8
43	60[b]	95	11.6
43	23[c]	89	11.2
9	60[b]	91	19.9
9	37[b]	68	12.5

Note: Monomer **43** is diethyloxybismethacrylate. Monomer **9** is triethylene glycol dimethacrylate.

[a]In solution.
[b]In bulk.
[c]Bulk photopolymerization.

tailored for specific applications. Properties such as resin viscosity and average molecular weight of the oligomeric component can also be controlled by adjusting the ratio of mono- to diacrylate used in the reaction.

Hydrophilic cyclopolymerizable monomers and oligomers were prepared (*47*) by procedures similar to those used to convert conventional alkyl mono- and diacrylates to the respective difunctional monomers and multifunctional oligomers. The hydrophilic monomers and oligomers have the same predominant 1,6-diene structure obtained for the alkyl diacrylates. Water-dispersible or water-soluble cyclopolymerizable monomers and resins are available through inclusion of hydrophilic functionalities such as tetrahydrofurfuryl groups and polyethylene oxide (PEO) segments. The PEO oligomers (**46**, Scheme VI) are novel hydro-

gels that should find application in composites, cements, and other resin-based materials. Their bulky size, cyclopolymerizability, and potential for hygroscopic expansion help moderate the effects of polymerization contraction.

Similarly, hydrophobic character can be introduced into the cyclopolymerizable monomers and oligomers by incorporation of siloxane- or fluorine-based segments. Like the alkyl analogs, the siloxane oligomers (**47**, Scheme VI) have mainly 1,6-diene arrangements of double bonds. However, the siloxane-oligomer synthesis requires the use of a solvent such as dimethyl sulfoxide to increase compatibility (*47*).

The use of dimethyl sulfoxide was also necessary to affect the synthesis of highly fluorinated difunctional monomers and oligomers from the corresponding fluoroacrylates (*48*). The presence of the electron-withdrawing fluorine groups on the acrylate results in a substantial contribution of an alternative pathway in the reaction with paraformaldehyde and DABCO; that is, 1,4-diene units (rather than the 1,6-diene units) predominate in the resulting structure (**48**). The 1,4-diene group does not effectively cyclize during free-radical polymerization. Formation of the 1,4-diene linkage with highly fluorinated materials can be circumvented by use of acrylates in which the fluorine substituents are located farther away from the ester carboxylate functionality (as demonstrated with the cyclopolymerizable monomer **49** and oligomer **50**) (*49*). The siloxane- and fluorine-containing oligomers, which combine large bulk, comparatively low viscosity, high flexibility, and cyclopolymerizability, offer a

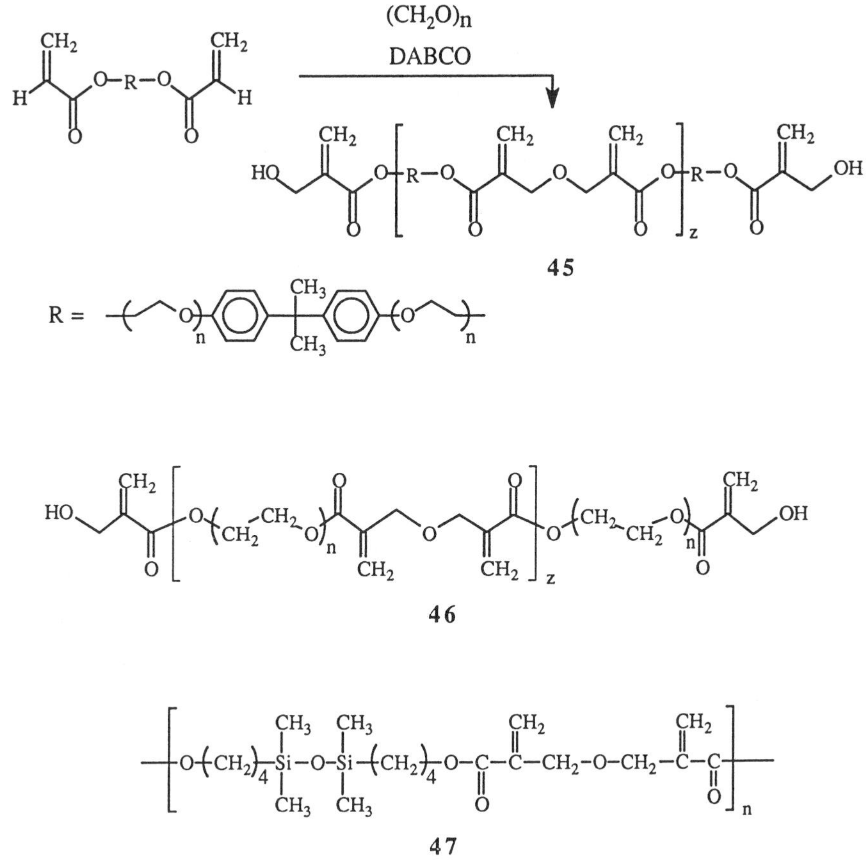

Scheme VI. Synthesis of a cyclopolymerizable multifunctional oligomer (**45**) by the chain-extension reaction of a diacrylate with paraformaldehyde. Examples of hydrophilic and hydrophobic cyclopolymerizable oligomers (**46** and **47**, respectively).

facile means of achieving hydrophobic, tough polymeric structures with low shrinkage that are suitable for use in a variety of dental and medical applications.

By adjusting the conditions of the DABCO-catalyzed formaldehyde condensation, the yield of α-hydroxymethylacrylate (**42**, Scheme V) can be maximized (*50*). As a structural isomer of HEMA **19**, this hydrophilic monomer is of interest in a number of dental applications including as a dentin primer to enhance bonding of composite restorations. Various polymers and copolymers of **42** (*51–53*) and those of its derivatives obtained by modifications of the hydroxyl functionality were investigated (*54–56*).

Nonacrylic Monomers and Oligomers

Single Ring-Opening Polymerization of Vinylene Acetals

Ring-opening monomers are potentially useful in a number of applications where low polymerization shrinkage or biodegradability is desirable (*57*). Compared with the common addition polymerization of vinyl monomers such as MMA or styrene, ring-opening polymerization of cyclic monomers results in sig-

Table VI. Properties of Photocured Composites Derived from BIS-GMA, Ethoxylated Bisphenol A Dimethacrylate (**16**), or Cyclopolymerizable Oligomer (**45**)

Parameter	BIS-GMA/9	16/9	45/9
Mole ratio	56:44	52:48	52:48
Cure (%)	65	77	74
DTS (MPa ± SD)	50.5 ± 1.3	51.6 ± 2.0	51.0 ± 0.9
TS (MPa ± SD)	91.4 ± 8.5	90.1 ± 8.0	86.0 ± 5.9
EB (mJ/mm³ ± SD)	0.101 ± 0.022	0.123 ± 0.036	0.133 ± 0.014

Note: DTS is diametral tensile strength; TS is transverse strength; EB is energy to break; SD is standard deviation.

nificantly less polymerization shrinkage, because bond formation and bond cleavage are balanced (*58*). Exomethylene-substituted (vinylene) cyclic acetals and cyclic ketene acetals are two general classes of cyclic monomers suitable for free-radical, ring-opening polymerization.

Vinylene cyclic acetals such as 4-methylene-2-phenyl-1,3-dioxolane **51** and *p*-phenylene-2-bis(4-methylene-1,3-dioxolane) **53** (Scheme VII) were evaluated in homo- and copolymerization (*59*) procedures and in formulations of photocured, dental-resin composites (*60*). Free-radical polymerizations of **51** gave

Structures 48–50. Fluorinated oligomer (**48**) with a predominance of 1,4-diene linkages and examples of fluorine-containing monomers (**49**) and oligomers (**50**) with efficiently cyclopolymerizable 1,6-diene groups.

complex polymers **52** comprised of mixtures of the ring-opened keto-ether structure together with units derived from 1,2-addition and elimination pathways. Virtually complete ring-opening polymerization of the cyclic acetal monomers such as **53** could be achieved by incorporating a photocationic initiator in the formulation (*61, 62*). Cyclic acetal compounds lacking the exo-methylene group were also used (*60*) with methacrylates in a rather unsuccessful attempt to achieve copolymerization through a hydrogen-abstraction, ring-opening pathway. Methacrylates bearing heterocyclic substituents produce cross-linked polymers as a result of hydrogen abstraction, but the efficiency of this process appears very low (*63*). Irrespective of the potential for ring-opening polymerization, the presence of a heterocyclic substituent in acrylic monomers appears to enhance the rate of polymerization (*64*).

Free-radical ring-opening polymerization occurs much more readily for cyclic ketene acetals, such as 5,6-benzo-2-methylene-1,3-dioxepane **55**, than for the vinylene cyclic acetals described previously. With the exo-vinylene group located at C_2 between two oxygen atoms (rather than at C_4 adjacent to a single oxygen), the double bond in the cyclic ketene acetals is more polar and more sterically accessible than in the vinylene cyclic acetals. Copolymerization of cyclic ketene

monomers and conventional vinyl monomers, such as MMA, can yield polymers with relatively high contents of ring-opened ester groups (**56**, Scheme VII) in the polymer backbone (*65, 66*). The mechanical strength of methacrylate-based, photocured composites is improved significantly by the addition of monomer **55** to the resin formulation (Table VII) (*67*). Ring size and substituents on the cyclic monomer and the polymerization conditions all affect the extent of ring-opening polymerization.

Double Ring-Opening Polymerization of Oxaspiro Systems

Vinyl monomers exhibit a volume contraction that reflects the conversion of the van der Waals distances between free monomer units to the covalent-bond distances linking these units in the polymer chain (*68*). In dental polymeric materials, as well as in some other polymer uses, polymerization shrinkage is responsible for the introduction of critical stresses that may ultimately result in diminished material performance or range of applications (*69–72*). A potential solution to this problem was devised by William J. Bailey (*73*) through the introduction of bicyclic oxaspiro monomers capable of double ring-opening

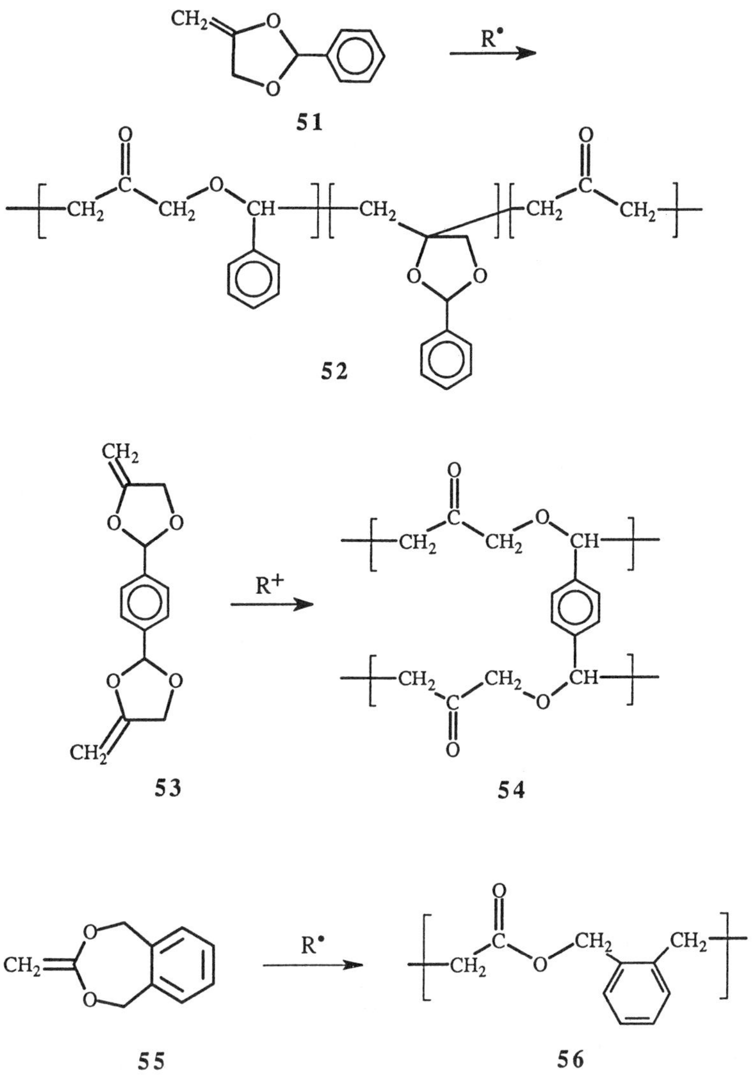

Scheme VII. Single ring-opening polymerization of vinylene cyclic acetals (**51** and **53**) and cyclic ketene acetals (**55**).

Table VII. Mechanical Properties of Photocured Composites of Ethoxylated Bisphenol A Dimethacrylate and Cyclic Ketene Acetal (**55**)

55 in Resin (wt%)	Diametral Tensile Strength (MPa ± SD)	Transverse Flexural Strength (MPa ± SD)
0	53.0 ± 2.1	90.8 ± 7.3
4.0	53.2 ± 2.0	94.0 ± 7.5
8.0	59.2 ± 1.9	102.0 ± 11.0
17.5	61.8 ± 1.9	118.1 ± 2.9
32.1	51.8 ± 2.1	89.9 ± 10.2

Note: SD is standard deviation.

polymerization. In these systems, two rings are opened for each monomer addition to the growing polymer chain (Scheme VIII). As a result, expansion occurs due to a net decrease in the number of covalent bonds in the polymer compared with the monomer. The polymerization of oxaspiro monomers (e.g., spiro orthocarbonates **57**) with either zero shrinkage

or a modest expansion in volume has attracted increasing attention for the development of dental and other polymeric materials (74).

One application where the potential advantages of ring-opening polymerization may be realized is that of dental-composite filling materials. Because the deleterious effects of shrinkage in conventional dental polymers are fairly well defined, the development of composites relying on zero-shrinkage resin systems appears highly desirable. This development would alleviate the serious problem of marginal gaps, which occur routinely in the placement of dental composites when the contractile stresses of polymerization exceed the adhesive bond strength of the composite-tooth interface. In addition, volume-neutral polymerization would eliminate the stress within the resin matrix and at the resin–filler interface, which may be responsible in part for the low durability of current dental composites (75). Dental composites provide a viable area for research on new ring-opening spirocyclic mono-

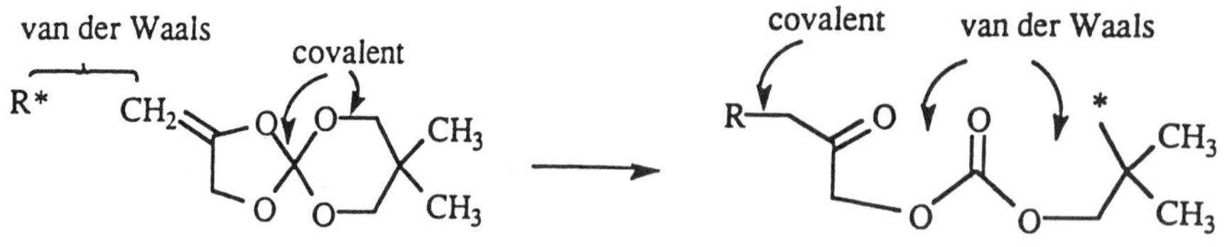

Scheme VIII. Source of expansion during the double ring-opening polymerization of a spiro orthocarbonate monomer (**57**) and examples of symmetric (**58** and **59**) and asymmetric (**60** and **61**) oxaspiro monomers.

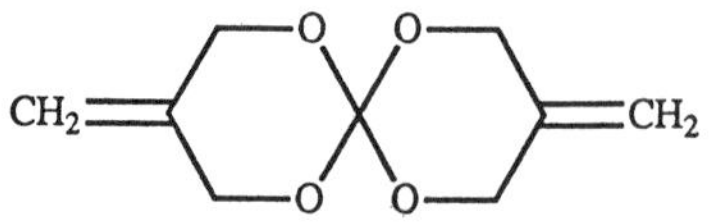

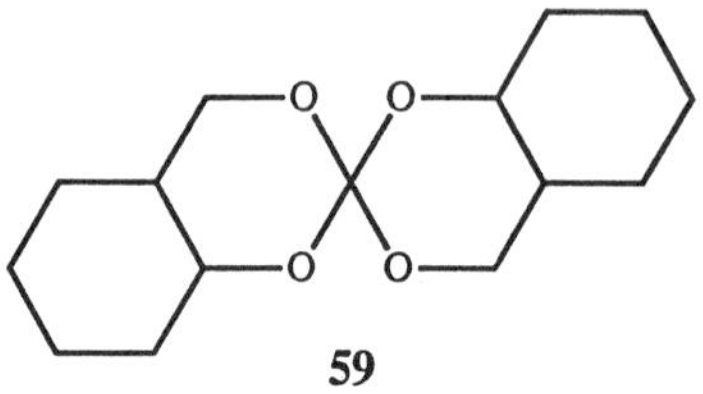

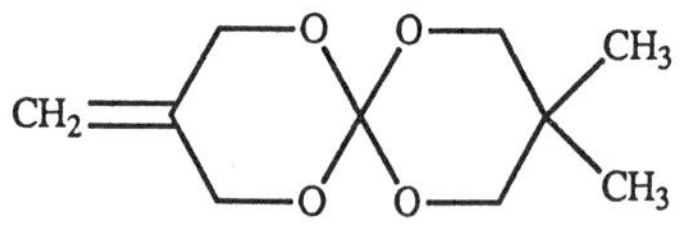

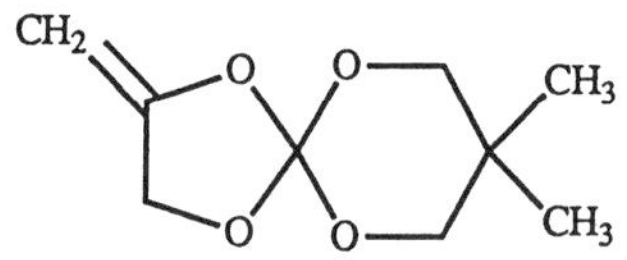

mers because resin cost is a minor contribution to the overall expense of dental restorations. Such an exercise would be highly cost-effective if the material performance can be improved significantly.

Soon after the development of free-radical polymerization of oxaspiro monomers, expanding monomers were introduced to dentistry and had promising results (*76*). Initially, a symmetric bis(methylene) spiro orthocarbonate (**58,** Scheme VIII) was used. However, this monomer has relatively limited solubility and reactivity. Recently, cationic, double ring-opening polymerization of new symmetric spiro orthocarbonate monomers such as 2,3,8,9-di(tetramethylene)-1,5,7,11-tetraoxaspiro[5.5]undecane **59** was studied (*77, 78*) for dental applications. The non-shrinking resin formulations used in this study contained relatively small quantities of **59** with a mixture of diepoxy comonomers, and the formulations were polymerized by photocationic initiators sensitized by visible light.

Other studies in this field used (*79*) free-radical polymerization of exo-methylene-substituted asymmetric spiro orthocarbonate monomers (oxaspiro-undecane **60** and oxaspiro-decane **61,** Scheme VIII) incorporated into dental composite formulations with BIS-GMA or an ethoxylated bisphenol A dimethacrylate. These experimental materials provided a good

correlation between decreased polymerization shrinkage and improved adhesive strength (*80*). Acrylate-substituted spiro orthoester and spiro orthocarbonate monomers (**62–64**) were prepared (*81–83*) as a means of achieving facile incorporation of oxaspiro monomers into polymers via more conventional free-radical-addition polymerization mechanisms that leave the pendant oxaspiro functionality intact. A subsequent cationic ring-opening polymerization of the oxaspiro groups results in cross-linked polymer networks and a slight expansion. In one study (*84*), the methacrylate-substituted spiro orthocarbonate **64** was polymerized by a simultaneous free-radical and cationic dual photocure process to obtain cross-linked, ring-opened polymers with low polymerization shrinkage in a single step.

Current studies of ring-opening polymerization for reducing shrinkage in dental composites address the problems of oxaspiro monomer reactivity and ring-opening proficiency, as well as the design of ring-opened polymers with higher glass-transition temperatures. Future work may also provide multicyclic oxaspiro monomers capable of opening three or more rings in a concerted fashion. This capability would result in dramatic expansions and thereby reduce the amount of oxaspiro monomer necessary to attain an overall volume-neutral polymerization.

Structures 62–64. Examples of methacrylate-substituted oxaspiro monomers.

Surface-Active Monomers, Oligomers, and Polymers

Polymer Adhesion to Tooth Structure

The ability to form strong, durable bonding between dental restorative materials and tooth structure has obvious benefits for both the patient and the dentist. Effective dental adhesives obviate the necessity for removing healthy dental tissue (mechanical undercuts for the retention of restorations) and can also reduce or eliminate microleakage and its attendant problems such as margin discoloration and secondary caries formation.

Current resin-based dental materials undergo significant polymerization shrinkage and show poor bonding to mineralized tissue. By contrast, zinc polycarboxylate and glass ionomer cements, which are polyelectrolyte-based composites produced from the interaction of basic fillers with aqueous solutions of poly(alkenoic acids), are noted for their intrinsic adhesion to both enamel and dentin (*1, 18, 85*). The mechanism of adhesion in this case seems to involve significant carboxylate salt formation via the interaction of these cements with the mildly basic calcium phosphate component of the tooth structure. A degree of micromechanical interlocking by virtue of acid etching as well as hydrogen bonding may also contribute to the adhesiveness of these materials. The bond strengths of glass ionomer cements to enamel are modest (3–6 MPa) and are even less to dentin, presumably because of the lower mineral content of dentin. Recently, strong composite-to-dentin bonding was achieved (*85–89*) by using vinyl resin-modified glass ionomer cements.

Adhesion of polymers to dental tissues involves a number of bonding mechanisms, including micromechanical interlocking; interfacial diffusion and entanglement; physicochemical interactions; and covalent, ionic, and coordinate bonding. Adhesion of acrylic composites and sealants to enamel was achieved by using mild acid-etching techniques, such as the use of aqueous phosphoric acid (*90*). By using this technique dental monomers diffuse into and form interlocking polymeric tags with the microporous enamel surfaces, thereby establishing micromechanical attachment. The procedure works well with both conventional dental monomer systems (MMA, **9**, BIS-GMA) and functional monomers having surface-active functionalities, such as carboxylic acid and phosphate groups (*91, 92*).

Early attempts to bond conventional monomers to dentin (a complex heterogeneous substrate consisting mainly of hydroxyapatite, collagen, and water; *see* Chapter 5.5) by the usual enamel-etching techniques proved ineffective. However, more recently surface-active monomers provided (*93, 94*) enhanced adhesion to etched enamel and variously conditioned dentin.

Adhesion-Promoting Acrylic Systems

It was soon recognized that adhesion to dentin would require surface modification techniques and the use of coupling agents or surface-active agents to mediate the bonding to the mineral or the organic components of dentin (*93, 94*). The following conditions appear to be necessary for effective bonding to dentin.

1. Removal or modification of the smear layer (a weak boundary layer formed during cavity preparation).
2. Diffusion of surface-active monomers and other agents throughout the conditioned dentinal surfaces.

3. A high degree of interfacial polymerization involving both the infused monomers and the main body of the sealant or restorative materials.

Other critical factors that may affect the long-term durability of dentin bonding are the stresses that develop at the interface due to polymerization shrinkage, the mismatch in the thermal expansion characteristics of the polymeric adhesive and the substrate, and the stability of the polymer-reinforced dentinal interface. The development of ring-opening monomers for nonshrinking or slightly expanding polymeric materials with coefficients of thermal expansion closer to dentin would certainly provide a route to more durable adhesive polymeric dental materials.

The design of adhesion-promoting monomers for dentinal bonding has historically focused on monomers having functional groups that are potentially capable of reacting with the calcium phosphate mineral or the organic component (collagen) of dentin. Early examples of potential coupling agents for enamel and dentin were the dimethacrylate of glyceryl phosphoric acid (65) (95) and the reaction product of N-phenylglycine and glycidyl methacrylate 68 (96, 97). The surface-active compound 68, originally designed to function as a monomer with chelating potential for calcium ions, also appears to promote adhesion by a recently proposed self-polymerization mechanism that involves intra- or intermolecular interaction of the carboxylic acid and amino functional groups (98, 99).

Structures 65–73. Examples of adhesion-promoting functional monomers for dental bonding.

Adhesion-promoting monomers, oligomers, and polymers with potential for interacting with the calcium ions of hydroxyapatite have surface-active functionalities such as phosphate, phosphonate, carboxylic acid (or anhydride), amino acid, or sulfonic acid (*91–107*). Adhesion-promoting resins with potential for specific chemical reactions with the collagenous phase of dentin have functional groups such as anhydride, acyl halide, isocyanate, or aldehyde capable of covalent-bond formation (*91–94, 100*). These types of coupling agents are designed to react primarily with amino, hydroxyl, carboxyl, and amide groups in collagen. Representative examples of these types of coupling agents are illustrated in **65–73**.

Many dentinal bonding systems contain monomers bearing one or more hydroxyl groups. The simplest member of this type of surface-active monomer is HEMA **19**, which is widely used in a number of commercial adhesive systems (*102*) (*see* Chapter 5.3). With these agents the predominant mode of interaction with dentin probably involves some hydrogen bonding. A recent interdiffusional approach (*92*) for bonding to dentin involves formulating monomer systems having solubility parameters similar to that of dentin.

Graft polymerization to the organic component of dentin by certain free-radical initiator systems, such as cerium(IV) salts, persulfates, and oxidized tri-*n*-butylborane, also have been used for bonding polymers to collagenous substrates (*100*). Effective dentin bonding by free-radical graft polymerization probably also benefits from the use of some surface-active monomers with affinity for dentin (*101–103*).

Alkyl 2-cyanoacrylates have attracted interest as potential dentin adhesives because of their ability to bond, presumably by anionic graft polymerization, to a variety of substrates. These monomers polymerize readily by contact with mild anionic initiators such as water, alcohol, and amines (*108*). However, because of their lack of durability in aqueous environments and concerns regarding their biocompatibility, cyanoacrylates are not currently used in dentin adhesive systems.

Bond formation between dentin and adhesion-promoting monomers of the type shown in **65–73** is assumed to involve primarily the formation of a hybrid, polymer-reinforced dentin layer created by resin infusion, aided by secondary bonding (e.g., hydrogen bonding) to the conditioned substrate, and subsequent in situ polymerization, which locks in the polymeric interphase by micromechanical attachment and by formation of an interpenetrating network structure (Scheme IX) (*101–103*). Given the limited sensitivities of Fourier transform IR and other analytical methods for detecting bond formation in the presence of complex interfering backgrounds arising from

dentin, primary bonding (ionic, covalent, and coordinate covalent) is not easy to measure, especially at low levels that result from adhesive-substrate reactions. However, direct evidence does exist for primary ionic bonding for polycarboxylic acid-based adhesives and cements (e.g., glass ionomer cements), poly(sulfonic acid)s, and phosphate- and phosphonate-containing resins that can react ionically with the calcium phosphate mineral of enamel and dentin (*106, 107*).

Important chemical factors that are critical to effective dentin bonding are the modes and quality of polymerization at the interface. For example, for strong bonding, efficient surface-generated interfacial polymerization is preferred to polymerization initiated only on the surface or in the bulk of the bonding resin. An effective dentin-bonding system involving a unique type of self-initiated interfacial polymerization is based on the interaction of carboxylic acid monomers (e.g., **71**) with amino acids such as *N*-phenyl-glycine and **68**. This mode of initiating polymerization appears to involve formation of an unstable acid–amine complex that subsequently decomposes to initiating radicals via electron transfer (*98, 99*). This redox acid–base adhesion mechanism seems to account for the observed "spontaneous" polymerization of **71** and other acidic monomers when placed in contact with dentin surfaces that have been infused with surface-active aryl amines (*109*). The general sequence of the reaction can be depicted as follows:

Acid + Aryl Amine ↔ [Salt-like Complex]

→ Initiating Radical

This mechanism also accounts for the instability toward polymerization of **68** and similar monomeric derivatives of *N*-aryl-α-amino acids that can form radicals by this mechanism by both intra- and intermolecular acid–amine interactions.

Conclusions and Outlook

A wide range of conventionally available polymers such as celluloid, phenol-formaldehyde resins, vinyl polymers, polyesters, polyamides, polycarbonates, polyurethanes, epoxides, polyethers, polysulfides, siloxanes, phosphazenes, and acrylics have been examined for a variety of dental applications. Acrylics have played a dominant role in preventive and restorative dentistry because of their unique combination of properties: facile free-radical polymerization and processing; excellent aesthetics; and acceptable mechanical, physicochemical, and toxicological properties. However, conventional acrylic-based dental materials suffer from a number of serious shortcomings such as high polymerization shrinkage, incom-

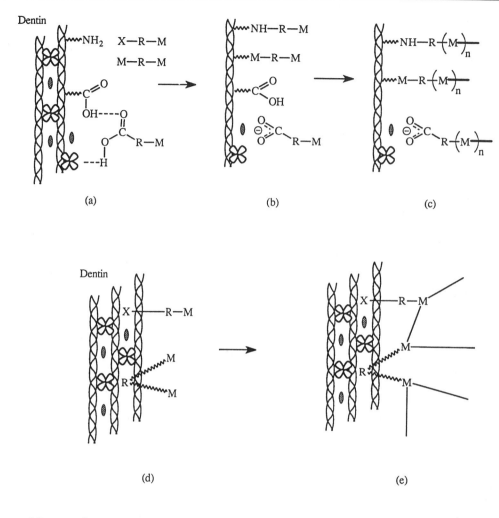

Scheme IX. Schematic representation of the various mechanisms for attaching bonding resins to dentin: (a) hydrogen bonding, (b) covalent and ionic bonding, followed by (c) subsequent graft polymerization of attached coupling agents. Interfacial diffusion of bonding resins (d) across the dentinal substrate leading to micromechanical polymer attachment and (e) interpenetrating network formation.

plete conversion of vinyl groups, susceptibility to retardation during cure, degradation in the oral environment, and poor adhesion to tooth structure.

Newly designed monomers, oligomers, and polymers, both acrylic and nonacrylic, are being developed in an effort to address these deficiencies. The new acrylics include hydrophobic, low-shrinking, multifunctional oligomers and cyclopolymerizable or cross-linking oligomers that combine low polymerization shrinkage with high conversion. Single- and multi-ring-opening monomers and oligomers are examples of nonacrylics that are designed to minimize (or eliminate) polymerization shrinkage, although some require new initiator systems (e.g., photocationic photopolymerization). Various functionalized monomers, oligomers, and polymers containing surface-active groups (carboxylate, phosphate, sulfonate, isocyanate,

and aldehyde) have also been developed to improve bonding to tooth structure.

Future research efforts are envisaged to focus on the design of highly sophisticated, ordered polymers such as two- and three-dimensional self-assemblies, dendrimers, liquid-crystalline systems, and organoceramics for more effective protection and restoration of both the function and aesthetics of natural tissues. For the development of such new monomers, oligomers, and polymers through molecular design, polymer science is likely to continue to play a significant role in the progress of dentistry in the foreseeable future.

Acknowledgments

This work was partially supported by the National Institute of Dental Research (Y01 DE30001). We are grateful for the assistance of Benjamin Reed in the preparation of the manuscript. Certain commercial materials and equipment are identified in this chapter for adequate definition of the experimental procedure. In no instance does such identification imply recommendation or endorsement by the National Institute of Standards and Technology, or that the material or equipment is necessarily the best available for the purpose.

References

1. Brauer, G. M.; Antonucci, J. M. In *Encyclopedia of Polymer Science and Engineering*, 2nd ed.; John Wiley & Sons: New York, 1987; Vol. 4, pp 698–719.
2. Bowen, R. L. *J. Dent. Res.* **1956**, *35*, 360–369.
3. Bowen, R. L. U.S. Patent 3 066 112, 1962.
4. Bowen, R. L. *J. Am. Dent. Assoc.* **1964**, *69*, 481–495.
5. Bowen, R. L. U.S. Patent 3 179 623, 1965.
6. Venz, S.; Antonucci, J. M. *J. Dent. Res.* **1984**, *64*, 229.
7. Atsuta, M.; Nakabayashi, N.; Masuhara, E. *J. Biomed. Mater. Res.* **1971**, *5*, 183–185.
8. Cowperthwaithe, G. F.; Foy, J. J.; Malloy, M. A. In *Biomedical and Dental Application of Polymers*; Gebelein, C. G.; Koblitz, F. F., Eds.; Plenum: New York, 1981; pp 379–985.
9. Ruyter, I. E. In *Posterior Composite Resin Dental Resorative Materials*; Vanherle, G.; Smith, D. C., Eds.; Peter Szulc: Utrecht, Netherlands, 1985; pp 109–135.
10. Waller, D. E. U.S. Patent 3 629 187, 1971.
11. Waller, D. E. German Patent 2 126 419, 1971.
12. Waller, D. E. U.S. Patent 3 709 866, 1973.
13. Foster, J.; Walker, R. J. U.S. Patent 3 825 518, 1974.
14. Foster, J. Br. Patent 1 430 303, 1976.
15. Dart, E. C.; Cantwell, J. B.; Traynor, J. R.; Jaworzyn, J. F. and Nemcek, J. U.S. Patent 4 110 184, 1978.
16. Antonucci, J. M. *J. Biomedical and Dental Applications of Polymers* Gebelein, C. G.; Koblitz, F. F., Eds; Plenum: New York, 1981; 357–371.
17. Antonucci, J. M.; Stansbury, J. W.; Liu, D-W. *J. Dent. Res.* **1993**, *72*, 369.
18. Antonucci, J. M. In *Polymers in Medicine II*; Chiellini, E.; Giusti, P.; Mgliaresi, E.; Nicolais, L., Ed.; Plenum: New York, 1986; pp 277–303.
19. Antonucci, J. M.; Stansbury, J. W.; Venz, S. In *Progress in Biomedical Polymers*; Gebelein, E.G.; Dunn, R. L., Ed.; Plenum: New York, 1990; pp 121–131.
20. Griffith, J. R.; O'Rear, J. G. In *Biomedical and Dental Applications of Polymers*; Gebelein, C. G.; Koblitz, F. F., Ed.; Plenum: New York, 1984; pp 357–371.
21. Kuo, J. S.; Antonucci, J. M.; Wu, W. *J. Dent. Res.* **1985**, *64*, 178.
22. Antonucci, J. M.; Stansbury, J. W.; Venz, S. *J. Dent. Res.* **1986**, *65*, 219.
23. Hirose, H.; Anzai, M.; Yoneyama, M.; Kawakami, T.; Watanabe, I.; Ohashi, M. *J. Nihon Univ. Sch. Dent.* **1987**, *29*, 287–297.
24. Ueda, M.; Takahashi, M.; Imai, Y.; Pittman, C. U. *J. Polym. Sci., Polym. Chem. Ed.* **1982**, *20*, 2819.
25. Akkapeddi, M. K. *Macromolecules* **1979**, *12*, 546.
26. Akkapeddi, M. K. *Polymer* **1979**, *20*, 1215.
27. Stansbury, J. W.; Antonucci, J. M. *Dent. Mater.* **1992**, *8*, 270.
28. Petragnani, N.; Ferraz, H. M. C.; Silva, G. V. J. *Synthesis* **1986**, 157.
29. Schlewer, G.; Stampf, J. L.; Benezra, C. *J. Med. Chem.* **1980**, *23*, 1031.
30. Howie, G. A.; Stamos, I. K.; Cassady, J. M. *J. Med. Chem.* **1976**, *19*, 309.
31. Butler, G. B. In *Encyclopedia of Polymer Science and Engineering*; John Wiley and Sons: New York, 1986; Vol. 4, pp 543–598.
32. Ruyter, I. E.; Rødsrud, P. G. *J. Dent. Res.* **1988**, *67*, 346 (abstract 1867).
33. Colletti, R. F.; Halley, R. J.; Mathias, L. J. *Macromolecules* **1991**, *24*, 2043.
34. Mathias, L. J.; Kusefoglu, S. H. *Macromolecules* **1987**, *20*, 2039.
35. Mathias, L. J.; Kusefoglu, S. H.; Ingram, J. E. *Macromolecules* **1988**, *21*, 545.
36. Stansbury, J. W. *J. Dent. Res.* **1989**, *68*, 248 (abstract 535).
37. Stansbury, J. W. *J. Dent. Res.* **1990**, *69*, 844.
38. Butler, G. B. In *Proceedings of the International Symposium on Macromolecules*; Mano, E. B., Ed.; Elsevier: New York, 1975; pp 57–76.
39. Hwa, J. C. H.; Fleming, W. A.; Miller, L. *J. Polym. Sci.: Part A, Polym. Chem.* **1964**, *2*, 2385.
40. Stansbury, J. W. *Macromolecules* **1991**, *24*, 2029.
41. Tsuda, T.; Mathias, L. J. *Macromolecules* **1993**, *26*, 4734.
42. Tsuda, T.; Mathias, L. J. *Polym. Prepr. (Am. Chem. Soc. Div. Polym. Chem.)* **1993**, *34(1)*, 499.
43. Stansbury, J. W. *J. Dent. Res.* **1992**, *71*, 434.
44. Mathias, L. J.; Dickerson, C. W. *J. Polym. Sci., Polym. Lett. Ed.* **1990**, *28*, 175.
45. Stansbury, J. W.; Liu, D. W. *J. Dent. Res.* **1993**, *72*, 369 (abstract 2128).
46. Mathias, L. J.; Dickerson, C. W. *Macromolecules* **1991**, *24*, 2048.
47. Antonucci, J. M.; Stansbury, J. W.; Cheng, G. W. *Polym. Prepr. (Am. Chem. Soc. Div. Polym. Chem.)* **1992**, *33(2)*, 522.
48. Antonucci, J. M.; Stansbury, J. W.; Cheng, G. W. *Polym. Prepr. (Am. Chem. Soc. Div. Polym. Chem.)* **1990**, *31(1)*, 320.
49. Antonucci, J. M.; Stansbury, J. W. *Polym. Prepr. (Am. Chem. Soc. Div. Polym. Chem.)* **1993**, *34(1)*, 403.
50. Stansbury, J. W. *Macromolecules* **1993**, *26*, 2981.
51. Mathias, L. J.; Kusefoglu, S. H.; Kress, A. O. *Macromolecules* **1987**, *20*, 2326.
52. Kress, A. O.; Mathias, L. J.; Cei, G. *Macromolecules* **1989**, *22*, 537.
53. Fernandez-Monreal, M. C.; Cuervo, R.; Madruga, E. L. *J. Polym. Sci., Polym. Chem. Ed.* **1992**, *30*, 2313.
54. Warren, S. C.; Mathias, L. J. *J. Polym. Sci., Polym. Chem. Ed.* **1990**, *28*, 1637.
55. Thompson, R. D.; Barclay, T. H.; Mathias, L. J. *Polym. Prepr. (Am. Chem. Soc. Div. Polym. Chem.)* **1993**, *34(1)*, 509.
56. Avci, D.; Kusefoglu, S. H. *J. Polym. Sci., Polym. Chem. Ed.* **1993**, *31*, 2941.
57. Bailey, W. J.; Chen, P. Y.; Chen, S. C.; Chiao, W. B.; Endo, T; Gapud, B.; Kuruganti, V.; Lin, Y. N.; Ni, Z.; Pan, C. Y.; Shaffer, S. E.; Sidney, L.; Wu, S. R.; Yamamoto, N.; Yamazaki, N.; Yonezawa, K.; Zhou, L. L. *Makromol. Chem., Macromol. Symp.* **1986**, *6*, 81.

58. Brady, R. F. *J. Macromol. Sci., Rev. Macromol. Chem. Phys.* **1992**, *32*, 135.

59. Gong, M. S.; Chang, S. I. *Makromol. Chem., Rapid Commun.* **1989**, *10*, 210.

60. Reed, B. B.; Stansbury, J. W.; Antonucci, J. M. *Polym. Prepr. (Am. Chem. Soc. Div. Polym. Chem.)* **1992**, *33(2)*, 520.

61. Stansbury, J. W.; Antonucci, J. M.; Reed, B. B. *J. Dent. Res.* **1993**, *72*, 385 (abstract 2253).

62. Park, J.; Kihara, N.; Ikeda, T.; Endo, T. *J. Polym. Sci., Polym. Chem. Ed.* **1993**, *31*, 1083.

63. Patel, M. P.; Braden, M. *Biomaterials* **1989**, *10*, 277.

64. Moussa, K.; Decker, C. *J. Polym. Sci., Polym. Chem. Ed.* **1993**, *31*, 2197.

65. Bailey, W. J.; Ni, Z.; Wu, S. R. *Macromolecules* **1982**, *15*, 711.

66. Hiraguri, Y.; Tokiwa, Y. *J. Polym. Sci., Polym. Chem. Ed.* **1993**, *31*, 3159.

67. Antonucci, J. M.; Stansbury, J. W.; Reed, B. B. *Polym. Prepr. (Am. Chem. Soc. Div. Polym. Chem.)* **1992**, *33(1)*, 507.

68. Patel, M. P.; Braden, M.; Davey, K. W. M. *Biomaterials* **1987**, *8*, 53.

69. Suliman, A. A.; Boyer, D. B.; Lakes, R. S. *J. Dent. Res.* **1993**, *72*, 1532.

70. Davidson, C. L.; De Gee, A. J.; Feilzer, A. J. *J. Dent. Res.* **1984**, *63*, 1396.

71. Feilzer, A. J.; De Gee, A. J.; Davidson, C. L. *J. Dent. Res.* **1987**, *66*, 1636.

72. Gross, J. D.; Retief, D. H.; Bradley, E. L. *Dent. Mater.* **1985**, *1*, 7.

73. Bailey, W. J.; Sun, R. L. *Polym. Prepr. (Am. Chem. Soc. Div. Polym. Chem.)* **1972**, *13(1)*, 281.

74. *Expanding Monomers: Synthesis, Characterization, and Applications*; Sadhir, R. K.; Luck, R. M., Eds.; CRC: Boca Raton, FL, 1992.

75. Jensen, M. E.; Chan, D. C. N. In *Posterior Composite Resin Dental Restorative Materials*; Vanherle, G.; Smith, D. C., Eds.; Peter Szulc: Utrecht, Netherlands, 1985; pp 243–262.

76. Thompson, V. P.; Williams, E. F.; Bailey, W. J. *J. Dent. Res.* **1979**, *58*, 1522.

77. Byerley, T. J.; Eick, J. D.; Chen, G. P.; Chappelow, C. C.; Millich, F. *Dent. Mater.* **1992**, *8*, 345.

78. Eick, J. D.; Byerley, T. J.; Chappell, R. P.; Chen, G. R.; Bowles, C. Q.; Chappelow, C. C. *Dent. Mater.* **1993**, *9*, 123.

79. Stansbury, J. W. *J. Dent. Res.* **1992**, *71*, 1408.

80. Stansbury, J. W.; Bailey, W. J. In *Progress in Biomedical Polymers*; Gebelein, C. G.; Dunn, R. L., Eds.; Plenum: New York, 1990; pp 133–139.

81. Endo, T.; Kitamura, N.; Takata, T.; Nishikubo, T. *J. Polym. Sci., Polym. Lett. Ed.*, **1988**, *26*, 517.

82. Kitamura, N.; Takata, T.; Endo, T.; Nishikubo, T. *J. Polym. Sci., Polym. Chem. Ed.* **1991**, *29*, 1151.

83. Haase, L.; Klemm, E. *Makromol. Chem.* **1990**, *191*, 549.

84. Stansbury, J. W. *Polym. Prepr. (Am. Chem. Soc. Div. Polym. Chem.)* **1992**, *33(2)*, 518.

85. Smith, D. C. *Oper. Dent.* **1992**, *Suppl. 5*, 177–183.

86. Antonucci, J. M.; McKinney, J. E.; Stansbury, J. W. *Transactions of the 13th Annual Meeting, Society for Biomaterials*; Society of Biomaterials: Minneapolis, MN, 1987; p 255.

87. Antonucci, J. M. *Trends Tech.* **1988**, *5*, 4.

88. Rusz, J. E.; Antonucci, J. M.; Eichmiller, F. C.; Anderson, M. H. *Dent. Mater.* **1992**, *8*, 31–36.

89. Mitra, S. B. *J. Dent. Res.* **1991**, *70*, 72–74.

90. Buonocore, M. G. *J. Dent. Res.* **1955**, *34*, 849–853.

91. Ruyter, J. E. *Oper. Dent.* **1992**, *Suppl. 5*, 32–43.

92. Asmussen, E.; Uno, S. *Oper. Dent.* **1992**, *Suppl. 5*, 68–74.

93. Bowen, R. L.; Marjenhoff, W. A. *Oper. Dent.* **1992**, *Suppl. 5*, 75–80.

94. Nakabayashi, N. In *Multiphase Biomedical Materials*; Tsuruta, T.; Nakajima, A., Eds.; VSP: Utrecht, Netherlands 1989; pp 89–104.

95. Buonocore, M. G.; Wileman, W.; Brudevold, F. *J. Dent. Res.* **1956**, *35*, 846–851.

96. Bowen, R. L. *J. Dent. Res.* **1965**, *44*, 895–902.

97. Bowen, R. L. *J. Dent. Res.* **1965**, *44*, 903–905.

98. Antonucci, J. M.; Stansbury, J. W.; Farahani, M. *J. Dent. Res.* **1992**, *71*, 239.

99. Schumacher, G. E.; Eichmiller, F. C.; Antonucci, J. M. *Dent. Mater.* **1992**, *8*, 278–282.

100. Brauer, G. M. In *Scientific Aspects of Dental Materials*; von Fraunhofer, J. A., Ed.; Butterworths: London, 1975; pp 49–96.

101. Nakabayashi, N.; Kojima, K.; Masuhara, E. *J. Biomed. Mater. Res.* **1982**, *16*, 265–273.

102. Erickson, R. L. *Oper. Dent.* **1992**, *Suppl. 5*, 81–94.

103. Nakabayashi, N. *Oper. Dent.* **1992**, *Suppl. 5*, 125–130.

104. Venz, S.; Dickens, B. *J. Dent. Res.* **1993**, *72*, 582–586.

105. Cabasso, I.; Sahni, S. *J. Biomed. Mater. Res.* **1990**, *24*, 705–720.

106. Beech, D. R. *Arch. Oral Biol.* **1972**, *17*, 907–911.

107. Wilson, A. D.; Prosser, H. J. *Br. Dent. J.* **1984**, *157*, 449–454.

108. Brauer, G. M.; Jackson, J. A.; Termini, D. J. *J. Dent. Res.* **1979**, *58*, 1900–1907.

109. Bowen, R. L.; Cobb, E. N.; Misra, D. N. *Ind. Eng. Chem. Prod. Res. Dev.* **1984**, *23*, 78–81.

Surface Modification of Hydroxyapatite for Dental Plaque Inhibition

Krister Holmberg and Jan Olsson

Various types of surface treatments of hydroxyapatite have been investigated to obtain a material with low protein and bacterial adsorption characteristics. Because hydroxyapatite is seen as a model for tooth enamel, a procedure providing a surface with low tendency for biofouling would be of interest in vivo to prevent buildup of dental plaque. This chapter summarizes surface modification work in the area, including treatments with proteins and polyelectrolytes, use of hydrophilic noncharged polymers, modification with silicones, and fluorination. Particularly interesting in vitro results have been obtained with hydrophilic polymers such as poly(alkylene oxides) and polysaccharides carrying phosphate groups for anchoring to the mineral surface.

The hard, durable material that surrounds the crown of the tooth is known as the enamel (Figure 1). It is a very heterogeneous material both from a chemical and structural point of view. It is the most highly mineralized tissue known, consisting of 95% inorganic material, 1% organic material, and 4% water (1). The inorganic phase is a crystalline calcium phosphate, known as hydroxyapatite. The organic material, which is mainly of proteinaceous nature, forms a fine network between the hydroxyapatite crystals. As with all biological materials, the composition of the tooth enamel differs slightly from person to person. It is affected by food habits and oral hygiene regimens. Different teeth of one individual may also show differences in composition of the enamel. Nevertheless, tooth enamel is generally regarded to be a type of impure hydroxyapatite, and hence this mineral is normally used as tooth models in in vitro experiments.

Hydroxyapatite in pure form has the molecular formula of $Ca_{10}(PO_4)_6(OH)_2$. It is a white solid with an isoelectric point of 6.5–7.0. Its main industrial application is as a raw material for the production of fertilizers. It has found some specific niches in the biotechnology field, such as solid support in column chromatography with special relevance to proteins, reconstructive material in bone surgery, and implant materials. However, the main biological interest in hydroxyapatite relates to its role in the oral cavity. The two major dental diseases, caries and periodontitis, both emanate from growth of bacteria on the tooth surface. Colonization of the oral bacteria is mediated by a proteinaceous film that is rapidly formed on a cleaned tooth surface when contacted by saliva.

Much current interest exists in interfering with this biofouling process on the teeth. One approach is to change the surface characteristics of the tooth enamel so as to reduce the tendency for protein adsorption. Alternatively, a modification of the hydroxyapatite surface may lead not to a reduction in the amount of salivary proteins adsorbed but to an altered adsorption pattern that may also result in altered bacterial colonization. This chapter reviews the more important methods of surface modification of hydroxyapatite (tooth enamel) to reduce formation of dental plaque.

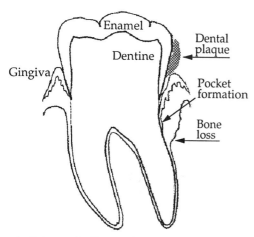

Figure 1. Schematic illustration of a tooth showing the location of the enamel and typical site of dental plaque. The dentine is the hard portion of the tooth surrounding the pulp, covered by enamel on the crown and cementum on the root. Gingiva is the mucous membrane with supporting fibrous tissue that surrounds the tooth.

Dental Plaque

Plaque Formation

Bacterial deposits (dental plaque) form continuously on the teeth. Plaque formation is rapid in sheltered areas, such as between teeth, along the gingival margin, and in pits and fissures on the tooth surface, where natural cleansing from saliva, tongue, lips, and cheeks does not reach. A tooth surface that has been cleaned by polishing is covered by a so-called salivary pellicle within seconds. This film consists primarily of salivary proteins, and forms the innermost layer of dental plaque. The pellicle may also contain serum products from the gingival crevice, as well as products from bacteria and epithelial cells.

Accumulation of bacteria on solid surfaces in the oral cavity is largely believed to be governed by the nature of the initially formed protein layer (2). The structure of this layer is likely to depend on the characteristics of the underlying surface. For instance, the proteinaceous film that develops on crowns and other artificial dental materials may be different from that formed on the enamel surface, and the amount and type of bacterial deposits will then also be different.

The role of salivary proteins in the formation of dental plaque is far from clear. In all biological environments, including the oral cavity, most exposed surfaces rapidly adsorb a proteinaceous layer to which microorganisms adhere (3, 4). The salivary pellicle is believed to change its composition with time, and this process of competitive adsorption of proteins goes on for at least two hours (5). Because the adhesion of oral streptococci to dental enamel is governed by the prop-

erties of the adsorbed salivary protein layer, bacterial adhesion in vivo is a very complex event.

The effect of saliva constituents on the binding of oral bacteria to the tooth surface is partly of a nonspecific nature and imparts charges or hydrophobicity to the surface (6). Christersson et al. demonstrated (7, 8) that the level of adsorption of oral bacteria in the presence of saliva depends on the critical surface tension of the substrate. Surfaces of medium critical surface tension (30–38 mN/m) retain higher numbers of microorganisms than both higher and lower energy surfaces.

Bacterial adherence and colonization of the tooth surface also seem to be dependent on more specific binding mechanisms. For instance, some bacteria attach by means of special surface appendages (so-called adhesins) to salivary components carrying specific receptors. One example of salivary components that are specific promoters of bacterial adherence are the proline-rich proteins (PRPs) that, when adsorbed at the dental enamel, serve as receptors promoting the adhesion of certain oral bacteria (9).

Species like *Actinomyces viscosus* and *Streptococcus sanguis* seem to use such specific receptors on the tooth surface for their colonization in humans. Although PRP molecules adsorbed on hydroxyapatite surfaces interact strongly with such cells, the same proteins in solution do not bind to the bacteria, nor do they affect their attachment to pellicles (9). The molecular segments of the PRPs that are responsible for specific bacterial binding are postulated to be hidden in solution and exposed only when the proteins become attached—via another segment—to the hydroxyapatite surface. Thus, the proteinaceous film that spontaneously develops on all dental surfaces in the oral cavity seems to influence bacterial binding of certain organisms by a highly specific mechanism.

The first bacteria colonize the tooth surface by binding directly to the pellicle. Further plaque formation results from growth of the early colonizers or from secondary colonizers. These secondary colonizers are bacteria that specialize in binding to other, already bound bacteria. If undisturbed, plaque may form thick deposits on the teeth in a few days. Oral bacteria also colonize salivary coatings on the mucosal surfaces. However, because of the continuous desquamation of epithelial cells, thick layers of bacteria do not form on the epithelium.

Because of the specific binding mechanisms, bacterial adhesion to different surfaces in humans and animals is very selective. This phenomenon, called tropism, is well demonstrated in the oral cavity. For instance, bacteria that colonize the tongue will not adhere to the tooth and vice versa. Thus, bacterial adherence is a critical factor in the microbial colonization of different surfaces in a host, and in fact, it deter-

concentration of 0.1 M, a bacterial cell with aver- characteristics (radius, surface charge, and surface rophobicity) has a shallow secondary minimum of w kT units (kT being the thermal energy per mole- value) units at around 8 nm distance from a flat, atively charged surface (surface potential, –30 mV) The secondary minimum is separated from the nary minimum by a large maximum of several dred kT at around 1 nm separation (Figure 2). This e constitutes a very high energy barrier. Thus, a erium approaching the hydroxyapatite surface ald normally not get beyond the secondary mini- n in which it will be loosely and reversibly ched (11).

One way for the bacterium to overcome the rgy maximum that hinders close contact with the ace is to allow proteins to first cover the surface. O calculations for proteins approaching a flat, atively charged surface give no secondary mini- n and only a relatively low energy barrier of a few units (10). Hence, for many proteins—and particu- y for the saliva proteins that are tailor-made for rption to enamel—a direct attachment to hydroxy- ite should present no problem from an energy t of view. The adsorbed protein may change the ace characteristics in terms of both hydrophobicity surface potential and distribution of charges, and is way the protein may lower the energy barrier a bacterium to arrive in the primary minimum. eins may also extend out from the surface in loops tails. Such segments may bridge the gap between

cell and surface, thus facilitating irreversible adhesion of the bacterium.

The issue of preventing adhesion of oral bacteria will then be intimately linked to that of reducing adsorption of salivary protein to the surface. It is well-known that solid materials can be made protein-repellent by specific modifications of their surfaces. Such a modification should eliminate the main driving forces for adsorption: that is, hydrophobic interactions and double-layer attractions (12, 13). Other surface characteristics that may be of importance and that should be considered are degree of hydration (in aqueous environment), surface mobility, and surface morphology.

Enamel Modification with Proteins and Polyelectrolytes

Because salivary proteins are known to play a key role in the adherence of oral bacteria, attempts have been made to manipulate the protein film on both pure hydroxyapatite and the enamel surface to reduce the formation of dental plaque. Rölla et al. demonstrated (14) that adherence of two important oral streptococci, S. mutans and S. sanguis, is reduced if hydroxyapatite is treated with an acidic protein such as albumin. If pretreatment of the mineral is made with a basic protein (e.g., protamine), on the other hand, more bacteria is bound. Pretreatment with saliva gives the same result as with albumin. In an earlier study it was shown that exposure of hydroxyapatite to saliva leads to a selective adsorption of acidic proteins (5). The results are shown in Figure 3. Evidently, these proteins bind to the amphoteric hydroxyapatite surface regardless of their net charge. The basic protein (protamine) is likely to give the mineral surface a net positive charge that will increase attraction of the negatively charged bacteria. The acidic protein (albumin) will reduce the zeta-potential of the solid surface, but the bacteria will still adhere, although to a smaller extent. Bridges by calcium and other multivalent cations

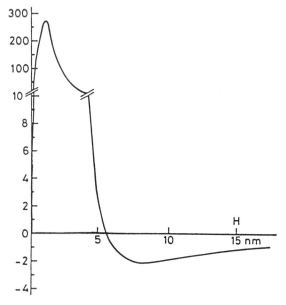

e 2. DLVO interaction curve for a bacterium mod- as a spherical particle with a radius of 5 × 10⁻⁷ m a flat surface: surface potentials, –30 mV; electrolyte ntration, 0.1 M; temperature, 25 °C. (Adapted from nce 10.)

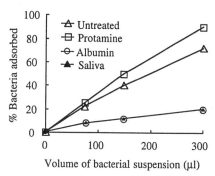

Figure 3. Effect of treatment of hydroxyapatite with different proteins on adhesion of S. mutans. (Reproduced with permission from reference 14. Copyright 1977.)

mines the type of microflora that will be acquired at different sites of the body.

Plaque-Related Diseases

The two major dental disorders are caries and periodontal disease. Both diseases are caused by the accumulation and growth of bacteria on the tooth surface. Dental caries result from demineralization of the tooth surface by acids that are produced during bacterial fermentation of carbohydrates, especially sucrose. Most oral bacteria have such acidogenic potential. However, for caries to develop on the tooth enamel, pH has to reach below a critical level of around 5.7. Some oral bacterial species, such as lactobacilli and streptococci, are aciduric; that is, they have the potential of producing acid in an acidic environment. Production of extracellular polymers, notably dextrans, is another characteristic feature of oral streptococci, which is considered important for the development of caries. Such polymers, which considerably increase the thickness and stickiness of the plaque film, are primarily produced by fermentation of the disaccharide sucrose. The bacteria use the energy-rich bond between glucose and fructose in sucrose to build the extracellular polymers. Thus, sucrose is a cariogenic substance on two accounts: it is the substrate for acid production, and it is the source of energy for building extracellular polymers. Among the oral streptococci, *Streptococcus mutans* is considered the most important cause of dental caries.

Periodontal disease is caused by bacteria that grow in the gingival pocket and cause inflammation in the soft tissue and subsequent breakdown of the supporting tissues: that is, the cementum, periodontal ligament, and bone. The disease starts as a result of colonization and growth of bacteria on the tooth surface adjacent to the gingiva. Over time noxious products from the bacteria will induce an inflammatory response in the gingiva with increased exsudation, swelling, and in later stages, formation of a pocket between the gingiva and the root surface. Because of these local changes of the environment there is a shift in the microflora toward more anaerobic and more Gram-negative. Some species associated with periodontal disease are *Actinobacillus actinomycetemcomitans*, *Porphyromonas gingivalis*, *Eikenella corrodens*, and *Fusobacterium nucleatum*. These bacteria are specialized in growing under the extremely anaerobic conditions prevailing in gingival pockets. The bacteria interfere in different ways with the inflammatory response, and they also have the capacity to break down host tissues by means of proteolytic enzymes. The inflammatory response per se has also been suggested to be important in the progression of the disease.

Ways of Plaque Inhibition

Dental plaque formation can be contr[...] ical and chemical methods. Brushing [...] tooth brush is a habit used for cent[...] even with modern day improvemen[...] and materials, mechanical plaque con[...] ticed by the patient is normally not su[...] tain a healthy status. Only profession[...] performed in a dental practice, has be[...] to have a clear effect on caries. In [...] other means of plaque control have [...] Chemical plaque control, which is use[...] to the mechanical measures, requires [...] Traditionally, chemical plaque control [...] istered via a dentifrice or a mouth was[...] chewing gums and lozenges have c[...] early days phenolic compounds (Liste[...] as active ingredients. Later, surfactants [...] ions, plant extracts, quaternary ar[...] pounds, and bisguanidines (chlor[...] employed. Except for chlorhexidine, tl[...] have been only marginally effective ag[...] mation. Chlorhexidine has a well-d[...] plaque effect, but the substance su[...] nounced side effects such as discolora[...] bitter taste, and mucous membrane ir[...] spread use of antimicrobial substances [...] of plaque control is also questionabl[...] potential toxicity of these products.

Enamel Modification as a Me[...] of Plaque Inhibition

Rationale of Surface Modificatio[...]

In a colloidal approach to bacterial a[...] surfaces, the microorganisms are cons[...] formable particles of fixed surface and [...] (10). The interaction between a bac[...] hydroxyapatite surface will then be de[...] der Waals, electrostatic, and steric force[...] distinction must be made between adh[...] mary minimum and in the secondary [...] sion in the primary minimum involv[...] between the mineral surface and the b[...] and the magnitude of attraction will b[...] specific interactions that differ from ca[...] sion in the secondary minimum, on th[...] mainly governed by van der Waals attr[...] ble layer repulsion. Whereas primary [...] sion is very strong and irreversible, [...] secondary minimum can be regarded [...] process.

By taking electrostatic and hydr[...] tions into account it can be shown th[...]

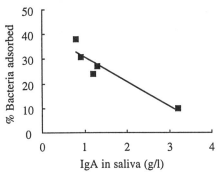

Figure 4. Adherence of *S. mutans* to hydroxyapatite pretreated with saliva containing varying amounts of IgA.

between negative sites are likely to contribute to the binding in the albumin case.

In another study directed toward *S. mutans* adherence, salivary fractions with varying activities of naturally occurring immunoglobulin A (IgA) antibodies were used for treatment of hydroxyapatite (*15*). The basis for this work was a previous observation that salivary IgA antibodies inhibited adherence of streptococci to buccal epithelial cells (*16*). As is seen from Figure 4, saliva containing IgA reduces the number of bacteria adhering to hydroxyapatite beads. Furthermore, a clear correlation exists between the concentration of IgA in the saliva and the adherence-inhibiting effect. The inhibiting effect of the antibodies attached to the mineral surface is probably of nonspecific nature, because no inhibition of adherence was observed when the bacteria were treated with the same IgA fractions before contacting the hydroxyapatite beads.

Reduced bacteria adherence to hydroxyapatite similar to those observed by pretreatment of the mineral with certain proteins can also be attained by treatment with polyelectrolytes. Olsson et al. demonstrated (*17*) that both heparin, which is rich in sulfate groups, and a phosphoprotein, which contains a high concentration of phosphate groups, reduce binding of *Streptococcus mutans* to hydroxyapatite. Both polymers are highly acidic. They both seem to adhere strongly to the hydroxyapatite surface and lower its zeta-potential. A basic polymer, cytochrome c, on the other hand, had negligible effect on *S. mutans* adherence, possibly due to little (or no) adhesion of this polymer to the hydroxyapatite beads.

Schematic structures of the three polymers used in this study are given in Chart I. and the results are shown in Figure 5. As can be seen, the effect exerted by the phosphoprotein is particularly striking, giving about 85% reduction in bacterial adherence already at very low concentrations. This result is probably due to a very strong interaction between phosphate groups on the polymer and so-called calcium sites at the mineral surface. Inorganic phosphate salts, particularly

pyrophosphate, also bind very strongly to the hydroxyapatite surface, and a pretreatment with such electrolytes drastically reduces adherence of oral bacteria (*18*). The pyrophosphate ions are believed to occupy the calcium sites of the mineral surface, thus preventing calcium bridges from being formed between the enamel and the approaching protein (Chart II).

Heparin contains sulfate groups that are believed to be responsible for the binding to the mineral surface. Also, the sulfated heteropolysaccharide chondroitin-4-sulfate (Chart I) is known to bind firmly to hydroxyapatite via the sulfate groups, as demonstrated by IR spectroscopy (*19*). Also, a low molecular weight sulfate, sodium dodecyl sulfate (SDS) (Chart I), was found to bind strongly to the enamel, probably with the sulfate groups occupying the calcium sites, giving a double layer of surfactant molecules on the hydroxyapatite surface (Chart III). Such a modified hydroxyapatite surface has very little driving force for protein adsorption (*20*, *21*).

Polyacrylates are also known to bind strongly to hydroxyapatite, and this good adhesion is taken advantage of in dental cements. On the basis of IR spectroscopy, adhesion is postulated to be achieved by the displacement of phosphate ions on the hydroxyapatite surface by the pendant carboxylate groups on the polymer chain (*22*). A calcium ion-bridging mechanism analogous to the one illustrated in Chart II cannot be ruled out, however.

Modification with Hydrophilic Noncharged Polymers

Several groups have demonstrated (*23–25*) that protein adsorption and bacterial adherence on plastic surfaces can be prevented by grafting of nonionic, water-soluble polymers. Derivatives of polyethylene glycol and polysaccharides have been widely used. The inert character of the surfaces is believed to be due to the solution properties of the polymer and its molecular conformation in aqueous solution, and to the fact that the polymer is completely noncharged (*26*). The ability of the grafted layer to prevent proteins and other biomolecules from approaching the surface can be seen as a steric stabilization effect.

Surface-active polyoxyalkylene derivatives such as ethoxylated and propoxylated fatty alcohols (Chart IV) prevent adhesion of bacteria to model apatite surfaces (*27*, *28*). However, because their interaction with the surface is weak, their attachment is probably reversible. To attain a more permanent surface modification, specific binding to the mineral surface is desirable.

Covalent grafting to the hydroxyapatite surface is hardly conceivable. However, from research on apatite ore flotation it is known that several functional groups interact strongly with hydroxyapatite (*29*, *30*). Such

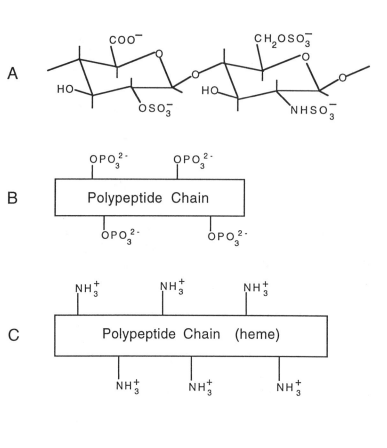

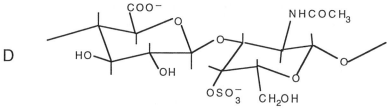

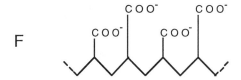

Chart I. Chemical or schematic structures of various polymers used for enamel modification: A, heparin; B, phosphoprotein; C, cytochrome c; D, chondroitin-4-sulfate; E, sodium dodecyl sulfate; F, polyacrylate.

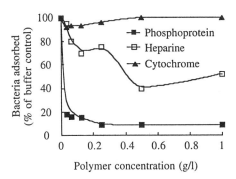

Figure 5. Adherence of *S. mutans* to hydroxyapatite pre-treated with varying concentrations of phosphoprotein, heparin, and cytochrome c.

groups may be used as anchoring groups for water-soluble substances to the mineral surface. Provided that all anchoring groups can be directed toward the hydroxyapatite surface, the grafted layer should contain no charged groups that could induce attractive double-layer interactions with oppositely charged groups of approaching proteins and cells.

In a series of papers, we reported (*31–34*) the use of homopolymers of ethylene oxide, block copolymers of ethylene oxide and propylene oxide, and ethoxylated polysaccharides as hydrophilic polymers for surface modification of hydroxyapatite beads. Phosphate, phosphonate, amino, and amino acid groups were attached to these polymers as anchoring moieties. The polysaccharide derivatives (*32*), and also some of the poly(alkylene oxide)s (*31*), were provided with more than one anchoring group to increase their interaction with the inorganic surface. In some cases the hydrophilic polymer layer on the surface was cross-linked by a diepoxide of a polyethylene glycol as a means of further reducing polymer detachment (*31*).

The surface-modified hydroxyapatite beads were investigated mainly in terms of adherence of *Strepto-*

coccus mutans. Some of the phosphate derivatives (compounds C, D, and E; Chart IV) were very effective in preventing bacterial adhesion in a buffer-treated system, as can be seen from Figure 6. However, there was a much less pronounced effect after saliva treatment. For comparison, adherence observed with non-phosphorylated polyglycerol alkoxylate (D*) is also included in the figure. It is clear that the phosphate groups play a vital role, most certainly as anchoring elements for the hydrophilic polymer to the mineral surface.

At least two possible explanations exist for the reduced adherence in the presence of saliva. The binding of the hydrophilizing agent to the hydroxyapatite surface may be so weak that it is readily replaced by salivary components, which are known to bind to hydroxyapatite with high affinity (*35*). The other possibility is that salivary components bind directly onto the grafted hydrophilic polymer, or possibly in combination with binding to isolated, exposed areas of hydroxyapatite.

The ^{14}C-labeled polymer D was used to distinguish between the two previously mentioned explanations (*31*). Figure 7 shows the results from treating hydroxyapatite beads modified by labeled compound D with two different samples of saliva: one of which is known to promote and one known to inhibit bacteria adherence. As can be seen, the amount of D remaining on the beads was almost the same after exposure to the two saliva samples. Furthermore, the salivas were not more effective than the aqueous buffer in desorbing compound D. Therefore, the adherence-enhancing effect by saliva is probably not related to saliva-induced desorption of the hydrophilic polymer. Compound D, when added to hydroxyapatite pre-treated with saliva, did not replace saliva components, and hence did not give a reduction of *S. mutans* adherence. Evidently, both the phosphated nonionic polymer and the saliva proteins become so strongly

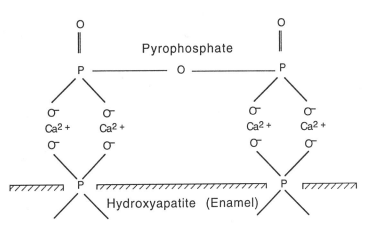

Chart II. Assumed blocking of calcium sites of hydroxyapatite by pyrophosphate preventing binding of acidic proteins. (Reproduced with permission from reference 18. Copyright 1988.)

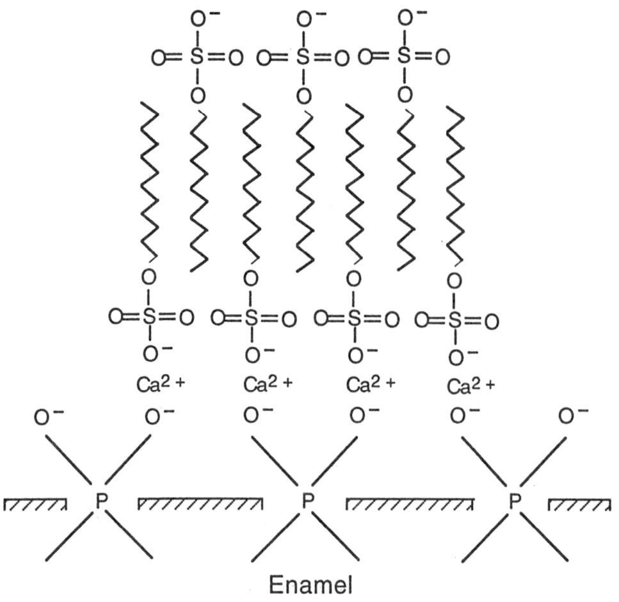

Chart III. Interaction between sodium dodecyl sulfate and calcium sites of hydroxyapatite.

attached to the mineral surface that they are difficult to remove by competitive adsorption.

Cross-linking of grafted D on the hydroxyapatite beads was performed to further improve its immobilization on the surface (31), but this procedure had no effect on the adhesion characteristics. This observation supports the view that adsorption of saliva constituents on the treated hydroxyapatite surface does not involve displacement of the hydrophilizing agent. Whether binding of salivary components to treated hydroxyapatite involves adhesion on top of the hydrophilic layer or to areas on the mineral surface not covered by the polymer cannot be deduced from these results.

The phosphorylated polyglycerol alkoxylate D, as well as the nonphosphorylated equivalent of E, were used in clinical trials to elucidate their in vivo effect on bacterial adherence (32). Neither compound gave any significant reduction of dental plaque formation.

Compounds C, D, and E all consist of a hydrophilic, noncharged, polymer carrying phosphate

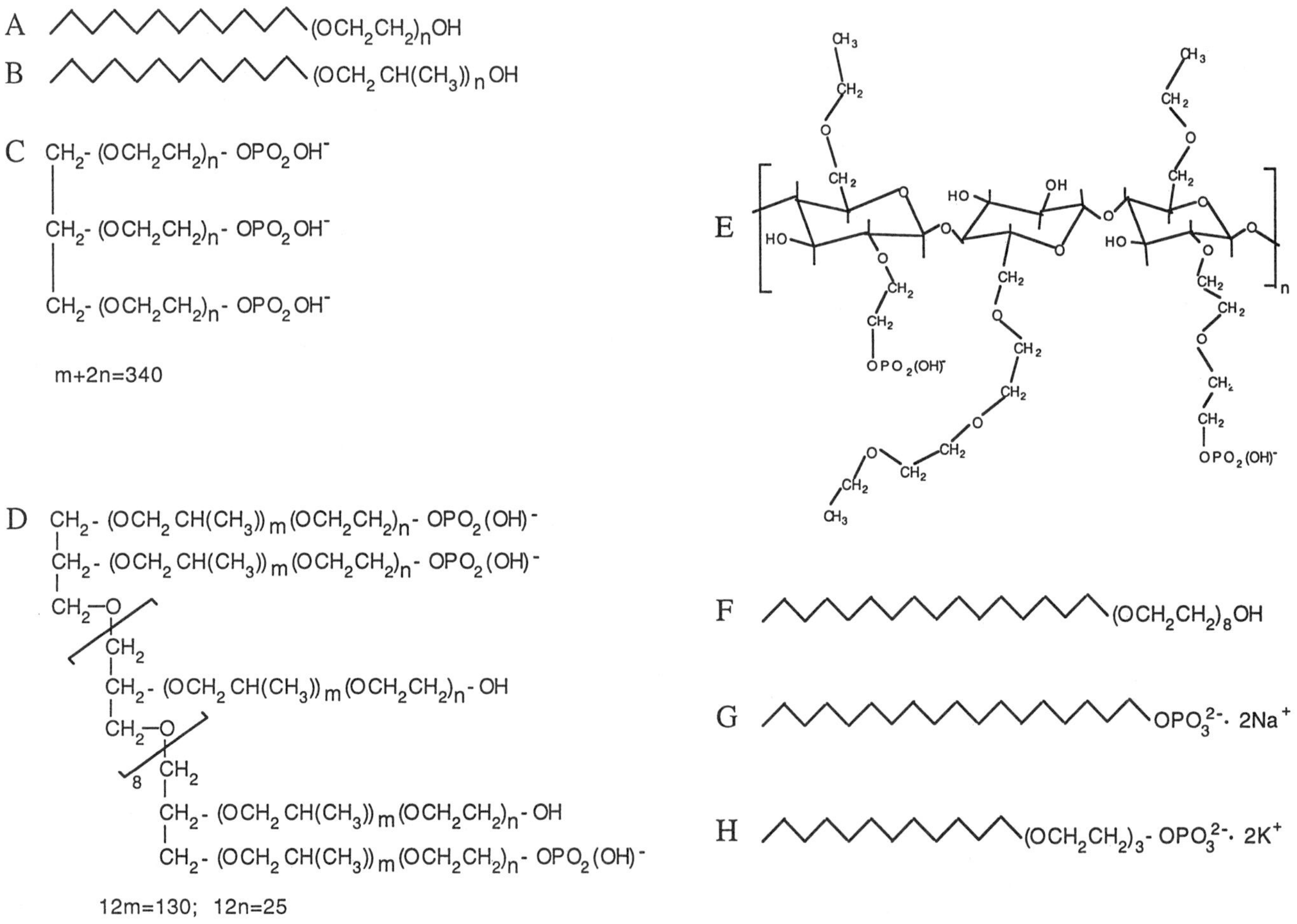

A $\diagdown\!\!\diagup\!\!\diagdown\!\!\diagup\!\!\diagdown\!\!\diagup\!\!\diagdown$(OCH$_2CH_2$)$_n$OH

B $\diagdown\!\!\diagup\!\!\diagdown\!\!\diagup\!\!\diagdown\!\!\diagup\!\!\diagdown$(OCH$_2$CH(CH$_3$))$_n$OH

C   CH$_2$- (OCH$_2$CH$_2$)$_n$- OPO$_2$OH$^-$
    |
    CH$_2$- (OCH$_2$CH$_2$)$_n$- OPO$_2$OH$^-$
    |
    CH$_2$- (OCH$_2$CH$_2$)$_n$- OPO$_2$OH$^-$

   m+2n=340

D   CH$_2$- (OCH$_2$CH(CH$_3$))$_m$(OCH$_2$CH$_2$)$_n$- OPO$_2$(OH)$^-$
    CH$_2$- (OCH$_2$CH(CH$_3$))$_m$(OCH$_2$CH$_2$)$_n$- OPO$_2$(OH)$^-$
    CH$_2$—O
      CH$_2$
      CH$_2$- (OCH$_2$CH(CH$_3$))$_m$(OCH$_2$CH$_2$)$_n$- OH
      CH$_2$—O
        CH$_2$
    8
        CH$_2$- (OCH$_2$CH(CH$_3$))$_m$(OCH$_2$CH$_2$)$_n$- OH
        CH$_2$- (OCH$_2$CH(CH$_3$))$_m$(OCH$_2$CH$_2$)$_n$- OPO$_2$(OH)$^-$

   12m=130; 12n=25

F $\diagdown\!\!\diagup\!\!\diagdown\!\!\diagup\!\!\diagdown\!\!\diagup\!\!\diagdown$(OCH$_2CH_2$)$_8$OH

G $\diagdown\!\!\diagup\!\!\diagdown\!\!\diagup\!\!\diagdown\!\!\diagup\!\!\diagdown$OPO$_3^{2-}$· 2Na$^+$

H $\diagdown\!\!\diagup\!\!\diagdown\!\!\diagup\!\!\diagdown$(OCH$_2CH_2$)$_3$- OPO$_3^{2-}$· 2K$^+$

Chart IV. Structure of A, ethoxylated fatty alcohol; B, propoxylated fatty alcohol; C, phosphorylated glycerol ethoxylate; D, phosphorylated decaglycerol ethoxylate–propoxylate; E, phosphorylated nonionic cellulose ether, ethylhydroxyethyl cellulose; F, octa(ethylene glycol) monohexadecyl ether; G, disodium phosphate of 1-octadecanol; H, dipotassium phosphate of tri(ethylene glycol) monododecyl ether.

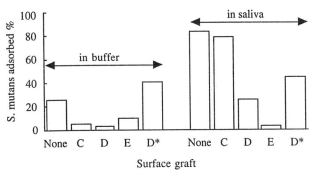

Figure 6. Adherence of *S. mutans* to hydroxyapatite modified with the phosphorylated polymers (C–E of Chart IV) and with the nonphosphorylated equivalent of compound D (D*). Experiments were carried out using either buffer or saliva.

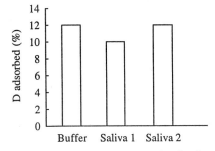

Figure 7. Amount of compound D adsorbed on hydroxyapatite. Tests were made after treatment either in buffer or in saliva from two individuals. Amount is given as percent of added activity.

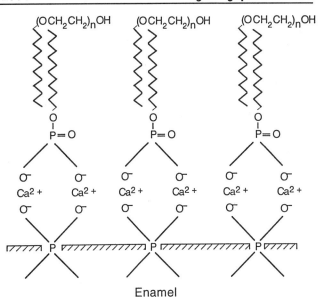

Chart V. Hypothetical unsymmetrical double layer of alkylphosphate and alcohol ethoxylate formed on the hydroxyapatite surface by simultaneous addition of the two components.

groups that are supposed to anchor the polymer to the mineral surface. An alternative approach to attachment of a hydrophilic polymer to hydroxyapatite is to use a nonionic surfactant together with an ionic surfactant, the polar group of which should have high affinity for hydroxyapatite. Under these conditions, the surface active agents may form an unsymmetrical double layer on the mineral surface (Chart V) (*33*).

In a study to find an optimal combination of anionic and nonionic surfactants, seven different types of alkylphosphates and three alcohol ethoxylates were examined for their ability to inhibit the adherence of ^{3}H-labeled cells of *Streptococcus mutans* both in buffer and in parotid saliva (*33*). Experiments with ^{14}C-labeled surfactants showed that the phosphate esters had a very strong affinity for the hydroxyapatite surface. At least some alkylphosphates bound even to surfaces that had been treated with saliva. Presumably, the alkylphosphate interaction with the surface (through ion pairing with lattice calcium ions and ion exchange) is strong enough to displace at least some salivary components bound to hydroxyapatite. However, the nonionic surfactants showed a much lower affinity for the mineral surface. The amount adsorbed was about 10% of the amount of bound alkylphos-

phate for buffer-treated surfaces, and the value was 5% for saliva-treated surfaces.

None of the compounds alone hindered binding of bacteria to buffer-treated or saliva-treated hydroxyapatite to any appreciable extent. A combination of certain of the alkylphosphates, notably the disodium phosphate of 1-octadecanol (G in Chart IV), together with the nonionic surfactant octa(ethylene glycol) monohexadecyl ether (F in Chart IV) (1:1 ratio) gave a strong inhibition of *S. mutans* adhesion in buffer and, interestingly, also on saliva-treated hydroxyapatite (Figure 8).

Adsorption experiments with radiolabeled surfactants also indicate that a symmetrical double layer of surface active agents is formed when an alkylphosphate is used alone, and an unsymmetrical double layer is formed when the alkylphosphate is employed together with a fatty alcohol ethoxylate. Synthesis difficulties prevented the use of labeled compound G. Instead, an alkylphosphate containing oxyethylene groups between the alkyl chain and the phosphate group was used (^{14}C was introduced via labeled ethylene oxide). The compound, dipotassium phosphate of tri(ethylene glycol) monododecyl ether (H in Chart IV), showed similar effects in terms of *S. mutans* rejection as compound G (*33*). The nonionic surfactant was the same as in the previous experiment: that is, compound F. The unsymmetrical double layer seems to consist of predominantly alkylphosphate in the inner layer and alcohol ethoxylate in the outer layer toward the water phase. The packing density of alcohol ethoxylate in the outer layer was calculated to be around

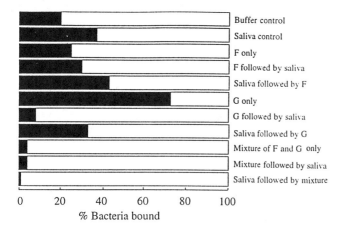

Figure 8. Adherence of *S. mutans* to hydroxyapatite treated with compounds F, G, and a 1:1 molar mixture of F and G. (Reproduced with permission from reference 34. Copyright 1991.)

Table 1. Electrophoretic Mobility (zeta potential in mV) of Hydroxyapatite Particles (<5 μm) After Treatment with Compound H or a 1:1 Molar Mixture of Compounds F and H, Before or After Addition of Saliva

Coating Conditions	No Coating	H Alone	F + H
Standard	−7 ± 1	−34 ± 2	−4 ± 1
Coated, then treated by saliva	1 ± 0.6	−5 ± 0.4	0.5 ± 1.8
Treated by saliva, then coated	1 ± 0.6	−20 ± 1.5	2 ± 0.1

Note: The structures for F and H are given in Figure 9.

Source: Adapted from reference 34.

50 Å^2/molecule. This value represents a rather dense packing of polyoxyethylene chains and means that the nonionic surfactant should be strongly bound to the surface. The inner alkyl phosphate layer, firmly attached to the hydroxyapatite surface by specific interaction of the phosphate group, constitutes a hydrophobic environment, which is capable of strong interaction with the hydrophobic tail of the alcohol ethoxylate. The driving force for adsorption of this mixed surfactant system seems to be strong, as judged from its ability to compete with salivary components.

Results from microelectrophoresis throw further light on the adsorption mechanism. As shown in Table I, compound H alone gives a strong negative charge on the hydroxyapatite surface, as expected. Post-treatment with saliva reduces the negative charge from −34 to −5 mV, indicating partial desorption of the alkylphosphate, as is indeed confirmed by experiments with radiolabeled surfactants (33). Addition of H to hydroxyapatite already treated with saliva changes the surface charge from 1 to −20 mV, indicative of strong alkylphosphate adsorption. The zeta-potential obtained with the alcohol ethoxylate–alkylphosphate mixture is particularly interesting. Very low values are obtained, both after pretreatment and post-treatment of the hydroxyapatite particles by saliva. The good results in prevention of bacterial adherence in vitro, obtained with the alkylphosphate–nonionic surfactant mixture of Chart IV, are likely to be related to low net surface charge, which should minimize attractive double-layer interactions.

The surface charge values obtained with the surfactant mixture are unexpected. Adsorption of one anionic and one nonionic species on a negatively charged surface would be expected to result in an increase in the negative surface charge. Instead, a decrease is observed. This finding strongly suggests that the inner alkylphosphate layer is bound to the apatite surface via calcium bridges (or ion exchange) and that the outer layer is virtually devoid of anionic species.

Modification with Silicones

Materials having very low surface free energy, such as perfluoropolymers and silicones, are known to have low tendency for biofouling. The mechanism by which these materials prevent adhesion of biomolecules and cells must clearly be very different from that of the hydrophilic noncharged surfaces. The protein-repelling character of low energy surfaces can be seen as a wetting effect: spreading on very low energy surfaces is energetically unfavorable for all molecules, including proteins.

Treatment of hydroxyapatite powder and of enamel with polydimethylsiloxane (Scheme I) gave a thin resistant film that on the enamel persisted for at least 3 h in the oral environment (36). Studies of albumin adsorption to the treated hydroxyapatite showed that the surface modification resulted in reduced protein binding (Figure 9). Similarly, reduced protein adsorption was also observed by scanning electron microscopic examination of enamel fragments that were implanted in the mouth to acquire pellicles. The enamel fragments treated with silicone oil exhibited a slower rate of pellicle formation, as compared with untreated enamel. Amino acid analysis also showed that the chemical composition of the pellicle material collected from the in vivo treated enamel was somewhat different from that of the untreated enamel. As can be seen from Figure 10, higher amounts of serine and glycine were obtained from the pellicle formed on the silicone-treated enamel. This result indicates that different salivary proteins may bind to the silicone-treated and the untreated surface, as would be expected from the large difference in surface characteristics of the two enamels. Thus, the pellicle formed

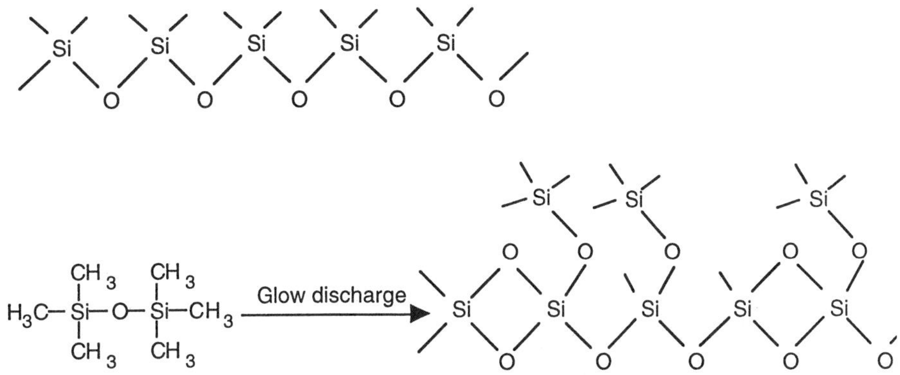

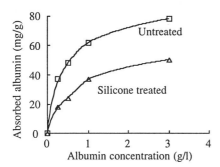

Scheme I. Structure of polydimethylsiloxane (top): plasma polymerization of hexamethyldisiloxane (bottom).

Figure 9. Adsorption isotherms of albumin on hydroxyapatite and silicon-treated hydroxyapatite. (Reproduced with permission from reference 36. Copyright 1990.)

on the silicone-treated surface is likely to expose functional groups different from those of the normal pellicle, and the bacteria-binding properties of the two surfaces are likely to differ. The effect of the silicone treatment on adhesion of oral bacteria was not investigated, however.

In another experiment different mineral surfaces were modified by plasma polymerization of hexamethyldisiloxane, producing a cross-linked silicone-like surface (schematically shown in Scheme I) (37). This modification did not result in any dramatic reduction of protein adsorption. However, in vitro adhesion of the oral bacterium S. mutans as well as plaque accumulation in vivo were small compared with untreated surfaces. A tentative explanation of this behavior is that the saliva proteins that bind to the modified surface adsorb in an unnatural conformation that does not favor bacteria adherence.

Fluorination

Fluoride ions can be incorporated into hydroxyapatite by simple adsorption from aqueous solution. When adsorption experiments were performed at neutral pH, no desorption of phosphate or hydroxyl ions were observed. For low fluoride concentrations, Langmuir-type adsorption of fluoride ions occurs with no evidence for the formation of definite phases. At higher

fluoride concentrations, once the solubility product of CaF_2 is exceeded, a CF_2 layer is deposited (38).

Most fluorinated surfaces give extremely low values of surface free energy. Poly(tetrafluoroethylene), for instance, has a critical surface tension of 18–19 mN/m, which is below the values obtained by silicone polymers. Even lower values can be attained with compounds containing trifluoromethyl groups. The adhesive properties of human enamel decrease after treatment with aqueous stannous fluoride, and the effect was ascribed to a decrease in surface free energy of the enamel (39, 40). An alternative explanation of the effect of fluorides is that they inhibit or retard pellicle formation by competing with acidic groups of salivary proteins for calcium sites on the crystal surface (41). Both in vitro (42) and in vivo (43) studies support the concept of fluoride as a competitive inhibitor of adherence. However, incorporation of fluoride into the hydroxyapatite crystal also lowers the solubility of the enamel, and this characteristic is an important aspect of the use of fluoro compounds for caries prevention.

A series of anionic perfluoroalkyl surfactants were tested as inhibitors of bacterial adherence both in vitro and in vivo (44). A pronounced anti-adhesion effect was obtained with a perfluorosulfonamidoalkyl ester of phosphorous acid. The effect is probably due to a strong interaction between the charged phosphate groups of the surfactant and calcium sites at the hydroxyapatite surface, as discussed previously. Electron spectroscopy for chemical analysis, as well as determination of total surface fluorine with an ion-specific electrode, show that the mineral surface is covered with a densely packed layer of perfluoroalkyl chains. The lower adhesion seen in this case is probably due to the marked reduction in surface free energy obtained by the surface treatment.

Conclusions and Future Prospects

Surface modification of tooth enamel is of considerable practical interest as a means of preventing the for-

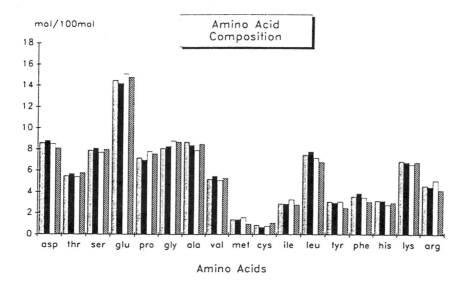

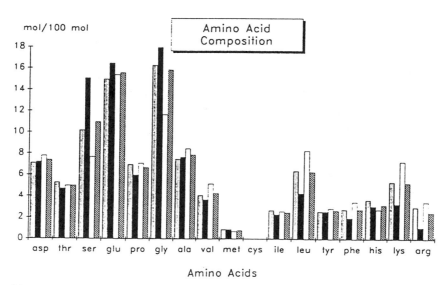

Figure 10. Amino acid composition of pellicle material collected in vivo from untreated enamel (top) and from silicone-oil-treated enamel (bottom). (Reproduced with permission from reference 36. Copyright 1990.)

mation of dental plaque. The two major dental disorders, caries and periodontal disease, are both caused by the accumulation and growth of bacteria on the tooth surface. Because hydroxyapatite is the main constituent of tooth enamel, most in vitro surface modifications have been made with this mineral, usually in the form of beads of well-defined size.

Plaque consists of bacterial deposits formed continuously on the teeth. Bacterial adherence is always preceded by the formation of a thin proteinaceous film, the salivary pellicle. Because formation of the pellicle is believed to be a prerequisite of plaque accumulation, the prime objective of surface modification has been to transform hydroxyapatite into a protein-repellent material.

Several routes of surface modification have been investigated. Promising in vitro results have been obtained by surface grafting of hydrophilic non-charged polymers, such as polyethylene glycol and polysaccharide derivatives. Phosphate groups are good anchoring groups for the hydrophilic polymers to the mineral surface. A combination of two surface active agents, an alkylphosphate and an alcohol ethoxylate, is of particular interest. The combination forms an unsymmetrical double layer on the hydroxyapatite surface: an inner layer of alkylphosphate with the phosphate groups bound to the mineral and an outer layer of alcohol ethoxylate with the oxyethylene chains pointing away from the mineral. In this way a completely noncharged surface is obtained that gives minimal bacterial adherence. Clinical trials using the same surface modification have been less successful, however, probably due to limited durability of the hydrophilic coating.

Plaque control by surface modification is an area that demands continued research in the future. The use of microbiocidal compounds is being questioned because of occasional irritation of the mucous membrane and the potential toxicity of the products. Hence, transforming the tooth enamel into a nonadhering material is increasingly attractive. Future research may focus on how to produce a sufficiently thick film of a hydrophilic polymer, applied via a dentifrice or a mouth wash. In this context, desquamating surfaces (i.e., surfaces that at a controlled rate release their outermost layer) may be of interest. This approach is being used to obtain nonfouling surfaces for other applications. An antiplaque treatment with a hydrophilic polymer giving a relatively thick coating with a controlled rate of leakage is an attractive concept. Research along this line is already on its way.

References

1. Jenkins, G. N. *The Physiology and Biochemistry of the Mouth*; Blackwell: Oxford, England, 1978; pp 54–112.
2. Busscher, H. J.; Cowan, M. M.; van der Mei, H. C. *FEMS Microbiol.* **1992**, *88*, 199–210.
3. Vassilakos, N.; Arnebrant, T.; Glantz, P.-O. *Scand. J. Dent. Res.* **1993**, *101*, 133–137.
4. Vassilakos, N.; Glantz, P.-O.; Arnebrant, T. *Scand. J. Dent. Res.* **1993**, *101*, 339–343.
5. Sönju, T.; Rölla, G. *Caries Res.* **1973**, *7*, 30–38.
6. Pratt-Terpstra, I. H.; Weerkamp, A. H.; Busscher, H. J. *J. Dent. Res.* **1989**, *68*, 463–467.
7. Christersson, C. E.; Dunford, R. G.; Glantz, P.-O. J.; Baier, R. E. *Scand. J. Dent. Res.* **1989** 97 247–256.
8. Christersson, C. E.; Glantz, P.-O. J. *Scand. J. Dent. Res.* **1992**, *100*, 98–103.
9. Gibbons, R. J.; Hay, D. I. *Infect. Immunity* **1988**, *56*, 439–445.
10. Norde, W.; Lyklema, J. *Colloids Surf.* **1989**, *38*, 1–13.
11. Lyklema, J.; Norde, W.; van Loosdrecht, M. C. M.; Zehnder, A. J. B. *Colloids Surf.* **1989**, *39*, 175–187.
12. Merrill, E. W. In *Poly(ethylene Glycol) Chemistry*; Harris, J. M., Ed.; Plenum: New York, 1992; pp 199–220.
13. Lee, J. H.; Kopecek, J.; Andrade, J. D. *J. Biomed. Mater. Res.* **1989**, *23*, 351–368.
14. Rölla, G.; Robrish, S. A.; Bowen, W. H. *Acta Path. Microbiol. Scand.* **1977**, *85*, 341–346.
15. Gahnberg, L.; Olsson, J.; Krasse, B.; Carlén, A. *Infect. Immunity* **1982**, *37*, 401–406.
16. Williams, R. C.; Gibbons, R. J. *Science (Washington, D.C.)* **1972**, *177*, 697–698.
17. Olsson, J.; Jontell, M.; Krasse, B. *Scand. J. Infect. Dis.* **1980**, *24*, 173178.
18. Rykke, M.; Rölla, G.; Sönju, T. *Scand. J. Dent. Res.* **1988**, *96*, 517–522.
19. Embery G.; Rölla, G. *Acta Odontol. Scand.* **1980**, *38*, 105–108.
20. Barkvoll, P.; Embery G.; Rölla, G. *J. Biol. Buccale* **1988**, *16*, 75–79.
21. Rykke, M.; Rölla, G.; Sönju, T. *Scand. J. Dent. Res.* **1990**, *98*, 135–143.
22. Wilson, A. D.; Prosser, H. J.; Powis, D. M. *J. Dent. Res.* **1983**, *62*, 590–592.
23. Merril, E. W.; Salzman, E. W. *ASAIO J.* **1983**, *6*, 60–64.
24. Lee, J. H.; Andrade, J. D. In *Polymer Surface Dynamics*; Andrade, J. D., Ed.; Plenum: New York, 1988; pp 119–136.
25. Österberg, E.; Bergström, K.; Holmberg, K.; Riggs, J. A.; Van Alstine, J. M.; Schuman, T. P.; Burns, N. L.; Harris, J. M. *Colloids Surf. A* **1993**, *77*, 159–169.
26. Malmsten, M.; Tiberg, F.; Lindman, B.; Holmberg, K. *Colloids Surf. A* **1993**, *77*, 91–100.
27. Humphries, M.; Jaworzyn, J. F.; Cantwell, J. B. *FEMS Microbiol. Ecol.* **1986**, *38*, 299–308.
28. Humphries, M.; Jaworzyn, J. F.; Cantwell, J. B.; Eakin, A. *FEMS Microbiol. Ecol.* **1987**, *42*, 91–101.
29. Rawls, R. H.; Cabasso, I. In *Adorption on and Surface Chemisty of Hydroxyapatite*; Misra, D. N., Ed.; Plenum: New York, 1984; pp 115–238.
30. Pradip *Minerals Metallurg. Processing* **1988**, *5*, 80–89.
31. Olsson, J.; Carlén, A.; Holmberg, K. *J. Dent. Res.* **1990**, *69*, 1586–1591.
32. Olsson, J.; Carlén, A.; Holmberg, K. *Arch. Oral Biol.* **1990**, *35*, 137–140.
33. Olsson, J.; Carlén, A.; Holmberg, K. *Caries Res.* **1991**, *25*, 51–57.
34. Olsson, J.; Hellsten, M.; Holmberg, K. *Colloid Polym. Sci.* **1991**, *269*, 1295–1302.
35. Ericson, T. *Acta Odont. Scand.* **1968**, *26*, 3–21.
36. Rykke, M.; Rölla, G. *Scand. J. Dent. Res.* **1990**, *98*, 401–411.
37. Lassen, B.; Holmberg, K.; Brink, C.; Carlén, A.; Olsson, J. *J. Colloid Polym. Sci.* **1994**, *272*, 1143–1150.
38. Gasser, P.; Voegel, J.C.; Gramain, Ph. *Colloids Surf. A* **1993**, *74*, 275–286.
39. Glantz, P. O. *Odontol. Revy* **1969**, *20*, *Suppl. 17*, 1–132.
40. van Pelt, A. W. J.; de Jong, H. P.; Busscher, H. J.; Arends, J. In *Surface and Colloidal Phenomena in the Oral Cavity: Methodological Aspects*; Frank, R. M.; Leach, S. A., Eds.; IRL Press: London, 1981; pp 111–118.
41. Rölla, G. *Caries Res.* **1977**, *11*, *Suppl. 1*, 243–261.
42. Rölla, G.; Melsen, B. *Caries Res.* **1975**, *9*, 66–73.
43. Tinanoff, N.; Brady, J. M.; Gross, A. *Caries Res.* **1976**, *10*, 415–426.
44. Gaffar, A.; Esposito, A.; Bahl, M.; Steinberg, L.; Mandel, I. *Colloids Surf.* **1987**, *26*, 109–121.

Polymer–Drug Conjugates

Hiroshi Maeda and Yuichiro Kojima

Advantages and rationales for the polymer-conjugated drugs of small molecular size proteins are discussed: namely, increased plasma half-life, increased targeting efficacy to the site of inflammation or cancer, and decreased toxicity of most of the cytotoxic drugs to normal tissues and organs. In addition to the primary functions of the parent drugs the secondary functions may be incorporated to the carrier polymers such as metal chelation, antithrombus formation, lipid formulation, and immunopotentiation. Several examples of such biocompatible carrier polymers and their conjugates are discussed. Some drawbacks such as immunogenicity of proteinaceous drugs will be eliminated while increasing therapeutic efficacy. A new drug in this category has been approved by the Ministry of Health and Welfare, Japan. These and other aspects of polymer–drug conjugates are discussed in this chapter.

Many intact proteins such as immunoglobulins, clotting factors, albumin, porcine insulin, growth hormones, kallikrein, aprotinin (Torasylol), urokinase, and streptokinase have been used in traditional medical practice. The prime objective of using these proteins is essentially supplementation of deficient endogenous components in a disease state. There are also a large number of proteins used clinically that are obtained from exogenous sources such as porcine insulin, neocarzinostatin (NCS), streptokinase, L-asparaginase, bovine superoxide dismutase (SD), cytochrome c, adenosine deaminase (ADA), microbial proteases. Furthermore, there are now many more proteins available from recombinant gene technology. These proteins include interferons, erythropoietin, human growth hormone, insulin, granulocyte colony stimulating factor, tissue plasminogen activating factor, interleukines (ILs), tumor necrosis factor, and other cytokines. Some of these proteins are greatly beneficial, whereas others have so far found only limited practical use. Protein drugs used in native form are classified here as first generation protein drugs (Table I).

The common problem for all of the first generation protein drugs is limited plasma half-life ($t_{1/2}$) in vivo, limited stability under harsh physical conditions

(pH and temperature), possible susceptibility against proteolytic enzymes, and immunogenicity.

Protein–Polymer Conjugates

Second generation protein drugs are conjugates of proteins with polymers such as polyethylene glycol (PEG), or PEG–proteins, which show improved $t_{1/2}$ and reduced immunogenicity (increased biocompatibility) (1, 2). These improvements have been observed for PEG-L-asparaginase, PEG–ADA, PEG–BOX (bilirubin oxidase), etc., but not for other proteins such as SD (see subsequent discussion).

In the third generation of polymer–drug conjugates, additional pharmacological functions that do not exist in the native protein are observed. Examples of such additional effects include interferon inducing capacity or immunopotentiating activity in anticancer drugs, anticlot formation or thrombolytic activity, or metal-chelating function, which is beneficial for hemosiderosis or excessive hemolysis. Other functions such as increased internalization rate into the cells, specific organ tropism or specific organ targeting, and combination of two different enzymes or hormones such as SD–catalase or heparin–enzyme conjugates are of future interest (Table I).

Table I. Three Generations of Protein–Polymer Drugs

Drug Generation	Definition (and Examples)	Remarks
First generation (protein drugs)	Native proteins such as enzymes, hormones, proteins (L-asparaginase, neocarzinostatin, IgG, erythropoietin, interferons, etc.)	Exhibits only primary function Limitations in in vivo activity (e.g., short half-life) Possible immunogenicity
Second generation (polymer drugs)	Polymer–protein conjugates (PEG–ADA, PEG–L-asparaginase, PEG–bilirubin oxidase, PVA–SD, etc.)	Enhanced primary function in vivo Prolonged half-life in vivo Disease site targeting Protease resistance Decreased immunogenicity Decreased toxicity
Third generation (polymer drugs)	Polymer–protein conjugates with primary function and additional functions (GEL–SD, SMANCS, etc.)	Addition of secondary functions (e.g., metal chelating, interferon induction, enhanced intracellular internalization, organ tropism–disease targeting, oxyradical resistance, clot–anticlot forming) Prolonged half-life in vivo Enhanced primary function Decreased toxicity, decreased immunogenicity

Biocompatibility and Immunogenicity

A number of problems must be considered before using polymers as pharmaceutical carriers. Polymers must be free from immunogenicity: that is, showing no antibody or cellular immunity (3–7). They must be devoid of contact activation of blood clotting factor (XII or Hageman factor), and free from hemolysis or any other cytotoxicity. Endotoxin shock-like syndrome, including shock, hypotension, and hemorrhage, and even activation of complements and release of anaphylatoxin need to be carefully examined (8). Some of these effects are the other side of the coin of biological response modifiers that are known to enhance nonspecific defense response in the host against cancer and infection.

In general, neutral and anionic polymers exhibit longer plasma $t_{1/2}$, whereas cationic polymers disappear from the circulation more rapidly (9). This phenomenon is attributed to the property of endothelial surface that is covered by highly negatively charged compounds such as heparin and glycocalyx. Cationic polymers are most likely to be captured by the first pass. Polycations are also known to have increased antigenicity for anionic proteins (10, 11). Polycations such as poly-L-lysine and poly-L-ornithine may have cytotoxicity. Among the neutral polymers, PEG is the most extensively studied (1, 2), and it is highly biocompatible. Another polymer known to possess high biocompatibility is poly(vinyl alcohol) (PVA) as proven by medical devices such as extracorporeal filtration systems. We prepared PVA–SD conjugates and confirmed this notion (12). Dextrans were examined to some extent, but none so far seems beneficial, and they may involve potential immunogenicity (unpublished results).

In animal experiments, we found that rats show significantly different immunological responses to a poly(styrene-co-maleic acid) conjugated NCS (SMANCS), whether animals were bred in conventional environments or in specific pathogen free or gnotobiotic conditions. Animals housed in conventional conditions showed much weaker responses against SMANCS, as judged by antibody production and passive cutaneous anaphylaxis reaction (unpublished results). Similarly, different immune responses were observed among two different strains of mice; A/J mice exhibited higher immune response than ddY mice against PVA–SD and succinyl gelatin–SD (12, 13) (Table II). Therefore, it is dangerous to extrapolate a group of experiments using a single strain of mice or rats to human setting. Naturally, difference between mouse and human is no smaller than that between different strains of mice.

Improved Plasma Half-Life of Macromolecules

Plasma concentration of a drug has important pharmacological and therapeutic consequences. In many cases high local tissue concentrations in the target organs are desirable for therapeutic effectiveness and reduced systemic side effects. To attain a high local tissue concentration, however, a high plasma concentration is needed for many small molecular weight drugs at one time or another in any systemic administration. All small proteins less than 50,000 daltons are now well known to be cleared into the urine very rapidly, within 5–10 min in mice (5, 6).

Tailoring of small proteins with biocompatible polymers greatly enhances their $t_{1/2}$ (see Table III).

Table II. Comparison of Passive Cutaneous Anaphylaxis (PCA) Reaction in Two Different Strains of Mice

Drug	Mouse Strain	Degree of PCA Reaction–Percentage of Responses[a]			
		−	+	++	+++
Experiment 1					
Native SD[b]	A/J	0	0	0	100
	ddY	60	20	0	20
Suc-gel–SD[c]	A/J	25	25	0	50
	ddY	100	0	0	0
Experiment 2					
Native SD	A/J	0	40	20	40
	ddY	50	25	25	0
PVA(L)–SD[d]	A/J	100	0	0	0
	ddY	100	0	0	0
PVA(H)–SD[e]	A/J	100	0	0	0
	ddY	100	0	0	0

Note: Sensitization doses of each SD derivative for two different strains of mice (A/J and ddY) were 0.1 and 1.0 mg SD equivalent per animal on day 0 and day 14, respectively. The induction dose for anaphylaxis of each SD derivative for Wistar rats was 1.0 mg SD equivalent per animal.

[a]Diameter of vascular permeability shown by Evans blue dye in the skin: −, <5 mm; +, 5–10 mm; ++, 11–15 mm; +++, 16–20 mm.
[b]Human recombinant Cu,Zn-SD (4390 U/mg protein).
[c]Succinyl-gelatin fragment (MW 23,000)–Cu,Zn-SD.
[d]Low molecular weight PVA(L) (MW 4500)–Cu,Zn-SD.
[e]High molecular weight PVA(H) (MW 11,200)–Cu,Zn-SD.

Source: Reproduced with permission from references 12 and 13. Copyright 1993.

Most of the polymer–protein conjugates examined showed increased $t_{1/2}$, and the area under the concentration curve was greatly improved. In PEG–BOX, for example, $t_{1/2}$ increased 26-fold (Table III and Figure 1).

Such protein tailoring could provide the following three main advantages, in addition to solving the immunological problems described previously (6).

1. Proteins that are ineffective due to their extremely short $t_{1/2}$ (e.g., BOX, SD) become effective drugs.
2. More selective targeting to the sites of lesion is achieved (e.g., at tumor).
3. Side effects at the same effective dose are reduced.

For instance, in our preliminary experiments, we found pyran–NCS conjugate (conjugate molecular weight (MW), 25,000, vs. 12,000 of native NCS) exhibited bone marrow toxicity to only about 30% of the native protein. Thus, a dose that is three times more effective can be given with the same toxicity, and therapeutic efficacy was much improved (14). This is interpreted to mean that by conjugating with polymers, the protein drug becomes less permeable through the blood capillary in the bone marrow tissue, as discussed subsequently.

In the case of SMANCS, despite its much prolonged $t_{1/2}$, its toxicity to the bone marrow, kidney, and other organs and tissues did not increase when compared with native NCS. The $t_{1/2}$ of NCS is extremely short, but its toxicity to the bone marrow and other organs is very severe because of its extremely high cytotoxicity (below 10^{-9} molar range).

Plasma concentration is not, however, solely a function of molecular size. For example, α_2-macroglobulin (α_2M), a plasma glycoprotein with a molecular weight of 720,000 (as tetrameric structure), is an important broad-spectrum protease inhibitor. It can combine with a number of protease molecules (trapping hypothesis) involving cleavage of peptide bonds in the so called *bait region*. After entrapping protease, α_2M undergoes a drastic conformational change, and its $t_{1/2}$ decreases drastically (from 140 h to less than 5 min) (15), a result indicating the importance of protein conformation. Similarly, when plasma albumin was modified with formaldehyde, its $t_{1/2}$ decreased (16). A more recent study indicated that acetylated low-density lipoprotein (LDL) is also more rapidly cleared as compared with native LDL (17, 18). Thus, it appears that denatured or modified proteins may generally have substantially shorter $t_{1/2}$.

When low molecular weight drugs or dyes are bound noncovalently to albumin (or other plasma proteins) or biocompatible synthetic polymers, their $t_{1/2}$ become almost as long as that of the polymer. However, the pharmacological activity is largely dependent on the association constant of ligands to the polymer.

Drug Targeting to Tumor and Lesion

Targeting to Tumor

The uniqueness of tumor vasculature has not been fully used for cancer chemotherapy, and little attention has been paid to this subject in drug design or tumor targeting. During our search for macromolecular anticancer agents, it became apparent that vascular properties of tumor tissues are greatly different from those of normal tissues (19, 20). There are several factors involved.

1. Tumor angiogenesis results in hypervasculature (21, 22).
2. Tumor derived factors facilitate enhanced permeability (23–26).
3. Vascular architecture and function of tumor are defective (27–29). Thus tumor blood vessels are leaky.
4. Activity of the lymphatic drainage system is low, and hence accumulation of macromolecules and lipids results in tumor tissue: they are usually recovered via this system (4, 19, 20, 30, 31).

Table III. Plasma Clearance Time for Various Proteins and Their Polymer Conjugates or Derivatives

Protein or Its Polymer Conjugate	Test Animal	Polymer Type or Modification	$MW \times 10^{-3}$	$t_{1/2}{}^a$	$t_{1/10}{}^b$	Ref.
NCS	mouse	none	12	1.8 min	15 min	4, 5
SMANCS^c	mouse	poly(styrene-co-maleic acid)	16	19 min	5 h	4, 5
Bovine ribonuclease	mouse	none	13.7	5 min	30 min	101
Bovine ribonuclease dimer	mouse	none	27	18 min	5 h	101
Bovine SD	rat	none	32	4 min.	30 min.	49
SMA^d–SD	rat	poly(styrene-co-maleic acid)	40	>5 h	>10 h	49
DIVEMA^e–SD	rat	Divinyl ether–maleic acid copolymer	42	30 min	>5 h	4, 14
Soybean trypsin inhibitor (SBTI, Kunitz type)	rabbit	none	20	<2 min	3 min	9
Dextran–SBTI	rabbit	dextran	127	~20 min	>80 min	9
Bilirubin oxidase	rat	none	50	15 min	74 min	62, 63
PEG–bilirubin oxidase	rat	PEG	70	5 h	48 h	62, 63
Chick ovomucoid	mouse	DTPA^f/^{51}Cr	29	5 min	34 min	5
Mouse serum albumin	mouse	none	68	3–4 days^g	—	5
	mouse	Evans blue–dye binding	—	2 h	30 h	5
	mouse	DTPA/^{51}Cr	—	6 h	30 h	5, 16
Bovine serum albumin	mouse	DTPA/^{51}Cr	—	1 h	24 h	5
	rat	iodination/^{125}I	—	4.5 h	65 h	16
Formaldehyde-modified human serum albumin	rat	formaldehyde/^{125}I	—	25 min	4 h	16
Human transferrin	rat	iodination/^{125}I	87	8 days	—	103
E. coli L-asparaginase	rat	none	65 × (2–8)	1.5–3.4 h	—	104
	rat	PEG$_2$-linked^h	—	56 h	11 days	104
Mouse immunoglobulin G	mouse	DTPA	150	60 h	—	5
	mouse	iodination/^{131}I	150	45.6 h	—	105
Human α_2-macroglobulin	mouse	iodination/^{125}I	180 × 4	140 h	22 days	15
Human α_2-macroglobulin (half-molecule)	mouse	iodination/^{125}I	180 × 2	36 h	—	15
Human α_2-macroglobulin–plasmin complex	mouse	iodination/^{125}I	180 × 2	2.5 min	20 min	15
Human α_2-macroglobulin–piasmin complex	mouse	iodination/^{125}I	180 × 4	5 min	—	15

^{a} $t_{1/2}$ indicates an initial decline, α phase in pharmacokinetics.
^{b} $t_{1/10}$ is required to reach to one-tenth of concentration of time zero by intrapolation.
^{c} SMANCS, poly(styrene-co-maleic acid) conjugated neocarzinostatin.
^{d} SMA, poly(styrene-co-maleic acid).
^{e} DIVEMA, copoly(divinyl ether–maleic acid), also called pyran copolymer.
^{f} DTPA, diethylenetriaminepentaacetic acid; (pK_a of ^{51}Cr = 24).
^{g} Human albumin in human, 19 days.
^{h} PEG$_2$, polyethylene glycol bis-substituted (two chains attached).

In the normal healthy state macromolecules such as plasma proteins are retained within the blood circulation without leaking out of the vascular endothelial compartment into the interstitial space. Thus, a majority of substances with molecular size above 50,000 daltons will not leak out into the normal tissue (3–5, 19, 20, 32). However, this characteristic is not the case in the vasculatures of inflammatory tissues (4, 32–36), and more importantly in tumor tissues. Radiolabelled fibrinogen, albumin, and α_1-acid glycoprotein, when injected in rats, were retained more in both solid and granulomas tumors, as compared with normal organs or tissues (37, 38). Similarly, macromolecular anticancer agent SMANCS (MW 16,000), as well as albumin and immunoglobulin G (IgG), accumulated preferentially in tumor compared with normal tissues (19, 20). The small protein NCS (MW 12,000) and small glycoprotein ovomucoid (MW 29,000) did not do so at all. SMANCS, which has a small size as a protein (MW 15,500) itself, can bind noncovalently to albumin and behave like a large protein of about 83,000 (39). Thus, it accumulates effectively in tumors (3, 5, 19, 20).

In the case of inflammatory tissues, proteins and lipids leak out of the blood vessels (at the postcapillary venule) into the interstitial space, and then they are recovered via the lymphatics. The proteins and lipids then return to the general circulation via the thoracic duct (3–6, 19, 20).

In tumor tissues, SMANCS, plasma albumin, IgG, or other polymers such as poly(N-2-hydroxypropyl-

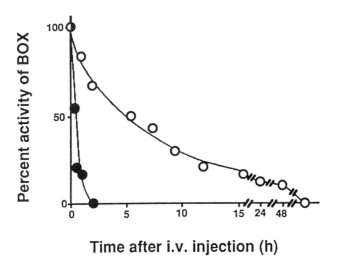

Figure 1. Residual plasma activities of bilirubin oxidase (●) and PEG-conjugated bilirubin oxidase (○) in rats after intravenous injection. (Reproduced with permission from reference 63. Copyright 1988.)

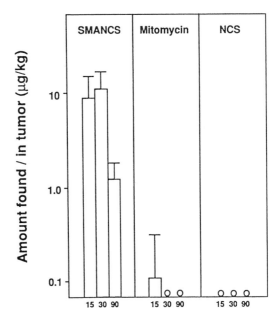

Figure 2. Intratumor concentration of SMANCS, myto-mycin C, and NCS in rabbits after intravenous injection of 10 mg/kg of each drug. VX-2 tumor was inoculated in the liver. Assay data were based on growth inhibition of *Micro-coccus luteus* of the tissue homogenates. (Reproduced with permission from reference 3. Copyright 1984.)

methacrylamide) (HPMA) can penetrate between the vascular endothelial cells into the interstitial space and accumulate in the tumor. An important finding is that such polymers are much less likely to be recovered into the lymphatic system in the tumor, as is observed in normal tissues. This characteristic is in contrast to many small molecular weight drugs. Thus, macromolecules and lipids are extravasated and retained in the tumor tissue more effectively. This phenomenon is called enhanced permeability and retention (EPR) effect (*6, 7, 19, 20*). Intratumor concentrations of SMANCS, NCS, and mitomycin C are compared in Figure 2 for VX-2 tumor implanted in the liver of a rabbit. In this study, SMANCS showed the highest concentration, and also the longest retention period in the tumor (EPR effect). Other studies also have shown that proteins and lipids were not cleared readily from the tumor, a result indicating low lymphatic drainage activity, which is again in contrast with normal tissue (*3, 19, 20, 30, 31*).

Targeting to Inflammation

In the case of inflammation caused by a bacterial infection, we have demonstrated that the bacterial proteases activate Hageman factor and prekallikrein, which subsequently leads to generation of kinin from kininogen, and ultimately results in extravasation (*4, 33–36*). Evans blue solution was injected intravenously in guinea pigs (1.0 mL 2.5% solution per kg body weight). Evans blue forms a complex with albumin immediately. The leaked dye complex was measured after 30 min. A similar effect was also found in heat burns. A common feature in these pathological lesions is that blood vessels become highly permeable to

Table IV. Enhancement of Vascular Permeability by Inflammation-Inducing Substances in Guinea Pig

Acting Substances	*Injection Dose–Site (μg) in Back Skin*	*Blood Vessel Leakage for Evans Blue-Albumin (μg)*
Histamine	3.0	122
	1.0	41
	0.3	23
	0.1	20
	0	0
Bradykinin	10.0	46
	1.0	36
	0.1	16
	0	0
Serratial protease	10.0	127
	3.0	41
	0.3	5
	0	0

macromolecules. Once leaked out of the blood vessels they are retained in there, and only slowly recovered via the lymphatic system. Thus, the EPR effect for macromolecules and lipids seems to be common to solid tumors and inflammatory sites (Table IV). Therefore, the EPR effect is the most important factor when one wants to direct and deliver proteins or macromolecular drugs to lesion sites.

An interesting example is the accumulation of Evans-blue complexed albumin in the guinea pig skin after intradermal injection of bacterial proteases (Table IV) (4, 33–39). When influenza virus infects the lung, a pathological lesion similar to bacterial protease becomes apparent. Thermally injured skin also exhibits an EPR effect (4). All inflammations are associated with the generation of vascular permeability enhancing factors such as kinins, histamines, leukotriens, and prostaglandins. Perhaps oxygen radicals and nitric oxide are also involved. Therefore, this phenomenon can be used for macromolecular therapeutics for selective targeting to inflammatory sites.

Promising Examples of Polymer–Drug Conjugates

Poly(styrene-co-maleic acid) Conjugated NCS

SMANCS, which was approved by Japanese Ministry of Health and Welfare in 1993 as an oily arterial injection against hepatoma, is a derivative of NCS, in which two amino groups, one at amino terminal alanine-1 and the other at lysine-20, are conjugated with poly-(styrene-co-maleic anhydride) [PSMA] (Chart I and Scheme I). The PSMA used for this preparation has a molecular weight of 1500 and a ratio of number-aver-

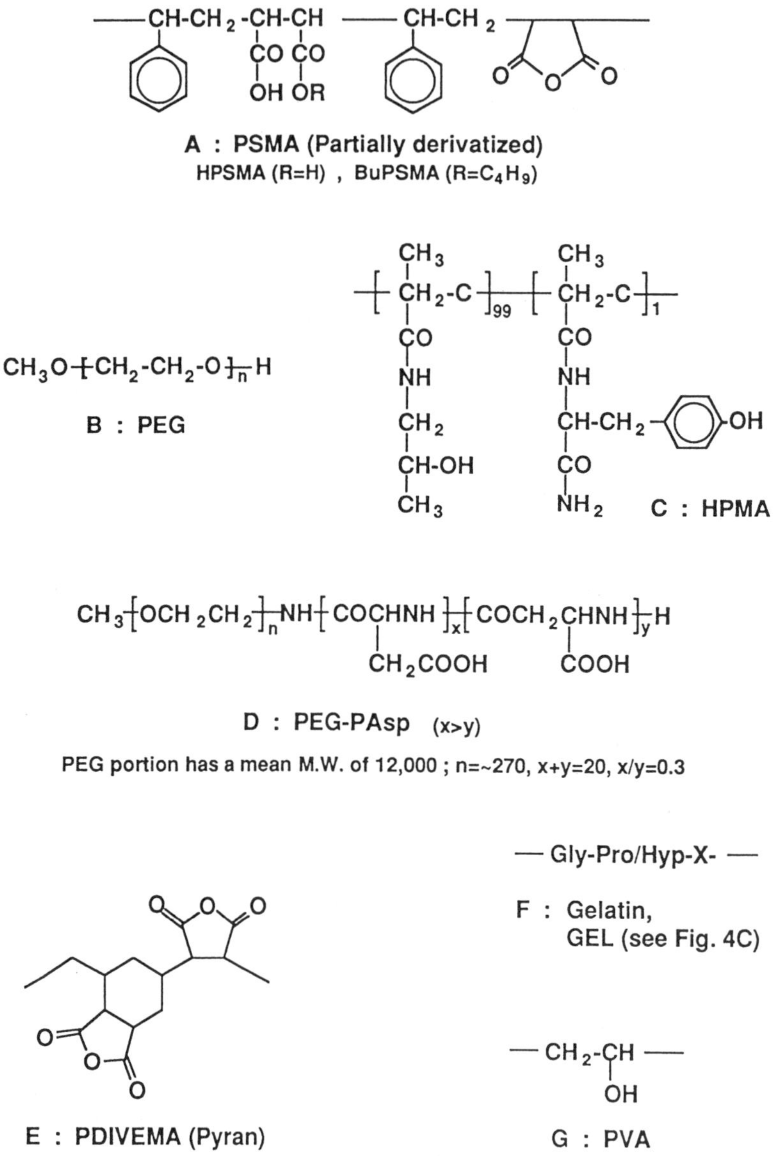

A : PSMA (Partially derivatized)
HPSMA (R=H) , BuPSMA (R=C_4H_9)

B : PEG $CH_3O + CH_2-CH_2-O +_n H$

C : HPMA

D : PEG-PAsp (x>y)

PEG portion has a mean M.W. of 12,000 ; n=~270, x+y=20, x/y=0.3

E : PDIVEMA (Pyran)

F : Gelatin, GEL (see Fig. 4C)
— Gly-Pro/Hyp-X- —

G : PVA
— CH_2-CH —
        OH

Chart I. Chemical structure of polymers: A, poly(styrene-co-maleic anhydride) [partially converted to acid (PSMA) or half ester, n-butyl ester, BuPSMA]; B, polyethylene glycol, PEG; C, N-(2-hydroxypropyl) methacrylamide carrying 1% specified amino acid (L-tyrosine) or peptides, HPMA; D, polyethylene glycol$_n$-poly(aspartic acid)$_{(x+y)}$, in which $n > (x+y)$; E, poly(divinylether-maleic anhydride), pyran; F, succinyl gelatin where gelatin contains about 26% of glycin, 32% of proline–hydroxyproline, and the reminder is any amino acids that composes about 40%; G, poly(vinyl alcohol).

age to weight-average molecular weight of 1.2 or less (*40*). The mechanism of action of SMANCS is similar to NCS: degradation of DNA and inhibition of DNA synthesis (*41*). Induction of interferon (mainly γ-type) (*42*) and activation of macrophage and killer T cell were described for SMANCS (*43–45*). These activation effects are acquired after polymer conjugation, particularly with polyanions. Thus, SMANCS is a third generation polymer drug. SMANCS with molecular weight of 15,000 binds with albumin, and hence its effective molecular weight in vivo or in plasma is about 83,000 (*39*). The greatly improved stability of SMANCS in aqueous and lipid media was described already (*46*) and seems to be a common feature of many polymer–protein conjugates (*47*). Proteolytic resistance of SMANCS and PSMA-conjugated SD also was improved compared with native form NCS. (*48, 49*).

The most remarkable antitumor effect was observed when the oily SMANCS formulation was given arterially with the oily contrast medium Lipiodol (*31, 50–53*). The response rate of previously incurable primary liver cancer has thus been greatly increased. The survival rates at 5 years posttreatment ranged from 30% to 90% depending on the extent of liver cirrhosis and the stage and grade of tumor in hepatocellular carcinoma patients (all tumors were highly advanced and inoperable). Efficacy of this protocol for other tumors is presently being evaluated. Tumors showing an effective response include lung cancer (adenocarcinoma or squamous cell carcinoma); renal cell carcinoma; pleural or ascitic carcinomatosis; some pancreatic, bile duct, and ovarian cancers; and leiomyosarcoma (*51, 53*, unpublished data).

Intravenous administration of aqueous SMANCS formulation is also effective. A pilot study by our group has shown some effect on tumors of the brain, lung, esophagus, stomach, kidney, urinary bladder, adrenal gland, pancreas, colon, and the ovary. More pronounced drug delivery can be achieved when SMANCS is given intravenously under angiotensin II induced hypertension (*54*). But in this case, ultimate target tumors and potential uses remain to be established.

PEG-Conjugated Adenosine Deaminase

Children born with a congenital deficiency of adenosine deaminase (ADA) are known to be very prone to microbial infections because of insufficient immune functions, and thus longevity may not be expected.

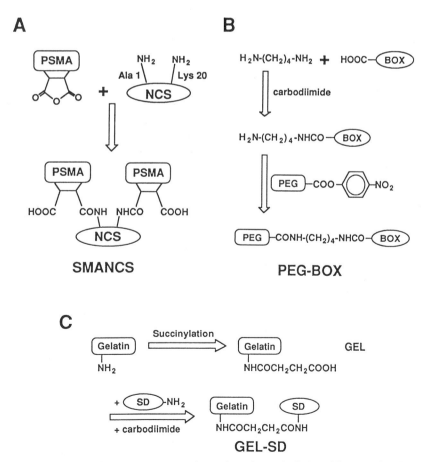

Scheme I. Reaction schemes for polymer–protein conjugations: A, SMANCS; B, PEG–BOX; C, GEL–SD.

Hershfield et al. (55) prepared a PEG-conjugated ADA ($t_{1/2}$ in man, 48–72 h) and used it for human trials. A weekly injection of 15 units/kg led to significant reduction in adenosine levels in erythrocytes and full restoration of lymphocyte functions (immune response) and bone marrow cells. No anaphylaxis was observed during the trial period of 10 months. This polymer–protein conjugate drug was approved by the U.S. Food and Drug Administration (FDA) in 1992 as an orphan drug.

PEG-Conjugated Interleukin-2

Interleukin-2 (IL-2) is a glycoprotein with a molecular weight of 15,000, and it is one of many lymphokines produced by lymphocytes as a sign of intercellular immune response. Therapeutic efficacy of IL-2 against various cancers and infectious diseases was reported (56, 57), and its large-scale production by genetic manipulation is established. However, such proteins obtained by recombinant DNA (rDNA) are lacking carbohydrate side chains and are very unstable in vivo. Furthermore, their solubility may be low, and some rDNA proteins may only be soluble in the presence of detergents.

Katre et al. (58) prepared a PEG-conjugate of human IL-2 produced from genetically modified *Escherichia coli*. The PEG used in this study had a molecular weight of about 5000. One or more PEG chains were attached to one IL-2 molecule. In the conjugate (PEG–IL-2), many of the problems associated with parent rIL-2 were overcome. In parallel with prolongation of $t_{1/2}$, the therapeutic efficacy against experimental mouse tumor (Meth A) was greatly improved. When rIL-2 is injected into the general circulation, it is cleared very rapidly, as observed for neocarzinostatin or SD; whereas PEG–rIL-2, with its increased Stokes radius, is cleared less rapidly. A derivative of rIL-2 with the polysaccharide pullulan also exhibited a behavior similar to the PEG conjugate (59).

PEG-Conjugated Bilirubin Oxidase

Bilirubin is a toxic yellow pigment formed as a catabolic end product of hemoglobin metabolism. It is lipid-soluble and is normally processed in the liver and excreted into the bile duct and then into the feces. When the liver function drops, the level of bilirubin in the blood plasma rises, and clinical symptoms appear. Thus, hyperbilirubinemia, which is associated with various liver disorders and hemolytic jaundice, is treated by various methods such as plasma exchange and steroid and phototherapy. More recently, a bilirubin-immobilized column (60) was clinically applied to remove bilirubin from plasma. However, none of these has proved to have therapeutic value as a first choice regimen.

We recently proposed a new tactic for the treatment of this disease by the bilirubin degrading enzyme, bilirubin oxidase (BOX). This enzyme is highly specific for BOX, and it has a molecular weight of about 50,000. BOX is produced by a microorganism, *Myrothecium verrucaria* (MT-1) (61).

BOX was conjugated with PEG (MW 5000) by using a spacer (diaminobutane) and carbodiimide (62) (Scheme IB). PEG–BOX exhibited a $t_{1/2}$ of about 20 times longerthan native BOX in rat (63) (see Figure 1). Therapeutic efficacy of PEG–BOX against experimentally prepared jaundice in rats and rabbits showed a remarkable decrease in bilirubin concentration. The toxicity of bilirubin, which was confirmed in culture, could be eliminated in the presence of PEG–BOX. The area under the curve for the activity of PEG–BOX was 26 times greater than that of native BOX. Antigenicity of BOX was reduced by PEG-conjugation, and anti-BOX antibody could not neutralize the enzyme activity of PEG–BOX. Anti-PEG–BOX antibody prepared rigorously with Freund's complete adjuvant could inhibit the activity of PEG–BOX only to the extent of 3.7%. No antibody development was observed during intravenous injection of PEG–BOX during several months of observation (63, 64). The potential value of this tailored polymer–drug conjugate for the treatment of hyperbilirubinemia such as fulminant hepatitis, the neonatal bilirubin encephalopathy, and liver insufficiency is apparent.

HPMA–Peptide Conjugated with Doxorubicin

Copolymers based on HPMA (Chart I) were investigated extensively by Kopecek et al. (65) as soluble drug carriers, particularly for tumor-targeted delivery of antitumor drugs. This polymer is biologically well tolerated and is essentially nonimmunogenic (66) and without observable toxicity both in vitro and in vivo (67). The versatile chemistry of HPMA means that a range of side chains and pendant drugs or targeting moieties can be introduced into the polymer structure, and the resulting conjugates are usually well tolerated by biological systems (68–70). Indeed, HPMA copolymers are capable of decreasing the immunogenicity of proteins conjugated to them (71).

Many different drugs including various anticancer agents have been linked covalently to the HPMA. Doxorubicin (DOX) has been incorporated using peptide sequences linked through the amino function of the daunosamine sugar residue. Peptide spacers such as Gly–Phe–Leu–Gly were used, which were designed to be stable extracellularly but cleavable following pinocytic internalization. Spacers longer than tetrapeptides proved particularly useful for this purpose, because the cleavage of the conjugate by lysosomal enzymes released free DOX in a quantitative fashion

(72). In addition targeting groups such as antibodies (68), hormones (73), or simple sugars (74) were incorporated into the conjugate. The incorporation of galactose is particularly interesting, because it affords the possibility of targeting therapeutic doses of DOX into the liver for organ-specific therapy of hepatic metastases of colorectal (and other) tumors, as well as for the targeted treatment of the primary hepatoma (75).

Of particular interest has been the mechanism of action and antitumor efficacy of untargeted HPMA–DOX conjugate. Free DOX has a very short $t_{1/2}$ (ca. 2 min in mice), and it diffuses rapidly through cell membrane resulting in high cardiac concentrations that closely correlate with a cumulative cardiotoxicity. DOX also shows relatively little tumor-selective accumulation (76). HPMA–DOX conjugate, on the other hand, has an extended $t_{1/2}$ (ca. 60 min) (77) and shows substantial accumulation in peripheral tumors. Following intravenous administration to mice bearing solid subcutaneous B16 F10 melanomas, free DOX (5 mg/kg body wt.) produced drug levels of only 0.55 μg/g tumor, whereas the same dose of DOX equivalent administered as polymer conjugate achieved levels of up to 7.5 μg free DOX equivalents/g tumor. The decreased cardiotoxicity of polymer–DOX conjugate permits the use of larger doses, and administration of doses of 18 mg DOX conjugate kg^{-1} g^{-1} results in drug levels up to 22 μg DOX/g tumor (53).

This passive targeting of HPMA–DOX conjugate to solid tumors is thought to result from its prolonged circulation in vivo and nonspecific capture by tumors as a result of the EPR effect described above. In studies of anticancer efficacy, this drug accumulation in tumor resulted in an impressive therapeutic effect in mice (78, 79). These conjugates are currently under clinical evaluation in the United Kingdom (personal communication from R. Duncan).

Micelle-Forming PEG–PAsp–ADR Conjugate

Yokoyama et al. (80, 81) prepared an adriamycin (ADR) conjugate of PEG–poly(aspartic acid) (PEG–PAsp [Chart I]). ADR was linked to the PAsp block of the copolymer by amide bond formation between an amino group of ADR and the carboxyl group on the polymer chain. This conjugate (PEG–PAsp–ADR) was observed to form micelles with a mean diameter of 50 nm and a narrow size distribution in phosphate-buffered saline. It also showed high water solubility despite the large amount of ADR contained (81). The PAsp, drug-binding segment forms the hydrophobic core of the micelle, whereas the PEG segment surrounds this core as a hydrated outer shell.

In vivo antitumor activity of PEG–PAsp–ADR was studied against several murine and human solid tumors (82). This conjugate showed higher antitumor

activity than free ADR against all six experimental mouse tumors examined except MKN-45 in mice. Especially, the use of this conjugate against mouse colon adenocarcinoma-26 led to complete disappearance of tumor growth in five of the six mice (at a dose of 150 mg ADR equivalents/kg), and considerably prolonged life span (treated group–control, >177%). The conjugate exhibited a toxicity of approximately 5% of free ADR in terms of body weight loss.

Polymer Conjugates of Cu,Zn-Superoxide Dismutase

Superoxide dismutase (SD) catalyzes the dismutation of highly reactive superoxide anion radical ($O_2^{\bullet-}$) to molecular oxygen and hydrogen peroxide. The hydrogen peroxide is then converted to molecular oxygen and water by catalase or glutathione peroxidase. SD was suggested as a possible therapeutic agent for many diseases associated with $O_2^{\bullet-}$ such as allergy (83), ischemic myocardial damage (84), Crohn's disease (85), and virus infection (86–88). We have shown in a model experiment that mice infected with influenza virus are cured to a great extent with the use of SD conjugated with a copolymer of divinyl ether-maleic anhydride (PDVEMA or pyran) (Chart I), whereas free SD has only poor therapeutic effect. The better therapeutic effect of the pyran–SD conjugate is partly attributed to its improved pharmacokinetics (86–88). Perhaps more importantly, this conjugate may possess metal chelating activity (*see* later).

Cell-lubricating activity observed in gelatin–SD conjugate (GEL–SD) (Scheme IC) may facilitate peripheral circulation of plasma cells. This activity also may be an important addition to SD activity.

Toxic Aspects of SD

In recent reports, Yim et al. (89) and our group (90) have shown that hydroxyl radicals ($^{\bullet}OH$) are generated during the reaction of Cu,Zn–SD with H_2O_2. Hydroxyl radical is known to be a toxic species in vivo generating many reactive intermediates (including lipid radicals) and causing direct damage to proteins, nucleic acids, cell membrane, and cells. Euphoria of SD as a panacea for numbers of diseases caused by oxygen radicals should take into account possible hazardous effects of $^{\bullet}OH$. Also, Cu,Zn–SD itself was inactivated by its own reaction product, H_2O_2, before inactivation of α_1-protease inhibitor (α_1-PI) (90–92). Hydroxyl radical thus generated is a potent inactivator of α_1-PI, a major endogenous elastase inhibitor and anti-inflammatory component (91–93). This inactivation of α_1-PI by $^{\bullet}OH$ results in deleterious effect to the infected or diseased hosts. If one could eliminated this SD-generated toxicosis, SD will be a useful therapeutic agent, though not a panacea.

Improved Polymer–SD Conjugates

To circumvent the previously mentioned SD toxicity, we prepared various SD conjugates with eight different polymers (MW 1500–30,000) and evaluated the stability of the polymer–SD conjugates against H_2O_2 (Table V). The effect of the polymer–SD conjugates on α_1-PI in the presence of H_2O_2 was also examined. The results showed widely different effects for different polymers against inactivation of SD and α_1-PI and generation of ·OH radicals (91).

Stability of several polymer–SD conjugates against externally added H_2O_2 was tested. To test the resistance of SD against its own enzyme-reaction product (H_2O_2), various kinds of polymer–SD conjugates (3.7 μM) were treated with H_2O_2 at 1 mM in 25 mM phosphate buffer at pH 7.4, and inhibitory effect against the inactivation by the polymer–conjugation was quantified as a function of incubation time after 15 min (Figure 3). Figure 3 shows most polymer–SD conjugates, except HPSMA–SD and BuPSMA–SD, became more resistant than native SD. In particular, the activity of high MW pyran–SD conjugate [pyran(H)–SD] was 78% of the initial activity after incubation with 0.1 mM H_2O_2 for 3 h at 37 °C, as compared with 39% for native SD under similar conditions.

In the same setting α_1-PI (α_1 protease inhibitor) was also added as a target molecule of the ·OH radical generated by Fenton reaction (H_2O_2 + Cu^{2+} of SD), and effect on elastase inhibitory activity of α_1-PI (at 20 μg/mL) was examined, where all reaction mixtures contained 7.4 μM Cu^{2+} at 37 °C and reacted for 3 h. We found that pyran(H)–SD in the presence of 1 mM H_2O_2 neither lost its SD activity nor α_1-PI activity, whereas native SD or SM(H)–SD was inactivated to a great extent and α_1-PI activity was lost in parallel to SD activity. From these data different polymers exhibited different degrees of protection or inhibition against inactivation by H_2O_2 (Table VII; Figure 4) (91,92).

We found that inactivation of α_1-PI was mediated by the generation of ·OH radicals produced by free

Cu^{2+} (Cu^{2+} is released from oxidatively damaged SD) (90). Therefore, the inhibitory or suppressive activity of ·OH radical generation of various polymers was tested. Electron spin resonance with 5,5'-dimethyl-l-pyrroline-N-oxide as a spin trap was employed to study ·OH generation in the reaction of Cu^{2+} with H_2O_2. Pyran(H) copolymer showed the highest inhibitory activity among all eight polymers tested (Table VI). Inhibiting effect against ·OH generation from

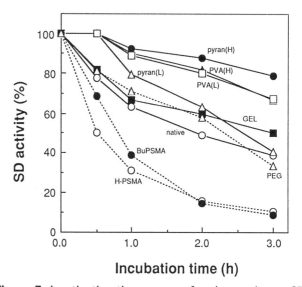

Figure 3. Inactivation time course of various polymer–SD conjugates with 0.1 mM H_2O_2 (see text for details): —○—, native SD; —●—, pyran(H)–SD; —△—, pyran(L)–SD; —▲—, PVA(H)–SD; —□—, PVA(L)–SD; —■—, GEL–SD; --△--, PEG–SD; --○--, H-PSMA–SD; --●--, BuPSMA–SD. Assay of SD is based on reference 107.

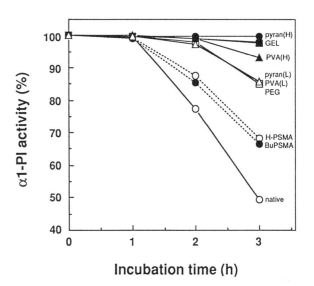

Figure 4. Inactivation time course of α_1-protease inhibitor (α_1-PI; 20 μg/mL) in the presence of various polymer–SD conjugates and 1 mM H_2O_2. See Figure 3 for symbols.

Table V. Properties of Various Polymer–SD Conjugates

Polymer–SD Conjugates	Polymer MW	SD Content (w/w%)	Polymer Chains Conjugated to One SD	Percent SD Activity in Conjugate
Native SD	—	100.0	0	100.0
Pyran(H)–SD	30,000	49.1	1.0	50.2
Pyran(L)–SD	5000	87.9	1.0	60.2
PVA(H)–SD	11,200	30.1	6.6	92.1
PVA(L)–SD	4500	55.2	5.7	96.2
GEL–SD	23,000	33.3	2.9	100.0
H-SMA–SD	1500	88.8	2.7	24.0
BuSMA–SD	1600	91.0	2.0	50.6
PEG–SD	5000	56.3	4.9	60.7

Note: Human recombinant Cu,Zn-SD (4390 U/mg protein). SD activity is on molar basis.

Table VI. Inhibitory Effect of Various Polymers and Diethylenetriaminopentaacetic acid (DTPA) for ˙OH Radical Generation in (Cu²⁺ + H₂O₂)

Polymer	Polymer MW	IC_{50} (M)[a]
Pyran copolymer	30,000	2.6×10^{-7}
Pyran copolymer	5000	2.4×10^{-6}
NEM–PVA[b]	4600	2.8×10^{-5}
NEM–PVA	11,300	2.4×10^{-5}
Succinyl-gelatin	23,000	6.6×10^{-5}
BuSMA	1500	2.7×10^{-3}
H-SMA	1500	2.9×10^{-3}
MPEG[c]	5000	$>1.0 \times 10^{-2}$
DTPA	393	1.6×10^{-6}

[a]Polymer concentration (M) for 50% inhibition.
[b]*N*-Ethylsuccinimidyl-*S*-poly(vinyl alcohol).
[c]Monomethoxy polyethylene glycol.

Table VII. Suppressive Effect of Various Polymers Against Inactivation of α₁-PI by ˙OH Radicals Produced by (Cu²⁺ + H₂O₂)

Polymer	Polymer MW	IC_{50} (M)[a]
Pyran copolymer	30,000	4.8×10^{-7}
Succinyl-gelatin	23,000	1.6×10^{-5}
NEM–PVA	11,300	3.7×10^{-5}
BuSMA	1600	$>1.0 \times 10^{-3}$
MPEG	5000	$>1.0 \times 10^{-3}$

[a]Polymer concentration (M) for 50% inhibition.

Cu^{2+} plus H_2O_2 system was well correlated with the rate of inactivation of α₁-PI. Pyran(H) copolymer alone showed the most suppressive effect on α₁-PI inactivation by Cu^{2+} plus H_2O_2, whereas either BuPSMA or PEG polymer alone showed much weaker suppression of ˙OH generation and hence more rapid inactivation of α₁-PI than native SD (Table VII). However, none of the polymers scavenged ˙OH radicals generated by H_2O_2 by UV irradiation. Thus, the suppressive effect of various polymers on ˙OH generation could be attributed to their chelating effect by some polymers against Cu^{2+} and hence inhibition of Fenton reaction.

Therefore, Cu^{2+} binding capacities of various polymers were examined, and the results are shown in Figure 5. Pyran copolymer had the highest chelating effect among all of the polymers tested. Despite low Cu^{2+} binding capacity of *N*-ethylsuccinimidyl–*S*-PVA (NEM–PVA), this polymer exhibited high ˙OH inhibitory activity, perhaps due to direct scavenging activity of the radical by sulfur atom (Table VI). Cu^{2+} chelating capacity and inhibitory activity of various polymers against inactivation of SD and α₁-PI were comparable for all the polymers except for sulfur-containing PVA (Table VII, Figures 3 and 4).

These results are interpreted to indicate that polymers such as pyran copolymer can capture loosely bound Cu^{2+} during the inactivation reaction of SD by H_2O_2, resulting in inhibition of ˙OH generation; therefore, inactivation of SD and α₁-PI does not take place in the presence of some of these polymers with high chelating capacity. PEG showed little Cu^{2+} chelating capacity and hence no protection against inactivations of SD or α₁-PI.

Cell-Lubricating Activity

Maeda et al. have shown (*94–96*) that α₁-acid glycoprotein and gelatin have a lubricating effect on red blood cells (i.e., facilitate cell passage) and reduce

hemolysis during cell passage under hydrostatic pressure. Tissue damage is thought to be associated with $O_2^{\bullet-}$ generation during ischemic reperfusion (*93*), and ischemia may result from increased vascular resistance at the capillary level. Thus, it may be desirable to decrease the resistance at the capillary level and facilitate plasma cell passage by lubrication and simultaneously to remove $O_2^{\bullet-}$ by polymer–SD conjugates such as GEL–SD.

To make the cross-linking reaction more selective between amino group of SD and carboxy group of gelatin, all free lysyl amino groups of gelatin fragment (Seiwakasei Co., Ltd., Osaka) were completely succinylated by using succinic anhydride. The preparation thus contained no more than 0.1 mole free amino group per mole gelatin (*12*), which we refer to here as GEL.

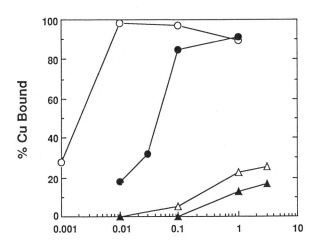

Figure 5. Binding of Cu^{2+} to various polymers as determined by ultrafiltration by using Centricon 3 (3000 MW cut-off) (*105*): —○—, pyran copolymer (MW 30,000); —●—, succinyl-gelatin (MW 23,000); —△—, NEM–PVA (MW 11,300); —▲—, PEG (MW 5000). Assay of free Cu^{2+} is based on the method using disodium bathocuproine disulfonic acid (*108*). Concentration of each polymer is based on molecular weight of polymer.

To this end, effect of GEL–SD on the passage of human red blood cells through membrane filters with 3-μm pores was examined (*13*). Red blood cells suspended in 10 mM phosphate buffered 0.15 M saline (used as control) did not pass through the membrane, and extensive hemolysis took place. When GEL–SD was added at 1.4 mg/mL to the cell suspension, about 7% of the cells penetrated the membrane and hemolysis was reduced from 19% to 6% (*13*). GEL–SD also possesses relatively high chelating activity, perhaps due to succinyl residues.

In disease states, highly elevated coagulating activity, antigen–antibody complex formation, and enhanced vascular permeability are known to occur. These may impede blood circulation by either physical obstruction, decreased fluidity, or slugging at the capillary level. Under these conditions, plasma flow can be facilitated in the presence of GEL–SD at the microcirculation level. The previously mentioned results indicate that a third function can be added to SD, and GEL–SD conjugates may be highly beneficial to improve the peripheral circulation.

Therapeutic Effect of Polymer–SD Conjugates

We tested pharmacological activities of pyran–SD, PVA–SD, GEL–SD, and other polymer–SD conjugates against disease models in mice: influenza virus infection, ischemic paw edema, and collagen-induced arthritis. In the case of mice infected with the influenza virus, the activities of adenosine deaminase and xanthine oxidase increased considerably in the supernatant of bronchoalveolar lavage fluid, lung tissue homogenate, and serum (plasma) (*86–88*). We also found that $O_2^{\bullet-}$ generation from xanthine oxidase in the bronchoalveolar lavage fluid, on days 6 and 8 after infection, was elevated 200–600-fold. Treatment of the infected mice with pyran–SD greatly improved the survival rate (0% of control vs. 90% of treated group) (*86–88*). These results indicate that generation of oxygen radicals by xanthine oxidase, coupled with catabolic supply of hypoxanthine from adenosine catabolism, is a pathogenic principle in influenza virus infection in mice, and thus a therapeutic elimination of oxygen radicals by pyran–SD seems possible.

Cerebral ischemia was produced in gerbils by bilateral common carotid occlusion for 5 min, which consistently resulted in delayed neuronal death in the CA1 region of the hippocampus. The effect of native SD and pyran–SD on early ischemic damage was investigated. Preischemic treatment by pyran–SD showed clear protective effect against both neuronal damage detected by immunohistochemistry after 5 min ischemia and delayed neuronal necrosis after one week of recovery. However, no beneficial effect was observed when this drug or native SD was administered just before the recirculation (*97*).

Ischemic reperfusion is known to involve $O_2^{\bullet-}$ generation, which results in edema and tissue damage in the case of artificially induced paw ischemia in mice (*97, 98*). GEL–SD was highly effective against paw edema in ischemic reperfusion, but native SD was ineffective (*13*). A similar result compared with GEL–SD was obtained for PVA–SD (*12*).

The effects of various polymer–SD conjugates and gelatin on murine collagen-induced arthritis, an animal model for human rheumatoid arthritis (*99, 100*), were reported recently (*101*). However, only GEL–SD suppressed the disease. Native SD and gelatin carrier alone were also ineffective. Interestingly, pyran–SD, which also has a long $t_{1/2}$ (about 30 min in mice), showed no effect on this disease model (*101*). This result may be attributed to the route of drug injection (intraperitoneal). The high molecular weight pyran–SD could not traverse the peritoneal membrane into the vascular blood vessels, and hence no significant plasma level could have been attained.

These results suggest that oxygen radicals may have an important role in the effector phase of immune response to manifest this chronic autoimmune polyarthritis, and the therapeutic effect of GEL–SD may involve an immunological mechanisms at the cellular level.

All five polymer–SD conjugates (i.e., pyran–, GEL–, PVA–, buSMA–, and PEG–) provide some clues for clarifying the pathogenic mechanism of these completely different diseases caused by superoxide. The possibility of using these polymer–SD conjugates as radical scavengers may prove useful also for the treatment of related diseases.

Concluding Remarks and Future Outlook

We have presented the results of our research on an emerging class of polymer–drug conjugates. The new polymer–drug conjugates are shown to be pharmacologically effective. They also show increased $t_{1/2}$ in vivo, targeting potential, and decreased toxicity and immunogenicity as compared with free drugs. Some of the polymer–drug conjugates also show added functions that do not exist in the native drugs, and hence they are medicinally more useful than the native compounds. Although officially approved polymer–drug conjugates are only a few, such drugs will increase in number, because many problems associated with recombinant proteins can be overcome by conjugation of these proteins with suitable polymers.

One reason for the slow use of the great potential of polymer–drug conjugates is the risk of development envisaged by entrepreneurs. Another obstacle relates to the inhomogeneous nature of these complex molecules (as compared with small molecules) such as molecular weight distribution, physical state, and difficulties of purification. Also, regulatory agencies are reluctant to approve their use. This negative attitude is now being gradually overcome by using modern technologies in purification and processing. The hallmark of this technology is the development of SMANCS, a highly purified product with consistent specifications, which was approved by Japanese Ministry of Health and Welfare in the fall of 1993. PEG-L-Asparaginase is another example being approved by the U.S. FDA. A third, polymer–drug candidate [*N*-(2-hydroxy-propyl)methacrylamide–doxorubicin conjugate] probably will be approved in the United Kingdom before long. It is hoped more drugs may be moving along this avenue and the full potential of polymer–drug conjugates will be gradually realized.

References

1. *Poly(ethlene glycol) Chemistry: Biotechnical and Biomedical Applications*; Harris, J. M., Ed.; Plenum: New York, 1992.
2. Abuchowski, A.; van Es, T.; Palczuk, N. C.; Davis, F. F. *J. Biol. Chem.* **1977**, *252*, 3578–3581.
3. Maeda, H.; Matsumoto, T.; Konno, T.; Iwai, K.; Ueda, M. *J. Protein Chem.* **1984**, *3*, 181–193.
4. Maeda, H.; Oda, T.; Matsumura, Y.; Kimura, M. *J. Bioact. Compat. Polym.* **1988**, *3*, 27–43.
5. Maeda, H.; Matsumura, Y.; Oda, T.; Sasamoto, K. In *Protein Tailoring for Food and Medical Uses*; Feeny, R. E.; Whitaker, J. R., Eds.; Marcel Dekker Inc.: New York, 1986; pp 353–382.
6. Maeda, H. In *High Performance Biomaterials. A Comprehensive Guide to Medical and Pharmaceutical Applications*; Szycher, M., Ed.; Technomic: Lancaster, England, 1991; pp 765–778.
7. Maeda, H. *Adv. Drug Delivery Rev.* **1991**, *6*, 181–202.
8. Regelson, W. In *International Encyclopedia Pharmacology Therapy*; Mitchell, M. S., Ed.; Pergamon: Oxford, England, 1985; pp 429–472.
9. Takakura, Y.; Takagi, A.; Hashida, M.; Sezaki, H. *Pharmaceut. Res.* **1987**, *4*, 293–300.
10. Sela, M. *Science (Washington, D.C.)* **1969**, *166*, 1365–1374.
11. Atassi, M. Z. In *Immunochemistry of Proteins*; Atassi, M. Z., Ed.; Plemun: New York, 1977; Vol. 2, Chapter 3, and other chapters; pp 77–176.
12. Kojima, Y.; Maeda, H. *J. Bioact. Compat. Polym.* **1993**, *8*, 115–131.
13. Kojima, Y.; Haruta, A.; Imai, T.; Otagiri, M.; Maeda, H. *Bioconjugate Chem.* **1993**, *4*, 490–498.
14. Yamamoto, H.; Miki, T.; Oda, T.; Hirano, T.; Sera, Y.; Akagi, M.; Maeda, H. *Eur. J. Cancer* **1990**, *26*, 253–260.
15. Gonias, S. L.; Pizzo, S. V. *Biochemistry* **1983**, *22*, 4933–4940.
16. Buys, C. H. C. M.; Dejong, A. S. H.; Bouma, J. M. W.; Gruber, M. *Biochim. Biophys. Acta* **1975**, *392*, 95–100.
17. Brown, M. S.; Goldstein, J. L. *Ann. Rev. Biochem.* **1983**, *52*, 223–261.
18. Murakami, M.; Horiuchi, S.; Takata, K.; Morino, Y. *J. Biochem.* **1987**, *101*, 729–741.
19. Matsumura, Y.; Maeda, H. *Cancer Res.* **1986**, *46*, 6387–6392.

20. Maeda, H.; Matsumura, Y. *CRC Crit. Rev. Ther. Drug Carrier Syst.* **1989**, *6*, 193–210.
21. Folkman, J. *Adv. Cancer Res.* **1974**, *19*, 331–358.
22. Folkman, J.; Klagsburn, M. *Science (Washington, D.C.)* **1987**, *235*, 442–447.
23. Senger, D. R.; Galli, S. J.; Dvorak, A. M.; Perruzzi, C. A.; Harrey, V. S.; Dvorak, H. F. *Science (Washington, D.C.)* **1983**, *219*, 983–985.
24. Senger, D. R.; Perruzzi, C. A.; Feder, J.; Dvorak, H. F. *Cancer Res.* **1986**, *46*, 5629–5632.
25. Maeda, H.; Matsumura, Y.; Kato, H. *J. Biol. Chem.* **1988**, *263*, 16051–16056.
26. Matsumura, Y.; Kimura, M.; Yamamoto, T.; Maeda, H. *Jpn. J. Cancer Res.* **1988**, *79*, 1327–1334.
27. Suzuki, M.; Takahashi, T.; Sato, T. *Cancer* **1987**, *59*, 444–450.
28. Suzuki, M.; Hori, K.; Abe, I.; Saito, S.; Sato, H. *J. Natl. Cancer Inst.* **1981**, *67*, 663–669.
29. Skinner, S. A.; Tutton, P. J.; O'Brien, P. E. *Cancer Res.* **1990**, *50*, 2411–2417.
30. Iwai, K.; Maeda, H.; Konno, T. *Cancer Res.* **1984**, *44*, 2114–2121.
31. Konno, T.; Maeda, H.; Iwai, K.; Tashiro, S.; Maki, S.; Morinaga, T.; Mochinaga, M.; Hiraoka, T.; Yokoyama, I. *Eur. J. Cancer Clin. Oncol.* **1983**, *19*, 1053–1065.
32. Simionescu, M.; Simionescu, N.; Silbert, J. E.; Palade, G. E. *J. Cell Biol.* **1981**, *90*, 614–621.
33. Matsumoto, K.; Yamamoto, T.; Kamata, R.; Maeda, H. *J. Biochem.* **1984**, *96*, 739–749.
34. Maeda, H.; Molla, A. *Clin. Chim. Acta* **1989**, *185*, 357–368.
35. Molla, A.; Yamamoto, T.; Akaike, T.; Miyoshi, S.; Maeda, H. *J. Biol. Chem.* **1989**, *264*, 10589–10594.
36. Maruo, K.; Akaike, T.; Inada, Y.; Ohkubo, I.; Ono, T.; Maeda, H. *J. Biol. Chem.* **1993**, *268*, 17711–17715.
37. Peterson, H.-I.; Appelgren, K. L. *Eur. J. Cancer* **1973**, *9*, 543–547.
38. Shibata, K.; Okubo, H.; Ishibashi, H.; Tsuda-Kawamura, K.; Yanase. T. *Br. J. Exp. Pathol.* **1978**, *59*, 601–608.
39. Kobayashi, A.; Oda, T.; Maeda, H. *J. Bioact. Compat. Polym.* **1988**, *3*, 319–328.
40. Maeda, H.; Ueda, M.; Morinaga, T.; Matsumoto, T. *J. Med. Chem.* **1985**, *29*, 455–461.
41. Oda, T.; Sato, F.; Yamamoto, H.; Akagi, M.; Maeda, H. *Anticancer Res.* **1987**, *9*, 261–266.
42. Suzuki, F.; Munakata, T.; Maeda, H. *Anticancer Res.* **1988**, *8*, 97–104.
43. Oda, T.; Morinaga, T.; Maeda, H. *Proc. Soc. Exp. Biol. Med.* **1986**, *181*, 9–17.
44. Suzuki, F.; Pollard, R. B.; Maeda, H. *Cancer Immunol. Immunother.* **1989**, *30*, 97–104.
45. Suzuki, F.; Pollard, R. B.; Uchimura, S.; Munakata, T.; Maeda, H. *Cancer Res.* **1990**, *50*, 3897–3904.
46. Hirayama, S.; Sato, F.; Oda, T.; Maeda, H. *Jpn. J. Antibiot.* **1986**, *39*, 815–822.
47. Inada, Y. *Protein Hybrids: Future Prospects of Chemical Modification*; Inada, Y.; Maeda, H., Eds.; Kyoritsu Shuppan: Tokyo, Japan, 1989; pp 1–40 (in Japanese).
48. Maeda, H.; Takeshita, J.; Kanamaru, R. *Int. J. Pept. Protein Res.* **1979**, *14*, 81–87.
49. Ogino, T.; Inoue, M.; Ando, Y.; Awai, M.; Maeda, H.; Morino, Y. *Int. J. Pept. Protein Res.* **1988**, *32*, 153–159.

50. Konno, T.; Maeda, H. In *Neoplasms of the Liver*; Okuda, K.; Ishak, K. G., Eds.; Springer-Verlag: New York, 1987; Chapter 27, pp 343–352.

51. Konno, T.; Maeda, H.; Iwai, K.; Maki, S.; Tashiro, S.; Uchida, M.; Miyauchi, Y. *Cancer* 1984, 54, 2367–2374.

52. Konno, T. *Eur. J. Cancer* 1992, 28, 403–409.

53. Maeda, H.; Seymour, L. W.; Miyamoto, Y. *Bioconjugate Chem.* 1992, 3, 351–362.

54. Li, C. J.; Miyamoto, Y.; Kojima, Y.; Maeda, H. *Br. J. Cancer* 1993, 67, 975–980.

55. Hershfield, M. S.; Buckley, R. H.; Greenberg, M. L.; Melton, A. L.; Schiff, R.; Hatem, C.; Kurtzberg, J.; Martert, M. L.; Kobayashi, R. H.; Kobayashi, A. L.; Abuchowski, A. *N. Engl. J. Med.* 1987, 316, 589–596.

56. Mule, J. J.; Shu, S.; Rosenberg, S. A. *J. Immunol.* 1985, 135, 646–652.

57. Ueno, Y.; Miyawaki, T.; Seki, H.; Hara, K.; Sato, T.; Taniguchi, N.; Takahashi, H.; Kondo, N. *Clin. Immunol. Immunopathol.* 1985, 35, 226–233.

58. Katre, N. V.; Knauf, M. J.; Laird, W. J. *Proc. Natl. Acad. Sci. U.S.A.* 1987, 84, 1487–1491.

59. Morikawa, K.; Okada, F.; Hosokawa, M.; Kobayashi, H. *Cancer Res.* 1987, 47, 37–41.

60. Lavin, A.; Sung, C.; Klibanov, A. M.; Langer, R. *Science (Washington D.C.)* 1985, 230, 543–545.

61. Murao, S.; Tanaka, N. *Agric. Biol. Chem.* 1981, 45, 2383–2384.

62. Maeda, H.; Kimura, M.; Sasaki, I.; Hirose, Y.; Konno, T. In *Poly(ethlene glycol) Chemistry: Biotechnical and Biomedical Applications*; Harris, J. M., Ed.; Plenum: New York, 1992; pp 153–169.

63. Kimura, M.; Matsumura, Y.; Miyauchi, Y.; Maeda, H. *Proc. Soc. Exp. Biol. Med.* 1988, 188, 364–369.

64. Kimura, M.; Matsumura, Y.; Konno, T.; Miyauchi, Y.; Maeda, H. *Proc. Soc. Exp. Biol. Med.* 1990, 195, 64–69.

65. Kopecek, J.; Rejmanova, P.; Duncan, R.; Lloyd, J. B. *Ann. N. Y. Acad. Sci.* 1985, 446, 93–104.

66. Rihova, B.; Kopecek, J.; Ulblich, K.; Chytry, V. *Makromol. Chem., Suppl.* 1985, 9, 13–24.

67. Duncan, R. *CRC Crit. Rev. Ther. Drug Carrier Syst.* 1985, 4, 281–310.

68. Seymour, L. W.; Flanagan, P. A.; Al-Shamkhani, A.; Subr, V.; Ulbrich, K.; Cassidy, J. A.; Duncan, R. *Sel. Cancer Ther.* 1991, 7, 59–73.

69. Rihova, B.; Ulbrich, K.; Strohalm, J.; Vetvicka, V.; Bilej, M.; Duncan, R.; Kopecek, J. *Biomaterials* 1989, 10, 335–342.

70. Flanagan, P. A.; Kopeckova, P.; Kopecek, J.; Duncan, R. *Biochim. Biophys. Acta* 1989, 993, 83–91.

71. Flanagan, P. A.; Rihova, B.; Subr, V.; Kopecek, J.; Duncan, R. *J. Bioact. Compat. Polym.* 1990, 5, 151–166.

72. Duncan, R.; Seymour, L. W.; O'Hare, K. B.; Flanagan, P. A.; Wedge, S.; Ulbrich, K.; Strohalm, J.; Subr, V.; Spreafico, F.; Grandi, M.; Ripamonti, M.; Farao, M.; Suarato, A. *J. Controlled Release* 1992, 18, 123–132.

73. O'Hare, K. B.; Duncan, R.; Strohalm, J.; Ulbrich, K.; Kopeckova, P. *J. Drug Targeting* 1993, 1, 217–229.

74. Duncan, R.; Seymour, L. W.; Scarlett, L.; Lloyd, J. B.; Rejmanova, P.; Kopecek, J. *Biochim. Biophys. Acta* 1986, 880, 62–71.

75. Seymour, L. W.; Ulbrich, K.; Strohalm, J.; Duncan, R. *Br. J. Cancer* 1991, 63, 859–866.

76. Stallard, S.; Morrison, J. G.; George, W. D.; Kaye, S. B. *Cancer Chemother. Pharmacol.* 1990, 25, 286–290.

77. Seymour, L. W.; Ulbrich, K.; Strohalm, J.; Kopecek, J.; Duncan, R. *Biochem. Pharmacol.* 1990, 39, 1125–1131.

78. Duncan, R.; Kopeckova-Rejmanova, P.; Strohalm, J.; Hume, I.; Lloyd, J. B.; Kopecek, J. *Br. J. Cancer* 1988, 57, 147–156.

79. Duncan, R.; Hume, I. C.; Kopeckova, P.; Ulbrich, K.; Strohalm, J.; Kopecek, J. *J. Controlled Release* 1989, 10, 51–63.

80. Yokoyama, M.; Miyauchi, M.; Yamada, N.; Okano, T.; Sakurai, Y.; Kataoka, K.; Inoue, S. *Cancer Res.* 1990, 50, 1693–1700.

81. Yokoyama, M; Kwon, G. S.; Okano, T.; Sakurai, Y.; Seto, T.; Kataoka, K. *Bioconjugate Chem.* 1992, 3, 295–301.

82. Yokoyama, M.; Okano, T.; Sakurai, Y.; Ekimoto, H.; Shibazaki, C.; Kataoka, K. *Cancer Res.* 1991, 51, 3229–3236.

83. Brigham, K. L. *Chest* 1986, 6, 859–863.

84. Gardner, T. J.; Stewart, J. R.; Casale, A. S.; Downey, J. M.; Chambers, D. E. *Surgery* 1983, 94, 423–427.

85. Suematsu, M.; Suzuki, M.; Kitahora, T.; Miura, S.; Suzuki, K.; Hibi, T.; Watanabe, M.; Nagata, H.; Asakura, H.; Tsuchiya, M. *J. Clin. Lab. Immunol.* 1987, 24, 125–128.

86. Oda, T.; Akaike, T.; Hamamoto, T.; Suzuki, F.; Hirano T.; Maeda, H. *Science (Washington, D.C.)* 1989, 244, 974–976.

87. Akaike, T.; Ando, M.; Oda, T.; Doi, T.; Ijiri, S.; Araki S.; Maeda, H. *J. Clin. Invest.* 1990, 85, 739–745.

88. Maeda, H.; Akaike, T. *Proc. Soc. Exp. Biol. Med.* 1991, 198, 721–727.

89. Yim, M. B.; Chock, P. B.; Stadtman, E. R. *Proc. Natl. Acad. Sci. U.S.A.* 1990, 87, 5006–5010.

90. Sato, K.; Akaike, T.; Kohno, M.; Ando, M.; Maeda, H. *J. Biol. Chem.* 1992, 267, 25371–25377.

91. Kojima, Y. Akaike, T.; Hirono, T.; Maeda H. *J. Bioact. Compat. Polym.*, in press.

92. Noda, J.; Otagiri, M.; Maeda, H. *J. Pharmacol. Exp. Therapeut.*, in press.

93. Halliwell, B.; Gutteridge, J. M. C. *Biochem. J.* 1984, 219, 1–14.

94. Maeda, H.; Nishi, K.; Mori, I. *Life Sci.* 1980, 27, 156–161.

95. Maeda, H.; Morinaga, T.; Mori, I.; Nishi, K. *Cell Struct. Funct.* 1984, 9, 279–290.

96. Maeda, H.; Nishi, K. U.S. Patent 4,362,718, 1982; U.S. Patent 4,446,136, 1984; Jpn. Patent 1,344,829, 1986; Eur. Patent 0,035,038, 1986; Jpn. Patent 1,483,537, 1989, and others.

97. Kitagawa, K.; Matsumoto, M.; Oda, T.; Niinobe, M.; Hata, R.; Handa, N.; Fukunaga, R.; Isaka, Y.; Kimura, K.; Maeda, H.; Mikoshiba, K.; Kamada, T. *Neuroscience* 1990, 35, 551–558.

98. Grisham, M. B.; Hernandez, L. A.; Granger, D. N. *Am. J. Physiol.* 1986, 251, G251–G267.

99. Trentham, D. E.; Townes, A. S.; Kang, A. H. *J. Exp. Med.* 1977, 146, 857–862.

100. Courtenay, J. S.; Dallman, A. D.; Dayan, A. D.; Martin, A.; Mosdale, B. *Nature (London)* 1980, 283, 666–669.

101. Kakimoto, K.; Kojima, Y.; Ishii, K.; Onoue, K.; Maeda, H. *Clin. Exp. Immunol.* 1993, 94, 241–246.

102. Bartholeyns, J.; Moore, S. *Science (Washington, D.C.)* 1974, 186, 444–445.

103. Morgan, E. H.; Peter, H., Jr. *J. Biol. Chem.* **1971**, *246*, 3508–3511.

104. Kamisaki, Y.; Wada, H.; Yagura, H.; Matsushima, A.; Inada, Y. *J. Pharmacol. Exp. Ther.* **1981**, *216*, 410–414.

105. Dixon, F. J.; Talmage, D. W.; Maurer, P. H.; Deichmiller, M. *J. Exp. Med.* **1952**, *96*, 313–318.

106. McCord, J.; Fridovich, I. *J. Biol. Chem.* **1969**, *244*, 6049–6055.

107. Ueno, K.; Imamura, T. In *Handbook of Organic Analytical Reagents,* 2nd ed.; Cheng, K. L.; Ueno, K.; Imamura, T, Eds.; CRC: Boca Raton, FL, 1992; pp 369–428.

Drug Targeting by Functional Polymers: Targeting of Polymer-Coated Liposomes

Vladimir P. Torchilin

The general problem of drug targeting is discussed. Also, advantages and limitations of individual drug targeting systems, from direct drug administration to complex multicomponent systems consisting of carrier, drug, and a targeting device, are discussed. A brief description of liposome use as a drug delivery vehicle is presented, together with the discussion of the main problems arising when liposomes are used as drug carriers. The use of polymers for the liposome surface modification is described. Among such polymers are polyelectrolytes capable of controlled permeation of liposomes and water-soluble flexible synthetic polymers, including polyethylene glycol (PEG). These polymers are able to drastically increase the residence time of liposomes in the blood. The mechanism of PEG action on the liposome blood stability is discussed in terms of statistical thermodynamics of polymer solutions. The possibility of combining long circulation time and targetability in liposomes is also discussed. PEG–immunoliposomes and their in vitro and in vivo properties are described. Several examples of in vivo application of PEG–liposomes and PEG–immunoliposomes for the delivery of diagnostic agents are also given.

For the majority of pharmaceuticals currently in use, specificity toward diseased sites is not based on their ability to accumulate selectively in the target organ or tissue. Usually, the drug is more or less evenly distributed within the body, and drug activity is maximally expressed in the affected zone. Moreover, to reach this zone, the drug has to cross many biological barriers—other organs, cells, and even intracellular compartments—where it can express undesirable influence on nontarget areas or can be partially inactivated. As a consequence, to achieve sufficiently high local therapeutic concentration of a given drug, one has to administer the drug in unnecessarily large quantities, the great part of which are just wasted by metabolizing in normal tissues. In addition, when cytotoxic agents, antigenic protein preparations, and some other drugs are used, a general negative effect of a drug on the organism can gradually develop.

The best solution of the problem may be drug targeting. We can define drug targeting as the ability of the drug to accumulate in the target organ or tissue selectively and quantitatively, independent of the place and the method of its administration. (This is an ideal case, and in reality we can only approach it by degrees.) Under ideal conditions, the local concentration of the drug in the target area should be high, and its concentration in other organs and tissues should be negligible, or at least low enough to prevent undesirable reactions. Some advantages of drug targeting are instantly clear. First, in many cases drug administration methods might be simplified. Second, drug quantity required for the successful therapy can be drastically reduced (resulting also in reduced therapy cost). Third, drug targeting makes it possible to increase drug concentration in the target area without any harmful side effects in normal organs or any systemic side effects.

From the very beginning of the concept of drug targeting, suggested by Paul Erlich almost a century ago, it was evident that the hypothetical *magic bullet* has to consist of two principal components: the first one should recognize and bind the target, while the second one should perform therapeutic action at the site of delivery. Thus, drug targeting in the organism includes the coordinated behavior of three elements—the target, the transporting and recognizing device, and the drug itself.

The proper choice of the target can be of crucial importance for successful drug delivery. The recognition of a target (an appropriate organ or tissue) can proceed on different levels: on the level of a whole organ, on the level of specific cells characteristic of the target organ, or on the level of individual components characteristic of these cells (i.e., cell surface antigens). So, if we are dealing with a high molecular weight synthetic polymer that concentrates in the kidneys, or macroparticles that are captured by the cells of the reticuloendothelial system (RES), we can say that the target is recognized by the organ. If the microcapsule or microparticle is modified in such a way that it acquires the ability to distinguish between Kupffer cells and hepatocytes in the liver, the target is recognized on the cellular level. This recognition can be very important if the drug has to be delivered into cells of a definite type, for example, during the treatment of some inherited metabolic diseases, when the metabolic error is located inside appropriate cells. However, the most universal form of target recognition is the recognition on the molecular level. In this particular case, the approach is based on the fact that every organ or tissue is characterized by a unique set of chemical compounds, among which components (antigens) can be found that are specific only to that organ or tissue. For successful targeting, another compound can be used as a transporting unit that is capable of the specific interaction with the target component (i.e., an antibody forming a complex with a target antigen).

Among many carriers suggested for targeted delivery of pharmaceuticals, liposomes have attracted special attention during the past 2 decades (1–3). These phospholipid-derived spherical vesicles are biocompatible and biodegradable, they can be easily prepared by different methods on a large scale, and they can effectively entrap both water-soluble and water-insoluble drugs (water-soluble drugs are entrapped into the interior water space of liposomes, and water-insoluble drugs are incorporated into the inner hydrophobic part of the bilayer membrane). However, there still are some unsolved problems with liposome stability and shelf life. The main drawback of liposomes as drug carriers in vivo is their fast elimination from the blood flow by cells of RES. Recently new long-circulating liposomes were described that can be obtained by incorporation of some natural or synthetic compounds into the liposomal membrane (4, 5).

In this chapter, I discuss the main approaches to targeted drug delivery, the systems studied and used so far, and the most important characteristics of liposomes and antibody-bearing liposomes (immunoliposomes) as drug delivery vehicles. Properties and possible use of a new generation of liposomes and immunoliposomes, the surfaces of which are modified with functionalized polyethylene glycol (PEG) to impart them increased stability in the biological milieu, are also discussed.

Principal Schemes of Targeted Drug Delivery

Direct Application of Drugs

In some cases drug targeting may be achieved by a very simple way—an appropriate drug in its native form can be administered directly into the affected zone. The intra-articular administration of hormonal preparations in the therapy of arthritis (6) or intracoronary infusion of thrombolytic enzymes in the therapy of thrombus-induced myocardial infarction (7) are successful examples of this process. The injection of microparticles containing covalently immobilized or incorporated antitumor drugs and carrying chemically reactive groups also was suggested (8). These reactive groups interact with cell surface proteins in the site of injection (tumors) and keep the particles there via the formation of multiple chemical bonds. The particles slowly degrade in the tumor tissue and continuously release active drug (adriamycin in this particular case). The experiments were performed with periodate-activated Sephadex granulae containing covalently and noncovalently bound drug. Free aldehyde groups in the oxidized Sephadex were used for adriamycin chemical immobilization and for the expected anchoring to the cell surface proteins via protein amino groups (Scheme I).

Physical Targeting of Drugs

Physical factors that can mediate targeted drug delivery can be of endogenous and exogenous origin. In the first case, the targeting effect can be based on the fact that the affected organ differs, for example, in temperature or pH, from the surrounding normal tissues. Thus, inflammation or neoplasia zones usually demonstrate some acidification and elevated temperature. This characteristic makes it possible to use drug-loaded stimuli-responsive carriers that can degrade under lower pH or higher temperature to release the entrapped drug. Even if the appropriate carrier is evenly distributed within the circulation, it will be degraded only in the target area, where the drug is released and accumulated. Strictly speaking, we can-

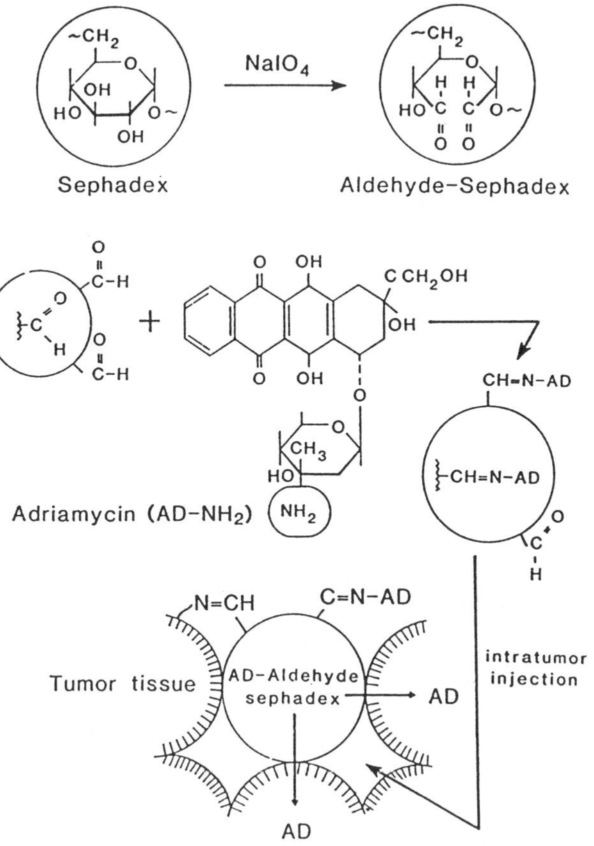

Scheme I. Scheme of preparation and intratumor administration of activated Sephadex microspheres containing covalently and noncovalently bound adriamycin. Free aldehyde groups of activated Sephadex interact with cell surface proteins providing firm binding of the preparation in the injection site. Free drugs release by diffusion and the solubilization of Sephadex fragments. (Adapted from reference 8.)

not consider this process as an *active* targeting, but the final goal can still be achieved.

The most typical studies of this approach have been performed with liposomes, which can easily be made temperature- or pH-sensitive from stimuli-responsive lipids. Thus, methotrexate incorporated into temperature-sensitive liposomes and injected intravenously into mice with inoculated tumors liberated the drug in the tumor several-fold faster if mice were subjected to localized external heat (*9*). Also, by using pH-sensitive liposomes (containing in the membrane pH-sensitive palmitoyl homocystein), fast and quantitative release of the intraliposomal compound can be induced by lowering pH from 7.4 to 6.0 (*10*). pH-sensitive liposomes also were used successfully for the intracellular delivery of drugs and genetic material (for review, *see* reference 11).

Another interesting example of targeted drug delivery by external physical influence is magnetic drug transport. For this purpose, the drug is immobi-

lized on a ferromagnetic microparticulate carrier, injected intravenously, and concentrated within the area to which the external magnetic field is applied. The ability of magnetic particles to concentrate depends on both the intensity of magnetic field and the blood flow rate (*12*). Thus, the particles can be more easily collected in small vessels where the linear flow velocity of the blood is about 0.05 cm/s, rather than in large vessels that have a blood flow velocity of up to 120 cm/s. In our studies (*13*), magnetic carriers were prepared by sedimentation of the ferromagnetic material within Sephadex granules or by coating Fe_3O_4 particles with a dextran layer. After the oxidative activation of polysaccharide carriers, different proteins were immobilized on their surface via Schiff base formation. After intravenous administration, the protein-containing magnetic carriers can be concentrated easily under a magnet field, when a magnet is placed, for example, in the vicinity of the marginal ear vein of a rabbit. Moreover, magnetic carriers with the immobilized thrombolytic agent streptokinase were used successfully for the lysis of the experimental thrombus in the carotid artery of a dog if the Sm–Co magnet was implanted near the supposed site of the thrombus formation (Figure 1) (*14*).

Drug Immobilization and Slow-Release Systems

The normal drug biodistribution can be affected in the desired fashion by its immobilization on different soluble or insoluble polymeric carriers. Polymer–drug conjugates represent an important and well-developed field discussed by Maeda and Kojima in Chapter 5.6. I would just like to mention that several different approaches can be used. First, some polymers possess definite affinity to some cells or organs. For example, polymers with molecular weight above 50,000 can accumulate in kidney tubules or be adsorbed on the surface of some cells with subsequent endocytosis (*15*). Thus, drug immobilization on such polymers may lead to preferential accumulation in specific areas. On the other hand, a drug can be chemically bound or entrapped into (micro)particles of a biocompatible polymer. After implantation or injection in the target area, these particles will create a long-acting drug depot in their vicinity. By varying the rate of drug diffusion from the carrier or the rate of the carrier biodegradation, one can achieve necessary therapeutic concentration of a given drug in the target organ. Thus, it was suggested that physiologically active compounds (proteins) should be included into pills or granules made of biocompatible synthetic polymers such as poly(vinylpyrrolidone) or poly(vinyl alcohol) in the process of granule formation (*16*). The rate of protein release from the granule can be con-

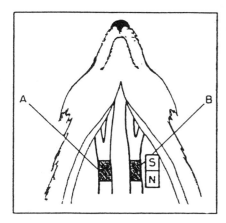

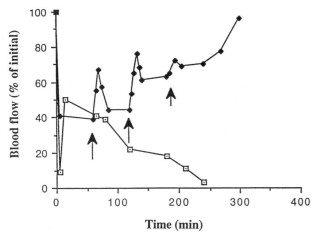

Figure 1. Magnetic targeting of streptokinase. Upper panel: scheme of the experiment; both carotid arteries (A and B) of the experimental dog were embolized with equal quantity of thrombotic mass prepared from the blood of the same animal, then Sm–Co magnet was implanted near one artery (B) at the occlusion site, "magnetic" streptokinase was injected repeatedly, and the blood flow in both arteries was registered. Lower panel: blood flow in both arteries of the experimental dog: the flow gradually decreases in the control artery (empty squares) but is completely restored in the artery with magnet (filled diamonds) after triple injection of "magnetic" thrombolytic enzyme. Arrows indicate time of injection. (Adapted from reference 14.)

trolled by controlling polymer-to-drug ratio or by coating granules with an additional layer of a polymer.

In still another case (17), proteins were covalently bound to a slowly biodegrading polysaccharide carrier. As a result, even after the solubilization, the drug remained covalently bound to a solubilized fragment of the carrier, which additionally stabilized the drug. Thrombolytic enzymes immobilized in this way effectively lysed the experimental thrombus in the femoral artery of dogs; the enzymes were delivered to the thrombus surface by catheterization. The approach

permitted the use of a sharply decreased thrombolytic enzyme dose in comparison with the routine intravenous drug administration (17). Implantable slow-release systems are now an active area of drug design.

Vector Molecules

Different approaches to drug targeting mentioned so far are not universal: sometimes direct administration into an affected organ is very difficult for technical reasons. Often, the affected area does not differ much in pH and temperature from normal tissues, slow release systems can demonstrate some adjuvant properties especially in the case of antigenic drugs (proteins), and magnetic delivery also has some theoretical limitations connected with the intensity of the blood flow through the target (18). The most natural and universal way to impart target affinity to a nonspecific drug is the binding of this drug with another molecule (usually referred to as a *vector molecule*) capable of recognizing and binding to a target organ or site. Antibodies; proteins; peptides; hormones; and mono-, oligo-, and polysaccharides are examples of vector molecules being studied. Monoclonal antibodies against characteristic components of target organs or tissues are of primary importance.

The direct coupling of a drug to a vector molecule seems to be the simplest way to prepare the targeted drug. However, only a few molecules of a drug (especially a high molecular weight drug) can be bound to a single vector molecule without affecting its specific properties. As a result, large quantity of a vector compound has to be used to create the necessary concentration of a drug in the target zone. This quantity may lead to both the waste of vector material and the development of an undesirable response of the organism if the vector compound possesses toxic or antigenic properties. Nevertheless, in some cases even the simplest approach can provide sufficient results. Thus, in the experimental targeted chemotherapy of malignancy, the direct coupling of some plant toxins with antibody against neoplastic cells (vector) results in the formation of immunotoxin (19). Immunotoxins, with their 1:1 drug-to-vector ratio, are effective because of the extreme activity of the toxin itself. Even a single toxin molecule entering a cell can kill it. The general scheme of immunotoxin synthesis (Scheme II) consists of splitting the toxin molecule into catalytic (active) and binding (with broad specificity) subunits, and subsequent chemical coupling of the catalytic subunit with the appropriate monoclonal antibody. As a result, the toxin acquires the ability to selectively kill the cells (cancer cells) that are recognized by the antibody.

An example of direct binding of a drug to a vector molecule is the coupling of chemotherapeutic agents to estrogen molecule (vector) for targeted human

TOXIN (e.g. ricin)

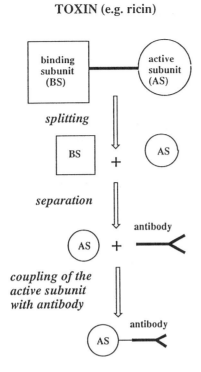

IMMUNOTOXIN

Scheme II. General scheme for immunotoxin synthesis by splitting toxin into active and cell-binding subunits, and subsequent coupling of the active subunit with target-specific monoclonal antibody.

breast cancer therapy via the recognition of specific estrogen receptors on the cancer cells (20). Another example is the coupling of some antiviral drugs (such as adeninearabinoside used for hepatitis B treatment) to glycoproteins containing terminal galactosyl groups for targeting to galactose receptors on hepatocyte surface (21).

Multifunctional Pharmacologically Active Polymers

Functional pharmacologically active polymers were described by Ringsdorf (22). Such polymers consists of a biocompatible and optionally biodegradable soluble backbone to which a drug is attached with stable or biodegradable bonds (Scheme III). The affinity molecule (antibody) and possibly some other effector groups that can influence conjugate solubility or biodistribution are also bound to the same polymeric chain. Thus, the whole complex combines multiple drug molecules with a single vector molecule via a polymeric chain. Such a structure of the multifunctional pharmacologically active polymer opens wide opportunities in the synthesis of new targeted polymeric drugs. Some of these drugs already were described in the literature. Goldberg suggested (23) treating atherosclerotic deposits with a drug consisting of a soluble polymer to which cholesterol oxidase, trypsin, and fibrinolysin are attached along with antibody against the protein component of atherogenic low-density lipoproteins. The same author also described a polymeric drug consisting of a soluble polymer with covalently attached hyaluronidase, proteases, and antibody against the protein component of urinary bladder stone for targeted treatment of urolithiasis.

The same principle was used for the synthesis of targeted polymeric antitumor drugs (24). With this in mind antitumor drugs such as methotrexate, bleomycin, and daunomycin were attached together with antineoplastic cell antibodies to a polysaccharide. As a result, the efficacy and targetability of antitumor drugs were improved, whereas their systemic toxicity was noticeably decreased. A similar chemical approach was used to couple ricin A-chain to dextran with subse-

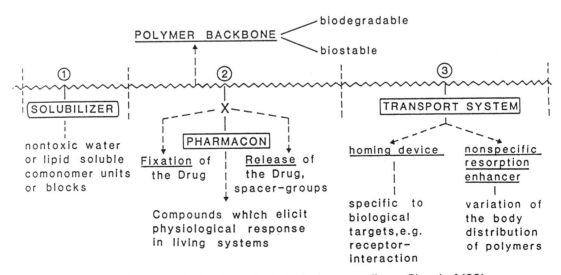

Scheme III. Schematic structure of polymeric drug according to Ringsdorf (22).

quent attachment of monoclonal antibody against smooth-muscle cell-surface antigen (*25*). This preparation demonstrated increased targetability and enhanced ability to kill target cells in culture.

These systems look more promising than those prepared by direct drug-to-antibody conjugation, and they permit the use of one vector molecule to transport numerous drug molecules attached to the same polymeric matrix. However, these systems require the existence of a set of biocompatible and biodegradable polymers that can be cleared from the body after the therapeutic function is completed. Moreover, sophisticated methods have to be used for the coupling of drug and vector molecules to the same polymeric chain without affecting both therapeutical and transporting properties of the conjugate components. Also, the polymer may create steric hindrances for normal drug–target interaction.

Microreservoir Systems for Targeted Drug Delivery

The previous considerations led to a conclusion that the optimal system for the targeted drug delivery has to consist of some microreservoir (microcapsule, cell ghost, liposome, etc.) filled with a drug, and vector molecules have to be attached to the outer surface of this microreservoir. In this case few surface-immobilized vector molecules can transport to the target a very large quantity of entrapped drug molecules (which means very high efficacy of the system). The drug can be released from the reservoir by simple diffusion or by the destruction of a reservoir under specific conditions (e.g., by the complement-dependent lysis). The most popular microreservoir systems include various forms of liposomes and microcapsules. Use of microcapsules for drug delivery is a separate and large area of research (*see*, for example, review by Arshady in *Trends in Polymer Science*, 1993, **1**, 86–93) and will not be considered here.

Liposomes as Targetable Drug Carriers

Liposomes (artificial phospholipid vesicles) are being studied as carriers for drugs (*1–3*). They are biologically inert and completely biocompatible, and they do not cause toxic or antigenic reactions. Drugs included into liposomes are protected from the destructive action of the external media, and liposomes are able to deliver their content inside cells and even inside different cell compartments.

Liposomes can be prepared easily by using different methods of dispersing of phospholipid mixtures in aqueous media. Among those methods are ultrasonication, freezing-thawing, and nuclepore filtration of lipid suspensions. Detergent dialysis from mixed micelles of phospholipid and detergent and reversed-phase evaporation of lipid solution in organic solvent also are used. Water-soluble drugs can be entrapped into inner water space of liposomes, whereas lipophilic compounds can be incorporated into the liposomal membrane.

Unfortunately, *plain* liposomes are, after introduced into the circulation, very quickly (usually within 15 to 30 min) sequestered by cells of RES, primarily by liver cells (*26*). From this point of view the use of targeted liposomes (i.e., liposomes with a specific affinity for the affected organ or tissue) must not only increase the efficacy of liposomal pharmaceutical agent but also decrease the loss of liposomes and their contents resulting from liposome destruction by blood components and their capture by cells. One more factor that has to be considered is the adjuvant property of liposomes (*27*) that can complicate the administration of immunogenic drugs (proteins and enzymes) in liposomes, especially by subcutaneous or intramuscular injection. On the other hand, this property opens new ways for liposome use as potent immunological adjuvants.

To increase the liposomal drug accumulation in the target organ for real clinical application of liposomal drug-targeting systems, one has to increase the efficacy of drug entrapment into liposome, decrease the liposome capture by RES, and increase the liposome affinity toward the target. These problems are analyzed within the rather simple kinetic model in reference 3. In addition, the longer the liposome circulation time, the higher the probability of interacting with the target, which improves targeting.

The improvement of drug incorporation into liposomes relates to liposome preparation. At present there already exist a variety of highly efficient methods of loading liposome with almost any drug, including genetic material (*1*). Numerous more-or-less successful attempts were made to protect liposomes from RES (we do not consider here the case when RES itself serves as the target for the liposomal drug delivery). These include presaturation of RES with empty liposomes before the administration of therapeutical ones (*28*) or liposome coating with certain plasma proteins (*29*). However, the most promising approach to the liposome targeting involves the use of long-circulating liposomes carrying a vector molecules on their surface.

To obtain targeted liposomes, different methods have been developed to bind corresponding vectors to the liposome surface. Immunoglobulins (usually monoclonal antibodies), primarily of IgG class, are the most promising and widely used targeting moieties. The strategy of antibody immobilization on the liposome surface may be crucial for successful targeting. Moreover, in some cases (the target with a limited blood supply or low concentration of the target antigen) even immunoliposome cannot provide good target accumulation before liposomes are eliminated from the blood flow by RES. To increase immunoliposome longevity in the blood (and to increase the probability of its interaction with a poor blood-supplied target) different methods are suggested, including liposome surface coating with inert and biocompatible PEG and similar polymers (*see* subsequent discussion).

Antibody Coupling to Liposomes

I reviewed (3, 30) numerous methods for antibody coupling to liposomes. In general, these methods have to meet the following important requirements:

1. Antibody specificity and affinity should not change on the binding to the liposome.
2. Sufficient transport molecules should be firmly bound to the liposome surface.
3. Liposomal integrity has to be preserved during binding.
4. The binding procedure should be simple and with high yield.

At present as much as 50 to 1000 antibody molecules can be bound to a single 200–250-nm liposome (3, 30). The most often used methods for protein (antibody) coupling to liposomes include covalent binding to a reactive group on the liposome membrane and hydrophobic interaction of proteins specifically modified by hydrophobic residues with the membrane (which, to some extent, resembles the binding of membrane proteins with cell membrane) (Scheme IV).

For chemical methods of protein binding to liposomes, both the liposome surface and the protein mol-

ecule should be chemically activated. Binding can be performed directly or via a spacer group. The main disadvantage of this method (especially in the case of bifunctional reagents) is cross-link formation between liposomes or proteins with the formation of large aggregates. This problem can be overcome by using heterobifunctional reagents, such as *N*-hydroxysuccinimidyl-3-(2-pyridyldithio)propionate (SPDP), which permits activation and coupling under different conditions (31) (Scheme V). Martin et al. (32) prepared large unilamellar liposomes containing a dithiopyridyl derivative of phoshatidylethanolamine (PE), and then they incubated the liposomes with Fab' fragments of antibodies against human red blood cells. The resulting proteoliposomes contained about 600 Fab' molecules per vesicle and were stable in 25% serum. The problem with SPDP is that S–S bonds formed between antibody and liposome can be reduced in vivo, and this reduction leads to the separation of antibody from the liposome. To form stable protein–liposome bonds, other heterobifunctional reagents such as succin-

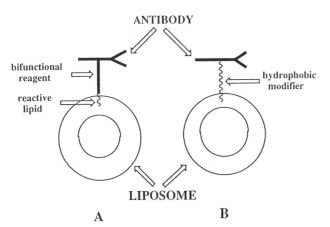

Scheme IV. Possible ways for the binding of target-specific monoclonal antibody to liposome surface: A, antibody covalent binding via reactive lipid and bifunctional reagent; B, antibody incorporation into the liposome membrane via the anchoring residue of hydrophobic modifier.

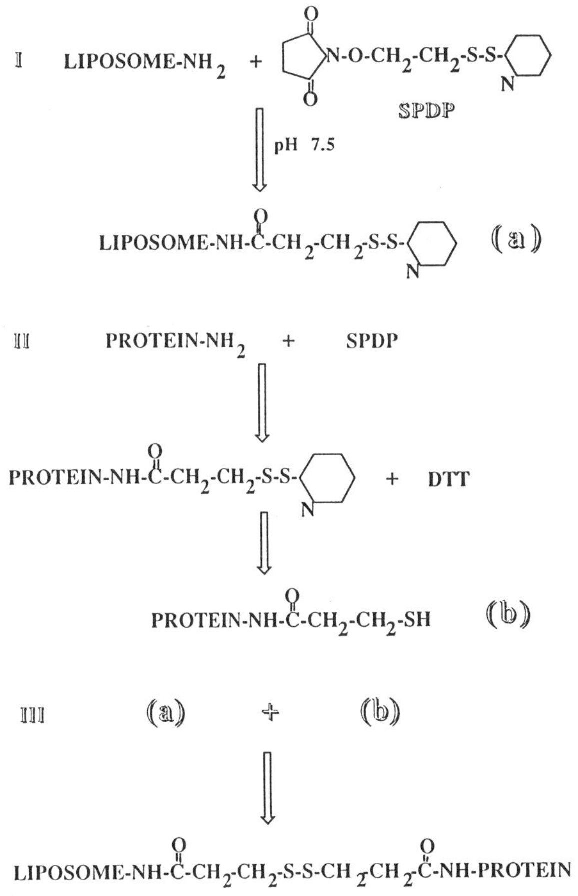

Scheme V. Covalent binding of a protein (antibody) to the liposome surface via phospholipid containing amino group and *N*-hydroxysuccinimidyl-3-(2-pyridyldithio)propionate (SPDP).

imidyl-4-(*n*-maleimidophenyl)butyrate (*33*) and succinimidyl-*S*-acetylthioacetate (*34*) were suggested.

Another approach for antibody immobilization on liposome surface is based on the fact that integral membrane proteins can be bound easily to liposomes because of the presence of large hydrophobic domains in these proteins (*35*). This characteristic makes protein association with the membrane advantageous from a thermodynamic point of view (*36*). Therefore, modification of a nonmembrane hydrophilic protein (such as antibody) with a hydrophobic reagent increases the affinity of the modified protein toward liposome. Protein binding via hydrophobic anchor must be firm: the change in Gibbs free energy (ΔG) of the transfer of a single CH_2 group from water into the organic phase is 0.7 kcal/mol, and the equivalent number of anchoring CH_2 groups in a hydrophobic tail can reach dozens. Hydrophobic groups can be incorporated easily into proteins by treating them with chloroanhydride of a long-chain fatty acid (*37*), activated phospholipids (e.g., oxidized phosphatidyl inositol) (*38*), or *N*-glutaryl phosphatidyl ethanolamine (*39*) (Scheme VI). Some other methods of antibody hydrophobization are also available (*3*).

The most important requirements for successful binding of immunoliposome to a target include the following:

- high specific affinity of immunoliposome toward the target
- accessibility of a specific binding site in the target for immunoliposome
- low nonspecific immunoliposome binding with the target

Studies on immunoliposome interaction with antigen monolayers have shown that the affinity binding of immunoliposomes to antigen is multipoint if a sufficient number of antibody molecules are present on the liposome surface (*40*). This affinity binding results in lower dissociation constant values for antigen:liposome–antibody conjugate complex than for antigen–antibody complex. To improve immunoliposome binding to the target, coupling of several antibodies against several specific target antigens with the liposome surface was suggested, and this complex can provide cooperative multipoint liposome-target binding (*41*). Numerous examples of in vitro immunoliposome

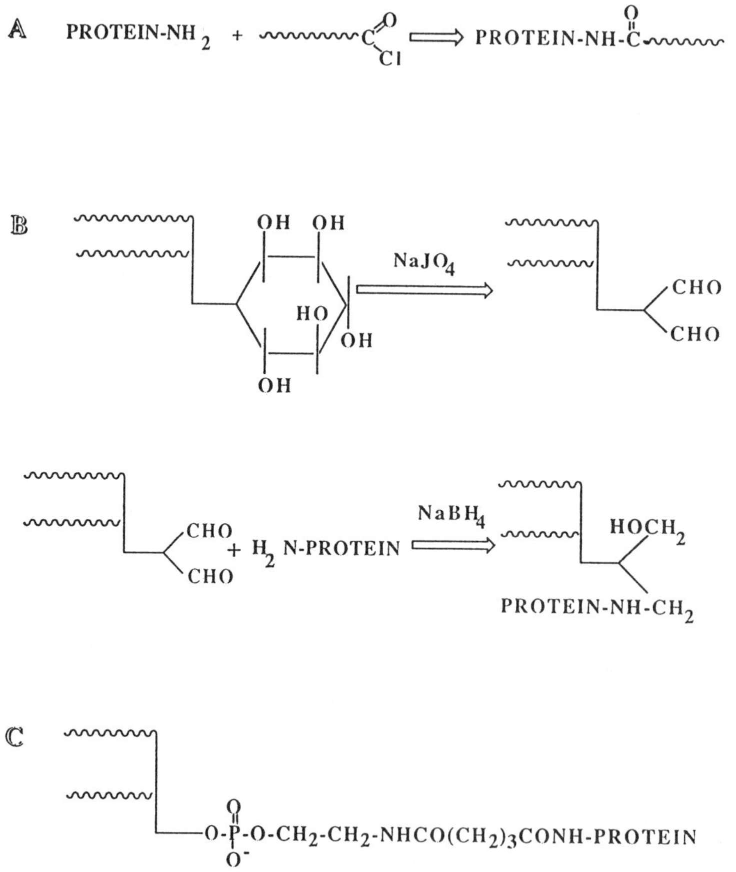

Scheme VI. Hydrophobization of a protein (antibody) for subsequent liposomal membrane incorporation via derivatization with A, long-chain fatty acid chloride; B, oxidized phosphatidyl inositol; and C, *N*-glutaryl phosphatidyl ehtanolamine.

binding to target antigens and cells have been reported (*3, 30*), but much fewer examples of immunoliposome targeting in vivo have been.

Immunoliposomes In Vivo

The most important problem for the targeting of any microparticulate carrier (including liposomes) is the inability of the carrier to extravasate and reach the targets inside tissues. Both plain liposomes and immunoliposomes extravasate very poorly (*42*). Despite limiting possible liposome use as drug carries, this fact should not affect liposome and immunoliposome ability to interact with cells and noncellular components within the circulation system, such as blood components, endothelial cells, subendothelial structures, and ischemic regions of the heart. Here, I will consider several such examples, keeping in mind that targeted delivery of immunoliposomes into RES organs or into tumors follows the same criteria.

It is commonly accepted that the initial stage of many vascular injuries, including thrombosis, is the disruption of the integrity of the vessel wall endothelial cover, leading to subendothelium exposure followed by platelet activation and adhesion (*43*). Naturally, it is tempting to think of early detection of such disruptions of endothelium and direct action at these sites to promote endothelium growth or prevent platelet adhesion onto the exposed collagen. To prove the possibility of using targeted immunoliposomes as specific drug carriers to these areas, conjugates were obtained between liposomes and antibodies against extracellular matrix antigens—collagen, laminin, and fibronectin (*44–46*). Human umbilical endothelial cells were grown on fibrillar collagen, which gave an experimental model including partial reconstitution of the luminal surface of *normal* and *injured* vessel wall. This surface was imitated by a bilayer structure of confluent (normal vessel wall) or preconfluent (injured vessel wall) endothelial cell cultures on collagen. Specific recognition of collagen-coated gaps in preconfluent culture was achieved by using ^{14}C-labeled liposome conjugates with anti-collagen antibody. The liposomes were prepared from a mixture of lecithin, cholesterol, and phosphatidyl ethanolamine in 6:2:2 molar ratio. For the coupling, liposomes were activated with glutaraldehyde, and after separation of unreacted glutaraldehyde, antibodies were allowed to react with activated liposomes via lysine amino groups. As a result, liposomes were obtained with the size of ca. 100 nm and 10–20 antibody molecules per liposome.

The data presented in Figure 2 clearly show that, opposite to plain liposomes and liposome–nonspecific-antibody conjugate, anti-collagen liposomes can specifically recognize and bind collagen gaps between endothelial cells in the preconfluent endothelial cell

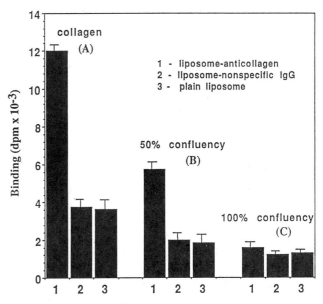

Figure 2. Binding of ^{14}C-labeled liposome-carrier conjugates by fibrillar collagen in endothelial cell cultures. (Adapted from reference 44.)

cultures grown on fibrillar collagen. Binding of all three preparations is equally low in the confluent cell culture (A), where the antigen is not exposed. The lower the confluency (i.e., the more collagen is exposed between cells) the higher the binding of the specific conjugate (*compare* cases B and C in Figure 2). The use of liposomes for directed transport of drugs to the injured area of the vessel wall in vivo may require the use of antibodies to several components of the subendothelium and intimal basal membrane. Liposome conjugates with antibodies against laminin and fibronectin (which are also the components of extracellular matrix) also were studied (*46*).

For the incorporation into liposomes, corresponding antibodies were modified with palmitic acid residues. Palmitoylated antibodies were added to the phospholipid suspension in the detergent and incorporated into liposomes during the detergent dialysis. By using ^{125}I-labeled immunoglobulins, it was established that the method enables the binding of 30–40 protein molecules randomly distributed between inner and outer monolayers of the liposomal membrane to a 100-nm liposome. The incubation of antibody–liposome conjugates with substrate-coated matrices demonstrated that antibodies on liposomes preserve their affinity, maintain their specificity, and target liposomes to an appropriate antigen. For example, anti-laminin-bearing liposomes specifically recognized laminin monolayers with up to 64 times higher binding to laminin than to either fibronectin or albumin monolayers. Binding of control antibody-free liposomes was nearly the same as nonspecific binding of antibody–liposomes. The dissociation constant for liposome–antibody conjugate binding to the target was estimated in the range of 1×10^{-8} to 1×10^{-9} M liposomes, which is well within the binding constants observed for the reaction of antigens with free antibodies.

To verify the applicability of the systems to conditions more close to the in vivo situations, we prepared ^{14}C-cholesterol oleate-containing liposomes conjugated with bovine or human anti-collagen antibodies and perfused them in situ through segments of bovine, rabbit, or human arteries partially denuded with a balloon catheter before perfusion (47). The results of experiments with bovine mesenteric arterial segments, rabbit aortal segments, and human carotid artery segments demonstrated that perfusion with ^{14}C-liposomes results in approximately equal association of liposomes with control and denuded areas in any artery. The same is true for liposomes with immobilized nonspecific rabbit IgG. At the same time, in case of anti-collagen antibody-targeted liposomes, their association with the denuded area is about 4 times higher. Thus, radiolabeled liposomes can be effectively targeted to certain areas of pathological luminal vessel wall, and these liposomes open certain opportunities for the targeted delivery of diagnostic (and possibly therapeutic) agents to these areas. The approach can permit fast and efficient visualization and treatment of affected regions that have increased risk of thrombus or atheroma formation.

Another example of possible liposome targeting within the vascular bed is that of highly specific and effective liposome targeting to pulmonary endothelium (48). In these studies two rat monoclonal antibodies, 34A and 201B, which specifically bind to a surface glycoprotein gp112 of the pulmonary endothelial cell surface, were coupled to unilamellar liposomes 250-nm in diameter. To couple antibodies, liposomes containing 10 mol% of N-glutaryl-PE were activated by 1-ethyl-3-(3-dimethyl aminopropyl) carbodiimide and N-hydroxysulfosuccinimide (39). More than 900 antibody molecules were bound per liposome. The 34A- and 201B-immunoliposomes bound effectively and specifically to mouse pulmonary artery endothelial cell culture at 4 °C. Moreover, specific immunoliposomes accumulated efficiently (30–75% of injected dose) in the mice lung. The targeting effect was shown by using both ^{125}I-labeled lipid marker and liposome-entrapped, labeled inulin derivative. The accumulation effect could be completely blocked by preadministration of free specific antibody, but the effect was insensitive to a nonspecific one. The binding efficacy increased with increasing antibody:lipid ratio in immunoliposomes, and maximum binding was a 1:8 M ratio. Time-course studies demonstrated that 34A–immunoliposomes bound to lung antigens within 1 min after injection, and liposomes remaining

unbound after several passages through the lung capillary bed were rapidly taken up by liver and spleen.

Another important target for the delivery of diagnostic and therapeutic agents is the infarcted myocardium. As usual, imaging agents [usually γ- or magnetic resonance (MR)-active heavy metals] have to be bound with some specific agent (antibody) able to interact specifically with some characteristic component of the affected tissue. The existence of such a targeted delivery system may pave the way for the targeted delivery of various therapeutic agents, such as thrombolytic enzymes, proteolytic drugs, and antioxidants, which have been proposed for treatment of infarcted myocardium. But these agents do not possess any specificity for the damaged myocardium. A targeting system based on liposome with anti-cardiac myosin antibody was reported for this purpose (49). The idea is based on the fact that normal myocardial cells with intact cell membrane will not permit extracellular macromolecules (e.g., anti-myosin antibody) to penetrate the cell membrane.

However, necrotic cardiomyocytes with membrane disruptions can no longer keep anti-myosin antibody from interacting with myosin (50). Thus, antibody to canine cardiac myosin was covalently coupled (via glutaraldehyde) to sonicated liposomes prepared from a mixture of egg lecithin, cholesterol, and phosphatidyl ethanolamine in 6:2:2 molar ratio (49). The preservation of anti-myosin activity after coupling to liposomes was proved by in vitro binding of anti-myosin–liposomes to ^{125}I-labeled canine cardiac myosin. In vivo studies were performed on dogs with experimental myocardial infarction developed in anesthetized animals by temporary occlusion of left anterior descending coronary artery. After reperfusion, anti-myosin–liposomes containing inside ^{111}InCl$_3$ were administered intravenously, and infarct imaging was performed on a γ-camera. Good accumulation of intraliposomal radioactive marker in the infarct was demonstrated and proved the possibility of immunoliposome targeting into necrotic myocardium and applicability of γ-emitting isotope-loaded liposomes for radioscintigraphy of myocardial infarction.

Protective Effect of PEG on Liposomes

Despite some promising results with immunoliposomes as carriers for radioisotopes and imaging agents, the whole approach is limited because of short lifetime of liposomes and immunoliposomes in the circulation. The fast elimination of liposomes from the blood and liposome capture by the cells of RES (mainly in the liver and spleen) are usually believed to be the result of their fast opsonization (26). To make liposomes capable of delivering pharmaceutical agents to targets other than RES, numerous attempts were made to prolong their lifetime in the circulation (*3*). The real breakthrough, however, was achieved with the discovery of surface-modified long-circulating liposomes. Originally (*51*), the longevity effect was achieved by incorporating ganglioside GM1 into the liposome membrane. Another approach involved liposome coating with PEG (*52–54*) (Scheme VII). The explanations of the PEG phenomenon involve the role of surface charge and hydrophilicity of PEG-coated liposomes (*55*), the participation of PEG in repulsive interactions between PEG-grafted membranes and other particles (*56*), and more generally, the decreased rate of plasma protein adsorption on the hydrophilic surface of liposomes incorporated with PEG (*57*).

We hypothesized (*58*) that the molecular mechanism of PEG action is determined by the behavior of a polymer molecule in solution and includes the formation of impermeable polymeric *cloud* over the liposome surface even at relatively low polymer concentration. Liposome elimination from the blood proceeds mainly via liposome recognition by phagocytic cells, mediated by plasma proteins adsorption onto the liposome surface. The most evident approach to slow down liposome clearance, therefore, should include the prevention of liposome–protein contact, for example, by coating the membrane with some inert polymer. This polymer should not contain any cell-specific fragments, and its contacts with plasma proteins should not result in opsonization.

We introduce some quantitative criteria for the characterization of liposome protection with a polymer (*59*). Considering the diffusional movement of a protein molecule toward the liposome surface as the initial step of a protein–liposome interaction, we can express the degree of liposome protection as the probability of protein collision with a polymer (P_{pol}) instead of liposome (P_{lip}). Thus, if the protein is totally unable to reach the liposome surface ($P_{lip} = 0$), then $P_{pol} = 1 - P_{lip} = 1$. When $P_{pol} = 0$, no protection can be achieved. In terms of chemical kinetics, the following is true:

$$d[LO]/dt = k[L][O] = P_{lip}k'[L][O] \qquad (1)$$

where [LO] is the concentration of liposome–opsonin complex, k and k' are rate constants of LO formation, [L] and [O] are concentrations of outer liposome monolayer lipid only and opsonin, respectively, and t is time.

To estimate possible P values, we have to consider the behavior of a liposome-attached polymer molecule within a simplified model of a polymer solution (*60*). Thus, the polymer solution can be described as a 3-D network, where each cell is occupied either by a polymer unit or a solvent molecule. According to this model, the more flexible the polymer (i.e., the more independent the motion of polymeric units relative to

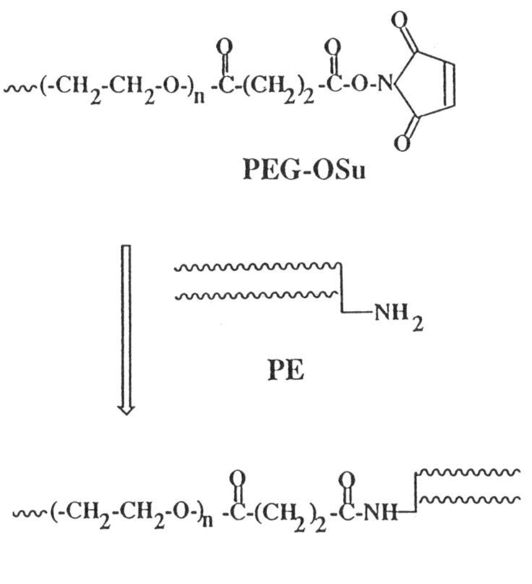

Scheme VII. Synthesis of PEG-PE for incorporation into liposome.

each other), the larger the total number of its possible conformations and the higher the transition rate from one conformation to another. Therefore, a water-soluble flexible polymer exists statistically as a distribution (or cloud) of probable conformations (Figure 3). The degree of water disturbance by the polymer correlates with polymer flexibility (i.e., its ability to occupy with high frequency many cells in solution by replacing water molecules). To reach the liposome surface, protein molecules have to penetrate the whole cloud formed by the liposome-attached polymer molecule. Thus, relatively small number of water-soluble and very flexible polymer molecules can create sufficiently dense conformational clouds over the liposome surface to protect it from being opsonized and recognized by RES cells. These molecules form protective umbrellas on the liposome surface, and the P_{lip} value depends on their amount, effective area, and reliability.

A simple calculation can give us the following expression:

$$P_{pol} = \gamma (S_P/S_1)P^* \qquad (2)$$

where γ is the molar ratio of polymer–lipid in the outer monolayer, S_P is the effective area under the umbrella, S_1 is the average area occupied by a single lipid molecule (for the given liposome size and composition), and P^* is the average P_{pol} within the umbrella volume. At high γ values, the polymer will be stretched out of the liposome and will form a dense *brush* (*61*). Therefore, $\gamma(S_P/S_1)$ is always <1. The maximum protection can be achieved when $\gamma S_P \approx S_1$ (the polymer clouds are practically fused) and P^* is close to 1.

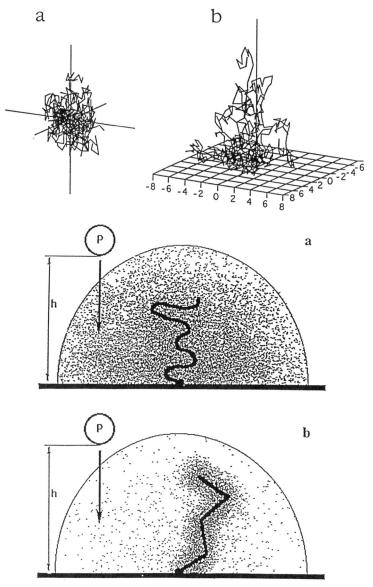

Figure 3. Top: Computer simulation of the mobility of a free (a) and surface-grafted (b) polymer molecule. Each conformational cloud includes 9 possible conformations, molecular length of 50 segments, and unrestricted segment motion assumed: scale, segments. The cloud shape in the solution is close to spherical, whereas surface-grafted polymer forms bell-shaped (or umbrella-like) cloud protecting larger area. Bottom: Schematic presentation of conformational mobility of a flexible (a) and rigid (b) polymer molecule, grafted to liposome surface. Pixel density represents the amount of the disturbed water, or the probability for protein molecule (P) to meet a polymer during the time of protein diffusion (bold arrow) from distance h to liposome surface. (Adapted from reference 59.)

If the immobilized polymer has a rigid chain (hindered unit motion), even its good water-solubility and hydrophilicity may not provide protection for the liposome. The number of possible conformations for such polymers is small and the conformational transitions proceed with a slower rate than those of a flexible polymer. Therefore, the density of the conformational cloud in this case will be very uneven during a single collision. In terms of the *cellular* model, sufficient water space may exist through which the normal diffusion of plasma proteins toward the liposome surface is possible. So, to protect liposome one has to bind a much larger number of rigid polymer molecules on the liposome surface. Only when the same polymer has both hydrophilicity and flexibility can it serve as an effective liposome protector even at low concentration at the surface. The most obvious example of such polymers is PEG; other possible candidates are poly(acrylamide) and poly(vinylpyrrolidone).

To prove our hypothesis experimentally we investigated the efficacy of the fluorescence quenching of the liposome-incorporated fluorescent phospholipid *N*-[7-

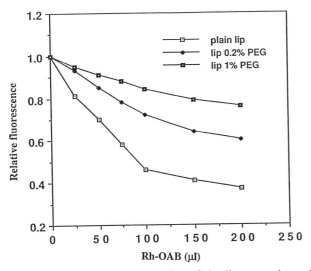

Figure 4. Fluorescence quenching of the liposome-bound NBD-PE with increasing quantities of free rhodamine-modified ovalbumin (Rh-OAB; µL/2 mL buffer) from solution at different concentrations of the liposomal PEG-PE. Quenching decreases with increasing PEG quantity on liposome surface. (Adapted from reference 58.)

nitrobenz-2-oxa-1,3-diazol-4-yl]dioleoyl-PE (NBD-PE) with soluble rhodamine-modified ovalbumin (Rh-OAB), depending on the type and quantity of the liposome-attached polymer. Liposomes were prepared by detergent (octyl glycoside) dialysis method from egg lecithin and cholesterol in 7:3 molar ratio with the addition of 1 mol% of NBD-PE and different quantities of PEG-PE (M_r 5000) or dextran-stearylamine (M_r 6000). The kinetics of NBD fluorescence quenching with Rh-OAB from solution were registered spectrofluorimetrically. The resulting data are presented in Figure 4. An increase in the liposomal PEG-PE concentration resulted in a decrease of the liposomal NBD quenching. The whole process is limited only by Rh-OAB diffusion from the solution to the liposome surface, and it is evident that the presence of PEG on the surface (even at such low concentration as 0.2 mol%) creates diffusional hindrances for this process. At PEG concentration of ca. 1 mol% these hindrances are more pronounced. At the same time, similar quantities of liposome-incorporated dextran have little influence on NBD quenching with Rh-OAB (data not shown).

To estimate the area on the liposome surface protected with a single polymer molecule of a given size and the number of polymer molecules needed to protect the liposome of a given size, we can use such parameters as the average end-to-end distance of a polymer random coil in solution, $R_{0,sol}$. A polymer molecule is located mainly in the volume between the ends (i.e., sphere for the molecule in solution and hemisphere for the surface-attached polymer). So, we can assume that the $R_{0,sol}$ value nearly corresponds to the radius of the dense cloud, which can hardly be penetrated with a protein molecule. The end-to-end distance for the surface-attached molecule should be at least 2 times longer than $R_{0,sol}$. Thus, we can assume that the radius of the protected area is about $2R_{0,sol}$ and is within this area, $P^* = 1$. Using this simple approach and $R_{0,sol}$ values for PEG (62), we can estimate the area that can be protected on the liposome surface with a single PEG molecule. Assuming about 4.25×10^4 lipid molecules in the outer monolayer of unilamellar 100-nm liposome (63), we can calculate the PEG-to-lipid molar ratio required for 100% protection of liposome surface (surface area, S, ca. 3.14×10^4 nm²) (Figure 5). This value matches well with published experimental data (54, 64).

Delivery of Imaging Agents

Another interesting approach in the use of PEG–liposomes is the targeted delivery of imaging agents. Plain liposomes are already widely used as imaging agents (65, 66) both for γ-scintigraphy and for MR tomography. For this purpose liposomes are loaded with corresponding imaging agents, such as ^{111}InCl$_3$ for γ-imaging (65) or Gd complex with the chelating agent diethylenetriamine pentaacetic acid (DTPA) (66) for MR imaging (MRI). Liposome loading can be performed by the entrapment of the imaging agent into aqueous interior of the liposome or by the attachment of the agent (usually heavy metal ion) to the liposome surface via a chelating group. For better incorporation into the liposomal membrane a chelator can be modified first with a fatty acid or phospholipid residue (67).

In many cases liposomes used for the delivery of diagnostic agents do not need to be administered via the circulation system. The major part of intramuscularly or subcutaneously injected liposomes is delivered through the lymphatic channels to the nearest lymph nodes and accumulates there (mainly taken' up by lymph node macrophages) (68). Imaging of lymph nodes plays a major role during the early detection of neoplastic involvement in cancer patients (69). Because lymph nodes are frequent sites for tumor metastases, this ability of liposomes is of great interest for the lymphatic delivery of diagnostic and therapeutic agents. In this particular case, the coating of the liposome surface aims to improve its accumulation in lymph nodes and to enhance the signal of the liposome-associated label.

In our experiments for enhanced lymph node visualization we used NMR spectroscopy with Gd-containing liposomes. The coating of MR-active Gd-liposomes with PEG changes the water content of Gd environment because of the presence of water molecules tightly associated with PEG molecule, and thus this coating enhances the NMR signal. Therefore, Gd-

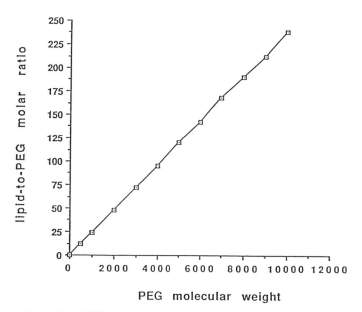

Figure 5. Dependence of calculated lipid:PEG molar ratio required for complete protection (coating) of 100-nm liposome surface on the molecular weight of PEG.

labeled PEG-modified liposomes (PEG M_r, 5000) were used for MRI of lymph nodes in rabbits and compared with nonmodified Gd-liposomes. For incorporation into the liposomal membrane, PEG was modified (hydrophobized) with PE (52). Liposomes (200 nm) were prepared from egg phosphatidyl choline, cholesterol, Gd-DTPA-PE, and PEG-PE in 60:25:10:5 molar ratio. Liposome relaxation parameters were measured by a 5-MHz NMR proton spin analyzer (RADX Corp.). In vivo imaging of axillary–subscapular lymph node area in rabbits was performed on a 1.5-T GE Signa MRI scanner (T_1 weighed pulse sequence, fat suppression mode) for 2 h starting from the moment of subcutaneous administration of a liposomal preparation into the paw of anesthetized rabbits ($n = 3$ for each group). The raw data from the MRI instrument were analyzed by image processing software to determine the relative target–nontarget (lymph node–muscle) pixel intensity.

In vivo imaging after subcutaneous injection of different preparations (20 mg of total lipid in 0.5 mL of saline) demonstrated that all Gd-liposomes are able to visualize axillar–subscapular lymph nodes within minutes (Figure 6). Relative signal intensity in the target area was noticeably higher for Gd-PEG–liposomes than for plain liposomes. Enhanced signal from Gd-PEG–liposomes in vivo is probably due to the increased relaxivity of this preparation because of the presence of PEG-bound water molecules.

Visualization of lymph nodes with Gd-liposomes is achieved within minutes after subcutaneous administration; hence, modification of liposome surface improves its characteristics as an MRI contrast agent. Coating of liposomes with PEG increases the *region of*

interest pixels intensity 1.5–2-fold compared with plain Gd-liposomes. Evidently, this example is only one of several for modified liposome utility for the delivery of diagnostic agents, including those for MRI. In addition, lymph nodes are natural targets for subcutaneously administered liposomes, in the same way as RES organs are targets for intravenously injected liposomes. The approach could be generalized using PEG–liposomes targeted to different areas of interest. However, we need to prove that the targeting moiety

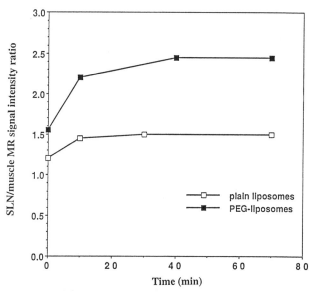

Figure 6. Ratio of MRI signal intensity in subscapular lymph node (SLN) to that in muscle in rabbit (typical pattern) following administration of Gd-liposomes into paw. (Adapted from reference 58.)

(antibody or its Fab fragment) can be coimmobilized on the liposome surface with PEG or other similar polymer, while preserving the functional properties of both antibody and polymer.

Coimmobilization of Antibody and PEG on Liposome Surface: Long-Circulating Immunoliposomes

If we can somehow combine the unique properties of long-circulating PEG-coated liposomes with target-ability of immunoliposomes in one preparation, the result could be very advantageous for the targeted delivery of imaging and therapeutic agents into affected areas with decreased blood flow or with low concentration of the target antigen. Under such circumstances the combination of liposome longevity and specificity should lead to the increase of productive collisions between the targeted liposomes and the target with time and result in liposome accumulation in the target.

One concern related to the use of PEG as a protection for immunoliposomes is that PEG can create steric hindrance for normal antibody–target interaction. To prove that under certain circumstances the peaceful coexistence of antibody and PEG is possible, we have to consider different cases for the coimmobilization of antibody and water-soluble flexible polymer on the same liposome (Figure 7). In the first case (Figure 7A), the polymer does not cover liposome completely. Antibody molecules are outside of the volume occupied by PEG, and two separate zones appear on the surface: umbrellas of PEG and antibodies. This structure should successfully interact with target antigens, so targeted liposome binding still occurs. At the same time the clearance rate (or, more exactly, the opsonization rate) depends on the surface area available for opsonin–membrane interactions. Because opsonization is often a cascade-type process and the binding of a single molecule such as a C-3 molecule can be amplified quickly, the clearance time of such a liposome may not differ much from that for normal immunoliposomes.

In the second (optimal) case (Figure 7B) the liposome surface is completely coated with overlapping umbrellas of PEG, but still some areas of loose conformational cloud exist, into which free antibodies (capable of lateral diffusion) can be squeezed. In this case opsonins have very limited opportunity for interaction with liposome, whereas antibodies are still able to recognize and to bind to the target.

The third case (Figure 7C) requires very high degree of surface modification in such a way that PEG forms conformationally stretched brushes (*61*). It is evident that in this case surface-immobilized antibody cannot overcome steric hindrance for binding to the

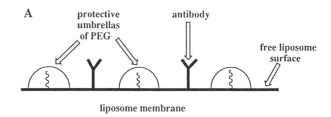

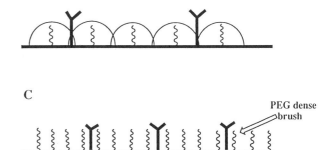

Figure 7. Possible schemes for antibody and PEG coimmobilization on the liposome surface: A, low PEG concentration; B, intermediate (optimal) PEG concentration; C, high PEG concentration in which polymer molecules form dense brushes hindering antibody mobility. (Adapted from reference 59.)

target. On the other hand, opsonins also cannot interact with the liposome surface, and very long circulation times are provided to such liposomes; half-clearance time in rabbits for immunoliposomes containing 10 mol% of PEG was more than 10 h (*54*). Brushes of this type, which should provide extremely long circulation time to liposomes, can probably be used for long-circulating immunoliposome preparation, if antibodies will be immobilized on the liposome surface via the long spacer group or even directly on termini of some PEG molecules (*70*).

To prove that long-circulating PEG-coated liposomes can be targeted by coincorporation of an antibody in the liposome surface, we studied the in vivo behavior of liposomes with anti-myosin antibody (*50*). This antibody, as we already mentioned, effectively binds myosin inside ischemic and necrotic cardiomyocytes with affected or destroyed cellular membranes, but it does not interact with normal cells (it is unable to penetrate the intact plasmic membrane) (*37, 71*). These characteristics form the basis for the targeted delivery of radiolabelled PEG-coated long-circulating liposomes in the region of ischemically damaged myocardium.

Infarcts in rabbits were generated as follows (*54*). Rabbits (New Zealand white rabbits, 3.3 kg) were anesthetized with ketamin–xylazine. A femoral artery cut-down was performed to establish a blood pressure line and for arterial blood sampling. An ear vein was catheterized to allow intravenous injections. An endotracheotomy was performed, followed by artificial ventilation. A left thoracotomy was performed and the mid-left anterior descending coronary artery was occluded with a silk suture. After 40 min, the snare was released and removed. Different radiolabelled liposome preparations were injected intravenously after 30 to 60 min of reperfusion. Blood samples were taken after specified time intervals to measure liposomal radioactivity. Five to six hours after liposome injection, animals were killed by an overdose of pentobarbital. The heart was excised and cut into 5-mm slices, stained with 2% triphenyl tetrasolium chloride, and each slice was further divided into smaller segments. Samples of normal and infarcted myocardium were weighed, and the radioactivity was counted in a γ-counter. The data for liposome accumulation in the heart were expressed as infarct:normal myocardium radioactivity ratio. Biodistribution of liposomes in sacrificed animals at 5-h postinjection was studied by measuring the liposome-associated radioactivity in liver, spleen, kidney, and lung.

Liposomes for this study were prepared by a detergent dialysis method from a mixture of phosphatidyl choline and cholesterol in 3:2 molar ratio. Additionally, liposomes contained 1 mol% of ^{111}In-labeled DTPA-stearylamine, and when necessary, 4 or 10 mol% of PEG-PE (52). Total liposome-associated radioactivity was usually between 100 and 300 µCi per 15–20 mg of total lipid. For the incorporation into the liposomal membrane, anti-myosin antibody was modified with a hydrophobic *anchor*, *N*-glutaryl-PE (Avanti Polar Lipids), as described previously (39). All liposomes were extruded through 0.4- and 0.2-µm nuclepore filters. According to measurements on Coulter Counter N-4, liposomes in all preparation were monodisperse and had a size between 150 and 190 nm.

The half-life of immunoliposomes in rabbit circulation was 40 min, which increased to 200 min with 4 mol% PEG and to about 1000 min with 10 mol% PEG. Antibody-free liposomes with 4 mol% of PEG had a half-life of about 300 min. Half-life of antibody-free liposomes with 10 mol% of PEG did not differ from that of immunoliposomes with 10 mol% of PEG. Two important conclusions can be drawn from these results. First, the increasing quantity of PEG increasingly protects liposomes from the clearance. Second, coimmobilization of an antibody and PEG decreases the half-life of liposomes only at the lower PEG concentration. High PEG concentration blocks the recognition of antibody by liver cells. The results agree well with our hypothesis (Figure 7).

The data on liposome biodistribution and infarct accumulation clearly demonstrate that PEG-coated anti-myosin–liposomes effectively accumulate in the infarct zone. Tissue radioactivity for such liposomes (expressed as a percent of injected dose per gram of tissue) is almost twice as high as for antibody-free PEG–liposomes or PEG-free immunoliposomes (0.25 against 0.13 and 0.14, respectively), and 12-times higher than for plain liposomes (0.018). Interestingly, liposomes modified only with monoclonal antibody or only with PEG show about the same infarct accumulation. Infarct–normal ratio is still much higher for anti-myosin modified liposomes (25 against 5) than for PEG–liposomes, because of lower nonspecific accumulation of the targeted liposomes in the normal myocardium. Therefore, target accumulation (at least in this case) can proceed via specific recognition and via nonspecific decreased filtration rate (liposomes can enter necrotic area via permeabilized capillaries, but cannot leave it because of affected drainage). The nonspecific mechanism requires prolonged accumulation times and can be realized only for long-circulating liposomes. For liposomes with 10 mol% PEG, infarct accumulation does not differ for the three preparations: antibody–liposomes, PEG–liposomes, and antibody–PEG–liposomes. This phenomenon can be explained by the lack of antibody participation in targeting in this case (Figure 7C).

These results show that liposomes can be made to be targetable and long-circulating at the same time, if proper PEG–monoclonal antibody ratio is maintained on the liposome surface. Prolonged circulation facilitates the accumulation of nontargetable (antibody-free) liposomes in areas with affected vasculature because of impaired filtration rate. Myocardial infarction seems to be a good target for the liposomal delivery of imaging and therapeutic agents (radiometals, thrombolytics, superoxide dismutase, and proteases). The data on experimental visualization of the myocardial infarction in dog using PEG–anti-myosin ^{111}In-labeled liposomes are presented in Figure 8. In all experiments, the infarct was also confirmed histologically by staining with triphenyl tetrasolium chloride. Stained areas coincided with areas of increased radioactivity accumulation (72, 73). Generally speaking, the principal behavior of PEG-coated liposomes in vivo agrees well with the hypothetical model of PEG action on liposomes.

Effect of Liposome Size, Antibody, and PEG on Liposome Biodistribution

To investigate the relative importance of size, antibody, and PEG presence on the liposome surface for liposome biodistribution and targeting, we studied the behavior of different liposome preparations in rabbits with experimental myocardial infarction (74). Liposomes of two sizes (small liposomes, 110–150 nm, and large liposomes, 330–400 nm) were used in our experiments.

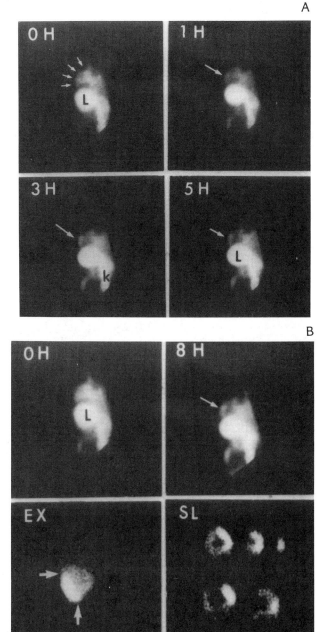

Figure 8. Radioimmunoscintigraphy of experimental myocardial infarction in dog with ^{111}In-PEG-anti-myosin-liposomes: A, time-course of infarct visualization. Images obtained in increasing time intervals gradually reveal the infarct localization, which can be seen as early as 1-h postinjection (single arrow). Multiple arrows show the area of the initial blood flow pattern in the heart area: L, liver; K, kidney. Figure 8B. Infarct localization in the excised heart (EX; between two arrows) and on postmortem heart slices (SL).

Blood Clearance

In case of small liposomes, the fastest clearance was observed for plain liposomes. Antibody slightly increases circulation time, probably by making part of the surface less accessible for opsonins, which interact mostly with the empty surface of liposome (*26*). No immunoliposome capture via Fc fragment can proceed when using Fab, even though this capture usually occurs in the case of whole IgG. PEG (4 mol%) sharply increased circulation time: Liposome half-life increased from 40 min for plain liposomes to 400 min for PEG–liposomes. Simultaneous incorporation of Fab and PEG somewhat decreased the circulation time of PEG–liposomes (probably because of more pronounced interaction between Fab and plasma proteins), but the liposomes still circulated long enough (half-life ca. 300 min) to perform effective binding to target. Large liposomes, in principle, behaved in a similar fashion to that for small ones, even though the half-life of large PEG–liposomes is somewhat shorter and Fab incorporation has little influence on circulation time of PEG–liposomes. This difference can be explained by size—the surface area of large liposome is about 6 times larger than that of small liposome. Possible irregularities in PEG location or PEG clusters formation can expose some part of the liposome surface for opsonization. Therefore, the circulation time for large PEG–liposomes is shorter. Additional incorporation of Fab onto the surface of large liposomes does not change the whole picture much. Thus, the circulation time is strongly influenced by all three factors studied: liposome size, the presence of Fab or PEG, or the presence of both on the liposome surface.

Infarct Accumulation

Small PEG–liposomes and small Fab–liposomes accumulate in the necrotic area almost identically if expressed in absolute quantities (0.13% and 0.14% dose/g, respectively). This similarity indicates two different ways for liposomes accumulation. The specific route requires the presence of antibodies on the surface of short-circulating liposomes to enhance their binding (*49*); the nonspecific route proceeds via impaired filtration mechanism in affected tissues and requires many passages of liposomes through the target (i.e., prolonged circulation). At the same time, infarct:normal ratio (or relative targeting) is much higher for Fab–liposomes than for PEG–liposomes (22.5 to 7.5, respectively). The reason for this interesting phenomenon is that the accumulation of Fab–liposomes in normal tissues is very low (the time of Fab–liposomes residence in the blood is too low for any nonspecific accumulation), whereas long-circulating PEG–liposomes can be accumulated slowly in any vascular defects that are present even in normal tissues. The combination of Fab and PEG on the liposome surface in relative terms gives less accumulation of liposomes than Fab–liposomes (again because of nonspecific capture in normal tissue). At the same time, in absolute terms (% dose/g), this combination gives excellent results (0.25% dose/g, or twice as high as for Fab–liposomes!), because both accumulation mechanisms are working in this particular case.

The increase in liposome size should definitely affect the ability for nonspecific accumulation in the necrotic tissues via impaired filtration. It can also affect the efficacy of Fab–liposome interaction with the target, for example, when a single Fab–antigen bond is not sufficient to anchor large liposome to the target and multiple bonds cannot be formed for any immunoliposome collision with the target. In addition, repeated passages of these unsuccessful liposomes through the target are very improbable because of their fast clearance (the number of productive collisions may be less than for small immunoliposomes). These considerations permit us to understand the results observed. Plain large liposomes do not give any noticeable accumulation in the infarction similar to small liposome; the residence time is both cases is too short to reveal any differences. The accumulation of both Fab–liposomes and PEG–liposomes is quite efficient. Moreover, PEG–liposomes show absolute accumulation even slightly better than Fab–liposomes (0.14% and 0.1% dose/g, respectively). Evidently, prolonged circulation can be more efficient for gradual accumulation than short-term specific interaction, part of which can be nonproductive because of large liposome size.

For infarct:normal ratio we can observe the same picture as for small liposomes. Very low nonspecific capture of Fab–liposomes makes their relative accumulation much higher than that for long-circulating PEG–liposomes (16 vs. 7.5 times for infarct:normal ratio). Coincorporation of Fab and PEG into the same liposome does not improve absolute accumulation of large liposomes (nonspecific accumulation via impaired filtration mechanism is the limiting step of the whole process, which does not leave room for additive effect with Fab). Infarct:normal ratio for this particular case is somewhat in between those for Fab–liposomes and PEG–liposomes. Maximal accumulation of large liposomes is somewhat less than that for small liposomes. Thus, the maximum infarct:normal ratios can be achieved for small Fab–liposomes, and these ratios make these conjugates attractive for the delivery of, for example, imaging agents, when the maximal difference between the area of interest and normal tissues is desired. At the same time maximum absolute delivery can be achieved in the case of small liposomes with coimmobilized Fab and PEG, and this

characteristic makes them suitable for the delivery of pharmaceutical agents.

Conclusion

Multicomponent polymer-based systems for targeted drug delivery have been widely investigated for almost 20 years. Recently, sterically protected polymer-coated liposomes have attracted increased attention as drug carriers (75). Numerous studies on long-circulating PEG–liposomes demonstrated that they can be used for nonspecific and targeted drug delivery, as well as for increasing the duration of drug action by maintaining drug levels in the body. They also demonstrate the ability to accumulate nonspecifically in the tissues with permeabilized vasculature.

We suggested a hypothetical model of PEG behavior on the liposome surface (59, 76) based on the statistical properties of polymer molecule in solution, and these properties include the formation of a dense "statistical (random) cloud" of possible polymer conformations over the liposome surface even at relatively low polymer concentrations. This cloud isolates the liposome surface from the plasma proteins and opsonins (26). The model postulates that polymers for liposome steric protection should be soluble and hydrophilic, have a highly flexible main chain, and be able to associate with the liposome surface. Poly(acryl amide), poly(vinylpyrrolidone), and poly(vinyl alcohol) were named as the most appropriate candidates among alternative liposome protectors (59, 76, 77). Our recent data clearly demonstrate that these polymers, which are grafted at one terminus with a hydrophobic residue such as a long-chain fatty acyl or phospholipid moiety, easily associate with liposomes and sharply increase in vivo liposome stability and circulation time (77).

The further development of the concept of long-circulating liposomes involves the attempt to combine the properties of long-circulating liposomes and immunoliposomes in one preparation (54, 70). This goal is especially important for efficient targeted delivery of liposomal drugs to targets with diminished blood supply or low antigen concentration when increased circulation times for liposome-to-target interaction are required.

References

1. *Liposome Technology;* Gregoriadis, G., Ed.; CRC: Boca Raton, FL, 1993; Vols. 1–3.
2. *Liposomes as Drug Carriers;* Gregoriadis, G., Ed.; John Wiley and Sons: Avon, England, 1988.
3. Torchilin, V. P. *CRC Crit. Rev. Ther. Drug Carriers Syst.* **1985,** *2,* 65–115.
4. Allen, T. M. In *Liposomes in the Therapy of Infectious Diseases and Cancer;* Lopez-Berestein, G.; Fidler, I. J., Eds.; Alan R. Liss: New York, 1989; pp 405–415.

5. Klibanov, A. L.; Maruyama, K.; Torchilin, V. P.; Huang, L. *FEBS Lett.* **1990**, *268*, 235–237.

6. Shaw, I. H.; Dingle, J. T. In *Liposomes in Biological Systems*; Gregoriadis, G.; Allison, A. C., Eds.; John Wiley & Sons: Chichester, England, 1980; Chapter 11.

7. Chazov, E. I.; Matveeva, L. S.; Mazaev, A. V.; Sargin, K. E.; Sadovskaya, G. V.; Ruda, M. Ya. *Ther. Arch. (Russ.)* **1976**, *48*, 8–12.

8. McLaughlin, C. A.; Goldberg, E. P.; In *Targeted Drugs*; Goldberg, E. P., Ed.; John Wiley and Sons: New York, 1983; pp 231–268.

9. Weinstein, J. N.; Magin, R. L.; Yatvin, M. B.; Saharko, D. S. *Science (Washington, D.C.)* **1979**, *204*, 188–191.

10. Yatvin, M. B.; Kreutz, W.; Horwitz, B. A.; Shinitzky, M. *Science (Washington, D.C.)* **1980**, *210*, 1253–1255.

11. Torchilin, V. P.; Fan Zhou; Leaf Huang *J. Liposome Res.* **1993**, *3*, 201–255.

12. Widder, K. J.; Marino, P. A.; Morris, R. M.; Senyei, A. E. In *Targeted Drugs*; Goldberg, E. P., Ed.; John Wiley & Sons: New York, 1983; pp 201–230.

13. Torchilin, V. P.; Papisov, M. I.; Smirnov, V. N. *J. Biomed. Mater. Res.* **1985**, *19*, 461–466.

14. Torchilin, V. P.; Papisov, M. I.; Orekhova, N. M.; Belyaev, A. A.; Petrov, A. D.; Ragimov, S. E. *Haemostasis* **1988**, *18*, 113–116.

15. Duncan, R.; Kopecek, J. *Adv. Polym. Sci.* **1984**, *57*, 53–101.

16. Langer, R.; Folkman, J. *Nature (London)* **1976**, *263*, 797–800.

17. Torchilin, V. P.; Tischenko, E. G.; Smirnov, V. N.; Chazov, E. I. *J. Biomed. Mater. Res.* **1977**, *11*, 223–235.

18. Papisov, M. I.; Torchilin, V. P. *Int. J. Pharm.* **1988**, *40*, 201–214.

19. Vitetta, E. S.; Krolick, K. A.; Miyama-Inaba, M.; Cushley, W.; Uhr, J. W. *Science (Washington, D.C.)* **1983**, *219*, 644–650.

20. Wall, M. E.; Abernathy, A. S.; Carroll, F. I.; Taylor, D. J. *J. Med. Chem.* **1969**, *12*, 810–813.

21. Fiume, L.; Busi, C.; Mattioli, A. *FEBS Lett.* **1983**, *153*, 6–10.

22. Ringsdorf, H. *J. Polym. Sci.* **1975**, *51*, 135–153.

23. Goldberg, E. P. In *Polymeric Drugs*; Donamura, L. G.; Vogel, O., Eds.; Acadenic:Orlando, FL, 1978; pp 239–254.

24. Arnon, R.; Hurwitz, E. In *Targeted Drugs*; Goldberg, E. P., Ed.; John Wiley & Sons: New York, 1983; pp 23–56.

25. Printseva, O. Yu.; Faerman, A. I.; Maksimenko, A. V.; Tonevitsky, A. G.; Ilynsky, O. B.; Torchilin, V. P. *Experientia* **1985**, *41*, 1342–1344.

26. Senior, J. H. *CRC Crit. Rev. Ther. Drug Carriers Syst.* **1987**, *3*, 123–193.

27. Shek, P. N.; Heath, T. D. *Immunology* **1983**, *50*, 101–106.

28. Kao, Y. J.; Juliano, R. L. *Biochim. Biophys. Acta* **1981**, *677*, 453–461.

29. Torchilin, V. P.; Berdichevsky, V. R.; Barsukov, A. A.; Smirnov, V. N. *FEBS Lett.* **1980**, *111*, 184–188.

30. Torchilin, V. P. In *Liposome Technology*; Gregoriadis, G., Ed.; CRC: Boca Raton, FL, 1984; Vol. 3, pp 75–94.

31. Leserman, L. D.; Barbet, J.; Kourilsky, F.; Weinstein, J. N. *Nature (London)* **1980**, *288*, 602–604.

32. Martin, F. J.; Hubbell, W. L.; Papahadjopoulos, D. *Biochemistry* **1981**, *20*, 4229–4238.

33. Martin, F. J.; Papahadjopoulos, D. *J. Biol. Chem.* **1982**, *257*, 286–288.

34. Derksen, J. T. P.; Scherphof, G. L. *Biochim. Biophys. Acta* **1985**, *814*, 151–155.

35. Singer, S. J. *Ann. Rev. Biochem.* **1974**, *43*, 805–833.

36. Tanford, C. *Science (Washington, D.C.)* **1978**, *200*, 1012–1018.

37. Torchilin, V. P.; Omelyanenko, V. G.; Klibanov, A. L.; Mikhailov, A. V.; Goldansky, V. I.; Smirnov, V. N. *Biochim. Biophys. Acta* **1981**, *602*, 511–521.

38. Torchilin, V. P.; Klibanov, A. L.; Smirnov, V. N. *FEBS Lett.* **1982**, *138*, 117–120.

39. Weissig, V.; Lasch, J.; Klibanov, A. L.; Torchilin, V. P. *FEBS Lett.* **1986**, *202*, 86–90.

40. Klibanov, A. L.; Muzykantov, V. R.; Ivanov, N. N.; Torchilin, V. P. *Anal. Biochem.* **1985**, *150*, 251–257.

41. Trubetskoy, V. S.; Berdichevsky, V. R.; Efremov, E. E.; Torchilin, V. P. *Biochem. Pharmacol.* **1987**, *36*, 839–842.

42. Poste, G.; Kirsh, R. *Biotechnology* **1983**, *1*, 869–878.

43. Ross, R. *Nature (London)* **1993**, *362*, 801–809.

44. Chazov, E. I.; Alexeev, A. V.; Antonov, A. V.; Koteliansky, V. E.; Leytin, V. L.; Lyubimova, E. V.; Repin, V. S.; Sviridov, D. D.; Torchilin, V. P.; Smirnov, V. N. *Proc. Natl. Acad. Sci. U.S.A.* **1981**, *78*, 5603–5607.

45. Smirnov, V. N.; Berdichevsky, V. R.; Sviridov, D. D.; Torchilin, V. P. In *Vessel Wall in Athero-and Thrombogenesis*; Chazov, E. I.; Smirnov, V. N., Eds.; Springer-Verlag: New York, 1982; pp 195–201.

46. Torchilin, V. P.; Klibanov, A. L.; Ivanov, N. N.; Glukhova, M. A.; Koteliansky, V. E.; Kleinman, H. K.; Martin, G. P. *J. Cell. Biochem.* **1985**, *28*, 23–29.

47. Smirnov, V. N.; Domogatsky, S. P.; Dolgov, V. V.; Hvatov, V. B.; Klibanov, A. L.; Koteliansky, V. E.; Muzykantov, V. R.; Repin, V. S.; Samokhin, G. P.; Shekhonin, B. V.; Smirnov, M. D.; Sviridov, D. D.; Torchilin, V. P.; Chazov, E. I. *Proc. Natl. Acad. Sci. U.S.A.* **1986**, *83*, 6603–6607.

48. Maruyama, K.; Holmberg, E.; Kennel, S. J.; Klibanov, A.; Torchilin, V. P.; Huang, L. *J. Pharm. Sci.* **1990**, *79*, 978–984.

49. Torchilin, V. P.; Khaw, B. A.; Smirnov, V. N.; Haber, E. *Biochem. Biophys. Res. Commun.* **1979**, *89*, 1114–1119.

50. Khaw, B. A.; Mattis, J. A.; Melnicoff, G.; Strauss, H. W.; Gold, H. K.; Haber, E. *Hybridoma* **1984**, *3*, 11–23.

51. Allen, T. M.; Chonn, A. *FEBS Lett.* **1987**, *223*, 42–46.

52. Klibanov, A. L.; Maruyama, K.; Beckerleg, A. M.; Torchilin, V. P.; Huang, L. *Biochim. Biophys. Acta* **1991**, *1062*, 142–148.

53. Mori, A.; Klibanov, A. L.; Torchilin, V. P.; Huang, L. *FEBS Lett.* **1991**, *284*, 263–266.

54. Torchilin, V. P.; Klibanov, A. L.; Huang, L.; O'Donnell, S.; Nossiff, N. D.; Khaw, B. A. *FASEB J.* **1992**, *6*, 2716–2719.

55. Gabizon, A.; Papahadjopoulos, D. *Biochim. Biophys. Acta* **1992**, *1103*, 94–100.

56. Needham, D.; McIntosh, T. J.; Lasic, D. D. *Biochim. Biophys. Acta* **1992**, *1108*, 40–48.

57. Lasic, D. D.; Martin, F. G.; Gabizon, A.; Huang, S. K.; Papahadjopoulos, D. *Biochim. Biophys. Acta* **1991**, *1070*, 187–192.

58. Torchilin, V. P.; Trubetskoy, V. S.; Milshteyn, A. M.; Canillo, J.; Wolf, G. L.; Papisov, M. I.; Bogdanov, A. A., Jr.; Narula, J.; Khaw, B. A.; Omelyanenko, V. G. *J. Contr. Release* **1994**, *28*, 45–58.

59. Torchilin, V. P.; Papisov, M. I. *J. Liposome Res.* **1994**, *4*, 725–739.

60. des Cloizeaux, J.; Jannink, G. *Polymers in Solution: Their Modelling and Structure;* Clarendon: Oxford, England, 1990; pp 63–108, 280–327, 539–649.

61. Milner, S.T. *Science (Washington, D.C.)* **1991**, *251*, 905–914.

62. Kurata, M.; Tsunashima, Y. In *Polymer Handbook;* Brandup, J.; Himmelgut, E. H., Eds.; John Wiley & Sons: New York, 1989; pp VII/1–VII/52.

63. Huang, C; Mason, J. T. *Proc. Natl. Acad. Sci. U.S.A.* **1978**, *75*, 308–310.

64. Allen, T. M.; Hansen, C. *Biochim. Biophys. Acta* **1991**, *1068*, 133–141.

65. Mauk, M. R.; Gamble, R. C. *Anal. Biochem.* **1979**, *94*, 302–307.

66. Kabalka, G.; Buonocore, E.; Hubner, K.; Moss, T.; Norley, N.; Huang, L. *Radiology* **1987**, *163*, 255–258.

67. Kabalka, G. W.; Davis, M. A.; Moss, T. H.; Buonocore, E.; Hubner, K.; Holmberg, E.; Maruyama, K.; Huang, L. *Magn. Reson. Med.* **1991**, *19*, 406–415.

68. Takakura, Y.; Hashida, M.; Sezaki, H. In *Lymphatic Transport of Drugs;* Charman W. N.; Stella, V. J., Eds.; CRC: Boca Raton, FL, 1992; pp 256–277.

69. *Lymphatic Transport of Drugs;* Charman, W. N.; Stella, V. J., Eds.; CRC: Boca Raton, FL, 1992.

70. Blume, G.; Cevc, G.; Crommelin, M. D. J. A.; Bakker-Woudenberg, I. A. J. M.; Kluft, C.; Storm, G. *Biochim. Biophys. Acta* **1993**, *1149*, 180–184.

71. Khaw, B. A.; Yasuda, T.; Gold, H. K.; Strauss, H. W.; Haber, E. *J. Nucl. Med.* **1987**, *28*, 1671–1678.

72. Khaw, B. A.; Klibanov, A.; O'Donnell, S. M.; Saito, T.; Nossif, N.; Slinkin, M. A.; Newell, J. B.; Strauss, H. W.; Torchilin, V. P. *J. Nucl. Med.* **1991**, *32*, 1742–1751.

73. Torchilin, V. P. *Immunomethods* **1994**, *4*, 244–258.

74. Torchilin, V. P.; Narula, J.; Khaw, B. A. In *Proceedings of the 21st International Symposium on Controlled Release of Bioactive Materials;* Controlled Release Society: Milwaukee, WI, 1994; pp 222–223.

75. *Stealth® Liposomes;* Martin, F.; Lasic, D., Eds.; CRC: Boca Raton, FL, 1995.

76. Torchilin, V. P.; Omelyanenko, V. G.; Papisov, M. I.; Bogdanov, A. A., Jr.; Trubetskoy, V. S.; Herron, J. N.; Gentry, C. A. *Biochim. Biophys. Acta* **1994**, *1195*, 11–20.

77. Torchilin, V. P.; Shtilman, M. I.; Trubetskoy, V. S.; Whiteman, K.; Milstein, A. M. *Biochim. Biophys. Acta* **1994**, *1195*, 181–184.

Index

P

Copyediting: Scott Hofmann-Reardon
Production: Margaret J. Brown and Kimberly N. Lassair
Acquisitions: Cheryl J. Shanks and Anne Wilson
Indexing: Colleen P. Stamm

Designed and Typeset by Betsy Kulamer, Washington, DC
Printed and bound by Maple Press, York, PA

DATE DUE

			PRINTED IN U.S.A.

Desk reference of functional
polymers